INTRODUCTION TO OPERATIONS RESEARCH

Ninth Edition

FREDERICK S. HILLIER

Stanford University

GERALD J. LIEBERMAN

Late of Stanford University

Mc
Graw
Hill

Boston   Burr Ridge, IL   Dubuque, IA   Madison, WI   New York   San Francisco   St. Louis
Bangkok   Bogotá   Caracas   Kuala Lumpur   Lisbon   London   Madrid   Mexico City
Milan   Montreal   New Delhi   Santiago   Seoul   Singapore   Sydney   Taipei   Toronto

The McGraw·Hill Companies

INTRODUCTION TO OPERATIONS RESEARCH, NINTH EDITION
International Edition 2010

10 09 08 07 06 05 04
15 14 13 12 11 10
CTP BJE

When ordering this title, use ISBN 978-007-126767-0 or MHID 007-126767-0

Printed in Singapore

www.mhhe.com

C2 AYL
<H>

ABOUT THE AUTHORS

Frederick S. Hillier was born and raised in Aberdeen, Washington, where he was an award winner in statewide high school contests in essay writing, mathematics, debate, and music. As an undergraduate at Stanford University he ranked first in his engineering class of over 300 students. He also won the McKinsey Prize for technical writing, won the Outstanding Sophomore Debater award, played in the Stanford Woodwind Quintet, and won the Hamilton Award for combining excellence in engineering with notable achievements in the humanities and social sciences. Upon his graduation with a B.S. degree in Industrial Engineering, he was awarded three national fellowships (National Science Foundation, Tau Beta Pi, and Danforth) for graduate study at Stanford with specialization in operations research. After receiving his PhD degree, he joined the faculty of Stanford University, where he earned tenure at the age of 28 and the rank of full professor at 32. He also received visiting appointments at Cornell University, Carnegie-Mellon University, the Technical University of Denmark, the University of Canterbury (New Zealand), and the University of Cambridge (England). After 35 years on the Stanford faculty, he took early retirement from his faculty responsibilities in 1996 in order to focus full time on textbook writing, and now is Professor Emeritus of Operations Research at Stanford.

Dr. Hillier's research has extended into a variety of areas, including integer programming, queueing theory and its application, statistical quality control, and the application of operations research to the design of production systems and to capital budgeting. He has published widely, and his seminal papers have been selected for republication in books of selected readings at least 10 times. He was the first-prize winner of a research contest on "Capital Budgeting of Interrelated Projects" sponsored by The Institute of Management Sciences (TIMS) and the U.S. Office of Naval Research. He and Dr. Lieberman also received the honorable mention award for the 1995 Lanchester Prize (best English-language publication of any kind in the field of operations research), which was awarded by the Institute of Operations Research and the Management Sciences (INFORMS) for the 6th edition of this book. In addition, he was the recipient of the prestigious 2004 INFORMS Expository Writing Award for the 8th edition of this book.

Dr. Hillier has held many leadership positions with the professional societies in his field. For example, he has served as Treasurer of the Operations Research Society of America (ORSA), Vice President for Meetings of TIMS, Co-General Chairman of the 1989 TIMS International Meeting in Osaka, Japan, Chair of the TIMS Publications Committee, Chair of the ORSA Search Committee for Editor of *Operations Research,* Chair of the ORSA Resources Planning Committee, Chair of the ORSA/TIMS Combined Meetings Committee, and Chair of the John von Neumann Theory Prize Selection Committee for INFORMS. He continues to serve as the Series Editor for Springer's International Series in Operations Research and Management Science, a particularly prominent book series that he founded in 1993.

In addition to *Introduction to Operations Research* and two companion volumes, *Introduction to Mathematical Programming* (2nd ed., 1995) and *Introduction to Stochastic Models in Operations Research* (1990), his books are *The Evaluation of Risky Interrelated Investments* (North-Holland, 1969), *Queueing Tables and Graphs* (Elsevier North-Holland, 1981, co-authored by O. S. Yu, with D. M. Avis, L. D. Fossett, F. D. Lo, and M. I. Reiman), and *Introduction to Management Science: A Modeling and Case Studies Approach with Spreadsheets* (3rd ed., McGraw-Hill/Irwin, 2008, co-authored by M. S. Hillier).

The late **Gerald J. Lieberman** sadly passed away in 1999. He had been Professor Emeritus of Operations Research and Statistics at Stanford University, where he was the founding chair of the Department of Operations Research. He was both an engineer (having received an undergraduate degree in mechanical engineering from Cooper Union) and an operations research statistician (with an AM from Columbia University in mathematical statistics, and a PhD from Stanford University in statistics).

Dr. Lieberman was one of Stanford's most eminent leaders in recent decades. After chairing the Department of Operations Research, he served as Associate Dean of the School of Humanities and Sciences, Vice Provost and Dean of Research, Vice Provost and Dean of Graduate Studies, Chair of the Faculty Senate, member of the University Advisory Board, and Chair of the Centennial Celebration Committee. He also served as Provost or Acting Provost under three different Stanford presidents.

Throughout these years of university leadership, he also remained active professionally. His research was in the stochastic areas of operations research, often at the interface of applied probability and statistics. He published extensively in the areas of reliability and quality control, and in the modeling of complex systems, including their optimal design, when resources are limited.

Highly respected as a senior statesman of the field of operations research, Dr. Lieberman served in numerous leadership roles, including as the elected president of The Institute of Management Sciences. His professional honors included being elected to the National Academy of Engineering, receiving the Shewhart Medal of the American Society for Quality Control, receiving the Cuthbertson Award for exceptional service to Stanford University, and serving as a fellow at the Center for Advanced Study in the Behavioral Sciences. In addition, the Institute of Operations Research and the Management Sciences (INFORMS) awarded him and Dr. Hillier the honorable mention award for the 1995 Lanchester Prize for the 6th edition of this book. In 1996, INFORMS also awarded him the prestigious Kimball Medal for his exceptional contributions to the field of operations research and management science.

In addition to *Introduction to Operations Research* and two companion volumes, *Introduction to Mathematical Programming* (2nd ed., 1995) and *Introduction to Stochastic Models in Operations Research* (1990), his books are *Handbook of Industrial Statistics* (Prentice-Hall, 1955, co-authored by A. H. Bowker), *Tables of the Non-Central t-Distribution* (Stanford University Press, 1957, co-authored by G. J. Resnikoff), *Tables of the Hypergeometric Probability Distribution* (Stanford University Press, 1961, co-authored by D. Owen), *Engineering Statistics*, Second Edition (Prentice-Hall, 1972, co-authored by A. H. Bowker), and *Introduction to Management Science: A Modeling and Case Studies Approach with Spreadsheets* (McGraw-Hill/Irwin, 2000, co-authored by F. S. Hillier and M. S. Hillier).

ABOUT THE CASE WRITERS

Karl Schmedders is an associate professor in the Department of Managerial Economics and Decision Sciences at the Kellogg Graduate School of Management (Northwestern University), where he teaches quantitative methods for managerial decision making. His research interests include applications of operations research in economic theory, general equilibrium theory with incomplete markets, asset pricing, and computational economics. Dr. Schmedders received his doctorate in operations research from Stanford University, where he taught both undergraduate and graduate classes in operations research. Among the classes taught was a case studies course in operations research, and he subsequently was invited to speak at a conference sponsored by the Institute of Operations Research and the Management Sciences (INFORMS) about his successful experience with this course. He received several teaching awards at Stanford, including the university's prestigious Walter J. Gores Teaching Award. He also has received several teaching awards, including the L. G. Lavengood Professor of the Year at the Kellogg School of Management. While serving as a visiting professor at WHU Koblenz (a leading German business school), he won teaching awards there as well.

Molly Stephens is an associate in the Los Angeles office of Quinn, Emanuel, Urquhart, Oliver & Hedges, LLP. She graduated from Stanford University with a B.S. degree in Industrial Engineering and an M.S. degree in Operations Research. Ms. Stephens taught public speaking in Stanford's School of Engineering and served as a teaching assistant for a case studies course in operations research. As a teaching assistant, she analyzed operations research problems encountered in the real world and the transformation of these problems into classroom case studies. Her research was rewarded when she won an undergraduate research grant from Stanford to continue her work and was invited to speak at an INFORMS conference to present her conclusions regarding successful classroom case studies. Following graduation, Ms. Stephens worked at Andersen Consulting as a systems integrator, experiencing real cases from the inside, before resuming her graduate studies to earn a JD degree (with honors) from the University of Texas Law School at Austin.

DEDICATION

To the memory of our parents

and

To the memory of my beloved mentor,
Gerald J. Lieberman, who was one of the true
giants of our field

TABLE OF CONTENTS

CHAPTER 13
Metaheuristics   607

CHAPTER 14
Game Theory   651

CHAPTER 15
Decision Analysis   672

CHAPTER 16
Markov Chains   723

SUPPLEMENTS AVAILABLE ON THE TEXT WEBSITE
www.mhhe.com/hillier

PREFACE

When Jerry Lieberman and I started working on the first edition of this book 45 years ago, our goal was to develop a pathbreaking textbook that would help establish the future direction of education in what was then the emerging field of operations research. Following publication, it was unclear how well this particular goal was met, but what did become clear was that the demand for the book was far larger than either of us had anticipated. Neither of us could have imagined that this extensive worldwide demand would continue at such a high level for such an extended period of time.

The enthusiastic response to our first eight editions has been most gratifying. It was a particular pleasure to have the field's leading professional society, the international Institute for Operations Research and the Management Sciences (INFORMS), award the 6th edition honorable mention for the 1995 INFORMS Lanchester Prize (the prize awarded for the year's most outstanding English-language publication of any kind in the field of operations research).

Then, just after the publication of the eighth edition, it was especially gratifying to be the recipient of the prestigious 2004 INFORMS Expository Writing Award for this book, including receiving the following citation:

> Over 37 years, successive editions of this book have introduced more than one-half million students to the field and have attracted many people to enter the field for academic activity and professional practice. Many leaders in the field and many current instructors first learned about the field via an edition of this book. The extensive use of international student editions and translations into 15 other languages has contributed to spreading the field around the world. The book remains preeminent even after 37 years. Although the eighth edition just appeared, the seventh edition had 46 percent of the market for books of its kind, and it ranked second in international sales among all McGraw-Hill publications in engineering.
>
> Two features account for this success. First, the editions have been outstanding from students' points of view due to excellent motivation, clear and intuitive explanations, good examples of professional practice, excellent organization of material, very useful supporting software, and appropriate but not excessive mathematics. Second, the editions have been attractive from instructors' points of view because they repeatedly infuse state-of-the-art material with remarkable lucidity and plain language. For example, a wonderful chapter on metaheuristics was created for the eighth edition.

When we began work on the book 45 years ago, Jerry already was a prominent member of the field, a successful textbook writer, and the chairman of a renowned operations research program at Stanford University. I was a very young assistant professor just starting my career. It was a wonderful opportunity for me to work with and to learn from the master. I will be forever indebted to Jerry for giving me this opportunity.

Now, sadly, Jerry is no longer with us. During the progressive illness that led to his death nine years ago, I resolved that I would pick up the torch and devote myself to subsequent editions of this book, maintaining a standard that would fully honor Jerry. Therefore, I took early retirement from my faculty responsibilities at Stanford in order to work full time on textbook writing for the foreseeable future. This has enabled me to spend far more than the usual amount of time in preparing each new edition. It also has enabled me to closely monitor new trends and developments in the field in order to bring this edition completely up to date. This monitoring has led to the choice of the major revisions outlined below.

THE MAJOR REVISIONS

- **A Greatly Increased Emphasis on Real Applications.** Unbeknownst to the general public, the field of operations research is continuing to have an increasingly dramatic impact on the success of numerous companies and organizations around the world. Therefore, a special goal of this edition has been to tell this story much more forcefully, thereby exciting students about the great relevance of the material they are studying. We have pursued this goal in four ways. One is the *addition of 29 application vignettes* separated from the regular textual material that describe in a few paragraphs how an actual application of operations research had a powerful impact on a company or organization by using techniques like those being studied in that portion of the book. A second is the *addition of 71 selected references of award winning OR applications* given at the end of various chapters. A third is the *addition of a link to the journal articles that fully describe these 100 applications*, through a special arrangement with INFORMS. The final way is the *addition of many problems that require reading one or more of these articles*. Thus, the instructor now can motivate his or her lectures by having the students delve into real applications that dramatically demonstrate the relevance of the material being covered in the lectures.

  We are particularly excited about our new partnership with INFORMS, our field's preeminent professional society, to provide a link to these 100 articles describing dramatic OR applications. The Institute for Operations Research and the Management Sciences (INFORMS®) is a learned professional society for students, academics, and practitioners in quantitative and analytical fields. Information about INFORMS® journals, meetings, job bank, scholarships, awards, and teaching materials is at www.informs.org.

- **Approximately 200 New or Revised Problems.** The new problems include the ones involving real applications mentioned above. Other new problems also have been added, including a considerable number that support the new or revised topics mentioned later. Two new cases have been added for the chapter on decision analysis that are less complex than the two that already were there. In addition, many of the problems from the eighth edition have been revised. Therefore, an instructor who does not wish to assign problems that were assigned in previous classes has a substantial number from which to choose.

- **An Updating of the Software Accompanying the Book.** The next section will outline the wealth of software options that are provided with this new edition. The main difference from the eighth edition is that new, improved versions of several of the software packages now are available. For example, *Excel 2007* represents by far the most major revision of Excel and its user interface in many, many years, so this new version of Excel and its Solver has been fully integrated into the book (while pointing out differences for those still using old versions). Another important example is that, for the first time in 10 years, new versions of *TreePlan* and *SensIt* have just now become available and have been fully integrated into the decision analysis chapter. The latest versions of all the other software packages also are being provided with this new edition.

- **A New Section on Revenue Management.** A hallmark of new editions of this book has been the addition of substantial coverage of dramatic, recent developments that are beginning to revolutionize how certain areas of operations research are being practiced. For example, the eighth edition added a new chapter on metaheuristics, a new section on the incorporation of constraint programming, and a new section on multiechelon inventory models for supply chain management. This edition is adding another key new topic with the *addition of a complete section on revenue management in the chapter on inventory theory*. This is a timely addition because of the dramatic impact that revenue management has been having in the airline industry and now is beginning to have in several other industries.

- **A Reorganization of the Chapter on the Theory of the Simplex Method.** Some instructors do not wish to take the time to cover the revised simplex method but may still want to introduce the matrix form of the simplex method and may still want to cover what we call the "fundamental insight" regarding the simplex method. Therefore, rather than covering the revised simplex method in Section 5.2 before turning to the fundamental insight in Section 5.3—as in the eighth edition—we now simply introduce the matrix form of the simplex method in Section 5.2, which flows directly into the fundamental insight in Section 5.3, after which we focus on the revised simplex method as an optional topic in Section 5.4.

- **A Simplified Method for Determining Utilities.** Among the various other smaller revisions throughout the book, perhaps the most noteworthy is a simplified presentation in Section 15.6 of how to determine utilities. This is done through outlining a simple "equivalent lottery method."

- **A Reorganization to Reduce the Size of the Book.** An unfortunate trend with early editions of this book was that each new edition was significantly larger than the previous one. This continued until the seventh edition had become considerably larger than is desirable for an introductory survey textbook. Therefore, I worked hard to substantially reduce the size of the eighth edition and adopted the goal of avoiding any growth in subsequent editions. The goal has been achieved for the current edition. This was accomplished through a variety of means. One was being careful not to add too much new material. Another was deleting two sections on real applications that had been in the eighth edition but no longer were needed because of the addition of application vignettes. Another was moving both the long Appendix 3.1 on the LINGO modeling language and the section on optimizing with OptQuest to the supplements on the book's website. (This decision regarding OptQuest was made easy by the fact that a new version is due out momentarily, but not in time for this edition, so it will be added later as a supplement.) Finally, a considerable number of sections were shortened. Otherwise, I have stuck closely to what I hope has become the familiar organization of the eighth edition after having made major changes for that edition.

- **Updating to Reflect the Current State of the Art.** A special effort has been made to keep the book completely up to date. This has included carefully updating both the selected references at the end of each chapter and the various footnotes referencing the latest research on the topics being covered.

◼ A WEALTH OF SOFTWARE OPTIONS

A wealth of software options is being provided on the book's website www.mhhe .com/hillier as outlined below.

- Excel spreadsheets: state-of-the-art spreadsheet formulations are displayed in Excel files for all relevant examples throughout the book.
- Several Excel add-ins, including Premium Solver for Education (an enhancement of the basic Excel Solver), TreePlan (for decision analysis), SensIt (for probabilistic sensitivity analysis), RiskSim (for simulation), and Solver Table (for sensitivity analysis).
- A number of Excel templates for solving basic models.
- Student versions of LINDO (a traditional optimizer) and LINGO (a popular algebraic modeling language), along with formulations and solutions for all relevant examples throughout the book.

- Student versions of MPL (a leading algebraic modeling language) and its prime solver CPLEX (the most widely used state-of-the-art optimizer), along with an MPL Tutorial and MPL/CPLEX formulations and solutions for all relevant examples throughout the book.
- Student versions of several additional MPL solvers, including CONOPT (for convex programming), LGO (for global optimization), LINDO (for mathematical programming), CoinMP (for linear and integer programming), and BendX (for some stochastic models).
- Queueing Simulator (for the simulation of queueing systems).
- OR Tutor for illustrating various algorithms in action.
- Interactive Operations Research (IOR) Tutorial for efficiently learning and executing algorithms interactively, implemented in Java 2 in order to be platform independent.

Numerous students have found OR Tutor and IOR Tutorial very helpful for learning algorithms of operations research. When moving to the next stage of solving OR models automatically, surveys have found instructors almost equally split in preferring one of the following options for their students' use: (1) Excel spreadsheets, including the Excel Solver and other add-ins, (2) convenient traditional software (LINDO and LINGO), and (3) state-of-the-art OR software (MPL and CPLEX). For this edition, therefore, I have retained the philosophy of the last couple of editions of providing enough introduction in the book to enable the basic use of any of the three options without distracting those using another, while also providing ample supporting material for each option on the book's website.

We have elected to no longer include the Crystal Ball software package that was bundled with the eighth edition. Fortunately, many universities now have a site license for Crystal Ball and the package currently can also be downloaded for a free 30-day trial period, so it still is feasible to have students use this software, at least for a limited time. Therefore, this edition continues to use Crystal Ball in Section 20.6 and certain supplements to illustrate the exciting functionality that is now available for analyzing simulation models.

Additional Online Resources

- Several examples for nearly every book chapter are included in a *Worked Examples section* of the book's website to provide additional help to occasional students who need it without disrupting the flow of the text and adding unneeded pages for others. (The book uses boldface to highlight whenever an additional example on the current topic is available.)
- A *glossary* for every book chapter.
- *Data files* for various cases are included to enable students to focus on analysis rather than inputting large data sets.
- An abundance of supplementary textual material (including eight complete chapters).
- A *test bank* featuring moderately difficult questions that require students to show their work is being provided to instructors. Most of the questions in this test bank have previously been used successfully as test questions by the authors.
- Also available to instructors are a solutions manual and image files.

Electronic Textbook Option

This text is offered through CourseSmart for both instructors and students. CourseSmart is an online resource where students can purchase access to this and other McGraw-Hill textbooks in a digital format. Through their browser, students can access the complete text online at almost half the cost of a traditional text. Purchasing the eTextbook also allows students to take advantage of CourseSmart's web tools for learning, which include full text search, notes and highlighting, and e-mail tools for sharing notes between classmates.

To learn more about CourseSmart options, contact your sales representative or visit www.CourseSmart.com.

THE USE OF THE BOOK

The overall thrust of all the revision efforts has been to build upon the strengths of previous editions to more fully meet the needs of today's students. These revisions make the book even more suitable for use in a modern course that reflects contemporary practice in the field. The use of software is integral to the practice of operations research, so the wealth of software options accompanying the book provides great flexibility to the instructor in choosing the preferred types of software for student use. All the educational resources accompanying the book further enhance the learning experience. Therefore, the book and its website should fit a course where the instructor wants the students to have a single self-contained textbook that complements and supports what happens in the classroom.

The McGraw-Hill editorial team and I think that the net effect of the revision has been to make this edition even more of a "student's book"—clear, interesting, and well-organized with lots of helpful examples and illustrations, good motivation and perspective, easy-to-find important material, and enjoyable homework, without too much notation, terminology, and dense mathematics. We believe and trust that the numerous instructors who have used previous editions will agree that this is the best edition yet.

The prerequisites for a course using this book can be relatively modest. As with previous editions, the mathematics has been kept at a relatively elementary level. Most of Chaps. 1 to 14 (introduction, linear programming, and mathematical programming) require no mathematics beyond high school algebra. Calculus is used only in Chaps. 12 (Nonlinear Programming) and in one example in Chap. 10 (Dynamic Programming). Matrix notation is used in Chap. 5 (The Theory of the Simplex Method), Chap. 6 (Duality Theory and Sensitivity Analysis), Sec. 7.4 (An Interior-Point Algorithm), and Chap. 12, but the only background needed for this is presented in Appendix 4. For Chaps. 15 to 20 (probabilistic models), a previous introduction to probability theory is assumed, and calculus is used in a few places. In general terms, the mathematical maturity that a student achieves through taking an elementary calculus course is useful throughout Chaps. 15 to 20 and for the more advanced material in the preceding chapters.

The content of the book is aimed largely at the upper-division undergraduate level (including well-prepared sophomores) and at first-year (master's level) graduate students. Because of the book's great flexibility, there are many ways to package the material into a course. Chapters 1 and 2 give an introduction to the subject of operations research. Chapters 3 to 14 (on linear programming and on mathematical programming) may essentially be covered independently of Chaps. 15 to 20 (on probabilistic models), and vice-versa. Furthermore, the individual chapters among Chaps. 3 to 14 are almost independent, except that they all use basic material presented in Chap. 3 and perhaps in Chap. 4. Chapter 6 and Sec. 7.2 also draw upon Chap. 5. Sections 7.1 and 7.2 use parts of Chap. 6. Section 9.6 assumes an acquaintance with the problem formulations in Secs. 8.1 and 8.3, while prior exposure to Secs. 7.3 and 8.2 is helpful (but not essential) in Sec. 9.7. Within Chaps. 15 to 20, there is considerable flexibility of coverage, although some integration of the material is available.

An elementary survey course covering linear programming, mathematical programming, and some probabilistic models can be presented in a quarter (40 hours) or semester by selectively drawing from material throughout the book. For example, a good survey of the field can be obtained from Chaps. 1, 2, 3, 4, 15, 17, 18, and 20, along with parts of

Chaps. 9 to 13. A more extensive elementary survey course can be completed in two quarters (60 to 80 hours) by excluding just a few chapters, for example, Chaps. 7, 14, and 19. Chapters 1 to 8 (and perhaps part of Chap. 9) form an excellent basis for a (one-quarter) course in linear programming. The material in Chaps. 9 to 14 covers topics for another (one-quarter) course in other deterministic models. Finally, the material in Chaps. 15 to 20 covers the probabilistic (stochastic) models of operations research suitable for presentation in a (one-quarter) course. In fact, these latter three courses (the material in the entire text) can be viewed as a basic one-year sequence in the techniques of operations research, forming the core of a master's degree program. Each course outlined has been presented at either the undergraduate or graduate level at Stanford University, and this text has been used in the manner suggested.

The book's website will provide updates about the book, including an errata. To access this site, visit www.mhhe.com/hillier.

ACKNOWLEDGMENTS

I am indebted to an excellent group of reviewers who provided sage advice for the revision process. This group included

Chun-Hung Chen, George Mason University
Mary Court, University of Oklahoma
Todd Easton, Kansas State University
Samuel H. Huang, University of Cincinnati
Ronald Giachetti, Florida International University
Mary E. Kurz, Clemson University
Wooseung Jang, University of Missouri-Columbia
Shafiu Jibrin, Northern Arizona University
Roger Johnson, South Dakota School of Mines & Technology
Emanuel Melachrinoudis, Northeastern University
Clark A. Mount-Campbell, The Ohio State University
Jose A. Ventura, Pennsylvania State University
John Wu, Kansas State University

I also am grateful to Garrett Van Ryzin for his expert advice regarding the new section on revenue management, to Charles McCallum, Jr., for providing lists of typos in the 8th edition three times, and to Bjarni Kristjansson for providing up-to-date information on the sizes of problems being solved successfully by the latest optimization software. In addition, thanks go to those instructors and students who sent email messages to provide their feedback on the 8th edition.

This edition was very much of a team effort. Our case writers, Karl Schmedders and Molly Stephens (both graduates of our department), wrote 24 elaborate cases for the 7th edition, and all of these cases continue to accompany this new edition. One of our department's current PhD students, Pelin Canbolat, did an excellent job in preparing the solutions manual. She went above and beyond the call of duty by typing nearly all of the solutions that had been handwritten for preceding editions, as well as providing helpful input for this edition. One of our former PhD students, Michael O'Sullivan, developed OR Tutor for the 7th edition (and continued here), based on part of the software that my son Mark Hillier had developed for the 5th and 6th editions. Mark (who was born the same year as the first edition, earned his PhD at Stanford, and now is a tenured Associate Professor of Quantitative Methods at the University of Washington) provided both the spreadsheets and the Excel files (including many Excel templates) for this edition, as well as the Solver Table and Queueing Simulator. He also gave helpful advice on both the textual material and software for this edition, and contributed greatly to Chapters 21 and 28 on the book's website. Another Stanford PhD graduate, William Sun (CEO of the software company Accelet Corporation), and his team did a brilliant job of starting with much of Mark's earlier software and implementing it anew in Java 2 as IOR Tutorial for the 7th edition. They again did a masterful job of further enhancing IOR Tutorial for the 8th and subsequent editions. Linus Schrage of the University of Chicago and LINDO Systems (and who took an introductory operations research course from me 45 years ago) provided LINGO and LINDO for the book's website. He also supervised the further development of LINGO/LINDO files for the various chapters as well as providing tutorial material for the book's website. Another long-time friend, Bjarni Kristjansson (who heads Maximal Software), did the same thing for the MPL/CPLEX files and MPL tutorial material, as well as arranging to provide student versions of MPL, CPLEX, and various other solvers for the book's website. My wife, Ann Hillier, devoted numerous long days and nights to sitting with a Macintosh, doing word processing and constructing many figures and tables. They all were vital members of the team.

In addition to Accelet Corporation, LINDO Systems, and Maximal Software, we are deeply indebted to several other companies for providing software to accompany this edition. These include Frontline Systems (for providing Premium Solver for Education), ILOG (for providing the CPLEX solver used with the MPL Student Edition), ARKI Corporation (for providing the CONOPT convex programming solver used with the MPL Student Edition), and PCS Inc. (for providing the LGO global optimization solver used with the MPL Student Edition). We also are grateful to Professor Michael Middleton for providing newly improved versions of TreePlan and SensIt, as well as RiskSim. Finally, we appreciate the cooperation of INFORMS in providing a link to the articles in *Interfaces* that describe the applications of OR that are summarized in the application vignettes and other selected references of award winning OR applications provided in the book.

It was a real pleasure working with McGraw-Hill's thoroughly professional editorial and production staff, including Debra Hash (Sponsoring Editor) and Lora Kalb-Neyens (Developmental Editor).

Just as so many individuals made important contributions to this edition, I would like to invite each of you to start contributing to the next edition by using my email address below to send me your comments, suggestions, and errata to help me improve the book in the future. In giving my email address, let me also assure instructors that I will continue to follow the policy of not providing solutions to problems and cases in the book to anybody (including your students) who contacts me.

Enjoy the book.

Frederick S. Hillier
Stanford University (fhillier@stanford.edu)

May 2008

CHAPTER 1

Introduction

■ 1.1 THE ORIGINS OF OPERATIONS RESEARCH

Since the advent of the industrial revolution, the world has seen a remarkable growth in the size and complexity of organizations. The artisans' small shops of an earlier era have evolved into the billion-dollar corporations of today. An integral part of this revolutionary change has been a tremendous increase in the division of labor and segmentation of management responsibilities in these organizations. The results have been spectacular. However, along with its blessings, this increasing specialization has created new problems, problems that are still occurring in many organizations. One problem is a tendency for the many components of an organization to grow into relatively autonomous empires with their own goals and value systems, thereby losing sight of how their activities and objectives mesh with those of the overall organization. What is best for one component frequently is detrimental to another, so the components may end up working at cross purposes. A related problem is that as the complexity and specialization in an organization increase, it becomes more and more difficult to allocate the available resources to the various activities in a way that is most effective for the organization as a whole. These kinds of problems and the need to find a better way to solve them provided the environment for the emergence of **operations research** (commonly referred to as **OR**).

The roots of OR can be traced back many decades,[1] when early attempts were made to use a scientific approach in the management of organizations. However, the beginning of the activity called *operations research* has generally been attributed to the military services early in World War II. Because of the war effort, there was an urgent need to allocate scarce resources to the various military operations and to the activities within each operation in an effective manner. Therefore, the British and then the U.S. military management called upon a large number of scientists to apply a scientific approach to dealing with this and other strategic and tactical problems. In effect, they were asked to do *research on* (military) *operations*. These teams of scientists were the first OR teams. By developing effective methods of using the new tool of radar, these teams were instrumental in winning the Air Battle of Britain. Through their research on how to better manage convoy and antisubmarine

---

[1]Selected Reference 2 provides an entertaining history of operations research that traces its roots as far back as 1564 by describing a considerable number of scientific contributions from 1564 to 1935 that influenced the subsequent development of OR.

operations, they also played a major role in winning the Battle of the North Atlantic. Similar efforts assisted the Island Campaign in the Pacific.

When the war ended, the success of OR in the war effort spurred interest in applying OR outside the military as well. As the industrial boom following the war was running its course, the problems caused by the increasing complexity and specialization in organizations were again coming to the forefront. It was becoming apparent to a growing number of people, including business consultants who had served on or with the OR teams during the war, that these were basically the same problems that had been faced by the military but in a different context. By the early 1950s, these individuals had introduced the use of OR to a variety of organizations in business, industry, and government. The rapid spread of OR soon followed.

At least two other factors that played a key role in the rapid growth of OR during this period can be identified. One was the substantial progress that was made early in improving the techniques of OR. After the war, many of the scientists who had participated on OR teams or who had heard about this work were motivated to pursue research relevant to the field; important advancements in the state of the art resulted. A prime example is the *simplex method* for solving linear programming problems, developed by George Dantzig in 1947. Many of the standard tools of OR, such as linear programming, dynamic programming, queueing theory, and inventory theory, were relatively well developed before the end of the 1950s.

A second factor that gave great impetus to the growth of the field was the onslaught of the *computer revolution*. A large amount of computation is usually required to deal most effectively with the complex problems typically considered by OR. Doing this by hand would often be out of the question. Therefore, the development of electronic digital computers, with their ability to perform arithmetic calculations millions of times faster than a human being can, was a tremendous boon to OR. A further boost came in the 1980s with the development of increasingly powerful personal computers accompanied by good software packages for doing OR. This brought the use of OR within the easy reach of much larger numbers of people, and this progress further accelerated in the 1990s and into the 21st century. Today, literally millions of individuals have ready access to OR software. Consequently, a whole range of computers from mainframes to laptops now are being routinely used to solve OR problems, including some of enormous size.

■ 1.2   THE NATURE OF OPERATIONS RESEARCH

As its name implies, operations research involves "research on operations." Thus, operations research is applied to problems that concern how to conduct and coordinate the *operations* (i.e., the *activities*) within an organization. The nature of the organization is essentially immaterial, and, in fact, OR has been applied extensively in such diverse areas as manufacturing, transportation, construction, telecommunications, financial planning, health care, the military, and public services, to name just a few. Therefore, the breadth of application is unusually wide.

The *research* part of the name means that operations research uses an approach that resembles the way research is conducted in established scientific fields. To a considerable extent, the *scientific method* is used to investigate the problem of concern. (In fact, the term *management science* sometimes is used as a synonym for operations research.) In particular, the process begins by carefully observing and formulating the problem, including gathering all relevant data. The next step is to construct a scientific (typically mathematical) model that attempts to abstract the essence of the real problem. It is then hypothesized that this model is a sufficiently precise representation of the essential features of the situation that the conclusions (solutions) obtained from the model are also valid for the real problem. Next, suitable experiments are conducted to test this hypothesis, modify it as needed, and eventually verify some form of the hypothesis. (This step is frequently referred to as *model validation.*) Thus, in a certain sense, operations research involves creative scientific research into the fundamental properties of operations. However, there is more to it than this. Specifically, OR is also concerned with the practical management of the organization. Therefore, to be successful, OR must also provide positive, understandable conclusions to the decision maker(s) when they are needed.

Still another characteristic of OR is its broad viewpoint. As implied in the preceding section, OR adopts an organizational point of view. Thus, it attempts to resolve the conflicts of interest among the components of the organization in a way that is best for the organization as a whole. This does not imply that the study of each problem must give explicit consideration to all aspects of the organization; rather, the objectives being sought must be consistent with those of the overall organization.

An additional characteristic is that OR frequently attempts to search for a *best* solution (referred to as an *optimal* solution) for the model that represents the problem under consideration. (We say *a* best instead of *the* best solution because there may be multiple solutions tied as best.) Rather than simply improving the status quo, the goal is to identify a best possible course of action. Although it must be interpreted carefully in terms of the practical needs of management, this "search for optimality" is an important theme in OR.

All these characteristics lead quite naturally to still another one. It is evident that no single individual should be expected to be an expert on all the many aspects of OR work or the problems typically considered; this would require a group of individuals having diverse backgrounds and skills. Therefore, when a full-fledged OR study of a new problem is undertaken, it is usually necessary to use a *team approach.* Such an OR team typically needs to include individuals who collectively are highly trained in mathematics, statistics and probability theory, economics, business administration, computer science, engineering and the physical sciences, the behavioral sciences, and the special techniques of OR. The team also needs to have the necessary experience and variety of skills to give appropriate consideration to the many ramifications of the problem throughout the organization.

■ 1.3   THE IMPACT OF OPERATIONS RESEARCH

Operations research has had an impressive impact on improving the efficiency of numerous organizations around the world. In the process, OR has made a significant contribution to increasing the productivity of the economies of various countries. There now are a few dozen member countries in the International Federation of Operational Research Societies (IFORS), with each country having a national OR society. Both Europe and Asia have federations of OR societies to coordinate holding international conferences and publishing international journals in those continents. In addition, the Institute for Operations Research and the Management Sciences (INFORMS) is an international OR society. Among its various journals is one called *Interfaces* that regularly publishes articles describing major OR studies and the impact they had on their organizations.

To give you a better notion of the wide applicability of OR, we list some actual applications in Table 1.1. Note the diversity of organizations and applications in the first two columns. The third column identifies the section where an "application vignette" devotes several paragraphs to describing the application and also references an article that provides full details. (You can see the first of these application vignettes in this section.) The last column indicates that these applications typically resulted in annual savings in the many millions of dollars. Furthermore, additional benefits not recorded in the table

■ **TABLE 1.1** Applications of operations research to be described in application vignettes

Organization	Area of Application	Section	Annual Savings
Federal Express	Logistical planning of shipments	1.3	Not estimated
Continental Airlines	Reassign crews to flights when schedule disruptions occur	2.2	$40 million
Swift & Company	Improve sales and manufacturing performance	3.1	$12 million
Memorial Sloan-Kettering Cancer Center	Design of radiation therapy	3.4	$459 million
United Airlines	Plan employee work schedules at airports and reservations offices	3.4	$6 million
Welch's	Optimize use and movement of raw materials	3.6	$150,000
Samsung Electronics	Reduce manufacturing times and inventory levels	4.3	$200 million more revenue
Pacific Lumber Company	Long-term forest ecosystem management	6.7	$398 million NPV
Procter & Gamble	Redesign the production and distribution system	8.1	$200 million
Canadian Pacific Railway	Plan routing of rail freight	9.3	$100 million
United Airlines	Reassign airplanes to flights when disruptions occur	9.6	Not estimated
U.S. Military	Logistical planning of Operations Desert Storm	10.3	Not estimated
Air New Zealand	Airline crew scheduling	11.2	$6.7 million
Taco Bell	Plan employee work schedules at restaurants	11.5	$13 million
Waste Management	Develop a route-management system for trash collection and disposal	11.7	$100 million
Bank Hapoalim Group	Develop a decision-support system for investment advisors	12.1	$31 million more revenue
Sears	Vehicle routing and scheduling for home services and deliveries	13.2	$42 million
Conoco-Phillips	Evaluate petroleum exploration projects	15.2	Not estimated
Workers' Compensation Board	Manage high-risk disability claims and rehabilitation	15.3	$4 million
Westinghouse	Evaluate research-and-development projects	15.4	Not estimated
Merrill Lynch	Manage liquidity risk for revolving credit lines	16.2	$4 billion more liquidity
PSA Peugeot Citroën	Guide the design process for efficient car assembly plants	16.8	$130 million more profit
KeyCorp	Improve efficiency of bank teller service	17.6	$20 million
General Motors	Improve efficiency of production lines	17.9	$90 million
Deere & Company	Management of inventories throughout a supply chain	18.5	$1 billion less inventory
Time Inc.	Management of distribution channels for magazines	18.7	$3.5 million more profit
Bank One Corporation	Management of credit lines and interest rates for credit cards	19.2	$75 million more profit
Merrill Lynch	Pricing analysis for providing financial services	20.2	$50 million more revenue
AT&T	Design and operation of call centers	20.5	$750 million more profit

(e.g., improved service to customers and better managerial control) sometimes were considered to be even more important than these financial benefits. (You will have an opportunity to investigate these less tangible benefits further in Probs. 1.3-1, 1.3-2, and 1.3-3.) A link to the articles that describe these applications in detail is included on our website, www.mhhe.com/hillier.

Although most routine OR studies provide considerably more modest benefits than the applications summarized in Table 1.1, the figures in the rightmost column of this table do accurately reflect the dramatic impact that large, well-designed OR studies occasionally can have.

Federal Express (FedEx) is the world's largest express transportation company. Every working day, it delivers more than 6.5 million documents, packages, and other items throughout the United States and more than 220 countries and territories around the world. In some cases, these shipments can be guaranteed overnight delivery by 10:30 A.M. the next morning.

The logistical challenges involved in providing this service are staggering. These millions of daily shipments must be individually sorted and routed to the correct general location (usually by aircraft) and then delivered to the exact destination (usually by motorized vehicle) in an amazingly short period of time. How is all this possible?

Operations research (OR) is the technological engine that drives this company. Ever since its founding in 1973, OR has helped make its major business decisions, including equipment investment, route structure, scheduling, finances, and location of facilities. After OR was credited with literally saving the company during its early years, it became the custom to have OR represented at the weekly senior management meetings and, indeed, several of the senior corporate vice presidents have come up from the outstanding FedEx OR group.

FedEx has come to be acknowledged as a world-class company. It routinely ranks among the top companies on *Fortune Magazine*'s annual listing of the "World's Most Admired Companies." It also was the first winner (in 1991) of the prestigious prize now known as the INFORMS Prize, which is awarded annually for the effective and repeated integration of OR into organizational decision making in pioneering, varied, novel, and lasting ways.

Source: R. O. Mason, J. L. McKenney, W. Carlson, and D. Copeland, "Absolutely, Positively Operations Research: The Federal Express Story," *Interfaces*, **27**(2): 17–36, March-April 1997. (A link to this article is provided on our website, www.mhhe.com/hillier.)

1.4 ALGORITHMS AND OR COURSEWARE

An important part of this book is the presentation of the major **algorithms** (systematic solution procedures) of OR for solving certain types of problems. Some of these algorithms are amazingly efficient and are routinely used on problems involving hundreds or thousands of variables. You will be introduced to how these algorithms work and what makes them so efficient. You then will use these algorithms to solve a variety of problems on a computer. The **OR Courseware** contained on the book's website (www.mhhe.com/hillier) will be a key tool for doing all this.

One special feature in your OR Courseware is a program called **OR Tutor.** This program is intended to be your personal tutor to help you learn the algorithms. It consists of many *demonstration examples* that display and explain the algorithms in action. These "demos" supplement the examples in the book.

In addition, your OR Courseware includes a special software package called **Interactive Operations Research Tutorial,** or **IOR Tutorial** for short. Implemented in Java, this innovative package is designed specifically to enhance the learning experience of students using this book. IOR Tutorial includes many *interactive procedures* for executing the algorithms interactively in a convenient format. The computer does all the routine calculations while you focus on learning and executing the logic of the algorithm. You should find these interactive procedures a very efficient and enlightening way of doing many of your homework problems. IOR Tutorial also includes a number of other helpful procedures, including some *automatic procedures* for executing algorithms automatically and several procedures that provide graphical displays of how the solution provided by an algorithm varies with the data of the problem.

In practice, the algorithms normally are executed by commercial software packages. We feel that it is important to acquaint students with the nature of these packages that they will be using after graduation. Therefore, your OR Courseware includes a wealth of material to introduce you to three particularly popular software packages described next.

Together, these packages will enable you to solve nearly all the OR models encountered in this book very efficiently. We have added our own *automatic procedures* to IOR Tutorial in a few cases where these packages are not applicable.

A very popular approach now is to use today's premier spreadsheet package, *Microsoft Excel,* to formulate small OR models in a spreadsheet format. The **Excel Solver** (or an enhanced version of this add-in, such as **Premium Solver for Education** included in your OR Courseware) then is used to solve the models. Your OR Courseware includes separate Excel files, based on the relatively new Excel 2007, for nearly every chapter in this book. Each time a chapter presents an example that can be solved using Excel, the complete spreadsheet formulation and solution is given in that chapter's Excel files. For many of the models in the book, an *Excel template* also is provided that already includes all the equations necessary to solve the model. Some *Excel add-ins* also are included on the book's website.

After many years, **LINDO** (and its companion modeling language **LINGO**) continues to be a popular OR software package. Student versions of LINDO and LINGO now can be downloaded free from the Web. This student version also is provided in your OR Courseware. As for Excel, each time an example can be solved with this package, all the details are given in a LINGO/LINDO file for that chapter in your OR Courseware.

CPLEX is an elite state-of-the-art software package that is widely used for solving large and challenging OR problems. When dealing with such problems, it is common to also use a *modeling system* to efficiently formulate the mathematical model and enter it into the computer. **MPL** is a user-friendly modeling system that uses CPLEX as its main solver, but also has several other solvers, including LINDO, CoinMP (introduced in Sec. 4.8), CONOPT (introduced in Sec. 12.9), LGO (introduced in Sec. 12.10), and BendX (useful for solving some stochastic models). A student version of MPL, along with the latest student version of CPLEX and its other solvers, is available free by downloading it from the Web. For your convenience, we also have included this student version (including all the solvers just mentioned) in your OR Courseware. Once again, all the examples that can be solved with this package are detailed in MPL/CPLEX files for the corresponding chapters in your OR Courseware.

We will further describe these three software packages and how to use them later (especially near the end of Chaps. 3 and 4). Appendix 1 also provides documentation for the OR Courseware, including OR Tutor and IOR Tutorial.

To alert you to relevant material in OR Courseware, the end of each chapter from Chap. 3 onward has a list entitled *Learning Aids for This Chapter on our Website.* As explained at the beginning of the problem section for each of these chapters, symbols also are placed to the left of each problem number or part where any of this material (including demonstration examples and interactive procedures) can be helpful.

Another learning aid provided on our website is a set of **Worked Examples** for each chapter (from Chap. 3 onward). These complete examples supplement the examples in the book for your use as needed, but without interrupting the flow of the material on those many occasions when you don't need to see an additional example. You also might find these supplementary examples helpful when preparing for an examination. We always will mention whenever a supplementary example on the current topic is included in the Worked Examples section of the book's website. To make sure you don't overlook this mention, we will boldface the words **additional example** (or something similar) each time.

The website also includes a glossary for each chapter.

■ SELECTED REFERENCES

1. Bell, P. C., C. K. Anderson, and S. P. Kaiser: "Strategic Operations Research and the Edelman Prize Finalist Applications 1989–1998," *Operations Research*, **51**(1): 17–31, January–February 2003.
2. Gass, S. I., and A. A. Assad: *An Annotated Timeline of Operations Research: An Informal History*, Kluwer Academic Publishers (now Springer), Boston, 2005.
3. Gass, S. I., and C. M. Harris (eds.): *Encyclopedia of Operations Research and Management Science*, 2d ed., Kluwer Academic Publishers (now Springer), Boston, 2001.
4. Horner, P.: "History in the Making," *OR/MS Today*, **29**(5): 30–39, October 2002.
5. Horner, P. (ed.): "Special Issue: Executive's Guide to Operations Research," *OR/MS Today*, Institute for Operations Research and the Management Sciences, **27**(3), June 2000.
6. Kirby, M. W.: "Operations Research Trajectories: The Anglo-American Experience from the 1940s to the 1990s," *Operations Research*, **48**(5): 661–670, September–October 2000.
7. Miser, H. J.: "The Easy Chair: What OR/MS Workers Should Know About the Early Formative Years of Their Profession," *Interfaces*, **30**(2): 99–111, March–April 2000.
8. Wein, L. M. (ed.): "50th Anniversary Issue," *Operations Research* (a special issue featuring personalized accounts of some of the key early theoretical and practical developments in the field), **50**(1), January–February 2002.

■ PROBLEMS

1.3-1. Select one of the applications of operations research listed in Table 1.1. Read the article that is referenced in the application vignette presented in the section shown in the third column. (A link to all these articles is provided on our website, www.mhhe.com/hillier.) Write a two-page summary of the application and the benefits (including nonfinancial benefits) it provided.

1.3-2. Select three of the applications of operations research listed in Table 1.1. For each one, read the article that is referenced in the application vignette presented in the section shown in the third column. (A link to all these articles is provided on our website, www.mhhe.com/hillier.) For each one, write a one-page summary of the application and the benefits (including nonfinancial benefits) it provided.

1.3-3. Read the referenced article that fully describes the OR study summarized in the application vignette presented in Sec. 1.3. List the various financial and nonfinancial benefits that resulted from this study.

CHAPTER 2

Overview of the Operations Research Modeling Approach

The bulk of this book is devoted to the mathematical methods of operations research (OR). This is quite appropriate because these quantitative techniques form the main part of what is known about OR. However, it does not imply that practical OR studies are primarily mathematical exercises. As a matter of fact, the mathematical analysis often represents only a relatively small part of the total effort required. The purpose of this chapter is to place things into better perspective by describing all the major phases of a typical OR study.

One way of summarizing the usual (overlapping) phases of an OR study is the following:

1. Define the problem of interest and gather relevant data.
2. Formulate a mathematical model to represent the problem.
3. Develop a computer-based procedure for deriving solutions to the problem from the model.
4. Test the model and refine it as needed.
5. Prepare for the ongoing application of the model as prescribed by management.
6. Implement.

Each of these phases will be discussed in turn in the following sections.

The selected references at the end of the chapter include some award-winning OR studies that provide excellent examples of how to execute these phases well. We will intersperse snippets from some of these examples throughout the chapter. If you decide that you would like to learn more about these award-winning applications of operations research, a link to the articles that describe these OR studies in detail is included on the book's website, www.mhhe.com/hillier.

■ 2.1 DEFINING THE PROBLEM AND GATHERING DATA

In contrast to textbook examples, most practical problems encountered by OR teams are initially described to them in a vague, imprecise way. Therefore, the first order of business is to study the relevant system and develop a well-defined statement of the problem to be considered. This includes determining such things as the appropriate objectives, constraints on what can be done, interrelationships between the area to be studied and other areas of the organization, possible alternative courses of action, time limits for making a decision, and so on. This process of problem definition is a crucial one because it greatly affects how relevant the conclusions of the study will be. It is difficult to extract a "right" answer from the "wrong" problem!

The first thing to recognize is that an OR team normally works in an *advisory capacity*. The team members are not just given a problem and told to solve it however they see fit. Instead, they advise management (often one key decision maker). The team performs a detailed technical analysis of the problem and then presents recommendations to management. Frequently, the report to management will identify a number of alternatives that are particularly attractive under different assumptions or over a different range of values of some policy parameter that can be evaluated only by management (e.g., the trade-off between *cost* and *benefits*). Management evaluates the study and its recommendations, takes into account a variety of intangible factors, and makes the final decision based on its best judgment. Consequently, it is vital for the OR team to get on the same wavelength as management, including identifying the "right" problem from management's viewpoint, and to build the support of management for the course that the study is taking.

Ascertaining the *appropriate objectives* is a very important aspect of problem definition. To do this, it is necessary first to identify the member (or members) of management who actually will be making the decisions concerning the system under study and then to probe into this individual's thinking regarding the pertinent objectives. (Involving the decision maker from the outset also is essential to build her or his support for the implementation of the study.)

By its nature, OR is concerned with the welfare of the *entire organization* rather than that of only certain of its components. An OR study seeks solutions that are optimal for the overall organization rather than suboptimal solutions that are best for only one component. Therefore, the objectives that are formulated ideally should be those of the entire organization. However, this is not always convenient. Many problems primarily concern only a portion of the organization, so the analysis would become unwieldy if the stated objectives were too general and if explicit consideration were given to all side effects on the rest of the organization. Instead, the objectives used in the study should be as specific as they can be while still encompassing the main goals of the decision maker and maintaining a reasonable degree of consistency with the higher-level objectives of the organization.

For profit-making organizations, one possible approach to circumventing the problem of suboptimization is to use *long-run profit maximization* (considering the time value of money) as the sole objective. The adjective *long-run* indicates that this objective provides the flexibility to consider activities that do not translate into profits *immediately* (e.g., research and development projects) but need to do so *eventually* in order to be worthwhile. This approach has considerable merit. This objective is specific enough to be used conveniently, and yet it seems to be broad enough to encompass the basic goal of profit-making organizations. In fact, some people believe that all other legitimate objectives can be translated into this one.

However, in actual practice, many profit-making organizations do not use this approach. A number of studies of U.S. corporations have found that management tends to adopt the goal of *satisfactory profits,* combined with *other objectives,* instead of focusing on long-run profit maximization. Typically, some of these *other* objectives might be to maintain stable profits, increase (or maintain) one's share of the market, provide for product diversification, maintain stable prices, improve worker morale, maintain family control of the business, and increase company prestige. Fulfilling these objectives might achieve long-run profit maximization, but the relationship may be sufficiently obscure that it may not be convenient to incorporate them all into this one objective.

Furthermore, there are additional considerations involving social responsibilities that are distinct from the profit motive. The five parties generally affected by a business firm located in a single country are (1) the *owners* (stockholders, etc.), who desire profits (dividends, stock appreciation, and so on); (2) the *employees,* who desire steady employment at reasonable wages; (3) the *customers,* who desire a reliable product at a reasonable price; (4) the *suppliers,* who desire integrity and a reasonable selling price for their goods; and (5) the *government* and hence the *nation,* which desire payment of fair taxes and consideration of the national interest. All five parties make essential contributions to the firm, and the firm should not be viewed as the exclusive servant of any one party for the exploitation of others. By the same token, international corporations acquire additional obligations to follow socially responsible practices. Therefore, while granting that management's prime responsibility is to make profits (which ultimately benefits all five parties), we note that its broader social responsibilities also must be recognized.

OR teams typically spend a surprisingly large amount of time *gathering relevant data* about the problem. Much data usually are needed both to gain an accurate understanding of the problem and to provide the needed input for the mathematical model being formulated in the next phase of study. Frequently, much of the needed data will not be available when the study begins, either because the information never has been kept or because what was kept is outdated or in the wrong form. Therefore, it often is necessary to install a new computer-based *management information system* to collect the necessary data on an ongoing basis and in the needed form. The OR team normally needs to enlist the assistance of various other key individuals in the organization, including *information technology* (IT) specialists, to track down all the vital data. Even with this effort, much of the data may be quite "soft," i.e., rough estimates based only on educated guesses. Typically, an OR team will spend considerable time trying to improve the precision of the data and then will make do with the best that can be obtained.

With the widespread use of databases and the explosive growth in their sizes in recent years, OR teams now frequently find that their biggest data problem is not that too little is available but that there is too much data. There may be thousands of sources of data, and the total amount of data may be measured in gigabytes or even terabytes. In this environment, locating the particularly relevant data and identifying the interesting patterns in these data can become an overwhelming task. One of the newer tools of OR teams is a technique called **data mining** that addresses this problem. Data mining methods search large databases for interesting patterns that may lead to useful decisions. (Selected Reference 2 at the end of the chapter provides further background about data mining.)

Example.   In the late 1990s, full-service financial services firms came under assault from electronic brokerage firms offering extremely low trading costs. **Merrill Lynch** responded by conducting a major OR study that led to a complete overhaul in how it charged for its services, ranging from a full-service asset-based option (charge a fixed percentage of the value of the assets held rather than for individual trades) to a low-cost option for clients wishing to invest online directly. *Data collection and processing* played a key role in the study. To analyze the impact of individual client behavior in response to different options, the team needed to assemble a comprehensive 200 gigabyte client database involving 5 million clients, 10 million accounts, 100 million trade records, and 250 million ledger records. This required merging, reconciling, filtering, and cleaning data from numerous production databases. The adoption of the recommendations of the study led to a one-year increase of nearly $50 billion in client assets held and nearly $80 million more revenue. (Selected Reference A2 describes this study in detail.)

■ 2.2 FORMULATING A MATHEMATICAL MODEL

After the decision maker's problem is defined, the next phase is to reformulate this problem in a form that is convenient for analysis. The conventional OR approach for doing this is to construct a mathematical model that represents the essence of the problem. Before discussing how to formulate such a model, we first explore the nature of models in general and of mathematical models in particular.

Models, or idealized representations, are an integral part of everyday life. Common examples include model airplanes, portraits, globes, and so on. Similarly, models play an important role in science and business, as illustrated by models of the atom, models of genetic structure, mathematical equations describing physical laws of motion or chemical reactions, graphs, organizational charts, and industrial accounting systems. Such models are invaluable for abstracting the essence of the subject of inquiry, showing interrelationships, and facilitating analysis.

Mathematical models are also idealized representations, but they are expressed in terms of mathematical symbols and expressions. Such laws of physics as $F = ma$ and $E = mc^2$ are familiar examples. Similarly, the mathematical model of a business problem is the system of equations and related mathematical expressions that describe the essence of the problem. Thus, if there are n related quantifiable decisions to be made, they are represented as **decision variables** (say, $x_1, x_2, \ldots, x_n$) whose respective values are to be determined. The appropriate measure of performance (e.g., profit) is then expressed as a mathematical function of these decision variables (for example, $P = 3x_1 + 2x_2 + \ldots + 5x_n$). This function is called the **objective function.** Any restrictions on the values that can be assigned to these decision variables are also expressed mathematically, typically by means of inequalities or equations (for example, $x_1 + 3x_1x_2 + 2x_2 \leq 10$). Such mathematical expressions for the restrictions often are called **constraints.** The constants (namely, the coefficients and right-hand sides) in the constraints and the objective function are called the **parameters** of the model. The mathematical model might then say that the problem is to choose the values of the decision variables so as to maximize the objective function, subject to the specified constraints. Such a model, and minor variations of it, typifies the models used in OR.

Determining the appropriate values to assign to the parameters of the model (one value per parameter) is both a critical and a challenging part of the model-building process. In contrast to textbook problems where the numbers are given to you, determining parameter values for real problems requires *gathering relevant data.* As discussed in the preceding section, gathering accurate data frequently is difficult. Therefore, the value assigned to a parameter often is, of necessity, only a rough estimate. Because of the uncertainty about the true value of the parameter, it is important to analyze how the solution derived from the model would change (if at all) if the value assigned to the parameter were changed to other plausible values. This process is referred to as **sensitivity analysis,** as discussed further in the next section (and much of Chap. 6).

Although we refer to "the" mathematical model of a business problem, real problems normally don't have just a single "right" model. Section 2.4 will describe how the process of testing a model typically leads to a succession of models that provide better and better representations of the problem. It is even possible that two or more completely different types of models may be developed to help analyze the same problem.

You will see numerous examples of mathematical models throughout the remainder of this book. One particularly important type that is studied in the next several chapters is the **linear programming model,** where the mathematical functions appearing in both the objective function and the constraints are all linear functions. In Chap. 3, specific linear programming models are constructed to fit such diverse problems as determining (1) the mix of products that maximizes profit, (2) the design of radiation therapy that effectively attacks a tumor while minimizing the damage to nearby healthy tissue, (3) the allocation of acreage to crops that maximizes total net return, and (4) the combination of pollution abatement methods that achieves air quality standards at minimum cost.

Mathematical models have many advantages over a verbal description of the problem. One advantage is that a mathematical model describes a problem much more concisely. This tends to make the overall structure of the problem more comprehensible, and it helps to reveal important cause-and-effect relationships. In this way, it indicates more clearly what additional data are relevant to the analysis. It also facilitates dealing with the problem in its entirety and considering all its interrelationships simultaneously. Finally, a mathematical model forms a bridge to the use of high-powered mathematical techniques and computers to analyze the problem. Indeed, packaged software for both personal computers and main-frame computers has become widely available for solving many mathematical models.

However, there are pitfalls to be avoided when you use mathematical models. Such a model is necessarily an abstract idealization of the problem, so approximations and sim-plifying assumptions generally are required if the model is to be *tractable* (capable of being solved). Therefore, care must be taken to ensure that the model remains a valid rep-resentation of the problem. The proper criterion for judging the validity of a model is whether the model predicts the relative effects of the alternative courses of action with sufficient accuracy to permit a sound decision. Consequently, it is not necessary to include unimportant details or factors that have approximately the same effect for all the alternative courses of action considered. It is not even necessary that the absolute magni-tude of the measure of performance be approximately correct for the various alternatives, provided that their relative values (i.e., the differences between their values) are suffi-ciently precise. Thus, all that is required is that there be a high *correlation* between the prediction by the model and what would actually happen in the real world. To ascertain whether this requirement is satisfied, it is important to do considerable *testing* and conse-quent modifying of the model, which will be the subject of Sec. 2.4. Although this testing phase is placed later in the chapter, much of this *model validation* work actually is con-ducted during the model-building phase of the study to help guide the construction of the mathematical model.

In developing the model, a good approach is to begin with a very simple version and then move in evolutionary fashion toward more elaborate models that more nearly reflect the complexity of the real problem. This process of *model enrichment* continues only as long as the model remains tractable. The basic trade-off under constant consideration is between the *precision* and the *tractability* of the model. (See Selected Reference 8 for a detailed description of this process.)

A crucial step in formulating an OR model is the construction of the objective func-tion. This requires developing a quantitative measure of performance relative to each of the decision maker's ultimate objectives that were identified while the problem was being defined. If there are multiple objectives, their respective measures commonly are then transformed and combined into a composite measure, called the **overall measure of per-formance.** This overall measure might be something tangible (e.g., profit) corresponding to a higher goal of the organization, or it might be abstract (e.g., utility). In the latter case, the task of developing this measure tends to be a complex one requiring a careful compar-ison of the objectives and their relative importance. After the overall measure of perfor-mance is developed, the objective function is then obtained by expressing this measure as a mathematical function of the decision variables. Alternatively, there also are methods for explicitly considering multiple objectives simultaneously, and one of these (goal program-ming) is discussed in the supplement to Chap. 7.

Continental Airlines is a major U.S. air carrier that transports passengers, cargo, and mail. It operates more than 2,000 daily departures to well over 100 domestic destinations and nearly 100 foreign destinations.

Airlines like Continental face schedule disruptions daily because of unexpected events, including inclement weather, aircraft mechanical problems, and crew unavailability. These disruptions can cause flight delays and cancellations. As a result, crews may not be in position to service their remaining scheduled flights. Airlines must reassign crews quickly to cover open flights and to return them to their original schedules in a cost-effective manner while honoring all government regulations, contractual obligations, and quality-of-life requirements.

To address such problems, an OR team at Continental Airlines developed a detailed *mathematical model* for reassigning crews to flights as soon as such emergencies arise. Because the airline has thousands of crews and daily flights, the model needed to be huge to consider all possible pairings of crews with flights. Therefore, the model has *millions of decision variables* and *many thousands of constraints.* In

its first year of use (mainly in 2001), the model was applied four times to recover from major schedule disruptions (two snowstorms, a flood, and the September 11 terrorist attacks). This led to *savings of approximately* **$40 million.** Subsequent applications extended to many daily minor disruptions as well.

Although other airlines subsequently scrambled to apply operations research in a similar way, this initial advantage over other airlines in being able to recover more quickly from schedule disruptions with fewer delays and cancelled flights left Continental Airlines in a relatively strong position as the airline industry struggled through a difficult period during the initial years of the 21st century. This initiative led to Continental winning the prestigious First Prize in the 2002 international competition for the Franz Edelman Award for Achievement in Operations Research and the Management Sciences.

Source: G. Yu, M. Argüello, C. Song, S. M. McGowan, and A. White, "A New Era for Crew Recovery at Continental Airlines," *Interfaces,* 33(1): 5–22, Jan.–Feb. 2003. (A link to this article is provided on our website, www.mhhe.com/hillier.)

Example. The Netherlands government agency responsible for water control and public works, the **Rijkswaterstaat,** commissioned a major OR study to guide the development of a new national water management policy. The new policy saved hundreds of millions of dollars in investment expenditures and reduced agricultural damage by about $15 million per year, while decreasing thermal and algae pollution. Rather than formulating *one* mathematical model, this OR study developed a comprehensive, integrated system of 50 models! Furthermore, for some of the models, both simple and complex versions were developed. The simple version was used to gain basic insights, including trade-off analyses. The complex version then was used in the final rounds of the analysis or whenever greater accuracy or more detailed outputs were desired. The overall OR study directly involved over 125 person-years of effort (more than one-third in data gathering), created several dozen computer programs, and structured an enormous amount of data. (Selected Reference A7 describes this study in detail.)

■ 2.3 DERIVING SOLUTIONS FROM THE MODEL

After a mathematical model is formulated for the problem under consideration, the next phase in an OR study is to develop a procedure (usually a computer-based procedure) for deriving solutions to the problem from this model. You might think that this must be the major part of the study, but actually it is not in most cases. Sometimes, in fact, it is a relatively simple step, in which one of the standard **algorithms** (systematic solution procedures) of OR is applied on a computer by using one of a number of readily available software packages. For experienced OR practitioners, finding a solution is the fun part, whereas the real work comes in the preceding and following steps, including the *postoptimality analysis* discussed later in this section.

Since much of this book is devoted to the subject of how to obtain solutions for various important types of mathematical models, little needs to be said about it here. However, we do need to discuss the nature of such solutions.

A common theme in OR is the search for an **optimal,** or best, **solution.** Indeed, many procedures have been developed, and are presented in this book, for finding such solutions for certain kinds of problems. However, it needs to be recognized that these solutions are optimal only with respect to the model being used. Since the model necessarily is an idealized rather than an exact representation of the real problem, there cannot be any utopian guarantee that the optimal solution for the model will prove to be the best possible solution that could have been implemented for the real problem. There just are too many imponderables and uncertainties associated with real problems. However, if the model is well formulated and tested, the resulting solution should tend to be a good approximation to an ideal course of action for the real problem. Therefore, rather than be deluded into demanding the impossible, you should make the test of the practical success of an OR study hinge on whether it provides a better guide for action than can be obtained by other means.

Eminent management scientist and Nobel Laureate in economics Herbert Simon points out that **satisficing** is much more prevalent than optimizing in actual practice. In coining the term *satisficing* as a combination of the words *satisfactory* and *optimizing,* Simon is describing the tendency of managers to seek a solution that is "good enough" for the problem at hand. Rather than trying to develop an overall measure of performance to optimally reconcile conflicts between various desirable objectives (including well-established criteria for judging the performance of different segments of the organization), a more pragmatic approach may be used. Goals may be set to establish minimum satisfactory levels of performance in various areas, based perhaps on past levels of performance or on what the competition is achieving. If a solution is found that enables all these goals to be met, it is likely to be adopted without further ado. Such is the nature of satisficing.

The distinction between optimizing and satisficing reflects the difference between theory and the realities frequently faced in trying to implement that theory in practice. In the words of one of England's pioneering OR leaders, Samuel Eilon, "Optimizing is the science of the ultimate; satisficing is the art of the feasible."[1]

OR teams attempt to bring as much of the "science of the ultimate" as possible to the decision-making process. However, the successful team does so in full recognition of the overriding need of the decision maker to obtain a satisfactory guide for action in a reasonable period of time. Therefore, the goal of an OR study should be to conduct the study in an optimal manner, regardless of whether this involves finding an optimal solution for the model. Thus, in addition to pursuing the science of the ultimate, the team should also consider the cost of the study and the disadvantages of delaying its completion, and then attempt to maximize the net benefits resulting from the study. In recognition of this concept, OR teams occasionally use only **heuristic procedures** (i.e., intuitively designed procedures that do not guarantee an optimal solution) to find a good **suboptimal solution.** This is most often the case when the time or cost required to find an optimal solution for an adequate model of the problem would be very large. In recent years, great progress has been made in developing efficient and effective **metaheuristics** that provide both a general structure and strategy guidelines for designing a specific heuristic procedure to fit a particular kind of problem. The use of metaheuristics (the subject of Chap. 13) is continuing to grow.

---

[1]S. Eilon, "Goals and Constraints in Decision-making," *Operational Research Quarterly,* **23:** 3–15, 1972. Address given at the 1971 annual conference of the Canadian Operational Research Society.

The discussion thus far has implied that an OR study seeks to find only one solution, which may or may not be required to be optimal. In fact, this usually is not the case. An optimal solution for the original model may be far from ideal for the real problem, so additional analysis is needed. Therefore, **postoptimality analysis** (analysis done after finding an optimal solution) is a very important part of most OR studies. This analysis also is sometimes referred to as **what-if analysis** because it involves addressing some questions about *what* would happen to the optimal solution *if* different assumptions are made about future conditions. These questions often are raised by the managers who will be making the ultimate decisions rather than by the OR team.

The advent of powerful spreadsheet software now has frequently given spreadsheets a central role in conducting postoptimality analysis. One of the great strengths of a spreadsheet is the ease with which it can be used interactively by anyone, including managers, to see what happens to the optimal solution when changes are made to the model. This process of experimenting with changes in the model also can be very helpful in providing understanding of the behavior of the model and increasing confidence in its validity.

In part, postoptimality analysis involves conducting **sensitivity analysis** to determine which parameters of the model are most critical (the "sensitive parameters") in determining the solution. A common definition of *sensitive parameter* (used throughout this book) is the following.

> For a mathematical model with specified values for all its parameters, the model's **sensitive parameters** are the parameters whose value cannot be changed without changing the optimal solution.

Identifying the sensitive parameters is important, because this identifies the parameters whose value must be assigned with special care to avoid distorting the output of the model.

The value assigned to a parameter commonly is just an *estimate* of some quantity (e.g., unit profit) whose exact value will become known only after the solution has been implemented. Therefore, after the sensitive parameters are identified, special attention is given to estimating each one more closely, or at least its range of likely values. One then seeks a solution that remains a particularly good one for all the various combinations of likely values of the sensitive parameters.

If the solution is implemented on an ongoing basis, any later change in the value of a sensitive parameter immediately signals a need to change the solution.

In some cases, certain parameters of the model represent policy decisions (e.g., resource allocations). If so, there frequently is some flexibility in the values assigned to these parameters. Perhaps some can be increased by decreasing others. Postoptimality analysis includes the investigation of such trade-offs.

In conjunction with the study phase discussed in Sec. 2.4 (testing the model), postoptimality analysis also involves obtaining a sequence of solutions that comprises a series of improving approximations to the ideal course of action. Thus, the apparent weaknesses in the initial solution are used to suggest improvements in the model, its input data, and perhaps the solution procedure. A new solution is then obtained, and the cycle is repeated. This process continues until the improvements in the succeeding solutions become too small to warrant continuation. Even then, a number of alternative solutions (perhaps solutions that are optimal for one of several plausible versions of the model and its input data) may be presented to management for the final selection. As suggested in Sec. 2.1, this presentation of alternative solutions would normally be done whenever the final choice among these alternatives should be based on considerations that are best left to the judgment of management.

Example.   Consider again the **Rijkswaterstaat** OR study of national water management policy for the Netherlands, introduced at the end of Sec. 2.2. This study did not conclude by recommending just a single solution. Instead, a number of attractive alternatives were identified, analyzed, and compared. The final choice was left to the Dutch political process, culminating with approval by Parliament. *Sensitivity analysis* played a major role in this study. For example, certain parameters of the models represented environmental standards. Sensitivity analysis included assessing the impact on water management problems if the values of these parameters were changed from the current environmental standards to other reasonable values. Sensitivity analysis also was used to assess the impact of changing the assumptions of the models, e.g., the assumption on the effect of future international treaties on the amount of pollution entering the Netherlands. A variety of *scenarios* (e.g., an extremely dry year and an extremely wet year) also were analyzed, with appropriate probabilities assigned.

■ 2.4   TESTING THE MODEL

Developing a large mathematical model is analogous in some ways to developing a large computer program. When the first version of the computer program is completed, it inevitably contains many bugs. The program must be thoroughly tested to try to find and correct as many bugs as possible. Eventually, after a long succession of improved programs, the programmer (or programming team) concludes that the current program now is generally giving reasonably valid results. Although some minor bugs undoubtedly remain hidden in the program (and may never be detected), the major bugs have been sufficiently eliminated that the program now can be reliably used.

Similarly, the first version of a large mathematical model inevitably contains many flaws. Some relevant factors or interrelationships undoubtedly have not been incorporated into the model, and some parameters undoubtedly have not been estimated correctly. This is inevitable, given the difficulty of communicating and understanding all the aspects and subtleties of a complex operational problem as well as the difficulty of collecting reliable data. Therefore, before you use the model, it must be thoroughly tested to try to identify and correct as many flaws as possible. Eventually, after a long succession of improved models, the OR team concludes that the current model now is giving reasonably valid results. Although some minor flaws undoubtedly remain hidden in the model (and may never be detected), the major flaws have been sufficiently eliminated so that the model now can be reliably used.

This process of testing and improving a model to increase its validity is commonly referred to as **model validation.**

It is difficult to describe how model validation is done, because the process depends greatly on the nature of the problem being considered and the model being used. However, we make a few general comments, and then we give an example. (See Selected Reference 3 for a detailed discussion.)

Since the OR team may spend months developing all the detailed pieces of the model, it is easy to "lose the forest for the trees." Therefore, after the details ("the trees") of the initial version of the model are completed, a good way to begin model validation is to take a fresh look at the overall model ("the forest") to check for obvious errors or oversights. The group doing this review preferably should include at least one individual who did not participate in the formulation of the model. Reexamining the definition of the problem and comparing it with the model may help to reveal mistakes. It is also useful to make sure that all the mathematical expressions are *dimensionally consistent* in the units used. Additional insight into the validity of the model can sometimes be obtained by varying the values of the parameters and/or the decision variables and checking to see whether the output from the model behaves in a plausible manner. This is often especially revealing when the parameters or variables are assigned extreme values near their maxima or minima.

A more systematic approach to testing the model is to use a **retrospective test.** When it is applicable, this test involves using historical data to reconstruct the past and then determining how well the model and the resulting solution would have performed if they had been used. Comparing the effectiveness of this hypothetical performance with what actually happened then indicates whether using this model tends to yield a significant improvement over current practice. It may also indicate areas where the model has short-comings and requires modifications. Furthermore, by using alternative solutions from the model and estimating their hypothetical historical performances, considerable evidence can be gathered regarding how well the model predicts the relative effects of alternative courses of actions.

On the other hand, a disadvantage of retrospective testing is that it uses the same data that guided the formulation of the model. The crucial question is whether the past is truly representative of the future. If it is not, then the model might perform quite differently in the future than it would have in the past.

To circumvent this disadvantage of retrospective testing, it is sometimes useful to con-tinue the status quo temporarily. This provides new data that were not available when the model was constructed. These data are then used in the same ways as those described here to evaluate the model.

Documenting the process used for model validation is important. This helps to increase confidence in the model for subsequent users. Furthermore, if concerns arise in the future about the model, this documentation will be helpful in diagnosing where prob-lems may lie.

Example.   Consider an OR study done for **IBM** to integrate its national network of spare-parts inventories to improve service support for IBM's customers. This study resulted in a new inventory system that improved customer service while reducing the value of IBM's inventories by over $250 million and saving an additional $20 million per year through improved operational efficiency. A particularly interesting aspect of the model validation phase of this study was the way that *future users* of the inventory system were incorporated into the testing process. Because these future users (IBM managers in functional areas responsible for implementation of the inventory system) were skeptical about the system being developed, representatives were appointed to a *user team* to serve as advisers to the OR team. After a preliminary version of the new system had been developed (based on a multiechelon inventory model), a *preimplementation test* of the system was conducted. Extensive feedback from the user team led to major improvements in the proposed system. (Selected Reference A5 describes this study in detail.)

■ 2.5   PREPARING TO APPLY THE MODEL

What happens after the testing phase has been completed and an acceptable model has been developed? If the model is to be used repeatedly, the next step is to install a well-documented *system* for applying the model as prescribed by management. This system will include the model, solution procedure (including postoptimality analysis), and operating procedures for implementation. Then, even as personnel changes, the system can be called on at regu-lar intervals to provide a specific numerical solution.

This system usually is *computer-based*. In fact, a considerable number of computer programs often need to be used and integrated. *Databases* and *management information systems* may provide up-to-date input for the model each time it is used, in which case interface programs are needed. After a solution procedure (another program) is applied to the model, additional computer programs may trigger the implementation of the results automatically. In other cases, an *interactive* computer-based system called a **decision support system** is installed to help managers use data and models to support (rather than replace) their decision making as needed. Another program may generate *managerial reports* (in the language of management) that interpret the output of the model and its implications for application.

In major OR studies, several months (or longer) may be required to develop, test, and install this computer system. Part of this effort involves developing and implementing a process for maintaining the system throughout its future use. As conditions change over time, this process should modify the computer system (including the model) accordingly.

Example.   The application vignette in Sec. 2.2 described an OR study done for **Continental Airlines** that led to the formulation of a huge mathematical model for reassigning crews to flights when schedule disruptions occur. Because the model needs to be applied immediately when a disruption occurs, a *decision support system* called *CrewSolver* was developed to incorporate both the model and a huge in-memory data store representing current operations. CrewSolver enables a crew coordinator to input data about the schedule disruption and then to use a graphical user interface to request an immediate solution for how to reassign crews to flights.

2.6   IMPLEMENTATION

After a system is developed for applying the model, the last phase of an OR study is to implement this system as prescribed by management. This phase is a critical one because it is here, and only here, that the benefits of the study are reaped. Therefore, it is important for the OR team to participate in launching this phase, both to make sure that model solutions are accurately translated to an operating procedure and to rectify any flaws in the solutions that are then uncovered.

The success of the implementation phase depends a great deal upon the support of both top management and operating management. The OR team is much more likely to gain this support if it has kept management well informed and encouraged management's active guidance throughout the course of the study. Good communications help to ensure that the study accomplishes what management wanted, and also give management a greater sense of ownership of the study, which encourages their support for implementation.

The implementation phase involves several steps. First, the OR team gives operating management a careful explanation of the new system to be adopted and how it relates to operating realities. Next, these two parties share the responsibility for developing the procedures required to put this system into operation. Operating management then sees that a detailed indoctrination is given to the personnel involved, and the new course of action is initiated. If successful, the new system may be used for years to come. With this in mind, the OR team monitors the initial experience with the course of action taken and seeks to identify any modifications that should be made in the future.

Throughout the entire period during which the new system is being used, it is important to continue to obtain feedback on how well the system is working and whether the assumptions of the model continue to be satisfied. When significant deviations from the original assumptions occur, the model should be revisited to determine if any modifications should be made in the system. The postoptimality analysis done earlier (as described in Sec. 2.3) can be helpful in guiding this review process.

Upon culmination of a study, it is appropriate for the OR team to *document* its methodology clearly and accurately enough so that the work is *reproducible. Replicability* should be part of the professional ethical code of the operations researcher. This condition is especially crucial when controversial public policy issues are being studied.

Example. This example illustrates how a successful implementation phase might need to involve thousands of employees before undertaking the new procedures. **Samsung Electronics Corp.** initiated a major OR study in March 1996 to develop new methodologies and scheduling applications that would streamline the entire semiconductor manufacturing process and reduce work-in-progress inventories. The study continued for over five years, culminating in June 2001, largely because of the extensive effort required for the implementation phase. The OR team needed to gain the support of numerous managers, manufacturing staff, and engineering staff by training them in the principles and logic of the new manufacturing procedures. Ultimately, more than 3,000 people attended training sessions. The new procedures then were phased in gradually to build confidence. However, this patient implementation process paid huge dividends. The new procedures transformed the company from being the least efficient manufacturer in the semiconductor industry to becoming the most efficient. This resulted in increased revenues of over $1 billion by the time the implementation of the OR study was completed. (Selected Reference A11 describes this study in detail.)

■ 2.7 CONCLUSIONS

Although the remainder of this book focuses primarily on *constructing* and *solving* mathematical models, in this chapter we have tried to emphasize that this constitutes only a portion of the overall process involved in conducting a typical OR study. The other phases described here also are very important to the success of the study. Try to keep in perspective the role of the model and the solution procedure in the overall process as you move through the subsequent chapters. Then, after gaining a deeper understanding of mathematical models, we suggest that you plan to return to review this chapter again in order to further sharpen this perspective.

OR is closely intertwined with the use of computers. In the early years, these generally were mainframe computers, but now personal computers and workstations are being widely used to solve OR models.

In concluding this discussion of the major phases of an OR study, it should be emphasized that there are many exceptions to the "rules" prescribed in this chapter. By its very nature, OR requires considerable ingenuity and innovation, so it is impossible to write down any standard procedure that should always be followed by OR teams. Rather, the preceding description may be viewed as a model that roughly represents how successful OR studies are conducted.

■ SELECTED REFERENCES

1. Board, J., C. Sutcliffe, and W. T. Ziemba: "Applying Operations Research Techniques to Financial Markets," *Interfaces,* **33**(2): 12–24, March–April 2003.
2. Bradley, P. S., U. M. Fayyad, and O. L. Mangasarian: "Mathematical Programming for Data Mining: Formulations and Challenges," *INFORMS Journal on Computing,* **11**(3): 217–238, Summer 1999.
3. Gass, S. I.: "Decision-Aiding Models: Validation, Assessment, and Related Issues for Policy Analysis," *Operations Research,* **31:** 603–631, 1983.
4. Gass, S. I.: "Model World: Danger, Beware the User as Modeler," *Interfaces,* **20**(3): 60–64, May–June 1990.
5. Hall, R. W.: "What's So Scientific about MS/OR?" *Interfaces,* **15**(2): 40–45, March–April 1985.
6. Howard, R. A.: "The Ethical OR/MS Professional," *Interfaces,* **31**(6): 69–82, November–December 2001.
7. Miser, H. J.: "The Easy Chair: Observation and Experimentation," *Interfaces,* **19**(5): 23–30, September–October 1989.
8. Morris, W. T.: "On the Art of Modeling," *Management Science,* **13:** B707–717, 1967.

9. Murphy, F. H.: "The Occasional Observer: Some Simple Precepts for Project Success," *Interfaces,* **28**(5): 25–28, September–October 1998.

10. Murphy, F. H.: "ASP, The Art and Science of Practice: Elements of the Practice of Operations Research: A Framework," *Interfaces,* **35**(2): 154–163, March–April 2005.

11. Pidd, M.: "Just Modeling Through: A Rough Guide to Modeling," *Interfaces,* **29**(2): 118–132, March–April 1999.

12. Williams, H. P.: *Model Building in Mathematical Programming,* 4th ed., Wiley, New York, 1999.

13. Wright, P. D., M. J. Liberatore, and R. L. Nydick: "A Survey of Operations Research Models and Applications in Homeland Security," *Interfaces,* **36**(6): 514–529, November–December 2006.

Some Award-Winning Applications of the OR Modeling Approach:

(A link to all these articles is provided on our website, www.mhhe.com/hillier.)

A1. Alden, J. M., L. D. Burns, T. Costy, R. D. Hutton, C. A. Jackson, D. S. Kim, K. A. Kohls, J. H. Owen, M. A. Turnquist, and D. J. V. Veen: "General Motors Increases Its Production Throughput," *Interfaces,* **36**(1): 6–25, January–February 2006.

A2. Altschuler, S., D. Batavia, J. Bennett, R. Labe, B. Liao, R. Nigam, and J. Oh: "Pricing Analysis for Merrill Lynch Integrated Choice," *Interfaces,* **32**(1): 5–19, January–February 2002.

A3. Bixby, A., B. Downs, and M. Self: "A Scheduling and Capable-to-Promise Application for Swift & Company," *Interfaces,* **36**(1): 69–86, January–February 2006.

A4. Braklow, J. W., W. W. Graham, S. M. Hassler, K. E. Peck, and W. B. Powell: "Interactive Optimization Improves Service and Performance for Yellow Freight System," *Interfaces,* **22**(1): 147–172, January–February 1992.

A5. Cohen, M., P. V. Kamesam, P. Kleindorfer, H. Lee, and A. Tekerian: "Optimizer: IBM's Multi-Echelon Inventory System for Managing Service Logistics," *Interfaces,* **20**(1): 65–82, January–February 1990.

A6. DeWitt, C. W., L. S. Lasdon, A. D. Waren, D. A. Brenner, and S. A. Melhem: "OMEGA: An Improved Gasoline Blending System for Texaco," *Interfaces,* **19**(1): 85–101, January–February 1990.

A7. Goeller, B. F., and the PAWN team: "Planning the Netherlands' Water Resources," *Interfaces,* **15**(1): 3–33, January–February 1985.

A8. Hicks, R., R. Madrid, C. Milligan, R. Pruneau, M. Kanaley, Y. Dumas, B. Lacroix, J. Desrosiers, and F. Soumis: "Bombardier Flexjet Significantly Improves Its Fractional Aircraft Ownership Operations," *Interfaces,* **35**(1): 49–60, January–February 2005.

A9. Kaplan, E. H., and E. O'Keefe: "Let the Needles Do the Talking! Evaluating the New Haven Needle Exchange," *Interfaces,* **23**(1): 7–26, January–February 1993.

A10. Kok, T. de, F. Janssen, J. van Doremalen, E. van Wachem, M. Clerkx, and W. Peeters: "Philips Electronics Synchronizes Its Supply Chain to End the Bullwhip Effect," *Interfaces,* **35**(1): 37–48, January–February 2005.

A11. Leachman, R. C., J. Kang, and V. Lin: "SLIM: Short Cycle Time and Low Inventory in Manufacturing at Samsung Electronics," *Interfaces,* **32**(1): 61–77, January–February 2002.

A12. Taylor, P. E., and S. J. Huxley: "A Break from Tradition for the San Francisco Police: Patrol Officer Scheduling Using an Optimization-Based Decision Support System," *Interfaces,* **19**(1): 4–24, January–February 1989.

■ PROBLEMS

2.1-1. The example in Sec. 2.1 summarizes an award-winning OR study done for Merrill Lynch. Read Selected Reference A2 that describes this study in detail.

(a) Summarize the background that led to undertaking this study.

(b) Quote the one-sentence statement of the general mission of the OR group (called the management science group) that conducted this study.

(c) Identify the type of data that the management science group obtained for each client.

(d) Identify the new pricing options that were provided to the company's clients as a result of this study.

(e) What was the resulting impact on Merrill Lynch's competitive position?

2.1-2. Read Selected Reference A1 that describes an award-winning OR study done for General Motors.

(a) Summarize the background that led to undertaking this study.

(b) What was the goal of this study?

(c) Describe how software was used to automate the collection of the needed data.

(d) The improved production throughput that resulted from this study yielded how much in documented savings and increased revenue?

2.1-3. Read Selected Reference A12 that describes an OR study done for the San Francisco Police Department.

(a) Summarize the background that led to undertaking this study.

(b) Define part of the problem being addressed by identifying the six directives for the scheduling system to be developed.

(c) Describe how the needed data were gathered.

(d) List the various tangible and intangible benefits that resulted from the study.

2.1-4. Read Selected Reference A9 that describes an OR study done for the Health Department of New Haven, Connecticut.

(a) Summarize the background that led to undertaking this study.

(b) Outline the system developed to track and test each needle and syringe in order to gather the needed data.

(c) Summarize the initial results from this tracking and testing system.

(d) Describe the impact and potential impact of this study on public policy.

2.2-1. Read the referenced article that fully describes the OR study summarized in the application vignette presented in Sec. 2.2. List the various financial and nonfinancial benefits that resulted from this study.

2.2-2. Read Selected Reference A3 that describes an OR study done for Swift & Company.

(a) Summarize the background that led to undertaking this study.

(b) Describe the purpose of each of the three general types of models formulated during this study.

(c) How many specific models does the company now use as a result of this study?

(d) List the various financial and nonfinancial benefits that resulted from this study.

2.2-3. Read Selected Reference A7 that describes an OR study done for the Rijkswaterstaat of the Netherlands. (Focus especially on pp. 3–20 and 30–32.)

(a) Summarize the background that led to undertaking this study.

(b) Summarize the purpose of each of the five mathematical models described on pp. 10–18.

(c) Summarize the "impact measures" (measures of performance) for comparing policies that are described on pp. 6–7 of this article.

(d) List the various tangible and intangible benefits that resulted from the study.

2.2-4. Read Selected Reference 5.

(a) Identify the author's example of a model in the natural sciences and of a model in OR.

(b) Describe the author's viewpoint about how basic precepts of using models to do research in the natural sciences can also be used to guide *research on operations* (OR).

2.3-1. Read Selected Reference A10 that describes an OR study done for Philips Electronics.

(a) Summarize the background that led to undertaking this study.

(b) What was the purpose of this study?

(c) What were the benefits of developing software to support problem solving speedily?

(d) List the four steps in the collaborative-planning process that resulted from this study.

(e) List the various financial and nonfinancial benefits that resulted from this study.

2.3-2. Refer to Selected Reference 5.

(a) Describe the author's viewpoint about whether the sole goal in using a model should be to find its optimal solution.

(b) Summarize the author's viewpoint about the complementary roles of modeling, evaluating information from the model, and then applying the decision maker's judgment when deciding on a course of action.

2.4-1. Refer to pp. 18–20 of Selected Reference A7 that describes an OR study done for the Rijkswaterstaat of the Netherlands. Describe an important lesson that was gained from model validation in this study.

2.4-2. Read Selected Reference 7. Summarize the author's viewpoint about the roles of observation and experimentation in the model validation process.

2.4-3. Read pp. 603–617 of Selected Reference 3.

(a) What does the author say about whether a model can be completely validated?

(b) Summarize the distinctions made between *model validity, data validity, logical/mathematical validity, predictive validity, operational validity,* and *dynamic validity.*

(c) Describe the role of *sensitivity analysis* in testing the *operational validity* of a model.

(d) What does the author say about whether there is a validation methodology that is appropriate for all models?

(e) Cite the page in the article that lists basic validation steps.

2.5-1. Read Selected Reference A6 that describes an OR study done for Texaco.

(a) Summarize the background that led to undertaking this study.

(b) Briefly describe the user interface with the decision support system OMEGA that was developed as a result of this study.

(c) OMEGA is constantly being updated and extended to reflect changes in the operating environment. Briefly describe the various kinds of changes involved.

(d) Summarize how OMEGA is used.

(e) List the various tangible and intangible benefits that resulted from the study.

2.5-2. Refer to Selected Reference A4 that describes an OR study done for Yellow Freight System, Inc.

(a) Referring to pp. 147–149 of this article, summarize the background that led to undertaking this study.

(b) Referring to p. 150, briefly describe the computer system SYSNET that was developed as a result of this study. Also summarize the applications of SYSNET.

(c) Referring to pp. 162–163, describe why the *interactive* aspects of SYSNET proved important.

(d) Referring to p. 163, summarize the outputs from SYSNET.

(e) Referring to pp. 168–172, summarize the various benefits that have resulted from using SYSNET.

2.6-1. Refer to pp. 163–167 of Selected Reference A4 that describes an OR study done for Yellow Freight System, Inc., and the resulting computer system SYSNET.

(a) Briefly describe how the OR team gained the support of upper management for implementing SYSNET.

(b) Briefly describe the implementation strategy that was developed.

(c) Briefly describe the field implementation.

(d) Briefly describe how management incentives and enforcement were used in implementing SYSNET.

2.6-2. Read Selected Reference A5 that describes an OR study done for IBM and the resulting computer system Optimizer.

(a) Summarize the background that led to undertaking this study.

(b) List the complicating factors that the OR team members faced when they started developing a model and a solution algorithm.

(c) Briefly describe the preimplementation test of Optimizer.

(d) Briefly describe the field implementation test.

(e) Briefly describe national implementation.

(f) List the various tangible and intangible benefits that resulted from the study.

2.7-1. From the bottom part of the selected references given at the end of the chapter, select one of these award-winning applications of the OR modeling approach (excluding any that have been assigned for other problems). Read this article and then write a two-page summary of the application and the benefits (including nonfinancial benefits) it provided.

2.7-2. From the bottom part of the selected references given at the end of the chapter, select three of these award-winning applications of the OR modeling approach (excluding any that have been assigned for other problems). For each one, read this article and write a one-page summary of the application and the benefits (including nonfinancial benefits) it provided.

2.7-3. Read Selected Reference 4. The author describes 13 detailed phases of any OR study that develops and applies a computer-based model, whereas this chapter describes six broader phases. For each of these broader phases, list the detailed phases that fall partially or primarily within the broader phase.

CHAPTER

3

Introduction to Linear Programming

The development of linear programming has been ranked among the most important scientific advances of the mid-20th century, and we must agree with this assessment. Its impact since just 1950 has been extraordinary. Today it is a standard tool that has saved many thousands or millions of dollars for many companies or businesses of even moderate size in the various industrialized countries of the world, and its use in other sectors of society has been spreading rapidly. A major proportion of all scientific computation on computers is devoted to the use of linear programming. Dozens of textbooks have been written about linear programming, and *published* articles describing important applications now number in the hundreds.

What is the nature of this remarkable tool, and what kinds of problems does it address? You will gain insight into this topic as you work through subsequent examples. However, a verbal summary may help provide perspective. Briefly, the most common type of application involves the general problem of allocating *limited resources* among *competing activities* in a best possible (i.e., *optimal*) way. More precisely, this problem involves selecting the level of certain activities that compete for scarce resources that are necessary to perform those activities. The choice of activity levels then dictates how much of each resource will be consumed by each activity. The variety of situations to which this description applies is diverse, indeed, ranging from the allocation of production facilities to products to the allocation of national resources to domestic needs, from portfolio selection to the selection of shipping patterns, from agricultural planning to the design of radiation therapy, and so on. However, the one common ingredient in each of these situations is the necessity for allocating resources to activities by choosing the levels of those activities.

Linear programming uses a mathematical model to describe the problem of concern. The adjective *linear* means that all the mathematical functions in this model are required to be *linear functions*. The word *programming* does not refer here to computer programming; rather, it is essentially a synonym for *planning*. Thus, linear programming involves the *planning of activities* to obtain an optimal result, i.e., a result that reaches the specified goal best (according to the mathematical model) among all feasible alternatives.

Although allocating resources to activities is the most common type of application, linear programming has numerous other important applications as well. In fact, *any* problem whose mathematical model fits the very general format for the linear programming model is a linear programming problem. (For this reason, a linear programming problem and its model often are referred to interchangeably as simply a *linear program,* or even as just an *LP.*) Furthermore, a remarkably efficient solution procedure, called the **simplex method,** is available for solving linear programming problems of even enormous size. These are some of the reasons for the tremendous impact of linear programming in recent decades.

Because of its great importance, we devote this and the next six chapters specifically to linear programming. After this chapter introduces the general features of linear programming, Chaps. 4 and 5 focus on the simplex method. Chapter 6 discusses the further analysis of linear programming problems *after* the simplex method has been initially applied. Chapter 7 presents several widely used extensions of the simplex method and introduces an *interior-point algorithm* that sometimes can be used to solve even larger linear programming problems than the simplex method can handle. Chapters 8 and 9 consider some special types of linear programming problems whose importance warrants individual study.

You also can look forward to seeing applications of linear programming to other areas of operations research (OR) in several later chapters.

We begin this chapter by developing a miniature prototype example of a linear programming problem. This example is small enough to be solved graphically in a straightforward way. Sections 3.2 and 3.3 present the general *linear programming model* and its basic assumptions. Section 3.4 gives some additional examples of linear programming applications. Section 3.5 describes how linear programming models of modest size can be conveniently displayed and solved on a spreadsheet. However, some linear programming problems encountered in practice require truly *massive* models. Section 3.6 illustrates how a massive model can arise and how it can still be formulated successfully with the help of a special modeling language such as MPL (its formulation is described in this section) or LINGO (its formulation of this model is presented in Supplement 2 to this chapter on the book's website).

■ 3.1 PROTOTYPE EXAMPLE

The WYNDOR GLASS CO. produces high-quality glass products, including windows and glass doors. It has three plants. Aluminum frames and hardware are made in Plant 1, wood frames are made in Plant 2, and Plant 3 produces the glass and assembles the products.

Because of declining earnings, top management has decided to revamp the company's product line. Unprofitable products are being discontinued, releasing production capacity to launch two new products having large sales potential:

Product 1: An 8-foot glass door with aluminum framing
Product 2: A 4 × 6 foot double-hung wood-framed window

Product 1 requires some of the production capacity in Plants 1 and 3, but none in Plant 2. Product 2 needs only Plants 2 and 3. The marketing division has concluded that the company could sell as much of either product as could be produced by these plants. However, because both products would be competing for the same production capacity in Plant 3, it is not clear which *mix* of the two products would be *most profitable.* Therefore, an OR team has been formed to study this question.

The OR team began by having discussions with upper management to identify management's objectives for the study. These discussions led to developing the following definition of the problem:

Determine what the *production rates* should be for the two products in order to *maximize their total profit,* subject to the restrictions imposed by the limited production capacities available in the three plants. (Each product will be produced in batches of 20, so the *production rate* is defined as the number of batches produced per week.) *Any* combination of production rates that satisfies these restrictions is permitted, including producing none of one product and as much as possible of the other.

The OR team also identified the data that needed to be gathered:

1. Number of hours of production time available per week in each plant for these new products. (Most of the time in these plants already is committed to current products, so the available capacity for the new products is quite limited.)
2. Number of hours of production time used in each plant for each batch produced of each new product.
3. Profit per batch produced of each new product. (*Profit per batch produced* was chosen as an appropriate measure after the team concluded that the incremental profit from each additional batch produced would be roughly *constant* regardless of the total number of batches produced. Because no substantial costs will be incurred to initiate the production and marketing of these new products, the total profit from each one is approximately this *profit per batch produced* times *the number of batches produced.*)

Obtaining reasonable estimates of these quantities required enlisting the help of key personnel in various units of the company. Staff in the manufacturing division provided the data in the first category above. Developing estimates for the second category of data required some analysis by the manufacturing engineers involved in designing the production processes for the new products. By analyzing cost data from these same engineers and the marketing division, along with a pricing decision from the marketing division, the accounting department developed estimates for the third category.

Table 3.1 summarizes the data gathered.

The OR team immediately recognized that this was a linear programming problem of the classic **product mix** type, and the team next undertook the formulation of the corresponding mathematical model.

■ **TABLE 3.1** Data for the Wyndor Glass Co. problem

| | Production Time per Batch, Hours | | |
| | Product | | Production Time |
Plant	1	2	Available per Week, Hours
1	1	0	4
2	0	2	12
3	3	2	18
Profit per batch	$3,000	$5,000	

Formulation as a Linear Programming Problem

The definition of the problem given above indicates that the decisions to be made are the number of batches of the respective products to be produced per week so as to maximize their total profit. Therefore, to formulate the mathematical (linear programming) model for this problem, let

x_1 = number of batches of product 1 produced per week

x_2 = number of batches of product 2 produced per week

Z = total profit per week (in thousands of dollars) from producing these
two products

Thus, x_1 and x_2 are the *decision variables* for the model. Using the bottom row of Table 3.1, we obtain

$$Z = 3x_1 + 5x_2.$$

The objective is to choose the values of x_1 and x_2 so as to *maximize* $Z = 3x_1 + 5x_2$, subject to the restrictions imposed on their values by the limited production capacities available in the three plants. Table 3.1 indicates that each batch of product 1 produced per week uses 1 hour of production time per week in Plant 1, whereas only 4 hours per week are available. This restriction is expressed mathematically by the inequality $x_1 \leq 4$. Similarly, Plant 2 imposes the restriction that $2x_2 \leq 12$. The number of hours of production time used per week in Plant 3 by choosing x_1 and x_2 as the new products' production rates would be $3x_1 + 2x_2$. Therefore, the mathematical statement of the Plant 3 restriction is $3x_1 + 2x_2 \leq 18$. Finally, since production rates cannot be negative, it is necessary to restrict the decision variables to be nonnegative: $x_1 \geq 0$ and $x_2 \geq 0$.

To summarize, in the mathematical language of linear programming, the problem is to choose values of x_1 and x_2 so as to

$$\text{Maximize} \quad Z = 3x_1 + 5x_2,$$

subject to the restrictions

$$x_1 \qquad\quad \leq 4$$
$$2x_2 \leq 12$$
$$3x_1 + 2x_2 \leq 18$$

and

$$x_1 \geq 0, \quad x_2 \geq 0.$$

(Notice how the layout of the coefficients of x_1 and x_2 in this linear programming model essentially duplicates the information summarized in Table 3.1.)

Graphical Solution

This very small problem has only two decision variables and therefore only two dimensions, so a graphical procedure can be used to solve it. This procedure involves constructing a two-dimensional graph with x_1 and x_2 as the axes. The first step is to identify the values of (x_1, x_2) that are permitted by the restrictions. This is done by drawing each line that borders the range of permissible values for one restriction. To begin, note that the nonnegativity restrictions $x_1 \geq 0$ and $x_2 \geq 0$ require (x_1, x_2) to lie on the *positive* side of the axes (including actually *on* either axis), i.e., in the first quadrant. Next, observe that the restriction $x_1 \leq 4$ means that (x_1, x_2) cannot lie to the right of the line $x_1 = 4$. These results are shown in Fig. 3.1, where the shaded area contains the only values of (x_1, x_2) that are still allowed.

In a similar fashion, the restriction $2x_2 \leq 12$ (or, equivalently, $x_2 \leq 6$) implies that the line $2x_2 = 12$ should be added to the boundary of the permissible region. The final restriction, $3x_1 + 2x_2 \leq 18$, requires plotting the points (x_1, x_2) such that $3x_1 + 2x_2 = 18$ (another line) to complete the boundary. (Note that the points such that $3x_1 + 2x_2 \leq 18$ are those that lie either underneath or on the line $3x_1 + 2x_2 = 18$, so this is the limiting line above which points do not satisfy the inequality.) The resulting region of permissible values of (x_1, x_2), called the **feasible region,** is shown in Fig. 3.2. (The demo called *Graphical Method* in your OR Tutor provides a more detailed example of constructing a feasible region.)

The final step is to pick out the point in this feasible region that maximizes the value of $Z = 3x_1 + 5x_2$. To discover how to perform this step efficiently, begin by trial and error. Try, for example, $Z = 10 = 3x_1 + 5x_2$ to see if there are in the permissible region any values of (x_1, x_2) that yield a value of Z as large as 10. By drawing the line $3x_1 + 5x_2 = 10$ (see Fig. 3.3), you can see that there are many points on this line that lie within the region. Having gained perspective by trying this arbitrarily chosen value of $Z = 10$, you should next try a larger arbitrary value of Z, say, $Z = 20 = 3x_1 + 5x_2$. Again, Fig. 3.3 reveals that a segment of the line $3x_1 + 5x_2 = 20$ lies within the region, so that the maximum permissible value of Z must be at least 20.

▓ **FIGURE 3.1**
Shaded area shows values of (x_1, x_2) allowed by $x_1 \geq 0$, $x_2 \geq 0$, $x_1 \leq 4$.

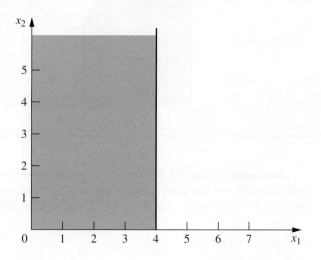

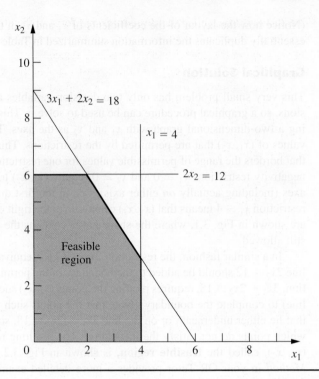

■ FIGURE 3.2
Shaded area shows the set of permissible values of (x_1, x_2), called the feasible region.

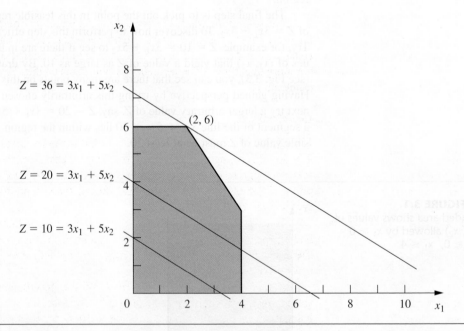

■ FIGURE 3.3
The value of (x_1, x_2) that maximizes $3x_1 + 5x_2$ is $(2, 6)$.

Now notice in Fig. 3.3 that the two lines just constructed are parallel. This is no coincidence, since *any* line constructed in this way has the form $Z = 3x_1 + 5x_2$ for the chosen value of Z, which implies that $5x_2 = -3x_1 + Z$ or, equivalently,

$$x_2 = -\frac{3}{5}x_1 + \frac{1}{5}Z$$

This last equation, called the **slope-intercept form** of the objective function, demonstrates that the *slope* of the line is $-\frac{3}{5}$ (since each unit increase in x_1 changes x_2 by $-\frac{3}{5}$), whereas the *intercept* of the line with the x_2 axis is $\frac{1}{5}Z$ (since $x_2 = \frac{1}{5}Z$ when $x_1 = 0$). The fact that the slope is fixed at $-\frac{3}{5}$ means that *all* lines constructed in this way are parallel.

Again, comparing the $10 = 3x_1 + 5x_2$ and $20 = 3x_1 + 5x_2$ lines in Fig. 3.3, we note that the line giving a larger value of Z ($Z = 20$) is farther up and away from the origin than the other line ($Z = 10$). This fact also is implied by the slope-intercept form of the objective function, which indicates that the intercept with the x_1 axis $(\frac{1}{5}Z)$ increases when the value chosen for Z is increased.

These observations imply that our trial-and-error procedure for constructing lines in Fig. 3.3 involves nothing more than drawing a family of parallel lines containing at least one point in the feasible region and selecting the line that corresponds to the largest value of Z. Figure 3.3 shows that this line passes through the point $(2, 6)$, indicating that the **optimal solution** is $x_1 = 2$ and $x_2 = 6$. The equation of this line is $3x_1 + 5x_2 = 3(2) + 5(6) = 36 = Z$, indicating that the optimal value of Z is $Z = 36$. The point $(2, 6)$ lies at the intersection of the two lines $2x_2 = 12$ and $3x_1 + 2x_2 = 18$, shown in Fig. 3.2, so that this point can be calculated algebraically as the simultaneous solution of these two equations.

Having seen the trial-and-error procedure for finding the optimal point $(2, 6)$, you now can streamline this approach for other problems. Rather than draw several parallel lines, it is sufficient to form a single line with a ruler to establish the slope. Then move the ruler with fixed slope through the feasible region in the direction of improving Z. (When the objective is to *minimize Z,* move the ruler in the direction that *decreases Z.*) Stop moving the ruler at the last instant that it still passes through a point in this region. This point is the desired *optimal solution.*

This procedure often is referred to as the **graphical method** for linear programming. It can be used to solve any linear programming problem with two decision variables. With considerable difficulty, it is possible to extend the method to three decision variables but not more than three. (The next chapter will focus on the *simplex method* for solving larger problems.)

Conclusions

The OR team used this approach to find that the optimal solution is $x_1 = 2$, $x_2 = 6$, with $Z = 36$. This solution indicates that the Wyndor Glass Co. should produce products 1 and 2 at the rate of 2 batches per week and 6 batches per week, respectively, with a resulting total profit of \$36,000 per week. No other mix of the two products would be so profitable— *according to the model.*

However, we emphasized in Chap. 2 that well-conducted OR studies do not simply find *one* solution for the *initial* model formulated and then stop. All six phases described in Chap. 2 are important, including thorough testing of the model (see Sec. 2.4) and postoptimality analysis (see Sec. 2.3).

In full recognition of these practical realities, the OR team now is ready to evaluate the validity of the model more critically (to be continued in Sec. 3.3) and to perform sensitivity analysis on the effect of the estimates in Table 3.1 being different because of inaccurate estimation, changes of circumstances, etc. (to be continued in Sec. 6.7).

Continuing the Learning Process with Your OR Courseware

This is the first of many points in the book where you may find it helpful to use your *OR Courseware* on the book's website. A key part of this courseware is a program called **OR Tutor.** This program includes a complete demonstration example of the *graphical method* introduced in this section. To provide you with **another example** of a model formulation as well, this demonstration begins by introducing a problem and formulating a linear programming model for the problem before then applying the graphical method step by step to solve the model. Like the many other demonstration examples accompanying other sections of the book, this computer demonstration highlights concepts that are difficult to convey on the printed page. You may refer to Appendix 1 for documentation of the software.

If you would like to see still **more examples,** you can go to the **Worked Examples** section of the book's website. This section includes a few examples with complete solutions for almost every chapter as a supplement to the examples in the book and in OR Tutor. The examples for the current chapter begin with a relatively straightforward problem that involves formulating a small linear programming model and applying the graphical method. The subsequent examples become progressively more challenging.

Another key part of your OR Courseware is a program called **IOR Tutorial.** This program features many interactive procedures for interactively executing various solution methods presented in the book, which enables you to focus on learning and executing the logic of the method efficiently while the computer does the number crunching. Included is an interactive procedure for applying the graphical method for linear programming. Once you get the hang of it, a second procedure enables you to quickly apply the graphical method for performing sensitivity analysis on the effect of revising the data of the problem. You then can print out your work and results for your homework. Like the other procedures in IOR Tutorial, these procedures are designed specifically to provide you with an efficient, enjoyable, and enlightening learning experience while you do your homework.

When you formulate a linear programming model with more than two decision variables (so the graphical method cannot be used), the *simplex method* described in Chap. 4 enables you to still find an optimal solution immediately. Doing so also is helpful for *model validation,* since finding a *nonsensical* optimal solution signals that you have made a mistake in formulating the model.

We mentioned in Sec. 1.4 that your OR Courseware introduces you to three particularly popular commercial software packages—the Excel Solver, LINGO/LINDO, and MPL/CPLEX—for solving a variety of OR models. All three packages include the simplex method for solving linear programming models. Section 3.5 describes how to use Excel to formulate and solve linear programming models in a spreadsheet format. Descriptions of the other packages are provided in Sec. 3.6 (MPL and LINGO), Supplements 1 and 2 to this chapter on the book's website (LINGO), Sec. 4.8 (CPLEX and LINDO), and Appendix 4.1 (LINGO and LINDO). MPL, LINGO, and LINDO tutorials also are provided on the book's website. In addition, your OR Courseware includes a file for each of the three packages showing how it can be used to solve each of the examples in this chapter.

■ 3.2   THE LINEAR PROGRAMMING MODEL

The Wyndor Glass Co. problem is intended to illustrate a typical linear programming problem (miniature version). However, linear programming is too versatile to be completely characterized by a single example. In this section we discuss the general characteristics of linear programming problems, including the various legitimate forms of the mathematical model for linear programming.

Let us begin with some basic terminology and notation. The first column of Table 3.2 summarizes the components of the Wyndor Glass Co. problem. The second column then introduces more general terms for these same components that will fit many linear programming problems. The key terms are *resources* and *activities,* where m denotes the number of different kinds of resources that can be used and n denotes the number of activities being considered. Some typical resources are money and particular kinds of machines,

TABLE 3.2 Common terminology for linear programming

Prototype Example	General Problem
Production capacities of plants 3 plants	Resources m resources
Production of products 2 products Production rate of product j, x_j	Activities n activities Level of activity j, x_j
Profit Z	Overall measure of performance Z

equipment, vehicles, and personnel. Examples of activities include investing in particular projects, advertising in particular media, and shipping goods from a particular source to a particular destination. In any application of linear programming, all the activities may be of one general kind (such as any one of these three examples), and then the individual activities would be particular alternatives within this general category.

As described in the introduction to this chapter, the most common type of application of linear programming involves allocating resources to activities. The amount available of each resource is limited, so a careful allocation of resources to activities must be made. Determining this allocation involves choosing the *levels* of the activities that achieve the best possible value of the *overall measure of performance.*

Certain symbols are commonly used to denote the various components of a linear programming model. These symbols are listed below, along with their interpretation for the general problem of allocating resources to activities.

Z = value of overall measure of performance.

x_j = level of activity j (for $j = 1, 2, \ldots, n$).

c_j = increase in Z that would result from each unit increase in level of activity j.

b_i = amount of resource i that is available for allocation to activities (for $i = 1, 2, \ldots, m$).

a_{ij} = amount of resource i consumed by each unit of activity j.

The model poses the problem in terms of making decisions about the levels of the activities, so $x_1, x_2, \ldots, x_n$ are called the **decision variables.** As summarized in Table 3.3, the

TABLE 3.3 Data needed for a linear programming model involving the allocation of resources to activities

Resource	Resource Usage per Unit of Activity				Amount of Resource Available
	Activity				
	1	2	...	n	
1	a_{11}	a_{12}	...	a_{1n}	b_1
2	a_{21}	a_{22}	...	a_{2n}	b_2
.					.
.	...	...	...	...	.
.					.
m	a_{m1}	a_{m2}	...	a_{mn}	b_m
Contribution to Z per unit of activity	c_1	c_2	...	c_n	

values of c_j, b_i, and a_{ij} (for $i = 1, 2, \ldots, m$ and $j = 1, 2, \ldots, n$) are the *input constants* for the model. The c_j, b_i, and a_{ij} are also referred to as the **parameters** of the model.

Notice the correspondence between Table 3.3 and Table 3.1.

A Standard Form of the Model

Proceeding as for the Wyndor Glass Co. problem, we can now formulate the mathematical model for this general problem of allocating resources to activities. In particular, this model is to select the values for $x_1, x_2, \ldots, x_n$ so as to

$$\text{Maximize} \quad Z = c_1 x_1 + c_2 x_2 + \cdots + c_n x_n,$$

subject to the restrictions

$$a_{11} x_1 + a_{12} x_2 + \cdots + a_{1n} x_n \leq b_1$$
$$a_{21} x_1 + a_{22} x_2 + \cdots + a_{2n} x_n \leq b_2$$
$$\vdots$$
$$a_{m1} x_1 + a_{m2} x_2 + \cdots + a_{mn} x_n \leq b_m,$$

and

$$x_1 \geq 0, \quad x_2 \geq 0, \quad \ldots, \quad x_n \geq 0.$$

We call this *our standard form*[1] for the linear programming problem. Any situation whose mathematical formulation fits this model is a linear programming problem.

Notice that the model for the Wyndor Glass Co. problem fits our standard form, with $m = 3$ and $n = 2$.

Common terminology for the linear programming model can now be summarized. The function being maximized, $c_1 x_1 + c_2 x_2 + \cdots + c_n x_n$, is called the **objective function.** The restrictions normally are referred to as **constraints.** The first m constraints (those with a *function* of all the variables $a_{i1} x_1 + a_{i2} x_2 + \cdots + a_{in} x_n$ on the left-hand side) are sometimes called **functional constraints** (or *structural constraints*). Similarly, the $x_j \geq 0$ restrictions are called **nonnegativity constraints** (or *nonnegativity conditions*).

Other Forms

We now hasten to add that the preceding model does not actually fit the natural form of some linear programming problems. The other *legitimate forms* are the following:

1. Minimizing rather than maximizing the objective function:

$$\text{Minimize} \quad Z = c_1 x_1 + c_2 x_2 + \cdots + c_n x_n.$$

2. Some functional constraints with a greater-than-or-equal-to inequality:

$$a_{i1} x_1 + a_{i2} x_2 + \cdots + a_{in} x_n \geq b_i \quad \text{for some values of } i.$$

3. Some functional constraints in equation form:

$$a_{i1} x_1 + a_{i2} x_2 + \cdots + a_{in} x_n = b_i \quad \text{for some values of } i.$$

4. Deleting the nonnegativity constraints for some decision variables:

$$x_j \text{ unrestricted in sign} \quad \text{for some values of } j.$$

Any problem that mixes some of or all these forms with the remaining parts of the preceding model is still a linear programming problem. Our interpretation of the words *allocating*

---

[1]This is called *our* standard form rather than *the* standard form because some textbooks adopt other forms.

limited resources among competing activities may no longer apply very well, if at all; but regardless of the interpretation or context, all that is required is that the mathematical statement of the problem fit the allowable forms. Thus, the concise definition of a linear programming problem is that each component of its model fits either the standard form or one of the other legitimate forms listed above.

Terminology for Solutions of the Model

You may be used to having the term *solution* mean the final answer to a problem, but the convention in linear programming (and its extensions) is quite different. Here, *any* specification of values for the decision variables $(x_1, x_2, \ldots, x_n)$ is called a **solution,** regardless of whether it is a desirable or even an allowable choice. Different types of solutions are then identified by using an appropriate adjective.

> A **feasible solution** is a solution for which *all* the constraints are *satisfied.*
> An **infeasible solution** is a solution for which *at least one* constraint is *violated.*

In the example, the points (2, 3) and (4, 1) in Fig. 3.2 are *feasible solutions,* while the points (-1, 3) and (4, 4) are *infeasible solutions.*

> The **feasible region** is the collection of all feasible solutions.

The feasible region in the example is the entire shaded area in Fig. 3.2.

It is possible for a problem to have **no feasible solutions.** This would have happened in the example if the new products had been required to return a net profit of at least $50,000 per week to justify discontinuing part of the current product line. The corresponding constraint, $3x_1 + 5x_2 \geq 50$, would eliminate the entire feasible region, so no mix of new products would be superior to the status quo. This case is illustrated in Fig. 3.4.

Given that there are feasible solutions, the goal of linear programming is to find a best feasible solution, as measured by the value of the objective function in the model.

◼ FIGURE 3.4
The Wyndor Glass Co. problem would have no feasible solutions if the constraint $3x_1 + 5x_2 \geq 50$ were added to the problem.

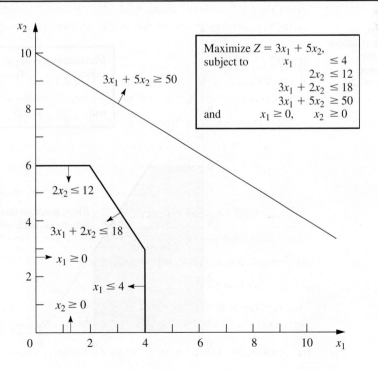

Maximize $Z = 3x_1 + 5x_2$,
subject to
$$x_1 \leq 4$$
$$2x_2 \leq 12$$
$$3x_1 + 2x_2 \leq 18$$
$$3x_1 + 5x_2 \geq 50$$
and $x_1 \geq 0, \quad x_2 \geq 0$

An **optimal solution** is a feasible solution that has the *most favorable value* of the objective function.

The **most favorable value** is the *largest value* if the objective function is to be *maximized,* whereas it is the *smallest value* if the objective function is to be *minimized.*

Most problems will have just one optimal solution. However, it is possible to have more than one. This would occur in the example if the *profit per batch produced* of product 2 were changed to $2,000. This changes the objective function to $Z = 3x_1 + 2x_2$, so that all the points on the line segment connecting (2, 6) and (4, 3) would be optimal. This case is illustrated in Fig. 3.5. As in this case, *any* problem having **multiple optimal solutions** will have an *infinite* number of them, each with the same optimal value of the objective function.

Another possibility is that a problem has **no optimal solutions.** This occurs only if (1) it has no feasible solutions or (2) the constraints do not prevent improving the value of the objective function (Z) indefinitely in the favorable direction (positive or negative). The latter case is referred to as having an **unbounded Z** or an *unbounded objective.* To illustrate, this case would result if the last two functional constraints were mistakenly deleted in the example, as illustrated in Fig. 3.6.

We next introduce a special type of feasible solution that plays the key role when the simplex method searches for an optimal solution.

A **corner-point feasible (CPF) solution** is a solution that lies at a corner of the feasible region.

(CPF solutions also are commonly referred to as *extreme points* or *vertices,* but we prefer the more suggestive *corner-point* terminology.) Figure 3.7 highlights the five CPF solutions for the example.

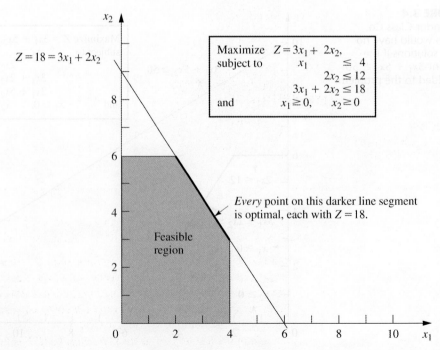

FIGURE 3.5
The Wyndor Glass Co. problem would have multiple optimal solutions if the objective function were changed to $Z = 3x_1 + 2x_2$.

$Z = 18 = 3x_1 + 2x_2$

Maximize $Z = 3x_1 + 2x_2$,
subject to $x_1 \leq 4$
                          $2x_2 \leq 12$
                          $3x_1 + 2x_2 \leq 18$
and $x_1 \geq 0, \quad x_2 \geq 0$

Every point on this darker line segment is optimal, each with $Z = 18$.

Feasible region

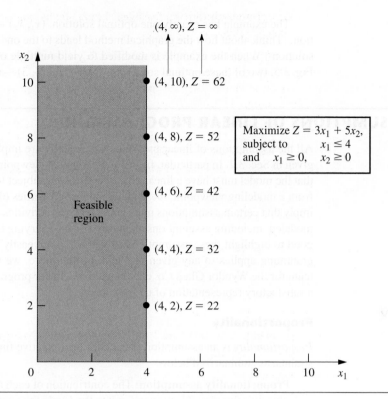

FIGURE 3.6
The Wyndor Glass Co. problem would have no optimal solutions if the only functional constraint were $x_1 \leq 4$, because x_2 then could be increased indefinitely in the feasible region without ever reaching the maximum value of $Z = 3x_1 + 5x_2$.

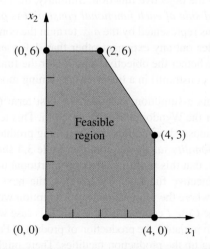

FIGURE 3.7
The five dots are the five CPF solutions for the Wyndor Glass Co. problem.

Sections 4.1 and 5.1 will delve into the various useful properties of CPF solutions for problems of any size, including the following relationship with optimal solutions.

Relationship between optimal solutions and CPF solutions: Consider any linear programming problem with feasible solutions and a bounded feasible region. The problem must possess CPF solutions and at least one optimal solution. Furthermore, the best CPF solution *must* be an optimal solution. Thus, if a problem has exactly one optimal solution, it *must* be a CPF solution. If the problem has multiple optimal solutions, at least two *must* be CPF solutions.

The example has exactly one optimal solution, $(x_1, x_2) = (2, 6)$, which is a CPF solution. (Think about how the graphical method leads to the one optimal solution being a CPF solution.) When the example is modified to yield multiple optimal solutions, as shown in Fig. 3.5, two of these optimal solutions—(2, 6) and (4, 3)—are CPF solutions.

3.3 ASSUMPTIONS OF LINEAR PROGRAMMING

All the assumptions of linear programming actually are implicit in the model formulation given in Sec. 3.2. In particular, from a mathematical viewpoint, the assumptions simply are that the model must have a linear objective function subject to linear constraints. However, from a modeling viewpoint, these mathematical properties of a linear programming model imply that certain assumptions must hold about the activities and data of the problem being modeled, including assumptions about the effect of varying the levels of the activities. It is good to highlight these assumptions so you can more easily evaluate how well linear programming applies to any given problem. Furthermore, we still need to see why the OR team for the Wyndor Glass Co. concluded that a linear programming formulation provided a satisfactory representation of the problem.

Proportionality

Proportionality is an assumption about both the objective function and the functional constraints, as summarized below.

> **Proportionality assumption:** The contribution of each activity to the *value of the objective function Z* is *proportional* to the *level of the activity x_j*, as represented by the $c_j x_j$ term in the objective function. Similarly, the contribution of each activity to the *left-hand side of each functional constraint* is *proportional* to the *level of the activity x_j*, as represented by the $a_{ij} x_j$ term in the constraint. Consequently, this assumption rules out any exponent other than 1 for any variable in any term of any function (whether the objective function or the function on the left-hand side of a functional constraint) in a linear programming model.[2]

To illustrate this assumption, consider the first term $(3x_1)$ in the objective function $(Z = 3x_1 + 5x_2)$ for the Wyndor Glass Co. problem. This term represents the profit generated per week (in thousands of dollars) by producing product 1 at the rate of x_1 batches per week. The *proportionality satisfied* column of Table 3.4 shows the case that was assumed in Sec. 3.1, namely, that this profit is indeed proportional to x_1 so that $3x_1$ is the appropriate term for the objective function. By contrast, the next three columns show different hypothetical cases where the proportionality assumption would be violated.

Refer first to the *Case 1* column in Table 3.4. This case would arise if there were *start-up costs* associated with initiating the production of product 1. For example, there might be costs involved with setting up the production facilities. There might also be costs associated with arranging the distribution of the new product. Because these are one-time costs, they would need to be amortized on a per-week basis to be commensurable with Z (profit in thousands of dollars per week). Suppose that this amortization were done and that the total start-up cost amounted to reducing Z by 1, but that the profit without considering the start-up cost would be $3x_1$. This would mean that the contribution from product 1 to Z should be $3x_1 - 1$ for $x_1 > 0$,

---

[2]When the function includes any *cross-product terms,* proportionality should be interpreted to mean that *changes* in the function value are proportional to *changes* in each variable (x_j) individually, given any fixed values for all the other variables. Therefore, a cross-product term satisfies proportionality as long as each variable in the term has an exponent of 1 (However, any cross-product term violates the *additivity assumption*, discussed next.)

■ **TABLE 3.4** Examples of satisfying or violating proportionality

	Profit from Product 1 ($000 per Week)			
		Proportionality Violated		
x_1	Proportionality Satisfied	Case 1	Case 2	Case 3
0	0	0	0	0
1	3	2	3	3
2	6	5	7	5
3	9	8	12	6
4	12	11	18	6

whereas the contribution would be $3x_1 = 0$ when $x_1 = 0$ (no start-up cost). This profit function,[3] which is given by the solid curve in Fig. 3.8, certainly is *not* proportional to x_1.

At first glance, it might appear that *Case 2* in Table 3.4 is quite similar to Case 1. However, Case 2 actually arises in a very different way. There no longer is a start-up cost, and the profit from the first unit of product 1 per week is indeed 3, as originally assumed. However, there now is an *increasing marginal return;* i.e., the *slope* of the *profit function* for product 1 (see the solid curve in Fig. 3.9) keeps increasing as x_1 is increased. This violation of proportionality might occur because of economies of scale that can sometimes be achieved at higher levels of production, e.g., through the use of more efficient high-volume machinery, longer production runs, quantity discounts for large purchases of raw materials, and the learning-curve effect whereby workers become more efficient as they gain experience with a particular mode of production. As the incremental cost goes down, the incremental profit will go up (assuming constant marginal revenue).

■ **FIGURE 3.8**
The solid curve violates the proportionality assumption because of the start-up cost that is incurred when x_1 is increased from 0. The values at the dots are given by the Case 1 column of Table 3.4.

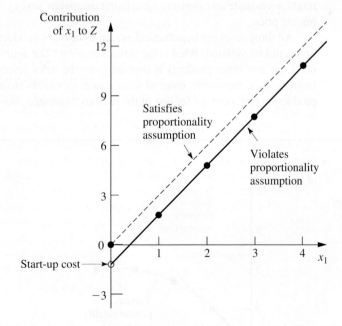

[3] If the contribution from product 1 to Z were $3x_1 - 1$ for *all* $x_1 \geq 0$, including $x_1 = 0$, then the fixed constant, -1, could be deleted from the objective function without changing the optimal solution and proportionality would be restored. However, this "fix" does not work here because the -1 constant does not apply when $x_1 = 0$.

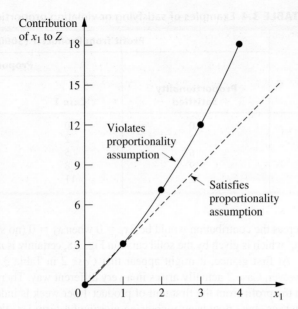

■ FIGURE 3.9
The solid curve violates the proportionality assumption because its slope (the *marginal return* from product 1) keeps increasing as x_1 is increased. The values at the dots are given by the Case 2 column of Table 3.4.

Referring again to Table 3.4, the reverse of Case 2 is *Case 3,* where there is a *decreasing marginal return.* In this case, the *slope* of the *profit function* for product 1 (given by the solid curve in Fig. 3.10) keeps decreasing as x_1 is increased. This violation of proportionality might occur because the *marketing costs* need to go up more than proportionally to attain increases in the level of sales. For example, it might be possible to sell product 1 at the rate of 1 per week ($x_1 = 1$) with no advertising, whereas attaining sales to sustain a production rate of $x_1 = 2$ might require a moderate amount of advertising, $x_1 = 3$ might necessitate an extensive advertising campaign, and $x_1 = 4$ might require also lowering the price.

All three cases are hypothetical examples of ways in which the proportionality assumption could be violated. What is the actual situation? The actual profit from producing product 1 (or any other product) is derived from the sales revenue minus various direct and indirect costs. Inevitably, some of these cost components are not strictly proportional to the production rate, perhaps for one of the reasons illustrated above. However, the real question

■ FIGURE 3.10
The solid curve violates the proportionality assumption because its slope (the *marginal return* from product 1) keeps decreasing as x_1 is increased. The values at the dots are given by the Case 3 column in Table 3.4.

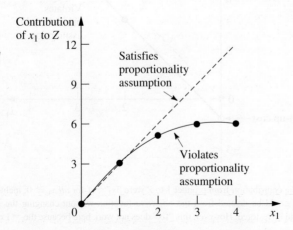

is whether, after all the components of profit have been accumulated, proportionality is a reasonable approximation for practical modeling purposes. For the Wyndor Glass Co. problem, the OR team checked both the objective function and the functional constraints. The conclusion was that proportionality could indeed be assumed without serious distortion.

For other problems, what happens when the proportionality assumption does not hold even as a reasonable approximation? In most cases, this means you must use *nonlinear programming* instead (presented in Chap. 12). However, we do point out in Sec. 12.8 that a certain important kind of nonproportionality can still be handled by linear programming by reformulating the problem appropriately. Furthermore, if the assumption is violated only because of start-up costs, there is an extension of linear programming (*mixed integer programming*) that can be used, as discussed in Sec. 11.3 (the fixed-charge problem).

Additivity

Although the proportionality assumption rules out exponents other than 1, it does not prohibit *cross-product terms* (terms involving the product of two or more variables). The additivity assumption does rule out this latter possibility, as summarized below.

> **Additivity assumption:** *Every* function in a linear programming model (whether the objective function or the function on the left-hand side of a functional constraint) is the *sum* of the *individual contributions* of the respective activities.

To make this definition more concrete and clarify why we need to worry about this assumption, let us look at some examples. Table 3.5 shows some possible cases for the objective function for the Wyndor Glass Co. problem. In each case, the *individual contributions* from the products are just as assumed in Sec. 3.1, namely, $3x_1$ for product 1 and $5x_2$ for product 2. The difference lies in the last row, which gives the *function value* for Z when the two products are produced jointly. The *additivity satisfied* column shows the case where this *function value* is obtained simply by adding the first two rows ($3 + 5 = 8$), so that $Z = 3x_1 + 5x_2$ as previously assumed. By contrast, the next two columns show hypothetical cases where the additivity assumption would be violated (but not the proportionality assumption).

Referring to the *Case 1* column of Table 3.5, this case corresponds to an objective function of $Z = 3x_1 + 5x_2 + x_1x_2$, so that $Z = 3 + 5 + 1 = 9$ for $(x_1, x_2) = (1, 1)$, thereby violating the additivity assumption that $Z = 3 + 5$. (The proportionality assumption still is satisfied since after the value of one variable is fixed, the increment in Z from the other variable is proportional to the value of that variable.) This case would arise if the two products were *complementary* in some way that *increases* profit. For example, suppose that a major advertising campaign would be required to market either new product produced by itself, but that the same single campaign can effectively promote both products if the decision is made to produce both. Because a major cost is saved for the second

TABLE 3.5 Examples of satisfying or violating additivity for the objective function

		Value of Z	
		Additivity Violated	
(x_1, x_2)	**Additivity Satisfied**	**Case 1**	**Case 2**
(1, 0)	3	3	3
(0, 1)	5	5	5
(1, 1)	8	9	7

product, their joint profit is somewhat more than the *sum* of their individual profits when each is produced by itself.

Case 2 in Table 3.5 also violates the additivity assumption because of the extra term in the corresponding objective function, $Z = 3x_1 + 5x_2 - x_1x_2$, so that $Z = 3 + 5 - 1 = 7$ for $(x_1, x_2) = (1, 1)$. As the reverse of the first case, Case 2 would arise if the two products were *competitive* in some way that *decreased* their joint profit. For example, suppose that both products need to use the same machinery and equipment. If either product were produced by itself, this machinery and equipment would be dedicated to this one use. However, producing both products would require switching the production processes back and forth, with substantial time and cost involved in temporarily shutting down the production of one product and setting up for the other. Because of this major extra cost, their joint profit is somewhat less than the *sum* of their individual profits when each is produced by itself.

The same kinds of interaction between activities can affect the additivity of the constraint functions. For example, consider the third functional constraint of the Wyndor Glass Co. problem: $3x_1 + 2x_2 \leq 18$. (This is the only constraint involving both products.) This constraint concerns the production capacity of Plant 3, where 18 hours of production time per week is available for the two new products, and the function on the left-hand side $(3x_1 + 2x_2)$ represents the number of hours of production time per week that would be used by these products. The *additivity satisfied* column of Table 3.6 shows this case as is, whereas the next two columns display cases where the function has an extra cross-product term that violates additivity. For all three columns, the *individual contributions* from the products toward using the capacity of Plant 3 are just as assumed previously, namely, $3x_1$ for product 1 and $2x_2$ for product 2, or $3(2) = 6$ for $x_1 = 2$ and $2(3) = 6$ for $x_2 = 3$. As was true for Table 3.5, the difference lies in the last row, which now gives the *total function value* for production time used when the two products are produced jointly.

For Case 3 (see Table 3.6), the production time used by the two products is given by the function $3x_1 + 2x_2 + 0.5x_1x_2$, so the *total function value* is $6 + 6 + 3 = 15$ when $(x_1, x_2) = (2, 3)$, which violates the additivity assumption that the value is just $6 + 6 = 12$. This case can arise in exactly the same way as described for Case 2 in Table 3.5; namely, extra time is wasted switching the production processes back and forth between the two products. The extra cross-product term $(0.5x_1x_2)$ would give the production time wasted in this way. (Note that wasting time switching between products leads to a positive cross-product term here, where the total function is measuring production time used, whereas it led to a negative cross-product term for Case 2 because the total function there measures profit.)

For Case 4 in Table 3.6, the function for production time used is $3x_1 + 2x_2 - 0.1x_1^2x_2$, so the *function value* for $(x_1, x_2) = (2, 3)$ is $6 + 6 - 1.2 = 10.8$. This case could arise in the following way. As in Case 3, suppose that the two products require the same type of machinery and equipment. But suppose now that the time required to switch from one

■ **TABLE 3.6 Examples of satisfying or violating additivity for a functional constraint**

| | **Amount of Resource Used** | | |
| | | **Additivity Violated** | |
(x_1, x_2)	**Additivity Satisfied**	**Case 3**	**Case 4**
(2, 0)	6	6	6
(0, 3)	6	6	6
(2, 3)	12	15	10.8

product to the other would be relatively small. Because each product goes through a sequence of production operations, individual production facilities normally dedicated to that product would incur occasional idle periods. During these otherwise idle periods, these facilities can be used by the other product. Consequently, the total production time used (including idle periods) when the two products are produced jointly would be less than the *sum* of the production times used by the individual products when each is produced by itself.

After analyzing the possible kinds of interaction between the two products illustrated by these four cases, the OR team concluded that none played a major role in the actual Wyndor Glass Co. problem. Therefore, the additivity assumption was adopted as a reasonable approximation.

For other problems, if additivity is not a reasonable assumption, so that some of or all the mathematical functions of the model need to be *nonlinear* (because of the cross-product terms), you definitely enter the realm of nonlinear programming (Chap. 12).

Divisibility

Our next assumption concerns the values allowed for the decision variables.

> **Divisibility assumption:** Decision variables in a linear programming model are allowed to have *any* values, including *noninteger* values, that satisfy the functional and nonnegativity constraints. Thus, these variables are *not* restricted to just integer values. Since each decision variable represents the level of some activity, it is being assumed that the activities can be run at *fractional levels*.

For the Wyndor Glass Co. problem, the decision variables represent production rates (the number of batches of a product produced per week). Since these production rates can have *any* fractional values within the feasible region, the divisibility assumption does hold.

In certain situations, the divisibility assumption does not hold because some of or all the decision variables must be restricted to *integer values*. Mathematical models with this restriction are called *integer programming* models, and they are discussed in Chap. 11.

Certainty

Our last assumption concerns the *parameters* of the model, namely, the coefficients in the objective function c_j, the coefficients in the functional constraints a_{ij}, and the right-hand sides of the functional constraints b_i.

> **Certainty assumption:** The value assigned to each parameter of a linear programming model is assumed to be a *known constant*.

In real applications, the certainty assumption is seldom satisfied precisely. Linear programming models usually are formulated to select some future course of action. Therefore, the parameter values used would be based on a prediction of future conditions, which inevitably introduces some degree of uncertainty.

For this reason it is usually important to conduct **sensitivity analysis** after a solution is found that is optimal under the assumed parameter values. As discussed in Sec. 2.3, one purpose is to identify the *sensitive* parameters (those whose value cannot be changed without changing the optimal solution), since any later change in the value of a sensitive parameter immediately signals a need to change the solution being used.

Sensitivity analysis plays an important role in the analysis of the Wyndor Glass Co. problem, as you will see in Sec. 6.7. However, it is necessary to acquire some more background before we finish that story.

Occasionally, the degree of uncertainty in the parameters is too great to be amenable to sensitivity analysis. In this case, it is necessary to treat the parameters explicitly as *random variables.* Formulations of this kind have been developed, as discussed in Secs. 23.6 and 23.7 on the book's website.

The Assumptions in Perspective

We emphasized in Sec. 2.2 that a mathematical model is intended to be only an idealized representation of the real problem. Approximations and simplifying assumptions generally are required in order for the model to be tractable. Adding too much detail and precision can make the model too unwieldy for useful analysis of the problem. All that is really needed is that there be a reasonably high correlation between the prediction of the model and what would actually happen in the real problem.

This advice certainly is applicable to linear programming. It is very common in real applications of linear programming that almost *none* of the four assumptions hold completely. Except perhaps for the *divisibility assumption,* minor disparities are to be expected. This is especially true for the *certainty assumption,* so sensitivity analysis normally is a must to compensate for the violation of this assumption.

However, it is important for the OR team to examine the four assumptions for the problem under study and to analyze just how large the disparities are. If any of the assumptions are violated in a major way, then a number of useful alternative models are available, as presented in later chapters of the book. A disadvantage of these other models is that the algorithms available for solving them are not nearly as powerful as those for linear programming, but this gap has been closing in some cases. For some applications, the powerful linear programming approach is used for the initial analysis, and then a more complicated model is used to refine this analysis.

As you work through the examples in Sec. 3.4, you will find it good practice to analyze how well each of the four assumptions of linear programming applies.

▪ 3.4 ADDITIONAL EXAMPLES

The Wyndor Glass Co. problem is a prototype example of linear programming in several respects: It involves allocating limited resources among competing activities, its model fits our standard form, and its context is the traditional one of improved business planning. However, the applicability of linear programming is much wider. In this section we begin broadening our horizons. As you study the following examples, note that it is their underlying mathematical model rather than their context that characterizes them as linear programming problems. Then give some thought to how the same mathematical model could arise in many other contexts by merely changing the names of the activities and so forth.

These examples are scaled-down versions of actual applications. Like the Wyndor problem and the demonstration example for the graphical method in OR Tutor, the first of these examples has only two decision variables and so can be solved by the graphical method. The new features are that it is a minimization problem and has a mixture of forms for the functional constraints. (This example considerably simplifies the real situation when designing radiation therapy, but the first application vignette in this section describes the exciting impact that OR actually is having in this area.) The subsequent examples have considerably more than two decision variables and so are more challenging to formulate. Although we will mention their optimal solutions that are obtained by the simplex method, the focus here is on how to formulate the linear programming model for these larger problems. Subsequent sections and the next chapter will turn to the question of the software tools and the algorithm (the simplex method) that are used to solve such problems.

If you find that you need **additional examples** of formulating small and relatively straightforward linear programming models before dealing with these more challenging formulation examples, we suggest that you go back to the demonstration example for the graphical method in OR Tutor and to the examples in the Worked Examples section for this chapter on the book's website.

Design of Radiation Therapy

MARY has just been diagnosed as having a cancer at a fairly advanced stage. Specifically, she has a large malignant tumor in the bladder area (a "whole bladder lesion").

Mary is to receive the most advanced medical care available to give her every possible chance for survival. This care will include extensive *radiation therapy.*

Radiation therapy involves using an external beam treatment machine to pass ionizing radiation through the patient's body, damaging both cancerous and healthy tissues. Normally, several beams are precisely administered from different angles in a two-dimensional plane. Due to attenuation, each beam delivers more radiation to the tissue near the entry point than to the tissue near the exit point. Scatter also causes some delivery of radiation to tissue outside the direct path of the beam. Because tumor cells are typically microscopically interspersed among healthy cells, the radiation dosage throughout the tumor region must be large enough to kill the malignant cells, which are slightly more radiosensitive, yet small enough to spare the healthy cells. At the same time, the aggregate dose to critical tissues must not exceed established tolerance levels, in order to prevent complications that can be more serious than the disease itself. For the same reason, the total dose to the entire healthy anatomy must be minimized.

Because of the need to carefully balance all these factors, the design of radiation therapy is a very delicate process. The goal of the design is to select the combination of beams to be used, and the intensity of each one, to generate the best possible dose distribution. (The dose strength at any point in the body is measured in units called *kilorads*.) Once the treatment design has been developed, it is administered in many installments, spread over several weeks.

In Mary's case, the size and location of her tumor make the design of her treatment an even more delicate process than usual. Figure 3.11 shows a diagram of a cross section of the tumor viewed from above, as well as nearby critical tissues to avoid. These tissues include critical organs (e.g., the rectum) as well as bony structures (e.g., the femurs and pelvis) that will attenuate the radiation. Also shown are the entry point and direction for the only two beams that can be used with any modicum of safety in this case. (Actually, we are simplifying the example at this point, because normally dozens of possible beams must be considered.)

For any proposed beam of given intensity, the analysis of what the resulting radiation absorption by various parts of the body would be requires a complicated process. In brief, based on careful anatomical analysis, the energy distribution within the two-dimensional cross section of the tissue can be plotted on an isodose map, where the contour lines represent the dose strength as a percentage of the dose strength at the entry point. A fine grid then is placed over the isodose map. By summing the radiation absorbed in the squares containing each type of tissue, the average dose that is absorbed by the tumor, healthy anatomy, and critical tissues can be calculated. With more than one beam (administered sequentially), the radiation absorption is additive.

After thorough analysis of this type, the medical team has carefully estimated the data needed to design Mary's treatment, as summarized in Table 3.7. The first column lists the areas of the body that must be considered, and then the next two columns give the fraction of the radiation dose at the entry point for each beam that is absorbed by the

■ FIGURE 3.11
Cross section of Mary's tumor (viewed from above), nearby critical tissues, and the radiation beams being used.

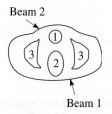

1. Bladder and
   tumor
2. Rectum, coccyx,
   etc.
3. Femur, part of
   pelvis, etc.

Prostate cancer is the most common form of cancer diagnosed in men. There were an estimated 220,000 new cases in just the United States alone in 2007. Like many other forms of cancer, *radiation therapy* is a common method of treatment for prostate cancer, where the goal is to have a sufficiently high radiation dosage in the tumor region to kill the malignant cells while minimizing the radiation exposure to critical healthy structures near the tumor. This treatment can be applied through either *external beam* radiation therapy (as illustrated by the first example in this section) or *brachytherapy,* which involves placing approximately 100 radioactive "seeds" within the tumor region. The challenge is to determine the most effective three-dimensional geometric pattern for placing these seeds.

Memorial Sloan-Kettering Cancer Center (MSKCC) in New York City is the world's oldest private cancer center. An OR team from the *Center for Operations Research in Medicine and HealthCare* at Georgia Institute of Technology worked with physicians at MSKCC to develop a highly sophisticated *next-generation method* of optimizing the application of brachytherapy to prostrate cancer. The underlying model fits the structure for linear programming with one exception. In addition to having the usual continuous variables that fit linear programming, the model also has some *binary variables* (variables whose only possible values are 0 and 1). (This kind of extension of linear programming to what is called *mixed-integer programming* will be discussed in

Chap. 11.) The optimization is done in a matter of minutes by an automated computerized planning system that can be operated readily by medical personnel when beginning the procedure of inserting the seeds into the patient's prostrate.

This breakthrough in optimizing the application of brachytherapy to prostrate cancer is having a profound impact on both health care costs and quality of life for treated patients because of its much greater effectiveness and the substantial reduction in side effects. When all U.S. clinics adopt this procedure, it is estimated that the annual cost savings will approximate **$500 million** due to eliminating the need for a pretreatment planning meeting and a postoperation CT scan, as well as providing a more efficient surgical procedure and reducing the need to treat subsequent side effects. It also is anticipated that this approach can be extended to other forms of brachytherapy, such as treatment of breast, cervix, esophagus, biliary tract, pancreas, head and neck, and eye.

This application of linear programming and its extensions led to the OR team winning the prestigious First Prize in the 2007 international competition for the Franz Edelman Award for Achievement in Operations Research and the Management Sciences.

Source: E. K. Lee and M. Zaider, "Operations Research Advances Cancer Therapeutics," *Interfaces,* **38**(1): 5–25, Jan.–Feb. 2008. (A link to this article is provided on our website, www.mhhe.com/hillier.)

respective areas on average. For example, if the dose level at the entry point for beam 1 is 1 kilorad, then an average of 0.4 kilorad will be absorbed by the entire healthy anatomy in the two-dimensional plane, an average of 0.3 kilorad will be absorbed by nearby critical tissues, an average of 0.5 kilorad will be absorbed by the various parts of the tumor, and 0.6 kilorad will be absorbed by the center of the tumor. The last column gives the restrictions on the total dosage from both beams that is absorbed on average by the respective areas of the body. In particular, the average dosage absorption for the

■ **TABLE 3.7** Data for the design of Mary's radiation therapy

Area	Fraction of Entry Dose Absorbed by Area (Average)		Restriction on Total Average Dosage, Kilorads
	Beam 1	Beam 2	
Healthy anatomy	0.4	0.5	Minimize
Critical tissues	0.3	0.1	≤ 2.7
Tumor region	0.5	0.5	= 6
Center of tumor	0.6	0.4	≥ 6

healthy anatomy must be *as small as possible,* the critical tissues must *not exceed* 2.7 kilorads, the average over the entire tumor must *equal* 6 kilorads, and the center of the tumor must be *at least* 6 kilorads.

Formulation as a Linear Programming Problem.   The decisions that need to be made are the dosages of radiation at the two entry points. Therefore, the two decision variables x_1 and x_2 represent the dose (in kilorads) at the entry point for beam 1 and beam 2, respectively. Because the total dosage reaching the healthy anatomy is to be minimized, let Z denote this quantity. The data from Table 3.7 can then be used directly to formulate the following linear programming model.[4]

$$\text{Minimize} \quad Z = 0.4x_1 + 0.5x_2,$$

subject to

$$0.3x_1 + 0.1x_2 \leq 2.7$$
$$0.5x_1 + 0.5x_2 = 6$$
$$0.6x_1 + 0.4x_2 \geq 6$$

and

$$x_1 \geq 0, \quad x_2 \geq 0.$$

Notice the differences between this model and the one in Sec. 3.1 for the Wyndor Glass Co. problem. The latter model involved *maximizing Z,* and all the functional constraints were in $\leq$ form. This new model does not fit this same standard form, but it does incorporate three other *legitimate* forms described in Sec. 3.2, namely, *minimizing Z,* functional constraints in $=$ form, and functional constraints in $\geq$ form.

However, both models have only two variables, so this new problem also can be solved by the *graphical method* illustrated in Sec. 3.1. Figure 3.12 shows the graphical solution. The *feasible region* consists of just the dark line segment between (6, 6) and (7.5, 4.5), because the points on this segment are the only ones that simultaneously satisfy all the constraints. (Note that the equality constraint limits the feasible region to the line containing this line segment, and then the other two functional constraints determine the two endpoints of the line segment.) The dashed line is the objective function line that passes through the optimal solution $(x_1, x_2) = (7.5, 4.5)$ with $Z = 5.25$. This solution is optimal rather than the point (6, 6) because *decreasing Z* (for positive values of Z) pushes the objective function line toward the origin (where $Z = 0$). And $Z = 5.25$ for (7.5, 4.5) is less than $Z = 5.4$ for (6, 6).

Thus, the optimal design is to use a total dose at the entry point of 7.5 kilorads for beam 1 and 4.5 kilorads for beam 2.

Regional Planning

The SOUTHERN CONFEDERATION OF KIBBUTZIM is a group of three kibbutzim (communal farming communities) in Israel. Overall planning for this group is done in its Coordinating Technical Office. This office currently is planning agricultural production for the coming year.

---

[4]This model is *much* smaller than normally would be needed for actual applications. For the best results, a realistic model might even need many tens of thousands of decision variables and constraints. For example, see H. E. Romeijn, R. K. Ahuja, J. F. Dempsey, and A. Kumar, "A New Linear Programming Approach to Radiation Therapy Treatment Planning Problems," *Operations Research,* **54**(2): 201–216, March–April 2006. For alternative approaches that combine linear programming with other OR techniques (like the first application vignette in this section), also see G. J. Lim, M. C. Ferris, S. J. Wright, D. M. Shepard, and M. A. Earl, "An Optimization Framework for Conformal Radiation Treatment Planning," *INFORMS Journal on Computing,* **19**(3): 366–380, Summer 2007.

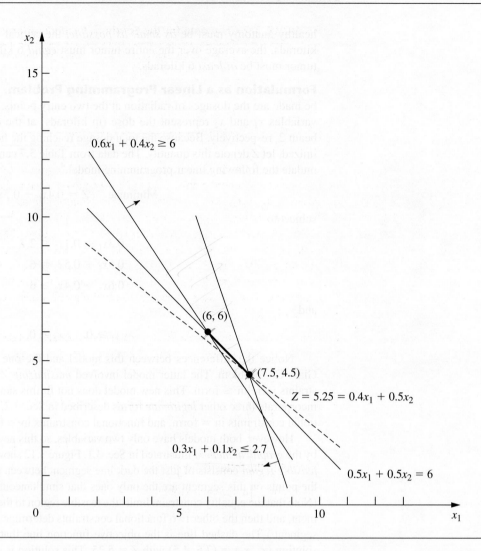

■ FIGURE 3.12
Graphical solution for the
design of Mary's radiation
therapy.

The agricultural output of each kibbutz is limited by both the amount of available irrigable land and the quantity of water allocated for irrigation by the Water Commissioner (a national government official). These data are given in Table 3.8.

The crops suited for this region include sugar beets, cotton, and sorghum, and these are the three being considered for the upcoming season. These crops differ primarily in their expected net return per acre and their consumption of water. In addition, the Ministry of Agriculture has set a maximum quota for the total acreage that can be devoted to each of these crops by the Southern Confederation of Kibbutzim, as shown in Table 3.9.

■ TABLE 3.8 Resource data for the Southern Confederation of Kibbutzim

Kibbutz	Usable Land (Acres)	Water Allocation (Acre Feet)
1	400	600
2	600	800
3	300	375

■ **TABLE 3.9** Crop data for the Southern Confederation of Kibbutzim

Crop	Maximum Quota (Acres)	Water Consumption (Acre Feet/Acre)	Net Return ($/Acre)
Sugar beets	600	3	1,000
Cotton	500	2	750
Sorghum	325	1	250

Because of the limited water available for irrigation, the Southern Confederation of Kibbutzim will not be able to use all its irrigable land for planting crops in the upcoming season. To ensure equity between the three kibbutzim, it has been agreed that every kibbutz will plant the same proportion of its available irrigable land. For example, if kibbutz 1 plants 200 of its available 400 acres, then kibbutz 2 must plant 300 of its 600 acres, while kibbutz 3 plants 150 acres of its 300 acres. However, any combination of the crops may be grown at any of the kibbutzim. The job facing the Coordinating Technical Office is to plan how many acres to devote to each crop at the respective kibbutzim while satisfying the given restrictions. The objective is to maximize the total net return to the Southern Confederation of Kibbutzim as a whole.

Formulation as a Linear Programming Problem. The quantities to be decided upon are the number of acres to devote to each of the three crops at each of the three kibbutzim. The decision variables x_j ($j = 1, 2, \ldots, 9$) represent these nine quantities, as shown in Table 3.10.

Since the measure of effectiveness Z is the total net return, the resulting linear programming model for this problem is

Maximize $Z = 1,000(x_1 + x_2 + x_3) + 750(x_4 + x_5 + x_6) + 250(x_7 + x_8 + x_9)$,

subject to the following constraints:

1. Usable land for each kibbutz:

$$x_1 + x_4 + x_7 \le 400$$
$$x_2 + x_5 + x_8 \le 600$$
$$x_3 + x_6 + x_9 \le 300$$

2. Water allocation for each kibbutz:

$$3x_1 + 2x_4 + x_7 \le 600$$
$$3x_2 + 2x_5 + x_8 \le 800$$
$$3x_3 + 2x_6 + x_9 \le 375$$

■ **TABLE 3.10** Decision variables for the Southern Confederation of Kibbutzim problem

	Allocation (Acres)		
	Kibbutz		
Crop	1	2	3
Sugar beets	x_1	x_2	x_3
Cotton	x_4	x_5	x_6
Sorghum	x_7	x_8	x_9

3. Total acreage for each crop:

$$x_1 + x_2 + x_3 \leq 600$$
$$x_4 + x_5 + x_6 \leq 500$$
$$x_7 + x_8 + x_9 \leq 325$$

4. Equal proportion of land planted:

$$\frac{x_1 + x_4 + x_7}{400} = \frac{x_2 + x_5 + x_8}{600}$$

$$\frac{x_2 + x_5 + x_8}{600} = \frac{x_3 + x_6 + x_9}{300}$$

$$\frac{x_3 + x_6 + x_9}{300} = \frac{x_1 + x_4 + x_7}{400}$$

5. Nonnegativity:

$$x_j \geq 0, \quad \text{for } j = 1, 2, \dots, 9.$$

This completes the model, except that the equality constraints are not yet in an appropriate form for a linear programming model because some of the variables are on the right-hand side. Hence, their final form[5] is

$$3(x_1 + x_4 + x_7) - 2(x_2 + x_5 + x_8) = 0$$
$$(x_2 + x_5 + x_8) - 2(x_3 + x_6 + x_9) = 0$$
$$4(x_3 + x_6 + x_9) - 3(x_1 + x_4 + x_7) = 0$$

The Coordinating Technical Office formulated this model and then applied the simplex method (developed in Chap. 4) to find an optimal solution

$$(x_1, x_2, x_3, x_4, x_5, x_6, x_7, x_8, x_9) = \left(133\frac{1}{3}, 100, 25, 100, 250, 150, 0, 0, 0\right),$$

as shown in Table 3.11. The resulting optimal value of the objective function is Z=633, $333\frac{1}{3}$, that is, a total net return of \$633,333.33.

■ **TABLE 3.11** Optimal solution for the Southern Confederation of Kibbutzim problem

	Best Allocation (Acres)		
	Kibbutz		
Crop	**1**	**2**	**3**
Sugar beets	$133\frac{1}{3}$	100	25
Cotton	100	250	150
Sorghum	0	0	0

[5]Actually, any one of these equations is redundant and can be deleted if desired. Also, because of these equations, any two of the usable land constraints also could be deleted because they automatically would be satisfied when both the remaining usable land constraint and these equations are satisfied. However, no harm is done (except a little more computational effort) by including unnecessary constraints, so you don't need to worry about identifying and deleting them in models you formulate.

Controlling Air Pollution

The NORI & LEETS CO., one of the major producers of steel in its part of the world, is located in the city of Steeltown and is the only large employer there. Steeltown has grown and prospered along with the company, which now employs nearly 50,000 residents. Therefore, the attitude of the townspeople always has been, What's good for Nori & Leets is good for the town. However, this attitude is now changing; uncontrolled air pollution from the company's furnaces is ruining the appearance of the city and endangering the health of its residents.

A recent stockholders' revolt resulted in the election of a new enlightened board of directors for the company. These directors are determined to follow socially responsible policies, and they have been discussing with Steeltown city officials and citizens' groups what to do about the air pollution problem. Together they have worked out stringent air quality standards for the Steeltown airshed.

The three main types of pollutants in this airshed are particulate matter, sulfur oxides, and hydrocarbons. The new standards require that the company reduce its annual emission of these pollutants by the amounts shown in Table 3.12. The board of directors has instructed management to have the engineering staff determine how to achieve these reductions in the most economical way.

The steelworks has two primary sources of pollution, namely, the blast furnaces for making pig iron and the open-hearth furnaces for changing iron into steel. In both cases the engineers have decided that the most effective types of abatement methods are (1) increasing the height of the smokestacks,[6] (2) using filter devices (including gas traps) in the smokestacks, and (3) including cleaner, high-grade materials among the fuels for the furnaces. Each of these methods has a technological limit on how heavily it can be used (e.g., a maximum feasible increase in the height of the smokestacks), but there also is considerable flexibility for using the method at a fraction of its technological limit.

Table 3.13 shows how much emission (in millions of pounds per year) can be eliminated from each type of furnace by fully using any abatement method to its technological limit. For purposes of analysis, it is assumed that each method also can be used less fully to achieve any fraction of the emission-rate reductions shown in this table. Furthermore, the fractions can be different for blast furnaces and for open-hearth furnaces. For either type of furnace, the emission reduction achieved by each method is not substantially affected by whether the other methods also are used.

After these data were developed, it became clear that no single method by itself could achieve all the required reductions. On the other hand, combining all three methods at full capacity on both types of furnaces (which would be prohibitively expensive if the company's

TABLE 3.12 Clean air standards for the Nori & Leets Co.

Pollutant	Required Reduction in Annual Emission Rate (Million Pounds)
Particulates	60
Sulfur oxides	150
Hydrocarbons	125

[6]Subsequent to this study, this particular abatement method has become a controversial one. Because its effect is to reduce ground-level pollution by spreading emissions over a greater distance, environmental groups contend that this creates more acid rain by keeping sulfur oxides in the air longer. Consequently, the U.S. Environmental Protection Agency adopted new rules in 1985 to remove incentives for using tall smokestacks.

■ **TABLE 3.13** Reduction in emission rate (in millions of pounds per year) from the maximum feasible use of an abatement method for Nori & Leets Co.

Pollutant	Taller Smokestacks		Filters		Better Fuels	
	Blast Furnaces	Open-Hearth Furnaces	Blast Furnaces	Open-Hearth Furnaces	Blast Furnaces	Open-Hearth Furnaces
Particulates	12	9	25	20	17	13
Sulfur oxides	35	42	18	31	56	49
Hydrocarbons	37	53	28	24	29	20

products are to remain competitively priced) is much more than adequate. Therefore, the engineers concluded that they would have to use some combination of the methods, perhaps with fractional capacities, based upon the relative costs. Furthermore, because of the differences between the blast and the open-hearth furnaces, the two types probably should not use the same combination.

An analysis was conducted to estimate the total annual cost that would be incurred by each abatement method. A method's annual cost includes increased operating and maintenance expenses as well as reduced revenue due to any loss in the efficiency of the production process caused by using the method. The other major cost is the *start-up cost* (the initial capital outlay) required to install the method. To make this one-time cost commensurable with the ongoing annual costs, the time value of money was used to calculate the annual expenditure (over the expected life of the method) that would be equivalent in value to this start-up cost.

This analysis led to the total annual cost estimates (in millions of dollars) given in Table 3.14 for using the methods at their full abatement capacities. It also was determined that the cost of a method being used at a lower level is roughly proportional to the fraction of the abatement capacity given in Table 3.13 that is achieved. Thus, for any given fraction achieved, the total annual cost would be roughly that fraction of the corresponding quantity in Table 3.14.

The stage now was set to develop the general framework of the company's plan for pollution abatement. This plan specifies which types of abatement methods will be used and at what fractions of their abatement capacities for (1) the blast furnaces and (2) the open-hearth furnaces. Because of the combinatorial nature of the problem of finding a plan that satisfies the requirements with the smallest possible cost, an OR team was formed to solve the problem. The team adopted a linear programming approach, formulating the model summarized next.

Formulation as a Linear Programming Problem. This problem has six decision variables x_j, $j = 1, 2, \ldots, 6$, each representing the use of one of the three abatement methods for one of the two types of furnaces, expressed as a *fraction of the abatement capacity* (so x_j cannot exceed 1). The ordering of these variables is shown in Table 3.15. Because the

■ **TABLE 3.14** Total annual cost from the maximum feasible use of an abatement method for Nori & Leets Co. ($ millions)

Abatement Method	Blast Furnaces	Open-Hearth Furnaces
Taller smokestacks	8	10
Filters	7	6
Better fuels	11	9

■ **TABLE 3.15** Decision variables (fraction of the maximum feasible use of an abatement method) for Nori & Leets Co.

Abatement Method	Blast Furnaces	Open-Hearth Furnaces
Taller smokestacks	x_1	x_2
Filters	x_3	x_4
Better fuels	x_5	x_6

objective is to minimize total cost while satisfying the emission reduction requirements, the data in Tables 3.12, 3.13, and 3.14 yield the following model:

$$\text{Minimize} \quad Z = 8x_1 + 10x_2 + 7x_3 + 6x_4 + 11x_5 + 9x_6,$$

subject to the following constraints:

1. Emission reduction:

$$12x_1 + 9x_2 + 25x_3 + 20x_4 + 17x_5 + 13x_6 \geq 60$$
$$35x_1 + 42x_2 + 18x_3 + 31x_4 + 56x_5 + 49x_6 \geq 150$$
$$37x_1 + 53x_2 + 28x_3 + 24x_4 + 29x_5 + 20x_6 \geq 125$$

2. Technological limit:

$$x_j \leq 1, \quad \text{for } j = 1, 2, \ldots, 6$$

3. Nonnegativity:

$$x_j \geq 0, \quad \text{for } j = 1, 2, \ldots, 6.$$

The OR team used this model[7] to find a minimum-cost plan

$$(x_1, x_2, x_3, x_4, x_5, x_6) = (1, 0.623, 0.343, 1, 0.048, 1),$$

with $Z = 32.16$ (total annual cost of $32.16 million). Sensitivity analysis then was conducted to explore the effect of making possible adjustments in the air standards given in Table 3.12, as well as to check on the effect of any inaccuracies in the cost data given in Table 3.14. (This story is continued in Case 6.1 at the end of Chap. 6.) Next came detailed planning and managerial review. Soon after, this program for controlling air pollution was fully implemented by the company, and the citizens of Steeltown breathed deep (cleaner) sighs of relief.

Reclaiming Solid Wastes

The SAVE-IT COMPANY operates a reclamation center that collects four types of solid waste materials and treats them so that they can be amalgamated into a salable product. (Treating and amalgamating are separate processes.) Three different grades of this product can be made (see the first column of Table 3.16), depending upon the mix of the materials used. Although there is some flexibility in the mix for each grade, quality standards may specify the minimum or maximum amount allowed for the proportion of a material in the product grade. (This proportion is the weight of the material expressed as a percentage of the total weight for the product grade.) For each of the two higher grades, a fixed percentage

---

[7]An equivalent formulation can express each decision variable in natural units for its abatement method; for example, x_1 and x_2 could represent the number of *feet* that the heights of the smokestacks are increased.

TABLE 3.16 Product data for Save-It Co.

Grade	Specification	Amalgamation Cost per Pound ($)	Selling Price per Pound ($)
A	Material 1: Not more than 30% of total Material 2: Not less than 40% of total Material 3: Not more than 50% of total Material 4: Exactly 20% of total	3.00	8.50
B	Material 1: Not more than 50% of total Material 2: Not less than 10% of total Material 4: Exactly 10% of total	2.50	7.00
C	Material 1: Not more than 70% of total	2.00	5.50

is specified for one of the materials. These specifications are given in Table 3.16 along with the cost of amalgamation and the selling price for each grade.

The reclamation center collects its solid waste materials from regular sources and so is normally able to maintain a steady rate for treating them. Table 3.17 gives the quantities available for collection and treatment each week, as well as the cost of treatment, for each type of material.

The Save-It Co. is solely owned by Green Earth, an organization devoted to dealing with environmental issues, so Save-It's profits are used to help support Green Earth's activities. Green Earth has raised contributions and grants, amounting to $30,000 per week, to be used exclusively to cover the entire treatment cost for the solid waste materials. The board of directors of Green Earth has instructed the management of Save-It to divide this money among the materials in such a way that *at least half* of the amount available of each material is actually collected and treated. These additional restrictions are listed in Table 3.17.

Within the restrictions specified in Tables 3.16 and 3.17, management wants to determine the *amount* of each product grade to produce *and* the exact *mix* of materials to be used for each grade. The objective is to maximize the net weekly profit (total sales income *minus* total amalgamation cost), exclusive of the fixed treatment cost of $30,000 per week that is being covered by gifts and grants.

Formulation as a Linear Programming Problem.   Before attempting to construct a linear programming model, we must give careful consideration to the proper definition of the decision variables. Although this definition is often obvious, it sometimes becomes the crux of the entire formulation. After clearly identifying what information is really desired and the most convenient form for conveying this information by means of decision variables, we can develop the objective function and the constraints on the values of these decision variables.

TABLE 3.17 Solid waste materials data for the Save-It Co.

Material	Pounds per Week Available	Treatment Cost per Pound ($)	Additional Restrictions
1	3,000	3.00	1. For each material, at least half of the pounds per week available should be collected and treated.
2	2,000	6.00	
3	4,000	4.00	
4	1,000	5.00	2. $30,000 per week should be used to treat these materials.

In this particular problem, the decisions to be made are well defined, but the appropriate means of conveying this information may require some thought. (Try it and see if you first obtain the following *inappropriate* choice of decision variables.)

Because one set of decisions is the *amount* of each product grade to produce, it would seem natural to define one set of decision variables accordingly. Proceeding tentatively along this line, we define

$$y_i = \text{number of pounds of product grade } i \text{ produced per week} \quad (i = A, B, C).$$

The other set of decisions is the *mix* of materials for each product grade. This mix is identified by the proportion of each material in the product grade, which would suggest defining the other set of decision variables as

$$z_{ij} = \text{proportion of material } j \text{ in product grade } i \quad (i = A, B, C; j = 1, 2, 3, 4).$$

However, Table 3.17 gives both the treatment cost and the availability of the materials by *quantity* (pounds) rather than *proportion,* so it is this *quantity* information that needs to be recorded in some of the constraints. For material j ($j = 1, 2, 3, 4$),

$$\text{Number of pounds of material } j \text{ used per week} = z_{Aj}y_A + z_{Bj}y_B + z_{Cj}y_C.$$

For example, since Table 3.17 indicates that 3,000 pounds of material 1 is available per week, one constraint in the model would be

$$z_{A1}y_A + z_{B1}y_B + z_{C1}y_C \leq 3{,}000.$$

Unfortunately, this is not a legitimate linear programming constraint. The expression on the left-hand side is *not* a linear function because it involves products of variables. Therefore, a linear programming model cannot be constructed with these decision variables.

Fortunately, there is another way of defining the decision variables that will fit the linear programming format. (Do you see how to do it?) It is accomplished by merely replacing each *product* of the old decision variables by a single variable! In other words, define

$$x_{ij} = z_{ij}y_i \quad (\text{for } i = A, B, C; j = 1, 2, 3, 4)$$
$$= \text{number of pounds of material } j \text{ allocated to product grade } i \text{ per week,}$$

and then we let the x_{ij} be the decision variables. Combining the x_{ij} in different ways yields the following quantities needed in the model (for $i = A, B, C; j = 1, 2, 3, 4$).

$$x_{i1} + x_{i2} + x_{i3} + x_{i4} = \text{number of pounds of product grade } i \text{ produced per week.}$$
$$x_{Aj} + x_{Bj} + x_{Cj} = \text{number of pounds of material } j \text{ used per week.}$$
$$\frac{x_{ij}}{x_{i1} + x_{i2} + x_{i3} + x_{i4}} = \text{proportion of material } j \text{ in product grade } i.$$

The fact that this last expression is a *nonlinear* function does not cause a complication. For example, consider the first specification for product grade A in Table 3.16 (the proportion of material 1 should not exceed 30 percent). This restriction gives the nonlinear constraint

$$\frac{x_{A1}}{x_{A1} + x_{A2} + x_{A3} + x_{A4}} \leq 0.3.$$

However, multiplying through both sides of this inequality by the denominator yields an *equivalent* constraint

$$x_{A1} \leq 0.3(x_{A1} + x_{A2} + x_{A3} + x_{A4}),$$

so

$$0.7x_{A1} - 0.3x_{A2} - 0.3x_{A3} - 0.3x_{A4} \leq 0,$$

which is a legitimate linear programming constraint.

With this adjustment, the three quantities given above lead directly to all the functional constraints of the model. The objective function is based on management's objective of <u>maximizing net weekly profit</u> (total sales income *minus* total amalgamation cost) from the three product grades. Thus, for each product grade, the profit per pound is obtained by subtracting the amalgamation cost given in the third column of Table 3.16 from the selling price in the fourth column. These *differences* provide the coefficients for the objective function.

Therefore, the complete linear programming model is

$$\text{Maximize} \quad Z = 5.5(x_{A1} + x_{A2} + x_{A3} + x_{A4}) + 4.5(x_{B1} + x_{B2} + x_{B3} + x_{B4})$$
$$+ 3.5(x_{C1} + x_{C2} + x_{C3} + x_{C4}),$$

subject to the following constraints:

1. Mixture specifications (second column of Table 3.16):

$$x_{A1} \leq 0.3(x_{A1} + x_{A2} + x_{A3} + x_{A4}) \quad \text{(grade } A, \text{ material 1)}$$
$$x_{A2} \geq 0.4(x_{A1} + x_{A2} + x_{A3} + x_{A4}) \quad \text{(grade } A, \text{ material 2)}$$
$$x_{A3} \leq 0.5(x_{A1} + x_{A2} + x_{A3} + x_{A4}) \quad \text{(grade } A, \text{ material 3)}$$
$$x_{A4} = 0.2(x_{A1} + x_{A2} + x_{A3} + x_{A4}) \quad \text{(grade } A, \text{ material 4)}$$
$$x_{B1} \leq 0.5(x_{B1} + x_{B2} + x_{B3} + x_{B4}) \quad \text{(grade } B, \text{ material 1)}$$
$$x_{B2} \geq 0.1(x_{B1} + x_{B2} + x_{B3} + x_{B4}) \quad \text{(grade } B, \text{ material 2)}$$
$$x_{B4} = 0.1(x_{B1} + x_{B2} + x_{B3} + x_{B4}) \quad \text{(grade } B, \text{ material 4)}$$
$$x_{C1} \leq 0.7(x_{C1} + x_{C2} + x_{C3} + x_{C4}) \quad \text{(grade } C, \text{ material 1).}$$

2. Availability of materials (second column of Table 3.17):

$$x_{A1} + x_{B1} + x_{C1} \leq 3{,}000 \quad \text{(material 1)}$$
$$x_{A2} + x_{B2} + x_{C2} \leq 2{,}000 \quad \text{(material 2)}$$
$$x_{A3} + x_{B3} + x_{C3} \leq 4{,}000 \quad \text{(material 3)}$$
$$x_{A4} + x_{B4} + x_{C4} \leq 1{,}000 \quad \text{(material 4).}$$

3. Restrictions on amounts treated (right side of Table 3.17):

$$x_{A1} + x_{B1} + x_{C1} \geq 1{,}500 \quad \text{(material 1)}$$
$$x_{A2} + x_{B2} + x_{C2} \geq 1{,}000 \quad \text{(material 2)}$$
$$x_{A3} + x_{B3} + x_{C3} \geq 2{,}000 \quad \text{(material 3)}$$
$$x_{A4} + x_{B4} + x_{C4} \geq 500 \quad \text{(material 4).}$$

4. Restriction on treatment cost (right side of Table 3.17):

$$3(x_{A1} + x_{B1} + x_{C1}) + 6(x_{A2} + x_{B2} + x_{C2}) + 4(x_{A3} + x_{B3} + x_{C3})$$
$$+ 5(x_{A4} + x_{B4} + x_{C4}) = 30{,}000.$$

5. Nonnegativity constraints:

$$x_{A1} \geq 0, \quad x_{A2} \geq 0, \quad \ldots, \quad x_{C4} \geq 0.$$

■ **TABLE 3.18** Optimal solution for the Save-It Co. problem

| | Pounds Used per Week | | | | |
| | Material | | | | Number of Pounds |
Grade	1	2	3	4	Produced per Week
A	412.3	859.6	447.4	429.8	2149
	(19.2%)	(40%)	(20.8%)	(20%)	
B	2587.7	517.5	1552.6	517.5	5175
	(50%)	(10%)	(30%)	(10%)	
C	0	0	0	0	0
Total	3000	1377	2000	947	

This formulation completes the model, except that the constraints for the mixture specifications need to be rewritten in the proper form for a linear programming model by bringing all variables to the left-hand side and combining terms, as follows:

Mixture specifications:

$$0.7x_{A1} - 0.3x_{A2} - 0.3x_{A3} - 0.3x_{A4} \leq 0 \quad \text{(grade } A, \text{material 1)}$$
$$-0.4x_{A1} + 0.6x_{A2} - 0.4x_{A3} - 0.4x_{A4} \geq 0 \quad \text{(grade } A, \text{material 2)}$$
$$-0.5x_{A1} - 0.5x_{A2} + 0.5x_{A3} - 0.5x_{A4} \leq 0 \quad \text{(grade } A, \text{material 3)}$$
$$-0.2x_{A1} - 0.2x_{A2} - 0.2x_{A3} + 0.8x_{A4} = 0 \quad \text{(grade } A, \text{material 4)}$$
$$0.5x_{B1} - 0.5x_{B2} - 0.5x_{B3} - 0.5x_{B4} \leq 0 \quad \text{(grade } B, \text{material 1)}$$
$$-0.1x_{B1} + 0.9x_{B2} - 0.1x_{B3} - 0.1x_{B4} \geq 0 \quad \text{(grade } B, \text{material 2)}$$
$$-0.1x_{B1} - 0.1x_{B2} - 0.1x_{B3} + 0.9x_{B4} = 0 \quad \text{(grade } B, \text{material 4)}$$
$$0.3x_{C1} - 0.7x_{C2} - 0.7x_{C3} - 0.7x_{C4} \leq 0 \quad \text{(grade } C, \text{material 1)}.$$

An optimal solution for this model is shown in Table 3.18, and then these x_{ij} values are used to calculate the other quantities of interest given in the table. The resulting optimal value of the objective function is $Z = 35,109.65$ (a total weekly profit of \$35,109.65).

The Save-It Co. problem is an example of a **blending problem.** The objective for a blending problem is to find the best blend of ingredients into final products to meet certain specifications. Some of the earliest applications of linear programming were for *gasoline blending,* where petroleum ingredients were blended to obtain various grades of gasoline. Other blending problems involve such final products as steel, fertilizer, and animal feed.

Personnel Scheduling

UNION AIRWAYS is adding more flights to and from its hub airport, and so it needs to hire additional customer service agents. However, it is not clear just how many more should be hired. Management recognizes the need for cost control while also consistently providing a satisfactory level of service to customers. Therefore, an OR team is studying how to schedule the agents to provide satisfactory service with the smallest personnel cost.

Based on the new schedule of flights, an analysis has been made of the *minimum* number of customer service agents that need to be on duty at different times of the day to provide a satisfactory level of service. The rightmost column of Table 3.19 shows the number of agents needed for the time periods given in the first column. The other entries in this table reflect one of the provisions in the company's current contract with the union that

Cost control is essential for survival in the airline industry. Therefore, upper management of **United Airlines** initiated an operations research study to improve the utilization of personnel at the airline's reservations offices and airports by matching work schedules to customer needs more closely. The number of employees needed at each location to provide the required level of service varies greatly during the 24-hour day and might fluctuate considerably from one half-hour to the next.

Trying to design the work schedules for all the employees at a given location to meet these service requirements most efficiently is a nightmare of combinatorial considerations. Once an employee arrives, he or she will be there continuously for the entire shift (2 to 10 hours, depending on the employee), *except* for either a meal break or short rest breaks every two hours. Given the *minimum* number of employees needed on duty for *each* half-hour interval over a 24-hour day (this minimum changes from day to day over a seven-day week), *how many* employees of *each shift length* should begin work at *what start time* over *each* 24-hour day of a seven-day week? Fortunately, linear programming thrives on such combinatorial nightmares. The linear programming model for some of the locations scheduled involves over 20,000 decisions!

This application of linear programming was credited with *saving United Airlines more than* **$6 million** *annually* in just direct salary and benefit costs. Other benefits included improved customer service and reduced workloads for support staff.

Source: T. J. Holloran and J. E. Bryne, "United Airlines Station Manpower Planning System," *Interfaces,* **16**(1): 39–50, Jan.–Feb. 1986. (A link to this article is provided on our website, www.mhhe.com/hillier.)

represents the customer service agents. The provision is that each agent work an 8-hour shift 5 days per week, and the authorized shifts are

Shift 1: 6:00 A.M. to 2:00 P.M.
Shift 2: 8:00 A.M. to 4:00 P.M.
Shift 3: Noon to 8:00 P.M.
Shift 4: 4:00 P.M. to midnight
Shift 5: 10:00 P.M. to 6:00 A.M.

Checkmarks in the main body of Table 3.19 show the hours covered by the respective shifts. Because some shifts are less desirable than others, the wages specified in the contract differ by shift. For each shift, the daily compensation (including benefits) for each agent is shown in the bottom row. The problem is to determine how many agents should be

■ **TABLE 3.19** Data for the Union Airways personnel scheduling problem

Time Period	Time Periods Covered					Minimum Number of Agents Needed
	Shift					
	1	2	3	4	5	
6:00 A.M. to 8:00 A.M.	✔					48
8:00 A.M. to 10:00 A.M.	✔	✔				79
10:00 A.M. to noon	✔	✔				65
Noon to 2:00 P.M.	✔	✔	✔			87
2:00 P.M. to 4:00 P.M.		✔	✔			64
4:00 P.M. to 6:00 P.M.			✔	✔		73
6:00 P.M. to 8:00 P.M.			✔	✔		82
8:00 P.M. to 10:00 P.M.				✔		43
10:00 P.M. to midnight				✔	✔	52
Midnight to 6:00 A.M.					✔	15
Daily cost per agent	$170	$160	$175	$180	$195	

assigned to the respective shifts each day to minimize the *total* personnel cost for agents, based on this bottom row, while meeting (or surpassing) the service requirements given in the rightmost column.

Formulation as a Linear Programming Problem.   Linear programming problems always involve finding the best *mix of activity levels.* The key to formulating this particular problem is to recognize the nature of the activities.

Activities correspond to shifts, where the *level* of each activity is the number of agents assigned to that shift. Thus, this problem involves finding the *best mix of shift sizes.* Since the decision variables always are the levels of the activities, the five decision variables here are

x_j = number of agents assigned to shift j,   for $j = 1, 2, 3, 4, 5$.

The main restrictions on the values of these decision variables are that the number of agents working during each time period must satisfy the minimum requirement given in the rightmost column of Table 3.19. For example, for 2:00 P.M. to 4:00 P.M., the total number of agents assigned to the shifts that cover this time period (shifts 2 and 3) must be at least 64, so

$x_2 + x_3 \geq 64$

is the functional constraint for this time period.

Because the objective is to minimize the total cost of the agents assigned to the five shifts, the coefficients in the objective function are given by the last row of Table 3.19.

Therefore, the complete linear programming model is

Minimize $Z = 170x_1 + 160x_2 + 175x_3 + 180x_4 + 195x_5$,

subject to

$$
\begin{aligned}
x_1 && \geq 48 && \text{(6–8 A.M.)} \\
x_1 + x_2 && \geq 79 && \text{(8–10 A.M.)} \\
x_1 + x_2 && \geq 65 && \text{(10 A.M. to noon)} \\
x_1 + x_2 + x_3 && \geq 87 && \text{(Noon–2 P.M.)} \\
x_2 + x_3 && \geq 64 && \text{(2–4 P.M.)} \\
x_3 + x_4 && \geq 73 && \text{(4–6 P.M.)} \\
x_3 + x_4 && \geq 82 && \text{(6–8 P.M.)} \\
x_4 && \geq 43 && \text{(8–10 P.M.)} \\
x_4 + x_5 && \geq 52 && \text{(10 P.M.–midnight)} \\
x_5 && \geq 15 && \text{(Midnight–6 A.M.)}
\end{aligned}
$$

and

$x_j \geq 0$,   for $j = 1, 2, 3, 4, 5$.

With a keen eye, you might have noticed that the third constraint, $x_1 + x_2 \geq 65$, actually is not necessary because the second constraint, $x_1 + x_2 \geq 79$, ensures that $x_1 + x_2$ will be larger than 65. Thus, $x_1 + x_2 \geq 65$ is a *redundant* constraint that can be deleted. Similarly, the sixth constraint, $x_3 + x_4 \geq 73$, also is a *redundant* constraint because the seventh constraint is $x_3 + x_4 \geq 82$. (In fact, three of the nonnegativity constraints— $x_1 \geq 0$, $x_4 \geq 0$, $x_5 \geq 0$—also are redundant constraints because of the first, eighth, and tenth functional constraints: $x_1 \geq 48$, $x_4 \geq 43$, and $x_5 \geq 15$. However, no computational advantage is gained by deleting these three nonnegativity constraints.)

The optimal solution for this model is $(x_1, x_2, x_3, x_4, x_5) = (48, 31, 39, 43, 15)$. This yields $Z = 30,610$, that is, a total daily personnel cost of \$30,610.

This problem is an example where the divisibility assumption of linear programming actually is not satisfied. The number of agents assigned to each shift needs to be an integer. Strictly speaking, the model should have an additional constraint for each decision variable specifying that the variable must have an integer value. Adding these constraints would convert the linear programming model to an integer programming model (the topic of Chap. 11).

Without these constraints, the optimal solution given above turned out to have integer values anyway, so no harm was done by not including the constraints. (The form of the functional constraints made this outcome a likely one.) If some of the variables had turned out to be noninteger, the easiest approach would have been to *round up* to integer values. (Rounding up is feasible for this example because all the functional constraints are in $\geq$ form with nonnegative coefficients.) Rounding up does not ensure obtaining an optimal solution for the integer programming model, but the error introduced by rounding up such large numbers would be negligible for most practical situations. Alternatively, integer programming techniques described in Chap. 11 could be used to solve exactly for an optimal solution with integer values.

The second application vignette in this section describes how United Airlines used linear programming to develop a personnel scheduling system on a vastly larger scale than this example.

Distributing Goods through a Distribution Network

The Problem.   The DISTRIBUTION UNLIMITED CO. will be producing the same new product at two different factories, and then the product must be shipped to two warehouses, where either factory can supply either warehouse. The distribution network available for shipping this product is shown in Fig. 3.13, where F1 and F2 are the two factories, W1 and W2 are the two warehouses, and DC is a distribution center. The amounts to be shipped from F1 and F2 are shown to their left, and the amounts to be received at W1 and W2 are shown to their right. Each arrow represents a feasible shipping lane. Thus, F1 can ship directly to W1 and has three possible routes (F1 → DC → W2, F1 → F2 → DC → W2, and F1 → W1 → W2) for shipping to W2. Factory F2 has just one route to W2 (F2 → DC → W2) and one to W1 (F2 → DC → W2 → W1). The cost per unit shipped through each shipping lane is shown next to the arrow. Also shown next to F1 → F2 and DC → W2 are the maximum amounts that can be shipped through these lanes. The other lanes have sufficient shipping capacity to handle everything these factories can send.

The decision to be made concerns how much to ship through each shipping lane. The objective is to minimize the total shipping cost.

Formulation as a Linear Programming Problem.   With seven shipping lanes, we need seven decision variables ($x_{\text{F1-F2}}$, $x_{\text{F1-DC}}$, $x_{\text{F1-W1}}$, $x_{\text{F2-DC}}$, $x_{\text{DC-W2}}$, $x_{\text{W1-W2}}$, $x_{\text{W2-W1}}$) to represent the amounts shipped through the respective lanes.

There are several restrictions on the values of these variables. In addition to the usual nonnegativity constraints, there are two *upper-bound constraints,* $x_{\text{F1-F2}} \leq 10$ and $x_{\text{DC-W2}} \leq 80$, imposed by the limited shipping capacities for the two lanes, F1 → F2 and DC → W2. All the other restrictions arise from five *net flow constraints,* one for each of the five locations. These constraints have the following form.

Net flow constraint for each location:

Amount shipped out − amount shipped in = required amount.

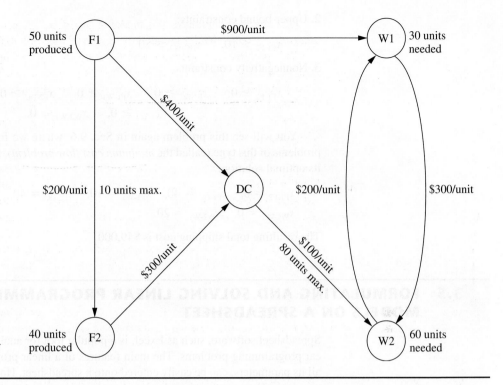

■ FIGURE 3.13
The distribution network for
Distribution Unlimited Co.

As indicated in Fig. 3.13, these required amounts are 50 for F1, 40 for F2, -30 for W1, and -60 for W2.

What is the required amount for DC? All the units produced at the factories are ultimately needed at the warehouses, so any units shipped from the factories to the distribution center should be forwarded to the warehouses. Therefore, the total amount shipped from the distribution center to the warehouses should *equal* the total amount shipped from the factories to the distribution center. In other words, the *difference* of these two shipping amounts (the required amount for the net flow constraint) should be *zero*.

Since the objective is to minimize the total shipping cost, the coefficients for the objective function come directly from the unit shipping costs given in Fig. 3.13. Therefore, by using money units of hundreds of dollars in this objective function, the complete linear programming model is

$$\text{Minimize} \quad Z = 2x_{\text{F1-F2}} + 4x_{\text{F1-DC}} + 9x_{\text{F1-W1}} + 3x_{\text{F2-DC}} + x_{\text{DC-W2}}$$
$$+ 3x_{\text{W1-W2}} + 2x_{\text{W2-W1}},$$

subject to the following constraints:

1. Net flow constraints:

$$
\begin{aligned}
x_{\text{F1-F2}} + x_{\text{F1-DC}} + x_{\text{F1-W1}} &= 50 \,(\text{factory } 1)\\
-x_{\text{F1-F2}} \qquad\qquad\quad + x_{\text{F2-DC}} &= 40 \,(\text{factory } 2)\\
- x_{\text{F1-DC}} \qquad\quad - x_{\text{F2-DC}} + x_{\text{DC-W2}} &= 0 \,(\text{distribution}\\
&\qquad\quad\text{center})\\
- x_{\text{F1-W1}} \qquad\qquad + x_{\text{W1-W2}} - x_{\text{W2-W1}} &= -30 \,(\text{warehouse } 1)\\
- x_{\text{DC-W2}} - x_{\text{W1-W2}} + x_{\text{W2-W1}} &= -60 \,(\text{warehouse } 2)
\end{aligned}
$$

2. Upper-bound constraints:

$$x_{F1-F2} \leq 10, \qquad x_{DC-W2} \leq 80$$

3. Nonnegativity constraints:

$$x_{F1-F2} \geq 0, \qquad x_{F1-DC} \geq 0, \qquad x_{F1-W1} \geq 0, \qquad x_{F2-DC} \geq 0, \qquad x_{DC-W2} \geq 0,$$
$$x_{W1-W2} \geq 0, \qquad x_{W2-W1} \geq 0.$$

You will see this problem again in Sec. 9.6, where we focus on linear programming problems of this type (called the *minimum cost flow problem*). In Sec. 9.7, we will solve for its optimal solution:

$$x_{F1-F2} = 0, \quad x_{F1-DC} = 40, \quad x_{F1-W1} = 10, \quad x_{F2-DC} = 40, \quad x_{DC-W2} = 80,$$
$$x_{W1-W2} = 0, \quad x_{W2-W1} = 20.$$

The resulting total shipping cost is $49,000.

■ 3.5   FORMULATING AND SOLVING LINEAR PROGRAMMING MODELS ON A SPREADSHEET

Spreadsheet software, such as Excel, is a popular tool for analyzing and solving small linear programming problems. The main features of a linear programming model, including all its parameters, can be easily entered onto a spreadsheet. However, spreadsheet software can do much more than just display data. If we include some additional information, the spreadsheet can be used to quickly analyze potential solutions. For example, a potential solution can be checked to see if it is feasible and what Z value (profit or cost) it achieves. Much of the power of the spreadsheet lies in its ability to immediately reveal the results of any changes made in the solution.

In addition, the Excel Solver can quickly apply the simplex method to find an optimal solution for the model. We will describe how this is done in the latter part of this section.

To illustrate this process of formulating and solving linear programming models on a spreadsheet, we now return to the Wyndor example introduced in Sec. 3.1.

Formulating the Model on a Spreadsheet

Figure 3.14 displays the Wyndor problem by transferring the data from Table 3.1 onto a spreadsheet. (Columns E and F are being reserved for later entries described below.) We will refer to the cells showing the data as **data cells.** These cells are lightly shaded to distinguish them from other cells in the spreadsheet.[8]

You will see later that the spreadsheet is made easier to interpret by using range names. **A range name** is a descriptive name given to a block of cells that immediately identifies what is there. Thus, the data cells in the Wyndor problem are given the range names UnitProfit (C4:D4), HoursUsedPerBatchProduced (C7:D9), and HoursAvailable (G7:G9). Note that no spaces are allowed in a range name so each new word begins with a capital letter. Although optional, the range of cells being given each range name can be specified in parentheses following the name. (For example, the range C7:D9 is Excel shorthand for the *range from* C7 *to* D9; that is, the entire block of cells in column C or D and in row 7, 8, or 9.) To enter a range name, first select the range of cells, then choose

---

[8]Borders and cell shading can be added either by using the borders button and the fill color button on the formatting toolbar or by choosing Cells from the Format menu and then selecting the Borders tab and/or the Patterns tab.

Welch's, Inc., is the world's largest processor of Concord and Niagara grapes, with annual sales surpassing $550 million per year. Such products as Welch's grape jelly and Welch's grape juice have been enjoyed by generations of American consumers.

Every September, growers begin delivering grapes to processing plants that then press the raw grapes into juice. Time must pass before the grape juice is ready for conversion into finished jams, jellies, juices, and concentrates.

Deciding how to use the grape crop is a complex task given changing demand and uncertain crop quality and quantity. Typical decisions include what recipes to use for major product groups, the transfer of grape juice between plants, and the mode of transportation for these transfers.

Because Welch's lacked a formal system for optimizing raw material movement and the recipes used for production, an OR team developed a preliminary linear programming model. This was a large model with 8,000 decision variables that focused on the component level of detail. Small-scale testing proved that the model worked.

To make the model more useful, the team then revised it by aggregating demand by product group rather than by component. This reduced its size to 324 decision variables and 361 functional constraints. *The model then was incorporated into a spreadsheet.*

The company has run the continually updated version of this *spreadsheet model* each month since 1994 to provide senior management with information on the optimal logistics plan generated by the Solver. *The savings* from using and optimizing this model *were approximately* **$150,000** *in the first year alone.* A major advantage of incorporating the linear programming model into a spreadsheet has been the ease of explaining the model to managers with differing levels of mathematical understanding. This has led to a widespread appreciation of the operations research approach for both this application and others.

Source: E. W. Schuster and S. J. Allen, "Raw Material Management at Welch's, Inc.," *Interfaces,* **28**(5): 13–24, Sept.–Oct. 1998. (A link to this article is provided on our website, www.mhhe.com/hillier.)

Name\Define from the Insert menu and type a range name (or click in the name box on the left of the formula bar above the spreadsheet and type a name).

Three questions need to be answered to begin the process of using the spreadsheet to formulate a linear programming model for the problem.

1. What are the *decisions* to be made? For this problem, the necessary decisions are the *production rates* (number of batches produced per week) for the two new products.
2. What are the *constraints* on these decisions? The constraints here are that the number of hours of production time used per week by the two products in the respective plants cannot exceed the number of hours available.
3. What is the overall *measure of performance* for these decisions? Wyndor's overall measure of performance is the *total profit* per week from the two products, so the *objective* is to *maximize* this quantity.

Figure 3.15 shows how these answers can be incorporated into the spreadsheet. Based on the first answer, the *production rates* of the two products are placed in cells C12 and

FIGURE 3.14
The initial spreadsheet for the Wyndor problem after transferring the data from Table 3.1 into data cells.

	A	B	C	D	E	F	G
1		**Wyndor Glass Co. Product-Mix Problem**					
2							
3			Doors	Windows			
4		Profit Per Batch	$3,000	$5,000			
5							Hours
6			Hours Used Per Batch Produced				Available
7		Plant 1	1	0			4
8		Plant 2	0	2			12
9		Plant 3	3	2			18

	A	B	C	D	E	F	G
1	**Wyndor Glass Co. Product-Mix Problem**						
2							
3			Doors	Windows			
4		Profit Per Batch	$3,000	$5,000			
5					Hours		Hours
6			Hours Used Per Batch Produced		Used		Available
7		Plant 1	1	0	0	<=	4
8		Plant 2	0	2	0	<=	12
9		Plant 3	3	2	0	<=	18
10							
11			Doors	Windows			Total Profit
12		Batches Produced	0	0			$0

■ **FIGURE 3.15**
The complete spreadsheet for the Wyndor problem with an initial trial solution (both production rates equal to zero) entered into the changing cells (C12 and D12).

D12 to locate them in the columns for these products just under the data cells. Since we don't know yet what these production rates should be, they are just entered as zeroes at this point. (Actually, any trial solution can be entered, although *negative* production rates should be excluded since they are impossible.) Later, these numbers will be changed while seeking the best mix of production rates. Therefore, these cells containing the decisions to be made are called **changing cells** (or *adjustable cells*). To highlight the changing cells, they are shaded and have a border. (In the spreadsheet files contained in OR Courseware, the changing cells appear in bright yellow on a color monitor.) The changing cells are given the range name BatchesProduced (C12:D12).

Using the answer to question 2, the total number of hours of production time used per week by the two products in the respective plants is entered in cells E7, E8, and E9, just to the right of the corresponding data cells. The Excel equations for these three cells are

$$E7 = C7*C12 + D7*D12$$

$$E8 = C8*C12 + D8*D12$$

$$E9 = C9*C12 + D9*D12$$

where each asterisk denotes multiplication. Since each of these cells provides output that depends on the changing cells (C12 and D12), they are called **output cells.**

Notice that each of the equations for the output cells involves the sum of two products. There is a function in Excel called SUMPRODUCT that will sum up the product of each of the individual terms in two different ranges of cells when the two ranges have the same number of rows and the same number of columns. Each product being summed is the product of a term in the first range and the term in the corresponding location in the second range. For example, consider the two ranges, C7:D7 and C12:D12, so that each range has one row and two columns. In this case, SUMPRODUCT (C7:D7, C12:D12) takes each of the individual terms in the range C7:D7, multiplies them by the corresponding term in the range C12:D12, and then sums up these individual products, as shown in the first equation above. Using the range name BatchesProduced (C12:D12), the formula becomes SUMPRODUCT (C7:D7, BatchesProduced). Although optional with such short equations, this function is especially handy as a shortcut for entering longer equations.

Next, ≤ signs are entered in cells F7, F8, and F9 to indicate that each total value to their left cannot be allowed to exceed the corresponding number in column G. The spreadsheet still will allow you to enter trial solutions that violate the ≤ signs. However, these ≤ signs serve as a reminder that such trial solutions need to be rejected if no changes are made in the numbers in column G.

Finally, since the answer to the third question is that the overall measure of performance is the total profit from the two products, this profit (per week) is entered in cell G12. Much like the numbers in column E, it is the sum of products,

G12 = SUMPRODUCT (C4:D4, C12:D12)

Utilizing range names of TotalProfit (G12), ProfitPerBatch (C4:D4), and BatchesProduced (C12:D12), this equation becomes

TotalProfit = SUMPRODUCT (ProfitPerBatch, BatchesProduced)

This is a good example of the benefit of using range names for making the resulting equation easier to interpret. Rather than needing to refer to the spreadsheet to see what is in cells G12, C4:D4, and C12:D12, the range names immediately reveal what the equation is doing.

TotalProfit (G12) is a special kind of output cell. It is the particular cell that is being targeted to be made as large as possible when making decisions regarding production rates. Therefore, TotalProfit (G12) is referred to as the **target cell** (or *objective cell*). The target cell is shaded darker than the changing cells and is further distinguished by having a heavy border. (In the spreadsheet files contained in OR Courseware, this cell appears in orange on a color monitor.)

The bottom of Fig. 3.16 summarizes all the formulas that need to be entered in the Hours Used column and in the Total Profit cell. Also shown is a summary of the range names (in alphabetical order) and the corresponding cell addresses.

This completes the formulation of the spreadsheet model for the Wyndor problem.

With this formulation, it becomes easy to analyze any trial solution for the production rates. Each time production rates are entered in cells C12 and D12, Excel immediately

FIGURE 3.16
The spreadsheet model for the Wyndor problem, including the formulas for the target cell TotalProfit (G12) and the other output cells in column E, where the objective is to maximize the target cell.

	A	B	C	D	E	F	G
1			**Wyndor Glass Co. Product-Mix Problem**				
2							
3			Doors	Windows			
4		Profit Per Batch	$3,000	$5,000			
5					Hours		Hours
6			Hours Used Per Batch Produced		Used		Available
7		Plant 1	1	0	0	<=	4
8		Plant 2	0	2	0	<=	12
9		Plant 3	3	2	0	<=	18
10							
11			Doors	Windows			Total Profit
12		Batches Produced	0	0			$0

Range Name	Cells
BatchesProduced	C12:D12
HoursAvailable	G7:G9
HoursUsed	E7:E9
HoursUsedPerBatchProduced	C7:D9
ProfitPerBatch	C4:D4
TotalProfit	G12

	E
5	Hours
6	Used
7	=SUMPRODUCT(C7:D7,BatchesProduced)
8	=SUMPRODUCT(C8:D8,BatchesProduced)
9	=SUMPRODUCT(C9:D9,BatchesProduced)

	G
11	Total Profit
12	=SUMPRODUCT(ProfitPerBatch,BatchesProduced)

calculates the output cells for hours used and total profit. However, it is not necessary to use trial and error. We shall describe next how the Excel Solver can be used to quickly find the optimal solution.

Using the Excel Solver to Solve the Model

Excel includes a tool called Solver that uses the simplex method to find an optimal solution. (A more powerful version of Solver, called *Premium Solver for Education,* also is available in your OR Courseware.)

To access Solver the first time, you need to install it by going to Excel's Add-in menu and adding Solver, after which you will find it on the Data tab (for Excel 2007) or in the Tools menu (for earlier versions of Excel).

To get started, an arbitrary trial solution has been entered in Fig. 3.16 by placing zeroes in the changing cells. The Solver will then change these to the optimal values after solving the problem.

This procedure is started by choosing Solver. The Solver dialogue box is shown in Fig. 3.17.

Before the Solver can start its work, it needs to know exactly where each component of the model is located on the spreadsheet. The Solver dialogue box is used to enter this information. You have the choice of typing the range names, typing in the cell addresses, or clicking on the cells in the spreadsheet.[9] Figure 3.17 shows the result of using the first choice, so TotalProfit (rather than G12) has been entered for the target cell and Batches Produced (rather than the range C12:D12) has been entered for the changing cells. Since the goal is to maximize the target cell, Max also has been selected.

Next, the cells containing the functional constraints need to be specified. This is done by clicking on the Add button on the Solver dialogue box. This brings up the Add Constraint dialogue box shown in Fig. 3.18. The ≤ signs in cells F7, F8, and F9 of Fig. 3.16 are a reminder that the cells in HoursUsed (E7:E9) all need to be less than or equal to the

■ FIGURE 3.17
This Solver dialogue box specifies which cells in Fig. 3.16 are the target cell and the changing cells. It also indicates that the target cell is to be maximized.

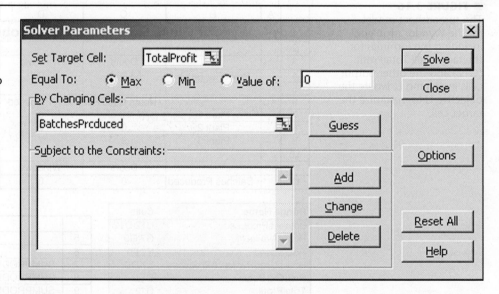

[9]If you select cells by clicking on them, they will first appear in the dialogue box with their cell addresses and with dollar signs (e.g., $C$9:$D$9). You can ignore the dollar signs. Solver will eventually replace both the cell addresses and the dollar signs with the corresponding range name (if a range name has been defined for the given cell addresses), but only after either adding a constraint or closing and reopening the Solver dialogue box.

■ FIGURE 3.18
The Add Constraint dialogue
box after entering the set of
constraints, HoursUsed
(E7:E9) ≤ HoursAvailable
(G7:G9), which specifies that
cells E7, E8, and E9 in
Fig. 3.16 are required to be
less than or equal to cells G7,
G8, and G9, respectively.

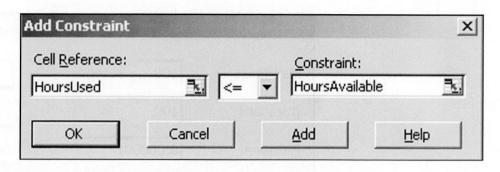

corresponding cells in HoursAvailable (G7:G9). These constraints are specified for the Solver by entering HoursUsed (or E7:E9) on the left-hand side of the Add Constraint dialogue box and HoursAvailable (or G7:G9) on the right-hand side. For the sign between these two sides, there is a menu to choose between <= (less than or equal), =, or >= (greater than or equal), so <= has been chosen. This choice is needed even though ≤ signs were previously entered in column F of the spreadsheet because the Solver only uses the functional constraints that are specified with the Add Constraint dialogue box.

If there were more functional constraints to add, you would click on Add to bring up a new Add Constraint dialogue box. However, since there are no more in this example, the next step is to click on OK to go back to the Solver dialogue box.

The Solver dialogue box now summarizes the complete model (see Fig. 3.19) in terms of the spreadsheet in Fig. 3.16. However, before asking Solver to solve the model, one more step should be taken. Clicking on the Options button brings up the dialogue box shown in Fig. 3.20. This box allows you to specify a number of options about how the problem will be solved. The most important of these are the Assume Linear Model option and the Assume Non-Negative option. Be sure that both options are checked as shown in the figure. This tells Solver that the problem is a *linear* programming problem and that nonnegativity constraints are needed for the changing cells to reject negative production rates. Regarding the other options, accepting the default values shown in the figure usually is fine for small problems. Clicking on the OK button then returns you to the Solver dialogue box.

■ FIGURE 3.19
The Solver dialogue box after specifying the entire model in terms of the spreadsheet.

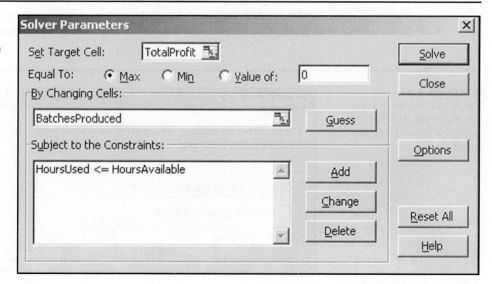

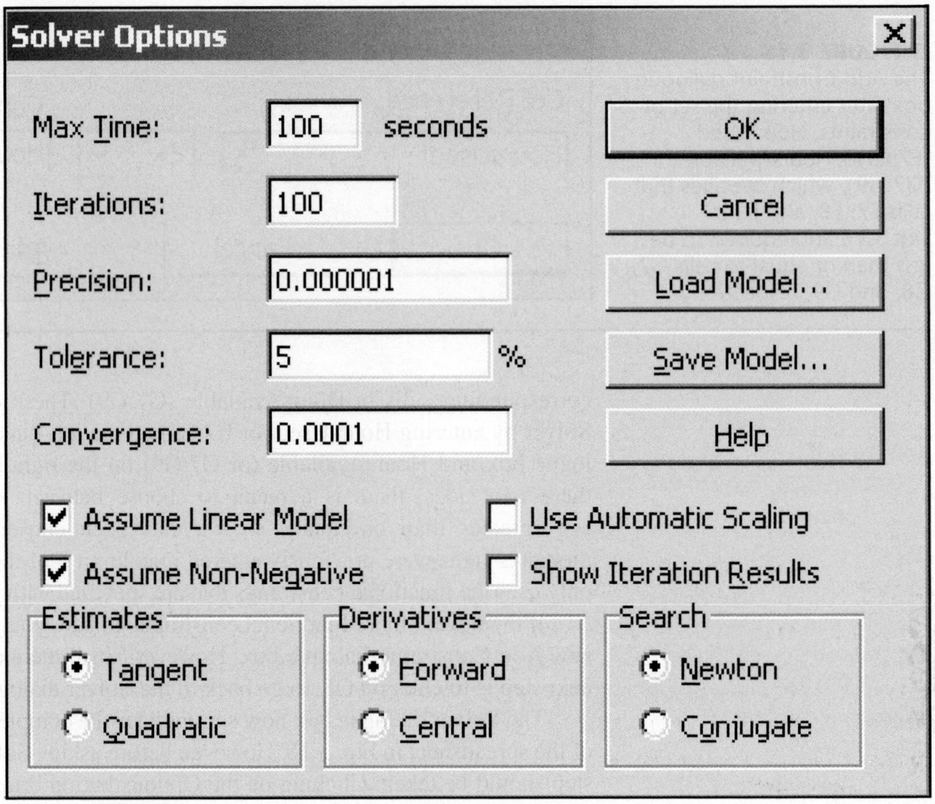

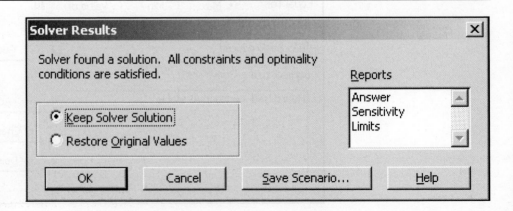

FIGURE 3.20
The Solver Options dialogue box after checking the Assume Linear Model and Assume Non-Negative options to indicate that we wish to solve a linear programming model that has nonnegativity constraints.

Now you are ready to click on Solve in the Solver dialogue box, which will start the process of solving the problem in the background. After a few seconds (for a small problem), Solver will then indicate the outcome. Typically, it will indicate that it has found an optimal solution, as specified in the Solver Results dialogue box shown in Fig. 3.21. If the model has no feasible solutions or no optimal solution, the dialogue box will indicate that instead by stating that "Solver could not find a feasible solution" or that "The Set Cell values do not converge." The dialogue box also presents the option of generating various reports. One of these (the Sensitivity Report) will be discussed later in Secs. 4.7 and 6.8.

FIGURE 3.21
The Solver Results dialogue box that indicates that an optimal solution has been found.

After solving the model, the Solver replaces the original numbers in the changing cells with the optimal numbers, as shown in Fig. 3.22. Thus, the optimal solution is to produce two batches of doors per week and six batches of windows per week, just as was found by the graphical method in Sec. 3.1. The spreadsheet also indicates the corresponding number in the target cell (a total profit of $36,000 per week), as well as the numbers in the output cells HoursUsed (E7:E9).

At this point, you might want to check what would happen to the optimal solution if any of the numbers in the data cells were changed to other possible values. This is easy to do because Solver saves all the addresses for the target cell, changing cells, constraints, and so on when you save the file. All you need to do is make the changes you want in the data cells and then click on Solve in the Solver dialogue box again. (Sections 4.7 and 6.8 will focus on this kind of *sensitivity analysis,* including how to use the Solver's Sensitivity Report to expedite this type of what-if analysis.)

To assist you with experimenting with these kinds of changes, your OR Courseware includes Excel files for this chapter (as for others) that provide a complete formulation and solution of the examples here (the Wyndor problem and the ones in Sec. 3.4) in a spreadsheet format. We encourage you to "play" with these examples to see what happens with different data, different solutions, and so forth. You might also find these spreadsheets useful as templates for solving homework problems.

In addition, we suggest that you use this chapter's Excel files to take a careful look at the spreadsheet formulations for some of the examples in Sec. 3.4. This will demonstrate

■ **FIGURE 3.22**
The spreadsheet obtained after solving the Wyndor problem.

	A	B	C	D	E	F	G
1		**Wyndor Glass Co. Product-Mix Problem**					
2							
3			Doors	Windows			
4		Profit Per Batch	$3,000	$5,000			
5					Hours		Hours
6			Hours Used Per Batch Produced		Used		Available
7		Plant 1	1	0	2	<=	4
8		Plant 2	0	2	12	<=	12
9		Plant 3	3	2	18	<=	18
10							
11			Doors	Windows			Total Profit
12		Batches Produced	2	6			$36,000

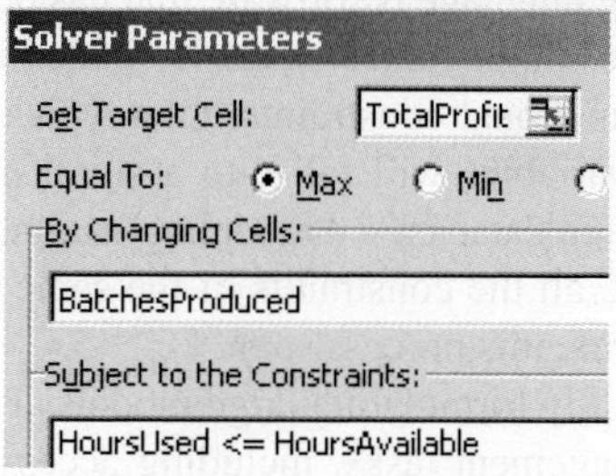

Solver Parameters

Set Target Cell: [TotalProfit]

Equal To: ● Max ○ Min ○

By Changing Cells:

[BatchesProduced]

Subject to the Constraints:

[HoursUsed <= HoursAvailable]

	E
5	Hours
6	Used
7	=SUMPRODUCT(C7:D7,BatchesProduced)
8	=SUMPRODUCT(C8:D8,BatchesProduced)
9	=SUMPRODUCT(C9:D9,BatchesProduced)

	G
11	Total Profit
12	=SUMPRODUCT(ProfitPerBatch,BatchesProduced)

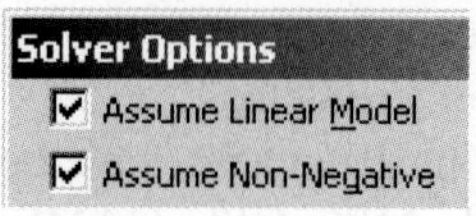

Solver Options

☑ Assume Linear Model
☑ Assume Non-Negative

Range Name	Cells
BatchesProduced	C12:D12
HoursAvailable	G7:G9
HoursUsed	E7:E9
HoursUsedPerBatchProduced	C7:D9
ProfitPerBatch	C4:D4
TotalProfit	G12

how to formulate linear programming models in a spreadsheet that are larger and more complicated than for the Wyndor problem.

You will see other examples of how to formulate and solve various kinds of OR models in a spreadsheet in later chapters. The supplementary chapters on the book's website also include a complete chapter (Chap. 21) that is devoted to the art of modeling in spreadsheets. That chapter describes in detail both the general process and the basic guidelines for building a spreadsheet model. It also presents some techniques for debugging such models.

■ 3.6   FORMULATING VERY LARGE LINEAR PROGRAMMING MODELS

Linear programming models come in many different sizes. For the examples in Secs. 3.1 and 3.4, the model sizes range from three functional constraints and two decision variables (for the Wyndor and radiation therapy problems) up to 17 functional constraints and 12 decision variables (for the Save-It Company problem). The latter case may seem like a rather large model. After all, it does take a substantial amount of time just to write down a model of this size. However, by contrast, the models for the application vignettes presented in this chapter are much, much larger. For example, the model for the United Airlines application in Sec. 3.4 often has over 20,000 decision variables.

Such model sizes are not at all unusual. Linear programming models in practice commonly have many hundreds or thousands of functional constraints. In fact, they occasionally will have even millions of functional constraints. The number of decision variables frequently is even larger than the number of functional constraints, and occasionally will range well into the millions.

Formulating such monstrously large models can be a daunting task. Even a "medium-sized" model with a thousand functional constraints and a thousand decision variables has over a million parameters (including the million coefficients in these constraints). It simply is not practical to write out the algebraic formulation, or even to fill in the parameters on a spreadsheet, for such a model.

So how are these very large models formulated in practice? It requires the use of a *modeling language.*

Modeling Languages

A mathematical modeling language is software that has been specifically designed for efficiently formulating large mathematical models, including linear programming models. Even with millions of functional constraints, they typically are of a relatively few types. Similarly, the decision variables will fall into a small number of categories. Therefore, using large blocks of data in databases, a modeling language will use a single expression to simultaneously formulate all the constraints of the same type in terms of the variables of each type. We will illustrate this process soon.

In addition to efficiently formulating large models, a modeling language will expedite a number of model management tasks, including accessing data, transforming data into model parameters, modifying the model whenever desired, and analyzing solutions from the model. It also may produce summary reports in the vernacular of the decision makers, as well as document the model's contents.

Several excellent modeling languages have been developed over the last couple of decades. These include AMPL, MPL, OPL, GAMS, and LINGO.

The student version of one of these, **MPL** (short for Mathematical Programming Language), is provided for you on the book's website along with extensive tutorial material. As subsequent versions are released in future years, the latest student version also can be downloaded from the website, maximalsoftware.com. MPL is a product of Maximal Software, Inc. One feature is extensive support for Excel in MPL. This includes both importing and exporting Excel ranges from MPL. Full support also is provided for the Excel VBA macro language through OptiMax 2000. (The student version of OptiMax 2000 is on the book's website as well.) This product allows the user to fully integrate MPL models into Excel and solve with any of the powerful solvers that MPL supports, including **CPLEX** (described in Sec. 4.8).

LINGO is a product of LINDO Systems, Inc., which also markets a spreadsheet-add-in optimizer called What's*Best!* that is designed for large industrial problems, as well as a callable subroutine library called the LINDO API. The LINGO software includes as a subset the LINDO interface that has been a popular introduction to linear programming for many people. The student version of LINGO with the LINDO interface is part of the software included on the book's website. All of the LINDO Systems products can also be downloaded from www.lindo.com. Like MPL, LINGO is a powerful general-purpose modeling language. A notable feature of LINGO is its great flexibility for dealing with a wide variety of OR problems in addition to linear programming. For example, when dealing with highly nonlinear models, it contains a global optimizer that will find a globally optimal solution. (More about this in Sec. 12.10.) New to the current edition of this book, the latest LINGO also has a built-in compatible programming language so that you can do things like solve several different optimization problems as part of one run, which is particularly useful when doing parametric analysis.

The book's website includes MPL, LINGO and LINDO formulations for essentially every example in this book to which these modeling languages and optimizers can be applied.

Now let us look at a simplified example that illustrates how a very large linear programming model can arise.

An Example of a Problem with a Huge Model

Management of the WORLDWIDE CORPORATION needs to address a *product-mix problem,* but one that is vastly more complex than the Wyndor product-mix problem introduced in Sec. 3.1. This corporation has 10 plants in various parts of the world. Each of these plants produces the same 10 products and then sells them within its region. The *demand* (sales potential) for each of these products from each plant is known for each of the next 10 months. Although the amount of a product sold by a plant in a given month cannot exceed the demand, the amount produced can be larger, where the excess amount would be stored in inventory (at some unit cost per month) for sale in a later month. Each unit of each product takes the same amount of space in inventory, and each plant has some upper limit on the total number of units that can be stored (the *inventory capacity*).

Each plant has the same 10 production processes (we'll refer to them as *machines*), each of which can be used to produce any of the 10 products. Both the production cost per unit of a product and the production rate of the product (number of units produced per day devoted to that product) depend on the combination of plant and machine involved (but not the month). The number of working days (*production days available*) varies somewhat from month to month.

Since some plants and machines can produce a particular product either less expensively or at a faster rate than other plants and machines, it is sometimes worthwhile to ship some units of the product from one plant to another for sale by the latter plant. For each combination of a plant being shipped from (the *fromplant*) and a plant being shipped to (the *toplant*), there is a certain cost per unit shipped of any product, where this unit shipping cost is the same for all the products.

Management now needs to determine how much of each product should be produced by each machine in each plant during each month, as well as how much each plant should sell of each product in each month and how much each plant should ship of each product in each month to each of the other plants. Considering the worldwide price for each product, the objective is to find the feasible plan that maximizes the total profit (total sales revenue *minus* the sum of the total production costs, inventory costs, and shipping costs).

We should note again that this is a simplified example in a number of ways. We have assumed that the number of plants, machines, products, and months are exactly the same (10). In most real situations, the number of products probably will be far larger and the planning horizon is likely to be considerably longer than 10 months, whereas the number of "machines" (types of production processes) may be less than 10. We also have assumed that every plant has all the same types of machines (production processes) and every machine type can produce every product. In reality, the plants may have some differences in terms of their machine types and the products they are capable of producing. The net result is that the corresponding model for some corporations may be smaller than the one for this example, but the model for other corporations may be considerably larger (perhaps even vastly larger) than this one.

The Structure of the Resulting Model

Because of the inventory costs and the limited inventory capacities, it is necessary to keep track of the amount of each product kept in inventory in each plant during each month. Consequently, the linear programming model has four types of decision variables: production quantities, inventory quantities, sales quantities, and shipping quantities. With 10 plants, 10 machines, 10 products, and 10 months, this gives a total of 21,000 decision variables, as outlined below.

Decision Variables.

10,000 production variables: one for each combination of a plant, machine, product, and
     month
1,000 inventory variables: one for each combination of a plant, product, and month
1,000 sales variables: one for each combination of a plant, product, and month
9,000 shipping variables: one for each combination of a product, month, plant (the fromplant),
     and another plant (the toplant)

Multiplying each of these decision variables by the corresponding unit cost or unit revenue, and then summing over each type, the following objective function can be calculated:

Objective Function.

Maximize   profit = total sales revenue − total cost,

where

     Total cost = total production cost + total inventory cost + total shipping cost.

When maximizing this objective function, the 21,000 decision variables need to satisfy nonnegativity constraints as well as four types of functional constraints—production capacity constraints, plant balance constraints (equality constraints that provide appropriate values to the inventory variables), maximum inventory constraints, and maximum sales constraints. As enumerated below, there are a total of 3,100 functional constraints, but all the constraints of each type follow the same pattern.

Functional Constraints.

1,000 production capacity constraints (one for each combination of a plant, machine, and month):

Production days used ≤ production days available,

where the left-hand side is the sum of 10 fractions, one for each product, where each fraction is that product's production quantity (a decision variable) *divided* by the product's production rate (a given constant).

1,000 plant balance constraints (one for each combination of a plant, product, and month):

Amount produced + inventory last month + amount shipped in = sales + current inventory + amount shipped out,

where the *amount produced* is the sum of the decision variables representing the production quantities at the machines, the *amount shipped in* is the sum of the decision variables representing the shipping quantities in from the other plants, and the *amount shipped out* is the sum of the decision variables representing the shipping quantities out to the other plants.

100 maximum inventory constraints (one for each combination of a plant and month):

Total inventory ≤ inventory capacity,

where the left-hand side is the sum of the decision variables representing the inventory quantities for the individual products.

1,000 maximum sales constraints (one for each combination of a plant, product, and month):

Sales ≤ demand.

Now let us see how the MPL Modeling Language can formulate this huge model very compactly.

Formulation of the Model in MPL

The modeler begins by assigning a title to the model and listing an *index* for each of the entities of the problem, as illustrated below.

```
TITLE
  Production_Planning;

INDEX
  product    := (A1, A2, A3, A4, A5, A6, A7, A8, A9, A10);
  month      := (Jan, Feb, Mar, Apr, May, Jun, Jul, Aug, Sep, Oct);
  plant      := (p1, p2, p3, p4, p5, p6, p7, p8, p9, p10);
  fromplant  := plant;
  toplant    := plant;
  machine    := (m1, m2, m3, m4, m5, m6, m7, m8, m9, m10);
```

Except for the months, the entries on the right-hand side are arbitrary labels for the respective products, plants, and machines, where these same labels are used in the data files. Note that a colon is placed after the name of each entry and a semicolon is placed at the end of each statement (but a statement is allowed to extend over more than one line).

A big job with any large model is collecting and organizing the various types of data into data files. A data file can be in either dense format or sparse format. In *dense format,* the file will contain an entry for every combination of all possible values of the respective indexes. For example, suppose that the data file contains the production rates for producing the various products with the various machines (production processes) in the various plants. In dense format, the file will contain an entry for every combination of a plant, a machine, and a product. However, the entry may need to be zero for most of the combinations because that particular plant may not have that particular machine or, even if it does, that particular machine may not be capable of producing that particular product in that particular plant. The percentage of the entries in dense format that are *nonzero* is referred to as the *density* of the data set. In practice, it is common for large data sets to have a density under 5 percent, and it frequently is under 1 percent. Data sets with such a low density are referred to as being *sparse*. In such situations, it is more efficient to use a data file in *sparse format*. In this format, only the nonzero values (and an identification of the index values they refer to) are entered into the data file. Generally, data are entered in sparse format either from a text file or from corporate databases. The ability to handle sparse data sets efficiently is one key for successfully formulating and solving large-scale optimization models. MPL can readily work with data in either dense format or sparse format.

In the Worldwide Corp. example, eight data files are needed to hold the product prices, demands, production costs, production rates, production days available, inventory costs, inventory capacities, and shipping costs. We assume that these data files are available in sparse format. The next step is to give a brief suggestive name to each one and to identify (inside square brackets) the index or indexes for that type of data, as shown below.

```
DATA
Price[product]             := SPARSEFILE("Price.dat");
Demand[plant,   product,   month] := SPARSEFILE("Demand.dat");
ProdCost[plant, machine, product] := SPARSEFILE("Produce.dat", 4);
ProdRate[plant, machine, product] := SPARSEFILE("Produce.dat", 5);
ProdDaysAvail[month]       := SPARSEFILE("ProdDays.dat");
InvtCost[plant, product]   := SPARSEFILE("InvtCost.dat");
InvtCapacity[plant]        := SPARSEFILE("InvtCap.dat");
ShipCost[fromplant, toplant]     := SPARSEFILE ("ShipCost.dat");
```

To illustrate the contents of these data files, consider the one that provides production costs and production rates. Here is a sample of the first few entries of SPARSEFILE produce.dat:

```
!
    ! Produce.dat - Production Cost and Rate
    !
    ! ProdCost[plant, machine, product]:
    ! ProdRate[plant, machine, product]:
    !
      p1, m11, A1, 73.30, 500,
      p1, m11, A2, 52.90, 450,
      p1, m12, A3, 65.40, 550,
      p1, m13, A3, 47.60, 350,
```

Next, the modeler gives a short name to each type of decision variable. Following the name, inside square brackets, is the index or indexes over which the subscripts run.

```
VARIABLES
    Produce[plant, machine, product, month]        -> Prod;
    Inventory[plant, product, month]               -> Invt;
    Sales[plant, product, month]                   -> Sale;
    Ship[product, month, fromplant, toplant]
        WHERE (fromplant <> toplant);
```

In the case of the decision variables with names longer than four letters, the arrows on the right point to four-letter abbreviations to fit the size limitations of many solvers. The last line indicates that the fromplant subscript and toplant subscript are not allowed to have the same value.

There is one more step before writing down the model. To make the model easier to read, it is useful first to introduce *macros* to represent the summations in the objective function.

```
MACROS
    Total Revenue    := SUM(plant, product, month: Price*Sales);
    TotalProdCost    := SUM(plant, machine, product, month:
                             ProdCost*Produce);
    TotalInvtCost    := SUM(plant, product, month:
                             InvtCost*Inventory);
    TotalShipCost    := SUM(product, month, fromplant, toplant:
                             ShipCost*Ship);
    TotalCost        := TotalProdCost + TotalInvtCost + TotalShipCost;
```

The first four macros use the MPL keyword SUM to execute the summation involved. Following each SUM keyword (inside the parentheses) is, first, the index or indexes over which the summation runs. Next (after the colon) is the vector product of a data vector (one of the data files) times a variable vector (one of the four types of decision variables).

Now this model with 3,100 functional constraints and 21,000 decision variables can be written down in the following compact form.

```
MODEL
MAX Profit = TotalRevenue - TotalCost;

SUBJECT TO
ProdCapacity[plant, machine, month] -> PCap:
    SUM(product: Produce/ProdRate) <= ProdDaysAvail;
PlantBal[plant, product, month] -> PBal:
        SUM(machine: Produce) + Inventory [month - 1]
    + SUM(fromplant: Ship[fromplant, toplant:= plant])

    =

        Sales + Inventory
    + SUM(toplant: Ship[fromplant:= plant, toplant]);

MaxInventory [plant, month] -> MaxI:
    SUM(product: Inventory) <= InvtCapacity;

BOUNDS
        Sales <= Demand;
END
```

For each of the four types of constraints, the first line gives the name for this type. There is one constraint of this type for each combination of values for the indexes inside the square brackets following the name. To the right of the brackets, the arrow points to a four-letter abbreviation of the name that a solver can use. Below the first line, the general form of constraints of this type is shown by using the SUM operator.

For each production capacity constraint, each term in the summation consists of a decision variable (the production quantity of that product on that machine in that plant during that month) divided by the corresponding production rate, which gives the number of production days being used. Summing over the products then gives the total number of production days being used on that machine in that plant during that month, so this number must not exceed the number of production days available.

The purpose of the plant balance constraint for each plant, product, and month is to give the correct value to the current inventory variable, given the values of all the other decision variables including the inventory level for the preceding month. Each of the SUM

A3. Chalermkraivuth, K. C., S. Bollapragada, M. C. Clark, J. Deaton, L. Kiaer, J. P. Murdzek, W. Neeves, B. J. Scholz, and D. Toledano: "GE Asset Management, Genworth Financial, and GE Insurance Use a Sequential-Linear-Programming Algorithm to Optimize Portfolios, *Interfaces,* **35**(5): 370–380, September–October 2005.

A4. Elimam, A. A., M. Girgis, and S. Kotob: "A Solution to Post Crash Debt Entanglements in Kuwait's al-Manakh Stock Market," *Interfaces,* **27**(1): 89–106, January–February 1997.

A5. Epstein, R., R. Morales, J. Serón, and A. Weintraub: "Use of OR Systems in the Chilean Forest Industries," *Interfaces,* **29**(1): 7–29, January–February 1999.

A6. Geraghty, M. K., and E. Johnson: "Revenue Management Saves National Car Rental," *Interfaces,* **27**(1): 107–127, January–February 1997.

A7. Leachman, R. C., R. F. Benson, C. Liu, and D. J. Raar: "IMPReSS: An Automated Production-Planning and Delivery-Quotation System at Harris Corporation—Semiconductor Sector," *Interfaces,* **26**(1): 6–37, January–February 1996.

A8. Mukuch, W. M., J. L. Dodge, J. G. Ecker, D. C. Granfors, and G. J. Hahn: "Managing Consumer Credit Delinquency in the U.S. Economy: A Multi-Billion Dollar Management Science Application," *Interfaces,* **22**(1): 90–109, January–February 1992.

A9. Murty, K. G., Y.-w. Wan, J. Liu, M. M. Tseng, E. Leung, K.-K. Lai, and H. W. C. Chiu: "Hongkong International Terminals Gains Elastic Capacity Using a Data-Intensive Decision-Support System," *Interfaces,* **35**(1): 61–75, January–February 2005.

A10. Yoshino, T., T. Sasaki, and T. Hasegawa: "The Traffic-Control System on the Hanshin Expressway," *Interfaces,* **25**(1): 94–108, January–February 1995.

■ LEARNING AIDS FOR THIS CHAPTER ON OUR WEBSITE (www.mhhe.com/hillier)

Worked Examples:

Examples for Chapter 3

A Demonstration Example in OR Tutor:

Graphical Method

Procedures in IOR Tutorial:

Interactive Graphical Method
Graphical Method and Sensitivity Analysis

An Excel Add-In:

Premium Solver for Education

"Ch. 3—Intro to LP" Files for Solving the Examples:

Excel Files
LINGO/LINDO File
MPL/CPLEX File

Glossary for Chapter 3

Supplements to This Chapter:

The LINGO Modeling Language
More About LINGO.

See Appendix 1 for documentation of the software.

■ PROBLEMS

The symbols to the left of some of the problems (or their parts) have the following meaning:

D: The demonstration example listed above may be helpful.

I: You may find it helpful to use the corresponding procedure in IOR Tutorial (the printout records your work).

C: Use the computer to solve the problem by applying the simplex method. The available software options for doing this include the Excel Solver or Premium Solver (Sec. 3.5), MPL/CPLEX (Sec. 3.6), LINGO (Supplements 1 and 2 to this chapter on the book's website and Appendix 4.1), and LINDO (Appendix 4.1), but follow any instructions given by your instructor regarding the option to use.

An asterisk on the problem number indicates that at least a partial answer is given in the back of the book.

3.1-1. Read the referenced article that fully describes the OR study summarized in the application vignette presented in Sec. 3.1. Briefly describe how linear programming was applied in this study. Then list the various financial and nonfinancial benefits that resulted from this study.

D **3.1-2.*** For each of the following constraints, draw a separate graph to show the nonnegative solutions that satisfy this constraint.
(a) $x_1 + 3x_2 \leq 6$
(b) $4x_1 + 3x_2 \leq 12$
(c) $4x_1 + x_2 \leq 8$
(d) Now combine these constraints into a single graph to show the feasible region for the entire set of functional constraints plus nonnegativity constraints.

D **3.1-3.** Consider the following objective function for a linear programming model:

Maximize $Z = 2x_1 + 3x_2$

(a) Draw a graph that shows the corresponding objective function lines for $Z = 6$, $Z = 12$, and $Z = 18$.
(b) Find the slope-intercept form of the equation for each of these three objective function lines. Compare the slope for these three lines. Also compare the intercept with the x_2 axis.

3.1-4. Consider the following equation of a line:

$60x_1 + 40x_2 = 600$

(a) Find the slope-intercept form of this equation.
(b) Use this form to identify the slope and the intercept with the x_2 axis for this line.
(c) Use the information from part (b) to draw a graph of this line.

D,I **3.1-5.*** Use the graphical method to solve the problem:

Maximize $Z = 2x_1 + x_2$,

subject to

$x_2 \leq 10$

$2x_1 + 5x_2 \leq 60$
$x_1 + x_2 \leq 18$
$3x_1 + x_2 \leq 44$

and

$x_1 \geq 0, \quad x_2 \geq 0.$

D,I **3.1-6.** Use the graphical method to solve the problem:

Maximize $Z = 10x_1 + 20x_2$,

subject to

$-x_1 + 2x_2 \leq 15$
$x_1 + x_2 \leq 12$
$5x_1 + 3x_2 \leq 45$

and

$x_1 \geq 0, \ x_2 \geq 0.$

3.1-7. The Whitt Window Company is a company with only three employees which makes two different kinds of hand-crafted windows: a wood-framed and an aluminum-framed window. They earn $180 profit for each wood-framed window and $90 profit for each aluminum-framed window. Doug makes the wood frames, and can make 6 per day. Linda makes the aluminum frames, and can make 4 per day. Bob forms and cuts the glass, and can make 48 square feet of glass per day. Each wood-framed window uses 6 square feet of glass and each aluminum-framed window uses 8 square feet of glass.

The company wishes to determine how many windows of each type to produce per day to maximize total profit.
(a) Describe the analogy between this problem and the Wyndor Glass Co. problem discussed in Sec. 3.1. Then construct and fill in a table like Table 3.1 for this problem, identifying both the activities and the resources.
(b) Formulate a linear programming model for this problem.
D,I (c) Use the graphical method to solve this model.
I (d) A new competitor in town has started making wood-framed windows as well. This may force the company to lower the price they charge and so lower the profit made for each wood-framed window. How would the optimal solution change (if at all) if the profit per wood-framed window decreases from $180 to $120? From $180 to $60? (You may find it helpful to use the Graphical Analysis and Sensitivity Analysis procedure in IOR Tutorial.)
I (e) Doug is considering lowering his working hours, which would decrease the number of wood frames he makes per day. How would the optimal solution change if he makes only 5 wood frames per day? (You may find it helpful to use the Graphical Analysis and Sensitivity Analysis procedure in IOR Tutorial.)

3.1-8. The WorldLight Company produces two light fixtures (products 1 and 2) that require both metal frame parts and electrical

D,I **3.2-4.** Use the graphical method to find all optimal solutions for the following model:

Maximize $Z = 500x_1 + 300x_2$,

subject to

$$15x_1 + 5x_2 \leq 300$$
$$10x_1 + 6x_2 \leq 240$$
$$8x_1 + 12x_2 \leq 450$$

and

$$x_1 \geq 0, \quad x_2 \geq 0.$$

D **3.2-5.** Use the graphical method to demonstrate that the following model has no feasible solutions.

Maximize $Z = 5x_1 + 7x_2$,

subject to

$$2x_1 - x_2 \leq -1$$
$$-x_1 + 2x_2 \leq -1$$

and

$$x_1 \geq 0, \quad x_2 \geq 0.$$

D **3.2-6.** Suppose that the following constraints have been provided for a linear programming model.

$$-x_1 + 2x_2 \leq 50$$
$$-2x_1 + x_2 \leq 50$$

and

$$x_1 \geq 0, \quad x_2 \geq 0.$$

(a) Demonstrate that the feasible region is unbounded.
(b) If the objective is to maximize $Z = -x_1 + x_2$, does the model have an optimal solution? If so, find it. If not, explain why not.
(c) Repeat part (b) when the objective is to maximize $Z = x_1 - x_2$.
(d) For objective functions where this model has no optimal solution, does this mean that there are no good solutions according to the model? Explain. What probably went wrong when formulating the model?

3.3-1. Reconsider Prob. 3.2-3. Indicate why each of the four assumptions of linear programming (Sec. 3.3) appears to be reasonably satisfied for this problem. Is one assumption more doubtful than the others? If so, what should be done to take this into account?

3.3-2. Consider a problem with two decision variables, x_1 and x_2, which represent the levels of activities 1 and 2, respectively. For each variable, the permissible values are 0, 1, and 2, where the feasible combinations of these values for the two variables are determined from a variety of constraints. The objective is to maximize a certain measure of performance denoted by Z. The values of Z for the possibly feasible values of (x_1, x_2) are estimated to be those given in the following table:

	x_2		
x_1	0	1	2
0	0	4	8
1	3	8	13
2	6	12	18

Based on this information, indicate whether this problem completely satisfies each of the four assumptions of linear programming. Justify your answers.

3.4-1. Read the referenced article that fully describes the OR study summarized in the first application vignette presented in Sec. 3.4. Briefly describe how linear programming was applied in this study. Then list the various financial and nonfinancial benefits that resulted from this study.

3.4-2. Read the referenced article that fully describes the OR study summarized in the second application vignette presented in Sec. 3.4. Briefly describe how linear programming was applied in this study. Then list the various financial and nonfinancial benefits that resulted from this study.

3.4-3.* For each of the four assumptions of linear programming discussed in Sec. 3.3, write a one-paragraph analysis of how well you feel it applies to each of the following examples given in Sec. 3.4:
(a) Design of radiation therapy (Mary).
(b) Regional planning (Southern Confederation of Kibbutzim).
(c) Controlling air pollution (Nori & Leets Co.).

3.4-4. For each of the four assumptions of linear programming discussed in Sec. 3.3, write a one-paragraph analysis of how well it applies to each of the following examples given in Sec. 3.4.
(a) Reclaiming solid wastes (Save-It Co.).
(b) Personnel scheduling (Union Airways).
(c) Distributing goods through a distribution network (Distribution Unlimited Co.).

D,I **3.4-5.** Use the graphical method to solve this problem:

Minimize $Z = 15x_1 + 20x_2$,

subject to

$$x_1 + 2x_2 \geq 10$$
$$2x_1 - 3x_2 \leq 6$$
$$x_1 + x_2 \geq 6$$

and

$$x_1 \geq 0, \quad x_2 \geq 0.$$

D,I **3.4-6.** Use the graphical method to solve this problem:

Minimize $Z = 3x_1 + 2x_2$,

subject to

$$x_1 + 2x_2 \leq 12$$

$$2x_1 + 3x_2 = 12$$
$$2x_1 + x_2 \geq 8$$

and

$$x_1 \geq 0, \quad x_2 \geq 0.$$

D **3.4-7.** Consider the following problem, where the value of c_1 has not yet been ascertained.

Maximize $Z = c_1x_1 + 2x_2,$

subject to

$$4x_1 + x_2 \leq 12$$
$$x_1 - x_2 \geq 2$$

and

$$x_1 \geq 0, \quad x_2 \geq 0.$$

Use graphical analysis to determine the optimal solution(s) for (x_1, x_2) for the various possible values of c_1.

D,I **3.4-8.** Consider the following model:

Minimize $Z = 40x_1 + 50x_2,$

subject to

$$2x_1 + 3x_2 \geq 30$$
$$x_1 + x_2 \geq 12$$
$$2x_1 + x_2 \geq 20$$

and

$$x_1 \geq 0, \quad x_2 \geq 0.$$

(a) Use the graphical method to solve this model.
(b) How does the optimal solution change if the objective function is changed to $Z = 40x_1 + 70x_2$? (You may find it helpful to use the Graphical Analysis and Sensitivity Analysis procedure in IOR Tutorial.)
(c) How does the optimal solution change if the third functional constraint is changed to $2x_1 + x_2 \geq 15$? (You may find it helpful to use the Graphical Analysis and Sensitivity Analysis procedure in IOR Tutorial.)

3.4-9. Ralph Edmund loves steaks and potatoes. Therefore, he has decided to go on a steady diet of only these two foods (plus some liquids and vitamin supplements) for all his meals. Ralph realizes that this isn't the healthiest diet, so he wants to make sure that he eats the right quantities of the two foods to satisfy some key nutritional requirements. He has obtained the nutritional and cost information shown at the top of the next column.

Ralph wishes to determine the number of daily servings (may be fractional) of steak and potatoes that will meet these requirements at a minimum cost.
(a) Formulate a linear programming model for this problem.
D,I (b) Use the graphical method to solve this model.
c (c) Use a computer to solve this model by the simplex method.

Ingredient	Grams of Ingredient per Serving		Daily Requirement (Grams)
	Steak	Potatoes	
Carbohydrates	5	15	≥ 50
Protein	20	5	≥ 40
Fat	15	2	≤ 60
Cost per serving	$4	$2	

3.4-10. Web Mercantile sells many household products through an online catalog. The company needs substantial warehouse space for storing its goods. Plans now are being made for leasing warehouse storage space over the next 5 months. Just how much space will be required in each of these months is known. However, since these space requirements are quite different, it may be most economical to lease only the amount needed each month on a month-by-month basis. On the other hand, the additional cost for leasing space for additional months is much less than for the first month, so it may be less expensive to lease the maximum amount needed for the entire 5 months. Another option is the intermediate approach of changing the total amount of space leased (by adding a new lease and/or having an old lease expire) at least once but not every month.

The space requirement and the leasing costs for the various leasing periods are as follows:

Month	Required Space (Sq. Ft.)	Leasing Period (Months)	Cost per Sq. Ft. Leased
1	30,000	1	$ 65
2	20,000	2	$100
3	40,000	3	$135
4	10,000	4	$160
5	50,000	5	$190

The objective is to minimize the total leasing cost for meeting the space requirements.
(a) Formulate a linear programming model for this problem.
c (b) Solve this model by the simplex method.

3.4-11. Larry Edison is the director of the Computer Center for Buckly College. He now needs to schedule the staffing of the center. It is open from 8 A.M. until midnight. Larry has monitored the usage of the center at various times of the day, and determined that the following number of computer consultants are required:

Time of Day	Minimum Number of Consultants Required to Be on Duty
8 A.M.–noon	4
Noon–4 P.M.	8
4 P.M.–8 P.M.	10
8 P.M.–midnight	6

Two types of computer consultants can be hired: full-time and part-time. The full-time consultants work for 8 consecutive hours in any of the following shifts: morning (8 A.M.–4 P.M.), afternoon (noon–8 P.M.), and evening (4 P.M.–midnight). Full-time consultants are paid $40 per hour.

Part-time consultants can be hired to work any of the four shifts listed in the above table. Part-time consultants are paid $30 per hour.

An additional requirement is that during every time period, there must be at least 2 full-time consultants on duty for every part-time consultant on duty.

Larry would like to determine how many full-time and how many part-time workers should work each shift to meet the above requirements at the minimum possible cost.
(a) Formulate a linear programming model for this problem.
C (b) Solve this model by the simplex method.

3.4-12.* The Medequip Company produces precision medical diagnostic equipment at two factories. Three medical centers have placed orders for this month's production output. The table below shows what the cost would be for shipping each unit from each factory to each of these customers. Also shown are the number of units that will be produced at each factory and the number of units ordered by each customer.

From \ To	Unit Shipping Cost			Output
	Customer 1	Customer 2	Customer 3	
Factory 1	$600	$800	$700	400 units
Factory 2	$400	$900	$600	500 units
Order size	300 units	200 units	400 units	

A decision now needs to be made about the shipping plan for how many units to ship from each factory to each customer.
(a) Formulate a linear programming model for this problem.
C (b) Solve this model by the simplex method.

3.4-13.* Al Ferris has $60,000 that he wishes to invest now in order to use the accumulation for purchasing a retirement annuity in 5 years. After consulting with his financial adviser, he has been offered four types of fixed-income investments, which we will label as investments A, B, C, D.

Investments A and B are available at the beginning of each of the next 5 years (call them years 1 to 5). Each dollar invested in A at the beginning of a year returns $1.40 (a profit of $0.40) 2 years later (in time for immediate reinvestment). Each dollar invested in B at the beginning of a year returns $1.70 three years later.

Investments C and D will each be available at one time in the future. Each dollar invested in C at the beginning of year 2 returns $1.90 at the end of year 5. Each dollar invested in D at the beginning of year 5 returns $1.30 at the end of year 5.

Al wishes to know which investment plan maximizes the amount of money that can be accumulated by the beginning of year 6.
(a) All the functional constraints for this problem can be expressed as equality constraints. To do this, let A_t, B_t, C_t, and D_t be the amount invested in investment A, B, C, and D, respectively, at the beginning of year t for each t where the investment is available and will mature by the end of year 5. Also let R_t be the number of available dollars *not* invested at the beginning of year t (and so available for investment in a later year). Thus, the amount invested at the beginning of year t *plus* R_t must equal the number of dollars available for investment at that time. Write such an equation in terms of the relevant variables above for the beginning of each of the 5 years to obtain the five functional constraints for this problem.
(b) Formulate a complete linear programming model for this problem.
C (c) Solve this model by the simplex model.

3.4-14. The Metalco Company desires to blend a new alloy of 40 percent tin, 35 percent zinc, and 25 percent lead from several available alloys having the following properties:

Property	Alloy				
	1	2	3	4	5
Percentage of tin	60	25	45	20	50
Percentage of zinc	10	15	45	50	40
Percentage of lead	30	60	10	30	10
Cost ($/lb)	77	70	88	84	94

The objective is to determine the proportions of these alloys that should be blended to produce the new alloy at a minimum cost.
(a) Formulate a linear programming model for this problem.
C (b) Solve this model by the simplex method.

3.4-15* A cargo plane has three compartments for storing cargo: front, center, and back. These compartments have capacity limits on both *weight* and *space*, as summarized below:

Compartment	Weight Capacity (Tons)	Space Capacity (Cubic Feet)
Front	12	7,000
Center	18	9,000
Back	10	5,000

Furthermore, the weight of the cargo in the respective compartments must be the same proportion of that compartment's weight capacity to maintain the balance of the airplane.

The following four cargoes have been offered for shipment on an upcoming flight as space is available:

Cargo	Weight (Tons)	Volume (Cubic Feet/Ton)	Profit ($/Ton)
1	20	500	320
2	16	700	400
3	25	600	360
4	13	400	290

Any portion of these cargoes can be accepted. The objective is to determine how much (if any) of each cargo should be accepted and how to distribute each among the compartments to maximize the total profit for the flight.

(a) Formulate a linear programming model for this problem.

c (b) Solve this model by the simplex method to find one of its multiple optimal solutions.

3.4-16. Oxbridge University maintains a powerful mainframe computer for research use by its faculty, Ph.D. students, and research associates. During all working hours, an operator must be available to operate and maintain the computer, as well as to perform some programming services. Beryl Ingram, the director of the computer facility, oversees the operation.

It is now the beginning of the fall semester, and Beryl is confronted with the problem of assigning different working hours to her operators. Because all the operators are currently enrolled in the university, they are available to work only a limited number of hours each day, as shown in the following table.

Operators	Wage Rate	Maximum Hours of Availability				
		Mon.	Tue.	Wed.	Thurs.	Fri.
K. C.	$25/hour	6	0	6	0	6
D. H.	$26/hour	0	6	0	6	0
H. B.	$24/hour	4	8	4	0	4
S. C.	$23/hour	5	5	5	0	5
K. S.	$28/hour	3	0	3	8	0
N. K.	$30/hour	0	0	0	6	2

There are six operators (four undergraduate students and two graduate students). They all have different wage rates because of differences in their experience with computers and in their programming ability. The above table shows their wage rates, along with the maximum number of hours that each can work each day.

Each operator is guaranteed a certain minimum number of hours per week that will maintain an adequate knowledge of the operation. This level is set arbitrarily at 8 hours per week for the undergraduate students (K. C., D. H., H. B., and S. C.) and 7 hours per week for the graduate students (K. S. and N. K.).

The computer facility is to be open for operation from 8 A.M. to 10 P.M. Monday through Friday with exactly one operator on duty during these hours. On Saturdays and Sundays, the computer is to be operated by other staff.

Because of a tight budget, Beryl has to minimize cost. She wishes to determine the number of hours she should assign to each operator on each day.

(a) Formulate a linear programming model for this problem.

c (b) Solve this model by the simplex method.

3.4-17. Joyce and Marvin run a day care for preschoolers. They are trying to decide what to feed the children for lunches. They would like to keep their costs down, but also need to meet the nutritional requirements of the children. They have already decided to go with peanut butter and jelly sandwiches, and some combination of graham crackers, milk, and orange juice. The nutritional content of each food choice and its cost are given in the table below.

Food Item	Calories from Fat	Total Calories	Vitamin C (mg)	Protein (g)	Cost (¢)
Bread (1 slice)	10	70	0	3	5
Peanut butter (1 tbsp)	75	100	0	4	4
Strawberry jelly (1 tbsp)	0	50	3	0	7
Graham cracker (1 cracker)	20	60	0	1	8
Milk (1 cup)	70	150	2	8	15
Juice (1 cup)	0	100	120	1	35

The nutritional requirements are as follows. Each child should receive between 400 and 600 calories. No more than 30 percent of the total calories should come from fat. Each child should consume at least 60 milligrams (mg) of vitamin C and 12 grams (g) of protein. Furthermore, for practical reasons, each child needs exactly 2 slices of bread (to make the sandwich), at least twice as much peanut butter as jelly, and at least 1 cup of liquid (milk and/or juice).

Joyce and Marvin would like to select the food choices for each child which minimize cost while meeting the above requirements.

(a) Formulate a linear programming model for this problem.

c (b) Solve this model by the simplex method.

3.5-1. Read the referenced article that fully describes the OR study summarized in the application vignette presented in Sec. 3.5. Briefly describe how linear programming was applied in this study. Then list the various financial and nonfinancial benefits that resulted from this study

3.5-2.* You are given the following data for a linear programming problem where the objective is to maximize the profit from allocating three resources to two nonnegative activities.

Resource	Resource Usage per Unit of Each Activity		Amount of Resource Available
	Activity 1	**Activity 2**	
1	2	1	10
2	3	3	20
3	2	4	20
Contribution per unit	$20	$30	

Contribution per unit = profit per unit of the activity.

(a) Formulate a linear programming model for this problem.
D,I (b) Use the graphical method to solve this model.
(c) Display the model on an Excel spreadsheet.
(d) Use the spreadsheet to check the following solutions: $(x_1, x_2) = (2, 2), (3, 3), (2, 4), (4, 2), (3, 4), (4, 3)$. Which of these solutions are feasible? Which of these feasible solutions has the best value of the objective function?
C (e) Use the Excel Solver to solve the model by the simplex method.

3.5-3. Ed Butler is the production manager for the Bilco Corporation, which produces three types of spare parts for automobiles. The manufacture of each part requires processing on each of two machines, with the following processing times (in hours):

Machine	Part		
	A	**B**	**C**
1	0.02	0.03	0.05
2	0.05	0.02	0.04

Each machine is available 40 hours per month. Each part manufactured will yield a unit profit as follows:

	Part		
	A	**B**	**C**
Profit	$300	$250	$200

Ed wants to determine the mix of spare parts to produce in order to maximize total profit.
(a) Formulate a linear programming model for this problem.
(b) Display the model on an Excel spreadsheet.
(c) Make three guesses of your own choosing for the optimal solution. Use the spreadsheet to check each one for feasibility and, if feasible, to find the value of the objective function.

Which feasible guess has the best objective function value?
(d) Use the Excel Solver to solve the model by the simplex method.

3.5-4. You are given the following data for a linear programming problem where the objective is to minimize the cost of conducting two nonnegative activities so as to achieve three benefits that do not fall below their minimum levels.

Benefit	Benefit Contribution per Unit of Each Activity		Minimum Acceptable Level
	Activity 1	**Activity 2**	
1	5	3	60
2	2	2	30
3	7	9	126
Unit cost	$60	$50	

(a) Formulate a linear programming model for this problem.
D,J (b) Use the graphical method to solve this model.
(c) Display the model on an Excel spreadsheet.
(d) Use the spreadsheet to check the following solutions: $(x_1, x_2) = (7, 7), (7, 8), (8, 7), (8, 8), (8, 9), (9, 8)$. Which of these solutions are feasible? Which of these feasible solutions has the best value of the objective function?
C (e) Use the Excel Solver to solve this model by the simplex method.

3.5-5.* Fred Jonasson manages a family-owned farm. To supplement several food products grown on the farm, Fred also raises pigs for market. He now wishes to determine the quantities of the available types of feed (corn, tankage, and alfalfa) that should be given to each pig. Since pigs will eat any mix of these feed types, the objective is to determine which mix will meet certain nutritional requirements at a *minimum cost*. The number of units of each type of basic nutritional ingredient contained within a kilogram of each feed type is given in the following table, along with the daily nutritional requirements and feed costs:

Nutritional Ingredient	Kilogram of Corn	Kilogram of Tankage	Kilogram of Alfalfa	Minimum Daily Requirement
Carbohydrates	90	20	40	200
Protein	30	80	60	180
Vitamins	10	20	60	150
Cost (¢)	84	72	60	

(a) Formulate a linear programming model for this problem.
(b) Display the model on an Excel spreadsheet.
(c) Use the spreadsheet to check if $(x_1, x_2, x_3) = (1, 2, 2)$ is a feasible solution and, if so, what the daily cost would be for this

diet. How many units of each nutritional ingredient would this diet provide daily?

(d) Take a few minutes to use a trial-and-error approach with the spreadsheet to develop your best guess for the optimal solution. What is the daily cost for your solution?

C **(e)** Use the Excel Solver to solve the model by the simplex method.

3.5-6. Maureen Laird is the chief financial officer for the Alva Electric Co., a major public utility in the midwest. The company has scheduled the construction of new hydroelectric plants 5, 10, and 20 years from now to meet the needs of the growing population in the region served by the company. To cover at least the construction costs, Maureen needs to invest some of the company's money now to meet these future cash-flow needs. Maureen may purchase only three kinds of financial assets, each of which costs $1 million per unit. Fractional units may be purchased. The assets produce income 5, 10, and 20 years from now, and that income is needed to cover at least minimum cash-flow requirements in those years. (Any excess income above the minimum requirement for each time period will be used to increase dividend payments to shareholders rather than saving it to help meet the minimum cash-flow requirement in the next time period.) The following table shows both the amount of income generated by each unit of each asset and the minimum amount of income needed for each of the future time periods when a new hydroelectric plant will be constructed.

Year	Income per Unit of Asset			Minimum Cash Flow Required
	Asset 1	**Asset 2**	**Asset 3**	
5	$2 million	$1 million	$0.5 million	$400 million
10	$0.5 million	$0.5 million	$1 million	$100 million
20	0	$1.5 million	$2 million	$300 million

Maureen wishes to determine the mix of investments in these assets that will cover the cash-flow requirements while minimizing the total amount invested.

(a) Formulate a linear programming model for this problem.

(b) Display the model on a spreadsheet.

(c) Use the spreadsheet to check the possibility of purchasing 100 units of Asset 1, 100 units of Asset 2, and 200 units of Asset 3. How much cash flow would this mix of investments generate 5, 10, and 20 years from now? What would be the total amount invested?

(d) Take a few minutes to use a trial-and-error approach with the spreadsheet to develop your best guess for the optimal solution. What is the total amount invested for your solution?

C **(e)** Use the Excel Solver to solve the model by the simplex method.

3.6-1. The Philbrick Company has two plants on opposite sides of the United States. Each of these plants produces the same two products and then sells them to wholesalers within its half of the country. The orders from wholesalers have already been received

for the next 2 months (February and March), where the number of units requested are shown below. (The company is not obligated to completely fill these orders but will do so if it can without decreasing its profits.)

Product	Plant 1		Plant 2	
	February	**March**	**February**	**March**
1	3,600	6,300	4,900	4,200
2	4,500	5,400	5,100	6,000

Each plant has 20 production days available in February and 23 production days available in March to produce and ship these products. Inventories are depleted at the end of January, but each plant has enough inventory capacity to hold 1,000 units total of the two products if an excess amount is produced in February for sale in March. In either plant, the cost of holding inventory in this way is $3 per unit of product 1 and $4 per unit of product 2.

Each plant has the same two production processes, each of which can be used to produce either of the two products. The production cost per unit produced of each product is shown below for each process in each plant.

Product	Plant 1		Plant 2	
	Process 1	**Process 2**	**Process 1**	**Process 2**
1	$62	$59	$61	$65
2	$78	$85	$89	$86

The production rate for each product (number of units produced per day devoted to that product) also is given for each process in each plant below.

Product	Plant 1		Plant 2	
	Process 1	**Process 2**	**Process 1**	**Process 2**
1	100	140	130	110
2	120	150	160	130

The net sales revenue (selling price minus normal shipping costs) the company receives when a plant sells the products to its own customers (the wholesalers in its half of the country) is $83 per unit of product 1 and $112 per unit of product 2. However, it also is possible (and occasionally desirable) for a plant to make a shipment to the other half of the country to help fill the sales of the other plant. When this happens, an extra shipping cost of $9 per unit of product 1 and $7 per unit of product 2 is incurred.

Management now needs to determine how much of each product should be produced by each production process in each plant during each month, as well as how much each plant should sell of each product in each month and how much each plant should ship of each product in each month to the other plant's customers. The objective is to determine which feasible plan would maximize the total profit (total net sales revenue minus the sum of the production costs, inventory costs, and extra shipping costs).

(a) Formulate a complete linear programming model in algebraic form that shows the individual constraints and decision variables for this problem.

c **(b)** Formulate this same model on an Excel spreadsheet instead. Then use the Excel Solver to solve the model.

c **(c)** Use MPL to formulate this model in a compact form. Then use the MPL solver CPLEX to solve the model.

c **(d)** Use LINGO to formulate this model in a compact form. Then use the LINGO solver to solve the model.

c **3.6-2.** Reconsider Prob. 3.1-11.

(a) Use MPL/CPLEX to formulate and solve the model for this problem.

(b) Use LINGO to formulate and solve this model.

c **3.6-3.** Reconsider Prob. 3.4-12.

(a) Use MPL/CPLEX to formulate and solve the model for this problem.

(b) Use LINGO to formulate and solve this model.

c **3.6-4.** Reconsider Prob. 3.4-16.

(a) Use MPL/CPLEX to formulate and solve the model for this problem.

(b) Use LINGO to formulate and solve this model.

c **3.6-5.** Reconsider Prob. 3.5-5.

(a) Use MPL/CPLEX to formulate and solve the model for this problem.

(b) Use LINGO to formulate and solve this model.

c **3.6-6.** Reconsider Prob. 3.5-6.

(a) Use MPL/CPLEX to formulate and solve the model for this problem.

(b) Use LINGO to formulate and solve this model.

3.6-7. A large paper manufacturing company, the Quality Paper Corporation, has 10 paper mills from which it needs to supply 1,000 customers. It uses three alternative types of machines and four types of raw materials to make five different types of paper. Therefore, the company needs to develop a detailed production distribution plan on a monthly basis, with an objective of minimizing the total cost of producing and distributing the paper during the month. Specifically, it is necessary to determine jointly the amount of each type of paper to be made at each paper mill on each type of machine *and* the amount of each type of paper to be shipped from each paper mill to each customer.

The relevant data can be expressed symbolically as follows:

D_{jk} = number of units of paper type k demanded by customer j,

r_{klm} = number of units of raw material m needed to produce 1 unit of paper type k on machine type l,

R_{im} = number of units of raw material m available at paper mill i,

c_{kl} = number of capacity units of machine type l that will produce 1 unit of paper type k,

C_{il} = number of capacity units of machine type l available at paper mill i,

P_{ikl} = production cost for each unit of paper type k produced on machine type l at paper mill i,

T_{ijk} = transportation cost for each unit of paper type k shipped from paper mill i to customer j.

(a) Using these symbols, formulate a linear programming model for this problem by hand.

(b) How many functional constraints and decision variables does this model have?

c **(c)** Use MPL to formulate this problem.

c **(d)** Use LINGO to formulate this problem.

3.7-1. From the bottom part of the selected references given at the end of the chapter, select one of these award-winning applications of linear programming. Read this article and then write a two-page summary of the application and the benefits (including nonfinancial benefits) it provided.

3.7-2. From the bottom part of the selected references given at the end of the chapter, select three of these award-winning applications of linear programming. For each one, read the article and then write a one-page summary of the application and the benefits (including nonfinancial benefits) it provided.

■ CASES

CASE 3.1 Auto Assembly

Automobile Alliance, a large automobile manufacturing company, organizes the vehicles it manufactures into three families: a family of trucks, a family of small cars, and a family of midsized and luxury cars. One plant outside Detroit, MI, assembles two models from the family of midsized and luxury cars. The first model, the Family Thrillseeker, is a four-door sedan with vinyl seats, plastic interior, standard features, and excellent gas mileage. It is marketed as a smart buy for middle-class families with tight budgets, and each Family Thrillseeker sold generates a modest profit of $3,600 for the company. The second model, the Classy Cruiser, is a two-door luxury sedan with leather seats, wooden interior, custom features, and navigational capabilities. It is marketed as a privilege of affluence for upper-middle-class families, and each Classy Cruiser sold generates a healthy profit of $5,400 for the company.

Rachel Rosencrantz, the manager of the assembly plant, is currently deciding the production schedule for the next month. Specifically, she must decide how many Family Thrillseekers and how many Classy Cruisers to assemble in the plant to maximize profit for the company. She knows that the plant possesses a capacity of 48,000 labor-hours during the month. She also knows that it takes 6 labor-hours to assemble one Family Thrillseeker and 10.5 labor-hours to assemble one Classy Cruiser.

Because the plant is simply an assembly plant, the parts required to assemble the two models are not produced at the plant. They are instead shipped from other plants around the Michigan area to the assembly plant. For example, tires, steering wheels, windows, seats, and doors all arrive from various supplier plants. For the next month, Rachel knows that she will be able to obtain only 20,000 doors (10,000 left-hand doors and 10,000 right-hand doors) from the door supplier. A recent labor strike forced the shutdown of that particular supplier plant for several days, and that plant will not be able to meet its production schedule for the next month. Both the Family Thrillseeker and the Classy Cruiser use the same door part.

In addition, a recent company forecast of the monthly demands for different automobile models suggests that the demand for the Classy Cruiser is limited to 3,500 cars. There is no limit on the demand for the Family Thrillseeker within the capacity limits of the assembly plant.

(a) Formulate and solve a linear programming problem to determine the number of Family Thrillseekers and the number of Classy Cruisers that should be assembled.

Before she makes her final production decisions, Rachel plans to explore the following questions independently except where otherwise indicated.

(b) The marketing department knows that it can pursue a targeted $500,000 advertising campaign that will raise the demand for the Classy Cruiser next month by 20 percent. Should the campaign be undertaken?

(c) Rachel knows that she can increase next month's plant capacity by using overtime labor. She can increase the plant's labor-hour capacity by 25 percent. With the new assembly plant capacity, how many Family Thrillseekers and how many Classy Cruisers should be assembled?

(d) Rachel knows that overtime labor does not come without an extra cost. What is the maximum amount she should be willing to

pay for all overtime labor beyond the cost of this labor at regular time rates? Express your answer as a lump sum.

(e) Rachel explores the option of using both the targeted advertising campaign and the overtime labor-hours. The advertising campaign raises the demand for the Classy Cruiser by 20 percent, and the overtime labor increases the plant's labor-hour capacity by 25 percent. How many Family Thrillseekers and how many Classy Cruisers should be assembled using the advertising campaign and overtime labor-hours if the profit from each Classy Cruiser sold continues to be 50 percent more than for each Family Thrillseeker sold?

(f) Knowing that the advertising campaign costs $500,000 and the maximum usage of overtime labor-hours costs $1,600,000 beyond regular time rates, is the solution found in part (e) a wise decision compared to the solution found in part (a)?

(g) Automobile Alliance has determined that dealerships are actually heavily discounting the price of the Family Thrillseekers to move them off the lot. Because of a profit-sharing agreement with its dealers, the company is therefore not making a profit of $3,600 on the Family Thrillseeker but is instead making a profit of $2,800. Determine the number of Family Thrillseekers and the number of Classy Cruisers that should be assembled given this new discounted price.

(h) The company has discovered quality problems with the Family Thrillseeker by randomly testing Thrillseekers at the end of the assembly line. Inspectors have discovered that in over 60 percent of the cases, two of the four doors on a Thrillseeker do not seal properly. Because the percentage of defective Thrillseekers determined by the random testing is so high, the floor supervisor has decided to perform quality control tests on every Thrillseeker at the end of the line. Because of the added tests, the time it takes to assemble one Family Thrillseeker has increased from 6 to 7.5 hours. Determine the number of units of each model that should be assembled given the new assembly time for the Family Thrillseeker.

(i) The board of directors of Automobile Alliance wishes to capture a larger share of the luxury sedan market and therefore would like to meet the full demand for Classy Cruisers. They ask Rachel to determine by how much the profit of her assembly plant would decrease as compared to the profit found in part (a). They then ask her to meet the full demand for Classy Cruisers if the decrease in profit is not more than $2,000,000.

(j) Rachel now makes her final decision by combining all the new considerations described in parts (f), (g), and (h). What are her final decisions on whether to undertake the advertising campaign, whether to use overtime labor, the number of Family Thrillseekers to assemble, and the number of Classy Cruisers to assemble?

■ PREVIEWS OF ADDED CASES ON OUR WEBSITE (www.mhhe.com/hillier)

CASE 3.2   Cutting Cafeteria Costs

This case focuses on a subject that is dear to the heart of many students. How should the manager of a college cafeteria choose the ingredients of a casserole dish to make it sufficiently tasty for the students while also minimizing costs? In this case, linear programming models with only two decision variables can be used to address seven specific issues being faced by the manager.

CASE 3.3   Staffing a Call Center

California Children's Hospital currently uses a confusing, decentralized appointment and registration process for its patients. Therefore, the decision has been made to centralize the process by establishing one call center devoted exclusively to appointments and registration. The hospital manager now needs to develop a plan for how many employees of each kind (full-time or part-time, English speaking, Spanish speaking, or bilingual) should be hired for each of several possible work shifts. Linear programming is needed to find a plan that minimizes the total cost of providing a satisfactory level of service throughout the 14 hours that the call center will be open each weekday. The model requires more than two decision variables, so a software

package such as described in Sec. 3.5 or Sec. 3.6 will be needed to solve the two versions of the model.

CASE 3.4   Promoting a Breakfast Cereal

The vice president for marketing of the Super Grain Corporation needs to develop a promotional campaign for the company's new breakfast cereal. Three advertising media have been chosen for the campaign, but decisions now need to be made regarding how much of each medium should be used. Constraints include a limited advertising budget, a limited planning budget, and a limited number of TV commercial spots available, as well as requirements for effectively reaching two special target audiences (young children and parents of young children) and for making full use of a rebate program. The corresponding linear programming model requires more than two decision variables, so a software package such as described in Sec. 3.5 or Sec. 3.6 will be needed to solve the model. This case also asks for an analysis of how well the four assumptions of linear programming are satisfied for this problem. Does linear programming actually provide a reasonable basis for managerial decision making in this situation? (Case 12.3 will provide a continuation of this case.)

4

Solving Linear Programming Problems: The Simplex Method

We now are ready to begin studying the *simplex method,* a general procedure for solving linear programming problems. Developed by the brilliant George Dantzig[1] in 1947, it has proved to be a remarkably efficient method that is used routinely to solve huge problems on today's computers. Except for its use on tiny problems, this method is always executed on a computer, and sophisticated software packages are widely available. Extensions and variations of the simplex method also are used to perform *postoptimality analysis* (including sensitivity analysis) on the model.

This chapter describes and illustrates the main features of the simplex method. The first section introduces its general nature, including its geometric interpretation. The following three sections then develop the procedure for solving any linear programming model that is in our standard form (maximization, all functional constraints in $\leq$ form, and nonnegativity constraints on all variables) and has only *nonnegative* right-hand sides b_i in the functional constraints. Certain details on resolving ties are deferred to Sec. 4.5, and Sec. 4.6 describes how to adapt the simplex method to other model forms. Next we discuss postoptimality analysis (Sec. 4.7), and describe the computer implementation of the simplex method (Sec. 4.8). Section 4.9 then introduces an alternative to the simplex method (the interior-point approach) for solving large linear programming problems.

■ 4.1   THE ESSENCE OF THE SIMPLEX METHOD

The simplex method is an *algebraic* procedure. However, its underlying concepts are *geometric.* Understanding these geometric concepts provides a strong intuitive feeling for how the simplex method operates and what makes it so efficient. Therefore, before delving into algebraic details, we focus in this section on the big picture from a geometric viewpoint.

---

[1]Widely revered as perhaps the most important pioneer of operations research, George Dantzig is commonly referred to as the *father of linear programming* because of the development of the simplex method and many key subsequent contributions. The authors had the privilege of being his faculty colleagues in the Department of Operations Research at Stanford University for nearly 30 years. Dr. Dantzig remained professionally active right up until he passed away in 2005 at the age of 90.

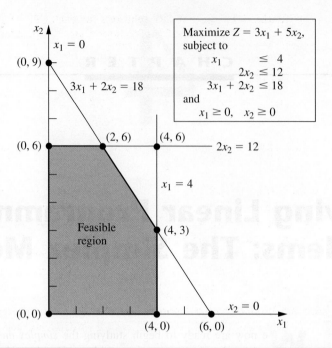

FIGURE 4.1
Constraint boundaries and corner-point solutions for the Wyndor Glass Co. problem.

To illustrate the general geometric concepts, we shall use the Wyndor Glass Co. example presented in Sec. 3.1. (Sections 4.2 and 4.3 use the *algebra* of the simplex method to solve this same example.) Section 5.1 will elaborate further on these geometric concepts for larger problems.

To refresh your memory, the model and graph for this example are repeated in Fig. 4.1. The five constraint boundaries and their points of intersection are highlighted in this figure because they are the keys to the analysis. Here, each **constraint boundary** is a line that forms the boundary of what is permitted by the corresponding constraint. The points of intersection are the **corner-point solutions** of the problem. The five that lie on the corners of the *feasible region*—(0, 0), (0, 6), (2, 6), (4, 3), and (4, 0)—are the *corner-point feasible solutions* (**CPF solutions**). [The other three—(0, 9), (4, 6), and (6, 0)—are called *corner-point infeasible solutions*.]

In this example, each corner-point solution lies at the intersection of *two* constraint boundaries. (For a linear programming problem with n decision variables, each of its corner-point solutions lies at the intersection of n constraint boundaries.[2]) Certain pairs of the CPF solutions in Fig. 4.1 share a constraint boundary, and other pairs do not. It will be important to distinguish between these cases by using the following general definitions.

> For any linear programming problem with n decision variables, two CPF solutions are **adjacent** to each other if they share $n - 1$ constraint boundaries. The two adjacent CPF solutions are connected by a line segment that lies on these same shared constraint boundaries. Such a line segment is referred to as an **edge** of the feasible region.

Since $n = 2$ in the example, two of its CPF solutions are adjacent if they share *one* constraint boundary; for example, (0, 0) and (0, 6) are adjacent because they share the $x_1 = 0$ constraint boundary. The feasible region in Fig. 4.1 has five edges, consisting of the five line segments forming the boundary of this region. Note that two edges emanate from

---

[2]Although a corner-point solution is defined in terms of n constraint boundaries whose intersection gives this solution, it also is possible that one or more *additional* constraint boundaries pass through this same point.

■ TABLE 4.1 Adjacent CPF solutions for each CPF solution of the Wyndor Glass Co. problem

CPF Solution	Its Adjacent CPF Solutions
(0, 0)	(0, 6) and (4, 0)
(0, 6)	(2, 6) and (0, 0)
(2, 6)	(4, 3) and (0, 6)
(4, 3)	(4, 0) and (2, 6)
(4, 0)	(0, 0) and (4, 3)

each CPF solution. Thus, each CPF solution has two adjacent CPF solutions (each lying at the other end of one of the two edges), as enumerated in Table 4.1. (In each row of this table, the CPF solution in the first column is adjacent to each of the two CPF solutions in the second column, but the two CPF solutions in the second column are *not* adjacent to each other.)

One reason for our interest in adjacent CPF solutions is the following general property about such solutions, which provides a very useful way of checking whether a CPF solution is an optimal solution.

> **Optimality test:** Consider any linear programming problem that possesses at least one optimal solution. If a CPF solution has no *adjacent* CPF solutions that are *better* (as measured by Z), then it *must* be an *optimal* solution.

Thus, for the example, (2, 6) must be optimal simply because its $Z = 36$ is larger than $Z = 30$ for (0, 6) and $Z = 27$ for (4, 3). (We will delve further into why this property holds in Sec. 5.1.) This optimality test is the one used by the simplex method for determining when an optimal solution has been reached.

Now we are ready to apply the simplex method to the example.

Solving the Example

Here is an outline of what the simplex method does (from a geometric viewpoint) to solve the Wyndor Glass Co. problem. At each step, first the conclusion is stated and then the reason is given in parentheses. (Refer to Fig. 4.1 for a visualization.)

Initialization: Choose (0, 0) as the *initial* CPF solution to examine. (This is a convenient choice because no calculations are required to identify this CPF solution.)

Optimality Test: Conclude that (0, 0) is *not* an optimal solution. (Adjacent CPF solutions are better.)

Iteration 1: Move to a better *adjacent* CPF solution, (0, 6), by performing the following three steps.

1. Considering the two edges of the feasible region that emanate from (0, 0), choose to move along the edge that leads up the x_2 axis. (With an objective function of $Z = 3x_1 + 5x_2$, moving up the x_2 axis increases Z at a faster rate than moving along the x_1 axis.)
2. Stop at the first new constraint boundary: $2x_2 = 12$. [Moving farther in the direction selected in step 1 leaves the feasible region; e.g., moving to the second new constraint boundary hit when moving in that direction gives (0, 9), which is a corner-point *infeasible* solution.]
3. Solve for the intersection of the new set of constraint boundaries: (0, 6). (The equations for these constraint boundaries, $x_1 = 0$ and $2x_2 = 12$, immediately yield this solution.)

Optimality Test: Conclude that (0, 6) is *not* an optimal solution. (An adjacent CPF solution is better.)

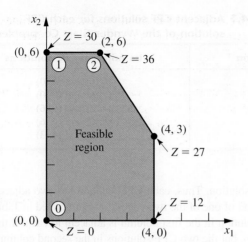

■ **FIGURE 4.2**
This graph shows the
sequence of CPF solutions
(⓪, ①, ②) examined by the
simplex method for the
Wyndor Glass Co. problem.
The optimal solution (2, 6) is
found after just three
solutions are examined.

Iteration 2: Move to a better adjacent CPF solution, (2, 6), by performing the following three steps.

1. Considering the two edges of the feasible region that emanate from (0, 6), choose to move along the edge that leads to the right. (Moving along this edge increases Z, whereas backtracking to move back down the x_2 axis decreases Z.)
2. Stop at the first new constraint boundary encountered when moving in that direction: $3x_1 + 2x_2 = 12$. (Moving farther in the direction selected in step 1 leaves the feasible region.)
3. Solve for the intersection of the new set of constraint boundaries: (2, 6). (The equations for these constraint boundaries, $3x_1 + 2x_2 = 18$ and $2x_2 = 12$, immediately yield this solution.)

Optimality Test: Conclude that (2, 6) *is* an optimal solution, so stop. (None of the adjacent CPF solutions are better.)

This sequence of CPF solutions examined is shown in Fig. 4.2, where each circled number identifies which iteration obtained that solution. (See the Worked Examples section on the book's website for **another example** of how the simplex method marches through a sequence of CPF solutions to reach the optimal solution.)

Now let us look at the six key solution concepts of the simplex method that provide the rationale behind the above steps. (Keep in mind that these concepts also apply for solving problems with more than two decision variables where a graph like Fig. 4.2 is not available to help quickly find an optimal solution.)

The Key Solution Concepts

The first solution concept is based directly on the relationship between optimal solutions and CPF solutions given at the end of Sec. 3.2.

Solution concept 1: The simplex method focuses solely on CPF solutions. For any problem with at least one optimal solution, finding one requires only finding a best CPF solution.[3]

---

[3]The only restriction is that the problem must possess CPF solutions. This is ensured if the feasible region is bounded.

Since the number of feasible solutions generally is infinite, reducing the number of solutions that need to be examined to a small finite number (just three in Fig. 4.2) is a tremendous simplification.

The next solution concept defines the flow of the simplex method.

Solution concept 2: The simplex method is an *iterative algorithm* (a systematic solution procedure that keeps repeating a fixed series of steps, called an *iteration*, until a desired result has been obtained) with the following structure.

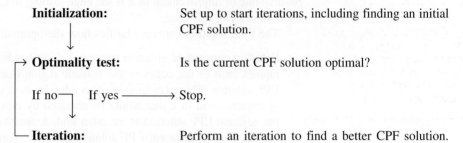

Initialization:	Set up to start iterations, including finding an initial CPF solution.
Optimality test:	Is the current CPF solution optimal?
If no ⌐ If yes ⟶ Stop.	
Iteration:	Perform an iteration to find a better CPF solution.

When the example was solved, note how this flow diagram was followed through two iterations until an optimal solution was found.

We next focus on how to get started.

Solution concept 3: Whenever possible, the initialization of the simplex method chooses the *origin* (all decision variables equal to zero) to be the initial CPF solution. When there are too many decision variables to find an initial CPF solution graphically, this choice eliminates the need to use algebraic procedures to find and solve for an initial CPF solution.

Choosing the origin commonly is possible when all the decision variables have nonnegativity constraints, because the intersection of these constraint boundaries yields the origin as a corner-point solution. This solution then is a CPF solution *unless* it is *infeasible* because it violates one or more of the functional constraints. If it is infeasible, special procedures described in Sec. 4.6 are needed to find the initial CPF solution.

The next solution concept concerns the choice of a better CPF solution at each iteration.

Solution concept 4: Given a CPF solution, it is much quicker computationally to gather information about its *adjacent* CPF solutions than about other CPF solutions. Therefore, each time the simplex method performs an iteration to move from the current CPF solution to a better one, it *always* chooses a CPF solution that is *adjacent* to the current one. No other CPF solutions are considered. Consequently, the entire path followed to eventually reach an optimal solution is along the *edges* of the feasible region.

The next focus is on which adjacent CPF solution to choose at each iteration.

Solution concept 5: After the current CPF solution is identified, the simplex method examines each of the edges of the feasible region that emanate from this CPF solution. Each of these edges leads to an *adjacent* CPF solution at the other end, but the simplex method does not even take the time to solve for the adjacent CPF solution. Instead, it simply identifies the *rate of improvement* in Z that would be obtained by moving along the edge. Among the edges with a *positive* rate of improvement in Z, it then chooses to move along the one with the *largest* rate of improvement in Z. The iteration is completed by first solving for the adjacent CPF solution at the other end of this one edge and then relabeling this adjacent

CPF solution as the *current* CPF solution for the optimality test and (if needed) the next iteration.

At the first iteration of the example, moving from (0, 0) along the edge on the x_1 axis would give a rate of improvement in Z of 3 (Z increases by 3 per unit increase in x_1), whereas moving along the edge on the x_2 axis would give a rate of improvement in Z of 5 (Z increases by 5 per unit increase in x_2), so the decision is made to move along the latter edge. At the second iteration, the only edge emanating from (0, 6) that would yield a *positive* rate of improvement in Z is the edge leading to (2, 6), so the decision is made to move next along this edge.

The final solution concept clarifies how the optimality test is performed efficiently.

Solution concept 6: Solution concept 5 describes how the simplex method examines each of the edges of the feasible region that emanate from the current CPF solution. This examination of an edge leads to quickly identifying the rate of improvement in Z that would be obtained by moving along the edge toward the adjacent CPF solution at the other end. A *positive* rate of improvement in Z implies that the adjacent CPF solution is *better* than the current CPF solution, whereas a *negative* rate of improvement in Z implies that the adjacent CPF solution is *worse*. Therefore, the optimality test consists simply of checking whether *any* of the edges give a *positive* rate of improvement in Z. If *none* do, then the current CPF solution is optimal.

In the example, moving along *either* edge from (2, 6) decreases Z. Since we want to maximize Z, this fact immediately gives the conclusion that (2, 6) is optimal.

■ 4.2   SETTING UP THE SIMPLEX METHOD

Section 4.1 stressed the geometric concepts that underlie the simplex method. However, this algorithm normally is run on a computer, which can follow only algebraic instructions. Therefore, it is necessary to translate the conceptually geometric procedure just described into a usable algebraic procedure. In this section, we introduce the *algebraic language* of the simplex method and relate it to the concepts of the preceding section.

The algebraic procedure is based on solving systems of equations. Therefore, the first step in setting up the simplex method is to convert the functional *inequality constraints* to equivalent *equality constraints*. (The nonnegativity constraints are left as inequalities because they are treated separately.) This conversion is accomplished by introducing **slack variables.** To illustrate, consider the first functional constraint in the Wyndor Glass Co. example of Sec. 3.1

$$x_1 \leq 4.$$

The slack variable for this constraint is defined to be

$$x_3 = 4 - x_1,$$

which is the amount of slack in the left-hand side of the inequality. Thus,

$$x_1 + x_3 = 4.$$

Given this equation, $x_1 \leq 4$ if and only if $4 - x_1 = x_3 \geq 0$. Therefore, the original constraint $x_1 \leq 4$ is entirely *equivalent* to the pair of constraints

$$x_1 + x_3 = 4 \qquad \text{and} \qquad x_3 \geq 0.$$

Upon the introduction of slack variables for the other functional constraints, the original linear programming model for the example (shown below on the left) can now be replaced by the equivalent model (called the *augmented form* of the model) shown below on the right:

<table>
<tr><td align="center">*Original Form of the Model*</td><td align="center">*Augmented Form of the Model*[4]</td></tr>
<tr><td>

Maximize $\quad Z = 3x_1 + 5x_2,$

subject to

$$x_1 \leq 4$$
$$2x_2 \leq 12$$
$$3x_1 + 2x_2 \leq 18$$

and

$$x_1 \geq 0, \quad x_2 \geq 0.$$

</td><td>

Maximize $\quad Z = 3x_1 + 5x_2,$

subject to

(1) $\quad x_1 \qquad\quad + x_3 \qquad\qquad = 4$
(2) $\qquad\quad 2x_2 \qquad + x_4 \qquad = 12$
(3) $\quad 3x_1 + 2x_2 \qquad\qquad + x_5 = 18$

and

$$x_j \geq 0, \qquad \text{for } j = 1, 2, 3, 4, 5.$$

</td></tr>
</table>

Although both forms of the model represent exactly the same problem, the new form is much more convenient for algebraic manipulation and for identification of CPF solutions. We call this the **augmented form** of the problem because the original form has been *augmented* by some supplementary variables needed to apply the simplex method.

If a slack variable equals 0 in the current solution, then this solution lies on the constraint boundary for the corresponding functional constraint. A value greater than 0 means that the solution lies on the *feasible* side of this constraint boundary, whereas a value less than 0 means that the solution lies on the *infeasible* side of this constraint boundary. A demonstration of these properties is provided by the **demonstration example** in your OR Tutor entitled *Interpretation of the Slack Variables.*

The terminology used in Section 4.1 (corner-point solutions, etc.) applies to the original form of the problem. We now introduce the corresponding terminology for the augmented form.

> An **augmented solution** is a solution for the original variables (the *decision variables*) that has been augmented by the corresponding values of the *slack variables*.

For example, augmenting the solution (3, 2) in the example yields the augmented solution (3, 2, 1, 8, 5) because the corresponding values of the slack variables are $x_3 = 1$, $x_4 = 8$, and $x_5 = 5$.

> A **basic solution** is an *augmented* corner-point solution.

To illustrate, consider the corner-point infeasible solution (4, 6) in Fig. 4.1. Augmenting it with the resulting values of the slack variables $x_3 = 0$, $x_4 = 0$, and $x_5 = -6$ yields the corresponding basic solution (4, 6, 0, 0, −6).

The fact that corner-point solutions (and so basic solutions) can be either feasible or infeasible implies the following definition:

> A **basic feasible (BF) solution** is an *augmented* CPF solution.

Thus, the CPF solution (0, 6) in the example is equivalent to the BF solution (0, 6, 4, 0, 6) for the problem in augmented form.

The only difference between basic solutions and corner-point solutions (or between BF solutions and CPF solutions) is whether the values of the slack variables are included.

---

[4]The slack variables are not shown in the objective function because the coefficients there are 0.

For any basic solution, the corresponding corner-point solution is obtained simply by deleting the slack variables. Therefore, the geometric and algebraic relationships between these two solutions are very close, as described in Sec. 5.1.

Because the terms *basic solution* and *basic feasible solution* are very important parts of the standard vocabulary of linear programming, we now need to clarify their algebraic properties. For the augmented form of the example, notice that the system of functional constraints has 5 variables and 3 equations, so

Number of variables − number of equations = 5 − 3 = 2.

This fact gives us 2 *degrees of freedom* in solving the system, since any two variables can be chosen to be set equal to any arbitrary value in order to solve the three equations in terms of the remaining three variables.[5] The simplex method uses zero for this arbitrary value. Thus, two of the variables (called the *nonbasic variables*) are set equal to zero, and then the simultaneous solution of the three equations for the other three variables (called the *basic variables*) is a *basic solution*. These properties are described in the following general definitions.

A **basic solution** has the following properties:

1. Each variable is designated as either a nonbasic variable or a basic variable.
2. The *number of basic variables* equals the number of functional constraints (now equations). Therefore, the *number of nonbasic variables* equals the total number of variables *minus* the number of functional constraints.
3. The **nonbasic variables** are set equal to zero.
4. The values of the **basic variables** are obtained as the simultaneous solution of the system of equations (functional constraints in augmented form). (The set of basic variables is often referred to as *the basis.*)
5. If the basic variables satisfy the *nonnegativity constraints,* the basic solution is a **BF solution.**

To illustrate these definitions, consider again the BF solution (0, 6, 4, 0, 6). This solution was obtained before by augmenting the CPF solution (0, 6). However, another way to obtain this same solution is to choose x_1 and x_4 to be the two nonbasic variables, and so the two variables are set equal to zero. The three equations then yield, respectively, $x_3 = 4$, $x_2 = 6$, and $x_5 = 6$ as the solution for the three basic variables, as shown below (with the basic variables in bold type):

$$x_1 = 0 \text{ and } x_4 = 0 \text{ so}$$

$$
\begin{array}{llll}
(1) & x_1 + \boldsymbol{x_3} = 4 & \boldsymbol{x_3} = 4 \\
(2) & 2\boldsymbol{x_2} + x_4 = 12 & \boldsymbol{x_2} = 6 \\
(3) & 3x_1 + 2\boldsymbol{x_2} + \boldsymbol{x_5} = 18 & \boldsymbol{x_5} = 6
\end{array}
$$

Because all three of these basic variables are nonnegative, this *basic solution* (0, 6, 4, 0, 6) is indeed a *BF solution.* The Worked Examples section of the book's website includes **another example** of the relationship between CPF solutions and BF solutions.

Just as certain pairs of CPF solutions are *adjacent,* the corresponding pairs of BF solutions also are said to be adjacent. Here is an easy way to tell when two BF solutions are adjacent.

Two BF solutions are **adjacent** if *all but one* of their *nonbasic variables* are the same. This implies that *all but one* of their *basic variables* also are the same, although perhaps with different numerical values.

---

[5]This method of determining the number of degrees of freedom for a system of equations is valid as long as the system does not include any redundant equations. This condition always holds for the system of equations formed from the functional constraints in the augmented form of a linear programming model.

Consequently, moving from the current BF solution to an adjacent one involves switching one variable from nonbasic to basic and vice versa for one other variable (and then adjusting the values of the basic variables to continue satisfying the system of equations).

To illustrate *adjacent BF solutions,* consider one pair of adjacent CPF solutions in Fig. 4.1: $(0, 0)$ and $(0, 6)$. Their augmented solutions, $(0, 0, 4, 12, 18)$ and $(0, 6, 4, 0, 6)$, automatically are adjacent BF solutions. However, you do not need to look at Fig. 4.1 to draw this conclusion. Another signpost is that their nonbasic variables, (x_1, x_2) and (x_1, x_4), are the same with just the one exception—x_2 has been replaced by x_4. Consequently, moving from $(0, 0, 4, 12, 18)$ to $(0, 6, 4, 0, 6)$ involves switching x_2 from nonbasic to basic and vice versa for x_4.

When we deal with the problem in augmented form, it is convenient to consider and manipulate the objective function equation at the same time as the new constraint equations. Therefore, before we start the simplex method, the problem needs to be rewritten once again in an equivalent way:

Maximize Z,

subject to

$$
\begin{array}{lllll}
(0) & Z - 3x_1 - 5x_2 & & & = 0 \\
(1) & x_1 & + x_3 & & = 4 \\
(2) & 2x_2 & + x_4 & & = 12 \\
(3) & 3x_1 + 2x_2 & & + x_5 & = 18
\end{array}
$$

and

$$x_j \geq 0, \quad \text{for } j = 1, 2, \ldots, 5.$$

It is just as if Eq. (0) actually were one of the original constraints; but because it already is in equality form, no slack variable is needed. While adding one more equation, we also have added one more unknown (Z) to the system of equations. Therefore, when using Eqs. (1) to (3) to obtain a basic solution as described above, we use Eq. (0) to solve for Z at the same time.

Somewhat fortuitously, the model for the Wyndor Glass Co. problem fits *our standard form,* and all its functional constraints have nonnegative right-hand sides b_i. If this had not been the case, then additional adjustments would have been needed at this point before the simplex method was applied. These details are deferred to Sec. 4.6, and we now focus on the simplex method itself.

■ 4.3 THE ALGEBRA OF THE SIMPLEX METHOD

We continue to use the prototype example of Sec. 3.1, as rewritten at the end of Sec. 4.2, for illustrative purposes. To start connecting the geometric and algebraic concepts of the simplex method, we begin by outlining side by side in Table 4.2 how the simplex method solves this example from both a geometric and an algebraic viewpoint. The geometric viewpoint (first presented in Sec. 4.1) is based on the *original form* of the model (no slack variables), so again refer to Fig. 4.1 for a visualization when you examine the second column of the table. Refer to the *augmented form* of the model presented at the end of Sec. 4.2 when you examine the third column of the table.

We now fill in the details for each step of the third column of Table 4.2.

■ **TABLE 4.2** Geometric and algebraic interpretations of how the simplex method solves the Wyndor Glass Co. problem

Method Sequence	Geometric Interpretation	Algebraic Interpretation
Initialization	Choose (0, 0) to be the initial CPF solution.	Choose x_1 and x_2 to be the nonbasic variables (= 0) for the initial BF solution: (0, 0, 4, 12, 18).
Optimality test	Not optimal, because moving along either edge from (0, 0) increases Z.	Not optimal, because increasing either nonbasic variable (x_1 or x_2) increases Z.
Iteration 1		
Step 1	Move up the edge lying on the x_2 axis.	Increase x_2 while adjusting other variable values to satisfy the system of equations.
Step 2	Stop when the first new constraint boundary ($2x_2 = 12$) is reached.	Stop when the first basic variable (x_3, x_4, or x_5) drops to zero (x_4).
Step 3	Find the intersection of the new pair of constraint boundaries: (0, 6) is the new CPF solution.	With x_2 now a basic variable and x_4 now a nonbasic variable, solve the system of equations: (0, 6, 4, 0, 6) is the new BF solution.
Optimality test	Not optimal, because moving along the edge from (0, 6) to the right increases Z.	Not optimal, because increasing one nonbasic variable (x_1) increases Z.
Iteration 2		
Step 1	Move along this edge to the right.	Increase x_1 while adjusting other variable values to satisfy the system of equations.
Step 2	Stop when the first new constraint boundary ($3x_1 + 2x_2 = 18$) is reached.	Stop when the first basic variable (x_2, x_3, or x_5) drops to zero (x_5).
Step 3	Find the intersection of the new pair of constraint boundaries: (2, 6) is the new CPF solution.	With x_1 now a basic variable and x_5 now a nonbasic variable, solve the system of equations: (2, 6, 2, 0, 0) is the new BF solution.
Optimality test	(2, 6) is optimal, because moving along either edge from (2, 6) decreases Z.	(2, 6, 2, 0, 0) is optimal, because increasing either nonbasic variable (x_4 or x_5) decreases Z.

Initialization

The choice of x_1 and x_2 to be the *nonbasic* variables (the variables set equal to zero) for the initial BF solution is based on solution concept 3 in Sec. 4.1. This choice eliminates the work required to solve for the *basic variables* (x_3, x_4, x_5) from the following system of equations (where the basic variables are shown in bold type):

$$x_1 = 0 \text{ and } x_2 = 0 \text{ so}$$

$$
\begin{array}{llll}
(1) & x_1 & + \boldsymbol{x_3} & = 4 & \boldsymbol{x_3} = 4 \\
(2) & 2x_2 & + \boldsymbol{x_4} & = 12 & \boldsymbol{x_4} = 12 \\
(3) & 3x_1 + 2x_2 & + \boldsymbol{x_5} & = 18 & \boldsymbol{x_5} = 18
\end{array}
$$

Thus, the **initial BF solution** is (0, 0, 4, 12, 18).

Notice that this solution can be read immediately because each equation has just one basic variable, which has a coefficient of 1, and this basic variable does not appear in any other equation. You will soon see that when the set of basic variables changes, the simplex method uses an algebraic procedure (Gaussian elimination) to convert the equations to this same convenient form for reading every subsequent BF solution as well. This form is called **proper form from Gaussian elimination**.

Samsung Electronics Corp., Ltd. (SEC) is a leading merchant of dynamic and static random access memory devices and other advanced digital integrated circuits. Its site at Kiheung, South Korea (probably the largest semiconductor fabrication site in the world) fabricates more than 300,000 silicon wafers per month and employs over 10,000 people.

Cycle time is the industry's term for the elapsed time from the release of a batch of blank silicon wafers into the fabrication process until completion of the devices that are fabricated on those wafers. Reducing cycle times is an ongoing goal since it both decreases costs and enables offering shorter lead times to potential customers, a real key to maintaining or increasing market share in a very competitive industry.

Three factors present particularly major challenges when striving to reduce cycle times. One is that the product mix changes continually. Another is that the company often needs to make substantial changes in the fab-out schedule inside the target cycle time as it revises forecasts of customer demand. The third is that the machines of a general type are not homogenous so only a small number of machines are qualified to perform each device-step.

An OR team developed *a huge linear programming model with tens of thousands of decision variables and functional constraints* to cope with these challenges. The objective function involved minimizing back-orders and finished-goods inventory. Despite the huge size of this model, it was readily solved in minutes whenever needed by using a highly sophisticated implementation of the simplex method (and related techniques) in the CPLEX optimization software. (CPLEX will be discussed further in Sec. 4.8.)

The ongoing implementation of this model enabled the company to reduce manufacturing cycle times to fabricate dynamic random access memory devices from more than 80 days to less than 30 days. This tremendous improvement and the resulting reduction in both manufacturing costs and sale prices enabled Samsung to capture *an additional* **$200 million** *in annual sales revenue*.

Source: R. C. Leachman, J. Kang, and Y. Lin: "SLIM: Short Cycle Time and Low Inventory in Manufacturing at Samsung Electronics," *Interfaces,* **32**(1): 61–77, Jan.–Feb. 2002. (A link to this article is provided on our website, www.mhhe.com/hillier.)

Optimality Test

The objective function is

$$Z = 3x_1 + 5x_2,$$

so $Z = 0$ for the initial BF solution. Because none of the basic variables (x_3, x_4, x_5) have a *nonzero* coefficient in this objective function, the coefficient of each nonbasic variable (x_1, x_2) gives the rate of improvement in Z if that variable were to be increased from zero (while the values of the basic variables are adjusted to continue satisfying the system of equations).[6] These rates of improvement (3 and 5) are *positive*. Therefore, based on solution concept 6 in Sec. 4.1, we conclude that (0, 0, 4, 12, 18) is not optimal.

For each BF solution examined after subsequent iterations, at least one basic variable has a nonzero coefficient in the objective function. Therefore, the optimality test then will use the new Eq. (0) to rewrite the objective function in terms of just the nonbasic variables, as you will see later.

Determining the Direction of Movement (Step 1 of an Iteration)

Increasing one nonbasic variable from zero (while adjusting the values of the basic variables to continue satisfying the system of equations) corresponds to moving along one edge emanating from the current CPF solution. Based on solution concepts 4 and 5 in Sec. 4.1, the choice of which nonbasic variable to increase is made as follows:

---

[6]Note that this interpretation of the coefficients of the x_j variables is based on these variables being on the right-hand side, $Z = 3x_1 + 5x_2$. When these variables are brought to the left-hand side for Eq. (0), $Z - 3x_1 - 5x_2 = 0$, the nonzero coefficients change their signs.

$$Z = 3x_1 + 5x_2$$

Increase x_1?   Rate of improvement in $Z = 3$.
Increase x_2?   Rate of improvement in $Z = 5$.
$5 > 3$, so choose x_2 to increase.

As indicated next, we call x_2 the *entering basic variable* for iteration 1.

> At any iteration of the simplex method, the purpose of step 1 is to choose one *nonbasic variable* to increase from zero (while the values of the basic variables are adjusted to continue satisfying the system of equations). Increasing this nonbasic variable from zero will convert it to a *basic variable* for the next BF solution. Therefore, this variable is called the **entering basic variable** for the current iteration (because it is entering the basis).

Determining Where to Stop (Step 2 of an Iteration)

Step 2 addresses the question of how far to increase the entering basic variable x_2 before stopping. Increasing x_2 increases Z, so we want to go as far as possible without leaving the feasible region. The requirement to satisfy the functional constraints in augmented form (shown below) means that increasing x_2 (while keeping the nonbasic variable $x_1 = 0$) changes the values of some of the basic variables as shown on the right.

$$x_1 = 0, \quad \text{so}$$

(1) $\quad x_1 \qquad + x_3 \qquad\qquad = 4 \qquad x_3 = 4$
(2) $\qquad\qquad 2x_2 \quad + x_4 \qquad = 12 \qquad x_4 = 12 - 2x_2$
(3) $\quad 3x_1 + 2x_2 \qquad\quad + x_5 = 18 \qquad x_5 = 18 - 2x_2$.

The other requirement for feasibility is that all the variables be *nonnegative*. The nonbasic variables (including the entering basic variable) are nonnegative, but we need to check how far x_2 can be increased without violating the nonnegativity constraints for the basic variables.

$x_3 = 4 \geq 0 \qquad \Rightarrow$ no upper bound on x_2.

$x_4 = 12 - 2x_2 \geq 0 \Rightarrow x_2 \leq \dfrac{12}{2} = 6 \quad \leftarrow$ minimum.

$x_5 = 18 - 2x_2 \geq 0 \Rightarrow x_2 \leq \dfrac{18}{2} = 9.$

Thus, x_2 can be increased just to 6, at which point x_4 has dropped to 0. Increasing x_2 beyond 6 would cause x_4 to become negative, which would violate feasibility.

These calculations are referred to as the **minimum ratio test.** The objective of this test is to determine which basic variable drops to zero first as the entering basic variable is increased. We can immediately rule out the basic variable in any equation where the coefficient of the entering basic variable is zero or negative, since such a basic variable would not decrease as the entering basic variable is increased. [This is what happened with x_3 in Eq. (1) of the example.] However, for each equation where the coefficient of the entering basic variable is *strictly positive* (> 0), this test calculates the *ratio* of the right-hand side to the coefficient of the entering basic variable. The basic variable in the equation with the *minimum ratio* is the one that drops to zero first as the entering basic variable is increased.

> At any iteration of the simplex method, step 2 uses the *minimum ratio test* to determine which basic variable drops to zero first as the entering basic variable is increased. Decreasing this basic variable to zero will convert it to a *nonbasic variable* for the next BF solution. Therefore, this variable is called the **leaving basic variable** for the current iteration (because it is leaving the basis).

Thus, x_4 is the leaving basic variable for iteration 1 of the example.

Solving for the New BF Solution (Step 3 of an Iteration)

Increasing $x_2 = 0$ to $x_2 = 6$ moves us from the *initial* BF solution on the left to the *new* BF solution on the right.

	Initial BF solution	New BF solution
Nonbasic variables:	$x_1 = 0$, $x_2 = 0$	$x_1 = 0$, $x_4 = 0$
Basic variables:	$x_3 = 4$, $x_4 = 12$, $x_5 = 18$	$x_3 = ?$, $x_2 = 6$, $x_5 = ?$

The purpose of step 3 is to convert the system of equations to a more convenient form (proper form from Gaussian elimination) for conducting the optimality test and (if needed) the next iteration with this new BF solution. In the process, this form also will identify the values of x_3 and x_5 for the new solution.

Here again is the complete original system of equations, where the *new* basic variables are shown in bold type (with Z playing the role of the basic variable in the objective function equation):

$$
\begin{aligned}
(0) \quad & Z - 3x_1 - 5\boldsymbol{x_2} && = 0 \\
(1) \quad & x_1 && + \boldsymbol{x_3} && = 4 \\
(2) \quad & 2\boldsymbol{x_2} && + x_4 && = 12 \\
(3) \quad & 3x_1 + 2\boldsymbol{x_2} && + \boldsymbol{x_5} && = 18.
\end{aligned}
$$

Thus, x_2 has replaced x_4 as the basic variable in Eq. (2). To solve this system of equations for Z, x_2, x_3, and x_5, we need to perform some **elementary algebraic operations** to reproduce the current pattern of coefficients of x_4 $(0, 0, 1, 0)$ as the new coefficients of x_2. We can use either of two types of elementary algebraic operations:

1. Multiply (or divide) an equation by a nonzero constant.
2. Add (or subtract) a multiple of one equation to (or from) another equation.

To prepare for performing these operations, note that the coefficients of x_2 in the above system of equations are -5, 0, 2, and 2, respectively, whereas we want these coefficients to become 0, 0, 1, and 0, respectively. To turn the coefficient of 2 in Eq. (2) into 1, we use the first type of elementary algebraic operation by dividing Eq. (2) by 2 to obtain

$$
(2) \quad x_2 + \frac{1}{2}x_4 = 6.
$$

To turn the coefficients of -5 and 2 into zeros, we need to use the second type of elementary algebraic operation. In particular, we add 5 times this new Eq. (2) to Eq. (0), and subtract 2 times this new Eq. (2) from Eq. (3). The resulting complete new system of equations is

$$
\begin{aligned}
(0) \quad & Z - 3x_1 && + \frac{5}{2}x_4 && = 30 \\
(1) \quad & x_1 && + \boldsymbol{x_3} && = 4 \\
(2) \quad & \boldsymbol{x_2} && + \frac{1}{2}x_4 && = 6 \\
(3) \quad & 3x_1 && - x_4 + \boldsymbol{x_5} && = 6.
\end{aligned}
$$

Since $x_1 = 0$ and $x_4 = 0$, the equations in this form immediately yield the new BF solution, $(x_1, x_2, x_3, x_4, x_5) = (0, 6, 4, 0, 6)$, which yields $Z = 30$.

This procedure for obtaining the simultaneous solution of a system of linear equations is called the *Gauss-Jordan method of elimination,* or **Gaussian elimination** for

■ **TABLE 4.3** Initial system of equations for the Wyndor Glass Co. problem

(a) Algebraic Form	(b) Tabular Form								
	Basic Variable	Eq.	Coefficient of:						Right Side
			Z	x_1	x_2	x_3	x_4	x_5	
(0) $Z - 3x_1 - 5x_2 = 0$	Z	(0)	1	−3	−5	0	0	0	0
(1) $x_1 \quad + x_3 = 4$	x_3	(1)	0	1	0	1	0	0	4
(2) $\quad 2x_2 \quad + x_4 = 12$	x_4	(2)	0	0	2	0	1	0	12
(3) $3x_1 + 2x_2 \quad + x_5 = 18$	x_5	(3)	0	3	2	0	0	1	18

on the left and in the first column of the simplex tableau on the right. [Although only the x_j variables are basic or nonbasic, Z plays the role of the basic variable for Eq. (0).] All variables *not* listed in this *basic variable* column (x_1, x_2) automatically are *nonbasic variables*. After we set $x_1 = 0$, $x_2 = 0$, the *right side* column gives the resulting solution for the basic variables, so that the initial BF solution is (x_1, x_2, x_3, x_4, x_5) = (0, 0, 4, 12, 18) which yields $Z = 0$.

The *tabular form* of the simplex method uses a **simplex tableau** to compactly display the system of equations yielding the current BF solution. For this solution, each variable in the leftmost column equals the corresponding number in the rightmost column (and variables not listed equal zero). When the optimality test or an iteration is performed, the only relevant numbers are those to the right of the Z column.[9] The term **row** refers to just a row of numbers to the right of the Z column (including the *right side* number), where row i corresponds to Eq. (i).

We summarize the tabular form of the simplex method below and, at the same time, briefly describe its application to the Wyndor Glass Co. problem. Keep in mind that the logic is identical to that for the algebraic form presented in the preceding section. Only the form for displaying both the current system of equations and the subsequent iteration has changed (plus we shall no longer bother to bring variables to the right-hand side of an equation before drawing our conclusions in the optimality test or in steps 1 and 2 of an iteration).

Summary of the Simplex Method (and Iteration 1 for the Example)

Initialization. Introduce slack variables. Select the *decision variables* to be the *initial nonbasic variables* (set equal to zero) and the *slack variables* to be the *initial basic variables*. (See Sec. 4.6 for the necessary adjustments if the model is not in our standard form—maximization, only ≤ functional constraints, and all nonnegativity constraints—or if any b_i values are negative.)

For the Example: This selection yields the initial simplex tableau shown in column (b) of Table 4.3, so the initial BF solution is (0, 0, 4, 12, 18).

Optimality Test. The current BF solution is *optimal* if and only if *every* coefficient in row 0 is nonnegative (≥ 0). If it is, stop; otherwise, go to an iteration to obtain the next BF solution, which involves changing one nonbasic variable to a basic variable (step 1) and vice versa (step 2) and then solving for the new solution (step 3).

For the Example: Just as $Z = 3x_1 + 5x_2$ indicates that increasing either x_1 or x_2 will increase Z, so the current BF solution is not optimal, the same conclusion is drawn from

---

[9]For this reason, it is permissible to delete the Eq. and Z columns to reduce the size of the simplex tableau. We prefer to retain these columns as a reminder that the simplex tableau is displaying the current system of equations and that Z is one of the variables in Eq. (0).

the equation $Z - 3x_1 - 5x_2 = 0$. These coefficients of -3 and -5 are shown in row 0 in column (b) of Table 4.3.

Iteration. *Step 1:* Determine the *entering basic variable* by selecting the variable (automatically a nonbasic variable) with the *negative coefficient* having the largest absolute value (i.e., the "most negative" coefficient) in Eq. (0). Put a box around the column below this coefficient, and call this the **pivot column.**

For the Example: The most negative coefficient is -5 for x_2 $(5 > 3)$, so x_2 is to be changed to a basic variable. (This change is indicated in Table 4.4 by the box around the x_2 column below -5.)

Step 2: Determine the *leaving basic variable* by applying the minimum ratio test.

Minimum Ratio Test

1. Pick out each coefficient in the pivot column that is strictly positive (> 0).
2. Divide each of these coefficients into the *right side* entry for the same row.
3. Identify the row that has the *smallest* of these ratios.
4. The basic variable for that row is the leaving basic variable, so replace that variable by the entering basic variable in the basic variable column of the next simplex tableau.

Put a box around this row and call it the **pivot row.** Also call the number that is in *both* boxes the **pivot number.**

For the Example: The calculations for the minimum ratio test are shown to the right of Table 4.4. Thus, row 2 is the pivot row (see the box around this row in the first simplex tableau of Table 4.5), and x_4 is the leaving basic variable. In the next simplex tableau (see the bottom of Table 4.5), x_2 replaces x_4 as the basic variable for row 2.

Step 3: Solve for the *new BF solution* by using **elementary row operations** (multiply or divide a row by a nonzero constant; add or subtract a multiple of one row to another row) to construct a new simplex tableau in proper form from Gaussian elimination below the current one, and then return to the optimality test. The specific elementary row operations that need to be performed are listed below.

1. Divide the pivot row by the pivot number. Use this *new* pivot row in steps 2 and 3.
2. For each other row (including row 0) that has a *negative* coefficient in the pivot column, *add* to this row the *product* of the absolute value of this coefficient and the new pivot row.
3. For each other row that has a *positive* coefficient in the pivot column, *subtract* from this row the *product* of this coefficient and the new pivot row.

TABLE 4.4 Applying the minimum ratio test to determine the first leaving basic variable for the Wyndor Glass Co. problem

Basic Variable	Eq.	Coefficient of:						Right Side	Ratio
		Z	x_1	x_2	x_3	x_4	x_5		
Z	(0)	1	-3	-5	0	0	0	0	
x_3	(1)	0	1	0	1	0	0	4	
x_4	(2)	0	0	2	0	1	0	$12 \rightarrow \dfrac{12}{2} = 6 \leftarrow$ minimum
x_5	(3)	0	3	2	0	0	1	$18 \rightarrow \dfrac{18}{2} = 9$

$x_2 \leftarrow x_4$

■ **TABLE 4.5** Simplex tableaux for the Wyndor Glass Co. problem after the first pivot row is divided by the first pivot number

Iteration	Basic Variable	Eq.	Coefficient of:							Right Side
			Z	x_1	x_2	x_3	x_4	x_5		
0	Z	(0)	1	-3	-5	0	0	0		0
	x_3	(1)	0	1	0	1	0	0		4
	x_4	(2)	0	0	2	0	1	0		12
	x_5	(3)	0	3	2	0	0	1		18
1	Z	(0)	1							
	x_3	(1)	0							
	x_2	(2)	0	0	1	0	$\frac{1}{2}$	0		6
	x_5	(3)	0							

For the Example: Since x_2 is replacing x_4 as a basic variable, we need to reproduce the first tableau's pattern of coefficients in the column of x_4 (0, 0, 1, 0) in the second tableau's column of x_2. To start, divide the pivot row (row 2) by the pivot number (2), which gives the new row 2 shown in Table 4.5. Next, we add to row 0 the product, 5 times the new row 2. Then we subtract from row 3 the product, 2 times the new row 2 (or equivalently, subtract from row 3 the *old* row 2). These calculations yield the new tableau shown in Table 4.6 for iteration 1. Thus, the new BF solution is (0, 6, 4, 0, 6), with $Z = 30$. We next return to the optimality test to check if the new BF solution is optimal. Since the new row 0 still has a negative coefficient (-3 for x_1), the solution is not optimal, and so at least one more iteration is needed.

Iteration 2 for the Example and the Resulting Optimal Solution

The second iteration starts anew from the second tableau of Table 4.6 to find the next BF solution. Following the instructions for steps 1 and 2, we find x_1 as the entering basic variable and x_5 as the leaving basic variable, as shown in Table 4.7.

For step 3, we start by dividing the pivot row (row 3) in Table 4.7 by the pivot number (3). Next, we add to row 0 the product, 3 times the new row 3. Then we subtract the new row 3 from row 1.

We now have the set of tableaux shown in Table 4.8. Therefore, the new BF solution is (2, 6, 2, 0, 0), with $Z = 36$. Going to the optimality test, we find that this solution is

■ **TABLE 4.6** First two simplex tableaux for the Wyndor Glass Co. problem

Iteration	Basic Variable	Eq.	Coefficient of:							Right Side
			Z	x_1	x_2	x_3	x_4	x_5		
0	Z	(0)	1	-3	-5	0	0	0		0
	x_3	(1)	0	1	0	1	0	0		4
	x_4	(2)	0	0	2	0	1	0		12
	x_5	(3)	0	3	2	0	0	1		18
1	Z	(0)	1	-3	0	0	$\frac{5}{2}$	0		30
	x_3	(1)	0	1	0	1	0	0		4
	x_2	(2)	0	0	1	0	$\frac{1}{2}$	0		6
	x_5	(3)	0	3	0	0	-1	1		6

■ TABLE 4.7 Steps 1 and 2 of iteration 2 for the Wyndor Glass Co. problem

Iteration	Basic Variable	Eq.	Z	x_1	x_2	x_3	x_4	x_5	Right Side	Ratio
	Z	(0)	1	-3	0	0	$\frac{5}{2}$	0	30	
	x_3	(1)	0	1	0	1	0	0	4	$\frac{4}{1}=4$
1	x_2	(2)	0	0	1	0	$\frac{1}{2}$	0	6	
	x_5	(3)	0	3	0	0	-1	1	6	$\frac{6}{3}=2 \leftarrow$ minimum

optimal because none of the coefficients in row 0 is negative, so the algorithm is finished. Consequently, the optimal solution for the Wyndor Glass Co. problem (before slack variables are introduced) is $x_1 = 2$, $x_2 = 6$.

Now compare Table 4.8 with the work done in Sec. 4.3 to verify that these two forms of the simplex method really are *equivalent*. Then note how the algebraic form is superior for learning the logic behind the simplex method, but the tabular form organizes the work being done in a considerably more convenient and compact form. We generally use the tabular form from now on.

An **additional example** of applying the simplex method in tabular form is available to you in the OR Tutor. See the demonstration entitled *Simplex Method—Tabular Form*. **Another example** also is included in the Worked Examples section of the book's website.

■ TABLE 4.8 Complete set of simplex tableaux for the Wyndor Glass Co. problem

Iteration	Basic Variable	Eq.	Z	x_1	x_2	x_3	x_4	x_5	Right Side
	Z	(0)	1	-3	-5	0	0	0	0
0	x_3	(1)	0	1	0	1	0	0	4
	x_4	(2)	0	0	2	0	1	0	12
	x_5	(3)	0	3	2	0	0	1	18
	Z	(0)	1	-3	0	0	$\frac{5}{2}$	0	30
1	x_3	(1)	0	1	0	1	0	0	4
	x_2	(2)	0	0	1	0	$\frac{1}{2}$	0	6
	x_5	(3)	0	3	0	0	-1	1	6
	Z	(0)	1	0	0	0	$\frac{3}{2}$	1	36
2	x_3	(1)	0	0	0	1	$\frac{1}{3}$	$-\frac{1}{3}$	2
	x_2	(2)	0	0	1	0	$\frac{1}{2}$	0	6
	x_1	(3)	0	1	0	0	$-\frac{1}{3}$	$\frac{1}{3}$	2

■ 4.5 TIE BREAKING IN THE SIMPLEX METHOD

You may have noticed in the preceding two sections that we never said what to do if the various choice rules of the simplex method do not lead to a clear-cut decision, because of either ties or other similar ambiguities. We discuss these details now.

Tie for the Entering Basic Variable

Step 1 of each iteration chooses the nonbasic variable having the *negative* coefficient with the *largest absolute value* in the current Eq. (0) as the entering basic variable. Now suppose that two or more nonbasic variables are tied for having the largest negative coefficient (in absolute terms). For example, this would occur in the first iteration for the Wyndor Glass Co. problem if its objective function were changed to $Z = 3x_1 + 3x_2$, so that the initial Eq. (0) became $Z - 3x_1 - 3x_2 = 0$. How should this tie be broken?

The answer is that the selection between these contenders may be made *arbitrarily*. The optimal solution will be reached eventually, regardless of the tied variable chosen, and there is no convenient method for predicting in advance which choice will lead there sooner. In this example, the simplex method happens to reach the optimal solution (2, 6) in three iterations with x_1 as the initial entering basic variable, versus two iterations if x_2 is chosen.

Tie for the Leaving Basic Variable—Degeneracy

Now suppose that two or more basic variables tie for being the leaving basic variable in step 2 of an iteration. Does it matter which one is chosen? Theoretically it does, and in a very critical way, because of the following sequence of events that could occur. First, all the tied basic variables reach zero simultaneously as the entering basic variable is increased. Therefore, the one or ones *not* chosen to be the leaving basic variable also will have a value of zero in the new BF solution. (Note that basic variables with a value of *zero* are called **degenerate,** and the same term is applied to the corresponding BF solution.) Second, if one of these degenerate basic variables retains its value of zero until it is chosen at a subsequent iteration to be a leaving basic variable, the corresponding entering basic variable also must remain zero (since it cannot be increased without making the leaving basic variable negative), so the value of Z must remain unchanged. Third, if Z may remain the same rather than increase at each iteration, the simplex method may then go around in a loop, repeating the same sequence of solutions periodically rather than eventually increasing Z toward an optimal solution. In fact, examples have been artificially constructed so that they do become entrapped in just such a perpetual loop.[10]

Fortunately, although a perpetual loop is theoretically possible, it has rarely been known to occur in practical problems. If a loop were to occur, one could always get out of it by changing the choice of the leaving basic variable. Furthermore, special rules[11] have been constructed for breaking ties so that such loops are always avoided. However, these rules frequently are ignored in actual application, and they will not be repeated here. For your purposes, just break this kind of tie arbitrarily and proceed without worrying about the degenerate basic variables that result.

---

[10]For further information about cycling around a perpetual loop, see J. A. J. Hall and K. I. M. McKinnon: "The Simplest Examples Where the Simplex Method Cycles and Conditions Where EXPAND Fails to Prevent Cycling," *Mathematical Programming,* Series B, **100**(1): 135–150, May 2004.

[11]See R. Bland: "New Finite Pivoting Rules for the Simplex Method," *Mathematics of Operations Research,* **2**: 103–107, 1977.

TABLE 4.9 Initial simplex tableau for the Wyndor Glass Co. problem without the last two functional constraints

Basic Variable	Eq.	Coefficient of:				Right Side	Ratio
		Z	x_1	x_2	x_3		
Z	(0)	1	-3	-5	0	0	
x_3	(1)	0	1	$\boxed{0}$	1	4	None

With $x_1 = 0$ and x_2 increasing, $x_3 = 4 - 1x_1 - 0x_2 = 4 > 0$.

No Leaving Basic Variable—Unbounded Z

In step 2 of an iteration, there is one other possible outcome that we have not yet discussed, namely, that *no* variable qualifies to be the leaving basic variable.[12] This outcome would occur if the entering basic variable could be increased *indefinitely* without giving negative values to *any* of the current basic variables. In tabular form, this means that *every* coefficient in the pivot column (excluding row 0) is either negative or zero.

As illustrated in Table 4.9, this situation arises in the example displayed in Fig. 3.6. In this example, the last two functional constraints of the Wyndor Glass Co. problem have been overlooked and so are not included in the model. Note in Fig. 3.6 how x_2 can be increased indefinitely (thereby increasing Z indefinitely) without ever leaving the feasible region. Then note in Table 4.9 that x_2 is the entering basic variable but the only coefficient in the pivot column is zero. Because the minimum ratio test uses only coefficients that are greater than zero, there is no ratio to provide a leaving basic variable.

The interpretation of a tableau like the one shown in Table 4.9 is that the constraints do not prevent the value of the objective function Z from increasing indefinitely, so the simplex method would stop with the message that Z is *unbounded*. Because even linear programming has not discovered a way of making infinite profits, the real message for practical problems is that a mistake has been made! The model probably has been misformulated, either by omitting relevant constraints or by stating them incorrectly. Alternatively, a computational mistake may have occurred.

Multiple Optimal Solutions

We mentioned in Sec. 3.2 (under the definition of **optimal solution**) that a problem can have more than one optimal solution. This fact was illustrated in Fig. 3.5 by changing the objective function in the Wyndor Glass Co. problem to $Z = 3x_1 + 2x_2$, so that every point on the line segment between (2, 6) and (4, 3) is optimal. Thus, all optimal solutions are a *weighted average* of these two optimal CPF solutions

$$(x_1, x_2) = w_1(2, 6) + w_2(4, 3),$$

where the weights w_1 and w_2 are numbers that satisfy the relationships

$$w_1 + w_2 = 1 \quad \text{and} \quad w_1 \geq 0, \quad w_2 \geq 0.$$

For example, $w_1 = \frac{1}{3}$ and $w_2 = \frac{2}{3}$ give

---

[12]Note that the analogous case (no *entering* basic variable) cannot occur in step 1 of an iteration, because the optimality test would stop the algorithm first by indicating that an optimal solution had been reached.

$$(x_1, x_2) = \frac{1}{3}(2, 6) + \frac{2}{3}(4, 3) = \left(\frac{2}{3} + \frac{8}{3}, \frac{6}{3} + \frac{6}{3}\right) = \left(\frac{10}{3}, 4\right)$$

as one optimal solution.

In general, any weighted average of two or more solutions (vectors) where the weights are nonnegative and sum to 1 is called a **convex combination** of these solutions. Thus, every optimal solution in the example is a convex combination of (2, 6) and (4, 3).

This example is typical of problems with multiple optimal solutions.

As indicated at the end of Sec. 3.2, *any* linear programming problem with multiple optimal solutions (and a bounded feasible region) has at least two CPF solutions that are optimal. *Every* optimal solution is a convex combination of these optimal CPF solutions. Consequently, in augmented form, every optimal solution is a convex combination of the optimal BF solutions.

(Problems 4.5-5 and 4.5-6 guide you through the reasoning behind this conclusion.)

The simplex method automatically stops after *one* optimal BF solution is found. However, for many applications of linear programming, there are intangible factors not incorporated into the model that can be used to make meaningful choices between alternative optimal solutions. In such cases, these other optimal solutions should be identified as well. As indicated above, this requires finding all the other optimal BF solutions, and then every optimal solution is a convex combination of the optimal BF solutions.

After the simplex method finds one optimal BF solution, you can detect if there are any others and, if so, find them as follows:

Whenever a problem has more than one optimal BF solution, at least one of the nonbasic variables has a coefficient of zero in the final row 0, so increasing any such variable will not change the value of Z. Therefore, these other optimal BF solutions can be identified (if desired) by performing additional iterations of the simplex method, each time choosing a nonbasic variable with a zero coefficient as the entering basic variable.[13]

To illustrate, consider again the case just mentioned, where the objective function in the Wyndor Glass Co. problem is changed to $Z = 3x_1 + 2x_2$. The simplex method obtains the first three tableaux shown in Table 4.10 and stops with an optimal BF solution. However, because a nonbasic variable (x_3) has a zero coefficient in row 0, we perform one more iteration in Table 4.10 to identify the other optimal BF solution. Thus, the two optimal BF solutions are (4, 3, 0, 6, 0) and (2, 6, 2, 0, 0), each yielding $Z = 18$. Notice that the last tableau also has a *nonbasic* variable (x_4) with a zero coefficient in row 0. This situation is inevitable because the extra iteration does not change row 0, so this leaving basic variable necessarily retains its zero coefficient. Making x_4 an entering basic variable now would only lead back to the third tableau. (Check this.) Therefore, these two are the only BF solutions that are optimal, and all *other* optimal solutions are a convex combination of these two.

$$(x_1, x_2, x_3, x_4, x_5) = w_1(2, 6, 2, 0, 0) + w_2(4, 3, 0, 6, 0),$$
$$w_1 + w_2 = 1, \qquad w_1 \geq 0, \qquad w_2 \geq 0.$$

---

[13]If such an iteration has no *leaving* basic variable, this indicates that the feasible region is unbounded and the entering basic variable can be increased indefinitely without changing the value of Z.

TABLE 4.10 Complete set of simplex tableaux to obtain all optimal BF solutions for the Wyndor Glass Co. problem with $c_2 = 2$

Iteration	Basic Variable	Eq.	Z	Coefficient of:						Right Side	Solution Optimal?
				x_1	x_2	x_3	x_4	x_5			
0	Z	(0)	1	-3	-2	0	0	0		0	No
	x_3	(1)	0	1	0	1	0	0		4	
	x_4	(2)	0	0	2	0	1	0		12	
	x_5	(3)	0	3	2	0	0	1		18	
1	Z	(0)	1	0	-2	3	0	0		12	No
	x_1	(1)	0	1	0	1	0	0		4	
	x_4	(2)	0	0	2	0	1	0		12	
	x_5	(3)	0	0	2	-3	0	1		6	
2	Z	(0)	1	0	0	0	0	1		18	Yes
	x_1	(1)	0	1	0	1	0	0		4	
	x_4	(2)	0	0	0	3	1	-1		6	
	x_2	(3)	0	0	1	$-\frac{3}{2}$	0	$\frac{1}{2}$		3	
Extra	Z	(0)	1	0	0	0	0	1		18	Yes
	x_1	(1)	0	1	0	0	$-\frac{1}{3}$	$\frac{1}{3}$		2	
	x_3	(2)	0	0	0	1	$\frac{1}{3}$	$-\frac{1}{3}$		2	
	x_2	(3)	0	0	1	0	$\frac{1}{2}$	0		6	

4.6 ADAPTING TO OTHER MODEL FORMS

Thus far we have presented the details of the simplex method under the assumptions that the problem is in our standard form (maximize Z subject to functional constraints in $\leq$ form and nonnegativity constraints on all variables) and that $b_i \geq 0$ for all $i = 1, 2, \ldots, m$. In this section we point out how to make the adjustments required for other legitimate forms of the linear programming model. You will see that all these adjustments can be made during the initialization, so the rest of the simplex method can then be applied just as you have learned it already.

The only serious problem introduced by the other forms for functional constraints (the = or $\geq$ forms, or having a negative right-hand side) lies in identifying an *initial BF solution*. Before, this initial solution was found very conveniently by letting the slack variables be the initial basic variables, so that each one just equals the *nonnegative* right-hand side of its equation. Now, something else must be done. The standard approach that is used for all these cases is the **artificial-variable technique.** This technique constructs a more convenient *artificial problem* by introducing a dummy variable (called an *artificial variable*) into each constraint that needs one. This new variable is introduced just for the purpose of being the initial basic variable for that equation. The usual nonnegativity constraints are placed on these variables, and the objective function also is modified to impose an exorbitant penalty on their having values larger than zero. The iterations of the simplex method then automatically force the artificial variables to disappear (become zero), one at a time, until they are all gone, after which the *real* problem is solved.

To illustrate the artificial-variable technique, first we consider the case where the only nonstandard form in the problem is the presence of one or more equality constraints.

Equality Constraints

Any equality constraint

$$a_{i1}x_1 + a_{i2}x_2 + \cdots + a_{in}x_n = b_i$$

actually is equivalent to a pair of inequality constraints:

$$a_{i1}x_1 + a_{i2}x_2 + \cdots + a_{in}x_n \le b_i$$
$$a_{i1}x_1 + a_{i2}x_2 + \cdots + a_{in}x_n \ge b_i.$$

However, rather than making this substitution and thereby increasing the number of constraints, it is more convenient to use the artificial-variable technique. We shall illustrate this technique with the following example.

Example.   Suppose that the Wyndor Glass Co. problem in Sec. 3.1 is modified to *require* that Plant 3 be used at full capacity. The only resulting change in the linear programming model is that the third constraint, $3x_1 + 2x_2 \le 18$, instead becomes an equality constraint

$$3x_1 + 2x_2 = 18,$$

so that the complete model becomes the one shown in the upper right-hand corner of Fig. 4.3. This figure also shows in darker ink the feasible region which now consists of *just* the line segment connecting $(2, 6)$ and $(4, 3)$.

After the slack variables still needed for the inequality constraints are introduced, the system of equations for the augmented form of the problem becomes

---

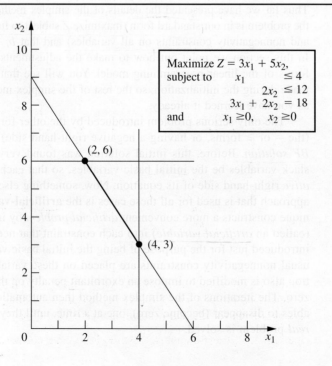

■ **FIGURE 4.3**
When the third functional constraint becomes an equality constraint, the feasible region for the Wyndor Glass Co. problem becomes the line segment between $(2, 6)$ and $(4, 3)$.

Maximize $Z = 3x_1 + 5x_2$,
subject to $x_1 \quad\quad \le 4$
                                    $2x_2 \le 12$
                           $3x_1 + 2x_2 = 18$
and $x_1 \ge 0, \quad x_2 \ge 0$

$$
\begin{aligned}
(0) \quad & Z - 3x_1 - 5x_2 && = 0 \\
(1) \quad & x_1 && + x_3 && = 4 \\
(2) \quad & 2x_2 && + x_4 = 12 \\
(3) \quad & 3x_1 + 2x_2 && = 18.
\end{aligned}
$$

Unfortunately, these equations do not have an obvious initial BF solution because there is no longer a slack variable to use as the initial basic variable for Eq. (3). It is necessary to find an initial BF solution to start the simplex method.

This difficulty can be circumvented in the following way.

Obtaining an Initial BF Solution. The procedure is to construct an **artificial problem** that has the same optimal solution as the real problem by making two modifications of the real problem.

1. Apply the **artificial-variable technique** by introducing a *nonnegative* **artificial variable** (call it $\bar{x}_5$)[14] into Eq. (3), just as if it were a slack variable

   $$(3) \quad 3x_1 + 2x_2 + \bar{x}_5 = 18.$$

2. Assign an *overwhelming penalty* to having $\bar{x}_5 > 0$ by changing the objective function $Z = 3x_1 + 5x_2$ to

   $$Z = 3x_1 + 5x_2 - M\bar{x}_5,$$

   where M symbolically represents a *huge* positive number. (This method of forcing $\bar{x}_5$ to be $\bar{x}_5 = 0$ in the optimal solution is called the **Big M method**.)

Now find the optimal solution for the real problem by applying the simplex method to the artificial problem, starting with the following initial BF solution:

Initial BF Solution
Nonbasic variables: $x_1 = 0, \quad x_2 = 0$
Basic variables: $x_3 = 4, \quad x_4 = 12, \quad \bar{x}_5 = 18.$

Because $\bar{x}_5$ plays the role of the slack variable for the third constraint in the artificial problem, this constraint is equivalent to $3x_1 + 2x_2 \le 18$ (just as for the original Wyndor Glass Co. problem in Sec. 3.1). We show below the resulting artificial problem (before augmenting) next to the real problem.

The Real Problem	*The Artificial Problem*
Maximize $Z = 3x_1 + 5x_2$,	Define $\bar{x}_5 = 18 - 3x_1 - 2x_2$. Maximize $Z = 3x_1 + 5x_2 - M\bar{x}_5$,
subject to	subject to
$x_1 \le 4$	$x_1 \le 4$
$2x_2 \le 12$	$2x_2 \le 12$
$3x_1 + 2x_2 = 18$	$3x_1 + 2x_2 \le 18$
and	(so $3x_1 + 2x_2 + \bar{x}_5 = 18$)
$x_1 \ge 0, \quad x_2 \ge 0.$	and
	$x_1 \ge 0, \quad x_2 \ge 0, \quad \bar{x}_5 \ge 0.$

Therefore, just as in Sec. 3.1, the feasible region for (x_1, x_2) for the artificial problem is the one shown in Fig. 4.4. The only portion of this feasible region that coincides with the feasible region for the real problem is where $\bar{x}_5 = 0$ (so $3x_1 + 2x_2 = 18$).

[14]We shall always label the artificial variables by putting a bar over them.

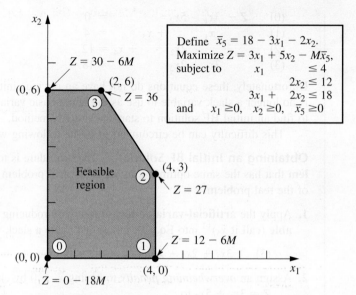

■ FIGURE 4.4
This graph shows the feasible region and the sequence of CPF solutions (⓪, ①, ②, ③) examined by the simplex method for the artificial problem that corresponds to the real problem of Fig. 4.3.

Figure 4.4 also shows the order in which the simplex method examines the CPF solutions (or BF solutions after augmenting), where each circled number identifies which iteration obtained that solution. Note that the simplex method moves counterclockwise here whereas it moved clockwise for the original Wyndor Glass Co. problem (see Fig. 4.2). The reason for this difference is the extra term $-M\bar{x}_5$ in the objective function for the artificial problem.

Before applying the simplex method and demonstrating that it follows the path shown in Fig. 4.4, the following preparatory step is needed.

Converting Equation (0) to Proper Form.   The system of equations after the artificial problem is augmented is

$$
\begin{array}{lllll}
(0) & Z - 3x_1 - 5x_2 & & + M\bar{x}_5 & = 0 \\
(1) & x_1 & + x_3 & & = 4 \\
(2) & 2x_2 & + x_4 & & = 12 \\
(3) & 3x_1 + 2x_2 & & + \bar{x}_5 & = 18
\end{array}
$$

where the initial basic variables $(x_3, x_4, \bar{x}_5)$ are shown in bold type. However, this system is not yet in proper form from Gaussian elimination because a basic variable $\bar{x}_5$ has a nonzero coefficient in Eq. (0). Recall that all basic variables must be algebraically eliminated from Eq. (0) before the simplex method can either apply the optimality test or find the entering basic variable. This elimination is necessary so that the negative of the coefficient of each nonbasic variable will give the rate at which Z would increase if that nonbasic variable were to be increased from 0 while adjusting the values of the basic variables accordingly.

To algebraically eliminate $\bar{x}_5$ from Eq. (0), we need to subtract from Eq. (0) the product, M times Eq. (3).

$$
\begin{array}{l}
 \quad Z - 3x_1 - 5x_2 + M\bar{x}_5 = 0 \\
 \quad \underline{-M(3x_1 + 2x_2 + \bar{x}_5 = 18)} \\
\text{New (0)} \quad Z - (3M + 3)x_1 - (2M + 5)x_2 = -18M.
\end{array}
$$

Application of the Simplex Method.   This new Eq. (0) gives Z in terms of *just* the nonbasic variables (x_1, x_2),

$$Z = -18M + (3M + 3)x_1 + (2M + 5)x_2.$$

Since $3M + 3 > 2M + 5$ (remember that M represents a huge number), increasing x_1 increases Z at a faster rate than increasing x_2 does, so x_1 is chosen as the entering basic variable. This leads to the move from $(0, 0)$ to $(4, 0)$ at iteration 1, shown in Fig. 4.4, thereby increasing Z by $4(3M + 3)$.

The quantities involving M never appear in the system of equations except for Eq. (0), so they need to be taken into account only in the optimality test and when an entering basic variable is determined. One way of dealing with these quantities is to assign some particular (huge) numerical value to M and use the resulting coefficients in Eq. (0) in the usual way. However, this approach may result in significant rounding errors that invalidate the optimality test. Therefore, it is better to do what we have just shown, namely, to express each coefficient in Eq. (0) as a linear function $aM + b$ of the *symbolic* quantity M by separately recording and updating the current numerical value of (1) the *multiplicative* factor a and (2) the *additive* term b. Because M is assumed to be so large that b always is negligible compared with M when $a \neq 0$, the decisions in the optimality test and the choice of the entering basic variable are made by using just the *multiplicative* factors in the usual way, except for breaking ties with the *additive* factors.

Using this approach on the example yields the simplex tableaux shown in Table 4.11. Note that the artificial variable $\bar{x}_5$ is a *basic variable* ($\bar{x}_5 > 0$) in the first two tableaux

■ TABLE 4.11 Complete set of simplex tableaux for the problem shown in Fig. 4.4

Iteration	Basic Variable	Eq.	Z	x_1	x_2	x_3	x_4	$\bar{x}_5$	Right Side
0	Z	(0)	1	$-3M - 3$	$-2M - 5$	0	0	0	$-18M$
	x_3	(1)	0	1	0	1	0	0	4
	x_4	(2)	0	0	2	0	1	0	12
	$\bar{x}_5$	(3)	0	3	2	0	0	1	18
1	Z	(0)	1	0	$-2M - 5$	$3M + 3$	0	0	$-6M + 12$
	x_1	(1)	0	1	0	1	0	0	4
	x_4	(2)	0	0	2	0	1	0	12
	$\bar{x}_5$	(3)	0	0	2	-3	0	1	6
2	Z	(0)	1	0	0	$-\dfrac{9}{2}$	0	$M + \dfrac{5}{2}$	27
	x_1	(1)	0	1	0	1	0	0	4
	x_4	(2)	0	0	0	3	1	-1	6
	x_2	(3)	0	0	1	$-\dfrac{3}{2}$	0	$\dfrac{1}{2}$	3
3	Z	(0)	1	0	0	0	$\dfrac{3}{2}$	$M + 1$	36
	x_1	(1)	0	1	0	0	$-\dfrac{1}{3}$	$\dfrac{1}{3}$	2
	x_3	(2)	0	0	0	1	$\dfrac{1}{3}$	$-\dfrac{1}{3}$	2
	x_2	(3)	0	0	1	0	$\dfrac{1}{2}$	0	6

and a *nonbasic variable* ($\bar{x}_5 = 0$) in the last two. Therefore, the first two BF solutions for this artificial problem are *infeasible* for the real problem whereas the last two also are BF solutions for the real problem.

This example involved only one equality constraint. If a linear programming model has more than one, each is handled in just the same way. (If the right-hand side is negative, multiply through both sides by -1 first.)

Negative Right-Hand Sides

The technique mentioned in the preceding sentence for dealing with an equality constraint with a negative right-hand side (namely, multiply through both sides by -1) also works for any inequality constraint with a negative right-hand side. Multiplying through both sides of an inequality by -1 also reverses the direction of the inequality; i.e., $\leq$ changes to $\geq$ or vice versa. For example, doing this to the constraint

$$x_1 - x_2 \leq -1 \qquad \text{(that is, } x_1 \leq x_2 - 1\text{)}$$

gives the equivalent constraint

$$-x_1 + x_2 \geq 1 \qquad \text{(that is, } x_2 - 1 \geq x_1\text{)}$$

but now the right-hand side is positive. Having nonnegative right-hand sides for all the functional constraints enables the simplex method to begin, because (after augmenting) these right-hand sides become the respective values of the *initial basic variables,* which must satisfy nonnegativity constraints.

We next focus on how to augment $\geq$ constraints, such as $-x_1 + x_2 \geq 1$, with the help of the artificial-variable technique.

Functional Constraints in $\geq$ Form

To illustrate how the artificial-variable technique deals with functional constraints in $\geq$ form, we will use the model for designing Mary's radiation therapy, as presented in Sec. 3.4. For your convenience, this model is repeated below, where we have placed a box around the constraint of special interest here.

Radiation Therapy Example

Minimize $Z = 0.4x_1 + 0.5x_2,$

subject to

$$0.3x_1 + 0.1x_2 \leq 2.7$$
$$0.5x_1 + 0.5x_2 = 6$$
$$\boxed{0.6x_1 + 0.4x_2 \geq 6}$$

and

$$x_1 \geq 0, \qquad x_2 \geq 0.$$

The graphical solution for this example (originally presented in Fig. 3.12) is repeated here in a slightly different form in Fig. 4.5. The three lines in the figure, along with the two axes, constitute the five constraint boundaries of the problem. The dots lying at the intersection of a pair of constraint boundaries are the *corner-point solutions.* The only two

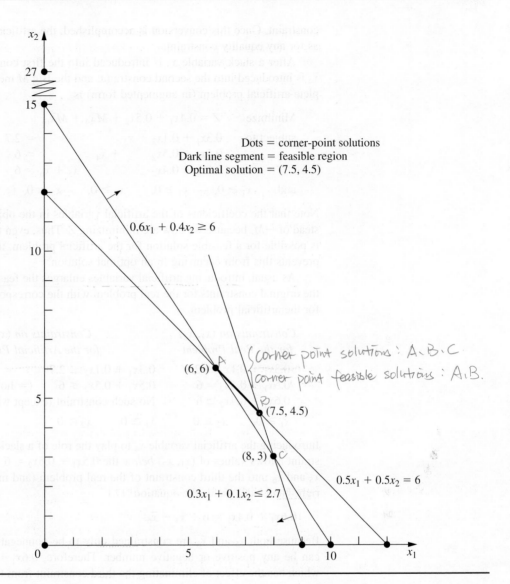

■ FIGURE 4.5
Graphical display of the radiation therapy example and its corner-point solutions.

corner-point *feasible* solutions are (6, 6) and (7.5, 4.5), and the feasible region is the line segment connecting these two points. The optimal solution is $(x_1, x_2) = (7.5, 4.5)$, with $Z = 5.25$.

We soon will show how the simplex method solves this problem by directly solving the corresponding artificial problem. However, first we must describe how to deal with the third constraint.

Our approach involves introducing *both* a surplus variable x_5 (defined as $x_5 = 0.6x_1 + 0.4x_2 - 6$) and an artificial variable $\bar{x}_6$, as shown next.

$$0.6x_1 + 0.4x_2 \qquad\qquad \geq 6$$
$$\rightarrow \quad 0.6x_1 + 0.4x_2 - x_5 \qquad = 6 \qquad (x_5 \geq 0)$$
$$\rightarrow \quad 0.6x_1 + 0.4x_2 - x_5 + \bar{x}_6 = 6 \qquad (x_5 \geq 0, \bar{x}_6 \geq 0).$$

Here x_5 is called a **surplus variable** because it subtracts the surplus of the left-hand side over the right-hand side to convert the inequality constraint to an equivalent equality constraint. Once this conversion is accomplished, the artificial variable is introduced just as for any equality constraint.

After a slack variable x_3 is introduced into the first constraint, an artificial variable $\bar{x}_4$ is introduced into the second constraint, and the Big M method is applied, so the complete artificial problem (in augmented form) is

Minimize $Z = 0.4x_1 + 0.5x_2 + M\bar{x}_4 + M\bar{x}_6,$

subject to $0.3x_1 + 0.1x_2 + x_3 \qquad\qquad\qquad = 2.7$

$\qquad\qquad\quad 0.5x_1 + 0.5x_2 \qquad + \bar{x}_4 \qquad\qquad = 6$

$\qquad\qquad\quad 0.6x_1 + 0.4x_2 \qquad\qquad\quad - x_5 + \bar{x}_6 = 6$

and $x_1 \geq 0, \qquad x_2 \geq 0, \qquad x_3 \geq 0, \qquad \bar{x}_4 \geq 0, \qquad x_5 \geq 0, \qquad \bar{x}_6 \geq 0.$

Note that the coefficients of the artificial variables in the objective function are $+M$, instead of $-M$, because we now are minimizing Z. Thus, even though $\bar{x}_4 > 0$ and/or $\bar{x}_6 > 0$ is possible for a feasible solution for the artificial problem, the huge unit penalty of $+M$ prevents this from occurring in an optimal solution.

As usual, introducing artificial variables enlarges the feasible region. Compare below the original constraints for the real problem with the corresponding constraints on (x_1, x_2) for the artificial problem.

Constraints on (x_1, x_2) for the Real Problem	*Constraints on (x_1, x_2) for the Artificial Problem*
$0.3x_1 + 0.1x_2 \leq 2.7$	$0.3x_1 + 0.1x_2 \leq 2.7$
$0.5x_1 + 0.5x_2 = 6$	$0.5x_1 + 0.5x_2 \leq 6$ (= holds when $\bar{x}_4 = 0$)
$0.6x_1 + 0.4x_2 \geq 6$	No such constraint (except when $\bar{x}_6 = 0$)
$x_1 \geq 0, \qquad x_2 \geq 0$	$x_1 \geq 0, \qquad x_2 \geq 0$

Introducing the artificial variable $\bar{x}_4$ to play the role of a slack variable in the second constraint allows values of (x_1, x_2) *below* the $0.5x_1 + 0.5x_2 = 6$ line in Fig. 4.5. Introducing x_5 and $\bar{x}_6$ into the third constraint of the real problem (and moving these variables to the right-hand side) yields the equation

$$0.6x_1 + 0.4x_2 = 6 + x_5 - \bar{x}_6.$$

Because both x_5 and $\bar{x}_6$ are constrained only to be nonnegative, their difference $x_5 - \bar{x}_6$ can be any positive or negative number. Therefore, $0.6x_1 + 0.4x_2$ can have any value, which has the effect of eliminating the third constraint from the artificial problem and allowing points on either side of the $0.6x_1 + 0.4x_2 = 6$ line in Fig. 4.5. (We keep the third constraint in the system of equations only because it will become relevant again later, after the Big M method forces $\bar{x}_6$ to be zero.) Consequently, the feasible region for the artificial problem is the entire polyhedron in Fig. 4.5 whose vertices are $(0, 0)$, $(9, 0)$, $(7.5, 4.5)$, and $(0, 12)$.

Since the origin now is feasible for the artificial problem, the simplex method starts with $(0, 0)$ as the initial CPF solution, i.e., with $(x_1, x_2, x_3, \bar{x}_4, x_5, \bar{x}_6) = (0, 0, 2.7, 6, 0, 6)$ as the initial BF solution. (Making the origin feasible as a convenient starting point for the simplex method is the whole point of creating the artificial problem.) We soon will trace the entire path followed by the simplex method from the origin to the optimal solution for both the artificial and real problems. But, first, how does the simplex method handle *minimization*?

Minimization

One straightforward way of minimizing Z with the simplex method is to exchange the roles of the positive and negative coefficients in row 0 for both the optimality test and step 1 of an iteration. However, rather than changing our instructions for the simplex method for this case, we present the following simple way of converting any minimization problem to an equivalent maximization problem:

$$Minimizing \quad Z = \sum_{j=1}^{n} c_j x_j$$

is equivalent to

$$maximizing \quad -Z = \sum_{j=1}^{n} (-c_j) x_j;$$

i.e., the two formulations yield the same optimal solution(s).

The two formulations are equivalent because the smaller Z is, the larger $-Z$ is, so the solution that gives the *smallest* value of Z in the entire feasible region must also give the *largest* value of $-Z$ in this region.

Therefore, in the radiation therapy example, we make the following change in the formulation:

$$\begin{aligned} &\text{Minimize} && Z = 0.4x_1 + 0.5x_2 \\ \rightarrow\ &\text{Maximize} && -Z = -0.4x_1 - 0.5x_2. \end{aligned}$$

After artificial variables $\bar{x}_4$ and $\bar{x}_6$ are introduced and then the Big M method is applied, the corresponding conversion is

$$\begin{aligned} &\text{Minimize} && Z = 0.4x_1 + 0.5x_2 + M\bar{x}_4 + M\bar{x}_6 \\ \rightarrow\ &\text{Maximize} && -Z = -0.4x_1 - 0.5x_2 - M\bar{x}_4 - M\bar{x}_6. \end{aligned}$$

Solving the Radiation Therapy Example

We now are nearly ready to apply the simplex method to the radiation therapy example. By using the maximization form just obtained, the entire system of equations is now

$$\begin{aligned} (0) \quad -Z + 0.4x_1 + 0.5x_2 + M\bar{x}_4 + M\bar{x}_6 &= 0 \\ (1) \quad 0.3x_1 + 0.1x_2 + x_3 &= 2.7 \\ (2) \quad 0.5x_1 + 0.5x_2 + \bar{x}_4 &= 6 \\ (3) \quad 0.6x_1 + 0.4x_2 - x_5 + \bar{x}_6 &= 6. \end{aligned}$$

The basic variables $(x_3, \bar{x}_4, \bar{x}_6)$ for the initial BF solution (for this artificial problem) are shown in bold type.

Note that this system of equations is not yet in proper form from Gaussian elimination, as required by the simplex method, since the basic variables $\bar{x}_4$ and $\bar{x}_6$ still need to be algebraically eliminated from Eq. (0). Because $\bar{x}_4$ and $\bar{x}_6$ both have a coefficient of M, Eq. (0) needs to have subtracted from it *both* M times Eq. (2) *and* M times Eq. (3). The calculations for all the coefficients (and the right-hand sides) are summarized below, where the vectors are the relevant rows of the simplex tableau corresponding to the above system of equations.

Row 0:

[0.4,	0.5,	0,	M,	0,	M,	0]
$-M$[0.5,	0.5,	0,	1,	0,	0,	6]
$-M$[0.6,	0.4,	0,	0,	-1,	1,	6]

New row 0 = $[-1.1M + 0.4, \quad -0.9M + 0.5, \quad 0, \quad 0, \quad M, \quad 0, \quad -12M]$

The resulting initial simplex tableau, ready to begin the simplex method, is shown at the top of Table 4.12. Applying the simplex method in just the usual way then yields the

■ **TABLE 4.12** The Big M method for the radiation therapy example

Iteration	Basic Variable	Eq.	Z	Coefficient of:						Right Side
				x_1	x_2	x_3	$\bar{x}_4$	x_5	$\bar{x}_6$	
0	Z	(0)	-1	$-1.1M + 0.4$	$-0.9M + 0.5$	0	0	M	0	$-12M$
	x_3	(1)	0	0.3	0.1	1	0	0	0	2.7
	$\bar{x}_4$	(2)	0	0.5	0.5	0	1	0	0	6
	$\bar{x}_6$	(3)	0	0.6	0.4	0	0	-1	1	6
1	Z	(0)	-1	0	$-\frac{16}{30}M + \frac{11}{30}$	$\frac{11}{3}M - \frac{4}{3}$	0	M	0	$-2.1M - 3.6$
	x_1	(1)	0	1	$\frac{1}{3}$	$\frac{10}{3}$	0	0	0	9
	$\bar{x}_4$	(2)	0	0	$\frac{1}{3}$	$-\frac{5}{3}$	1	0	0	1.5
	$\bar{x}_6$	(3)	0	0	0.2	-2	0	-1	1	0.6
2	Z	(0)	-1	0	0	$-\frac{5}{3}M + \frac{7}{3}$	0	$-\frac{5}{3}M + \frac{11}{6}$	$\frac{8}{3}M - \frac{11}{6}$	$-0.5M - 4.7$
	x_1	(1)	0	1	0	$\frac{20}{3}$	0	$\frac{5}{3}$	$-\frac{5}{3}$	8
	$\bar{x}_4$	(2)	0	0	0	$\frac{5}{3}$	1	$\frac{5}{3}$	$-\frac{5}{3}$	0.5
	x_2	(3)	0	0	1	-10	0	-5	5	3
3	Z	(0)	-1	0	0	0.5	$M - 1.1$	0	M	-5.25
	x_1	(1)	0	1	0	5	-1	0	0	7.5
	x_5	(2)	0	0	0	1	0.6	1	-1	0.3
	x_2	(3)	0	0	1	-5	3	0	0	4.5

sequence of simplex tableaux shown in the rest of Table 4.12. For the optimality test and the selection of the entering basic variable at each iteration, the quantities involving M are treated just as discussed in connection with Table 4.11. Specifically, whenever M is present, only its multiplicative factor is used, unless there is a tie, in which case the tie is broken by using the corresponding additive terms. Just such a tie occurs in the last selection of an entering basic variable (see the next-to-last tableau), where the coefficients of x_3 and x_5 in row 0 both have the same multiplicative factor of $-\frac{5}{3}$. Comparing the additive terms, $\frac{11}{6} < \frac{7}{3}$ leads to choosing x_5 as the entering basic variable.

Note in Table 4.12 the progression of values of the artificial variables $\bar{x}_4$ and $\bar{x}_6$ and of Z. We start with large values, $\bar{x}_4 = 6$ and $\bar{x}_6 = 6$, with $Z = 12M$ ($-Z = -12M$). The first iteration greatly reduces these values. The Big M method succeeds in driving $\bar{x}_6$ to zero (as a new nonbasic variable) at the second iteration and then in doing the same to $\bar{x}_4$ at the next iteration. With both $\bar{x}_4 = 0$ and $\bar{x}_6 = 0$, the basic solution given in the last tableau is guaranteed to be feasible for the real problem. Since it passes the optimality test, it also is optimal.

Now see what the Big M method has done graphically in Fig. 4.6. The feasible region for the artificial problem initially has four CPF solutions—(0, 0), (9, 0), (0, 12), and (7.5, 4.5)—and then replaces the first three with two new CPF solutions—(8, 3), (6, 6)—after $\bar{x}_6$ decreases to $\bar{x}_6 = 0$ so that $0.6x_1 + 0.4x_2 \geq 6$ becomes an additional constraint. (Note that the three replaced CPF solutions—(0, 0), (9, 0), and (0, 12)—actually were corner-point *infeasible* solutions for the real problem shown in Fig. 4.5.) Starting with the origin as the convenient initial CPF solution for the artificial problem, we move around

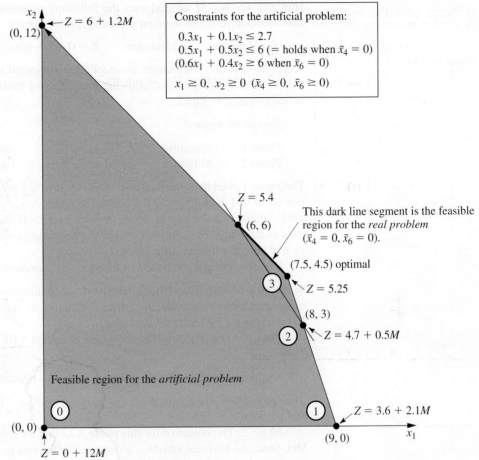

Constraints for the artificial problem:

$0.3x_1 + 0.1x_2 \leq 2.7$
$0.5x_1 + 0.5x_2 \leq 6$ (= holds when $\bar{x}_4 = 0$)
$(0.6x_1 + 0.4x_2 \geq 6$ when $\bar{x}_6 = 0)$

$x_1 \geq 0, \ x_2 \geq 0 \ (\bar{x}_4 \geq 0, \ \bar{x}_6 \geq 0)$

x_2
$\leftarrow Z = 6 + 1.2M$
(0, 12)

$Z = 5.4$
(6, 6)

This dark line segment is the feasible region for the *real problem* ($\bar{x}_4 = 0, \bar{x}_6 = 0$).

(7.5, 4.5) optimal

③ $\leftarrow Z = 5.25$

(8, 3)

② $\leftarrow Z = 4.7 + 0.5M$

Feasible region for the *artificial problem*

⓪

① $\leftarrow Z = 3.6 + 2.1M$

(0, 0)
(9, 0) x_1
$\uparrow$
$Z = 0 + 12M$

■ **FIGURE 4.6**
This graph shows the feasible region and the sequence of CPF solutions (⓪, ①, ②, ③) examined by the simplex method (with the Big M method) for the artificial problem that corresponds to the real problem of Fig. 4.5.

the boundary to three other CPF solutions—(9, 0), (8, 3), and (7.5, 4.5). The last of these is the first one that also is feasible for the real problem. Fortuitously, this first feasible solution also is optimal, so no additional iterations are needed.

For other problems with artificial variables, it may be necessary to perform additional iterations to reach an optimal solution after the first feasible solution is obtained for the real problem. (This was the case for the example solved in Table 4.11.) Thus, the Big M method can be thought of as having two phases. In the *first phase,* all the artificial variables are driven to zero (because of the penalty of M per unit for being greater than zero) in order to reach an initial BF solution for the *real* problem. In the *second phase,* all the artificial variables are kept at zero (because of this same penalty) while the simplex method generates a sequence of BF solutions for the real problem that leads to an optimal solution. The *two-phase method* described next is a streamlined procedure for performing these two phases directly, without even introducing M explicitly.

The Two-Phase Method

For the radiation therapy example just solved in Table 4.12, recall its real objective function

Real problem: Minimize $\quad Z = 0.4x_1 + 0.5x_2.$

However, the Big M method uses the following objective function (or its equivalent in maximization form) throughout the entire procedure:

    Big M method: Minimize $Z = 0.4x_1 + 0.5x_2 + M\bar{x}_4 + M\bar{x}_6.$

Since the first two coefficients are negligible compared to M, the two-phase method is able to drop M by using the following two objective functions with completely different definitions of Z in turn.

    Two-phase method:

Phase 1:	Minimize	$Z = \bar{x}_4 + \bar{x}_6$	(until $\bar{x}_4 = 0, \bar{x}_6 = 0$).
Phase 2:	Minimize	$Z = 0.4x_1 + 0.5x_2$	(with $\bar{x}_4 = 0, \bar{x}_6 = 0$).

The phase 1 objective function is obtained by dividing the Big M method objective function by M and then dropping the negligible terms. Since phase 1 concludes by obtaining a BF solution for the real problem (one where $\bar{x}_4 = 0$ and $\bar{x}_6 = 0$), this solution is then used as the *initial* BF solution for applying the simplex method to the real problem (with its real objective function) in phase 2.

Before solving the example in this way, we summarize the general method.

Summary of the Two-Phase Method.   *Initialization:* Revise the constraints of the original problem by introducing artificial variables as needed to obtain an obvious initial BF solution for the *artificial problem.*

    Phase 1: The objective for this phase is to find a BF solution for the *real problem.* To do this,

    Minimize $Z = \Sigma$ artificial variables, subject to revised constraints.

The optimal solution obtained for this problem (with $Z = 0$) will be a BF solution for the real problem.

    Phase 2: The objective for this phase is to find an *optimal solution* for the real problem. Since the artificial variables are not part of the real problem, these variables can now be dropped (they are all zero now anyway).[15] Starting from the BF solution obtained at the end of phase 1, use the simplex method to solve the real problem.

For the example, the problems to be solved by the simplex method in the respective phases are summarized below.

    Phase 1 Problem (Radiation Therapy Example):

    Minimize $Z = \bar{x}_4 + \bar{x}_6,$

subject to

$$0.3x_1 + 0.1x_2 + x_3 \qquad\qquad\qquad = 2.7$$
$$0.5x_1 + 0.5x_2 \qquad\quad + \bar{x}_4 \qquad\quad = 6$$
$$0.6x_1 + 0.4x_2 \qquad\qquad\quad - x_5 + \bar{x}_6 = 6$$

and

$$x_1 \geq 0, \qquad x_2 \geq 0, \qquad x_3 \geq 0, \qquad \bar{x}_4 \geq 0, \qquad x_5 \geq 0, \qquad \bar{x}_6 \geq 0.$$

    Phase 2 Problem (Radiation Therapy Example):

    Minimize $Z = 0.4x_1 + 0.5x_2,$

---

[15]We are skipping over three other possibilities here: (1) artificial variables > 0 (discussed in the next subsection), (2) artificial variables that are degenerate basic variables, and (3) retaining the artificial variables as nonbasic variables in phase 2 (and not allowing them to become basic) as an aid to subsequent postoptimality analysis. Your IOR Tutorial allows you to explore these possibilities.

subject to

$$0.3x_1 + 0.1x_2 + x_3 \quad\quad = 2.7$$
$$0.5x_1 + 0.5x_2 \quad\quad\quad = 6$$
$$0.6x_1 + 0.4x_2 \quad\quad - x_5 = 6$$

and

$$x_1 \geq 0, \quad x_2 \geq 0, \quad x_3 \geq 0, \quad x_5 \geq 0.$$

The only differences between these two problems are in the objective function and in the inclusion (phase 1) or exclusion (phase 2) of the artificial variables $\bar{x}_4$ and $\bar{x}_6$. Without the artificial variables, the phase 2 problem does not have an obvious *initial BF solution.* [The sole purpose of solving the phase 1 problem is to obtain a BF solution with $\bar{x}_4 = 0$ and $\bar{x}_6 = 0$ so that this solution (without the artificial variables) can be used as the initial BF solution for phase 2.]

Table 4.13 shows the result of applying the simplex method to this phase 1 problem. [Row 0 in the initial tableau is obtained by converting Minimize $Z = \bar{x}_4 + \bar{x}_6$ to Maximize $(-Z) = -\bar{x}_4 - \bar{x}_6$ and then using *elementary row operations* to eliminate the basic variables $\bar{x}_4$ and $\bar{x}_6$ from $-Z + \bar{x}_4 + \bar{x}_6 = 0$.] In the next-to-last tableau, there is a tie for the *entering basic variable* between x_3 and x_5, which is broken arbitrarily in favor of x_3. The solution obtained at the end of phase 1, then, is $(x_1, x_2, x_3, \bar{x}_4, x_5, \bar{x}_6) = (6, 6, 0.3, 0, 0, 0)$ or, after $\bar{x}_4$ and $\bar{x}_6$ are dropped, $(x_1, x_2, x_3, x_5) = (6, 6, 0.3, 0)$.

TABLE 4.13 Phase 1 of the two-phase method for the radiation therapy example

Iteration	Basic Variable	Eq.	Z	x_1	x_2	x_3	$\bar{x}_4$	x_5	$\bar{x}_6$	Right Side
0	Z	(0)	−1	−1.1	−0.9	0	0	1	0	−12
	x_3	(1)	0	0.3	0.1	1	0	0	0	2.7
	$\bar{x}_4$	(2)	0	0.5	0.5	0	1	0	0	6
	$\bar{x}_6$	(3)	0	0.6	0.4	0	0	−1	1	6
1	Z	(0)	−1	0	$-\frac{16}{30}$	$\frac{11}{3}$	0	1	0	−2.1
	x_1	(1)	0	1	$\frac{1}{3}$	$\frac{10}{3}$	0	0	0	9
	$\bar{x}_4$	(2)	0	0	$\frac{1}{3}$	$-\frac{5}{3}$	1	0	0	1.5
	$\bar{x}_6$	(3)	0	0	0.2	−2	0	−1	1	0.6
2	Z	(0)	−1	0	0	$-\frac{5}{3}$	0	$-\frac{5}{3}$	$\frac{8}{3}$	−0.5
	x_1	(1)	0	1	0	$\frac{20}{3}$	0	$\frac{5}{3}$	$-\frac{5}{3}$	8
	$\bar{x}_4$	(2)	0	0	0	$\frac{5}{3}$	1	$\frac{5}{3}$	$-\frac{5}{3}$	0.5
	x_2	(3)	0	0	1	−10	0	−5	5	3
3	Z	(0)	−1	0	0	0	1	0	1	0
	x_1	(1)	0	1	0	0	−4	−5	5	6
	x_3	(2)	0	0	0	1	$\frac{3}{5}$	1	−1	0.3
	x_2	(3)	0	0	1	0	6	5	−5	6

As claimed in the summary, this solution from phase 1 is indeed a BF solution for the *real* problem (the phase 2 problem) because it is the solution (after you set $x_5 = 0$) to the system of equations consisting of the three functional constraints for the phase 2 problem. In fact, after deleting the $\bar{x}_4$ and $\bar{x}_6$ columns as well as row 0 for each iteration, Table 4.13 shows one way of using Gaussian elimination to solve this system of equations by reducing the system to the form displayed in the final tableau.

Table 4.14 shows the preparations for beginning phase 2 after phase 1 is completed. Starting from the final tableau in Table 4.13, we drop the artificial variables ($\bar{x}_4$ and $\bar{x}_6$), substitute the phase 2 objective function ($-Z = -0.4x_1 - 0.5x_2$ in maximization form) into row 0, and then restore the proper form from Gaussian elimination (by algebraically eliminating the basic variables x_1 and x_2 from row 0). Thus, row 0 in the last tableau is obtained by performing the following *elementary row operations* in the next-to-last tableau: from row 0 subtract both the product, 0.4 times row 1, and the product, 0.5 times row 3. Except for the deletion of the two columns, note that rows 1 to 3 never change. The only adjustments occur in row 0 in order to replace the phase 1 objective function by the phase 2 objective function.

The last tableau in Table 4.14 is the initial tableau for applying the simplex method to the phase 2 problem, as shown at the top of Table 4.15. Just one iteration then leads to the optimal solution shown in the second tableau: $(x_1, x_2, x_3, x_5) = (7.5, 4.5, 0, 0.3)$. This solution is the desired optimal solution for the real problem of interest rather than the artificial problem constructed for phase 1.

Now we see what the two-phase method has done graphically in Fig. 4.7. Starting at the origin, phase 1 examines a total of four CPF solutions for the artificial problem. The first three actually were corner-point infeasible solutions for the real problem shown in Fig. 4.5. The fourth CPF solution, at (6, 6), is the first one that also is feasible for the real problem, so it becomes the initial CPF solution for phase 2. One iteration in phase 2 leads to the optimal CPF solution at (7.5, 4.5).

■ **TABLE 4.14** Preparing to begin phase 2 for the radiation therapy example

	Basic Variable	Eq.	Z	x_1	x_2	x_3	$\bar{x}_4$	x_5	$\bar{x}_6$	Right Side
							Coefficient of:			
Final Phase 1 tableau	Z	(0)	−1	0	0	0	1	0	1	0
	x_1	(1)	0	1	0	0	−4	−5	5	6
	x_3	(2)	0	0	0	1	$\frac{3}{5}$	1	−1	0.3
	x_2	(3)	0	0	1	0	6	5	−5	6
Drop $\bar{x}_4$ and $\bar{x}_6$	Z	(0)	−1	0	0	0		0		0
	x_1	(1)	0	1	0	0		−5		6
	x_3	(2)	0	0	0	1		1		0.3
	x_2	(3)	0	0	1	0		5		6
Substitute phase 2 objective function	Z	(0)	−1	0.4	0.5	0		0		0
	x_1	(1)	0	1	0	0		−5		6
	x_3	(2)	0	0	0	1		1		0.3
	x_2	(3)	0	0	1	0		5		6
Restore proper form from Gaussian elimination	Z	(0)	−1	0	0	0		−0.5		−5.4
	x_1	(1)	0	1	0	0		−5		6
	x_3	(2)	0	0	0	1		1		0.3
	x_2	(3)	0	0	1	0		5		6

■ **TABLE 4.15** Phase 2 of the two-phase method for the radiation therapy example

Iteration	Basic Variable	Eq.	Coefficient of:					Right Side
			Z	x_1	x_2	x_3	x_5	
0	Z	(0)	-1	0	0	0	-0.5	-5.4
	x_1	(1)	0	1	0	0	-5	6
	x_3	(2)	0	0	0	1	1	0.3
	x_2	(3)	0	0	1	0	5	6
1	Z	(0)	-1	0	0	0.5	0	-5.25
	x_1	(1)	0	1	0	5	0	7.5
	x_5	(2)	0	0	0	1	1	0.3
	x_2	(3)	0	0	1	-5	0	4.5

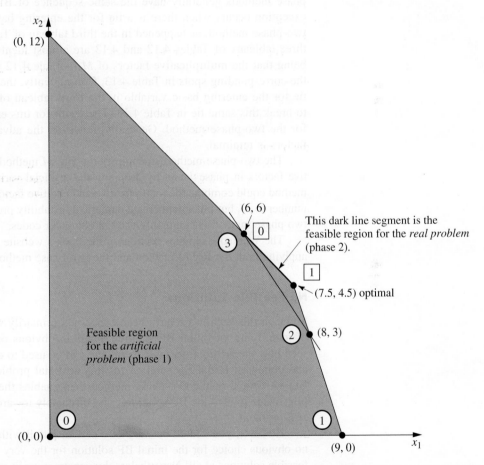

This dark line segment is the feasible region for the *real problem* (phase 2).

(7.5, 4.5) optimal

Feasible region for the *artificial problem* (phase 1)

■ **FIGURE 4.7**
This graph shows the sequence of CPF solutions for phase 1 (⓪, ①, ②, ③) and then for phase 2 (⓪, ①) when the two-phase method is applied to the radiation therapy example.

If the tie for the entering basic variable in the next-to-last tableau of Table 4.13 had been broken in the other way, then phase 1 would have gone directly from (8, 3) to (7.5, 4.5). After (7.5, 4.5) was used to set up the initial simplex tableau for phase 2, the *optimality test* would have revealed that this solution was optimal, so no iterations would be done.

It is interesting to compare the Big M and two-phase methods. Begin with their objective functions.

Big M Method:

Minimize $Z = 0.4x_1 + 0.5x_2 + M\bar{x}_4 + M\bar{x}_6.$

Two-Phase Method:

Phase 1:   Minimize $Z = \bar{x}_4 + \bar{x}_6.$
Phase 2:   Minimize $Z = 0.4x_1 + 0.5x_2.$

Because the $M\bar{x}_4$ and $M\bar{x}_6$ terms dominate the $0.4x_1$ and $0.5x_2$ terms in the objective function for the Big M method, this objective function is essentially equivalent to the phase 1 objective function as long as $\bar{x}_4$ and/or $\bar{x}_6$ is greater than zero. Then, when both $\bar{x}_4 = 0$ and $\bar{x}_6 = 0$, the objective function for the Big M method becomes completely equivalent to the phase 2 objective function.

Because of these virtual equivalencies in objective functions, the Big M and two-phase methods generally have the same sequence of BF solutions. The one possible exception occurs when there is a tie for the entering basic variable in phase 1 of the two-phase method, as happened in the third tableau of Table 4.13. Notice that the first three tableaux of Tables 4.12 and 4.13 are almost identical, with the only difference being that the multiplicative factors of M in Table 4.12 become the sole quantities in the corresponding spots in Table 4.13. Consequently, the additive terms that broke the tie for the entering basic variable in the third tableau of Table 4.12 were not present to break this same tie in Table 4.13. The result for this example was an extra iteration for the two-phase method. Generally, however, the advantage of having the additive factors is minimal.

The two-phase method streamlines the Big M method by using only the multiplicative factors in phase 1 and by dropping the artificial variables in phase 2. (The Big M method could combine the multiplicative and additive factors by assigning an actual huge number to M, but this might create numerical instability problems.) For these reasons, the two-phase method is commonly used in computer codes.

The Worked Examples section on the book's website provides **another example** of applying both the Big M method and the two-phase method to the same problem.

No Feasible Solutions

So far in this section we have been concerned primarily with the fundamental problem of identifying an initial BF solution when an obvious one is not available. You have seen how the artificial-variable technique can be used to construct an artificial problem and obtain an initial BF solution for this artificial problem instead. Use of either the Big M method or the two-phase method then enables the simplex method to begin its pilgrimage toward the BF solutions, and ultimately toward the optimal solution, for the *real* problem.

However, you should be wary of a certain pitfall with this approach. There may be no obvious choice for the initial BF solution for the very good reason that there are no feasible solutions at all! Nevertheless, by constructing an artificial feasible solution, there is nothing to prevent the simplex method from proceeding as usual and ultimately reporting a supposedly optimal solution.

Fortunately, the artificial-variable technique provides the following signpost to indicate when this has happened:

If the original problem has *no feasible solutions,* then either the Big M method or phase 1 of the two-phase method yields a final solution that has at least one artificial variable *greater* than zero. Otherwise, they *all* equal zero.

TABLE 4.16 The Big M method for the revision of the radiation therapy example that has no feasible solutions

| Iteration | Basic Variable | Eq. | Coefficient of: | | | | | | | Right Side |
			Z	x_1	x_2	x_3	$\bar{x}_4$	x_5	$\bar{x}_6$	
0	Z	(0)	-1	$-1.1M + 0.4$	$-0.9M + 0.5$	0	0	M	0	$-12M$
	x_3	(1)	0	0.3	0.1	1	0	0	0	1.8
	$\bar{x}_4$	(2)	0	0.5	0.5	0	1	0	0	6
	$\bar{x}_6$	(3)	0	0.6	0.4	0	0	-1	1	6
1	Z	(0)	-1	0	$-\frac{16}{30}M + \frac{11}{30}$	$\frac{11}{3}M - \frac{4}{3}$	0	M	0	$-5.4M - 2.4$
	x_1	(1)	0	1	$\frac{1}{3}$	$\frac{10}{3}$	0	0	0	6
	$\bar{x}_4$	(2)	0	0	$\frac{1}{3}$	$-\frac{5}{3}$	1	0	0	3
	$\bar{x}_6$	(3)	0	0	0.2	-2	0	-1	1	2.4
2	Z	(0)	-1	0	0	$M + 0.5$	$1.6M - 1.1$	M	0	$-0.6M - 5.7$
	x_1	(1)	0	1	0	5	-1	0	0	3
	x_2	(2)	0	0	1	-5	3	0	0	9
	$\bar{x}_6$	(3)	0	0	0	-1	-0.6	-1	1	0.6

To illustrate, let us change the first constraint in the radiation therapy example (see Fig. 4.5) as follows:

$$0.3x_1 + 0.1x_2 \leq 2.7 \quad \rightarrow \quad 0.3x_1 + 0.1x_2 \leq 1.8,$$

so that the problem no longer has any feasible solutions. Applying the Big M method just as before (see Table 4.12) yields the tableaux shown in Table 4.16. (Phase 1 of the two-phase method yields the same tableaux except that each expression involving M is replaced by just the multiplicative factor.) Hence, the Big M method normally would be indicating that the optimal solution is (3, 9, 0, 0, 0, 0.6). However, since an artificial variable $\bar{x}_6 = 0.6 > 0$, the real message here is that the problem has no feasible solutions.[16]

Variables Allowed to Be Negative

In most practical problems, negative values for the decision variables would have no physical meaning, so it is necessary to include nonnegativity constraints in the formulations of their linear programming models. However, this is not always the case. To illustrate, suppose that the Wyndor Glass Co. problem is changed so that product 1 already is in production, and the first decision variable x_1 represents the *increase* in its production rate. Therefore, a negative value of x_1 would indicate that product 1 is to be cut back by that amount. Such reductions might be desirable to allow a larger production rate for the new, more profitable product 2, so negative values should be allowed for x_1 in the model.

Since the procedure for determining the *leaving basic variable* requires that all the variables have nonnegativity constraints, any problem containing variables allowed to be negative must be converted to an *equivalent* problem involving only nonnegative variables before the simplex method is applied. Fortunately, this conversion can be done. The

---

[16]Techniques have been developed (and incorporated into linear programming software) to analyze what causes a large linear programming problem to have no feasible solutions so that any errors in the formulation can be corrected. For example, see J. W. Chinneck: *Feasibility and Infeasibility in Optimization: Algorithms and Computational Methods*, Springer Science + Business Media, New York, 2008.

modification required for each variable depends upon whether it has a (negative) lower bound on the values allowed. Each of these two cases is now discussed.

Variables with a Bound on the Negative Values Allowed. Consider any decision variable x_j that is allowed to have negative values which satisfy a constraint of the form

$$x_j \geq L_j,$$

where L_j is some negative constant. This constraint can be converted to a nonnegativity constraint by making the change of variables

$$x'_j = x_j - L_j, \qquad \text{so} \qquad x'_j \geq 0.$$

Thus, $x'_j + L_j$ would be substituted for x_j throughout the model, so that the redefined decision variable x'_j cannot be negative. (This same technique can be used when L_j is *positive* to convert a functional constraint $x_j \geq L_j$ to a nonnegativity constraint $x'_j \geq 0$.)

To illustrate, suppose that the current production rate for product 1 in the Wyndor Glass Co. problem is 10. With the definition of x_1 just given, the complete model at this point is the same as that given in Sec. 3.1 except that the nonnegativity constraint $x_1 \geq 0$ is replaced by

$$x_1 \geq -10.$$

To obtain the equivalent model needed for the simplex method, this decision variable would be redefined as the *total* production rate of product 1

$$x'_j = x_1 + 10,$$

which yields the changes in the objective function and constraints as shown:

$$
\begin{array}{l}
Z = 3x_1 + 5x_2 \\
\begin{array}{rl}
x_1 & \leq 4 \\
2x_2 & \leq 12 \\
3x_1 + 2x_2 & \leq 18 \\
x_1 \geq -10, & x_2 \geq 0
\end{array}
\end{array}
\;\rightarrow\;
\begin{array}{l}
Z = 3(x'_1 - 10) + 5x_2 \\
\begin{array}{rl}
x'_1 - 10 & \leq 4 \\
2x_2 & \leq 12 \\
3(x'_1 - 10) + 2x_2 & \leq 18 \\
x'_1 - 10 \geq -10, & x_2 \geq 0
\end{array}
\end{array}
\;\rightarrow\;
\begin{array}{l}
Z = -30 + 3x'_1 + 5x_2 \\
\begin{array}{rl}
x'_1 & \leq 14 \\
2x_2 & \leq 12 \\
3x'_1 + 2x_2 & \leq 48 \\
x'_1 \geq 0, & x_2 \geq 0
\end{array}
\end{array}
$$

Variables with No Bound on the Negative Values Allowed. In the case where x_j does *not* have a lower-bound constraint in the model formulated, another approach is required: x_j is replaced throughout the model by the *difference* of two new *nonnegative* variables

$$x_j = x_j^+ - x_j^-, \qquad \text{where } x_j^+ \geq 0, x_j^- \geq 0.$$

Since x_j^+ and x_j^- can have any nonnegative values, this difference $x_j^+ - x_j^-$ can have *any* value (positive or negative), so it is a legitimate substitute for x_j in the model. But after such substitutions, the simplex method can proceed with just nonnegative variables.

The new variables x_j^+ and x_j^- have a simple interpretation. As explained in the next paragraph, each BF solution for the new form of the model necessarily has the property that *either* $x_j^+ = 0$ or $x_j^- = 0$ (or both). Therefore, at the optimal solution obtained by the simplex method (a BF solution),

$$
x_j^+ = \begin{cases} x_j & \text{if } x_j \geq 0, \\ 0 & \text{otherwise;} \end{cases}
$$

$$
x_j^- = \begin{cases} |x_j| & \text{if } x_j \leq 0, \\ 0 & \text{otherwise;} \end{cases}
$$

so that x_j^+ represents the positive part of the decision variable x_j and x_j^- its negative part (as suggested by the superscripts).

For example, if $x_j = 10$, the above expressions give $x_j^+ = 10$ and $x_j^- = 0$. This same value of $x_j = x_j^+ - x_j^- = 10$ also would occur with larger values of x_j^+ and x_j^- such that $x_j^+ = x_j^- + 10$. Plotting these values of x_j^+ and x_j^- on a two-dimensional graph gives a line with an endpoint at $x_j^+ = 10$, $x_j^- = 0$ to avoid violating the nonnegativity constraints. This endpoint is the only corner-point solution on the line. Therefore, only this endpoint can be part of an overall CPF solution or BF solution involving all the variables of the model. This illustrates why each BF solution necessarily has either $x_j^+ = 0$ or $x_j^- = 0$ (or both).

To illustrate the use of the x_j^+ and x_j^-, let us return to the example on the preceding page where x_1 is redefined as the increase over the current production rate of 10 for product 1 in the Wyndor Glass Co. problem.

However, now suppose that the $x_1 \geq -10$ constraint was not included in the original model because it clearly would not change the optimal solution. (In some problems, certain variables do not need explicit lower-bound constraints because the functional constraints already prevent lower values.) Therefore, before the simplex method is applied, x_1 would be replaced by the difference

$$x_1 = x_1^+ - x_1^-, \qquad \text{where } x_1^+ \geq 0, \ x_1^- \geq 0,$$

as shown:

<table>
<tr><td>Maximize $Z = 3x_1 + 5x_2$,
subject to
   x_1 ≤ 4
          $2x_2 \leq 12$
   $3x_1 + 2x_2 \leq 18$
           $x_2 \geq 0$ (only)</td><td>$\rightarrow$</td><td>Maximize $Z = 3x_1^+ - 3x_1^- + 5x_2$,
subject to
   $x_1^+ - x_1^-$ ≤ 4
              $2x_2 \leq 12$
   $3x_1^+ - 3x_1^- + 2x_2 \leq 18$
   $x_1^+ \geq 0, \quad x_1^- \geq 0, \quad x_2 \geq 0$</td></tr>
</table>

From a computational viewpoint, this approach has the disadvantage that the new equivalent model to be used has more variables than the original model. In fact, if *all* the original variables lack lower-bound constraints, the new model will have *twice* as many variables. Fortunately, the approach can be modified slightly so that the number of variables is increased by only one, regardless of how many original variables need to be replaced. This modification is done by replacing each such variable x_j by

$$x_j = x_j' - x'', \qquad \text{where } x_j' \geq 0, \ x'' \geq 0,$$

instead, where x'' is the *same* variable for all relevant j. The interpretation of x'' in this case is that $-x''$ is the current value of the *largest* (in absolute terms) negative original variable, so that x_j' is the amount by which x_j exceeds this value. Thus, the simplex method now can make some of the x_j' variables larger than zero even when $x'' > 0$.

■ 4.7 POSTOPTIMALITY ANALYSIS

We stressed in Secs. 2.3, 2.4, and 2.5 that *postoptimality analysis*—the analysis done *after* an optimal solution is obtained for the initial version of the model—constitutes a very major and very important part of most operations research studies. The fact that postoptimality analysis is very important is particularly true for typical linear programming applications. In this section, we focus on the role of the simplex method in performing this analysis.

Table 4.17 summarizes the typical steps in postoptimality analysis for linear programming studies. The rightmost column identifies some algorithmic techniques that

■ **TABLE 4.17** Postoptimality analysis for linear programming

Task	Purpose	Technique
Model debugging	Find errors and weaknesses in model	Reoptimization
Model validation	Demonstrate validity of final model	See Sec. 2.4
Final managerial decisions on resource allocations (the b_i values)	Make appropriate division of organizational resources between activities under study and other important activities	Shadow prices
Evaluate estimates of model parameters	Determine crucial estimates that may affect optimal solution for further study	Sensitivity analysis
Evaluate trade-offs between model parameters	Determine best trade-off	Parametric linear programming

involve the simplex method. These techniques are introduced briefly here with the technical details deferred to later chapters.

Reoptimization

As discussed in Sec. 3.6, linear programming models that arise in practice commonly are very large, with hundreds, thousands, or even millions of functional constraints and decision variables. In such cases, many variations of the basic model may be of interest for considering different scenarios. Therefore, after having found an optimal solution for one version of a linear programming model, we frequently must solve again (often many times) for the solution of a slightly different version of the model. We nearly always have to solve again several times during the model debugging stage (described in Secs. 2.3 and 2.4), and we usually have to do so a large number of times during the later stages of postoptimality analysis as well.

One approach is simply to reapply the simplex method from scratch for each new version of the model, even though each run may require hundreds or even thousands of iterations for large problems. However, a *much more efficient* approach is to *reoptimize*. Reoptimization involves deducing how changes in the model get carried along to the *final* simplex tableau (as described in Secs. 5.3 and 6.6). This revised tableau and the optimal solution for the prior model are then used as the *initial tableau* and the *initial basic solution* for solving the new model. If this solution is feasible for the new model, then the simplex method is applied in the usual way, starting from this initial BF solution. If the solution is not feasible, a related algorithm called the *dual simplex method* (described in Sec. 7.1) probably can be applied to find the new optimal solution,[17] starting from this initial basic solution.

The big advantage of this **reoptimization technique** over re-solving from scratch is that an optimal solution for the revised model probably is going to be *much* closer to the prior optimal solution than to an initial BF solution constructed in the usual way for the simplex method. Therefore, assuming that the model revisions were modest, only a few iterations should be required to reoptimize instead of the hundreds or thousands that may be required when you start from scratch. In fact, the optimal solutions for the prior and revised models are frequently the same, in which case the reoptimization technique requires only one application of the optimality test and *no* iterations.

---

[17]The one requirement for using the dual simplex method here is that the *optimality test* is still passed when applied to row 0 of the *revised* final tableau. If not, then still another algorithm called the *primal-dual method* can be used instead.

Shadow Prices

Recall that linear programming problems often can be interpreted as allocating resources to activities. In particular, when the functional constraints are in $\leq$ form, we interpreted the b_i (the right-hand sides) as the amounts of the respective resources being made available for the activities under consideration. In many cases, there may be some latitude in the amounts that will be made available. If so, the b_i values used in the initial (validated) model actually may represent management's *tentative initial decision* on how much of the organization's resources will be provided to the activities considered in the model instead of to other important activities under the purview of management. From this broader perspective, some of the b_i values can be increased in a revised model, but only if a sufficiently strong case can be made to management that this revision would be beneficial.

Consequently, information on the economic contribution of the resources to the measure of performance (Z) for the current study often would be extremely useful. The simplex method provides this information in the form of *shadow prices* for the respective resources.

The **shadow price** for resource i (denoted by y_i^*) measures the *marginal value* of this resource, i.e., the rate at which Z could be increased by (slightly) increasing the amount of this resource (b_i) being made available.[18,19] The simplex method identifies this shadow price by $y_i^* = $ coefficient of the ith slack variable in row 0 of the final simplex tableau.

To illustrate, for the Wyndor Glass Co. problem,

Resource i = production capacity of Plant i ($i = 1, 2, 3$) being made available to the two new products under consideration,

b_i = hours of production time per week being made available in Plant i for these new products.

Providing a substantial amount of production time for the new products would require adjusting production times for the current products, so choosing the b_i value is a difficult managerial decision. The tentative initial decision has been

$$b_1 = 4, \qquad b_2 = 12, \qquad b_3 = 18,$$

as reflected in the basic model considered in Sec. 3.1 and in this chapter. However, management now wishes to evaluate the effect of changing any of the b_i values.

The shadow prices for these three resources provide just the information that management needs. The final tableau in Table 4.8 yields

$$y_1^* = 0 = \text{shadow price for resource 1,}$$

$$y_2^* = \frac{3}{2} = \text{shadow price for resource 2,}$$

$$y_3^* = 1 = \text{shadow price for resource 3.}$$

With just two decision variables, these numbers can be verified by checking graphically that individually increasing any b_i by 1 indeed would increase the optimal value of Z by y_i^*. For example, Fig. 4.8 demonstrates this increase for resource 2 by reapplying the

---

[18]The increase in b_i must be sufficiently small that the current set of basic variables remains optimal since this rate (marginal value) changes if the set of basic variables changes.

[19]In the case of a functional constraint in $\geq$ or $=$ form, its shadow price is again defined as the rate at which Z could be increased by (slightly) increasing the value of b_i, although the interpretation of b_i now would normally be something other than the amount of a resource being made available.

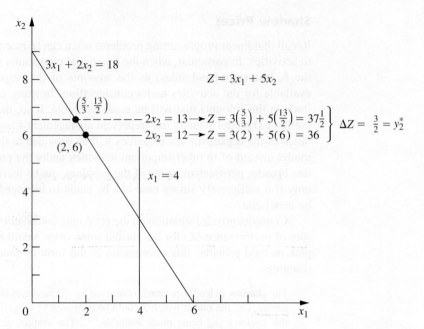

■ FIGURE 4.8
This graph shows that the shadow price is $y_2^* = \frac{3}{2}$ for resource 2 for the Wyndor Glass Co. problem. The two dots are the optimal solutions for $b_2 = 12$ or $b_2 = 13$, and plugging these solutions into the objective function reveals that increasing b_2 by 1 increases Z by $y_2^* = \frac{3}{2}$.

graphical method presented in Sec. 3.1. The optimal solution, $(2, 6)$ with $Z = 36$, changes to $\left(\frac{5}{3}, \frac{13}{2}\right)$ with $Z = 37\frac{1}{2}$ when b_2 is increased by 1 (from 12 to 13), so that

$$y_2^* = \Delta Z = 37\frac{1}{2} - 36 = \frac{3}{2}.$$

Since Z is expressed in thousands of dollars of profit per week, $y_2^* = \frac{3}{2}$ indicates that adding 1 more hour of production time per week in Plant 2 for these two new products would increase their total profit by \$1,500 per week. Should this actually be done? It depends on the marginal profitability of other products currently using this production time. If there is a current product that contributes less than \$1,500 of weekly profit per hour of weekly production time in Plant 2, then some shift of production time to the new products would be worthwhile.

We shall continue this story in Sec. 6.7, where the Wyndor OR team uses shadow prices as part of its *sensitivity analysis* of the model.

Figure 4.8 demonstrates that $y_2^* = \frac{3}{2}$ is the rate at which Z could be increased by increasing b_2 slightly. However, it also demonstrates the common phenomenon that this interpretation holds only for a small increase in b_2. Once b_2 is increased beyond 18, the optimal solution stays at $(0, 9)$ with no further increase in Z. (At that point, the set of basic variables in the optimal solution has changed, so a new final simplex tableau will be obtained with new shadow prices, including $y_2^* = 0$.)

Now note in Fig. 4.8 why $y_1^* = 0$. Because the constraint on resource 1, $x_1 \leq 4$, is *not binding* on the optimal solution $(2, 6)$, there is a *surplus* of this resource. Therefore, increasing b_1 beyond 4 cannot yield a new optimal solution with a larger value of Z.

By contrast, the constraints on resources 2 and 3, $2x_2 \leq 12$ and $3x_1 + 2x_2 \leq 18$, are **binding constraints** (constraints that hold with equality at the optimal solution). Because the limited supply of these resources ($b_2 = 12$, $b_3 = 18$) *binds* Z from being increased further, they have *positive* shadow prices. Economists refer to such resources as *scarce goods,* whereas resources available in surplus (such as resource 1) are *free goods* (resources with a zero shadow price).

The kind of information provided by shadow prices clearly is valuable to management when it considers reallocations of resources within the organization. It also is very helpful when an increase in b_i can be achieved only by going outside the organization to purchase more of the resource in the marketplace. For example, suppose that Z represents *profit* and that the unit profits of the activities (the c_j values) include the costs (at regular prices) of all the resources consumed. Then a *positive* shadow price of y_i^* for resource i means that the total profit Z can be increased by y_i^* by purchasing 1 more unit of this resource at its regular price. Alternatively, if a *premium* price must be paid for the resource in the marketplace, then y_i^* represents the *maximum* premium (excess over the regular price) that would be worth paying.[20]

The theoretical foundation for shadow prices is provided by the duality theory described in Chap. 6.

Sensitivity Analysis

When discussing the *certainty assumption* for linear programming at the end of Sec. 3.3, we pointed out that the values used for the model parameters (the a_{ij}, b_i, and c_j identified in Table 3.3) generally are just *estimates* of quantities whose true values will not become known until the linear programming study is implemented at some time in the future. A main purpose of sensitivity analysis is to identify the **sensitive parameters** (i.e., those that cannot be changed without changing the optimal solution). The sensitive parameters are the parameters that need to be estimated with special care to minimize the risk of obtaining an erroneous optimal solution. They also will need to be monitored particularly closely as the study is implemented. If it is discovered that the true value of a sensitive parameter differs from its estimated value in the model, this immediately signals a need to change the solution.

How are the sensitive parameters identified? In the case of the b_i, you have just seen that this information is given by the shadow prices provided by the simplex method. In particular, if $y_i^* > 0$, then the optimal solution changes if b_i is changed, so b_i is a sensitive parameter. However, $y_i^* = 0$ implies that the optimal solution is not sensitive to at least small changes in b_i. Consequently, if the value used for b_i is an estimate of the amount of the resource that will be available (rather than a managerial decision), then the b_i values that need to be monitored more closely are those with *positive* shadow prices—especially those with *large* shadow prices.

When there are just two variables, the sensitivity of the various parameters can be analyzed graphically. For example, in Fig. 4.9, $c_1 = 3$ can be changed to any other value from 0 to 7.5 without the optimal solution changing from (2, 6). (The reason is that any value of c_1 within this range keeps the slope of $Z = c_1 x_1 + 5 x_2$ between the slopes of the lines $2x_2 = 12$ and $3x_1 + 2x_2 = 18$.) Similarly, if $c_2 = 5$ is the only parameter changed, it can have any value greater than 2 without affecting the optimal solution. Hence, neither c_1 nor c_2 is a sensitive parameter. (The procedure called **Graphical Method and Sensitivity Analysis** in IOR Tutorial enables you to perform this kind of graphical analysis very efficiently.)

The easiest way to analyze the sensitivity of each of the a_{ij} parameters graphically is to check whether the corresponding constraint is *binding* at the optimal solution. Because $x_1 \leq 4$ is *not* a binding constraint, any sufficiently small change in its coefficients ($a_{11} = 1$, $a_{12} = 0$) is not going to change the optimal solution, so these are *not* sensitive parameters. On the other hand, both $2x_2 \leq 12$ and $3x_1 + 2x_2 \leq 18$ are *binding constraints*,

---

[20]If the unit profits do *not* include the costs of the resources consumed, then y_i^* represents the maximum *total* unit price that would be worth paying to increase b_i.

■ **FIGURE 4.9**
This graph demonstrates the sensitivity analysis of c_1 and c_2 for the Wyndor Glass Co. problem. Starting with the original objective function line [where $c_1 = 3$, $c_2 = 5$, and the optimal solution is $(2, 6)$], the other two lines show the extremes of how much the slope of the objective function line can change and still retain $(2, 6)$ as an optimal solution. Thus, with $c_2 = 5$, the allowable range for c_1 is $0 \leq c_1 \leq 7.5$. With $c_1 = 3$, the allowable range for c_2 is $c_2 \geq 2$.

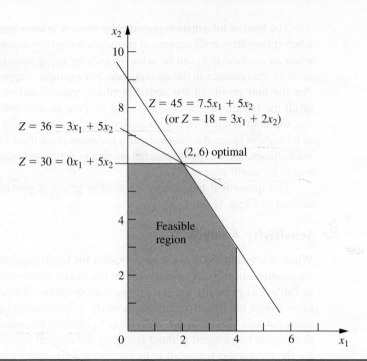

so changing *any* one of their coefficients ($a_{21} = 0$, $a_{22} = 2$, $a_{31} = 3$, $a_{32} = 2$) is going to change the optimal solution, and therefore these are sensitive parameters.

Typically, greater attention is given to performing sensitivity analysis on the b_i and c_j parameters than on the a_{ij} parameters. On real problems with hundreds or thousands of constraints and variables, the effect of changing one a_{ij} value is usually negligible, but changing one b_i or c_j value can have real impact. Furthermore, in many cases, the a_{ij} values are determined by the technology being used (the a_{ij} values are sometimes called *technological coefficients*), so there may be relatively little (or no) uncertainty about their final values. This is fortunate, because there are far more a_{ij} parameters than b_i and c_j parameters for large problems.

For problems with more than two (or possibly three) decision variables, you cannot analyze the sensitivity of the parameters graphically as was just done for the Wyndor Glass Co. problem. However, you can extract the same kind of information from the simplex method. Getting this information requires using the *fundamental insight* described in Sec. 5.3 to deduce the changes that get carried along to the final simplex tableau as a result of changing the value of a parameter in the original model. The rest of the procedure is described and illustrated in Secs. 6.6 and 6.7.

Using Excel to Generate Sensitivity Analysis Information

Sensitivity analysis normally is incorporated into software packages based on the simplex method. For example, the Excel Solver will generate sensitivity analysis information upon request. As was shown in Fig. 3.21, when the Solver gives the message that it has found a solution, it also gives on the right a list of three reports that can be provided. By selecting the second one (labeled "Sensitivity") after solving the Wyndor Glass Co. problem, you will obtain the *sensitivity report* shown in Fig. 4.10. The upper table in this report provides sensitivity analysis information about the decision variables and their coefficients in the objective function. The lower table does the same for the functional constraints and their right-hand sides.

Adjustable Cells

Cell	Name	Final Value	Reduced Cost	Objective Coefficient	Allowable Increase	Allowable Decrease
$C$12	Batches Produced Doors	2	0	3,000	4,500	3,000
$D$12	Batches Produced Windows	6	0	5,000	1E+30	3,000

Constraints

Cell	Name	Final Value	Shadow Price	Constraint R.H. Side	Allowable Increase	Allowable Decrease
$E$7	Plant 1 Used	2	0	4	1E+30	2
$E$8	Plant 2 Used	12	1,500	12	6	6
$E$9	Plant 3 Used	18	1,000	18	6	6

■ **FIGURE 4.10**
The sensitivity report provided by the Excel Solver for the Wyndor Glass Co. problem.

Look first at the upper table in this figure. The "Final Value" column indicates the optimal solution. The next column gives the *reduced costs*. (We will not discuss these reduced costs now because the information they provide can also be gleaned from the rest of the upper table.) The next three columns provide the information needed to identify the *allowable range* for each coefficient c_j in the objective function.

For any c_j, its **allowable range** is the range of values for this coefficient over which the current optimal solution remains optimal, assuming no change in the other coefficients.

The "Objective Coefficient" column gives the current value of each coefficient, and then the next two columns give the *allowable increase* and the *allowable decrease* from this value to remain within the allowable range. The spreadsheet model (Fig. 3.22) expresses the profits per batch in units of *dollars,* whereas the c_j in the algebraic version of the linear programming model uses units of *thousands of dollars,* so the quantities in all three of these columns need to be divided by 1000 to use the same units as the c_j. Therefore,

$$\frac{3,000 - 3,000}{1,000} \le c_1 \le \frac{3,000 + 4,500}{1,000}, \qquad \text{so} \qquad 0 \le c_1 \le 7.5$$

is the allowable range for c_1 over which the current optimal solution will stay optimal (assuming $c_2 = 5$), just as was found graphically in Fig. 4.9. Similarly, since Excel uses 1E + 30 (10^{30}) to represent infinity,

$$\frac{5,000 - 3,000}{1,000} \le c_2 \le \frac{5,000 + \infty}{1,000}, \qquad \text{so} \qquad 2 \le c_2$$

is the allowable range for c_2.

The fact that both the allowable increase and the allowable decrease are greater than zero for the coefficient of both decision variables provides another useful piece of information, as described below.

When the upper table in the sensitivity report generated by the Excel Solver indicates that both the allowable increase and the allowable decrease are greater than zero for every objective coefficient, this is a signpost that the optimal solution in the "Final Value" column is the only optimal solution. Conversely, having any allowable increase or allowable decrease equal to zero is a signpost that there are multiple optimal solutions. Changing the corresponding coefficient a tiny amount beyond the zero allowed and re-solving provides another optimal CPF solution for the original model.

Now consider the lower table in Fig. 4.10 that focuses on sensitivity analysis for the three functional constraints. The "Final Value" column gives the value of each constraint's left-hand side for the optimal solution. The next two columns give the shadow price and the current value of the right-hand side (b_i) for each constraint. (These shadow prices from the spreadsheet model use units of *dollars,* so they need to be divided by 1000 to use the same units of *thousands of dollars* as Z in the algebraic version of the linear programming model.) When just one b_i value is then changed, the last two columns give the *allowable increase* or *allowable decrease* in order to remain within its *allowable range.*

For any b_i, its **allowable range** is the range of values for this right-hand side over which the current optimal BF solution (with adjusted values[21] for the basic variables) remains feasible, assuming no change in the other right-hand sides. A key property of this range of values is that the current *shadow price* for b_i remains valid for evaluating the effect on Z of changing b_i only as long as b_i remains within this allowable range.

Thus, using the lower table in Fig. 4.10, combining the last two columns with the current values of the right-hand sides gives the following allowable ranges:

$$2 \leq b_1$$
$$6 \leq b_2 \leq 18$$
$$12 \leq b_3 \leq 24.$$

This sensitivity report generated by the Excel Solver is typical of the sensitivity analysis information provided by linear programming software packages. You will see in Appendix 4.1 that LINDO and LINGO provide essentially the same report. MPL/CPLEX does also when it is requested with the Solution File dialogue box. Once again, this information obtained algebraically also can be derived from graphical analysis for this two-variable problem. (See Prob. 4.7-1.) For example, when b_2 is increased from 12 in Fig. 4.8, the originally optimal CPF solution at the intersection of two constraint boundaries $2x_2 = b_2$ and $3x_1 + 2x_2 = 18$ will remain feasible (including $x_1 \geq 0$) only for $b_2 \leq 18$.

The Worked Examples section of the book's website includes **another example** of applying sensitivity analysis (using both graphical analysis and the sensitivity report). The latter part of Chap. 6 also will delve into this type of analysis more deeply.

Parametric Linear Programming

Sensitivity analysis involves changing one parameter at a time in the original model to check its effect on the optimal solution. By contrast, **parametric linear programming** (or **parametric programming** for short) involves the systematic study of how the optimal solution changes as *many* of the parameters change *simultaneously* over some range. This study can provide a very useful extension of sensitivity analysis, e.g., to check the effect of "correlated" parameters that change together due to exogenous factors such as the state of the economy. However, a more important application is the investigation of *trade-offs* in parameter values. For example, if the c_j values represent the unit profits of the respective activities, it may be possible to increase some of the c_j values at the expense of decreasing others by an appropriate shifting of personnel and equipment among activities. Similarly, if the

---

[21]Since the values of the basic variables are obtained as the simultaneous solution of a system of equations (the functional constraints in augmented form), at least some of these values change if one of the right-hand sides changes. However, the adjusted values of the current set of basic variables still will satisfy the nonnegativity constraints, and so still will be feasible, as long as the new value of this right-hand side remains within its allowable range. If the adjusted basic solution is still feasible, it also will still be optimal. We shall elaborate further in Sec. 6.7.

b_i values represent the amounts of the respective resources being made available, it may be possible to increase some of the b_i values by agreeing to accept decreases in some of the others. The analysis of such possibilities is discussed and illustrated at the end of Sec. 6.7.

In some applications, the main purpose of the study is to determine the most appropriate trade-off between two basic factors, such as *costs* and *benefits*. The usual approach is to express one of these factors in the objective function (e.g., minimize total cost) and incorporate the other into the constraints (e.g., benefits $\geq$ minimum acceptable level), as was done for the Nori & Leets Co. air pollution problem in Sec. 3.4. Parametric linear programming then enables systematic investigation of what happens when the initial tentative decision on the trade-off (e.g., the minimum acceptable level for the benefits) is changed by improving one factor at the expense of the other.

The algorithmic technique for parametric linear programming is a natural extension of that for sensitivity analysis, so it, too, is based on the simplex method. The procedure is described in Sec. 7.2.

4.8 COMPUTER IMPLEMENTATION

If the electronic computer had never been invented, you probably would have never heard of linear programming and the simplex method. Even though it is possible to apply the simplex method by hand (perhaps with the aid of a calculator) to solve tiny linear programming problems, the calculations involved are just too tedious to do this on a routine basis. However, the simplex method is ideally suited for execution on a computer. It is the computer revolution that has made possible the widespread application of linear programming in recent decades.

Implementation of the Simplex Method

Computer codes for the simplex method now are widely available for essentially all modern computer systems. These codes commonly are part of a sophisticated software package for mathematical programming that includes many of the procedures described in subsequent chapters (including those used for postoptimality analysis).

These production computer codes do not closely follow either the algebraic form or the tabular form of the simplex method presented in Secs. 4.3 and 4.4. These forms can be streamlined considerably for computer implementation. Therefore, the codes use instead a *matrix form* (usually called the *revised simplex method*) that is especially well suited for the computer. This form accomplishes exactly the same things as the algebraic or tabular form, but it does this while computing and storing only the numbers that are actually needed for the current iteration; and then it carries along the essential data in a more compact form. The revised simplex method is described in Secs. 5.2 and 5.4.

The simplex method is used routinely to solve surprisingly large linear programming problems. For example, powerful desktop computers (including workstations) commonly are used to solve problems with hundreds of thousands, or even millions, of functional constraints and a larger number of decision variables. Occasionally, successfully solved problems have even tens of millions of functional constraints and decision variables.[22] For certain *special types* of linear programming problems (such as the transportation,

---

[22]Do not try this at home. Attacking such a massive problem requires an especially sophisticated linear programming system that uses the latest techniques for exploiting sparcity in the coefficient matrix as well as other special techniques (e.g., *crashing techniques* for quickly finding an advanced initial BF solution). When problems are re-solved periodically after minor updating of the data, much time often is saved by using (or modifying) the last optimal solution to provide the initial BF solution for the new run.

assignment, and minimum cost flow problems to be described later in the book), even larger problems now can be solved by *specialized* versions of the simplex method.

Several factors affect how long it will take to solve a linear programming problem by the general simplex method. The most important one is the *number of ordinary functional constraints.* In fact, computation time tends to be roughly proportional to the cube of this number, so that doubling this number may multiply the computation time by a factor of approximately 8. By contrast, the number of variables is a relatively minor factor.[23] Thus, doubling the number of variables probably will not even double the computation time. A third factor of some importance is the *density* of the table of constraint coefficients (i.e., the *proportion* of the coefficients that are *not* zero), because this affects the computation time *per iteration.* (For large problems encountered in practice, it is common for the density to be under 5 percent, or even under 1 percent, and this much "sparsity" tends to greatly accelerate the simplex method.) One common rule of thumb for the *number of iterations* is that it tends to be roughly twice the number of functional constraints.

With large linear programming problems, it is inevitable that some mistakes and faulty decisions will be made initially in formulating the model and inputting it into the computer. Therefore, as discussed in Sec. 2.4, a thorough process of testing and refining the model (*model validation*) is needed. The usual end product is not a single static model that is solved once by the simplex method. Instead, the OR team and management typically consider a long series of variations on a basic model (sometimes even thousands of variations) to examine different scenarios as part of postoptimality analysis. This entire process is greatly accelerated when it can be carried out *interactively* on a *desktop computer.* And, with the help of both mathematical programming modeling languages and improving computer technology, this now is becoming common practice.

Until the mid-1980s, linear programming problems were solved almost exclusively on *mainframe computers.* Since then, there has been an explosion in the capability of doing linear programming on desktop computers, including personal computers as well as workstations. Workstations, including some with parallel processing capabilities, now are commonly used instead of mainframe computers to solve massive linear programming models. The fastest personal computers are not lagging far behind, although solving huge models usually requires additional memory.

Linear Programming Software Featured in This Book

A considerable number of excellent software packages for linear programming and its extensions now are available to fill a variety of needs. One leading package of this type is **Express-MP**, a product of Dash Optimization (which now has joined Fair Isaac). Another that is widely regarded to be a particularly powerful package for solving massive problems is **CPLEX**, a product of ILOG, Inc., located in Silicon Valley. Since 1988, CPLEX has helped to lead the way in solving larger and larger linear programming problems. An extensive research and development effort has enabled a series of upgrades with dramatic increases in efficiency. CPLEX 11 released in 2007 provided another major improvement. This software package frequently is capable of solving real linear programming problems arising in industry with tens of millions of functional constraints and decision variables! CPLEX often uses the simplex method and its variants (such as the dual simplex method presented in Sec. 7.1) to solve these massive problems. In addition to the simplex method, CPLEX also features some other powerful weapons for attacking linear programming problems. One is a lightning-fast algorithm (referred to as the *barrier algorithm*) that uses the *interior-point approach* introduced in Section 4.9. This algorithm can solve some huge

---

[23]This statement assumes that the revised simplex method described in Secs. 5.2 and 5.4 is being used.

general linear programming problems that the simplex method cannot (and vice versa). Another feature is the *network simplex method* (described in Sec. 9.7) that can solve even larger special types of linear programming problems. CPLEX 11 also extends beyond linear programming by including state-of-the-art algorithms for *integer programming* (Chap. 11) and *quadratic programming* (Sec. 12.7), as well as *integer quadratic programming*.

We anticipate that these major improvements in the state-of-the-art optimization software packages such as CPLEX will continue in the future as well. Continuing rapid improvements in the speed of computers also will further accelerate the speedup of these future software packages.

Because it often is used to solve really large problems, CPLEX normally is used in conjunction with a mathematical programming *modeling language.* As described in Sec. 3.7, modeling languages are designed for efficiently formulating large linear programming models (and related models) in a compact way, after which a solver is called upon to solve the model. Several of the prominent modeling languages support CPLEX as a solver. ILOG also has introduced its own modeling language, called the *Optimization Programming Language (OPL),* that can be used with CPLEX to form the *OPL-CPLEX Development System.* (A trial version of that product is available at ILOG's website, www.ilog.com.)

As we mentioned in Sec. 3.7, the student version of CPLEX is included in your OR Courseware as the main solver for the MPL modeling language. This version features the simplex method for solving linear programming problems.

The student version of MPL in your OR Courseware also includes two other solvers that are an alternative to CPLEX for solving both linear programming problems and integer programming problems (discussed in Chap. 11). One is **CoinMP,** an open source solver that can solve larger problems than the student version of CPLEX (which is limited to 300 constraints and variables). The other is LINDO.

LINDO (short for Linear, Interactive, and Discrete Optimizer) has an even longer history than CPLEX in the realm of applications of linear programming and its extensions. The easy-to-use LINDO interface is available as a subset of the **LINGO** optimization modeling package from LINDO Systems, www.lindo.com. The long-time popularity of LINDO is partially due to its ease of use. For "textbook-sized" problems, the model can be entered and solved in an intuitive, straightforward manner, so the LINDO interface provides a convenient tool for students to use. Although easy to use for small models, LINDO/LINGO can also solve large models, e.g., the largest version has solved real problems with 4 million variables and 2 million constraints.

The OR Courseware provided on this book's website contains a student version of LINDO/LINGO, accompanied by an extensive tutorial. Appendix 4.1 provides a quick introduction. Additionally, the software contains extensive online help. The OR Courseware also contains LINGO/LINDO formulations for the major examples used in the book.

Spreadsheet-based solvers are becoming increasingly popular for linear programming and its extensions. Leading the way are the solvers produced by Frontline Systems for Microsoft Excel and other spreadsheet packages. In addition to the basic solver shipped with these packages, more powerful *Premium Solver* products also are available. Because of the widespread use of spreadsheet packages such as Microsoft Excel today, these solvers are introducing large numbers of people to the potential of linear programming for the first time. For textbook-sized linear programming problems (and considerably larger problems as well), spreadsheets provide a convenient way to formulate and solve the model, as described in Sec. 3.5. The more powerful spreadsheet solvers can solve fairly large models with many thousand decision variables. However, when the spreadsheet grows to an unwieldy size, a good modeling language and its solver may provide a more efficient approach to formulating and solving the model.

Spreadsheets provide an excellent communication tool, especially when dealing with typical managers who are very comfortable with this format but not with the algebraic formulations of OR models. Therefore, optimization software packages and modeling languages now can commonly import and export data and results in a spreadsheet format. For example, the MPL modeling language now includes an enhancement (called the *OptiMax 2000 Component Library*) that enables the modeler to create the feel of a spreadsheet model for the user of the model while still using MPL to formulate the model very efficiently. (The student version of OptiMax 2000 is included in your OR Courseware.)

Premium Solver for Education is one of the Excel add-ins included on the book's website. You can install this add-in to obtain more functionality than with the standard Excel Solver.

Consequently, all the software, tutorials, and examples packed on the book's website are providing you with several attractive software options for linear programming.

Available Software Options for Linear Programming

1. Demonstration examples (in OR Tutor) and both interactive and automatic procedures in IOR Tutorial for efficiently learning the simplex method.
2. Excel and its Premium Solver for formulating and solving linear programming models in a spreadsheet format.
3. MPL/CPLEX for efficiently formulating and solving large linear programming models.
4. LINGO and its solver (shared with LINDO) for an alternative way of efficiently formulating and solving large linear programming models.

Your instructor may specify which software to use. Whatever the choice, you will be gaining experience with the kind of state-of-the-art software that is used by OR professionals.

■ 4.9   THE INTERIOR-POINT APPROACH TO SOLVING LINEAR PROGRAMMING PROBLEMS

The most dramatic new development in operations research during the 1980s was the discovery of the interior-point approach to solving linear programming problems. This discovery was made in 1984 by a young mathematician at AT&T Bell Laboratories, Narendra Karmarkar, when he successfully developed a new algorithm for linear programming with this kind of approach. Although this particular algorithm experienced only mixed success in competing with the simplex method, the key solution concept described below appeared to have great potential for solving *huge* linear programming problems beyond the reach of the simplex method. Many top researchers subsequently worked on modifying Karmarkar's algorithm to fully tap this potential. Much progress has been made (and continues to be made), and a number of powerful algorithms using the interior-point approach have been developed. Today, the more powerful software packages that are designed for solving really large linear programming problems (such as CPLEX) include at least one algorithm using the interior-point approach along with the simplex method and its variants. As research continues on these algorithms, their computer implementations continue to improve. This has spurred renewed research on the simplex method, and its computer implementations continue to improve as well. The competition between the two approaches for supremacy in solving huge problems is continuing.

Now let us look at the key idea behind Karmarkar's algorithm and its subsequent variants that use the interior-point approach.

The Key Solution Concept

Although radically different from the simplex method, Karmarkar's algorithm does share a few of the same characteristics. It is an *iterative* algorithm. It gets started by identifying a feasible *trial solution*. At each iteration, it moves from the current trial solution to a better trial solution in the feasible region. It then continues this process until it reaches a trial solution that is (essentially) optimal.

The big difference lies in the nature of these trial solutions. For the simplex method, the trial solutions are *CPF solutions* (or BF solutions after augmenting), so all movement is along edges on the *boundary* of the feasible region. For Karmarkar's algorithm, the trial solutions are **interior points,** i.e., points *inside* the boundary of the feasible region. For this reason, Karmarkar's algorithm and its variants are referred to as **interior-point algorithms.**

However, because of an early patent obtained on an early version of an interior-point algorithm, such an algorithm also is commonly referred to as a **barrier algorithm** (or *barrier method*). The term *barrier* is used because, from the perspective of a search whose trial solutions are *interior points,* each constraint boundary is treated as a barrier. Most optimization software packages now use the barrier terminology when referring to their solver option that is based on the interior-point approach. Both CPLEX and LINDO API include a "barrier algorithm" that can be used to solve either linear programming problems or quadratic programming problems (discussed in Sec. 12.7).

To illustrate the interior-point approach, Fig. 4.11 shows the path followed by the interior-point algorithm in your OR Courseware when it is applied to the Wyndor Glass Co. problem, starting from the initial trial solution (1, 2). Note how all the trial solutions (dots)

■ **FIGURE 4.11**
The curve from (1, 2) to (2, 6) shows a typical path followed by an interior-point algorithm, right through the *interior* of the feasible region for the Wyndor Glass Co. problem.

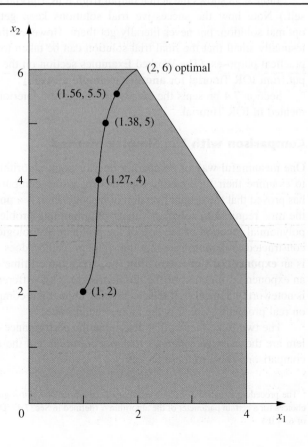

■ **TABLE 4.18** Output of interior-point algorithm in OR Courseware
for Wyndor Glass Co. problem

Iteration	x_1	x_2	Z
0	1	2	13
1	1.27298	4	23.8189
2	1.37744	5	29.1323
3	1.56291	5.5	32.1887
4	1.80268	5.71816	33.9989
5	1.92134	5.82908	34.9094
6	1.96639	5.90595	35.429
7	1.98385	5.95199	35.7115
8	1.99197	5.97594	35.8556
9	1.99599	5.98796	35.9278
10	1.99799	5.99398	35.9639
11	1.999	5.99699	35.9819
12	1.9995	5.9985	35.991
13	1.99975	5.99925	35.9955
14	1.99987	5.99962	35.9977
15	1.99994	5.99981	35.9989

shown on this path are inside the boundary of the feasible region as the path approaches the optimal solution (2, 6). (All the subsequent trial solutions not shown also are inside the boundary of the feasible region.) Contrast this path with the path followed by the simplex method around the boundary of the feasible region from (0, 0) to (0, 6) to (2, 6).

Table 4.18 shows the actual output from IOR Tutorial for this problem.[24] (Try it yourself.) Note how the successive trial solutions keep getting closer and closer to the optimal solution, but never literally get there. However, the deviation becomes so infinitesimally small that the final trial solution can be taken to be the optimal solution for all practical purposes. (The Worked Examples section on the book's website shows the output from IOR Tutorial for **another example** as well.)

Section 7.4 presents the details of the specific interior-point algorithm that is implemented in IOR Tutorial.

Comparison with the Simplex Method

One meaningful way of comparing interior-point algorithms with the simplex method is to examine their theoretical properties regarding computational complexity. Karmarkar has proved that the original version of his algorithm is a **polynomial time algorithm;** i.e., the time required to solve *any* linear programming problem can be bounded above by a polynomial function of the size of the problem. Pathological counterexamples have been constructed to demonstrate that the simplex method does not possess this property, so it is an **exponential time algorithm** (i.e., the required time can be bounded above only by an exponential function of the problem size). This difference in *worst-case performance* is noteworthy. However, it tells us nothing about their comparison in average performance on real problems, which is the more crucial issue.

The two basic factors that determine the performance of an algorithm on a real problem are the *average computer time per iteration* and the *number of iterations*. Our next comparisons concern these factors.

[24]The procedure is called *Solve Automatically by the Interior-Point Algorithm*. The option menu provides two choices for a certain parameter of the algorithm α (defined in Sec. 7.4). The choice used here is the default value of $\alpha = 0.5$.

Interior-point algorithms are far more complicated than the simplex method. Considerably more extensive computations are required for each iteration to find the next trial solution. Therefore, the computer time per iteration for an interior-point algorithm is many times longer than that for the simplex method.

For fairly small problems, the numbers of iterations needed by an interior-point algorithm and by the simplex method tend to be somewhat comparable. For example, on a problem with 10 functional constraints, roughly 20 iterations would be typical for either kind of algorithm. Consequently, on problems of similar size, the total computer time for an interior-point algorithm will tend to be many times longer than that for the simplex method.

On the other hand, a key advantage of interior-point algorithms is that large problems do not require many more iterations than small problems. For example, a problem with 10,000 functional constraints probably will require well under 100 iterations. Even considering the very substantial computer time per iteration needed for a problem of this size, such a small number of iterations makes the problem quite tractable. By contrast, the simplex method might need 20,000 iterations and so might not finish within a reasonable amount of computer time. Therefore, interior-point algorithms often are faster than the simplex method for such huge problems.

The reason for this very large difference in the number of iterations on huge problems is the difference in the paths followed. At each iteration, the simplex method moves from the current CPF solution to an adjacent CPF solution along an edge on the boundary of the feasible region. Huge problems have an astronomical number of CPF solutions. The path from the initial CPF solution to an optimal solution may be a very circuitous one around the boundary, taking only a small step each time to the next adjacent CPF solution, so a huge number of steps may be required to reach an optimal solution. By contrast, an interior-point algorithm bypasses all this by shooting through the interior of the feasible region toward an optimal solution. Adding more functional constraints adds more constraint boundaries to the feasible region, but has little effect on the number of trial solutions needed on this path through the interior. This makes it possible for interior-point algorithms to solve problems with a huge number of functional constraints.

A final key comparison concerns the ability to perform the various kinds of postoptimality analysis described in Sec. 4.7. The simplex method and its extensions are very well suited to and are widely used for this kind of analysis. For example, an ILOG product called *Optimization Decision Manager* makes full use of the simplex method in CPLEX to perform a wide variety of postoptimality analysis tasks in convenient ways. Unfortunately, the interior-point approach currently has limited capability in this area.[25] Given the great importance of postoptimality analysis, this is a crucial drawback of interior-point algorithms. However, we point out next how the simplex method can be combined with the interior-point approach to overcome this drawback.

The Complementary Roles of the Simplex Method and the Interior-Point Approach

Ongoing research is continuing to provide substantial improvements in computer implementations of both the simplex method (including its variants) and interior-point algorithms. Therefore, any predictions about their future roles are risky. However, we do summarize below our current assessment of their complementary roles.

---

[25]However, research aimed at increasing this capability continues to make progress. For example, see E. A. Yildirim and M. J. Todd: "Sensitivity Analysis in Linear Programming and Semidefinite Programming Using Interior-Point Methods," *Mathematical Programming*, Series A, **90**(2): 229–261, April 2001.

The simplex method (and its variants) continues to be the standard algorithm for the routine use of linear programming. It continues to be the most efficient algorithm for problems with less than, say, 10,000 functional constraints. It also is the most efficient for some (but not all) problems with up to, say, 100,000 functional constraints and nearly an unlimited number of decision variables, so most users are continuing to use the simplex method for such problems. However, as the number of functional constraints increases even further, it becomes increasingly likely that an interior-point approach will be the most efficient, so it often is now used instead. As the size grows into the hundreds of thousands, or even millions, of functional constraints, the interior-point approach may be the only one capable of solving the problem. However, this certainly is not always the case. As mentioned in the preceding section, the latest state-of-the-art software is successfully using the simplex method and its variants to solve some truly massive problems with millions, or even tens of millions, of functional constraints and decision variables.

These generalizations about how the interior-point approach and the simplex method should compare for various problem sizes will not hold across the board. The specific software packages and computer equipment being used have a major impact. The comparison also is affected considerably by the *specific type* of linear programming problem being solved. As time goes on, we should learn much more about how to identify specific types which are better suited for one kind of algorithm.

One of the by-products of the emergence of the interior-point approach has been a major renewal of efforts to improve the efficiency of computer implementations of the simplex method. As we indicated, impressive progress has been made in recent years, and more lies ahead. At the same time, ongoing research and development of the interior-point approach will further increase its power, and perhaps at a faster rate than for the simplex method.

Improving computer technology, such as massive parallel processing (a huge number of computer units operating in parallel on different parts of the same problem), also will substantially increase the size of problem that either kind of algorithm can solve. However, it now appears that the interior-point approach has much greater potential to take advantage of parallel processing than the simplex method does.

As discussed earlier, a key disadvantage of the interior-point approach is its limited capability for performing postoptimality analysis. To overcome this drawback, researchers have been developing procedures for switching over to the simplex method after an interior-point algorithm has finished. Recall that the trial solutions obtained by an interior-point algorithm keep getting closer and closer to an optimal solution (the best CPF solution), but never quite get there. Therefore, a switching procedure requires identifying a CPF solution (or BF solution after augmenting) that is very close to the final trial solution.

For example, by looking at Fig. 4.11, it is easy to see that the final trial solution in Table 4.18 is very near the CPF solution (2, 6). Unfortunately, on problems with thousands of decision variables (so no graph is available), identifying a nearby CPF (or BF) solution is a very challenging and time-consuming task. However, good progress has been made in developing procedures to do this. For example, the full-fledged professional version of CPLEX includes a *crossover algorithm* which converts the solutions obtained by its "barrier algorithm" into a BF solution.

Once this nearby BF solution has been found, the optimality test for the simplex method is applied to check whether this actually is the optimal BF solution. If it is not optimal, some iterations of the simplex method are conducted to move from this BF solution to an optimal solution. Generally, only a very few iterations (perhaps one) are needed because the interior-point algorithm has brought us so close to an optimal solution. Therefore, these iterations should be done quite quickly, even on problems that are too huge to be solved from scratch. After an optimal solution is actually reached, the simplex method and its variants are applied to help perform postoptimality analysis.

Because of the difficulties involved in applying a switching procedure (including the extra computer time), some practitioners prefer to just use the simplex method from the outset. This makes good sense when you only occasionally encounter problems that are large enough for an interior-point algorithm to be modestly faster (before switching) than the simplex method. This modest speed-up would not justify both the extra computer time for a switching procedure and the high cost of acquiring (and learning to use) a software package based on the interior-point approach. However, for organizations which frequently must deal with extremely large linear programming problems, acquiring a state-of-the-art software package of this kind (including a switching procedure) probably is worthwhile. For sufficiently huge problems, the only available way of solving them may be with such a package.

Applications of huge linear programming models sometimes lead to savings of millions of dollars. Just one such application can pay many times over for a state-of-the-art software package based on the interior-point approach plus switching over to the simplex method at the end.

4.10 CONCLUSIONS

The simplex method is an efficient and reliable algorithm for solving linear programming problems. It also provides the basis for performing the various parts of postoptimality analysis very efficiently.

Although it has a useful geometric interpretation, the simplex method is an algebraic procedure. At each iteration, it moves from the current BF solution to a better, adjacent BF solution by choosing both an entering basic variable and a leaving basic variable and then using Gaussian elimination to solve a system of linear equations. When the current solution has no adjacent BF solution that is better, the current solution is optimal and the algorithm stops.

We presented the full algebraic form of the simplex method to convey its logic, and then we streamlined the method to a more convenient tabular form. To set up for starting the simplex method, it is sometimes necessary to use artificial variables to obtain an initial BF solution for an artificial problem. If so, either the Big M method or the two-phase method is used to ensure that the simplex method obtains an optimal solution for the real problem.

Computer implementations of the simplex method and its variants have become so powerful that they now are frequently used to solve linear programming problems with many hundreds of thousands of functional constraints and decision variables, and occasionally vastly larger problems. Interior-point algorithms also provide a powerful tool for solving very large problems.

APPENDIX 4.1 AN INTRODUCTION TO USING LINDO AND LINGO

The LINGO software can accept optimization models in either of two styles or syntax: (a) LINDO syntax or (b) LINGO syntax. We will first describe LINDO syntax. The relative advantages of LINDO syntax are that it is very easy and natural for simple linear and integer programming problems. It has been in wide use since 1981.

The LINDO syntax allows you to enter a model in a natural form, essentially as presented in a textbook. For example, here is how the Wyndor Glass Co. example introduced in Sec. 3.1. is

These are the basics for getting started with LINGO/LINDO. You can turn on or turn off the generation of reports. For example, if the automatic generation of the standard solution report has been turned off (Terse mode), you can turn it back on by clicking on: LINGO | Options | Interface | Output level | Verbose | Apply. The ability to generate range reports can be turned on or off by clicking on: LINGO | Options | General solver | Dual computations | Prices & Ranges | Apply.

The second input style that LINGO supports is LINGO syntax. LINGO syntax is dramatically more powerful than LINDO syntax. The advantages to using LINGO syntax are: (a) it allows arbitrary mathematical expressions, including parentheses and all familiar mathematical operators such as division, multiplication, log, sin, etc., (b) the ability to solve not just linear programming problems but also nonlinear programming problems, (c) scalability to large applications using subscripted variables and sets, (d) the ability to read input data from a spreadsheet or database and send solution information back into a spreadsheet or database, (e) the ability to naturally represent sparse relationships, (f) programming ability so that you can solve a series of models automatically as when doing parametric analysis. A formulation of the Wyndor problem in LINGO, using the subscript/sets feature is:

```
! Wyndor Glass Co. Problem;

SETS:
  PRODUCT: PPB, X;          ! Each product has a profit/batch
and amount;
  RESOURCE: HOURSAVAILABLE;  ! Each resource has a capacity;
! Each resource product combination has an hours/batch;
  RXP(RESOURCE,PRODUCT): HPB;
ENDSETS
DATA:
  PRODUCT     = DOORS   WINDOWS;   ! The products;
  PPB         =   3       5;       ! Profit per batch;

  RESOURCE    = PLANT1 PLANT2 PLANT3;
  HOURSAVAILABLE =   4      12      18;

  HPB  =   1   0     ! Hours per batch;
           0   2
           3   2;
ENDDATA
! Sum over all products j the profit per batch times batches
produced;
MAX = @SUM( PRODUCT(j): PPB(j)*X(j));

@FOR( RESOURCE(i)):   ! For each resource i...;
    ! Sum over all products j of hours per batch time batches
produced...;
    @SUM(RXP(i,j): HPB(i,j)*X(j)) <= HOURSAVAILABLE(i);
    );
```

The original Wyndor problem has two products and three resources. If Wyndor expands to having four products and five resources, it is a trivial change to insert the appropriate new data into the DATA section. The formulation of the model adjusts automatically. The subscript/sets capability also allows one to naturally represent three dimensional or higher models. The large problem described in Sec. 3.6 has five dimensions: plants, machines, products, regions/customers, and time periods. This would be hard to fit into a two-dimensional spreadsheet but is easy to represent in a modeling language with sets and subscripts. In practice, for problems like that in Sec. 3.6, many of the 10(10)(10)(10)(10) = 100,000 possible combinations of relationships do not exist; e.g., not all plants can make all products, and not all customers demand all products. The subscript/sets capability in modeling languages make it easy to represent such sparse relationships.

For most models that you enter, LINGO will be able to detect automatically whether you are using LINDO syntax or LINGO syntax. You may choose your default syntax by clicking on: LINGO | Options | Interface | File format | lng (for LINGO) or ltx (for LINDO).

LINGO includes an extensive online Help menu to give more details and examples. Supplements 1 and 2 to Chapter 3 (shown on the book's website) provide a relatively complete introduction to LINGO. The LINGO tutorial on the website also provides additional details. The LINGO/LINDO files on the website for various chapters show LINDO/LINGO formulations for numerous examples from most of the chapters.

■ SELECTED REFERENCES

1. Bixby, R. E.: "Solving Real-World Linear Programs: A Decade and More of Progress," *Operations Research,* **50**(1): 3–15, January–February 2002.
2. Dantzig, G. B., and M. N. Thapa: *Linear Programming 1: Introduction,* Springer, New York, 1997.
3. Fourer, R.: "Software Survey: Linear Programming," *OR/MS Today,* June 2007, pp. 42–51.
4. Luenberger, D., and Y. Ye: *Linear and Nonlinear Programming,* 3rd ed., Springer, New York, 2008.
5. Maros, I.: *Computational Techniques of the Simplex Method,* Kluwer Academic Publishers (now Springer), Boston, MA, 2003.
6. Schrage, L.: *Optimization Modeling with LINGO,* LINDO Systems, Chicago, 2008.
7. Tretkoff, C., and I. Lustig: "New Age of Optimization Applications," *OR/MS Today*, December 2006, pp. 46–49.
8. Vanderbei, R. J.: *Linear Programming: Foundations and Extensions,* 3rd ed., Springer, New York, 2008.

■ LEARNING AIDS FOR THIS CHAPTER ON OUR WEBSITE (www.mhhe.com/hillier)

Worked Examples:

Examples for Chapter 4

Demonstration Examples in OR Tutor:

Interpretation of the Slack Variables
Simplex Method—Algebraic Form
Simplex Method—Tabular Form

Interactive Procedures in IOR Tutorial:

Enter or Revise a General Linear Programming Model
Set Up for the Simplex Method—Interactive Only
Solve Interactively by the Simplex Method
Interactive Graphical Method

Automatic Procedures in IOR Tutorial:

Solve Automatically by the Simplex Method
Solve Automatically by the Interior-Point Algorithm
Graphical Method and Sensitivity Analysis

An Excel Add-In:

Premium Solver for Education

D,I **(b)** Repeat part (a) with the corresponding interactive routine in your IOR Tutorial.

C **(c)** Verify the optimal solution you obtained by using a software package based on the simplex method.

D,I **4.3-4.*** Work through the simplex method (in algebraic form) step by step to solve the following problem.

Maximize $Z = 4x_1 + 3x_2 + 6x_3,$

subject to

$$3x_1 + x_2 + 3x_3 \leq 30$$
$$2x_1 + 2x_2 + 3x_3 \leq 40$$

and

$$x_1 \geq 0, \qquad x_2 \geq 0, \qquad x_3 \geq 0.$$

D,I **4.3-5.** Work through the simplex method (in algebraic form) step by step to solve the following problem.

Maximize $Z = 3x_1 + 4x_2 + 5x_3,$

subject to

$$3x_1 + x_2 + 5x_3 \leq 150$$
$$x_1 + 4x_2 + x_3 \leq 120$$
$$2x_1 + 2x_3 \leq 105$$

and

$$x_1 \geq 0, \qquad x_2 \geq 0, \qquad x_3 \geq 0.$$

4.3-6. Consider the following problem.

Maximize $Z = 5x_1 + 3x_2 + 4x_3,$

subject to

$$2x_1 + x_2 + x_3 \leq 20$$
$$3x_1 + x_2 + 2x_3 \leq 30$$

and

$$x_1 \geq 0, \qquad x_2 \geq 0, \qquad x_3 \geq 0.$$

You are given the information that the *nonzero* variables in the optimal solution are x_2 and x_3.

(a) Describe how you can use this information to adapt the simplex method to solve this problem in the minimum possible number of iterations (when you start from the usual initial BF solution). Do *not* actually perform any iterations.

(b) Use the procedure developed in part (a) to solve this problem by hand. (Do *not* use your OR Courseware.)

4.3-7. Consider the following problem.

Maximize $Z = 2x_1 + 4x_2 + 3x_3,$

subject to

$$x_1 + 3x_2 + 2x_3 \leq 30$$
$$x_1 + x_2 + x_3 \leq 24$$
$$3x_1 + 5x_2 + 3x_3 \leq 60$$

and

$$x_1 \geq 0, \qquad x_2 \geq 0, \qquad x_3 \geq 0.$$

You are given the information that $x_1 > 0$, $x_2 = 0$, and $x_3 > 0$ in the optimal solution.

(a) Describe how you can use this information to adapt the simplex method to solve this problem in the minimum possible number of iterations (when you start from the usual initial BF solution). Do *not* actually perform any iterations.

(b) Use the procedure developed in part (a) to solve this problem by hand. (Do *not* use your OR Courseware.)

4.3-8. Label each of the following statements as true or false, and then justify your answer by referring to specific statements in the chapter.

(a) The simplex method's rule for choosing the entering basic variable is used because it always leads to the *best* adjacent BF solution (largest Z).

(b) The simplex method's minimum ratio rule for choosing the leaving basic variable is used because making another choice with a larger ratio would yield a basic solution that is not feasible.

(c) When the simplex method solves for the next BF solution, elementary algebraic operations are used to eliminate each nonbasic variable from all but one equation (*its* equation) and to give it a coefficient of $+1$ in that one equation.

D,I **4.4-1.** Repeat Prob. 4.3-2, using the tabular form of the simplex method.

D,I,C **4.4-2.** Repeat Prob. 4.3-3, using the tabular form of the simplex method.

4.4-3. Consider the following problem.

Maximize $Z = 2x_1 + x_2,$

subject to

$$x_1 + x_2 \leq 40$$
$$4x_1 + x_2 \leq 100$$

and

$$x_1 \geq 0, \qquad x_2 \geq 0.$$

(a) Solve this problem graphically in a freehand manner. Also identify all the CPF solutions.

D,I **(b)** Now use IOR Tutorial to solve the problem graphically.

D **(c)** Use hand calculations to solve this problem by the simplex method in algebraic form.

D,I **(d)** Now use IOR Tutorial to solve this problem interactively by the simplex method in algebraic form.

D **(e)** Use hand calculations to solve this problem by the simplex method in tabular form.

D,I **(f)** Now use IOR Tutorial to solve this problem interactively by the simplex method in tabular form.

C **(g)** Use a software package based on the simplex method to solve the problem.

4.4-4. Repeat Prob. 4.4-3 for the following problem.

Maximize $Z = 2x_1 + 3x_2$,

subject to

$$x_1 + 2x_2 \leq 30$$
$$x_1 + x_2 \leq 20$$

and

$$x_1 \geq 0, \qquad x_2 \geq 0.$$

4.4-5. Consider the following problem.

Maximize $Z = 5x_1 + 9x_2 + 7x_3$,

subject to

$$x_1 + 3x_2 + 2x_3 \leq 10$$
$$3x_1 + 4x_2 + 2x_3 \leq 12$$
$$2x_1 + x_2 + 2x_3 \leq 8$$

and

$$x_1 \geq 0, \qquad x_2 \geq 0, \qquad x_3 \geq 0.$$

D,I **(a)** Work through the simplex method step by step in algebraic form.

D,I **(b)** Work through the simplex method step by step in tabular form.

C **(c)** Use a software package based on the simplex method to solve the problem.

4.4-6. Consider the following problem.

Maximize $Z = 3x_1 + 5x_2 + 6x_3$,

subject to

$$2x_1 + x_2 + x_3 \leq 4$$
$$x_1 + 2x_2 + x_3 \leq 4$$
$$x_1 + x_2 + 2x_3 \leq 4$$
$$x_1 + x_2 + x_3 \leq 3$$

and

$$x_1 \geq 0, \qquad x_2 \geq 0, \qquad x_3 \geq 0.$$

D,I **(a)** Work through the simplex method step by step in algebraic form.

D,I **(b)** Work through the simplex method in tabular form.

C **(c)** Use a computer package based on the simplex method to solve the problem.

D,I **4.4-7.** Work through the simplex method step by step (in tabular form) to solve the following problem.

Maximize $Z = 2x_1 - x_2 + x_3$,

subject to

$$3x_1 + x_2 + x_3 \leq 6$$
$$x_1 - x_2 + 2x_3 \leq 1$$
$$x_1 + x_2 - x_3 \leq 2$$

and

$$x_1 \geq 0, \qquad x_2 \geq 0, \qquad x_3 \geq 0.$$

D,I **4.4-8.** Work through the simplex method step by step to solve the following problem.

Maximize $Z = -x_1 + x_2 + 2x_3$,

subject to

$$x_1 + 2x_2 - x_3 \leq 20$$
$$-2x_1 + 4x_2 + 2x_3 \leq 60$$
$$2x_1 + 3x_2 + x_3 \leq 50$$

and

$$x_1 \geq 0, \qquad x_2 \geq 0, \qquad x_3 \geq 0.$$

4.5-1. Consider the following statements about linear programming and the simplex method. Label each statement as true or false, and then justify your answer.

(a) In a particular iteration of the simplex method, if there is a tie for which variable should be the leaving basic variable, then the next BF solution must have at least one basic variable equal to zero.

(b) If there is no leaving basic variable at some iteration, then the problem has no feasible solutions.

(c) If at least one of the basic variables has a coefficient of zero in row 0 of the final tableau, then the problem has multiple optimal solutions.

(d) If the problem has multiple optimal solutions, then the problem must have a bounded feasible region.

4.5-2. Suppose that the following constraints have been provided for a linear programming model with decision variables x_1 and x_2.

$$-2x_1 + 3x_2 \leq 12$$
$$-3x_1 + 2x_2 \leq 12$$

and

$$x_1 \geq 0, \qquad x_2 \geq 0.$$

(a) Demonstrate graphically that the feasible region is unbounded.

(b) If the objective is to maximize $Z = -x_1 + x_2$, does the model have an optimal solution? If so, find it. If not, explain why not.

(c) Repeat part (b) when the objective is to maximize $Z = x_1 - x_2$.

(d) For objective functions where this model has no optimal solution, does this mean that there are no good solutions according to the model? Explain. What probably went wrong when formulating the model?

D,I **(e)** Select an objective function for which this model has no optimal solution. Then work through the simplex method step by step to demonstrate that Z is unbounded.

C **(f)** For the objective function selected in part (e), use a software package based on the simplex method to determine that Z is unbounded.

4.5-3. Follow the instructions of Prob. 4.5-2 when the constraints are the following:

$$2x_1 - x_2 \leq 20$$
$$x_1 - 2x_2 \leq 20$$

I **(a)** Using the Big M method, work through the simplex method step by step to solve the problem.

I **(b)** Using the two-phase method, work through the simplex method step by step to solve the problem.

(c) Compare the sequence of BF solutions obtained in parts (a) and (b). Which of these solutions are feasible only for the artificial problem obtained by introducing artificial variables and which are actually feasible for the real problem?

C **(d)** Use a software package based on the simplex method to solve the problem.

4.6-10. Follow the instructions of Prob. 4.6-9 for the following problem.

Minimize $Z = 3x_1 + 2x_2 + 7x_3,$

subject to

$$-x_1 + x_2 \qquad = 10$$
$$2x_1 - x_2 + x_3 \geq 10$$

and

$$x_1 \geq 0, \qquad x_2 \geq 0, \qquad x_3 \geq 0.$$

4.6-11. Label each of the following statements as true or false, and then justify your answer.

(a) When a linear programming model has an equality constraint, an artificial variable is introduced into this constraint in order to start the simplex method with an obvious initial basic solution that is feasible for the original model.

(b) When an artificial problem is created by introducing artificial variables and using the Big M method, if all artificial variables in an optimal solution for the artificial problem are equal to zero, then the real problem has no feasible solutions.

(c) The two-phase method is commonly used in practice because it usually requires fewer iterations to reach an optimal solution than the Big M method does.

4.6-12. Consider the following problem.

Maximize $Z = 3x_1 + 7x_2 + 5x_3,$

subject to

$$3x_1 + x_2 + 2x_3 \leq 9$$
$$-2x_1 + x_2 + 3x_3 \leq 12$$

and

$$x_2 \geq 0, \qquad x_3 \geq 0$$

(no nonnegativity constraint for x_1).

(a) Reformulate this problem so all variables have nonnegativity constraints.

D,I **(b)** Work through the simplex method step by step to solve the problem.

C **(c)** Use a software package based on the simplex method to solve the problem.

4.6-13.* Consider the following problem.

Maximize $Z = -x_1 + 4x_2,$

subject to

$$-3x_1 + x_2 \leq 6$$

$$x_1 + 2x_2 \leq 4$$
$$x_2 \geq -3$$

(no lower bound constraint for x_1).

D,I **(a)** Solve this problem graphically.

(b) Reformulate this problem so that it has only two functional constraints and all variables have nonnegativity constraints.

D,I **(c)** Work through the simplex method step by step to solve the problem.

4.6-14. Consider the following problem.

Maximize $Z = -x_1 + 2x_2 + x_3,$

subject to

$$3x_2 + x_3 \leq 120$$
$$x_1 - x_2 - 4x_3 \leq 80$$
$$-3x_1 + x_2 + 2x_3 \leq 100$$

(no nonnegativity constraints).

(a) Reformulate this problem so that all variables have nonnegativity constraints.

D,I **(b)** Work through the simplex method step by step to solve the problem.

C **(c)** Use a computer package based on the simplex method to solve the problem.

4.6-15. This chapter has described the simplex method as applied to linear programming problems where the objective function is to be maximized. Section 4.6 then described how to convert a minimization problem to an equivalent maximization problem for applying the simplex method. Another option with minimization problems is to make a few modifications in the instructions for the simplex method given in the chapter in order to apply the algorithm directly.

(a) Describe what these modifications would need to be.

(b) Using the Big M method, apply the modified algorithm developed in part (a) to solve the following problem directly by hand. (Do not use your OR Courseware.)

Minimize $Z = 3x_1 + 8x_2 + 5x_3,$

subject to

$$3x_2 + 4x_3 \geq 70$$
$$3x_1 + 5x_2 + 2x_3 \geq 70$$

and

$$x_1 \geq 0, \qquad x_2 \geq 0, \qquad x_3 \geq 0.$$

4.6-16. Consider the following problem.

Maximize $Z = -2x_1 + x_2 - 4x_3 + 3x_4,$

subject to

$$x_1 + x_2 + 3x_3 + 2x_4 \leq 4$$
$$x_1 \qquad - x_3 + x_4 \geq -1$$
$$2x_1 + x_2 \qquad \leq 2$$
$$x_1 + 2x_2 + x_3 + 2x_4 = 2$$

and

$$x_2 \geq 0, \qquad x_3 \geq 0, \qquad x_4 \geq 0$$

(no nonnegativity constraint for x_1).

(a) Reformulate this problem to fit our standard form for a linear programming model presented in Sec. 3.2.

(b) Using the Big M method, construct the complete first simplex tableau for the simplex method and identify the corresponding initial (artificial) BF solution. Also identify the initial entering basic variable and the leaving basic variable.

(c) Using the two-phase method, construct row 0 of the first simplex tableau for phase 1.

C (d) Use a computer package based on the simplex method to solve the problem.

I **4.6-17.** Consider the following problem.

Maximize $Z = 4x_1 + 5x_2 + 3x_3,$

subject to

$$x_1 + x_2 + 2x_3 \geq 20$$
$$15x_1 + 6x_2 - 5x_3 \leq 50$$
$$x_1 + 3x_2 + 5x_3 \leq 30$$

and

$$x_1 \geq 0, \qquad x_2 \geq 0, \qquad x_3 \geq 0.$$

Work through the simplex method step by step to demonstrate that this problem does not possess any feasible solutions.

4.7-1. Refer to Fig. 4.10 and the resulting *allowable range* for the respective right-hand sides of the Wyndor Glass Co. problem given in Sec. 3.1. Use graphical analysis to demonstrate that each given allowable range is correct.

4.7-2. Reconsider the model in Prob. 4.1-5. Interpret the right-hand side of the respective functional constraints as the amount available of the respective resources.

I (a) Use graphical analysis as in Fig. 4.8 to determine the shadow prices for the respective resources.

I (b) Use graphical analysis to perform sensitivity analysis on this model. In particular, check each parameter of the model to determine whether it is a *sensitive* parameter (a parameter whose value cannot be changed without changing the optimal solution) by examining the graph that identifies the optimal solution.

I (c) Use graphical analysis as in Fig. 4.9 to determine the allowable range for each c_j value (coefficient of x_j in the objective function) over which the current optimal solution will remain optimal.

I (d) Changing just one b_i value (the right-hand side of functional constraint i) will shift the corresponding constraint boundary. If the current optimal CPF solution lies on this constraint boundary, this CPF solution also will shift. Use graphical analysis to determine the allowable range for each b_i value over which this CPF solution will remain feasible.

C (e) Verify your answers in parts (a), (c), and (d) by using a computer package based on the simplex method to solve the problem and then to generate sensitivity analysis information.

4.7-3. You are given the following linear programming problem.

Maximize $Z = 3x_1 + 2x_2,$

subject to

$$3x_1 \qquad \leq 60 \qquad \text{(resource 1)}$$

$$2x_1 + 3x_2 \leq 75 \qquad \text{(resource 2)}$$
$$2x_2 \leq 40 \qquad \text{(resource 3)}$$

and

$$x_1 \geq 0, \qquad x_2 \geq 0.$$

D,I (a) Solve this problem graphically.

(b) Use graphical analysis to find the shadow prices for the resources.

(c) Determine how many additional units of resource 1 would be needed to increase the optimal value of Z by 15.

4.7-4. Consider the following problem.

Maximize $Z = x_1 - 7x_2 + 3x_3,$

subject to

$$2x_1 + x_2 - x_3 \leq 4 \qquad \text{(resource 1)}$$
$$4x_1 - 3x_2 \qquad \leq 2 \qquad \text{(resource 2)}$$
$$-3x_1 + 2x_2 + x_3 \leq 3 \qquad \text{(resource 3)}$$

and

$$x_1 \geq 0, \qquad x_2 \geq 0, \qquad x_3 \geq 0.$$

D,I (a) Work through the simplex method step by step to solve the problem.

(b) Identify the shadow prices for the three resources and describe their significance.

C (c) Use a software package based on the simplex method to solve the problem and then to generate sensitivity information. Use this information to identify the shadow price for each resource, the allowable range for each objective function coefficient, and the allowable range for each right-hand side.

4.7-5.* Consider the following problem.

Maximize $Z = 2x_1 - 2x_2 + 3x_3,$

subject to

$$-x_1 + x_2 + x_3 \leq 4 \qquad \text{(resource 1)}$$
$$2x_1 - x_2 + x_3 \leq 2 \qquad \text{(resource 2)}$$
$$x_1 + x_2 + 3x_3 \leq 12 \qquad \text{(resource 3)}$$

and

$$x_1 \geq 0, \qquad x_2 \geq 0, \qquad x_3 \geq 0.$$

D,I (a) Work through the simplex method step by step to solve the problem.

(b) Identify the shadow prices for the three resources and describe their significance.

C (c) Use a software package based on the simplex method to solve the problem and then to generate sensitivity information. Use this information to identify the shadow price for each resource, the allowable range for each objective function coefficient and the allowable range for each right-hand side.

4.7-6. Consider the following problem.

Maximize $Z = 5x_1 + 4x_2 - x_3 + 3x_4,$

subject to

$$3x_1 + 2x_2 - 3x_3 + x_4 \leq 24 \qquad \text{(resource 1)}$$
$$3x_1 + 3x_2 + x_3 + 3x_4 \leq 36 \qquad \text{(resource 2)}$$

and

$$x_1 \geq 0, \qquad x_2 \geq 0, \qquad x_3 \geq 0, \qquad x_4 \geq 0.$$

D,I **(a)** Work through the simplex method step by step to solve the problem.

(b) Identify the shadow prices for the two resources and describe their significance.

C **(c)** Use a software package based on the simplex method to solve the problem and then to generate sensitivity information. Use this information to identify the shadow price for each resource, the allowable range for each objective function coefficient, and the allowable range for each right-hand side.

4.9.1. Use the interior-point algorithm in your IOR Tutorial to solve the model in Prob. 4.1-4. Choose $\alpha = 0.5$ from the Option menu, use $(x_1, x_2) = (0.1, 0.4)$ as the initial trial solution, and run 15 iterations. Draw a graph of the feasible region, and then plot the trajectory of the trial solutions through this feasible region.

4.9-2. Repeat Prob. 4.9-1 for the model in Prob. 4.1-5.

■ CASES

CASE 4.1   Fabrics and Fall Fashions

From the tenth floor of her office building, Katherine Rally watches the swarms of New Yorkers fight their way through the streets infested with yellow cabs and the sidewalks littered with hot dog stands. On this sweltering July day, she pays particular attention to the fashions worn by the various women and wonders what they will choose to wear in the fall. Her thoughts are not simply random musings; they are critical to her work since she owns and manages TrendLines, an elite women's clothing company.

Today is an especially important day because she must meet with Ted Lawson, the production manager, to decide upon next month's production plan for the fall line. Specifically, she must determine the quantity of each clothing item she should produce given the plant's production capacity, limited resources, and demand forecasts. Accurate planning for next month's production is critical to fall sales since the items produced next month will appear in stores during September, and women generally buy the majority of the fall fashions when they first appear in September.

She turns back to her sprawling glass desk and looks at the numerous papers covering it. Her eyes roam across the clothing patterns designed almost six months ago, the lists of materials requirements for each pattern, and the lists of demand forecasts for each pattern determined by customer surveys at fashion shows. She remembers the hectic and sometimes nightmarish days of designing the fall line and presenting it at fashion shows in New York, Milan, and Paris. Ultimately, she paid her team of six designers a total of $860,000 for their work on her fall line. With the cost of hiring runway models, hair stylists, and makeup artists, sewing and fitting clothes, building the set, choreographing and rehearsing the show, and renting the conference hall, each of the three fashion shows cost her an additional $2,700,000.

She studies the clothing patterns and material requirements. Her fall line consists of both professional and casual fashions. She determined the prices for each clothing item by taking into account the quality and cost of material, the cost of labor and machining, the demand for the item, and the prestige of the TrendLines brand name.

The fall professional fashions include:

Clothing Item	Materials Requirements	Price	Labor and Machine Cost
Tailored wool slacks	3 yards of wool 2 yards of acetate for lining	$300	$160
Cashmere sweater	1.5 yards of cashmere	$450	$150
Silk blouse	1.5 yards of silk	$180	$100
Silk camisole	0.5 yard of silk	$120	$ 60
Tailored skirt	2 yards of rayon 1.5 yards of acetate for lining	$270	$120
Wool blazer	2.5 yards of wool 1.5 yards of acetate for lining	$320	$140

The fall casual fashions include:

Clothing Item	Materials Requirements	Price	Labor and Machine Cost
Velvet pants	3 yards of velvet 2 yards of acetate for lining	$350	$175
Cotton sweater	1.5 yards of cotton	$130	$ 60
Cotton miniskirt	0.5 yard of cotton	$ 75	$ 40
Velvet shirt	1.5 yards of velvet	$200	$160
Button-down blouse	1.5 yards of rayon	$120	$ 90

She knows that for the next month, she has ordered 45,000 yards of wool, 28,000 yards of acetate, 9,000 yards of cashmere, 18,000 yards of silk, 30,000 yards of rayon, 20,000 yards of velvet, and 30,000 yards of cotton for production. The prices of the materials are as follows:

Material	Price per yard
Wool	$ 9.00
Acetate	$ 1.50
Cashmere	$60.00
Silk	$13.00
Rayon	$ 2.25
Velvet	$12.00
Cotton	$ 2.50

Any material that is not used in production can be sent back to the textile wholesaler for a full refund, although scrap material cannot be sent back to the wholesaler.

She knows that the production of both the silk blouse and cotton sweater leaves leftover scraps of material. Specifically, for the production of one silk blouse or one cotton sweater, 2 yards of silk and cotton, respectively, are needed. From these 2 yards, 1.5 yards are used for the silk blouse or the cotton sweater and 0.5 yard is left as scrap material. She does not want to waste the material, so she plans to use the rectangular scrap of silk or cotton to produce a silk camisole or cotton miniskirt, respectively. Therefore, whenever a silk blouse is produced, a silk camisole is also produced. Likewise, whenever a cotton sweater is produced, a cotton miniskirt is also produced. Note that it is possible to produce a silk camisole without producing a silk blouse and a cotton miniskirt without producing a cotton sweater.

The demand forecasts indicate that some items have limited demand. Specifically, because the velvet pants and velvet shirts are fashion fads, TrendLines has forecasted that it can sell only 5,500 pairs of velvet pants and 6,000 velvet

shirts. TrendLines does not want to produce more than the forecasted demand because once the pants and shirts go out of style, the company cannot sell them. TrendLines can produce less than the forecasted demand, however, since the company is not required to meet the demand. The cashmere sweater also has limited demand because it is quite expensive, and TrendLines knows it can sell at most 4,000 cashmere sweaters. The silk blouses and camisoles have limited demand because many women think silk is too hard to care for, and TrendLines projects that it can sell at most 12,000 silk blouses and 15,000 silk camisoles.

The demand forecasts also indicate that the wool slacks, tailored skirts, and wool blazers have a great demand because they are basic items needed in every professional wardrobe. Specifically, the demand for wool slacks is 7,000 pairs of slacks, and the demand for wool blazers is 5,000 blazers. Katherine wants to meet at least 60 percent of the demand for these two items in order to maintain her loyal customer base and not lose business in the future. Although the demand for tailored skirts could not be estimated, Katherine feels she should make at least 2,800 of them.

(a) Ted is trying to convince Katherine not to produce any velvet shirts since the demand for this fashion fad is quite low. He argues that this fashion fad alone accounts for $500,000 of the fixed design and other costs. The net contribution (price of clothing item − materials cost − labor cost) from selling the fashion fad should cover these fixed costs. Each velvet shirt generates a net contribution of $22. He argues that given the net contribution, even satisfying the maximum demand will not yield a profit. What do you think of Ted's argument?

(b) Formulate and solve a linear programming problem to maximize profit given the production, resource, and demand constraints.

Before she makes her final decision, Katherine plans to explore the following questions independently except where otherwise indicated.

(c) The textile wholesaler informs Katherine that the velvet cannot be sent back because the demand forecasts show that the demand for velvet will decrease in the future. Katherine can therefore get no refund for the velvet. How does this fact change the production plan?

(d) What is an intuitive economic explanation for the difference between the solutions found in parts (*b*) and (*c*)?

(e) The sewing staff encounters difficulties sewing the arms and lining into the wool blazers since the blazer pattern has an awkward shape and the heavy wool material is difficult to cut and sew. The increased labor time to sew a wool blazer increases the labor and machine cost for each blazer by $80. Given this new cost, how many of each clothing item should TrendLines produce to maximize profit?

(f) The textile wholesaler informs Katherine that since another textile customer canceled his order, she can obtain an extra 10,000 yards of acetate. How many of each clothing item should TrendLines now produce to maximize profit?

(g) TrendLines assumes that it can sell every item that was not sold during September and October in a big sale in November at 60 percent of the original price. Therefore, it can sell all items in unlimited quantity during the November sale. (The previously mentioned upper limits on demand concern only the sales during September and October.) What should the new production plan be to maximize profit?

■ PREVIEWS OF ADDED CASES ON OUR WEBSITE (www.mhhe.com/hillier)

CASE 4.2   New Frontiers

AmeriBank will soon begin offering Web banking to its customers. To guide its planning for the services to provide over the Internet, a survey will be conducted with four different age groups in three types of communities. AmeriBank is imposing a number of constraints on how extensively each age group and each community should be surveyed. Linear programming is needed to develop a plan for the survey that will minimize its total cost while meeting all the survey constraints under several different scenarios.

CASE 4.3   Assigning Students to Schools

After deciding to close one of its middle schools, the Springfield school board needs to reassign all of next year's middle school students to the three remaining middle schools. Many

of the students will be bussed, so minimizing the total bussing cost is one objective. Another is to minimize the inconvenience and safety concerns for the students who will walk or bicycle to school. Given the capacities of the three schools, as well as the need to roughly balance the number of students in the three grades at each school, how can linear programming be used to determine how many students from each of the city's six residential areas should be assigned to each school? What would happen if each entire residential area must be assigned to the same school? (This case will be continued in Cases 6.3 and 11.4.)

CHAPTER

5

The Theory of the Simplex Method

Chapter 4 introduced the basic mechanics of the simplex method. Now we shall delve a little more deeply into this algorithm by examining some of its underlying theory. The first section further develops the general geometric and algebraic properties that form the foundation of the simplex method. We then describe the *matrix form* of the simplex method, which streamlines the procedure considerably for computer implementation. Next we use this matrix form to present a fundamental insight about a property of the simplex method that enables us to deduce how changes that are made in the original model get carried along to the final simplex tableau. This insight will provide the key to the important topics of Chap. 6 (duality theory and sensitivity analysis). The chapter then concludes by presenting the *revised simplex method*, which further streamlines the matrix form of the simplex method. Commercial computer codes of the simplex method normally are based on the revised simplex method.

■ 5.1 FOUNDATIONS OF THE SIMPLEX METHOD

Section 4.1 introduced *corner-point feasible (CPF) solutions* and the key role they play in the simplex method. These geometric concepts were related to the algebra of the simplex method in Secs. 4.2 and 4.3. However, all this was done in the context of the Wyndor Glass Co. problem, which has only *two decision variables* and so has a straightforward geometric interpretation. How do these concepts generalize to higher dimensions when we deal with larger problems? We address this question in this section.

We begin by introducing some basic terminology for any linear programming problem with n decision variables. While we are doing this, you may find it helpful to refer to Fig. 5.1 (which repeats Fig. 4.1) to interpret these definitions in two dimensions ($n = 2$).

Terminology

It may seem intuitively clear that optimal solutions for any linear programming problem must lie on the boundary of the feasible region, and in fact this is a general property. Because boundary is a geometric concept, our initial definitions clarify how the boundary of the feasible region is identified algebraically.

The **constraint boundary equation** for any constraint is obtained by replacing its ≤, =, or ≥ sign with an = sign.

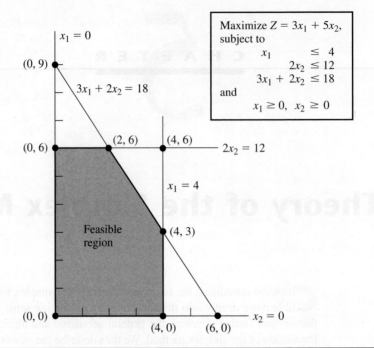

Maximize $Z = 3x_1 + 5x_2$,
subject to
$$x_1 \qquad \leq 4$$
$$2x_2 \leq 12$$
$$3x_1 + 2x_2 \leq 18$$
and
$$x_1 \geq 0, \quad x_2 \geq 0$$

■ **FIGURE 5.1**
Constraint boundaries,
constraint boundary
equations, and corner-point
solutions for the Wyndor
Glass Co. problem.

Consequently, the form of a constraint boundary equation is $a_{i1}x_1 + a_{i2}x_2 + \cdots + a_{in}x_n = b_i$ for functional constraints and $x_j = 0$ for nonnegativity constraints. Each such equation defines a "flat" geometric shape (called a **hyperplane**) in n-dimensional space, analogous to the line in two-dimensional space and the plane in three-dimensional space. This hyperplane forms the **constraint boundary** for the corresponding constraint. When the constraint has either a $\leq$ or a $\geq$ sign, this *constraint boundary* separates the points that satisfy the constraint (all the points on one side up to and including the constraint boundary) from the points that violate the constraint (all those on the other side of the constraint boundary). When the constraint has an $=$ sign, only the points on the constraint boundary satisfy the constraint.

For example, the Wyndor Glass Co. problem has five constraints (three functional constraints and two nonnegativity constraints), so it has the five *constraint boundary equations* shown in Fig. 5.1. Because $n = 2$, the hyperplanes defined by these constraint boundary equations are simply lines. Therefore, the constraint boundaries for the five constraints are the five lines shown in Fig. 5.1.

The **boundary** of the feasible region contains just those feasible solutions that satisfy one or more of the constraint boundary equations.

Geometrically, any point on the boundary of the feasible region lies on one or more of the hyperplanes defined by the respective constraint boundary equations. Thus, in Fig. 5.1, the boundary consists of the five darker line segments.

Next, we give a general definition of *CPF solution* in n-dimensional space.

A **corner-point feasible (CPF) solution** is a feasible solution that does not lie on *any* line segment[1] connecting two *other* feasible solutions.

As this definition implies, a feasible solution that *does* lie on a line segment connecting two other feasible solutions is *not* a CPF solution. To illustrate when $n = 2$, consider Fig. 5.1.

[1]An algebraic expression for a line segment is given in Appendix 2.

The point (2, 3) is *not* a CPF solution, because it lies on various such line segments, e.g., the line segment connecting (0, 3) and (4, 3). Similarly, (0, 3) is *not* a CPF solution, because it lies on the line segment connecting (0, 0) and (0, 6). However, (0, 0) *is* a CPF solution, because it is impossible to find two *other* feasible solutions that lie on completely opposite sides of (0, 0). (Try it.)

When the number of decision variables n is greater than 2 or 3, this definition for *CPF solution* is not a very convenient one for identifying such solutions. Therefore, it will prove most helpful to interpret these solutions algebraically. For the Wyndor Glass Co. example, each CPF solution in Fig. 5.1 lies at the intersection of two ($n = 2$) constraint lines; i.e., it is the *simultaneous solution* of a system of two constraint boundary equations. This situation is summarized in Table 5.1, where **defining equations** refer to the constraint boundary equations that yield (define) the indicated CPF solution.

For any linear programming problem with n decision variables, each CPF solution lies at the intersection of n constraint boundaries; i.e., it is the *simultaneous solution* of a system of n constraint boundary equations.

However, this is not to say that *every* set of n constraint boundary equations chosen from the $n + m$ constraints (n nonnegativity and m functional constraints) yields a CPF solution. In particular, the simultaneous solution of such a system of equations might violate one or more of the other m constraints not chosen, in which case it is a corner-point *infeasible* solution. The example has three such solutions, as summarized in Table 5.2. (Check to see why they are infeasible.)

■ **TABLE 5.1** Defining equations for each CPF solution for the Wyndor Glass Co. problem

CPF Solution	Defining Equations
(0, 0)	$x_1 = 0$ $x_2 = 0$
(0, 6)	$x_1 = 0$ $2x_2 = 12$
(2, 6)	$2x_2 = 12$ $3x_1 + 2x_2 = 18$
(4, 3)	$3x_1 + 2x_2 = 18$ $x_1 = 4$
(4, 0)	$x_1 = 4$ $x_2 = 0$

■ **TABLE 5.2** Defining equations for each corner-point infeasible solution for the Wyndor Glass Co. problem

Corner-Point Infeasible Solution	Defining Equations
(0, 9)	$x_1 = 0$ $3x_1 + 2x_2 = 18$
(4, 6)	$2x_2 = 12$ $x_1 = 4$
(6, 0)	$3x_1 + 2x_2 = 18$ $x_2 = 0$

Furthermore, a system of n constraint boundary equations might have no solution at all. This occurs twice in the example, with the pairs of equations (1) $x_1 = 0$ and $x_1 = 4$ and (2) $x_2 = 0$ and $2x_2 = 12$. Such systems are of no interest to us.

The final possibility (which never occurs in the example) is that a system of n constraint boundary equations has multiple solutions because of redundant equations. You need not be concerned with this case either, because the simplex method circumvents its difficulties.

We also should mention that it is possible for more than one system of n constraint boundary equations to yield the same CP solution. For example, if the $x_1 \leq 4$ constraint in the Wyndor Glass Co. problem were to be replaced by $x_1 \leq 2$, note in Fig. 5.1 how the CPF solution (2, 6) can be derived from any one of three pairs of constraint boundary equations. (This is an example of the *degeneracy* discussed in a different context in Sec. 4.5.)

To summarize for the example, with five constraints and two variables, there are 10 pairs of constraint boundary equations. Five of these pairs became defining equations for CPF solutions (Table 5.1), three became defining equations for corner-point infeasible solutions (Table 5.2), and each of the final two pairs had no solution.

Adjacent CPF Solutions

Section 4.1 introduced adjacent CPF solutions and their role in solving linear programming problems. We now elaborate.

Recall from Chap. 4 that (when we ignore slack, surplus, and artificial variables) each iteration of the simplex method moves from the current CPF solution to an *adjacent* one. What is the *path* followed in this process? What really is meant by *adjacent* CPF solution? First we address these questions from a geometric viewpoint, and then we turn to algebraic interpretations.

These questions are easy to answer when $n = 2$. In this case, the *boundary* of the feasible region consists of several connected *line segments* forming a *polygon,* as shown in Fig. 5.1 by the five darker line segments. These line segments are the *edges* of the feasible region. Emanating from each CPF solution are *two* such edges leading to an adjacent CPF solution at the other end. (Note in Fig. 5.1 how each CPF solution has two adjacent ones.) The path followed in an iteration is to move along one of these edges from one end to the other. In Fig. 5.1, the first iteration involves moving along the edge from (0, 0) to (0, 6), and then the next iteration moves along the edge from (0, 6) to (2, 6). As Table 5.1 illustrates, each of these moves to an adjacent CPF solution involves just one change in the set of defining equations (constraint boundaries on which the solution lies).

When $n = 3$, the answers are slightly more complicated. To help you visualize what is going on, Fig. 5.2 shows a three-dimensional drawing of a typical feasible region when $n = 3$, where the dots are the CPF solutions. This feasible region is a *polyhedron* rather than the polygon we had with $n = 2$ (Fig. 5.1), because the constraint boundaries now are *planes* rather than lines. The faces of the polyhedron form the *boundary* of the feasible region, where each face is the portion of a constraint boundary that satisfies the other constraints as well. Note that each CPF solution lies at the intersection of three constraint boundaries (sometimes including some of the $x_1 = 0$, $x_2 = 0$, and $x_3 = 0$ constraint boundaries for the nonnegativity constraints), and the solution also satisfies the other constraints. Such intersections that do not satisfy one or more of the other constraints yield corner-point *infeasible* solutions instead.

The darker line segment in Fig. 5.2 depicts the path of the simplex method on a typical iteration. The point (2, 4, 3) is the *current* CPF solution to begin the iteration, and the point (4, 2, 4) will be the new CPF solution at the end of the iteration. The point (2, 4, 3) lies at the intersection of the $x_2 = 4$, $x_1 + x_2 = 6$, and $-x_1 + 2x_3 = 4$ constraint boundaries, so these three equations are the *defining equations* for this CPF solution. If the $x_2 = 4$ defining equation were removed, the intersection of the other two constraint

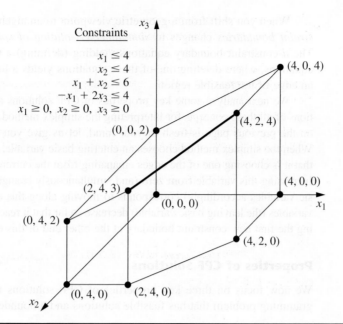

Constraints

$$x_1 \leq 4$$
$$x_2 \leq 4$$
$$x_1 + x_2 \leq 6$$
$$-x_1 + 2x_3 \leq 4$$
$$x_1 \geq 0, \ x_2 \geq 0, \ x_3 \geq 0$$

■ **FIGURE 5.2**
Feasible region and CPF
solutions for a three-variable
linear programming
problem.

boundaries (planes) would form a line. One segment of this line, shown as the dark line segment from (2, 4, 3) to (4, 2, 4) in Fig. 5.2, lies on the boundary of the feasible region, whereas the rest of the line is infeasible. This line segment is an edge of the feasible region, and its endpoints (2, 4, 3) and (4, 2, 4) are adjacent CPF solutions.

For $n = 3$, all the *edges* of the feasible region are formed in this way as the feasible segment of the line lying at the intersection of two constraint boundaries, and the two endpoints of an edge are *adjacent* CPF solutions. In Fig. 5.2 there are 15 edges of the feasible region, and so there are 15 pairs of adjacent CPF solutions. For the current CPF solution (2, 4, 3), there are three ways to remove one of its three defining equations to obtain an intersection of the other two constraint boundaries, so there are three edges emanating from (2, 4, 3). These edges lead to (4, 2, 4), (0, 4, 2), and (2, 4, 0), so these are the CPF solutions that are adjacent to (2, 4, 3).

For the next iteration, the simplex method chooses one of these three edges, say, the darker line segment in Fig. 5.2, and then moves along this edge away from (2, 4, 3) until it reaches the first new constraint boundary, $x_1 = 4$, at its other endpoint. [We cannot continue farther along this line to the next constraint boundary, $x_2 = 0$, because this leads to a corner-point infeasible solution—(6, 0, 5).] The intersection of this first new constraint boundary with the two constraint boundaries forming the edge yields the *new* CPF solution (4, 2, 4).

When $n > 3$, these same concepts generalize to higher dimensions, except the constraint boundaries now are *hyperplanes* instead of planes. Let us summarize.

Consider any linear programming problem with n decision variables and a bounded feasible region. A CPF solution lies at the intersection of n constraint boundaries (and satisfies the other constraints as well). An **edge** of the feasible region is a feasible line segment that lies at the intersection of $n - 1$ constraint boundaries, where each endpoint lies on one additional constraint boundary (so that these endpoints are CPF solutions). Two CPF solutions are **adjacent** if the line segment connecting them is an edge of the feasible region. Emanating from each CPF solution are n such edges, each one leading to one of the n adjacent CPF solutions. Each iteration of the simplex method moves from the current CPF solution to an adjacent one by moving along one of these n edges.

When you shift from a geometric viewpoint to an algebraic one, *intersection of constraint boundaries* changes to *simultaneous solution of constraint boundary equations.* The n constraint boundary equations yielding (defining) a CPF solution are its defining equations, where deleting one of these equations yields a line whose feasible segment is an edge of the feasible region.

We next analyze some key properties of CPF solutions and then describe the implications of all these concepts for interpreting the simplex method. However, while the summary on the previous page is fresh in your mind, let us give you a preview of its implications. When the simplex method chooses an entering basic variable, the geometric interpretation is that it is choosing one of the edges emanating from the current CPF solution to move along. Increasing this variable from zero (and simultaneously changing the values of the other basic variables accordingly) corresponds to moving along this edge. Having one of the basic variables (the leaving basic variable) decrease so far that it reaches zero corresponds to reaching the first new constraint boundary at the other end of this edge of the feasible region.

Properties of CPF Solutions

We now focus on three key properties of CPF solutions that hold for *any* linear programming problem that has feasible solutions and a bounded feasible region.

Property 1: (*a*) If there is exactly one optimal solution, then it must be a CPF solution. (*b*) If there are multiple optimal solutions (and a bounded feasible region), then at least two must be adjacent CPF solutions.

Property 1 is a rather intuitive one from a geometric viewpoint. First consider Case (*a*), which is illustrated by the Wyndor Glass Co. problem (see Fig. 5.1) where the one optimal solution (2, 6) is indeed a CPF solution. Note that there is nothing special about this example that led to this result. For any problem having just one optimal solution, it always is possible to keep raising the objective function line (hyperplane) until it just touches one point (the optimal solution) at a corner of the feasible region.

We now give an algebraic proof for this case.

Proof of Case (*a*) of Property 1: We set up a *proof by contradiction* by assuming that there is exactly one optimal solution and that it is *not* a CPF solution. We then show below that this assumption leads to a contradiction and so cannot be true. (The solution assumed to be optimal will be denoted by $\mathbf{x}^*$, and its objective function value by Z^*.)

Recall the definition of *CPF solution* (a feasible solution that does not lie on any line segment connecting two other feasible solutions). Since we have assumed that the optimal solution $\mathbf{x}^*$ is not a CPF solution, this implies that there must be two other feasible solutions such that the line segment connecting them contains the optimal solution. Let the vectors $\mathbf{x}'$ and $\mathbf{x}''$ denote these two other feasible solutions, and let Z_1 and Z_2 denote their respective objective function values. Like each other point on the line segment connecting $\mathbf{x}'$ and $\mathbf{x}''$,

$$\mathbf{x}^* = \alpha \mathbf{x}'' + (1 - \alpha)\mathbf{x}'$$

for some value of α such that $0 < \alpha < 1$. Thus, since the coefficients of the variables are identical for Z^*, Z_1, and Z_2, it follows that

$$Z^* = \alpha Z_2 + (1 - \alpha)Z_1.$$

Since the weights α and $1 - \alpha$ add to 1, the only possibilities for how Z^*, Z_1, and Z_2 compare are (1) $Z^* = Z_1 = Z_2$, (2) $Z_1 < Z^* < Z_2$, and (3) $Z_1 > Z^* > Z_2$. The first possibility implies that $\mathbf{x}'$ and $\mathbf{x}''$ also are optimal, which contradicts the assumption that there is exactly one optimal solution. Both the latter possibilities contradict the assumption that $\mathbf{x}*$ (not a CPF solution) is optimal. The resulting conclusion is that it is impossible to have a single optimal solution that is not a CPF solution.

Now consider Case (*b*), which was demonstrated in Sec. 3.2 under the definition of *optimal solution* by changing the objective function in the example to $Z = 3x_1 + 2x_2$ (see Fig. 3.5 in Sec. 3.2). What then happens when you are solving graphically is that the objective function line keeps getting raised until it contains the line segment connecting the two CPF solutions (2, 6) and (4, 3). The same thing would happen in higher dimensions except that an objective function *hyperplane* would keep getting raised until it contained the line segment(s) connecting two (or more) adjacent CPF solutions. As a consequence, *all* optimal solutions can be obtained as weighted averages of optimal CPF solutions. (This situation is described further in Probs. 4.5-5 and 4.5-6.)

The real significance of Property 1 is that it greatly simplifies the search for an optimal solution because now only CPF solutions need to be considered. The magnitude of this simplification is emphasized in Property 2.

Property 2: There are only a *finite* number of CPF solutions.

This property certainly holds in Figs. 5.1 and 5.2, where there are just 5 and 10 CPF solutions, respectively. To see why the number is finite in general, recall that each CPF solution is the simultaneous solution of a system of *n* out of the $m + n$ constraint boundary equations. The number of different combinations of $m + n$ equations taken *n* at a time is

$$\binom{m+n}{n} = \frac{(m+n)!}{m!n!},$$

which is a finite number. This number, in turn, in an *upper bound* on the number of CPF solutions. In Fig. 5.1, $m = 3$ and $n = 2$, so there are 10 different systems of two equations, but only half of them yield CPF solutions. In Fig. 5.2, $m = 4$ and $n = 3$, which gives 35 different systems of three equations, but only 10 yield CPF solutions.

Property 2 suggests that, in principle, an optimal solution can be obtained by exhaustive enumeration; i.e., find and compare all the finite number of CPF solutions. Unfortunately, there are finite numbers, and then there are finite numbers that (for all practical purposes) might as well be infinite. For example, a rather small linear programming problem with only $m = 50$ and $n = 50$ would have $100!/(50!)^2 \approx 10^{29}$ systems of equations to be solved! By contrast, the simplex method would need to examine only approximately 100 CPF solutions for a problem of this size. This tremendous savings can be obtained because of the optimality test given in Sec. 4.1 and restated here as Property 3.

Property 3: If a CPF solution has no *adjacent* CPF solutions that are *better* (as measured by *Z*), then there are no *better* CPF solutions anywhere. Therefore, such a CPF solution is guaranteed to be an *optimal* solution (by Property 1), assuming only that the problem possesses at least one optimal solution (guaranteed if the problem possesses feasible solutions and a bounded feasible region).

To illustrate Property 3, consider Fig. 5.1 for the Wyndor Glass Co. example. For the CPF solution (2, 6), its adjacent CPF solutions are (0, 6) and (4, 3), and neither has a better value of *Z* than (2, 6) does. This outcome implies that none of the other CPF solutions— (0, 0) and (4, 0)—can be better than (2, 6), so (2, 6) must be optimal.

By contrast, Fig. 5.3 shows a feasible region that can *never* occur for a linear programming problem (since the continuation of the constraint boundary lines that pass

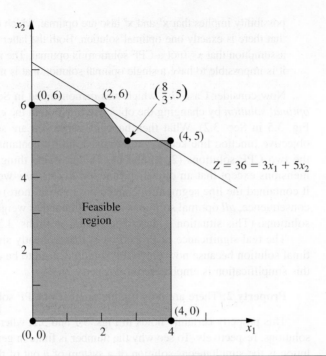

■ **FIGURE 5.3**
Modification of the Wyndor Glass Co. problem that violates both linear programming and Property 3 for CPF solutions in linear programming.

through $(\frac{8}{3}, 5)$ would chop off part of this region) but that does violate Property 3. The problem shown is identical to the Wyndor Glass Co. example (including the same objective function) *except* for the enlargement of the feasible region to the right of $(\frac{8}{3}, 5)$. Consequently, the adjacent CPF solutions for (2, 6) now are (0, 6) and $(\frac{8}{3}, 5)$, and again neither is better than (2, 6). However, another CPF solution (4, 5) now is better than (2, 6), thereby violating Property 3. The reason is that the boundary of the feasible region goes down from (2, 6) to $(\frac{8}{3}, 5)$ and then "bends outward" to (4, 5), beyond the objective function line passing through (2, 6).

The key point is that the kind of situation illustrated in Fig. 5.3 can never occur in linear programming. The feasible region in Fig. 5.3 implies that the $2x_2 \leq 12$ and $3x_1 + 2x_2 \leq 18$ constraints apply for $0 \leq x_1 \leq \frac{8}{3}$. However, under the condition that $\frac{8}{3} \leq x_1 \leq 4$, the $3x_1 + 2x_2 \leq 18$ constraint is dropped and replaced by $x_2 \leq 5$. Such "conditional constraints" just are not allowed in linear programming.

The basic reason that Property 3 holds for any linear programming problem is that the feasible region always has the property of being a *convex set*[2], as defined in Appendix 2 and illustrated in several figures there. For two-variable linear programming problems, this convex property means that the *angle* inside the feasible region at *every* CPF solution is less than 180°. This property is illustrated in Fig. 5.1, where the angles at (0, 0), (0, 6), and (4, 0) are 90° and those at (2, 6) and (4, 3) are between 90° and 180°. By contrast, the feasible region in Fig. 5.3 is *not* a convex set, because the angle at $(\frac{8}{3}, 5)$ is more than 180°. This is the kind of "bending outward" at an angle greater than 180° that can never occur in linear programming. In higher dimensions, the same intuitive notion of "never bending outward" (a basic property of a convex set) continues to apply.

[2]If you already are familiar with convex sets, note that the set of solutions that satisfy any linear programming constraint (whether it be an inequality or equality constraint) is a convex set. For any linear programming problem, its feasible region is the *intersection* of the sets of solutions that satisfy its individual constraints. Since the intersection of convex sets is a convex set, this feasible region necessarily is a convex set.

To clarify the significance of a convex feasible region, consider the objective function hyperplane that passes through a CPF solution that has no adjacent CPF solutions that are better. [In the original Wyndor Glass Co. example, this hyperplane is the objective function line passing through $(2, 6)$.] All these adjacent solutions [$(0, 6)$ and $(4, 3)$ in the example] must lie either on the hyperplane or on the unfavorable side (as measured by Z) of the hyperplane. The feasible region being convex means that its boundary cannot "bend outward" beyond an adjacent CPF solution to give another CPF solution that lies on the favorable side of the hyperplane. So Property 3 holds.

Extensions to the Augmented Form of the Problem

For any linear programming problem in our standard form (including functional constraints in $\leq$ form), the appearance of the functional constraints after slack variables are introduced is as follows:

$$(1) \quad a_{11}x_1 + a_{12}x_2 + \cdots + a_{1n}x_n + x_{n+1} \qquad\qquad = b_1$$
$$(2) \quad a_{21}x_1 + a_{22}x_2 + \cdots + a_{2n}x_n \qquad + x_{n+2} \qquad = b_2$$
$$\cdots\cdots\cdots\cdots\cdots\cdots\cdots\cdots\cdots\cdots\cdots\cdots\cdots\cdots\cdots\cdots\cdots$$
$$(m) \quad a_{m1}x_1 + a_{m2}x_2 + \cdots + a_{mn}x_n \qquad\qquad\quad + x_{n+m} = b_m,$$

where $x_{n+1}, x_{n+2}, \ldots, x_{n+m}$ are the slack variables. For other linear programming problems, Sec. 4.6 described how essentially this same appearance (proper form from Gaussian elimination) can be obtained by introducing artificial variables, etc. Thus, the original solutions $(x_1, x_2, \ldots, x_n)$ now are augmented by the corresponding values of the slack or artificial variables $(x_{n+1}, x_{n+2}, \ldots, x_{n+m})$ and perhaps some surplus variables as well. This augmentation led in Sec. 4.2 to defining **basic solutions** as *augmented corner-point solutions* and **basic feasible solutions (BF solutions)** as *augmented CPF solutions*. Consequently, the preceding three properties of CPF solutions also hold for BF solutions.

Now let us clarify the algebraic relationships between basic solutions and corner-point solutions. Recall that each corner-point solution is the simultaneous solution of a system of n constraint boundary equations, which we called its *defining equations*. The key question is: How do we tell whether a particular constraint boundary equation is one of the defining equations when the problem is in augmented form? The answer, fortunately, is a simple one. Each constraint has an **indicating variable** that completely indicates (by whether its value is zero) whether that constraint's boundary equation is satisfied by the current solution. A summary appears in Table 5.3. For the type of constraint in each row

TABLE 5.3 Indicating variables for constraint boundary equations*

Type of Constraint	Form of Constraint	Constraint in Augmented Form	Constraint Boundary Equation	Indicating Variable
Nonnegativity	$x_j \geq 0$	$x_j \geq 0$	$x_j = 0$	x_j
Functional ($\leq$)	$\sum_{j=1}^{n} a_{ij}x_j \leq b_i$	$\sum_{j=1}^{n} a_{ij}x_j + x_{n+i} = b_i$	$\sum_{j=1}^{n} a_{ij}x_j = b_i$	x_{n+i}
Functional ($=$)	$\sum_{j=1}^{n} a_{ij}x_j = b_i$	$\sum_{j=1}^{n} a_{ij}x_j + \bar{x}_{n+i} = b_i$	$\sum_{j=1}^{n} a_{ij}x_j = b_i$	$\bar{x}_{n+i}$
Functional ($\geq$)	$\sum_{j=1}^{n} a_{ij}x_j \geq b_i$	$\sum_{j=1}^{n} a_{ij}x_j + \bar{x}_{n+i} - x_{s_i} = b_i$	$\sum_{j=1}^{n} a_{ij}x_j = b_i$	$\bar{x}_{n+i} - x_{s_i}$

*Indicating variable $= 0 \Rightarrow$ constraint boundary equation satisfied;
indicating variable $\neq 0 \Rightarrow$ constraint boundary equation violated.

■ TABLE 5.7 Sequence of solutions obtained by the simplex method for the Wyndor Glass Co. problem

Iteration	CPF Solution	Defining Equations	BF Solution	Nonbasic Variables	Functional Constraints in Augmented Form
0	(0, 0)	$x_1 = 0$ $x_2 = 0$	(0, 0, 4, 12, 18)	$x_1 = 0$ $x_2 = 0$	$x_1 \quad\quad + \mathbf{x}_3 = 4$ $2x_2 + \mathbf{x}_4 = 12$ $3x_1 + 2x_2 + \mathbf{x}_5 = 18$
1	(0, 6)	$x_1 = 0$ $2x_2 = 12$	(0, 6, 4, 0, 6)	$x_1 = 0$ $x_4 = 0$	$x_1 \quad\quad + \mathbf{x}_3 = 4$ $2\mathbf{x}_2 + x_4 = 12$ $3x_1 + 2\mathbf{x}_2 + \mathbf{x}_5 = 18$
2	(2, 6)	$2x_2 = 12$ $3x_1 + 2x_2 = 18$	(2, 6, 2, 0, 0)	$x_4 = 0$ $x_5 = 0$	$\mathbf{x}_1 \quad\quad + \mathbf{x}_3 = 4$ $2\mathbf{x}_2 + x_4 = 12$ $3\mathbf{x}_1 + 2\mathbf{x}_2 + x_5 = 18$

Equivalently, in our new terminology, the simplex method reaches an adjacent BF solution from the current one by (1) deleting one variable (the entering basic variable) from the set of n nonbasic variables defining the current solution, (2) moving away from the current solution by *increasing* this one variable from zero (and adjusting the other basic variables to still satisfy the system of equations) while keeping the remaining $n - 1$ nonbasic variables at zero, and (3) stopping when the *first* of the basic variables (the leaving basic variable) reaches a value of zero (its constraint boundary). With either interpretation, the choice among the n alternatives in step 1 is made by selecting the one that would give the best rate of improvement in Z (per unit increase in the entering basic variable) during step 2.

Table 5.7 illustrates the close correspondence between these geometric and algebraic interpretations of the simplex method. Using the results already presented in Secs. 4.3 and 4.4, the fourth column summarizes the sequence of BF solutions found for the Wyndor Glass Co. problem, and the second column shows the corresponding CPF solutions. In the third column, note how each iteration results in deleting one constraint boundary (defining equation) and substituting a new one to obtain the new CPF solution. Similarly, note in the fifth column how each iteration results in deleting one nonbasic variable and substituting a new one to obtain the new BF solution. Furthermore, the nonbasic variables being deleted and added are the indicating variables for the defining equations being deleted and added in the third column. The last column displays the initial system of equations [excluding Eq. (0)] for the augmented form of the problem, with the current basic variables shown in bold type. In each case, note how setting the nonbasic variables equal to zero and then solving this system of equations for the basic variables must yield the same solution for (x_1, x_2) as the corresponding pair of defining equations in the third column.

The Worked Examples section of the book's website provides **another example** of developing the type of information given in Table 5.7 for a minimization problem.

■ 5.2 THE SIMPLEX METHOD IN MATRIX FORM

Chapter 4 describes the simplex method in both an algebraic form and a tabular form. Further insight into the theory and power of the simplex method can be obtained by examining its *matrix* form. We begin by introducing matrix notation to represent linear programming problems. (See Appendix 4 for a review of matrices.).

To help you distinguish between matrices, vectors, and scalars, we consistently use **BOLDFACE CAPITAL** letters to represent matrices, **boldface lowercase** letters to represent vectors, and *italicized* letters in ordinary print to represent scalars. We also use a boldface zero (**0**) to denote a *null vector* (a vector whose elements all are zero) in either column or row form (which one should be clear from the context), whereas a zero in ordinary print (0) continues to represent the number zero.

Using matrices, our standard form for the general linear programming model given in Sec. 3.2 becomes

$$
\begin{array}{ll}
\text{Maximize} & Z = \mathbf{c}\mathbf{x}, \\
\text{subject to} & \\
\mathbf{A}\mathbf{x} \leq \mathbf{b} & \text{and} \qquad \mathbf{x} \geq \mathbf{0},
\end{array}
$$

where **c** is the row vector

$$\mathbf{c} = [c_1, c_2, \ldots, c_n],$$

x, **b**, and **0** are the column vectors such that

$$
\mathbf{x} = \begin{bmatrix} x_1 \\ x_2 \\ \vdots \\ x_n \end{bmatrix}, \qquad
\mathbf{b} = \begin{bmatrix} b_1 \\ b_2 \\ \vdots \\ b_m \end{bmatrix}, \qquad
\mathbf{0} = \begin{bmatrix} 0 \\ 0 \\ \vdots \\ 0 \end{bmatrix},
$$

and **A** is the matrix

$$
\mathbf{A} = \begin{bmatrix}
a_{11} & a_{12} & \cdots & a_{1n} \\
a_{21} & a_{22} & \cdots & a_{2n} \\
\multicolumn{4}{c}{\dotfill} \\
a_{m1} & a_{m2} & \cdots & a_{mn}
\end{bmatrix}.
$$

To obtain the *augmented form* of the problem, introduce the column vector of slack variables

$$
\mathbf{x}_s = \begin{bmatrix} x_{n+1} \\ x_{n+2} \\ \vdots \\ x_{n+m} \end{bmatrix}
$$

so that the constraints become

$$
[\mathbf{A}, \mathbf{I}] \begin{bmatrix} \mathbf{x} \\ \mathbf{x}_s \end{bmatrix} = \mathbf{b} \qquad \text{and} \qquad \begin{bmatrix} \mathbf{x} \\ \mathbf{x}_s \end{bmatrix} \geq \mathbf{0},
$$

where **I** is the $m \times m$ identity matrix, and the null vector **0** now has $n + m$ elements. (We comment at the end of the section about how to deal with problems that are not in our standard form.)

Solving for a Basic Feasible Solution

Recall that the general approach of the simplex method is to obtain a sequence of *improving BF solutions* until an optimal solution is reached. One of the key features of the matrix form of the simplex method involves the way in which it solves for each new

Matrix Form of the Current Set of Equations

The last preliminary before we summarize the matrix form of the simplex method is to show the matrix form of the set of equations appearing in the simplex tableau for any iteration of the original simplex method.

For the *original* set of equations, the matrix form is

$$\begin{bmatrix} 1 & -\mathbf{c} & \mathbf{0} \\ 0 & \mathbf{A} & \mathbf{I} \end{bmatrix} \begin{bmatrix} Z \\ \mathbf{x} \\ \mathbf{x}_s \end{bmatrix} = \begin{bmatrix} 0 \\ \mathbf{b} \end{bmatrix}.$$

This set of equations also is exhibited in the first simplex tableau of Table 5.8.

The algebraic operations performed by the simplex method (multiply an equation by a constant and add a multiple of one equation to another equation) are expressed in matrix form by premultiplying both sides of the original set of equations by the appropriate matrix. This matrix would have the same elements as the identity matrix, *except* that each multiple for an algebraic operation would go into the spot needed to have the matrix multiplication perform this operation. Even after a series of algebraic operations over several iterations, we still can deduce what this matrix must be (symbolically) for the entire series by using what we already know about the right-hand sides of the new set of equations. In particular, after any iteration, $\mathbf{x}_B = \mathbf{B}^{-1}\mathbf{b}$ and $Z = \mathbf{c}_B\mathbf{B}^{-1}\mathbf{b}$, so the right-hand sides of the new set of equations have become

$$\begin{bmatrix} Z \\ \mathbf{x}_B \end{bmatrix} = \begin{bmatrix} 1 & \mathbf{c}_B\mathbf{B}^{-1} \\ 0 & \mathbf{B}^{-1} \end{bmatrix} \begin{bmatrix} 0 \\ \mathbf{b} \end{bmatrix} = \begin{bmatrix} \mathbf{c}_B\mathbf{B}^{-1}\mathbf{b} \\ \mathbf{B}^{-1}\mathbf{b} \end{bmatrix}.$$

Because we perform the same series of algebraic operations on *both* sides of the original set of equations, we use this same matrix that premultiplies the original right-hand side to premultiply the original left-hand side. Consequently, since

$$\begin{bmatrix} 1 & \mathbf{c}_B\mathbf{B}^{-1} \\ 0 & \mathbf{B}^{-1} \end{bmatrix} \begin{bmatrix} 1 & -\mathbf{c} & \mathbf{0} \\ 0 & \mathbf{A} & \mathbf{I} \end{bmatrix} = \begin{bmatrix} 1 & \mathbf{c}_B\mathbf{B}^{-1}\mathbf{A} - \mathbf{c} & \mathbf{c}_B\mathbf{B}^{-1} \\ 0 & \mathbf{B}^{-1}\mathbf{A} & \mathbf{B}^{-1} \end{bmatrix},$$

TABLE 5.8 Initial and later simplex tableaux in matrix form

Iteration	Basic Variable	Eq.	Z	Coefficient of: Original Variables	Coefficient of: Slack Variables	Right Side
0	Z	(0)	1	$-\mathbf{c}$	$\mathbf{0}$	0
	$\mathbf{x}_B$	$(1, 2, \ldots, m)$	0	$\mathbf{A}$	$\mathbf{I}$	$\mathbf{b}$
Any	Z	(0)	1	$\mathbf{c}_B\mathbf{B}^{-1}\mathbf{A} - \mathbf{c}$	$\mathbf{c}_B\mathbf{B}^{-1}$	$\mathbf{c}_B\mathbf{B}^{-1}\mathbf{b}$
	$\mathbf{x}_B$	$(1, 2, \ldots, m)$	0	$\mathbf{B}^{-1}\mathbf{A}$	$\mathbf{B}^{-1}$	$\mathbf{B}^{-1}\mathbf{b}$

the desired matrix form of the *set of equations after any iteration* is

$$\begin{bmatrix} 1 & c_B B^{-1} A - c & c_B B^{-1} \\ 0 & B^{-1} A & B^{-1} \end{bmatrix} \begin{bmatrix} Z \\ x \\ x_s \end{bmatrix} = \begin{bmatrix} c_B B^{-1} b \\ B^{-1} b \end{bmatrix}.$$

The second simplex tableau of Table 5.8 also exhibits this same set of equations.

Example. To illustrate this matrix form for the current set of equations, we will show how it yields the final set of equations resulting from iteration 2 for the Wyndor Glass Co. problem. Using the B^{-1} and c_B given for iteration 2 at the end of the preceding subsection, we have

$$B^{-1} A = \begin{bmatrix} 1 & \frac{1}{3} & -\frac{1}{3} \\ 0 & \frac{1}{2} & 0 \\ 0 & -\frac{1}{3} & \frac{1}{3} \end{bmatrix} \begin{bmatrix} 1 & 0 \\ 0 & 2 \\ 3 & 2 \end{bmatrix} = \begin{bmatrix} 0 & 0 \\ 0 & 1 \\ 1 & 0 \end{bmatrix},$$

$$c_B B^{-1} = [0, 5, 3] \begin{bmatrix} 1 & \frac{1}{3} & -\frac{1}{3} \\ 0 & \frac{1}{2} & 0 \\ 0 & -\frac{1}{3} & \frac{1}{3} \end{bmatrix} = [0, \tfrac{3}{2}, 1],$$

$$c_B B^{-1} A - c = [0, 5, 3] \begin{bmatrix} 0 & 0 \\ 0 & 1 \\ 1 & 0 \end{bmatrix} - [3, 5] = [0, 0].$$

Also, by using the values of $x_B = B^{-1} b$ and $Z = c_B B^{-1} b$ calculated at the end of the preceding subsection, these results give the following set of equations:

$$\begin{bmatrix} 1 & 0 & 0 & 0 & \frac{3}{2} & 1 \\ 0 & 0 & 0 & 1 & \frac{1}{3} & -\frac{1}{3} \\ 0 & 0 & 1 & 0 & \frac{1}{2} & 0 \\ 0 & 1 & 0 & 0 & -\frac{1}{3} & \frac{1}{3} \end{bmatrix} \begin{bmatrix} Z \\ x_1 \\ x_2 \\ x_3 \\ x_4 \\ x_5 \end{bmatrix} = \begin{bmatrix} 36 \\ 2 \\ 6 \\ 2 \end{bmatrix},$$

as shown in the final simplex tableau in Table 4.8.

The matrix form of the set of equations after any iteration (as shown in the box just before the above example) provides the key to the execution of the matrix form of the simplex method. The matrix expressions shown in these equations (or in the bottom part of Table 5.8) provide a direct way of calculating all the numbers that would appear in the current set of equations (for the algebraic form of the simplex method) or in the current simplex tableau (for the tableau form of the simplex method). The three forms of the simplex method make exactly the same decisions (entering basic variable, leaving basic variable, etc.) step after step and iteration after iteration. The only difference between these forms is in the methods used

Also using $\mathbf{x}_B$ obtained at the end of the preceding iteration, the minimum ratio test indicates that x_5 is the leaving basic variable since $6/3 < 4/1$. Iteration 2 for the first example in this section already shows the resulting updated $\mathbf{B}$, $\mathbf{B}^{-1}$, $\mathbf{x}_B$, and $\mathbf{c}_B$, namely,

$$\mathbf{B} = \begin{bmatrix} 1 & 0 & 1 \\ 0 & 2 & 0 \\ 0 & 2 & 3 \end{bmatrix}, \quad \mathbf{B}^{-1} = \begin{bmatrix} 1 & \frac{1}{3} & -\frac{1}{3} \\ 0 & \frac{1}{2} & 0 \\ 0 & -\frac{1}{3} & \frac{1}{3} \end{bmatrix}, \quad \mathbf{x}_B = \begin{bmatrix} x_3 \\ x_2 \\ x_1 \end{bmatrix} = \mathbf{B}^{-1}\mathbf{b} = \begin{bmatrix} 2 \\ 6 \\ 2 \end{bmatrix}, \quad \mathbf{c}_B = [0, 5, 3],$$

so x_1 has replaced x_5 in $\mathbf{x}_B$, in providing an element of $\mathbf{c}_B$ from $[3, 5, 0, 0, 0]$, and in providing a column from $[\mathbf{A},\mathbf{I}]$ in $\mathbf{B}$.

Optimality test

The nonbasic variables now are x_4 and x_5. Using the calculations already shown for the second example in this section, their coefficients in Eq. (0) are 3/2 and 1, respectively. Since neither of these coefficients are negative, the current BF solution ($x_1 = 2$, $x_2 = 6$, $x_3 = 2$, $x_4 = 0$, $x_5 = 0$) is optimal and the procedure terminates.

Final Observations

The above example illustrates that the matrix form of the simplex method uses just a few matrix expressions to perform all the needed calculations. These matrix expressions are summarized in the bottom part of Table 5.8. A fundamental insight from this table is that it is only necessary to know the current $\mathbf{B}^{-1}$ and $\mathbf{c}_B\mathbf{B}^{-1}$, which appear in the slack variables portion of the current simplex tableau, in order to calculate all the other numbers in this tableau in terms of the original parameters ($\mathbf{A}$, $\mathbf{b}$, and $\mathbf{c}$) of the model being solved. When dealing with the *final* simplex tableau, this insight proves to be a particularly valuable one, as will be described in the next section.

A drawback of the matrix form of the simplex method as it has been outlined in this section is that it is necessary to derive $\mathbf{B}^{-1}$, the inverse of the updated basis matrix, at the end of each iteration. Although routines are available for inverting small square (nonsingular) matrices (and this can even be done readily by hand for 2 x 2 or perhaps 3 x 3 matrices), the time required to invert matrices grows very rapidly with the size of the matrices. Fortunately, there is a much more efficient procedure available for updating $\mathbf{B}^{-1}$ from one iteration to the next rather than inverting the new basis matrix from scratch. When this procedure is incorporated into the matrix form of the simplex method, this improved version of the matrix form is conventionally called the **revised simplex method.** This is the version of the simplex method (along with further improvements) that normally is used in commercial software for linear programming. We will describe the procedure for updating $\mathbf{B}^{-1}$ in Sec. 5.4.

The Worked Examples section of the book's website gives **another example** of applying the matrix form of the simplex method. This example also incorporates the efficient procedure for updating $\mathbf{B}^{-1}$ at each iteration instead of inverting the updated basis matrix from scratch, so the full-fledged revised simplex method is applied.

Finally, we should remind you that the description of the matrix form of the simplex method throughout this section has assumed that the problem being solved fits *our standard form* for the general linear programming model given in Sec. 3.2. However, the modifications for other forms of the model are relatively straightforward. The initialization step would be conducted just as was described in Sec. 4.6 for either the algebraic form or tabular form of the simplex method. When this step involves introducing artificial variables to obtain an initial BF solution (and thereby to obtain an *identity matrix as the initial basis matrix*), these variables are included among the m elements of $\mathbf{x}_s$.

5.3 A FUNDAMENTAL INSIGHT

We shall now focus on a property of the simplex method (in any form) that has been revealed by the matrix form of the simplex method in Sec. 5.2. This fundamental insight provides the key to both duality theory and sensitivity analysis (Chap. 6), two very important parts of linear programming.

We shall first describe this insight when the problem being solved fits *our standard form* for linear programming models (Sec. 3.2) and then discuss how to adapt to other forms later. The insight is based directly on Table 5.8 in Sec. 5.2, as described below.

> **The insight provided by Table 5.8:** Using matrix notation, Table 5.8 gives the rows of the *initial* simplex tableau as $[-\mathbf{c}, \mathbf{0}, 0]$ for row 0 and $[\mathbf{A}, \mathbf{I}, \mathbf{b}]$ for the rest of the rows. After any iteration, the coefficients of the slack variables in the current simplex tableau become $\mathbf{c}_B \mathbf{B}^{-1}$ for row 0 and $\mathbf{B}^{-1}$ for the rest of the rows, where $\mathbf{B}$ is the current basis matrix. Examining the rest of the current simplex tableau, the insight is that these coefficients of the slack variables immediately reveal how the *entire* rows of the current simplex tableau have been obtained from the rows in the *initial* simplex tableau. In particular, after any iteration,
>
> $$\text{Row } 0 = [-\mathbf{c}, \mathbf{0}, 0] + \mathbf{c}_B \mathbf{B}^{-1}[\mathbf{A}, \mathbf{I}, \mathbf{b}]$$
> $$\text{Rows } 1 \text{ to } m = \mathbf{B}^{-1}[\mathbf{A}, \mathbf{I}, \mathbf{b}]$$

We shall describe the applications of this insight at the end of this section. These applications are particularly important only when we are dealing with the *final* simplex tableau after the optimal solution has been obtained. Therefore, we will focus hereafter on discussing the "fundamental insight" just in terms of the optimal solution.

To distinguish between the matrix notation used after *any* iteration ($\mathbf{B}^{-1}$, etc.) and the corresponding notation after just the *last* iteration, we now introduce the following notation for the latter case.

When $\mathbf{B}$ is the basis matrix for the *optimal solution* found by the simplex method, let

$\mathbf{S}^* = \mathbf{B}^{-1} =$ coefficients of the *slack* variables in rows 1 to m

$\mathbf{A}^* = \mathbf{B}^{-1}\mathbf{A} =$ coefficients of the *original* variables in rows 1 to m

$\mathbf{y}^* = \mathbf{c}_B \mathbf{B}^{-1} =$ coefficients of the *slack* variables in row 0

$\mathbf{z}^* = \mathbf{c}_B \mathbf{B}^{-1}\mathbf{A}$, so $\mathbf{z}^* - \mathbf{c} =$ coefficients of the *original* variables in row 0

$Z^* = \mathbf{c}_B \mathbf{B}^{-1}\mathbf{b} =$ optimal value of the objective function

$\mathbf{b}^* = \mathbf{B}^{-1}\mathbf{b} =$ optimal right-hand sides of rows 1 to m

The bottom half of Table 5.9 shows where each of these symbols fits in the final simplex tableau. To illustrate all the notation, the top half of Table 5.9 includes the initial tableau for the Wyndor Glass Co. problem and the bottom half includes the final tableau for this problem.

Referring to this again, suppose now that you are given the initial tableau, $\mathbf{t}$ and $\mathbf{T}$, and just $\mathbf{y}^*$ and $\mathbf{S}^*$ from the final tableau. How can this information alone be used to calculate the rest of the final tableau? The answer is provided by the fundamental insight summarized below.

Fundamental Insight

(1) $\mathbf{t}^* = \mathbf{t} + \mathbf{y}^*\mathbf{T} = [\mathbf{y}^*\mathbf{A} - \mathbf{c} \mid \mathbf{y}^* \mid \mathbf{y}^*\mathbf{b}].$

(2) $\mathbf{T}^* = \mathbf{S}^*\mathbf{T} = [\mathbf{S}^*\mathbf{A} \mid \mathbf{S}^* \mid \mathbf{S}^*\mathbf{b}].$

of these managerial decisions on resource allocations were to be changed in various ways. By using the formulas,

$$\mathbf{x}_B = \mathbf{S}^*\mathbf{b}$$
$$Z^* = \mathbf{y}^*\mathbf{b},$$

you can see exactly how the optimal BF solution changes (or whether it becomes infeasible because of negative variables), as well as how the optimal value of the objective function changes, as a function of $\mathbf{b}$. You do *not* have to reapply the simplex method over and over for each new $\mathbf{b}$, because the coefficients of the slack variables tell all!

For example, consider the change from $b_2 = 12$ to $b_2 = 13$ as illustrated in Fig. 4.8 for the Wyndor Glass Co. problem. It is not necessary to *solve* for the new optimal solution $(x_1, x_2) = (\frac{5}{3}, \frac{13}{2})$ because the values of the basic variables in the final tableau ($\mathbf{b}^*$) are immediately revealed by the fundamental insight:

$$\begin{bmatrix} x_3 \\ x_2 \\ x_1 \end{bmatrix} = \mathbf{b}^* = \mathbf{S}^*\mathbf{b} = \begin{bmatrix} 1 & \frac{1}{3} & -\frac{1}{3} \\ 0 & \frac{1}{2} & 0 \\ 0 & -\frac{1}{3} & \frac{1}{3} \end{bmatrix} \begin{bmatrix} 4 \\ 13 \\ 18 \end{bmatrix} = \begin{bmatrix} \frac{7}{3} \\ \frac{13}{2} \\ \frac{5}{3} \end{bmatrix}.$$

There is an even easier way to make this calculation. Since the only change is in the *second* component of $\mathbf{b}$ ($\Delta b_2 = 1$), which gets premultiplied by only the *second* column of $\mathbf{S}^*$, the *change* in $\mathbf{b}^*$ can be calculated as simply

$$\Delta \mathbf{b}^* = \begin{bmatrix} \frac{1}{3} \\ \frac{1}{2} \\ -\frac{1}{3} \end{bmatrix} \Delta b_2 = \begin{bmatrix} \frac{1}{3} \\ \frac{1}{2} \\ -\frac{1}{3} \end{bmatrix},$$

so the original values of the basic variables in the final tableau ($x_3 = 2$, $x_2 = 6$, $x_1 = 2$) now become

$$\begin{bmatrix} x_3 \\ x_2 \\ x_1 \end{bmatrix} = \begin{bmatrix} 2 \\ 6 \\ 2 \end{bmatrix} + \begin{bmatrix} \frac{1}{3} \\ \frac{1}{2} \\ -\frac{1}{3} \end{bmatrix} = \begin{bmatrix} \frac{7}{13} \\ \frac{13}{2} \\ \frac{5}{3} \end{bmatrix}.$$

(If any of these new values were *negative*, and thus infeasible, then the reoptimization technique described in Sec. 4.7 would be applied, starting from this revised final tableau.) Applying *incremental analysis* to the preceding equation for Z^* also immediately yields

$$\Delta Z^* = \frac{3}{2}\Delta b_2 = \frac{3}{2}.$$

The fundamental insight can be applied to investigating other kinds of changes in the original model in a very similar fashion; it is the crux of the sensitivity analysis procedure described in the latter part of Chap. 6.

You also will see in the next chapter that the fundamental insight plays a key role in the very useful duality theory for linear programming.

5.4 THE REVISED SIMPLEX METHOD

The revised simplex method is based directly on the matrix form of the simplex method presented in Sec. 5.2. However, as mentioned at the end of that section, the difference is that the revised simplex method incorporates a key improvement into the matrix form. Instead of needing to invert the new basis matrix $\mathbf{B}$ after each iteration, which is computationally expensive for large matrices, the revised simplex method uses a much more efficient procedure that simply updates $\mathbf{B}^{-1}$ from one iteration to the next. We focus on describing and illustrating this procedure in this section.

This procedure is based on two properties of the simplex method. One is described in *the insight provided by Table 5.8* at the beginning of Sec. 5.3. In particular, after any iteration, the coefficients of the *slack variables* for all the rows except row 0 in the current simplex tableau become $\mathbf{B}^{-1}$, where $\mathbf{B}$ is the current basis matrix. This property always holds as long as the problem being solved fits *our standard form* described in Sec. 3.2 for linear programming models. (For nonstandard forms where artificial variables need to be introduced, the only difference is that it is the set of appropriately ordered columns that form an identity matrix $\mathbf{I}$ below row 0 in the initial simplex tableau that then provides $\mathbf{B}^{-1}$ in any subsequent tableau.)

The other relevant property of the simplex method is that step 3 of an iteration changes the numbers in the simplex tableau, including the numbers giving $\mathbf{B}^{-1}$, only by performing the elementary algebraic operations (such as dividing an equation by a constant or subtracting a multiple of some equation from another equation) that are needed to restore proper form from Gaussian elimination. Therefore, all that is needed to update $\mathbf{B}^{-1}$ from one iteration to the next is to obtain the new $\mathbf{B}^{-1}$ (denote it by $\mathbf{B}_{\text{new}}^{-1}$) from the old $\mathbf{B}^{-1}$ (denote it by $\mathbf{B}_{\text{old}}^{-1}$) by performing the usual algebraic operations on $\mathbf{B}_{\text{old}}^{-1}$ that the algebraic form of the simplex method would perform on the entire system of equations (except Eq. (0)) for this iteration. Thus, given the choice of the entering basic variable and leaving basic variable from steps 1 and 2 of an iteration, the procedure is to apply step 3 of an iteration (as described in Secs. 4.3 and 4.4) to the $\mathbf{B}^{-1}$ portion of the current simplex tableau or system of equations.

To describe this procedure formally, let

x_k = entering basic variable,
a'_{ik} = coefficient of x_k in current Eq. (i), for $i = 1, 2, \ldots, m$ (identified in step 2 of an iteration),
r = number of equation containing the leaving basic variable.

Recall that the new set of equations [excluding Eq. (0)] can be obtained from the preceding set by subtracting a'_{ik}/a'_{rk} times Eq. (r) from Eq. (i), for all $i = 1, 2, \ldots, m$ except $i = r$, and then dividing Eq. (r) by a'_{rk}. Therefore, the element in row i and column j of $\mathbf{B}_{\text{new}}^{-1}$ is

$$(\mathbf{B}_{\text{new}}^{-1})_{ij} = \begin{cases} (\mathbf{B}_{\text{old}}^{-1})_{ij} - \dfrac{a'_{ik}}{a'_{rk}}(\mathbf{B}_{\text{old}}^{-1})_{rj} & \text{if } i \neq r, \\[2ex] \dfrac{1}{a'_{rk}}(\mathbf{B}_{\text{old}}^{-1})_{rj} & \text{if } i = r. \end{cases}$$

These formulas are expressed in matrix notation as

$$\mathbf{B}_{\text{new}}^{-1} = \mathbf{E}\mathbf{B}_{\text{old}}^{-1},$$

where matrix $\mathbf{E}$ is an identity matrix except that its rth column is replaced by the vector

$$\boldsymbol{\eta} = \begin{bmatrix} \eta_1 \\ \eta_2 \\ \vdots \\ \eta_m \end{bmatrix}, \qquad \text{where} \qquad \eta_i = \begin{cases} -\dfrac{a'_{ik}}{a'_{rk}} & \text{if } i \neq r, \\[2ex] \dfrac{1}{a'_{rk}} & \text{if } i = r. \end{cases}$$

Thus, $\mathbf{E} = [\mathbf{U}_1, \mathbf{U}_2, \ldots, \mathbf{U}_{r-1}, \boldsymbol{\eta}, \mathbf{U}_{r+1}, \ldots, \mathbf{U}_m]$, where the m elements of each of the $\mathbf{U}_i$ column vectors are 0 except for a 1 in the ith position.

■ LEARNING AIDS FOR THIS CHAPTER ON OUR WEBSITE (www.mhhe.com/hillier)

Worked Examples:

Examples for Chapter 5

A Demonstration Example in OR Tutor:

Fundamental Insight

Interactive Procedures in IOR Tutorial:

Interactive Graphical Method
Enter or Revise a General Linear Programming Model
Set Up for the Simplex Method—Interactive Only
Solve Interactively by the Simplex Method

Automatic Procedures in IOR Tutorial:

Solve Automatically by the Simplex Method
Graphical Method and Sensitivity Analysis

Files (Chapter 3) for Solving the Wyndor Example:

Excel Files
LINGO/LINDO File
MPL/CPLEX File

Glossary for Chapter 5

See Appendix 1 for documentation of the software.

■ PROBLEMS

The symbols to the left of some of the problems (or their parts) have the following meaning:

D: The demonstration example listed above may be helpful.
I: You can check some of your work by using procedures listed above.

An asterisk on the problem number indicates that at least a partial answer is given in the back of the book.

5.1-1.* Consider the following problem.

Maximize $Z = 3x_1 + 2x_2,$

subject to

$$2x_1 + x_2 \le 6$$
$$x_1 + 2x_2 \le 6$$

and

$$x_1 \ge 0, \qquad x_2 \ge 0.$$

I **(a)** Solve this problem graphically. Identify the CPF solutions by circling them on the graph.

(b) Identify all the sets of two defining equations for this problem. For each set, solve (if a solution exists) for the corresponding corner-point solution, and classify it as a CPF solution or corner-point infeasible solution.

(c) Introduce slack variables in order to write the functional constraints in augmented form. Use these slack variables to identify the basic solution that corresponds to each corner-point solution found in part (*b*).

(d) Do the following for *each* set of two defining equations from part (*b*): Identify the indicating variable for each defining equation. Display the set of equations from part (*c*) *after* deleting these two indicating (nonbasic) variables. Then use the latter set of equations to solve for the two remaining variables (the basic variables). Compare the resulting basic solution to the corresponding basic solution obtained in part (*c*).

(e) Without executing the simplex method, use its geometric interpretation (and the objective function) to identify the path (sequence of CPF solutions) it would follow to reach the optimal solution. For each of these CPF solutions in turn, identify the following decisions being made for the next iteration: (*i*) which defining equation is being deleted and which is being added; (*ii*) which indicating variable is being deleted (the entering basic variable) and which is being added (the leaving basic variable).

5.1-2. Repeat Prob. 5.1-1 for the model in Prob. 3.1-6.

5.1-3. Consider the following problem.

Maximize $Z = 5x_1 + 8x_2$,

subject to

$$4x_1 + 2x_2 \leq 80$$
$$-3x_1 + x_2 \leq 4$$
$$-x_1 + 2x_2 \leq 20$$
$$4x_1 - x_2 \leq 40$$

and

$$x_1 \geq 0, \qquad x_2 \geq 0.$$

I **(a)** Solve this problem graphically. Identify the CPF solutions by circling them on the graph.
(b) Develop a table giving each of the CPF solutions and the corresponding defining equations, BF solution, and nonbasic variables. Calculate Z for each of these solutions, and use just this information to identify the optimal solution.
(c) Develop the corresponding table for the corner-point infeasible solutions, etc. Also identify the sets of defining equations and nonbasic variables that do not yield a solution.

5.1-4. Consider the following problem.

Maximize $Z = 2x_1 - x_2 + x_3$,

subject to

$$3x_1 + x_2 + x_3 \leq 60$$
$$x_1 - x_2 + 2x_3 \leq 10$$
$$x_1 + x_2 - x_3 \leq 20$$

and

$$x_1 \geq 0, \qquad x_2 \geq 0, \qquad x_3 \geq 0.$$

After slack variables are introduced and then one complete iteration of the simplex method is performed, the following simplex tableau is obtained.

	Basic					Coefficient of:					Right
Iteration	Variable	Eq.	Z	x_1	x_2	x_3	x_4	x_5	x_6		Side
	Z	(0)	1	0	−1	3	0	2	0		20
1	x_4	(1)	0	0	4	−5	1	−3	0		30
	x_1	(2)	0	1	−1	2	0	1	0		10
	x_6	(3)	0	0	2	−3	0	−1	1		10

(a) Identify the CPF solution obtained at iteration 1.
(b) Identify the constraint boundary equations that define this CPF solution.

5.1-5. Consider the three-variable linear programming problem shown in Fig. 5.2.
(a) Construct a table like Table 5.1, giving the set of defining equations for each CPF solution.
(b) What are the defining equations for the corner-point infeasible solution (6, 0, 5)?
(c) Identify one of the systems of three constraint boundary equations that yields neither a CPF solution nor a corner-point infeasible solution. Explain why this occurs for this system.

5.1-6. Consider the following problem.

Minimize $Z = 8x_1 + 5x_2$,

subject to

$$-3x_1 + 2x_2 \leq 30$$
$$2x_1 + x_2 \geq 50$$
$$x_1 + x_2 \geq 30$$

and

$$x_1 \geq 0, \qquad x_2 \geq 0.$$

(a) Identify the 10 sets of defining equations for this problem. For each one, solve (if a solution exists) for the corresponding corner-point solution, and classify it as a CPF solution or a corner-point infeasible solution.
(b) For each corner-point solution, give the corresponding basic solution and its set of nonbasic variables.

5.1-7. Reconsider the model in Prob. 3.1-5.
(a) Identify the 15 sets of defining equations for this problem. For each one, solve (if a solution exists) for the corresponding corner-point solution, and classify it as a CPF solution or a corner-point infeasible solution.
(b) For each corner-point solution, give the corresponding basic solution and its set of nonbasic variables.

5.1-8. Each of the following statements is true under most circumstances, but not always. In each case, indicate when the statement will not be true and why.
(a) The best CPF solution is an optimal solution.
(b) An optimal solution is a CPF solution.
(c) A CPF solution is the only optimal solution if none of its adjacent CPF solutions are better (as measured by the value of the objective function).

5.1-9. Consider the original form (before augmenting) of a linear programming problem with n decision variables (each with a nonnegativity constraint) and m functional constraints. Label each of the following statements as true or false, and then justify your answer with specific references (including page citations) to material in the chapter.
(a) If a feasible solution is optimal, it must be a CPF solution.

and

$$x_j \geq 0, \quad j = 1, 2, 3, 4, 5.$$

5.2-3. Reconsider Prob. 5.1-1. For the sequence of CPF solutions identified in part (e), construct the basis matrix **B** for each of the corresponding BF solutions. For each one, invert **B** manually, use this $\mathbf{B}^{-1}$ to calculate the current solution, and then perform the next iteration (or demonstrate that the current solution is optimal).

I **5.2-4.** Work through the matrix form of the simplex method step by step to solve the model given in Prob. 4.1-5.

I **5.2-5.** Work through the matrix form of the simplex method step by step to solve the model given in Prob. 4.7-6.

D **5.3-1.*** Consider the following problem.

Maximize $Z = x_1 - x_2 + 2x_3,$

subject to

$$2x_1 - 2x_2 + 3x_3 \leq 5$$
$$x_1 + x_2 - x_3 \leq 3$$
$$x_1 - x_2 + x_3 \leq 2$$

and

$$x_1 \geq 0, \quad x_2 \geq 0, \quad x_3 \geq 0.$$

Let x_4, x_5, and x_6 denote the slack variables for the respective constraints. After you apply the simplex method, a portion of the final simplex tableau is as follows:

Basic Variable	Eq.	Z	x_1	x_2	x_3	x_4	x_5	x_6	Right Side
						Coefficient of:			
Z	(0)	1				1	1	0	
x_2	(1)	0				1	3	0	
x_6	(2)	0				0	1	1	
x_3	(3)	0				1	2	0	

(a) Use the fundamental insight presented in Sec. 5.3 to identify the missing numbers in the final simplex tableau. Show your calculations.
(b) Identify the defining equations of the CPF solution corresponding to the optimal BF solution in the final simplex tableau.

D **5.3-2.** Consider the following problem.

Maximize $Z = 4x_1 + 3x_2 + x_3 + 2x_4,$

subject to

$$4x_1 + 2x_2 + x_3 + x_4 \leq 5$$
$$3x_1 + x_2 + 2x_3 + x_4 \leq 4$$

and

$$x_1 \geq 0, \quad x_2 \geq 0, \quad x_3 \geq 0, \quad x_4 \geq 0.$$

Let x_5 and x_6 denote the slack variables for the respective constraints. After you apply the simplex method, a portion of the final simplex tableau is as follows:

Basic Variable	Eq.	Z	x_1	x_2	x_3	x_4	x_5	x_6	Right Side
						Coefficient of:			
Z	(0)	1					1	1	
x_2	(1)	0					1	-1	
x_4	(2)	0					-1	2	

(a) Use the fundamental insight presented in Sec. 5.3 to identify the missing numbers in the final simplex tableau. Show your calculations.
(b) Identify the defining equations of the CPF solution corresponding to the optimal BF solution in the final simplex tableau.

D **5.3-3.** Consider the following problem.

Maximize $Z = 6x_1 + x_2 + 2x_3,$

subject to

$$2x_1 + 2x_2 + \frac{1}{2}x_3 \leq 2$$
$$-4x_1 - 2x_2 - \frac{3}{2}x_3 \leq 3$$
$$x_1 + 2x_2 + \frac{1}{2}x_3 \leq 1$$

and

$$x_1 \geq 0, \quad x_2 \geq 0, \quad x_3 \geq 0.$$

Let x_4, x_5, and x_6 denote the slack variables for the respective constraints. After you apply the simplex method, a portion of the final simplex tableau is as follows:

Basic Variable	Eq.	Z	x_1	x_2	x_3	x_4	x_5	x_6	Right Side
						Coefficient of:			
Z	(0)	1				2	0	2	
x_5	(1)	0				1	1	2	
x_3	(2)	0				-2	0	4	
x_1	(3)	0				1	0	-1	

Use the fundamental insight presented in Sec. 5.3 to identify the missing numbers in the final simplex tableau. Show your calculations.

D **5.3-4.** Consider the following problem.

$$\text{Maximize} \quad Z = 20x_1 + 6x_2 + 8x_3,$$

subject to

$$
\begin{aligned}
8x_1 + 2x_2 + 3x_3 &\leq 200 \\
4x_1 + 3x_2 &\leq 100 \\
2x_1 \qquad + x_3 &\leq 50 \\
x_3 &\leq 20
\end{aligned}
$$

and

$$x_1 \geq 0, \qquad x_2 \geq 0, \qquad x_3 \geq 0.$$

Let x_4, x_5, x_6, and x_7 denote the slack variables for the first through fourth constraints, respectively. Suppose that after some number of iterations of the simplex method, a portion of the current simplex tableau is as follows:

Basic Variable	Eq.	Z	x_1	x_2	x_3	x_4	x_5	x_6	x_7	Right Side
Z	(0)	1				$\frac{9}{4}$	$\frac{1}{2}$	0	0	
x_1	(1)	0				$\frac{3}{16}$	$-\frac{1}{8}$	0	0	
x_2	(2)	0				$-\frac{1}{4}$	$\frac{1}{2}$	0	0	
x_6	(3)	0				$-\frac{3}{8}$	$\frac{1}{4}$	1	0	
x_7	(4)	0				0	0	0	1	

(a) Use the fundamental insight presented in Sec. 5.3 to identify the missing numbers in the current simplex tableau. Show your calculations.
(b) Indicate which of these missing numbers would be generated by the matrix form of the simplex method to perform the next iteration.
(c) Identify the defining equations of the CPF solution corresponding to the BF solution in the current simplex tableau.

D **5.3-5.** Consider the following problem.

$$\text{Maximize} \quad Z = c_1x_1 + c_2x_2 + c_3x_3,$$

subject to

$$
\begin{aligned}
x_1 + 2x_2 + x_3 &\leq b \\
2x_1 + x_2 + 3x_3 &\leq 2b
\end{aligned}
$$

and

$$x_1 \geq 0, \qquad x_2 \geq 0, \qquad x_3 \geq 0.$$

Note that values have not been assigned to the coefficients in the objective function (c_1, c_2, c_3), and that the only specification for the right-hand side of the functional constraints is that the second one ($2b$) be twice as large as the first (b).

Now suppose that your boss has inserted her best estimate of the values of c_1, c_2, c_3, and b without informing you and then has run the simplex method. You are given the resulting final simplex tableau below (where x_4 and x_5 are the slack variables for the respective functional constraints), but you are unable to read the value of Z^*.

Basic Variable	Eq.	Z	x_1	x_2	x_3	x_4	x_5	Right Side
					Coefficient of:			
Z	(0)	1	$\frac{7}{10}$	0	0	$\frac{3}{5}$	$\frac{4}{5}$	Z^*
x_2	(1)	0	$\frac{1}{5}$	1	0	$\frac{3}{5}$	$-\frac{1}{5}$	1
x_3	(2)	0	$\frac{3}{5}$	0	1	$-\frac{1}{5}$	$\frac{2}{5}$	3

(a) Use the fundamental insight presented in Sec. 5.3 to identify the value of (c_1, c_2, c_3) that was used.
(b) Use the fundamental insight presented in Sec. 5.3 to identify the value of b that was used.
(c) Calculate the value of Z^* in two ways, where one way uses your results from part (a) and the other way uses your result from part (b). Show your two methods for finding Z^*.

5.3-6. For iteration 2 of the example in Sec. 5.3, the following expression was shown:

$$\text{Final row } 0 = [-3, \quad -5 \,|\, 0, \quad 0, \quad 0 \,|\, 0]$$

$$
+ [0, \tfrac{3}{2}, 1] \begin{bmatrix} 1 & 0 & 1 & 0 & 0 & 4 \\ 0 & 2 & 0 & 1 & 0 & 12 \\ 3 & 2 & 0 & 0 & 1 & 18 \end{bmatrix}.
$$

Derive this expression by combining the algebraic operations (in matrix form) for iterations 1 and 2 that affect row 0.

5.3-7. Most of the description of the fundamental insight presented in Sec. 5.3 assumes that the problem is in our standard form. Now consider each of the following other forms, where the additional adjustments in the initialization step are those presented in Sec. 4.6, including the use of artificial variables and the Big M method where appropriate. Describe the resulting adjustments in the fundamental insight.
(a) Equality constraints
(b) Functional constraints in $\geq$ form
(c) Negative right-hand sides
(d) Variables allowed to be negative (with no lower bound)

■ 6.1   THE ESSENCE OF DUALITY THEORY

Given our standard form for the *primal problem* at the left (perhaps after conversion from another form), its *dual problem* has the form shown to the right.

Primal Problem	Dual Problem
Maximize $\quad Z = \sum_{j=1}^{n} c_j x_j,$	Minimize $\quad W = \sum_{i=1}^{m} b_i y_i,$
subject to	subject to
$\sum_{j=1}^{n} a_{ij} x_j \leq b_i, \quad$ for $i = 1, 2, \ldots, m$	$\sum_{i=1}^{m} a_{ij} y_i \geq c_j, \quad$ for $j = 1, 2, \ldots, n$
and	and
$x_j \geq 0, \quad$ for $j = 1, 2, \ldots, n.$	$y_i \geq 0, \quad$ for $i = 1, 2, \ldots, m.$

Thus, with the primal problem in *maximization* form, the dual problem is in *minimization* form instead. Furthermore, the dual problem uses exactly the same *parameters* as the primal problem, but in different locations, as summarized below.

1. The coefficients in the objective function of the primal problem are the *right-hand sides* of the functional constraints in the dual problem.
2. The right-hand sides of the functional constraints in the primal problem are the coefficients in the objective function of the dual problem.
3. The coefficients of a variable in the functional constraints of the primal problem are the coefficients in a functional constraint of the dual problem.

To highlight the comparison, now look at these same two problems in matrix notation (as introduced at the beginning of Sec. 5.2), where **c** and $\mathbf{y} = [y_1, y_2, \ldots, y_m]$ are row vectors but **b** and **x** are column vectors.

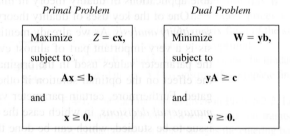

Primal Problem	Dual Problem
Maximize $\quad Z = \mathbf{cx},$	Minimize $\quad W = \mathbf{yb},$
subject to	subject to
$\mathbf{Ax} \leq \mathbf{b}$	$\mathbf{yA} \geq \mathbf{c}$
and	and
$\mathbf{x} \geq \mathbf{0}.$	$\mathbf{y} \geq \mathbf{0}.$

To illustrate, the primal and dual problems for the Wyndor Glass Co. example of Sec. 3.1 are shown in Table 6.1 in both algebraic and matrix form.

The **primal-dual table** for linear programming (Table 6.2) also helps to highlight the correspondence between the two problems. It shows all the linear programming parameters (the a_{ij}, b_i, and c_j) and how they are used to construct the two problems. All the headings for the primal problem are horizontal, whereas the headings for the dual problem are read by turning the book sideways. For the primal problem, each *column* (except the Right Side column) gives the coefficients of a single variable in the respective constraints and then in the objective function, whereas each *row* (except the bottom one) gives the parameters for a single contraint. For the dual problem, each *row* (except the Right Side row) gives the coefficients of a single variable in the respective constraints and then in the objective function, whereas each *column* (except the rightmost one) gives

■ **TABLE 6.1** Primal and dual problems for the Wyndor Glass Co. example

Primal Problem in Algebraic Form

Maximize $\quad Z = 3x_1 + 5x_2,$

subject to

$$
\begin{aligned}
x_1 \qquad &\leq 4 \\
2x_2 &\leq 12 \\
3x_1 + 2x_2 &\leq 18
\end{aligned}
$$

and $\quad x_1 \geq 0, \qquad x_2 \geq 0.$

Dual Problem in Algebraic Form

Minimize $\quad W = 4y_1 + 12y_2 + 18y_3,$

subject to

$$
\begin{aligned}
y_1 \qquad + 3y_3 &\geq 3 \\
2y_2 + 2y_3 &\geq 5
\end{aligned}
$$

and

$$y_1 \geq 0, \qquad y_2 \geq 0, \qquad y_3 \geq 0.$$

Primal Problem in Matrix Form

Maximize $\quad Z = [3, 5]\begin{bmatrix} x_1 \\ x_2 \end{bmatrix},$

subject to

$$
\begin{bmatrix} 1 & 0 \\ 0 & 2 \\ 3 & 2 \end{bmatrix} \begin{bmatrix} x_1 \\ x_2 \end{bmatrix} \leq \begin{bmatrix} 4 \\ 12 \\ 18 \end{bmatrix}
$$

and

$$
\begin{bmatrix} x_1 \\ x_2 \end{bmatrix} \geq \begin{bmatrix} 0 \\ 0 \end{bmatrix}.
$$

Dual Problem in Matrix Form

Minimize $\quad W = [y_1, y_2, y_3]\begin{bmatrix} 4 \\ 12 \\ 18 \end{bmatrix}$

subject to

$$
[y_1, y_2, y_3]\begin{bmatrix} 1 & 0 \\ 0 & 2 \\ 3 & 2 \end{bmatrix} \geq [3, 5]
$$

and

$$[y_1, y_2, y_3] \geq [0, 0, 0].$$

■ **TABLE 6.2** Primal-dual table for linear programming, illustrated by the Wyndor Glass Co. example

(a) General Case

			Primal Problem					
			Coefficient of:				Right Side	
			x_1	x_2	$\cdots$	x_n		
Dual Problem	Coefficient of:	y_1	a_{11}	a_{12}	$\cdots$	a_{1n}	$\leq b_1$	Coefficients for Objective Function (Minimize)
		y_2	a_{21}	a_{22}	$\cdots$	a_{2n}	$\leq b_2$	
		$\vdots$					$\vdots$	
		y_m	a_{m1}	a_{m2}	$\cdots$	a_{mn}	$\leq b_m$	
	Right Side		VI c_1	VI c_2	$\cdots$	VI c_n		
			Coefficients for Objective Function (Maximize)					

(b) Wyndor Glass Co. Example

	x_1	x_2	
y_1	1	0	≤ 4
y_2	0	2	≤ 12
y_3	3	2	≤ 18
	VI 3	VI 5	

To illustrate, the left-hand side of Table 6.5 shows row 0 for the respective iterations when the simplex method is applied to the Wyndor Glass Co. example. In each case, row 0 is partitioned into three parts: the coefficients of the decision variables (x_1, x_2), the coefficients of the slack variables (x_3, x_4, x_5), and the right-hand side (value of Z). Since the coefficients of the slack variables give the corresponding values of the dual variables (y_1, y_2, y_3), each row 0 identifies a corresponding solution for the dual problem, as shown in the y_1, y_2, and y_3 columns of Table 6.5. To interpret the next two columns, recall that $(z_1 - c_1)$ and $(z_2 - c_2)$ are the surplus variables for the functional constraints in the dual problem, so the full dual problem after augmenting with these surplus variables is

Minimize $W = 4y_1 + 12y_2 + 18y_3$,

subject to

$$y_1 \quad + 3y_3 - (z_1 - c_1) = 3$$
$$2y_2 + 2y_3 - (z_2 - c_2) = 5$$

and

$$y_1 \geq 0, \qquad y_2 \geq 0, \qquad y_3 \geq 0.$$

Therefore, by using the numbers in the y_1, y_2, and y_3 columns, the values of these surplus variables can be calculated as

$$z_1 - c_1 = y_1 + 3y_3 - 3,$$
$$z_2 - c_2 = 2y_2 + 2y_3 - 5.$$

Thus, a negative value for either surplus variable indicates that the corresponding constraint is violated. Also included in the rightmost column of the table is the calculated value of the dual objective function $W = 4y_1 + 12y_2 + 18y_3$.

As displayed in Table 6.4, *all* these quantities to the right of row 0 in Table 6.5 already are identified by row 0 without requiring any new calculations. In particular, note in Table 6.5 how *each* number obtained for the dual problem already appears in row 0 in the spot indicated by Table 6.4.

For the initial row 0, Table 6.5 shows that the corresponding dual solution $(y_1, y_2, y_3) = (0, 0, 0)$ is infeasible because both surplus variables are negative. The first iteration succeeds in eliminating one of these negative values, but not the other. After two iterations, the optimality test is satisfied for the primal problem because all the dual variables and surplus variables are nonnegative. This dual solution $(y_1^*, y_2^*, y_3^*) = (0, \frac{3}{2}, 1)$ is optimal (as could be verified by applying the simplex method directly to the dual problem), so the optimal value of Z and W is $Z^* = 36 = W^*$.

■ **TABLE 6.5** Row 0 and corresponding dual solution for each iteration
for the Wyndor Glass Co. example

Primal Problem			Dual Problem					
Iteration	**Row 0**		y_1	y_2	y_3	$z_1 - c_1$	$z_2 - c_2$	**W**
0	$[-3,$ -5 $0,$ $0,$ 0 $0]$		0	0	0	-3	-5	0
1	$[-3,$ 0 $0,$ $\frac{5}{2},$ 0 $30]$		0	$\frac{5}{2}$	0	-3	0	30
2	$[\;0,$ 0 $0,$ $\frac{3}{2},$ 1 $36]$		0	$\frac{3}{2}$	1	0	0	36

Summary of Primal-Dual Relationships

Now let us summarize the newly discovered key relationships between the primal and dual problems.

Weak duality property: If $\mathbf{x}$ is a feasible solution for the primal problem and $\mathbf{y}$ is a feasible solution for the dual problem, then

$$\mathbf{cx} \leq \mathbf{yb}.$$

For example, for the Wyndor Glass Co. problem, one feasible solution is $x_1 = 3$, $x_2 = 3$, which yields $Z = \mathbf{cx} = 24$, and one feasible solution for the dual problem is $y_1 = 1$, $y_2 = 1$, $y_3 = 2$, which yields a larger objective function value $W = \mathbf{yb} = 52$. These are just sample feasible solutions for the two problems. For *any* such pair of feasible solutions, this inequality must hold because the *maximum* feasible value of $Z = \mathbf{cx}$ (36) *equals* the *minimum* feasible value of the dual objective function $W = \mathbf{yb}$, which is our next property.

Strong duality property: If $\mathbf{x}^*$ is an optimal solution for the primal problem and $\mathbf{y}^*$ is an optimal solution for the dual problem, then

$$\mathbf{cx}^* = \mathbf{y}^*\mathbf{b}.$$

Thus, these two properties imply that $\mathbf{cx} < \mathbf{yb}$ for feasible solutions if one or both of them are *not optimal* for their respective problems, whereas equality holds when both are optimal.

The *weak duality property* describes the relationship between any pair of solutions for the primal and dual problems where *both* solutions are *feasible* for their respective problems. At each iteration, the simplex method finds a specific pair of solutions for the two problems, where the primal solution is feasible but the dual solution is *not feasible* (except at the final iteration). Our next property describes this situation and the relationship between this pair of solutions.

△ **Complementary solutions property:** At each iteration, the simplex method simultaneously identifies a CPF solution $\mathbf{x}$ for the primal problem and a **complementary solution** $\mathbf{y}$ for the dual problem (found in row 0, the coefficients of the slack variables), where

$$\mathbf{cx} = \mathbf{yb}.$$

If $\mathbf{x}$ is *not optimal* for the primal problem, then $\mathbf{y}$ is *not feasible* for the dual problem.

To illustrate, after one iteration for the Wyndor Glass Co. problem, $x_1 = 0$, $x_2 = 6$, and $y_1 = 0$, $y_2 = \frac{5}{2}$, $y_3 = 0$, with $\mathbf{cx} = 30 = \mathbf{yb}$. This $\mathbf{x}$ is feasible for the primal problem, but this $\mathbf{y}$ is not feasible for the dual problem (since it violates the constraint, $y_1 + 3y_3 \geq 3$).

The complementary solutions property also holds at the final iteration of the simplex method, where an optimal solution is found for the primal problem. However, more can be said about the complementary solution $\mathbf{y}$ in this case, as presented in the next property.

△ **Complementary optimal solutions property:** At the final iteration, the simplex method simultaneously identifies an optimal solution $\mathbf{x}^*$ for the primal problem and a **complementary optimal solution** $\mathbf{y}^*$ for the dual problem (found in row 0, the coefficients of the slack variables), where

$$\mathbf{cx}^* = \mathbf{y}^*\mathbf{b}.$$

The y_i^* are the shadow prices for the primal problem.

For the example, the final iteration yields $x_1^* = 2$, $x_2^* = 6$, and $y_1^* = 0$, $y_2^* = \frac{3}{2}$, $y_3^* = 1$, with $\mathbf{cx}^* = 36 = \mathbf{y}^*\mathbf{b}$.

We shall take a closer look at some of these properties in Sec. 6.3. There you will see that the complementary solutions property can be extended considerably further. In particular, after slack and surplus variables are introduced to augment the respective problems, every *basic* solution in the primal problem has a complementary *basic* solution in the dual problem. We already have noted that the simplex method identifies the values of the surplus variables for the dual problem as $z_j - c_j$ in Table 6.4. This result then leads to an additional *complementary slackness property* that relates the basic variables in one problem to the nonbasic variables in the other (Tables 6.7 and 6.8), but more about that later.

In Sec. 6.4, after describing how to construct the dual problem when the primal problem is *not* in our standard form, we discuss another very useful property, which is summarized as follows:

> **Symmetry property:** For *any* primal problem and its dual problem, all relationships between them must be *symmetric* because the dual of this dual problem is this primal problem.

Therefore, all the preceding properties hold regardless of which of the two problems is labeled as the primal problem. (The direction of the inequality for the weak duality property does require that the primal problem be expressed or reexpressed in maximization form and the dual problem in minimization form.) Consequently, the simplex method can be applied to either problem, and it simultaneously will identify complementary solutions (ultimately a complementary optimal solution) for the other problem.

So far, we have focused on the relationships between *feasible* or *optimal* solutions in the primal problem and corresponding solutions in the dual problem. However, it is possible that the primal (or dual) problem either has *no feasible solutions* or has feasible solutions but *no optimal solution* (because the objective function is unbounded). Our final property summarizes the primal-dual relationships under all these possibilities.

> **Duality theorem:** The following are the only possible relationships between the primal and dual problems.
>
> 1. If one problem has *feasible solutions* and a *bounded* objective function (and so has an optimal solution), then so does the other problem, so both the weak and strong duality properties are applicable.
> 2. If one problem has *feasible solutions* and an *unbounded* objective function (and so *no optimal solution*), then the other problem has *no feasible solutions.*
> 3. If one problem has *no feasible solutions,* then the other problem has either *no feasible solutions* or an *unbounded* objective function.

Applications

As we have just implied, one important application of duality theory is that the *dual* problem can be solved directly by the simplex method in order to identify an optimal solution for the primal problem. We discussed in Sec. 4.8 that the number of functional constraints affects the computational effort of the simplex method far more than the number of variables does. If $m > n$, so that the dual problem has fewer functional constraints (n) than the primal problem (m), then applying the simplex method directly to the dual problem instead of the primal problem probably will achieve a substantial reduction in computational effort.

The *weak* and *strong duality properties* describe key relationships between the primal and dual problems. One useful application is for evaluating a proposed solution for the primal problem. For example, suppose that **x** is a feasible solution that has been proposed for implementation and that a feasible solution **y** has been found by inspection for the dual problem such that $\mathbf{cx} = \mathbf{yb}$. In this case, $\mathbf{x}$ must be *optimal* without the simplex method even being applied! Even if $\mathbf{cx} < \mathbf{yb}$, then $\mathbf{yb}$ still provides an upper bound on the optimal value of Z, so if $\mathbf{yb} - \mathbf{cx}$ is small, intangible factors favoring $\mathbf{x}$ may lead to its selection without further ado.

One of the key applications of the complementary solutions property is its use in the dual simplex method presented in Sec. 7.1. This algorithm operates on the primal problem exactly as if the simplex method were being applied simultaneously to the dual problem, which can be done because of this property. Because the roles of row 0 and the right side in the simplex tableau have been reversed, the dual simplex method requires that row 0 *begin and remain nonnegative* while the right side *begins* with some *negative* values (subsequent iterations strive to reach a nonnegative right side). Consequently, this algorithm occasionally is used because it is more convenient to set up the initial tableau in this form than in the form required by the simplex method. Furthermore, it frequently is used for reoptimization (discussed in Sec. 4.7), because changes in the original model lead to the revised final tableau fitting this form. This situation is common for certain types of sensitivity analysis, as you will see later in the chapter.

In general terms, duality theory plays a central role in sensitivity analysis. This role is the topic of Sec. 6.5.

Another important application is its use in the economic interpretation of the dual problem and the resulting insights for analyzing the primal problem. You already have seen one example when we discussed shadow prices in Sec. 4.7. Section 6.2 describes how this interpretation extends to the entire dual problem and then to the simplex method.

6.2 ECONOMIC INTERPRETATION OF DUALITY

The economic interpretation of duality is based directly upon the typical interpretation for the primal problem (linear programming problem in our standard form) presented in Sec. 3.2. To refresh your memory, we have summarized this interpretation of the primal problem in Table 6.6.

Interpretation of the Dual Problem

To see how this interpretation of the primal problem leads to an economic interpretation for the dual problem,[1] note in Table 6.4 that W is the value of Z (total profit) at the current iteration. Because

$$W = b_1 y_1 + b_2 y_2 + \cdots + b_m y_m,$$

■ **TABLE 6.6** Economic interpretation of the primal problem

Quantity	Interpretation
x_j	Level of activity j $\quad (j = 1, 2, \ldots, n)$
c_j	Unit profit from activity j
Z	Total profit from all activities
b_i	Amount of resource i available $\quad (i = 1, 2, \ldots, m)$
a_{ij}	Amount of resource i consumed by each unit of activity j

[1]Actually, several slightly different interpretations have been proposed. The one presented here seems to us to be the most useful because it also directly interprets what the simplex method does in the primal problem.

each $b_i y_i$ can thereby be interpreted as the current *contribution to profit* by having b_i units of resource i available for the primal problem. Thus,

> The dual variable y_i is interpreted as the contribution to profit per unit of resource i $(i = 1, 2, \ldots, m)$, when the current set of basic variables is used to obtain the primal solution.

In other words, the y_i values (or y_i^* values in the optimal solution) are just the **shadow prices** discussed in Sec. 4.7.

For example, when iteration 2 of the simplex method finds the optimal solution for the Wyndor problem, it also finds the optimal values of the dual variables (as shown in the bottom row of Table 6.5) to be $y_1^* = 0$, $y_2^* = \frac{3}{2}$, and $y_3^* = 1$. These are precisely the shadow prices found in Sec. 4.7 for this problem through graphical analysis. Recall that the resources for the Wyndor problem are the production capacities of the three plants being made available to the two new products under consideration, so that b_i is the number of hours of production time per week being made available in Plant i for these new products, where $i = 1, 2, 3$. As discussed in Sec. 4.7, the shadow prices indicate that individually increasing any b_i by 1 would increase the optimal value of the objective function (total weekly profit in units of thousands of dollars) by y_i^*. Thus, y_i^* can be interpreted as the contribution to profit per unit of resource i when using the optimal solution.

This interpretation of the dual variables leads to our interpretation of the overall dual problem. Specifically, since each unit of activity j in the primal problem consumes a_{ij} units of resource i,

> $\Sigma_{i=1}^{m} a_{ij} y_i$ is interpreted as the current contribution to profit of the mix of resources that would be consumed if 1 unit of activity j were used $(j = 1, 2, \ldots, n)$.

For the Wyndor problem, 1 unit of activity j corresponds to producing 1 batch of product j per week, where $j = 1, 2$. The mix of resources consumed by producing 1 batch of product 1 is 1 hour of production time in Plant 1 and 3 hours in Plant 3. The corresponding mix per batch of product 2 is 2 hours each in Plants 2 and 3. Thus, $y_1 + 3y_3$ and $2y_2 + 2y_3$ are interpreted as the current contributions to profit (in thousands of dollars per week) of these respective mixes of resources per batch produced per week of the respective products.

For each activity j, this same mix of resources (and more) probably can be used in other ways as well, but no alternative use should be considered if it is less profitable than 1 unit of activity j. Since c_j is interpreted as the unit profit from activity j, each functional constraint in the dual problem is interpreted as follows:

> $\Sigma_{i=1}^{m} a_{ij} y_i \geq c_j$ says that the actual contribution to profit of the above mix of resources must be at least as much as if they were used by 1 unit of activity j; otherwise, we would not be making the best possible use of these resources.

For the Wyndor problem, the unit profits (in thousands of dollars per week) are $c_1 = 3$ and $c_2 = 5$, so the dual functional constraints with this interpretation are $y_1 + 3y_3 \geq 3$ and $2y_2 + 2y_3 \geq 5$. Similarly, the interpretation of the nonnegativity constraints is the following:

> $y_i \geq 0$ says that the contribution to profit of resource i $(i = 1, 2, \ldots, m)$ must be nonnegative: otherwise, it would be better not to use this resource at all.

The objective

$$\text{Minimize} \quad W = \sum_{i=1}^{m} b_i y_i$$

can be viewed as minimizing the total implicit value of the resources consumed by the activities. For the Wyndor problem, the total implicit value (in thousands of dollars per week) of the resources consumed by the two products is $W = 4y_1 + 12y_2 + 18y_3$.

This interpretation can be sharpened somewhat by differentiating between basic and nonbasic variables in the primal problem for any given BF solution $(x_1, x_2, \ldots, x_{n+m})$. Recall that the *basic* variables (the only variables whose values can be nonzero) *always* have a coefficient of *zero* in row 0. Therefore, referring again to Table 6.4 and the accompanying equation for z_j, we see that

$$\sum_{i=1}^{m} a_{ij}y_i = c_j, \quad \text{if } x_j > 0 \quad (j = 1, 2, \ldots, n),$$

$$y_i = 0, \quad \text{if } x_{n+i} > 0 \quad (i = 1, 2, \ldots, m).$$

(This is one version of the complementary slackness property discussed in Sec. 6.3.) The economic interpretation of the first statement is that whenever an activity j operates at a strictly positive level ($x_j > 0$), the marginal value of the resources it consumes *must equal* (as opposed to exceeding) the unit profit from this activity. The second statement implies that the marginal value of resource i is *zero* ($y_i = 0$) whenever the supply of this resource is not exhausted by the activities ($x_{n+i} > 0$). In economic terminology, such a resource is a "free good"; the price of goods that are oversupplied must drop to zero by the law of supply and demand. This fact is what justifies interpreting the objective for the dual problem as minimizing the total implicit value of the resources *consumed*, rather than the resources *allocated*.

To illustrate these two statements, consider the optimal BF solution (2, 6, 2, 0, 0) for the Wyndor problem. The basic variables are x_1, x_2, and x_3, so their coefficients in row 0 are zero, as shown in the bottom row of Table 6.5. This bottom row also gives the corresponding dual solution: $y_1^* = 0$, $y_2^* = \frac{3}{2}$, $y_3^* = 1$, with surplus variables $(z_1^* - c_1) = 0$ and $(z_2^* - c_2) = 0$. Since $x_1 > 0$ and $x_2 > 0$, both these surplus variables and direct calculations indicate that $y_1^* + 3y_3^* = c_1 = 3$ and $2y_2^* + 2y_3^* = c_2 = 5$. Therefore, the value of the resources consumed per batch of the respective products produced does indeed equal the respective unit profits. The slack variable for the constraint on the amount of Plant 1 capacity used is $x_3 > 0$, so the marginal value of adding any Plant 1 capacity would be zero ($y_1^* = 0$).

Interpretation of the Simplex Method

The interpretation of the dual problem also provides an economic interpretation of what the simplex method does in the primal problem. The *goal* of the simplex method is to find how to use the available resources in the most profitable feasible way. To attain this goal, we must reach a BF solution that satisfies all the *requirements* on profitable use of the resources (the constraints of the dual problem). These requirements comprise the *condition for optimality* for the algorithm. For any given BF solution, the requirements (dual constraints) associated with the basic variables are automatically satisfied (with equality). However, those associated with nonbasic variables may or may not be satisfied.

In particular, if an original variable x_j is nonbasic so that activity j is not used, then the current contribution to profit of the resources that would be required to undertake each unit of activity j

$$\sum_{i=1}^{m} a_{ij}y_i$$

may be smaller than, larger than, or equal to the unit profit c_j obtainable from the activity. If it is smaller, so that $z_j - c_j < 0$ in row 0 of the simplex tableau, then these resources can be used more profitably by initiating this activity. If it is larger ($z_j - c_j > 0$), then these resources already are being assigned elsewhere in a more profitable way, so they should not be diverted to activity j. If $z_j - c_j = 0$, there would be no change in profitability by initiating activity j.

Similarly, if a slack variable x_{n+i} is nonbasic so that the total allocation b_i of resource i is being used, then y_i is the current contribution to profit of this resource on a marginal basis. Hence, if $y_i < 0$, profit can be increased by cutting back on the use of this resource (i.e., increasing x_{n+i}). If $y_i > 0$, it is worthwhile to continue fully using this resource, whereas this decision does not affect profitability if $y_i = 0$.

Therefore, what the simplex method does is to examine all the nonbasic variables in the current BF solution to see which ones can provide a *more profitable use of the resources* by being increased. If *none* can, so that no feasible shifts or reductions in the current proposed use of the resources can increase profit, then the current solution must be optimal. If one or more can, the simplex method selects the variable that, if increased by 1, would *improve the profitability* of the use of the resources the most. It then actually increases this variable (the entering basic variable) as much as it can until the marginal values of the resources change. This increase results in a new BF solution with a new row 0 (dual solution), and the whole process is repeated.

The economic interpretation of the dual problem considerably expands our ability to analyze the primal problem. However, you already have seen in Sec. 6.1 that this interpretation is just one ramification of the relationships between the two problems. In Sec 6.3, we delve into these relationships more deeply.

6.3 PRIMAL–DUAL RELATIONSHIPS

Because the dual problem is a linear programming problem, it also has corner-point solutions. Furthermore, by using the augmented form of the problem, we can express these corner-point solutions as basic solutions. Because the functional constraints have the $\geq$ form, this augmented form is obtained by *subtracting* the surplus (rather than adding the slack) from the left-hand side of each constraint j ($j = 1, 2, \ldots, n$).[2] This surplus is

$$z_j - c_j = \sum_{i=1}^{m} a_{ij}y_i - c_j, \qquad \text{for } j = 1, 2, \ldots, n.$$

Thus, $z_j - c_j$ plays the role of the *surplus variable* for constraint j (or its slack variable if the constraint is multiplied through by -1). Therefore, augmenting each corner-point solution ($y_1, y_2, \ldots, y_m$) yields a basic solution ($y_1, y_2, \ldots, y_m, z_1 - c_1, z_2 - c_2, \ldots, z_n - c_n$) by using this expression for $z_j - c_j$. Since the augmented form of the dual problem has n functional constraints and $n + m$ variables, each basic solution has n basic variables and m nonbasic variables. (Note how m and n reverse their previous roles here because, as Table 6.3 indicates, dual constraints correspond to primal variables and dual variables correspond to primal constraints.)

---

[2]You might wonder why we do not also introduce *artificial variables* into these constraints as discussed in Sec. 4.6. The reason is that these variables have no purpose other than to change the feasible region temporarily as a convenience in starting the simplex method. We are not interested now in applying the simplex method to the dual problem, and we do not want to change its feasible region.

Complementary Basic Solutions

One of the important relationships between the primal and dual problems is a direct correspondence between their basic solutions. The key to this correspondence is row 0 of the simplex tableau for the primal basic solution, such as shown in Table 6.4 or 6.5. Such a row 0 can be obtained for *any* primal basic solution, feasible or not, by using the formulas given in the bottom part of Table 5.8.

Note again in Tables 6.4 and 6.5 how a complete solution for the dual problem (including the surplus variables) can be read directly from row 0. Thus, because of its coefficient in row 0, each variable in the primal problem has an associated variable in the dual problem, as summarized in Table 6.7, first for any problem and then for the Wyndor problem.

A key insight here is that the dual solution read from row 0 must also be a basic solution! The reason is that the m basic variables for the primal problem are required to have a coefficient of zero in row 0, which thereby requires the m associated dual variables to be zero, i.e., nonbasic variables for the dual problem. The values of the remaining n (basic) variables then will be the simultaneous solution to the system of equations given at the beginning of this section. In matrix form, this system of equations is $\mathbf{z} - \mathbf{c} = \mathbf{yA} - \mathbf{c}$, and the fundamental insight of Sec. 5.3 actually identifies its solution for $\mathbf{z} - \mathbf{c}$ and $\mathbf{y}$ as being the corresponding entries in row 0.

Because of the symmetry property quoted in Sec. 6.1 (and the direct association between variables shown in Table 6.7), the correspondence between basic solutions in the primal and dual problems is a symmetric one. Furthermore, a pair of complementary basic solutions has the same objective function value, shown as W in Table 6.4.

Let us now summarize our conclusions about the correspondence between primal and dual basic solutions, where the first property extends the complementary solutions property of Sec. 6.1 to the augmented forms of the two problems and then to any basic solution (feasible or not) in the primal problem.

△ **Complementary basic solutions property:** Each *basic* solution in the *primal problem* has a **complementary basic solution** in the *dual problem,* where their respective objective function values (Z and W) are equal. Given row 0 of the simplex tableau for the primal basic solution, the complementary dual basic solution $(\mathbf{y}, \mathbf{z} - \mathbf{c})$ is found as shown in Table 6.4.

The next property shows how to identify the basic and nonbasic variables in this complementary basic solution.

△ **Complementary slackness property:** Given the association between variables in Table 6.7, the variables in the primal basic solution and the complementary dual basic solution satisfy the **complementary slackness** relationship shown in Table 6.8. Furthermore, this relationship is a symmetric one, so that these two basic solutions are complementary to each other.

■ **TABLE 6.7** Association between variables in primal and dual problems

	Primal Variable	Associated Dual Variable
Any problem	(Decision variable) x_j (Slack variable) x_{n+i}	$z_j - c_j$ (surplus variable) $j = 1, 2, \ldots, n$ y_i (decision variable) $\quad i = 1, 2, \ldots, m$
Wyndor problem	Decision variables: x_1 $\quad\quad\quad\quad\quad\quad\quad x_2$ Slack variables: $\quad x_3$ $\quad\quad\quad\quad\quad\quad\quad x_4$ $\quad\quad\quad\quad\quad\quad\quad x_5$	$z_1 - c_1$ (surplus variables) $z_2 - c_2$ $y_1 \quad$ (decision variables) y_2 y_3

■ **TABLE 6.8** Complementary slackness relationship for complementary basic solutions

Primal Variable	Associated Dual Variable	
Basic	Nonbasic	(m variables)
Nonbasic	Basic	(n variables)

■ **TABLE 6.9** Complementary basic solutions for the Wyndor Glass Co. example

	Primal Problem			Dual Problem	
No.	Basic Solution	Feasible?	$Z = W$	Feasible?	Basic Solution
1	$(0, 0, 4, 12, 18)$	Yes	0	No	$(0, 0, 0, -3, -5)$
2	$(4, 0, 0, 12, 6)$	Yes	12	No	$(3, 0, 0, 0, -5)$
3	$(6, 0, -2, 12, 0)$	No	18	No	$(0, 0, 1, 0, -3)$
4	$(4, 3, 0, 6, 0)$	Yes	27	No	$\left(-\dfrac{9}{2}, 0, \dfrac{5}{2}, 0, 0\right)$
5	$(0, 6, 4, 0, 6)$	Yes	30	No	$\left(0, \dfrac{5}{2}, 0, -3, 0\right)$
6	$(2, 6, 2, 0, 0)$	Yes	36	Yes	$\left(0, \dfrac{3}{2}, 1, 0, 0\right)$
7	$(4, 6, 0, 0, -6)$	No	42	Yes	$\left(3, \dfrac{5}{2}, 0, 0, 0\right)$
8	$(0, 9, 4, -6, 0)$	No	45	Yes	$\left(0, 0, \dfrac{5}{2}, \dfrac{9}{2}, 0\right)$

The reason for using the name *complementary slackness* for this latter property is that it says (in part) that for each pair of associated variables, if one of them has *slack* in its nonnegativity constraint (a basic variable > 0), then the other one must have *no slack* (a nonbasic variable $= 0$). We mentioned in Sec. 6.2 that this property has a useful economic interpretation for linear programming problems.

Example.   To illustrate these two properties, again consider the Wyndor Glass Co. problem of Sec. 3.1. All eight of its basic solutions (five feasible and three infeasible) are shown in Table 6.9. Thus, its dual problem (see Table 6.1) also must have eight basic solutions, each complementary to one of these primal solutions, as shown in Table 6.9.

The three BF solutions obtained by the simplex method for the primal problem are the first, fifth, and sixth primal solutions shown in Table 6.9. You already saw in Table 6.5 how the complementary basic solutions for the dual problem can be read directly from row 0, starting with the coefficients of the slack variables and then the original variables. The other dual basic solutions also could be identified in this way by constructing row 0 for each of the other primal basic solutions, using the formulas given in the bottom part of Table 5.8.

Alternatively, for each primal basic solution, the complementary slackness property can be used to identify the basic and nonbasic variables for the complementary dual basic solution, so that the system of equations given at the beginning of the section can be solved directly to obtain this complementary solution. For example, consider the next-to-last primal basic solution in Table 6.9, $(4, 6, 0, 0, -6)$. Note that x_1, x_2, and x_5 are *basic variables*, since these variables are not equal to 0. Table 6.7 indicates that the associated dual variables are $(z_1 - c_1)$, $(z_2 - c_2)$, and y_3. Table 6.8 specifies that these associated dual variables are *nonbasic variables* in the complementary basic solution, so

$$z_1 - c_1 = 0, \qquad z_2 - c_2 = 0, \qquad y_3 = 0.$$

Consequently, the augmented form of the functional constraints in the dual problem,

$$y_1 \quad + 3y_3 - (z_1 - c_1) = 3$$
$$2y_2 + 2y_3 - (z_2 - c_2) = 5,$$

reduce to

$$y_1 \quad + 0 - 0 = 3$$
$$2y_2 + 0 - 0 = 5,$$

so that $y_1 = 3$ and $y_2 = \frac{5}{2}$. Combining these values with the values of 0 for the nonbasic variables gives the basic solution $(3, \frac{5}{2}, 0, 0, 0)$, shown in the rightmost column and next-to-last row of Table 6.9. Note that this dual solution is feasible for the dual problem because all five variables satisfy the nonnegativity constraints.

Finally, notice that Table 6.9 demonstrates that $(0, \frac{3}{2}, 1, 0, 0)$ is the optimal solution for the dual problem, because it is the basic *feasible* solution with minimal W (36).

Relationships between Complementary Basic Solutions

We now turn our attention to the relationships between complementary basic solutions, beginning with their *feasibility* relationships. The middle columns in Table 6.9 provide some valuable clues. For the pairs of complementary solutions, notice how the yes or no answers on feasibility also satisfy a complementary relationship in most cases. In particular, with one exception, whenever one solution is feasible, the other is not. (It also is possible for *neither* solution to be feasible, as happened with the third pair.) The one exception is the sixth pair, where the primal solution is known to be optimal. The explanation is suggested by the $Z = W$ column. Because the sixth dual solution also is optimal (by the complementary optimal solutions property), with $W = 36$, the first five dual solutions *cannot be feasible* because $W < 36$ (remember that the dual problem objective is to *minimize* W). By the same token, the last two primal solutions cannot be feasible because $Z > 36$.

This explanation is further supported by the strong duality property that optimal primal and dual solutions have $Z = W$.

Next, let us state the *extension* of the complementary optimal solutions property of Sec. 6.1 for the augmented forms of the two problems.

> **Complementary optimal basic solutions property:** An *optimal* basic solution in the *primal problem* has a **complementary optimal basic solution** in the dual problem, where their respective objective function values (Z and W) are equal. Given row 0 of the simplex tableau for the optimal primal solution, the complementary optimal dual solution $(\mathbf{y}^*, \mathbf{z}^* - \mathbf{c})$ is found as shown in Table 6.4.

To review the reasoning behind this property, note that the dual solution $(\mathbf{y}^*, \mathbf{z}^* - \mathbf{c})$ must be feasible for the dual problem because the condition for optimality for the primal problem requires that *all* these dual variables (including surplus variables) be *nonnegative*. Since this solution is *feasible,* it must be *optimal* for the dual problem by the weak duality property (since $W = Z$, so $\mathbf{y}^*\mathbf{b} = \mathbf{cx}^*$ where $\mathbf{x}^*$ is optimal for the primal problem).

Basic solutions can be classified according to whether they satisfy each of two conditions. One is the *condition for feasibility,* namely, whether *all* the variables (including slack variables) in the augmented solution are *nonnegative*. The other is the *condition for optimality,* namely, whether *all* the coefficients in row 0 (i.e., all the variables in the complementary basic solution) are *nonnegative*. Our names for the different types of basic solutions are summarized in Table 6.10. For example, in Table 6.9, primal basic

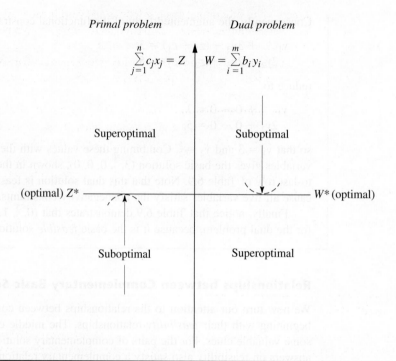

Primal problem   Dual problem

$$\sum_{j=1}^{n} c_j x_j = Z \qquad W = \sum_{i=1}^{m} b_i y_i$$

Superoptimal   Suboptimal

(optimal) Z^* ————————————————————— W^* (optimal)

Suboptimal   Superoptimal

■ **FIGURE 6.1**
Range of possible values of
$Z = W$ for certain types of
complementary basic
solutions.

■ **TABLE 6.10** Classification of basic solutions

		Satisfies Condition for Optimality?	
		Yes	**No**
Feasible?	**Yes**	Optimal	Suboptimal
	No	Superoptimal	Neither feasible nor superoptimal

■ **TABLE 6.11** Relationships between complementary basic solutions

Primal Basic Solution	Complementary Dual Basic Solution	Both Basic Solutions	
		Primal Feasible?	**Dual Feasible?**
Suboptimal	Superoptimal	Yes	No
Optimal	Optimal	Yes	Yes
Superoptimal	Suboptimal	No	Yes
Neither feasible nor superoptimal	Neither feasible nor superoptimal	No	No

solutions 1, 2, 4, and 5 are suboptimal, 6 is optimal, 7 and 8 are superoptimal, and 3 is neither feasible nor superoptimal.

Given these definitions, the general relationships between complementary basic solutions are summarized in Table 6.11. The resulting range of possible (common) values for the objective functions ($Z = W$) for the first three pairs given in Table 6.11 (the last pair can have any value) is shown in Fig. 6.1. Thus, while the simplex method is dealing directly with suboptimal basic solutions and working toward optimality in the primal problem, it is simultaneously dealing indirectly with complementary superoptimal solutions and working toward feasibility in the dual problem. Conversely, it sometimes is more convenient (or necessary) to work directly with superoptimal basic solutions and to move toward feasibility in the primal problem, which is the purpose of the dual simplex method described in Sec. 7.1.

The third and fourth columns of Table 6.11 introduce two other common terms that are used to describe a pair of complementary basic solutions. The two solutions are said to be **primal feasible** if the primal basic solution is feasible, whereas they are called **dual feasible** if the complementary dual basic solution is feasible for the dual problem. Using this terminology, the simplex method deals with primal feasible solutions and strives toward achieving dual feasibility as well. When this is achieved, the two complementary basic solutions are optimal for their respective problems.

These relationships prove very useful, particularly in sensitivity analysis, as you will see later in the chapter.

6.4 ADAPTING TO OTHER PRIMAL FORMS

Thus far it has been assumed that the model for the primal problem is in our standard form. However, we indicated at the beginning of the chapter that any linear programming problem, whether in our standard form or not, possesses a dual problem. Therefore, this section focuses on how the dual problem changes for other primal forms.

Each nonstandard form was discussed in Sec. 4.6, and we pointed out how it is possible to convert each one to an equivalent standard form if so desired. These conversions are summarized in Table 6.12. Hence, you always have the option of converting any model to our standard form and *then* constructing its dual problem in the usual way. To illustrate, we do this for our standard dual problem (it must have a dual also) in Table 6.13. Note that what we end up with is just our standard primal problem! Since any pair of primal and dual problems can be converted to these forms, this fact implies that the dual of the dual problem always is the primal problem. Therefore, for any primal problem and its dual problem, all relationships between them must be symmetric. This is just the symmetry property already stated in Sec. 6.1 (without proof), but now Table 6.13 demonstrates why it holds.

One consequence of the symmetry property is that all the statements made earlier in the chapter about the relationships of the dual problem to the primal problem also hold in reverse.

■ **TABLE 6.12** Conversions to standard form for linear programming models

Nonstandard Form	Equivalent Standard Form
Minimize $\quad Z$	Maximize $\quad (-Z)$
$\displaystyle\sum_{j=1}^{n} a_{ij}x_j \geq b_i$	$\displaystyle -\sum_{j=1}^{n} a_{ij}x_j \leq -b_i$
$\displaystyle\sum_{j=1}^{n} a_{ij}x_j = b_i$	$\displaystyle\sum_{j=1}^{n} a_{ij}x_j \leq b_i \quad$ and $\quad \displaystyle -\sum_{j=1}^{n} a_{ij}x_j \leq -b_i$
x_j unconstrained in sign	$x_j^+ - x_j^-, \quad x_j^+ \geq 0, \quad x_j^- \geq 0$

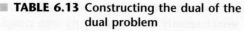

■ **TABLE 6.13** Constructing the dual of the dual problem

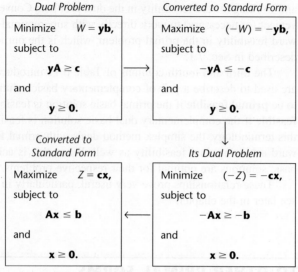

Another consequence is that it is immaterial which problem is called the primal and which is called the dual. In practice, you might see a linear programming problem fitting our standard form being referred to as the dual problem. The convention is that the model formulated to fit the actual problem is called the primal problem, regardless of its form.

Our illustration of how to construct the dual problem for a nonstandard primal problem did not involve either equality constraints or variables unconstrained in sign. Actually, for these two forms, a shortcut is available. It is possible to show (see Probs. 6.4-7 and 6.4-2a) that an *equality constraint* in the primal problem should be treated just like a $\leq$ constraint in constructing the dual problem except that the nonnegativity constraint for the corresponding dual variable should be *deleted* (i.e., this variable is unconstrained in sign). By the symmetry property, deleting a nonnegativity constraint in the primal problem affects the dual problem only by changing the corresponding inequality constraint to an equality constraint.

Another shortcut involves functional constraints in $\geq$ form for a maximization problem. The straightforward (but longer) approach would begin by converting each such constraint to $\leq$ form

$$\sum_{j=1}^{n} a_{ij}x_j \geq b_i \longrightarrow -\sum_{j=1}^{n} a_{ij}x_j \leq -b_i.$$

Constructing the dual problem in the usual way then gives $-a_{ij}$ as the coefficient of y_i in functional constraint j (which has $\geq$ form) and a coefficient of $-b_i$ in the objective function (which is to be minimized), where y_i also has a nonnegativity constraint $y_i \geq 0$. Now suppose we define a new variable $y_i' = -y_i$. The changes caused by expressing the dual problem in terms of y_i' instead of y_i are that (1) the coefficients of the variable become a_{ij} for functional constraint j and b_i for the objective function and (2) the constraint on the variable becomes $y_i' \leq 0$ (a *nonpositivity constraint*). The shortcut is to use y_i' instead of y_i as a dual variable so that the parameters in the original constraint (a_{ij} and b_i) immediately become the coefficients of this variable in the dual problem.

Here is a useful mnemonic device for remembering what the forms of dual constraints should be. With a maximization problem, it might seem *sensible* for a functional constraint to be in $\leq$ form, slightly *odd* to be in = form, and somewhat *bizarre* to be in $\geq$ form. Similarly, for a minimization problem, it might seem *sensible* to be in $\geq$ form, slightly *odd* to be in = form, and somewhat *bizarre* to be in $\leq$ form. For the constraint on an individual variable in either kind of problem, it might seem *sensible* to have a nonnegativity constraint, somewhat *odd* to have no constraint (so the variable is unconstrained in sign), and quite *bizarre* for the variable to be restricted to be *less* than or equal to zero. Now recall the correspondence between entities in the primal and dual problems indicated in Table 6.3; namely, functional constraint i in one problem corresponds to variable i in the other problem, and vice versa. The **sensible-odd-bizarre method,** or **SOB method** for short, says that the form of a functional constraint or the constraint on a variable in the dual problem should be sensible, odd, or bizarre, depending on whether the form for the corresponding entity in the primal problem is sensible, odd, or bizarre. Here is a summary.

The SOB Method for Determining the Form of Constraints in the Dual.[3]

1. Formulate the primal problem in either maximization form or minimization form, and then the dual problem automatically will be in the other form.
2. Label the different forms of functional constraints and of constraints on individual variables in the primal problem as being *sensible, odd,* or *bizarre* according to Table 6.14. The labeling of the functional constraints depends on whether the problem is a *maximization* problem (use the second column) or a *minimization* problem (use the third column).
3. For each constraint on an *individual variable* in the dual problem, use the form that has the same label as for the functional constraint in the primal problem that corresponds to this dual variable (as indicated by Table 6.3).
4. For each *functional constraint* in the dual problem, use the form that has the same label as for the constraint on the corresponding individual variable in the primal problem (as indicated by Table 6.3).

The arrows between the second and third columns of Table 6.14 spell out the correspondence between the forms of constraints in the primal and dual. Note that the correspondence always is between a functional constraint in one problem and a constraint on an individual variable in the other problem. Since the primal problem can be either a maximization or minimization problem, where the dual then will be of the opposite type, the second column of the table gives the form for whichever is the maximization problem and the third column gives the form for the other problem (a minimization problem).

To illustrate, consider the radiation therapy example presented at the beginning of Sec. 3.4. To show the conversion in both directions in Table 6.14, we begin with the maximization form of this model as the primal problem, before using the (original) minimization form.

The primal problem in maximization form is shown on the left side of Table 6.15. By using the second column of Table 6.14 to represent this problem, the arrows in this table indicate the form of the dual problem in the third column. These same arrows are used in Table 6.15 to show the resulting dual problem. (Because of these arrows, we have placed the functional constraints last in the dual problem rather than in their usual top position.)

---

[3]This particular mnemonic device (and a related one) for remembering what the forms of the dual constraints should be has been suggested by Arthur T. Benjamin, a mathematics professor at Harvey Mudd College. An interesting and wonderfully bizarre fact about Professor Benjamin himself is that he is one of the world's great human calculators who can perform such feats as quickly multiplying six-digit numbers in his head. For a further discussion and derivation of the SOB method, see A. T. Benjamin: "Sensible Rules for Remembering Duals — The S-O-B Method," *SIAM Review,* **37**(1): 85–87, 1995.

■ TABLE 6.14 Corresponding primal-dual forms

Label	Primal Problem (or Dual Problem) Maximize Z (or W)	Dual Problem (or Primal Problem) Minimize W (or Z)
	Constraint i:	Variable y_i (or x_i):
Sensible	≤ form ⟷	⟶ $y_i \geq 0$
Odd	= form ⟷	⟶ Unconstrained
Bizarre	≥ form ⟷	⟶ $y_i' \leq 0$
	Variable x_j (or y_j):	Constraint j:
Sensible	$x_j \geq 0$ ⟷	⟶ ≥ form
Odd	Unconstrained ⟷	⟶ = form
Bizarre	$x_j' \leq 0$ ⟷	⟶ ≤ form

■ TABLE 6.15 One primal-dual form for the radiation therapy example

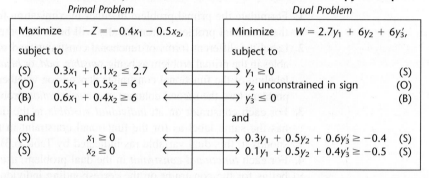

Primal Problem

Maximize $-Z = -0.4x_1 - 0.5x_2,$

subject to

(S) $0.3x_1 + 0.1x_2 \leq 2.7$
(O) $0.5x_1 + 0.5x_2 = 6$
(B) $0.6x_1 + 0.4x_2 \geq 6$

and

(S) $x_1 \geq 0$
(S) $x_2 \geq 0$

Dual Problem

Minimize $W = 2.7y_1 + 6y_2 + 6y_3',$

subject to

$y_1 \geq 0$ (S)
y_2 unconstrained in sign   (O)
$y_3' \leq 0$ (B)

and

$0.3y_1 + 0.5y_2 + 0.6y_3' \geq -0.4$ (S)
$0.1y_1 + 0.5y_2 + 0.4y_3' \geq -0.5$ (S)

Beside each constraint in both problems, we have inserted (in parentheses) an S, O, or B to label the form as sensible, odd, or bizarre. As prescribed by the SOB method, the label for each dual constraint always is the same as for the corresponding primal constraint.

However, there was no need (other than for illustrative purposes) to convert the primal problem to maximization form. Using the original minimization form, the equivalent primal problem is shown on the left side of Table 6.16. Now we use the *third column* of Table 6.14 to represent this primal problem, where the arrows indicate the form of the dual problem in the *second column*. These same arrows in Table 6.16 show the resulting dual problem on the right side. Again, the labels on the constraints show the application of the SOB method.

Just as the primal problems in Tables 6.15 and 6.16 are equivalent, the two dual problems also are completely equivalent. The key to recognizing this equivalency lies in the fact that the variables in each version of the dual problem are the negative of those in the other version ($y_1' = -y_1$, $y_2' = -y_2$, $y_3 = -y_3'$). Therefore, for each version, if the variables in the other version are used instead, and if both the objective function and the constraints are multiplied through by -1, then the other version is obtained. (Problem 6.4-5 asks you to verify this.)

If you would like to see **another example** of using the SOB method to construct a dual problem, one is given in the Worked Examples section of the book's website.

If the simplex method is to be applied to either a primal or a dual problem that has any variables constrained to be *nonpositive* (for example, $y_3' \leq 0$ in the dual problem of Table 6.15), this variable may be replaced by its *nonnegative* counterpart (for example, $y_3 = -y_3'$).

■ TABLE 6.16 The other primal-dual form for the radiation therapy example

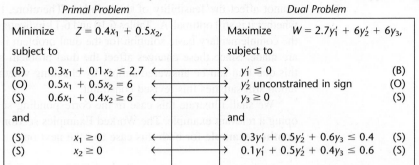

Primal Problem	Dual Problem
Minimize $\quad Z = 0.4x_1 + 0.5x_2,$	Maximize $\quad W = 2.7y_1' + 6y_2' + 6y_3,$
subject to	subject to
(B) $\quad 0.3x_1 + 0.1x_2 \leq 2.7 \quad \longleftrightarrow \quad$	$\longrightarrow \quad y_1' \leq 0 \quad$ (B)
(O) $\quad 0.5x_1 + 0.5x_2 = 6 \quad \longleftrightarrow \quad$	$\longrightarrow \quad y_2'$ unconstrained in sign $\quad$ (O)
(S) $\quad 0.6x_1 + 0.4x_2 \geq 6 \quad \longleftrightarrow \quad$	$\longrightarrow \quad y_3 \geq 0 \quad$ (S)
and	and
(S) $\qquad\qquad x_1 \geq 0 \quad \longleftrightarrow \quad$	$\longrightarrow \quad 0.3y_1' + 0.5y_2' + 0.6y_3 \leq 0.4 \quad$ (S)
(S) $\qquad\qquad x_2 \geq 0 \quad \longleftrightarrow \quad$	$\longrightarrow \quad 0.1y_1' + 0.5y_2' + 0.4y_3 \leq 0.6 \quad$ (S)

When artificial variables are used to help the simplex method solve a primal problem, the duality interpretation of row 0 of the simplex tableau is the following: Since artificial variables play the role of slack variables, their coefficients in row 0 now provide the values of the corresponding dual variables in the complementary basic solution for the dual problem. Since artificial variables are used to replace the real problem with a more convenient artificial problem, this dual problem actually is the dual of the artificial problem. However, after all the artificial variables become nonbasic, we are back to the real primal and dual problems. With the two-phase method, the artificial variables would need to be retained in phase 2 in order to read off the complete dual solution from row 0. With the Big M method, since M has been added initially to the coefficient of each artificial variable in row 0, the current value of each corresponding dual variable is the current coefficient of this artificial variable *minus M*.

For example, look at row 0 in the final simplex tableau for the radiation therapy example, given at the bottom of Table 4.12. After M is subtracted from the coefficients of the artificial variables $\bar{x}_4$ and $\bar{x}_6$, the optimal solution for the corresponding dual problem given in Table 6.15 is read from the coefficients of x_3, $\bar{x}_4$, and $\bar{x}_6$ as (y_1, y_2, y_3') = $(0.5, -1.1, 0)$. As usual, the surplus variables for the two functional constraints are read from the coefficients of x_1 and x_2 as $z_1 - c_1 = 0$ and $z_2 - c_2 = 0$.

■ 6.5 THE ROLE OF DUALITY THEORY IN SENSITIVITY ANALYSIS

As described further in the next three sections, sensitivity analysis basically involves investigating the effect on the optimal solution of making changes in the values of the model parameters a_{ij}, b_i, and c_j. However, changing parameter values in the primal problem also changes the corresponding values in the dual problem. Therefore, you have your choice of which problem to use to investigate each change. Because of the primal-dual relationships presented in Secs. 6.1 and 6.3 (especially the complementary basic solutions property), it is easy to move back and forth between the two problems as desired. In some cases, it is more convenient to analyze the dual problem directly in order to determine the complementary effect on the primal problem. We begin by considering two such cases.

Changes in the Coefficients of a Nonbasic Variable

Suppose that the changes made in the original model occur in the coefficients of a variable that was nonbasic in the original optimal solution. What is the effect of these changes on this solution? Is it still feasible? Is it still optimal?

Because the variable involved is nonbasic (value of zero), changing its coefficients cannot affect the feasibility of the solution. Therefore, the open question in this case is whether it is still optimal. As Tables 6.10 and 6.11 indicate, an equivalent question is whether the complementary basic solution for the dual problem is still feasible after these changes are made. Since these changes affect the dual problem by changing only one constraint, this question can be answered simply by checking whether this complementary basic solution still satisfies this revised constraint.

We shall illustrate this case in the corresponding subsection of Sec. 6.7 after developing a relevant example. The Worked Examples section of the book's website also gives **another example** for both this case and the next one.

Introduction of a New Variable

As indicated in Table 6.6, the decision variables in the model typically represent the levels of the various activities under consideration. In some situations, these activities were selected from a larger group of *possible* activities, where the remaining activities were not included in the original model because they seemed less attractive. Or perhaps these other activities did not come to light until after the original model was formulated and solved. Either way, the key question is whether any of these previously unconsidered activities are sufficiently worthwhile to warrant initiation. In other words, would adding any of these activities to the model change the original optimal solution?

Adding another activity amounts to introducing a new variable, with the appropriate coefficients in the functional constraints and objective function, into the model. The only resulting change in the dual problem is to add a *new constraint* (see Table 6.3).

After these changes are made, would the original optimal solution, along with the new variable equal to zero (nonbasic), still be optimal for the primal problem? As for the preceding case, an equivalent question is whether the complementary basic solution for the dual problem is still feasible. And, as before, this question can be answered simply by checking whether this complementary basic solution satisfies one constraint, which in this case is the new constraint for the dual problem.

To illustrate, suppose for the Wyndor Glass Co. problem of Sec. 3.1 that a possible third new product now is being considered for inclusion in the product line. Letting x_{new} represent the production rate for this product, we show the resulting revised model as follows:

$$\text{Maximize} \quad Z = 3x_1 + 5x_2 + 4x_{new},$$

subject to

$$x_1 \qquad\qquad + 2x_{new} \leq 4$$
$$2x_2 + 3x_{new} \leq 12$$
$$3x_1 + 2x_2 + \;\; x_{new} \leq 18$$

and

$$x_1 \geq 0, \qquad x_2 \geq 0, \qquad x_{new} \geq 0.$$

After we introduced slack variables, the original optimal solution for this problem without x_{new} (given by Table 4.8) was $(x_1, x_2, x_3, x_4, x_5) = (2, 6, 2, 0, 0)$. Is this solution, along with $x_{new} = 0$, still optimal?

To answer this question, we need to check the complementary basic solution for the dual problem. As indicated by the *complementary optimal basic solutions property* in Sec. 6.3, this solution is given in row 0 of the *final* simplex tableau for the primal problem, using the locations shown in Table 6.4 and illustrated in Table 6.5. Therefore, as given in both the bottom row of Table 6.5 and the sixth row of Table 6.9, the solution is

$$(y_1, y_2, y_3, z_1 - c_1, z_2 - c_2) = \left(0, \frac{3}{2}, 1, 0, 0\right).$$

(Alternatively, this complementary basic solution can be derived in the way that was illustrated in Sec. 6.3 for the complementary basic solution in the next-to-last row of Table 6.9.)

Since this solution was optimal for the original dual problem, it certainly satisfies the original dual constraints shown in Table 6.1. But does it satisfy this new dual constraint?

$$2y_1 + 3y_2 + y_3 \geq 4$$

Plugging in this solution, we see that

$$2(0) + 3\left(\frac{3}{2}\right) + (1) \geq 4$$

is satisfied, so this dual solution is still feasible (and thus still optimal). Consequently, the original primal solution (2, 6, 2, 0, 0), along with $x_{new} = 0$, is still optimal, so this third possible new product should *not* be added to the product line.

This approach also makes it very easy to conduct sensitivity analysis on the coefficients of the new variable added to the primal problem. By simply checking the new dual constraint, you can immediately see how far any of these parameter values can be changed before they affect the feasibility of the dual solution and so the optimality of the primal solution.

Other Applications

Already we have discussed two other key applications of duality theory to sensitivity analysis, namely, *shadow prices* and the *dual simplex method*. As described in Secs. 4.7 and 6.2, the optimal dual solution $(y_1^*, y_2^*, \ldots, y_m^*)$ provides the shadow prices for the respective resources that indicate how Z would change if (small) changes were made in the b_i (the resource amounts). The resulting analysis will be illustrated in some detail in Sec. 6.7.

In more general terms, the economic interpretation of the dual problem and of the simplex method presented in Sec. 6.2 provides some useful insights for sensitivity analysis.

When we investigate the effect of changing the b_i or the a_{ij} values (for basic variables), the original optimal solution may become a *superoptimal* basic solution (as defined in Table 6.10) instead. If we then want to *reoptimize* to identify the new optimal solution, the dual simplex method (discussed at the end of Secs. 6.1 and 6.3) should be applied, starting from this basic solution. (This important variant of the simplex method will be described in Sec. 7.1.)

We mentioned in Sec. 6.1 that sometimes it is more efficient to solve the dual problem directly by the simplex method in order to identify an optimal solution for the primal problem. When the solution has been found in this way, sensitivity analysis for the primal problem then is conducted by applying the procedure described in the next two sections directly to the dual problem and then inferring the complementary effects on the primal problem (e.g., see Table 6.11). This approach to sensitivity analysis is relatively straightforward because of the close primal-dual relationships described in Secs. 6.1 and 6.3. (See Prob. 6.6-3.)

■ 6.6 THE ESSENCE OF SENSITIVITY ANALYSIS

The work of the operations research team usually is not even nearly done when the simplex method has been successfully applied to identify an optimal solution for the model. As we pointed out at the end of Sec. 3.3, one assumption of linear programming is that all the parameters of the model (a_{ij}, b_i, and c_j) are *known constants*. Actually, the parameter values used in the model normally are just *estimates* based on a *prediction of future conditions*. The data obtained to develop these estimates often are rather crude or nonexistent, so that the parameters in the original formulation may represent little more than quick rules of thumb provided by busy line personnel. The data may even represent deliberate overestimates or underestimates to protect the interests of the estimators.

Thus, the successful manager and operations research staff will maintain a healthy skepticism about the original numbers coming out of the computer and will view them in many cases as only a starting point for further analysis of the problem. An "optimal" solution is optimal only with respect to the specific model being used to represent the real problem, and such a solution becomes a reliable guide for action only after it has been verified as performing well for other reasonable representations of the problem. Furthermore, the model parameters (particularly b_i) sometimes are set as a result of managerial policy decisions (e.g., the amount of certain resources to be made available to the activities), and these decisions should be reviewed after their potential consequences are recognized.

For these reasons it is important to perform **sensitivity analysis** to investigate the effect on the optimal solution provided by the simplex method if the parameters take on other possible values. Usually there will be some parameters that can be assigned any reasonable value without the optimality of this solution being affected. However, there may also be parameters with likely alternative values that would yield a new optimal solution. This situation is particularly serious if the original solution would then have a substantially inferior value of the objective function, or perhaps even be infeasible!

Therefore, one main purpose of sensitivity analysis is to identify the **sensitive parameters** (i.e., the parameters whose values cannot be changed without changing the optimal solution). For coefficients in the objective function that are not categorized as sensitive, it is also very helpful to determine the *range of values* of the coefficient over which the optimal solution will remain unchanged. (We call this range of values the *allowable range for that coefficient*.) In some cases, changing the right-hand side of a functional constraint can affect the *feasibility* of the optimal BF solution. For such parameters, it is useful to determine the range of values over which the optimal BF solution (with adjusted values for the basic variables) will remain feasible. (We call this range of values the *allowable range* for the right-hand side involved.) This range of values also is the range over which the current *shadow price* for the corresponding constraint remains valid. In the next section, we will describe the specific procedures for obtaining this kind of information.

Such information is invaluable in two ways. First, it identifies the more important parameters, so that special care can be taken to estimate them closely and to select a solution that performs well for most of their likely values. Second, it identifies the parameters that will need to be monitored particularly closely as the study is implemented. If it is discovered that the true value of a parameter lies outside its allowable range, this immediately signals a need to change the solution.

For small problems, it would be straightforward to check the effect of a variety of changes in parameter values simply by reapplying the simplex method each time to see if the optimal solution changes. This is particularly convenient when using a spreadsheet formulation. Once the Solver has been set up to obtain an optimal solution, all you have to do is make any desired change on the spreadsheet and then click on the Solve button again.

However, for larger problems of the size typically encountered in practice, sensitivity analysis would require an exorbitant computational effort if it were necessary to reapply the simplex method from the beginning to investigate each new change in a parameter value. Fortunately, the fundamental insight discussed in Sec. 5.3 virtually eliminates computational effort. The basic idea is that the fundamental insight *immediately* reveals just how any changes in the original model would change the numbers in the final simplex tableau (assuming that the *same* sequence of algebraic operations originally performed by the simplex method were to be *duplicated*). Therefore, after making a few simple calculations to revise this tableau, we can check easily whether the original optimal BF solution is now nonoptimal (or infeasible). If so, this solution would be used as the initial basic solution to restart the simplex method (or dual simplex method) to find the new optimal solution, if desired. If the changes in the model are not major, only a very few iterations should be required to reach the new optimal solution from this "advanced" initial basic solution.

To describe this procedure more specifically, consider the following situation. The simplex method already has been used to obtain an optimal solution for a linear programming model with specified values for the b_i, c_j, and a_{ij} parameters. To initiate sensitivity analysis, at least one of the parameters is changed. After the changes are made, let $\bar{b}_i$, $\bar{c}_j$, and $\bar{a}_{ij}$ denote the values of the various parameters. Thus, in matrix notation,

$$\mathbf{b} \to \bar{\mathbf{b}}, \qquad \mathbf{c} \to \bar{\mathbf{c}}, \qquad \mathbf{A} \to \bar{\mathbf{A}},$$

for the revised model.

The first step is to revise the final simplex tableau to reflect these changes. In particular, we want to find the revised final tableau that would result if *exactly* the same algebraic operations (including the same multiples of rows being added to or subtracted from other rows) that led from the initial tableau to the final tableau were repeated when starting from the new initial tableau. (This isn't necessarily the same as reapplying the simplex method since the changes in the initial tableau might cause the simplex method to change some of the algebraic operations being used.) Continuing to use the notation presented in Table 5.9, as well as the accompanying formulas for the fundamental insight [(1) $\mathbf{t}^* = \mathbf{t} + \mathbf{y}^*\mathbf{T}$ and (2) $\mathbf{T}^* = \mathbf{S}^*\mathbf{T}$], the revised final tableau is calculated from $\mathbf{y}^*$ and $\mathbf{S}^*$ (which have not changed) and the new initial tableau, as shown in Table 6.17. Note that $\mathbf{y}^*$ and $\mathbf{S}^*$ together are the coefficients of the *slack variables* in the final simplex tableau, where the vector $\mathbf{y}^*$ (the dual variables) equals these coefficients in row 0 and the matrix $\mathbf{S}^*$ gives these coefficients in the other rows of the tableau. Thus, simply by using $\mathbf{y}^*$, $\mathbf{S}^*$, and the revised numbers in the *initial* tableau, Table 6.17 reveals how the revised numbers in the rest of the *final* tableau are calculated immediately without having to repeat any algebraic operations.

Example (Variation 1 of the Wyndor Model). To illustrate, suppose that the first revision in the model for the Wyndor Glass Co. problem of Sec. 3.1 is the one shown in Table 6.18.

■ **TABLE 6.17** Revised final simplex tableau resulting from changes in original model

	Eq.	Z	Coefficient of: Original Variables	Coefficient of: Slack Variables	Right Side
New initial tableau	(0)	1	$-\bar{\mathbf{c}}$	**0**	0
	(1, 2, . . . , m)	**0**	$\bar{\mathbf{A}}$	**I**	$\bar{\mathbf{b}}$
Revised final tableau	(0)	1	$\mathbf{z}^* - \bar{\mathbf{c}} = \mathbf{y}^*\bar{\mathbf{A}} - \bar{\mathbf{c}}$	$\mathbf{y}^*$	$Z^* = \mathbf{y}^*\bar{\mathbf{b}}$
	(1, 2, . . . , m)	**0**	$\mathbf{A}^* = \mathbf{S}^*\bar{\mathbf{A}}$	$\mathbf{S}^*$	$\mathbf{b}^* = \mathbf{S}^*\bar{\mathbf{b}}$

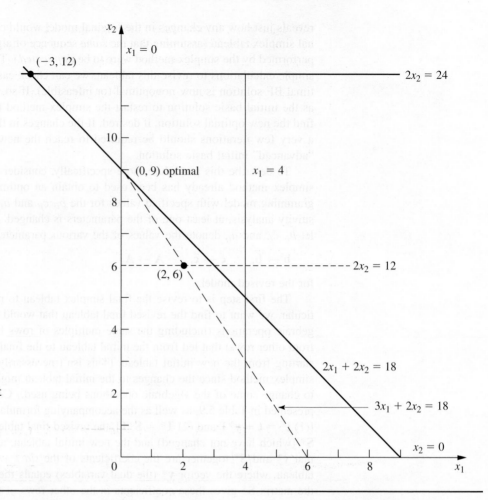

■ FIGURE 6.2
Shift of the final corner-point
solution from (2, 6) to
(−3, 12) for Variation 1 of
the Wyndor Glass Co. model
where $c_1 = 3 \rightarrow 4$,
$a_{31} = 3 \rightarrow 2$, and
$b_2 = 12 \rightarrow 24$.

■ TABLE 6.18 The original model and the first revised model (variation 1) for conducting sensitivity analysis on the Wyndor Glass Co. model

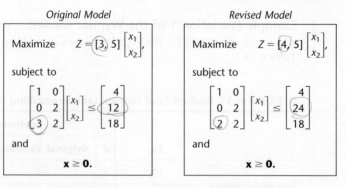

Thus, the changes from the original model are $c_1 = 3 \rightarrow 4$, $a_{31} = 3 \rightarrow 2$, and $b_2 = 12 \rightarrow 24$. Figure 6.2 shows the graphical effect of these changes. For the original model, the simplex method already has identified the optimal CPF solution as (2, 6), lying at the intersection of the two constraint boundaries, shown as dashed lines $2x_2 = 12$ and $3x_1 + 2x_2 = 18$. Now the revision of the model has shifted both of these constraint boundaries as shown by the dark lines $2x_2 = 24$ and $2x_1 + 2x_2 = 18$. Consequently, the previous

CPF solution (2, 6) now shifts to the new intersection $(-3, 12)$, which is a corner-point *infeasible* solution for the revised model. The procedure described in the preceding paragraphs finds this shift *algebraically* (in augmented form). Furthermore, it does so in a manner that is very efficient even for huge problems where graphical analysis is impossible.

To carry out this procedure, we begin by displaying the parameters of the revised model in matrix form:

$$\bar{c} = [4, 5], \qquad \overline{A} = \begin{bmatrix} 1 & 0 \\ 0 & 2 \\ 2 & 2 \end{bmatrix}, \qquad \overline{b} = \begin{bmatrix} 4 \\ 24 \\ 18 \end{bmatrix}.$$

The resulting new initial simplex tableau is shown at the top of Table 6.19. Below this tableau is the original final tableau (as first given in Table 4.8). We have drawn dark boxes around the portions of this final tableau that the changes in the model definitely *do not change*, namely, the coefficients of the slack variables in both row 0 (y^*) and the rest of the rows (S^*). Thus,

$$y^* = [0, \tfrac{3}{2}, 1], \qquad S^* = \begin{bmatrix} 1 & \tfrac{1}{3} & -\tfrac{1}{3} \\ 0 & \tfrac{1}{2} & 0 \\ 0 & -\tfrac{1}{3} & \tfrac{1}{3} \end{bmatrix}.$$

These coefficients of the slack variables necessarily are unchanged with the same algebraic operations originally performed by the simplex method because the coefficients of these same variables in the initial tableau are unchanged.

■ **TABLE 6.19** Obtaining the revised final simplex tableau for Variation 1 of the Wyndor Glass Co. model

	Basic Variable	Eq.	Z	x_1	x_2	x_3	x_4	x_5	Right Side
						Coefficient of:			
New initial tableau	Z	(0)	1	−4	−5	0	0	0	0
	x_3	(1)	0	1	0	1	0	0	4
	x_4	(2)	0	0	2	0	1	0	24
	x_5	(3)	0	2	2	0	0	1	18
Final tableau for original model	Z	(0)	1	0	0	0	$\tfrac{3}{2}$	1	36
	x_3	(1)	0	0	0	1	$\tfrac{1}{3}$	$-\tfrac{1}{3}$	2
	x_2	(2)	0	0	1	0	$\tfrac{1}{2}$	0	6
	x_1	(3)	0	1	0	0	$-\tfrac{1}{3}$	$\tfrac{1}{3}$	2
Revised final tableau	Z	(0)	1	−2	0	0	$\tfrac{3}{2}$	1	54
	x_3	(1)	0	$\tfrac{1}{3}$	0	1	$\tfrac{1}{3}$	$-\tfrac{1}{3}$	6
	x_2	(2)	0	0	1	0	$\tfrac{1}{2}$	0	12
	x_1	(3)	0	$\tfrac{2}{3}$	0	0	$-\tfrac{1}{3}$	$\tfrac{1}{3}$	−2

uses just one iteration to move from the corner-point solution $(-3, 12)$ to the optimal CPF solution $(0, 9)$. (It is often useful in sensitivity analysis to identify the solutions that are optimal for some set of likely values of the model parameters and then to determine which of these solutions most *consistently* performs well for the various likely parameter values.)

If the basic solution $(-3, 12, 7, 0, 0)$ had been *neither* primal feasible nor dual feasible (i.e., if the tableau had negative entries in *both* the *right side* column and row 0), artificial variables could have been introduced to convert the tableau to the proper form for an initial simplex tableau.[4]

The General Procedure.   When one is testing to see how *sensitive* the original optimal solution is to the various parameters of the model, the common approach is to check each parameter (or at least c_j and b_i) individually. In addition to finding allowable ranges as described in the next section, this check might include changing the value of the parameter from its initial estimate to other possibilities in the *range of likely values* (including the endpoints of this range). Then some combinations of simultaneous changes of parameter values (such as changing an entire functional constraint) may be investigated. *Each* time one (or more) of the parameters is changed, the procedure described and illustrated here would be applied. Let us now summarize this procedure.

Summary of Procedure for Sensitivity Analysis

1. *Revision of model:* Make the desired change or changes in the model to be investigated next.
2. *Revision of final tableau:* Use the fundamental insight (as summarized by the formulas on the bottom of Table 6.17) to determine the resulting changes in the final simplex tableau. (See Table 6.19 for an illustration.)
3. *Conversion to proper form from Gaussian elimination:* Convert this tableau to the proper form for identifying and evaluating the current basic solution by applying (as necessary) Gaussian elimination. (See Table 6.20 for an illustration.)
4. *Feasibility test:* Test this solution for feasibility by checking whether all its basic variable values in the right-side column of the tableau still are nonnegative.
5. *Optimality test:* Test this solution for optimality (if feasible) by checking whether all its nonbasic variable coefficients in row 0 of the tableau still are nonnegative.
6. *Reoptimization:* If this solution fails either test, the new optimal solution can be obtained (if desired) by using the current tableau as the initial simplex tableau (and making any necessary conversions) for the simplex method or dual simplex method.

The interactive routine entitled *sensitivity analysis* in IOR Tutorial will enable you to efficiently practice applying this procedure. In addition, a demonstration in OR Tutor (also entitled *sensitivity analysis*) provides you with **another example**.

For problems with only two decision variables, graphical analysis provides an alternative to the above algebraic procedure for performing sensitivity analysis. IOR Tutorial includes a procedure called *Graphical Method and Sensitivity Analysis* for performing such graphical analysis efficiently.

In the next section, we shall discuss and illustrate the application of the above algebraic procedure to each of the major categories of revisions in the original model. We also will use graphical analysis to illuminate what is being accomplished algebraically. This discussion will involve, in part, expanding upon the example introduced in this section for investigating changes in the Wyndor Glass Co. model. In fact, we shall begin by *individually* checking each of the preceding changes. At the same time, we shall integrate some of the applications of duality theory to sensitivity analysis discussed in Sec. 6.5.

---

[4]There also exists a primal-dual algorithm that can be directly applied to such a simplex tableau without any conversion.

6.7 APPLYING SENSITIVITY ANALYSIS

Sensitivity analysis often begins with the investigation of changes in the values of b_i, the amount of resource i ($i = 1, 2, \ldots, m$) being made available for the activities under consideration. The reason is that there generally is more flexibility in setting and adjusting these values than there is for the other parameters of the model. As already discussed in Secs. 4.7 and 6.2, the economic interpretation of the dual variables (the y_i) as shadow prices is extremely useful for deciding which changes should be considered.

Case 1—Changes in b_i

Suppose that the only changes in the current model are that one or more of the b_i parameters ($i = 1, 2, \ldots, m$) has been changed. In this case, the *only* resulting changes in the final simplex tableau are in the *right-side* column. Consequently, the tableau still will be in proper form from Gaussian elimination and all the nonbasic variable coefficients in row 0 still will be nonnegative. Therefore, both the *conversion to proper form from Gaussian elimination* and the *optimality test* steps of the general procedure can be skipped. After revising the right-side column of the tableau, the only question will be whether all the basic variable values in this column still are nonnegative (the feasibility test).

As shown in Table 6.17, when the vector of the b_i values is changed from $\mathbf{b}$ to $\bar{\mathbf{b}}$, the formulas for calculating the new *right-side* column in the final tableau are

Right side of final row 0: $\qquad\qquad Z^* = \mathbf{y}^*\bar{\mathbf{b}},$
Right side of final rows 1, 2, $\ldots$, m: $\qquad \mathbf{b}^* = \mathbf{S}^*\mathbf{b}.$

(See the bottom of Table 6.17 for the location of the unchanged vector $\mathbf{y}^*$ and matrix $\mathbf{S}^*$ in the final tableau.) The first equation has a natural economic interpretation that relates to the economic interpretation of the dual variables presented at the beginning of Sec. 6.2. The vector $\mathbf{y}^*$ gives the optimal values of the dual variables, where these values are interpreted as the *shadow prices* of the respective resources. In particular, when Z^* represents the profit from using the optimal primal solution $\mathbf{x}^*$ and each b_i represents the amount of resource i being made available, y_i^* indicates how much the profit could be increased per unit increase in b_i (for small increases in b_i).

Example (Variation 2 of the Wyndor Model). Sensitivity analysis is begun for the original Wyndor Glass Co. problem of Sec. 3.1 by examining the optimal values of the y_i dual variables ($y_1^* = 0$, $y_2^* = \frac{3}{2}$, $y_3^* = 1$). These *shadow prices* give the marginal value of each resource i (the available production capacity of Plant i) for the activities (two new products) under consideration, where marginal value is expressed in the units of Z (thousands of dollars of profit per week). As discussed in Sec. 4.7 (see Fig. 4.8), the total profit from these activities can be increased $1,500 per week ($y_2^*$ times $1,000 per week) for each additional unit of resource 2 (hour of production time per week in Plant 2) that is made available. This increase in profit holds for relatively small changes that do not affect the feasibility of the current basic solution (and so do not affect the y_i^* values).

Consequently, the OR team has investigated the marginal profitability from the other current uses of this resource to determine if any are less than $1,500 per week. This investigation reveals that one old product is far less profitable. The production rate for this product already has been reduced to the minimum amount that would justify its marketing expenses. However, it can be discontinued altogether, which would provide an additional 12 units of resource 2 for the new products. Thus, the next step is to determine the profit that could be obtained from the new products if this shift were made. This shift changes b_2 from 12 to 24 in the linear programming model. Figure 6.3 shows the graphical effect of this change, including the shift in the final corner-point solution from (2, 6) to (−2, 12).

The **Pacific Lumber Company (PALCO)** is a large timber-holding company with headquarters in Scotia, California. The company has over 200,000 acres of highly productive forest lands that support five mills located in Humboldt County in northern California. The lands include some of the most spectacular redwood groves in the world that have been given or sold at low cost to be preserved as parks. PALCO manages the remaining lands intensively for sustained timber production, subject to strong forest practice laws. Since PALCO's forests are home to many species of wildlife, including endangered species such as spotted owls and marbled murrelets, the provisions of the federal Endangered Species Act also need to be carefully observed.

To obtain a sustained yield plan for the entire landholding, PALCO management contracted with a team of OR consultants to develop a 120-year, 12-period, long-term forest ecosystem management plan. The OR team performed this task by formulating and applying a linear programming model to optimize the company's overall timberland operations and profitability after satisfying the various constraints. The model was a huge one with approximately 8,500 functional constraints and 353,000 decision variables.

A major challenge in applying the linear programming model was the many uncertainties in estimating what the parameters of the model should be. The major factors causing these uncertainties were the continuing fluctuations in market supply and demand, logging costs, and environmental regulations. Therefore, the OR team made extensive use of *detailed sensitivity analysis*. The resulting sustained yield plan *increased the company's present net worth by over* **$398 million** while also generating a better mix of wildlife habitat acres.

Source: L. R. Fletcher, H. Alden, S. P. Holmen, D. P. Angelis, and M. J. Etzenhouser: "Long-Term Forest Ecosystem Planning at Pacific Lumber," *Interfaces*, **29**(1): 90–112, Jan–Feb. 1999. (A link to this article is provided on our website, www.mhhe.com/hillier.)

Based on the results with $b_2 = 24$, the relatively unprofitable old product will be discontinued and the unused 6 units of resource 2 will be saved for some future use. Since y_3^* still is positive, a similar study is made of the possibility of changing the allocation of resource 3, but the resulting decision is to retain the current allocation. Therefore, the current linear programming model at this point (Variation 2) has the parameter values and optimal solution shown in Table 6.21. This model will be used as the starting point for investigating other types of changes in the model later in this section. However, before turning to these other cases, let us take a broader look at the current case.

The Allowable Range for a Right-Hand Side. Although $\Delta b_2 = 12$ proved to be too large an increase in b_2 to retain feasibility (and so optimality) with the basic solution where x_1, x_2, and x_3 are the basic variables (middle of Table 6.19), the above incremental analysis shows immediately just how large an increase is feasible. In particular, note that

$$b_1^* = 2 + \frac{1}{3} \Delta b_2,$$

$$b_2^* = 6 + \frac{1}{2} \Delta b_2,$$

$$b_3^* = 2 - \frac{1}{3} \Delta b_2,$$

where these three quantities are the values of x_3, x_2, and x_1, respectively, for this basic solution. The solution remains feasible, and so optimal, as long as all three quantities remain nonnegative.

$$2 + \frac{1}{3} \Delta b_2 \geq 0 \quad \Rightarrow \quad \frac{1}{3} \Delta b_2 \geq -2 \quad \Rightarrow \quad \Delta b_2 \geq -6,$$

$$6 + \frac{1}{2} \Delta b_2 \geq 0 \quad \Rightarrow \quad \frac{1}{2} \Delta b_2 \geq -6 \quad \Rightarrow \quad \Delta b_2 \geq -12,$$

$$2 - \frac{1}{3} \Delta b_2 \geq 0 \quad \Rightarrow \quad 2 \geq \frac{1}{3} \Delta b_2 \quad \Rightarrow \quad \Delta b_2 \leq 6.$$

Therefore, since $b_2 = 12 + \Delta b_2$, the solution remains feasible only if

$$-6 \leq \Delta b_2 \leq 6, \qquad \text{that is,} \qquad 6 \leq b_2 \leq 18.$$

(Verify this graphically in Fig. 6.3.) As introduced in Sec. 4.7, this range of values for b_2 is referred to as its *allowable range*.

> For any b_i, recall from Sec. 4.7 that its **allowable range** is the range of values over which the current optimal BF solution[5] (with adjusted values for the basic variables) remains feasible. Thus, the *shadow price* for b_i remains valid for evaluating the effect on Z of changing b_i only as long as b_i remains within this allowable range. (It is assumed that the change in this one b_i value is the only change in the model.) The adjusted values for the basic variables are obtained from the formula $\mathbf{b}^* = \mathbf{S}^*\mathbf{b}$. The calculation of the allowable range then is based on finding the range of values of b_i such that $\mathbf{b}^* \geq \mathbf{0}$.

Many linear programming software packages use this same technique for automatically generating the allowable range for each b_i. (A similar technique, discussed under Cases 2a and 3, also is used to generate an *allowable range* for each c_j.) In Chap. 4, we showed the corresponding output for the Excel Solver and LINDO in Figs. 4.10 and A4.2, respectively. Table 6.22 summarizes this same output with respect to the b_i for the original Wyndor Glass Co. model. For example, both the *allowable increase* and *allowable decrease* for b_2 are 6, that is, $-6 \leq \Delta b_2 \leq 6$. The analysis in the preceding paragraph shows how these quantities were calculated.

Analyzing Simultaneous Changes in Right-Hand Sides. When multiple b_i values are changed simultaneously, the formula $\mathbf{b}^* = \mathbf{S}^*\mathbf{b}$ can again be used to see how the right-hand sides change in the final tableau. If all these right-hand sides still are nonnegative, the feasibility test will indicate that the revised solution provided by this tableau still is feasible. Since row 0 has not changed, being feasible implies that this solution also is optimal.

Although this approach works fine for checking the effect of a *specific* set of changes in the b_i, it does not give much insight into how far the b_i can be simultaneously changed from their original values before the revised solution will no longer be feasible. As part of postoptimality analysis, the management of an organization often is interested in investigating the effect of various changes in policy decisions (e.g., the amounts of resources being made available to the activities under consideration) that determine the right-hand sides. Rather than considering just one specific set of changes, management may want to explore *directions* of changes where some right-hand sides increase while others decrease. Shadow

TABLE 6.22 Typical software output for sensitivity analysis of the right-hand sides for the original Wyndor Glass Co. model

Constraint	Shadow Price	Current RHS	Allowable Increase	Allowable Decrease
Plant 1	0	4	∞	2
Plant 2	1.5	12	6	6
Plant 3	1	18	6	6

[5]When there is more than one optimal BF solution for the current model (before changing b_i), we are referring here to the one obtained by the simplex method.

prices are invaluable for this kind of exploration. However, shadow prices remain valid for evaluating the effect of such changes on Z only within certain ranges of changes. For each b_i, the *allowable range* gives this range if *none* of the other b_i are changing at the same time. What do these *allowable ranges* become when some of the b_i are changing simultaneously?

A partial answer to this question is provided by the following 100 percent rule, which combines the *allowable changes* (increase or decrease) for the individual b_i that are given by the last two columns of a table like Table 6.22.

The 100 Percent Rule for Simultaneous Changes in Right-Hand Sides: The shadow prices remain valid for predicting the effect of simultaneously changing the right-hand sides of some of the functional constraints as long as the changes are not too large. To check whether the changes are small enough, calculate for each change the percentage of the allowable change (increase or decrease) for that right-hand side to remain within its allowable range. If the *sum* of the percentage changes does *not* exceed 100 percent, the shadow prices definitely will still be valid. (If the sum *does* exceed 100 percent, then we cannot be sure.)

Example (Variation 3 of the Wyndor Model).   To illustrate this rule, consider *Variation* 3 of the Wyndor Glass Co. model, which revises the original model by changing the right-hand side vector as follows:

$$\mathbf{b} = \begin{bmatrix} 4 \\ 12 \\ 18 \end{bmatrix} \rightarrow \overline{\mathbf{b}} = \begin{bmatrix} 4 \\ 15 \\ 15 \end{bmatrix}.$$

The calculations for the 100 percent rule in this case are

$$b_2\colon 12 \rightarrow 15.\quad \text{Percentage of allowable increase} = 100\left(\frac{15 - 12}{6}\right) = 50\%$$

$$b_3\colon 18 \rightarrow 15.\quad \text{Percentage of allowable decrease} = 100\left(\frac{18 - 15}{6}\right) = 50\%$$

$$\text{Sum} = 100\%$$

Since the sum of 100 percent barely does *not* exceed 100 percent, the shadow prices definitely are valid for predicting the effect of these changes on Z. In particular, since the shadow prices of b_2 and b_3 are 1.5 and 1, respectively, the resulting change in Z would be

$$\Delta Z = 1.5(3) + 1(-3) = 1.5,$$

so Z^* would increase from 36 to 37.5.

Figure 6.4 shows the feasible region for this revised model. (The dashed lines show the original locations of the revised constraint boundary lines.) The optimal solution now is the CPF solution (0, 7.5), which gives

$$Z = 3x_1 + 5x_2 = 0 + 5(7.5) = 37.5,$$

just as predicted by the shadow prices. However, note what would happen if either b_2 were further increased above 15 or b_3 were further decreased below 15, so that the sum of the percentages of allowable changes would exceed 100 percent. This would cause the previously optimal corner-point solution to slide to the left of the x_2 axis ($x_1 < 0$), so this *infeasible* solution would no longer be optimal. Consequently, the old shadow prices would no longer be valid for predicting the new value of Z^*.

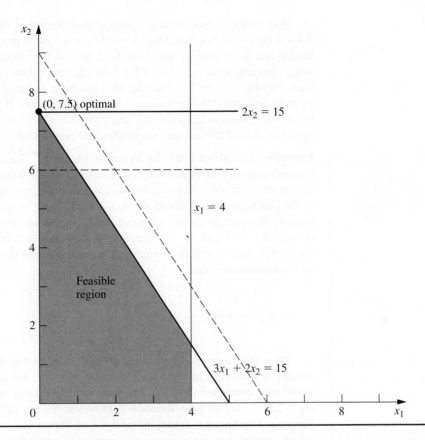

FIGURE 6.4
Feasible region for Variation 3 of the Wyndor Glass Co. model where $b_2 = 12 \rightarrow 15$ and $b_3 = 18 \rightarrow 15$.

Case 2a—Changes in the Coefficients of a Nonbasic Variable

Consider a particular variable x_j (fixed j) that is a nonbasic variable in the optimal solution shown by the final simplex tableau. In Case 2a, the only change in the current model is that one or more of the coefficients of this variable—$c_j, a_{1j}, a_{2j}, \ldots, a_{mj}$—have been changed. Thus, letting $\bar{c}_j$ and $\bar{a}_{ij}$ denote the new values of these parameters, with $\overline{\mathbf{A}}_j$ (column j of matrix $\mathbf{A}$) as the vector containing the $\bar{a}_{ij}$, we have

$$c_j \longrightarrow \bar{c}_j, \qquad \mathbf{A}_j \longrightarrow \overline{\mathbf{A}}_j$$

for the revised model.

As described at the beginning of Sec. 6.5, duality theory provides a very convenient way of checking these changes. In particular, if the *complementary* basic solution $\mathbf{y}^*$ in the dual problem still satisfies the single dual constraint that has changed, then the original optimal solution in the primal problem *remains optimal* as is. Conversely, if $\mathbf{y}^*$ violates this dual constraint, then this primal solution is *no longer optimal*.

If the optimal solution has changed and you wish to find the new one, you can do so rather easily. Simply apply the fundamental insight to revise the x_j column (the only one that has changed) in the final simplex tableau. Specifically, the formulas in Table 6.17 reduce to the following:

Coefficient of x_j in final row 0: $z_j^* - \bar{c}_j = \mathbf{y}^*\overline{\mathbf{A}}_j - \bar{c}_j,$

Coefficient of x_j in final rows 1 to m: $\mathbf{A}_j^* = \mathbf{S}^*\overline{\mathbf{A}}_j.$

With the current basic solution no longer optimal, the new value of $z_j^* - c_j$ now will be the one negative coefficient in row 0, so restart the simplex method with x_j as the initial entering basic variable.

Note that this procedure is a streamlined version of the general procedure summarized at the end of Sec. 6.6. Steps 3 and 4 (conversion to proper form from Gaussian elimination and the feasibility test) have been deleted as irrelevant, because the only column being changed in the revision of the final tableau (before reoptimization) is for the nonbasic variable x_j. Step 5 (optimality test) has been replaced by a quicker test of optimality to be performed right after step 1 (revision of model). It is only if this test reveals that the optimal solution has changed, and you wish to find the new one, that steps 2 and 6 (revision of final tableau and reoptimization) are needed.

Example (Variation 4 of the Wyndor Model).   Since x_1 is nonbasic in the current optimal solution (see Table 6.21) for Variation 2 of the Wyndor Glass Co. model, the next step in its sensitivity analysis is to check whether any reasonable changes in the estimates of the coefficients of x_1 could still make it advisable to introduce product 1. The set of changes that goes as far as realistically possible to make product 1 more attractive would be to reset $c_1 = 4$ and $a_{31} = 2$. Rather than exploring each of these changes independently (as is often done in sensitivity analysis), we will consider them together. Thus, the changes under consideration are

$$c_1 = 3 \longrightarrow \bar{c}_1 = 4, \qquad \mathbf{A}_1 = \begin{bmatrix} 1 \\ 0 \\ 3 \end{bmatrix} \longrightarrow \bar{\mathbf{A}}_1 = \begin{bmatrix} 1 \\ 0 \\ 2 \end{bmatrix}.$$

These two changes in Variation 2 give us *Variation 4* of the Wyndor model. Variation 4 actually is equivalent to Variation 1 considered in Sec. 6.6 and depicted in Fig. 6.2, since Variation 1 combined these two changes with the change in the original Wyndor model ($b_2 = 12 \rightarrow 24$) that gave Variation 2. However, the key difference from the treatment of Variation 1 in Sec. 6.6 is that the analysis of Variation 4 treats Variation 2 as being the original model, so our starting point is the final simplex tableau given in Table 6.21 where x_1 now is a nonbasic variable.

The change in a_{31} revises the feasible region from that shown in Fig. 6.3 to the corresponding region in Fig. 6.5. The change in c_1 revises the objective function from $Z = 3x_1 + 5x_2$ to $Z = 4x_1 + 5x_2$. Figure 6.5 shows that the optimal objective function line $Z = 45 = 4x_1 + 5x_2$ still passes through the current optimal solution (0, 9), so this solution remains optimal after these changes in a_{31} and c_1.

To use duality theory to draw this same conclusion, observe that the changes in c_1 and a_{31} lead to a single revised constraint for the dual problem, namely, the constraint that $a_{11}y_1 + a_{21}y_2 + a_{31}y_3 \geq c_1$. Both this revised constraint and the current $\mathbf{y}^*$ (coefficients of the slack variables in row 0 of Table 6.21) are shown below.

$$y_1^* = 0, \qquad y_2^* = 0, \qquad y_3^* = \frac{5}{2},$$

$$y_1 + 3y_3 \geq 3 \longrightarrow y_1 + 2y_3 \geq 4,$$

$$0 + 2\left(\frac{5}{2}\right) \geq 4.$$

Since $\mathbf{y}^*$ *still* satisfies the revised constraint, the current primal solution (Table 6.21) is still optimal.

Because this solution is still optimal, there is no need to revise the x_j column in the final tableau (step 2). Nevertheless, we do so below for illustrative purposes.

$$z_1^* - \bar{c}_1 = \mathbf{y}^* \bar{\mathbf{A}}_1 - c_1 = [0, 0, \tfrac{5}{2}] \begin{bmatrix} 1 \\ 0 \\ 2 \end{bmatrix} - 4 = 1.$$

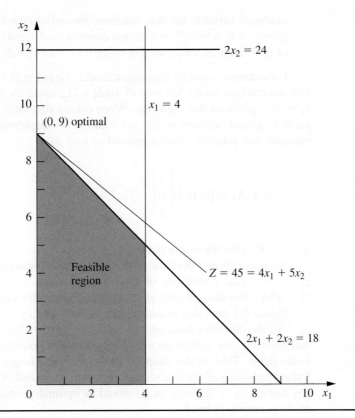

■ FIGURE 6.5
Feasible region for Variation 4 of the Wyndor model where Variation 2 (Fig. 6.3) has been revised so $a_{31} = 3 \rightarrow 2$ and $c_1 = 3 \rightarrow 4$.

$$\mathbf{A}_1^* = \mathbf{S}^* \overline{\mathbf{A}}_1 = \begin{bmatrix} 1 & 0 & 0 \\ 0 & 0 & \frac{1}{2} \\ 0 & 1 & -1 \end{bmatrix} \begin{bmatrix} 1 \\ 0 \\ 2 \end{bmatrix} = \begin{bmatrix} 1 \\ 1 \\ -2 \end{bmatrix}.$$

The fact that $z_1^* - \overline{c}_1 \geq 0$ again confirms the optimality of the current solution. Since $z_1^* - c_1$ is the surplus variable for the revised constraint in the dual problem, this way of testing for optimality is equivalent to the one used above.

This completes the analysis of the effect of changing the current model (Variation 2) to Variation 4. Because any larger changes in the original estimates of the coefficients of x_1 would be unrealistic, the OR team concludes that these coefficients are *insensitive* parameters in the current model. Therefore, they will be kept fixed at their best estimates shown in Table 6.21—$c_1 = 3$ and $a_{31} = 3$—for the remainder of the sensitivity analysis.

The Allowable Range for an Objective Function Coefficient of a Nonbasic Variable.

We have just described and illustrated how to analyze *simultaneous* changes in the coefficients of a nonbasic variable x_j. It is common practice in sensitivity analysis to also focus on the effect of changing just *one* parameter, c_j. As introduced in Sec. 4.7, this involves streamlining the above approach to find the *allowable range* for c_j.

For any c_j, recall from Sec. 4.7 that its **allowable range** is the range of values over which the current optimal solution (as obtained by the simplex method for the current model before c_j is changed) remains optimal. (It is assumed that the change in this one c_j is the only change in the current model.) When x_j is a nonbasic variable for this solution, the solution remains optimal as long as $z_j^* - c_j \geq 0$, where $z_j^* = \mathbf{y}^*\mathbf{A}_j$ is a constant unaffected by any change in the value of c_j. Therefore, the allowable range for c_j can be calculated as $c_j \leq \mathbf{y}^*\mathbf{A}_j$.

For example, consider the current model (Variation 2) for the Wyndor Glass Co. problem summarized on the left side of Table 6.21, where the current optimal solution (with $c_1 = 3$) is given on the right side. When considering only the decision variables, x_1 and x_2, this optimal solution is $(x_1, x_2) = (0, 9)$, as displayed in Fig. 6.3. When just c_1 is changed, this solution remains optimal as long as

$$c_1 \leq \mathbf{y}^*\mathbf{A}_1 = [0, 0, \tfrac{5}{2}] \begin{bmatrix} 1 \\ 0 \\ 3 \end{bmatrix} = 7\tfrac{1}{2},$$

so $c_1 \leq 7\tfrac{1}{2}$ is the allowable range.

An alternative to performing this vector multiplication is to note in Table 6.21 that $z_1^* - c_1 = \tfrac{9}{2}$ (the coefficient of x_1 in row 0) when $c_1 = 3$, so $z_1^* = 3 + \tfrac{9}{2} = 7\tfrac{1}{2}$. Since $z_1^* = \mathbf{y}^*\mathbf{A}_1$, this immediately yields the same allowable range.

Figure 6.3 provides graphical insight into why $c_1 \leq 7\tfrac{1}{2}$ is the allowable range. At $c_1 = 7\tfrac{1}{2}$, the objective function becomes $Z = 7.5x_1 + 5x_2 = 2.5(3x_1 + 2x_2)$, so the optimal objective line will lie on top of the constraint boundary line $3x_1 + 2x_2 = 18$ shown in the figure. Thus, at this endpoint of the allowable range, we have multiple optimal solutions consisting of the line segment between $(0, 9)$ and $(4, 3)$. If c_1 were to be increased any further $(c_1 > 7\tfrac{1}{2})$, only $(4, 3)$ would be optimal. Consequently, we need $c_1 \leq 7\tfrac{1}{2}$ for $(0, 9)$ to remain optimal.

IOR Tutorial includes a procedure called *Graphical Method and Sensitivity Analysis* that enables you to perform this kind of graphical analysis very efficiently.

For any nonbasic decision variable x_j, the value of $z_j^* - c_j$ sometimes is referred to as the **reduced cost** for x_j, because it is the minimum amount by which the unit *cost* of activity j would have to be *reduced* to make it worthwhile to undertake activity j (increase x_j from zero). Interpreting c_j as the unit profit of activity j (so reducing the unit cost increases c_j by the same amount), the value of $z_j^* - c_j$ thereby is the maximum allowable increase in c_j to keep the current BF solution optimal.

The sensitivity analysis information generated by linear programming software packages normally includes both the reduced cost and the allowable range for each coefficient in the objective function (along with the types of information displayed in Table 6.22). This was illustrated in Fig. 4.10 for the Excel Solver and in Figs. A4.1 and A4.2 for LINGO and LINDO. Table 6.23 displays this information in a typical form for our current model (Variation 2 of the Wyndor Glass Co. model). The last three columns are used to calculate the allowable range for each coefficient, so these allowable ranges are

$$c_1 \leq 3 + 4.5 = 7.5,$$
$$c_2 \geq 5 - 3 = 2.$$

■ **TABLE 6.23** Typical software output for sensitivity analysis of the objective function coefficients for Variation 2 of the Wyndor Glass Co. model

Variable	Value	Reduced Cost	Current Coefficient	Allowable Increase	Allowable Decrease
x_1	0	4.5	3	4.5	∞
x_2	9	0	5	∞	3

As was discussed in Sec. 4.7, if any of the allowable increases or decreases had turned out to be zero, this would have been a signpost that the optimal solution given in the table is only one of multiple optimal solutions. In this case, changing the corresponding coefficient a tiny amount beyond the zero allowed and re-solving would provide another optimal CPF solution for the original model.

Thus far, we have described how to calculate the type of information in Table 6.23 for only nonbasic variables. For a basic variable like x_2, the reduced cost automatically is 0. We will discuss how to obtain the allowable range for c_j when x_j is a basic variable under Case 3.

Analyzing Simultaneous Changes in Objective Function Coefficients. Regardless of whether x_j is a basic or nonbasic variable, the allowable range for c_j is valid only if this objective function coefficient is the only one being changed. However, when simultaneous changes are made in the coefficients of the objective function, a 100 percent rule is available for checking whether the original solution must still be optimal. Much like the 100 percent rule for simultaneous changes in right-hand sides, this 100 percent rule combines the *allowable changes* (increase or decrease) for the individual c_j that are given by the last two columns of a table like Table 6.23, as described below.

> **The 100 Percent Rule for Simultaneous Changes in Objective Function Coefficients:** If simultaneous changes are made in the coefficients of the objective function, calculate for each change the percentage of the allowable change (increase or decrease) for that coefficient to remain within its allowable range. If the *sum* of the percentage changes does *not* exceed 100 percent, the original optimal solution definitely will still be optimal. (If the sum *does* exceed 100 percent, then we cannot be sure.)

Using Table 6.23 (and referring to Fig. 6.3 for visualization), this 100 percent rule says that (0, 9) will remain optimal for Variation 2 of the Wyndor Glass Co. model even if we simultaneously increase c_1 from 3 and decrease c_2 from 5 as long as these changes are not too large. For example, if c_1 is increased by 1.5 ($33\frac{1}{3}$ percent of the allowable change), then c_2 can be decreased by as much as 2 ($66\frac{2}{3}$ percent of the allowable change). Similarly, if c_1 is increased by 3 ($66\frac{2}{3}$ percent of the allowable change), then c_2 can only be decreased by as much as 1 ($33\frac{1}{3}$ percent of the allowable change). These maximum changes revise the objective function to either $Z = 4.5x_1 + 3x_2$ or $Z = 6x_1 + 4x_2$, which causes the optimal objective function line in Fig. 6.3 to rotate clockwise until it coincides with the constraint boundary equation $3x_1 + 2x_2 = 18$.

In general, when objective function coefficients change in the *same* direction, it is possible for the percentages of allowable changes to sum to more than 100 percent without changing the optimal solution. We will give an example at the end of the discussion of Case 3.

Case 2b—Introduction of a New Variable

After solving for the optimal solution, we may discover that the linear programming formulation did not consider all the attractive alternative activities. Considering a new activity requires introducing a new variable with the appropriate coefficients into the objective function and constraints of the current model—which is Case 2b.

The convenient way to deal with this case is to treat it just as if it were Case 2a! This is done by pretending that the new variable x_j actually was in the original model with all its coefficients equal to zero (so that they still are zero in the final simplex tableau) and that x_j is a nonbasic variable in the current BF solution. Therefore, if we change these zero coefficients to their actual values for the new variable, the procedure (including any reoptimization) does indeed become identical to that for Case 2a.

In particular, all you have to do to check whether the current solution still is optimal is to check whether the complementary basic solution $\mathbf{y}^*$ satisfies the one new dual constraint that corresponds to the new variable in the primal problem. We already have described this approach and then illustrated it for the Wyndor Glass Co. problem in Sec. 6.5.

Case 3—Changes in the Coefficients of a Basic Variable

Now suppose that the variable x_j (fixed j) under consideration is a *basic* variable in the optimal solution shown by the final simplex tableau. Case 3 assumes that the only changes in the current model are made to the coefficients of this variable.

Case 3 differs from Case 2a because of the requirement that a simplex tableau be in proper form from Gaussian elimination. This requirement allows the column for a nonbasic variable to be anything, so it does not affect Case 2a. However, for Case 3, the basic variable x_j must have a coefficient of 1 in its row of the simplex tableau and a coefficient of 0 in every other row (including row 0). Therefore, after the changes in the x_j column of the final simplex tableau have been calculated,[6] it probably will be necessary to apply Gaussian elimination to restore this form, as illustrated in Table 6.20. In turn, this step probably will change the value of the current basic solution and may make it either infeasible or nonoptimal (so reoptimization may be needed). Consequently, all the steps of the overall procedure summarized at the end of Sec. 6.6 are required for Case 3.

Before Gaussian elimination is applied, the formulas for revising the x_j column are the same as for Case 2a, as summarized below.

Coefficient of x_j in final row 0: $\qquad z_j^* - \overline{c}_j = \mathbf{y}^*\mathbf{A}_j - \overline{c}_j.$

Coefficient of x_j in final rows 1 to m: $\qquad \mathbf{A}_j^* = \mathbf{S}^*\mathbf{A}_j.$

Example (Variation 5 of the Wyndor Model).

Because x_2 is a basic variable in Table 6.21 for Variation 2 of the Wyndor Glass Co. model, sensitivity analysis of its coefficients fits Case 3. Given the current optimal solution ($x_1 = 0$, $x_2 = 9$), product 2 is the *only* new product that should be introduced, and its production rate should be relatively large. Therefore, the key question now is whether the initial estimates that led to the coefficients of x_2 in the current model (Variation 2) could have *overestimated* the attractiveness of product 2 so much as to invalidate this conclusion. This question can be tested by checking the *most pessimistic* set of reasonable estimates for these coefficients, which turns out to be $c_2 = 3$, $a_{22} = 3$, and $a_{32} = 4$. Consequently, the changes to be investigated (Variation 5 of the Wyndor model) are

$$c_2 = 5 \longrightarrow \overline{c}_2 = 3, \qquad \mathbf{A}_2 = \begin{bmatrix} 0 \\ 2 \\ 2 \end{bmatrix} \longrightarrow \overline{\mathbf{A}}_2 = \begin{bmatrix} 0 \\ 3 \\ 4 \end{bmatrix}.$$

The graphical effect of these changes is that the feasible region changes from the one shown in Fig. 6.3 to the one in Fig. 6.6. The optimal solution in Fig. 6.3 is $(x_1, x_2) = (0, 9)$, which is the corner-point solution lying at the intersection of the $x_1 = 0$ and $3x_1 + 2x_2 = 18$ constraint boundaries. With the revision of the constraints, the corresponding corner-point solution in Fig. 6.6 is $(0, \frac{9}{2})$. However, this solution no longer is optimal, because the revised objective function of $Z = 3x_1 + 3x_2$ now yields a new optimal solution of $(x_1, x_2) = (4, \frac{3}{2})$.

---

[6]For the relatively sophisticated reader, we should point out a possible pitfall for Case 3 that would be discovered at this point. Specifically, the changes in the initial tableau can destroy the linear independence of the columns of coefficients of basic variables. This event occurs only if the unit coefficient of the basic variable x_j in the final tableau has been changed to zero at this point, in which case more extensive simplex method calculations must be used for Case 3.

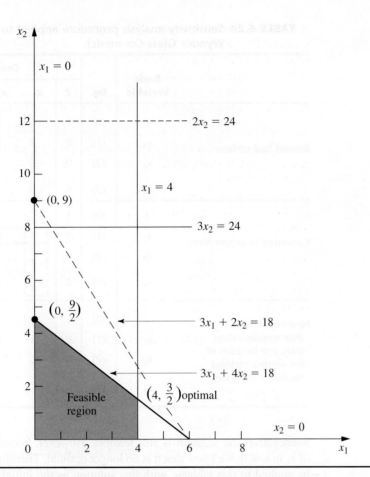

FIGURE 6.6
Feasible region for Variation 5 of the Wyndor model where Variation 2 (Fig. 6.3) has been revised so $c_2 = 5 \rightarrow 3$, $a_{22} = 2 \rightarrow 3$, and $a_{32} = 2 \rightarrow 4$.

Analysis of Variation 5. Now let us see how we draw these same conclusions algebraically. Because the only changes in the model are in the coefficients of x_2, the *only* resulting changes in the final simplex tableau (Table 6.21) are in the x_2 column. Therefore, the above formulas are used to recompute just this column.

$$z_2 - \overline{c}_2 = \mathbf{y^*\overline{A}}_2 - \overline{c}_2 = [0, 0, \tfrac{5}{2}] \begin{bmatrix} 0 \\ 3 \\ 4 \end{bmatrix} - 3 = 7.$$

$$\mathbf{A}_2^* = \mathbf{S^*\overline{A}}_2 = \begin{bmatrix} 1 & 0 & 0 \\ 0 & 0 & \tfrac{1}{2} \\ 0 & 1 & -1 \end{bmatrix} \begin{bmatrix} 0 \\ 3 \\ 4 \end{bmatrix} = \begin{bmatrix} 0 \\ 2 \\ -1 \end{bmatrix}.$$

(Equivalently, incremental analysis with $\Delta c_2 = -2$, $\Delta a_{22} = 1$, and $\Delta a_{32} = 2$ can be used in the same way to obtain this column.)

The resulting revised final tableau is shown at the top of Table 6.24. Note that the new coefficients of the basic variable x_2 do not have the required values, so the conversion to proper form from Gaussian elimination must be applied next. This step involves dividing row 2 by 2, subtracting 7 times the new row 2 from row 0, and adding the new row 2 to row 3.

The resulting second tableau in Table 6.24 gives the new value of the current basic solution, namely, $x_3 = 4$, $x_2 = \tfrac{9}{2}$, $x_4 = \tfrac{21}{2}$ ($x_1 = 0$, $x_5 = 0$). Since all these variables are

■ **TABLE 6.24** Sensitivity analysis procedure applied to Variation 5 of the Wyndor Glass Co. model

	Basic Variable	Eq.	Z	Coefficient of:					Right Side
				x_1	x_2	x_3	x_4	x_5	
Revised final tableau	Z	(0)	1	$\frac{9}{2}$	7	0	0	$\frac{5}{2}$	45
	x_3	(1)	0	1	0	1	0	0	4
	x_2	(2)	0	$\frac{3}{2}$	2	0	0	$\frac{1}{2}$	9
	x_4	(3)	0	-3	-1	0	1	-1	6
Converted to proper form	Z	(0)	1	$-\frac{3}{4}$	0	0	0	$\frac{3}{4}$	$\frac{27}{2}$
	x_3	(1)	0	1	0	1	0	0	4
	x_2	(2)	0	$\frac{3}{4}$	1	0	0	$\frac{1}{4}$	$\frac{9}{2}$
	x_4	(3)	0	$-\frac{9}{4}$	0	0	1	$-\frac{3}{4}$	$\frac{21}{2}$
New final tableau after reoptimization (only one iteration of the simplex method needed in this case)	Z	(0)	1	0	0	$\frac{3}{4}$	0	$\frac{3}{4}$	$\frac{33}{2}$
	x_1	(1)	0	1	0	1	0	0	4
	x_2	(2)	0	0	1	$-\frac{3}{4}$	0	$\frac{1}{4}$	$\frac{3}{2}$
	x_4	(3)	0	0	0	$\frac{9}{4}$	1	$-\frac{3}{4}$	$\frac{39}{2}$

nonnegative, the solution is still feasible. However, because of the negative coefficient of x_1 in row 0, we know that it is no longer optimal. Therefore, the simplex method would be applied to this tableau, with this solution as the initial BF solution, to find the new optimal solution. The initial entering basic variable is x_1, with x_3 as the leaving basic variable. Just one iteration is needed in this case to reach the new optimal solution $x_1 = 4$, $x_2 = \frac{3}{2}$, $x_4 = \frac{39}{2}$ ($x_3 = 0$, $x_5 = 0$), as shown in the last tableau of Table 6.24.

All this analysis suggests that c_2, a_{22}, and a_{32} are relatively sensitive parameters. However, additional data for estimating them more closely can be obtained only by conducting a pilot run. Therefore, the OR team recommends that production of product 2 be initiated immediately on a small scale ($x_2 = \frac{3}{2}$) and that this experience be used to guide the decision on whether the remaining production capacity should be allocated to product 2 or product 1.

The Allowable Range for an Objective Function Coefficient of a Basic Variable. For Case 2a, we described how to find the allowable range for any c_j such that x_j is a nonbasic variable for the current optimal solution (before c_j is changed). When x_j is a basic variable instead, the procedure is somewhat more involved because of the need to convert to proper form from Gaussian elimination before testing for optimality.

To illustrate the procedure, consider Variation 5 of the Wyndor Glass Co. model (with $c_2 = 3$, $a_{22} = 3$, $a_{23} = 4$) that is graphed in Fig. 6.6 and solved in Table 6.24. Since x_2 is a basic variable for the optimal solution (with $c_2 = 3$) given at the bottom of this table, the steps needed to find the allowable range for c_2 are the following:

1. Since x_2 is a basic variable, note that its coefficient in the new final row 0 (see the bottom tableau in Table 6.24) is automatically $z_2^* - c_2 = 0$ before c_2 is changed from its current value of 3.

2. Now increment $c_2 = 3$ by Δc_2 (so $c_2 = 3 + \Delta c_2$). This changes the coefficient noted in step 1 to $z_2^* - c_2 = -\Delta c_2$, which changes row 0 to

$$\text{Row } 0 = \left[0, -\Delta c_2, \frac{3}{4}, 0, \frac{3}{4} \;\middle|\; \frac{33}{2} \right].$$

3. With this coefficient now not zero, we must perform elementary row operations to restore proper form from Gaussian elimination. In particular, add to row 0 the product, Δc_2 times row 2, to obtain the new row 0, as shown below.

$$\left[0, -\Delta c_2, \frac{3}{4}, \quad 0, \quad \frac{3}{4} \;\middle|\; \frac{33}{2} \right]$$

$$+ \left[0, \quad \Delta c_2, -\frac{3}{4}\Delta c_2, 0, \frac{1}{4}\Delta c_2 \;\middle|\; \frac{3}{2}\Delta c_2 \right]$$

$$\text{New row } 0 = \left[0, \quad 0, \frac{3}{4} - \frac{3}{4}\Delta c_2, 0, \frac{3}{4} + \frac{1}{4}\Delta c_2 \;\middle|\; \frac{33}{2} + \frac{3}{2}\Delta c_2 \right]$$

4. Using this new row 0, solve for the range of values of Δc_2 that keeps the coefficients of the nonbasic variables (x_3 and x_5) nonnegative.

$$\frac{3}{4} - \frac{3}{4}\Delta c_2 \geq 0 \quad \Rightarrow \quad \frac{3}{4} \geq \frac{3}{4}\Delta c_2 \quad \Rightarrow \quad \Delta c_2 \leq 1.$$

$$\frac{3}{4} + \frac{1}{4}\Delta c_2 \geq 0 \quad \Rightarrow \quad \frac{1}{4}\Delta c_2 \geq -\frac{3}{4} \quad \Rightarrow \quad \Delta c_2 \geq -3.$$

Thus, the range of values is $-3 \leq \Delta c_2 \leq 1$.

5. Since $c_2 = 3 + \Delta c_2$, add 3 to this range of values, which yields

$$0 \leq c_2 \leq 4$$

as the allowable range for c_2.

   With just two decision variables, this allowable range can be verified graphically by using Fig. 6.6 with an objective function of $Z = 3x_1 + c_2x_2$. With the current value of $c_2 = 3$, the optimal solution is $(4, \frac{3}{2})$. When c_2 is increased, this solution remains optimal only for $c_2 \leq 4$. For $c_2 \geq 4$, $(0, \frac{9}{2})$ becomes optimal (with a tie at $c_2 = 4$), because of the constraint boundary $3x_1 + 4x_2 = 18$. When c_2 is decreased instead, $(4, \frac{3}{2})$ remains optimal only for $c_2 \geq 0$. For $c_2 \leq 0$, $(4, 0)$ becomes optimal because of the constraint boundary $x_1 = 4$.

   In a similar manner, the allowable range for c_1 (with c_2 fixed at 3) can be derived either algebraically or graphically to be $c_1 \geq \frac{9}{4}$. (Problem 6.7-10 asks you to verify this both ways.)

   Thus, the *allowable decrease* for c_1 from its current value of 3 is only $\frac{3}{4}$. However, it is possible to decrease c_1 by a larger amount without changing the optimal solution if c_2 also decreases sufficiently. For example, suppose that *both* c_1 and c_2 are decreased by 1 from their current value of 3, so that the objective function changes from $Z = 3x_1 + 3x_2$ to $Z = 2x_1 + 2x_2$. According to the 100 percent rule for simultaneous changes in objective function coefficients, the percentages of allowable changes are $133\frac{1}{3}$ percent and $33\frac{1}{3}$ percent, respectively, which sum to far over 100 percent. However, the slope of the objective function line has not changed at all, so $(4, \frac{3}{2})$ still is optimal.

Case 4—Introduction of a New Constraint

In this case, a new constraint must be introduced to the model after it has already been solved. This case may occur because the constraint was overlooked initially or because new considerations have arisen since the model was formulated. Another possibility is that the constraint was deleted purposely to decrease computational effort because it appeared to be less restrictive than other constraints already in the model, but now this impression needs to be checked with the optimal solution actually obtained.

To see if the current optimal solution would be affected by a new constraint, all you have to do is to check directly whether the optimal solution satisfies the constraint. If it does, then it would still be the *best feasible solution* (i.e., the optimal solution), even if the constraint were added to the model. The reason is that a new constraint can only eliminate some previously feasible solutions without adding any new ones.

If the new constraint does eliminate the current optimal solution, and if you want to find the new solution, then introduce this constraint into the final simplex tableau (as an additional row) *just* as if this were the initial tableau, where the usual additional variable (slack variable or artificial variable) is designated to be the basic variable for this new row. Because the new row probably will have *nonzero* coefficients for some of the other basic variables, the conversion to proper form from Gaussian elimination is applied next, and then the reoptimization step is applied in the usual way.

Just as for some of the preceding cases, this procedure for Case 4 is a streamlined version of the general procedure summarized at the end of Sec. 6.6. The only question to be addressed for this case is whether the previously optimal solution still is *feasible,* so step 5 (optimality test) has been deleted. Step 4 (feasibility test) has been replaced by a much quicker test of feasibility (does the previously optimal solution satisfy the new constraint?) to be performed right after step 1 (revision of model). It is only if this test provides a negative answer, and you wish to reoptimize, that steps 2, 3, and 6 are used (revision of final tableau, conversion to proper form from Gaussian elimination, and reoptimization).

Example (Variation 6 of the Wyndor Model).   To illustrate this case, we consider Variation 6 of the Wyndor Glass Co. model, which simply introduces the new constraint

$$2x_1 + 3x_2 \leq 24$$

into the Variation 2 model given in Table 6.21. The graphical effect is shown in Fig. 6.7. The previous optimal solution $(0, 9)$ violates the new constraint, so the optimal solution changes to $(0, 8)$.

To analyze this example algebraically, note that $(0, 9)$ yields $2x_1 + 3x_2 = 27 > 24$, so this previous optimal solution is no longer feasible. To find the new optimal solution, add the new constraint to the current final simplex tableau as just described, with the slack variable x_6 as its initial basic variable. This step yields the first tableau shown in Table 6.25. The conversion to proper form from Gaussian elimination then requires subtracting from the new row the product, 3 times row 2, which identifies the current basic solution $x_3 = 4$, $x_2 = 9$, $x_4 = 6$, $x_6 = -3$ $(x_1 = 0, x_5 = 0)$, as shown in the second tableau. Applying the dual simplex method (described in Sec. 7.1) to this tableau then leads in just one iteration (more are sometimes needed) to the new optimal solution in the last tableau of Table 6.25.

Systematic Sensitivity Analysis—Parametric Programming

So far we have described how to test specific changes in the model parameters. Another common approach to sensitivity analysis is to vary one or more parameters continuously over some interval(s) to see when the optimal solution changes.

For example, with Variation 2 of the Wyndor Glass Co. model, rather than beginning by testing the specific change from $b_2 = 12$ to $\bar{b}_2 = 24$, we might instead set

$$\bar{b}_2 = 12 + \theta$$

and then vary θ continuously from 0 to 12 (the maximum value of interest). The geometric interpretation in Fig. 6.3 is that the $2x_2 = 12$ constraint line is being shifted upward to

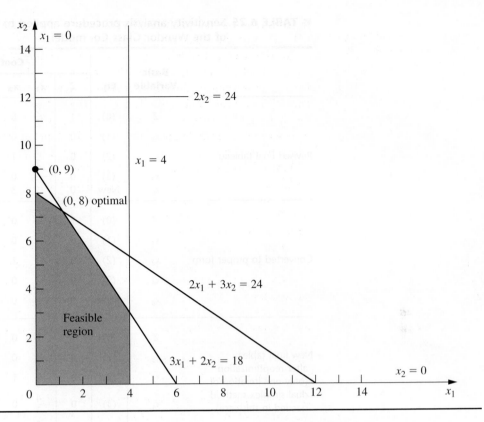

■ FIGURE 6.7
Feasible region for Variation 6 of the Wyndor model where Variation 2 (Fig. 6.3) has been revised by adding the new constraint, $2x_1 + 3x_2 \leq 24$.

$2x_2 = 12 + \theta$, with θ being increased from 0 to 12. The result is that the original optimal CPF solution (2, 6) shifts up the $3x_1 + 2x_2 = 18$ constraint line toward $(-2, 12)$. This corner-point solution remains optimal as long as it is still feasible ($x_1 \geq 0$), after which (0, 9) becomes the optimal solution.

The algebraic calculations of the effect of having $\Delta b_2 = \theta$ are directly analogous to those for the Case 1 example where $\Delta b_2 = 12$. In particular, we use the expressions for Z^* and $\mathbf{b}^*$ given for Case 1,

$$Z^* = \mathbf{y}^* \overline{\mathbf{b}}$$
$$\mathbf{b}^* = \mathbf{S}^* \overline{\mathbf{b}}$$

where $\overline{\mathbf{b}}$ now is

$$\overline{\mathbf{b}} = \begin{bmatrix} 4 \\ 12+\theta \\ 18 \end{bmatrix}$$

and where $\mathbf{y}^*$ and $\mathbf{S}^*$ are given in the boxes in the middle tableau in Table 6.19. These equations indicate that the optimal solution is

$$Z^* = 36 + \frac{3}{2}\theta$$
$$x_3 = 2 + \frac{1}{3}\theta$$
$$x_2 = 6 + \frac{1}{2}\theta \qquad (x_4 = 0, \; x_5 = 0)$$
$$x_1 = 2 - \frac{1}{3}\theta$$

■ **TABLE 6.25** Sensitivity analysis procedure applied to Variation 6 of the Wyndor Glass Co. model

	Basic Variable	Eq.	Z	Coefficient of:						Right Side
				x_1	x_2	x_3	x_4	x_5	x_6	
Revised final tableau	Z	(0)	1	$\frac{9}{2}$	0	0	0	$\frac{5}{2}$	0	45
	x_3	(1)	0	1	0	1	0	0	0	4
	x_2	(2)	0	$\frac{3}{2}$	1	0	0	$\frac{1}{2}$	0	9
	x_4	(3)	0	-3	0	0	1	-1	0	6
	x_6	New	0	2	3	0	0	0	1	24
Converted to proper form	Z	(0)	1	$\frac{9}{2}$	0	0	0	$\frac{5}{2}$	0	45
	x_3	(1)	0	1	0	1	0	0	0	4
	x_2	(2)	0	$\frac{3}{2}$	1	0	0	$\frac{1}{2}$	0	9
	x_4	(3)	0	-3	0	0	1	-1	0	6
	x_6	New	0	$-\frac{5}{2}$	0	0	0	$-\frac{3}{2}$	1	-3
New final tableau after reoptimization (only one iteration of dual simplex method needed in this case)	Z	(0)	1	$\frac{1}{3}$	0	0	0	0	$\frac{5}{3}$	40
	x_3	(1)	0	1	0	1	0	0	0	4
	x_2	(2)	0	$\frac{2}{3}$	1	0	0	0	$\frac{1}{3}$	8
	x_4	(3)	0	$-\frac{4}{3}$	0	0	1	0	$-\frac{2}{3}$	8
	x_5	New	0	$\frac{5}{3}$	0	0	0	1	$-\frac{2}{3}$	2

for θ small enough that this solution still is feasible, i.e., for $\theta \leq 6$. For $\theta > 6$, the dual simplex method (described in Sec. 7.1) yields the tableau shown in Table 6.21 except for the value of x_4. Thus, $Z = 45$, $x_3 = 4$, $x_2 = 9$ (along with $x_1 = 0$, $x_5 = 0$), and the expression for $\mathbf{b}^*$ yields

$$x_4 = b_3^* = 0(4) + 1(12 + \theta) - 1(18) = -6 + \theta.$$

This information can then be used (along with other data not incorporated into the model on the effect of increasing b_2) to decide whether to retain the original optimal solution and, if not, how much to increase b_2.

In a similar way, we can investigate the effect on the optimal solution of varying several parameters simultaneously. When we vary just the b_i parameters, we express the new value $\bar{b}_i$ in terms of the original value b_i as follows:

$$\bar{b}_i = b_i + \alpha_i\theta, \qquad \text{for } i = 1, 2, \ldots, m,$$

where the α_i values are input constants specifying the desired rate of increase (positive or negative) of the corresponding right-hand side as θ is increased.

For example, suppose that it is possible to shift some of the production of a current Wyndor Glass Co. product from Plant 2 to Plant 3, thereby increasing b_2 by decreasing b_3. Also suppose that b_3 decreases twice as fast as b_2 increases. Then

$$\bar{b}_2 = 12 + \theta$$
$$\bar{b}_3 = 18 - 2\theta,$$

where the (nonnegative) value of θ measures the amount of production shifted. (Thus, $\alpha_1 = 0$, $\alpha_2 = 1$, and $\alpha_3 = -2$ in this case.) In Fig. 6.3, the geometric interpretation is that as θ is increased from 0, the $2x_2 = 12$ constraint line is being pushed up to $2x_2 = 12 + \theta$ (ignore the $2x_2 = 24$ line) and simultaneously the $3x_1 + 2x_2 = 18$ constraint line is being pushed down to $3x_1 + 2x_2 = 18 - 2\theta$. The original optimal CPF solution $(2, 6)$ lies at the intersection of the $2x_2 = 12$ and $3x_1 + 2x_2 = 18$ lines, so shifting these lines causes this corner-point solution to shift. However, with the objective function of $Z = 3x_1 + 5x_2$, this corner-point solution will remain optimal as long as it is still feasible ($x_1 \geq 0$).

An algebraic investigation of simultaneously changing b_2 and b_3 in this way again involves using the formulas for Case 1 (treating θ as representing an unknown number) to calculate the resulting changes in the final tableau (middle of Table 6.19), namely,

$$Z^* = \mathbf{y}^* \overline{\mathbf{b}} = [0, \tfrac{3}{2}, 1] \begin{bmatrix} 4 \\ 12 + \theta \\ 18 - 2\theta \end{bmatrix} = 36 - \tfrac{1}{2}\theta,$$

$$\mathbf{b}^* = \mathbf{S}^* \overline{\mathbf{b}} = \begin{bmatrix} 1 & \tfrac{1}{3} & -\tfrac{1}{3} \\ 0 & \tfrac{1}{2} & 0 \\ 0 & -\tfrac{1}{3} & \tfrac{1}{3} \end{bmatrix} \begin{bmatrix} 4 \\ 12 + \theta \\ 18 - 2\theta \end{bmatrix} = \begin{bmatrix} 2 + \theta \\ 6 + \tfrac{1}{2}\theta \\ 2 - \theta \end{bmatrix}.$$

Therefore, the optimal solution becomes

$$Z^* = 36 - \frac{1}{2}\theta$$

$$x_3 = 2 + \theta$$

$$x_2 = 6 + \frac{1}{2}\theta \qquad (x_4 = 0, \qquad x_5 = 0)$$

$$x_1 = 2 - \theta$$

for θ small enough that this solution still is feasible, i.e., for $\theta \leq 2$. (Check this conclusion in Fig. 6.3.) However, the fact that Z decreases as θ increases from 0 indicates that the best choice for θ is $\theta = 0$, so none of the possible shifting of production should be done.

The approach to varying several c_j parameters simultaneously is similar. In this case, we express the new value $\overline{c}_j$ in terms of the original value of c_j as

$$\overline{c}_j = c_j + \alpha_j \theta, \qquad \text{for } j = 1, 2, \ldots, n,$$

where the α_j are input constants specifying the desired rate of increase (positive or negative) of c_j as θ is increased.

To illustrate this case, reconsider the sensitivity analysis of c_1 and c_2 for the Wyndor Glass Co. problem that was performed earlier in this section. Starting with Variation 2 of the Wyndor model presented in Table 6.21 and Fig. 6.3, we separately considered the effect of changing c_1 from 3 to 4 (its most optimistic estimate) and c_2 from 5 to 3 (its most pessimistic estimate). Now we can simultaneously consider both changes, as well as various intermediate cases with smaller changes, by setting

$$\overline{c}_1 = 3 + \theta \qquad \text{and} \qquad \overline{c}_2 = 5 - 2\theta,$$

where the value of θ measures the *fraction* of the maximum possible change that is made. The result is to replace the original objective function $Z = 3x_1 + 5x_2$ by a *function* of θ

$$Z(\theta) = (3 + \theta)x_1 + (5 - 2\theta)x_2,$$

so the optimization now can be performed for any desired (fixed) value of θ between 0 and 1. By checking the effect as θ increases from 0 to 1, we can determine just when and how the optimal solution changes as the error in the original estimates of these parameters increases.

Considering these changes simultaneously is especially appropriate if there are factors that cause the parameters to change together. Are the two products competitive in some sense, so that a larger-than-expected unit profit for one implies a smaller-than-expected unit profit for the other? Are they both affected by some exogenous factor, such as the advertising emphasis of a competitor? Is it possible to simultaneously change both unit profits through appropriate shifting of personnel and equipment?

In the feasible region shown in Fig. 6.3, the geometric interpretation of changing the objective function from $Z = 3x_1 + 5x_2$ to $Z(\theta) = (3 + \theta)x_1 + (5 - 2\theta)x_2$ is that we are changing the *slope* of the original objective function line ($Z = 45 = 3x_1 + 5x_2$) that passes through the optimal solution $(0, 9)$. If θ is increased enough, this slope will change sufficiently that the optimal solution will switch from $(0, 9)$ to another CPF solution $(4, 3)$. (Check graphically whether this occurs for $\theta \leq 1$.)

The algebraic procedure for dealing simultaneously with these two changes ($\Delta c_1 = \theta$ and $\Delta c_2 = -2\theta$) is shown in Table 6.26. Although the changes now are expressed in terms of θ rather than specific numerical amounts, θ is treated just as an unknown number. The table displays just the relevant rows of the tableaux involved (row 0 and the row for the basic variable x_2). The first tableau shown is just the final tableau for the current version of the model (before c_1 and c_2 are changed) as given in Table 6.21. Refer to the formulas in Table 6.17. The only changes in the *revised* final tableau shown next are that Δc_1 and Δc_2 are subtracted from the row 0 coefficients of x_1 and x_2, respectively. To convert this tableau to proper form from Gaussian elimination, we subtract 2θ times row 2 from row 0, which yields the last tableau shown. The expressions in terms of θ for the coefficients of nonbasic variables x_1 and x_5 in row 0 of this tableau show that the current BF solution remains optimal for $\theta \leq \frac{9}{8}$. Because $\theta = 1$ is the maximum realistic value of θ, this indicates that c_1 and c_2 together are insensitive parameters with respect to the Variation 2 model in Table 6.21. There is no need to try to estimate these parameters more closely unless other parameters change (as occurred for Variation 5 of the Wyndor model).

■ **TABLE 6.26** Dealing with $\Delta c_1 = \theta$ and $\Delta c_2 = -2\theta$ for Variation 2 of the Wyndor model as given in Table 6.21

	Basic Variable	Eq.	Z	Coefficient of:					Right Side
				x_1	x_2	x_3	x_4	x_5	
Final tableau	Z	(0)	1	$\frac{9}{2}$	0	0	0	$\frac{5}{2}$	45
	x_2	(2)	0	$\frac{3}{2}$	1	0	0	$\frac{1}{2}$	9
Revised final tableau when $\Delta c_1 = \theta$ and $\Delta c_2 = -2\theta$	$Z(\theta)$	(0)	1	$\frac{9}{2} - \theta$	2θ	0	0	$\frac{5}{2}$	45
	x_2	(2)	0	$\frac{3}{2}$	1	0	0	$\frac{1}{2}$	9
Converted to proper form	$Z(\theta)$	(0)	1	$\frac{9}{2} - 4\theta$	0	0	0	$\frac{5}{2} - \theta$	$45 - 18\theta$
	x_2	(2)	0	$\frac{3}{2}$	1	0	0	$\frac{1}{2}$	9

As we discussed in Sec. 4.7, this way of continuously varying several parameters simultaneously is referred to as *parametric linear programming.* Section 7.2 presents the complete parametric linear programming procedure (including identifying new optimal solutions for larger values of θ) when just the c_j parameters are being varied and then when just the b_i parameters are being varied. Some linear programming software packages also include routines for varying just the coefficients of a single variable or just the parameters of a single constraint. In addition to the other applications discussed in Sec. 4.7, these procedures provide a convenient way of conducting sensitivity analysis systematically.

■ 6.8   PERFORMING SENSITIVITY ANALYSIS ON A SPREADSHEET[7]

With the help of the Excel Solver, spreadsheets provide an alternative, relatively straightforward way of performing much of the sensitivity analysis described in Secs. 6.5–6.7. The spreadsheet approach is basically the same for each of the cases considered in Sec. 6.7 for the types of changes made in the original model. Therefore, we will focus on only the effect of changes in the coefficients of the variables in the objective function (Cases 2*a* and 3 in Sec. 6.7). We will illustrate this effect by making changes in the *original* Wyndor model formulated in Sec. 3.1, where the coefficients of x_1 (number of batches of the new door produced per week) and x_2 (number of batches of the new window produced per week) in the objective function are

$c_1 = 3 =$ profit (in thousands of dollars) per batch of the new type of door,
$c_2 = 5 =$ profit (in thousands of dollars) per batch of the new type of window.

For your convenience, the spreadsheet formulation of this model (Fig. 3.22) is repeated here as Fig. 6.8. Note that the cells containing the quantities to be changed are ProfitPerBatch (C4:D4). Since the profits in these cells are expressed in dollars, whereas c_1 and c_2 are in units of thousands of dollars, we hereafter will discuss the sensitivity analysis in terms of the changes in the profits shown in these cells instead of changes in c_1 and c_2. To this end, we will denote these profits by

$P_D =$ profit per batch of doors currently entered in cell C4,
$P_W =$ profit per batch of windows currently entered in cell D4.

Spreadsheets actually provide three methods of performing sensitivity analysis. One is to check the effect of an individual change in the model by simply making the change on the spreadsheet and re-solving. A second is to systematically generate a table on a single spreadsheet that shows the effect of a series of changes in one or two parameters of the model. A third is to obtain and apply Excel's sensitivity report. We describe each of these methods in turn below.

Checking Individual Changes in the Model

One of the great strengths of a spreadsheet is the ease with which it can be used interactively to perform various kinds of sensitivity analysis. Once the Solver has been set up to obtain an optimal solution, you can immediately find out what would happen if one of the parameters of the model were changed to some other value. All you have to do is make this change on the spreadsheet and then click on the Solve button again.

---

[7]We have written this section in a way that can be understood without first reading any of the preceding sections in this chapter. However, Sec. 4.7 is important background for the latter part of this section.

FIGURE 6.8
The spreadsheet model and the optimal solution obtained for the original Wyndor problem before performing sensitivity analysis.

	A	B	C	D	E	F	G
1		**Wyndor Glass Co. Product-Mix Problem**					
2							
3			Doors	Windows			
4		Profit Per Batch	$3,000	$5,000			
5					Hours		Hours
6			Hours Used Per Batch Produced		Used		Available
7		Plant 1	1	0	2	<=	4
8		Plant 2	0	2	12	<=	12
9		Plant 3	3	2	18	<=	18
10							
11			Doors	Windows			Total Profit
12		Batches Produced	2	6			$36,000

Solver Parameters

Set Target Cell: TotalProfit
Equal To: ● Max ○ Min
By Changing Cells:
BatchesProduced
Subject to the Constraints:
HoursUsed <= HoursAvailable

Solver Options
☑ Assume Linear Model
☑ Assume Non-Negative

	E
5	Hours
6	Used
7	=SUMPRODUCT(C7:D7,BatchesProduced)
8	=SUMPRODUCT(C8:D8,BatchesProduced)
9	=SUMPRODUCT(C9:D9,BatchesProduced)

	G
11	Total Profit
12	=SUMPRODUCT(ProfitPerBatch,BatchesProduced)

Range Name	Cells
BatchesProduced	C12:D12
HoursAvailable	G7:G9
HoursUsed	E7:E9
HoursUsedPerBatchProduced	C7:D9
ProfitPerBatch	C4:D4
TotalProfit	G12

FIGURE 6.9
The revised Wyndor problem where the estimate of the profit per batch of doors has been decreased from $P_D = \$3,000$ to $P_D = \$2,000$, but no change occurs in the optimal solution for the product mix.

	A	B	C	D	E	F	G
1		**Wyndor Glass Co. Product-Mix Problem**					
2							
3			Doors	Windows			
4		Profit Per Batch	$2,000	$5,000			
5					Hours		Hours
6			Hours Used Per Batch Produced		Used		Available
7		Plant 1	1	0	2	<=	4
8		Plant 2	0	2	12	<=	12
9		Plant 3	3	2	18	<=	18
10							
11			Doors	Windows			Total Profit
12		Batches Produced	2	6			$34,000

To illustrate, suppose that Wyndor management is quite uncertain about what the profit per batch of doors (P_D) will turn out to be. Although the figure of $3,000 given in Fig. 6.8 is considered to be a reasonable initial estimate, management feels that the true profit could end up deviating substantially from this figure in either direction. However, the range between $P_D = \$2,000$ and $P_D = \$5,000$ is considered fairly likely.

Figure 6.9 shows what would happen if the profit per batch of doors were to drop from $P_D = \$3,000$ to $P_D = \$2,000$. Comparing with Fig. 6.8, there is no change at all in

■ **FIGURE 6.10**
The revised Wyndor problem where the estimate of the profit per batch of doors has been increased from $P_D = \$3,000$ to $P_D = \$5,000$, but no change occurs in the optimal solution for the product mix.

	A	B	C	D	E	F	G
1		**Wyndor Glass Co. Product-Mix Problem**					
2							
3			Doors	Windows			
4		Profit Per Batch	$5,000	$5,000			
5					Hours		Hours
6			Hours Used Per Batch Produced		Used		Available
7		Plant 1	1	0	2	<=	4
8		Plant 2	0	2	12	<=	12
9		Plant 3	3	2	18	<=	18
10							
11			Doors	Windows			Total Profit
12		Batches Produced	2	6			$40,000

the optimal solution for the product mix. In fact, the *only* changes in the new spreadsheet are the new value of P_D in cell C4 and a decrease of $2,000 in the total profit shown in cell G12 (because each of the two batches of doors produced per week provides $1,000 less profit). Because the optimal solution does not change, we now know that the original estimate of $P_D = \$3,000$ can be considerably *too high* without invalidating the model's optimal solution.

But what happens if this estimate is *too low* instead? Figure 6.10 shows what would happen if P_D were increased to $P_D = \$5,000$. Again, there is no change in the optimal solution. Therefore, we now know that the range of values of P_D over which the current optimal solution remains optimal (i.e., the *allowable range* discussed in Sec. 6.7) includes the range from $2,000 to $5,000 and may extend further.

Because the original value of $P_D = \$3,000$ can be changed considerably in either direction without changing the optimal solution, P_D is a relatively insensitive parameter. It is not necessary to pin down this estimate with great accuracy in order to have confidence that the model is providing the correct optimal solution.

This may be all the information that is needed about P_D. However, if there is a good possibility that the true value of P_D will turn out to be even outside this broad range from $2,000 to $5,000, further investigation would be desirable. How much higher or lower can P_D be before the optimal solution would change?

Figure 6.11 demonstrates that the optimal solution would indeed change if P_D is increased all the way up to $P_D = \$10,000$. Thus, we now know that this change occurs somewhere between $5,000 and $10,000 during the process of increasing P_D.

■ **FIGURE 6.11**
The revised Wyndor problem where the estimate of the profit per batch of doors has been increased from $P_D = \$3,000$ to $P_D = \$10,000$, which results in a change in the optimal solution for the product mix.

	A	B	C	D	E	F	G
1		**Wyndor Glass Co. Product-Mix Problem**					
2							
3			Doors	Windows			
4		Profit Per Batch	$10,000	$5,000			
5					Hours		Hours
6			Hours Used Per Batch Produced		Used		Available
7		Plant 1	1	0	4	<=	4
8		Plant 2	0	2	6	<=	12
9		Plant 3	3	2	18	<=	18
10							
11			Doors	Windows			Total Profit
12		Batches Produced	4	3			$55,000

Using the Solver Table to Do Sensitivity Analysis Systematically

To pin down just when the optimal solution will change, we could continue selecting new values of P_D at random. However, a better approach is to systematically consider a range of values of P_D. An Excel add-in developed by Professor Mark Hillier, called the *Solver Table,* is designed to perform just this sort of analysis. It is available to you in your OR Courseware on the book's website. To install it, you need simply to open the Solver Table file in OR Courseware.

The Solver Table is used to show the results in the changing cells and/or certain output cells for various trial values in a data cell. For each trial value in the data cell, Solver is called on to re-solve the problem. Therefore, the Solver Table (or any comparable Excel add-in) provides a systematic way of performing sensitivity analysis and then displaying the results to managers and others who are not familiar with the more technical aspects of sensitivity analysis.

To use the Solver Table, first expand the original spreadsheet (Fig. 6.8) to make a table with headings as shown in Fig. 6.12. In the first column of the table (cells B19:B28), list the trial values for the data cell (the profit per batch of doors), except leave the first row (cell B18) blank. The headings of the next columns specify which output will be evaluated. For each of these columns, use the first row of the table (cells C18:E18) to write an equation that sets the value in each of these cells equal to the relevant changing cell or output cell. In this case, the cells of interest are DoorBatchesProduced (C12), WindowBatchesProduced (D12), and TotalProfit (G12), so the equations for C18:E18 are those shown just below the spreadsheet in Fig. 6.12.

Next, select the entire table by clicking and dragging from cells B18 through E28, and then choose Solver Table from the Add-Ins tab (for Excel 2007) or Tools menu (for earlier versions of Excel), after having installed this Excel add-in provided in your OR Courseware. In the Solver Table dialogue box (as shown at the bottom of Fig. 6.12), indicate the column input cell (C4), which refers to the data cell that is being changed in the first column of the table. Nothing is entered for the row input cell because no row is being used to list the trial values of a data cell in this case.

The Solver Table shown in Fig. 6.13 is then generated automatically by clicking on the OK button. For each trial value listed in the first column of the table for the data cell of interest, Excel re-solves the problem using Solver and then fills in the corresponding values in the other columns of the tables. (The numbers in the first row of the table come from the original solution in the spreadsheet before the original value in the data cell was changed.)

The table reveals that the optimal solution remains the same all the way from $P_D = \$1,000$ (and perhaps lower) to $P_D = \$7,000$, but that a change occurs somewhere between $7,000 and $8,000. We next could systematically consider values of P_D between $7,000 and $8,000 to determine more closely where the optimal solution changes. However, this is not necessary since, as discussed a little later, a shortcut is to use the Excel sensitivity report to determine exactly where the optimal solution changes.

Thus far, we have illustrated how to systematically investigate the effect of changing only P_D (cell C4 in Fig. 6.8). The approach is the same for P_W (cell D4). In fact, the Solver Table can be used in this way to investigate the effect of changing *any* single data cell in the model, including any cell in HoursAvailable (G7:G9) or HoursUsedPerBatchProduced (C7:D9).

We next will illustrate how to investigate simultaneous changes in two data cells with a spreadsheet, first by itself and then with the help of the Solver Table.

Checking Two-Way Changes in the Model

When using the original estimates for P_D ($3,000) and P_W ($5,000), the optimal solution indicated by the model (Fig. 6.8) is heavily weighted toward producing the windows (6 batches per week) rather than the doors (only 2 batches per week). Suppose that

	A	B	C	D	E	F	G
1		**Wyndor Glass Co. Product-Mix Problem**					
2							
3			Doors	Windows			
4		Profit Per Batch	$3,000	$5,000			
5					Hours		Hours
6			Hours Used Per Batch Produced		Used		Available
7		Plant 1	1	0	2	<=	4
8		Plant 2	0	2	12	<=	12
9		Plant 3	3	2	18	<=	18
10							
11			Doors	Windows			Total Profit
12		Batches Produced	2	6			$36,000
13							
14							
15							
16		Profit Per Batch	Optimal Batches Produced		Total		
17		for Doors	Doors	Windows	Profit		Select these
18			2	6	$36,000		cells
19		$1,000					(B18:E28),
20		$2,000					before
21		$3,000					choosing the
22		$4,000					Solver Table.
23		$5,000					
24		$6,000					
25		$7,000					
26		$8,000					
27		$9,000					
28		$10,000					

	C	D	E
16	Optimal Batches Produced		Total
17	Doors	Windows	Profit
18	=DoorBatchesProduced	=WindowBatchesProduced	=TotalProfit

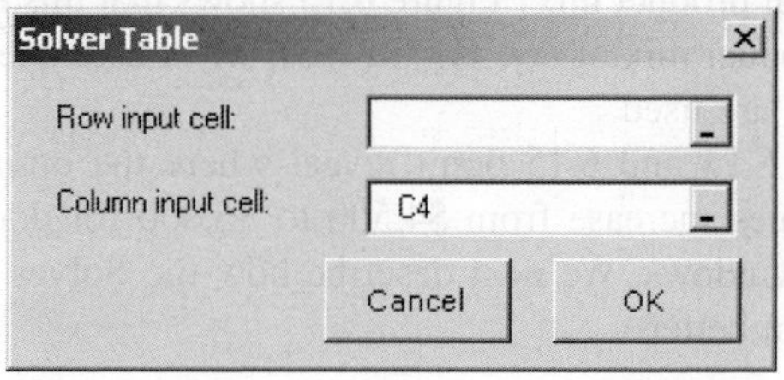

FIGURE 6.12
Expansion of the spreadsheet in Fig. 6.8 to prepare for using the Solver Table to show the effect of systematically varying the estimate of the profit per batch of doors in the Wyndor problem.

Solver Table

Row input cell:

Column input cell: C4

Cancel   OK

Range Name	Cells
DoorBatchesProduced	C12
TotalProfit	G12
WindowBatchesProduced	D12

Wyndor management is concerned about this imbalance and feels that the problem may be that the estimate for P_D is too low and the estimate for P_W is too high. This raises the question: If the estimates are indeed off in these directions, would this lead to a more balanced product mix being the most profitable one? (Keep in mind that it is the *ratio* of P_D to P_W that is relevant for determining the optimal product mix, so having their estimates be off in the *same* direction with little change in this ratio is unlikely to change the optimal product mix).

This question can be answered in a matter of seconds simply by substituting new estimates of the profits per batch in the original spreadsheet in Fig. 6.8 and clicking on the Solve button. Figure 6.14 shows that new estimates of $4,500 for doors and $4,000 for windows causes no change at all in the solution for the optimal product mix. (The total profit does change, but this occurs only because of the changes in the profits per batch.) Would even larger changes in the estimates of profits per batch finally lead to a change

	A	B	C	D	E	F	G
1		**Wyndor Glass Co. Product-Mix Problem**					
2							
3			Doors	Windows			
4		Profit Per Batch	$3,000	$5,000			
5					Hours		Hours
6			Hours Used Per Batch Produced		Used		Available
7		Plant 1	1	0	2	<=	4
8		Plant 2	0	2	12	<=	12
9		Plant 3	3	2	18	<=	18
10							
11			Doors	Windows			Total Profit
12		Batches Produced	2	6			$36,000
13							
14							
15							
16		Profit Per Batch	Optimal Batches Produced		Total		
17		for Doors	Doors	Windows	Profit		
18			2	6	$36,000		
19		$1,000	2	6	$32,000		
20		$2,000	2	6	$34,000		
21		$3,000	2	6	$36,000		
22		$4,000	2	6	$38,000		
23		$5,000	2	6	$40,000		
24		$6,000	2	6	$42,000		
25		$7,000	2	6	$44,000		
26		$8,000	4	3	$47,000		
27		$9,000	4	3	$51,000		
28		$10,000	4	3	$55,000		

■ **FIGURE 6.13**
An application of the Solver Table that shows the effect of systematically varying the estimate of the profit per batch for doors in the Wyndor problem.

in the optimal product mix? Figure 6.15 shows that this does happen, yielding a relatively balanced product mix of $(x_1, x_2) = (4, 3)$, when estimates of $6,000 for doors and $3,000 for windows are used.

Figures 6.14 and 6.15 don't reveal where the optimal product mix changes as the profit estimates increase from $4,500 to $6,000 for doors and decrease from $4,000 to $3,000 for windows. We next describe how the Solver Table can systematically help to pin this down better.

■ **FIGURE 6.14**
The revised Wyndor problem where the estimates of the profits per batch of doors and windows have been changed to $P_D = \$4,500$ and $P_W = \$4,000$, respectively, but no change occurs in the optimal product mix.

	A	B	C	D	E	F	G
1		**Wyndor Glass Co. Product-Mix Problem**					
2							
3			Doors	Windows			
4		Profit Per Batch	$4,500	$4,000			
5					Hours		Hours
6			Hours Used Per Batch Produced		Used		Available
7		Plant 1	1	0	2	<=	4
8		Plant 2	0	2	12	<=	12
9		Plant 3	3	2	18	<=	18
10							
11			Doors	Windows			Total Profit
12		Batches Produced	2	6			$33,000

	A	B	C	D	E	F	G
1		**Wyndor Glass Co. Product-Mix Problem**					
2							
3			Doors	Windows			
4		Profit Per Batch	$6,000	$3,000			
5					Hours		Hours
6			Hours Used Per Batch Produced		Used		Available
7		Plant 1	1	0	4	<=	4
8		Plant 2	0	2	6	<=	12
9		Plant 3	3	2	18	<=	18
10							
11			Doors	Windows			Total Profit
12		Batches Produced	4	3			$33,000

■ **FIGURE 6.15**
The revised Wyndor problem where the estimates of the profits per batch of doors and windows have been changed to $6,000 and $3,000, respectively, which results in a change in the optimal product mix.

Using the Solver Table for Two-Way Sensitivity Analysis

A two-way version of the Solver Table provides a way of systematically investigating the effect if the estimates entered into *two* data cells are inaccurate simultaneously. (However, two is the maximum number of data cells that can be considered simultaneously by the Solver Table.) In this case, the Solver Table shows the results in a single output cell for various trial values in the two data cells.

To illustrate this approach, we again will investigate the effect of increasing P_D and decreasing P_W simultaneously. Before considering the effect on the optimal product mix, we will look at the effect on the total profit. To do this, the Solver Table will be used to show how TotalProfit (G12) in Fig. 6.8 varies over a range of trial values in the two data cells, ProfitPerBatch (C4:D4). For each pair of trial values in these data cells, Solver will be called on to re-solve the problem.

To create a two-way Solver Table for the Wyndor problem, expand the original spreadsheet (Fig. 6.8) to make a table with column and row headings as shown in rows 16–21 of the spreadsheet in Fig. 6.16. In the upper left-hand corner of the table (C17), write an equation (=TotalProfit) that refers to the target cell. In the first column of the table (column C, below the equation in cell C17), insert various trial values for the first data cell of interest (the profit per batch of the doors). In the first row of the table (row 17, to the right of the equation in cell C17), insert various trial values for the second data cell of interest (the profit per batch of the windows).

Next, select the entire table (C17:H21) and choose Solver Table from the Add-Ins tab (for Excel 2007) or Tools menu (for earlier versions of Excel), after having installed this Excel add-in provided in your OR Courseware. In the Solver Table dialogue box (shown at the bottom of Fig. 6.16), indicate which data cells are being changed simultaneously. The column input cell C4 refers to the data cell whose various trial values are listed in the first column of the table (C18:C21), while the row input cell D4 refers to the data cell whose various trial values are listed in the first row of the table (D17:H17).

The Solver Table shown in Fig. 6.17 is then generated automatically by clicking on the OK button. For each pair of trial values for the two data cells, Excel re-solves the problem using Solver and then fills in the total profit in the corresponding spot in the table. (The number in C17 comes from the target cell in the original spreadsheet before the original values in the two data cells are changed.)

Unlike a one-way Solver Table that can show the results of *multiple* changing cells and/or output cells for various trial values of a single data cell, a two-way Solver Table is limited to showing the results in a *single* cell for each pair of trial values in the two data cells of interest.

	A	B	C	D	E	F	G	H	I
1		**Wyndor Glass Co. Product-Mix Problem**							
2									
3			Doors	Windows					
4		Profit Per Batch	$3,000	$5,000					
5					Hours		Hours		
6			Hours Used Per Batch Produced		Used		Available		
7		Plant 1	1	0	2	<=	4		
8		Plant 2	0	2	12	<=	12		
9		Plant 3	3	2	18	<=	18		Select these
10									cells
11			Doors	Windows			Total Profit		(C17:H21),
12		Batches Produced	2	6			$36,000		before
13									choosing the
14									Solver Table.
15									
16		**Total Profit**			Profit Per Batch for Windows				
17			$36,000	$1,000	$2,000	$3,000	$4,000	$5,000	
18			$3,000						
19		Profit Per Batch	$4,000						
20		for Doors	$5,000						
21			$6,000						

■ **FIGURE 6.16**

Expansion of the spreadsheet in Fig. 6.8 to prepare for using a two-dimensional Solver Table to show the effect on total profits of systematically varying the estimates of the profits per batch of doors and windows for the Wyndor problem.

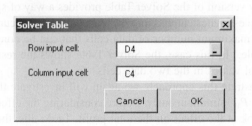

	Solver Table	
Row input cell:	D4	
Column input cell:	C4	

		C
17		=TotalProfit

Range Name	Cell
TotalProfit	G12

■ **FIGURE 6.17**

A two-dimensional application of the Solver Table that shows the effect on the optimal total profit of systematically varying the estimates of the profits per batch of doors and windows for the Wyndor problem.

	B	C	D	E	F	G	H
16	**Total Profit**		Profit Per Batch for Windows				
17		$36,000	$1,000	$2,000	$3,000	$4,000	$5,000
18		$3,000	$15,000	$18,000	$24,000	$30,000	$36,000
19	Profit Per Batch	$4,000	$19,000	$22,000	$26,000	$32,000	$38,000
20	for Doors	$5,000	$23,000	$26,000	$29,000	$34,000	$40,000
21		$6,000	$27,000	$30,000	$33,000	$36,000	$42,000

However, there is a trick using the & symbol that enables Solver Table to show the results from multiple changing cells and/or output cells within a single cell of the table. We utilize this trick in the Solver Table shown in Fig. 6.18 to show the results for *both* changing cells, DoorBatchesProduced (C12) and WindowBatchesProduced (D12), for each pair of trial values for ProfitPerBatch (C4:D4). The key formula is in cell C25:

C25 = "("& DoorBatchesProduced &", "& WindowBatchesProduced &")"

The & character tells Excel to concatenate, so the result will be a left parenthesis, followed by the value in DoorBatchesProduced (C12), then a comma and the contents in Window-BatchesProduced (D12), and finally a right parenthesis. If DoorBatchesProduced = 2 and WindowBatchesProduced = 6, the result is (2, 6). Thus, the results from *both* changing cells are displayed within a *single* cell of the table.

After the usual preliminaries in entering the information shown in rows 24–25 and columns B-C of Fig. 6.18, along with the formula in C25, clicking on the OK button

	B	C	D	E	F	G	H
24	Batches Produced (Doors, Windows)			Profit Per Batch for Windows			
25		(2,6)	$1,000	$2,000	$3,000	$4,000	$5,000
26		$3,000	(4,3)	(4,3)	(2,6)	(2,6)	(2,6)
27	Profit Per Batch	$4,000	(4,3)	(4,3)	(2,6)	(2,6)	(2,6)
28	for Doors	$5,000	(4,3)	(4,3)	(4,3)	(2,6)	(2,6)
29		$6,000	(4,3)	(4,3)	(4,3)	(4,3)	(4,3)

	C
25	="(" & DoorBatchesProduced & "," & WindowBatchesProduced & ")"

■ **FIGURE 6.18**
A two-dimensional application of the Solver Table that shows the effect on the optimal product mix of systematically varying the estimates of the profits per batch of doors and windows for the Wyndor problem.

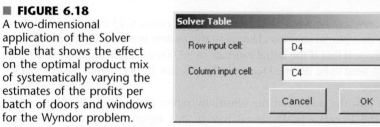

Range Name	Cells
DoorBatchesProduced	C12
WindowBatchesProduced	D12

automatically generates the entire Solver Table. Cells D26:H29 show the optimal solution for the various combinations of trial values for the profits per batch of the doors and windows. The upper right-hand corner (cell H26) of this Solver Table gives the optimal solution of $(x_1, x_2) = (2, 6)$ when using the original profit estimates of $3,000 per batch of doors and $5,000 per batch of windows. Moving down from this cell corresponds to increasing this estimate for doors while moving to the left amounts to decreasing the estimate for windows. (The cells when moving up or to the right of H26 are not shown because these changes would only increase the attractiveness of $(x_1, x_2) = (2, 6)$ as the optimal solution.) Note that $(x_1, x_2) = (2, 6)$ continues to be the optimal solution for all the cells near H26. This indicates that the original estimates of profit per batch would need to be very inaccurate indeed before the optimal product mix would change.

Using the Sensitivity Report to Perform Sensitivity Analysis

You now have seen how some sensitivity analysis can be performed readily on a spreadsheet either by interactively making changes in data cells and re-solving or by using the Solver Table to generate similar information systematically. However, there is a shortcut. Some of the same information (and more) can be obtained more quickly and precisely by simply using the sensitivity report provided by the Excel Solver. (Essentially the same sensitivity report is a standard part of the output available from other linear programming software packages as well, including MPL/CPLEX, LINDO, and LINGO.)

Section 4.7 already has discussed the sensitivity report and how it is used to perform sensitivity analysis. Figure 4.10 in that section shows the sensitivity report for the Wyndor problem. Part of this report is shown here in Fig. 6.19. Rather than repeating Sec. 4.7, we will focus here on illustrating how the sensitivity report can efficiently address the specific questions raised in the preceding subsections for the Wyndor problem.

The question considered in the first two subsections was how far the initial estimate of $3,000 for P_D could be off before the current optimal solution, $(x_1, x_2) = (2, 6)$, would change. Figures 6.10 and 6.11 showed that the optimal solution would not change until

■ **FIGURE 6.19**
Part of the sensitivity report
generated by the Excel Solver
for the original Wyndor
problem (Fig. 6.8), where the
last three columns identify
the allowable ranges for the
profits per batch of doors
and windows.

Adjustable Cells

Cell	Name	Final Value	Reduced Cost	Objective Coefficient	Allowable Increase	Allowable Decrease
$C$12	DoorBatchesProduced	2	0	3000	4500	3000
$D$12	WindowBatchesProduced	6	0	5000	1E+30	3000

P_D is raised to somewhere between \$5,000 and \$10,000. Figure 6.13 then narrowed down the gap for where the optimal solution changes to somewhere between \$7,000 and \$8,000. This figure also showed that if the initial estimate of \$3,000 for P_D is too high rather than too low, P_D would need to be dropped to somewhere below \$1,000 before the optimal solution would change.

Now look at how the portion of the sensitivity report in Figure 6.19 addresses this same question. The DoorBatchesProduced row in this report provides the following information (without the dollar signs) about P_D.

Current value of P_D:	3,000.	
Allowable increase in P_D:	4,500.	So $P_D \leq 3,000 + 4,500 = 7,500$
Allowable decrease in P_D:	3,000.	So $P_D \geq 3,000 - 3,000 = 0$.
Allowable range for P_D:		$0 \leq P_D \leq 7,500$.

Therefore, if P_D is changed from its current value (without making any other change in the model), the current solution $(x_1, x_2) = (2, 6)$ will remain optimal so long as the new value of P_D is within this *allowable range*, $0 \leq P_D \leq \$7,500$.

Figure 6.20 provides graphical insight into this allowable range. For the original value of $P_D = 3,000$, the solid line in the figure shows the slope of the objective function line passing through (2, 6). At the lower end of the allowable range, where $P_D = 0$, the objective function line that passes through (2, 6) now is line B in the figure, so every point on the line segment between (0, 6) and (2, 6) is an optimal solution. For any value of $P_D < 0$, the objective function line will have rotated even further so that (0, 6) becomes the only optimal solution. At the upper end of the allowable range, when $P_D = 7,500$, the objective function line that passes through (2, 6) becomes line C, so every point on the line segment between (2, 6) and (4, 3) becomes an optimal solution. For any value of $P_D > 7,500$, the objective function line is even steeper than line C, so (4, 3) becomes the only optimal solution. Consequently, the original optimal solution, $(x_1, x_2) = (2, 6)$ remains optimal only as long as $0 \leq P_D \leq \$7,500$.

The procedure called *Graphical Method and Sensitivity Analysis* in IOR Tutorial is designed to help you perform this kind of graphical analysis. After you enter the model for the original Wyndor problem, the module provides you with the graph shown in Fig. 6.20 (without the dashed lines). You then can simply drag one end of the objective line up or down to see how far you can increase or decrease P_D before $(x_1, x_2) = (2, 6)$ will no longer be optimal.

Conclusion: The allowable range for P_D is $0 \leq P_D \leq \$7,500$, because $(x_1, x_2) = (2, 6)$ remains optimal over this range but not beyond. (When $P_D = 0$ or $P_D = \$7,500$, there are multiple optimal solutions, but $(x_1, x_2) = (2, 6)$ still is one of them.) With the range this wide around the original estimate of \$3,000

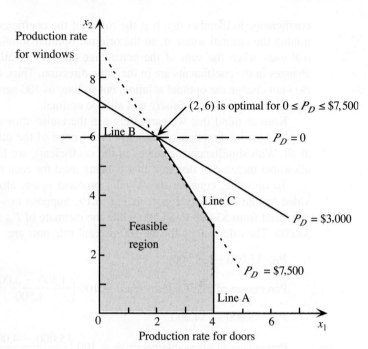

■ FIGURE 6.20
The two dashed lines that pass through solid constraint boundary lines are the objective function lines when P_D (the profit per batch of doors) is at an endpoint of its allowable range, $0 \leq P_D \leq \$7,500$, since either line or any objective function line in between still yields $(x_1, x_2) = (2, 6)$ as an optimal solution for the Wyndor problem.

($P_D = \$3,000$) for the profit per batch of doors, we can be quite confident of obtaining the correct optimal solution for the true profit.

Now let us turn to the question considered in the preceding two subsections. What would happen if the estimate of P_D ($\$3,000$) were too low and the estimate of P_W ($\$5,000$) were too high simultaneously? Specifically, how far can the estimates be off in these directions before the current optimal solution, $(x_1, x_2) = (2, 6)$, would change?

Figure 6.14 showed that if P_D were increased by $\$1,500$ (from $\$3,000$ to $\$4,500$) and P_W were decreased by $\$1,000$ (from $\$5,000$ to $\$4,000$), the optimal solution would remain the same. Figure 6.15 then indicated that doubling these changes would result in a change in the optimal solution. However, it is unclear where the change in the optimal solution occurs. Figure 6.18 provided further information, but not a definitive answer to this question.

Fortunately, additional information can be gleaned from the sensitivity report (Fig. 6.19) by using its allowable increases and allowable decreases in P_D and P_W. The key is to apply the following rule (as first stated in Sec. 6.7):

The 100 Percent Rule for Simultaneous Changes in Objective Function Coefficients: If simultaneous changes are made in the coefficients of the objective function, calculate for each change the percentage of the allowable change (increase or decrease) for that coefficient to remain within its allowable range. If the *sum* of the percentage changes does *not* exceed 100 percent, the original optimal solution definitely will still be optimal. (If the sum *does* exceed 100 percent, then we cannot be sure.)

This rule does not spell out what happens if the sum of the percentage changes *does* exceed 100 percent. The consequence depends on the directions of the changes in the coefficients. Remember that it is the *ratios* of the coefficients that are relevant for determining the optimal solution, so the original optimal solution might indeed remain optimal even when the sum of the percentage changes greatly exceeds 100 percent if the changes in the coefficients are in the same direction. Thus, exceeding 100 percent may or may not change the optimal solution, but so long as 100 percent is not exceeded, the original optimal solution *definitely* will still be optimal.

Keep in mind that we can safely use the entire allowable increase or decrease in a single objective function coefficient only if none of the other coefficients have changed at all. With simultaneous changes in the coefficients, we focus on the *percentage* of the allowable increase or decrease that is being used for each coefficient.

To illustrate, consider the Wyndor problem again, along with the information provided by the sensitivity report in Fig. 6.19. Suppose now that the estimate of P_D has increased from \$3,000 to \$4,500 while the estimate of P_W has decreased from \$5,000 to \$4,000. The calculations for the 100 percent rule now are

$$P_D: \$3,000 \rightarrow \$4,500.$$

$$\text{Percentage of allowable increase} = 100 \left(\frac{4,500 - 3,000}{4,500} \right)\% = 33\frac{1}{3}\%$$

$$P_W: \$5,000 \rightarrow \$4,000.$$

$$\text{Percentage of allowable decrease} = 100 \left(\frac{5,000 - 4,000}{3,000} \right)\% = 33\frac{1}{3}\%$$

$$\text{Sum} = 66\frac{2}{3}\%.$$

Since the sum of the percentages does not exceed 100 percent, the original optimal solution $(x_1, x_2) = (2, 6)$ definitely is still optimal, just as we found earlier in Fig. 6.14.

Now suppose that the estimate of P_D has increased from \$3,000 to \$6,000 while the estimate P_W has decreased from \$5,000 to \$3,000. The calculations for the 100 percent rule now are

$$P_D: \$3,000 \rightarrow \$6,000.$$

$$\text{Percentage of allowable increase} = 100 \left(\frac{6,000 - 3,000}{4,500} \right)\% = 66\frac{2}{3}\%$$

$$P_W: \$5,000 \rightarrow \$3,000.$$

$$\text{Percentage of allowable decrease} = 100 \left(\frac{5,000 - 3,000}{3,000} \right)\% = 66\frac{2}{3}\%$$

$$\text{Sum} = 133\frac{1}{3}\%.$$

Since the sum of the percentages now exceeds 100 percent, the 100 percent rule says that we can no longer guarantee that $(x_1, x_2) = (2, 6)$ is still optimal. In fact, we found earlier in both Figs. 6.15 and 6.18 that the optimal solution has changed to $(x_1, x_2) = (4, 3)$.

These results suggest how to find just where the optimal solution changes while P_D is being increased and P_W is being decreased by these relative amounts. Since 100 percent is midway between $66\frac{2}{3}$ percent and $133\frac{1}{3}$ percent, the sum of the percentage changes will equal 100 percent when the values of P_D and P_W are midway between their values in the above cases. In particular, $P_D = \$5,250$ is midway between \$4,500 and \$6,000 and

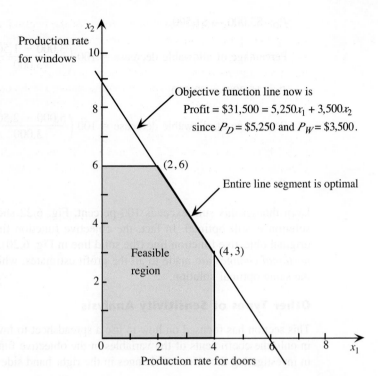

x_2

Production rate
for windows

Objective function line now is
   Profit = \$31,500 = $5,250x_1 + 3,500x_2$
   since P_D = \$5,250 and P_W = \$3,500.

(2, 6)

Entire line segment is optimal

Feasible
region

(4, 3)

Production rate for doors

x_1

■ **FIGURE 6.21**
When the estimates of the
profits per batch of doors
and windows change to
P_D = \$5,250 and
P_W = \$3,500, which lies at
the edge of what is allowed
by the 100 percent rule, the
graphical method shows that
(x_1, x_2) = (2, 6) still is an
optimal solution, but now
every other point on the line
segment between this
solution and (4, 3) also is
optimal.

P_W = \$3,500 is midway between \$4,000 and \$3,000. The corresponding calculations for the 100 percent rule are

P_D: \$3,000 → \$5,250.

$$\text{Percentage of allowable increase} = 100 \left(\frac{5,250 - 3,000}{4,500} \right) \% = 50\%$$

P_W: \$5,000 → \$3,500.

$$\text{Percentage of allowable decrease} = 100 \left(\frac{5,000 - 3,500}{3,000} \right) \% = 50\%$$

Sum = 100%.

Although the sum of the percentages equals 100 percent, the fact that it does not *exceed* 100 percent guarantees that (x_1, x_2) = (2, 6) is still optimal. Figure 6.21 shows graphically that *both* (2, 6) and (4, 3) are now optimal, as well as all the points on the line segment connecting these two points. However, if P_D and P_W were to be changed any further from their original values (so that the sum of the percentages exceeds 100 percent), the objective function line would be rotated so far toward the vertical that (x_1, x_2) = (4, 3) would become the only optimal solution.

At the same time, keep in mind that having the sum of the percentages of allowable changes exceed 100 percent does not automatically mean that the optimal solution will change. For example, suppose that the estimates of both unit profits are halved. The resulting calculations for the 100 percent rule are

P_D: \$3,000 → \$1,500.

$$\text{Percentage of allowable decrease} = 100\left(\frac{3{,}000 - 1{,}500}{3{,}000}\right)\% = \quad 50\%$$

P_W: \$5,000 → \$2,500.

$$\text{Percentage of allowable decrease} = 100\left(\frac{5{,}000 - 2{,}500}{3{,}000}\right)\% = \quad 83\tfrac{1}{3}\%$$

$$\text{Sum} = 133\tfrac{1}{3}\%.$$

Even though this sum exceeds 100 percent, Fig. 6.22 shows that the original optimal solution is still optimal. In fact, the objective function line has the same slope as the original objective function line (the solid line in Fig. 6.20). This happens whenever *proportional changes* are made to all the profit estimates, which will automatically lead to the same optimal solution.

Other Types of Sensitivity Analysis

This section has focused on how to use a spreadsheet to investigate the effect of changes in only the coefficients of the variables in the objective function. One often is interested in investigating the effect of changes in the right-hand sides of the functional constraints

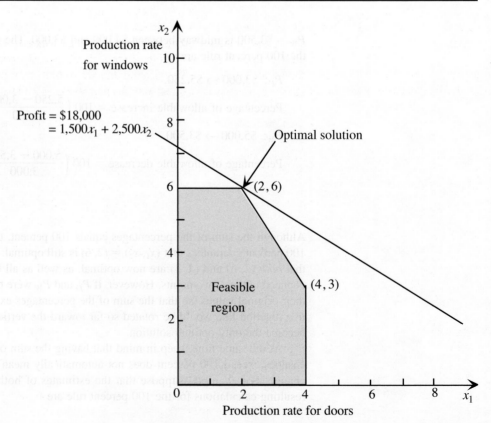

■ **FIGURE 6.22**
When the estimates of the profits per batch of doors and windows change to $P_D = \$1,500$ and $P_W = \$2,500$ (half of their original values), the graphical method shows that the optimal solution still is $(x_1, x_2) = (2, 6)$, even though the 100 percent rule says that the optimal solution might change.

as well. Occasionally you might even want to check whether the optimal solution would change if changes need to be made in some coefficients in the functional constraints.

The spreadsheet approach for investigating these other kinds of changes in the model is virtually the same as for the coefficients in the objective function. Once again, you can try out any changes in the data cells by simply making these changes on the spreadsheet and using the Excel Solver to re-solve the model. And once again, you can systematically check the effect of a series of changes in any one or two data cells by using the Solver Table. As already described in Sec. 4.7, the sensitivity report generated by the Excel Solver (or any other linear programming software package) also provides some valuable information, including the shadow prices, regarding the effect of changing the right-hand side of any single functional constraint. When changing a number of right-hand sides simultaneously, there also is a "100 percent rule" for this case that is analogous to the 100 percent rule for simultaneous changes in objective function constraints. (See the Case 1 portion of Sec. 6.7 for details about how to investigate the effect of changes in right-hand sides, including the application of the 100 percent rule for simultaneous changes in right-hand sides.)

The Worked Examples section of the book's website includes **another example** of using a spreadsheet to investigate the effect of changing individual right-hand sides.

■ 6.9 CONCLUSIONS

Every linear programming problem has associated with it a dual linear programming problem. There are a number of very useful relationships between the original (primal) problem and its dual problem that enhance our ability to analyze the primal problem. For example, the economic interpretation of the dual problem gives shadow prices that measure the marginal value of the resources in the primal problem and provides an interpretation of the simplex method. Because the simplex method can be applied directly to either problem in order to solve both of them simultaneously, considerable computational effort sometimes can be saved by dealing directly with the dual problem. Duality theory, including the dual simplex method (Sec. 7.1) for working with superoptimal basic solutions, also plays a major role in sensitivity analysis.

The values used for the parameters of a linear programming model generally are just estimates. Therefore, sensitivity analysis needs to be performed to investigate what happens if these estimates are wrong. The fundamental insight of Sec. 5.3 provides the key to performing this investigation efficiently. The general objectives are to identify the sensitive parameters that affect the optimal solution, to try to estimate these sensitive parameters more closely, and then to select a solution that remains good over the range of likely values of the sensitive parameters. This analysis is a very important part of most linear programming studies.

With the help of the Excel Solver, spreadsheets also provide some useful methods of performing sensitivity analysis. One method is to repeatedly enter changes in one or more parameters of the model into the spreadsheet and then click on the Solve button to see immediately if the optimal solution changes. A second is to use the Solver Table to systematically check on the effect of making a series of changes in one or two parameters of the model. A third is to use the sensitivity report provided by the Excel Solver to identify the allowable range for the coefficients in the objective function, the shadow prices for the functional constraints, and the allowable range for each right-hand side over which its shadow price remains valid. (Other software that applies the simplex method—including MPL/CPLEX, LINDO, and LINGO—also provides such a sensitivity report upon request.)

■ SELECTED REFERENCES

1. Bertsimas, D., and M. Sim: "The Price of Robustness," *Operations Research*, **52**(1): 35–53, January–February 2004.
2. Dantzig, G. B., and M. N. Thapa: *Linear Programming* 1: *Introduction*, Springer, New York, 1997.
3. Gal, T., and H. Greenberg (eds.): *Advances in Sensitivity Analysis and Parametric Programming*, Kluwer Academic Publishers (now Springer), Boston, MA, 1997.
4. Higle, J. L., and S. W. Wallace: "Sensitivity Analysis and Uncertainty in Linear Programming," *Interfaces*, **33**(4): 53–60, July–August 2003.
5. Hillier, F. S., and M. S. Hillier: *Introduction to Management Science*: *A Modeling and Case Studies Approach with Spreadsheets*, 3rd ed., McGraw-Hill/Irwin, Burr Ridge, IL, 2008, chap. 5.
6. Nazareth, J. L.: *An Optimization Primer: On Models, Algorithms, and Duality*, Springer-Verlag, New York, 2004.
7. Vanderbei, R. J.: *Linear Programming*: *Foundations and Extensions*, 3rd ed., Springer, New York, 2008.
8. Wendell, R. E.: "Tolerance Sensitivity and Optimality Bounds in Linear Programming," *Management Science*, **50**(6): 797–803, June 2004.

■ LEARNING AIDS FOR THIS CHAPTER ON OUR WEBSITE (www.mhhe.com/hillier)

Worked Examples:

Examples for Chapter 6

A Demonstration Example in OR Tutor:

Sensitivity Analysis

Interactive Procedures in IOR Tutorial:

Interactive Graphical Method
Enter or Revise a General Linear Programming Model
Solve Interactively by the Simplex Method
Sensitivity Analysis

Automatic Procedures in IOR Tutorial:

Solve Automatically by the Simplex Method
Graphical Method and Sensitivity Analysis

Excel Add-Ins:

Premium Solver for Education
Solver Table

Files (Chapter 3) for Solving the Wyndor Example:

Excel Files
LINGO/LINDO File
MPL/CPLEX File

Glossary for Chapter 6

See Appendix 1 for documentation of the software.

PROBLEMS

The symbols to the left of some of the problems (or their parts) have the following meaning:

D: The demonstration example just listed may be helpful.
I: We suggest that you use the corresponding interactive procedure just listed (the printout records your work).
C: Use the computer with any of the software options available to you (or as instructed by your instructor) to solve the problem automatically.
E*: Use Excel.

An asterisk on the problem number indicates that at least a partial answer is given in the back of the book.

6.1-1.* Construct the dual problem for each of the following linear programming models fitting our standard form.
(a) Model in Prob. 3.1-6
(b) Model in Prob. 4.7-5

6.1-2. Consider the linear programming model in Prob. 4.5-4.
(a) Construct the primal-dual table and the dual problem for this model.
(b) What does the fact that Z is unbounded for this model imply about its dual problem?

6.1-3. For each of the following linear programming models, give your recommendation on which is the more efficient way (probably) to obtain an optimal solution: by applying the simplex method directly to this primal problem or by applying the simplex method directly to the dual problem instead. Explain.
(a) Maximize $Z = 10x_1 - 4x_2 + 7x_3$,

subject to

$$3x_1 - x_2 + 2x_3 \leq 25$$
$$x_1 - 2x_2 + 3x_3 \leq 25$$
$$5x_1 + x_2 + 2x_3 \leq 40$$
$$x_1 + x_2 + x_3 \leq 90$$
$$2x_1 - x_2 + x_3 \leq 20$$

and

$$x_1 \geq 0, \qquad x_2 \geq 0, \qquad x_3 \geq 0.$$

(b) Maximize $Z = 2x_1 + 5x_2 + 3x_3 + 4x_4 + x_5$,

subject to

$$x_1 + 3x_2 + 2x_3 + 3x_4 + x_5 \leq 6$$
$$4x_1 + 6x_2 + 5x_3 + 7x_4 + x_5 \leq 15$$

and

$$x_j \geq 0, \qquad \text{for } j = 1, 2, 3, 4, 5.$$

6.1-4. Consider the following problem.

Maximize $Z = -x_1 - 2x_2 - x_3$,

subject to

$$x_1 + x_2 + 2x_3 \leq 12$$
$$x_1 + x_2 - x_3 \leq 1$$

and

$$x_1 \geq 0, \qquad x_2 \geq 0, \qquad x_3 \geq 0.$$

(a) Construct the dual problem.
(b) Use duality theory to show that the optimal solution for the primal problem has $Z \leq 0$.

6.1-5. Consider the following problem.

Maximize $Z = 5x_1 + 4x_2 + 3x_3$,

subject to

$$x_1 + x_3 \leq 15 \quad \text{(resource 1)}$$
$$x_2 + 2x_3 \leq 25 \quad \text{(resource 2)}$$

and

$$x_1 \geq 0, \qquad x_2 \geq 0, \qquad x_3 \geq 0.$$

(a) Construct the dual problem for this primal problem.
I **(b)** Solve the dual problem graphically. Use this solution to identify the shadow prices for the resources in the primal problem.
C **(c)** Confirm your results from part (b) by solving the primal problem automatically by the simplex method and then identifying the shadow prices.

6.1-6. Follow the instructions of Prob. 6.1-5 for the following problem.

Maximize $Z = x_1 - 3x_2 + 2x_3$,

subject to

$$2x_1 + 2x_2 - 2x_3 \leq 6 \quad \text{(resource 1)}$$
$$-x_2 + 2x_3 \leq 4 \quad \text{(resource 2)}$$

and

$$x_1 \geq 0, \qquad x_2 \geq 0, \qquad x_3 \geq 0.$$

6.1-7. Consider the following problem.

Maximize $Z = 2x_1 + 3x_2$,

subject to

$$4x_1 + x_2 \leq 20$$
$$-x_1 + x_2 \leq 10$$

and

$$x_1 \geq 0, \qquad x_2 \geq 0.$$

I **(a)** Demonstrate graphically that this problem has no feasible solutions.
(b) Construct the dual problem.

I **(c)** Demonstrate graphically that the dual problem has an unbounded objective function.

I **6.1-8.** Construct and graph a primal problem with two decision variables and two functional constraints that has feasible solutions and an unbounded objective function. Then construct the dual problem and demonstrate graphically that it has no feasible solutions.

I **6.1-9.** Construct a pair of primal and dual problems, each with two decision variables and two functional constraints, such that both problems have no feasible solutions. Demonstrate this property graphically.

6.1-10. Construct a pair of primal and dual problems, each with two decision variables and two functional constraints, such that the primal problem has no feasible solutions and the dual problem has an unbounded objective function.

6.1-11. Use the weak duality property to prove that if both the primal and the dual problem have feasible solutions, then both must have an optimal solution.

6.1-12. Consider the primal and dual problems in our standard form presented in matrix notation at the beginning of Sec. 6.1. Use only this definition of the dual problem for a primal problem in this form to prove each of the following results.
(a) The weak duality property presented in Sec. 6.1.
(b) If the primal problem has an unbounded feasible region that permits increasing Z indefinitely, then the dual problem has no feasible solutions.

6.1-13. Consider the primal and dual problems in our standard form presented in matrix notation at the beginning of Sec. 6.1. Let $\mathbf{y}^*$ denote the optimal solution for this dual problem. Suppose that $\mathbf{b}$ is then replaced by $\bar{\mathbf{b}}$. Let $\bar{\mathbf{x}}$ denote the optimal solution for the new primal problem. Prove that

$$\mathbf{c}\bar{\mathbf{x}} \leq \mathbf{y}^*\bar{\mathbf{b}}.$$

6.1-14. For any linear programming problem in our standard form and its dual problem, label each of the following statements as true or false and then justify your answer.
(a) The sum of the number of functional constraints and the number of variables (before augmenting) is the same for both the primal and the dual problems.
(b) At each iteration, the simplex method simultaneously identifies a CPF solution for the primal problem and a CPF solution for the dual problem such that their objective function values are the same.
(c) If the primal problem has an unbounded objective function, then the optimal value of the objective function for the dual problem must be zero.

6.2-1. Consider the simplex tableaux for the Wyndor Glass Co. problem given in Table 4.8. For each tableau, give the economic interpretation of the following items:
(a) Each of the coefficients of the slack variables (x_3, x_4, x_5) in row 0
(b) Each of the coefficients of the decision variables (x_1, x_2) in row 0

(c) The resulting choice for the entering basic variable (or the decision to stop after the final tableau)

6.3-1.* Consider the following problem.

Maximize $Z = 6x_1 + 8x_2,$

subject to

$$5x_1 + 2x_2 \leq 20$$
$$x_1 + 2x_2 \leq 10$$

and

$$x_1 \geq 0, \qquad x_2 \geq 0.$$

(a) Construct the dual problem for this primal problem.
(b) Solve both the primal problem and the dual problem graphically. Identify the CPF solutions and corner-point infeasible solutions for both problems. Calculate the objective function values for all these solutions.
(c) Use the information obtained in part (b) to construct a table listing the complementary basic solutions for these problems. (Use the same column headings as for Table 6.9.)
I **(d)** Work through the simplex method step by step to solve the primal problem. After each iteration (including iteration 0), identify the BF solution for this problem and the complementary basic solution for the dual problem. Also identify the corresponding corner-point solutions.

6.3-2. Consider the model with two functional constraints and two variables given in Prob. 4.1-5. Follow the instructions of Prob. 6.3-1 for this model.

6.3-3. Consider the primal and dual problems for the Wyndor Glass Co. example given in Table 6.1. Using Tables 5.5, 5.6, 6.8, and 6.9, construct a new table showing the eight sets of nonbasic variables for the primal problem in column 1, the corresponding sets of associated variables for the dual problem in column 2, and the set of nonbasic variables for each complementary basic solution in the dual problem in column 3. Explain why this table demonstrates the complementary slackness property for this example.

6.3-4. Suppose that a primal problem has a *degenerate* BF solution (one or more basic variables equal to zero) as its optimal solution. What does this degeneracy imply about the dual problem? Why? Is the converse also true?

6.3-5. Consider the following problem.

Maximize $Z = 3x_1 - 8x_2,$

subject to

$$x_1 - 2x_2 \leq 10$$

and

$$x_1 \geq 0, \qquad x_2 \geq 0.$$

(a) Construct the dual problem, and then find its optimal solution by inspection.
(b) Use the complementary slackness property and the optimal

solution for the dual problem to find the optimal solution for the primal problem.

(c) Suppose that c_1, the coefficient of x_1 in the primal objective function, actually can have any value in the model. For what values of c_1 does the dual problem have no feasible solutions? For these values, what does duality theory then imply about the primal problem?

6.3-6. Consider the following problem.

$$\text{Maximize} \quad Z = 2x_1 + 7x_2 + 4x_3,$$

subject to

$$x_1 + 2x_2 + x_3 \leq 10$$
$$3x_1 + 3x_2 + 2x_3 \leq 10$$

and

$$x_1 \geq 0, \quad x_2 \geq 0, \quad x_3 \geq 0.$$

(a) Construct the dual problem for this primal problem.
(b) Use the dual problem to demonstrate that the optimal value of Z for the primal problem cannot exceed 25.
(c) It has been conjectured that x_2 and x_3 should be the basic variables for the optimal solution of the primal problem. Directly derive this basic solution (and Z) by using Gaussian elimination. Simultaneously derive and identify the complementary basic solution for the dual problem by using Eq. (0) for the primal problem. Then draw your conclusions about whether these two basic solutions are optimal for their respective problems.
I **(d)** Solve the dual problem graphically. Use this solution to identify the basic variables and the nonbasic variables for the optimal solution of the primal problem. Directly derive this solution, using Gaussian elimination.

6.3-7.* Reconsider the model of Prob. 6.1-3b.
(a) Construct its dual problem.
I **(b)** Solve this dual problem graphically.
(c) Use the result from part (b) to identify the nonbasic variables and basic variables for the optimal BF solution for the primal problem.
(d) Use the results from part (c) to obtain the optimal solution for the primal problem directly by using Gaussian elimination to solve for its basic variables, starting from the initial system of equations [excluding Eq. (0)] constructed for the simplex method and setting the nonbasic variables to zero.
(e) Use the results from part (c) to identify the defining equations (see Sec. 5.1) for the optimal CPF solution for the primal problem, and then use these equations to find this solution.

6.3-8. Consider the model given in Prob. 5.3-10.
(a) Construct the dual problem.
(b) Use the given information about the basic variables in the optimal primal solution to identify the nonbasic variables and basic variables for the optimal dual solution.
(c) Use the results from part (b) to identify the defining equations (see Sec. 5.1) for the optimal CPF solution for the dual problem, and then use these equations to find this solution.

I **(d)** Solve the dual problem graphically to verify your results from part (c).

6.3-9. Consider the model given in Prob. 3.1-5.
(a) Construct the dual problem for this model.
(b) Use the fact that $(x_1, x_2) = (13, 5)$ is optimal for the primal problem to identify the nonbasic variables and basic variables for the optimal BF solution for the dual problem.
(c) Identify this optimal solution for the dual problem by directly deriving Eq. (0) corresponding to the optimal primal solution identified in part (b). Derive this equation by using Gaussian elimination.
(d) Use the results from part (b) to identify the defining equations (see Sec. 5.1) for the optimal CPF solution for the dual problem. Verify your optimal dual solution from part (c) by checking to see that it satisfies this system of equations.

6.3-10. Suppose that you also want information about the dual problem when you apply the matrix form of the simplex method (see Sec. 5.2) to the primal problem in our standard form.
(a) How would you identify the optimal solution for the dual problem?
(b) After obtaining the BF solution at each iteration, how would you identify the complementary basic solution in the dual problem?

6.4-1. Consider the following problem.

$$\text{Maximize} \quad Z = 5x_1 + 4x_2,$$

subject to

$$2x_1 + 3x_2 \geq 10$$
$$x_1 + 2x_2 = 20$$

and

$$x_2 \geq 0 \qquad (x_1 \text{ unconstrained in sign}).$$

(a) Use the SOB method to construct the dual problem.
(b) Use Table 6.12 to convert the primal problem to our standard form given at the beginning of Sec. 6.1, and construct the corresponding dual problem. Then show that this dual problem is equivalent to the one obtained in part (a).

6.4-2. Consider the primal and dual problems in our standard form presented in matrix notation at the beginning of Sec. 6.1. Use only this definition of the dual problem for a primal problem in this form to prove each of the following results.
(a) If the functional constraints for the primal problem $\mathbf{Ax} \leq \mathbf{b}$ are changed to $\mathbf{Ax} = \mathbf{b}$, the only resulting change in the dual problem is to *delete* the nonnegativity constraints, $\mathbf{y} \geq \mathbf{0}$. (*Hint:* The constraints $\mathbf{Ax} = \mathbf{b}$ are equivalent to the set of constraints $\mathbf{Ax} \leq \mathbf{b}$ *and* $\mathbf{Ax} \geq \mathbf{b}$.)
(b) If the functional constraints for the primal problem $\mathbf{Ax} \leq \mathbf{b}$ are changed to $\mathbf{Ax} \geq \mathbf{b}$, the only resulting change in the dual problem is that the nonnegativity constraints $\mathbf{y} \geq \mathbf{0}$ are replaced by nonpositivity constraints $\mathbf{y} \leq \mathbf{0}$, where the current dual variables are interpreted as the negative of the original dual variables. (*Hint:* The constraints $\mathbf{Ax} \geq \mathbf{b}$ are equivalent to $-\mathbf{Ax} \leq -\mathbf{b}$.)

(c) If the nonnegativity constraints for the primal problem $\mathbf{x} \geq \mathbf{0}$ are deleted, the only resulting change in the dual problem is to replace the functional constraints $\mathbf{yA} \geq \mathbf{c}$ by $\mathbf{yA} = \mathbf{c}$. (*Hint:* A variable unconstrained in sign can be replaced by the difference of two nonnegative variables.)

6.4-3.* Construct the dual problem for the linear programming problem given in Prob. 4.6-3.

6.4-4. Consider the following problem.

$$\text{Minimize} \qquad Z = 5x_1 + 10x_2,$$

subject to

$$
\begin{aligned}
-4x_1 + 2x_2 &\geq 4 \\
5x_1 - 10x_2 &\geq 10
\end{aligned}
$$

and

$$x_1 \geq 0, \qquad x_2 \geq 0.$$

(a) Construct the dual problem.
I (b) Use graphical analysis of the dual problem to determine whether the primal problem has feasible solutions and, if so, whether its objective function is bounded.

6.4-5. Consider the two versions of the dual problem for the radiation therapy example that are given in Tables 6.15 and 6.16. Review in Sec. 6.4 the general discussion of why these two versions are completely equivalent. Then fill in the details to verify this equivalency by proceeding step by step to convert the version in Table 6.15 to equivalent forms until the version in Table 6.16 is obtained.

6.4-6. For each of the following linear programming models, use the SOB method to construct its dual problem.
(a) Model in Prob. 4.6-7
(b) Model in Prob. 4.6-16

6.4-7. Consider the model with equality constraints given in Prob. 4.6-2.
(a) Construct its dual problem.
(b) Demonstrate that the answer in part (a) is correct (i.e., equality constraints yield dual variables without nonnegativity constraints) by first converting the primal problem to our standard form (see Table 6.12), then constructing its dual problem, and next converting this dual problem to the form obtained in part (a).

6.4-8.* Consider the model without nonnegativity constraints given in Prob. 4.6-14.
(a) Construct its dual problem.
(b) Demonstrate that the answer in part (a) is correct (i.e., variables without nonnegativity constraints yield equality constraints in the dual problem) by first converting the primal problem to our standard form (see Table 6.12), then constructing its dual problem, and finally converting this dual problem to the form obtained in part (a).

6.4-9. Consider the dual problem for the Wyndor Glass Co. example given in Table 6.1. Demonstrate that *its* dual problem is the primal problem given in Table 6.1 by going through the conversion steps given in Table 6.13.

6.4-10. Consider the following problem.

$$\text{Minimize} \qquad Z = -5x_1 - 15x_2,$$

subject to

$$
\begin{aligned}
2x_1 - 4x_2 &\leq 8 \\
-3x_1 + 3x_2 &\leq 24
\end{aligned}
$$

and

$$x_1 \geq 0, \qquad x_2 \geq 0.$$

I (a) Demonstrate graphically that this problem has an unbounded objective function.
(b) Construct the dual problem.
I (c) Demonstrate graphically that the dual problem has no feasible solutions.

6.5-1. Consider the model of Prob. 6.7-2. Use duality theory directly to determine whether the current basic solution remains optimal after each of the following independent changes.
(a) The change in part (e) of Prob. 6.7-2
(b) The change in part (g) of Prob. 6.7-2

6.5-2. Consider the model of Prob. 6.7-4. Use duality theory directly to determine whether the current basic solution remains optimal after each of the following independent changes.
(a) The change in part (b) of Prob. 6.7-4
(b) The change in part (d) of Prob. 6.7-4

6.5-3. Reconsider part (d) of Prob. 6.7-6. Use duality theory directly to determine whether the original optimal solution is still optimal.

6.6-1.* Consider the following problem.

$$\text{Maximize} \qquad Z = 3x_1 + x_2 + 4x_3,$$

subject to

$$
\begin{aligned}
6x_1 + 3x_2 + 5x_3 &\leq 25 \\
3x_1 + 4x_2 + 5x_3 &\leq 20
\end{aligned}
$$

and

$$x_1 \geq 0, \qquad x_2 \geq 0, \qquad x_3 \geq 0.$$

The corresponding final set of equations yielding the optimal solution is

$$(0) \qquad Z \qquad + 2x_2 \qquad + \frac{1}{5}x_4 + \frac{3}{5}x_5 = 17$$

$$(1) \qquad x_1 - \frac{1}{3}x_2 \qquad + \frac{1}{3}x_4 - \frac{1}{3}x_5 = \frac{5}{3}$$

$$(2) \qquad x_2 + x_3 - \frac{1}{5}x_4 + \frac{2}{5}x_5 = 3.$$

(a) Identify the optimal solution from this set of equations.
(b) Construct the dual problem.

I **(c)** Identify the optimal solution for the dual problem from the final set of equations. Verify this solution by solving the dual problem graphically.

(d) Suppose that the original problem is changed to

Maximize $Z = 3x_1 + 3x_2 + 4x_3$,

subject to

$$6x_1 + 2x_2 + 5x_3 \le 25$$
$$3x_1 + 3x_2 + 5x_3 \le 20$$

and

$$x_1 \ge 0, \qquad x_2 \ge 0, \qquad x_3 \ge 0.$$

Use duality theory to determine whether the previous optimal solution is still optimal.

(e) Use the fundamental insight presented in Sec. 5.3 to identify the new coefficients of x_2 in the final set of equations after it has been adjusted for the changes in the original problem given in part (d).

(f) Now suppose that the only change in the original problem is that a new variable x_{new} has been introduced into the model as follows:

Maximize $Z = 3x_1 + x_2 + 4x_3 + 2x_{new}$,

subject to

$$6x_1 + 3x_2 + 5x_3 + 3x_{new} \le 25$$
$$3x_1 + 4x_2 + 5x_3 + 2x_{new} \le 20$$

and

$$x_1 \ge 0, \qquad x_2 \ge 0, \qquad x_3 \ge 0, \qquad x_{new} \ge 0.$$

Use duality theory to determine whether the previous optimal solution, along with $x_{new} = 0$, is still optimal.

(g) Use the fundamental insight presented in Sec. 5.3 to identify the coefficients of x_{new} as a nonbasic variable in the final set of equations resulting from the introduction of x_{new} into the original model as shown in part (f).

D,I **6.6-2.** Reconsider the model of Prob. 6.6-1. You are now to conduct sensitivity analysis by *independently* investigating each of the following six changes in the original model. For each change, use the sensitivity analysis procedure to revise the given final set of equations (in tableau form) and convert it to proper form from Gaussian elimination. Then test this solution for feasibility and for optimality. (Do not reoptimize.)

(a) Change the right-hand side of constraint 1 to $b_1 = 10$.
(b) Change the right-hand side of constraint 2 to $b_2 = 10$.
(c) Change the coefficient of x_2 in the objective function to $c_2 = 3$.
(d) Change the coefficient of x_3 in the objective function to $c_3 = 2$.
(e) Change the coefficient of x_2 in constraint 2 to $a_{22} = 2$.
(f) Change the coefficient of x_1 in constraint 1 to $a_{11} = 8$.

D,I **6.6-3.** Consider the following problem.

Minimize $W = 5y_1 + 4y_2$,

subject to

$$4y_1 + 3y_2 \ge 4$$
$$2y_1 + y_2 \ge 3$$
$$y_1 + 2y_2 \ge 1$$
$$y_1 + y_2 \ge 2$$

and

$$y_1 \ge 0, \qquad y_2 \ge 0.$$

Because this primal problem has more functional constraints than variables, suppose that the simplex method has been applied directly to its dual problem. If we let x_5 and x_6 denote the slack variables for this dual problem, the resulting final simplex tableau is

Basic Variable	Eq.	Coefficient of:							Right Side
		Z	x_1	x_2	x_3	x_4	x_5	x_6	
Z	(0)	1	3	0	2	0	1	1	9
x_2	(1)	0	1	1	−1	0	1	−1	1
x_4	(2)	0	2	0	3	1	−1	2	3

For each of the following independent changes in the original primal model, you now are to conduct sensitivity analysis by directly investigating the effect on the dual problem and then inferring the complementary effect on the primal problem. For each change, apply the procedure for sensitivity analysis summarized at the end of Sec. 6.6 to the dual problem (do *not* reoptimize), and then give your conclusions as to whether the current basic solution for the primal problem still is feasible and whether it still is optimal. Then check your conclusions by a direct graphical analysis of the primal problem.

(a) Change the objective function to $W = 3y_1 + 5y_2$.
(b) Change the right-hand sides of the functional constraints to 3, 5, 2, and 3, respectively.
(c) Change the first constraint to $2y_1 + 4y_2 \ge 7$.
(d) Change the second constraint to $5y_1 + 2y_2 \ge 10$.

6.7-1. Read the referenced article that fully describes the OR study summarized in the application vignette presented in Sec. 6.7. Briefly describe how sensitivity analysis was applied in this study. Then list the various financial and nonfinancial benefits that resulted from the study.

D,I **6.7-2.*** Consider the following problem.

Maximize $Z = -5x_1 + 5x_2 + 13x_3$,

subject to

$$-x_1 + x_2 + 3x_3 \le 20$$
$$12x_1 + 4x_2 + 10x_3 \le 90$$

and

$$x_j \ge 0 \qquad (j = 1, 2, 3).$$

If we let x_4 and x_5 be the slack variables for the respective constraints, the simplex method yields the following final set of equations:

(0) $Z \qquad + 2x_3 + 5x_4 \qquad = 100$
(1) $\quad -x_1 + x_2 + 3x_3 + \ x_4 \qquad = 20$
(2) $\quad 16x_1 \qquad - 2x_3 - 4x_4 + x_5 = \ 10.$

Now you are to conduct sensitivity analysis by *independently* investigating each of the following nine changes in the original model. For each change, use the sensitivity analysis procedure to revise this set of equations (in tableau form) and convert it to proper form from Gaussian elimination for identifying and evaluating the current basic solution. Then test this solution for feasibility and for optimality. (Do not reoptimize.)

(a) Change the right-hand side of constraint 1 to

$b_1 = 30.$

(b) Change the right-hand side of constraint 2 to

$b_2 = 70.$

(c) Change the right-hand sides to

$$\begin{bmatrix} b_1 \\ b_2 \end{bmatrix} = \begin{bmatrix} 10 \\ 100 \end{bmatrix}.$$

(d) Change the coefficient of x_3 in the objective function to

$c_3 = 8.$

(e) Change the coefficients of x_1 to

$$\begin{bmatrix} c_1 \\ a_{11} \\ a_{21} \end{bmatrix} = \begin{bmatrix} -2 \\ 0 \\ 5 \end{bmatrix}.$$

(f) Change the coefficients of x_2 to

$$\begin{bmatrix} c_2 \\ a_{12} \\ a_{22} \end{bmatrix} = \begin{bmatrix} 6 \\ 2 \\ 5 \end{bmatrix}.$$

(g) Introduce a new variable x_6 with coefficients

$$\begin{bmatrix} c_6 \\ a_{16} \\ a_{26} \end{bmatrix} = \begin{bmatrix} 10 \\ 3 \\ 5 \end{bmatrix}.$$

(h) Introduce a new constraint $2x_1 + 3x_2 + 5x_3 \le 50$. (Denote its slack variable by x_6.)

(i) Change constraint 2 to

$10x_1 + 5x_2 + 10x_3 \le 100.$

6.7-3.* Reconsider the model of Prob. 6.7-2. Suppose that we now want to apply parametric linear programming analysis to this problem. Specifically, the right-hand sides of the functional constraints are changed to

$20 + 2\theta \qquad$ (for constraint 1)

and

$90 - \theta \qquad$ (for constraint 2),

where θ can be assigned any positive or negative values.

Express the basic solution (and Z) corresponding to the original optimal solution as a function of θ. Determine the lower and upper bounds on θ before this solution would become infeasible.

D,I **6.7-4.** Consider the following problem.

Maximize $\qquad Z = 2x_1 + 7x_2 - 3x_3,$

subject to

$x_1 + 3x_2 + 4x_3 \le 30$
$x_1 + 4x_2 - \ x_3 \le 10$

and

$\qquad x_1 \ge 0, \qquad x_2 \ge 0, \qquad x_3 \ge 0.$

By letting x_4 and x_5 be the slack variables for the respective constraints, the simplex method yields the following final set of equations:

(0) $Z \ + \ x_2 + \ x_3 \qquad + 2x_5 = 20$
(1) $\qquad - \ x_2 + 5x_3 + x_4 - \ x_5 = 20$
(2) $\qquad x_1 + 4x_2 - \ x_3 \qquad + \ x_5 = 10.$

Now you are to conduct sensitivity analysis by *independently* investigating each of the following seven changes in the original model. For each change, use the sensitivity analysis procedure to revise this set of equations (in tableau form) and convert it to proper form from Gaussian elimination for identifying and evaluating the current basic solution. Then test this solution for feasibility and for optimality. If either test fails, reoptimize to find a new optimal solution.

(a) Change the right-hand sides to

$$\begin{bmatrix} b_1 \\ b_2 \end{bmatrix} = \begin{bmatrix} 20 \\ 30 \end{bmatrix}.$$

(b) Change the coefficients of x_3 to

$$\begin{bmatrix} c_3 \\ a_{13} \\ a_{23} \end{bmatrix} = \begin{bmatrix} -2 \\ 3 \\ -2 \end{bmatrix}.$$

(c) Change the coefficients of x_1 to

$$\begin{bmatrix} c_1 \\ a_{11} \\ a_{21} \end{bmatrix} = \begin{bmatrix} 4 \\ 3 \\ 2 \end{bmatrix}.$$

(d) Introduce a new variable x_6 with coefficients

$$\begin{bmatrix} c_6 \\ a_{16} \\ a_{26} \end{bmatrix} = \begin{bmatrix} -3 \\ 1 \\ 2 \end{bmatrix}.$$

(e) Change the objective function to $Z = x_1 + 5x_2 - 2x_3.$

(f) Introduce a new constraint $3x_1 + 2x_2 + 3x_3 \le 25$.
(g) Change constraint 2 to $x_1 + 2x_2 + 2x_3 \le 35$.

6.7-5. Reconsider the model of Prob. 6.7-4. Suppose that we now want to apply parametric linear programming analysis to this problem. Specifically, the right-hand sides of the functional constraints are changed to

$$30 + 3\theta \qquad \text{(for constraint 1)}$$

and

$$10 - \theta \qquad \text{(for constraint 2)},$$

where θ can be assigned any positive or negative values.

Express the basic solution (and Z) corresponding to the original optimal solution as a function of θ. Determine the lower and upper bounds on θ before this solution would become infeasible.

D,I **6.7-6.** Consider the following problem.

$$\text{Maximize} \qquad Z = 2x_1 - x_2 + x_3,$$

subject to

$$
\begin{array}{rcl}
3x_1 - 2x_2 + 2x_3 & \le & 15 \\
-x_1 + x_2 + x_3 & \le & 3 \\
x_1 - x_2 + x_3 & \le & 4
\end{array}
$$

and

$$x_1 \ge 0, \qquad x_2 \ge 0, \qquad x_3 \ge 0.$$

If we let x_4, x_5, and x_6 be the slack variables for the respective constraints, the simplex method yields the following final set of equations:

$$
\begin{array}{lrcl}
(0) & Z \qquad\quad + 2x_3 + x_4 + x_5 & = & 18 \\
(1) & x_2 + 5x_3 + x_4 + 3x_5 & = & 24 \\
(2) & 2x_3 \qquad + x_5 + x_6 & = & 7 \\
(3) & x_1 + 4x_3 + x_4 + 2x_5 & = & 21.
\end{array}
$$

Now you are to conduct sensitivity analysis by *independently* investigating each of the following eight changes in the original model. For each change, use the sensitivity analysis procedure to revise this set of equations (in tableau form) and convert it to proper form from Gaussian elimination for identifying and evaluating the current basic solution. Then test this solution for feasibility and for optimality. If either test fails, reoptimize to find a new optimal solution.
(a) Change the right-hand sides to

$$
\begin{bmatrix} b_1 \\ b_2 \\ b_3 \end{bmatrix} = \begin{bmatrix} 10 \\ 4 \\ 2 \end{bmatrix}.
$$

(b) Change the coefficient of x_3 in the objective function to $c_3 = 2$.
(c) Change the coefficient of x_1 in the objective function to $c_1 = 3$.
(d) Change the coefficients of x_3 to

$$
\begin{bmatrix} c_3 \\ a_{13} \\ a_{23} \\ a_{33} \end{bmatrix} = \begin{bmatrix} 4 \\ 3 \\ 2 \\ 1 \end{bmatrix}.
$$

(e) Change the coefficients of x_1 and x_2 to

$$
\begin{bmatrix} c_1 \\ a_{11} \\ a_{21} \\ a_{31} \end{bmatrix} = \begin{bmatrix} 1 \\ 1 \\ -2 \\ 3 \end{bmatrix} \quad \text{and} \quad \begin{bmatrix} c_2 \\ a_{12} \\ a_{22} \\ a_{32} \end{bmatrix} = \begin{bmatrix} -2 \\ -2 \\ 3 \\ 2 \end{bmatrix},
$$

respectively.
(f) Change the objective function to $Z = 5x_1 + x_2 + 3x_3$.
(g) Change constraint 1 to $2x_1 - x_2 + 4x_3 \le 12$.
(h) Introduce a new constraint $2x_1 + x_2 + 3x_3 \le 60$.

C **6.7-7** Consider the Distribution Unlimited Co. problem presented in Sec. 3.4 and summarized in Fig. 3.13.

Although Fig. 3.13 gives estimated unit costs for shipping through the various shipping lanes, there actually is some uncertainty about what these unit costs will turn out to be. Therefore, before adopting the optimal solution given at the end of Sec. 3.4, management wants additional information about the effect of inaccuracies in estimating these unit costs.

Use a computer package based on the simplex method to generate sensitivity analysis information preparatory to addressing the following questions.
(a) Which of the unit shipping costs given in Fig. 3.13 has the smallest margin for error without invalidating the optimal solution given in Sec. 3.4? Where should the greatest effort be placed in estimating the unit shipping costs?
(b) What is the allowable range for each of the unit shipping costs?
(c) How should these allowable ranges be interpreted to management?
(d) If the estimates change for more than one of the unit shipping costs, how can you use the generated sensitivity analysis information to determine whether the optimal solution might change?

6.7-8. Consider the following problem.

$$\text{Maximize} \qquad Z = c_1x_1 + c_2x_2,$$

subject to

$$
\begin{array}{rcl}
2x_1 - x_2 & \le & b_1 \\
x_1 - x_2 & \le & b_2
\end{array}
$$

and

$$x_1 \ge 0, \qquad x_2 \ge 0.$$

Let x_3 and x_4 denote the slack variables for the respective functional constraints. When $c_1 = 3$, $c_2 = -2$, $b_1 = 30$, and $b_2 = 10$, the simplex method yields the following final simplex tableau.

Basic Variable	Eq.	Coefficient of:					Right Side
		Z	x_1	x_2	x_3	x_4	
Z	(0)	1	0	0	1	1	40
x_2	(1)	0	0	1	1	-2	10
x_1	(2)	0	1	0	1	-1	20

I **(a)** Use graphical analysis to determine the allowable range for c_1 and c_2.

(b) Use algebraic analysis to derive and verify your answers in part (a).

I **(c)** Use graphical analysis to determine the allowable range for b_1 and b_2.

(d) Use algebraic analysis to derive and verify your answers in part (c)

C **(e)** Use a software package based on the simplex method to find these allowable ranges.

I **6.7-9.** Consider Variation 5 of the Wyndor Glass Co. model (see Fig. 6.6 and Table 6.24), where the changes in the parameter values given in Table 6.21 are $\bar{c}_2 = 3$, $\bar{a}_{22} = 3$, and $\bar{a}_{32} = 4$. Use the formula $\mathbf{b}^* = \mathbf{S}^*\bar{\mathbf{b}}$ to find the allowable range for each b_i. Then interpret each allowable range graphically.

I **6.7-10.** Consider Variation 5 of the Wyndor Glass Co. model (see Fig. 6.6 and Table 6.24), where the changes in the parameter values given in Table 6.21 are $\bar{c}_2 = 3$, $\bar{a}_{22} = 3$, and $\bar{a}_{32} = 4$. Verify both algebraically and graphically that the allowable range for c_1 is $c_1 \geq \frac{9}{4}$.

6.7-11. For the problem given in Table 6.21, find the allowable range for c_2. Show your work algebraically, using the tableau given in Table 6.21. Then justify your answer from a geometric viewpoint, referring to Fig. 6.3.

6.7-12.* For the original Wyndor Glass Co. problem, use the last tableau in Table 4.8 to do the following.
(a) Find the allowable range for each b_i.
(b) Find the allowable range for c_1 and c_2.
C **(c)** Use a software package based on the simplex method to find these allowable ranges.

6.7-13. For Variation 6 of the Wyndor Glass Co. model presented in Sec. 6.7, use the last tableau in Table 6.25 to do the following.
(a) Find the allowable range for each b_i.
(b) Find the allowable range for c_1 and c_2.
C **(c)** Use a software package based on the simplex method to find these allowable ranges.

6.7-14. Consider Variation 5 of the Wyndor Glass Co. model presented in Sec. 6.7, where $\bar{c}_2 = 3$, $\bar{a}_{22} = 3$, $\bar{a}_{32} = 4$, and where the other parameters are given in Table 6.21. Starting from the resulting final tableau given at the bottom of Table 6.24, construct a table like Table 6.26 to perform parametric linear programming analysis, where

$$c_1 = 3 + \theta \quad \text{and} \quad c_2 = 3 + 2\theta.$$

How far can θ be increased above 0 before the current basic solution is no longer optimal?

6.7-15. Reconsider the model of Prob. 6.7-6. Suppose that you now have the option of making trade-offs in the profitability of the first two activities, whereby the objective function coefficient of x_1 can be increased by any amount by simultaneously decreasing the objective function coefficient of x_2 by the same amount. Thus, the alternative choices of the objective function are

$$Z(\theta) = (2 + \theta)x_1 - (1 + \theta)x_2 + x_3,$$

where any nonnegative value of θ can be chosen.

Construct a table like Table 6.26 to perform parametric linear programming analysis on this problem. Determine the upper bound on θ before the original optimal solution would become nonoptimal. Then determine the best choice of θ over this range.

6.7-16. Consider the following parametric linear programming problem.

Maximize $Z(\theta) = (10 - 4\theta)x_1 + (4 - \theta)x_2 + (7 + \theta)x_3,$

subject to

$$3x_1 + x_2 + 2x_3 \leq 7 \quad \text{(resource 1)},$$
$$2x_1 + x_2 + 3x_3 \leq 5 \quad \text{(resource 2)},$$

and

$$x_1 \geq 0, \qquad x_2 \geq 0, \qquad x_3 \geq 0,$$

where θ can be assigned any positive or negative values. Let x_4 and x_5 be the slack variables for the respective constraints. After we apply the simplex method with $\theta = 0$, the final simplex tableau is

Basic Variable	Eq.	\multicolumn{6}{c	}{Coefficient of:}	Right Side
		Z	x_1	x_2	x_3	x_4	x_5	
Z	(0)	1	0	0	3	2	2	24
x_1	(1)	0	1	0	-1	1	-1	2
x_2	(2)	0	0	1	5	-2	3	1

(a) Determine the range of values of θ over which the above BF solution will remain optimal. Then find the best choice of θ within this range.
(b) Given that θ is within the range of values found in part (a), find the allowable range for b_1 (the available amount of resource 1). Then do the same for b_2 (the available amount of resource 2).
(c) Given that θ is within the range of values found in part (a), identify the shadow prices (as a function of θ) for the two resources. Use this information to determine how the optimal value of the objective function would change (as a function of θ) if the available amount of resource 1 were decreased by 1 and the available amount of resource 2 simultaneously were increased by 1.
(d) Construct the dual of this parametric linear programming problem. Set $\theta = 0$ and solve this dual problem graphically to find the corresponding shadow prices for the two resources of the primal problem. Then find these shadow prices as a function of θ [within the range of values found in part (a)] by algebraically solving for this same optimal CPF solution for the dual problem as a function of θ.

6.7-17. Consider the following parametric linear programming problem.

Maximize $Z(\theta) = 2x_1 + 4x_2 + 5x_3,$

subject to

$$x_1 + 3x_2 + 2x_3 \leq 5 + \theta$$
$$x_1 + 2x_2 + 3x_3 \leq 6 + 2\theta$$

and

$$x_1 \geq 0, \qquad x_2 \geq 0, \qquad x_3 \geq 0,$$

where θ can be assigned any positive or negative values. Let x_4 and x_5 be the slack variables for the respective functional constraints. After we apply the simplex method with $\theta = 0$, the final simplex tableau is

Basic Variable	Eq.	Z	x_1	x_2	x_3	x_4	x_5	Right Side
Z	(0)	0	0	1	0	1	1	11
x_1	(1)	1	1	5	0	3	-2	3
x_3	(2)	2	0	-1	1	-1	1	1

(a) Express the BF solution (and Z) given in this tableau as a function of θ. Determine the lower and upper bounds on θ before this optimal BF solution would become infeasible. Then determine the best choice of θ between these bounds.

(b) Given that θ is between the bounds found in part (a), determine the allowable range for c_1 (the coefficient of x_1 in the objective function).

6.7-18. Consider the following problem.

Maximize $\quad Z = 10x_1 + 4x_2,$

subject to

$$3x_1 + x_2 \leq 30$$
$$2x_1 + x_2 \leq 25$$

and

$$x_1 \geq 0, \qquad x_2 \geq 0.$$

Let x_3 and x_4 denote the slack variables for the respective functional constraints. After we apply the simplex method, the final simplex tableau is

Basic Variable	Eq.	Z	x_1	x_2	x_3	x_4	Right Side
Z	(0)	1	0	0	2	2	110
x_2	(1)	0	0	1	-2	3	15
x_1	(2)	0	1	0	1	-1	5

Now suppose that both of the following changes are made simultaneously in the original model:

1. The first constraint is changed to $4x_1 + x_2 \leq 40$.

2. Parametric programming is introduced to change the objective function to the alternative choices of

$$Z(\theta) = (10 - 2\theta)x_1 + (4 + \theta)x_2,$$

where any nonnegative value of θ can be chosen.

(a) Construct the resulting revised final tableau (as a function of θ), and then convert this tableau to proper form from Gaussian elimination. Use this tableau to identify the new optimal solution that applies for either $\theta = 0$ or sufficiently small values of θ.

(b) What is the upper bound on θ before this optimal solution would become nonoptimal?

(c) Over the range of θ from zero to this upper bound, which choice of θ gives the largest value of the objective function?

6.7-19. Consider the following problem.

Maximize $\quad Z = 9x_1 + 8x_2 + 5x_3,$

subject to

$$2x_1 + 3x_2 + x_3 \leq 4$$
$$5x_1 + 4x_2 + 3x_3 \leq 11$$

and

$$x_1 \geq 0, \qquad x_2 \geq 0, \qquad x_3 \geq 0.$$

Let x_4 and x_5 denote the slack variables for the respective functional constraints. After we apply the simplex method, the final simplex tableau is

Basic Variable	Eq.	Z	x_1	x_2	x_3	x_4	x_5	Right Side
Z	(0)	1	0	2	0	2	1	19
x_1	(1)	0	1	5	0	3	-1	1
x_3	(2)	0	0	-7	1	-5	2	2

D,I **(a)** Suppose that a new technology has become available for conducting the first activity considered in this problem. If the new technology were adopted to replace the existing one, the coefficients of x_1 in the model would change

from $\quad \begin{bmatrix} c_1 \\ a_{11} \\ a_{21} \end{bmatrix} = \begin{bmatrix} 9 \\ 2 \\ 5 \end{bmatrix} \quad$ to $\quad \begin{bmatrix} c_1 \\ a_{11} \\ a_{21} \end{bmatrix} = \begin{bmatrix} 18 \\ 3 \\ 6 \end{bmatrix}.$

Use the sensitivity analysis procedure to investigate the potential effect and desirability of adopting the new technology. Specifically, assuming it were adopted, construct the resulting revised final tableau, convert this tableau to proper form from Gaussian elimination, and then reoptimize (if necessary) to find the new optimal solution.

(b) Now suppose that you have the option of mixing the old and new technologies for conducting the first activity. Let θ denote the fraction of the technology used that is from the new technology, so $0 \leq \theta \leq 1$. Given θ, the coefficients of x_1 in the model become

$$\begin{bmatrix} c_1 \\ a_{11} \\ a_{21} \end{bmatrix} = \begin{bmatrix} 9 + 9\theta \\ 2 + \theta \\ 5 + \theta \end{bmatrix}.$$

Construct the resulting revised final tableau (as a function of θ), and convert this tableau to proper form from Gaussian elimination. Use this tableau to identify the current basic solution as a function of θ. Over the allowable values of $0 \leq \theta \leq 1$, give the range of values of θ for which this solution is both feasible and optimal. What is the best choice of θ within this range?

6.7-20. Consider the following problem.

Maximize $Z = 3x_1 + 5x_2 + 2x_3,$

subject to

$$-2x_1 + 2x_2 + x_3 \leq 5$$
$$3x_1 + x_2 - x_3 \leq 10$$

and

$$x_1 \geq 0, \qquad x_2 \geq 0, \qquad x_3 \geq 0.$$

Let x_4 and x_5 be the slack variables for the respective functional constraints. After we apply the simplex method, the final simplex tableau is

Basic Variable	Eq.	Coefficient of:						Right Side
		Z	x_1	x_2	x_3	x_4	x_5	
Z	(0)	1	0	20	0	9	7	115
x_1	(1)	0	1	3	0	1	1	15
x_3	(2)	0	0	8	1	3	2	35

Parametric linear programming analysis now is to be applied simultaneously to the objective function and right-hand sides, where the model in terms of the new parameter is the following:

Maximize $Z(\theta) = (3 + 2\theta)x_1 + (5 + \theta)x_2 + (2 - \theta)x_3,$

subject to

$$-2x_1 + 2x_2 + x_3 \leq 5 + 6\theta$$
$$3x_1 + x_2 - x_3 \leq 10 - 8\theta$$

and

$$x_1 \geq 0, \qquad x_2 \geq 0, \qquad x_3 \geq 0.$$

Construct the resulting revised final tableau (as a function of θ), and convert this tableau to proper form from Gaussian elimination. Use this tableau to identify the current basic solution as a function of θ. For $\theta \geq 0$, give the range of values of θ for which this solution is both feasible and optimal. What is the best choice of θ within this range?

6.7-21. Consider the Wyndor Glass Co. problem described in Sec. 3.1. Suppose that, in addition to considering the introduction of two new products, management now is considering changing the production rate of a certain old product that is still profitable. Refer to Table 3.1. The number of production hours per week used per unit production rate of this old product is 1, 4, and 3 for Plants 1, 2, and 3, respectively. Therefore, if we let θ denote the *change* (positive or negative) in the production rate of this old product, the right-hand sides of the three functional constraints in Sec. 3.1 become $4 - \theta$, $12 - 4\theta$, and $18 - 3\theta$, respectively. Thus, choosing a negative value of θ would free additional capacity for producing more of the two new products, whereas a positive value would have the opposite effect.

(a) Use a parametric linear programming formulation to determine the effect of different choices of θ on the optimal solution for the product mix of the two new products given in the final tableau of Table 4.8. In particular, use the fundamental insight of Sec. 5.3 to obtain expressions for Z and the basic variables x_3, x_2, and x_1 in terms of θ, assuming that θ is sufficiently close to zero that this "final" basic solution still is feasible and thus optimal for the given value of θ.

(b) Now consider the broader question of the choice of θ along with the product mix for the two new products. What is the breakeven unit profit for the old product (in comparison with the two new products) below which its production rate should be decreased ($\theta < 0$) in favor of the new products and above which its production rate should be increased ($\theta > 0$)?

(c) If the unit profit is above this breakeven point, how much can the old product's production rate be increased before the final BF solution would become infeasible?

(d) If the unit profit is below this breakeven point, how much can the old product's production rate be decreased (assuming its previous rate was larger than this decrease) before the final BF solution would become infeasible?

6.7-22. Consider the following problem.

Maximize $Z = 2x_1 - x_2 + 3x_3,$

subject to

$$x_1 + x_2 + x_3 = 3$$
$$x_1 - 2x_2 + x_3 \geq 1$$
$$2x_2 + x_3 \leq 2$$

and

$$x_1 \geq 0, \qquad x_2 \geq 0, \qquad x_3 \geq 0.$$

Suppose that the Big M method (see Sec. 4.6) is used to obtain the initial (artificial) BF solution. Let $\bar{x}_4$ be the artificial slack variable for the first constraint, x_5 the surplus variable for the second constraint, $\bar{x}_6$ the artificial variable for the second constraint, and x_7 the slack variable for the third constraint. The corresponding final set of equations yielding the optimal solution is

(0) $Z + 5x_2 \qquad\qquad + (M + 2)\bar{x}_4 \qquad + M\bar{x}_6 + x_7 = 8$
(1) $x_1 - x_2 \qquad + \qquad \bar{x}_4 \qquad\qquad - x_7 = 1$
(2) $2x_2 + x_3 \qquad\qquad\qquad\qquad\qquad + x_7 = 2$
(3) $3x_2 \qquad\qquad + \qquad \bar{x}_4 + x_5 - \bar{x}_6 \qquad = 2.$

Suppose that the original objective function is changed to $Z = 2x_1 + 3x_2 + 4x_3$ and that the original third constraint is changed to $2x_2 + x_3 \leq 1$. Use the sensitivity analysis procedure to revise the final set of equations (in tableau form) and convert it to proper form from Gaussian elimination for identifying and evaluating the current basic solution. Then test this solution for feasibility and for optimality. (Do not reoptimize.)

6.8-1. Consider the following problem.

Maximize $\quad Z = 2x_1 + 5x_2,$

subject to

$$x_1 + 2x_2 \leq 10 \text{ (resource 1)}$$
$$x_1 + 3x_2 \leq 12 \text{ (resource 2)}$$

and

$$x_1 \geq 0, \qquad x_2 \geq 0,$$

where Z measures the profit in dollars from the two activities.

While doing sensitivity analysis, you learn that the estimates of the unit profits are accurate only to within ± 50 percent. In other words, the ranges of *likely values* for these unit profits are \$1 to \$3 for activity 1 and \$2.50 to \$7.50 for activity 2.

E* **(a)** Formulate a spreadsheet model for this problem based on the original estimates of the unit profits. Then use the Solver to find an optimal solution and to generate the sensitivity report.

E* **(b)** Use the spreadsheet and Solver to check whether this optimal solution remains optimal if the unit profit for activity 1 changes from \$2 to \$1. From \$2 to \$3.

E* **(c)** Also check whether the optimal solution remains optimal if the unit profit for activity 1 still is \$2 but the unit profit for activity 2 changes from \$5 to \$2.50. From \$5 to \$7.50.

E* **(d)** Use the Solver Table to systematically generate the optimal solution and total profit as the unit profit of activity 1 increases in 20¢ increments from \$1 to \$3 (without changing the unit profit of activity 2). Then do the same as the unit profit of activity 2 increases in 50¢ increments from \$2.50 to \$7.50 (without changing the unit profit of activity 1). Use these results to estimate the allowable range for the unit profit of each activity.

I **(e)** Use the Graphical Method and Sensitivity Analysis procedure in IOR Tutorial to estimate the allowable range for the unit profit of each activity.

E* **(f)** Use the sensitivity report provided by the Excel Solver to find the allowable range for the unit profit of each activity. Then use these ranges to check your results in parts (b–e).

E* **(g)** Use a two-way Solver Table to systematically generate the optimal solution as the unit profits of the two activities are changed simultaneously as described in part (d).

I **(h)** Use the Graphical Method and Sensitivity Analysis procedure in IOR Tutorial to interpret the results in part (g) graphically.

E* **6.8-2.** Reconsider the model given in Prob. 6.8-1. While doing sensitivity analysis, you learn that the estimates of the right-hand sides of the two functional constraints are accurate only to within ± 50 percent. In other words, the ranges of *likely values* for these parameters are 5 to 15 for the first right-hand side and 6 to 18 for the second right-hand side.

(a) After solving the original spreadsheet model, determine the shadow price for the first functional constraint by increasing its right-hand side by 1 and solving again.

(b) Use the Solver Table to generate the optimal solution and total profit as the right-hand side of the first functional constraint is incremented by 1 from 5 to 15. Use this table to estimate the allowable range for this right-hand side, i.e., the range over which the shadow price obtained in part (a) is valid.

(c) Repeat part (a) for the second functional constraint.

(d) Repeat part (b) for the second functional constraint where its right-hand side is incremented by 1 from 6 to 18.

(e) Use the Solver's sensitivity report to determine the shadow price for each functional constraint and the allowable range for the right-hand side of each of these constraints.

6.8-3. Consider the following problem.

Maximize $\quad Z = x_1 + 2x_2,$

subject to

$$x_1 + 3x_2 \leq 8 \text{ (resource 1)}$$
$$x_1 + x_2 \leq 4 \text{ (resource 2)}$$

and

$$x_1 \geq 0, \qquad x_2 \geq 0,$$

where Z measures the profit in dollars from the two activities and the right-hand sides are the number of units available of the respective resources.

I **(a)** Use the graphical method to solve this model.

I **(b)** Use graphical analysis to determine the shadow price for each of these resources by solving again after increasing the amount of the resource available by 1.

E* **(c)** Use the spreadsheet model and the Solver instead to do parts (a) and (b).

E* **(d)** For each resource in turn, use the Solver Table to systematically generate the optimal solution and the total profit when the only change is that the amount of that resource available increases in increments of 1 from 4 less than the original value up to 6 more than the current value. Use these results to estimate the allowable range for the amount available of each resource.

(e) Use the Solver's sensitivity report to obtain the shadow prices. Also use this report to find the range for the amount of each resource available over which the corresponding shadow price remains valid.

(f) Describe why these shadow prices are useful when management has the flexibility to change the amounts of the resources being made available.

6.8-4.* One of the products of the G.A. Tanner Company is a special kind of toy that provides an estimated unit profit of \$3. Because of a large demand for this toy, management would like to increase its production rate from the current level of 1,000 per day.

However, a limited supply of two subassemblies (A and B) from vendors makes this difficult. Each toy requires two subassemblies of type A, but the vendor providing these subassemblies would only be able to increase its supply rate from the current 2,000 per day to a maximum of 3,000 per day. Each toy requires only one subassembly of type B, but the vendor providing these subassemblies would be unable to increase its supply rate above the current level of 1,000 per day. Because no other vendors currently are available to provide these subassemblies, management is considering initiating a new production process internally that would simultaneously produce an equal number of subassemblies of the two types to supplement the supply from the two vendors. It is estimated that the company's cost for producing one subassembly of each type would be $2.50 more than the cost of purchasing these subassemblies from the two vendors. Management wants to determine both the production rate of the toy and the production rate of each pair of subassemblies (one A and one B) that would maximize the total profit.

The following table summarizes the data for the problem.

| Resource | Resource Usage per Unit of Each Activity | | Amount of Resource Available |
| | Activity | | |
Resource	Produce Toys	Produce Subassemblies	Amount of Resource Available
Subassembly A	2	−1	3,000
Subassembly B	1	−1	1,000
Unit profit	$3	−$2.50	

E* (a) Formulate and solve a spreadsheet model for this problem.

E* (b) Since the stated unit profits for the two activities are only estimates, management wants to know how much each of these estimates can be off before the optimal solution would change. Begin exploring this question for the first activity (producing toys) by using the spreadsheet and Solver to manually generate a table that gives the optimal solution and total profit as the unit profit for this activity increases in 50¢ increments from $2 to $4. What conclusion can be drawn about how much the estimate of this unit profit can differ in each direction from its original value of $3 before the optimal solution would change?

E* (c) Repeat part (b) for the second activity (producing subassemblies) by generating a table as the unit profit for this activity increases in 50¢ increments from −$3.50 to −$1.50 (with the unit profit for the first activity fixed at $3).

E* (d) Use the Solver Table to systematically generate all the data requested in parts (b) and (c), except use 25¢ increments instead of 50¢ increments. Use these data to refine your conclusions in parts (b) and (c).

I (e) Use the Graphical Method and Sensitivity Analysis procedure in IOR Tutorial to determine how much the unit profit of each activity can change in either direction (without changing the unit profit of the other activity) before the optimal solution would change. Use this information to specify the allowable range for the unit profit of each activity.

E* (f) Use Excel's sensitivity report to find the allowable range for the unit profit of each activity.

E* (g) Use a two-way Solver Table to systematically generate the optimal solution as the unit profits of the two activities are changed simultaneously as described in parts (b) and (c).

(h) Use the information provided by Excel's sensitivity report to describe how far the unit profits of the two activities can change simultaneously before the optimal solution might change.

E* 6.8-5. Reconsider Prob. 6.8-4. After further negotiations with each vendor, management of the G.A. Tanner Co. has learned that either of them would be willing to consider increasing their supply of their respective subassemblies over the previously stated maxima (3,000 subassemblies of type A per day and 1,000 of type B per day) if the company would pay a small premium over the regular price for the extra subassemblies. The size of the premium for each type of subassembly remains to be negotiated. The demand for the toy being produced is sufficiently high so that 2,500 per day could be sold if the supply of subassemblies could be increased enough to support this production rate. Assume that the original estimates of unit profits given in Prob. 6.8-4 are accurate.

(a) Formulate and solve a spreadsheet model for this problem with the original maximum supply levels and the additional constraint that no more than 2,500 toys should be produced per day.

(b) Without considering the premium, use the spreadsheet and Solver to determine the shadow price for the subassembly A constraint by solving the model again after increasing the maximum supply by 1. Use this shadow price to determine the maximum premium that the company should be willing to pay for each subassembly of this type.

(c) Repeat part (b) for the subassembly B constraint.

(d) Estimate how much the maximum supply of subassemblies of type A could be increased before the shadow price (and the corresponding premium) found in part (b) would no longer be valid by using the Solver Table to generate the optimal solution and total profit (excluding the premium) as the maximum supply increases in increments of 100 from 3,000 to 4,000.

(e) Repeat part (d) for subassemblies of type B by using the Solver Table as the maximum supply increases in increments of 100 from 1,000 to 2,000.

(f) Use the Solver's sensitivity report to determine the shadow price for each of the subassembly constraints and the allowable range for the right-hand side of each of these constraints.

E* 6.8-6.* Consider the Union Airways problem presented in Sec. 3.4, including the data given in Table 3.19. The Excel files for Chap. 3 include a spreadsheet that shows the formulation and optimal solution for this problem. You are to use this spreadsheet and the Excel Solver to do parts (a) to (g) below.

Management is about to begin negotiations on a new contract with the union that represents the company's customer service agents. This might result in some small changes in the daily costs per agent given in Table 3.19 for the various shifts. Several possible changes listed below are being considered separately. In each case, management would like to know whether the change might result in the solution in the spreadsheet no longer being optimal. Answer this question in parts (a) to (e) by using the spreadsheet and Solver directly. If the optimal solution changes, record the new solution.

(a) The daily cost per agent for Shift 2 changes from $160 to $165.
(b) The daily cost per agent for Shift 4 changes from $180 to $170.
(c) The changes in parts (a) and (b) both occur.
(d) The daily cost per agent increases by $4 for shifts 2, 4, and 5, but decreases by $4 for shifts 1 and 3.
(e) The daily cost per agent increases by 2 percent for each shift.
(f) Use the Solver to generate the sensitivity report for this problem. Suppose that the above changes are being considered later without having the spreadsheet model immediately available on a computer. Show in each case how the sensitivity report can be used to check whether the original optimal solution must still be optimal.
(g) For each of the five shifts in turn, use the Solver Table to systematically generate the optimal solution and total cost when the only change is that the daily cost per agent on that shift increases in $3 increments from $15 less than the current cost up to $15 more than the current cost.

E* **6.8-7.** Reconsider the Union Airways problem and its spreadsheet model that was dealt with in Prob. 6.8-6.

Management now is considering increasing the level of service provided to customers by increasing one or more of the numbers in the rightmost column of Table 3.19 for the minimum number of agents needed in the various time periods. To guide them in making this decision, they would like to know what impact this change would have on total cost.

Use the Excel Solver to generate the sensitivity report in preparation for addressing the following questions.

(a) Which of the numbers in the rightmost column of Table 3.19 can be increased without increasing total cost? In each case, indicate how much it can be increased (if it is the only one being changed) without increasing total cost.
(b) For each of the other numbers, how much would the total cost increase per increase of 1 in the number? For each answer, indicate how much the number can be increased (if it is the only one being changed) before the answer is no longer valid.
(c) Do your answers in part (b) definitely remain valid if all the numbers considered in part (b) are simultaneously increased by one?
(d) Do your answers in part (b) definitely remain valid if all 10 numbers are simultaneously increased by one?
(e) How far can all 10 numbers be simultaneously increased by the same amount before your answers in part (b) may no longer be valid?

6.8-8. David, LaDeana, and Lydia are the sole partners and workers in a company which produces fine clocks. David and LaDeana each are available to work a maximum of 40 hours per week at the company, while Lydia is available to work a maximum of 20 hours per week.

The company makes two different types of clocks: a grandfather clock and a wall clock. To make a clock, David (a mechanical

engineer) assembles the inside mechanical parts of the clock while LaDeana (a woodworker) produces the handcarved wood casings. Lydia is responsible for taking orders and shipping the clocks. The amount of time required for each of these tasks is shown below.

	Time Required	
Task	**Grandfather Clock**	**Wall Clock**
Assemble clock mechanism	6 hours	4 hours
Carve wood casing	8 hours	4 hours
Shipping	3 hours	3 hours

Each grandfather clock built and shipped yields a profit of $300, while each wall clock yields a profit of $200.

The three partners now want to determine how many clocks of each type should be produced per week to maximize the total profit.

(a) Formulate a linear programming model in algebraic form for this problem.
I (b) Use the Graphical Method and Sensitivity Analysis procedure in IOR Tutorial to solve the model. Then use this procedure to check if the optimal solution would change if the unit profit for grandfather clocks is changed from $300 to $375 (with no other changes in the model). Then check if the optimal solution would change if, in addition to this change in the unit profit for grandfather clocks, the estimated unit profit for wall clocks also changes from $200 to $175.
E* (c) Formulate and solve this model on a spreadsheet.
E* (d) Use the Excel Solver to check the effect of the changes specified in part (b).
E* (e) Use the Solver Table to systematically generate the optimal solution and total profit as the unit profit for grandfather clocks is increased in $20 increments from $150 to $450 (with no change in the unit profit for wall clocks). Then do the same as the unit profit for wall clocks is increased in $20 increments from $50 to $350 (with no change in the unit profit for grandfather clocks). Use this information to estimate the allowable range for the unit profit of each type of clock.
E* (f) Use a two-way Solver Table to systematically generate the optimal solution as the unit profits for the two types of clocks are changed simultaneously as specified in part (e), except use $50 increments instead of $20 increments.
E* (g) For each of the three partners in turn, use the Excel Solver to determine the effect on the optimal solution and the total profit if that partner alone were to increase the maximum number of hours available to work per week by 5 hours.
E* (h) Use the Solver Table to systematically generate the optimal solution and the total profit when the only change is that David's maximum number of hours available to work per week changes to each of the following values: 35, 37, 39, 41, 43, 45. Then do the same when the only change is that LaDeana's number changes in the same way. Then do the

same when the only change is that Lydia's number changes to each of the following values: 15, 17, 19, 21, 23, 25.

E* (i) Generate the Excel sensitivity report and use it to determine the allowable range for the unit profit for each type of clock and the allowable range for the maximum number of hours each partner is available to work per week.

(j) To increase the total profit, the three partners have agreed that one of them will slightly increase the maximum number of hours available to work per week. The choice of which one will be based on which one would increase the total profit the most. Use the sensitivity report to make this choice. (Assume no change in the original estimates of the unit profits.)

(k) Explain why one of the shadow prices is equal to zero.

(l) Can the shadow prices in the sensitivity report be validly used to determine the effect if Lydia were to change her maximum number of hours available to work per week from 20 to 25? If so, what would be the increase in the total profit?

(m) Repeat part (l) if, in addition to the change for Lydia, David also were to change his maximum number of hours available to work per week from 40 to 35.

I (n) Use graphical analysis to verify your answer in part (m).

■ CASES

CASE 6.1　Controlling Air Pollution

Refer to Sec. 3.4 (subsection entitled "Controlling Air Pollution") for the Nori & Leets Co. problem. After the OR team obtained an optimal solution, we mentioned that the team then conducted sensitivity analysis. We now continue this story by having you retrace the steps taken by the OR team, after we provide some additional background.

The values of the various parameters in the original formulation of the model are given in Tables 3.12, 3.13, and 3.14. Since the company does not have much prior experience with the pollution abatement methods under consideration, the cost estimates given in Table 3.14 are fairly rough, and each one could easily be off by as much as 10 percent in either direction. There also is some uncertainty about the parameter values given in Table 3.13, but less so than for Table 3.14. By contrast, the values in Table 3.12 are policy standards, and so are prescribed constants.

However, there still is considerable debate about where to set these policy standards on the required reductions in the emission rates of the various pollutants. The numbers in Table 3.12 actually are preliminary values tentatively agreed upon before learning what the total cost would be to meet these standards. Both the city and company officials agree that the final decision on these policy standards should be based on the *trade-off* between costs and benefits. With this in mind, the city has concluded that each 10 percent increase in the policy standards over the current values (all the numbers in Table 3.12) would be worth $3.5 million to the city. Therefore, the city has agreed to reduce the company's tax payments to the city by $3.5 million for *each* 10 percent reduction in the policy standards (up to 50 percent) that is accepted by the company.

Finally, there has been some debate about the *relative* values of the policy standards for the three pollutants. As indicated in Table 3.12, the required reduction for particulates now is less than half of that for either sulfur oxides or

hydrocarbons. Some have argued for decreasing this disparity. Others contend that an even greater disparity is justified because sulfur oxides and hydrocarbons cause considerably more damage than particulates. Agreement has been reached that this issue will be reexamined after information is obtained about which trade-offs in policy standards (increasing one while decreasing another) are available without increasing the total cost.

(a) Use any available linear programming software to solve the model for this problem as formulated in Sec. 3.4. In addition to the optimal solution, obtain the additional output provided for performing postoptimality analysis (e.g., the Sensitivity Report when using Excel). This output provides the basis for the following steps.

(b) Ignoring the constraints with no uncertainty about their parameter values (namely, $x_j \leq 1$ for $j = 1, 2, \ldots, 6$), identify the parameters of the model that should be classified as *sensitive parameters*. (*Hint:* See the subsection "Sensitivity Analysis" in Sec. 4.7.) Make a resulting recommendation about which parameters should be estimated more closely, if possible.

(c) Analyze the effect of an inaccuracy in estimating each cost parameter given in Table 3.14. If the true value is 10 percent *less* than the estimated value, would this alter the optimal solution? Would it change if the true value were 10 percent *more* than the estimated value? Make a resulting recommendation about where to focus further work in estimating the cost parameters more closely.

(d) Consider the case where your model has been converted to maximization form before applying the simplex method. Use Table 6.14 to construct the corresponding dual problem, and use the output from applying the simplex method to the primal problem to identify an optimal solution for this dual problem. If the primal problem had been left in minimization form, how would this affect the form of the dual problem and the sign of the optimal dual variables?

(e) For each pollutant, use your results from part (d) to specify the rate at which the total cost of an optimal solution would change with any small change in the required reduction in the annual emission rate of the pollutant. Also specify how much this required reduction can be changed (up or down) without affecting the rate of change in the total cost.

(f) For each unit change in the policy standard for particulates given in Table 3.12, determine the change in the opposite direction for sulfur oxides that would keep the total cost of an optimal solution unchanged. Repeat this for hydrocarbons instead of sulfur oxides. Then do it for a simultaneous and equal change for both sulfur oxides and hydrocarbons in the opposite direction from particulates.

(g) Letting θ denote the percentage increase in all the policy standards given in Table 3.12, formulate the problem of analyzing the effect of simultaneous proportional increases in these standards as a parametric linear programming problem. Then use your results from part (e) to determine the rate at which the total cost of an optimal solution would increase with a small increase in θ from zero.

(h) Use the simplex method to find an optimal solution for the parametric linear programming problem formulated in part (g) for each $\theta = 10, 20, 30, 40, 50$. Considering the tax incentive offered by the city, use these results to determine which value of θ (including the option of $\theta = 0$) should be chosen to minimize the company's total cost of both pollution abatement and taxes.

(i) For the value of θ chosen in part (h), repeat parts (e) and (f) so that the decision makers can make a final decision on the *relative* values of the policy standards for the three pollutants.

■ PREVIEWS OF ADDED CASES ON OUR WEBSITE (www.mhhe.com/hillier)

CASE 6.2 Farm Management

The Ploughman family has owned and operated a 640-acre farm for several generations. The family now needs to make a decision about the mix of livestock and crops for the coming year. By assuming that normal weather conditions will prevail next year, a linear programming model can be formulated and solved to guide this decision. However, adverse weather conditions would harm the crops and greatly reduce the resulting value. Therefore, considerable postoptimality analysis is needed to explore the effect of several possible scenarios for the weather next year and the implications for the family's decision.

CASE 6.3 Assigning Students to Schools, Revisited

This case is a continuation of Case 4.3, which involved the Springfield School Board assigning students from six residential areas to the city's three remaining middle schools. After solving a linear programming model for the problem with any software package, that package's sensitivity analysis report now needs to be used for two purposes. One is to check on the effect of an increase in certain bussing costs because of ongoing road construction in one of the residential areas. The other is to explore the advisability of adding portable classrooms to increase the capacity of one or more of the middle schools for a few years.

CASE 6.4 Writing a Nontechnical Memo

After setting goals for how much the sales of three products should increase as a result of an upcoming advertising campaign, the management of the Profit & Gambit Co. now wants to explore the trade-off between advertising cost and increased sales. Your first task is to perform the associated sensitivity analysis. Your main task then is to write a nontechnical memo to Profit & Gambit management presenting your results in the language of management.

Other Algorithms for Linear Programming

The key to the extremely widespread use of linear programming is the availability of an exceptionally efficient algorithm—the simplex method—that will routinely solve the large-size problems that typically arise in practice. However, the simplex method is only part of the arsenal of algorithms regularly used by linear programming practitioners. We now turn to these other algorithms.

This chapter begins with three algorithms that are, in fact, *variants* of the simplex method. In particular, the next three sections introduce the *dual simplex method* (a modification particularly useful for sensitivity analysis), *parametric linear programming* (an extension for systematic sensitivity analysis), and the *upper bound technique* (a streamlined version of the simplex method for dealing with variables having upper bounds). We will not go into the kind of detail with these algorithms that we did with the simplex method in Chaps. 4 and 5. The goal instead will be to briefly introduce their main ideas.

Section 4.9 introduced another algorithmic approach to linear programming—a type of algorithm that moves through the interior of the feasible region. We describe this *interior-point approach* further in Sec. 7.4.

A supplement to this chapter on the book's website also introduces *linear goal programming*. In this case, rather than having a *single objective* (maximize or minimize Z) as for linear programming, the problem instead has *several goals* toward which we must strive simultaneously. Certain formulation techniques enable converting a linear goal programming problem back into a linear programming problem so that solution procedures based on the simplex method can still be used. The supplement describes these techniques and procedures.

■ 7.1 THE DUAL SIMPLEX METHOD

The *dual simplex method* is based on the duality theory presented in the first part of Chap. 6. To describe the basic idea behind this method, it is helpful to use some terminology introduced in Tables 6.10 and 6.11 of Sec. 6.3 for describing any pair of complementary basic solutions in the primal and dual problems. In particular, recall that both solutions are said to be *primal feasible* if the primal basic solution is feasible, whereas they are called *dual feasible* if the complementary dual basic solution is feasible for the dual problem. Also recall (as indicated on the right side of Table 6.11) that each complementary basic solution is optimal for its problem only if it is *both* primal feasible and dual feasible.

The dual simplex method can be thought of as the *mirror image* of the simplex method. The simplex method deals directly with basic solutions in the primal problem that are *primal feasible* but not dual feasible. It then moves toward an optimal solution by striving to achieve dual feasibility as well (the optimality test for the simplex method). By contrast, the dual simplex method deals with basic solutions in the primal problem that are *dual feasible* but not primal feasible. It then moves toward an optimal solution by striving to achieve primal feasibility as well.

Furthermore, the dual simplex method deals with a problem as if the simplex method were being applied simultaneously to its dual problem. If we make their *initial* basic solutions *complementary*, the two methods move in complete sequence, obtaining *complementary* basic solutions with each iteration.

The dual simplex method is very useful in certain special types of situations. Ordinarily it is easier to find an initial basic solution that is feasible than one that is dual feasible. However, it is occasionally necessary to introduce many *artificial* variables to construct an initial BF solution artificially. In such cases it may be easier to begin with a dual feasible basic solution and use the dual simplex method. Furthermore, fewer iterations may be required when it is not necessary to drive many artificial variables to zero.

When dealing with a problem whose initial basic solutions (without artificial variables) are *neither* primal feasible nor dual feasible, it also is possible to combine the ideas of the simplex method and dual simplex method into a *primal-dual algorithm* that strives toward both primal feasibility and dual feasibility.

As we mentioned several times in Chap. 6 as well as in Sec. 4.7, another important primary application of the dual simplex method is its use in conjunction with sensitivity analysis. Suppose that an optimal solution has been obtained by the simplex method but that it becomes necessary (or of interest for sensitivity analysis) to make minor changes in the model. If the formerly optimal basic solution is *no longer primal feasible* (but still satisfies the optimality test), you can immediately apply the dual simplex method by starting with this *dual feasible* basic solution. (We will illustrate this at the end of this section.) Applying the dual simplex method in this way usually leads to the new optimal solution much more quickly than would solving the new problem from the beginning with the simplex method.

The dual simplex method also can be useful in solving certain huge linear programming problems from scratch because it is such an efficient algorithm. Computational experience with the most powerful versions of CPLEX indicates that the dual simplex method often is more efficient than the simplex method for solving particularly massive problems encountered in practice.

The rules for the dual simplex method are very similar to those for the simplex method. In fact, once the methods are started, the only difference between them is in the criteria used for selecting the entering and leaving basic variables and for stopping the algorithm.

To start the dual simplex method (for a maximization problem), we must have all the coefficients in Eq. (0) *nonnegative* (so that the basic solution is dual feasible). The basic solutions will be infeasible (except for the last one) only because some of the variables are negative. The method continues to decrease the value of the objective function, always retaining *nonnegative coefficients* in Eq. (0), until all the *variables* are nonnegative. Such a basic solution is feasible (it satisfies all the equations) and is, therefore, optimal by the simplex method criterion of nonnegative coefficients in Eq. (0).

The details of the dual simplex method are summarized next.

Summary of the Dual Simplex Method

1. *Initialization:*   After converting any functional constraints in $\geq$ form to $\leq$ form (by multiplying through both sides by -1), introduce slack variables as needed to construct a set of equations describing the problem. Find a basic solution such that the coefficients in Eq. (0) are zero for basic variables and nonnegative for nonbasic variables (so the solution is optimal if it is feasible). Go to the feasibility test.
2. *Feasibility test:*   Check to see whether all the basic variables are *nonnegative.* If they are, then this solution is feasible, and therefore optimal, so stop. Otherwise, go to an iteration.
3. *Iteration:*

   Step 1   Determine the *leaving basic variable:* Select the *negative* basic variable that has the largest absolute value.

   Step 2   Determine the *entering basic variable:* Select the nonbasic variable whose coefficient in Eq. (0) reaches zero first as an increasing multiple of the equation containing the leaving basic variable is added to Eq. (0). This selection is made by checking the nonbasic variables with *negative coefficients* in that equation (the one containing the leaving basic variable) and selecting the one with the smallest absolute value of the ratio of the Eq. (0) coefficient to the coefficient in that equation.

   Step 3   Determine the *new basic solution:* Starting from the current set of equations, solve for the basic variables in terms of the nonbasic variables by Gaussian elimination. When we set the nonbasic variables equal to zero, each basic variable (and Z) equals the new right-hand side of the one equation in which it appears (with a coefficient of $+1$). Return to the feasibility test.

To fully understand the dual simplex method, you must realize that the method proceeds just as if the *simplex method* were being applied to the complementary basic solutions in the *dual problem.* (In fact, this interpretation was the motivation for constructing the method as it is.) Step 1 of an iteration, determining the leaving basic variable, is equivalent to determining the entering basic variable in the dual problem. The negative variable with the largest absolute value corresponds to the negative coefficient with the largest absolute value in Eq. (0) of the dual problem (see Table 6.3). Step 2, determining the entering basic variable, is equivalent to determining the leaving basic variable in the dual problem. The coefficient in Eq. (0) that reaches zero first corresponds to the variable in the dual problem that reaches zero first. The two criteria for stopping the algorithm are also complementary.

An Example

We shall now illustrate the dual simplex method by applying it to the *dual problem* for the Wyndor Glass Co. (see Table 6.1). Normally this method is applied directly to the problem of concern (a primal problem). However, we have chosen this problem because you have already seen the simplex method applied to *its* dual problem (namely, the primal problem[1]) in Table 4.8 so you can compare the two. To facilitate the comparison, we shall continue to denote the decision variables in the problem being solved by y_i rather than x_j.

In *maximization* form, the problem to be solved is

$$\text{Maximize} \quad Z = -4y_1 - 12y_2 - 18y_3,$$

subject to

$$y_1 \quad + 3y_3 \geq 3$$
$$2y_2 + 2y_3 \geq 5$$

---

[1] Recall that the symmetry property in Sec. 6.1 points out that the dual of a dual problem is the original primal problem.

■ TABLE 7.1 Dual simplex method applied to the Wyndor Glass Co. dual problem

Iteration	Basic Variable	Eq.	Z	Coefficient of:					Right Side
				y_1	y_2	y_3	y_4	y_5	
0	Z	(0)	1	4	12	18	0	0	0
	y_4	(1)	0	-1	0	-3	1	0	-3
	y_5	(2)	0	0	-2	-2	0	1	-5
1	Z	(0)	1	4	0	6	0	6	-30
	y_4	(1)	0	-1	0	-3	1	0	-3
	y_2	(2)	0	0	1	1	0	$-\frac{1}{2}$	$\frac{5}{2}$
2	Z	(0)	1	2	0	0	2	6	-36
	y_3	(1)	0	$\frac{1}{3}$	0	1	$-\frac{1}{3}$	0	1
	y_2	(2)	0	$-\frac{1}{3}$	1	0	$\frac{1}{3}$	$-\frac{1}{2}$	$\frac{3}{2}$

and

$$y_1 \geq 0, \qquad y_2 \geq 0, \qquad y_3 \geq 0.$$

Since negative right-hand sides are now allowed, we do not need to introduce artificial variables to be the initial basic variables. Instead, we simply convert the functional constraints to $\leq$ form and introduce slack variables to play this role. The resulting initial set of equations is that shown for iteration 0 in Table 7.1. Notice that all the coefficients in Eq. (0) are nonnegative, so the solution is optimal if it is feasible.

The initial basic solution is $y_1 = 0$, $y_2 = 0$, $y_3 = 0$, $y_4 = -3$, $y_5 = -5$, with $Z = 0$, which is not feasible because of the negative values. The leaving basic variable is y_5 ($5 > 3$), and the entering basic variable is y_2 ($12/2 < 18/2$), which leads to the second set of equations, labeled as iteration 1 in Table 7.1. The corresponding basic solution is $y_1 = 0$, $y_2 = \frac{5}{2}$, $y_3 = 0$, $y_4 = -3$, $y_5 = 0$, with $Z = -30$, which is not feasible.

The next leaving basic variable is y_4, and the entering basic variable is y_3 ($6/3 < 4/1$), which leads to the final set of equations in Table 7.1. The corresponding basic solution is $y_1 = 0$, $y_2 = \frac{3}{2}$, $y_3 = 1$, $y_4 = 0$, $y_5 = 0$, with $Z = -36$, which is feasible and therefore optimal.

Notice that the optimal solution for the dual of this problem[2] is $x_1^* = 2$, $x_2^* = 6$, $x_3^* = 2$, $x_4^* = 0$, $x_5^* = 0$, as was obtained in Table 4.8 by the simplex method. We suggest that you now trace through Tables 7.1 and 4.8 simultaneously and compare the complementary steps for the two mirror-image methods.

As mentioned earlier, an important primary application of the dual simplex method is that it frequently can be used to quickly re-solve a problem when sensitivity analysis results in making small changes in the original model. In particular, if the formerly optimal basic solution is no longer primal feasible (one or more right-hand sides now are negative) but still satisfies the optimality test (no negative coefficients in Row 0), you can immediately apply the dual simplex method by starting with this dual feasible basic solution. For example, this situation arises when a new constraint that violates the formerly optimal solution is added to the original model. To illustrate, suppose that the problem solved in Table 7.1 originally did not include its first functional constraint ($y_1 + 3y_3 \geq 3$).

---

[2]The *complementary optimal basic solutions property* presented in Sec. 6.3 indicates how to read the optimal solution for the dual problem from row 0 of the final simplex tableau for the primal problem. This same conclusion holds regardless of whether the simplex method or the dual simplex method is used to obtain the final tableau.

After deleting Row 1, the iteration 1 tableau in Table 7.1 shows that the resulting optimal solution is $y_1 = 0$, $y_2 = \frac{5}{2}$, $y_3 = 0$, $y_5 = 0$, with $Z = -30$. Now suppose that sensitivity analysis leads to adding the originally omitted constraint, $y_1 + 3y_3 \geq 3$, which is violated by the original optimal solution since both $y_1 = 0$ and $y_3 = 0$. To find the new optimal solution, this constraint (including its slack variable y_4) now would be added as Row 1 of the middle tableau in Table 7.1. Regardless of whether this tableau had been obtained by applying the simplex method or the dual simplex method to obtain the original optimal solution (perhaps after many iterations), applying the dual simplex method to this tableau leads to the new optimal solution in just one iteration.

If you would like to see **another example** of applying the dual simplex method, one is provided in the Worked Examples section of the book's website.

■ 7.2 PARAMETRIC LINEAR PROGRAMMING

At the end of Sec. 6.7 we described *parametric linear programming* and its use for conducting sensitivity analysis systematically by gradually changing various model parameters simultaneously. We shall now present the algorithmic procedure, first for the case where the c_j parameters are being changed and then where the b_i parameters are varied.

Systematic Changes in the c_j Parameters

For the case where the c_j parameters are being changed, the *objective function* of the ordinary linear programming model

$$Z = \sum_{j=1}^{n} c_j x_j$$

is replaced by

$$Z(\theta) = \sum_{j=1}^{n} (c_j + \alpha_j \theta) x_j,$$

where the α_j are given input constants representing the *relative* rates at which the coefficients are to be changed. Therefore, gradually increasing θ from zero changes the coefficients at these relative rates.

The values assigned to the α_j may represent interesting simultaneous changes of the c_j for systematic sensitivity analysis of the effect of increasing the magnitude of these changes. They may also be based on how the coefficients (e.g., unit profits) would change together with respect to some factor measured by θ. This factor might be uncontrollable, e.g., the state of the economy. However, it may also be under the control of the decision maker, e.g., the amount of personnel and equipment to shift from some of the activities to others.

For any given value of θ, the optimal solution of the corresponding linear programming problem can be obtained by the simplex method. This solution may have been obtained already for the original problem where $\theta = 0$. However, the objective is to *find the optimal solution* of the modified linear programming problem [maximize $Z(\theta)$ subject to the original constraints] *as a function of θ*. Therefore, in the solution procedure you need to be able to determine when and how the optimal solution changes (if it does) as θ increases from zero to any specified positive number.

Figure 7.1 illustrates how $Z^*(\theta)$, the objective function value for the optimal solution (given θ), changes as θ increases. In fact, $Z^*(\theta)$ always has this *piecewise linear* and *convex*[3] form (see Prob. 7.2-7). The corresponding optimal solution changes (as θ increases) *just*

---

[3]See Appendix 2 for a definition and discussion of convex functions.

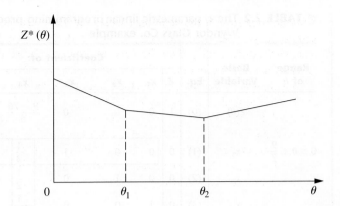

■ FIGURE 7.1
The objective function value for an optimal solution as a function of θ for parametric linear programming with systematic changes in the c_j parameters.

at the values of θ where the slope of the $Z^*(\theta)$ function changes. Thus, Fig. 7.1 depicts a problem where three different solutions are optimal for different values of θ, the first for $0 \le \theta \le \theta_1$, the second for $\theta_1 \le \theta \le \theta_2$, and the third for $\theta \ge \theta_2$. Because the value of each x_j remains the same within each of these intervals for θ, the value of $Z^*(\theta)$ varies with θ only because the *coefficients* of the x_j are changing as a linear function of θ. The solution procedure is based directly upon the sensitivity analysis procedure for investigating changes in the c_j parameters (Cases 2a and 3, Sec. 6.7). As described in the last subsection of Sec. 6.7, the only basic difference with parametric linear programming is that the changes now are expressed in terms of θ rather than as specific numbers.

Example. To illustrate the solution procedure, suppose that $\alpha_1 = 2$ and $\alpha_2 = -1$ for the original Wyndor Glass Co. problem presented in Sec. 3.1, so that

$$Z(\theta) = (3 + 2\theta)x_1 + (5 - \theta)x_2.$$

Beginning with the final simplex tableau for $\theta = 0$ (Table 4.8), we see that its Eq. (0)

$$(0) \quad Z + \frac{3}{2}x_4 + x_5 = 36$$

would first have these changes from the original ($\theta = 0$) coefficients added into it on the left-hand side:

$$(0) \quad Z - 2\theta x_1 + \theta x_2 + \frac{3}{2}x_4 + x_5 = 36.$$

Because both x_1 and x_2 are basic variables [appearing in Eqs. (3) and (2), respectively], they both need to be eliminated algebraically from Eq. (0):

$$Z - 2\theta x_1 + \theta x_2 + \frac{3}{2}x_4 + x_5 = 36$$
$$+ 2\theta \text{ times Eq. (3)}$$
$$- \theta \text{ times Eq. (2)}$$

$$(0) \quad Z + \left(\frac{3}{2} - \frac{7}{6}\theta\right)x_4 + \left(1 + \frac{2}{3}\theta\right)x_5 = 36 - 2\theta.$$

The optimality test says that the current BF solution will remain optimal as long as these coefficients of the nonbasic variables remain nonnegative:

$$\frac{3}{2} - \frac{7}{6}\theta \ge 0, \quad \text{for } 0 \le \theta \le \frac{9}{7},$$

$$1 + \frac{2}{3}\theta \ge 0, \quad \text{for all } \theta \ge 0.$$

■ **TABLE 7.2** The c_j parametric linear programming procedure applied to the Wyndor Glass Co. example

Range of θ	Basic Variable	Eq.	Z	x_1	x_2	x_3	x_4	x_5	Right Side	Optimal Solution
	$Z(\theta)$	(0)	1	0	0	0	$\dfrac{9-7\theta}{6}$	$\dfrac{3+2\theta}{3}$	$36-2\theta$	$x_4 = 0$
										$x_5 = 0$
$0 \le \theta \le \dfrac{9}{7}$	x_3	(1)	0	0	0	1	$\dfrac{1}{3}$	$-\dfrac{1}{3}$	2	$x_3 = 2$
	x_2	(2)	0	0	1	0	$\dfrac{1}{2}$	0	6	$x_2 = 6$
	x_1	(3)	0	1	0	0	$-\dfrac{1}{3}$	$\dfrac{1}{3}$	2	$x_1 = 2$
	$Z(\theta)$	(0)	1	0	0	$\dfrac{-9+7\theta}{2}$	0	$\dfrac{5-\theta}{2}$	$27+5\theta$	$x_3 = 0$
										$x_5 = 0$
$\dfrac{9}{7} \le \theta \le 5$	x_4	(1)	0	0	0	3	1	-1	6	$x_4 = 6$
	x_2	(2)	0	0	1	$-\dfrac{3}{2}$	0	$\dfrac{1}{2}$	3	$x_2 = 3$
	x_1	(3)	0	1	0	1	0	0	4	$x_1 = 4$
	$Z(\theta)$	(0)	1	0	$-5+\theta$	$3+2\theta$	0	0	$12+8\theta$	$x_2 = 0$
										$x_3 = 0$
$\theta \ge 5$	x_4	(1)	0	0	2	0	1	0	12	$x_4 = 12$
	x_5	(2)	0	0	2	-3	0	1	6	$x_5 = 6$
	x_1	(3)	0	1	0	1	0	0	4	$x_1 = 4$

Therefore, after θ is increased past $\theta = \frac{9}{7}$, x_4 would need to be the entering basic variable for another iteration of the simplex method to find the new optimal solution. Then θ would be increased further until another coefficient goes negative, and so on until θ has been increased as far as desired.

This entire procedure is now summarized, and the example is completed in Table 7.2.

Summary of the Parametric Linear Programming Procedure for Systematic Changes in the c_j Parameters

1. Solve the problem with $\theta = 0$ by the simplex method.
2. Use the sensitivity analysis procedure (Cases 2a and 3, Sec. 6.7) to introduce the $\Delta c_j = \alpha_j \theta$ changes into Eq. (0).
3. Increase θ until one of the nonbasic variables has its coefficient in Eq. (0) go negative (or until θ has been increased as far as desired).
4. Use this variable as the entering basic variable for an iteration of the simplex method to find the new optimal solution. Return to step 3.

Systematic Changes in the b_i Parameters

For the case where the b_i parameters change systematically, the one modification made in the original linear programming model is that b_i is replaced by $b_i + \alpha_i\theta$, for $i = 1$, $2, \ldots, m$, where the α_i are given input constants. Thus, the problem becomes

$$\text{Maximize} \quad Z(\theta) = \sum_{j=1}^{n} c_j x_j,$$

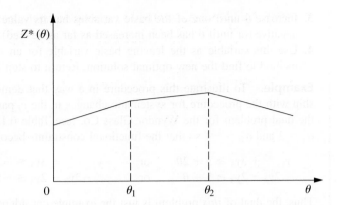

FIGURE 7.2
The objective function value for an optimal solution as a function of θ for parametric linear programming with systematic changes in the b_i parameters.

subject to

$$\sum_{j=1}^{n} a_{ij}x_j \leq b_i + \alpha_i\theta \qquad \text{for } i = 1, 2, \ldots, m$$

and

$$x_j \geq 0 \qquad \text{for } j = 1, 2, \ldots, n.$$

The goal is to identify the optimal solution as a function of θ.

With this formulation, the corresponding objective function value $Z^*(\theta)$ always has the *piecewise linear* and *concave*[4] form shown in Fig. 7.2. (See Prob. 7.2-8.) The set of basic variables in the optimal solution still changes (as θ increases) *only* where the slope of $Z^*(\theta)$ changes. However, in contrast to the preceding case, the values of these variables now change as a (linear) function of θ between the slope changes. The reason is that increasing θ changes the right-hand sides in the initial set of equations, which then causes changes in the right-hand sides in the final set of equations, i.e., in the values of the final set of basic variables. Figure 7.2 depicts a problem with three sets of basic variables that are optimal for different values of θ, the first for $0 \leq \theta \leq \theta_1$, the second for $\theta_1 \leq \theta \leq \theta_2$, and the third for $\theta \geq \theta_2$. Within each of these intervals of θ, the value of $Z^*(\theta)$ varies with θ despite the fixed coefficients c_j because the x_j values are changing.

The following solution procedure summary is very similar to that just presented for systematic changes in the c_j parameters. The reason is that changing the b_i values is equivalent to changing the coefficients in the objective function of the *dual* model. Therefore, the procedure for the primal problem is exactly *complementary* to applying simultaneously the procedure for systematic changes in the c_j parameters to the *dual* problem. Consequently, the *dual simplex method* (see Sec. 7.1) now would be used to obtain each new optimal solution, and the applicable sensitivity analysis case (see Sec. 6.7) now is Case 1, but these differences are the only major differences.

Summary of the Parametric Linear Programming Procedure for Systematic Changes in the b_i Parameters

1. Solve the problem with $\theta = 0$ by the simplex method.
2. Use the sensitivity analysis procedure (Case 1, Sec. 6.7) to introduce the $\Delta b_i = \alpha_i\theta$ changes to the *right side* column.

---

[4]See Appendix 2 for a definition and discussion of concave functions.

3. Increase θ until one of the basic variables has its value in the *right side* column go negative (or until θ has been increased as far as desired).

4. Use this variable as the leaving basic variable for an iteration of the dual simplex method to find the new optimal solution. Return to step 3.

Example.   To illustrate this procedure in a way that demonstrates its *duality* relationship with the procedure for systematic changes in the c_j parameters, we now apply it to the dual problem for the Wyndor Glass Co. (see Table 6.1). In particular, suppose that $\alpha_1 = 2$ and $\alpha_2 = -1$ so that the functional constraints become

$$y_1 \quad + 3y_3 \geq 3 + 2\theta \qquad \text{or} \qquad -y_1 \quad - 3y_3 \leq -3 - 2\theta$$
$$2y_2 + 2y_3 \geq 5 - \theta \qquad \text{or} \qquad -2y_2 - 2y_3 \leq -5 + \theta.$$

Thus, the dual of *this* problem is just the example considered in Table 7.2.

This problem with $\theta = 0$ has already been solved in Table 7.1, so we begin with the final simplex tableau given there. Using the sensitivity analysis procedure for Case 1, Sec. 6.7, we find that the entries in the *right side* column of the tableau change to the values given below.

$$Z^* = \mathbf{y}^*\overline{\mathbf{b}} = [2, 6] \begin{bmatrix} -3 - 2\theta \\ -5 + \theta \end{bmatrix} = -36 + 2\theta,$$

$$\mathbf{b}^* = \mathbf{S}^*\overline{\mathbf{b}} = \begin{bmatrix} -\frac{1}{3} & 0 \\ \frac{1}{3} & -\frac{1}{2} \end{bmatrix} \begin{bmatrix} -3 - 2\theta \\ -5 + \theta \end{bmatrix} = \begin{bmatrix} 1 + \frac{2\theta}{3} \\ \frac{3}{2} - \frac{7\theta}{6} \end{bmatrix}.$$

Therefore, the two basic variables in this tableau

$$y_3 = \frac{3 + 2\theta}{3} \qquad \text{and} \qquad y_2 = \frac{9 - 7\theta}{6}$$

remain nonnegative for $0 \leq \theta \leq \frac{9}{7}$. Increasing θ past $\theta = \frac{9}{7}$ requires making y_2 a leaving basic variable for another iteration of the dual simplex method, and so on, as summarized in Table 7.3.

■ **TABLE 7.3** The b_i parametric linear programming procedure applied to the dual of the Wyndor Glass Co. example

Range of θ	Basic Variable	Eq.	Z	y_1	y_2	y_3	y_4	y_5	Right Side	Optimal Solution
	$Z(\theta)$	(0)	1	2	0	0	2	6	$-36 + 2\theta$	$y_1 = y_4 = y_5 = 0$
$0 \leq \theta \leq \frac{9}{7}$	y_3	(1)	0	$\frac{1}{3}$	0	1	$-\frac{1}{3}$	0	$\frac{3 + 2\theta}{3}$	$y_3 = \frac{3 + 2\theta}{3}$
	y_2	(2)	0	$-\frac{1}{3}$	1	0	$\frac{1}{3}$	$-\frac{1}{2}$	$\frac{9 - 7\theta}{6}$	$y_2 = \frac{9 - 7\theta}{6}$
	$Z(\theta)$	(0)	1	0	6	0	4	3	$-27 - 5\theta$	$y_2 = y_4 = y_5 = 0$
$\frac{9}{7} \leq \theta \leq 5$	y_3	(1)	0	0	1	1	0	$-\frac{1}{2}$	$\frac{5 - \theta}{2}$	$y_3 = \frac{5 - \theta}{2}$
	y_1	(2)	0	1	-3	0	-1	$\frac{3}{2}$	$\frac{-9 + 7\theta}{2}$	$y_1 = \frac{-9 + 7\theta}{2}$
	$Z(\theta)$	(0)	1	0	12	6	4	0	$-12 - 8\theta$	$y_2 = y_3 = y_4 = 0$
$\theta \geq 5$	y_5	(1)	0	0	-2	-2	0	1	$-5 + \theta$	$y_5 = -5 + \theta$
	y_1	(2)	0	1	0	3	-1	0	$3 + 2\theta$	$y_1 = 3 + 2\theta$

We suggest that you now trace through Tables 7.2 and 7.3 simultaneously to note the duality relationship between the two procedures.

The Worked Examples section of the book's website includes **another example** of the procedure for systematic changes in the b_i parameters.

7.3 THE UPPER BOUND TECHNIQUE

It is fairly common in linear programming problems for some of or all the *individual* x_j variables to have *upper bound constraints*

$$x_j \le u_j,$$

where u_j is a positive constant representing the maximum *feasible* value of x_j. We pointed out in Sec. 4.8 that the most important determinant of computation time for the simplex method is the *number of functional constraints*, whereas the number of *nonnegativity* constraints is relatively unimportant. Therefore, having a large number of upper bound constraints among the functional constraints greatly increases the computational effort required.

The *upper bound technique* avoids this increased effort by removing the upper bound constraints from the functional constraints and treating them separately, essentially like nonnegativity constraints.[5] Removing the upper bound constraints in this way causes no problems as long as none of the variables gets increased over its upper bound. The only time the simplex method increases some of the variables is when the entering basic variable is increased to obtain a new BF solution. Therefore, the upper bound technique simply applies the simplex method in the usual way to the *remainder* of the problem (i.e., without the upper bound constraints) but with the one additional restriction that each new BF solution must satisfy the upper bound constraints in addition to the usual lower bound (nonnegativity) constraints.

To implement this idea, note that a decision variable x_j with an upper bound constraint $x_j \le u_j$ can always be replaced by

$$x_j = u_j - y_j,$$

where y_j would then be the decision variable. In other words, you have a choice between letting the decision variable be the *amount above zero* (x_j) or the *amount below* u_j ($y_j = u_j - x_j$). (We shall refer to x_j and y_j as *complementary* decision variables.) Because

$$0 \le x_j \le u_j$$

it also follows that

$$0 \le y_j \le u_j.$$

Thus, at any point during the simplex method, you can either

1. Use x_j, where $0 \le x_j \le u_j$, or
2. Replace x_j by $u_j - y_j$, where $0 \le y_j \le u_j$.

The upper bound technique uses the following rule to make this choice:

Rule: Begin with choice 1.
Whenever $x_j = 0$, use choice 1, so x_j is *nonbasic*.

---

[5]The upper bound technique assumes that the variables have the usual nonnegativity constraints in addition to the upper bound constraints. If a variable has a lower bound other than 0, say, $x_j \ge L_j$, then this constraint can be converted into a nonnegativity constraint by making the change of variables, $x'_j = x_j - L_j$, so $x'_j \ge 0$.

Whenever $x_j = u_j$, use choice 2, so $y_j = 0$ is *nonbasic.*
Switch choices only when the other extreme value of x_j is reached.

Therefore, whenever a basic variable reaches its upper bound, you should switch choices and use its complementary decision variable as the new nonbasic variable (the leaving basic variable) for identifying the new BF solution. Thus, the one substantive modification being made in the simplex method is in the rule for selecting the leaving basic variable.

Recall that the simplex method selects as the leaving basic variable the one that would be the first to become infeasible by going negative as the entering basic variable is increased. The modification now made is to select instead the variable that would be the first to become infeasible *in any way,* either by going negative or by going over the upper bound, as the entering basic variable is increased. (Notice that one possibility is that the entering basic variable may become infeasible first by going over its upper bound, so that its complementary decision variable becomes the leaving basic variable.) If the leaving basic variable reaches zero, then proceed as usual with the simplex method. However, if it reaches its upper bound instead, then switch choices and make its complementary decision variable the leaving basic variable.

An Example

To illustrate the upper bound technique, consider this problem:

Maximize $Z = 2x_1 + x_2 + 2x_3,$

subject to

$$4x_1 + x_2 \quad\quad = 12$$
$$-2x_1 \quad\quad + x_3 = 4$$

and

$$0 \le x_1 \le 4, \quad\quad 0 \le x_2 \le 15, \quad\quad 0 \le x_3 \le 6.$$

Thus, all three variables have upper bound constraints ($u_1 = 4$, $u_2 = 15$, $u_3 = 6$).

The two equality constraints are already in proper form from Gaussian elimination for identifying the initial BF solution ($x_1 = 0$, $x_2 = 12$, $x_3 = 4$), and none of the variables in this solution exceeds its upper bound, so x_2 and x_3 can be used as the initial basic variables without artificial variables being introduced. However, these variables then need to be eliminated algebraically from the objective function to obtain the initial Eq. (0), as follows:

$$
\begin{array}{l}
Z \quad - 2x_1 - x_2 - 2x_3 = \ 0 \\
\quad\quad + (4x_1 + x_2 \quad\quad = 12) \\
\quad\quad + 2(- \ 2x_1 \quad\quad + x_3 = \ 4) \\
\hline
(0) \quad Z \quad - 2x_1 \quad\quad\quad\quad = 20.
\end{array}
$$

To start the first iteration, this initial Eq. (0) indicates that the initial *entering* basic variable is x_1. Since the upper bound constraints are not to be included, the entire initial set of equations and the corresponding calculations for selecting the leaving basic variables are those shown in Table 7.4. The second column shows how much the entering basic variable x_1 can be *increased* from zero before some basic variable (including x_1) becomes infeasible. The maximum value given next to Eq. (0) is just the upper bound constraint for x_1. For Eq. (1), since the coefficient of x_1 is *positive, increasing* x_1 to 3 decreases the basic variable in this equation (x_2) from 12 to its *lower* bound of *zero.* For Eq. (2), since the coefficient of x_1 is *negative, increasing* x_1 to 1 *increases* the basic variable in this equation (x_3) from 4 to its *upper bound* of 6.

■ **TABLE 7.4** Equations and calculations for the initial leaving basic variable in the example for the upper bound technique

Initial Set of Equations	Maximum Feasible Value of x_1
(0) $Z - 2x_1 \qquad\qquad = 20$	$x_1 \leq 4 \quad$ (since $u_1 = 4$)
(1) $\qquad 4x_1 + x_2 \qquad = 12$	$x_1 \leq \dfrac{12}{4} = 3$
(2) $\qquad -2x_1 \qquad + x_3 = 4$	$x_1 \leq \dfrac{6-4}{2} = 1 \leftarrow$ minimum (because $u_3 = 6$)

Because Eq. (2) has the *smallest* maximum feasible value of x_1 in Table 7.4, the basic variable in this equation (x_3) provides the *leaving* basic variable. However, because x_3 reached its *upper* bound, replace x_3 by $6 - y_3$, so that $y_3 = 0$ becomes the new nonbasic variable for the next BF solution and x_1 becomes the new basic variable in Eq. (2). This replacement leads to the following changes in this equation:

$$
\begin{aligned}
(2) \qquad -2x_1 + \; x_3 \; &= \; 4 \\
\rightarrow -2x_1 + 6 - y_3 &= \; 4 \\
\rightarrow -2x_1 - \; y_3 \; &= -2 \\
\rightarrow \qquad x_1 + \tfrac{1}{2}y_3 &= \; 1
\end{aligned}
$$

Therefore, after we eliminate x_1 algebraically from the other equations, the *second* complete set of equations becomes

$$
\begin{aligned}
(0) \quad Z \qquad\quad + \; y_3 &= 22 \\
(1) \qquad x_2 - 2y_3 &= 8 \\
(2) \qquad x_1 + \tfrac{1}{2}y_3 &= 1.
\end{aligned}
$$

The resulting BF solution is $x_1 = 1$, $x_2 = 8$, $y_3 = 0$. By the optimality test, it also is an optimal solution, so $x_1 = 1$, $x_2 = 8$, $x_3 = 6 - y_3 = 6$ is the desired solution for the original problem.

If you would like to see **another example** of the upper bound technique, the Worked Examples section of the book's website includes one.

■ 7.4 AN INTERIOR-POINT ALGORITHM

In Sec. 4.9 we discussed a dramatic development in linear programming that occurred in 1984, namely, the invention by Narendra Karmarkar of AT&T Bell Laboratories of a powerful algorithm for solving huge linear programming problems with an approach very different from the simplex method. We now introduce the nature of Karmarkar's approach by describing a relatively elementary variant (the "affine" or "affine-scaling" variant) of his algorithm.[6] (Your IOR Tutorial also includes this variant under the title, *Solve Automatically by the Interior-Point Algorithm.*)

Throughout this section we shall focus on Karmarkar's main ideas on an intuitive level while avoiding mathematical details. In particular, we shall bypass certain details

---

[6]The basic approach for this variant actually was proposed in 1967 by a Russian mathematician I. I. Dikin and then rediscovered soon after the appearance of Karmarkar's work by a number of researchers, including E. R. Barnes, T. M. Cavalier, and A. L. Soyster. Also see R. J. Vanderbei, M. S. Meketon, and B. A. Freedman, "A Modification of Karmarkar's Linear Programming Algorithm," *Algorithmica,* **1**(4) (Special Issue on New Approaches to Linear Programming): 395–407, 1986.

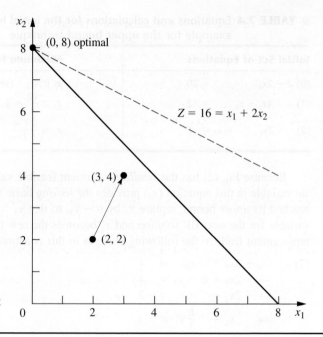

■ FIGURE 7.3
Example for the interior-point algorithm.

that are needed for the full implementation of the algorithm (e.g., how to find an initial feasible trial solution) but are not central to a basic conceptual understanding. The ideas to be described can be summarized as follows:

Concept 1: Shoot through the *interior* of the feasible region toward an optimal solution.
Concept 2: Move in a direction that improves the objective function value at the fastest possible rate.
Concept 3: Transform the feasible region to place the current trial solution near its center, thereby enabling a large improvement when concept 2 is implemented.

To illustrate these ideas throughout the section, we shall use the following example:

$$\text{Maximize} \quad Z = x_1 + 2x_2,$$

subject to

$$x_1 + x_2 \le 8$$

and

$$x_1 \ge 0, \qquad x_2 \ge 0.$$

This problem is depicted graphically in Fig. 7.3, where the optimal solution is seen to be $(x_1, x_2) = (0, 8)$ with $Z = 16$. (We will describe the significance of the arrow in the figure shortly.)

You will see that our interior-point algorithm requires a considerable amount of work to solve this tiny example. The reason is that the algorithm is designed to solve *huge* problems efficiently, but is much less efficient than the simplex method (or the graphical method in this case) for small problems. However, having an example with only two variables will allow us to depict graphically what the algorithm is doing.

The Relevance of the Gradient for Concepts 1 and 2

The algorithm begins with an initial trial solution that (like all subsequent trial solutions) lies in the *interior* of the feasible region, i.e., *inside the boundary* of the feasible region. Thus, for the example, the solution must not lie on any of the three lines ($x_1 = 0$, $x_2 = 0$,

$x_1 + x_2 = 8$) that form the boundary of this region in Fig. 7.3. (A trial solution that lies on the boundary cannot be used because this would lead to the undefined mathematical operation of division by zero at one point in the algorithm.) We have arbitrarily chosen $(x_1, x_2) = (2, 2)$ to be the initial trial solution.

To begin implementing concepts 1 and 2, note in Fig. 7.3 that the direction of movement from $(2, 2)$ that increases Z at the fastest possible rate is *perpendicular* to (and toward) the objective function line $Z = 16 = x_1 + 2x_2$. We have shown this direction by the arrow from $(2, 2)$ to $(3, 4)$. Using vector addition, we have

$$(3, 4) = (2, 2) + (1, 2),$$

where the vector $(1, 2)$ is the **gradient** of the objective function. (We will discuss gradients further in Sec. 12.5 in the broader context of *nonlinear programming,* where algorithms similar to Karmarkar's have long been used.) The components of $(1, 2)$ are just the coefficients in the objective function. Thus, with one subsequent modification, the gradient $(1, 2)$ defines the ideal direction to which to move, where the question of the *distance to move* will be considered later.

The algorithm actually operates on linear programming problems after they have been rewritten in augmented form. Letting x_3 be the slack variable for the functional constraint of the example, we see that this form is

Maximize $Z = x_1 + 2x_2,$

subject to

$$x_1 + x_2 + x_3 = 8$$

and

$$x_1 \geq 0, \qquad x_2 \geq 0, \qquad x_3 \geq 0.$$

In matrix notation (slightly different from Chap. 5 because the slack variable now is incorporated into the notation), the augmented form can be written in general as

Maximize $Z = \mathbf{c}^T\mathbf{x},$

subject to

$$\mathbf{Ax} = \mathbf{b}$$

and

$$\mathbf{x} \geq \mathbf{0},$$

where

$$\mathbf{c} = \begin{bmatrix} 1 \\ 2 \\ 0 \end{bmatrix}, \qquad \mathbf{x} = \begin{bmatrix} x_1 \\ x_2 \\ x_3 \end{bmatrix}, \qquad \mathbf{A} = [1, \quad 1, \quad 1], \qquad \mathbf{b} = [8], \qquad \mathbf{0} = \begin{bmatrix} 0 \\ 0 \\ 0 \end{bmatrix}$$

for the example. Note that $\mathbf{c}^T = [1, 2, 0]$ now is the gradient of the objective function.

The augmented form of the example is depicted graphically in Fig. 7.4. The feasible region now consists of the triangle with vertices $(8, 0, 0)$, $(0, 8, 0)$, and $(0, 0, 8)$. Points in the interior of this feasible region are those where $x_1 > 0$, $x_2 > 0$, and $x_3 > 0$. Each of these three $x_j > 0$ conditions has the effect of forcing (x_1, x_2) away from one of the three lines forming the boundary of the feasible region in Fig. 7.3.

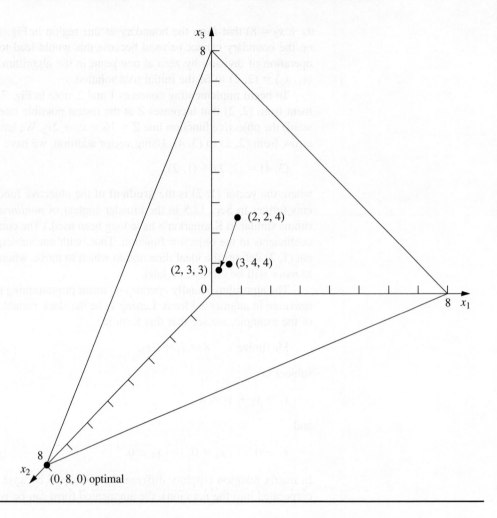

■ FIGURE 7.4
Example in augmented form
for the interior-point
algorithm.

Using the Projected Gradient to Implement Concepts 1 and 2

In augmented form, the initial trial solution for the example is $(x_1, x_2, x_3) = (2, 2, 4)$. Adding the gradient $(1, 2, 0)$ leads to

$$(3, 4, 4) = (2, 2, 4) + (1, 2, 0).$$

However, now there is a complication. The algorithm cannot move from $(2, 2, 4)$ to $(3, 4, 4)$, because $(3, 4, 4)$ is infeasible! When $x_1 = 3$ and $x_2 = 4$, then $x_3 = 8 - x_1 - x_2 = 1$ instead of 4. The point $(3, 4, 4)$ lies on the near side as you look down on the feasible triangle in Fig. 7.4. Therefore, to remain feasible, the algorithm (indirectly) *projects* the point $(3, 4, 4)$ down onto the feasible triangle by dropping a line that is *perpendicular* to this triangle. A vector from $(0, 0, 0)$ to $(1, 1, 1)$ is perpendicular to this triangle, so the perpendicular line through $(3, 4, 4)$ is given by the equation

$$(x_1, x_2, x_3) = (3, 4, 4) - \theta(1, 1, 1),$$

where θ is a scalar. Since the triangle satisfies the equation $x_1 + x_2 + x_3 = 8$, this perpendicular line intersects the triangle at $(2, 3, 3)$. Because

$$(2, 3, 3) = (2, 2, 4) + (0, 1, -1),$$

the **projected gradient** of the objective function (the gradient projected onto the feasible region) is $(0, 1, -1)$. It is this projected gradient that defines the direction of movement from $(2, 2, 4)$ for the algorithm, as shown by the arrow in Fig. 7.4.

A formula is available for computing the projected gradient directly. By defining the *projection matrix* $\mathbf{P}$ as

$$\mathbf{P} = \mathbf{I} - \mathbf{A}^T(\mathbf{A}\mathbf{A}^T)^{-1}\mathbf{A},$$

the *projected gradient* (in column form) is

$$\mathbf{c}_p = \mathbf{P}\mathbf{c}.$$

Thus, for the example,

$$
\mathbf{P} = \begin{bmatrix} 1 & 0 & 0 \\ 0 & 1 & 0 \\ 0 & 0 & 1 \end{bmatrix} - \begin{bmatrix} 1 \\ 1 \\ 1 \end{bmatrix} \left(\begin{bmatrix} 1 & 1 & 1 \end{bmatrix} \begin{bmatrix} 1 \\ 1 \\ 1 \end{bmatrix} \right)^{-1} \begin{bmatrix} 1 & 1 & 1 \end{bmatrix}
$$

$$
= \begin{bmatrix} 1 & 0 & 0 \\ 0 & 1 & 0 \\ 0 & 0 & 1 \end{bmatrix} - \frac{1}{3} \begin{bmatrix} 1 \\ 1 \\ 1 \end{bmatrix} \begin{bmatrix} 1 & 1 & 1 \end{bmatrix}
$$

$$
= \begin{bmatrix} 1 & 0 & 0 \\ 0 & 1 & 0 \\ 0 & 0 & 1 \end{bmatrix} - \frac{1}{3} \begin{bmatrix} 1 & 1 & 1 \\ 1 & 1 & 1 \\ 1 & 1 & 1 \end{bmatrix} = \begin{bmatrix} \frac{2}{3} & -\frac{1}{3} & -\frac{1}{3} \\ -\frac{1}{3} & \frac{2}{3} & -\frac{1}{3} \\ -\frac{1}{3} & -\frac{1}{3} & \frac{2}{3} \end{bmatrix},
$$

so

$$
\mathbf{c}_p = \begin{bmatrix} \frac{2}{3} & -\frac{1}{3} & -\frac{1}{3} \\ -\frac{1}{3} & \frac{2}{3} & -\frac{1}{3} \\ -\frac{1}{3} & -\frac{1}{3} & \frac{2}{3} \end{bmatrix} \begin{bmatrix} 1 \\ 2 \\ 0 \end{bmatrix} = \begin{bmatrix} 0 \\ 1 \\ -1 \end{bmatrix}.
$$

Moving from $(2, 2, 4)$ in the direction of the projected gradient $(0, 1, -1)$ involves increasing α from zero in the formula

$$
\mathbf{x} = \begin{bmatrix} 2 \\ 2 \\ 4 \end{bmatrix} + 4\alpha\mathbf{c}_p = \begin{bmatrix} 2 \\ 2 \\ 4 \end{bmatrix} + 4\alpha \begin{bmatrix} 0 \\ 1 \\ -1 \end{bmatrix},
$$

where the coefficient 4 is used simply to give an upper bound of 1 for α to maintain feasibility (all $x_j \geq 0$). Note that increasing α to $\alpha = 1$ would cause x_3 to decrease to $x_3 = 4 + 4(1)(-1) = 0$, where $\alpha > 1$ yields $x_3 < 0$. Thus, α measures the fraction used of the distance that could be moved before the feasible region is left.

How large should α be made for moving to the next trial solution? Because the increase in Z is proportional to α, a value close to the upper bound of 1 is good for giving a relatively large step toward optimality on the current iteration. However, the problem with a value too close to 1 is that the next trial solution then is jammed against a constraint boundary, thereby making it difficult to take large improving steps during subsequent iterations. Therefore, it is very helpful for trial solutions to be near the center of the feasible region (or at least near the center of the portion of the feasible region in the vicinity of an optimal solution), and not too close to any constraint boundary. With this in mind, Karmarkar has stated for his algorithm that a value as large as $\alpha = 0.25$ should be "safe." In practice, much larger values (for example, $\alpha = 0.9$) sometimes are used. For the purposes of this example (and the problems at the end of the chapter), we have chosen $\alpha = 0.5$. (Your IOR Tutorial uses $\alpha = 0.5$ as the default value, but also has $\alpha = 0.9$ available.)

A Centering Scheme for Implementing Concept 3

We now have just one more step to complete the description of the algorithm, namely, a special scheme for transforming the feasible region to place the current trial solution near its center. We have just described the benefit of having the trial solution near the center, but another important benefit of this centering scheme is that it keeps turning the direction of the projected gradient to point more nearly toward an optimal solution as the algorithm converges toward this solution.

The basic idea of the centering scheme is straightforward—simply change the scale (units) for each of the variables so that the trial solution becomes equidistant from the constraint boundaries in the new coordinate system. (Karmarkar's original algorithm uses a more sophisticated centering scheme.)

For the example, there are three constraint boundaries in Fig. 7.3, each one corresponding to a zero value for one of the three variables of the problem in augmented form, namely, $x_1 = 0$, $x_2 = 0$, and $x_3 = 0$. In Fig. 7.4, see how these three constraint boundaries intersect the $\mathbf{Ax} = \mathbf{b}$ ($x_1 + x_2 + x_3 = 8$) plane to form the boundary of the feasible region. The initial trial solution is $(x_1, x_2, x_3) = (2, 2, 4)$, so this solution is 2 units away from the $x_1 = 0$ and $x_2 = 0$ constraint boundaries and 4 units away from the $x_3 = 0$ constraint boundary, when the units of the respective variables are used. However, whatever these units are in each case, they are quite arbitrary and can be changed as desired without changing the problem. Therefore, let us rescale the variables as follows:

$$\tilde{x}_1 = \frac{x_1}{2}, \qquad \tilde{x}_2 = \frac{x_2}{2}, \qquad \tilde{x}_3 = \frac{x_3}{4}$$

in order to make the current trial solution of $(x_1, x_2, x_3) = (2, 2, 4)$ become

$$(\tilde{x}_1, \tilde{x}_2, \tilde{x}_3) = (1, 1, 1).$$

In these new coordinates (substituting $2\tilde{x}_1$ for x_1, $2\tilde{x}_2$ for x_2, and $4\tilde{x}_3$ for x_3), the problem becomes

Maximize $Z = 2\tilde{x}_1 + 4\tilde{x}_2,$

subject to

$$2\tilde{x}_1 + 2\tilde{x}_2 + 4\tilde{x}_3 = 8$$

and

$$\tilde{x}_1 \geq 0, \qquad \tilde{x}_2 \geq 0, \qquad \tilde{x}_3 \geq 0,$$

as depicted graphically in Fig. 7.5.

Note that the trial solution (1, 1, 1) in Fig. 7.5 is equidistant from the three constraint boundaries $\tilde{x}_1 = 0$, $\tilde{x}_2 = 0$, $\tilde{x}_3 = 0$. For each subsequent iteration as well, the problem is rescaled again to achieve this same property, so that the current trial solution always is (1, 1, 1) in the current coordinates.

Summary and Illustration of the Algorithm

Now let us summarize and illustrate the algorithm by going through the first iteration for the example, then giving a summary of the general procedure, and finally applying this summary to a second iteration.

Iteration 1. Given the initial trial solution $(x_1, x_2, x_3) = (2, 2, 4)$, let $\mathbf{D}$ be the corresponding *diagonal matrix* such that $\mathbf{x} = \mathbf{D\tilde{x}}$, so that

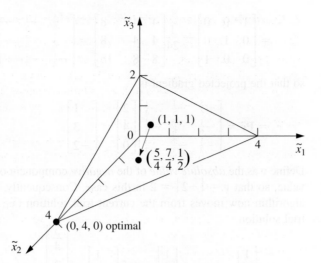

■ **FIGURE 7.5**
Example after rescaling for
iteration 1.

$$
\mathbf{D} = \begin{bmatrix} 2 & 0 & 0 \\ 0 & 2 & 0 \\ 0 & 0 & 4 \end{bmatrix}.
$$

The rescaled variables then are the components of

$$
\tilde{\mathbf{x}} = \mathbf{D}^{-1}\mathbf{x} = \begin{bmatrix} \dfrac{1}{2} & 0 & 0 \\ 0 & \dfrac{1}{2} & 0 \\ 0 & 0 & \dfrac{1}{4} \end{bmatrix} \begin{bmatrix} x_1 \\ x_2 \\ x_3 \end{bmatrix} = \begin{bmatrix} \dfrac{x_1}{2} \\ \dfrac{x_2}{2} \\ \dfrac{x_3}{4} \end{bmatrix}.
$$

In these new coordinates, $\mathbf{A}$ and $\mathbf{c}$ have become

$$
\tilde{\mathbf{A}} = \mathbf{A}\mathbf{D} = \begin{bmatrix} 1 & 1 & 1 \end{bmatrix} \begin{bmatrix} 2 & 0 & 0 \\ 0 & 2 & 0 \\ 0 & 0 & 4 \end{bmatrix} = \begin{bmatrix} 2 & 2 & 4 \end{bmatrix},
$$

$$
\tilde{\mathbf{c}} = \mathbf{D}\mathbf{c} = \begin{bmatrix} 2 & 0 & 0 \\ 0 & 2 & 0 \\ 0 & 0 & 4 \end{bmatrix} \begin{bmatrix} 1 \\ 2 \\ 0 \end{bmatrix} = \begin{bmatrix} 2 \\ 4 \\ 0 \end{bmatrix}.
$$

Therefore, the projection matrix is

$$
\mathbf{P} = \mathbf{I} - \tilde{\mathbf{A}}^T(\tilde{\mathbf{A}}\tilde{\mathbf{A}}^T)^{-1}\tilde{\mathbf{A}}
$$

$$
= \begin{bmatrix} 1 & 0 & 0 \\ 0 & 1 & 0 \\ 0 & 0 & 1 \end{bmatrix} - \begin{bmatrix} 2 \\ 2 \\ 4 \end{bmatrix} \left(\begin{bmatrix} 2 & 2 & 4 \end{bmatrix} \begin{bmatrix} 2 \\ 2 \\ 4 \end{bmatrix} \right)^{-1} \begin{bmatrix} 2 & 2 & 4 \end{bmatrix}
$$

$$
= \begin{bmatrix} 1 & 0 & 0 \\ 0 & 1 & 0 \\ 0 & 0 & 1 \end{bmatrix} - \frac{1}{24} \begin{bmatrix} 4 & 4 & 8 \\ 4 & 4 & 8 \\ 8 & 8 & 16 \end{bmatrix} = \begin{bmatrix} \frac{5}{6} & -\frac{1}{6} & -\frac{1}{3} \\ -\frac{1}{6} & \frac{5}{6} & -\frac{1}{3} \\ -\frac{1}{3} & -\frac{1}{3} & \frac{1}{3} \end{bmatrix},
$$

so that the projected gradient is

$$
\mathbf{c}_p = \mathbf{P}\tilde{\mathbf{c}} = \begin{bmatrix} \frac{5}{6} & -\frac{1}{6} & -\frac{1}{3} \\ -\frac{1}{6} & \frac{5}{6} & -\frac{1}{3} \\ -\frac{1}{3} & -\frac{1}{3} & \frac{1}{3} \end{bmatrix} \begin{bmatrix} 2 \\ 4 \\ 0 \end{bmatrix} = \begin{bmatrix} 1 \\ 3 \\ -2 \end{bmatrix}.
$$

Define v as the *absolute value* of the *negative* component of $\mathbf{c}_p$ having the *largest* absolute value, so that $v = |-2| = 2$ in this case. Consequently, in the current coordinates, the algorithm now moves from the current trial solution $(\tilde{x}_1, \tilde{x}_2, \tilde{x}_3) = (1, 1, 1)$ to the next trial solution

$$
\tilde{\mathbf{x}} = \begin{bmatrix} 1 \\ 1 \\ 1 \end{bmatrix} + \frac{\alpha}{v} \mathbf{c}_p = \begin{bmatrix} 1 \\ 1 \\ 1 \end{bmatrix} + \frac{0.5}{2} \begin{bmatrix} 1 \\ 3 \\ -2 \end{bmatrix} = \begin{bmatrix} \frac{5}{4} \\ \frac{7}{4} \\ \frac{1}{2} \end{bmatrix},
$$

as shown in Fig. 7.5. (The definition of v has been chosen to make the smallest component of $\tilde{\mathbf{x}}$ equal to zero when $\alpha = 1$ in this equation for the next trial solution.) In the original coordinates, this solution is

$$
\begin{bmatrix} x_1 \\ x_2 \\ x_3 \end{bmatrix} = \mathbf{D}\tilde{\mathbf{x}} = \begin{bmatrix} 2 & 0 & 0 \\ 0 & 2 & 0 \\ 0 & 0 & 4 \end{bmatrix} \begin{bmatrix} \frac{5}{4} \\ \frac{7}{4} \\ \frac{1}{2} \end{bmatrix} = \begin{bmatrix} \frac{5}{2} \\ \frac{7}{2} \\ 2 \end{bmatrix}.
$$

This completes the iteration, and this new solution will be used to start the next iteration.

These steps can be summarized as follows for any iteration.

Summary of the Interior-Point Algorithm

1. Given the current trial solution $(x_1, x_2, \ldots, x_n)$, set

$$
\mathbf{D} = \begin{bmatrix} x_1 & 0 & 0 & \cdots & 0 \\ 0 & x_2 & 0 & \cdots & 0 \\ 0 & 0 & x_3 & \cdots & 0 \\ \vdots & & & & \vdots \\ 0 & 0 & 0 & \cdots & x_n \end{bmatrix}
$$

2. Calculate $\tilde{\mathbf{A}} = \mathbf{A}\mathbf{D}$ and $\tilde{\mathbf{c}} = \mathbf{D}\mathbf{c}.$

3. Calculate $\mathbf{P} = \mathbf{I} - \tilde{\mathbf{A}}^T(\tilde{\mathbf{A}}\tilde{\mathbf{A}}^T)^{-1}\tilde{\mathbf{A}}$ and $\mathbf{c}_p = \mathbf{P}\tilde{\mathbf{c}}.$

4. Identify the negative component of $\mathbf{c}_p$ having the largest absolute value, and set v equal to this absolute value. Then calculate

$$
\tilde{\mathbf{x}} = \begin{bmatrix} 1 \\ 1 \\ \vdots \\ 1 \end{bmatrix} + \frac{\alpha}{v} \mathbf{c}_p,
$$

where α is a selected constant between 0 and 1 (for example, $\alpha = 0.5$).

5. Calculate $\mathbf{x} = \mathbf{D}\tilde{\mathbf{x}}$ as the trial solution for the next iteration (step 1). (If this trial solution is virtually unchanged from the preceding one, then the algorithm has virtually converged to an optimal solution, so stop.)

Now let us apply this summary to iteration 2 for the example.

Iteration 2

Step 1:

Given the current trial solution $(x_1, x_2, x_3) = (\frac{5}{2}, \frac{7}{2}, 2)$, set

$$\mathbf{D} = \begin{bmatrix} \frac{5}{2} & 0 & 0 \\ 0 & \frac{7}{2} & 0 \\ 0 & 0 & 2 \end{bmatrix}.$$

(Note that the rescaled variables are

$$\begin{bmatrix} \tilde{x}_1 \\ \tilde{x}_2 \\ \tilde{x}_3 \end{bmatrix} = \mathbf{D}^{-1}\mathbf{x} = \begin{bmatrix} \frac{2}{5} & 0 & 0 \\ 0 & \frac{2}{7} & 0 \\ 0 & 0 & \frac{1}{2} \end{bmatrix}\begin{bmatrix} x_1 \\ x_2 \\ x_3 \end{bmatrix} = \begin{bmatrix} \frac{2}{5}x_1 \\ \frac{2}{7}x_2 \\ \frac{1}{2}x_3 \end{bmatrix},$$

so that the BF solutions in these new coordinates are

$$\tilde{\mathbf{x}} = \mathbf{D}^{-1}\begin{bmatrix} 8 \\ 0 \\ 0 \end{bmatrix} = \begin{bmatrix} \frac{16}{5} \\ 0 \\ 0 \end{bmatrix}, \qquad \tilde{\mathbf{x}} = \mathbf{D}^{-1}\begin{bmatrix} 0 \\ 8 \\ 0 \end{bmatrix} = \begin{bmatrix} 0 \\ \frac{16}{7} \\ 0 \end{bmatrix},$$

and

$$\tilde{\mathbf{x}} = \mathbf{D}^{-1}\begin{bmatrix} 0 \\ 0 \\ 8 \end{bmatrix} = \begin{bmatrix} 0 \\ 0 \\ 4 \end{bmatrix},$$

as depicted in Fig. 7.6.)

Step 2:

$$\tilde{\mathbf{A}} = \mathbf{A}\mathbf{D} = [\tfrac{5}{2}, \tfrac{7}{2}, 2] \qquad \text{and} \qquad \tilde{\mathbf{c}} = \mathbf{D}\mathbf{c} = \begin{bmatrix} \frac{5}{2} \\ 7 \\ 0 \end{bmatrix}.$$

■ **FIGURE 7.6**

Example after rescaling for iteration 2.

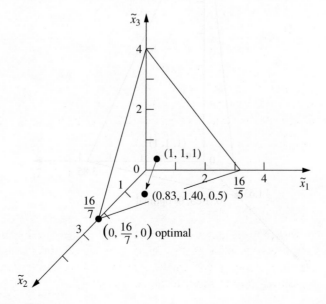

Step 3:

$$\mathbf{P} = \begin{bmatrix} \frac{13}{18} & -\frac{7}{18} & -\frac{2}{9} \\ -\frac{7}{18} & \frac{41}{90} & -\frac{14}{45} \\ -\frac{2}{9} & -\frac{14}{45} & \frac{37}{45} \end{bmatrix} \quad \text{and} \quad \mathbf{c}_p = \begin{bmatrix} -\frac{11}{12} \\ \frac{133}{60} \\ -\frac{41}{15} \end{bmatrix}.$$

Step 4:

$\left| -\frac{41}{15} \right| > \left| -\frac{11}{12} \right|$, so $v = \frac{41}{15}$ and

$$\tilde{\mathbf{x}} = \begin{bmatrix} 1 \\ 1 \\ 1 \end{bmatrix} + \frac{0.5}{\frac{41}{15}} \begin{bmatrix} -\frac{11}{12} \\ \frac{133}{60} \\ -\frac{41}{15} \end{bmatrix} = \begin{bmatrix} \frac{273}{328} \\ \frac{461}{328} \\ \frac{1}{2} \end{bmatrix} \approx \begin{bmatrix} 0.83 \\ 1.40 \\ 0.50 \end{bmatrix}.$$

Step 5:

$$\mathbf{x} = \mathbf{D}\tilde{\mathbf{x}} = \begin{bmatrix} \frac{1365}{656} \\ \frac{3227}{656} \\ 1 \end{bmatrix} \approx \begin{bmatrix} 2.08 \\ 4.92 \\ 1.00 \end{bmatrix}$$

is the trial solution for iteration 3.

Since there is little to be learned by repeating these calculations for additional iterations, we shall stop here. However, we do show in Fig. 7.7 the reconfigured feasible region after rescaling based on the trial solution just obtained for iteration 3. As always,

■ **FIGURE 7.7**
Example after rescaling for
iteration 3.

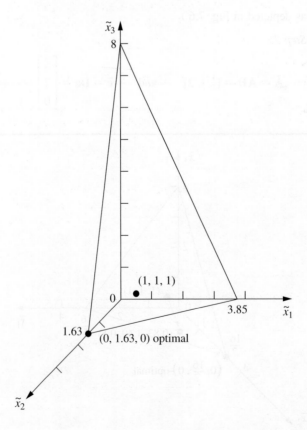

the rescaling has placed the trial solution at $(\tilde{x}_1, \tilde{x}_2, \tilde{x}_3) = (1, 1, 1)$, equidistant from the $\tilde{x}_1 = 0$, $\tilde{x}_2 = 0$, and $\tilde{x}_3 = 0$ constraint boundaries. Note in Figs. 7.5, 7.6, and 7.7 how the sequence of iterations and rescaling have the effect of "sliding" the optimal solution toward $(1, 1, 1)$ while the other BF solutions tend to slide away. Eventually, after enough iterations, the optimal solution will lie very near $(\tilde{x}_1, \tilde{x}_2, \tilde{x}_3) = (0, 1, 0)$ after rescaling, while the other two BF solutions will be *very* far from the origin on the $\tilde{x}_1$ and $\tilde{x}_3$ axes. Step 5 of that iteration then will yield a solution in the original coordinates very near the optimal solution of $(x_1, x_2, x_3) = (0, 8, 0)$.

Figure 7.8 shows the progress of the algorithm in the original $x_1 = x_2$ coordinate system before the problem is augmented. The three points—$(x_1, x_2) = (2, 2)$, $(2.5, 3.5)$, and $(2.08, 4.92)$—are the trial solutions for initiating iterations 1, 2, and 3, respectively. We then have drawn a smooth curve through and beyond these points to show the trajectory of the algorithm in subsequent iterations as it approaches $(x_1, x_2) = (0, 8)$.

The functional constraint for this particular example happened to be an inequality constraint. However, equality constraints cause no difficulty for the algorithm, since it deals with the constraints only after any necessary augmenting has been done to convert them to equality form ($\mathbf{Ax} = \mathbf{b}$) anyway. To illustrate, suppose that the only change in the example is that the constraint $x_1 + x_2 \leq 8$ is changed to $x_1 + x_2 = 8$. Thus, the feasible region in Fig. 7.3 changes to just the line segment between $(8, 0)$ and $(0, 8)$. Given an initial feasible trial solution in the interior ($x_1 > 0$ and $x_2 > 0$) of this line segment—say, $(x_1, x_2) = (4, 4)$—the algorithm can proceed just as presented in the five-step summary with just the two variables and $\mathbf{A} = [1, 1]$. For each iteration, the projected gradient points along this line segment in the direction of $(0, 8)$. With $\alpha = \frac{1}{2}$, iteration 1 leads from $(4, 4)$ to $(2, 6)$, iteration 2 leads from $(2, 6)$ to $(1, 7)$, etc. (Problem 7.4-3 asks you to verify these results.)

Although either version of the example has only one functional constraint, having more than one leads to just one change in the procedure as already illustrated (other than more extensive calculations). Having a single functional constraint in the example meant that **A**

■ FIGURE 7.8

Trajectory of the interior-point algorithm for the example in the original x_1-x_2 coordinate system.

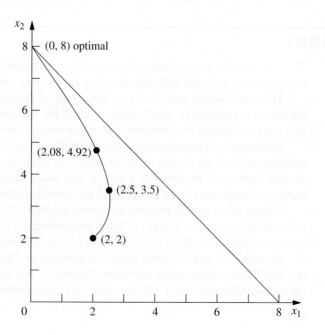

had only a single row, so the $(\tilde{\mathbf{A}}\tilde{\mathbf{A}}^T)^{-1}$ term in step 3 only involved taking the reciprocal of the number obtained from the vector product $\tilde{\mathbf{A}}\tilde{\mathbf{A}}^T$. Multiple functional constraints mean that $\mathbf{A}$ has multiple rows, so then the $(\tilde{\mathbf{A}}\tilde{\mathbf{A}}^T)^{-1}$ term involves finding the *inverse* of the matrix obtained from the matrix product $\tilde{\mathbf{A}}\tilde{\mathbf{A}}^T$.

To conclude, we need to add a comment to place the algorithm into better perspective. For our extremely small example, the algorithm requires relatively extensive calculations and then, after many iterations, obtains only an approximation of the optimal solution. By contrast, the graphical procedure of Sec. 3.1 finds the optimal solution in Fig. 7.3 immediately, and the simplex method requires only one quick iteration. However, do not let this contrast fool you into downgrading the efficiency of the interior-point algorithm. This algorithm is designed for dealing with *big* problems that may have many thousands of functional constraints. The simplex method typically requires thousands of iterations on such problems. By "shooting" through the interior of the feasible region, the interior-point algorithm tends to require a substantially smaller number of iterations (although with considerably more work per iteration). This sometimes enables an interior-point algorithm to efficiently solve huge linear programming problems that might even be beyond the reach of either the simplex method or the dual simplex method. Therefore, interior-point algorithms similar to the one presented here should play an important role in the future of linear programming.

See Sec. 4.9 for a comparison of the interior-point approach with the simplex method. Section 4.9 also discusses the complementary roles of the interior-point approach and the simplex method, including how they can even be combined into a hybrid algorithm.

Finally, we should emphasize that this section has provided only a conceptual introduction to the interior-point approach to linear programming by describing an elementary variant of Karmakar's path-breaking 1984 algorithm. Over the many subsequent years, a number of top-notch researchers have developed many key advances in the interior-point approach. Further coverage of this advanced topic is beyond the scope of this book. However, the interested reader can find many details in the selected references listed at the end of this chapter.

7.5   CONCLUSIONS

The *dual simplex method* and *parametric linear programming* are especially valuable for postoptimality analysis, although they also can be very useful in other contexts.

The *upper bound technique* provides a way of streamlining the simplex method for the common situation in which many or all of the variables have explicit upper bounds. It can greatly reduce the computational effort for large problems.

Mathematical-programming computer packages usually include all three of these procedures, and they are widely used. Because their basic structure is based largely upon the simplex method as presented in Chap. 4, they retain the exceptional computational efficiency to handle very large problems of the sizes described in Sec. 4.8.

Various other special-purpose algorithms also have been developed to exploit the special structure of particular types of linear programming problems (such as those to be discussed in Chaps. 8 and 9). Much research is currently being done in this area.

Karmarkar's interior-point algorithm initiated another key line of research into how to solve linear programming problems. Variants of this algorithm now provide a powerful approach for efficiently solving some very large problems.

■ SELECTED REFERENCES

1. Hooker, J. N.: "Karmarkar's Linear Programming Algorithm," *Interfaces,* **16:** 75–90, July–August 1986.

2. Luenberger, D., and Y. Ye: *Linear and Nonlinear Programming*, 3rd ed., Springer, New York, 2008.

3. Marsten, R., R. Subramanian, M. Saltzman, I. Lustig, and D. Shanno: "Interior-Point Methods for Linear Programming: Just Call Newton, Lagrange, and Fiacco and McCormick!," *Interfaces,* **20:** 105–116, July–August 1990.

4. Vanderbei, R. J.: "Affine-Scaling for Linear Programs with Free Variables," *Mathematical Programming,* **43:** 31–44, 1989.

5. Vanderbei, R. J.: *Linear Programming: Foundations and Extensions,* 3rd ed., Springer, New York, 2008.

6. Ye, Y.: *Interior-Point Algorithms: Theory and Analysis,* Wiley, New York, 1997.

■ LEARNING AIDS FOR THIS CHAPTER ON OUR WEBSITE (www.mhhe.com/hillier)

Worked Examples:

Examples for Chapter 7

Interactive Procedures in IOR Tutorial:

Enter or Revise a General Linear Programming Model
Set Up for the Simplex Method—Interactive Only
Solve Interactively by the Simplex Method
Interactive Graphical Method

Automatic Procedures in IOR Tutorial:

Solve Automatically by the Simplex Method
Solve Automatically by the Interior-Point Algorithm
Graphical Method and Sensitivity Analysis

An Excel Add-In:

Premium Solver for Education

"Ch. 7—Other Algorithms for LP" Files for Solving the Examples:

Excel Files
LINGO/LINDO File
MPL/CPLEX File

Glossary for Chapter 7

Supplement to This Chapter:

Linear Goal Programming and Its Solution Procedures (includes two accompanying cases: A Cure for Cuba and Airport Security)

See Appendix 1 for documentation of the software.

■ PROBLEMS

The symbols to the left of some of the problems (or their parts) have the following meaning:

  I: We suggest that you use one of the procedures in IOR Tutorial (the print-out records your work). For parametric linear programming, this only applies to $\theta = 0$, after which you should proceed manually.

  C: Use the computer to solve the problem by using the automatic procedure for the interior-point algorithm in IOR Tutorial.

An asterisk on the problem number indicates that at least a partial answer is given in the back of the book.

7.1-1. Consider the following problem.

   Maximize $Z = -x_1 - 2x_2,$

subject to

$$2x_1 + x_2 \leq 40$$
$$x_2 \geq 15$$
$$-2x_1 + x_2 \leq 10$$

and

   $x_1 \geq 0,$ $x_2 \geq 0.$

I **(a)** Solve this problem graphically.
(b) Use the *dual simplex method* manually to solve this problem.
(c) Trace graphically the path taken by the dual simplex method.

7.1-2.* Use the *dual simplex method* manually to solve the following problem.

   Minimize $Z = 5x_1 + 2x_2 + 4x_3,$

subject to

$$3x_1 + x_2 + 2x_3 \geq 4$$
$$6x_1 + 3x_2 + 5x_3 \geq 10$$

and

   $x_1 \geq 0,$ $x_2 \geq 0,$ $x_3 \geq 0.$

7.1-3. Use the *dual simplex method* manually to solve the following problem.

   Minimize $Z = 7x_1 + 2x_2 + 5x_3 + 4x_4,$

subject to

$$2x_1 + 4x_2 + 7x_3 + x_4 \geq 5$$
$$8x_1 + 4x_2 + 6x_3 + 4x_4 \geq 8$$
$$3x_1 + 8x_2 + x_3 + 4x_4 \geq 4$$

and

   $x_j \geq 0,$ for $j = 1, 2, 3, 4.$

7.1-4. Consider the following problem.

   Maximize $Z = 5x_1 + 10x_2,$

subject to

$$3x_1 + x_2 \leq 40$$
$$x_1 + x_2 \leq 20$$
$$5x_1 + 3x_2 \leq 90$$

and

   $x_1 \geq 0,$ $x_2 \geq 0.$

I **(a)** Solve by the *original simplex method* (in tabular form). Identify the *complementary* basic solution for the dual problem obtained at each iteration.
(b) Solve the *dual* of this problem manually by the *dual simplex method*. Compare the resulting sequence of basic solutions with the complementary basic solutions obtained in part (a).

7.1-5. Consider the example for case 1 of sensitivity analysis given in Sec. 6.7, where the initial simplex tableau of Table 4.8 is modified by changing b_2 from 12 to 24, thereby changing the respective entries in the right-side column of the *final* simplex tableau to 54, 6, 12, and -2. Starting from this revised final simplex tableau, use the *dual simplex method* to obtain the new optimal solution shown in Table 6.21. Show your work.

7.1-6.* Consider part (a) of Prob. 6.7-2. Use the *dual simplex method* manually to reoptimize, starting from the revised final tableau.

7.2-1.* Consider the following problem.

   Maximize $Z = 8x_1 + 24x_2,$

subject to

$$x_1 + 2x_2 \leq 10$$
$$2x_1 + x_2 \leq 10$$

and

   $x_1 \geq 0,$ $x_2 \geq 0.$

Suppose that Z represents profit and that it is possible to modify the objective function somewhat by an appropriate shifting of key personnel between the two activities. In particular, suppose that the unit profit of activity 1 can be increased above 8 (to a maximum of 18) at the expense of decreasing the unit profit of activity 2 below 24 by twice the amount. Thus, Z can actually be represented as

$$Z(\theta) = (8 + \theta)x_1 + (24 - 2\theta)x_2,$$

where θ is also a decision variable such that $0 \leq \theta \leq 10.$

I **(a)** Solve the original form of this problem graphically. Then extend this graphical procedure to solve the parametric extension of the problem; i.e., find the optimal solution and the optimal value of $Z(\theta)$ as a function of θ, for $0 \leq \theta \leq 10.$
I **(b)** Find an optimal solution for the original form of the problem by the simplex method. Then use *parametric linear programming* to find an optimal solution and the optimal value of $Z(\theta)$ as a function of θ, for $0 \leq \theta \leq 10.$ Plot $Z(\theta)$.

(c) Determine the optimal value of θ. Then indicate how this optimal value could have been identified directly by solving only two ordinary linear programming problems. (*Hint:* A convex function achieves its maximum at an endpoint.)

I **7.2-2.** Use *parametric linear programming* to find the optimal solution for the following problem as a function of θ, for $0 \leq \theta \leq 20$.

$$\text{Maximize} \quad Z(\theta) = (20 + 4\theta)x_1 + (30 - 3\theta)x_2 + 5x_3,$$

subject to

$$3x_1 + 3x_2 + x_3 \leq 10$$
$$8x_1 + 6x_2 + 4x_3 \leq 25$$
$$6x_1 + x_2 + x_3 \leq 15$$

and

$$x_1 \geq 0, \quad x_2 \geq 0, \quad x_3 \geq 0.$$

I **7.2-3.** Consider the following problem.

$$\text{Maximize} \quad Z(\theta) = (10 - \theta)x_1 + (12 + \theta)x_2 + (7 + 2\theta)x_3,$$

subject to

$$x_1 + 2x_2 + 2x_3 \leq 30$$
$$x_1 + x_2 + x_3 \leq 20$$

and

$$x_1 \geq 0, \quad x_2 \geq 0, \quad x_3 \geq 0.$$

(a) Use *parametric linear programming* to find an optimal solution for this problem as a function of θ, for $\theta \geq 0$.
(b) Construct the dual model for this problem. Then find an optimal solution for this dual problem as a function of θ, for $\theta \geq 0$, by the method described in the latter part of Sec. 7.2. Indicate graphically what this algebraic procedure is doing. Compare the basic solutions obtained with the complementary basic solutions obtained in part (*a*).

I **7.2-4.*** Use the *parametric linear programming* procedure for making systematic changes in the b_i parameters to find an optimal solution for the following problem as a function of θ, for $0 \leq \theta \leq 25$.

$$\text{Maximize} \quad Z(\theta) = 2x_1 + x_2,$$

subject to

$$x_1 \qquad\quad \leq 10 + 2\theta$$
$$x_1 + x_2 \leq 25 - \theta$$
$$\qquad\quad x_2 \leq 10 + 2\theta$$

and

$$x_1 \geq 0, \quad x_2 \geq 0.$$

Indicate graphically what this algebraic procedure is doing.

I **7.2-5.** Use *parametric linear programming* to find an optimal solution for the following problem as a function of θ, for $0 \leq \theta \leq 30$.

$$\text{Maximize} \quad Z(\theta) = 5x_1 + 42x_2 + 28x_3 + 49x_4,$$

subject to

$$3x_1 - 2x_2 + x_3 + 3x_4 \leq 135 - 2\theta$$
$$2x_1 + 4x_2 - x_3 + 2x_4 \leq 78 - \theta$$
$$x_1 + 2x_2 + x_3 + 2x_4 \leq 30 + \theta$$

and

$$x_j \geq 0, \qquad \text{for } j = 1, 2, 3, 4.$$

Then identify the value of θ that gives the largest optimal value of $Z(\theta)$.

7.2-6. Consider Prob. 6.7-3. Use *parametric linear programming* to find an optimal solution as a function of θ for $-20 \leq \theta \leq 0$. (*Hint:* Substitute $-\theta'$ for θ, and then increase θ' from zero.)

7.2-7. Consider the $Z^*(\theta)$ function shown in Fig. 7.1 for *parametric linear programming* with systematic changes in the c_j parameters.
(a) Explain why this function is piecewise linear.
(b) Show that this function must be convex.

7.2-8. Consider the $Z^*(\theta)$ function shown in Fig. 7.2 for *parametric linear programming* with systematic changes in the b_i parameters.
(a) Explain why this function is piecewise linear.
(b) Show that this function must be concave.

7.2-9. Let

$$Z^* = \max \left\{ \sum_{j=1}^{n} c_j x_j \right\},$$

subject to

$$\sum_{j=1}^{n} a_{ij} x_j \leq b_i, \qquad \text{for } i = 1, 2, \ldots, m,$$

and

$$x_j \geq 0, \qquad \text{for } j = 1, 2, \ldots, n$$

(where the a_{ij}, b_i, and c_j are fixed constants), and let $(y_1^*, y_2^*, \ldots, y_m^*)$ be the corresponding optimal dual solution. Then let

$$Z^{**} = \max \left\{ \sum_{j=1}^{n} c_j x_j \right\},$$

subject to

$$\sum_{j=1}^{n} a_{ij} x_j \leq b_i + k_i, \qquad \text{for } i = 1, 2, \ldots, m,$$

and

$$x_j \geq 0, \qquad \text{for } j = 1, 2, \ldots, n,$$

where $k_1, k_2, \ldots, k_m$ are given constants. Show that

$$Z^{**} \leq Z^* + \sum_{i=1}^{m} k_i y_i^*.$$

7.3-1. Consider the following problem.

$$\text{Maximize} \quad Z = 2x_1 + 3x_2,$$

subject to

$$3x_1 - 9x_2 \leq 20$$
$$3x_1 \qquad \leq 40$$
$$9x_2 \leq 40$$

and

$$x_1 \geq 0, \qquad x_2 \geq 0.$$

I **(a)** Solve this problem graphically.
(b) Use the *upper bound technique* manually to solve this problem.
(c) Trace graphically the path taken by the upper bound technique.

7.3-2.* Use the *upper bound technique* manually to solve the following problem.

Maximize $Z = x_1 + 3x_2 - 2x_3,$

subject to

$$x_2 - 2x_3 \leq 1$$
$$2x_1 + x_2 + 2x_3 \leq 8$$
$$x_1 \qquad\qquad \leq 1$$
$$x_2 \qquad \leq 3$$
$$x_3 \leq 2$$

and

$$x_1 \geq 0, \qquad x_2 \geq 0, \qquad x_3 \geq 0.$$

7.3-3. Use the *upper bound technique* manually to solve the following problem.

Maximize $Z = 2x_1 + 3x_2 - 2x_3 + 5x_4,$

subject to

$$2x_1 + 2x_2 + x_3 + 2x_4 \leq 5$$
$$x_1 + 2x_2 - 3x_3 + 4x_4 \leq 5$$

and

$$0 \leq x_j \leq 1, \qquad \text{for } j = 1, 2, 3, 4.$$

7.3-4. Use the *upper bound technique* manually to solve the following problem.

Maximize $Z = 2x_1 + 5x_2 + 3x_3 + 4x_4 + x_5,$

subject to

$$x_1 + 3x_2 + 2x_3 + 3x_4 + x_5 \leq 6$$
$$4x_1 + 6x_2 + 5x_3 + 7x_4 + x_5 \leq 15$$

and

$$0 \leq x_j \leq 1, \qquad \text{for } j = 1, 2, 3, 4, 5.$$

7.3-5. Simultaneously use the *upper bound technique* and the *dual simplex method* manually to solve the following problem.

Minimize $Z = 3x_1 + 4x_2 + 2x_3,$

subject to

$$x_1 + x_2 + x_3 \geq 15$$
$$x_2 + x_3 \geq 10$$

and

$$0 \leq x_1 \leq 25, \qquad 0 \leq x_2 \leq 5, \qquad 0 \leq x_3 \leq 15.$$

C **7.4-1.** Reconsider the example used to illustrate the interior-point algorithm in Sec. 7.4. Suppose that $(x_1, x_2) = (1, 3)$ were used instead as the initial feasible trial solution. Perform two iterations manually, starting from this solution. Then use the automatic procedure in your IOR Tutorial to check your work.

7.4-2. Consider the following problem.

Maximize $Z = 3x_1 + x_2,$

subject to

$$x_1 + x_2 \leq 4$$

and

$$x_1 \geq 0, \qquad x_2 \geq 0.$$

I **(a)** Solve this problem graphically. Also identify all CPF solutions.
C **(b)** Starting from the initial trial solution $(x_1, x_2) = (1, 1)$, perform four iterations of the interior-point algorithm presented in Sec. 7.4 manually. Then use the automatic procedure in your IOR Tutorial to check your work.
(c) Draw figures corresponding to Figs. 7.4, 7.5, 7.6, 7.7, and 7.8 for this problem. In each case, identify the basic (or corner-point) feasible solutions in the current coordinate system. (Trial solutions can be used to determine projected gradients.)

7.4-3. Consider the following problem.

Maximize $Z = x_1 + 2x_2,$

subject to

$$x_1 + x_2 = 8$$

and

$$x_1 \geq 0, \qquad x_2 \geq 0.$$

C **(a)** Near the end of Sec. 7.4, there is a discussion of what the interior-point algorithm does on this problem when starting from the initial feasible trial solution $(x_1, x_2) = (4, 4)$. Verify the results presented there by performing two iterations manually. Then use the automatic procedure in your IOR Tutorial to check your work.
(b) Use these results to predict what subsequent trial solutions would be if additional iterations were to be performed.
(c) Suppose that the stopping rule adopted for the algorithm in this application is that the algorithm stops when two successive trial solutions differ by no more than 0.01 in any component. Use your predictions from part (*b*) to predict the final trial solution and the total number of iterations required to get there. How close would this solution be to the optimal solution $(x_1, x_2) = (0, 8)$?

7.4-4. Consider the following problem.

Maximize $Z = 3x_1 + x_2,$

subject to

$$3x_1 + 2x_2 \leq 45$$
$$6x_1 + x_2 \leq 45$$

and

$$x_1 \geq 0, \qquad x_2 \geq 0.$$

I **(a)** Solve the problem graphically.

(b) Find the *gradient* of the objective function in the original x_1-x_2 coordinate system. If you move from the origin in the direction of the gradient until you reach the boundary of the feasible region, where does it lead relative to the optimal solution?

C **(c)** Starting from the initial trial solution $(x_1, x_2) = (1, 1)$, use your IOR Tutorial to perform 10 iterations of the interior-point algorithm presented in Sec. 7.4.

C **(d)** Repeat part (c) with $\alpha = 0.9$.

7.4-5. Consider the following problem.

Maximize $Z = 2x_1 + 5x_2 + 7x_3,$

subject to

$$x_1 + 2x_2 + 3x_3 = 6$$

and

$$x_1 \geq 0, \qquad x_2 \geq 0, \qquad x_3 \geq 0.$$

I **(a)** Graph the feasible region.

(b) Find the *gradient* of the objective function, and then find the *projected gradient* onto the feasible region.

(c) Starting from the initial trial solution $(x_1, x_2, x_3) = (1, 1, 1)$, perform two iterations of the interior-point algorithm presented in Sec. 7.4 manually.

C **(d)** Starting from this same initial trial solution, use your IOR Tutorial to perform 10 iterations of this algorithm.

C **7.4-6.** Starting from the initial trial solution $(x_1, x_2) = (2, 2)$, use your IOR Tutorial to apply 15 iterations of the interior-point algorithm presented in Sec. 7.4 to the Wyndor Glass Co. problem presented in Sec. 3.1. Also draw a figure like Fig. 7.8 to show the trajectory of the algorithm in the original x_1-x_2 coordinate system.

8

The Transportation and Assignment Problems

Chapter 3 emphasized the wide applicability of linear programming. We continue to broaden our horizons in this chapter by discussing two particularly important (and related) types of linear programming problems. One type, called the *transportation problem,* received this name because many of its applications involve determining how to optimally transport goods. However, some of its important applications (e.g., production scheduling) actually have nothing to do with transportation.

The second type, called the *assignment problem,* involves such applications as assigning people to tasks. Although its applications appear to be quite different from those for the transportation problem, we shall see that the assignment problem can be viewed as a special type of transportation problem.

The next chapter will introduce additional special types of linear programming problems involving *networks,* including the *minimum cost flow problem* (Sec. 9.6). There we shall see that both the transportation and assignment problems actually are special cases of the minimum cost flow problem. We introduce the network representation of the transportation and assignment problems in this chapter.

Applications of the transportation and assignment problems tend to require a very large number of constraints and variables, so a straightforward computer application of the simplex method may require an exorbitant computational effort. Fortunately, a key characteristic of these problems is that most of the a_{ij} coefficients in the constraints are zeros, and the relatively few nonzero coefficients appear in a distinctive pattern. As a result, it has been possible to develop special *streamlined* algorithms that achieve dramatic computational savings by exploiting this special structure of the problem. Therefore, it is important to become sufficiently familiar with these special types of problems that you can recognize them when they arise and apply the proper computational procedure.

To describe special structures, we shall introduce the table (matrix) of constraint coefficients shown in Table 8.1, where a_{ij} is the coefficient of the jth variable in the ith functional constraint. Later, portions of the table containing only coefficients equal to zero will be indicated by leaving them blank, whereas blocks containing nonzero coefficients will be shaded.

After presenting a prototype example for the transportation problem, we describe the special structure in its model and give additional examples of its applications. Section 8.2 presents the *transportation simplex method,* a special streamlined version of the simplex

■ **TABLE 8.1** Table of
constraint coefficients
for linear programming

$$
\mathbf{A} = \begin{bmatrix} a_{11} & a_{12} & \cdots & a_{1n} \\ a_{21} & a_{22} & \cdots & a_{2n} \\ \cdots\cdots\cdots\cdots\cdots\cdots\cdots \\ a_{m1} & a_{m2} & \cdots & a_{mn} \end{bmatrix}
$$

method for efficiently solving transportation problems. (You will see in Sec. 9.7 that this algorithm is related to the *network simplex method,* another streamlined version of the simplex method for efficiently solving any minimum cost flow problem, including both transportation and assignment problems.) Section 8.3 focuses on the assignment problem. Section 8.4 then presents a specialized algorithm, called the *Hungarian algorithm,* for solving only assignment problems very efficiently.

The book's website also provides a supplement to this chapter. It is a complete case study (including the analysis) that illustrates how a corporate decision regarding where to locate a new facility (an oil refinery in this case) may require solving many transportation problems. (One of the cases for this chapter asks you to continue the analysis for an extension of this case study.)

8.1 THE TRANSPORTATION PROBLEM

Prototype Example

One of the main products of the P & T COMPANY is canned peas. The peas are prepared at three canneries (near Bellingham, Washington; Eugene, Oregon; and Albert Lea, Minnesota) and then shipped by truck to four distributing warehouses in the western United States (Sacramento, California; Salt Lake City, Utah; Rapid City, South Dakota; and Albuquerque, New Mexico), as shown in Fig. 8.1. Because the shipping costs are a major expense, management is initiating a study to reduce them as much as possible. For the upcoming season, an estimate has been made of the output from each cannery, and each warehouse has been allocated a certain amount from the total supply of peas. This information (in units of truckloads), along with the shipping cost per truckload for each cannery-warehouse combination, is given in Table 8.2. Thus, there are a total of 300 truckloads to be shipped. The problem now is to determine which plan for assigning these shipments to the various cannery-warehouse combinations would *minimize the total shipping cost.*

By ignoring the geographical layout of the canneries and warehouses, we can provide a *network representation* of this problem in a simple way by lining up all the canneries in one column on the left and all the warehouses in one column on the right. This representation is shown in Fig. 8.2. The arrows show the possible routes for the truckloads, where the number next to each arrow is the shipping cost per truckload for that route. A square bracket next to each location gives the number of truckloads to be shipped *out* of that location (so that the allocation into each warehouse is given as a negative number).

The problem depicted in Fig. 8.2 is actually a linear programming problem of the *transportation problem type.* To formulate the model, let Z denote total shipping cost, and let x_{ij} ($i = 1, 2, 3$; $j = 1, 2, 3, 4$) be the number of truckloads to be shipped from cannery

An Application Vignette

Procter & Gamble (P & G) makes and markets over 300 brands of consumer goods worldwide. The company has grown continuously over its long history tracing back to the 1830s. To maintain and accelerate that growth, a major OR study was undertaken to strengthen P & G's global effectiveness. Prior to the study, the company's supply chain consisted of hundreds of suppliers, over 50 product categories, over 60 plants, 15 distribution centers, and over 1,000 customer zones. However, as the company moved toward global brands, management realized that it needed to consolidate plants to reduce manufacturing expenses, improve speed to market, and reduce capital investment. Therefore, the study focused on redesigning the company's production and distribution system for its North American operations. The result was a reduction in the number of North American

plants by almost 20 percent, *saving over* **$200 million** in pretax costs *per year*.

A major part of the study revolved around *formulating and solving transportation problems* for individual product categories. For each option regarding the plants to keep open, and so forth, solving the corresponding transportation problem for a product category showed what the distribution cost would be for shipping the product category from those plants to the distribution centers and customer zones.

Source: J. D. Camm, T. E. Chorman, F. A. Dill, J. R. Evans, D. J. Sweeney, and G. W. Wegryn: "Blending OR/MS, Judgment, and GIS: Restructuring P & G's Supply Chain," *Interfaces,* **27**(1): 128–142, Jan.–Feb. 1997. (A link to this article is provided on our website, www.mhhe.com/hillier.)

FIGURE 8.1

Location of canneries and warehouses for the P & T Co. problem.

■ **TABLE 8.2** Shipping data for P & T Co.

		Shipping Cost ($) per Truckload				
		Warehouse				
		1	2	3	4	Output
Cannery	1	464	513	654	867	75
	2	352	416	690	791	125
	3	995	682	388	685	100
Allocation		80	65	70	85	

i to warehouse j. Thus, the objective is to choose the values of these 12 decision variables (the x_{ij}) so as to

Minimize $Z = 464x_{11} + 513x_{12} + 654x_{13} + 867x_{14} + 352x_{21} + 416x_{22}$
$+ 690x_{23} + 791x_{24} + 995x_{31} + 682x_{32} + 388x_{33} + 685x_{34},$

subject to the constraints

$$x_{11} + x_{12} + x_{13} + x_{14} \qquad\qquad\qquad\qquad\qquad = 75$$
$$x_{21} + x_{22} + x_{23} + x_{24} \qquad\qquad\qquad = 125$$
$$x_{31} + x_{32} + x_{33} + x_{34} = 100$$
$$x_{11} \qquad\qquad + x_{21} \qquad\qquad + x_{31} \qquad\qquad = 80$$
$$x_{12} \qquad\qquad + x_{22} \qquad\qquad + x_{32} \qquad = 65$$
$$x_{13} \qquad\qquad + x_{23} \qquad\qquad + x_{33} = 70$$
$$x_{14} \qquad\qquad + x_{24} \qquad\qquad + x_{34} = 85$$

and

$$x_{ij} \geq 0 \qquad (i = 1, 2, 3; j = 1, 2, 3, 4).$$

■ **FIGURE 8.2**
Network representation of the P & T Co. problem.

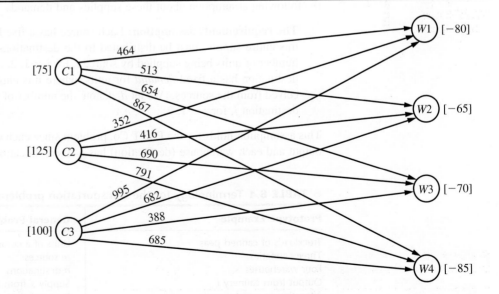

■ TABLE 8.3 Constraint coefficients for P & T Co.

					Coefficient of:							
	x_{11}	x_{12}	x_{13}	x_{14}	x_{21}	x_{22}	x_{23}	x_{24}	x_{31}	x_{32}	x_{33}	x_{34}
	1	1	1	1								
					1	1	1	1				
									1	1	1	1
A =	1				1				1			
		1				1				1		
			1				1				1	
				1				1				1

Rows 1–3: Cannery constraints. Rows 4–7: Warehouse constraints.

Table 8.3 shows the constraint coefficients. As you will see later in this section, it is the special structure in the pattern of these coefficients that distinguishes this problem as a transportation problem, not its context. However, we first will describe the various other characteristics of the transportation problem model.

The Transportation Problem Model

To describe the general model for the transportation problem, we need to use terms that are considerably less specific than those for the components of the prototype example. In particular, the general transportation problem is concerned (literally or figuratively) with distributing *any* commodity from *any* group of supply centers, called **sources,** to *any* group of receiving centers, called **destinations,** in such a way as to minimize the total distribution cost. The correspondence in terminology between the prototype example and the general problem is summarized in Table 8.4.

As indicated by the fourth and fifth rows of the table, each source has a certain **supply** of units to distribute to the destinations, and each destination has a certain **demand** for units to be received from the sources. The model for a transportation problem makes the following assumption about these supplies and demands.

The requirements assumption: Each source has a fixed *supply* of units, where this entire supply must be distributed to the destinations. (We let s_i denote the number of units being supplied by source i, for $i = 1, 2, \ldots, m$.) Similarly, each destination has a fixed *demand* for units, where this entire demand must be received from the sources. (We let d_j denote the number of units being received by destination j, for $j = 1, 2, \ldots, n$.)

This assumption holds for the P & T Co. problem since each cannery (source) has a fixed output and each warehouse (destination) has a fixed allocation.

■ TABLE 8.4 Terminology for the transportation problem

Prototype Example	General Problem
Truckloads of canned peas	Units of a commodity
Three canneries	m sources
Four warehouses	n destinations
Output from cannery i	Supply s_i from source i
Allocation to warehouse j	Demand d_j at destination j
Shipping cost per truckload from cannery i to warehouse j	Cost c_{ij} per unit distributed from source i to destination j

This assumption that there is no leeway in the amounts to be sent or received means that there needs to be a balance between the total supply from all sources and the total demand at all destinations.

> **The feasible solutions property:** A transportation problem will have feasible solutions if and only if
>
> $$\sum_{i=1}^{m} s_i = \sum_{j=1}^{n} d_j.$$

Fortunately, these sums are equal for the P & T Co. since Table 8.2 indicates that the supplies (outputs) sum to 300 truckloads and so do the demands (allocations).

In some real problems, the supplies actually represent *maximum* amounts (rather than fixed amounts) to be distributed. Similarly, in other cases, the demands represent maximum amounts (rather than fixed amounts) to be received. Such problems do not quite fit the model for a transportation problem because they violate the *requirements assumption.* However, it is possible to *reformulate* the problem so that they then fit this model by introducing a *dummy destination* or a *dummy source* to take up the slack between the actual amounts and maximum amounts being distributed. We will illustrate how this is done with two examples at the end of this section.

The last row of Table 8.4 refers to a cost per unit distributed. This reference to a *unit cost* implies the following basic assumption for any transportation problem.

> **The cost assumption:** The cost of distributing units from any particular source to any particular destination is *directly proportional* to the number of units distributed. Therefore, this cost is just the *unit cost* of distribution *times* the *number of units distributed.* (We let c_{ij} denote this unit cost for source i and destination j.)

This assumption holds for the P & T Co. problem since the cost of shipping peas from any cannery to any warehouse is directly proportional to the number of truckloads being shipped.

The only data needed for a transportation problem model are the supplies, demands, and unit costs. These are the *parameters of the model.* All these parameters can be summarized conveniently in a single *parameter table* as shown in Table 8.5.

> **The model:** Any problem (whether involving transportation or not) fits the model for a transportation problem if it can be described completely in terms of a *parameter table* like Table 8.5 and it satisfies both the *requirements assumption* and the *cost assumption.* The objective is to minimize the total cost of distributing the units. All the parameters of the model are included in this parameter table.

■ **TABLE 8.5** Parameter table for the transportation problem

		Cost per Unit Distributed				
		Destination				
		1	**2**	...	**n**	**Supply**
Source	1	c_{11}	c_{12}	...	c_{1n}	s_1
	2	c_{21}	c_{22}	...	c_{2n}	s_2
	⋮					⋮
	m	c_{m1}	c_{m2}	...	c_{mn}	s_m
Demand		d_1	d_2	...	d_n	

Therefore, formulating a problem as a transportation problem only requires filling out a parameter table in the format of Table 8.5. (The parameter table for the P & T Co. problem is shown in Table 8.2.) Alternatively, the same information can be provided by using the network representation of the problem shown in Fig. 8.3 (as was done in Fig. 8.2 for the P & T Co. problem). Some problems that have nothing to do with transportation also can be formulated as a transportation problem in either of these two ways. The Worked Examples section of the book's website includes **another example** of such a problem.

Since a transportation problem can be formulated simply by either filling out a parameter table or drawing its network representation, it is not necessary to write out a formal mathematical model for the problem. However, we will go ahead and show you this model once for the general transportation problem just to emphasize that it is indeed a special type of linear programming problem.

Letting Z be the total distribution cost and x_{ij} ($i = 1, 2, \ldots, m; j = 1, 2, \ldots, n$) be the number of units to be distributed from source i to destination j, the linear programming formulation of this problem is

$$\text{Minimize} \quad Z = \sum_{i=1}^{m} \sum_{j=1}^{n} c_{ij} x_{ij},$$

subject to

$$\sum_{j=1}^{n} x_{ij} = s_i \quad \text{for } i = 1, 2, \ldots, m,$$

■ **FIGURE 8.3**
Network representation of the transportation problem.

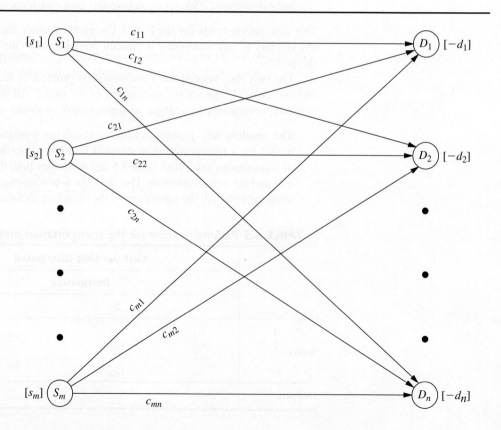

$$\sum_{i=1}^{m} x_{ij} = d_j \qquad \text{for } j = 1, 2, \ldots, n,$$

and

$$x_{ij} \geq 0, \qquad \text{for all } i \text{ and } j.$$

Note that the resulting table of constraint coefficients has the special structure shown in Table 8.6. *Any* linear programming problem that fits this special formulation is of the transportation problem type, regardless of its physical context. In fact, there have been numerous applications unrelated to transportation that have been fitted to this special structure, as we shall illustrate in the next example later in this section. (The assignment problem described in Sec. 8.3 is an additional example.) This is one of the reasons why the transportation problem is considered such an important special type of linear programming problem.

For many applications, the supply and demand quantities in the model (the s_i and d_j) have integer values, and implementation will require that the distribution quantities (the x_{ij}) also have integer values. Fortunately, because of the special structure shown in Table 8.6, all such problems have the following property.

Integer solutions property: For transportation problems where every s_i and d_j have an integer value, all the basic variables (allocations) in *every* basic feasible (BF) solution (including an optimal one) also have *integer* values.

The solution procedure described in Sec. 8.2 deals only with BF solutions, so it automatically will obtain an *integer* optimal solution for this case. (You will be able to see why this solution procedure actually gives a proof of the integer solutions property after you learn the procedure; Prob. 8.2-20 guides you through the reasoning involved.) Therefore, it is unnecessary to add a constraint to the model that the x_{ij} must have integer values.

As with other linear programming problems, the usual software options (Excel, LINGO/LINDO, MPL/CPLEX) are available to you for setting up and solving transportation problems (and assignment problems), as demonstrated in the files for this chapter in your OR Courseware. However, because the Excel approach now is somewhat different from what you have seen previously, we next describe this approach.

Using Excel to Formulate and Solve Transportation Problems

As described in Sec. 3.6, the process of using a spreadsheet to formulate a linear programming model for a problem begins by developing answers to three questions. What are the *decisions* to be made? What are the *constraints* on these decisions? What is the *overall measure of performance* for these decisions? Since a transportation problem is a special type of

■ **TABLE 8.6** Constraint coefficients for the transportation problem

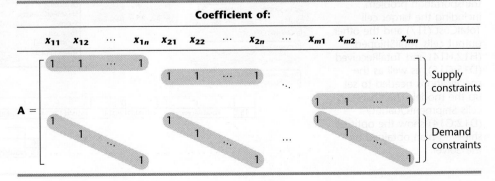

linear programming problem, addressing these questions also is a suitable starting point for formulating this kind of problem on a spreadsheet. The design of the spreadsheet then revolves around laying out this information and the associated data in a logical way.

To illustrate, consider the P & T Co. problem again. The decisions to be made are the number of truckloads of peas to ship from each cannery to each warehouse. The constraints on these decisions are that the total amount shipped from each cannery must equal its output (the supply) and the total amount received at each warehouse must equal its allocation (the demand). The overall measure of performance is the total shipping cost, so the objective is to minimize this quantity.

This information leads to the spreadsheet model shown in Fig. 8.4. All the data provided in Table 8.2 are displayed in the following data cells: UnitCost (D5:G7), Supply (J12:J14), and Demand (D17:G17). The decisions on shipping quantities are given by the changing cells, ShipmentQuantity (D12:G14). The output cells are TotalShipped (H12:H14) and TotalReceived (D15:G15), where the SUM functions entered into these cells are shown near the bottom of Fig. 8.4. The constraints, TotalShipped (H12:H14) = Supply (J12:J14) and TotalReceived (D15:G15) = Demand (D17:G17), have been specified on the spreadsheet and entered into the Solver dialogue box. The target cell is TotalCost (J17), where its SUMPRODUCT function is shown in the lower right-hand corner of Fig. 8.4. The Solver dialogue box specifies that the objective is to minimize this target cell. One of the selected Solver options (Assume Non-Negative) specifies that all shipment quantities must be nonnegative. The other one (Assume Linear Model) indicates that this transportation problem is also a linear programming problem.

To begin the process of solving the problem, any value (such as 0) can be entered in each of the changing cells. After clicking on the Solve button, the Solver will use the simplex method to solve the transportation problem and determine the best value for each of

	A	B	C	D	E	F	G	H	I	J
1	P&T Co. Distribution Problem									
2										
3		Unit Cost		Destination (Warehouse)						
4				Sacramento	Salt Lake City	Rapid City	Albuquerque			
5		Source	Bellingham	$464	$513	$654	$867			
6		(Cannery)	Eugene	$352	$416	$690	$791			
7			Albert Lea	$995	$682	$388	$685			
8										
9										
10		Shipment Quantity		Destination (Warehouse)						
11		(Truckloads)		Sacramento	Salt Lake City	Rapid City	Albuquerque	Total Shipped		Supply
12		Source	Bellingham	0	20	0	55	75	=	75
13		(Cannery)	Eugene	80	45	0	0	125	=	125
14			Albert Lea	0	0	70	30	100	=	100
15			Total Received	80	65	70	85			
16				=	=	=	=			Total Cost
17			Demand	80	65	70	85			$ 152,535

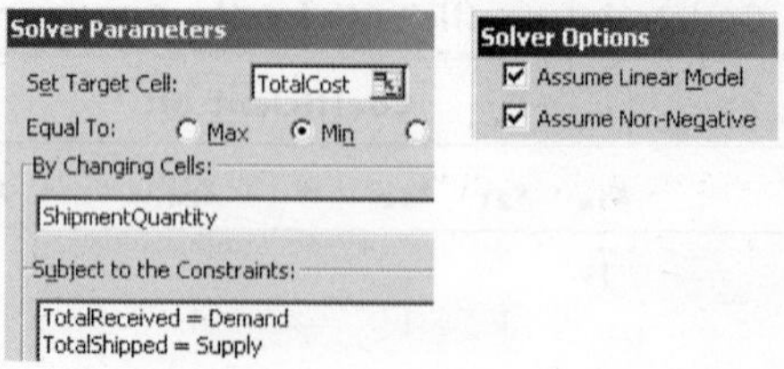

Range Name	Cells
Demand	D17:G17
ShipmentQuantity	D12:G14
Supply	J12:J14
TotalCost	J17
TotalReceived	D15:G15
TotalShipped	H12:H14
UnitCost	D5:G7

	H
11	Total Shipped
12	=SUM(D12:G12)
13	=SUM(D13:G13)
14	=SUM(D14:G14)

	C	D	E	F	G
15	Total Received	=SUM(D12:D14)	=SUM(E12:E14)	=SUM(F12:F14)	=SUM(G12:G14)

	J
16	Total Cost
17	=SUMPRODUCT(UnitCost,ShipmentQuantity)

■ **FIGURE 8.4**
A spreadsheet formulation of the P & T Co. problem as a transportation problem, including the target cell TotalCost (J17) and the other output cells TotalShipped (H12:H14) and TotalReceived (D15:G15), as well as the specifications needed to set up the model. The changing cells ShipmentQuantity (D12:G14) show the optimal shipping plan obtained by the Solver.

the decision variables. This optimal solution is shown in ShipmentQuantity (D12:G14) in Fig. 8.4, along with the resulting value $152,535 in the target cell TotalCost (J17).

Note that the Solver simply uses the general simplex method to solve a transportation problem rather than a streamlined version that is specially designed for solving transportation problems very efficiently, such as the transportation simplex method presented in the next section. Therefore, a software package that includes such a streamlined version should solve a large transportation problem much faster than the Excel Solver.

We mentioned earlier that some problems do not quite fit the model for a transportation problem because they violate the requirements assumption, but that it is possible to reformulate such a problem to fit this model by introducing a dummy destination or a dummy source. When using the Excel Solver, it is not necessary to do this reformulation since the simplex method can solve the original model where the supply constraints are in $\leq$ form or the demand constraints are in $\geq$ form. (The Excel files for the next two examples in your OR Courseware illustrate spreadsheet formulations that retain either the supply constraints or the demand constraints in their original inequality form.) However, the larger the problem, the more worthwhile it becomes to do the reformulation and use the transportation simplex method (or equivalent) instead with another software package.

The next two examples illustrate how to do this kind of reformulation.

An Example with a Dummy Destination

The NORTHERN AIRPLANE COMPANY builds commercial airplanes for various airline companies around the world. The last stage in the production process is to produce the jet engines and then to install them (a very fast operation) in the completed airplane frame. The company has been working under some contracts to deliver a considerable number of airplanes in the near future, and the production of the jet engines for these planes must now be scheduled for the next four months.

To meet the contracted dates for delivery, the company must supply engines for installation in the quantities indicated in the second column of Table 8.7. Thus, the cumulative number of engines produced by the end of months 1, 2, 3, and 4 must be at least 10, 25, 50, and 70, respectively.

The facilities that will be available for producing the engines vary according to other production, maintenance, and renovation work scheduled during this period. The resulting monthly differences in the maximum number that can be produced and the cost (in millions of dollars) of producing each one are given in the third and fourth columns of Table 8.7.

Because of the variations in production costs, it may well be worthwhile to produce some of the engines a month or more before they are scheduled for installation, and this possibility is being considered. The drawback is that such engines must be stored until the scheduled installation (the airplane frames will not be ready early) at a storage cost of $15,000 per month (including interest on expended capital) for each engine,[1] as shown in the rightmost column of Table 8.7.

The production manager wants a schedule developed for the number of engines to be produced in each of the four months so that the total of the production and storage costs will be minimized.

Formulation. One way to formulate a mathematical model for this problem is to let x_j be the number of jet engines to be produced in month j, for $j = 1, 2, 3, 4$. By using only

---

[1]For modeling purposes, assume that this storage cost is incurred at the *end of the month* for just those engines that are being held over into the next month. Thus, engines that are produced in a given month for installation in the same month are assumed to incur no storage cost.

■ **TABLE 8.7** Production scheduling data for Northern Airplane Co.

Month	Scheduled Installations	Maximum Production	Unit Cost* of Production	Unit Cost* of Storage
1	10	25	1.08	0.015
2	15	35	1.11	0.015
3	25	30	1.10	0.015
4	20	10	1.13	

*Cost is expressed in millions of dollars.

these four decision variables, the problem can be formulated as a linear programming problem that does *not* fit the transportation problem type. (See Prob. 8.2-18.)

On the other hand, by adopting a different viewpoint, we can instead formulate the problem as a transportation problem that requires *much* less effort to solve. This viewpoint will describe the problem in terms of sources and destinations and then identify the corresponding x_{ij}, c_{ij}, s_i, and d_j. (See if you can do this before reading further.)

Because the units being distributed are jet engines, each of which is to be scheduled for production in a particular month and then installed in a particular (perhaps different) month,

$$\text{Source } i = \text{production of jet engines in month } i \ (i = 1, 2, 3, 4)$$

$$\text{Destination } j = \text{installation of jet engines in month } j \ (j = 1, 2, 3, 4)$$

$$x_{ij} = \text{number of engines produced in month } i \text{ for installation in month } j$$

$$c_{ij} = \text{cost associated with each unit of } x_{ij}$$

$$= \begin{cases} \text{cost per unit for production and any storage} & \text{if } i \leq j \\ ? & \text{if } i > j \end{cases}$$

$$s_i = ?$$

$$d_j = \text{number of scheduled installations in month } j.$$

The corresponding (incomplete) parameter table is given in Table 8.8. Thus, it remains to identify the missing costs and the supplies.

Since it is impossible to produce engines in one month for installation in an earlier month, x_{ij} must be zero if $i > j$. Therefore, there is no real cost that can be associated with such x_{ij}. Nevertheless, in order to have a well-defined transportation problem to which the solution procedure of Sec. 8.2 can be applied, it is necessary to assign some value for the unidentified costs. Fortunately, we can use the Big M method introduced in Sec. 4.6 to

■ **TABLE 8.8** Incomplete parameter table for Northern Airplane Co.

		Cost per Unit Distributed				
		Destination				
		1	2	3	4	Supply
Source	1	1.080	1.095	1.110	1.125	?
	2	?	1.110	1.125	1.140	?
	3	?	?	1.100	1.115	?
	4	?	?	?	1.130	?
Demand		10	15	25	20	

assign this value. Thus, we assign a *very* large number (denoted by M for convenience) to the unidentified cost entries in Table 8.8 to force the corresponding values of x_{ij} to be zero in the final solution.

The numbers that need to be inserted into the supply column of Table 8.8 are not obvious because the "supplies," the amounts produced in the respective months, are not fixed quantities. In fact, the objective is to solve for the most desirable values of these production quantities. Nevertheless, it is necessary to assign some fixed number to every entry in the table, including those in the supply column, to have a transportation problem. A clue is provided by the fact that although the supply constraints are not present in the usual form, these constraints do exist in the form of upper bounds on the amount that can be supplied, namely,

$$x_{11} + x_{12} + x_{13} + x_{14} \leq 25,$$
$$x_{21} + x_{22} + x_{23} + x_{24} \leq 35,$$
$$x_{31} + x_{32} + x_{33} + x_{34} \leq 30,$$
$$x_{41} + x_{42} + x_{43} + x_{44} \leq 10.$$

The only change from the standard model for the transportation problem is that these constraints are in the form of inequalities instead of equalities.

To convert these inequalities to equations in order to fit the transportation problem model, we use the familiar device of *slack variables,* introduced in Sec. 4.2. In this context, the slack variables are allocations to a single **dummy destination** that represent the *unused production capacity* in the respective months. This change permits the supply in the transportation problem formulation to be the total production capacity in the given month. Furthermore, because the demand for the dummy destination is the total unused capacity, this demand is

$$(25 + 35 + 30 + 10) - (10 + 15 + 25 + 20) = 30.$$

With this demand included, the sum of the supplies now equals the sum of the demands, which is the condition given by the *feasible solutions property* for having feasible solutions.

The cost entries associated with the dummy destination should be zero because there is no cost incurred by a fictional allocation. (Cost entries of M would be *inappropriate* for this column because we do not want to force the corresponding values of x_{ij} to be zero. In fact, these values need to sum to 30.)

The resulting final parameter table is given in Table 8.9, with the dummy destination labeled as destination 5(D). By using this formulation, it is quite easy to find the optimal production schedule by the solution procedure described in Sec. 8.2. (See Prob. 8.2-10 and its answer in the back of the book.)

■ **TABLE 8.9** Complete parameter table for Northern Airplane Co.

		Cost per Unit Distributed					
		Destination					
		1	2	3	4	5(*D*)	Supply
	1	1.080	1.095	1.110	1.125	0	25
Source	2	M	1.110	1.125	1.140	0	35
	3	M	M	1.100	1.115	0	30
	4	M	M	M	1.130	0	10
Demand		10	15	25	20	30	

An Example with a Dummy Source

METRO WATER DISTRICT is an agency that administers water distribution in a large geographic region. The region is fairly arid, so the district must purchase and bring in water from outside the region. The sources of this imported water are the Colombo, Sacron, and Calorie rivers. The district then resells the water to users in the region. Its main customers are the water departments of the cities of Berdoo, Los Devils, San Go, and Hollyglass.

It is possible to supply any of these cities with water brought in from any of the three rivers, with the exception that no provision has been made to supply Hollyglass with Calorie River water. However, because of the geographic layouts of the aqueducts and the cities in the region, the cost to the district of supplying water depends upon both the source of the water and the city being supplied. The variable cost per acre foot of water (in tens of dollars) for each combination of river and city is given in Table 8.10. Despite these variations, the price per acre foot charged by the district is independent of the source of the water and is the same for all cities.

The management of the district is now faced with the problem of how to allocate the available water during the upcoming summer season. In units of 1 million acre feet, the amounts available from the three rivers are given in the rightmost column of Table 8.10. The district is committed to providing a certain minimum amount to meet the essential needs of each city (with the exception of San Go, which has an independent source of water), as shown in the *minimum needed* row of the table. The *requested* row indicates that Los Devils desires no more than the minimum amount, but that Berdoo would like to buy as much as 20 more, San Go would buy up to 30 more, and Hollyglass will take as much as it can get.

Management wishes to allocate *all* the available water from the three rivers to the four cities in such a way as to at least meet the essential needs of each city while minimizing the total cost to the district.

Formulation. Table 8.10 already is close to the proper form for a parameter table, with the rivers being the sources and the cities being the destinations. However, the one basic difficulty is that it is not clear what the demands at the destinations should be. The amount to be received at each destination (except Los Devils) actually is a decision variable, with both a lower bound and an upper bound. This upper bound is the amount requested unless the request exceeds the total supply remaining after the minimum needs of the other cities are met, in which case this *remaining supply* becomes the upper bound. Thus, insatiably thirsty Hollyglass has an upper bound of

$$(50 + 60 + 50) - (30 + 70 + 0) = 60.$$

Unfortunately, just like the other numbers in the parameter table of a transportation problem, the demand quantities must be *constants*, not bounded decision variables. To

■ **TABLE 8.10** Water resources data for Metro Water District

	Cost (Tens of Dollars) per Acre Foot				
	Berdoo	**Los Devils**	**San Go**	**Hollyglass**	**Supply**
Colombo River	16	13	22	17	50
Sacron River	14	13	19	15	60
Calorie River	19	20	23	—	50
Minimum needed	30	70	0	10	(in units of 1
Requested	50	70	30	∞	million acre feet)

begin resolving this difficulty, temporarily suppose that it is not necessary to satisfy the minimum needs, so that the upper bounds are the only constraints on amounts to be allocated to the cities. In this circumstance, can the requested allocations be viewed as the demand quantities for a transportation problem formulation? After one adjustment, yes! (Do you see already what the needed adjustment is?)

The situation is analogous to Northern Airplane Co.'s production scheduling problem, where there was *excess supply capacity.* Now there is *excess demand capacity.* Consequently, rather than introducing a *dummy destination* to "receive" the unused supply capacity, the adjustment needed here is to introduce a **dummy source** to "send" the *unused demand capacity.* The imaginary supply quantity for this dummy source would be the amount by which the sum of the demands exceeds the sum of the real supplies:

$$(50 + 70 + 30 + 60) - (50 + 60 + 50) = 50.$$

This formulation yields the parameter table shown in Table 8.11, which uses units of million acre feet and tens of millions of dollars. The cost entries in the *dummy* row are zero because there is no cost incurred by the fictional allocations from this dummy source. On the other hand, a huge unit cost of M is assigned to the Calorie River–Hollyglass spot. The reason is that Calorie River water cannot be used to supply Hollyglass, and assigning a cost of M will prevent any such allocation.

Now let us see how we can take each city's minimum needs into account in this kind of formulation. Because San Go has no minimum need, it is all set. Similarly, the formulation for Hollyglass does not require any adjustments because its demand (60) exceeds the dummy source's supply (50) by 10, so the amount supplied to Hollyglass from the *real* sources will be *at least 10* in any feasible solution. Consequently, its minimum need of 10 from the rivers is guaranteed. (If this coincidence had not occurred, Hollyglass would need the same adjustments that we shall have to make for Berdoo.)

Los Devils' minimum need equals its requested allocation, so its *entire* demand of 70 must be filled from the real sources rather than the dummy source. This requirement calls for the Big M method! Assigning a huge unit cost of M to the allocation from the dummy source to Los Devils ensures that this allocation will be zero in an optimal solution.

Finally, consider Berdoo. In contrast to Hollyglass, the dummy source has an adequate (fictional) supply to "provide" at least some of Berdoo's minimum need in addition to its extra requested amount. Therefore, since Berdoo's minimum need is 30, adjustments must be made to prevent the dummy source from contributing more than 20 to Berdoo's total demand of 50. This adjustment is accomplished by splitting Berdoo into two destinations, one having a demand of 30 with a unit cost of M for any allocation from the dummy source and the other having a demand of 20 with a unit cost of zero for the dummy source allocation. This formulation gives the final parameter table shown in Table 8.12.

TABLE 8.11 Parameter table without minimum needs for Metro Water District

		Cost (Tens of Millions of Dollars) per Unit Distributed				
		Destination				
		Berdoo	**Los Devils**	**San Go**	**Hollyglass**	**Supply**
Source	Colombo River	16	13	22	17	50
	Sacron River	14	13	19	15	60
	Calorie River	19	20	23	M	50
	Dummy	0	0	0	0	50
Demand		50	70	30	60	

■ **TABLE 8.12** Parameter table for Metro Water District

			Cost (Tens of Millions of Dollars) per Unit Distributed					
			Destination					
			Berdoo (min.) 1	Berdoo (extra) 2	Los Devils 3	San Go 4	Hollyglass 5	Supply
	Colombo River	1	16	16	13	22	17	50
Source	Sacron River	2	14	14	13	19	15	60
	Calorie River	3	19	19	20	23	M	50
	Dummy	4(D)	M	0	M	0	0	50
Demand			30	20	70	30	60	

This problem will be solved in Sec. 8.2 to illustrate the solution procedure presented there.

Generalizations of the Transportation Problem

Even after the kinds of reformulations illustrated by the two preceding examples, some problems involving the distribution of units from sources to destinations fail to satisfy the model for the transportation problem. One reason may be that the distribution does not go directly from the sources to the destinations but instead passes through transfer points along the way. The Distribution Unlimited Co example in Sec. 3.4 (See Fig. 3.13) illustrates such a problem. In this case, the sources are the two factories and the destinations are the two warehouses. However, a shipment from a particular factory to a particular warehouse may first get transferred at a distribution center, or even at the other factory or the other warehouse, before reaching its destination. The unit shipping costs differ for these different shipping lanes. Furthermore, there are upper limits on how much can be shipped through some of the shipping lanes. Although it is not a transportation problem, this kind of problem still is a special type of linear programming problem, called the *minimum cost flow problem*, that will be discussed in Sec. 9.6. The *network simplex method* described in Sec. 9.7 provides an efficient way of solving minimum cost flow problems. A minimum cost flow problem that does not impose any upper limits on how much can be shipped through the shipping lanes is referred to as a *transshipment problem*. Section 23.1 on the book's website is devoted to discussing transshipment problems.

In other cases, the distribution may go directly from sources to destinations, but other assumptions of the transportation problem may be violated. The *cost assumption* will be violated if the cost of distributing units from any particular source to any particular destination is a nonlinear function of the number of units distributed. The *requirements assumption* will be violated if either the supplies from the sources or the demands at the destinations are not fixed. For example, the final demand at a destination may not become known until after the units have arrived and then a nonlinear cost is incurred if the amount received deviates from the final demand. If the supply at a source is not fixed, the cost of producing the amount supplied may be a nonlinear function of this amount. For example, a fixed cost may be part of the cost associated with a decision to open up a new source. Considerable research has been done to generalize the transportation problem and its solution procedure in these kinds of directions.[2]

[2]For example, see K. Holmberg and H. Tuy: "A Production-Transportation Problem with Stochastic Demand and Concave Production Costs," *Mathematical Programming Series A,* **85:** 157–179, 1999.

8.2   A STREAMLINED SIMPLEX METHOD FOR THE TRANSPORTATION PROBLEM

Because the transportation problem is just a special type of linear programming problem, it can be solved by applying the simplex method as described in Chap. 4. However, you will see in this section that some tremendous computational shortcuts can be taken in this method by exploiting the special structure shown in Table 8.6. We shall refer to this streamlined procedure as the **transportation simplex method.**

As you read on, note particularly how the special structure is exploited to achieve great computational savings. This will illustrate an important OR technique—streamlining an algorithm to exploit the special structure in the problem at hand.

Setting Up the Transportation Simplex Method

To highlight the streamlining achieved by the transportation simplex method, let us first review how the general (unstreamlined) simplex method would set up a transportation problem in tabular form. After constructing the table of constraint coefficients (see Table 8.6), converting the objective function to maximization form, and using the Big M method to introduce artificial variables $z_1, z_2, \ldots, z_{m+n}$ into the $m + n$ respective equality constraints (see Sec. 4.6), typical columns of the simplex tableau would have the form shown in Table 8.13, where all entries *not shown* in these columns are *zeros*. [The one remaining adjustment to be made before the first iteration of the simplex method is to algebraically eliminate the nonzero coefficients of the initial (artificial) basic variables in row 0.]

After any subsequent iteration, row 0 then would have the form shown in Table 8.14. Because of the pattern of 0s and 1s for the coefficients in Table 8.13, by the *fundamental insight* presented in Sec. 5.3, u_i and v_j would have the following interpretation:

u_i = multiple of *original* row i that has been subtracted (directly or indirectly) from *original* row 0 by the simplex method during all iterations leading to the current simplex tableau.

v_j = multiple of *original* row $m + j$ that has been subtracted (directly or indirectly) from *original* row 0 by the simplex method during all iterations leading to the current simplex tableau.

TABLE 8.13 Original simplex tableau before simplex method is applied to transportation problem

Basic Variable	Eq.	Coefficient of:								Right side
		Z	$\cdots$	x_{ij}	$\cdots$	z_i	$\cdots$	z_{m+j}	$\cdots$	
Z	(0)	-1		c_{ij}		M		M		0
	(1)									
	$\vdots$									
z_i	(i)	0		1		1				s_i
	$\vdots$									
z_{m+j}	($m + j$)	0		1				1		d_j
	$\vdots$									
	($m + n$)									

■ **TABLE 8.14** Row 0 of simplex tableau when simplex method is applied to transportation problem

Basic Variable	Eq.	Coefficient of:						Right Side
		Z	$\cdots$	x_{ij}	$\cdots$	z_i	$\cdots$	z_{m+j} $\cdots$	
Z	(0)	-1		$c_{ij} - u_i - v_j$		$M - u_i$		$M - v_j$	$-\sum\limits_{i=1}^{m} s_i u_i - \sum\limits_{j=1}^{n} d_j v_j$

Using the duality theory introduced in Chap. 6, another property of the u_i and v_j is that they are the *dual variables*.[3] If x_{ij} is a nonbasic variable, $c_{ij} - u_i - v_j$ is interpreted as the rate at which Z will change as x_{ij} is increased.

The Needed Information. To lay the groundwork for simplifying this setup, recall what information is needed by the simplex method. In the initialization, an initial BF solution must be obtained, which is done artificially by introducing artificial variables as the initial basic variables and setting them equal to s_i and d_j. The optimality test and step 1 of an iteration (selecting an entering basic variable) require knowing the current row 0, which is obtained by subtracting a certain multiple of another row from the preceding row 0. Step 2 (determining the leaving basic variable) must identify the basic variable that reaches zero first as the entering basic variable is increased, which is done by comparing the current coefficients of the entering basic variable and the corresponding right side. Step 3 must determine the new BF solution, which is found by subtracting certain multiples of one row from the other rows in the current simplex tableau.

Greatly Streamlined Ways of Obtaining This Information. Now, how does the *transportation simplex method* obtain the same information in much simpler ways? This story will unfold fully in the coming pages, but here are some preliminary answers.

First, *no artificial variables* are needed, because a simple and convenient procedure (with several variations) is available for constructing an initial BF solution.

Second, the current row 0 can be obtained *without using any other row* simply by calculating the current values of u_i and v_j directly. Since each basic variable must have a coefficient of zero in row 0, the current u_i and v_j are obtained by solving the set of equations

$$c_{ij} - u_i - v_j = 0 \qquad \text{for each } i \text{ and } j \text{ such that } x_{ij} \text{ is a basic variable.}$$

(We will illustrate this straightforward procedure later when discussing the optimality test for the transportation simplex method.) The special structure in Table 8.13 makes this convenient way of obtaining row 0 possible by yielding $c_{ij} - u_i - v_j$ as the coefficient of x_{ij} in Table 8.14.

Third, the leaving basic variable can be identified in a simple way without (explicitly) using the coefficients of the entering basic variable. The reason is that the special structure of the problem makes it easy to see how the solution must change as the entering basic variable is increased. As a result, the new BF solution also can be identified immediately *without any algebraic manipulations* on the rows of the simplex tableau. (You will see the details when we describe how the transportation simplex method performs an iteration.)

The grand conclusion is that *almost the entire simplex tableau* (and the work of maintaining it) *can be eliminated*! Besides the input data (the c_{ij}, s_i, and d_j values), the only

---

[3]It would be easier to recognize these variables as dual variables by relabeling all these variables as y_i and then changing all the signs in row 0 of Table 8.14 by converting the objective function back to its original minimization form.

information needed by the transportation simplex method is the current BF solution,[4] the current values of u_i and v_j, and the resulting values of $c_{ij} - u_i - v_j$ for nonbasic variables x_{ij}. When you solve a problem by hand, it is convenient to record this information for each iteration in a **transportation simplex tableau**, such as shown in Table 8.15. (Note carefully that the values of x_{ij} and $c_{ij} - u_i - v_j$ are distinguished in these tableaux by circling the former but not the latter.)

The Resulting Great Improvement in Efficiency. You can gain a fuller appreciation for the great difference in efficiency and convenience between the simplex and the transportation simplex methods by applying both to the same small problem (see Prob. 8.2-17). However, the difference becomes even more pronounced for large problems that must be solved on a computer. This pronounced difference is suggested somewhat by comparing the sizes of the simplex and the transportation simplex tableaux. Thus, for a transportation problem having m sources and n destinations, the simplex tableau would have $m + n + 1$ rows and $(m + 1)(n + 1)$ columns (excluding those to the left of the x_{ij} columns), and the transportation simplex tableau would have m rows and n columns (excluding the two extra informational rows and columns). Now try plugging in various values for m and n (for example, $m = 10$ and $n = 100$ would be a rather typical medium-size transportation problem), and note how the ratio of the number of cells in the simplex tableau to the number in the transportation simplex tableau increases as m and n increase.

Initialization

Recall that the objective of the initialization is to obtain an initial BF solution. Because all the functional constraints in the transportation problem are *equality* constraints, the simplex method would obtain this solution by introducing artificial variables and using them as the initial basic variables, as described in Sec. 4.6. The resulting basic solution

■ **TABLE 8.15** Format of a transportation simplex tableau

		Destination				Supply	u_i
		1	**2**	...	**n**		
Source	1	c_{11}	c_{12}	...	c_{1n}	s_1	
	2	c_{21}	c_{22}	...	c_{2n}	s_2	
	⋮	...			...	⋮	
	m	c_{m1}	c_{m2}	...	c_{mn}	s_m	
Demand		d_1	d_2	...	d_n	$Z =$	
v_j							

Additional information to be added to each cell:

If x_{ij} is a basic variable	If x_{ij} is a nonbasic variable
c_{ij} $\;\;\;$ (x_{ij})	c_{ij} $\;\;\;$ $c_{ij} - u_i - v_j$

[4]Since nonbasic variables are automatically zero, the current BF solution is fully identified by recording just the values of the basic variables. We shall use this convention from now on.

actually is feasible only for a revised version of the problem, so a number of iterations are needed to drive these artificial variables to zero in order to reach the real BF solutions. The transportation simplex method bypasses all this by instead using a simpler procedure to directly construct a real BF solution on a transportation simplex tableau.

Before outlining this procedure, we need to point out that the number of basic variables in any basic solution of a transportation problem is one fewer than you might expect. Ordinarily, there is one basic variable for each functional constraint in a linear programming problem. For transportation problems with m sources and n destinations, the number of functional constraints is $m + n$. However,

Number of basic variables = $m + n - 1$.

The reason is that the functional constraints are equality constraints, and this set of $m + n$ equations has one *extra* (or *redundant*) equation that can be deleted without changing the feasible region; i.e., any one of the constraints is automatically satisfied whenever the other $m + n - 1$ constraints are satisfied. (This fact can be verified by showing that any supply constraint exactly equals the sum of the demand constraints minus the sum of the *other* supply constraints, and that any demand equation also can be reproduced by summing the supply equations and subtracting the other demand equations. See Prob. 8.2-19.) Therefore, any *BF solution* appears on a transportation simplex tableau with exactly $m + n - 1$ circled *nonnegative* allocations, where the sum of the allocations for each row or column equals its supply or demand.[5]

The procedure for constructing an initial BF solution selects the $m + n - 1$ basic variables one at a time. After each selection, a value that will satisfy one additional constraint (thereby eliminating that constraint's row or column from further consideration for providing allocations) is assigned to that variable. Thus, after $m + n - 1$ selections, an entire basic solution has been constructed in such a way as to satisfy all the constraints. A number of different criteria have been proposed for selecting the basic variables. We present and illustrate three of these criteria here, after outlining the general procedure.

General Procedure[6] for Constructing an Initial BF Solution.

To begin, all source rows and destination columns of the transportation simplex tableau are initially under consideration for providing a basic variable (allocation).

1. From the rows and columns still under consideration, select the next basic variable (allocation) according to some criterion.
2. Make that allocation large enough to exactly use up the remaining supply in its row or the remaining demand in its column (whichever is smaller).
3. Eliminate that row or column (whichever had the smaller remaining supply or demand) from further consideration. (If the row and column have the same remaining supply and demand, then arbitrarily select the *row* as the one to be eliminated. The column will be used later to provide a *degenerate* basic variable, i.e., a circled allocation of zero.)
4. If only one row or only one column remains under consideration, then the procedure is completed by selecting every *remaining* variable (i.e., those variables that were neither previously selected to be basic nor eliminated from consideration by eliminating

---

[5]However, note that any feasible solution with $m + n - 1$ nonzero variables is *not necessarily* a basic solution because it might be the weighted average of two or more degenerate BF solutions (i.e., BF solutions having some basic variables equal to zero). We need not be concerned about mislabeling such solutions as being basic, however, because the transportation simplex method constructs only legitimate BF solutions.

[6]In Sec. 4.1 we pointed out that the simplex method is an example of the algorithms (systematic solution procedures) so prevalent in OR work. Note that this procedure also is an algorithm, where each successive execution of the (four) steps constitutes an iteration.

their row or column) associated with that row or column to be basic with the only feasible allocation. Otherwise, return to step 1.

Alternative Criteria for Step 1

1. *Northwest corner rule:* Begin by selecting x_{11} (that is, start in the northwest corner of the transportation simplex tableau). Thereafter, if x_{ij} was the last basic variable selected, then next select $x_{i,j+1}$ (that is, move one column to the *right*) if source i has any supply remaining. Otherwise, next select $x_{i+1,j}$ (that is, move one row *down*).

Example. To make this description more concrete, we now illustrate the general procedure on the Metro Water District problem (see Table 8.12) with the northwest corner rule being used in step 1. Because $m = 4$ and $n = 5$ in this case, the procedure would find an initial BF solution having $m + n - 1 = 8$ basic variables.

As shown in Table 8.16, the first allocation is $x_{11} = 30$, which exactly uses up the demand in column 1 (and eliminates this column from further consideration). This first iteration leaves a supply of 20 remaining in row 1, so next select $x_{1,1+1} = x_{12}$ to be a basic variable. Because this supply is no larger than the demand of 20 in column 2, all of it is allocated, $x_{12} = 20$, and this row is eliminated from further consideration. (Row 1 is chosen for elimination rather than column 2 because of the parenthetical instruction in step 3.) Therefore, select $x_{1+1,2} = x_{22}$ next. Because the remaining demand of 0 in column 2 is less than the supply of 60 in row 2, allocate $x_{22} = 0$ and eliminate column 2.

Continuing in this manner, we eventually obtain the entire *initial BF solution* shown in Table 8.16, where the circled numbers are the values of the basic variables ($x_{11} = 30, \ldots, x_{45} = 50$) and all the other variables (x_{13}, etc.) are nonbasic variables equal to zero. Arrows have been added to show the order in which the basic variables (allocations) were selected. The value of Z for this solution is

$$Z = 16(30) + 16(20) + \cdots + 0(50) = 2{,}470 + 10M.$$

2. *Vogel's approximation method:* For each row and column remaining under consideration, calculate its **difference,** which is defined as *the arithmetic difference between the smallest and next-to-the-smallest unit cost c_{ij} still remaining in that row or column.* (If two unit costs tie for being the smallest remaining in a row or column, then

■ **TABLE 8.16** Initial BF solution from the Northwest Corner Rule

		Destination					Supply	u_i
		1	2	3	4	5		
Source	1	16 ⓈⓄ(30) → 16 (20)		13	22	17	50	
	2	14	14 (0) →	13 (60)	19	15	60	
	3	19	19	20 (10) →	23 (30) →	M (10) →	50	
	4(D)	M	0	M	0	0 (50)	50	
Demand		30	20	70	30	60	$Z = 2{,}470 + 10M$	
v_j								

the *difference* is 0.) In that row or column having the *largest difference,* select the variable having the *smallest remaining unit cost.* (Ties for the largest difference, or for the smallest remaining unit cost, may be broken arbitrarily.)

Example.   Now let us apply the general procedure to the Metro Water District problem by using the criterion for Vogel's approximation method to select the next basic variable in step 1. With this criterion, it is more convenient to work with parameter tables (rather than with complete transportation simplex tableaux), beginning with the one shown in Table 8.12. At each iteration, after the difference for every row and column remaining under consideration is calculated and displayed, the largest difference is circled and the smallest unit cost in its row or column is enclosed in a box. The resulting selection (and value) of the variable having this unit cost as the next basic variable is indicated in the lower right-hand corner of the current table, along with the row or column thereby being eliminated from further consideration (see steps 2 and 3 of the general procedure). The table for the next iteration is exactly the same except for deleting this row or column and subtracting the last allocation from its supply or demand (whichever remains).

Applying this procedure to the Metro Water District problem yields the sequence of parameter tables shown in Table 8.17, where the resulting initial BF solution consists of the eight basic variables (allocations) given in the lower right-hand corner of the respective parameter tables.

This example illustrates two relatively subtle features of the general procedure that warrant special attention. First, note that the final iteration selects *three* variables (x_{31}, x_{32}, and x_{33}) to become basic instead of the single selection made at the other iterations. The reason is that only *one* row (row 3) remains under consideration at this point. Therefore, step 4 of the general procedure says to select *every* remaining variable associated with row 3 to be basic.

Second, note that the allocation of $x_{23} = 20$ at the next-to-last iteration exhausts *both* the remaining supply in its row *and* the remaining demand in its column. However, rather than eliminate both the row and column from further consideration, step 3 says to eliminate *only the row,* saving the column to provide a *degenerate* basic variable later. Column 3 is, in fact, used for just this purpose at the final iteration when $x_{33} = 0$ is selected as one of the basic variables. For another illustration of this same phenomenon, see Table 8.16 where the allocation of $x_{12} = 20$ results in eliminating only row 1, so that column 2 is saved to provide a degenerate basic variable, $x_{22} = 0$, at the next iteration.

Although a zero allocation might seem irrelevant, it actually plays an important role. You will see soon that the transportation simplex method must know *all* $m + n - 1$ basic variables, including those with value zero, in the current BF solution.

3. *Russell's approximation method:* For each source row i remaining under consideration, determine its $\bar{u}_i$, which is the largest unit cost c_{ij} still remaining in that row. For each destination column j remaining under consideration, determine its $\bar{v}_j$, which is the largest unit cost c_{ij} still remaining in that column. For each variable x_{ij} not previously selected in these rows and columns, calculate $\Delta_{ij} = c_{ij} - \bar{u}_i - \bar{v}_j$. Select the variable having the *largest* (in absolute terms) *negative* value of Δ_{ij}. (Ties may be broken arbitrarily.)

Example.   Using the criterion for Russell's approximation method in step 1, we again apply the general procedure to the Metro Water District problem (see Table 8.12). The results, including the sequence of basic variables (allocations), are shown in Table 8.18.

At iteration 1, the largest unit cost in row 1 is $\bar{u}_1 = 22$, the largest in column 1 is $\bar{v}_1 = M$, and so forth. Thus,

$$\Delta_{11} = c_{11} - \bar{u}_1 - \bar{v}_1 = 16 - 22 - M = -6 - M.$$

■ **TABLE 8.17** Initial BF solution from Vogel's approximation method

| | | Destination | | | | | | Row |
		1	2	3	4	5	Supply	Difference
	1	16	16	13	22	17	50	3
Source	2	14	14	13	19	15	60	1
	3	19	19	20	23	M	50	0
	4(D)	M	0	M	[0]	0	50	0
Demand		30	20	70	30	60	Select $x_{44} = 30$	
Column difference		2	14	0	(19)	15	Eliminate column 4	

| | | Destination | | | | | Row |
		1	2	3	5	Supply	Difference
	1	16	16	13	17	50	3
Source	2	14	14	13	15	60	1
	3	19	19	20	M	50	0
	4(D)	M	0	M	[0]	20	0
Demand		30	20	70	60	Select $x_{45} = 20$	
Column difference		2	14	0	(15)	Eliminate row 4(D)	

| | | Destination | | | | | Row |
		1	2	3	5	Supply	Difference
	1	16	16	[13]	17	50	(3)
Source	2	14	14	13	15	60	1
	3	19	19	20	M	50	0
Demand		30	20	70	40	Select $x_{13} = 50$	
Column difference		2	2	0	2	Eliminate row 1	

| | | Destination | | | | Row |
		1	2	3	5	Supply	Difference
	2	14	14	13	[15]	60	1
Source	3	19	19	20	M	50	0
Demand		30	20	20	40	Select $x_{25} = 40$	
Column difference		5	5	7	(M − 15)	Eliminate column 5	

| | | Destination | | | | Row |
		1	2	3	Supply	Difference
	2	14	14	[13]	20	1
Source	3	19	19	20	50	0
Demand		30	20	20	Select $x_{23} = 20$	
Column difference		5	5	(7)	Eliminate row 2	

| | | Destination | | | |
		1	2	3	Supply
Source	3	19	19	20	50
Demand		30	20	0	Select $x_{31} = 30$
					$x_{32} = 20$
					$x_{33} = 0$

$Z = 2,460$

Calculating all the Δ_{ij} values for $i = 1, 2, 3, 4$ and $j = 1, 2, 3, 4, 5$ shows that $\Delta_{45} = 0 - 2M$ has the largest negative value, so $x_{45} = 50$ is selected as the first basic variable (allocation). This allocation exactly uses up the supply in row 4, so this row is eliminated from further consideration.

Note that eliminating this row changes $\bar{v}_1$ and $\bar{v}_3$ for the next iteration. Therefore, the second iteration requires recalculating the Δ_{ij} with $j = 1, 3$ as well as eliminating $i = 4$. The largest negative value now is

$$\Delta_{15} = 17 - 22 - M = -5 - M,$$

so $x_{15} = 10$ becomes the second basic variable (allocation), eliminating column 5 from further consideration.

The subsequent iterations proceed similarly, but you may want to test your understanding by verifying the remaining allocations given in Table 8.18. As with the other procedures in this (and other) section(s), you should find your IOR Tutorial useful for doing the calculations involved and illuminating the approach. (See the interactive procedure for finding an initial BF solution.)

Comparison of Alternative Criteria for Step 1.
Now let us compare these three criteria for selecting the next basic variable. The main virtue of the northwest corner rule is that it is quick and easy. However, because it pays no attention to unit costs c_{ij}, usually the solution obtained will be far from optimal. (Note in Table 8.16 that $x_{35} = 10$ even though $c_{35} = M$.) Expending a little more effort to find a good initial BF solution might greatly reduce the number of iterations then required by the transportation simplex method to reach an optimal solution (see Probs. 8.2-7 and 8.2-9). Finding such a solution is the objective of the other two criteria.

Vogel's approximation method has been a popular criterion for many years,[7] partially because it is relatively easy to implement by hand. Because the *difference* represents the minimum extra unit cost incurred by failing to make an allocation to the cell having the smallest unit cost in that row or column, this criterion does take costs into account in an effective way.

Russell's approximation method provides another excellent criterion[8] that is still quick to implement on a computer (but not manually). Although it is unclear as to which is more

TABLE 8.18 Initial BF solution from Russell's approximation method

Iteration	$\bar{u}_1$	$\bar{u}_2$	$\bar{u}_3$	$\bar{u}_4$	$\bar{v}_1$	$\bar{v}_2$	$\bar{v}_3$	$\bar{v}_4$	$\bar{v}_5$	Largest Negative Δ_{ij}	Allocation
1	22	19	M	M	M	19	M	23	M	$\Delta_{45} = -2M$	$x_{45} = 50$
2	22	19	M		19	19	20	23	M	$\Delta_{15} = -5 - M$	$x_{15} = 10$
3	22	19	23		19	19	20	23		$\Delta_{13} = -29$	$x_{13} = 40$
4		19	23		19	19	20	23		$\Delta_{23} = -26$	$x_{23} = 30$
5		19	23		19	19		23		$\Delta_{21} = -24^*$	$x_{21} = 30$
6										Irrelevant	$x_{31} = 0$
											$x_{32} = 20$
											$x_{34} = 30$
											$Z = 2,570$

*Tie with $\Delta_{22} = -24$ broken arbitrarily.

[7]N. V. Reinfeld and W. R. Vogel: *Mathematical Programming,* Prentice-Hall, Englewood Cliffs, NJ, 1958.

[8]E. J. Russell: "Extension of Dantzig's Algorithm to Finding an Initial Near-Optimal Basis for the Transportation Problem," *Operations Research,* **17**: 187–191, 1969.

effective *on average,* this criterion *frequently* does obtain a better solution than Vogel's. (For the example, Vogel's approximation method happened to find the optimal solution with $Z = 2,460$, whereas Russell's misses slightly with $Z = 2,570$.) For a large problem, it may be worthwhile to apply both criteria and then use the better solution to start the iterations of the transportation simplex method.

One distinct advantage of Russell's approximation method is that it is patterned directly after step 1 for the transportation simplex method (as you will see soon), which somewhat simplifies the overall computer code. In particular, the $\bar{u}_i$ and $\bar{v}_j$ values have been defined in such a way that the relative values of the $c_{ij} - \bar{u}_i - \bar{v}_j$ *estimate* the relative values of $c_{ij} - u_i - v_j$ that will be obtained when the transportation simplex method reaches an optimal solution.

We now shall use the initial BF solution obtained in Table 8.18 by Russell's approximation method to illustrate the remainder of the transportation simplex method. Thus, our *initial transportation simplex tableau* (before we solve for u_i and v_j) is shown in Table 8.19.

The next step is to check whether this initial solution is optimal by applying the *optimality test*.

Optimality Test

Using the notation of Table 8.14, we can reduce the standard optimality test for the simplex method (see Sec. 4.3) to the following for the transportation problem:

> **Optimality test:** A BF solution is optimal if and only if $c_{ij} - u_i - v_j \geq 0$ for every (i, j) such that x_{ij} is nonbasic.[9]

Thus, the only work required by the optimality test is the derivation of the values of u_i and v_j for the current BF solution and then the calculation of these $c_{ij} - u_i - v_j$, as described below.

■ **TABLE 8.19** Initial transportation simplex tableau (before we obtain $c_{ij} - u_i - v_j$) from Russell's approximation method

Iteration 0		Destination					Supply	u_i
		1	2	3	4	5		
Source	1	16	16	13 (40)	22	17 (10)	50	
	2	14 (30)	14	13 (30)	19	15	60	
	3	19 (0)	19 (20)	20	23 (30)	M	50	
	4(D)	M	0	M	0	0 (50)	50	
Demand		30	20	70	30	60	$Z = 2,570$	
v_j								

[9]The one exception is that two or more equivalent degenerate BF solutions (i.e., identical solutions having different degenerate basic variables equal to zero) can be optimal with only some of these basic solutions satisfying the optimality test. This exception is illustrated later in the example (see the identical solutions in the last two tableaux of Table 8.23, where only the latter solution satisfies the criterion for optimality).

Since $c_{ij} - u_i - v_j$ is required to be zero if x_{ij} is a basic variable, u_i and v_j satisfy the set of equations

$$c_{ij} = u_i + v_j \qquad \text{for each } (i, j) \text{ such that } x_{ij} \text{ is basic.}$$

There are $m + n - 1$ basic variables, and so there are $m + n - 1$ of these equations. Since the number of unknowns (the u_i and v_j) is $m + n$, one of these variables can be assigned a value arbitrarily without violating the equations. The choice of this one variable and its value does not affect the value of any $c_{ij} - u_i - v_j$, even when x_{ij} is nonbasic, so the only (minor) difference it makes is in the ease of solving these equations. A convenient choice for this purpose is to select the u_i that has the *largest number of allocations in its row* (break any tie arbitrarily) and to assign to it the value zero. Because of the simple structure of these equations, it is then very simple to solve for the remaining variables algebraically.

To demonstrate, we give each equation that corresponds to a basic variable in our initial BF solution.

x_{31}:	$19 = u_3 + v_1$.	Set $u_3 = 0$, so $v_1 = 19$,
x_{32}:	$19 = u_3 + v_2$.	$v_2 = 19$,
x_{34}:	$23 = u_3 + v_4$.	$v_4 = 23$.
x_{21}:	$14 = u_2 + v_1$.	Know $v_1 = 19$, so $u_2 = -5$.
x_{23}:	$13 = u_2 + v_3$.	Know $u_2 = -5$, so $v_3 = 18$.
x_{13}:	$13 = u_1 + v_3$.	Know $v_3 = 18$, so $u_1 = -5$.
x_{15}:	$17 = u_1 + v_5$.	Know $u_1 = -5$, so $v_5 = 22$.
x_{45}:	$0 = u_4 + v_5$.	Know $v_5 = 22$, so $u_4 = -22$.

Setting $u_3 = 0$ (since row 3 of Table 8.19 has the largest number of allocations—3) and moving down the equations one at a time immediately give the derivation of values for the unknowns shown to the right of the equations. (Note that this derivation of the u_i and v_j values depends on which x_{ij} variables are *basic variables* in the current BF solution, so this derivation will need to be repeated each time a new BF solution is obtained.)

Once you get the hang of it, you probably will find it even more convenient to solve these equations without writing them down by working directly on the transportation simplex tableau. Thus, in Table 8.19 you begin by writing in the value $u_3 = 0$ and then picking out the circled allocations (x_{31}, x_{32}, x_{34}) in that row. For each one you set $v_j = c_{3j}$ and then look for circled allocations (except in row 3) in these columns (x_{21}). Mentally calculate $u_2 = c_{21} - v_1$, pick out x_{23}, set $v_3 = c_{23} - u_2$, and so on until you have filled in all the values for u_i and v_j. (Try it.) Then calculate and fill in the value of $c_{ij} - u_i - v_j$ for each nonbasic variable x_{ij} (that is, for each cell without a circled allocation), and you will have the completed initial transportation simplex tableau shown in Table 8.20.

We are now in a position to apply the optimality test by checking the values of $c_{ij} - u_i - v_j$ given in Table 8.20. Because two of these values ($c_{25} - u_2 - v_5 = -2$ and $c_{44} - u_4 - v_4 = -1$) are negative, we conclude that the current BF solution is not optimal. Therefore, the transportation simplex method must next go to an iteration to find a better BF solution.

An Iteration

As with the full-fledged simplex method, an iteration for this streamlined version must determine an entering basic variable (step 1), a leaving basic variable (step 2), and then identify the resulting new BF solution (step 3).

■ **TABLE 8.20** Completed initial transportation simplex tableau

Iteration 0	Destination					Supply	u_i
	1	**2**	**3**	**4**	**5**		
1	16 +2	16 +2	13 (40)	22 +4	17 (10)	50	−5
2	14 (30)	14 0	13 (30)	19 +1	15 −2	60	−5
Source 3	19 (0)	19 (20)	20 +2	23 (30)	M M − 22	50	0
4(D)	M M + 3	0 +3	M M + 4	0 −1	0 (50)	50	−22
Demand	30	20	70	30	60	Z = 2,570	
v_j	19	19	18	23	22		

Step 1: Find the Entering Basic Variable.   Since $c_{ij} - u_i - v_j$ represents the rate at which the objective function will change as the nonbasic variable x_{ij} is increased, the entering basic variable must have a *negative* $c_{ij} - u_i - v_j$ value to decrease the total cost Z. Thus, the candidates in Table 8.20 are x_{25} and x_{44}. To choose between the candidates, select the one having the larger (in absolute terms) negative value of $c_{ij} - u_i - v_j$ to be the entering basic variable, which is x_{25} in this case.

Step 2: Find the Leaving Basic Variable.   Increasing the entering basic variable from zero sets off a *chain reaction* of compensating changes in other basic variables (allocations), in order to continue satisfying the supply and demand constraints. The first basic variable to be decreased to zero then becomes the leaving basic variable.

With x_{25} as the entering basic variable, the chain reaction in Table 8.20 is the relatively simple one summarized in Table 8.21. (We shall always indicate the entering basic variable by placing a boxed plus sign in the center of its cell while leaving the corresponding value of $c_{ij} - u_i - v_j$ in the lower right-hand corner of this cell.) Increasing x_{25} by some amount requires decreasing x_{15} by the same amount to restore the demand of 60 in column 5. This change then requires increasing x_{13} by this same amount to restore the

■ **TABLE 8.21** Part of initial transportation simplex tableau showing the chain reaction caused by increasing the entering basic variable x_{25}

	Destination			Supply
	3	**4**	**5**	
Source 1	13 (40)+	22 +4	17 (10)−	50
2	13 (30)−	19 +1	15 [+] −2	60
	...	...	...	
Demand	70	30	60	

supply of 50 in row 1. This change then requires decreasing x_{23} by this amount to restore the demand of 70 in column 3. This decrease in x_{23} successfully completes the chain reaction because it also restores the supply of 60 in row 2. (Equivalently, we could have started the chain reaction by restoring this supply in row 2 with the decrease in x_{23}, and then the chain reaction would continue with the increase in x_{13} and decrease in x_{15}.)

The net result is that cells (2, 5) and (1, 3) become **recipient cells,** each receiving its additional allocation from one of the **donor cells,** (1, 5) and (2, 3). (These cells are indicated in Table 8.21 by the plus and minus signs.) Note that cell (1, 5) had to be the donor cell for column 5 rather than cell (4, 5), because cell (4, 5) would have no recipient cell in row 4 to continue the chain reaction. [Similarly, if the chain reaction had been started in row 2 instead, cell (2, 1) could not be the donor cell for this row because the chain reaction could not then be completed successfully after necessarily choosing cell (3, 1) as the next recipient cell and either cell (3, 2) or (3, 4) as its donor cell.] Also note that, except for the entering basic variable, *all* recipient cells and donor cells in the chain reaction must correspond to *basic* variables in the current BF solution.

Each donor cell decreases its allocation by exactly the same amount as the entering basic variable (and other recipient cells) is increased. Therefore, the donor cell that starts with the smallest allocation—cell (1, 5) in this case (since $10 < 30$ in Table 8.21)—must reach a zero allocation first as the entering basic variable x_{25} is increased. Thus, x_{15} becomes the leaving basic variable.

In general, there always is just *one* chain reaction (in either direction) that can be completed successfully to maintain feasibility when the entering basic variable is increased from zero. This chain reaction can be identified by selecting from the cells having a basic variable: first the donor cell in the *column* having the entering basic variable, then the recipient cell in the row having this donor cell, then the donor cell in the column having this recipient cell, and so on until the chain reaction yields a donor cell in the *row* having the entering basic variable. When a column or row has more than one additional basic variable cell, it may be necessary to trace them all further to see which one must be selected to be the donor or recipient cell. (All but this one eventually will reach a dead end in a row or column having no additional basic variable cell.) After the chain reaction is identified, *the donor cell having the smallest allocation automatically provides the leaving basic variable.* (In the case of a tie for the donor cell having the smallest allocation, any one can be chosen arbitrarily to provide the leaving basic variable.)

Step 3: Find the New BF Solution.

The *new BF solution* is identified simply by adding the value of the leaving basic variable (before any change) to the allocation for each recipient cell and subtracting *this same amount* from the allocation for each donor cell. In Table 8.21 the value of the leaving basic variable x_{15} is 10, so the portion of the transportation simplex tableau in this table changes as shown in Table 8.22 for the new solution. (Since x_{15} is nonbasic in the new solution, its new allocation of zero is no longer shown in this new tableau.)

We can now highlight a useful interpretation of the $c_{ij} - u_i - v_j$ quantities derived during the optimality test. Because of the shift of 10 allocation units from the donor cells to the recipient cells (shown in Tables 8.21 and 8.22), the total cost changes by

$$\Delta Z = 10(15 - 17 + 13 - 13) = 10(-2) = 10(c_{25} - u_2 - v_5).$$

Thus, the effect of increasing the entering basic variable x_{25} from zero has been a cost change at the rate of -2 per unit increase in x_{25}. This is precisely what the value of $c_{25} - u_2 - v_5 = -2$ in Table 8.20 indicates would happen. In fact, another (but less efficient) way of deriving $c_{ij} - u_i - v_j$ for each nonbasic variable x_{ij} is to identify the chain reaction caused by increasing this variable from 0 to 1 and then to calculate the resulting

■ **TABLE 8.22** Part of second transportation simplex tableau showing the changes in the BF solution

		Destination			Supply
		3	4	5	Supply
	1	...	13 (50)	22	17	50
Source	2	...	13 (20)	19	15 (10)	60
		...	...	...	...	
Demand		70	30	60	

cost change. This intuitive interpretation sometimes is useful for checking calculations during the optimality test.

Before completing the solution of the Metro Water District problem, we now summarize the rules for the transportation simplex method.

Summary of the Transportation Simplex Method

Initialization: Construct an initial BF solution by the procedure outlined earlier in this section. Go to the optimality test.

Optimality test: Derive u_i and v_j by selecting the row having the largest number of allocations, setting its $u_i = 0$, and then solving the set of equations $c_{ij} = u_i + v_j$ for each (i, j) such that x_{ij} is basic. If $c_{ij} - u_i - v_j \geq 0$ for every (i, j) such that x_{ij} is *nonbasic,* then the current solution is optimal, so stop. Otherwise, go to an iteration.

Iteration:

1. Determine the entering basic variable: Select the nonbasic variable x_{ij} having the *largest* (in absolute terms) *negative* value of $c_{ij} - u_i - v_j$.
2. Determine the leaving basic variable: Identify the chain reaction required to retain feasibility when the entering basic variable is increased. From the donor cells, select the basic variable having the *smallest* value.
3. Determine the new BF solution: Add the value of the leaving basic variable to the allocation for each recipient cell. Subtract this value from the allocation for each donor cell.

Continuing to apply this procedure to the Metro Water District problem yields the complete set of transportation simplex tableaux shown in Table 8.23. Since all the $c_{ij} - u_i - v_j$ values are nonnegative in the fourth tableau, the optimality test identifies the set of allocations in this tableau as being optimal, which concludes the algorithm.

It would be good practice for you to derive the values of u_i and v_j given in the second, third, and fourth tableaux. Try doing this by working directly on the tableaux. Also check out the chain reactions in the second and third tableaux, which are somewhat more complicated than the one you have seen in Table 8.21.

Special Features of This Example

Note three special points that are illustrated by this example. First, the initial BF solution is *degenerate* because the basic variable $x_{31} = 0$. However, this degenerate basic variable causes no complication, because cell (3, 1) becomes a *recipient cell* in the second tableau, which increases x_{31} to a value greater than zero.

■ **TABLE 8.23** Complete set of transportation simplex tableaux for the Metro Water District problem

Iteration 0

Source	Destination 1	Destination 2	Destination 3	Destination 4	Destination 5	Supply	u_i
1	16 (+2)	16 (+2)	13 (40) +	22 (+4)	17 (10) −	50	−5
2	14 (30)	14 (0)	13 (30) −	19 (+1)	15 (+) −2	60	−5
3	19 (0)	19 (20)	20 (+2)	23 (30)	M (M − 22)	50	0
4(D)	M (M + 3)	0 (+3)	M (M + 4)	0 (−1)	0 (50)	50	−22
Demand	30	20	70	30	60	Z = 2,570	
v_j	19	19	18	23	22		

Iteration 1

Source	Destination 1	Destination 2	Destination 3	Destination 4	Destination 5	Supply	u_i
1	16 (+2)	16 (+2)	13 (50)	22 (+4)	17 (+2)	50	−5
2	14 (30) −	14 (0)	13 (20)	19 (+1)	15 (10) +	60	−5
3	19 (0) +	19 (20)	20 (+2)	23 (30) −	M (M − 20)	50	0
4(D)	M (M + 1)	0 (+1)	M (M + 2)	0 (+) −3	0 (50) −	50	−20
Demand	30	20	70	30	60	Z = 2,550	
v_j	19	19	18	23	20		

Iteration 2

Source	Destination 1	Destination 2	Destination 3	Destination 4	Destination 5	Supply	u_i
1	16 (+5)	16 (+5)	13 (50)	22 (+7)	17 (+2)	50	−8
2	14 (+3)	14 (+3)	13 (20) −	19 (+4)	15 (40) +	60	−8
3	19 (30)	19 (20)	20 (+) −1	23 (0) −	M (M − 23)	50	0
4(D)	M (M + 4)	0 (+4)	M (M + 2)	0 (30) +	0 (20) −	50	−23
Demand	30	20	70	30	60	Z = 2,460	
v_j	19	19	21	23	23		

■ **TABLE 8.23** (*Continued*)

Iteration 3		Destination					Supply	u_i
		1	2	3	4	5		
Source	1	16 +4	16 +4	13 (50)	22 +7	17 +2	50	−7
	2	14 +2	14 +2	13 (20)	19 +4	15 (40)	60	−7
	3	19 (30)	19 (20)	20 (0)	23 +1	M M−22	50	0
	4(D)	M M+3	0 +3	M M+2	0 (30)	0 (20)	50	−22
Demand		30	20	70	30	60	Z = 2,460	
v_j		19	19	20	22	22		

Second, another degenerate basic variable (x_{34}) arises in the third tableau because the basic variables for *two* donor cells in the second tableau, cells (2, 1) and (3, 4), *tie* for having the smallest value (30). (This tie is broken arbitrarily by selecting x_{21} as the leaving basic variable; if x_{34} had been selected instead, then x_{21} would have become the degenerate basic variable.) This degenerate basic variable does appear to create a complication subsequently, because cell (3, 4) becomes a *donor cell* in the third tableau but has nothing to donate! Fortunately, such an event actually gives no cause for concern. Since zero is the amount to be added to or subtracted from the allocations for the recipient and donor cells, these allocations do not change. However, the degenerate basic variable does become the leaving basic variable, so it is replaced by the entering basic variable as the circled allocation of zero in the fourth tableau. This change in the set of basic variables changes the values of u_i and v_j. Therefore, if any of the $c_{ij} − u_i − v_j$ had been negative in the fourth tableau, the algorithm would have gone on to make *real* changes in the allocations (whenever all donor cells have nondegenerate basic variables).

Third, because none of the $c_{ij} − u_i − v_j$ turned out to be negative in the fourth tableau, the equivalent set of allocations in the third tableau is optimal also. Thus, the algorithm executed one more iteration than was necessary. This extra iteration is a flaw that occasionally arises in both the transportation simplex method and the simplex method because of degeneracy, but it is not sufficiently serious to warrant any adjustments to these algorithms.

If you would like to see additional (smaller) examples of the application of the transportation simplex method, two are available. One is the demonstration provided for the transportation problem area in your OR Tutor. In addition, the Worked Examples section of the book's website includes **another example** of this type. Also provided in your IOR Tutorial are both an interactive procedure and an automatic procedure for the transportation simplex method.

Now that you have studied the transportation simplex method, you are in a position to check for yourself how the algorithm actually provides a proof of the *integer solutions property* presented in Sec. 8.1. Problem 8.2-20 helps to guide you through the reasoning.

■ 8.3 THE ASSIGNMENT PROBLEM

The **assignment problem** is a special type of linear programming problem where **assignees** are being assigned to perform **tasks.** For example, the assignees might be employees who need to be given work assignments. Assigning people to jobs is a common application of the assignment problem.[10] However, the assignees need not be people. They also could be machines, or vehicles, or plants, or even time slots to be assigned tasks. The first example below involves machines being assigned to locations, so the tasks in this case simply involve holding a machine. A subsequent example involves plants being assigned products to be produced.

To fit the definition of an assignment problem, these kinds of applications need to be formulated in a way that satisfies the following assumptions.

1. The number of assignees and the number of tasks are the same. (This number is denoted by n.)
2. Each assignee is to be assigned to exactly *one* task.
3. Each task is to be performed by exactly *one* assignee.
4. There is a cost c_{ij} associated with assignee i ($i = 1, 2, \ldots, n$) performing task j ($j = 1, 2, \ldots, n$).
5. The objective is to determine how all n assignments should be made to minimize the total cost.

Any problem satisfying all these assumptions can be solved extremely efficiently by algorithms designed specifically for assignment problems.

The first three assumptions are fairly restrictive. Many potential applications do not quite satisfy these assumptions. However, it often is possible to reformulate the problem to make it fit. For example, *dummy assignees* or *dummy tasks* frequently can be used for this purpose. We illustrate these formulation techniques in the examples.

Prototype Example

The JOB SHOP COMPANY has purchased three new machines of different types. There are four available locations in the shop where a machine could be installed. Some of these locations are more desirable than others for particular machines because of their proximity to work centers that will have a heavy work flow to and from these machines. (There will be no work flow *between* the new machines.) Therefore, the objective is to assign the new machines to the available locations to minimize the total cost of materials handling. The estimated cost in dollars per hour of materials handling involving each of the machines is given in Table 8.24 for the respective locations. Location 2 is not considered suitable for machine 2, so no cost is given for this case.

To formulate this problem as an assignment problem, we must introduce a *dummy machine* for the extra location. Also, an extremely large cost M should be attached to the assignment of machine 2 to location 2 to prevent this assignment in the optimal solution. The resulting assignment problem *cost table* is shown in Table 8.25. This cost table contains all the necessary data for solving the problem. The optimal solution is to assign machine 1 to location 4, machine 2 to location 3, and machine 3 to location 1, for a total cost of $29 per hour. The dummy machine is assigned to location 2, so this location is available for some future real machine.

---

[10]For example, see L. J. LeBlanc, D. Randels, Jr., and T. K. Swann: "Heery International's Spreadsheet Optimization Model for Assigning Managers to Construction Projects," *Interfaces*, **30**(6): 95–106, Nov.–Dec. 2000. Page 98 of this article also cites seven other applications of the assignment problem.

TABLE 8.24 Materials-handling cost data ($) for Job Shop Co.

		Location			
		1	**2**	**3**	**4**
	1	13	16	12	11
Machine	2	15	—	13	20
	3	5	7	10	6

TABLE 8.25 Cost table for the Job Shop Co. assignment problem

		Task (Location)			
		1	**2**	**3**	**4**
	1	13	16	12	11
Assignee	2	15	M	13	20
(Machine)	3	5	7	10	6
	4(D)	0	0	0	0

We shall discuss how this solution is obtained after we formulate the mathematical model for the general assignment problem.

The Assignment Problem Model

The mathematical model for the assignment problem uses the following decision variables:

$$x_{ij} = \begin{cases} 1 & \text{if assignee } i \text{ performs task } j, \\ 0 & \text{if not,} \end{cases}$$

for $i = 1, 2, \ldots, n$ and $j = 1, 2, \ldots, n$. Thus, each x_{ij} is a *binary variable* (it has value 0 or 1). As discussed at length in the chapter on integer programming (Chap. 11), binary variables are important in OR for representing *yes/no decisions*. In this case, the yes/no decision is: Should assignee i perform task j?

By letting Z denote the total cost, the assignment problem model is

$$\text{Minimize} \quad Z = \sum_{i=1}^{n} \sum_{j=1}^{n} c_{ij} x_{ij},$$

subject to

$$\sum_{j=1}^{n} x_{ij} = 1 \quad \text{for } i = 1, 2, \ldots, n,$$

$$\sum_{i=1}^{n} x_{ij} = 1 \quad \text{for } j = 1, 2, \ldots, n,$$

and

$$x_{ij} \geq 0, \quad \text{for all } i \text{ and } j$$
$$(x_{ij} \text{ binary,} \quad \text{for all } i \text{ and } j).$$

The first set of functional constraints specifies that each assignee is to perform exactly one task, whereas the second set requires each task to be performed by exactly one assignee. If we delete the parenthetical restriction that the x_{ij} be binary, the model clearly is a special type of linear programming problem and so can be readily solved. Fortunately, for reasons about to unfold, we *can* delete this restriction. (This deletion is the reason that the assignment problem appears in this chapter rather than in the integer programming chapter.)

Now compare this model (without the binary restriction) with the transportation problem model presented in the third subsection of Sec. 8.1 (including Table 8.6). Note how similar their structures are. In fact, the assignment problem is just a special type of transportation problem where the *sources* now are *assignees* and the *destinations* now are *tasks* and where

Number of sources m = number of destinations n,

Every supply $s_i = 1$,

Every demand $d_j = 1$.

Now focus on the **integer solutions property** in the subsection on the transportation problem model. Because s_i and d_j are integers (= 1) now, this property implies that *every BF solution* (including an optimal one) is an *integer* solution for an assignment problem. The functional constraints of the assignment problem model prevent any variable from being greater than 1, and the nonnegativity constraints prevent values less than 0. Therefore, by deleting the binary restriction to enable us to solve an assignment problem as a linear programming problem, the resulting BF solutions obtained (including the final optimal solution) *automatically* will satisfy the binary restriction anyway.

Just as the transportation problem has a network representation (see Fig. 8.3), the assignment problem can be depicted in a very similar way, as shown in Fig. 8.5. The first column now lists the n assignees and the second column the n tasks. Each number in a square bracket indicates the number of assignees being provided at that location in the network, so the values are automatically 1 on the left, whereas the values of -1 on the right indicate that each task is using up one assignee.

For any particular assignment problem, practitioners normally do not bother writing out the full mathematical model. It is simpler to formulate the problem by filling out a cost table (e.g., Table 8.25), including identifying the assignees and tasks, since this table contains all the essential data in a far more compact form.

Problems occasionally arise that do not quite fit the model for an assignment problem because certain assignees will be assigned to more than one task. In this case, the problem can be reformulated to fit the model by splitting each such assignee into separate (but identical) new assignees where each new assignee will be assigned to exactly one task. (Table 8.29 will illustrate this for a subsequent example.) Similarly, if a task is to be performed by multiple assignees, that task can be split into separate (but identical) new tasks where each new task is to be performed by exactly one assignee according to the reformulated model. The Worked Examples section of the book's website provides **another example** that illustrates both cases and the resulting reformulation to fit the model for an assignment problem. An alternative formulation as a transportation problem also is shown.

Solution Procedures for Assignment Problems

Alternative solution procedures are available for solving assignment problems. Problems that aren't much larger than the Job Shop Co. example can be solved very quickly by the

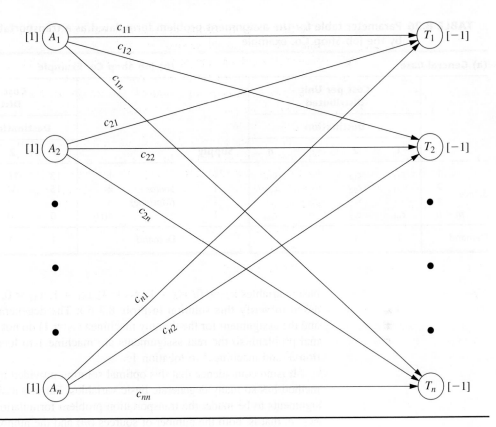

■ **FIGURE 8.5**
Network representation of
the assignment problem.

general simplex method, so it may be convenient to simply use a basic software package (such as Excel and its Solver) that only employs this method. If this were done for the Job Shop Co. problem, it would not have been necessary to add the dummy machine to Table 8.25 to make it fit the assignment problem model. The constraints on the number of machines assigned to each location would be expressed instead as

$$\sum_{i=1}^{3} x_{ij} \leq 1 \qquad \text{for } j = 1, 2, 3, 4.$$

As shown in the Excel files for this chapter, a spreadsheet formulation for this example would be very similar to the formulation for a transportation problem displayed in Fig. 8.4 except now all the supplies and demands would be 1 and the demand constraints would be ≤ 1 instead of $= 1$.

However, large assignment problems can be solved much faster by using more specialized solution procedures, so we recommend using such a procedure instead of the general simplex method for big problems.

Because the assignment problem is a special type of transportation problem, one convenient and relatively fast way to solve any particular assignment problem is to apply the transportation simplex method described in Sec. 8.2. This approach requires converting the cost table to a parameter table for the equivalent transportation problem, as shown in Table 8.26a.

For example, Table 8.26b shows the parameter table for the Job Shop Co. problem that is obtained from the cost table of Table 8.25. When the transportation simplex method is applied to this transportation problem formulation, the resulting optimal solution has

■ **TABLE 8.26** Parameter table for the assignment problem formulated as a transportation problem, illustrated by the Job Shop Co. example

(a) General Case

		Cost per Unit Distributed				
		Destination				
		1	**2**	$\cdots$	**n**	**Supply**
Source	1	c_{11}	c_{12}	$\cdots$	c_{1n}	1
	2	c_{21}	c_{22}	$\cdots$	c_{2n}	1
	$\vdots$	$\cdots$	$\cdots$	$\cdots$	$\cdots$	$\vdots$
	$m = n$	c_{n1}	c_{n2}	$\cdots$	c_{nn}	1
Demand		1	1	$\cdots$	1	

(b) Job Shop Co. Example

		Cost per Unit Distributed				
		Destination (Location)				
		1	**2**	**3**	**4**	**Supply**
Source (Machine)	1	13	16	12	11	1
	2	15	M	13	20	1
	3	5	7	10	6	1
	4(D)	0	0	0	0	1
Demand		1	1	1	1	

basic variables $x_{13} = 0$, $x_{14} = 1$, $x_{23} = 1$, $x_{31} = 1$, $x_{41} = 0$, $x_{42} = 1$, $x_{43} = 0$. (You are asked to verify this solution in Prob. 8.3-6.). The degenerate basic variables ($x_{ij} = 0$) and the assignment for the dummy machine ($x_{42} = 1$) do not mean anything for the original problem, so the real assignments are machine 1 to location 4, machine 2 to location 3, and machine 3 to location 1.

It is no coincidence that this optimal solution provided by the transportation simplex method has so many degenerate basic variables. For any assignment problem with n assignments to be made, the transportation problem formulation shown in Table 8.26a has $m = n$, that is, both the number of sources (m) and the number of destinations (n) in this formulation equal the number of assignments (n). Transportation problems in general have $m + n - 1$ basic variables (allocations), so every BF solution for this particular kind of transportation problem has $2n - 1$ basic variables, but exactly n of these x_{ij} equal 1 (corresponding to the n assignments being made). Therefore, since all the variables are binary variables, there always are $n - 1$ degenerate basic variables ($x_{ij} = 0$). As discussed at the end of Sec. 8.2, degenerate basic variables do not cause any major complication in the execution of the algorithm. However, they do frequently cause *wasted iterations,* where nothing changes (same allocations) except for the labeling of which allocations of zero correspond to degenerate basic variables rather than nonbasic variables. These wasted iterations are a major drawback to applying the transportation simplex method in this kind of situation, where there *always* are so many degenerate basic variables.

Another drawback of the transportation simplex method here is that it is purely a *general-purpose* algorithm for solving all transportation problems. Therefore, it does nothing to exploit the additional special structure in this special type of transportation problem ($m = n$, every $s_i = 1$, and every $d_j = 1$). Fortunately, specialized algorithms have been developed to fully streamline the procedure for solving just assignment problems. These algorithms operate directly on the cost table and do not bother with degenerate basic variables. When a computer code is available for one of these algorithms, it generally should be used in preference to the transportation simplex method, especially for really big problems.[11]

Section 8.4 describes one of these specialized algorithms (called the *Hungarian algorithm*) for solving only assignment problems very efficiently.

---

[11]For an article comparing various algorithms for the assignment problem, see J. L. Kennington and Z. Wang:"An Empirical Analysis of the Dense Assignment Problem: Sequential and Parallel Implementations," *ORSA Journal on Computing,* **3:** 299–306, 1991.

Your IOR Tutorial includes both an interactive procedure and an automatic procedure for applying this algorithm.

Example—Assigning Products to Plants

The BETTER PRODUCTS COMPANY has decided to initiate the production of four new products, using three plants that currently have excess production capacity. The products require a comparable production effort per unit, so the available production capacity of the plants is measured by the number of units of any product that can be produced per day, as given in the rightmost column of Table 8.27. The bottom row gives the required production rate per day to meet projected sales. Each plant can produce any of these products, *except* that Plant 2 *cannot* produce product 3. However, the variable costs per unit of each product differ from plant to plant, as shown in the main body of Table 8.27.

Management now needs to make a decision on how to split up the production of the products among plants. Two kinds of options are available.

Option 1: Permit *product splitting,* where the same product is produced in more than one plant.
Option 2: Prohibit *product splitting.*

This second option imposes a constraint that can only increase the cost of an optimal solution based on Table 8.27. On the other hand, the key advantage of Option 2 is that it eliminates some *hidden costs* associated with product splitting that are not reflected in Table 8.27, including extra setup, distribution, and administration costs. Therefore, management wants both options analyzed before a final decision is made. For Option 2, management further specifies that every plant should be assigned at least one of the products.

We will formulate and solve the model for each option in turn, where Option 1 leads to a transportation problem and Option 2 leads to an assignment problem.

Formulation of Option 1.   With product splitting permitted, Table 8.27 can be converted directly to a parameter table for a transportation problem. The plants become the sources, and the products become the destinations (or vice versa), so the supplies are the available production capacities and the demands are the required production rates. Only two changes need to be made in Table 8.27. First, because Plant 2 cannot produce product 3, such an allocation is prevented by assigning to it a huge unit cost of M. Second, the total capacity ($75 + 75 + 45 = 195$) exceeds the total required production ($20 + 30 + 30 + 40 = 120$), so a dummy destination with a demand of 75 is needed to balance these two quantities. The resulting parameter table is shown in Table 8.28.

The optimal solution for this transportation problem has basic variables (allocations) $x_{12} = 30$, $x_{13} = 30$, $x_{15} = 15$, $x_{24} = 15$, $x_{25} = 60$, $x_{31} = 20$, and $x_{34} = 25$, so

TABLE 8.27 Data for the Better Products Co. problem

		\multicolumn{4}{c}{Unit Cost ($) for Product}				Capacity Available
		1	**2**	**3**	**4**	
Plant	1	41	27	28	24	75
	2	40	29	—	23	75
	3	37	30	27	21	45
Production rate		20	30	30	40	

■ **TABLE 8.28** Parameter table for the transportation problem formulation of Option 1 for the Better Products Co. problem

| | | Cost per Unit Distributed | | | | | |
| | | Destination (Product) | | | | | |
		1	2	3	4	5(D)	Supply
Source	1	41	27	28	24	0	75
(Plant)	2	40	29	M	23	0	75
	3	37	30	27	21	0	45
Demand		20	30	30	40	75	

Plant 1 produces all of products 2 and 3.
Plant 2 produces 37.5 percent of product 4.
Plant 3 produces 62.5 percent of product 4 and all of product 1.

The total cost is $Z = \$3,260$ per day.

Formulation of Option 2.   Without product splitting, each product must be assigned to just one plant. Therefore, producing the products can be interpreted as the tasks for an assignment problem, where the plants are the assignees.

Management has specified that every plant should be assigned at least one of the products. There are more products (four) than plants (three), so one of the plants will need to be assigned two products. Plant 3 has only enough excess capacity to produce one product (see Table 8.27), so *either* Plant 1 or Plant 2 will take the extra product.

To make this assignment of an extra product possible within an assignment problem formulation, Plants 1 and 2 each are split into two assignees, as shown in Table 8.29.

The number of assignees (now five) must equal the number of tasks (now four), so a *dummy task* (product) is introduced into Table 8.29 as 5(D). The role of this dummy task is to provide the fictional second product to either Plant 1 or Plant 2, whichever one receives only one real product. There is no cost for producing a fictional product so, as usual, the cost entries for the dummy task are zero. The one exception is the entry of M in the last row of Table 8.29. The reason for M here is that Plant 3 must be assigned a real product (a choice of product 1, 2, 3, or 4), so the Big M method is needed to prevent the assignment of the fictional product to Plant 3 instead. (As in Table 8.28, M also is used to prevent the infeasible assignment of product 3 to Plant 2.)

The remaining cost entries in Table 8.29 are *not* the unit costs shown in Tables 8.27 or 8.28. Table 8.28 gives a transportation problem formulation (for Option 1), so unit costs

■ **TABLE 8.29** Cost table for the assignment problem formulation of Option 2 for the Better Products Co. problem

| | | Task (Product) | | | | |
		1	2	3	4	5(D)
Assignee	1a	820	810	840	960	0
(Plant)	1b	820	810	840	960	0
	2a	800	870	M	920	0
	2b	800	870	M	920	0
	3	740	900	810	840	M

are appropriate there, but now we are formulating an assignment problem (for Option 2). For an assignment problem, the cost c_{ij} is the *total* cost associated with assignee i performing task j. For Table 8.29, the *total cost* (per day) for Plant i to produce product j is the unit cost of production *times* the number of units produced (per day), where these two quantities for the multiplication are given separately in Table 8.27. For example, consider the assignment of Plant 1 to product 1. By using the corresponding unit cost in Table 8.28 ($41) and the corresponding demand (number of units produced per day) in Table 8.28 (20), we obtain

Cost of Plant 1 producing one unit of product 1   = $41

Required (daily) production of product 1   = 20 units

Total (daily) cost of assigning plant 1 to product 1   = 20 ($41)

               = $820

so 820 is entered into Table 8.29 for the cost of either Assignee 1a or 1b performing Task 1.

The optimal solution for this assignment problem is as follows:

Plant 1 produces products 2 and 3.
Plant 2 produces product 1.
Plant 3 produces product 4.

Here the dummy assignment is given to Plant 2. The total cost is $Z = \$3,290$ per day.

As usual, one way to obtain this optimal solution is to convert the cost table of Table 8.29 to a parameter table for the equivalent transportation problem (see Table 8.26) and then apply the transportation simplex method. Because of the identical rows in Table 8.29, this approach can be streamlined by combining the five assignees into three sources with supplies 2, 2, and 1, respectively. (See Prob. 8.3-5.) This streamlining also decreases by two the number of degenerate basic variables in every BF solution. Therefore, even though this streamlined formulation no longer fits the format presented in Table 8.26a for an assignment problem, it is a more efficient formulation for applying the transportation simplex method.

Figure 8.6 shows how Excel and its Solver can be used to obtain this optimal solution, which is displayed in the changing cells Assignment (C19:F21) of the spreadsheet. Since the general simplex method is being used, there is no need to fit this formulation into the format for either the assignment problem or transportation problem model. Therefore, the formulation does not bother to split Plants 1 and 2 into two assignees each, or to add a dummy task. Instead, Plants 1 and 2 are given a supply of 2 each, and then $\leq$ signs are entered into cells H19 and H20 as well as into the corresponding constraints in the Solver dialogue box. There also is no need to include the Big M method to prohibit assigning product 3 to Plant 2 in cell E20, since this dialogue box includes the constraint that E20 = 0. The target cell TotalCost (I24) shows the total cost of $3,290 per day.

Now look back and compare this solution to the one obtained for Option 1, which included the splitting of product 4 between Plants 2 and 3. The allocations are somewhat different for the two solutions, but the total daily costs are virtually the same ($3,260 for Option 1 versus $3,290 for Option 2). However, there are hidden costs associated with product splitting (including the cost of extra setup, distribution, and administration) that are not included in the objective function for Option 1. As with any application of OR, the mathematical model used can provide only an approximate representation of the total problem, so management needs to consider factors that cannot be incorporated into the model before it makes a final decision. In this case, after evaluating the disadvantages of product splitting, management decided to adopt the Option 2 solution.

	A	B	C	D	E	F	G	H	I
1		**Better Products Co. Production Planning Problem (Option 2)**							
2									
3		**Unit Cost**	Product 1	Product 2	Product 3	Product 4			
4		Plant 1	$41	$27	$28	$24			
5		Plant 2	$40	$29	-	$23			
6		Plant 3	$37	$30	$27	$21			
7									
8		Required Production	20	30	30	40			
9									
10									
11		**Cost ($/day)**	Product 1	Product 2	Product 3	Product 4			
12		Plant 1	$820	$810	$840	$960			
13		Plant 2	$800	$870	-	$920			
14		Plant 3	$740	$900	$810	$840			
15									
16									
17							Total		
18		**Assignment**	Product 1	Product 2	Product 3	Product 4	Assignments		Supply
19		Plant 1	0	1	1	0	2	<=	2
20		Plant 2	1	0	0	0	1	<=	2
21		Plant 3	0	0	0	1	1	=	1
22		Total Assigned	1	1	1	1			
23			=	=	=	=			Total Cost
24		Demand	1	1	1	1			$3,290

■ FIGURE 8.6

A spreadsheet formulation of Option 2 for the Better Products Co. problem as a variant of an assignment problem. The target cell is TotalCost (I24) and the other output cells are Cost (C12:F14), TotalAssignments (G19:G21), and TotalAssigned (C22:F22), where the equations entered into these cells are shown below the spreadsheet. The values of 1 in the changing cells Assignment (C19:F21) display the optimal production plan obtained by the Solver.

Solver Parameters

Set Target Cell: [TotalCost]

Equal To: ○ Max   ● Min   ○

By Changing Cells:

[Assignment]

Subject to the Constraints:

```
$E$20 = 0
$G$19:$G$20 <= $I$19:$I$20
$G$21 = $I$21
TotalAssigned = Demand
```

Solver Options

☑ Assume Linear Model
☑ Assume Non-Negative

	B	C	D	E	F
11	**Cost ($/day)**	Product 1	Product 2	Product 3	Product 4
12	Plant 1	=C4*C$8	=D4*D$8	=E4*E$8	=F4*F$8
13	Plant 2	=C5*C$8	=D5*D$8	-	=F5*F$8
14	Plant 3	=C6*C$8	=D6*D$8	=E6*E$8	=F6*F$8

	G
17	Total
18	Assignments
19	=SUM(C19:F19)
20	=SUM(C20:F20)
21	=SUM(C21:F21)

	B	C	D	E	F
22	Total Assigned	=SUM(C19:C21)	=SUM(D19:D21)	=SUM(E19:E21)	=SUM(F19:F21)

Range Name	Cells
Assignment	C19:F21
Cost	C12:F14
Demand	C24:F24
RequiredProduction	C8:F8
Supply	I19:I21
TotalAssigned	C22:F22
TotalAssignments	G19:G21
TotalCost	I24
UnitCost	C4:F6

	I
23	Total Cost
24	=SUMPRODUCT(Cost,Assignment)

8.4   A SPECIAL ALGORITHM FOR THE ASSIGNMENT PROBLEM

In Sec. 8.3, we pointed out that the transportation simplex method can be used to solve assignment problems but that a *specialized* algorithm designed for such problems should be more efficient. We now will describe a classic algorithm of this type. It is called the **Hungarian algorithm** (or *Hungarian method*) because it was developed by Hungarian mathematicians. We will focus just on the key ideas without filling in all the details needed for a complete computer implementation.

The Role of Equivalent Cost Tables

The algorithm operates directly on the *cost table* for the problem. More precisely, it converts the original cost table into a series of *equivalent* cost tables until it reaches one where an optimal solution is obvious. This final equivalent cost table is one consisting of only *positive* or *zero* elements where all the assignments can be made to the zero element positions. Since the total cost cannot be negative, this set of assignments with a zero total cost is clearly optimal. The question remaining is how to convert the original cost table into this form.

The key to this conversion is the fact that one can add or subtract any constant from every element of a row or column of the cost table without really changing the problem. That is, an optimal solution for the new cost table must also be optimal for the old one, and conversely.

Therefore, the algorithm begins by subtracting the smallest number in each row from every number in the row. This *row reduction* process will create an equivalent cost table that has a zero element in every row. If this cost table has any columns without a zero element, the next step is to perform a *column reduction* process by subtracting the smallest number in each such column from every number in the column.[12] The new equivalent cost table will have a zero element in every row and every column. If these zero elements provide a complete set of assignments, these assignments constitute an optimal solution and the algorithm if finished.

To illustrate, consider the cost table for the Job Shop Co. problem given in Table 8.25. To convert this cost table into an equivalent cost table, suppose that we begin the row reduction process by subtracting 11 from every element in row 1, which yields:

	1	2	3	4
1	2	5	1	0
2	15	M	13	20
3	5	7	10	6
4(D)	0	0	0	0

Since any feasible solution must have exactly one assignment in row 1, the total cost for the new table must always be exactly 11 less than for the old table. Hence, the solution which minimizes total cost for one table must also minimize total cost for the other.

Notice that, whereas the original cost table had only strictly positive elements in the first three rows, the new table has a zero element in row 1. Since the objective is to obtain enough strategically located zero elements to yield a complete set of assignments, this process should be continued on the other rows and columns. Negative elements are to be avoided, so the constant to be subtracted should be the minimum element in the row or column. Doing this for rows 2 and 3 yields the following equivalent cost table:

	1	2	3	4
1	2	5	1	[0]
2	2	M	[0]	7
3	[0]	2	5	1
4(D)	0	[0]	0	0

This cost table has all the zero elements required for a complete set of assignments, as shown by the four boxes, so these four assignments constitute an *optimal solution* (as

---

[12]The individual rows and columns actually can be reduced in any order, but starting with all the rows and then doing all the columns provides one systematic way of executing the algorithm.

claimed in Sec. 8.3 for this problem). The total cost for this optimal solution is seen in Table 8.25 to be $Z = 29$, which is just the sum of the numbers that have been subtracted from rows 1, 2, and 3.

Unfortunately, an optimal solution is not always obtained quite so easily, as we now illustrate with the assignment problem formulation of Option 2 for the Better Products Co. problem shown in Table 8.29.

Because this problem's cost table already has zero elements in every row but the last one, suppose we begin the process of converting to equivalent cost tables by subtracting the minimum element in each column from every entry in that column. The result is shown below.

	1	2	3	4	5(D)
1a	80	0	30	120	0
1b	80	0	30	120	0
2a	60	60	M	80	0
2b	60	60	M	80	0
3	0	90	0	0	M

Now *every* row and column has at least one zero element, but a complete set of assignments with zero elements is *not* possible this time. In fact, the maximum number of assignments that can be made in zero element positions in only 3. (Try it.) Therefore, one more idea must be implemented to finish solving this problem that was not needed for the first example.

The Creation of Additional Zero Elements

This idea involves a new way of creating *additional* positions with zero elements without creating any negative elements. Rather than subtracting a constant from a *single* row or column, we now add or subtract a constant from a *combination* of rows and columns.

This procedure begins by drawing a set of lines through some of the rows and columns in such a way as to *cover all the zeros*. This is done with a *minimum* number of lines, as shown in the next cost table.

	1	2	3	4	5(D)
1a	80	0	30	120	0
1b	80	0	30	120	0
2a	60	60	M	80	0
2b	60	60	M	80	0
3	0	90	0	0	M

Notice that the minimum element not crossed out is 30 in the two top positions in column 3. Therefore, subtracting 30 from every element in the entire table, i.e., from every row or from every column, will create a new zero element in these two positions. Then, in order to restore the previous zero elements and eliminate negative elements, we add 30 to each row or column with a line covering it—row 3 and columns 2 and 5(*D*). This yields the following equivalent cost table.

	1	2	3	4	5(*D*)
1*a*	50	0	0	90	0
1*b*	50	0	0	90	0
2*a*	30	60	*M*	50	0
2*b*	30	60	*M*	50	0
3	0	120	0	0	*M*

A shortcut for obtaining this cost table from the preceding one is to subtract 30 from just the elements without a line through them and then add 30 to every element that lies at the intersection of two lines.

Note that columns 1 and 4 in this new cost table have only a single zero element and they both are in the same row (row 3). Consequently, it now is possible to make four assignments to zero element positions, but still not five. (Try it.) In general, the minimum number of lines needed to cover all zeros equals the maximum number of assignments that can be made to zero element positions. Therefore, we repeat the above procedure, where four lines (the same number as the maximum number of assignments) now are the minimum needed to cover all zeros. One way of doing this is shown below.

	1	2	3	4	5(*D*)
1*a*	~~50~~	~~0~~	~~0~~	~~90~~	~~0~~
1*b*	~~50~~	~~0~~	~~0~~	~~90~~	~~0~~
2*a*	30	60	*M*	50	0
2*b*	30	60	*M*	50	0
3	~~0~~	~~120~~	~~0~~	~~0~~	~~M~~

The minimum element not covered by a line is again 30, where this number now appears in the first position in both rows 2*a* and 2*b*. Therefore, we subtract 30 from every *uncovered* element and add 30 to every *doubly covered* element (except for ignoring elements of *M*), which gives the following equivalent cost table.

	1	2	3	4	5(*D*)
1*a*	50	[0]	0	90	30
1*b*	50	0	[0]	90	30
2*a*	[0]	30	*M*	20	0
2*b*	0	30	*M*	20	[0]
3	0	120	0	[0]	*M*

■ CASES

CASE 8.1   Shipping Wood to Market

Alabama Atlantic is a lumber company that has three sources of wood and five markets to be supplied. The annual availability of wood at sources 1, 2, and 3 is 15, 20, and 15 million board feet, respectively. The amount that can be sold annually at markets 1, 2, 3, 4, and 5 is 11, 12, 9, 10, and 8 million board feet, respectively.

In the past the company has shipped the wood by train. However, because shipping costs have been increasing, the alternative of using ships to make some of the deliveries is being investigated. This alternative would require the company to invest in some ships. Except for these investment costs, the shipping costs in thousands of dollars per million board feet by rail and by water (when feasible) would be the following for each route:

	Unit Cost by Rail ($1,000's) Market					Unit Cost by Ship ($1,000's) Market				
Source	1	2	3	4	5	1	2	3	4	5
1	61	72	45	55	66	31	38	24	—	35
2	69	78	60	49	56	36	43	28	24	31
3	59	66	63	61	47	—	33	36	32	26

The capital investment (in thousands of dollars) in ships required for each million board feet to be transported annually by ship along each route is given as follows:

	Investment for Ships ($1,000's) Market				
Source	1	2	3	4	5
1	275	303	238	—	285
2	293	318	270	250	265
3	—	283	275	268	240

Considering the expected useful life of the ships and the time value of money, the equivalent uniform annual cost of these investments is one-tenth the amount given in the table. The objective is to determine the overall shipping plan that minimizes the total equivalent uniform annual cost (including shipping costs).

You are the head of the OR team that has been assigned the task of determining this shipping plan for each of the following three options.

Option 1: Continue shipping exclusively by rail.
Option 2: Switch to shipping exclusively by water (except where only rail is feasible).

Option 3: Ship by either rail or water, depending on which is less expensive for the particular route.

Present your results for each option. Compare.

Finally, consider the fact that these results are based on *current* shipping and investment costs, so the decision on the option to adopt now should take into account management's projection of how these costs are likely to change in the future. For each option, describe a scenario of future cost changes that would justify adopting that option now.

(*Note:* Data files for this case are provided on the book's website for your convenience.)

■ PREVIEWS OF ADDED CASES ON OUR WEBSITE (www.mhhe.com/hillier)

CASE 8.2 Continuation of the Texago Case Study

The supplement to this chapter on the book's website presents a case study of how the Texago Corp. solved many transportation problems to help make its decision regarding where to locate its new oil refinery. Management now needs to address the question of whether the capacity of the new refinery should be made somewhat larger than originally planned. This will require formulating and solving some additional transportation problems. A key part of the analysis then will involve combining two transportation problems into a single linear programming model that simultaneously considers the shipping of crude oil from the oil fields to the refineries and the shipping of final product from the refineries

to the distribution centers. A memo to management summarizing your results and recommendations also needs to be written.

CASE 8.3 Project Pickings

This case focuses on a series of applications of the assignment problem for a pharmaceutical manufacturing company. The decision has been made to undertake five research and development projects to attempt to develop new drugs that will treat five specific types of medical ailments. Five senior scientists are available to lead these projects as project directors. The problem now is to decide on how to assign these scientists to the projects on a one-to-one basis. A variety of likely scenarios need to be considered.

C H A P T E R

9

Network Optimization Models

Networks arise in numerous settings and in a variety of guises. Transportation, electrical, and communication networks pervade our daily lives. Network representations also are widely used for problems in such diverse areas as production, distribution, project planning, facilities location, resource management, and financial planning—to name just a few examples. In fact, a network representation provides such a powerful visual and conceptual aid for portraying the relationships between the components of systems that it is used in virtually every field of scientific, social, and economic endeavor.

One of the most exciting developments in operations research (OR) in recent years has been the unusually rapid advance in both the methodology and application of network optimization models. A number of algorithmic breakthroughs have had a major impact, as have ideas from computer science concerning data structures and efficient data manipulation. Consequently, algorithms and software now are available *and are being used* to solve huge problems on a routine basis that would have been completely intractable two or three decades ago.

Many network optimization models actually are special types of *linear programming* problems. For example, both the transportation problem and the assignment problem discussed in the preceding chapter fall into this category because of their network representations presented in Figs. 8.3 and 8.5.

One of the linear programming examples presented in Sec. 3.4 also is a network optimization problem. This is the Distribution Unlimited Co. problem of how to distribute its goods through the distribution network shown in Fig. 3.13. This special type of linear programming problem, called the *minimum cost flow* problem, is presented in Sec. 9.6. We shall return to this specific example in that section and then solve it with network methodology in the following section.

In this one chapter we only scratch the surface of the current state of the art of network methodology. However, we shall introduce you to five important kinds of network problems and some basic ideas of how to solve them (without delving into issues of data structures that are so vital to successful large-scale implementations). Each of the first three problem types—the *shortest-path problem,* the *minimum spanning tree problem,* and the *maximum flow problem*—has a very specific structure that arises frequently in applications.

The fourth type—the *minimum cost flow problem*—provides a unified approach to many other applications because of its far more general structure. In fact, this structure is so general that it includes as special cases both the shortest-path problem and the maximum flow problem as well as the transportation problem and the assignment problem from

Chap. 8. Because the minimum cost flow problem is a special type of linear programming problem, it can be solved extremely efficiently by a streamlined version of the simplex method called the *network simplex method.* (We shall not discuss even more general network problems that are more difficult to solve.)

The fifth kind of network problem considered here involves determining the most economical way to conduct a project so that it can be completed by its deadline. A technique called the *CPM method of time-cost trade-offs* is used to formulate a network model of the project and the time-cost trade-offs for its activities. Either marginal cost analysis or linear programming then is used to solve for the optimal project plan.

The first section introduces a prototype example that will be used subsequently to illustrate the approach to the first three of these problems. Section 9.2 presents some basic terminology for networks. The next four sections deal with the first four problems in turn, and Sec. 9.7 then is devoted to the network simplex method. Section 9.8 presents the CPM method of time-cost trade-offs.

■ 9.1 PROTOTYPE EXAMPLE

SEERVADA PARK has recently been set aside for a limited amount of sightseeing and backpack hiking. Cars are not allowed into the park, but there is a narrow, winding road system for trams and for jeeps driven by the park rangers. This road system is shown (without the curves) in Fig. 9.1, where location O is the entrance into the park; other letters designate the locations of ranger stations (and other limited facilities). The numbers give the distances of these winding roads in miles.

The park contains a scenic wonder at station T. A small number of trams are used to transport sightseers from the park entrance to station T and back.

The park management currently faces three problems. One is to determine which route from the park entrance to station T has the *smallest total distance* for the operation of the trams. (This is an example of the shortest-path problem to be discussed in Sec. 9.3.)

A second problem is that telephone lines must be installed under the roads to establish telephone communication among all the stations (including the park entrance). Because the installation is both expensive and disruptive to the natural environment, lines will be installed under just enough roads to provide some connection between every pair of stations. The question is where the lines should be laid to accomplish this with a *minimum* total number of miles of line installed. (This is an example of the minimum spanning tree problem to be discussed in Sec. 9.4.)

■ **FIGURE 9.1**
The road system for Seervada Park.

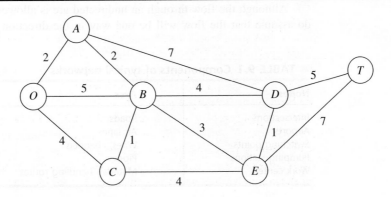

The third problem is that more people want to take the tram ride from the park entrance to station T than can be accommodated during the peak season. To avoid unduly disturbing the ecology and wildlife of the region, a strict ration has been placed on the number of tram trips that can be made on each of the roads per day. (These limits differ for the different roads, as we shall describe in detail in Sec. 9.5.) Therefore, during the peak season, various routes might be followed regardless of distance to increase the number of tram trips that can be made each day. The question pertains to how to route the various trips to *maximize* the number of trips that can be made per day without violating the limits on any individual road. (This is an example of the maximum flow problem to be discussed in Sec. 9.5.)

■ 9.2   THE TERMINOLOGY OF NETWORKS

A relatively extensive terminology has been developed to describe the various kinds of networks and their components. Although we have avoided as much of this special vocabulary as we could, we still need to introduce a considerable number of terms for use throughout the chapter. We suggest that you read through this section once at the outset to understand the definitions and then plan to return to refresh your memory as the terms are used in subsequent sections. To assist you, each term is highlighted in **boldface** at the point where it is defined.

A network consists of a set of *points* and a set of *lines* connecting certain pairs of the points. The points are called **nodes** (or vertices); e.g., the network in Fig. 9.1 has seven nodes designated by the seven circles. The lines are called **arcs** (or links or edges or branches); e.g., the network in Fig. 9.1 has 12 arcs corresponding to the 12 roads in the road system. Arcs are labeled by naming the nodes at either end; for example, AB is the arc between nodes A and B in Fig. 9.1.

The arcs of a network may have a flow of some type through them, e.g., the flow of trams on the roads of Seervada Park in Sec. 9.1. Table 9.1 gives several examples of flow in typical networks. If flow through an arc is allowed in only one direction (e.g., a one-way street), the arc is said to be a **directed arc.** The direction is indicated by adding an arrowhead at the end of the line representing the arc. When a directed arc is labeled by listing two nodes it connects, the *from* node always is given before the *to* node; e.g., an arc that is directed *from* node A *to* node B must be labeled as AB rather than BA. Alternatively, this arc may be labeled as $A \rightarrow B$.

If flow through an arc is allowed in either direction (e.g., a pipeline that can be used to pump fluid in either direction), the arc is said to be an **undirected arc.** To help you distinguish between the two kinds of arcs, we shall frequently refer to undirected arcs by the suggestive name of **links.**

Although the flow through an undirected arc is allowed to be in either direction, we do assume that the flow will be one way in the direction of choice rather than having

■ **TABLE 9.1** Components of typical networks

Nodes	Arcs	Flow
Intersections	Roads	Vehicles
Airports	Air lanes	Aircraft
Switching points	Wires, channels	Messages
Pumping stations	Pipes	Fluids
Work centers	Materials-handling routes	Jobs

simultaneous flows in opposite directions. (The latter case requires the use of a *pair of directed arcs* in opposite directions.) However, in the process of making the decision on the flow through an undirected arc, it is permissible to make a sequence of assignments of flows in opposite directions, but with the understanding that the actual flow will be the *net flow* (the difference of the assigned flows in the two directions). For example, if a flow of 10 has been assigned in one direction and then a flow of 4 is assigned in the opposite direction, the actual effect is to *cancel* 4 units of the original assignment by reducing the flow in the original direction from 10 to 6. Even for a directed arc, the same technique sometimes is used as a convenient device to reduce a previously assigned flow. In particular, you are allowed to make a fictional assignment of flow in the "wrong" direction through a directed arc to record a reduction of that amount in the flow in the "right" direction.

A network that has only directed arcs is called a **directed network.** Similarly, if all its arcs are undirected, the network is said to be an **undirected network.** A network with a mixture of directed and undirected arcs (or even all undirected arcs) can be converted to a directed network, if desired, by replacing each undirected arc by a pair of directed arcs in opposite directions. (You then have the choice of interpreting the flows through each pair of directed arcs as being simultaneous flows in opposite directions or providing a net flow in one direction, depending on which fits your application.)

When two nodes are not connected by an arc, a natural question is whether they are connected by a series of arcs. A **path** between two nodes is a *sequence of distinct arcs* connecting these nodes. For example, one of the paths connecting nodes O and T in Fig. 9.1 is the sequence of arcs OB–BD–DT ($O \rightarrow B \rightarrow D \rightarrow T$), or vice versa. When some of or all the arcs in the network are directed arcs, we then distinguish between directed paths and undirected paths. A **directed path** from node i to node j is a sequence of connecting arcs whose direction (if any) is *toward* node j, so that flow from node i to node j along this path is feasible. An **undirected path** from node i to node j is a sequence of connecting arcs whose direction (if any) can be *either* toward or away from node j. (Notice that a directed path also satisfies the definition of an undirected path, but not vice versa.) Frequently, an undirected path will have some arcs directed toward node j but others directed away (i.e., toward node i). You will see in Secs. 9.5 and 9.7 that, perhaps surprisingly, *undirected* paths play a major role in the analysis of *directed* networks.

To illustrate these definitions, Fig. 9.2 shows a typical directed network. (Its nodes and arcs are the same as in Fig. 3.13, where nodes A and B represent two factories, nodes D and E represent two warehouses, node C represents a distribution center, and the arcs represent shipping lanes.) The sequence of arcs AB–BC–CE ($A \rightarrow B \rightarrow C \rightarrow E$) is a directed path from node A to E, since flow toward node E along this entire path is feasible. On the other hand, BC–AC–AD ($B \rightarrow C \rightarrow A \rightarrow D$) is *not* a directed path from node B to node D, because the direction of arc AC is away from node D (on this path). However, $B \rightarrow C \rightarrow A \rightarrow D$ is an undirected path from node B to node D, because the sequence of arcs BC–AC–AD does *connect* these two nodes (even though the direction of arc AC prevents flow through this path).

As an example of the relevance of undirected paths, suppose that 2 units of flow from node A to node C had previously been assigned to arc AC. Given this previous assignment, it now is feasible to assign a smaller flow, say, 1 unit, to the entire undirected path $B \rightarrow C \rightarrow A \rightarrow D$, even though the direction of arc AC prevents positive flow through $C \rightarrow A$. The reason is that this assignment of flow in the "wrong" direction for arc AC actually just *reduces* the flow in the "right" direction by 1 unit. Sections 9.5 and 9.7 make heavy use of this technique of assigning a flow through an undirected path that includes arcs whose direction is opposite to this flow, where the real effect for these arcs is to reduce previously assigned positive flows in the "right" direction.

A path that begins and ends at the same node is called a **cycle.** In a *directed* network, a cycle is either a directed or an undirected cycle, depending on whether the

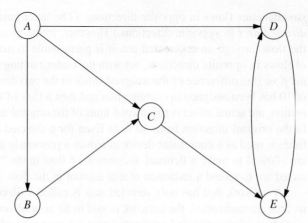

■ FIGURE 9.2
The distribution network for
Distribution Unlimited Co.,
first shown in Fig. 3.13,
illustrates a directed network.

path involved is a directed or an undirected path. (Since a directed path also is an undirected path, a directed cycle is an undirected cycle, but not vice versa in general.) In Fig. 9.2, for example, *DE–ED* is a directed cycle. By contrast, *AB–BC–AC* is *not* a directed cycle, because the direction of arc *AC* opposes the direction of arcs *AB* and *BC*. On the other hand, *AB–BC–AC* is an undirected cycle, because $A \rightarrow B \rightarrow C \rightarrow A$ is an undirected path. In the undirected network shown in Fig. 9.1, there are many cycles, for example, *OA–AB–BC–CO*. However, note that the definition of *path* (a sequence of *distinct* arcs) rules out retracing one's steps in forming a cycle. For example, *OB–BO* in Fig. 9.1 does not qualify as a cycle, because *OB* and *BO* are two labels for the *same* arc (link). On the other hand, *DE–ED* is a (directed) cycle in Fig. 9.2, because *DE* and *ED* are distinct arcs.

Two nodes are said to be **connected** if the network contains at least one *undirected* path between them. (Note that the path does not need to be directed even if the network is directed.) A **connected network** is a network where every pair of nodes is connected. Thus, the networks in Figs. 9.1 and 9.2 are both connected. However, the latter network would not be connected if arcs *AD* and *CE* were removed.

Consider a connected network with *n* nodes (e.g., the $n = 5$ nodes in Fig. 9.2) where all the arcs have been deleted. A "tree" can then be "grown" by adding one arc (or "branch") at a time from the original network in a certain way. The first arc can go anywhere to connect some pair of nodes. Thereafter, each new arc should be between a node that already is connected to other nodes and a new node not previously connected to any other nodes. Adding an arc in this way avoids creating a cycle and ensures that the number of connected nodes is 1 greater than the number of arcs. Each new arc creates a larger **tree,** which is a *connected network* (for some subset of the *n* nodes) that contains *no undirected cycles*. Once the $(n - 1)$st arc has been added, the process stops because the resulting tree *spans* (connects) all *n* nodes. This tree is called a **spanning tree,** i.e., a *connected network* for all *n* nodes that contains *no undirected cycles*. Every spanning tree has exactly $n - 1$ arcs, since this is the *minimum* number of arcs needed to have a connected network and the *maximum* number possible without having undirected cycles.

Figure 9.3 uses the five nodes and some of the arcs of Fig. 9.2 to illustrate this process of growing a tree one arc (branch) at a time until a spanning tree has been obtained. There are several alternative choices for the new arc at each stage of the process, so Fig. 9.3 shows only one of many ways to construct a spanning tree in this case. Note, however, how each new added arc satisfies the conditions specified in the preceding paragraph. We shall discuss and illustrate spanning trees further in Sec. 9.4.

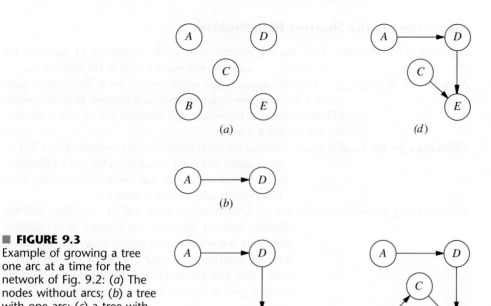

FIGURE 9.3
Example of growing a tree one arc at a time for the network of Fig. 9.2: (a) The nodes without arcs; (b) a tree with one arc; (c) a tree with two arcs; (d) a tree with three arcs; (e) a spanning tree.

Spanning trees play a key role in the analysis of many networks. For example, they form the basis for the *minimum spanning tree problem* discussed in Sec. 9.4. Another prime example is that (feasible) spanning trees correspond to the BF solutions for the *network simplex method* discussed in Sec. 9.7.

Finally, we shall need a little additional terminology about *flows* in networks. The maximum amount of flow (possibly infinity) that can be carried on a directed arc is referred to as the **arc capacity.** For nodes, a distinction is made among those that are net generators of flow, net absorbers of flow, or neither. A **supply node** (or source node or source) has the property that the flow *out* of the node exceeds the flow *into* the node. The reverse case is a **demand node** (or sink node or sink), where the flow *into* the node exceeds the flow *out* of the node. A **transshipment node** (or intermediate node) satisfies *conservation of flow,* so flow in equals flow out.

9.3 THE SHORTEST-PATH PROBLEM

Although several other versions of the shortest-path problem (including some for directed networks) are mentioned at the end of the section, we shall focus on the following simple version. Consider an *undirected* and *connected* network with two special nodes called the *origin* and the *destination.* Associated with each of the *links* (undirected arcs) is a nonnegative *distance.* The objective is to find the shortest path (the path with the minimum total distance) from the origin to the destination.

A relatively straightforward algorithm is available for this problem. The essence of this procedure is that it fans out from the origin, successively identifying the shortest path to each of the nodes of the network in the ascending order of their (shortest) distances from the origin, thereby solving the problem when the destination node is reached. We shall first outline the method and then illustrate it by solving the shortest-path problem encountered by the Seervada Park management in Sec. 9.1.

Algorithm for the Shortest-Path Problem

Objective of nth *iteration:* Find the nth nearest node to the origin (to be repeated for $n = 1, 2, \ldots$ until the nth nearest node is the destination.

Input for nth *iteration:* $n - 1$ nearest nodes to the origin (solved for at the previous iterations), including their shortest path and distance from the origin. (These nodes, plus the origin, will be called *solved nodes;* the others are *unsolved nodes.*)

Candidates for nth *nearest node:* Each solved node that is directly connected by a link to one or more unsolved nodes provides *one* candidate— the unsolved node with the *shortest* connecting link. (Ties provide additional candidates.)

Calculation of nth *nearest node:* For each such solved node and its candidate, add the distance between them and the distance of the shortest path from the origin to this solved node. The candidate with the smallest such total distance is the nth nearest node (ties provide additional solved nodes), and its shortest path is the one generating this distance.

Applying This Algorithm to the Seervada Park Shortest-Path Problem

The Seervada Park management needs to find the shortest path from the park entrance (node O) to the scenic wonder (node T) through the road system shown in Fig. 9.1. Applying the above algorithm to this problem yields the results shown in Table 9.2 (where the tie for the second nearest node allows skipping directly to seeking the fourth nearest node next). The first column (n) indicates the iteration count. The second column simply lists the *solved nodes* for beginning the current iteration after deleting the irrelevant ones (those not connected directly to any unsolved node). The third column then gives the *candidates* for the nth nearest node (the unsolved nodes with the *shortest* connecting link to a solved node). The fourth column calculates the distance of the shortest path from the origin to each of these candidates (namely, the distance to the solved node plus the link

■ **TABLE 9.2** Applying the shortest-path algorithm to the Seervada Park problem

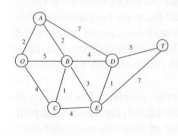

n	Solved Nodes Directly Connected to Unsolved Nodes	Closest Connected Unsolved Node	Total Distance Involved	nth Nearest Node	Minimum Distance	Last Connection
1	O	A	2	A	2	OA
2, 3	O	C	4	C	4	OC
	A	B	$2 + 2 = 4$	B	4	AB
4	A	D	$2 + 7 = 9$			
	B	E	$4 + 3 = 7$	E	7	BE
	C	E	$4 + 4 = 8$			
5	A	D	$2 + 7 = 9$			
	B	D	$4 + 4 = 8$	D	8	BD
	E	D	$7 + 1 = 8$	D	8	ED
6	D	T	$8 + 5 = 13$	T	13	DT
	E	T	$7 + 7 = 14$			

distance to the candidate). The candidate with the smallest such distance is the nth nearest node to the origin, as listed in the fifth column. The last two columns summarize the information for this *newest solved node* that is needed to proceed to subsequent iterations (namely, the distance of the shortest path from the origin to this node and the last link on this shortest path).

Now let us relate these columns directly to the outline given for the algorithm. The *input for nth iteration* is provided by the fifth and sixth columns for the preceding iterations, where the solved nodes in the fifth column are then listed in the second column for the current iteration after deleting those that are no longer directly connected to unsolved nodes. The *candidates for nth nearest node* next are listed in the third column for the current iteration. The *calculation of nth nearest node* is performed in the fourth column, and the results are recorded in the last three columns for the current iteration.

After the work shown in Table 9.2 is completed, the shortest path *from the destination to the origin* can be traced back through the last column of Table 9.2 as *either* $T \to D \to E \to B \to A \to O$ or $T \to D \to B \to A \to O$. Therefore, the two alternates for the shortest path *from the origin to the destination* have been identified as $O \to A \to B \to E \to D \to T$ and $O \to A \to B \to D \to T$, with a total distance of 13 miles on either path.

Using Excel to Formulate and Solve Shortest-Path Problems

This algorithm provides a particularly efficient way of solving large shortest-path problems. However, some mathematical programming software packages do not include this algorithm. If not, they often will include the *network simplex method* described in Sec. 9.7, which is another good option for these problems.

Since the shortest-path problem is a special type of linear programming problem, the general simplex method also can be used when better options are not readily available. Although not nearly as efficient as these specialized algorithms on large shortest-path problems, it is quite adequate for problems of even very substantial size (much larger than the Seervada Park problem). Excel, which relies on the general simplex method, provides a convenient way of formulating and solving shortest-path problems with dozens of arcs and nodes.

Figure 9.4 shows an appropriate spreadsheet formulation for the Seervada Park shortest-path problem. Rather than using the kind of formulation presented in Sec. 3.6 that uses a separate row for each functional constraint of the linear programming model, this formulation exploits the special structure by listing the *nodes* in column G and the *arcs* in columns B and C, as well as the distance (in miles) along each arc in column E. Since each *link* in the network is an *undirected arc*, whereas travel through the shortest path is in one direction, each link can be replaced by a pair of *directed* arcs in opposite directions. Thus, columns B and C together list both of the nearly vertical links in Fig. 9.1 (B–C and D–E) twice, once as a downward arc and once as an upward arc, since either direction might be on the chosen path. However, the other links are only listed as left-to-right arcs, since this is the only direction of interest for choosing a shortest path from the origin to the destination.

A trip from the origin to the destination is interpreted to be a "flow" of 1 on the chosen path through the network. The decisions to be made are which arcs should be included in the path to be traversed. A flow of 1 is assigned to an arc if it is included, whereas the flow is 0 if it is not included. Thus, the decision variables are

$$x_{ij} = \begin{cases} 0 & \text{if arc } i \to j \text{ is not included} \\ 1 & \text{if arc } i \to j \text{ is included} \end{cases}$$

Many applications require finding the shortest *directed* path from the origin to the destination through a *directed* network. The algorithm already presented can be easily modified to deal just with directed paths at each iteration. In particular, when candidates for the *n*th nearest node are identified, only directed arcs *from* a solved node *to* an unsolved node are considered.

Another version of the shortest-path problem is to find the shortest paths from the origin to *all* the other nodes of the network. Notice that the algorithm already solves for the shortest path to each node that is closer to the origin than the destination. Therefore, when all nodes are potential destinations, the only modification needed in the algorithm is that it does not stop until all nodes are solved nodes.

An even more general version of the shortest-path problem is to find the shortest paths from *every* node to every other node. Another option is to drop the restriction that "distances" (arc values) be nonnegative. Constraints also can be imposed on the paths that can be followed. All these variations occasionally arise in applications and so have been studied by researchers.

The algorithms for a wide variety of combinatorial optimization problems, such as certain vehicle routing or network design problems, often call for the solution of a large number of shortest-path problems as subroutines. Although we lack the space to pursue this topic further, this use may now be the most important kind of application of the shortest-path problem.

9.4   THE MINIMUM SPANNING TREE PROBLEM

The minimum spanning tree problem bears some similarities to the main version of the shortest-path problem presented in the preceding section. In both cases, an *undirected* and *connected* network is being considered, where the given information includes some measure of the positive *length* (distance, cost, time, etc.) associated with each link. Both problems also involve choosing a set of links that have the *shortest total length* among all sets of links that satisfy a certain property. For the shortest-path problem, this property is that the chosen links must provide a path between the origin and the destination. For the minimum spanning tree problem, the required property is that the chosen links must provide a path between *each* pair of nodes.

The minimum spanning tree problem can be summarized as follows.

1. You are given the *nodes* of a network but *not* the links. Instead, you are given the *potential links* and the positive *length* for each if it is inserted into the network. (Alternative measures for the length of a link include distance, cost, and time.)
2. You wish to design the network by inserting enough links to satisfy the requirement that there be a path between *every* pair of nodes.
3. The objective is to satisfy this requirement in a way that minimizes the total length of the links inserted into the network.

A network with *n* nodes requires only ($n - 1$) links to provide a path between each pair of nodes. No extra links should be used, since this would needlessly increase the total length of the chosen links. The ($n - 1$) links need to be chosen in such a way that the resulting network (with just the chosen links) forms a *spanning tree* (as defined in Sec. 9.2). Therefore, the problem is to find the spanning tree with a minimum total length of the links.

Figure 9.5 illustrates this concept of a spanning tree for the Seervada Park problem (see Sec. 9.1). Thus, Fig. 9.5*a* is *not* a spanning tree because nodes *O*, *A*, *B*, and *C* are not connected with nodes *D*, *E*, and *T*. It needs another link to make this connection. This network actually consists of two trees, one for each of these two sets of nodes. The links in Fig. 9.5*b* do *span* the network (i.e., the network is connected as defined in Sec. 9.2), but it is *not* a tree because there are two *cycles* (*O–A–B–C–O* and *D–T–E–D*). It has too

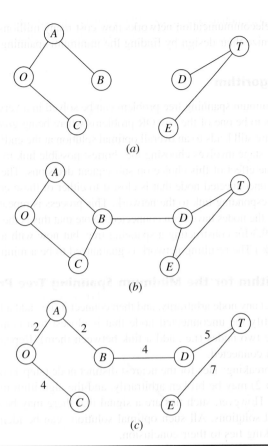

■ FIGURE 9.5
Illustrations of the spanning
tree concept for the
Seervada Park problem:
(*a*) Not a spanning tree;
(*b*) not a spanning tree;
(*c*) a spanning tree.

many links. Because the Seervada Park problem has $n = 7$ nodes, Sec. 9.2 indicates that
the network must have exactly $n - 1 = 6$ links, with *no cycles,* to qualify as a spanning
tree. This condition is achieved in Fig. 9.5*c,* so this network is a *feasible* solution (with a
value of 24 miles for the total length of the links) for the minimum spanning tree problem. (You soon will see that this solution is not *optimal* because it is possible to construct
a spanning tree with only 14 miles of links.)

Some Applications

Here is a list of some key types of applications of the minimum spanning tree problem.

1. Design of telecommunication networks (fiber-optic networks, computer networks,
   leased-line telephone networks, cable television networks, etc.)
2. Design of a lightly used transportation network to minimize the total cost of providing the links (rail lines, roads, etc.)
3. Design of a network of high-voltage electrical power transmission lines
4. Design of a network of wiring on electrical equipment (e.g., a digital computer system) to minimize the total length of the wire
5. Design of a network of pipelines to connect a number of locations

   In this age of the information superhighway, applications of this first type have
become particularly important. In a telecommunication network, it is only necessary
to insert enough links to provide a path between every pair of nodes, so designing such
a network is a classic application of the minimum spanning tree problem. Because

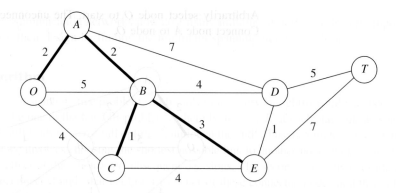

The unconnected node closest to node O, A, B, C, or E is node D (closest to E). Connect node D to node E.

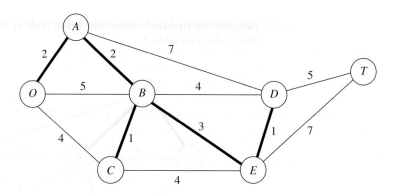

The only remaining unconnected node is node T. It is closest to node D. Connect node T to node D.

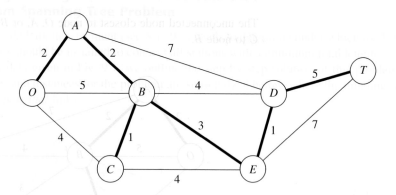

All nodes are now connected, so this solution to the problem is the desired (optimal) one. The total length of the links is 14 miles.

Although it may appear at first glance that the choice of the initial node will affect the resulting final solution (and its total link length) with this procedure, it really does not. We suggest you verify this fact for the example by reapplying the algorithm, starting with nodes other than node O.

The minimum spanning tree problem is the one problem we consider in this chapter that falls into the broad category of *network design*. In this category, the objective is to design the most appropriate network for the given application (frequently involving transportation systems) rather than analyzing an already designed network. Selected Reference 6 provides a survey of this important area.

9.5 THE MAXIMUM FLOW PROBLEM

Now recall that the third problem facing the Seervada Park management (see Sec. 9.1) during the peak season is to determine how to route the various tram trips from the park entrance (station O in Fig. 9.1) to the scenic wonder (station T) to maximize the number of trips per day. (Each tram will return by the same route it took on the outgoing trip, so the analysis focuses on outgoing trips only.) To avoid unduly disturbing the ecology and wildlife of the region, strict upper limits have been imposed on the number of outgoing trips allowed per day in the outbound direction on each individual road. For each road, the direction of travel for outgoing trips is indicated by an arrow in Fig. 9.6. The number at the base of the arrow gives the upper limit on the number of outgoing trips allowed per day. Given the limits, one *feasible solution* is to send 7 trams per day, with 5 using the route $O \rightarrow B \rightarrow E \rightarrow T$, 1 using $O \rightarrow B \rightarrow C \rightarrow E \rightarrow T$, and 1 using $O \rightarrow B \rightarrow C \rightarrow E \rightarrow D \rightarrow T$. However, because this solution blocks the use of any routes starting with $O \rightarrow C$ (because the $E \rightarrow T$ and $E \rightarrow D$ capacities are fully used), it is easy to find better feasible solutions. Many *combinations* of routes (and the number of trips to assign to each one) need to be considered to find the one(s) maximizing the number of trips made per day. This kind of problem is called a *maximum flow problem*.

In general terms, the maximum flow problem can be described as follows.

1. All flow through a directed and connected network originates at one node, called the **source,** and terminates at one other node, called the **sink.** (The source and sink in the Seervada Park problem are the park entrance at node O and the scenic wonder at node T, respectively.)
2. All the remaining nodes are *transshipment nodes.* (These are nodes A, B, C, D, and E in the Seervada Park problem.)
3. Flow through an arc is allowed only in the direction indicated by the arrowhead, where the maximum amount of flow is given by the *capacity* of that arc. At the *source,* all arcs point away from the node. At the *sink,* all arcs point into the node.

■ **FIGURE 9.6**
The Seervada Park maximum flow problem.

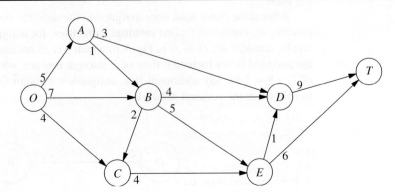

4. The objective is to maximize the total amount of flow from the source to the sink. This amount is measured in either of two equivalent ways, namely, either the amount *leaving the source* or the amount *entering the sink*.

Some Applications

Here are some typical kinds of applications of the maximum flow problem.

1. Maximize the flow through a company's distribution network from its factories to its customers.
2. Maximize the flow through a company's supply network from its vendors to its factories.
3. Maximize the flow of oil through a system of pipelines.
4. Maximize the flow of water through a system of aqueducts.
5. Maximize the flow of vehicles through a transportation network.

For some of these applications, the flow through the network may originate at more than one node and may also terminate at more than one node, even though a maximum flow problem is allowed to have only a single source and a single sink. For example, a company's distribution network commonly has multiple factories and multiple customers. A clever reformulation is used to make such a situation fit the maximum flow problem. This reformulation involves expanding the original network to include a *dummy source,* a *dummy sink,* and some new arcs. The dummy source is treated as the node that originates all the flow that, in reality, originates from some of the other nodes. For each of these other nodes, a new arc is inserted that leads from the dummy source to this node, where the capacity of this arc equals the maximum flow that, in reality, can originate from this node. Similarly, the dummy sink is treated as the node that absorbs all the flow that, in reality, terminates at some of the other nodes. Therefore, a new arc is inserted from each of these other nodes to the dummy sink, where the capacity of this arc equals the maximum flow that, in reality, can terminate at this node. Because of all these changes, all the nodes in the original network now are transshipment nodes, so the expanded network has the required single source (the dummy source) and single sink (the dummy sink) to fit the maximum flow problem.

An Algorithm

Because the maximum flow problem can be formulated as a *linear programming problem* (see Prob. 9.5-2), it can be solved by the simplex method, so any of the linear programming software packages introduced in Chaps. 3 and 4 can be used. However, an even more efficient *augmenting path algorithm* is available for solving this problem. This algorithm is based on two intuitive concepts, a *residual network* and an *augmenting path.*

After some flows have been assigned to the arcs, the **residual network** shows the *remaining* arc capacities (called **residual capacities**) for assigning *additional* flows. For example, consider arc $O \rightarrow B$ in Fig. 9.6, which has an arc capacity of 7. Now suppose that the assigned flows include a flow of 5 through this arc, which leaves a residual capacity of $7 - 5 = 2$ for any additional flow assignment through $O \rightarrow B$. This status is depicted as follows in the residual network.

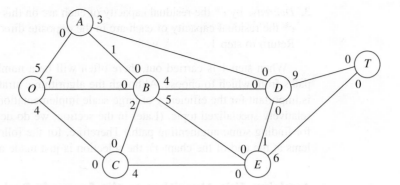

■ FIGURE 9.7
The initial residual network
for the Seervada Park
maximum flow problem.

The number on an arc next to a node gives the residual capacity for flow *from* that node *to* the other node. Therefore, in addition to the residual capacity of 2 for flow from O to B, the 5 on the right indicates a residual capacity of 5 for assigning some flow from B to O (that is, for canceling some previously assigned flow from O to B).

Initially, before any flows have been assigned, the residual network for the Seervada Park problem has the appearance shown in Fig. 9.7. Every arc in the original network (Fig. 9.6) has been changed from a *directed* arc to an *undirected* arc. However, the arc capacity in the original direction remains the same and the arc capacity in the opposite direction is *zero*, so the constraints on flows are unchanged.

Subsequently, whenever some amount of flow is assigned to an arc, that amount is *subtracted* from the residual capacity in the same direction and *added* to the residual capacity in the opposite direction.

An **augmenting path** is a directed path from the source to the sink in the residual network such that *every* arc on this path has *strictly positive* residual capacity. The *minimum* of these residual capacities is called the *residual capacity of the augmenting path* because it represents the amount of flow that can feasibly be added to the entire path. Therefore, each augmenting path provides an opportunity to further augment the flow through the original network.

The augmenting path algorithm repeatedly selects some augmenting path and adds a flow equal to its residual capacity to that path in the original network. This process continues until there are no more augmenting paths, so the flow from the source to the sink cannot be increased further. The key to ensuring that the final solution necessarily is optimal is the fact that augmenting paths can cancel some previously assigned flows in the original network, so an indiscriminate selection of paths for assigning flows cannot prevent the use of a better combination of flow assignments.

To summarize, each *iteration* of the algorithm consists of the following three steps.

The Augmenting Path Algorithm for the Maximum Flow Problem[1]

1. Identify an augmenting path by finding some directed path from the source to the sink in the residual network such that every arc on this path has strictly positive residual capacity. (If no augmenting path exists, the net flows already assigned constitute an optimal flow pattern.)
2. Identify the residual capacity c^* of this augmenting path by finding the *minimum* of the residual capacities of the arcs on this path. *Increase* the flow in this path by c^*.

[1]It is assumed that the arc capacities are either integers or rational numbers.

3. *Decrease* by c^* the residual capacity of each arc on this augmenting path. *Increase* by c^* the residual capacity of each arc in the opposite direction on this augmenting path. Return to step 1.

When step 1 is carried out, there often will be a number of alternative augmenting paths from which to choose. Although the algorithmic strategy for making this selection is important for the efficiency of large-scale implementations, we shall not delve into this relatively specialized topic. (Later in the section, we do describe a systematic procedure for finding some augmenting path.) Therefore, for the following example (and the problems at the end of the chapter), the selection is just made arbitrarily.

Applying This Algorithm to the Seervada Park Maximum Flow Problem

Applying this algorithm to the Seervada Park problem (see Fig. 9.6 for the original network) yields the results summarized next. (Also see the Worked Examples section of the book's website for **another example** of the application of this algorithm.) Starting with the initial residual network given in Fig. 9.7, we give the new residual network after each one or two iterations, where the total amount of flow from O to T achieved thus far is shown in **boldface** (next to nodes O and T).

Iteration 1: In Fig. 9.7, one of several augmenting paths is $O \rightarrow B \rightarrow E \rightarrow T$, which has a residual capacity of $\min\{7, 5, 6\} = 5$. By assigning a flow of 5 to this path, the resulting residual network is

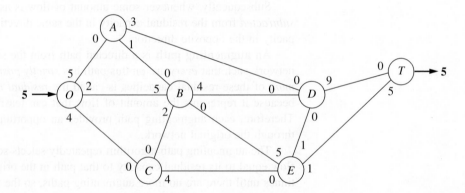

Iteration 2: Assign a flow of 3 to the augmenting path $O \rightarrow A \rightarrow D \rightarrow T$. The resulting residual network is

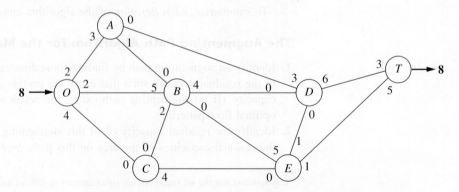

Iteration 3: Assign a flow of 1 to the augmenting path $O \rightarrow A \rightarrow B \rightarrow D \rightarrow T$.

Iteration 4: Assign a flow of 2 to the augmenting path $O \rightarrow B \rightarrow D \rightarrow T$. The resulting residual network is

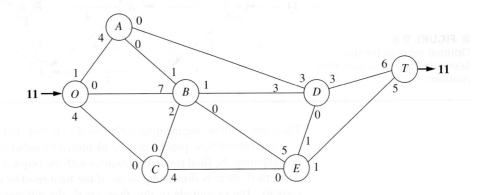

Iteration 5: Assign a flow of 1 to the augmenting path $O \rightarrow C \rightarrow E \rightarrow D \rightarrow T$.

Iteration 6: Assign a flow of 1 to the augmenting path $O \rightarrow C \rightarrow E \rightarrow T$. The resulting residual network is

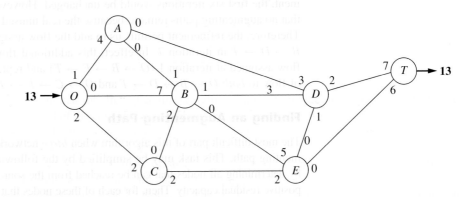

Iteration 7: Assign a flow of 1 to the augmenting path $O \rightarrow C \rightarrow E \rightarrow B \rightarrow D \rightarrow T$. The resulting residual network is

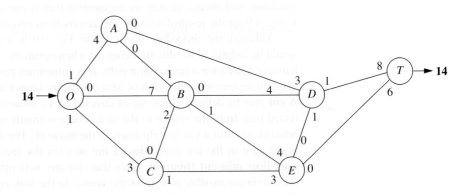

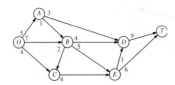

	A	B	C	D	E	F	G	H	I	J	K
1		Seervada Park Maximum Flow Problem									
2											
3		From	To	Flow		Capacity		Nodes	Net Flow		Supply/Demand
4		O	A	4	<=	5		O	14		
5		O	B	7	<=	7		A	0	=	0
6		O	C	3	<=	4		B	0	=	0
7		A	B	1	<=	1		C	0	=	0
8		A	D	3	<=	3		D	0	=	0
9		B	C	0	<=	2		E	0	=	0
10		B	D	4	<=	4		T	-14		
11		B	E	4	<=	5					
12		C	E	3	<=	4					
13		D	T	8	<=	9					
14		E	D	1	<=	1					
15		E	T	6	<=	6					
16											
17		**Maximum Flow**		14							

■ **FIGURE 9.11**

A spreadsheet formulation for the Seervada Park maximum flow problem, where the changing cells Flow (D4:D15) show the optimal solution obtained by the Excel Solver and the target cell MaxFlow (D17) gives the resulting maximum flow through the network. The network next to the spreadsheet shows the Seervada Park maximum flow problem as it was originally depicted in Fig. 9.6.

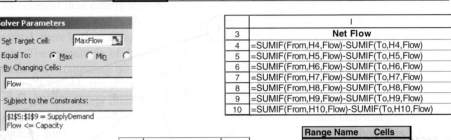

	I
3	**Net Flow**
4	=SUMIF(From,H4,Flow)-SUMIF(To,H4,Flow)
5	=SUMIF(From,H5,Flow)-SUMIF(To,H5,Flow)
6	=SUMIF(From,H6,Flow)-SUMIF(To,H6,Flow)
7	=SUMIF(From,H7,Flow)-SUMIF(To,H7,Flow)
8	=SUMIF(From,H8,Flow)-SUMIF(To,H8,Flow)
9	=SUMIF(From,H9,Flow)-SUMIF(To,H9,Flow)
10	=SUMIF(From,H10,Flow)-SUMIF(To,H10,Flow)

	C	D
17	Maximum Flow	=I4

Range Name	Cells
Capacity	F4:F15
Flow	D4:D15
From	B4:B15
MaxFlow	D17
NetFlow	I4:I10
Nodes	H4:H10
SupplyDemand	K5:K9
To	C4:C15

these quantities are entered in the changing cells Flow (D4:D15). Employing the equations given in the bottom right-hand corner of the figure, these flows then are used to calculate the net flow generated at each of the nodes (see columns H and I). These net flows are required to be 0 for the transshipment nodes (A, B, C, D, and E), as indicated by the first set of constraints (I5:I9 = SupplyDemand) in the Solver dialogue box. The second set of constraints (Flow ≤ Capacity) specifies the arc capacity constraints. The total amount of flow from the source (node O) to the sink (node T) equals the flow generated at the source (cell I4), so the target cell MaxFlow (D17) is set equal to I4. After specifying *maximization* of the target cell in the Solver dialogue box and then clicking on the Solve button, the optimal solution shown in Flow (D4:D15) is obtained.

9.6   THE MINIMUM COST FLOW PROBLEM

The minimum cost flow problem holds a central position among network optimization models, both because it encompasses such a broad class of applications and because it can be solved extremely efficiently. Like the maximum flow problem, it considers flow through a network with limited arc capacities. Like the shortest-path problem, it considers a cost (or distance) for flow through an arc. Like the transportation problem or assignment problem of Chap. 8, it can consider multiple sources (supply nodes) and multiple destinations (demand nodes) for the flow, again with associated costs. In fact, all four of these previously studied problems are special cases of the minimum cost flow problem, as we will demonstrate shortly.

The reason that the minimum cost flow problem can be solved so efficiently is that it can be formulated as a linear programming problem so it can be solved by a streamlined version of the simplex method called the *network simplex method*. We describe this algorithm in the next section.

The minimum cost flow problem is described below.

1. The network is a *directed* and *connected* network.
2. *At least one* of the nodes is a *supply node*.
3. *At least one* of the other nodes is a *demand node*.
4. All the remaining nodes are *transshipment nodes*.
5. Flow through an arc is allowed only in the direction indicated by the arrowhead, where the maximum amount of flow is given by the *capacity* of that arc. (If flow can occur in both directions, this would be represented by a pair of arcs pointing in opposite directions.)
6. The network has enough arcs with sufficient capacity to enable all the flow generated at the *supply nodes* to reach all the *demand nodes*.
7. The cost of the flow through each arc is *proportional* to the amount of that flow, where the cost per unit flow is known.
8. The objective is to minimize the total cost of sending the available supply through the network to satisfy the given demand. (An alternative objective is to maximize the total profit from doing this.)

Some Applications

Probably the most important kind of application of minimum cost flow problems is to the operation of a company's distribution network. As summarized in the first row of Table 9.3, this kind of application always involves determining a plan for shipping goods from its *sources* (factories, etc.) to *intermediate storage facilities* (as needed) and then on to the *customers*.

For some applications of minimum cost flow problems, all the transshipment nodes are *processing facilities* rather than intermediate storage facilities. This is the case for

■ **TABLE 9.3 Typical kinds of applications of minimum cost flow problems**

Kind of Application	Supply Nodes	Transshipment Nodes	Demand Nodes
Operation of a distribution network	Sources of goods	Intermediate storage facilities	Customers
Solid waste management	Sources of solid waste	Processing facilities	Landfill locations
Operation of a supply network	Vendors	Intermediate warehouses	Processing facilities
Coordinating product mixes at plants	Plants	Production of a specific product	Market for a specific product
Cash flow management	Sources of cash at a specific time	Short-term investment options	Needs for cash at a specific time

An especially challenging problem encountered daily by any major airline company is how to compensate effectively for disruptions in the airline's flight schedules. Bad weather can disrupt flight arrivals and departures; so can mechanical problems. Each delay or cancellation involving a particular airplane can then cause subsequent delays or cancellations because that airplane is not available on time for its next scheduled flights.

Such delays or cancellations may require both reassigning crews to flights and readjusting the plans for which airplanes will be used to fly the respective flights. The application vignette in Sec. 2.2 describes how Continental Airlines led the way in applying operations research to the problem of quickly reassigning crews to flights in the most cost-effective manner. However, a different approach is needed to address the problem of quickly reassigning airplanes to flights.

An airline has two primary ways of reassigning airplanes to flights to compensate for delays or cancellations. One is to swap aircraft so that an airplane scheduled for a later flight can take the place of the delayed or cancelled airplane. The other is to use a spare airplane (often after flying it in) to replace the delayed or cancelled airplane. However, it is a real challenge to quickly make good decisions of these types when a considerable number of delays or cancellations occur throughout the day.

United Airlines has led the way in applying operations research to this problem. This is done by formulating and solving the problem as a *minimum-cost flow problem* where each node in the network represents an airport and each arc represents the route of a flight. The objective of the model then is to keep the airplanes flowing through the network in a way that minimizes the cost incurred by having delays or cancellations. When a status monitor subsystem alerts an operations controller of impending delays or cancellations, the controller provides the necessary input into the model and then solves it in order to provide the updated operating plan in a matter of minutes. This application of the minimum-cost flow problem has resulted in *reducing passenger delays by about 50 percent.*

Source: A. Rakshit, N. Krishnamurthy, and G. Yu: "System Operations Advisor: A Real-Time Decision Support System for Managing Airline Operations at United Airlines," *Interfaces,* **26**(2): 50–58, Mar.–Apr. 1996. (A link to this article is provided on our website, www.mhhe.com/hillier.)

---

solid waste management, as indicated in the second row of Table 9.3. Here, the flow of materials through the network begins at the sources of the solid waste, then goes to the facilities for processing these waste materials into a form suitable for landfill, and then sends them on to the various landfill locations. However, the objective still is to determine the flow plan that minimizes the total cost, where the cost now is for both shipping and processing.

In other applications, the *demand nodes* might be processing facilities. For example, in the third row of Table 9.3, the objective is to find the minimum cost plan for obtaining supplies from various possible vendors, storing these goods in warehouses (as needed), and then shipping the supplies to the company's processing facilities (factories, etc.). Since the total amount that could be supplied by all the vendors is more than the company needs, the network includes a *dummy demand node* that receives (at zero cost) all the unused supply capacity at the vendors.

The next kind of application in Table 9.3 (coordinating product mixes at plants) illustrates that arcs can represent something other than a shipping lane for a physical flow of materials. This application involves a company with several plants (the supply nodes) that can produce the same products but at different costs. Each arc from a supply node represents the production of one of the possible products at that plant, where this arc leads to the transshipment node that corresponds to this product. Thus, this transshipment node has an arc coming in from each plant capable of producing this product, and then the arcs leading out of this node go to the respective customers (the demand nodes) for this product. The objective is to determine how to divide each plant's production capacity among the products so as to minimize the total cost of meeting the demand for the various products.

The last application in Table 9.3 (cash flow management) illustrates that different nodes can represent some event that occurs at different times. In this case, each supply node represents a specific time (or time period) when some cash will become available to the company (through maturing accounts, notes receivable, sales of securities, borrowing, etc.). The supply at each of these nodes is the amount of cash that will become available then. Similarly, each demand node represents a specific time (or time period) when the company will need to draw on its cash reserves. The demand at each such node is the amount of cash that will be needed then. The objective is to maximize the company's income from investing the cash between each time it becomes available and when it will be used. Therefore, each transshipment node represents the choice of a specific short-term investment option (e.g., purchasing a certificate of deposit from a bank) over a specific time interval. The resulting network will have a succession of flows representing a schedule for cash becoming available, being invested, and then being used after the maturing of the investment.

Formulation of the Model

Consider a directed and connected network where the n nodes include at least one supply node and at least one demand node. The decision variables are

x_{ij} = flow through arc $i \rightarrow j$,

and the given information includes

c_{ij} = cost per unit flow through arc $i \rightarrow j$,
u_{ij} = arc capacity for arc $i \rightarrow j$,
b_i = net flow generated at node i.

The value of b_i depends on the nature of node i, where

$b_i > 0$ if node i is a supply node,
$b_i < 0$ if node i is a demand node,
$b_i = 0$ if node i is a transshipment node.

The objective is to minimize the total cost of sending the available supply through the network to satisfy the given demand.

By using the convention that summations are taken only over existing arcs, the linear programming formulation of this problem is

$$\text{Minimize} \quad Z = \sum_{i=1}^{n} \sum_{j=1}^{n} c_{ij} x_{ij},$$

subject to

$$\sum_{j=1}^{n} x_{ij} - \sum_{j=1}^{n} x_{ji} = b_i, \quad \text{for each node } i,$$

and

$$0 \leq x_{ij} \leq u_{ij}, \quad \text{for each arc } i \rightarrow j.$$

The first summation in the *node constraints* represents the total flow *out* of node i, whereas the second summation represents the total flow *into* node i, so the difference is the net flow generated at this node.

The pattern of the coefficients in these node constraints is a key characteristic of minimum cost flow problems. It is not always easy to recognize a minimum cost flow problem, but formulating (or reformulating) a problem so that its constraint coefficients have this pattern is a good way of doing so. This then enables solving the problem extremely efficiently by the network simplex method.

In some applications, it is necessary to have a lower bound $L_{ij} > 0$ for the flow through each arc $i \rightarrow j$. When this occurs, use a translation of variables $x'_{ij} = x_{ij} - L_{ij}$, with $x'_{ij} + L_{ij}$ substituted for x_{ij} throughout the model, to convert the model back to the above format with nonnegativity constraints.

It is not guaranteed that the problem actually will possess *feasible* solutions, depending partially upon which arcs are present in the network and their arc capacities. However, for a reasonably designed network, the main condition needed is the following.

Feasible solutions property: A necessary condition for a minimum cost flow problem to have any feasible solutions is that

$$\sum_{i=1}^{n} b_i = 0.$$

That is, the total flow being generated at the supply nodes equals the total flow being absorbed at the demand nodes.

If the values of b_i provided for some application violate this condition, the usual interpretation is that either the supplies or the demands (whichever are in excess) actually represent upper bounds rather than exact amounts. When this situation arose for the transportation problem in Sec. 8.1, either a dummy destination was added to receive the excess supply or a dummy source was added to send the excess demand. The analogous step now is that either a dummy demand node should be added to absorb the excess supply (with $c_{ij} = 0$ arcs added from every supply node to this node) or a dummy supply node should be added to generate the flow for the excess demand (with $c_{ij} = 0$ arcs added from this node to every demand node).

For many applications, b_i and u_{ij} will have *integer* values, and implementation will require that the flow quantities x_{ij} also be integer. Fortunately, just as for the transportation problem, this outcome is guaranteed without explicitly imposing integer constraints on the variables because of the following property.

Integer solutions property: For minimum cost flow problems where every b_i and u_{ij} have integer values, all the basic variables in *every* basic feasible (BF) solution (including an optimal one) also have integer values.

An Example

Figure 9.12 shows an example of a minimum cost flow problem. This network actually is the *distribution network* for the Distribution Unlimited Co. problem presented in Sec. 3.4 (see Fig. 3.13). The quantities given in Fig. 3.13 provide the values of the b_i, c_{ij}, and u_{ij} shown here. The b_i values in Fig. 9.12 are shown in square brackets by the nodes, so the supply nodes ($b_i > 0$) are A and B (the company's two factories), the demand nodes ($b_i < 0$) are D and E (two warehouses), and the one transshipment node ($b_i = 0$) is C (a distribution center). The c_{ij} values are shown next to the arcs. In this example, all but two of the arcs have arc capacities exceeding the total flow generated (90), so $u_{ij} = \infty$ for all practical purposes. The two exceptions are arc $A \rightarrow B$, where $u_{AB} = 10$, and arc $C \rightarrow E$, which has $u_{CE} = 80$.

The linear programming model for this example is

Minimize $Z = 2x_{AB} + 4x_{AC} + 9x_{AD} + 3x_{BC} + x_{CE} + 3x_{DE} + 2x_{ED},$

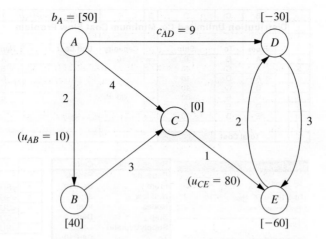

■ **FIGURE 9.12**
The Distribution Unlimited
Co. problem formulated as a
minimum cost flow problem.

subject to

$$
\begin{array}{rcl}
x_{AB} + x_{AC} + x_{AD} & = & 50 \\
-x_{AB} \qquad\qquad + x_{BC} & = & 40 \\
- x_{AC} \qquad - x_{BC} + x_{CE} & = & 0 \\
- x_{AD} \qquad\qquad + x_{DE} - x_{ED} & = & -30 \\
- x_{CE} - x_{DE} + x_{ED} & = & -60
\end{array}
$$

and

$$
x_{AB} \le 10, \qquad x_{CE} \le 80, \qquad \text{all } x_{ij} \ge 0.
$$

Now note the pattern of coefficients for each variable in the set of five *node constraints* (the equality constraints). Each variable has exactly *two* nonzero coefficients, where one is $+1$ and the other is -1. This pattern recurs in *every* minimum cost flow problem, and it is this special structure that leads to the integer solutions property.

Another implication of this special structure is that (any) one of the node constraints is *redundant*. The reason is that summing all these constraint equations yields nothing but zeros on both sides (assuming feasible solutions exist, so the b_i values sum to zero), so the negative of any one of these equations equals the sum of the rest of the equations. With just $n - 1$ nonredundant node constraints, these equations provide just $n - 1$ basic variables for a BF solution. In the next section, you will see that the network simplex method treats the $x_{ij} \le u_{ij}$ constraints as mirror images of the nonnegativity constraints, so the *total* number of basic variables is $n - 1$. This leads to a direct correspondence between the $n - 1$ arcs of a *spanning tree* and the $n - 1$ basic variables—but more about that story later.

Using Excel to Formulate and Solve Minimum Cost Flow Problems

Excel provides a convenient way of formulating and solving small minimum cost flow problems like this one, as well as somewhat larger problems. Figure 9.13 shows how this can be done. The format is almost the same as displayed in Fig. 9.11 for a maximum flow problem. One difference is that the unit costs (c_{ij}) now need to be included (in column G). Because b_i values are specified for every node, net flow constraints are needed for all the nodes. However, only two of the arcs happen to need arc capacity

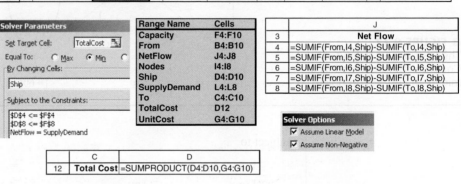

FIGURE 9.13
A spreadsheet formulation for the Distribution Unlimited Co. minimum cost flow problem, where the changing cells Ship (D4:D10) show the optimal solution obtained by the Excel Solver and the target cell TotalCost (D12) gives the resulting total cost of the flow of shipments through the network.

constraints. The target cell TotalCost (D12) now gives the total cost of the flow (shipments) through the network (see its equation at the bottom of the figure), so the objective specified in the Solver dialogue box is to *minimize* this quantity. The changing cells Ship (D4:D10) in this spreadsheet show the optimal solution obtained after clicking on the Solve button.

For much larger minimum cost flow problems, the *network simplex method* described in the next section provides a considerably more efficient solution procedure. It also is an attractive option for solving various special cases of the minimum cost flow problem outlined below. This algorithm is commonly included in mathematical programming software packages. For example, it is one of the options with CPLEX.

We shall soon solve this same example by the network simplex method. However, let us first see how some special cases fit into the network format of the minimum cost flow problem.

Special Cases

The Transportation Problem. To formulate the transportation problem presented in Sec. 8.1 as a minimum cost flow problem, a *supply node* is provided for each *source,* as well as a *demand node* for each *destination,* but no transshipment nodes are included in the network. All the arcs are directed from a supply node to a demand node, where distributing x_{ij} units from source i to destination j corresponds to a flow of x_{ij} through arc $i \rightarrow j$. The cost c_{ij} per unit distributed becomes the cost c_{ij} per unit of flow. Since the transportation problem does not impose upper bound constraints on individual x_{ij}, all the $u_{ij} = \infty$.

Using this formulation for the P & T Co. transportation problem presented in Table 8.2 yields the network shown in Fig. 8.2. The corresponding network for the general transportation problem is shown in Fig. 8.3.

The Assignment Problem. Since the assignment problem discussed in Sec. 8.3 is a special type of transportation problem, its formulation as a minimum cost flow problem fits into the same format. The additional factors are that (1) the number of supply nodes equals the number of demand nodes, (2) $b_i = 1$ for each supply node, and (3) $b_i = -1$ for each demand node.

Figure 8.5 shows this formulation for the general assignment problem.

The Transshipment Problem.   This special case actually includes all the general features of the minimum cost flow problem except for not having (finite) arc capacities. Thus, any minimum cost flow problem where each arc can carry any desired amount of flow is also called a transshipment problem.

For example, the Distribution Unlimited Co. problem shown in Fig. 9.13 would be a transshipment problem if the upper bounds on the flow through arcs $A \rightarrow B$ and $C \rightarrow E$ were removed.

Transshipment problems frequently arise as generalizations of transportation problems where units being distributed from each source to each destination can first pass through intermediate points. These intermediate points may include other sources and destinations, as well as additional transfer points that would be represented by transshipment nodes in the network representation of the problem. For example, the Distribution Unlimited Co. problem can be viewed as a generalization of a transportation problem with two sources (the two factories represented by nodes A and B in Fig. 9.13), two destinations (the two warehouses represented by nodes D and E), and one additional intermediate transfer point (the distribution center represented by node C).

(Chapter 23 on the book's website includes a further discussion of the transshipment problem.)

The Shortest-Path Problem.   Now consider the main version of the shortest-path problem presented in Sec. 9.3 (finding the shortest path from one origin to one destination through an *undirected* network). To formulate this problem as a minimum cost flow problem, one supply node with a supply of 1 is provided for the origin, one demand node with a demand of 1 is provided for the destination, and the rest of the nodes are transshipment nodes. Because the network of our shortest-path problem is undirected, whereas the minimum cost flow problem is assumed to have a directed network, we replace each link with a pair of directed arcs in opposite directions (depicted by a single line with arrowheads at both ends). The only exceptions are that there is no need to bother with arcs *into* the supply node or *out of* the demand node. The distance between nodes i and j becomes the unit cost c_{ij} or c_{ji} for flow in either direction between these nodes. As with the preceding special cases, no arc capacities are imposed, so all $u_{ij} = \infty$.

Figure 9.14 depicts this formulation for the Seervada Park shortest-path problem shown in Fig. 9.1, where the numbers next to the lines now represent the unit cost of flow in either direction.

The Maximum Flow Problem.   The last special case we shall consider is the maximum flow problem described in Sec. 9.5. In this case a network already is provided with one supply node (the source), one demand node (the sink), and various transshipment nodes, as well as the various arcs and arc capacities. Only three adjustments are needed to fit this problem into the format for the minimum cost flow problem. First, set $c_{ij} = 0$ for all existing arcs to reflect the absence of costs in the maximum flow problem. Second, select a quantity $\overline{F}$, which is a safe upper bound on the maximum feasible flow through the network, and then assign a supply and a demand of $\overline{F}$ to the supply node and the demand node, respectively. (Because all *other* nodes are transshipment nodes, they automatically have $b_i = 0$.) Third, add an arc going directly from the supply node to the demand node and assign it an arbitrarily large unit cost of $c_{ij} = M$ as well as an unlimited arc capacity ($u_{ij} = \infty$). Because of this positive unit cost for this arc and the zero unit cost

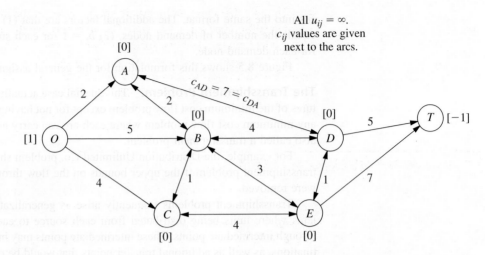

FIGURE 9.14
Formulation of the Seervada Park shortest-path problem as a minimum cost flow problem.

for all the *other* arcs, the minimum cost flow problem will send the maximum feasible flow through the *other* arcs, which achieves the objective of the maximum flow problem.

Applying this formulation to the Seervada Park maximum flow problem shown in Fig. 9.6 yields the network given in Fig. 9.15, where the numbers given next to the original arcs are the arc capacities.

Final Comments.   Except for the transshipment problem, each of these special cases has been the focus of a previous section in either this chapter or Chap. 8. When each was first presented, we talked about a special-purpose algorithm for solving it very efficiently. Therefore, it certainly is not necessary to reformulate these special cases to fit the format of the minimum cost flow problem in order to solve them. However, when a computer code is not readily available for the special-purpose algorithm, it is very reasonable to use the network simplex method instead. In fact, recent implementations of the network simplex method have become so powerful that it now provides an excellent alternative to the special-purpose algorithm.

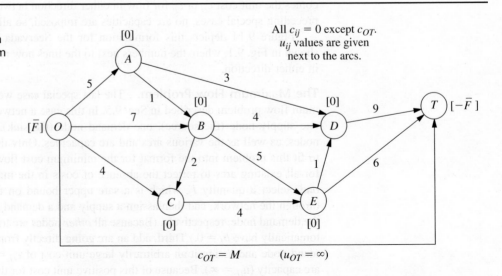

FIGURE 9.15
Formulation of the Seervada Park maximum flow problem as a minimum cost flow problem.

The fact that these problems are special cases of the minimum cost flow problem is of interest for other reasons as well. One reason is that the underlying theory for the minimum cost flow problem and for the network simplex method provides a unifying theory for all these special cases. Another reason is that some of the many applications of the minimum cost flow problem include features of one or more of the special cases, so it is important to know how to reformulate these features into the broader framework of the general problem.

■ 9.7   THE NETWORK SIMPLEX METHOD

The network simplex method is a highly streamlined version of the simplex method for solving minimum cost flow problems. As such, it goes through the same basic steps at each iteration—finding the entering basic variable, determining the leaving basic variable, and solving for the new BF solution—in order to move from the current BF solution to a better adjacent one. However, it executes these steps in ways that exploit the special network structure of the problem without ever needing a simplex tableau.

You may note some similarities between the network simplex method and the transportation simplex method presented in Sec. 8.2. In fact, both are streamlined versions of the simplex method that provide alternative algorithms for solving transportation problems in similar ways. The network simplex method extends these ideas to solving other types of minimum cost flow problems as well.

In this section, we provide a somewhat abbreviated description of the network simplex method that focuses just on the main concepts. We omit certain details needed for a full computer implementation, including how to construct an initial BF solution and how to perform certain calculations (such as for finding the entering basic variable) in the most efficient manner. These details are provided in various more specialized textbooks, such as Selected References 1 and 3.

Incorporating the Upper Bound Technique

The first concept is to incorporate the upper bound technique described in Sec. 7.3 to deal efficiently with the arc capacity constraints $x_{ij} \leq u_{ij}$. Thus, rather than these constraints being treated as *functional* constraints, they are handled just as *nonnegativity* constraints are. Therefore, they are considered only when the leaving basic variable is determined. In particular, as the entering basic variable is increased from zero, the leaving basic variable is the *first* basic variable that reaches either its lower bound (0) or its upper bound (u_{ij}). A nonbasic variable at its upper bound $x_{ij} = u_{ij}$ is replaced with $x_{ij} = u_{ij} - y_{ij}$, so $y_{ij} = 0$ becomes the nonbasic variable. See Sec. 7.3 for further details.

In our current context, y_{ij} has an interesting network interpretation. Whenever y_{ij} becomes a basic variable with a strictly positive value ($\leq u_{ij}$), this value can be thought of as flow from node j to node i (so in the "wrong" direction through arc $i \to j$) that, in actuality, is *canceling* that amount of the previously assigned flow ($x_{ij} = u_{ij}$) from node i to node j. Thus, when $x_{ij} = u_{ij}$ is replaced with $x_{ij} = u_{ij} - y_{ij}$, we also replace the *real* arc $i \to j$ with the **reverse arc** $j \to i$, where this new arc has arc capacity u_{ij} (the maximum amount of the $x_{ij} = u_{ij}$ flow that can be canceled) and unit cost $-c_{ij}$ (since each unit of flow canceled saves c_{ij}). To reflect the flow of $x_{ij} = u_{ij}$ through the deleted arc, we shift this amount of net flow generated from node i to node j by *decreasing* b_i by u_{ij} and *increasing* b_j by u_{ij}. Later, if y_{ij} becomes the leaving basic variable by reaching its upper bound, then $y_{ij} = u_{ij}$ is replaced with $y_{ij} = u_{ij} - x_{ij}$ with $x_{ij} = 0$ as the new

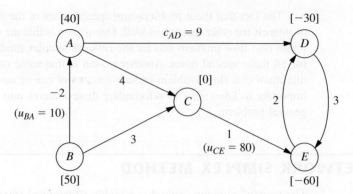

■ FIGURE 9.16
The adjusted network for the example when the upper-bound technique leads to replacing $x_{AB} = 10$ with $x_{AB} = 10 - y_{AB}$.

nonbasic variable, so the above process would be reversed (replace arc $j \rightarrow i$ by arc $i \rightarrow j$, etc.) to the original configuration.

To illustrate this process, consider the minimum cost flow problem shown in Fig. 9.12. While the network simplex method is generating a sequence of BF solutions, suppose that x_{AB} has become the leaving basic variable for some iteration by reaching its upper bound of 10. Consequently, $x_{AB} = 10$ is replaced with $x_{AB} = 10 - y_{AB}$, so $y_{AB} = 0$ becomes the new nonbasic variable. At the same time, we replace arc $A \rightarrow B$ with arc $B \rightarrow A$ (with y_{AB} as its flow quantity), and we assign this new arc a capacity of 10 and a unit cost of -2. To take $x_{AB} = 10$ into account, we also decrease b_A from 50 to 40 and increase b_B from 40 to 50. The resulting adjusted network is shown in Fig. 9.16.

We shall soon illustrate the entire network simplex method with this same example, starting with $y_{AB} = 0$ ($x_{AB} = 10$) as a nonbasic variable and so using Fig. 9.16. A later iteration will show x_{CE} reaching its upper bound of 80 and so being replaced with $x_{CE} = 80 - y_{CE}$, and so on, and then the next iteration has y_{AB} reaching its upper bound of 10. You will see that all these operations are performed directly on the network, so we will not need to use the x_{ij} or y_{ij} labels for arc flows or even to keep track of which arcs are *real* arcs and which are *reverse* arcs (except when we record the final solution). Using the upper bound technique leaves the *node constraints* (flow out minus flow in = b_i) as the only functional constraints. Minimum cost flow problems tend to have far more arcs than nodes, so the resulting number of functional constraints generally is only a small fraction of what it would have been if the arc capacity constraints had been included. The computation time for the simplex method goes up relatively rapidly with the number of functional constraints, but only slowly with the number of variables (or the number of bounding constraints on these variables). Therefore, incorporating the upper bound technique here tends to provide a tremendous saving in computation time.

However, this technique is not needed for *uncapacitated* minimum cost flow problems (including all but the last special case considered in the preceding section), where there are no arc capacity constraints.

Correspondence between BF Solutions and Feasible Spanning Trees

The most important concept underlying the network simplex method is its network representation of *BF solutions*. Recall from Sec. 9.6 that with n nodes, every BF solution has $(n - 1)$ basic variables, where each basic variable x_{ij} represents the flow through arc $i \rightarrow j$. These $(n - 1)$ arcs are referred to as **basic arcs**. (Similarly, the arcs corresponding to the *nonbasic* variables $x_{ij} = 0$ or $y_{ij} = 0$ are called **nonbasic arcs**.)

A key property of basic arcs is that they never form undirected *cycles*. (This property prevents the resulting solution from being a weighted average of another pair of feasible solutions, which would violate one of the general properties of BF solutions.) However, *any* set of $n - 1$ arcs that contains no undirected cycles forms a *spanning tree*. Therefore, any complete set of $n - 1$ basic arcs forms a spanning tree.

Thus, BF solutions can be obtained by "solving" spanning trees, as summarized below.

A **spanning tree solution** is obtained as follows:

1. For the arcs *not* in the spanning tree (the nonbasic arcs), set the corresponding variables (x_{ij} or y_{ij}) equal to zero.
2. For the arcs that are in the spanning tree (the basic arcs), solve for the corresponding variables (x_{ij} or y_{ij}) in the system of linear equations provided by the node constraints.

(The network simplex method actually solves for the new BF solution from the current one much more efficiently, without solving this system of equations from scratch.) Note that this solution process does not consider either the nonnegativity constraints or the arc capacity constraints for the basic variables, so the resulting spanning tree solution may or may not be feasible with respect to these constraints—which leads to our next definition.

A **feasible spanning tree** is a spanning tree whose solution from the node constraints also satisfies all the other constraints ($0 \leq x_{ij} \leq u_{ij}$ or $0 \leq y_{ij} \leq u_{ij}$).

With these definitions, we now can summarize our key conclusion as follows:

The **fundamental theorem for the network simplex method** says that basic solutions are *spanning tree solutions* (and conversely) and that BF solutions are solutions for *feasible spanning trees* (and conversely).

To begin illustrating the application of this fundamental theorem, consider the network shown in Fig. 9.16 that results from replacing $x_{AB} = 10$ with $x_{AB} = 10 - y_{AB}$ for our example in Fig. 9.12. One spanning tree for this network is the one shown in Fig. 9.3e, where the arcs are $A \rightarrow D$, $D \rightarrow E$, $C \rightarrow E$, and $B \rightarrow C$. With these as the *basic arcs,* the process of finding the spanning tree solution is shown below. On the left is the set of node constraints given in Sec. 9.6 after $10 - y_{AB}$ is substituted for x_{AB}, where the *basic* variables are shown in **boldface.** On the right, starting at the top and moving down, is the sequence of steps for setting or calculating the values of the variables.

$$
\begin{array}{rclcl}
 & & y_{AB} = 0,\ x_{AC} = 0,\ x_{ED} = 0 \\
-y_{AB} + x_{AC} + \mathbf{x_{AD}} & = & 40 & & x_{AD} = 40. \\
y_{AB} \qquad\qquad + \mathbf{x_{BC}} & = & 50 & & x_{BC} = 50. \\
-x_{AC} \quad - \mathbf{x_{BC}} + \mathbf{x_{CE}} & = & 0 & \text{so} & x_{CE} = 50. \\
-\mathbf{x_{AD}} \qquad\quad + \mathbf{x_{DE}} - x_{ED} & = & -30 & \text{so} & x_{DE} = 10. \\
-\mathbf{x_{CE}} - \mathbf{x_{DE}} + x_{ED} & = & -60 & & \text{Redundant.}
\end{array}
$$

Since the values of all these basic variables satisfy the nonnegativity constraints and the one relevant arc capacity constraint ($x_{CE} \leq 80$), the spanning tree is a *feasible spanning tree,* so we have a *BF solution.*

We shall use this solution as the initial BF solution for demonstrating the network simplex method. Figure 9.17 shows its network representation, namely, the feasible spanning tree and its solution. Thus, the numbers given next to the arcs now represent *flows* (values of x_{ij}) rather than the unit costs c_{ij} previously given. (To help you distinguish, we shall always put parentheses around flows but not around costs.)

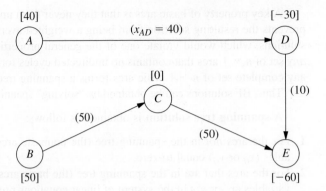

■ FIGURE 9.17
The initial feasible spanning
tree and its solution for the
example.

Selecting the Entering Basic Variable

To begin an iteration of the network simplex method, recall that the standard simplex method criterion for selecting the entering basic variable is to choose the nonbasic variable which, when increased from zero, will *improve Z at the fastest rate.* Now let us see how this is done without having a simplex tableau.

To illustrate, consider the nonbasic variable x_{AC} in our initial BF solution, i.e., the nonbasic arc $A \rightarrow C$. Increasing x_{AC} from zero to some value θ means that the arc $A \rightarrow C$ with flow θ must be added to the network shown in Fig. 9.17. Adding a nonbasic arc to a spanning tree *always* creates a unique undirected *cycle,* where the cycle in this case is seen in Fig. 9.18 to be *AC–CE–DE–AD.* Figure 9.18 also shows the effect of adding the flow θ to arc $A \rightarrow C$ on the other flows in the network. Specifically, the flow is thereby *increased* by θ for other arcs that have the *same* direction as $A \rightarrow C$ in the cycle (arc $C \rightarrow E$), whereas the *net* flow is *decreased* by θ for other arcs whose direction is *opposite* to $A \rightarrow C$ in the cycle (arcs $D \rightarrow E$ and $A \rightarrow D$). In the latter case, the new flow is, in effect, canceling a flow of θ in the opposite direction. Arcs not in the cycle (arc $B \rightarrow C$) are unaffected by the new flow. (Check these conclusions by noting the effect of the change in x_{AC} on the values of the other variables in the solution just derived for the initial feasible spanning tree.)

Now what is the incremental effect on Z (total flow cost) from adding the flow θ to arc $A \rightarrow C$? Figure 9.19 shows most of the answer by giving the unit cost times the change in the flow for each arc of Fig. 9.18. Therefore, the overall increment in Z is

$$\Delta Z = c_{AC}\theta + c_{CE}\theta + c_{DE}(-\theta) + c_{AD}(-\theta)$$
$$= 4\theta + \theta - 3\theta - 9\theta$$
$$= -7\theta.$$

Setting $\theta = 1$ then gives the *rate* of change of Z as x_{AC} is increased, namely,

$$\Delta Z = -7, \qquad \text{when } \theta = 1.$$

Because the objective is to *minimize* Z, this large rate of decrease in Z by increasing x_{AC} is very desirable, so x_{AC} becomes a prime candidate to be the entering basic variable.

We now need to perform the same analysis for the other nonbasic variables before we make the final selection of the entering basic variable. The only other nonbasic variables are y_{AB} and x_{ED}, corresponding to the two other nonbasic arcs $B \rightarrow A$ and $E \rightarrow D$ in Fig. 9.16.

Figure 9.20 shows the incremental effect on costs of adding arc $B \rightarrow A$ with flow θ to the initial feasible spanning tree given in Fig. 9.17. Adding this arc creates the undirected cycle *BA–AD–DE–CE–BC,* so the flow increases by θ for arcs $A \rightarrow D$ and $D \rightarrow E$

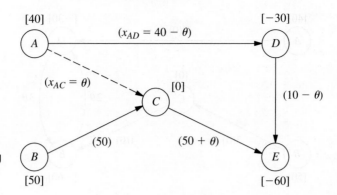

■ **FIGURE 9.18**
The effect on flows of adding arc $A \rightarrow C$ with flow θ to the initial feasible spanning tree.

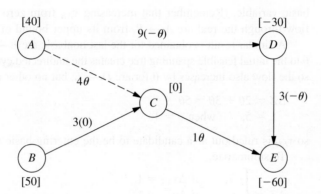

■ **FIGURE 9.19**
The incremental effect on costs of adding arc $A \rightarrow C$ with flow θ to the initial feasible spanning tree.

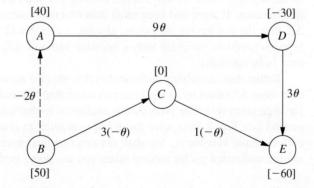

■ **FIGURE 9.20**
The incremental effect on costs of adding arc $B \rightarrow A$ with flow θ to the initial feasible spanning tree.

but decreases by θ for the two arcs in the opposite direction on this cycle, $C \rightarrow E$ and $B \rightarrow C$. These flow increments, θ and $-\theta$, are the multiplicands for the c_{ij} values in the figure. Therefore,

$$\Delta Z = -2\theta + 9\theta + 3\theta + 1(-\theta) + 3(-\theta) = 6\theta$$
$$= 6, \quad \text{when } \theta = 1.$$

The fact that Z *increases* rather than decreases when y_{AB} (flow through the reverse arc $B \rightarrow A$) is increased from zero rules out this variable as a candidate to be the entering

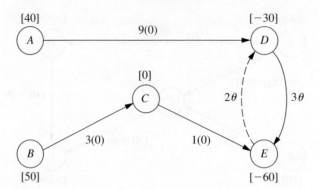

■ FIGURE 9.21
The incremental effect on costs of adding arc $E \rightarrow D$ with flow θ to the initial feasible spanning tree.

basic variable. (Remember that increasing y_{AB} from zero really means decreasing x_{AB}, flow through the real arc $A \rightarrow B$, from its upper bound of 10.)

A similar result is obtained for the last nonbasic arc $E \rightarrow D$. Adding this arc with flow θ to the initial feasible spanning tree creates the undirected cycle ED–DE shown in Fig. 9.21, so the flow also increases by θ for arc $D \rightarrow E$, but no other arcs are affected. Therefore,

$$\Delta Z = 2\theta + 3\theta = 5\theta$$
$$= 5, \qquad \text{when } \theta = 1,$$

so x_{ED} is ruled out as a candidate to be the entering basic variable.

To summarize,

$$\Delta Z = \begin{cases} -7, & \text{if } \Delta x_{AC} = 1 \\ 6, & \text{if } \Delta y_{AB} = 1 \\ 5, & \text{if } \Delta x_{ED} = 1 \end{cases}$$

so the negative value for x_{AC} implies that x_{AC} becomes the entering basic variable for the first iteration. If there had been more than one nonbasic variable with a *negative* value of ΔZ, then the one having the *largest* absolute value would have been chosen. (If there had been no nonbasic variables with a negative value of ΔZ, the current BF solution would have been optimal.)

Rather than identifying undirected cycles, etc., the network simplex method actually obtains these ΔZ values by an algebraic procedure that is considerably more efficient (especially for large networks). The procedure is analogous to that used by the transportation simplex method (see Sec. 8.2) to solve for u_i and v_j in order to obtain the value of $c_{ij} - u_i - v_j$ for each nonbasic variable x_{ij}. We shall not describe this procedure further, so you should just use the undirected cycles method when you are doing problems at the end of the chapter.

Finding the Leaving Basic Variable and the Next BF Solution

After selection of the entering basic variable, only one more quick step is needed to simultaneously determine the leaving basic variable and solve for the next BF solution. For the first iteration of the example, the key is Fig. 9.18. Since x_{AC} is the entering basic variable, the flow θ through arc $A \rightarrow C$ is to be increased from zero as far as possible until one of the basic variables reaches *either* its lower bound (0) or its upper bound (u_{ij}). For those arcs whose flow *increases* with θ in Fig. 9.18 (arcs $A \rightarrow C$ and $C \rightarrow E$), only the *upper* bounds ($u_{AC} = \infty$ and $u_{CE} = 80$) need to be considered:

$$x_{AC} = \theta \leq \infty.$$
$$x_{CE} = 50 + \theta \leq 80, \qquad \text{so} \qquad \theta \leq 30.$$

For those arcs whose flow *decreases* with θ (arcs $D \to E$ and $A \to D$), only the *lower* bound of 0 needs to be considered:

$$x_{DE} = 10 - \theta \geq 0, \qquad \text{so} \qquad \theta \leq 10.$$
$$x_{AD} = 40 - \theta \geq 0, \qquad \text{so} \qquad \theta \leq 40.$$

Arcs whose flow is unchanged by θ (i.e., those not part of the undirected cycle), which is just arc $B \to C$ in Fig. 9.18, can be ignored since no bound will be reached as θ is increased.

For the five arcs in Fig. 9.18, the conclusion is that x_{DE} must be the leaving basic variable because it reaches a bound for the smallest value of θ (10). Setting $\theta = 10$ in this figure thereby yields the flows through the basic arcs in the next BF solution:

$$x_{AC} = \theta = 10,$$
$$x_{CE} = 50 + \theta = 60,$$
$$x_{AD} = 40 - \theta = 30,$$
$$x_{BC} = 50.$$

The corresponding feasible spanning tree is shown in Fig. 9.22.

If the leaving basic variable had reached its upper bound, then the adjustments discussed for the upper bound technique would have been needed at this point (as you will see illustrated during the next two iterations). However, because it was the lower bound of 0 that was reached, nothing more needs to be done.

Completing the Example.   For the two remaining iterations needed to reach the optimal solution, the primary focus will be on some features of the upper bound technique they illustrate. The pattern for finding the entering basic variable, the leaving basic variable, and the next BF solution will be very similar to that described for the first iteration, so we only summarize these steps briefly.

Iteration 2: Starting with the feasible spanning tree shown in Fig. 9.22 and referring to Fig. 9.16 for the unit costs c_{ij}, we arrive at the calculations for selecting the entering basic variable in Table 9.4. The second column identifies the unique undirected cycle that is created by adding the nonbasic arc in the first column to this spanning tree, and the third column shows the incremental effect on costs because of the changes in flows on this cycle caused by adding a flow of $\theta = 1$ to the nonbasic arc. Arc $E \to D$ has the largest (in absolute terms) negative value of ΔZ, so x_{ED} is the entering basic variable.

We now make the flow θ through arc $E \to D$ as large as possible, while satisfying the following flow bounds:

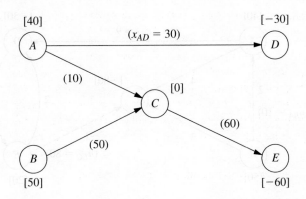

■ **FIGURE 9.22**
The second feasible spanning tree and its solution for the example.

■ **TABLE 9.4** Calculations for selecting the entering basic variable for iteration 2

Nonbasic Arc	Cycle Created	ΔZ When $\theta = 1$	
$B \rightarrow A$	BA–AC–BC	$-2 + 4 - 3 = -1$	
$D \rightarrow E$	DE–CE–AC–AD	$3 - 1 - 4 + 9 = 7$	
$E \rightarrow D$	ED–AD–AC–CE	$2 - 9 + 4 + 1 = -2$	← Minimum

$$x_{ED} = \theta \leq u_{ED} = \infty, \qquad \text{so} \qquad \theta \leq \infty.$$
$$x_{AD} = 30 - \theta \geq 0, \qquad \text{so} \qquad \theta \leq 30.$$
$$x_{AC} = 10 + \theta \leq u_{AC} = \infty, \qquad \text{so} \qquad \theta \leq \infty.$$
$$x_{CE} = 60 + \theta \leq u_{CE} = 80, \qquad \text{so} \qquad \theta \leq 20. \qquad \leftarrow \text{Minimum}$$

Because x_{CE} imposes the smallest upper bound (20) on θ, x_{CE} becomes the leaving basic variable. Setting $\theta = 20$ in the above expressions for x_{ED}, x_{AD}, and x_{AC} then yields the flow through the basic arcs for the next BF solution (with $x_{BC} = 50$ unaffected by θ), as shown in Fig. 9.23.

What is of special interest here is that the leaving basic variable x_{CE} was obtained by the variable reaching its upper bound (80). Therefore, by using the upper bound technique, x_{CE} is replaced with $80 - y_{CE}$, where $y_{CE} = 0$ is the new nonbasic variable. At the same time, the original arc $C \rightarrow E$ with $c_{CE} = 1$ and $u_{CE} = 80$ is replaced with the reverse arc $E \rightarrow C$ with $c_{EC} = -1$ and $u_{EC} = 80$. The values of b_E and b_C also are adjusted by adding 80 to b_E and subtracting 80 from b_C. The resulting adjusted network is shown in Fig. 9.24, where the nonbasic arcs are shown as dashed lines and the numbers by all the arcs are unit costs.

■ **FIGURE 9.23**
The third feasible spanning tree and its solution for the example.

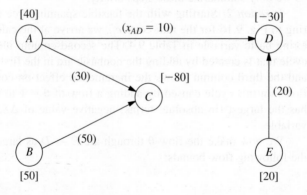

■ **FIGURE 9.24**
The adjusted network with unit costs at the completion of iteration 2.

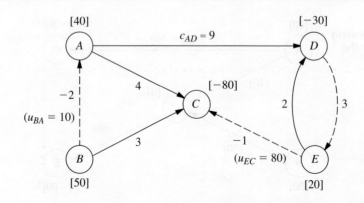

■ **TABLE 9.5** Calculations for selecting the entering basic variable for iteration 3

Nonbasic Arc	Cycle Created	ΔZ When $\theta = 1$	
$B \rightarrow A$	BA–AC–BC	$-2 + 4 - 3 = -1$	← Minimum
$D \rightarrow E$	DE–ED	$3 + 2 = 5$	
$E \rightarrow C$	EC–AC–AD–ED	$-1 - 4 + 9 - 2 = 2$	

Iteration 3: If Figs. 9.23 and 9.24 are used to initiate the next iteration, Table 9.5 shows the calculations that lead to selecting y_{AB} (reverse arc $B \rightarrow A$) as the entering basic variable. We then add as much flow θ through arc $B \rightarrow A$ as possible while satisfying the flow bounds below:

$$y_{AB} = \theta \le u_{BA} = 10, \qquad \text{so} \quad \theta \le 10. \quad \leftarrow \text{Minimum}$$
$$x_{AC} = 30 + \theta \le u_{AC} = \infty, \qquad \text{so} \quad \theta \le \infty.$$
$$x_{BC} = 50 - \theta \ge 0, \qquad \text{so} \quad \theta \le 50.$$

The smallest upper bound (10) on θ is imposed by y_{AB}, so this variable becomes the leaving basic variable. Setting $\theta = 10$ in these expressions for x_{AC} and x_{BC} (along with the unchanged values of $x_{AC} = 10$ and $x_{ED} = 20$) then yields the next BF solution, as shown in Fig. 9.25.

As with iteration 2, the leaving basic variable (y_{AB}) was obtained here by the variable reaching its upper bound. In addition, there are two other points of special interest concerning this particular choice. One is that the *entering* basic variable y_{AB} also became the *leaving* basic variable on the same iteration! This event occurs occasionally with the upper bound technique whenever increasing the entering basic variable from zero causes *its* upper bound to be reached first before any of the other basic variables reach a bound.

The other interesting point is that the arc $B \rightarrow A$ that now needs to be replaced by a *reverse* arc $A \rightarrow B$ (because of the leaving basic variable reaching an upper bound) already is a reverse arc! This is no problem, because the reverse arc for a reverse arc is simply the original *real* arc. Therefore, the arc $B \rightarrow A$ (with $c_{BA} = -2$ and $u_{BA} = 10$) in Fig. 9.24 now is replaced by arc $A \rightarrow B$ (with $c_{AB} = 2$ and $u_{AB} = 10$), which is the arc between nodes A and B in the original network shown in Fig. 9.12, and a generated net flow of 10 is shifted from node B ($b_B = 50 \rightarrow 40$) to node A ($b_A = 40 \rightarrow 50$). Simultaneously, the variable $y_{AB} = 10$ is replaced by $10 - x_{AB}$, with $x_{AB} = 0$ as the new nonbasic variable. The resulting adjusted network is shown in Fig. 9.26.

Passing the Optimality Test: At this point, the algorithm would attempt to use Figs. 9.25 and 9.26 to find the next entering basic variable with the usual calculations shown in Table 9.6. However, *none* of the nonbasic arcs gives a *negative* value of ΔZ,

■ **FIGURE 9.25**
The fourth (and final) feasible spanning tree and its solution for the example.

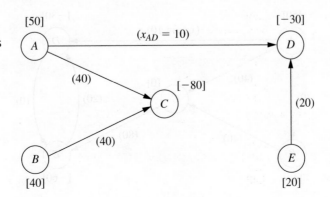

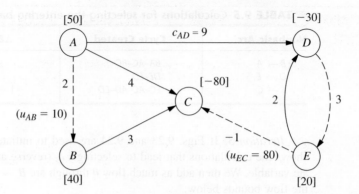

■ **FIGURE 9.26**
The adjusted network with unit costs at the completion of iteration 3.

■ **TABLE 9.6** Calculations for the optimality test at the end of iteration 3

Nonbasic Arc	Cycle Created	ΔZ When $\theta = 1$
$A \to B$	AB–BC–AC	$2 + 3 - 4 = 1$
$D \to E$	DE–ED	$3 + 2 = 5$
$E \to C$	EC–AC–AD–ED	$-1 - 4 + 9 - 2 = 2$

so an improvement in Z *cannot* be achieved by introducing flow through any of them. This means that the current BF solution shown in Fig. 9.25 has *passed* the optimality test, so the algorithm stops.

To identify the flows through real arcs rather than reverse arcs for this optimal solution, the current adjusted network (Fig. 9.26) should be compared with the original network (Fig. 9.12). Note that each of the arcs has the same direction in the two networks with the one exception of the arc between nodes C and E. This means that the only reverse arc in Fig. 9.26 is arc $E \to C$, where its flow is given by the variable y_{CE}. Therefore, calculate $x_{CE} = u_{CE} - y_{CE} = 80 - y_{CE}$. Arc $E \to C$ happens to be a nonbasic arc, so $y_{CE} = 0$ and $x_{CE} = 80$ is the flow through the real arc $C \to E$. All the other flows through real arcs are the flows given in Fig. 9.25. Therefore, the optimal solution is the one shown in Fig. 9.27.

Another complete example of applying the network simplex method is provided by the demonstration in the *Network Analysis Area* of your OR Tutor. **An additional example** is given in the Worked Examples section of the book's website as well. Also included in your IOR Tutorial is an interactive procedure for the network simplex method.

■ **FIGURE 9.27**
The optimal flow pattern in the original network for the Distribution Unlimited Co. example.

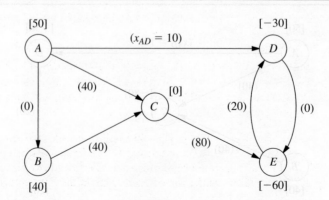

9.8 A NETWORK MODEL FOR OPTIMIZING A PROJECT'S TIME-COST TRADE-OFF

Networks provide a natural way of graphically displaying the flow of activities in a major project, such as a construction project or a research-and-development project. Therefore, one of the most important applications of network theory is in aiding the management of such projects.

In the late 1950s, two network-based OR techniques—**PERT** (program evaluation and review technique) and **CPM** (critical path method)—were developed independently to assist project managers in carrying out their responsibilities. These techniques were designed to help plan how to coordinate a project's various activities, develop a realistic schedule for the project, and then monitor the progress of the project after it is under way. Over the years, the better features of these two techniques have tended to be merged into what is now commonly referred to as the PERT/CPM technique. This network approach to project management continues to be widely used today.

One of the supplementary chapters on the book's website, Chap. 22 (Project Management with PERT/CPM), provides a complete description of the various features of PERT/CPM. We now will highlight one of these features for two reasons. First, it is a network optimization model and so fits into the theme of the current chapter. Second, it illustrates the kind of important applications that such models can have.

The feature we will highlight is referred to as the *CPM method of time-cost trade-offs* because it was a key part of the original CPM technique. It addresses the following problem for a project that needs to be completed by a specific deadline. Suppose that this deadline would not be met if all the activities are performed in the normal manner, but that there are various ways of meeting the deadline by spending more money to expedite some of the activities. What is the optimal plan for expediting some activities so as to minimize the total cost of performing the project within the deadline?

The general approach begins by using a network to display the various activities and the order in which they need to be performed. An optimization model then is formulated that can be solved by using either marginal analysis or linear programming. As with the other network optimization models considered earlier in this chapter, the special structure of the problem makes it relatively easy to solve efficiently.

This approach is illustrated below by using the same prototype example that is carried through Chap. 22.

A Prototype Example—the Reliable Construction Co. Problem

The RELIABLE CONSTRUCTION COMPANY has just made the winning bid of $5.4 million to construct a new plant for a major manufacturer. The manufacturer needs the plant to go into operation within 40 weeks.

Reliable is assigning its best construction manager, David Perty, to this project to help ensure that it stays on schedule. Mr. Perty will need to arrange for a number of crews to perform the various construction activities at different times. Table 9.7 shows his list of the various activities. The third column provides important additional information for coordinating the scheduling of the crews.

For any given activity, its **immediate predecessors** (as given in the third column of Table 9.7) are those activities that must be completed by no later than the starting time of the given activity. (Similarly, the given activity is called an **immediate successor** of each of its immediate predecessors.)

■ **TABLE 9.7** Activity list for the Reliable Construction Co. project

Activity	Activity Description	Immediate Predecessors	Estimated Duration
A	Excavate	—	2 weeks
B	Lay the foundation	A	4 weeks
C	Put up the rough wall	B	10 weeks
D	Put up the roof	C	6 weeks
E	Install the exterior plumbing	C	4 weeks
F	Install the interior plumbing	E	5 weeks
G	Put up the exterior siding	D	7 weeks
H	Do the exterior painting	E, G	9 weeks
I	Do the electrical work	C	7 weeks
J	Put up the wallboard	F, I	8 weeks
K	Install the flooring	J	4 weeks
L	Do the interior painting	J	5 weeks
M	Install the exterior fixtures	H	2 weeks
N	Install the interior fixtures	K, L	6 weeks

For example, the top entries in this column indicate that

1. Excavation does not need to wait for any other activities.
2. Excavation must be completed before starting to lay the foundation.
3. The foundation must be completely laid before starting to put up the rough wall, and so on.

When a given activity has *more than one* immediate predecessor, all must be finished before the activity can begin.

In order to schedule the activities, Mr. Perty consults with each of the crew supervisors to develop an estimate of how long each activity should take when it is done in the normal way. These estimates are given in the rightmost column of Table 9.7.

Adding up these times gives a grand total of 79 weeks, which is far beyond the deadline of 40 weeks for the project. Fortunately, some of the activities can be done in parallel, which substantially reduces the project completion time. We will see next how the project can be displayed graphically to better visualize the flow of the activities and to determine the total time required to complete the project if no delays occur.

We have seen in this chapter how valuable *networks* can be to represent and help analyze many kinds of problems. In much the same way, networks play a key role in dealing with projects. They enable showing the relationships between the activities and succinctly displaying the overall plan for the project. They also are helpful for analyzing the project.

Project Networks

A network used to represent a project is called a **project network.** A project network consists of a number of *nodes* (typically shown as small circles or rectangles) and a number of *arcs* (shown as arrows) that connect two different nodes.

As Table 9.7 indicates, three types of information are needed to describe a project.

1. Activity information: Break down the project into its individual *activities* (at the desired level of detail).
2. Precedence relationships: Identify the *immediate predecessor(s)* for each activity.
3. Time information: Estimate the *duration* of each activity.

The project network should convey all this information. Two alternative types of project networks are available for doing this.

One type is the **activity-on-arc (AOA)** project network, where each activity is represented by an *arc*. A node is used to separate an activity (an outgoing arc) from each of its immediate predecessors (an incoming arc). The sequencing of the arcs thereby shows the precedence relationships between the activities.

The second type is the **activity-on-node (AON)** project network, where each activity is represented by a *node*. Then the arcs are used just to show the precedence relationships that exist between the activities. In particular, the node for each activity with immediate predecessors has an arc coming in from each of these predecessors.

The original versions of PERT and CPM used AOA project networks, so this was the conventional type for some years. However, AON project networks have some important advantages over AOA project networks for conveying the same information.

1. AON project networks are considerably easier to construct than AOA project networks.
2. AON project networks are easier to understand than AOA project networks for inexperienced users, including many managers.
3. AON project networks are easier to revise than AOA project networks when there are changes in the project.

For these reasons, AON project networks have become increasingly popular with practitioners. It appears that they may become the standard format for project networks. Therefore, we will focus solely on AON project networks, and will drop the adjective AON.

Figure 9.28 shows the project network for Reliable's project.[2] Referring also to the third column of Table 9.7, note how there is an arc leading to each activity from each of its immediate predecessors. Because activity A has no immediate predecessors, there is an arc leading from the start node to this activity. Similarly, since activities M and N have no immediate successors, arcs lead from these activities to the finish node. Therefore, the project network nicely displays at a glance all the precedence relationships between all the activities (plus the start and finish of the project). Based on the rightmost column of Table 9.7, the number next to the node for each activity then records the estimated duration (in weeks) of that activity.

The Critical Path

How long should the project take? We noted earlier that summing the durations of all the activities gives a grand total of 79 weeks. However, this isn't the answer to the question because some of the activities can be performed (roughly) simultaneously.

What is relevant instead is the *length* of each *path* through the network.

A **path** through a project network is one of the routes following the arcs from the START node to the FINISH node. The **length** of a path is the *sum* of the (estimated) *durations* of the activities on the path.

The six paths through the project network in Fig. 9.28 are given in Table 9.8, along with the calculations of the lengths of these paths. The path lengths range from 31 weeks up to 44 weeks for the longest path (the fourth one in the table).

So given these path lengths, what should be the (estimated) **project duration** (the total time required for the project)? Let us reason it out.

Since the activities on any given path must be done in sequence with no overlap, the project duration cannot be *shorter* than the path length. However, the project duration can be *longer* because some activity on the path with multiple immediate predecessors might

---

[2]Although project networks often are drawn from left to right, we go from top to bottom to better fit on the printed page.

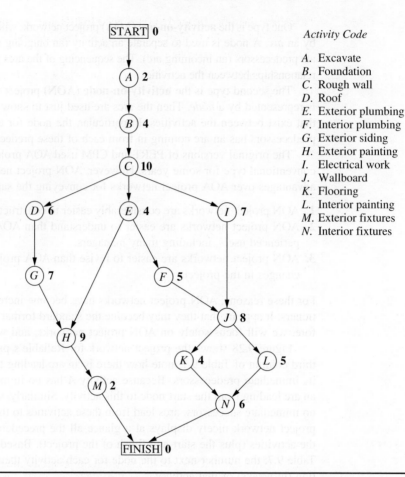

| START | 0 |

Activity Code

A. Excavate
B. Foundation
C. Rough wall
D. Roof
E. Exterior plumbing
F. Interior plumbing
G. Exterior siding
H. Exterior painting
I. Electrical work
J. Wallboard
K. Flooring
L. Interior painting
M. Exterior fixtures
N. Interior fixtures

■ **FIGURE 9.28**
The project network for the
Reliable Construction Co.
project.

■ **TABLE 9.8** The paths and path lengths through Reliable's project network

Path	Length	
START →A→B→C→D→G→H→M→ FINISH	2 + 4 + 10 + 6 + 7 + 9 + 2	= 40 weeks
START →A→B→C→E→H→M→ FINISH	2 + 4 + 10 + 4 + 9 + 2	= 31 weeks
START →A→B→C→E→F→J→K→N→ FINISH	2 + 4 + 10 + 4 + 5 + 8 + 4 + 6	= 43 weeks
START →A→B→C→E→F→J→L→N→ FINISH	2 + 4 + 10 + 4 + 5 + 8 + 5 + 6	= 44 weeks
START →A→B→C→I→J→K→N→ FINISH	2 + 4 + 10 + 7 + 8 + 4 + 6	= 41 weeks
START →A→B→C→I→J→L→N→ FINISH	2 + 4 + 10 + 7 + 8 + 5 + 6	= 42 weeks

have to wait longer for an immediate predecessor *not* on the path to finish than for the one on the path. For example, consider the second path in Table 9.8 and focus on activity *H*. This activity has two immediate predecessors, one (activity *G*) *not* on the path and one (activity *E*) that is. After activity *C* finishes, only 4 more weeks are required for activity *E* but 13 weeks will be needed for activity *D* and then activity *G* to finish. Therefore, the project duration must be considerably longer than the length of the second path in the table.

However, the project duration will not be longer than one particular path. This is the *longest path* through the project network. The activities on this path can be performed sequentially without interruption. (Otherwise, this would not be the longest path.)

Therefore, the time required to reach the FINISH node equals the length of this path. Furthermore, all the shorter paths will reach the FINISH node no later than this.

Here is the key conclusion.

The (estimated) *project duration* equals the *length of the longest path* through the project network. This longest path is called the **critical path.**[3] (If more than one path tie for the longest, they all are critical paths.)

Thus, for the Reliable Construction Co. project, we have

Critical path: START $\to A \to B \to C \to E \to F \to J \to L \to N \to$ FINISH
(Estimated) project duration = 44 weeks.

Therefore, if no delays occur, the total time required to complete the project should be about 44 weeks. Furthermore, the activities on this critical path are the critical bottleneck activities where any delays in their completion must be avoided to prevent delaying project completion. This is valuable information for Mr. Perty, since he now knows that he should focus most of his attention on keeping these particular activities on schedule in striving to keep the overall project on schedule. Furthermore, to reduce the duration of the project (remember that the deadline for completion is 40 weeks), these are the main activities where changes should be made to reduce their durations.

Mr. Perty now needs to determine specifically which activites should have their durations reduced, and by how much, in order to meet the deadline of 40 weeks in the least expensive way. He remembers that CPM provides an excellent procedure for investigating such *time-cost trade-offs,* so he will use this approach to address this question.

We begin with some background.

Time-Cost Trade-Offs for Individual Activities

The first key concept for this approach is that of *crashing.*

Crashing an activity refers to taking special costly measures to reduce the duration of an activity below its normal value. These special measures might include using overtime, hiring additional temporary help, using special time-saving materials, obtaining special equipment, etc. **Crashing the project** refers to crashing a number of activities in order to reduce the duration of the project below its normal value.

The **CPM method of time-cost trade-offs** is concerned with determining how much (if any) to crash each of the activities in order to reduce the anticipated duration of the project to a desired value.

The data necessary for determining how much to crash a particular activity are given by the *time-cost graph* for the activity. Figure 9.29 shows a typical time-cost graph. Note the two key points on this graph labeled *Normal* and *Crash.*

The **normal point** on the time-cost graph for an activity shows the time (duration) and cost of the activity when it is performed in the normal way. The **crash point** shows the time and cost when the activity is *fully crashed,* i.e., it is fully expedited with no cost spared to reduce its duration as much as possible. As an approximation, CPM assumes that these times and costs can be reliably predicted without significant uncertainty.

For most applications, it is assumed that *partially crashing* the activity at any level will give a combination of time and cost that will lie somewhere on the line segment between

---

[3]Although Table 9.8 illustrates how the enumeration of paths and path lengths can be used to find the critical path for small projects, Chap. 22 describes how PERT/CPM normally uses a considerably more efficient procedure to obtain a variety of useful information, including the critical path.

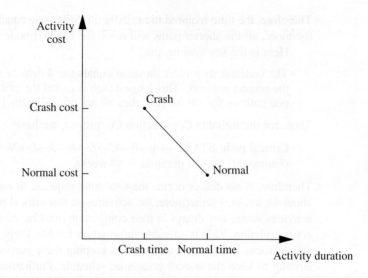

FIGURE 9.29
A typical time-cost graph for
an activity.

these two points.[4] (For example, this assumption says that *half* of a full crash will give a point on this line segment that is midway between the normal and crash points.) This simplifying approximation reduces the necessary data gathering to estimating the time and cost for just two situations: *normal conditions* (to obtain the normal point) and a *full crash* (to obtain the crash point).

Using this approach, Mr. Perty has his staff and crew supervisors working on developing these data for each of the activities of Reliable's project. For example, the supervisor of the crew responsible for putting up the wallboard indicates that adding two temporary employees and using overtime would enable him to reduce the duration of this activity from 8 weeks to 6 weeks, which is the minimum possible. Mr. Perty's staff then estimates the cost of fully crashing the activity in this way as compared to following the normal 8-week schedule, as shown below.

Activity *J* (put up the wallboard):

Normal point: time = 8 weeks, cost = \$430,000.
Crash point: time = 6 weeks, cost = \$490,000.
Maximum reduction in time = 8 − 6 = 2 weeks.

$$\text{Crash cost per week saved} = \frac{\$490,000 - \$430,000}{2}$$

$$= \$30,000.$$

After investigating the time-cost trade-off for each of the other activities in the same way, Table 9.9 gives the data obtained for all the activities.

Which Activities Should Be Crashed?

Summing the *normal cost* and *crash cost* columns of Table 9.9 gives

Sum of normal costs = \$4.55 million,
Sum of crash costs = \$6.15 million.

---

[4]This is a convenient assumption, but it often is only a rough approximation since the underlying assumptions of proportionality and divisibility may not hold completely. If the true time-cost graph is convex, linear programming can still be employed by using a piecewise linear approximation and then applying the separable programming technique described in Sec. 12.8.

TABLE 9.9 Time-cost trade-off data for the activities of Reliable's project

Activity	Time Normal	Time Crash	Cost Normal	Cost Crash	Maximum Reduction in Time	Crash Cost per Week Saved
A	2 weeks	1 week	$180,000	$ 280,000	1 week	$100,000
B	4 weeks	2 weeks	$320,000	$ 420,000	2 weeks	$ 50,000
C	10 weeks	7 weeks	$620,000	$ 860,000	3 weeks	$ 80,000
D	6 weeks	4 weeks	$260,000	$ 340,000	2 weeks	$ 40,000
E	4 weeks	3 weeks	$410,000	$ 570,000	1 week	$160,000
F	5 weeks	3 weeks	$180,000	$ 260,000	2 weeks	$ 40,000
G	7 weeks	4 weeks	$900,000	$1,020,000	3 weeks	$ 40,000
H	9 weeks	6 weeks	$200,000	$ 380,000	3 weeks	$ 60,000
I	7 weeks	5 weeks	$210,000	$ 270,000	2 weeks	$ 30,000
J	8 weeks	6 weeks	$430,000	$ 490,000	2 weeks	$ 30,000
K	4 weeks	3 weeks	$160,000	$ 200,000	1 week	$ 40,000
L	5 weeks	3 weeks	$250,000	$ 350,000	2 weeks	$ 50,000
M	2 weeks	1 week	$100,000	$ 200,000	1 week	$100,000
N	6 weeks	3 weeks	$330,000	$ 510,000	3 weeks	$ 60,000

Recall that the company will be paid $5.4 million for doing this project. This payment needs to cover some *overhead costs* in addition to the costs of the activities listed in the table, as well as provide a reasonable profit to the company. When developing the winning bid of $5.4 million, Reliable's management felt that this amount would provide a reasonable profit as long as the total cost of the activities could be held fairly close to the normal level of about $4.55 million. Mr. Perty understands very well that it is his responsibility to keep the project as close to both budget and schedule as possible.

As found previously in Table 9.8, if all the activities are performed in the normal way, the anticipated duration of the project would be 44 weeks (if delays can be avoided). If *all* the activities were to be *fully crashed* instead, then a similar calculation would find that this duration would be reduced to only 28 weeks. But look at the prohibitive cost ($6.15 million) of doing this! Fully crashing all activities clearly is not a viable option.

However, Mr. Perty still wants to investigate the possibility of partially or fully crashing just a few activities to reduce the anticipated duration of the project to 40 weeks.

The problem: What is the least expensive way of crashing some activities to reduce the (estimated) project duration to the specified level (40 weeks)?

One way of solving this problem is **marginal cost analysis,** which uses the last column of Table 9.9 (along with Table 9.8) to determine the least expensive way to reduce project duration 1 week at a time. The easiest way to conduct this kind of analysis is to set up a table like Table 9.10 that lists all the paths through the project network and the current length of each of these paths. To get started, this information can be copied directly from Table 9.8.

Since the fourth path listed in Table 9.10 has the longest length (44 weeks), the only way to reduce project duration by a week is to reduce the duration of the activities on this particular path by a week. Comparing the crash cost per week saved given in the last

TABLE 9.10 The initial table for starting marginal cost analysis of Reliable's project

Activity to Crash	Crash Cost	Length of Path ABCDGHM	ABCEHM	ABCEFJKN	ABCEFJLN	ABCIJKN	ABCIJLN
		40	31	43	44	41	42

■ **TABLE 9.11** The final table for performing marginal cost analysis on Reliable's project

Activity to Crash	Crash Cost	Length of Path					
		ABCDGHM	**ABCEHM**	**ABCEFJKN**	**ABCEFJLN**	**ABCIJKN**	**ABCIJLN**
		40	31	43	44	41	42
J	$30,000	40	31	42	43	40	41
J	$30,000	40	31	41	42	39	40
F	$40,000	40	31	40	41	39	40
F	$40,000	40	31	39	40	39	40

column of Table 9.9 for these activities, the smallest cost is $30,000 for activity J. (Note that activity I with this same cost is not on this path.) Therefore, the first change is to crash activity J enough to reduce its duration by a week.

This change results in reducing the length of each path that includes activity J (the third, fourth, fifth, and sixth paths in Table 9.10) by a week, as shown in the second row of Table 9.11. Because the fourth path still is the longest (43 weeks), the same process is repeated to find the least expensive activity to shorten on this path. This again is activity J, since the next-to-last column in Table 9.9 indicates that a maximum reduction of 2 weeks is allowed for this activity. This second reduction of a week for activity J leads to the third row of Table 9.11.

At this point, the fourth path still is the longest (42 weeks), but activity J cannot be shortened any further. Among the other activities on this path, activity F now is the least expensive to shorten ($40,000 per week) according to the last column of Table 9.9. Therefore, this activity is shortened by a week to obtain the fourth row of Table 9.11, and then (because a maximum reduction of 2 weeks is allowed) is shortened by another week to obtain the last row of this table.

The longest path (a tie between the first, fourth, and sixth paths) now has the desired length of 40 weeks, so we don't need to do any more crashing. (If we did need to go further, the next step would require looking at the activities on all three paths to find the least expensive way of shortening all three paths by a week.) The total cost of crashing activities J and F to get down to this project duration of 40 weeks is calculated by adding the costs in the second column of Table 9.11—a total of $140,000. Figure 9.30 shows the resulting project network, where the darker arrows show the critical paths.

Figure 9.30 shows that reducing the durations of activities F and J to their crash times has led to now having *three* critical paths through the network. The reason is that, as we found earlier from the last row of Table 9.11, the three paths tie for being the longest, each with a length of 40 weeks.

With larger networks, marginal cost analysis can become quite unwieldy. A more efficient procedure would be desirable for large projects. For this reason, the standard CPM procedure is to apply *linear programming* instead (commonly with a customized software package that exploits the special structure of this network optimization model).

Using Linear Programming to Make Crashing Decisions

The problem of finding the least expensive way of crashing activities can be rephrased in a form more familiar to linear programming as follows.

> **Restatement of the problem:** Let Z be the total cost of crashing activities. The problem then is to minimize Z, subject to the constraint that project duration must be less than or equal to the time desired by the project manager.

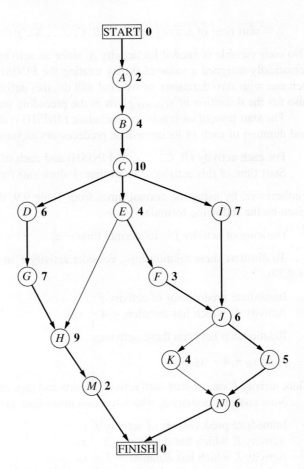

■ FIGURE 9.30
The project network if activities J and F are fully crashed (with all other activities normal) for Reliable's project. The darker arrows show the various critical paths through the project network.

The natural decision variables are

x_j = reduction in the duration of activity j due to crashing this activity,
for $j = A, B, \ldots, N$.

By using the last column of Table 9.9, the objective function to be minimized then is

$$Z = 100{,}000x_A + 50{,}000x_B + \cdots + 60{,}000x_N.$$

Each of the 14 decision variables on the right-hand side needs to be restricted to nonnegative values that do not exceed the maximum given in the next-to-last column of Table 9.9.

To impose the constraint that project duration must be less than or equal to the desired value (40 weeks), let

y_{FINISH} = project duration, i.e., the time at which the FINISH node in the project network is reached.

The constraint then is . . .

$$y_{\text{FINISH}} \leq 40.$$

To help the linear programming model assign the appropriate value to y_{FINISH}, given the values of $x_A, x_B, \ldots, x_N$, it is convenient to introduce into the model the following additional variables.

y_j = start time of activity j (for $j = B, C, \ldots, N$), given the values of $x_A, x_B, \ldots, x_N$.

(No such variable is needed for activity A, since an activity that begins the project is automatically assigned a value of 0.) By treating the FINISH node as another activity (albeit one with zero duration), as we now will do, this definition of y_j for activity FINISH also fits the definition of y_{FINISH} given in the preceding paragraph.

The start time of each activity (including FINISH) is directly related to the start time and duration of each of its immediate predecessors as summarized below.

> For each activity ($B, C, \ldots, N$, FINISH) and each of its immediate predecessors,
> Start time of this activity $\geq$ (start time + duration) for this immediate predecessor.

Furthermore, by using the normal times from Table 9.9, the duration of each activity is given by the following formula:

> Duration of activity j = its normal time $- x_j$,

To illustrate these relationships, consider activity F in the project network (Fig. 9.28 or 9.30).

> Immediate predecessor of activity F:
> Activity E, which has duration = $4 - x_E$.

Relationship between these activities:

$$y_F \geq y_E + 4 - x_E.$$

Thus, activity F cannot start until activity E starts and then completes its duration of $4 - x_E$. Now consider activity J, which has two immediate predecessors.

> Immediate predecessors of activity J:
> Activity F, which has duration = $5 - x_F$.
> Activity I, which has duration = $7 - x_I$.

Relationships between these activities:

$$y_J \geq y_F + 5 - x_F,$$
$$y_J \geq y_I + 7 - x_I.$$

These inequalities together say that activity j cannot start until both of its predecessors finish.

By including these relationships for all the activities as constraints, we obtain the complete linear programming model given below.

> Minimize $Z = 100{,}000x_A + 50{,}000x_B + \cdots + 60{,}000x_N$,

subject to the following constraints:

1. Maximum reduction constraints:
  Using the next-to-last column of Table 9.9,

$$x_A \leq 1, \ x_B \leq 2, \ \ldots, \ x_N \leq 3.$$

2. Nonnegativity constraints:

$$x_A \geq 0, \ x_B \geq 0, \ \ldots, \ x_N \geq 0$$
$$y_B \geq 0, \ y_C \geq 0, \ \ldots, \ y_N \geq 0, \ y_{\text{FINISH}} \geq 0.$$

3. Start-time constraints:
  As described above the objective function, with the exception of activity A (which starts the project), there is one start-time constraint for each activity with a single immediate

predecessor (activities *B, C, D, E, F, G, I, K, L, M*) and two constraints for each activity with two immediate predecessors (activities *H, J, N,* FINISH), as listed below.

One immediate predecessor	Two immediate predecessors
$y_B \geq 0 + 2 - x_A$	$y_H \geq y_G + 7 - x_G$
$y_C \geq y_B + 4 - x_B$	$y_H \geq y_E + 4 - x_E$
$y_D \geq y_C + 10 - x_C$	$\vdots$
$\vdots$	$y_{\text{FINISH}} \geq y_M + 2 - x_M$
$y_M \geq y_H + 9 - x_H$	$y_{\text{FINISH}} \geq y_N + 6 - x_N$

(In general, the number of start-time constraints for an activity equals its number of immediate predecessors since each immediate predecessor contributes one start-time constraint.)

4. Project duration constraint:

$$y_{\text{FINISH}} \leq 40.$$

Figure 9.31 shows how this problem can be formulated as a linear programming model on a spreadsheet. The decisions to be made are shown in the changing cells, StartTime (I6:I19), TimeReduction (J6:J19), and ProjectFinishTime (I22). Columns B to H correspond to the columns in Table 9.9. As the equations in the bottom half of the figure indicate, columns G and H are calculated in a straightforward way. The equations for column K express the fact that the finish time for each activity is its start time *plus* its normal time *minus* its time reduction due to crashing. The equation entered into the target cell TotalCost (I24) adds all the normal costs plus the extra costs due to crashing to obtain the total cost.

The last set of constraints in the Solver dialogue box, TimeReduction (J6:J19) ≤ MaxTimeReduction (G6:G19), specifies that the time reduction for each activity cannot exceed its maximum time reduction given in column G. The two preceding constraints, ProjectFinishTime (I22) ≥ MFinish (K18) and ProjectFinishTime (I22) ≥ NFINISH (K19), indicate that the project cannot finish until each of the two immediate predecessors (activities *M* and *N*) finish. The constraint that ProjectFinishTime (I22) ≤ MaxTime (K22) is a key one that specifies that the project must finish within 40 weeks.

The constraints involving StartTime (I6:I19) all are *start-time constraints* that specify that an activity cannot start until each of its immediate predecessors has finished. For example, the first constraint shown, BStart (I7) ≥ AFinish (K6), says that activity *B* cannot start until activity *A* (its immediate predecessor) finishes. When an activity has more than one immediate predecessor, there is one such constraint for each of them. To illustrate, activity *H* has both activities *E* and *G* as immediate predecessors. Consequently, activity *H* has two start-time constraints, HStart (I13) ≥ EFinish (K10) and HStart (I13) ≥ GFinish (K12).

You may have noticed that the ≥ form of the *start-time constraints* allows a delay in starting an activity after all its immediate predecessors have finished. Although such a delay is feasible in the model, it cannot be optimal for any activity on a critical path, since this needless delay would increase the total cost (by necessitating additional crashing to meet the project duration constraint). Therefore, an optimal solution for the model will not have any such delays, except possibly for activities not on a critical path.

Columns I and J in Fig. 9.31 show the optimal solution obtained after having clicked on the Solve button. (Note that this solution involves one delay—activity *K* starts at 30 even though its only immediate predecessor, activity *J*, finishes at 29—but this doesn't matter since activity *K* is not on a critical path.) This solution corresponds to the one displayed in Fig. 9.30 that was obtained by marginal cost analysis.

If you would like to see **another example** that illustrates both the marginal cost analysis approach and the linear programming approach to applying the CPM method of time-cost trade-offs, the Worked Examples section of the book's website provides one.

An Interactive Procedure in IOR Tutorial:

Network Simplex Method—Interactive

An Excel Add-in:

Premium Solver for Education

"Ch. 9—Network Opt Models" Files for Solving the Examples:

Excel Files
LINGO/LINDO File
MPL/CPLEX File

Glossary for Chapter 9

See Appendix 1 for documentation of the software.

■ PROBLEMS

The symbols to the left of some of the problems (or their parts) have the following meaning:

D: The demonstration example just listed in Learning Aids may be helpful.

I: We suggest that you use the interactive procedure just listed (the printout records your work).

C: Use the computer with any of the software options available to you (or as instructed by your instructor) to solve the problem.

An asterisk on the problem number indicates that at least a partial answer is given in the back of the book.

9.2-1. Consider the following directed network.

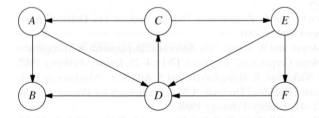

(a) Find a directed path from node A to node F, and then identify three other undirected paths from node A to node F.

(b) Find three directed cycles. Then identify an undirected cycle that includes every node.

(c) Identify a set of arcs that forms a spanning tree.

(d) Use the process illustrated in Fig. 9.3 to grow a tree one arc at a time until a spanning tree has been formed. Then repeat this process to obtain another spanning tree. [Do not duplicate the spanning tree identified in part (c).]

9.3-1. Read the referenced article that fully describes the OR study summarized in the application vignette presented in Sec. 9.3. Briefly describe how network optimization models were applied in this study. Then list the various financial and nonfinancial benefits that resulted from this study.

9.3-2. You need to take a trip by car to another town that you have never visited before. Therefore, you are studying a map to determine the shortest route to your destination. Depending on which route you choose, there are five other towns (call them A, B, C, D, E) that you might pass through on the way. The map shows the mileage along each road that directly connects two towns without any intervening towns. These numbers are summarized in the following table, where a dash indicates that there is no road directly connecting these two towns without going through any other towns.

Town	Miles between Adjacent Towns					
	A	**B**	**C**	**D**	**E**	**Destination**
Origin	40	60	50	—	—	—
A		10	—	70	—	—
B			20	55	40	—
C				—	50	—
D					10	60
E						80

(a) Formulate this problem as a shortest-path problem by drawing a network where nodes represent towns, links represent roads, and numbers indicate the length of each link in miles.

(b) Use the algorithm described in Sec. 9.3 to solve this shortest-path problem.

C **(c)** Formulate and solve a spreadsheet model for this problem.

(d) If each number in the table represented your *cost* (in dollars) for driving your car from one town to the next, would the answer in part (b) or (c) now give your minimum cost route?

(e) If each number in the table represented your *time* (in minutes) for driving your car from one town to the next, would the answer in part (b) or (c) now give your minimum time route?

9.3-3. At a small but growing airport, the local airline company is purchasing a new tractor for a tractor-trailer train to bring luggage to and from the airplanes. A new mechanized luggage system will be installed in 3 years, so the tractor will not be needed after that. However, because it will receive heavy use, so that the running and maintenance costs will increase rapidly as the tractor ages, it may still be more economical to replace the tractor after 1 or 2 years. The following table gives the total net discounted cost associated with purchasing a tractor (purchase price minus trade-in allowance, plus running and maintenance costs) at the end of year i and trading it in at the end of year j (where year 0 is now).

		j	
	1	**2**	**3**
0	$13,000	$28,000	$48,000
i **1**		$17,000	$33,000
2			$20,000

The problem is to determine at what times (if any) the tractor should be replaced to minimize the total cost for the tractors over 3 years.
(a) Formulate this problem as a shortest-path problem.
(b) Use the algorithm described in Sec. 9.3 to solve this shortest-path problem.
C (c) Formulate and solve a spreadsheet model for this problem.

9.3-4.* Use the algorithm described in Sec. 9.3 to find the *shortest path* through each of the following networks, where the numbers represent actual distances between the corresponding nodes.

(a)

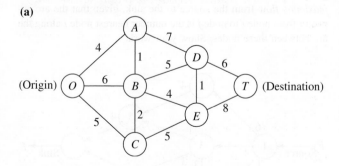

(b)

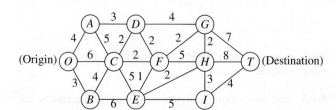

9.3-5. Formulate the shortest-path problem as a linear programming problem.

9.3-6. One of Speedy Airlines' flights is about to take off from Seattle for a nonstop flight to London. There is some flexibility in choosing the precise route to be taken, depending upon weather conditions. The following network depicts the possible routes under consideration, where SE and LN are Seattle and London, respectively, and the other nodes represent various intermediate locations.

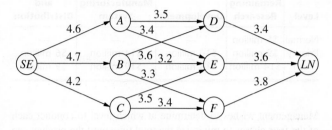

The winds along each arc greatly affect the flying time (and so the fuel consumption). Based on current meteorological reports, the flying times (in hours) for this particular flight are shown next to the arcs. Because the fuel consumed is so expensive, the management of Speedy Airlines has established a policy of choosing the route that minimizes the total flight time.
(a) What plays the role of "distances" in interpreting this problem to be a shortest-path problem?
(b) Use the algorithm described in Sec. 9.3 to solve this shortest-path problem.
C (c) Formulate and solve a spreadsheet model for this problem.

9.3-7. The Quick Company has learned that a competitor is planning to come out with a new kind of product with a great sales potential. Quick has been working on a similar product that had been scheduled to come to market in 20 months. However, research is nearly complete and Quick's management now wishes to rush the product out to meet the competition.

There are four nonoverlapping phases left to be accomplished, including the remaining research that currently is being conducted at a normal pace. However, each phase can instead be conducted at a priority or crash level to expedite completion, and these are the only levels that will be considered for the last three phases. The times required at these levels are given in the following table. (The times in parentheses at the normal level have been ruled out as too long.)

	Time			
Level	**Remaining Research**	**Development**	**Design of Manufacturing System**	**Initiate Production and Distribution**
Normal	5 months	(4 months)	(7 months)	(4 months)
Priority	4 months	3 months	5 months	2 months
Crash	2 months	2 months	3 months	1 month

Management has allocated $50 million for these four phases. The cost of each phase at the different levels under consideration is as follows:

	Cost			
Level	**Remaining Research**	**Development**	**Design of Manufacturing System**	**Initiate Production and Distribution**
Normal	$5 million	—	—	—
Priority	$9 million	$10 million	$14 million	$6 million
Crash	$14 million	$15 million	$19 million	$9 million

Management wishes to determine at which level to conduct each of the four phases to minimize the total time until the product can be marketed subject to the budget restriction of $50 million.

(a) Formulate this problem as a shortest-path problem.

(b) Use the algorithm described in Sec. 9.3 to solve this shortest-path problem.

9.4-1.* Reconsider the networks shown in Prob. 9.3-4. Use the algorithm described in Sec. 9.4 to find the *minimum spanning tree* for each of these networks.

9.4-2. The Wirehouse Lumber Company will soon begin logging eight groves of trees in the same general area. Therefore, it must develop a system of dirt roads that makes each grove accessible from every other grove. The distance (in miles) between every pair of groves is as follows:

	Distance between Pairs of Groves							
	1	**2**	**3**	**4**	**5**	**6**	**7**	**8**
1	—	1.3	2.1	0.9	0.7	1.8	2.0	1.5
2	1.3	—	0.9	1.8	1.2	2.6	2.3	1.1
3	2.1	0.9	—	2.6	1.7	2.5	1.9	1.0
Grove 4	0.9	1.8	2.6	—	0.7	1.6	1.5	0.9
5	0.7	1.2	1.7	0.7	—	0.9	1.1	0.8
6	1.8	2.6	2.5	1.6	0.9	—	0.6	1.0
7	2.0	2.3	1.9	1.5	1.1	0.6	—	0.5
8	1.5	1.1	1.0	0.9	0.8	1.0	0.5	—

Management now wishes to determine between which pairs of groves the roads should be constructed to connect all groves with a minimum total length of road.

(a) Describe how this problem fits the network description of the minimum spanning tree problem.

(b) Use the algorithm described in Sec. 9.4 to solve the problem.

9.4-3. The Premiere Bank soon will be hooking up computer terminals at each of its branch offices to the computer at its main office using special phone lines with telecommunications devices.

The phone line from a branch office need not be connected directly to the main office. It can be connected indirectly by being connected to another branch office that is connected (directly or indirectly) to the main office. The only requirement is that every branch office be connected by some route to the main office.

The charge for the special phone lines is $100 times the number of miles involved, where the distance (in miles) between every pair of offices is as follows:

	Distance between Pairs of Offices					
	Main	**B.1**	**B.2**	**B.3**	**B.4**	**B.5**
Main office	—	190	70	115	270	160
Branch 1	190	—	100	110	215	50
Branch 2	70	100	—	140	120	220
Branch 3	115	110	140	—	175	80
Branch 4	270	215	120	175	—	310
Branch 5	160	50	220	80	310	—

Management wishes to determine which pairs of offices should be directly connected by special phone lines in order to connect every branch office (directly or indirectly) to the main office at a minimum total cost.

(a) Describe how this problem fits the network description of the minimum spanning tree problem.

(b) Use the algorithm described in Sec. 9.4 to solve the problem.

9.5-1.* For the network shown below, use the augmenting path algorithm described in Sec. 9.5 to find the flow pattern giving the *maximum flow* from the source to the sink, given that the arc capacity from node i to node j is the number nearest node i along the arc between these nodes. Show your work.

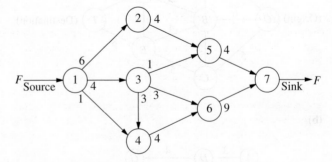

9.5-2. Formulate the maximum flow problem as a linear programming problem.

9.5-3. The next diagram depicts a system of aqueducts that originate at three rivers (nodes R1, R2, and R3) and terminate at a major city (node T), where the other nodes are junction points in the system.

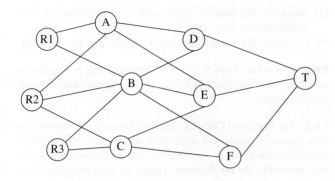

Using units of thousands of acre feet, the tables below the diagram show the maximum amount of water that can be pumped through each aqueduct per day.

To From	A	B	C
R1	130	115	—
R2	70	90	110
R3	—	140	120

To From	D	E	F
A	110	85	—
B	130	95	85
C	—	130	160

To From	T
D	220
E	330
F	240

The city water manager wants to determine a flow plan that will maximize the flow of water to the city.

(a) Formulate this problem as a maximum flow problem by identifying a source, a sink, and the transshipment nodes, and then drawing the complete network that shows the capacity of each arc.

(b) Use the augmenting path algorithm described in Sec. 9.5 to solve this problem.

c (c) Formulate and solve a spreadsheet model for this problem.

9.5-4. The Texago Corporation has four oil fields, four refineries, and four distribution centers. A major strike involving the transportation industries now has sharply curtailed Texago's capacity to ship oil from the oil fields to the refineries and to ship petroleum products from the refineries to the distribution centers. Using units of thousands of barrels of crude oil (and its equivalent in refined products), the following tables show the maximum number of units that can be shipped per day from each oil field to each refinery, and from each refinery to each distribution center.

Oil Field	Refinery			
	New Orleans	Charleston	Seattle	St. Louis
Texas	11	7	2	8
California	5	4	8	7
Alaska	7	3	12	6
Middle East	8	9	4	15

Refinery	Distribution Center			
	Pittsburgh	Atlanta	Kansas City	San Francisco
New Orleans	5	9	6	4
Charleston	8	7	9	5
Seattle	4	6	7	8
St. Louis	12	11	9	7

The Texago management now wants to determine a plan for how many units to ship from each oil field to each refinery and from each refinery to each distribution center that will maximize the total number of units reaching the distribution centers.

(a) Draw a rough map that shows the location of Texago's oil fields, refineries, and distribution centers. Add arrows to show the flow of crude oil and then petroleum products through this distribution network.

(b) Redraw this distribution network by lining up all the nodes representing oil fields in one column, all the nodes representing refineries in a second column, and all the nodes representing distribution centers in a third column. Then add arcs to show the possible flow.

(c) Modify the network in part (b) as needed to formulate this problem as a maximum flow problem with a single source, a single sink, and a capacity for each arc.

(d) Use the augmenting path algorithm described in Sec. 9.5 to solve this maximum flow problem.

c (e) Formulate and solve a spreadsheet model for this problem.

9.5-5. One track of the Eura Railroad system runs from the major industrial city of Faireparc to the major port city of Portstown. This track is heavily used by both express passenger and freight trains. The passenger trains are carefully scheduled and have priority over the slow freight trains (this is a European railroad), so that the freight trains must pull over onto a siding whenever a passenger train is scheduled to pass them soon. It is now necessary to increase the freight service, so the problem is to schedule the freight trains so as to maximize the number that can be sent each day without interfering with the fixed schedule for passenger trains.

Consecutive freight trains must maintain a schedule differential of at least 0.1 hour, and this is the time unit used for scheduling them (so that the daily schedule indicates the status of each freight train at times $0.0, 0.1, 0.2, \ldots, 23.9$). There are S sidings between Faireparc and Portstown, where siding i is long enough to hold n_i freight trains $(i = 1, \ldots, S)$. It requires t_i time units (rounded up to an integer) for a freight train to travel from siding i to siding $i + 1$ (where t_0 is the time from the Faireparc station to siding 1 and t_s is the time from siding S to the Portstown station). A freight train is allowed to pass or leave siding i $(i = 0, 1, \ldots, S)$ at time j $(j = 0.0, 0.1, \ldots, 23.9)$ only if it would not be overtaken by a scheduled passenger train before reaching siding $i + 1$ (let $\delta_{ij} = 1$ if it would not be overtaken, and let $\delta_{ij} = 0$ if it would be). A freight train also is required to stop at a siding if there will not be room for it at all subsequent sidings that it would reach before being overtaken by a passenger train.

Formulate this problem as a maximum flow problem by identifying each node (including the supply node and the demand node) as well as each arc and its arc capacity for the network representation of the problem. (*Hint:* Use a different set of nodes for each of the 240 times.)

9.5-6. Consider the maximum flow problem shown below, where the source is node *A*, the sink is node *F*, and the arc capacities are the numbers shown next to these directed arcs.

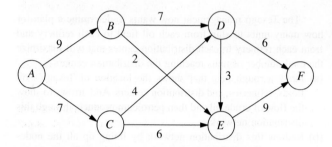

(a) Use the augmenting path algorithm described in Sec. 9.5 to solve this problem.

C **(b)** Formulate and solve a spreadsheet model for this problem.

9.6-1. Read the referenced article that fully describes the OR study summarized in the application vignette presented in Sec. 9.6. Briefly describe how the model for the minimum cost flow problem was applied in this study. Then list the various financial and nonfinancial benefits that resulted from this study.

9.6-2. Reconsider the maximum flow problem shown in Prob. 9.5-6. Formulate this problem as a minimum cost flow problem, including adding the arc $A \rightarrow F$. Use $\overline{F} = 20$.

9.6-3. A company will be producing the same new product at two different factories, and then the product must be shipped to two warehouses. Factory 1 can send an unlimited amount by rail to warehouse 1 only, whereas factory 2 can send an unlimited amount by rail to warehouse 2 only. However, independent truckers can be used to ship up to 50 units from each factory to a distribution center, from which up to 50 units can be shipped to each warehouse. The shipping cost per unit for each alternative is shown in the following table, along with the amounts to be produced at the factories and the amounts needed at the warehouses.

From \ To	Unit Shipping Cost Distribution Center	Warehouse 1	Warehouse 2	Output
Factory 1	3	7	—	80
Factory 2	4	—	9	70
Distribution center		2	4	
Allocation		60	90	

(a) Formulate the network representation of this problem as a minimum cost flow problem.

(b) Formulate the linear programming model for this problem.

9.6-4. Reconsider Prob. 9.3-3. Now formulate this problem as a minimum cost flow problem by showing the appropriate network representation.

9.6-5. The Makonsel Company is a fully integrated company that both produces goods and sells them at its retail outlets. After production, the goods are stored in the company's two warehouses until needed by the retail outlets. Trucks are used to transport the goods from the two plants to the warehouses, and then from the warehouses to the three retail outlets.

Using units of full truckloads, the following table shows each plant's monthly output, its shipping cost per truckload sent to each warehouse, and the maximum amount that it can ship per month to each warehouse.

From \ To	Unit Shipping Cost Warehouse 1	Warehouse 2	Shipping Capacity Warehouse 1	Warehouse 2	Output
Plant 1	$1175	$1580	375	450	600
Plant 2	$1430	$1700	525	600	900

For each retail outlet (RO), the next table shows its monthly demand, its shipping cost per truckload from each warehouse, and the maximum amount that can be shipped per month from each warehouse.

From \ To	Unit Shipping Cost RO1	RO2	RO3	Shipping Capacity RO1	RO2	RO3
Warehouse 1	$1370	$1505	$1490	300	450	300
Warehouse 2	$1190	$1210	$1240	375	450	225
Demand	450	600	450	450	600	450

Management now wants to determine a distribution plan (number of truckloads shipped per month from each plant to each warehouse and from each warehouse to each retail outlet) that will minimize the total shipping cost.

(a) Draw a network that depicts the company's distribution network. Identify the supply nodes, transshipment nodes, and demand nodes in this network.

(b) Formulate this problem as a minimum cost flow problem by inserting all the necessary data into this network.

C **(c)** Formulate and solve a spreadsheet model for this problem.

C **(d)** Use the computer to solve this problem without using Excel.

9.6-6. The Audiofile Company produces boomboxes. However, management has decided to subcontract out the production of the speakers needed for the boomboxes. Three vendors are available to supply the speakers. Their price for each shipment of 1,000 speakers is shown below.

Vendor	Price
1	$22,500
2	$22,700
3	$22,300

In addition, each vendor would charge a shipping cost. Each shipment would go to one of the company's two warehouses. Each vendor has its own formula for calculating this shipping cost based on the mileage to the warehouse. These formulas and the mileage data are shown below.

Vendor	Charge per Shipment
1	$300 + 40¢/mile
2	$200 + 50¢/mile
3	$500 + 20¢/mile

Vendor	Warehouse 1	Warehouse 2
1	1,600 miles	400 miles
2	500 miles	600 miles
3	2,000 miles	1,000 miles

Whenever one of the company's two factories needs a shipment of speakers to assemble into the boomboxes, the company hires a trucker to bring the shipment in from one of the warehouses. The cost per shipment is given next, along with the number of shipments needed per month at each factory.

	Unit Shipping Cost	
	Factory 1	Factory 2
Warehouse 1	$200	$700
Warehouse 2	$400	$500
Monthly demand	10	6

Each vendor is able to supply as many as 10 shipments per month. However, because of shipping limitations, each vendor is able to send a maximum of only 6 shipments per month to each

warehouse. Similarly, each warehouse is able to send a maximum of only 6 shipments per month to each factory.

Management now wants to develop a plan for each month regarding how many shipments (if any) to order from each vendor, how many of those shipments should go to each warehouse, and then how many shipments each warehouse should send to each factory. The objective is to minimize the sum of the purchase costs (including the shipping charge) and the shipping costs from the warehouses to the factories.

(a) Draw a network that depicts the company's supply network. Identify the supply nodes, transshipment nodes, and demand nodes in this network.

(b) Formulate this problem as a minimum cost flow problem by inserting all the necessary data into this network. Also include a dummy demand node that receives (at zero cost) all the unused supply capacity at the vendors.

C (c) Formulate and solve a spreadsheet model for this problem.

C (d) Use the computer to solve this problem without using - Excel.

D **9.7-1.** Consider the minimum cost flow problem shown below, where the b_i values (net flows generated) are given by the nodes, the c_{ij} values (costs per unit flow) are given by the arcs, and the u_{ij} values (arc capacities) are given between nodes C and D. Do the following work manually.

(a) Obtain an initial BF solution by solving the feasible spanning tree with basic arcs $A \rightarrow B$, $C \rightarrow E$, $D \rightarrow E$, and $C \rightarrow A$

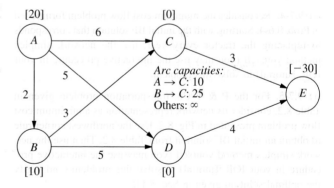

(a reverse arc), where one of the nonbasic arcs ($C \rightarrow B$) also is a reverse arc. Show the resulting network (including b_i, c_{ij}, and u_{ij}) in the same format as the above one (except use dashed lines to draw the nonbasic arcs), and add the flows in parentheses next to the basic arcs.

(b) Use the optimality test to verify that this initial BF solution is optimal and that there are multiple optimal solutions. Apply one iteration of the network simplex method to find the other optimal BF solution, and then use these results to identify the other optimal solutions that are not BF solutions.

(c) Now consider the following BF solution.

PREVIEWS OF ADDED CASES ON OUR WEBSITE (www.mhhe.com/hillier)

CASE 9.2   Aiding Allies

CHAPTER 10

Dynamic Programming

Dynamic programming is a useful mathematical technique for making a sequence of interrelated decisions. It provides a systematic procedure for determining the optimal combination of decisions.

In contrast to linear programming, there does not exist a standard mathematical formulation of "the" dynamic programming problem. Rather, dynamic programming is a general type of approach to problem solving, and the particular equations used must be developed to fit each situation. Therefore, a certain degree of ingenuity and insight into the general structure of dynamic programming problems is required to recognize when and how a problem can be solved by dynamic programming procedures. These abilities can best be developed by an exposure to a wide variety of dynamic programming applications and a study of the characteristics that are common to all these situations. A large number of illustrative examples are presented for this purpose.

■ 10.1   A PROTOTYPE EXAMPLE FOR DYNAMIC PROGRAMMING

EXAMPLE 1   The Stagecoach Problem

The STAGECOACH PROBLEM is a problem specially constructed[1] to illustrate the features and to introduce the terminology of dynamic programming. It concerns a mythical fortune seeker in Missouri who decided to go west to join the gold rush in California during the mid-19th century. The journey would require traveling by stagecoach through unsettled country where there was serious danger of attack by marauders. Although his starting point and destination were fixed, he had considerable choice as to which states (or territories that subsequently became states) to travel through en route. The possible routes are shown in Fig. 10.1, where each state is represented by a circled letter and the direction of travel is always from left to right in the diagram. Thus, four stages (stagecoach runs) were required to travel from his point of embarkation in state *A* (Missouri) to his destination in state *J* (California).

This fortune seeker was a prudent man who was quite concerned about his safety. After some thought, he came up with a rather clever way of determining the safest route. Life insurance policies were offered to stagecoach passengers. Because the cost of the policy

[1]This problem was developed by Professor Harvey M. Wagner while he was at Stanford University.

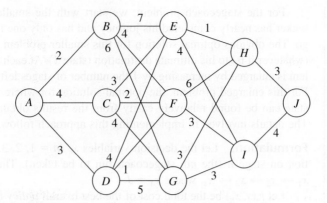

■ FIGURE 10.1
The road system and costs
for the stagecoach problem.

for taking any given stagecoach run was based on a careful evaluation of the safety of that run, the safest route should be the one with the cheapest total life insurance policy.

The cost for the standard policy on the stagecoach run from state i to state j, which will be denoted by c_{ij}, is

	B	C	D			E	F	G			H	I			J
A	2	4	3		B	7	4	6		E	1	4		H	3
					C	3	2	4		F	6	3		I	4
					D	4	1	5		G	3	3			

These costs are also shown in Fig. 10.1.

We shall now focus on the question of which route minimizes the total cost of the policy.

Solving the Problem

First note that the shortsighted approach of selecting the cheapest run offered by each successive stage need not yield an overall optimal decision. Following this strategy would give the route $A \rightarrow B \rightarrow F \rightarrow I \rightarrow J$, at a total cost of 13. However, sacrificing a little on one stage may permit greater savings thereafter. For example, $A \rightarrow D \rightarrow F$ is cheaper overall than $A \rightarrow B \rightarrow F$.

One possible approach to solving this problem is to use trial and error.[2] However, the number of possible routes is large (18), and having to calculate the total cost for each route is not an appealing task.

Fortunately, dynamic programming provides a solution with much less effort than exhaustive enumeration. (The computational savings are enormous for larger versions of this problem.) Dynamic programming starts with a small portion of the original problem and finds the optimal solution for this smaller problem. It then gradually enlarges the problem, finding the current optimal solution from the preceding one, until the original problem is solved in its entirety.

[2]This problem also can be formulated as a *shortest-path problem* (see Sec. 9.3), where *costs* here play the role of *distances* in the shortest-path problem. The algorithm presented in Sec. 9.3 actually uses the philosophy of dynamic programming. However, because the present problem has a fixed number of stages, the dynamic programming approach presented here is even better.

For the stagecoach problem, we start with the smaller problem where the fortune seeker has nearly completed his journey and has only one more stage (stagecoach run) to go. The obvious optimal solution for this smaller problem is to go from his current state (whatever it is) to his ultimate destination (state J). At each subsequent iteration, the problem is enlarged by increasing by 1 the number of stages left to go to complete the journey. For this enlarged problem, the optimal solution for where to go next from each possible state can be found relatively easily from the results obtained at the preceding iteration. The details involved in implementing this approach follow.

Formulation.   Let the decision variables x_n ($n = 1, 2, 3, 4$) be the immediate destination on stage n (the nth stagecoach run to be taken). Thus, the route selected is $A \rightarrow x_1 \rightarrow x_2 \rightarrow x_3 \rightarrow x_4$, where $x_4 = J$.

Let $f_n(s, x_n)$ be the total cost of the best overall *policy* for the *remaining* stages, given that the fortune seeker is in state s, ready to start stage n, and selects x_n as the immediate destination. Given s and n, let x_n^* denote any value of x_n (not necessarily unique) that minimizes $f_n(s, x_n)$, and let $f_n^*(s)$ be the corresponding minimum value of $f_n(s, x_n)$. Thus,

$$f_n^*(s) = \min_{x_n} f_n(s, x_n) = f_n(s, x_n^*),$$

where

$$f_n(s, x_n) = \text{immediate cost (stage } n) + \text{minimum future cost (stages } n+1 \text{ onward)}$$
$$= c_{sx_n} + f_{n+1}^*(x_n).$$

The value of c_{sx_n} is given by the preceding tables for c_{ij} by setting $i = s$ (the current state) and $j = x_n$ (the immediate destination). Because the ultimate destination (state J) is reached at the end of stage 4, $f_5^*(J) = 0$.

The objective is to find $f_1^*(A)$ and the corresponding route. Dynamic programming finds it by successively finding $f_4^*(s)$, $f_3^*(s)$, $f_2^*(s)$, for each of the possible states s and then using $f_2^*(s)$ to solve for $f_1^*(A)$.[3]

Solution Procedure.   When the fortune seeker has only one more stage to go ($n = 4$), his route thereafter is determined entirely by his current state s (either H or I) and his final destination $x_4 = J$, so the route for this final stagecoach run is $s \rightarrow J$. Therefore, since $f_4^*(s) = f_4(s, J) = c_{s,J}$, the immediate solution to the $n = 4$ problem is

$n = 4$:	s	$f_4^*(s)$	x_4^*
	H	3	J
	I	4	J

When the fortune seeker has two more stages to go ($n = 3$), the solution procedure requires a few calculations. For example, suppose that the fortune seeker is in state F. Then, as depicted below, he must next go to either state H or I at an immediate cost of $c_{F,H} = 6$ or $c_{F,I} = 3$, respectively. If he chooses state H, the minimum additional cost after he reaches there is given in the preceding table as $f_4^*(H) = 3$, as shown above the H node in the diagram. Therefore, the total cost for this decision is $6 + 3 = 9$. If he chooses state I instead, the total cost is $3 + 4 = 7$, which is smaller. Therefore, the optimal choice is this latter one, $x_3^* = I$, because it gives the minimum cost $f_3^*(F) = 7$.

---

[3]Because this procedure involves moving *backward* stage by stage, some writers also count n backward to denote the number of *remaining stages* to the destination. We use the more natural *forward counting* for greater simplicity.

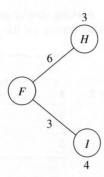

Similar calculations need to be made when you start from the other two possible states $s = E$ and $s = G$ with two stages to go. Try it, proceeding both graphically (Fig. 10.1) and algebraically [combining c_{ij} and $f_4^*(s)$ values], to verify the following complete results for the $n = 3$ problem.

		$f_3(s, x_3) = c_{sx_3} + f_4^*(x_3)$			
$n = 3$:	s $\diagdown$ x_3	H	I	$f_3^*(s)$	x_3^*
	E	4	8	4	H
	F	9	7	7	I
	G	6	7	6	H

The solution for the second-stage problem ($n = 2$), where there are three stages to go, is obtained in a similar fashion. In this case, $f_2(s, x_2) = c_{sx_2} + f_3^*(x_2)$. For example, suppose that the fortune seeker is in state C, as depicted below.

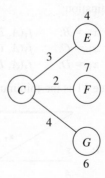

He must next go to state E, F, or G at an immediate cost of $c_{C,E} = 3$, $c_{C,F} = 2$, or $c_{C,G} = 4$, respectively. After getting there, the minimum additional cost for stage 3 to the end is given by the $n = 3$ table as $f_3^*(E) = 4$, $f_3^*(F) = 7$, or $f_3^*(G) = 6$, respectively, as shown above the E and F nodes and below the G node in the preceding diagram. The resulting calculations for the three alternatives are summarized below.

$$x_2 = E: \quad f_2(C, E) = c_{C,E} + f_3^*(E) = 3 + 4 = 7.$$
$$x_2 = F: \quad f_2(C, F) = c_{C,F} + f_3^*(F) = 2 + 7 = 9.$$
$$x_2 = G: \quad f_2(C, G) = c_{C,G} + f_3^*(G) = 4 + 6 = 10.$$

The minimum of these three numbers is 7, so the minimum total cost from state C to the end is $f_2^*(C) = 7$, and the immediate destination should be $x_2^* = E$.

Making similar calculations when you start from state B or D (try it) yields the following results for the $n = 2$ problem:

$n = 2$:	s	$f_2(s, x_2) = c_{sx_2} + f_3^*(x_2)$			$f_2^*(s)$	x_2^*
		E	F	G		
	B	11	11	12	11	E or F
	C	7	9	10	7	E
	D	8	8	11	8	E or F

In the first and third rows of this table, note that E and F tie as the minimizing value of x_2, so the immediate destination from either state B or D should be $x_2^* = E$ or F.

Moving to the first-stage problem ($n = 1$), with all four stages to go, we see that the calculations are similar to those just shown for the second-stage problem ($n = 2$), except now there is just *one* possible starting state $s = A$, as depicted below.

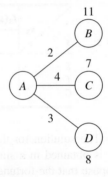

These calculations are summarized next for the three alternatives for the immediate destination:

$x_1 = B$: $f_1(A, B) = c_{A,B} + f_2^*(B) = 2 + 11 = 13.$
$x_1 = C$: $f_1(A, C) = c_{A,C} + f_2^*(C) = 4 + 7 = 11.$
$x_1 = D$: $f_1(A, D) = c_{A,D} + f_2^*(D) = 3 + 8 = 11.$

Since 11 is the minimum, $f_1^*(A) = 11$ and $x_1^* = C$ or D, as shown in the following table.

$n = 1$:	s	$f_1(s, x_1) = c_{sx_1} + f_2^*(x_1)$			$f_1^*(s)$	x_1^*
		B	C	D		
	A	13	11	11	11	C or D

An optimal solution for the entire problem can now be identified from the four tables. Results for the $n = 1$ problem indicate that the fortune seeker should go initially to either state C or state D. Suppose that he chooses $x_1^* = C$. For $n = 2$, the result for $s = C$ is $x_2^* = E$. This result leads to the $n = 3$ problem, which gives $x_3^* = H$ for $s = E$, and the $n = 4$ problem yields $x_4^* = J$ for $s = H$. Hence, one optimal route is $A \rightarrow C \rightarrow E \rightarrow H \rightarrow J$. Choosing $x_1^* = D$ leads to the other two optimal routes $A \rightarrow D \rightarrow E \rightarrow H \rightarrow J$ and $A \rightarrow D \rightarrow F \rightarrow I \rightarrow J$. They all yield a total cost of $f_1^*(A) = 11$.

These results of the dynamic programming analysis also are summarized in Fig. 10.2. Note how the two arrows for stage 1 come from the first and last columns of the $n = 1$ table and the resulting cost comes from the next-to-last column. Each of the other

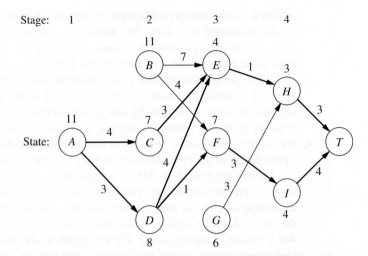

■ FIGURE 10.2
Graphical display of the dynamic programming solution of the stagecoach problem. Each arrow shows an optimal policy decision (the best immediate destination) from that state, where the number by the state is the resulting cost from there to the end. Following the boldface arrows from A to T gives the three optimal solutions (the three routes giving the minimum total cost of 11).

arrows (and the resulting cost) comes from one row in one of the other tables in just the same way.

You will see in the next section that the special terms describing the particular context of this problem—*stage, state,* and *policy*—actually are part of the general terminology of dynamic programming with an analogous interpretation in other contexts.

■ 10.2   CHARACTERISTICS OF DYNAMIC PROGRAMMING PROBLEMS

The stagecoach problem is a literal prototype of dynamic programming problems. In fact, this example was purposely designed to provide a literal physical interpretation of the rather abstract structure of such problems. Therefore, one way to recognize a situation that can be formulated as a dynamic programming problem is to notice that its basic structure is analogous to the stagecoach problem.

These basic features that characterize dynamic programming problems are presented and discussed here.

1. The problem can be divided into **stages**, with a **policy decision** required at each stage.

   The stagecoach problem was literally divided into its four stages (stagecoaches) that correspond to the four legs of the journey. The policy decision at each stage was which life insurance policy to choose (i.e., which destination to select for the next stagecoach ride). Similarly, other dynamic programming problems require making a *sequence of interrelated decisions,* where each decision corresponds to one stage of the problem.

2. Each stage has a number of **states** associated with the beginning of that stage.

   The states associated with each stage in the stagecoach problem were the states (or territories) in which the fortune seeker could be located when embarking on that particular leg of the journey. In general, the states are the various *possible conditions* in which the system might be at that stage of the problem. The number of states may be either finite (as in the stagecoach problem) or infinite (as in some subsequent examples).

3. The effect of the policy decision at each stage is to *transform the current state to a state associated with the beginning of the next stage* (possibly according to a probability distribution).

   The fortune seeker's decision as to his next destination led him from his current state to the next state on his journey. This procedure suggests that dynamic programming problems can be interpreted in terms of the *networks* described in Chap. 9. Each *node* would correspond to a *state.* The network would consist of columns of nodes, with each *column* corresponding to a *stage,* so that the flow from a node can go only to a node in the next column to the right. The links from a node to nodes in the next column correspond to the possible policy decisions on which state to go to next. The value assigned to each link usually can be interpreted as the *immediate contribution* to the objective function from making that policy decision. In most cases, the objective corresponds to finding either the *shortest* or the *longest path* through the network.

4. The solution procedure is designed to find an **optimal policy** for the overall problem, i.e., a prescription of the optimal policy decision at each stage for *each* of the possible states.

   For the stagecoach problem, the solution procedure constructed a table for each stage (n) that prescribed the optimal decision (x_n^*) for *each* possible state (s). Thus, in addition to identifying three *optimal solutions* (optimal routes) for the overall problem, the results show the fortune seeker how he should proceed if he gets detoured to a state that is not on an optimal route. For any problem, dynamic programming provides this kind of *policy* prescription of what to do under every possible circumstance (which is why the actual decision made upon reaching a particular state at a given stage is referred to as a *policy* decision). Providing this additional information beyond simply specifying an optimal solution (optimal sequence of decisions) can be helpful in a variety of ways, including sensitivity analysis.

5. Given the current state, an *optimal policy for the remaining stages* is *independent* of the policy decisions adopted in *previous stages.* Therefore, the optimal immediate decision depends on only the current state and not on how you got there. This is the **principle of optimality** for dynamic programming.

   Given the state in which the fortune seeker is currently located, the optimal life insurance policy (and its associated route) from this point onward is independent of how he got there. For dynamic programming problems in general, knowledge of the current state of the system conveys all the information about its previous behavior necessary for determining the optimal policy henceforth. (This property is the *Markovian property,* discussed in Sec. 16.2.) Any problem lacking this property cannot be formulated as a dynamic programming problem.

6. The solution procedure begins by finding the *optimal policy for the last stage.*

   The optimal policy for the last stage prescribes the optimal policy decision for *each* of the possible states at that stage. The solution of this one-stage problem is usually trivial, as it was for the stagecoach problem.

7. A **recursive relationship** that identifies the optimal policy for stage n, given the optimal policy for stage $n + 1$, is available.

   For the stagecoach problem, this recursive relationship was

$$f_n^*(s) = \min_{x_n} \{c_{sx_n} + f_{n+1}^*(x_n)\}.$$

Therefore, finding the *optimal policy decision* when you start in state s at stage n requires finding the minimizing value of x_n. For this particular problem, the corresponding minimum cost is achieved by using this value of x_n and then following the optimal policy when you start in state x_n at stage $n + 1$.

   The precise form of the recursive relationship differs somewhat among dynamic programming problems. However, notation analogous to that introduced in the preceding section will continue to be used here, as summarized below.

> N = number of stages.
>
> n = label for current stage ($n = 1, 2, \ldots, N$).
>
> s_n = current *state* for stage n.

$$x_n = \text{decision variable for stage } n.$$

$$x_n^* = \text{optimal value of } x_n \text{ (given } s_n).$$

$f_n(s_n, x_n) = $ contribution of stages $n, n + 1, \ldots, N$ to objective function if system starts in state s_n at stage n, immediate decision is x_n, and optimal decisions are made thereafter.

$$f_n^*(s_n) = f_n(s_n, x_n^*).$$

The recursive relationship will always be of the form

$$f_n^*(s_n) = \max_{x_n} \{f_n(s_n, x_n)\} \qquad \text{or} \qquad f_n^*(s_n) = \min_{x_n} \{f_n(s_n, x_n)\},$$

where $f_n(s_n, x_n)$ would be written in terms of s_n, x_n, $f_{n+1}^*(s_{n+1})$, and probably some measure of the immediate contribution of x_n to the objective function. It is the inclusion of $f_{n+1}^*(s_{n+1})$ on the right-hand side, so that $f_n^*(s_n)$ is defined in terms of $f_{n+1}^*(s_{n+1})$, that makes the expression for $f_n^*(s_n)$ a recursive relationship.

The recursive relationship keeps recurring as we move backward stage by stage. When the current stage number n is decreased by 1, the new $f_n^*(s_n)$ function is derived by using the $f_{n+1}^*(s_{n+1})$ function that was just derived during the preceding iteration, and then this process keeps repeating. This property is emphasized in the next (and final) characteristic of dynamic programming.

8. When we use this recursive relationship, the solution procedure starts at the end and moves *backward* stage by stage—each time finding the optimal policy for that stage—until it finds the optimal policy starting at the *initial* stage. This optimal policy immediately yields an optimal solution for the entire problem, namely, x_1^* for the initial state s_1, then x_2^* for the resulting state s_2, then x_3^* for the resulting state s_3, and so forth to x_N^* for the resulting stage s_N.

This backward movement was demonstrated by the stagecoach problem, where the optimal policy was found successively beginning in each state at stages 4, 3, 2, and 1, respectively.[4] For all dynamic programming problems, a table such as the following would be obtained for each stage $(n = N, N - 1, \ldots, 1)$.

s_n	x_n	$f_n(s_n, x_n)$		
			$f_n^*(s_n)$	x_n^*

When this table is finally obtained for the initial stage $(n = 1)$, the problem of interest is solved. Because the initial state is known, the initial decision is specified by x_1^* in this table. The optimal value of the other decision variables is then specified by the other tables in turn according to the state of the system that results from the preceding decisions.

■ 10.3 DETERMINISTIC DYNAMIC PROGRAMMING

This section further elaborates upon the dynamic programming approach to *deterministic* problems, where the *state* at the *next stage* is *completely determined* by the *state* and *policy decision* at the *current stage*. The *probabilistic* case, where there is a probability distribution for what the next state will be, is discussed in the next section.

---

[4]Actually, for this problem the solution procedure can move *either* backward or forward. However, for many problems (especially when the stages correspond to *time periods*), the solution procedure *must* move backward.

Deterministic dynamic programming can be described diagrammatically as shown in Fig. 10.3. Thus, at stage n the process will be in some state s_n. Making policy decision x_n then moves the process to some state s_{n+1} at stage $n + 1$. The contribution *thereafter* to the objective function under an optimal policy has been previously calculated to be $f_{n+1}^*(s_{n+1})$. The policy decision x_n also makes some contribution to the objective function. Combining these two quantities in an appropriate way provides $f_n(s_n, x_n)$, the contribution of stages n onward to the objective function. Optimizing with respect to x_n then gives $f_n^*(s_n) = f_n(s_n, x_n^*)$. After x_n^* and $f_n^*(s_n)$ are found for each possible value of s_n, the solution procedure is ready to move back one stage.

One way of categorizing deterministic dynamic programming problems is by the *form of the objective function.* For example, the objective might be to minimize the sum of the contributions from the individual stages (as for the stagecoach problem), or to maximize such a sum, or to minimize a product of such terms, and so on. Another categorization is in terms of the nature of the *set of states* for the respective stages. In particular, states s_n might be representable by a *discrete* state variable (as for the stagecoach problem) or by a *continuous* state variable, or perhaps a state *vector* (more than one variable) is required. Similarly, the decision variables $(x_1, x_2, \ldots, x_N)$ also can be either discrete or continuous.

Several examples are presented to illustrate these various possibilities. More importantly, they illustrate that these apparently major differences are actually quite inconsequential (except in terms of computational difficulty) because the underlying basic structure shown in Fig. 10.3 always remains the same.

The first new example arises in a much different context from the stagecoach problem, but it has the same *mathematical formulation* except that the objective is to *maximize* rather than to minimize a sum.

EXAMPLE 2   Distributing Medical Teams to Countries

The WORLD HEALTH COUNCIL is devoted to improving health care in the underdeveloped countries of the world. It now has five medical teams available to allocate among three such countries to improve their medical care, health education, and training programs. Therefore, the council needs to determine how many teams (if any) to allocate to each of these countries to maximize the total effectiveness of the five teams. The teams must be kept intact, so the number allocated to each country must be an integer.

The measure of performance being used is *additional person-years of life.* (For a particular country, this measure equals the *increased life expectancy* in years times the country's population.) Table 10.1 gives the estimated additional person-years of life (in multiples of 1,000) for each country for each possible allocation of medical teams.

Which allocation maximizes the measure of performance?

Formulation. This problem requires making three *interrelated decisions,* namely, how many medical teams to allocate to each of the three countries. Therefore, even though there is no fixed sequence, these three countries can be considered as the three stages in

---

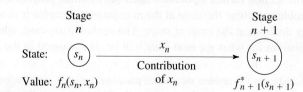

Stage n Stage $n + 1$

State: s_n $\xrightarrow[\text{of } x_n]{x_n}$ s_{n+1}

Contribution

Value: $f_n(s_n, x_n)$ $f_{n+1}^*(s_{n+1})$

Six days after Saddam Hussein ordered his Iraqi military forces to invade Kuwait on August 2, 1990, the United States began the long process of deploying many of its own military units and cargo to the region. After developing a coalition force from 35 nations led by the United States, the military operation called **Operation Desert Storm** was launched on January 17, 1991, to expel the Iraqi troops from Kuwait. This led to a decisive victory for the coalition forces, which liberated Kuwait and penetrated Iraq.

The logistical challenge involved in quickly transporting the needed troops and cargo to the war zone was a daunting one. A typical airlift mission carrying troops and cargo from the United States to the Persian Gulf required a three-day round-trip, visited seven or more different airfields, burned almost one million pounds of fuel, and cost $280,000. During Operation Desert Storm, the Military Airlift Command (MAC) averaged more than 100 such missions daily as it managed the largest airlift in history.

To meet this challenge, operations research was applied to develop the decision support systems needed to schedule and route each airlift mission. The OR technique used to drive this process was *dynamic programming*. The stages in the dynamic programming formulation correspond to the airfields in the network of flight legs relevant to the mission. For a given airfield, the states are characterized by the departure time from the airfield and the remaining available duty for the current crew. The objective function to be minimized is a weighted sum of several measures of performance: the lateness of deliveries, the flying time of the mission, the ground time, and the number of crew changes. The constraints include a lower bound on the load carried by the mission and upper bounds on the availability of crew and ground-support resources at airfields.

This application of dynamic programming had a dramatic impact on the ability to deliver the necessary cargo and personnel to the Persian gulf quickly to support Operation Desert Storm. For example, when speaking to the developers of this approach, MAC's deputy chief of staff for operations and transportation is quoted as saying, "I guarantee you that we could not have done that (the deployment to the Persian Gulf) without your help and the contributions you made to (the decision support systems)—we absolutely could not have done that."

Source: M. C. Hilliard, R. S. Solanki, C. Liu, I. K. Busch, G. Harrison, and R. D. Kraemer: "Scheduling the Operation Desert Storm Airlift: An Advanced Automated Scheduling Support System," *Interfaces*, **22**(1): 131–146, Jan.–Feb. 1992.

■ **TABLE 10.1** Data for the World Health Council problem

Medical Teams	Thousands of Additional Person-Years of Life		
	Country		
	1	2	3
0	0	0	0
1	45	20	50
2	70	45	70
3	90	75	80
4	105	110	100
5	120	150	130

a dynamic programming formulation. The decision variables x_n ($n = 1, 2, 3$) are the number of teams to allocate to stage (country) n.

The identification of the states may not be readily apparent. To determine the states, we ask questions such as the following. What is it that changes from one stage to the next? Given that the decisions have been made at the previous stages, how can the status of the situation at the current stage be described? What information about the current state of affairs is necessary to determine the optimal policy hereafter? On these bases, an appropriate choice for the "state of the system" is

s_n = number of medical teams still available for allocation to remaining countries $(n, \ldots, 3)$.

Thus, at stage 1 (country 1), where all three countries remain under consideration for allocations, $s_1 = 5$. However, at stage 2 or 3 (country 2 or 3), s_n is just 5 minus the number of teams allocated at preceding stages, so that the sequence of states is

$$s_1 = 5, \qquad s_2 = 5 - x_1, \qquad s_3 = s_2 - x_2.$$

With the dynamic programming procedure of solving backward stage by stage, when we are solving at stage 2 or 3, we shall not yet have solved for the allocations at the preceding stages. Therefore, we shall consider every possible state we could be in at stage 2 or 3, namely, $s_n = 0, 1, 2, 3, 4,$ or 5.

Figure 10.4 shows the states to be considered at each stage. The links (line segments) show the possible transitions in states from one stage to the next from making a feasible allocation of medical teams to the country involved. The numbers shown next to the links are the corresponding contributions to the measure of performance, where these numbers

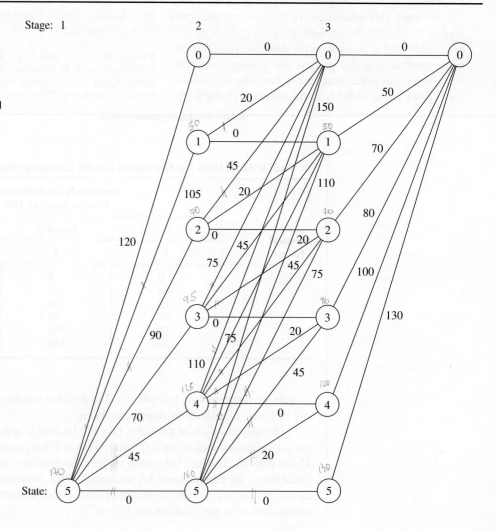

■ **FIGURE 10.4**
Graphical display of the World Health Council problem, showing the possible states at each stage, the possible transitions in states, and the corresponding contributions to the measure of performance.

come from Table 10.1. From the perspective of this figure, the overall problem is to find the path from the initial state 5 (beginning stage 1) to the final state 0 (after stage 3) that maximizes the sum of the numbers along the path.

To state the overall problem mathematically, let $p_i(x_i)$ be the measure of performance from allocating x_i medical teams to country i, as given in Table 10.1. Thus, the objective is to choose x_1, x_2, x_3 so as to

$$\text{Maximize} \quad \sum_{i=1}^{3} p_i(x_i),$$

subject to

$$\sum_{i=1}^{3} x_i = 5,$$

and

x_i are nonnegative integers.

Using the notation presented in Sec. 10.2, we see that $f_n(s_n, x_n)$ is

$$f_n(s_n, x_n) = p_n(x_n) + \max \sum_{i=n+1}^{3} p_i(x_i),$$

where the maximum is taken over $x_{n+1}, \ldots, x_3$ such that

$$\sum_{i=n}^{3} x_i = s_n$$

and the x_i are nonnegative integers, for $n = 1, 2, 3$. In addition,

$$f_n^*(s_n) = \max_{x_n=0,1,\ldots,s_n} f_n(s_n, x_n).$$

Therefore,

$$f_n(s_n, x_n) = p_n(x_n) + f_{n+1}^*(s_n - x_n)$$

(with f_4^* defined to be zero). These basic relationships are summarized in Fig. 10.5.

Consequently, the *recursive relationship* relating functions f_1^*, f_2^*, and f_3^* for this problem is

$$f_n^*(s_n) = \max_{x_n=0,1,\ldots,s_n} \{p_n(x_n) + f_{n+1}^*(s_n - x_n)\}, \qquad \text{for } n = 1, 2.$$

For the last stage ($n = 3$),

$$f_3^*(s_3) = \max_{x_3=0,1,\ldots,s_3} p_3(x_3).$$

The resulting dynamic programming calculations are given next.

Solution Procedure. Beginning with the last stage ($n = 3$), we note that the values of $p_3(x_3)$ are given in the last column of Table 10.1 and these values keep increasing as we move down the column. Therefore, with s_3 medical teams still available for allocation to country 3, the maximum of $p_3(x_3)$ is automatically achieved by allocating all s_3 teams; so $x_3^* = s_3$ and $f_3^*(s_3) = p_3(s_3)$, as shown in the following table.

$n = 3$:	s_3	$f_3^*(s_3)$	x_3^*
	0	0	0
	1	50	1
	2	70	2
	3	80	3
	4	100	4
	5	130	5

We now move backward to start from the next-to-last stage ($n = 2$). Here, finding x_2^* requires calculating and comparing $f_2(s_2, x_2)$ for the alternative values of x_2, namely, $x_2 = 0, 1, \ldots, s_2$. To illustrate, we depict this situation when $s_2 = 2$ graphically:

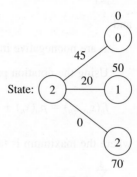

This diagram corresponds to Fig. 10.5 except that all three possible states at stage 3 are shown. Thus, if $x_2 = 0$, the resulting state at stage 3 will be $s_2 - x_2 = 2 - 0 = 2$, whereas $x_2 = 1$ leads to state 1 and $x_2 = 2$ leads to state 0. The corresponding values of $p_2(x_2)$ from the country 2 column of Table 10.1 are shown along the links, and the values of $f_3^*(s_2 - x_2)$ from the $n = 3$ table are given next to the stage 3 nodes. The required calculations for this case of $s_2 = 2$ are summarized below.

Formula: $f_2(2, x_2) = p_2(x_2) + f_3^*(2 - x_2)$.
$p_2(x_2)$ is given in the country 2 column of Table 10.1.
$f_3^*(2 - x_2)$ is given in the $n = 3$ table above.

$x_2 = 0$: $\quad f_2(2, 0) = p_2(0) + f_3^*(2) = \quad 0 + 70 = 70.$
$x_2 = 1$: $\quad f_2(2, 1) = p_2(1) + f_3^*(1) = 20 + 50 = 70.$
$x_2 = 2$: $\quad f_2(2, 2) = p_2(2) + f_3^*(0) = 45 + \quad 0 = 45.$

Because the objective is *maximization*, $x_2^* = 0$ or 1 with $f_2^*(2) = 70$.

■ **FIGURE 10.5**
The basic structure for the World Health Council problem.

$$\text{State:} \quad \overset{\text{Stage}}{\underset{n}{\boxed{s_n}}} \xrightarrow{\quad x_n \quad} \overset{\text{Stage}}{\underset{n+1}{\boxed{s_n - x_n}}}$$

$$\begin{array}{ccc} \text{Value: } f_n(s_n, x_n) & p_n(x_n) & f_{n+1}^*(s_n - x_n) \\ \quad = p_n(x_n) + f_{n+1}^*(s_n - x_n) & & \end{array}$$

Proceeding in a similar way with the other possible values of s_2 (try it) yields the following table.

| $n = 2$:
s_2 | $f_2(s_2, x_2) = p_2(x_2) + f_3^*(s_2 - x_2)$ | | | | | | $f_2^*(s_2)$ | x_2^* |
	x_2 0	1	2	3	4	5		
0	0						0	0
1	50	20					50	0
2	70	70	45				70	0 or 1
3	80	90	95	75			95	2
4	100	100	115	125	110		125	3
5	130	120	125	145	160	150	160	4

We now are ready to move backward to solve the original problem where we are starting from stage 1 ($n = 1$). In this case, the only state to be considered is the starting state of $s_1 = 5$, as depicted below.

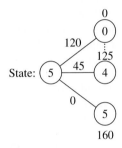

Since allocating x_1 medical teams to country 1 leads to a state of $5 - x_1$ at stage 2, a choice of $x_1 = 0$ leads to the bottom node on the right, $x_1 = 1$ leads to the next node up, and so forth up to the top node with $x_1 = 5$. The corresponding $p_1(x_1)$ values from Table 10.1 are shown next to the links. The numbers next to the nodes are obtained from the $f_2^*(s_2)$ column of the $n = 2$ table. As with $n = 2$, the calculation needed for each alternative value of the decision variable involves adding the corresponding link value and node value, as summarized below.

Formula: $f_1(5, x_1) = p_1(x_1) + f_2^*(5 - x_1)$.
          $p_1(x_1)$ is given in the country 1 column of Table 10.1.
          $f_2^*(5 - x_1)$ is given in the $n = 2$ table.

$x_1 = 0$: $f_1(5, 0) = p_1(0) + f_2^*(5) =   0 + 160 = 160$.
$x_1 = 1$: $f_1(5, 1) = p_1(1) + f_2^*(4) =   45 + 125 = 170$.
⋮
$x_1 = 5$: $f_1(5, 5) = p_1(5) + f_2^*(0) = 120 +   0 = 120$.

The similar calculations for $x_1 = 2, 3, 4$ (try it) verify that $x_1^* = 1$ with $f_1^*(5) = 170$, as shown in the following table.

| $n = 1$:
s_1 | $f_1(s_1, x_1) = p_1(x_1) + f_2^*(s_1 - x_1)$ | | | | | | $f_1^*(s_1)$ | x_1^* |
	x_2 0	1	2	3	4	5		
5	160	170	165	160	155	120	170	1

The structure of the next example is similar to the one for the World Health Council because it, too, is a distribution of effort problem. However, its recursive relationship differs in that its objective is to minimize a product of terms for the respective stages.

At first glance, this example may appear *not* to be a deterministic dynamic programming problem because probabilities are involved. However, it does indeed fit our definition because the state at the next stage is completely determined by the state and policy decision at the current stage.

EXAMPLE 3   Distributing Scientists to Research Teams

A government space project is conducting research on a certain engineering problem that must be solved before people can fly safely to Mars. Three research teams are currently trying three different approaches for solving this problem. The estimate has been made that, under present circumstances, the probability that the respective teams—call them 1, 2, and 3—will not succeed is 0.40, 0.60, and 0.80, respectively. Thus, the current probability that all three teams will fail is $(0.40)(0.60)(0.80) = 0.192$. Because the objective is to minimize the probability of failure, two more top scientists have been assigned to the project.

Table 10.2 gives the estimated probability that the respective teams will fail when 0, 1, or 2 additional scientists are added to that team. Only integer numbers of scientists are considered because each new scientist will need to devote full attention to one team. The problem is to determine how to allocate the two additional scientists to minimize the probability that all three teams will fail.

Formulation. Because both Examples 2 and 3 are distribution of effort problems, their underlying structure is actually very similar. In this case, scientists replace medical teams as the kind of resource involved, and research teams replace countries as the activities. Therefore, instead of medical teams being allocated to countries, scientists are being allocated to research teams. The only basic difference between the two problems is in their objective functions.

With so few scientists and teams involved, this problem could be solved very easily by a process of exhaustive enumeration. However, the dynamic programming solution is presented for illustrative purposes.

In this case, stage n ($n = 1, 2, 3$) corresponds to research team n, and the state s_n is the number of new scientists *still available* for allocation to the remaining teams. The decision variables x_n ($n = 1, 2, 3$) are the number of additional scientists allocated to team n.

Let $p_i(x_i)$ denote the probability of failure for team i if it is assigned x_i additional scientists, as given by Table 10.2. If we let Π denote multiplication, the government's objective is to choose x_1, x_2, x_3 so as to

$$\text{Minimize} \quad \prod_{i=1}^{3} p_i(x_i) = p_1(x_1)p_2(x_2)p_3(x_3),$$

■ **TABLE 10.2** Data for the Government Space Project problem

| | Probability of Failure | | |
| | Team | | |
New Scientists	1	2	3
0	0.40	0.60	0.80
1	0.20	0.40	0.50
2	0.15	0.20	0.30

subject to

$$\sum_{i=1}^{3} x_i = 2$$

and

x_i are nonnegative integers.

Consequently, $f_n(s_n, x_n)$ for this problem is

$$f_n(s_n, x_n) = p_n(x_n) \cdot \min \prod_{i=n+1}^{3} p_i(x_i),$$

where the minimum is taken over $x_{n+1}, \ldots, x_3$ such that

$$\sum_{i=n}^{3} x_i = s_n$$

and

x_i are nonnegative integers,

for $n = 1, 2, 3$. Thus,

$$f_n^*(s_n) = \min_{x_n = 0, 1, \ldots, s_n} f_n(s_n, x_n),$$

where

$$f_n(s_n, x_n) = p_n(x_n) \cdot f_{n+1}^*(s_n - x_n)$$

(with f_4^* defined to be 1). Figure 10.7 summarizes these basic relationships.

Thus, the *recursive relationship* relating the f_1^*, f_2^*, and f_3^* functions in this case is

$$f_n^*(s_n) = \min_{x_n = 0, 1, \ldots, s_n} \{ p_n(x_n) \cdot f_{n+1}^*(s_n - x_n) \}, \qquad \text{for } n = 1, 2,$$

and, when $n = 3$,

$$f_3^*(s_3) = \min_{x_3 = 0, 1, \ldots, s_3} p_3(x_3).$$

Solution Procedure.   The resulting dynamic programming calculations are as follows:

$n = 3$:	s_3	$f_3^*(s_3)$	x_3^*
	0	0.80	0
	1	0.50	1
	2	0.30	2

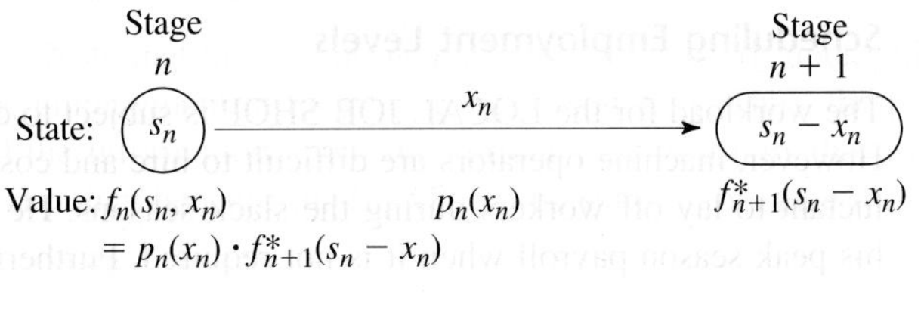

■ **FIGURE 10.7**
The basic structure for the government space project problem.

Stage n

Stage $n + 1$

State: $\left(s_n \right)$ $\xrightarrow{\quad x_n \quad}$ $\left(s_n - x_n \right)$

Value: $f_n(s_n, x_n)$ $p_n(x_n)$ $f_{n+1}^*(s_n - x_n)$
$= p_n(x_n) \cdot f_{n+1}^*(s_n - x_n)$

When $n = 1$, $s_1 = x_0 = x_4 = 255$.

For your ease of reference while working through the problem, a summary of the data is given in Table 10.3 for each of the four stages.

The objective for the problem is to choose x_1, x_2, x_3 (with $x_0 = x_4 = 255$) so as to

$$\text{Minimize} \quad \sum_{i=1}^{4} [200(x_i - x_{i-1})^2 + 2{,}000(x_i - r_i)],$$

subject to

$$r_i \leq x_i \leq 255, \quad \text{for } i = 1, 2, 3, 4.$$

Thus, for stage n onward ($n = 1, 2, 3, 4$), since $s_n = x_{n-1}$

$$f_n(s_n, x_n) = 200(x_n - s_n)^2 + 2{,}000(x_n - r_n)$$
$$+ \min_{r_i \leq x_i \leq 255} \sum_{i=n+1}^{4} [200(x_i - x_{i-1})^2 + 2{,}000(x_i - r_i)],$$

where this summation equals zero when $n = 4$ (because it has no terms). Also,

$$f_n^*(s_n) = \min_{r_n \leq x_n \leq 255} f_n(s_n, x_n).$$

Hence,

$$f_n(s_n, x_n) = 200(x_n - s_n)^2 + 2{,}000(x_n - r_n) + f_{n+1}^*(x_n)$$

(with f_5^* defined to be zero because costs after stage 4 are irrelevant to the analysis). A summary of these basic relationships is given in Fig. 10.8.

Consequently, the recursive relationship relating the f_n^* functions is

$$f_n^*(s_n) = \min_{r_n \leq x_n \leq 255} \{200(x_n - s_n)^2 + 2{,}000(x_n - r_n) + f_{n+1}^*(x_n)\}.$$

The dynamic programming approach uses this relationship to identify successively these functions—$f_4^*(s_4)$, $f_3^*(s_3)$, $f_2^*(s_2)$, $f_1^*(255)$—and the corresponding minimizing x_n.

■ **TABLE 10.3** Data for the Local Job Shop problem

n	r_n	Feasible x_n	Possible $s_n = x_{n-1}$	Cost
1	220	$220 \leq x_1 \leq 255$	$s_1 = 255$	$200(x_1 - 255)^2 + 2{,}000(x_1 - 220)$
2	240	$240 \leq x_2 \leq 255$	$220 \leq s_2 \leq 255$	$200(x_2 - x_1)^2 + 2{,}000(x_2 - 240)$
3	200	$200 \leq x_3 \leq 255$	$240 \leq s_3 \leq 255$	$200(x_3 - x_2)^2 + 2{,}000(x_3 - 200)$
4	255	$x_4 = 255$	$200 \leq s_4 \leq 255$	$200(255 - x_3)^2$

■ **FIGURE 10.8**
The basic structure for the Local Job Shop problem.

Stage n — Stage $n + 1$

State: $s_n \longrightarrow x_n$

Value: $f_n(s_n, x_n)$ $200(x_n - s_n)^2 + 2{,}000(x_n - r_n)$ $f_{n+1}^*(x_n)$
= sum

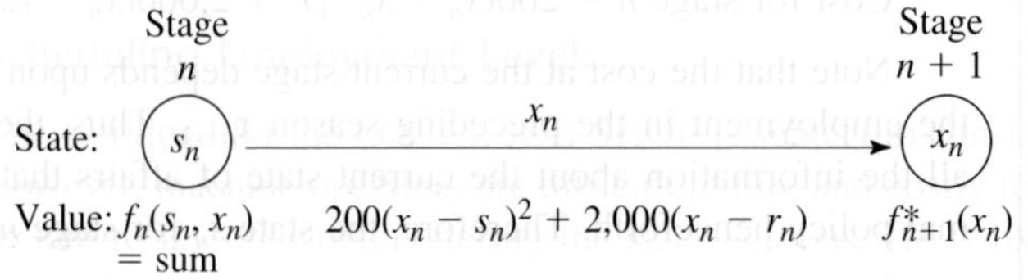

Solution Procedure. *Stage 4:* Beginning at the last stage ($n = 4$), we already know that $x_4^* = 255$, so the necessary results are

$n = 4$:	s_4	$f_4^*(s_4)$	x_4^*
	$200 \le s_4 \le 255$	$200(255 - s_4)^2$	255

Stage 3: For the problem consisting of just the last *two* stages ($n = 3$), the recursive relationship reduces to

$$f_3^*(s_3) = \min_{200 \le x_3 \le 255} \{200(x_3 - s_3)^2 + 2{,}000(x_3 - 200) + f_4^*(x_3)\}$$
$$= \min_{200 \le x_3 \le 255} \{200(x_3 - s_3)^2 + 2{,}000(x_3 - 200) + 200(255 - x_3)^2\},$$

where the possible values of s_3 are $240 \le s_3 \le 255$.

One way to solve for the value of x_3 that minimizes $f_3(s_3, x_3)$ for any particular value of s_3 is the graphical approach illustrated in Fig. 10.9.

However, a faster way is to use *calculus*. We want to solve for the minimizing x_3 in terms of s_3 by considering s_3 to have some fixed (but unknown) value. Therefore, set the first (partial) derivative of $f_3(s_3, x_3)$ with respect to x_3 equal to zero:

$$\frac{\partial}{\partial x_3} f_3(s_3, x_3) = 400(x_3 - s_3) + 2{,}000 - 400(255 - x_3)$$
$$= 400(2x_3 - s_3 - 250)$$
$$= 0,$$

which yields

$$x_3^* = \frac{s_3 + 250}{2}.$$

Because the second derivative is positive, and because this solution lies in the feasible interval for x_3 ($200 \le x_3 \le 255$) for all possible s_3 ($240 \le s_3 \le 255$), it is indeed the desired minimum.

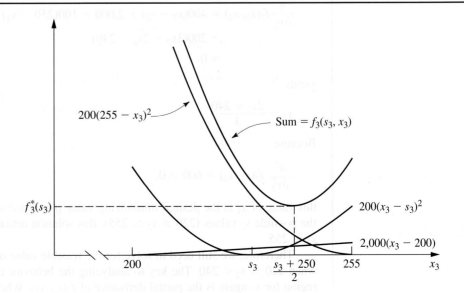

■ **FIGURE 10.9**
Graphical solution for $f_3^*(s_3)$
for the Local Job Shop
problem.

set

$$\frac{\partial}{\partial x_1} f_1(s_1, x_1) = 0,$$

which yields

$$x_1 = \frac{3s_1 + 225}{4}.$$

Because $s_1 = 255$, it follows that $x_1 = 247.5$ minimizes $f_1(s_1, x_1)$ over the region $240 \leq x_1 \leq 255$.

Note that this region ($240 \leq x_1 \leq 255$) includes $x_1 = 240$, so that $f_1(s_1, 240) > f_1(s_1, 247.5)$. In the next-to-last paragraph, we found that $x_1 = 240$ minimizes $f_1(s_1, x_1)$ over the region $220 \leq x_1 \leq 240$. Consequently, we now can conclude that $x_1 = 247.5$ also minimizes $f_1(s_1, x_1)$ over the *entire* feasible region $220 \leq x_1 \leq 255$.

Our final calculation is to find $f_1^*(s_1)$ for $s_1 = 255$ by plugging $x_1 = 247.5$ into the expression for $f_1(255, x_1)$ that holds for $240 \leq x_1 \leq 255$. Hence,

$$f_1^*(255) = 200(247.5 - 255)^2 + 2,000(247.5 - 220)$$

$$+ \frac{200}{9} [2(250 - 247.5)^2 + (265 - 247.5)^2 + 30(742.5 - 575)]$$

$$= 185,000.$$

These results are summarized as follows:

$n = 1$:	s_1	$f_1^*(s_1)$	x_1^*
	255	185,000	247.5

Therefore, by tracing back through the tables for $n = 2$, $n = 3$, and $n = 4$, respectively, and setting $s_n = x_{n-1}^*$ each time, the resulting optimal solution is $x_1^* = 247.5$, $x_2^* = 245$, $x_3^* = 247.5$, $x_4^* = 255$, with a total estimated cost per cycle of $185,000.

To conclude our illustrations of deterministic dynamic programming, we give one example that requires *more than one* variable to describe the state at each stage.

EXAMPLE 5   Wyndor Glass Company Problem

Consider the following linear programming problem:

Maximize $Z = 3x_1 + 5x_2,$

subject to

$$\begin{aligned}
x_1 &\leq 4 \\
2x_2 &\leq 12 \\
3x_1 + 2x_2 &\leq 18
\end{aligned}$$

and

$$x_1 \geq 0, \qquad x_2 \geq 0.$$

(You might recognize this as being the model for the Wyndor Glass Co. problem—introduced in Sec. 3.1.) One way of solving small linear (or nonlinear) programming problems like this one is by dynamic programming, which is illustrated below.

Formulation.   This problem requires making two interrelated decisions, namely, the level of activity 1, denoted by x_1, and the level of activity 2, denoted by x_2. Therefore, these two activities can be interpreted as the two stages in a dynamic programming formulation. Although they can be taken in either order, let stage $n =$ activity n ($n = 1, 2$). Thus, x_n is the decision variable at stage n.

What are the states? In other words, given that the decision had been made at prior stages (if any), what information is needed about the current state of affairs before the decision can be made at stage n? Reflection might suggest that the required information is the *amount of slack* left in the functional constraints. Interpret the right-hand side of these constraints (4, 12, and 18) as the total available amount of resources 1, 2, and 3, respectively (as described in Sec. 3.1). Then state s_n can be defined as

State $s_n =$ amount of respective resources still available for allocation to remaining activities.

(Note that the definition of the state is analogous to that for distribution of effort problems, including Examples 2 and 3, except that there are now three resources to be allocated instead of just one.) Thus,

$$s_n = (R_1, R_2, R_3),$$

where R_i is the amount of resource i remaining to be allocated ($i = 1, 2, 3$). Therefore,

$$s_1 = (4, 12, 18),$$
$$s_2 = (4 - x_1, 12, 18 - 3x_1).$$

However, when we begin by solving for stage 2, we do not yet know the value of x_1, and so we use $s_2 = (R_1, R_2, R_3)$ at that point.

Therefore, in contrast to the preceding examples, this problem has *three* state variables (i.e., a *state vector* with three components) at each stage rather than one. From a theoretical standpoint, this difference is not particularly serious. It only means that, instead of considering all possible values of the one state variable, we must consider all possible *combinations* of values of the several state variables. However, from the standpoint of computational efficiency, this difference tends to be a very serious complication. Because the number of combinations, in general, can be as large as the *product* of the number of possible values of the respective variables, the number of required calculations tends to "blow up" rapidly when additional state variables are introduced. This phenomenon has been given the apt name of the **curse of dimensionality.**

Each of the three state variables is *continuous.* Therefore, rather than consider each possible combination of values separately, we must use the approach introduced in Example 4 of solving for the required information as a *function* of the state of the system.

Despite these complications, this problem is small enough that it can still be solved without great difficulty. To solve it, we need to introduce the usual dynamic programming notation. Thus,

$$f_2(R_1, R_2, R_3, x_2) = \text{contribution of activity 2 to } Z \text{ if system starts in state}$$
$$(R_1, R_2, R_3) \text{ at stage 2 and decision is } x_2$$
$$= 5x_2,$$

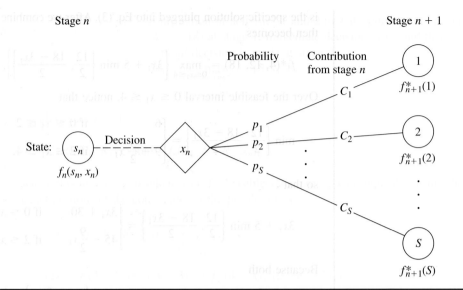

■ FIGURE 10.11
The basic structure for probabilistic dynamic programming.

and policy decision at the current stage. The resulting basic structure for probabilistic dynamic programming is described diagrammatically in Fig. 10.11.

For the purposes of this diagram, we let S denote the number of possible states at stage $n + 1$ and label these states on the right side as $1, 2, \ldots , S$. The system goes to state i with probability p_i $(i = 1, 2, \ldots , S)$ given state s_n and decision x_n at stage n. If the system goes to state i, C_i is the contribution of stage n to the objective function.

When Fig. 10.11 is expanded to include all the possible states and decisions at all the stages, it is sometimes referred to as a **decision tree.** If the decision tree is not too large, it provides a useful way of summarizing the various possibilities.

Because of the probabilistic structure, the relationship between $f_n(s_n, x_n)$ and the $f_{n+1}^*(s_{n+1})$ necessarily is somewhat more complicated than that for deterministic dynamic programming. The precise form of this relationship will depend upon the form of the overall objective function.

To illustrate, suppose that the objective is to *minimize* the *expected sum* of the contributions from the individual stages. In this case, $f_n(s_n, x_n)$ represents the minimum expected sum from stage n onward, *given* that the state and policy decision at stage n are s_n and x_n, respectively. Consequently,

$$f_n(s_n, x_n) = \sum_{i=1}^{S} p_i [C_i + f_{n+1}^*(i)],$$

with

$$f_{n+1}^*(i) = \min_{x_{n+1}} f_{n+1}(i, x_{n+1}),$$

where this minimization is taken over the *feasible* values of x_{n+1}.

Example 6 has this same form. Example 7 will illustrate another form.

EXAMPLE 6   Determining Reject Allowances

The HIT-AND-MISS MANUFACTURING COMPANY has received an order to supply one item of a particular type. However, the customer has specified such stringent quality requirements that the manufacturer may have to produce more than one item to obtain an item that is acceptable. The number of *extra* items produced in a production run is called the *reject allowance*. Including a reject allowance is common practice when producing for a custom order, and it seems advisable in this case.

The manufacturer estimates that each item of this type that is produced will be *acceptable* with probability $\frac{1}{2}$ and *defective* (without possibility for rework) with probability $\frac{1}{2}$. Thus, the number of acceptable items produced in a lot of size L will have a *binomial distribution;* i.e., the probability of producing no acceptable items in such a lot is $(\frac{1}{2})^L$.

Marginal production costs for this product are estimated to be $100 per item (even if defective), and excess items are worthless. In addition, a setup cost of $300 must be incurred whenever the production process is set up for this product, and a completely new setup at this same cost is required for each subsequent production run if a lengthy inspection procedure reveals that a completed lot has not yielded an acceptable item. The manufacturer has time to make no more than three production runs. If an acceptable item has not been obtained by the end of the third production run, the cost to the manufacturer in lost sales income and penalty costs will be $1,600.

The objective is to determine the policy regarding the lot size (1 + reject allowance) for the required production run(s) that minimizes total expected cost for the manufacturer.

Formulation. A dynamic programming formulation for this problem is

Stage n = production run n ($n = 1, 2, 3$),
x_n = lot size for stage n,
State s_n = number of acceptable items still needed (1 or 0) at beginning of stage n.

Thus, at stage 1, state $s_1 = 1$. If at least one acceptable item is obtained subsequently, the state changes to $s_n = 0$, after which no additional costs need to be incurred.

Because of the stated objective for the problem,

$f_n(s_n, x_n)$ = total expected cost for stages $n, \ldots, 3$ if system starts in state s_n at stage n, immediate decision is x_n, and optimal decisions are made thereafter,

$$f_n^*(s_n) = \min_{x_n = 0, 1, \ldots} f_n(s_n, x_n),$$

where $f_n^*(0) = 0$. Using $100 as the unit of money, the contribution to cost from stage n is $[K(x_n) + x_n]$ regardless of the next state, where $K(x_n)$ is a function of x_n such that

$$K(x_n) = \begin{cases} 0, & \text{if } x_n = 0 \\ 3, & \text{if } x_n > 0. \end{cases}$$

Therefore, for $s_n = 1$,

$$f_n(1, x_n) = K(x_n) + x_n + \left(\frac{1}{2}\right)^{x_n} f_{n+1}^*(1) + \left[1 - \left(\frac{1}{2}\right)^{x_n}\right] f_{n+1}^*(0)$$

$$= K(x_n) + x_n + \left(\frac{1}{2}\right)^{x_n} f_{n+1}^*(1)$$

[where $f_4^*(1)$ is defined to be 16, the terminal cost if no acceptable items have been obtained]. A summary of these basic relationships is given in Fig. 10.12.

Consequently, the recursive relationship for the dynamic programming calculations is

$$f_n^*(1) = \min_{x_n = 0, 1, \ldots} \left\{ K(x_n) + x_n + \left(\frac{1}{2}\right)^{x_n} f_{n+1}^*(1) \right\}$$

for $n = 1, 2, 3$.

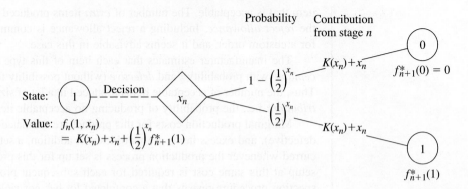

FIGURE 10.12
The basic structure for the Hit-and-Miss Manufacturing Co. problem.

Solution Procedure.   The calculations using this recursive relationship are summarized as follows.

	x_3	$f_3(1, x_3) = K(x_3) + x_3 + 16\left(\frac{1}{2}\right)^{x_3}$							
$n = 3$:	s_3	**0**	**1**	**2**	**3**	**4**	**5**	$f_3^*(s_3)$	x_3^*
	0	0						0	0
	1	16	12	9	8	8	$8\frac{1}{2}$	8	3 or 4

	x_2	$f_2(1, x_2) = K(x_2) + x_2 + \left(\frac{1}{2}\right)^{x_2} f_3^*(1)$						
$n = 2$:	s_2	**0**	**1**	**2**	**3**	**4**	$f_2^*(s_2)$	x_2^*
	0	0					0	0
	1	8	8	7	7	$7\frac{1}{2}$	7	2 or 3

	x_1	$f_1(1, x_1) = K(x_1) + x_1 + \left(\frac{1}{2}\right)^{x_1} f_2^*(1)$						
$n = 1$:	s_1	**0**	**1**	**2**	**3**	**4**	$f_1^*(s_1)$	x_1^*
	1	7	$7\frac{1}{2}$	$6\frac{3}{4}$	$6\frac{7}{8}$	$7\frac{7}{16}$	$6\frac{3}{4}$	2

Thus, the optimal policy is to produce two items on the first production run; if none is acceptable, then produce either two or three items on the second production run; if none is acceptable, then produce either three or four items on the third production run. The total expected cost for this policy is $675.

EXAMPLE 7   **Winning in Las Vegas**

An enterprising young statistician believes that she has developed a system for winning a popular Las Vegas game. Her colleagues do not believe that her system works, so they have made a large bet with her that if she starts with three chips, she will not have at least five chips after three plays of the game. Each play of the game involves betting any desired number of available chips and then either winning or losing this number of chips. The statistician believes that her system will give her a probability of $\frac{2}{3}$ of winning a given play of the game.

Assuming the statistician is correct, we now use dynamic programming to determine her optimal policy regarding how many chips to bet (if any) at each of the three plays of the game. The decision at each play should take into account the results of earlier plays. The objective is to maximize the probability of winning her bet with her colleagues.

Formulation.   The dynamic programming formulation for this problem is

Stage n = nth play of game ($n = 1, 2, 3$),
  x_n = number of chips to bet at stage n,
State s_n = number of chips in hand to begin stage n.

This definition of the state is chosen because it provides the needed information about the current situation for making an optimal decision on how many chips to bet next.

Because the objective is to maximize the probability that the statistician will win her bet, the objective function to be maximized at each stage must be the probability of finishing the three plays with at least five chips. (Note that the value of ending with more than five chips is just the same as ending with exactly five, since the bet is won either way.) Therefore,

$f_n(s_n, x_n)$ = probability of finishing three plays with at least five chips, given that the statistician starts stage n in state s_n, makes immediate decision x_n, and makes optimal decisions thereafter,

$$f_n^*(s_n) = \max_{x_n=0, 1, \ldots, s_n} f_n(s_n, x_n).$$

The expression for $f_n(s_n, x_n)$ must reflect the fact that it may still be possible to accumulate five chips eventually even if the statistician should lose the next play. If she loses, the state at the next stage will be $s_n - x_n$, and the probability of finishing with at least five chips will then be $f_{n+1}^*(s_n - x_n)$. If she wins the next play instead, the state will become $s_n + x_n$, and the corresponding probability will be $f_{n+1}^*(s_n + x_n)$. Because the assumed probability of winning a given play is $\frac{2}{3}$, it now follows that

$$f_n(s_n, x_n) = \frac{1}{3} f_{n+1}^*(s_n - x_n) + \frac{2}{3} f_{n+1}^*(s_n + x_n)$$

[where $f_4^*(s_4)$ is defined to be 0 for $s_4 < 5$ and 1 for $s_4 \geq 5$]. Thus, there is no direct contribution to the objective function from stage n other than the effect of then being in the next state. These basic relationships are summarized in Fig. 10.13.

Therefore, the recursive relationship for this problem is

$$f_n^*(s_n) = \max_{x_n=0, 1, \ldots, s_n} \left\{ \frac{1}{3} f_{n+1}^*(s_n - x_n) + \frac{2}{3} f_{n+1}^*(s_n + x_n) \right\},$$

for $n = 1, 2, 3$, with $f_4^*(s_4)$ as just defined.

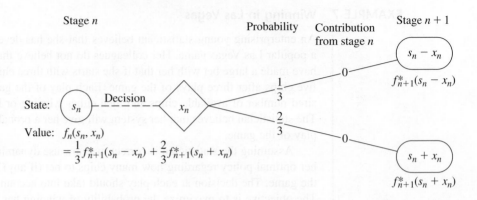

■ FIGURE 10.13
The basic structure for the
Las Vegas problem.

Solution Procedure.   This recursive relationship leads to the following computational
results.

n = 3:

s_3	$f_3^*(s_3)$	x_3^*
0	0	—
1	0	—
2	0	—
3	$\frac{2}{3}$	2 (or more)
4	$\frac{2}{3}$	1 (or more)
≥ 5	1	0 (or $\leq s_3 - 5$)

n = 2:

$$f_2(s_2, x_2) = \frac{1}{3}f_3^*(s_2 - x_2) + \frac{2}{3}f_3^*(s_2 + x_2)$$

s_2 \ x_2	0	1	2	3	4	$f_2^*(s_2)$	x_2^*
0	0					0	—
1	0	0				0	—
2	0	$\frac{4}{9}$	$\frac{4}{9}$			$\frac{4}{9}$	1 or 2
3	$\frac{2}{3}$	$\frac{4}{9}$	$\frac{2}{3}$	$\frac{2}{3}$		$\frac{2}{3}$	0, 2, or 3
4	$\frac{2}{3}$	$\frac{8}{9}$	$\frac{2}{3}$	$\frac{2}{3}$	$\frac{2}{3}$	$\frac{8}{9}$	1
≥ 5	1					1	0 (or $\leq s_2 - 5$)

n = 1:

$$f_1(s_1, x_1) = \frac{1}{3}f_2^*(s_1 - x_1) + \frac{2}{3}f_2^*(s_1 + x_1)$$

s_1 \ x_1	0	1	2	3	$f_1^*(s_1)$	x_1^*
3	$\frac{2}{3}$	$\frac{20}{27}$	$\frac{2}{3}$	$\frac{2}{3}$	$\frac{20}{27}$	1

Therefore, the optimal policy is

$$
x_1^* = 1 \begin{cases} \text{if win,} & x_2^* = 1 \begin{cases} \text{if win,} & x_3^* = 0 \\ \text{if lose,} & x_3^* = 2 \text{ or } 3. \end{cases} \\[2em] \text{if lose,} & x_2^* = 1 \text{ or } 2 \begin{cases} \text{if win,} & x_3^* = \begin{cases} 2 \text{ or } 3 & (\text{for } x_2^* = 1) \\ 1, 2, 3, \text{ or } 4 & (\text{for } x_2^* = 2) \end{cases} \\ \text{if lose,} & \text{bet is lost} \end{cases} \end{cases}
$$

This policy gives the statistician a probability of $\frac{20}{27}$ of winning her bet with her colleagues.

10.5 CONCLUSIONS

Dynamic programming is a very useful technique for making a *sequence of interrelated decisions*. It requires formulating an appropriate *recursive relationship* for each individual problem. However, it provides a great computational savings over using exhaustive enumeration to find the best combination of decisions, especially for large problems. For example, if a problem has 10 stages with 10 states and 10 possible decisions at each stage, then exhaustive enumeration must consider up to 10 billion combinations, whereas dynamic programming need make no more than a thousand calculations (10 for each state at each stage).

This chapter has considered only dynamic programming with a *finite* number of stages. Chapter 19 is devoted to a general kind of model for probabilistic dynamic programming where the stages continue to recur indefinitely, namely, Markov decision processes.

SELECTED REFERENCES

1. Bertsekas, D. P.: *Dynamic Programming: Deterministic and Stochastic Models,* Prentice-Hall, Englewood Cliffs, NJ, 1987.
2. Denardo, E. V.: *Dynamic Programming Theory and Applications,* Prentice-Hall, Englewood Cliffs, NJ, 1982.
3. Howard, R. A.: "Dynamic Programming," *Management Science,* **12:** 317–345, 1966.
4. Lew, A., and H. Mauch: *Dynamic Programming: A Computational Tool,* Springer, New York, 2007.
5. Smith, D. K.: *Dynamic Programming: A Practical Introduction,* Ellis Horwood, London, 1991.
6. Sniedovich, M.: *Dynamic Programming,* Marcel Dekker, New York, 1991.

■ LEARNING AIDS FOR THIS CHAPTER ON OUR WEBSITE (www.mhhe.com/hillier)

Worked Examples:

Examples for Chapter 10

"Ch. 10—Dynamic Programming" LINGO File

Glossary for Chapter 10

See Appendix 1 for documentation of the software.

■ PROBLEMS

An asterisk on the problem number indicates that at least a partial answer is given in the back of the book.

10.2-1. Consider the following network, where each number along a link represents the actual distance between the pair of nodes connected by that link. The objective is to find the shortest path from the origin to the destination.

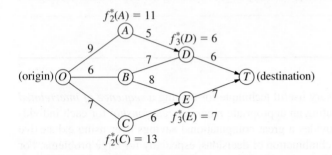

(a) What are the stages and states for the dynamic programming formulation of this problem?

(b) Use dynamic programming to solve this problem. However, instead of using the usual tables, show your work graphically (similar to Fig. 10.2). In particular, start with the given network, where the answers already are given for $f_n^*(s_n)$ for four of the nodes; then solve for and fill in $f_2^*(B)$ and $f_1^*(O)$. Draw an arrowhead that shows the optimal link to traverse out of each of the latter two nodes. Finally, identify the optimal path by following the arrows from node O onward to node T.

(c) Use dynamic programming to solve this problem by manually constructing the usual tables for $n = 3$, $n = 2$, and $n = 1$.

(d) Use the shortest-path algorithm presented in Sec. 9.3 to solve this problem. Compare and contrast this approach with the one in parts (b) and (c).

10.2-2. The sales manager for a publisher of college textbooks has six traveling salespeople to assign to three different regions of the country. She has decided that each region should be assigned at least one salesperson and that each individual salesperson should be restricted to one of the regions, but now she wants to determine how many salespeople should be assigned to the respective regions in order to maximize sales.

The next table gives the estimated increase in sales (in appropriate units) in each region if it were allocated various numbers of salespeople:

	Region		
Salespersons	**1**	**2**	**3**
1	40	24	32
2	54	47	46
3	78	63	70
4	99	78	84

(a) Use dynamic programming to solve this problem. Instead of using the usual tables, show your work graphically by constructing and filling in a network such as the one shown for Prob. 10.2-1. Proceed as in Prob. 10.2-1b by solving for $f_n^*(s_n)$ for each node (except the terminal node) and writing its value by the node. Draw an arrowhead to show the optimal link (or links in case of a tie) to take out of each node. Finally, identify the resulting optimal path (or paths) through the network and the corresponding optimal solution (or solutions).

(b) Use dynamic programming to solve this problem by constructing the usual tables for $n = 3$, $n = 2$, and $n = 1$.

10.2-3. Consider the following project network (as described in Sec. 9.8), where the number over each node is the time required for the corresponding activity. Consider the problem of finding the *longest path* (the largest total time) through this network from start to finish, since the longest path is the critical path.

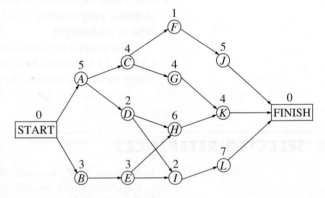

(a) What are the stages and states for the dynamic programming formulation of this problem?

(b) Use dynamic programming to solve this problem. However, instead of using the usual tables, show your work graphically. In particular, fill in the values of the various $f_n^*(s_n)$ under the corresponding nodes, and show the resulting optimal arc to traverse out of each node by drawing an arrowhead near the beginning of the arc. Then identify the optimal path (the longest path) by following these arrowheads from the Start node to the Finish node. If there is more than one optimal path, identify them all.

(c) Use dynamic programming to solve this problem by constructing the usual tables for $n = 4$, $n = 3$, $n = 2$, and $n = 1$.

10.2-4. Consider the following statements about solving dynamic programming problems. Label each statement as true or false, and then justify your answer by referring to specific statements in the chapter.

(a) The solution procedure uses a recursive relationship that enables solving for the optimal policy for stage $(n + 1)$ given the optimal policy for stage n.

(b) After completing the solution procedure, if a nonoptimal decision is made by mistake at some stage, the solution procedure will need to be reapplied to determine the new optimal decisions (given this nonoptimal decision) at the subsequent stages.

(c) Once an optimal policy has been found for the overall problem, the information needed to specify the optimal decision at a particular stage is the state at that stage and the decisions made at preceding stages.

10.3-1. Read the referenced article that fully describes the OR study summarized in the application vignette presented in Sec. 10.3. Briefly describe how dynamic programming was applied in this study. Then list the various financial and nonfinancial benefits that resulted from this study.

10.3-2.* The owner of a chain of three grocery stores has purchased five crates of fresh strawberries. The estimated probability distribution of potential sales of the strawberries before spoilage differs among the three stores. Therefore, the owner wants to know how to allocate five crates to the three stores to maximize expected profit.

For administrative reasons, the owner does not wish to split crates between stores. However, he is willing to distribute no crates to any of his stores.

The following table gives the estimated expected profit at each store when it is allocated various numbers of crates:

| | Store | | |
Crates	1	2	3
0	0	0	0
1	5	6	4
2	9	11	9
3	14	15	13
4	17	19	18
5	21	22	20

Use dynamic programming to determine how many of the five crates should be assigned to each of the three stores to maximize the total expected profit.

10.3-3. A college student has 7 days remaining before final examinations begin in her four courses, and she wants to allocate this study time as effectively as possible. She needs at least 1 day on each course, and she likes to concentrate on just one course each day, so she wants to allocate 1, 2, 3, or 4 days to each course. Having recently taken an OR course, she decides to use dynamic programming to make these allocations to maximize the total grade points to be obtained from the four courses. She estimates

that the alternative allocations for each course would yield the number of grade points shown in the following table:

| | Estimated Grade Points | | | |
| | Course | | | |
Study Days	1	2	3	4
1	1	5	4	4
2	3	6	6	4
3	6	8	7	5
4	8	8	9	8

Solve this problem by dynamic programming.

10.3-4. A political campaign is entering its final stage, and polls indicate a very close election. One of the candidates has enough funds left to purchase TV time for a total of five prime-time commercials on TV stations located in four different areas. Based on polling information, an estimate has been made of the number of additional votes that can be won in the different broadcasting areas depending upon the number of commercials run. These estimates are given in the following table in thousands of votes:

| | Area | | | |
Commercials	1	2	3	4
0	0	0	0	0
1	4	6	5	3
2	7	8	9	7
3	9	10	11	12
4	12	11	10	14
5	15	12	9	16

Use dynamic programming to determine how the five commercials should be distributed among the four areas in order to maximize the estimated number of votes won.

10.3-5. A county chairwoman of a certain political party is making plans for an upcoming presidential election. She has received the services of six volunteer workers for precinct work, and she wants to assign them to four precincts in such a way as to maximize their effectiveness. She feels that it would be inefficient to assign a worker to more than one precinct, but she is willing to assign no workers to any one of the precincts if they can accomplish more in other precincts.

The following table gives the estimated increase in the number of votes for the party's candidate in each precinct if it were allocated various numbers of workers:

	Precinct			
Workers	**1**	**2**	**3**	**4**
0	0	0	0	0
1	4	7	5	6
2	9	11	10	11
3	15	16	15	14
4	18	18	18	16
5	22	20	21	17
6	24	21	22	18

This problem has several optimal solutions for how many of the six workers should be assigned to each of the four precincts to maximize the total estimated increase in the plurality of the party's candidate. Use dynamic programming to find all of them so the chairwoman can make the final selection based on other factors.

10.3-6. Use dynamic programming to solve the Northern Airplane Co. production scheduling problem presented in Sec. 8.1 (see Table 8.7). Assume that production quantities must be integer multiples of 5.

10.3-7.* A company will soon be introducing a new product into a very competitive market and is currently planning its marketing strategy. The decision has been made to introduce the product in three phases. Phase 1 will feature making a special introductory offer of the product to the public at a greatly reduced price to attract first-time buyers. Phase 2 will involve an intensive advertising campaign to persuade these first-time buyers to continue purchasing the product at a regular price. It is known that another company will be introducing a new competitive product at about the time that phase 2 will end. Therefore, phase 3 will involve a follow-up advertising and promotion campaign to try to keep the regular purchasers from switching to the competitive product.

A total of $4 million has been budgeted for this marketing campaign. The problem now is to determine how to allocate this money most effectively to the three phases. Let m denote the initial share of the market (expressed as a percentage) attained in phase 1, f_2 the fraction of this market share that is retained in phase 2, and f_3 the fraction of the remaining market share that is retained in phase 3. Use dynamic programming to determine how to allocate the $4 million to maximize the final share of the market for the new product, i.e., to maximize mf_2f_3.

(a) Assume that the money must be spent in integer multiples of $1 million in each phase, where the minimum permissible multiple is 1 for phase 1 and 0 for phases 2 and 3. The following table gives the estimated effect of expenditures in each phase:

Millions of Dollars Expended	Effect on Market Share		
	m	**f_2**	**f_3**
0	—	0.2	0.3
1	20	0.4	0.5
2	30	0.5	0.6
3	40	0.6	0.7
4	50	—	—

(b) Now assume that *any* amount within the total budget can be spent in each phase, where the estimated effect of spending an amount x_i (in units of *millions* of dollars) in phase i ($i = 1$, 2, 3) is

$$m = 10x_1 - x_1^2$$
$$f_2 = 0.40 + 0.10x_2$$
$$f_3 = 0.60 + 0.07x_3.$$

[*Hint:* After solving for the $f_2^*(s)$ and $f_3^*(s)$ functions analytically, solve for x_1^* graphically.]

10.3-8. Consider an electronic system consisting of four components, each of which must work for the system to function. The reliability of the system can be improved by installing several parallel units in one or more of the components. The following table gives the probability that the respective components (labeled as Comp. 1, 2, 3, and 4) will function if they consist of one, two, or three parallel units:

	Probability of Functioning			
Parallel Units	**Comp. 1**	**Comp. 2**	**Comp. 3**	**Comp. 4**
1	0.5	0.6	0.7	0.5
2	0.6	0.7	0.8	0.7
3	0.8	0.8	0.9	0.9

The probability that the system will function is the product of the probabilities that the respective components will function.

The cost (in hundreds of dollars) of installing one, two, or three parallel units in the respective components (labeled as Comp. 1, 2, 3, and 4) is given by the following table:

	Cost			
Parallel Units	**Comp. 1**	**Comp. 2**	**Comp. 3**	**Comp. 4**
1	1	2	1	2
2	2	4	3	3
3	3	5	4	4

Because of budget limitations, a maximum of $1,000 can be expended.

Use dynamic programming to determine how many parallel units should be installed in each of the four components to maximize the probability that the system will function.

10.3-9. Consider the following integer nonlinear programming problem.

$$\text{Maximize} \qquad Z = 3x_1^2 - x_1^3 + 5x_2^2 - x_2^3,$$

subject to

$$x_1 + 2x_2 \le 4$$

and

$$x_1 \ge 0, \qquad x_2 \ge 0$$
$$x_1, x_2 \text{ are integers.}$$

Use dynamic programming to solve this problem.

10.3-10. Consider the following integer nonlinear programming problem.

$$\text{Maximize} \qquad Z = 32x_1 - 2x_1^2 + 30x_2 + 20x_3,$$

subject to

$$3x_1 + 7x_2 + 5x_3 \le 20$$

and

$$x_1, x_2, x_3 \text{ are nonnegative integers.}$$

Use dynamic programming to solve this problem.

10.3-11.* Consider the following nonlinear programming problem.

$$\text{Maximize} \qquad Z = 36x_1 + 9x_1^2 - 6x_1^3$$
$$+ 36x_2 - 3x_2^3,$$

subject to

$$x_1 + x_2 \le 3$$

and

$$x_1 \ge 0, \qquad x_2 \ge 0.$$

Use dynamic programming to solve this problem.

10.3-12. Re-solve the Local Job Shop employment scheduling problem (Example 4) when the total cost of changing the level of employment from one season to the next is changed to $100 times the square of the difference in employment levels.

10.3-13. Consider the following nonlinear programming problem.

$$\text{Maximize} \qquad Z = 2x_1^2 + 2x_2 + 4x_3 - x_3^2$$

subject to

$$2x_1 + x_2 + x_3 \le 4$$

and

$$x_1 \ge 0, \qquad x_2 \ge 0, \qquad x_3 \ge 0.$$

Use dynamic programming to solve this problem.

10.3-14. Consider the following nonlinear programming problem.

$$\text{Minimize} \qquad Z = x_1^4 + 2x_2^2$$

subject to

$$x_1^2 + x_2^2 \ge 2.$$

(There are no nonnegativity constraints.) Use dynamic programming to solve this problem.

10.3-15. Consider the following nonlinear programming problem.

$$\text{Maximize} \qquad Z = x_1^3 + 4x_2^2 + 16x_3,$$

subject to

$$x_1 x_2 x_3 = 4$$

and

$$x_1 \ge 1, \qquad x_2 \ge 1, \qquad x_3 \ge 1.$$

(a) Solve by dynamic programming when, in addition to the given constraints, all three variables also are required to be integer.
(b) Use dynamic programming to solve the problem as given (continuous variables).

10.3-16. Consider the following nonlinear programming problem.

$$\text{Maximize} \qquad Z = x_1(1 - x_2)x_3,$$

subject to

$$x_1 - x_2 + x_3 \le 1$$

and

$$x_1 \ge 0, \qquad x_2 \ge 0, \qquad x_3 \ge 0.$$

Use dynamic programming to solve this problem.

10.3-17. Consider the following linear programming problem.

$$\text{Maximize} \qquad Z = 15x_1 + 10x_2,$$

subject to

$$x_1 + 2x_2 \le 6$$
$$3x_1 + x_2 \le 8$$

and

$$x_1 \ge 0, \qquad x_2 \ge 0.$$

Use dynamic programming to solve this problem.

10.3-18. Consider the following "fixed-charge" problem.

$$\text{Maximize} \qquad Z = 3x_1 + 7x_2 + 6f(x_3),$$

subject to

$$x_1 + 3x_2 + 2x_3 \le 6$$
$$x_1 + x_2 \le 5$$

and

$$x_1 \ge 0, \qquad x_2 \ge 0, \qquad x_3 \ge 0,$$

where

$$f(x_3) = \begin{cases} 0 & \text{if } x_3 = 0 \\ -1 + x_3 & \text{if } x_3 > 0. \end{cases}$$

Use dynamic programming to solve this problem.

10.4-1. A backgammon player will be playing three consecutive matches with friends tonight. For each match, he will have the opportunity to place an even bet that he will win; the amount bet can be *any* quantity of his choice between zero and the amount of money he still has left after the bets on the preceding matches. For each match, the probability is $\frac{1}{2}$ that he will win the match and thus win the amount bet, whereas the probability is $\frac{1}{2}$ that he will lose the match and thus lose the amount bet. He will begin with $75, and his goal is to have $100 at the end. (Because these are friendly matches, he does not want to end up with more than $100.) Therefore, he wants to find the optimal betting policy (including all ties) that maximizes the probability that he will have exactly $100 after the three matches.

Use dynamic programming to solve this problem.

10.4-2. Imagine that you have $10,000 to invest and that you will have an opportunity to invest that amount in either of two investments (A or B) at the beginning of each of the next 3 years. Both investments have uncertain returns. For investment A you will either lose your money entirely or (with higher probability) get back $20,000 (a profit of $10,000) at the end of the year. For investment B you will get back either just your $10,000 or (with low probability) $20,000 at the end of the year. The probabilities for these events are as follows:

Investment	Amount Returned ($)	Probability
A	0	0.25
	20,000	0.75
B	10,000	0.9
	20,000	0.1

You are allowed to make only (at most) *one* investment each year, and you can invest only $10,000 each time. (Any additional money accumulated is left idle.)

(a) Use dynamic programming to find the investment policy that maximizes the expected amount of money you will have after 3 years.

(b) Use dynamic programming to find the investment policy that maximizes the probability that you will have at least $20,000 after 3 years.

10.4-3.* Suppose that the situation for the Hit-and-Miss Manufacturing Co. problem (Example 6) has changed somewhat. After a more careful analysis, you now estimate that each item produced will be acceptable with probability $\frac{2}{3}$, rather than $\frac{1}{2}$, so that the probability of producing *zero* acceptable items in a lot of size L is $(\frac{1}{3})^L$. Furthermore, there now is only enough time available to make two production runs. Use dynamic programming to determine the new optimal policy for this problem.

10.4-4. Reconsider Example 7. Suppose that the bet is changed as follows: "Starting with two chips, she will not have at least five chips after five plays of the game." By referring to the previous computational results, make additional calculations to determine the new optimal policy for the enterprising young statistician.

10.4-5. The Profit & Gambit Co. has a major product that has been losing money recently because of declining sales. In fact, during the current quarter of the year, sales will be 4 million units below the break-even point. Because the marginal revenue for each unit sold exceeds the marginal cost by $5, this amounts to a loss of $20 million for the quarter. Therefore, management must take action quickly to rectify this situation. Two alternative courses of action are being considered. One is to abandon the product immediately, incurring a cost of $20 million for shutting down. The other alternative is to undertake an intensive advertising campaign to increase sales and then abandon the product (at the cost of $20 million) only if the campaign is not sufficiently successful. Tentative plans for this advertising campaign have been developed and analyzed. It would extend over the next three quarters (subject to early cancellation), and the cost would be $30 million in each of the three quarters. It is estimated that the increase in sales would be approximately 3 million units in the first quarter, another 2 million units in the second quarter, and another 1 million units in the third quarter. However, because of a number of unpredictable market variables, there is considerable uncertainty as to what impact the advertising actually would have; and careful analysis indicates that the estimates for each quarter could turn out to be off by as much as 2 million units in either direction. (To quantify this uncertainty, assume that the additional increases in sales in the three quarters are independent random variables having a uniform distribution with a range from 1 to 5 million, from 0 to 4 million, and from −1 to 3 million, respectively.) If the actual increases are too small, the advertising campaign can be discontinued and the product abandoned at the end of either of the next two quarters.

If the intensive advertising campaign were initiated and continued to its completion, it is estimated that the sales for some time thereafter would continue to be at about the same level as in the third (last) quarter of the campaign. Therefore, if the sales in that quarter still were below the break-even point, the product would be abandoned. Otherwise, it is estimated that the expected discounted profit thereafter would be $40 for each unit sold over the break-even point in the third quarter.

Use dynamic programming to determine the optimal policy maximizing the expected profit.

11

Integer Programming

In Chap. 3 you saw several examples of the numerous and diverse applications of linear programming. However, one key limitation that prevents many more applications is the assumption of divisibility (see Sec. 3.3), which requires that noninteger values be permissible for decision variables. In many practical problems, the decision variables actually make sense only if they have integer values. For example, it is often necessary to assign people, machines, and vehicles to activities in integer quantities. If requiring integer values is the only way in which a problem deviates from a linear programming formulation, then it is an *integer programming* (**IP**) problem. (The more complete name is *integer linear programming*, but the adjective *linear* normally is dropped except when this problem is contrasted with the more esoteric integer nonlinear programming problem, which is beyond the scope of this book.)

The mathematical model for integer programming is the linear programming model (see Sec. 3.2) with the one additional restriction that the variables must have integer values. If only *some* of the variables are required to have integer values (so the divisibility assumption holds for the rest), this model is referred to as **mixed integer programming (MIP).** When distinguishing the all-integer problem from this mixed case, we call the former *pure* integer programming.

For example, the Wyndor Glass Co. problem presented in Sec. 3.1 actually would have been an IP problem if the two decision variables x_1 and x_2 had represented the total number of units to be produced of products 1 and 2, respectively, instead of the production rates. Because both products (glass doors and wood-framed windows) necessarily come in whole units, x_1 and x_2 would have to be restricted to integer values.

There have been numerous applications of integer programming that involve a direct extension of linear programming where the divisibility assumption must be dropped. However, another area of application may be of even greater importance, namely, problems involving a number of interrelated "yes-or-no decisions." In such decisions, the only two possible choices are *yes* and *no*. For example, should we undertake a particular fixed project? Should we make a particular fixed investment? Should we locate a facility in a particular site?

With just two choices, we can represent such decisions by decision variables that are restricted to just two values, say 0 and 1. Thus, the jth yes-or-no decision would be represented by, say, x_j such that

$$x_j = \begin{cases} 1 & \text{if decision } j \text{ is yes} \\ 0 & \text{if decision } j \text{ is no.} \end{cases}$$

Such variables are called **binary variables** (or 0–1 variables). Consequently, IP problems that contain only binary variables sometimes are called **binary integer programming (BIP)** problems (or 0–1 integer programming problems).

Section 11.1 presents a miniature version of a typical BIP problem and Sec. 11.2 surveys a variety of other BIP applications. Additional formulation possibilities with binary variables are discussed in Sec. 11.3, and Sec. 11.4 presents a series of formulation examples. Sections 11.5–11.8 then deal with ways to solve IP problems, including both BIP and MIP problems. The chapter concludes in Sec. 11.9 by introducing an exciting recent development *(constraint programming)* that promises to greatly expand our ability to formulate and solve integer programming models.

11.1 PROTOTYPE EXAMPLE

The CALIFORNIA MANUFACTURING COMPANY is considering expansion by building a new factory in either Los Angeles or San Francisco, or perhaps even in both cities. It also is considering building at most one new warehouse, but the choice of location is restricted to a city where a new factory is being built. The *net present value* (total profitability considering the time value of money) of each of these alternatives is shown in the fourth column of Table 11.1. The rightmost column gives the capital required (already included in the net present value) for the respective investments, where the total capital available is $10 million. The objective is to find the feasible combination of alternatives that maximizes the total net present value.

The BIP Model

Although this problem is small enough that it can be solved very quickly by inspection (build factories in both cities but no warehouse), let us formulate the IP model for illustrative purposes. All the decision variables have the *binary* form

$$x_j = \begin{cases} 1 & \text{if decision } j \text{ is yes,} \\ 0 & \text{if decision } j \text{ is no,} \end{cases} \quad (j = 1, 2, 3, 4).$$

Let

Z = total net present value of these decisions.

If the investment is made to build a particular facility (so that the corresponding decision variable has a value of 1), the estimated net present value from that investment is given in the fourth column of Table 11.1. If the investment is not made (so the decision variable equals 0), the net present value is 0. Therefore, using units of millions of dollars,

$$Z = 9x_1 + 5x_2 + 6x_3 + 4x_4.$$

TABLE 11.1 Data for the California Manufacturing Co. example

Decision Number	Yes-or-No Question	Decision Variable	Net Present Value	Capital Required
1	Build factory in Los Angeles?	x_1	$9 million	$6 million
2	Build factory in San Francisco?	x_2	$5 million	$3 million
3	Build warehouse in Los Angeles?	x_3	$6 million	$5 million
4	Build warehouse in San Francisco?	x_4	$4 million	$2 million

Capital available: $10 million

The rightmost column of Table 11.1 indicates that the amount of capital expended on the four facilities cannot exceed $10 million. Consequently, continuing to use units of millions of dollars, one constraint in the model is

$$6x_1 + 3x_2 + 5x_3 + 2x_4 \leq 10.$$

Because the last two decisions represent *mutually exclusive alternatives* (the company wants *at most* one new warehouse), we also need the constraint

$$x_3 + x_4 \leq 1.$$

Furthermore, decisions 3 and 4 are *contingent decisions,* because they are contingent on decisions 1 and 2, respectively (the company would consider building a warehouse in a city only if a new factory also were going there). Thus, in the case of decision 3, we require that $x_3 = 0$ if $x_1 = 0$. This restriction on x_3 (when $x_1 = 0$) is imposed by adding the constraint

$$x_3 \leq x_1.$$

Similarly, the requirement that $x_4 = 0$ if $x_2 = 0$ is imposed by adding the constraint

$$x_4 \leq x_2.$$

Therefore, after we rewrite these two constraints to bring all variables to the left-hand side, the complete BIP model is

Maximize $Z = 9x_1 + 5x_2 + 6x_3 + 4x_4,$

subject to

$$
\begin{aligned}
6x_1 + 3x_2 + 5x_3 + 2x_4 &\leq 10 \\
x_3 + x_4 &\leq 1 \\
-x_1 \qquad\quad + x_3 \qquad\;\; &\leq 0 \\
-x_2 \qquad\quad + x_4 &\leq 0 \\
x_j &\leq 1 \\
x_j &\geq 0
\end{aligned}
$$

and

     x_j is integer,   for $j = 1, 2, 3, 4.$

Equivalently, the last three lines of this model can be replaced by the single restriction

     x_j is binary,   for $j = 1, 2, 3, 4.$

Except for its small size, this example is typical of many real applications of integer programming where the basic decisions to be made are of the yes-or-no type. Like the second pair of decisions for this example, groups of yes-or-no decisions often constitute groups of **mutually exclusive alternatives** such that *only one* decision in the group can be yes. Each group requires a constraint that the sum of the corresponding binary variables must be equal to 1 (if *exactly one* decision in the group must be yes) or less than or equal to 1 (if *at most one* decision in the group can be yes). Occasionally, decisions of the yes-or-no type are **contingent decisions**, i.e., decisions that depend upon previous decisions. For example, one decision is said to be *contingent* on another decision if it is allowed to be yes *only if* the other is yes. This situation occurs when the contingent decision involves a follow-up action that would become irrelevant, or even impossible, if the other decision were no. The form that the resulting constraint takes always is that illustrated by the third and fourth constraints in the example.

Software Options for Solving Such Models

All the software packages featured in your OR Courseware (Excel, LINGO/LINDO, and MPL/CPLEX) include an algorithm for solving (pure or mixed) BIP models, as well as an algorithm for solving general (pure or mixed) IP models where variables need to be integer but not binary. However, since binary variables are considerably easier to deal with than general integer variables, the former algorithm generally can solve substantially larger problems than the latter algorithm.

When using the Excel Solver, the procedure is basically the same as for linear programming. The one difference arises when you click on the "Add" button on the Solver dialogue box to add the constraints. In addition to the constraints that fit linear programming, you also need to add the integer constraints. In the case of integer variables that are not binary, this is accomplished in the Add Constraint dialogue box by choosing the range of integer-restricted variables on the left-hand side and then choosing "int" from the pop-up menu. In the case of binary variables, choose "bin" from the pop-up menu instead.

One of the Excel files for this chapter shows the complete spreadsheet formulation and solution for the California Manufacturing Co. example. The Worked Examples section of the book's website also includes **a small minimization example** with two integer-restricted variables. This example illustrates the formulation of the IP model and its graphical solution, along with a spreadsheet formulation and solution.

A LINGO model uses the function @BIN() to specify that the variable named inside the parentheses is a binary variable. For a *general* integer variable (one restricted to integer values but not just binary values), the function @GIN() is used in the same way. In either case, the function can be embedded inside an @FOR statement to impose this binary or integer constraint on an entire set of variables.

In a LINDO syntax model, the binary or integer constraints are inserted after the END statement. A variable X is specified to be a general integer variable by entering GIN X. Alternatively, for any positive integer value of *n,* the statement GIN *n* specifies that the first *n* variables are general integer variables. Binary variables are handled in the same way except for substituting the word INTEGER for GIN.

For an MPL model, the keyword INTEGER is used to designate general integer variables, whereas BINARY is used for binary variables. In the variables section of an MPL model, all you need to do is add the appropriate adjective (INTEGER or BINARY) in front of the label VARIABLES to specify that the set of variables listed below the label is of that type. Alternatively, you can ignore this specification in the variables section and instead place the integer or binary constraints in the model section anywhere after the other constraints. In this case, the label over the set of variables becomes just INTEGER or BINARY.

The prime MPL solver CPLEX includes state-of-the-art algorithms for solving pure or mixed IP or BIP models. By selecting the *MIP Strategy* tab from the *CPLEX Parameters* dialogue box in the *Options* menu, an experienced practitioner can even choose from a wide variety of options for exactly how to execute the algorithm to best fit the particular problem.

These instructions for how to use the various software packages become clearer when you see them applied to examples. The Excel, LINGO/LINDO, and MPL/CPLEX files for this chapter in your OR Courseware show how each of these software options would be applied to the prototype example introduced in this section, as well as to the subsequent IP examples.

The latter part of the chapter will focus on IP algorithms that are similar to those used in these software packages. Section 11.6 will use the prototype example to illustrate the application of the pure BIP algorithm presented there.

■ 11.2 SOME BIP APPLICATIONS

Just as in the California Manufacturing Co. example, managers frequently must face *yes-or-no decisions*. Therefore, *binary integer programming* (BIP) is widely used to aid in these decisions.

We now will introduce various types of yes-or-no decisions. We also will mention some examples of actual applications where BIP was used to address these decisions.

Each of these applications is fully described in an article in the journal called *Interfaces*. We will point out several of these articles that are included in the selected references of award-winning applications cited at the end of the chapter, since a link to these articles is provided on the book's website. For the other articles, we still will mention the specific issue of *Interfaces* in which the article appears.

Investment Analysis

Linear programming sometimes is used to make capital budgeting decisions about how much to invest in various projects. However, as the California Manufacturing Co. example demonstrates, some capital budgeting decisions do not involve *how much* to invest, but rather, *whether* to invest a fixed amount. Specifically, the four decisions in the example were whether to invest the fixed amount of capital required to build a certain kind of facility (factory or warehouse) in a certain location (Los Angeles or San Francisco).

Management often must face decisions about whether to make fixed investments (those where the amount of capital required has been fixed in advance). Should we acquire a certain subsidiary being spun off by another company? Should we purchase a certain source of raw materials? Should we add a new production line to produce a certain input item ourselves rather than continuing to obtain it from a supplier?

In general, capital budgeting decisions about fixed investments are yes-or-no decisions of the following type.

Each yes-or-no decision:
Should we make a certain fixed investment?

Its decision variable $= \begin{cases} 1 & \text{if yes} \\ 0 & \text{if no.} \end{cases}$

The July–August 1990 issue of *Interfaces* describes how the *Turkish Petroleum Refineries Corporation* used BIP to analyze capital investments worth tens of millions of dollars to expand refinery capacity and conserve energy.

A rather different example that still falls somewhat into this category is described in Selected Reference A7. A major OR study was conducted for the *South African National Defense Force* to upgrade its capabilities with a smaller budget. The "investments" under consideration in this case were acquisition costs and ongoing expenses that would be required to provide specific types of military capabilities. A mixed BIP model was formulated to choose those specific capabilities that would maximize the overall effectiveness of the Defense Force while satisfying a budget constraint. The model had over 16,000 variables (including 256 binary variables) and over 5,000 functional constraints. The resulting optimization of the size and shape of the defense force provided savings of over $1.1 billion per year as well as vital nonmonetary benefits. The impact of this study won it the prestigious *first prize* among the 1996 Franz Edelman Awards for Management Science Achievement.

In a somewhat similar military application, the United States Air Force Space Command spends many billions of dollars each year acquiring and developing launch vehicles and space systems. The July–August 2003 issue of *Interfaces* describes how Space Command uses integer programming to optimize these long-term investments over a 24-year time horizon.

Selected Reference A3 presents another award-winning application of a mixed BIP model to investment analysis. This particular model has been used by the investment firm *Grantham, Mayo, Van Otterloo and Company* to construct many quantitatively managed portfolios representing over $8 billion in assets. In each case, a portfolio has been constructed that is close (in terms of sector and security exposure) to a target portfolio but with a far smaller and more manageable number of distinct stocks. A binary variable is used to represent each yes-or-no decision as to whether a particular stock should be included in the portfolio and then a separate continuous variable represents the amount of the stock to include. Given a current portfolio that needs to be rebalanced, it is desirable to reduce transaction costs by minimizing the number of transactions needed to obtain the final portfolio, so binary variables also are included to represent the yes-or-no decisions as to whether to make the transactions to change the amounts of individual stocks being held. The inclusion of this consideration in the model has reduced the annual cost of trading the portfolios being managed by at least $4 million.

Site Selection

In this global economy, many corporations are opening up new plants in various parts of the world to take advantage of lower labor costs, etc. Before selecting a site for a new plant, many potential sites may need to be analyzed and compared. (The California Manufacturing Co. example had just two potential sites for each of two kinds of facilities.) Each of the potential sites involves a yes-or-no decision of the following type.

Each yes-or-no decision:
Should a certain site be selected for the location of a certain new facility?

$$\text{Its decision variable} = \begin{cases} 1 & \text{if yes} \\ 0 & \text{if no.} \end{cases}$$

In many cases, the objective is to select the sites so as to minimize the total cost of the new facilities that will provide the required output.

As described in Selected Reference A11, AT&T used a BIP model to help dozens of their customers select the sites for their telemarketing centers. The model minimizes labor, communications, and real estate costs while providing the desired level of coverage by the centers. In one year alone (1988), this approach enabled 46 AT&T customers to make their yes-or-no decisions on site locations swiftly and confidently, while committing to $375 million in annual network services and $31 million in equipment sales from AT&T.

We next describe an important type of problem for many corporations where site selection plays a key role.

Designing a Production and Distribution Network

Manufacturers today face great competitive pressure to get their products to market more quickly as well as to reduce their production and distribution costs. Therefore, any corporation that distributes its products over a wide geographical area (or even worldwide) must pay continuing attention to the design of its production and distribution network.

This design involves addressing the following kinds of yes-or-no decisions.

Should a certain plant remain open?
Should a certain site be selected for a new plant?
Should a certain distribution center remain open?
Should a certain site be selected for a new distribution center?

If each market area is to be served by a single distribution center, then we also have another kind of yes-or-no decision for each combination of a market area and a distribution center.

Should a certain distribution center be assigned to serve a certain market area?

For each of the yes-or-no decisions of any of these kinds,

$$\text{Its decision variable} = \begin{cases} 1 & \text{if yes} \\ 0 & \text{if no.} \end{cases}$$

Ault Foods Limited (July–August 1994 issue of *Interfaces*) used this approach to design its production and distribution center. Management considered 10 sites for plants, 13 sites for distribution centers, and 48 market areas. This application of BIP was credited with saving the company $200,000 per year.

Digital Equipment Corporation (January–February 1995 issue of *Interfaces*) provides another example of an application of this kind. At the time, this large multinational corporation was serving one-quarter million customer sites, with more than half of its $14 billion annual revenues coming from 81 countries outside the United States. Therefore, this application involved restructuring the corporation's entire *global supply chain,* consisting of its suppliers, plants, distribution centers, potential sites, and market areas all around the world. The restructuring generated annual cost reductions of $500 million in manufacturing and $300 million in logistics, as well as a reduction of over $400 million in required capital assets.

Dispatching Shipments

Once a production and distribution network has been designed and put into operation, daily operating decisions need to be made about how to send the shipments. Some of these decisions again are yes-or-no decisions.

For example, suppose that trucks are being used to transport the shipments and each truck typically makes deliveries to several customers during each trip. It then becomes necessary to select a route (sequence of customers) for each truck, so each candidate for a route leads to the following yes-or-no decision.

Should a certain route be selected for one of the trucks?

$$\text{Its decision variable} = \begin{cases} 1 & \text{if yes} \\ 0 & \text{if no.} \end{cases}$$

The objective would be to select the routes that would minimize the total cost of making all the deliveries.

Various complications also can be considered. For example, if different truck sizes are available, each candidate for selection would include both a certain route and a certain truck size. Similarly, if timing is an issue, a time period for the departure also can be specified as part of the yes-or-no decision. With both factors, each yes-or-no decision would have the form shown next.

Should all the following be selected simultaneously for a delivery run:

1. A certain route,
2. A certain size of truck, and
3. A certain time period for the departure?

$$\text{Its decision variable} = \begin{cases} 1 & \text{if yes} \\ 0 & \text{if no.} \end{cases}$$

For example, *Sears, Roebuck and Company* (January–February 1999 issue of *Interfaces*) achieved annual savings of over $42 million by using a vehicle-routing-and-scheduling system based on BIP and a geographic information system to run its delivery and home service fleets more efficiently.

Scheduling Interrelated Activities

We all schedule interrelated activities in our everyday lives, even if it is just scheduling when to begin our various homework assignments. So too, managers must schedule various kinds of interrelated activities. When should we begin production for various new orders? When should we begin marketing various new products? When should we make various capital investments to expand our production capacity?

For any such activity, the decision about when to begin can be expressed in terms of a series of yes-or-no decisions, with one of these decisions for each of the possible time periods in which to begin, as shown below.

Should a certain activity begin in a certain time period?

$$\text{Its decision variable} = \begin{cases} 1 & \text{if yes} \\ 0 & \text{if no.} \end{cases}$$

Since a particular activity can begin in only one time period, the choice of the various time periods provides a group of *mutually exclusive alternatives,* so the decision variable for only one time period can have a value of 1.

For example, consider the following application that occurred in *China* (January–February 1995 issue of *Interfaces*). China was facing at least $240 billion in new investments over a 15-year horizon to meet the energy needs of its rapidly growing economy. Shortages of coal and electricity required developing new infrastructure for transporting coal and transmitting electricity, as well as building new dams and plants for generating thermal, hydro, and nuclear power. Therefore, the Chinese State Planning Commission and the World Bank collaborated in developing a huge mixed BIP model to guide the decisions on which projects to approve and when to undertake them over the 15-year planning period to minimize the total discounted cost. It is estimated that this OR application is saving China about $6.4 billion over the 15 years.

Airline Applications

The airline industry is an especially heavy user of OR throughout its operations. Many hundreds of OR professionals now work in this area. Major airline companies typically have a large in-house department that works on OR applications. In addition, there are some prominent consulting firms that focus solely on the problems of companies involved with transportation, including especially airlines. We will mention here just two of the applications which specifically use BIP.

One is the *fleet assignment problem*. Given several different types of airplanes available, the problem is to assign a specific type to each flight leg in the schedule so as to maximize the total profit from meeting the schedule. The basic trade-off is that if the airline uses an airplane that is too small on a particular flight leg, it will leave potential customers behind, while if it uses an airplane that is too large, it will suffer the greater expense of the larger airplane to fly empty seats.

For each combination of an airplane type and a flight leg, we have the following yes-or-no decision.

Should a certain type of airplane be assigned to a certain flight leg?

$$\text{Its decision variable} = \begin{cases} 1 & \text{if yes} \\ 0 & \text{if no.} \end{cases}$$

Commercial airlines must solve two difficult scheduling problems to ensure that aircrews are available for all scheduled flights. One, called the *tours-of-duty planning problem*, involves constructing sequences of flights with interspersed rest periods that will comprise tours of duty over perhaps many days for individual crews. The second one, called the *rostering problem*, involves allocating these tours of duty to individual crew members. Management seeks minimum-cost or maximum-productivity solutions for these problems that also satisfy labor agreements and consider the preferences of crew members.

Many major airlines around the world have achieved impressive savings in recent years by using *BIP models* to obtain optimal solutions for these problems. One of these airlines is **Air New Zealand**, which is the largest national and international airline based in New Zealand. It employs over 2,000 crew members and operates flights to Australia, Asia, North America, and Europe, as well as between the major centers within New Zealand.

The BIP models used by Air New Zealand typically have *hundreds of functional constraints and many thousands of binary variables*, where advanced techniques are then used to solve these models. A conservative estimate of the *savings* resulting from the use of these models is **US$6.7 million** *per year*, which accounted for 11 % of the company's operating profit in one recent year. There also are many intangible benefits, including quick implementations, efficiently accommodating late schedule changes, and improved passenger service.

Source: E. R. Butchers, P. R. Day, A. P. Goldie, S. Miller, J. A. Meyer, D. M. Ryan, A. C. Scott, and C. A. Wallace: "Optimized Crew Scheduling at Air New Zealand," *Interfaces*, **31**(1): 30–56, Jan.–Feb. 2001. (A link to this article is provided on our website, www.mhhe.com/hillier.)

Delta Air Lines flies over 2,500 domestic flight legs every day, using about 450 airplanes of 10 different types. As described in Selected Reference A12, they have used a huge integer programming model (about 40,000 functional constraints, 20,000 binary variables, and 40,000 general integer variables) to solve their fleet assignment problem each time a change is needed. This application has saved Delta approximately $100 million per year.

A fairly similar application is the *crew scheduling problem*. Here, rather than assigning airplane types to flight legs, we are instead assigning sequences of flight legs to crews of pilots and flight attendants. Thus, for each feasible sequence of flight legs that leaves from a crew base and returns to the same base, the following yes-or-no decision must be made.

Should a certain sequence of flight legs be assigned to a crew?

$$\text{Its decision variable} = \begin{cases} 1 & \text{if yes} \\ 0 & \text{if no.} \end{cases}$$

The objective is to minimize the total cost of providing crews that cover each flight leg in the schedule.

American Airlines (July–August 1989 and January–February 1991 issues of *Interfaces*) achieved annual savings of over $20 million by using BIP to solve its crew scheduling problem on a monthly basis. This approach also is being used extensively by airline companies headquartered outside the United States. For example, *Air New Zealand* (January–February 2001 issue of *Interfaces*) saves approximately $6.7 million per year by using BIP to optimize crew scheduling, as described further in an application vignette in this section.

A full-fledged formulation example of this type will be presented at the end of Sec. 11.4.

A related problem for airline companies is that their crew schedules occasionally need to be revised quickly when flight delays or cancellations occur because of inclement weather, aircraft mechanical problems, or crew unavailability. As described in an application vignette in Sec. 2.2 (as well as in Selected Reference A14), *Continental Airlines* achieved savings of $40 million in the first year of using an elaborate decision support system based on BIP for optimizing the *reassignment* of crews to flights when such emergencies occur. (Continental Airlines won first prize among the 2002 Franz Edelman Awards for Management Science Achievement for this innovative application.)

Many of the problems that face airline companies also arise in other segments of the transportation industry. Therefore, some of the airline applications of OR are being extended to these other segments, including railroad travel. For example, three of the first-prize winners of the Franz Edelman Award for Management Science Achievement in recent years have been for railroad applications by the national railroad of France, the Canadian Pacific Railway, and Netherlands Railways that provided dramatic financial benefits. (See the January–February 1998, January–February 2004, and January–February 2009 issues of *Interfaces*.) Selected Reference A1 also describes how Netherlands Railways (NS Reizigers) now is saving approximately $4.8 million per year by using BIP to optimize its crew scheduling.

11.3   INNOVATIVE USES OF BINARY VARIABLES IN MODEL FORMULATION

You have just seen a number of examples where the *basic decisions* of the problem are of the *yes-or-no type*, so that *binary variables* are introduced to represent these decisions. We now will look at some other ways in which binary variables can be very useful. In particular, we will see that these variables sometimes enable us to take a problem whose natural formulation is intractable and *reformulate* it as a pure or mixed IP problem.

This kind of situation arises when the original formulation of the problem fits either an IP or a linear programming format *except* for minor disparities involving combinatorial relationships in the model. By expressing these combinatorial relationships in terms of questions that must be answered yes or no, **auxiliary binary variables** can be introduced to the model to represent these yes-or-no decisions. (Rather than being a decision variable for the original problem under consideration, an *auxiliary* binary variable is a binary variable that is introduced into the model of the problem simply to help formulate the model as a pure or mixed BIP model.) Introducing these variables reduces the problem to an MIP problem (or a *pure* IP problem if all the original variables also are required to have integer values).

Some cases that can be handled by this approach are discussed next, where the x_j denote the *original* variables of the problem (they may be either continuous or integer variables) and the y_i denote the *auxiliary* binary variables that are introduced for the reformulation.

Either-Or Constraints

Consider the important case where a choice can be made between two constraints, so that *only one* (either one) must hold (whereas the other one can hold but is not required to do so). For example, there may be a choice as to which of two resources to use for a certain purpose, so that it is necessary for only one of the two resource availability constraints to hold mathematically. To illustrate the approach to such situations, suppose that one of the requirements in the overall problem is that

$$\text{Either} \quad 3x_1 + 2x_2 \leq 18$$
$$\text{or} \quad x_1 + 4x_2 \leq 16,$$

i.e., at least one of these two inequalities must hold but not necessarily both. This requirement must be reformulated to fit it into the linear programming format where *all*

and

$$y_i \text{ is binary,} \quad \text{for } i = 1, 2, \ldots, N.$$

so this new set of constraints would replace this requirement in the statement of the overall problem. This set of constraints provides an *equivalent* formulation because exactly one y_i must equal 1 and the others must equal 0, so exactly one d_i is being chosen as the value of the function. In this case, there are N yes-or-no questions being asked, namely, should d_i be the value chosen ($i = 1, 2, \ldots, N$)? Because the y_i respectively represent these *yes-or-no decisions,* the second constraint makes them *mutually exclusive alternatives.*

To illustrate how this case can arise, reconsider the Wyndor Glass Co. problem presented in Sec. 3.1. Eighteen hours of production time per week in Plant 3 currently is unused and available for the two new products *or* for certain future products that will be ready for production soon. In order to leave any remaining capacity in usable blocks for these future products, management now wants to impose the restriction that the production time used by the two current new products be 6 *or* 12 *or* 18 hours per week. Thus, the third constraint of the original model ($3x_1 + 2x_2 \leq 18$) now becomes

$$3x_1 + 2x_2 = 6 \quad \text{or} \quad 12 \quad \text{or} \quad 18.$$

In the preceding notation, $N = 3$ with $d_1 = 6$, $d_2 = 12$, and $d_3 = 18$. Consequently, management's new requirement should be formulated as follows:

$$3x_1 + 2x_2 = 6y_1 + 12y_2 + 18y_3$$
$$y_1 + y_2 + y_3 = 1$$

and

$$y_1, y_2, y_3 \text{ are binary.}$$

The overall model for this new version of the problem then consists of the original model (see Sec. 3.1) plus this new set of constraints that replaces the original third constraint. This replacement yields a very tractable MIP formulation.

The Fixed-Charge Problem

It is quite common to incur a fixed charge or setup cost when undertaking an activity. For example, such a charge occurs when a production run to produce a batch of a particular product is undertaken and the required production facilities must be set up to initiate the run. In such cases, the total cost of the activity is the sum of a variable cost related to the level of the activity and the setup cost required to initiate the activity. Frequently the variable cost will be at least roughly proportional to the level of the activity. If this is the case, the *total cost* of the activity (say, activity j) can be represented by a function of the form

$$f_j(x_j) = \begin{cases} k_j + c_j x_j & \text{if } x_j > 0 \\ 0 & \text{if } x_j = 0, \end{cases}$$

where x_j denotes the level of activity j ($x_j \geq 0$), k_j denotes the setup cost, and c_j denotes the cost for each incremental unit. Were it not for the setup cost k_j, this cost structure would suggest the possibility of a *linear programming* formulation to determine the optimal levels of the competing activities. Fortunately, even with the k_j, MIP can still be used.

To formulate the overall model, suppose that there are n activities, each with the preceding cost structure (with $k_j \geq 0$ in every case and $k_j > 0$ for some $j = 1, 2, \ldots, n$), and that the problem is to

Minimize $\quad Z = f_1(x_1) + f_2(x_2) + \cdots + f_n(x_n),$

subject to

given linear programming constraints.

To convert this problem to an MIP format, we begin by posing n questions that must be answered yes or no; namely, for each $j = 1, 2, \ldots, n$, should activity j be undertaken ($x_j > 0$)? Each of these *yes-or-no decisions* is then represented by an auxiliary *binary variable* y_j, so that

$$Z = \sum_{j=1}^{n} (c_j x_j + k_j y_j),$$

where

$$y_j = \begin{cases} 1 & \text{if } x_j > 0 \\ 0 & \text{if } x_j = 0. \end{cases}$$

Therefore, the y_j can be viewed as *contingent decisions* similar to (but not identical to) the type considered in Sec. 11.1. Let M be an extremely large positive number that exceeds the maximum feasible value of any x_j ($j = 1, 2, \ldots, n$). Then the constraints

$$x_j \leq M y_j \qquad \text{for } j = 1, 2, \ldots, n$$

will ensure that $y_j = 1$ rather than 0 whenever $x_j > 0$. The one difficulty remaining is that these constraints leave y_j free to be either 0 or 1 when $x_j = 0$. Fortunately, this difficulty is automatically resolved because of the nature of the objective function. The case where $k_j = 0$ can be ignored because y_j can then be deleted from the formulation. So we consider the only other case, namely, where $k_j > 0$. When $x_j = 0$, so that the constraints permit a choice between $y_j = 0$ and $y_j = 1$, $y_j = 0$ must yield a smaller value of Z than $y_j = 1$. Therefore, because the objective is to minimize Z, an algorithm yielding an optimal solution would always choose $y_j = 0$ when $x_j = 0$.

To summarize, the MIP formulation of the fixed-charge problem is

$$\text{Minimize} \qquad Z = \sum_{j=1}^{n} (c_j x_j + k_j y_j),$$

subject to

the original constraints, plus
$$x_j - M y_j \leq 0$$
and

y_j is binary,   for $j = 1, 2, \ldots, n$.

If the x_j also had been restricted to be integer, then this would be a *pure* IP problem.

To illustrate this approach, look again at the Nori & Leets Co. air pollution problem described in Sec. 3.4. The first of the abatement methods considered—increasing the height of the smokestacks—actually would involve a substantial *fixed charge* to get ready for *any* increase in addition to a variable cost that would be roughly proportional to the amount of increase. After conversion to the equivalent annual costs used in the formulation, this fixed charge would be $2 million each for the blast furnaces and the open-hearth furnaces, whereas the variable costs are those identified in Table 3.14. Thus, in the preceding notation, $k_1 = 2$, $k_2 = 2$, $c_1 = 8$, and $c_2 = 10$, where the objective function is expressed in units of *millions* of dollars. Because the other abatement methods do not involve any fixed charges, $k_j = 0$ for $j = 3, 4, 5, 6$. Consequently, the new MIP formulation of this problem is

$$\text{Minimize} \qquad Z = 8x_1 + 10x_2 + 7x_3 + 6x_4 + 11x_5 + 9x_6 + 2y_1 + 2y_2,$$

subject to

> the constraints given in Sec. 3.4, plus
>
> $$x_1 - My_1 \leq 0,$$
> $$x_2 - My_2 \leq 0,$$

and

> y_1, y_2 are binary.

Binary Representation of General Integer Variables

Suppose that you have a pure IP problem where most of the variables are *binary* variables, but the presence of a few *general* integer variables prevents you from solving the problem by one of the very efficient BIP algorithms now available. A nice way to circumvent this difficulty is to use the *binary representation* for each of these general integer variables. Specifically, if the bounds on an integer variable x are

$$0 \leq x \leq u$$

and if N is defined as the integer such that

$$2^N \leq u < 2^{N+1},$$

then the **binary representation** of x is

$$x = \sum_{i=0}^{N} 2^i y_i,$$

where the y_i variables are (auxiliary) binary variables. Substituting this binary representation for each of the general integer variables (with a different set of auxiliary binary variables for each) thereby reduces the entire problem to a BIP model.

For example, suppose that an IP problem has just two general integer variables x_1 and x_2 along with many binary variables. Also suppose that the problem has nonnegativity constraints for both x_1 and x_2 and that the functional constraints include

$$
\begin{aligned}
x_1 &\leq 5 \\
2x_1 + 3x_2 &\leq 30.
\end{aligned}
$$

These constraints imply that $u = 5$ for x_1 and $u = 10$ for x_2, so the above definition of N gives $N = 2$ for x_1 (since $2^2 \leq 5 < 2^3$) and $N = 3$ for x_2 (since $2^3 \leq 10 < 2^4$). Therefore, the binary representations of these variables are

$$
\begin{aligned}
x_1 &= y_0 + 2y_1 + 4y_2 \\
x_2 &= y_3 + 2y_4 + 4y_5 + 8y_6.
\end{aligned}
$$

After we substitute these expressions for the respective variables throughout all the functional constraints and the objective function, the two functional constraints noted above become

$$
\begin{aligned}
y_0 + 2y_1 + 4y_2 &\leq 5 \\
2y_0 + 4y_1 + 8y_2 + 3y_3 + 6y_4 + 12y_5 + 24y_6 &\leq 30.
\end{aligned}
$$

Observe that each feasible value of x_1 corresponds to one of the feasible values of the vector (y_0, y_1, y_2), and similarly for x_2 and (y_3, y_4, y_5, y_6). For example, $x_1 = 3$ corresponds to $(y_0, y_1, y_2) = (1, 1, 0)$, and $x_2 = 5$ corresponds to $(y_3, y_4, y_5, y_6) = (1, 0, 1, 0)$.

For an IP problem where *all* the variables are (bounded) general integer variables, it is possible to use this same technique to reduce the problem to a BIP model. However, this is not advisable for most cases because of the explosion in the number of variables involved. Applying a good IP algorithm to the original IP model generally should be more efficient than applying a good BIP algorithm to the much larger BIP model.[1]

In general terms, for *all* the formulation possibilities with auxiliary binary variables discussed in this section, we need to strike the same note of caution. This approach sometimes requires adding a relatively large number of such variables, which can make the model *computationally infeasible*. (Section 11.5 will provide some perspective on the sizes of IP problems that can be solved.)

■ 11.4 SOME FORMULATION EXAMPLES

We now present a series of examples that illustrate a variety of formulation techniques with binary variables, including those discussed in the preceding sections. For the sake of clarity, these examples have been kept very small. (**A somewhat larger formulation example**, with dozens of binary variables and constraints, is included in the Worked Examples section of the book's website.) In actual applications, these formulations typically would be just a small part of a vastly larger model.

EXAMPLE 1   Making Choices When the Decision Variables Are Continuous

The Research and Development Division of the GOOD PRODUCTS COMPANY has developed three possible new products. However, to avoid undue diversification of the company's product line, management has imposed the following restriction.

> **Restriction 1:** From the three possible new products, *at most two* should be chosen to be produced.

Each of these products can be produced in either of two plants. For administrative reasons, management has imposed a second restriction in this regard.

> **Restriction 2:** Just one of the two plants should be chosen to be the sole producer of the new products.

The production cost per unit of each product would be essentially the same in the two plants. However, because of differences in their production facilities, the number of hours of production time needed per unit of each product might differ between the two plants. These data are given in Table 11.2, along with other relevant information, including marketing

■ **TABLE 11.2** Data for Example 1 (the Good Products Co. problem)

	Production Time Used for Each Unit Produced			Production Time Available per Week
	Product 1	**Product 2**	**Product 3**	
Plant 1	3 hours	4 hours	2 hours	30 hours
Plant 2	4 hours	6 hours	2 hours	40 hours
Unit profit	5	7	3	(thousands of dollars)
Sales potential	7	5	9	(units per week)

[1]For evidence supporting this conclusion, see J. H. Owen and S. Mehrotra, "On the Value of Binary Expansions for General Mixed Integer Linear Programs," *Operations Research,* **50:** 810–819, 2002.

estimates of the number of units of each product that could be sold per week if it is produced. The objective is to choose the products, the plant, and the production rates of the chosen products so as to maximize total profit.

In some ways, this problem resembles a standard *product mix problem* such as the Wyndor Glass Co. example described in Sec. 3.1. In fact, if we changed the problem by dropping the two restrictions *and* by requiring each unit of a product to use the production hours given in Table 11.2 in *both plants* (so the two plants now perform different operations needed by the products), it would become just such a problem. In particular, if we let x_1, x_2, x_3 be the production rates of the respective products, the model then becomes

$$\text{Maximize} \quad Z = 5x_1 + 7x_2 + 3x_3,$$

subject to

$$
\begin{aligned}
3x_1 + 4x_2 + 2x_3 &\le 30 \\
4x_1 + 6x_2 + 2x_3 &\le 40 \\
x_1 &\le 7 \\
x_2 &\le 5 \\
x_3 &\le 9
\end{aligned}
$$

and

$$x_1 \ge 0, \qquad x_2 \ge 0, \qquad x_3 \ge 0.$$

For the real problem, however, restriction 1 necessitates adding to the model the constraint

The number of strictly positive decision variables (x_1, x_2, x_3) must be ≤ 2.

This constraint does not fit into a linear or an integer programming format, so the key question is how to convert it to such a format so that a corresponding algorithm can be used to solve the overall model. If the decision variables were binary variables, then the constraint would be expressed in this format as $x_1 + x_2 + x_3 \le 2$. However, with *continuous* decision variables, a more complicated approach involving the introduction of auxiliary binary variables is needed.

Requirement 2 necessitates replacing the first two functional constraints ($3x_1 + 4x_2 + 2x_3 \le 30$ and $4x_1 + 6x_2 + 2x_3 \le 40$) by the restriction

Either $3x_1 + 4x_2 + 2x_3 \le 30$
or $4x_1 + 6x_2 + 2x_3 \le 40$

must hold, where the choice of which constraint must hold corresponds to the choice of which plant will be used to produce the new products. We discussed in the preceding section how such an either-or constraint can be converted to a linear or an integer programming format, again with the help of an auxiliary binary variable.

Formulation with Auxiliary Binary Variables. To deal with requirement 1, we introduce three auxiliary binary variables (y_1, y_2, y_3) with the interpretation

$$
y_j = \begin{cases} 1 & \text{if } x_j > 0 \text{ can hold (can produce product } j) \\ 0 & \text{if } x_j = 0 \text{ must hold (cannot produce product } j), \end{cases}
$$

for $j = 1, 2, 3$. To enforce this interpretation in the model with the help of M (an extremely large positive number), we add the constraints

$$x_1 \leq My_1$$
$$x_2 \leq My_2$$
$$x_3 \leq My_3$$
$$y_1 + y_2 + y_3 \leq 2$$
$$y_j \text{ is binary}, \qquad \text{for } j = 1, 2, 3.$$

The either-or constraint and nonnegativity constraints give a *bounded* feasible region for the decision variables (so each $x_j \leq M$ throughout this region). Therefore, in each $x_j \leq My_j$ constraint, $y_j = 1$ allows any value of x_j in the feasible region, whereas $y_j = 0$ forces $x_j = 0$. (Conversely, $x_j > 0$ forces $y_j = 1$, whereas $x_j = 0$ allows either value of y_j.) Consequently, when the fourth constraint forces choosing at most two of the y_j to equal 1, this amounts to choosing at most two of the new products as the ones that can be produced.

To deal with requirement 2, we introduce another auxiliary binary variable y_4 with the interpretation

$$y_4 = \begin{cases} 1 & \text{if } 4x_1 + 6x_2 + 2x_3 \leq 40 \text{ must hold (choose Plant 2)} \\ 0 & \text{if } 3x_1 + 4x_2 + 2x_3 \leq 30 \text{ must hold (choose Plant 1).} \end{cases}$$

As discussed in Sec. 11.3, this interpretation is enforced by adding the constraints,

$$3x_1 + 4x_2 + 2x_3 \leq 30 + My_4$$
$$4x_1 + 6x_2 + 2x_3 \leq 40 + M(1 - y_4)$$
$$y_4 \text{ is binary}.$$

Consequently, after we move all variables to the left-hand side of the constraints, the complete model is

Maximize $Z = 5x_1 + 7x_2 + 3x_3,$

subject to

$$x_1 \leq 7$$
$$x_2 \leq 5$$
$$x_3 \leq 9$$
$$x_1 - My_1 \leq 0$$
$$x_2 - My_2 \leq 0$$
$$x_3 - My_3 \leq 0$$
$$y_1 + y_2 + y_3 \leq 2$$
$$3x_1 + 4x_2 + 2x_3 - My_4 \leq 30$$
$$4x_1 + 6x_2 + 2x_3 + My_4 \leq 40 + M$$

and

$$x_1 \geq 0, \qquad x_2 \geq 0, \qquad x_3 \geq 0$$
$$y_j \text{ is binary}, \qquad \text{for } j = 1, 2, 3, 4.$$

This now is an MIP model, with three variables (the x_j) not required to be integer and four binary variables, so an MIP algorithm can be used to solve the model. When this is done (after substituting a large numerical value for M),[2] the optimal solution is $y_1 = 1$,

---

[2]In practice, some care is taken to choose a value for M that definitely is large enough to avoid eliminating any feasible solutions, but as small as possible otherwise in order to avoid unduly enlarging the feasible region for the LP relaxation (described in the next section) and to avoid numerical instability. For this example, a careful examination of the constraints reveals that the minimum feasible value of M is $M = 9$.

$y_2 = 0$, $y_3 = 1$, $y_4 = 1$, $x_1 = 5\frac{1}{2}$, $x_2 = 0$, and $x_3 = 9$; that is, choose products 1 and 3 to produce, choose Plant 2 for the production, and choose the production rates of $5\frac{1}{2}$ units per week for product 1 and 9 units per week for product 3. The resulting total profit is $54,500 per week.

EXAMPLE 2   **Violating Proportionality**

The SUPERSUDS CORPORATION is developing its marketing plans for next year's new products. For three of these products, the decision has been made to purchase a total of five TV spots for commercials on national television networks. The problem we will focus on is how to allocate the five spots to these three products, with a maximum of three spots (and a minimum of zero) for each product.

Table 11.3 shows the estimated impact of allocating zero, one, two, or three spots to each product. This impact is measured in terms of the *profit* (in units of millions of dollars) from the *additional sales* that would result from the spots, considering also the cost of producing the commercial and purchasing the spots. The objective is to allocate five spots to the products so as to maximize the total profit.

This small problem can be solved easily by dynamic programming (Chap. 10) or even by inspection. (The optimal solution is to allocate two spots to product 1, no spots to product 2, and three spots to product 3.) However, we will show two different BIP formulations for illustrative purposes. Such a formulation would become necessary if this small problem needed to be incorporated into a larger IP model involving the allocation of resources to marketing activities for all the corporation's new products.

One Formulation with Auxiliary Binary Variables.   A natural formulation would be to let x_1, x_2, x_3 be the number of TV spots allocated to the respective products. The contribution of each x_j to the objective function then would be given by the corresponding column in Table 11.3. However, each of these columns violates the assumption of proportionality described in Sec. 3.3. Therefore, we cannot write a *linear* objective function in terms of these integer decision variables.

Now see what happens when we introduce an *auxiliary binary variable* y_{ij} for each positive integer value of $x_i = j$ ($j = 1, 2, 3$), where y_{ij} has the interpretation

$$y_{ij} = \begin{cases} 1 & \text{if } x_i = j \\ 0 & \text{otherwise.} \end{cases}$$

(For example, $y_{21} = 0$, $y_{22} = 0$, and $y_{23} = 1$ mean that $x_2 = 3$.) The resulting *linear* BIP model is

Maximize $Z = y_{11} + 3y_{12} + 3y_{13} + 2y_{22} + 3y_{23} - y_{31} + 2y_{32} + 4y_{33}$,

TABLE 11.3 Data for Example 2 (the Supersuds Corp. problem)

Number of TV Spots	Profit		
	Product		
	1	2	3
0	0	0	0
1	1	0	−1
2	3	2	2
3	3	3	4

subject to

$$y_{11} + y_{12} + y_{13} \leq 1$$
$$y_{21} + y_{22} + y_{23} \leq 1$$
$$y_{31} + y_{32} + y_{33} \leq 1$$
$$y_{11} + 2y_{12} + 3y_{13} + y_{21} + 2y_{22} + 3y_{23} + y_{31} + 2y_{32} + 3y_{33} = 5$$

and

each y_{ij} is binary.

Note that the first three functional constraints ensure that each x_i will be assigned just one of its possible values. (Here $y_{i1} + y_{i2} + y_{i3} = 0$ corresponds to $x_i = 0$, which contributes nothing to the objective function.) The last functional constraint ensures that $x_1 + x_2 + x_3 = 5$. The *linear* objective function then gives the total profit according to Table 11.3.

Solving this BIP model gives an optimal solution of

$$
\begin{array}{llllll}
y_{11} = 0, & y_{12} = 1, & y_{13} = 0, & \text{so} & x_1 = 2 \\
y_{21} = 0, & y_{22} = 0, & y_{23} = 0, & \text{so} & x_2 = 0 \\
y_{31} = 0, & y_{32} = 0, & y_{33} = 1, & \text{so} & x_3 = 3.
\end{array}
$$

Another Formulation with Auxiliary Binary Variables.   We now redefine the above auxiliary binary variables y_{ij} as follows:

$$
y_{ij} = \begin{cases} 1 & \text{if } x_i \geq j \\ 0 & \text{otherwise.} \end{cases}
$$

Thus, the difference is that $y_{ij} = 1$ now if $x_i \geq j$ instead of $x_i = j$. Therefore,

$$
\begin{array}{lllll}
x_i = 0 & \Rightarrow & y_{i1} = 0, & y_{i2} = 0, & y_{i3} = 0, \\
x_i = 1 & \Rightarrow & y_{i1} = 1, & y_{i2} = 0, & y_{i3} = 0, \\
x_i = 2 & \Rightarrow & y_{i1} = 1, & y_{i2} = 1, & y_{i3} = 0, \\
x_i = 3 & \Rightarrow & y_{i1} = 1, & y_{i2} = 1, & y_{i3} = 1, \\
\end{array}
$$
$$\text{so } x_i = y_{i1} + y_{i2} + y_{i3}$$

for $i = 1, 2, 3$. Because allowing $y_{i2} = 1$ is contingent upon $y_{i1} = 1$ and allowing $y_{i3} = 1$ is contingent upon $y_{i2} = 1$, these definitions are enforced by adding the constraints

$$y_{i2} \leq y_{i1} \quad \text{and} \quad y_{i3} \leq y_{i2}, \quad \text{for } i = 1, 2, 3.$$

The new definition of the y_{ij} also changes the objective function, as illustrated in Fig. 11.1 for the product 1 portion of the objective function. Since y_{11}, y_{12}, y_{13} provide the successive increments (if any) in the value of x_1 (starting from a value of 0), the coefficients of y_{11}, y_{12}, y_{13} are given by the respective *increments* in the product 1 column of Table 11.3 ($1 - 0 = 1, 3 - 1 = 2, 3 - 3 = 0$). These *increments* are the *slopes* in Fig. 11.1, yielding $1y_{11} + 2y_{12} + 0y_{13}$ for the product 1 portion of the objective function. Note that applying this approach to all three products still must lead to a *linear* objective function.

After we bring all variables to the left-hand side of the constraints, the resulting complete BIP model is

Maximize $Z = y_{11} + 2y_{12} + 2y_{22} + y_{23} - y_{31} + 3y_{32} + 2y_{33},$

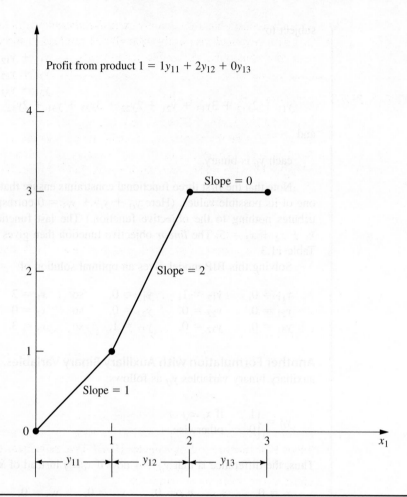

Profit from product $1 = 1y_{11} + 2y_{12} + 0y_{13}$

Slope = 0

Slope = 2

Slope = 1

■ **FIGURE 11.1**
The profit from the additional sales of product 1 that would result from x_1 TV spots, where the slopes give the corresponding coefficients in the objective function for the second BIP formulation for Example 2 (the Supersuds Corp. problem).

subject to

$$y_{12} - y_{11} \leq 0$$
$$y_{13} - y_{12} \leq 0$$
$$y_{22} - y_{21} \leq 0$$
$$y_{23} - y_{22} \leq 0$$
$$y_{32} - y_{31} \leq 0$$
$$y_{33} - y_{32} \leq 0$$
$$y_{11} + y_{12} + y_{13} + y_{21} + y_{22} + y_{23} + y_{31} + y_{32} + y_{33} = 5$$

and

each y_{ij} is binary.

Solving this BIP model gives an optimal solution of

$$y_{11} = 1, \qquad y_{12} = 1, \qquad y_{13} = 0, \qquad \text{so} \qquad x_1 = 2$$
$$y_{21} = 0, \qquad y_{22} = 0, \qquad y_{23} = 0, \qquad \text{so} \qquad x_2 = 0$$
$$y_{31} = 1, \qquad y_{32} = 1, \qquad y_{33} = 1, \qquad \text{so} \qquad x_3 = 3.$$

There is little to choose between this BIP model and the preceding one other than personal taste. They have the same number of binary variables (the prime consideration in determining computational effort for BIP problems). They also both have some *special structure* (constraints for *mutually exclusive alternatives* in the first model and constraints for *contingent decisions* in the second) that can lead to speedup. The second model does have more functional constraints than the first.

EXAMPLE 3 Covering All Characteristics

SOUTHWESTERN AIRWAYS needs to assign its crews to cover all its upcoming flights. We will focus on the problem of assigning three crews based in San Francisco to the flights listed in the first column of Table 11.4. The other 12 columns show the 12 feasible sequences of flights for a crew. (The numbers in each column indicate the order of the flights.) Exactly three of the sequences need to be chosen (one per crew) in such a way that every flight is covered. (It is permissible to have more than one crew on a flight, where the extra crews would fly as passengers, but union contracts require that the extra crews would still need to be paid for their time as if they were working.) The cost of assigning a crew to a particular sequence of flights is given (in thousands of dollars) in the bottom row of the table. The objective is to minimize the total cost of the three crew assignments that cover all the flights.

Formulation with Binary Variables.   With 12 feasible sequences of flights, we have 12 yes-or-no decisions:

Should sequence j be assigned to a crew? $(j = 1, 2, \ldots, 12)$

Therefore, we use 12 binary variables to represent these respective decisions:

$$x_j = \begin{cases} 1 & \text{if sequence } j \text{ is assigned to a crew} \\ 0 & \text{otherwise.} \end{cases}$$

The most interesting part of this formulation is the nature of each constraint that ensures that a corresponding flight is covered. For example, consider the last flight in Table 11.4 [Seattle to Los Angeles (LA)]. Five sequences (namely, sequences 6, 9, 10, 11, and 12) include this flight. Therefore, at least one of these five sequences must be chosen. The resulting constraint is

$$x_6 + x_9 + x_{10} + x_{11} + x_{12} \geq 1.$$

Using similar constraints for the other 10 flights, the complete BIP model is

Minimize $Z = 2x_1 + 3x_2 + 4x_3 + 6x_4 + 7x_5 + 5x_6 + 7x_7 + 8x_8 + 9x_9$
$+ 9x_{10} + 8x_{11} + 9x_{12},$

▬ **TABLE 11.4** Data for Example 3 (the Southwestern Airways problem)

Flight	\multicolumn{12}{c}{Feasible Sequence of Flights}											
	1	2	3	4	5	6	7	8	9	10	11	12
1. San Francisco to Los Angeles	1			1			1			1		
2. San Francisco to Denver		1			1			1			1	
3. San Francisco to Seattle			1			1			1			1
4. Los Angeles to Chicago				2			2		3	2		3
5. Los Angeles to San Francisco	2					3			5	5		
6. Chicago to Denver				3	3				4			
7. Chicago to Seattle							3	3		3	3	4
8. Denver to San Francisco		2		4	4				5			
9. Denver to Chicago					2			2			2	
10. Seattle to San Francisco			2				4	4				5
11. Seattle to Los Angeles						2			2	4	4	2
Cost, $1,000's	2	3	4	6	7	5	7	8	9	9	8	9

subject to

$$x_1 + x_4 + x_7 + x_{10} \geq 1 \qquad \text{(SF to LA)}$$
$$x_2 + x_5 + x_8 + x_{11} \geq 1 \qquad \text{(SF to Denver)}$$
$$x_3 + x_6 + x_9 + x_{12} \geq 1 \qquad \text{(SF to Seattle)}$$
$$x_4 + x_7 + x_9 + x_{10} + x_{12} \geq 1 \qquad \text{(LA to Chicago)}$$
$$x_1 + x_6 + x_{10} + x_{11} \geq 1 \qquad \text{(LA to SF)}$$
$$x_4 + x_5 + x_9 \geq 1 \qquad \text{(Chicago to Denver)}$$
$$x_7 + x_8 + x_{10} + x_{11} + x_{12} \geq 1 \qquad \text{(Chicago to Seattle)}$$
$$x_2 + x_4 + x_5 + x_9 \geq 1 \qquad \text{(Denver to SF)}$$
$$x_5 + x_8 + x_{11} \geq 1 \qquad \text{(Denver to Chicago)}$$
$$x_3 + x_7 + x_8 + x_{12} \geq 1 \qquad \text{(Seattle to SF)}$$
$$x_6 + x_9 + x_{10} + x_{11} + x_{12} \geq 1 \qquad \text{(Seattle to LA)}$$

$$\sum_{j=1}^{12} x_j = 3 \qquad \text{(assign three crews)}$$

and

$$x_j \text{ is binary}, \qquad \text{for } j = 1, 2, \ldots, 12.$$

One optimal solution for this BIP model is

$$x_3 = 1 \qquad \text{(assign sequence 3 to a crew)}$$
$$x_4 = 1 \qquad \text{(assign sequence 4 to a crew)}$$
$$x_{11} = 1 \qquad \text{(assign sequence 11 to a crew)}$$

and all other $x_j = 0$, for a total cost of \$18,000. (Another optimal solution is $x_1 = 1$, $x_5 = 1$, $x_{12} = 1$, and all other $x_j = 0$.)

This example illustrates a broader class of problems called **set covering problems**.[3] Any set covering problem can be described in general terms as involving a number of potential *activities* (such as flight sequences) and *characteristics* (such as flights). Each activity possesses some but not all of the characteristics. The objective is to determine the least costly combination of activities that collectively possess (cover) each characteristic at least once. Thus, let S_i be the set of all activities that possess characteristic i. At least one member of the set S_i must be included among the chosen activities, so a constraint,

$$\sum_{j \in S_i} x_j \geq 1,$$

is included for each characteristic i.

A related class of problems, called **set partitioning problems,** changes each such constraint to

$$\sum_{j \in S_i} x_j = 1,$$

so now *exactly* one member of each set S_i must be included among the chosen activities. For the crew scheduling example, this means that each flight must be included *exactly* once among the chosen flight sequences, which rules out having extra crews (as passengers) on any flight.

---

[3]Strictly speaking, a set covering problem does not include any *other* functional constraints such as the last functional constraint in the above crew scheduling example. It also is sometimes assumed that every coefficient in the objective function being minimized equals *one*, and then the name *weighted set covering problem* is used when this assumption does not hold.

■ 11.5   SOME PERSPECTIVES ON SOLVING INTEGER PROGRAMMING PROBLEMS

It may seem that IP problems should be relatively easy to solve. After all, *linear programming* problems can be solved extremely efficiently, and the only difference is that IP problems have far fewer solutions to be considered. In fact, *pure* IP problems with a bounded feasible region are guaranteed to have just a *finite* number of feasible solutions.

Unfortunately, there are two fallacies in this line of reasoning. One is that having a finite number of feasible solutions ensures that the problem is readily solvable. Finite numbers can be astronomically large. For example, consider the simple case of BIP problems. With n variables, there are 2^n solutions to be considered (where some of these solutions can subsequently be discarded because they violate the functional constraints). Thus, each time n is increased by 1, the number of solutions is *doubled*. This pattern is referred to as the **exponential growth** of the difficulty of the problem. With $n = 10$, there are more than 1,000 solutions (1,024); with $n = 20$, there are more than 1,000,000; with $n = 30$, there are more than 1 billion; and so forth. Therefore, even the fastest computers are incapable of performing exhaustive enumeration (checking each solution for feasibility and, if it is feasible, calculating the value of the objective value) for BIP problems with more than a few dozen variables, let alone for *general* IP problems with the same number of integer variables. Fortunately, by starting with the ideas described in subsequent sections, today's best IP algorithms are vastly superior to exhaustive enumeration. The improvement over just the past two decades has been dramatic. BIP problems that would have required years of computing time to solve 20 years ago now can be solved in seconds with today's best commercial software (such as CPLEX). This huge speedup is due to great progress in three areas—dramatic improvements in BIP algorithms (as well as other IP algorithms), striking improvements in linear programming algorithms that are heavily used within the integer programming algorithms, and the great speedup in computers (including desktop computers). As a result, vastly larger BIP problems now are sometimes being solved than would have been possible in past decades. The best algorithms today are capable of solving *some* pure BIP problems with over a hundred thousand variables. Nevertheless, because of *exponential growth*, even the best algorithms cannot be guaranteed to solve every relatively small problem (less than a few hundred binary variables). Depending on their characteristics, certain relatively small problems can be much more difficult to solve than some much larger ones.

When dealing with general integer variables instead of binary variables, the size of the problems that can be solved tend to be substantially smaller. However, there are exceptions. For example, several years ago, the professional version of CPLEX 8.0 successfully solved an IP problem with 215,000 general integer variables, 75,000 functional constraints, and 6,000,000 nonzero constraint coefficients, and current versions of CPLEX have become far more powerful.

The second fallacy is that removing some feasible solutions (the noninteger ones) from a linear programming problem will make it easier to solve. To the contrary, it is only because all these feasible solutions are there that the guarantee usually can be given (see Sec. 5.1) that there will be a corner-point feasible (CPF) solution [and so a corresponding basic feasible (BF) solution] that is optimal for the overall problem. This guarantee is the key to the remarkable efficiency of the simplex method. As a result, linear programming problems generally are considerably easier to solve than IP problems.

Consequently, most successful algorithms for integer programming incorporate a linear programming algorithm, such as the simplex method (or dual simplex method), as much as they can by relating portions of the IP problem under consideration to the corresponding linear programming problem (i.e., the same problem except that the integer restriction is deleted). For any given IP problem, this corresponding linear programming

Taco Bell Corporation has over 6,500 quick-service restaurants in the United States and a growing international market. It serves approximately 2 billion meals per year, generating about $5.4 billion in annual sales income.

At each Taco Bell restaurant, the amount of business is highly variable throughout the day (and from day to day), with a heavy concentration during the normal meal times. Therefore, determining how many employees should be scheduled to perform what functions in the restaurant at any given time is a complex and vexing problem.

To attack this problem, Taco Bell management instructed an OR team (including several consultants) to develop a new labor-management system. The team concluded that the system needed three major components: (1) *a forecasting model* for predicting customer transactions at any time, (2) *a simulation model* (such as those described in Chap. 20) to translate customer transactions to labor requirements, and (3) *an integer programming model* to schedule employees to satisfy labor requirements and minimize payroll.

The integer decision variables for this integer programming model for any restaurant are the number of employees assigned to each of the shifts that begin at various specified times. The lengths of these shifts also are decision variables (constrained to be between minimum and maximum permissible shift lengths), but *continuous* decision variables in this case, so the model is a mixed IP model. The main constraints specify that the number of employees working during each 15-minute time interval must be greater than or equal to the minimum number required during that interval (according to the forecasting model).

This MIP model is similar to the *linear programming* model for assigning employees to shifts at United Airlines facilities that is described in an application vignette in Sec. 3.4. However, the key difference is that the number of employees working shifts at Taco Bell restaurants is much smaller than the number at United Airlines facilities, so it is necessary to restrict these decision variables to integer values for the Taco Bell model (whereas noninteger values in a solution for the United Airlines can readily be rounded to integer values with little loss of accuracy).

The implementation of this MIP model along with the other components of the labor-management system has provided Taco Bell with *documented savings of* **$13 million** *per year* in labor costs.

Source: J. Hueter and W. Swart: "An Integrated Labor-Management System for Taco Bell," *Interfaces,* **28**(1): 75–91, Jan.–Feb. 1998. (A link to this article is provided on our website, www.mhhe.com/hillier.)

problem commonly is referred to as its **LP relaxation.** The algorithms presented in the next two sections illustrate how a sequence of LP relaxations for portions of an IP problem can be used to solve the overall IP problem effectively.

There is one special situation where solving an IP problem is no more difficult than solving its LP relaxation once by the simplex method, namely, when the optimal solution to the latter problem turns out to satisfy the integer restriction of the IP problem. When this situation occurs, this solution *must* be optimal for the IP problem as well, because it is the best solution among all the feasible solutions for the LP relaxation, which includes all the feasible solutions for the IP problem. Therefore, it is common for an IP algorithm to begin by applying the simplex method to the LP relaxation to check whether this fortuitous outcome has occurred.

Although it generally is quite fortuitous indeed for the optimal solution to the LP relaxation to be integer as well, there actually exist several *special types* of IP problems for which this outcome is *guaranteed.* You already have seen the most prominent of these special types in Chaps. 8 and 9, namely, the *minimum cost flow problem* (with integer parameters) and its special cases (including the *transportation problem,* the *assignment problem,* the *shortest-path problem,* and the *maximum flow problem*). This guarantee can be given for these types of problems because they possess a certain *special structure* (e.g., see Table 8.6) that ensures that every BF solution is integer, as stated in the integer solutions property given in Secs. 8.1 and 9.6. Consequently, these special types of IP problems can be treated as linear programming problems, because they can be solved completely by a streamlined version of the simplex method.

Although this much simplification is somewhat unusual, in practice IP problems frequently have *some* special structure that can be exploited to simplify the problem. (Examples 2 and 3 in the preceding section fit into this category, because of their *mutually exclusive alternatives* constraints or *contingent decisions* constraints or *set-covering* constraints.) Sometimes, very large versions of these problems can be solved successfully. Special-purpose algorithms designed specifically to exploit certain kinds of special structures are becoming increasingly important in integer programming.

Thus, the three primary determinants of *computational difficulty* for an IP problem are (1) the *number of integer variables,* (2) whether these integer variables are *binary* variables or *general* integer variables, and (3) any *special structure* in the problem. This situation is in contrast to linear programming, where the number of (functional) constraints is much more important than the number of variables. In integer programming, the number of constraints is of *some* importance (especially if LP relaxations are being solved), but it is strictly secondary to the other three factors. In fact, there occasionally are cases where *increasing* the number of constraints *decreases* the computation time because the number of feasible solutions has been reduced. For MIP problems, it is the number of *integer* variables rather than the *total* number of variables that is important, because the continuous variables have almost no effect on the computational effort.

Because IP problems are, in general, much more difficult to solve than linear programming problems, sometimes it is tempting to use the approximate procedure of simply applying the simplex method to the LP relaxation and then *rounding* the noninteger values to integers in the resulting solution. This approach may be adequate for some applications, especially if the values of the variables are quite large so that rounding creates relatively little error. However, you should beware of two pitfalls involved in this approach.

One pitfall is that an optimal linear programming solution is *not necessarily feasible* after it is rounded. Often it is difficult to see in which way the rounding should be done to retain feasibility. It may even be necessary to change the value of some variables by one or more units after rounding. To illustrate, consider the following problem:

Maximize $Z = x_2,$

subject to

$$-x_1 + x_2 \leq \frac{1}{2}$$

$$x_1 + x_2 \leq 3\frac{1}{2}$$

and

$$x_1 \geq 0, \qquad x_2 \geq 0$$
$$x_1, x_2 \text{ are integers.}$$

As Fig. 11.2 shows, the optimal solution for the LP relaxation is $x_1 = 1\frac{1}{2}$, $x_2 = 2$, but it is impossible to round the noninteger variable x_1 to 1 or 2 (or any other integer) and retain feasibility. Feasibility can be retained only by also changing the integer value of x_2. It is easy to imagine how such difficulties can be compounded when there are tens or hundreds of constraints and variables.

Even if an optimal solution for the LP relaxation is rounded successfully, there remains another pitfall. There is no guarantee that this rounded solution will be the optimal

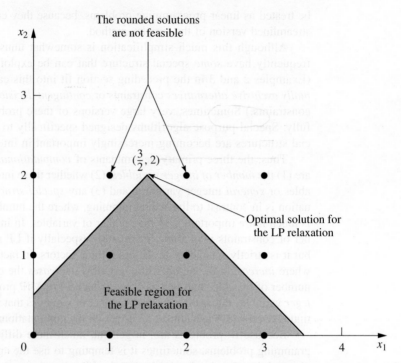

■ FIGURE 11.2
An example of an IP problem where the optimal solution for the LP relaxation cannot be rounded in any way that retains feasibility.

integer solution. In fact, it may even be far from optimal in terms of the value of the objective function. This fact is illustrated by the following problem:

Maximize $Z = x_1 + 5x_2$,

subject to

$$x_1 + 10x_2 \leq 20$$
$$x_1 \qquad\;\; \leq 2$$

and

$$x_1 \geq 0, \qquad x_2 \geq 0$$
x_1, x_2 are integers.

Because there are only two decision variables, this problem can be depicted graphically as shown in Fig. 11.3. Either the graph or the simplex method may be used to find that the optimal solution for the LP relaxation is $x_1 = 2$, $x_2 = \frac{9}{5}$, with $Z = 11$. If a graphical solution were not available (which would be the case with more decision variables), then the variable with the noninteger value $x_2 = \frac{9}{5}$ would normally be rounded in the feasible direction to $x_2 = 1$. The resulting integer solution is $x_1 = 2$, $x_2 = 1$, which yields $Z = 7$. Notice that this solution is far from the optimal solution $(x_1, x_2) = (0, 2)$, where $Z = 10$.

Because of these two pitfalls, a better approach for dealing with IP problems that are too large to be solved exactly is to use one of the available *heuristic algorithms*. These algorithms are extremely efficient for large problems, but they are not guaranteed to find an optimal solution. However, they do tend to be considerably more effective than the rounding approach just discussed in finding very good feasible solutions.

One of the particularly exciting developments in OR in recent years has been the rapid progress in developing very effective heuristic algorithms (commonly called *metaheuristics*)

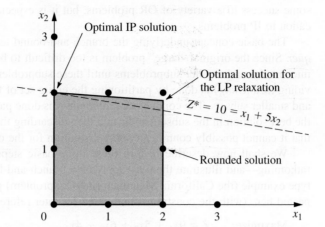

■ FIGURE 11.3
An example where rounding the optimal solution for the LP relaxation is far from optimal for the IP problem.

for various combinatorial problems such as IP problems. Three prominent types of metaheuristics (tabu search, simulated annealing, and genetic algorithms) will be described in Chap. 13. These sophisticated metaheuristics can even be applied to integer *nonlinear* programming problems that have locally optimal solutions that may be far removed from a globally optimal solution. They also can be applied to various *combinatorial optimization* problems, which frequently can be represented in a model that has integer variables but also has some constraints that are more complicated than for an IP model. (We'll discuss such applications further in Chap. 13.)

Returning to integer *linear* programming, for IP problems that are small enough to be solved to optimality, a considerable number of algorithms now are available. However, no IP algorithm possesses computational efficiency that is nearly comparable to the *simplex method* (except on special types of problems). Therefore, developing IP algorithms has continued to be an active area of research. Fortunately, some exciting algorithmic advances have been made and additional progress can be anticipated during the coming years. These advances are discussed further in Secs. 11.8 and 11.9.

The most popular traditional mode for IP algorithms is to use the *branch-and-bound technique* and related ideas to *implicitly enumerate* the feasible integer solutions, and we shall focus on this approach. The next section presents the branch-and-bound technique in a general context, and illustrates it with a basic branch-and-bound algorithm for BIP problems. Section 11.7 presents another algorithm of the same type for general MIP problems.

■ 11.6 THE BRANCH-AND-BOUND TECHNIQUE AND ITS APPLICATION TO BINARY INTEGER PROGRAMMING

Because any bounded *pure* IP problem has only a finite number of feasible solutions, it is natural to consider using some kind of *enumeration procedure* for finding an optimal solution. Unfortunately, as we discussed in the preceding section, this finite number can be, and usually is, very large. Therefore, it is imperative that any enumeration procedure be cleverly structured so that only a tiny fraction of the feasible solutions actually need be examined. For example, dynamic programming (see Chap. 10) provides one such kind of procedure for many problems having a finite number of feasible solutions (although it is not particularly efficient for most IP problems). Another such approach is provided by the *branch-and-bound technique.* This technique and variations of it have been applied with some success to a variety of OR problems, but it is especially well known for its application to IP problems.

The basic concept underlying the branch-and-bound technique is to *divide and conquer.* Since the original "large" problem is too difficult to be solved directly, it is divided into smaller and smaller subproblems until these subproblems can be conquered. The dividing (*branching*) is done by partitioning the entire set of feasible solutions into smaller and smaller subsets. The conquering (*fathoming*) is done partially by *bounding* how good the best solution in the subset can be and then discarding the subset if its bound indicates that it cannot possibly contain an optimal solution for the original problem.

We shall now describe in turn these three basic steps—branching, bounding, and fathoming—and illustrate them by applying a branch-and-bound algorithm to the prototype example (the California Manufacturing Co. problem) presented in Sec. 11.1 and repeated here (with the constraints numbered for later reference).

$$\text{Maximize} \quad Z = 9x_1 + 5x_2 + 6x_3 + 4x_4,$$

subject to

$$
\begin{aligned}
(1) \quad & 6x_1 + 3x_2 + 5x_3 + 2x_4 \le 10 \\
(2) \quad & x_3 + x_4 \le 1 \\
(3) \quad & -x_1 + x_3 \le 0 \\
(4) \quad & -x_2 + x_4 \le 0
\end{aligned}
$$

and

$$(5) \quad x_j \text{ is binary,} \quad \text{for } j = 1, 2, 3, 4.$$

Branching

When you are dealing with binary variables, the most straightforward way to partition the set of feasible solutions into subsets is to fix the value of one of the variables (say, x_1) at $x_1 = 0$ for one subset and at $x_1 = 1$ for the other subset. Doing this for the prototype example divides the whole problem into the two smaller subproblems shown next.

Subproblem 1:

Fix $x_1 = 0$ so the resulting subproblem reduces to

$$\text{Maximize} \quad Z = 5x_2 + 6x_3 + 4x_4,$$

subject to

$$
\begin{aligned}
(1) \quad & 3x_2 + 5x_3 + 2x_4 \le 10 \\
(2) \quad & x_3 + x_4 \le 1 \\
(3) \quad & x_3 \le 0 \\
(4) \quad & -x_2 + x_4 \le 0 \\
(5) \quad & x_j \text{ is binary,} \quad \text{for } j = 2, 3, 4.
\end{aligned}
$$

Subproblem 2:

Fix $x_1 = 1$ so the resulting subproblem reduces to

$$\text{Maximize} \quad Z = 9 + 5x_2 + 6x_3 + 4x_4,$$

subject to

$$
\begin{aligned}
(1) \quad & 3x_2 + 5x_3 + 2x_4 \le 4 \\
(2) \quad & x_3 + x_4 \le 1 \\
(3) \quad & x_3 \le 1 \\
(4) \quad & -x_2 + x_4 \le 0 \\
(5) \quad & x_j \text{ is binary,} \quad \text{for } j = 2, 3, 4.
\end{aligned}
$$

Variable: x_1

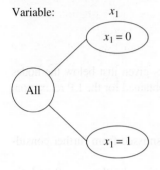

Figure 11.4 portrays this dividing (branching) into subproblems by a *tree* (defined in Sec. 9.2) with *branches* (arcs) from the *All* node (corresponding to the whole problem having *all* feasible solutions) to the two nodes corresponding to the two subproblems. This tree, which will continue "growing branches" iteration by iteration, is referred to as the **branching tree** (or *solution tree* or *enumeration tree*) for the algorithm. The variable used to do this branching at any iteration by assigning values to the variable (as with x_1 above) is called the **branching variable.** (Sophisticated methods for selecting branching variables are an important part of most branch-and-bound algorithms but, for simplicity, we always select them in their natural order—$x_1, x_2, \ldots, x_n$—throughout this section.)

Later in the section you will see that one of these subproblems can be conquered (fathomed) immediately, whereas the other subproblem will need to be divided further into smaller subproblems by setting $x_2 = 0$ or $x_2 = 1$.

For other IP problems where the integer variables have more than two possible values, the branching can still be done by setting the branching variable at its respective individual values, thereby creating more than two new subproblems. However, a good alternate approach is to specify a *range* of values (for example, $x_j \leq 2$ or $x_j \geq 3$) for the branching variable for each new subproblem. This is the approach used for the algorithm presented in Sec. 11.7.

Bounding

For each of these subproblems, we now need to obtain a *bound* on how good its best feasible solution can be. The standard way of doing this is to quickly solve a simpler *relaxation* of the subproblem. In most cases, a **relaxation** of a problem is obtained simply by *deleting* ("relaxing") one set of constraints that had made the problem difficult to solve. For IP problems, the most troublesome constraints are those requiring the respective variables to be integer. Therefore, the most widely used relaxation is the **LP relaxation** that deletes this set of constraints.

To illustrate for the example, consider first the whole problem given in Sec. 11.1 (and repeated at the beginning of this section). Its LP relaxation is obtained by replacing the last line of the model (x_j is binary, for $j = 1, 2, 3, 4$) by the following new (relaxed) version of this constraint (5).

$$(5) \qquad 0 \leq x_j \leq 1, \qquad \text{for } j = 1, 2, 3, 4.$$

Using the simplex method to quickly solve this LP relaxation yields its optimal solution

$$(x_1, x_2, x_3, x_4) = \left(\frac{5}{6}, 1, 0, 1\right), \qquad \text{with } Z = 16\frac{1}{2}.$$

Therefore, $Z \leq 16\frac{1}{2}$ for all feasible solutions for the original BIP problem (since these solutions are a subset of the feasible solutions for the LP relaxation). In fact, as indicated later in the summary of the algorithm, this *bound* of $16\frac{1}{2}$ can be rounded down to 16, because all coefficients in the objective function are integer, so all integer solutions must have an integer value for Z.

Bound for whole problem: $Z \leq 16$.

Now let us obtain the bounds for the two subproblems in the same way. For subproblem 1, where x_1 has been fixed at $x_1 = 0$, this can be conveniently expressed in its LP relaxation by adding the constraint that $x_1 \leq 0$ since combining this with the current constraint that $0 \leq x_1 \leq 1$ forces $x_1 = 0$. Similarly, fixing x_1 at $x_1 = 1$ for subproblem 2 leads to adding the constraint that $x_1 \geq 1$ for its LP relaxation. Applying the simplex method then yields the optimal solutions shown below for these LP relaxations.

LP relaxation of subproblem 1: (5) $x_1 \leq 0$ and $0 \leq x_j \leq 1$ for $j = 1, 2, 3, 4$.
Optimal solution: $(x_1, x_2, x_3, x_4) = (0, 1, 0, 1)$ with $Z = 9$.

LP relaxation of subproblem 2: (5) $x_1 \geq 1$ and $0 \leq x_j \leq 1$ for $j = 1, 2, 3, 4$.
Optimal solution: $(x_1, x_2, x_3, x_4) = \left(1, \dfrac{4}{5}, 0, \dfrac{4}{5}\right)$ with $Z = 16\dfrac{1}{5}$.

The resulting bounds for the subproblems then are

Bound for subproblem 1: $Z \leq 9$,
Bound for subproblem 2: $Z \leq 16$.

Figure 11.5 summarizes these results, where the numbers given just below the nodes are the bounds and below each bound is the optimal solution obtained for the LP relaxation.

Fathoming

A subproblem can be conquered (fathomed), and thereby dismissed from further consideration, in the three ways described below.

① One way is illustrated by the results for subproblem 1 given by the $x_1 = 0$ node in Fig. 11.5. Note that the (unique) optimal solution for its LP relaxation, $(x_1, x_2, x_3, x_4) = (0, 1, 0, 1)$, is an *integer* solution. Therefore, this solution must also be the optimal solution for subproblem 1 itself. This solution should be stored as the first **incumbent** (the best feasible solution found so far) for the whole problem, along with its value of Z. This value is denoted by

$Z^* = $ value of Z for current incumbent,

so $Z^* = 9$ at this point. Since this solution has been stored, there is no reason to consider subproblem 1 any further by branching from the $x_1 = 0$ node, etc. Doing so could only lead to other feasible solutions that are inferior to the incumbent, and we have no interest in such solutions. Because it has been solved, we **fathom** (dismiss) subproblem 1 now.

② The above results suggest a second key fathoming test. Since $Z^* = 9$, there is no reason to consider further any subproblem whose *bound* (after rounding down) ≤ 9, since such a subproblem cannot have a feasible solution better than the *incumbent*. Stated more generally, a subproblem is fathomed whenever its

Bound $\leq Z^*$.

This outcome does not occur in the current iteration of the example because subproblem 2 has a bound of 16 that is larger than 9. However, it might occur later for **descendants** of this subproblem (new smaller subproblems created by branching on this subproblem, and then perhaps branching further through subsequent "generations"). Furthermore, as new incumbents with larger values of Z^* are found, it will become easier to *fathom* in this way.

③ The third way of fathoming is quite straightforward. If the simplex method finds that a subproblem's LP relaxation has *no feasible solutions,* then the subproblem itself must have *no feasible solutions,* so it can be dismissed (fathomed).

In all three cases, we are conducting our search for an optimal solution by retaining for further investigation only those subproblems that could possibly have a feasible solution better than the current incumbent.

Summary of Fathoming Tests.   A subproblem is *fathomed* (dismissed from further consideration) if
Test 1: Its bound $\leq Z^*$,
or
Test 2: Its LP relaxation has no feasible solutions,

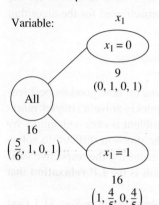

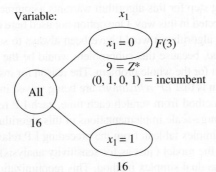

FIGURE 11.6
The branching tree after the first iteration of the BIP branch-and-bound algorithm for the example in Sec. 11.1.

or

Test 3: The optimal solution for its LP relaxation is *integer*. (If this solution is better than the incumbent, it becomes the new incumbent, and test 1 is reapplied to all unfathomed subproblems with the new larger Z^*.)

Figure 11.6 summarizes the results of applying these three tests to subproblems 1 and 2 by showing the current *branching tree*. Only subproblem 1 has been fathomed, by test 3, as indicated by $F(3)$ next to the $x_1 = 0$ node. The resulting incumbent also is identified below this node.

The subsequent iterations will illustrate successful applications of all three tests. However, before continuing the example, we summarize the algorithm being applied to this BIP problem. (This algorithm assumes that the objective function is to be *maximized*, that all coefficients in the objective function are integer and, for simplicity, that the ordering of the variables for branching is $x_1, x_2, \ldots, x_n$. As noted previously, most branch-and-bound algorithms use sophisticated methods for selecting branching variables instead.)

Summary of the BIP Branch-and-Bound Algorithm

Initialization: Set $Z^* = -\infty$. Apply the bounding step, fathoming step, and optimality test described below to the whole problem. If not fathomed, classify this problem as the one remaining "subproblem" for performing the first full iteration below.

Steps for each iteration:

1. *Branching:* Among the *remaining* (unfathomed) subproblems, select the one that was created *most recently*. (Break ties according to which has the *larger bound*.) Branch from the node for this subproblem to create two new subproblems by fixing the next variable (the branching variable) at either 0 or 1.

2. *Bounding:* For each new subproblem, apply the simplex method to its LP relaxation to obtain an optimal solution, including the value of Z, for this LP relaxation. If this value of Z is not an integer, round it down to an integer. (If it was already an integer, no change is needed.) This integer value of Z is the *bound* for the subproblem.

3. *Fathoming:* For each new subproblem, apply the three fathoming tests summarized above, and discard those subproblems that are fathomed by any of the tests.

Optimality test: Stop when there are *no remaining* subproblems; the current *incumbent* is optimal.[4] Otherwise, return to perform another iteration.

---

[4]If there is no incumbent, the conclusion is that the problem has no feasible solutions.

The branching step for this algorithm warrants a comment as to why the subproblem to branch from is selected in this way. One option not used here (but sometimes adopted in other branch-and-bound algorithms) would have been always to select the remaining subproblem with the *best bound,* because this subproblem would be the most promising one to contain an optimal solution for the whole problem. The reason for instead selecting the *most recently created* subproblem is that *LP relaxations* are being solved in the bounding step. Rather than start the simplex method from scratch each time, each LP relaxation generally is solved by *reoptimization* in large-scale implementations of this algorithm. This reoptimization involves revising the final simplex tableau from the preceding LP relaxation as needed because of the few differences in the model (just as for sensitivity analysis) and then applying a few iterations of perhaps the dual simplex method. This reoptimization tends to be *much* faster than starting from scratch, *provided* the preceding and current models are closely related. The models will tend to be closely related under the branching rule used, but *not* when you are skipping around in the branching tree by selecting the subproblem with the best bound.

Completing the Example

The pattern for the remaining iterations will be quite similar to that for the first iteration described above except for the ways in which fathoming occurs. Therefore, we shall summarize the branching and bounding steps fairly briefly and then focus on the fathoming step.

Iteration 2. The only remaining subproblem corresponds to the $x_1 = 1$ node in Fig. 11.6, so we shall branch from this node to create the two new subproblems given below.

Subproblem 3:
Fix $x_1 = 1$, $x_2 = 0$ so the resulting subproblem reduces to

Maximize $Z = 9 + 6x_3 + 4x_4$,

subject to

$$\begin{align}
(1) \quad & 5x_3 + 2x_4 \le 4 \\
(2) \quad & x_3 + x_4 \le 1 \\
(3) \quad & x_3 \le 1 \\
(4) \quad & x_4 \le 0 \\
(5) \quad & x_j \text{ is binary,} \quad \text{for } j = 3, 4.
\end{align}$$

Subproblem 4:
Fix $x_1 = 1$, $x_2 = 1$ so the resulting subproblem reduces to

Maximize $Z = 14 + 6x_3 + 4x_4$,

subject to

$$\begin{align}
(1) \quad & 5x_3 + 2x_4 \le 1 \\
(2) \quad & x_3 + x_4 \le 1 \\
(3) \quad & x_3 \le 1 \\
(4) \quad & x_4 \le 1 \\
(5) \quad & x_j \text{ is binary,} \quad \text{for } j = 3, 4.
\end{align}$$

The LP relaxations of these subproblems are obtained by adding the additional constraint shown below into the relaxed version of constraint (5). Their optimal solutions also are shown below.

LP relaxation of subproblem 3:   (5) $x_1 \ge 1$, $x_2 \le 0$, and $0 \le x_j \le 1$
for $j = 1, 2, 3, 4$.

Optimal solution: $(x_1, x_2, x_3, x_4) = \left(1, 0, \frac{4}{5}, 0\right)$ with $Z = 13\frac{4}{5}$,

LP relaxation of subproblem 4:   (5) $x_1 \geq 1$, $x_2 \geq 1$,   and $0 \leq x_j \leq 1$
for $j = 1, 2, 3, 4$.

Optimal solution: $(x_1, x_2, x_3, x_4) = \left(1, 1, 0, \frac{1}{2}\right)$ with $Z = 16$.

The resulting bounds for the subproblems are

Bound for subproblem 3: $Z \leq 13$,
Bound for subproblem 4: $Z \leq 16$.

Note that both these bounds are larger than $Z^* = 9$, so fathoming test 1 fails in both cases. Test 2 also fails, since both LP relaxations have feasible solutions (as indicated by the existence of an optimal solution). Alas, test 3 fails as well, because both optimal solutions include variables with noninteger values.

Figure 11.7 shows the resulting branching tree at this point. The lack of an *F* to the right of either new node indicates that both remain unfathomed.

Iteration 3.   So far, the algorithm has created four subproblems. Subproblem 1 has been fathomed, and subproblem 2 has been replaced by (separated into) subproblems 3 and 4, but these last two remain under consideration. Because they were created simultaneously, but subproblem 4 ($x_1 = 1$, $x_2 = 1$) has the larger *bound* ($16 > 13$), the next branching is done from the $(x_1, x_2) = (1, 1)$ node in the branching tree, which creates the following new subproblems (where constraint 3 disappears because it does not contain x_4).

Subproblem 5:

Fix $x_1 = 1$, $x_2 = 1$, $x_3 = 0$ so the resulting subproblem reduces to

Maximize $Z = 14 + 4x_4$,

subject to

(1) $2x_4 \leq 1$
(2), (4) $x_4 \leq 1$ (twice)
(5) x_4 is binary.

■ FIGURE 11.7
The branching tree after iteration 2 of the BIP branch-and-bound algorithm for the example in Sec. 11.1.

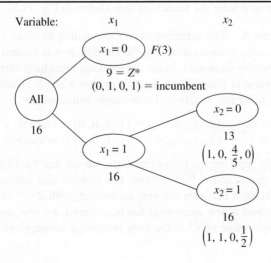

Subproblem 6:

Fix $x_1 = 1$, $x_2 = 1$, $x_3 = 1$ so the resulting subproblem reduces to

Maximize $Z = 20 + 4x_4$,

subject to

(1) $2x_4 \leq -4$
(2) $x_4 \leq 0$
(4) $x_4 \leq 1$
(5) x_4 is binary.

The corresponding LP relaxations have the relaxed version of constraint (5), the optimal solution, and the bound (when it exists) shown below.

LP relaxation of subproblem 5:

(5) $x_1 \geq 1$, $x_2 \geq 1$, $x_3 \leq 0$,   and $0 \leq x_j \leq 1$
        for $j = 1, 2, 3, 4$.

Optimal solution: $(x_1, x_2, x_3, x_4) = \left(1, 1, 0, \dfrac{1}{2}\right)$,   with $Z = 16$.

Bound: $Z \leq 16$.

LP relaxation of subproblem 6:

(5) $x_1 \geq 1$, $x_2 \geq 1$, $x_3 \geq 1$,   and $0 \leq x_j \leq 1$
        for $j = 1, 2, 3, 4$.

Optimal solution: None since there are no feasible solutions.
        Bound: None

For both of these subproblems, the relaxed version of constraint (5) has the effect of fixing the values of x_1, x_2, and x_3 at the desired values and then requiring that $0 \leq x_4 \leq 1$. Therefore, the LP relaxations for these subproblems reduce to the statements of the subproblems given above except for replacing constraint (5) by $0 \leq x_4 \leq 1$. Reducing these LP relaxations to one-variable problems (plus the fixed values of x_1, x_2, and x_3) make it easy to see that the optimal solution for the LP relaxation of subproblem 5 is indeed the one given above. Similarly, note how the combination of constraint 1 and $0 \leq x_4 \leq 1$ in the LP relaxation of subproblem 6 prevents any feasible solutions. Therefore, this subproblem is fathomed by test 2. However, subproblem 5 fails this test, as well as test 1 ($16 > 9$) and test 3 ($x_4 = \frac{1}{2}$ is not integer), so it remains under consideration.

We now have the branching tree shown in Fig. 11.8.

Iteration 4.   The subproblems corresponding to nodes (1, 0) and (1, 1, 0) in Fig. 11.8 remain under consideration, but the latter node was created more recently, so it is selected for branching from next. Since the resulting branching variable x_4 is the *last* variable, fixing its value at either 0 or 1 actually creates a *single solution* rather than subproblems requiring fuller investigation. These single solutions are

$x_4 = 0$: $(x_1, x_2, x_3, x_4) = (1, 1, 0, 0)$ is feasible, with $Z = 14$,
$x_4 = 1$: $(x_1, x_2, x_3, x_4) = (1, 1, 0, 1)$ is infeasible.

Formally applying the fathoming tests, we see that the first solution passes test 3 and the second passes test 2. Furthermore, this feasible first solution is better than the incumbent ($14 > 9$), so it becomes the new incumbent, with $Z^* = 14$.

Because a new incumbent has been found, we now reapply fathoming test 1 with the new larger value of Z^* to the only remaining subproblem, the one at node (1, 0).

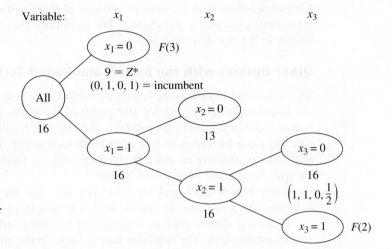

■ FIGURE 11.8
The branching tree after iteration 3 of the BIP branch-and-bound algorithm for the example in Sec. 11.1.

■ FIGURE 11.9
The branching tree after the final (fourth) iteration of the BIP branch-and-bound algorithm for the example in Sec. 11.1.

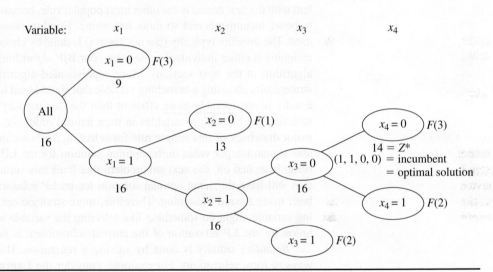

Subproblem 3:

Bound $= 13 \leq Z^* = 14$.

Therefore, this subproblem now is fathomed.

We now have the branching tree shown in Fig. 11.9. Note that there are *no remaining* (unfathomed) subproblems. Consequently, the optimality test indicates that the current incumbent

$$(x_1, x_2, x_3, x_4) = (1, 1, 0, 0)$$

is optimal, so we are done.

Your OR Tutor includes **another example** of applying this algorithm. Also included in the IOR Tutorial is an interactive procedure for executing this algorithm. As usual, the Excel, LINGO/LINDO, and MPL/CPLEX files for this chapter in your OR

Courseware show how the student versions of these software packages are applied to the various examples in the chapter. The algorithms they use for BIP problems all are similar to the one described above.[5]

Other Options with the Branch-and-Bound Technique

This section has illustrated the branch-and-bound technique by describing a basic branch-and-bound algorithm for solving BIP problems. However, the general framework of the branch-and-bound technique provides a great deal of flexibility in how to design a specific algorithm for any given type of problem such as BIP. There are many options available, and constructing an efficient algorithm requires tailoring the specific design to fit the specific structure of the problem type.

Every branch-and-bound algorithm has the same three basic steps of *branching, bounding,* and *fathoming.* The flexibility lies in how these steps are performed.

Branching always involves *selecting* one remaining subproblem and *dividing* it into smaller subproblems. The flexibility here is found in the rules for selecting and dividing. Our BIP algorithm selected the *most recently created* subproblem, because this is very efficient for *reoptimizing* each LP relaxation from the preceding one. Selecting the subproblem with the *best bound* is the other most popular rule, because it tends to lead more quickly to better incumbents and so more fathoming. Combinations of the two rules also can be used. The *dividing* typically (but not always) is done by choosing a *branching variable* and assigning it either individual values (e.g., our BIP algorithm) or ranges of values (e.g., the algorithm in the next section). More sophisticated algorithms generally use a rule for strategically choosing a branching variable that should tend to lead to early fathoming. This usually is considerably more efficient than the rule used by our BIP algorithm of simply selecting the branching variables in their natural order—$x_1, x_2, \ldots, x_n$. For example, a major drawback of this simple rule for selecting the branching variable is that if this variable has an integer value in the optimal solution for the LP relaxation of the subproblem being branched on, the next subproblem that fixes this variable at this same integer value also will have the same optimal solution for its LP relaxation, so no progress will have been made toward fathoming. Therefore, more strategic options for selecting the branching variable might do something like selecting the variable whose value in the optimal solution for the LP relaxation of the current subproblem is *furthest* from being an integer.

Bounding usually is done by solving a *relaxation.* However, there are a variety of ways to form relaxations. For example, consider the **Lagrangian relaxation,** where the entire set of functional constraints $\mathbf{Ax} \leq \mathbf{b}$ (in matrix notation) is *deleted* (except possibly for any "convenient" constraints) and then the objective function

Maximize $Z = \mathbf{cx},$

is replaced by

Maximize $Z_R = \mathbf{cx} - \boldsymbol{\lambda}(\mathbf{Ax} - \mathbf{b}),$

where the fixed vector $\boldsymbol{\lambda} \geq \mathbf{0}$. If $\mathbf{x}^*$ is an optimal solution for the original problem, its $Z \leq Z_R$, so solving the Lagrangian relaxation for the optimal value of Z_R provides a valid *bound.* If $\boldsymbol{\lambda}$ is chosen well, this bound tends to be a reasonably tight one (at least comparable to the bound from the LP relaxation). Without any functional constraints, this relaxation also can be solved extremely quickly. The drawbacks are that fathoming tests 2 and 3 (revised) are not as powerful as for the LP relaxation.

---

[5]In the professional version of LINGO, LINDO, and CPLEX, the BIP algorithm also uses a variety of sophisticated techniques along the lines described in Sec. 11.8.

In general terms, two features are sought in choosing a relaxation: it can be solved relatively quickly, and provides a relatively tight bound. Neither alone is adequate. The LP relaxation is popular because it provides an excellent trade-off between these two factors.

One option occasionally employed is to use a quickly solved relaxation and then, if fathoming is not achieved, to tighten the relaxation in some way to obtain a somewhat tighter bound.

Fathoming generally is done pretty much as described for the BIP algorithm. The three fathoming criteria can be stated in more general terms as follows.

Summary of Fathoming Criteria.   A subproblem is *fathomed* if an analysis of its *relaxation* reveals that

Criterion 1: Feasible solutions of the subproblem must have $Z \leq Z^*$, or
Criterion 2: The subproblem has no feasible solutions, or
Criterion 3: An optimal solution of the subproblem has been found.

Just as for the BIP algorithm, the first two criteria usually are applied by solving the relaxation to obtain a bound for the subproblem and then checking whether this bound is $\leq Z^*$ (test 1) or whether the relaxation has no feasible solutions (test 2). If the relaxation differs from the subproblem *only* by the deletion (or loosening) of some constraints, then the third criterion usually is applied by checking whether the optimal solution for the relaxation is *feasible* for the subproblem, in which case it must be *optimal* for the subproblem. For other relaxations (such as the Lagrangian relaxation), additional analysis is required to determine whether the optimal solution for the relaxation is also optimal for the subproblem.

If the original problem involves *minimization* rather than maximization, two options are available. One is to convert to maximization in the usual way (see Sec. 4.6). The other is to convert the branch-and-bound algorithm directly to minimization form, which requires changing the direction of the inequality for fathoming test 1 from

Is the subproblem's bound $\leq Z^*$?

to

Is the subproblem's bound $\geq Z^*$?

When using this latter inequality, if the value of Z for the optimal solution for the LP relaxation of the subproblem is not an integer, it now would be rounded *up* to an integer to obtain the subproblem's bound.

So far, we have described how to use the branch-and-bound technique to find only *one* optimal solution. However, in the case of ties for the optimal solution, it is sometimes desirable to identify *all* these optimal solutions so that the final choice among them can be made on the basis of intangible factors not incorporated into the mathematical model. To find them all, you need to make only a few slight alterations in the procedure. First, change the weak inequality for fathoming test 1 (Is the subproblem's bound $\leq Z^*$?) to a strict inequality (Is the subproblem's bound $< Z^*$?), so that fathoming will not occur if the subproblem can have a feasible solution *equal* to the incumbent. Second, if fathoming test 3 passes and the optimal solution for the subproblem has $Z = Z^*$, then store this solution as *another* (tied) incumbent. Third, if test 3 provides a new incumbent (tied or otherwise), then check whether the optimal solution obtained for the *relaxation* is *unique*. If it is not, then identify the other optimal solutions for the relaxation and check whether they are optimal for the subproblem as well, in which case they also become incumbents. Finally, when the *optimality test* finds that there are *no remaining* (unfathomed) subsets, *all* the current *incumbents* will be the *optimal* solutions.

Finally, note that rather than find an optimal solution, the branch-and-bound technique can be used to find a *nearly optimal* solution, generally with much less computational effort. For some applications, a solution is "good enough" if its Z is "close enough" to the value of Z for an optimal solution (call it Z^{**}). *Close enough* can be defined in either of two ways as either

$$Z^{**} - K \le Z \quad \text{or} \quad (1 - \alpha)Z^{**} \le Z$$

for a specified (positive) constant K or α. For example, if the second definition is chosen and $\alpha = 0.05$, then the solution is required to be within 5 percent of optimal. Consequently, if it were known that the value of Z for the current incumbent (Z^*) satisfies either

$$Z^{**} - K \le Z^* \quad \text{or} \quad (1 - \alpha)Z^{**} \le Z^*$$

then the procedure could be terminated immediately by choosing the incumbent as the desired nearly optimal solution. Although the procedure does not actually identify an optimal solution and the corresponding Z^{**}, if this (unknown) solution is feasible (and so optimal) for the subproblem currently under investigation, then fathoming test 1 finds an upper bound such that

$$Z^{**} \le \text{bound}$$

so that either

$$\text{Bound} - K \le Z^* \quad \text{or} \quad (1 - \alpha)\text{bound} \le Z^*$$

would imply that the corresponding inequality in the preceding sentence is satisfied. Even if this solution is not feasible for the current subproblem, a valid upper bound is still obtained for the value of Z for the subproblem's optimal solution. Thus, satisfying either of these last two inequalities is sufficient to fathom this subproblem because the incumbent must be "close enough" to the subproblem's optimal solution.

Therefore, to find a solution that is close enough to being optimal, only one change is needed in the usual branch-and-bound procedure. This change is to replace the usual fathoming test 1 for a subproblem

$$\text{Bound} \le Z^*?$$

by either

$$\text{Bound} - K \le Z^*?$$

or

$$(1 - \alpha)(\text{bound}) \le Z^*?$$

and then perform this test *after* test 3 (so that a feasible solution found with $Z > Z^*$ is still kept as the new incumbent). The reason this weaker test 1 suffices is that regardless of how close Z for the subproblem's (unknown) optimal solution is to the subproblem's bound, the incumbent is still close enough to this solution (if the new inequality holds) that the subproblem does not need to be considered further. When there are no remaining subproblems, the current incumbent will be the desired *nearly optimal* solution. However, it is much easier to fathom with this new fathoming test (in either form), so the algorithm should run much faster. For an extremely large problem, this acceleration may make the difference between finishing with a solution guaranteed to be close to optimal and never terminating. For many extremely large problems arising in practice, since the model provides only an idealized representation of the real problem anyway, finding a nearly optimal solution for the model in this way may be sufficient for all practical purposes. Therefore, this shortcut is used fairly frequently in practice.

■ 11.7   A BRANCH-AND-BOUND ALGORITHM FOR MIXED INTEGER PROGRAMMING

We shall now consider the general MIP problem, where *some* of the variables (say, I of them) are restricted to integer values (but not necessarily just 0 and 1) but the rest are ordinary continuous variables. For notational convenience, we shall order the variables so that the first I variables are the *integer-restricted* variables. Therefore, the general form of the problem being considered is

$$\text{Maximize} \quad Z = \sum_{j=1}^{n} c_j x_j,$$

subject to

$$\sum_{j=1}^{n} a_{ij} x_j \leq b_i, \quad \text{for } i = 1, 2, \ldots, m,$$

and

$$x_j \geq 0, \quad \text{for } j = 1, 2, \ldots, n,$$
$$x_j \text{ is integer}, \quad \text{for } j = 1, 2, \ldots, I; \ I \leq n.$$

(When $I = n$, this problem becomes the pure IP problem.)

We shall describe a basic branch-and-bound algorithm for solving this problem that, with a variety of refinements, has provided a standard approach to MIP. The structure of this algorithm was first developed by R. J. Dakin,[6] based on a pioneering branch-and-bound algorithm by A. H. Land and A. G. Doig.[7]

This algorithm is quite similar in structure to the BIP algorithm presented in the preceding section. Solving *LP relaxations* again provides the basis for both the *bounding* and *fathoming* steps. In fact, only four changes are needed in the BIP algorithm to deal with the generalizations from *binary* to *general* integer variables and from *pure* IP to *mixed* IP.

One change involves the choice of the *branching variable*. Before, the *next* variable in the natural ordering—$x_1, x_2, \ldots, x_n$—was chosen automatically. Now, the only variables considered are the *integer-restricted* variables that have a *noninteger* value in the optimal solution for the LP relaxation of the current subproblem. Our rule for choosing among these variables is to select the *first* one in the natural ordering. (Production codes generally use a more sophisticated rule.)

The second change involves the values assigned to the branching variable for creating the new smaller subproblems. Before, the *binary* variable was fixed at 0 and 1, respectively, for the two new subproblems. Now, the *general* integer-restricted variable could have a very large number of possible integer values, and it would be inefficient to create *and* analyze *many* subproblems by fixing the variable at its individual integer values. Therefore, what is done instead is to create just *two* new subproblems (as before) by specifying two *ranges* of values for the variable.

To spell out how this is done, let x_j be the current branching variable, and let x_j^* be its (noninteger) value in the optimal solution for the LP relaxation of the current subproblem. Using square brackets to denote

$$[x_j^*] = \text{greatest integer} \leq x_j^*,$$

[6]R. J. Dakin, "A Tree Search Algorithm for Mixed Integer Programming Problems," *Computer Journal,* **8**(3): 250–255, 1965.

[7]A. H. Land and A. G. Doig, "An Automatic Method of Solving Discrete Programming Problems," *Econometrica,* **28:** 497–520, 1960.

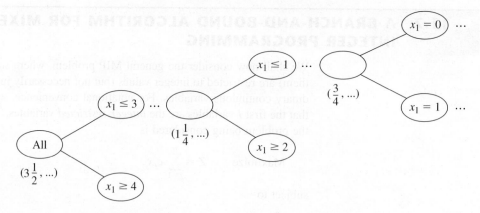

■ FIGURE 11.10
Illustration of the phenomenon of a *recurring branching variable*, where here x_1 becomes a branching variable three times because it has a noninteger value in the optimal solution for the LP relaxation at three nodes.

we have for the range of values for the two new subproblems

$$x_j \le [x_j^*] \qquad \text{and} \qquad x_j \ge [x_j^*] + 1,$$

respectively. Each inequality becomes an *additional constraint* for that new subproblem. For example, if $x_j^* = 3\frac{1}{2}$, then

$$x_j \le 3 \qquad \text{and} \qquad x_j \ge 4$$

are the respective additional constraints for the new subproblem.

When the two changes to the BIP algorithm described above are combined, an interesting phenomenon of a *recurring branching variable* can occur. To illustrate, as shown in Fig. 11.10, let $j = 1$ in the above example where $x_j^* = 3\frac{1}{2}$, and consider the new subproblem where $x_1 \le 3$. When the LP relaxation of a descendant of this subproblem is solved, suppose that $x_1^* = 1\frac{1}{4}$. Then x_1 *recurs* as the branching variable, and the two new subproblems created have the additional constraint $x_1 \le 1$ and $x_1 \ge 2$, respectively (as well as the previous additional constraint $x_1 \le 3$). Later, when the LP relaxation for a descendant of, say, the $x_1 \le 1$ subproblem is solved, suppose that $x_1^* = \frac{3}{4}$. Then x_1 *recurs* again as the branching variable, and the two new subproblems created have $x_1 = 0$ (because of the new $x_1 \le 0$ constraint and the nonnegativity constraint on x_1) and $x_1 = 1$ (because of the new $x_1 \ge 1$ constraint and the previous $x_1 \le 1$ constraint).

The third change involves the *bounding step*. Before, with a *pure* IP problem and integer coefficients in the objective function, the value of Z for the optimal solution for the subproblem's LP relaxation was *rounded down* to obtain the bound, because any feasible solution for the subproblem must have an *integer* Z. Now, with some of the variables *not* integer-restricted, the bound is the value of Z *without* rounding down.

The fourth (and final) change to the BIP algorithm to obtain our MIP algorithm involves fathoming test 3. Before, with a *pure* IP problem, the test was that the optimal solution for the subproblem's LP relaxation is *integer*, since this ensures that the solution is feasible, and therefore optimal, for the subproblem. Now, with a *mixed* IP problem, the test requires only that the *integer-restricted* variables be *integer* in the optimal solution for the subproblem's LP relaxation, because this suffices to ensure that the solution is feasible, and therefore optimal, for the subproblem.

Incorporating these four changes into the summary presented in the preceding section for the BIP algorithm yields the following summary for the new algorithm for MIP.

With headquarters in Houston, Texas, **Waste Management, Inc.** (a Fortune 100 company) is the leading provider of comprehensive waste-management services in North America. Its network of operations includes 293 active landfill disposal sites, 16 waste-to-energy plants, 72 landfill gas-to-energy facilities, 146 recycling plants, 346 transfer stations, and 435 collection operations (depots) to provide services to nearly 20 million residential customers and 2 million commercial customers throughout the United States and Canada.

The company's collection-and-transfer vehicles need to follow nearly 20,000 daily routes. With an annual operating cost of nearly $120,000 per vehicle, management wanted to have a comprehensive route-management system that would make every route as profitable and efficient as possible. Therefore, an OR team that included a number of consultants was formed to attack this problem.

The heart of the route-management system developed by this team is a *huge mixed BIP model* that optimizes the routes assigned to the respective collection-and-transfer

vehicles. Although the objective function takes several factors into account, the primary goal is the minimization of total travel time. The main decision variables are binary variables that equal 1 if the route assigned to a particular vehicle includes a particular possible leg and equal 0 otherwise. A geographical information system (GIS) provides the data about the distance and time required to go between any two points. All of this is imbedded within a Web-based Java application that is integrated with the company's other systems.

It is estimated that the recent implementation of this comprehensive route-management system will *increase the company's cash flow by* **$648 million** *over a 5-year period*, largely because of *savings of* **$498 million** in operational expenses over this same period. It also is providing better customer service.

Source: S. Sahoo, S. Kim, B.-I. Kim, B. Krass, and A. Popov, Jr.: "Routing Optimization for Waste Management," *Interfaces*, **35**(1): 24–36, Jan.–Feb. 2005. (A link to this article is provided on our website, www.mhhe.com/hillier.)

(As before, this summary assumes that the objective function is to be *maximized*, but the only change needed for minimization is to change the direction of the inequality for fathoming test 1.)

Summary of the MIP Branch-and-Bound Algorithm

Initialization: Set $Z^* = -\infty$. Apply the bounding step, fathoming step, and optimality test described below to the whole problem. If not fathomed, classify this problem as the one remaining subproblem for performing the first full iteration below.

Steps for each iteration:

1. *Branching:* Among the *remaining* (unfathomed) subproblems, select the one that was created *most recently*. (Break ties according to which has the *larger bound*.) Among the *integer-restricted* variables that have a *noninteger* value in the optimal solution for the LP relaxation of the subproblem, choose the *first one* in the natural ordering of the variables to be the *branching variable*. Let x_j be this variable and x_j^* its value in this solution. Branch from the node for the subproblem to create two new subproblems by adding the respective constraints $x_j \leq [x_j^*]$ and $x_j \geq [x_j^*] + 1$.

2. *Bounding:* For each new subproblem, obtain its bound by applying the simplex method (or the dual simplex method when reoptimizing) to its LP relaxation and using the value of Z for the resulting optimal solution.

3. *Fathoming:* For each new subproblem, apply the three fathoming tests given below, and discard those subproblems that are fathomed by any of the tests.

   Test 1: Its bound $\leq Z^*$, where Z^* is the value of Z for the current *incumbent*.

   Test 2: Its LP relaxation has no feasible solutions.

Test 3: The optimal solution for its LP relaxation has *integer* values for the *integer-restricted* variables. (If this solution is better than the incumbent, it becomes the new incumbent and test 1 is reapplied to all unfathomed subproblems with the new larger Z^*.)

Optimality test: Stop when there are no remaining subproblems; the current *incumbent* is optimal.[8] Otherwise, perform another iteration.

An MIP Example. We will now illustrate this algorithm by applying it to the following MIP problem:

$$\text{Maximize} \quad Z = 4x_1 - 2x_2 + 7x_3 - x_4,$$

subject to

$$
\begin{aligned}
x_1 \qquad\quad + 5x_3 \qquad\quad &\leq 10 \\
x_1 + x_2 - x_3 \qquad\quad &\leq 1 \\
6x_1 - 5x_2 \qquad\qquad\quad &\leq 0 \\
-x_1 \qquad\quad + 2x_3 - 2x_4 &\leq 3
\end{aligned}
$$

and

$$x_j \geq 0, \qquad \text{for } j = 1, 2, 3, 4$$
$$x_j \text{ is an integer}, \qquad \text{for } j = 1, 2, 3.$$

Note that the number of integer-restricted variables is $I = 3$, so x_4 is the only continuous variable.

Initialization. After setting $Z^* = -\infty$, we form the LP relaxation of this problem by *deleting* the set of constraints that x_j is an integer for $j = 1, 2, 3$. Applying the simplex method to this LP relaxation yields its optimal solution below.

$$\text{LP relaxation of whole problem:} \quad (x_1, x_2, x_3, x_4) = \left(\frac{5}{4}, \frac{3}{2}, \frac{7}{4}, 0\right), \qquad \text{with } Z = 14\frac{1}{4}.$$

Because it has *feasible* solutions and this optimal solution has *noninteger* values for its integer-restricted variables, the whole problem is not fathomed, so the algorithm continues with the first full iteration below.

Iteration 1. In this optimal solution for the LP relaxation, the *first* integer-restricted variable that has a noninteger value is $x_1 = \frac{5}{4}$, so x_1 becomes the branching variable. Branching from the *All* node (*all* feasible solutions) with this branching variable then creates the following two subproblems:

Subproblem 1:
Original problem plus additional constraint

$$x_1 \leq 1.$$

Subproblem 2:
Original problem plus additional constraint

$$x_1 \geq 2.$$

Deleting the set of integer constraints again and solving the resulting LP relaxations of these two subproblems yield the following results.

---

[8]If there is no incumbent, the conclusion is that the problem has no feasible solutions.

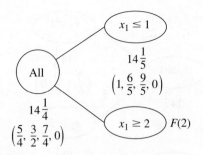

■ **FIGURES 11.11**
The branching tree after the first iteration of the MIP branch-and-bound algorithm for the MIP example.

Subproblem 1:
Optimal solution for LP relaxation: $(x_1, x_2, x_3, x_4) = \left(1, \frac{6}{5}, \frac{9}{5}, 0\right)$,   with $Z = 14\frac{1}{5}$.

Bound: $Z \le 14\frac{1}{5}$.

Subproblem 2:
LP relaxation: No feasible solutions.

This outcome for subproblem 2 means that it is fathomed by test 2. However, just as for the whole problem, subproblem 1 fails all fathoming tests.

These results are summarized in the branching tree shown in Fig. 11.11.

Iteration 2.   With only one remaining subproblem, corresponding to the $x_1 \le 1$ node in Fig. 11.11, the next branching is from this node. Examining its LP relaxation's optimal solution given above, we see that this node reveals that the *branching variable* is x_2, because $x_2 = \frac{6}{5}$ is the first integer-restricted variable that has a noninteger value. Adding one of the constraints $x_2 \le 1$ or $x_2 \ge 2$ then creates the following two new subproblems.

Subproblem 3:
Original problem plus additional constraints

   $x_1 \le 1,$ $x_2 \le 1.$

Subproblem 4:
Original problem plus additional constraints

   $x_1 \le 1,$ $x_2 \ge 2.$

Solving their LP relaxations gives the following results.

Subproblem 3:
Optimal solution for LP relaxation: $(x_1, x_2, x_3, x_4) = \left(\frac{5}{6}, 1, \frac{11}{6}, 0\right)$, with $Z = 14\frac{1}{6}$.

Bound: $Z \le 14\frac{1}{6}$.

Subproblem 4:
Optimal solution for LP relaxation: $(x_1, x_2, x_3, x_4) = \left(\frac{5}{6}, 2, \frac{11}{6}, 0\right)$, with $Z = 12\frac{1}{6}$.

Bound: $Z \le 12\frac{1}{6}$.

Because both solutions exist (feasible solutions) and have noninteger values for integer-restricted variables, neither subproblem is fathomed. (Test 1 still is not operational, since $Z^* = -\infty$ until the first incumbent is found.)

The branching tree at this point is given in Fig. 11.12.

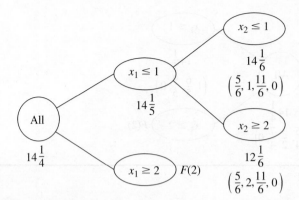

■ FIGURE 11.12
The branching tree after the second iteration of the MIP branch-and-bound algorithm for the MIP example.

Iteration 3. With two remaining subproblems (3 and 4) that were created simultaneously, the one with the larger bound (subproblem 3, with $14\frac{1}{6} > 12\frac{1}{6}$) is selected for the next branching. Because $x_1 = \frac{5}{6}$ has a noninteger value in the optimal solution for this subproblem's LP relaxation, x_1 becomes the branching variable. (Note that x_1 now is a *recurring* branching variable, since it also was chosen at iteration 1.) This leads to the following new subproblems.

Subproblem 5:
Original problem plus additional constraints

$$x_1 \leq 1$$
$$x_2 \leq 1$$
$$x_1 \leq 0 \quad \text{(so } x_1 = 0).$$

Subproblem 6:
Original problem plus additional constraints

$$x_1 \leq 1$$
$$x_2 \leq 1$$
$$x_1 \geq 1 \quad \text{(so } x_1 = 1).$$

The results from solving their LP relaxations are given below.

Subproblem 5:
　　Optimal solution for LP relaxation: $(x_1, x_2, x_3, x_4) = \left(0, 0, 2, \frac{1}{2}\right)$, with $Z = 13\frac{1}{2}$.

　　Bound: $Z \leq 13\frac{1}{2}$.

Subproblem 6:
　　LP relaxation:　　　　　　　No feasible solutions.

Subproblem 6 is immediately fathomed by test 2. However, note that subproblem 5 also can be fathomed. Test 3 passes because the optimal solution for its LP relaxation has integer values ($x_1 = 0$, $x_2 = 0$, $x_3 = 2$) for all three integer-restricted variables. (It does not matter that $x_4 = \frac{1}{2}$, since x_4 is not integer-restricted.) This *feasible* solution for the original problem becomes our first incumbent:

$$\text{Incumbent} = \left(0, 0, 2, \frac{1}{2}\right) \quad \text{with } Z^* = 13\frac{1}{2}.$$

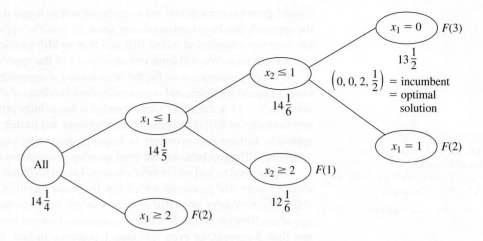

■ **FIGURE 11.13**
The branching tree after the final (third) iteration of the MIP branch-and-bound algorithm for the MIP example.

Using this Z^* to reapply fathoming test 1 to the only other subproblem (subproblem 4) is successful, because its bound $12\frac{1}{6} \leq Z^*$.

This iteration has succeeded in fathoming subproblems in all three possible ways. Furthermore, there now are no remaining subproblems, so the current incumbent is optimal.

$$\text{Optimal solution} = \left(0, 0, 2, \frac{1}{2}\right) \quad \text{with } Z = 13\frac{1}{2}.$$

These results are summarized by the final branching tree given in Fig. 11.13.

Another example of applying the MIP algorithm is presented in your OR Tutor. In addition, **a small example** (only two variables, both integer-restricted) that includes graphical displays is provided in the Worked Examples section of the book's website. The IOR Tutorial also includes an interactive procedure for executing the MIP algorithm.

■ 11.8 THE BRANCH-AND-CUT APPROACH TO SOLVING BIP PROBLEMS

Integer programming has been an especially exciting area of OR since the mid-1980s because of the dramatic progress being made in its solution methodology.

Background

To place this progress into perspective, consider the historical background. One big breakthrough had come in the 1960s and early 1970s with the development and refinement of the branch-and-bound approach. But then the state of the art seemed to hit a plateau. Relatively small problems (well under 100 variables) could be solved very efficiently, but even a modest increase in problem size might cause an explosion in computation time beyond feasible limits. Little progress was being made in overcoming this exponential growth in computation time as the problem size was increased. Many important problems arising in practice could not be solved.

Then came the next breakthrough in the mid-1980s, with the introduction of the *branch-and-cut approach* to solving BIP problems. There were early reports of very large problems with as many as a couple thousand variables being solved using this approach. This created great excitement and led to intensive research and development activities to refine the approach that have continued ever since. At first, the approach was limited to *pure* BIP, but soon was extended to *mixed* BIP, and then to MIP problems with some general integer variables as well. We will limit our description of the approach to the *pure* BIP case.

It is fairly common now for the branch-and-cut approach to solve some problems with many thousand variables, and occasionally even hundreds of thousands of variables. As mentioned in Sec. 11.4, this tremendous speedup is due to huge progress in three areas—dramatic improvements in BIP algorithms by incorporating and further developing the branch-and-cut approach, striking improvements in linear programming algorithms that are heavily used within the BIP algorithms, and the great speedup in computers (including desktop computers).

We do need to add one note of caution. This algorithmic approach cannot consistently solve *all* pure BIP problems with a few thousand variables, or even a few hundred variables. The very large pure BIP problems solved have *sparse* **A** matrices; i.e., the percentage of coefficients in the functional constraints that are *nonzeros* is quite small (perhaps less than 5 percent, or even less than 1 percent). In fact, the approach depends heavily upon this sparsity. (Fortunately, this kind of sparsity is typical in large practical problems.) Furthermore, there are other important factors besides sparsity and size that affect just how difficult a given IP problem will be to solve. IP formulations of fairly substantial size should still be approached with considerable caution.

Although it would be beyond the scope and level of this book to fully describe the algorithmic approach discussed above, we will now give a brief overview. Since this overview is limited to *pure* BIP, *all* variables introduced later in this section are *binary* variables.

The approach mainly uses a combination of three kinds[9] of techniques: *automatic problem preprocessing,* the *generation of cutting planes,* and clever *branch-and-bound* techniques. You already are familiar with branch-and-bound techniques, and we will not elaborate further on the more advanced versions incorporated here. An introduction to the other two kinds of techniques is given below.

Automatic Problem Preprocessing for Pure BIP

Automatic problem preprocessing involves a "computer inspection" of the user-supplied formulation of the IP problem in order to spot reformulations that make the problem quicker to solve without eliminating any feasible solutions. These reformulations fall into three categories:

1. *Fixing variables:* Identify variables that can be fixed at one of their possible values (either 0 or 1) because the other value cannot possibly be part of a solution that is both feasible and optimal.
2. *Eliminating redundant constraints:* Identify and eliminate *redundant constraints* (constraints that automatically are satisfied by solutions that satisfy all the other constraints).
3. *Tightening constraints:* Tighten some constraints in a way that reduces the feasible region for the LP relaxation without eliminating any feasible solutions for the BIP problem.

These categories are described in turn.

Fixing Variables.   One general principle for fixing variables is the following.

> If one value of a variable cannot satisfy a certain constraint, even when the other variables equal their best values for trying to satisfy the constraint, then that variable should be fixed at its other value.

---

[9]As discussed briefly in Sec. 11.4, still another technique that has played a significant role in the recent progress has been the use of *heuristics* for quickly finding good feasible solutions.

For example, *each* of the following $\leq$ constraints would enable us to fix x_1 at $x_1 = 0$, since $x_1 = 1$ with the best values of the other variables (0 with a nonnegative coefficient and 1 with a negative coefficient) would violate the constraint.

$$3x_1 \leq 2 \quad \Rightarrow \quad x_1 = 0, \quad \text{since} \quad 3(1) > 2.$$
$$3x_1 + x_2 \leq 2 \quad \Rightarrow \quad x_1 = 0, \quad \text{since} \quad 3(1) + 1(0) > 2.$$
$$5x_1 + x_2 - 2x_3 \leq 2 \quad \Rightarrow \quad x_1 = 0, \quad \text{since} \quad 5(1) + 1(0) - 2(1) > 2.$$

The general procedure for checking any $\leq$ constraint is to identify the variable with the *largest positive coefficient,* and if the *sum* of *that coefficient* and any *negative coefficients* exceeds the right-hand side, then that variable should be fixed at 0. (Once the variable has been fixed, the procedure can be repeated for the variable with the next largest positive coefficient, etc.)

An analogous procedure with $\geq$ constraints can enable us to fix a variable at 1 instead, as illustrated below three times.

$$3x_1 \geq 2 \quad \Rightarrow \quad x_1 = 1, \quad \text{since} \quad 3(0) < 2.$$
$$3x_1 + x_2 \geq 2 \quad \Rightarrow \quad x_1 = 1, \quad \text{since} \quad 3(0) + 1(1) < 2.$$
$$3x_1 + x_2 - 2x_3 \geq 2 \quad \Rightarrow \quad x_1 = 1, \quad \text{since} \quad 3(0) + 1(1) - 2(0) < 2.$$

A $\geq$ constraint also can enable us to fix a variable at 0, as illustrated next.

$$x_1 + x_2 - 2x_3 \geq 1 \quad \Rightarrow \quad x_3 = 0, \quad \text{since} \quad 1(1) + 1(1) - 2(1) < 1.$$

The next example shows a $\geq$ constraint fixing one variable at 1 and another at 0.

$$3x_1 + x_2 - 3x_3 \geq 2 \quad \Rightarrow \quad x_1 = 1, \quad \text{since} \quad 3(0) + 1(1) - 3(0) < 2$$
$$\textit{and} \quad \Rightarrow \quad x_3 = 0, \quad \text{since} \quad 3(1) + 1(1) - 3(1) < 2.$$

Similarly, a $\leq$ constraint with a *negative* right-hand side can result in either 0 or 1 becoming the fixed value of a variable. For example, both happen with the following constraint.

$$3x_1 - 2x_2 \leq -1 \quad \Rightarrow \quad x_1 = 0, \quad \text{since} \quad 3(1) - 2(1) > -1$$
$$\textit{and} \quad \Rightarrow \quad x_2 = 1, \quad \text{since} \quad 3(0) - 2(0) > -1.$$

Fixing a variable from one constraint can sometimes generate a chain reaction of then being able to fix other variables from other constraints. For example, look at what happens with the following three constraints.

$$3x_1 + x_2 - 2x_3 \geq 2 \quad \Rightarrow \quad x_1 = 1 \quad \text{(as above)}.$$

Then

$$x_1 + x_4 + x_5 \leq 1 \quad \Rightarrow \quad x_4 = 0, \quad x_5 = 0.$$

Then

$$-x_5 + x_6 \leq 0 \quad \Rightarrow \quad x_6 = 0.$$

In some cases, it is possible to combine one or more *mutually exclusive alternatives* constraints with another constraint to fix a variable, as illustrated below,

$$\left. \begin{array}{r} 8x_1 - 4x_2 - 5x_3 + 3x_4 \leq 2 \\ x_2 + x_3 \leq 1 \end{array} \right\} \quad \Rightarrow \quad x_1 = 0,$$

$$\text{since} \quad 8(1) - \max\{4, 5\}(1) + 3(0) > 2.$$

There are additional techniques for fixing variables, including some involving optimality considerations, but we will not delve further into this topic.

Fixing variables can have a dramatic impact on reducing the size of a problem. It is not unusual to eliminate over half of the problem's variables from further consideration.

Eliminating Redundant Constraints.   Here is one easy way to detect a redundant constraint.

> If a functional constraint satisfies even the most challenging binary solution, then it has been made redundant by the binary constraints and can be eliminated from further consideration. For a $\leq$ constraint, the most challenging binary solution has variables equal to 1 when they have nonnegative coefficients and other variables equal to 0. (Reverse these values for a $\geq$ constraint.)

Some examples are given below.

$$3x_1 + 2x_2 \leq 6 \qquad \text{is redundant, since } 3(1) + 2(1) \leq 6.$$
$$3x_1 - 2x_2 \leq 3 \qquad \text{is redundant, since } 3(1) - 2(0) \leq 3.$$
$$3x_1 - 2x_2 \geq -3 \qquad \text{is redundant, since } 3(0) - 2(1) \geq -3.$$

In most cases where a constraint has been identified as redundant, it was not redundant in the original model but became so after fixing some variables. Of the 11 examples of fixing variables given above, *all* but the last one left a constraint that then was redundant.

Tightening Constraints.[10]   Consider the following problem.

$$\text{Maximize} \qquad Z = 3x_1 + 2x_2,$$

subject to

$$2x_1 + 3x_2 \leq 4$$

and

$$x_1, x_2 \text{ binary.}$$

This BIP problem has just three feasible solutions—(0, 0), (1, 0), and (0, 1)—where the optimal solution is (1, 0) with $Z = 3$. The feasible region for the LP relaxation of this problem is shown in Fig. 11.14. The optimal solution for this LP relaxation is $(1, \frac{2}{3})$

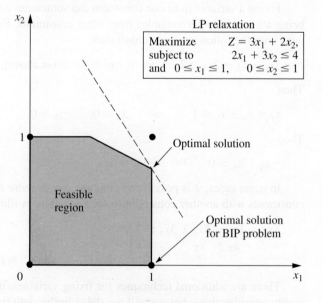

---

[10]Also commonly called *coefficient reduction*.

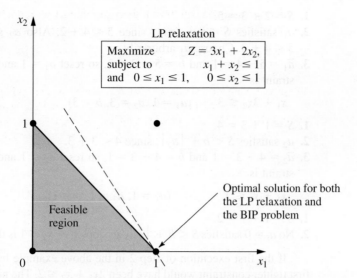

FIGURE 11.15
The LP relaxation after tightening the constraint, $2x_1 + 3x_2 \leq 4$, to $x_1 + x_2 \leq 1$ for the example of Fig. 11.14.

with $Z = 4\frac{1}{3}$, which is not very close to the optimal solution for the BIP problem. A branch-and-bound algorithm would have some work to do to identify the optimal BIP solution.

Now look what happens when the functional constraint $2x_1 + 3x_2 \leq 4$ is replaced by

$$x_1 + x_2 \leq 1.$$

The feasible solutions for the BIP problem remain exactly the same—(0, 0), (1, 0), and (0, 1)—so the optimal solution still is (1, 0). However, the feasible region for the LP relaxation has been greatly reduced, as shown in Fig. 11.15. In fact, this feasible region has been reduced so much that the optimal solution for the LP relaxation now is (1, 0), so the optimal solution for the BIP problem has been found without needing any additional work.

This is an example of tightening a constraint in a way that reduces the feasible region for the LP relaxation without eliminating any feasible solutions for the BIP problem. It was easy to do for this tiny two-variable problem that could be displayed graphically. However, with application of the same principles for tightening a constraint without eliminating any feasible BIP solutions, the following algebraic procedure can be used to do this for any $\leq$ constraint with any number of variables.

Procedure for Tightening a $\leq$ Constraint
Denote the constraint by $a_1x_1 + a_2x_2 + \cdots + a_nx_n \leq b$.

1. Calculate S = sum of the *positive* a_j.
2. Identify any $a_j \neq 0$ such that $S < b + \left| a_j \right|$.
   (a) If none, stop; the constraint cannot be tightened further.
   (b) If $a_j > 0$, go to step 3.
   (c) If $a_j < 0$, go to step 4.
3. ($a_j > 0$)   Calculate $\bar{a}_j = S - b$ and $\bar{b} = S - a_j$. Reset $a_j = \bar{a}_j$ *and* $b = \bar{b}$. Return to step 1.
4. ($a_j < 0$)   Increase a_j to $a_j = b - S$. Return to step 1.

Applying this procedure to the functional constraint in the above example flows as follows:
The constraint is $2x_1 + 3x_2 \leq 4$ ($a_1 = 2$, $a_2 = 3$, $b = 4$).

1. $S = 2 + 3 = 5$.
2. a_1 satisfies $S < b + |a_1|$, since $5 < 4 + 2$. Also a_2 satisfies $S < b + |a_2|$, since $5 < 4 + 3$. Choose a_1 arbitrarily.
3. $\bar{a}_1 = 5 - 4 = 1$ and $b = 5 - 2 = 3$, so reset $a_1 = 1$ and $b = 3$. The new tighter constraint is

$$x_1 + 3x_2 \leq 3 \qquad (a_1 = 1, a_2 = 3, b = 3).$$

1. $S = 1 + 3 = 4$.
2. a_2 satisfies $S < b + |a_2|$, since $4 < 3 + 3$.
3. $\bar{a}_2 = 4 - 3 = 1$ and $b = 4 - 3 = 1$, so reset $a_2 = 1$ and $b = 1$. The new tighter constraint is

$$x_1 + x_2 \leq 1 \qquad (a_1 = 1, a_2 = 1, b = 1).$$

1. $S = 1 + 1 = 2$.
2. No $a_j \neq 0$ satisfies $S < b + |a_j|$, so stop; $x_1 + x_2 \leq 1$ is the desired tightened constraint.

If the first execution of step 2 in the above example had chosen a_2 instead, then the first tighter constraint would have been $2x_1 + x_2 \leq 2$. The next series of steps again would have led to $x_1 + x_2 \leq 1$.

In the next example, the procedure tightens the constraint on the left to become the one on its right and then tightens further to become the second one on the right.

$$4x_1 - 3x_2 + x_3 + 2x_4 \leq 5 \quad \Rightarrow \quad 2x_1 - 3x_2 + x_3 + 2x_4 \leq 3$$
$$\Rightarrow \quad 2x_1 - 2x_2 + x_3 + 2x_4 \leq 3.$$

(Problem 11.8-5 asks you to apply the procedure to confirm these results.)

A constraint in $\geq$ form can be converted to $\leq$ form (by multiplying through both sides by -1) to apply this procedure directly.

Generating Cutting Planes for Pure BIP

A **cutting plane** (or **cut**) for any IP problem is a new functional constraint that reduces the feasible region for the LP relaxation without eliminating any feasible solutions for the IP problem. In fact, you have just seen one way of generating cutting planes for pure BIP problems, namely, apply the above procedure for tightening constraints. Thus, $x_1 + x_2 \leq 1$ is a cutting plane for the BIP problem considered in Fig. 11.14, which leads to the reduced feasible region for the LP relaxation shown in Fig. 11.15.

In addition to this procedure, a number of other techniques have been developed for generating cutting planes that will tend to accelerate how quickly a branch-and-bound algorithm can find an optimal solution for a pure BIP problem. We will focus on just one of these techniques.

To illustrate this technique, consider the California Manufacturing Co. pure BIP problem presented in Sec. 11.1 and used to illustrate the BIP branch-and-bound algorithm in Sec. 11.6. The optimal solution for its LP relaxation is given in Fig. 11.5 as $(x_1, x_2, x_3, x_4) = (\frac{5}{6}, 1, 0, 1)$. One of the functional constraints is

$$6x_1 + 3x_2 + 5x_3 + 2x_4 \leq 10.$$

Now note that the binary constraints and this constraint together imply that

$$x_1 + x_2 + x_4 \leq 2.$$

This new constraint is a *cutting plane*. It eliminates part of the feasible region for the LP relaxation, including what had been the optimal solution, $(\frac{5}{6}, 1, 0, 1)$, but it does not eliminate any feasible *integer* solutions. Adding just this one cutting plane to the original model would improve the performance of the BIP branch-and-bound algorithm in Sec. 11.6 (see Fig. 11.9) in two ways. First, the optimal solution for the new (tighter) LP relaxation would be $(1, 1, \frac{1}{5}, 0)$, with $Z = 15\frac{1}{5}$, so the bounds for the *All* node, $x_1 = 1$ node, and $(x_1, x_2) = (1, 1)$ node now would be 15 instead of 16. Second, one less iteration would be needed because the optimal solution for the LP relaxation at the $(x_1, x_2, x_3) = (1, 1, 0)$ node now would be $(1, 1, 0, 0)$, which provides a new *incumbent* with $Z^* = 14$. Therefore, on the *third* iteration (see Fig. 11.8), this node would be fathomed by test 3, and the $(x_1, x_2) = (1, 0)$ node would be fathomed by test 1, thereby revealing that this incumbent is the optimal solution for the original BIP problem.

Here is the general procedure used to generate this cutting plane.

A Procedure for Generating Cutting Planes

1. Consider any functional constraint in $\leq$ form with only nonnegative coefficients.
2. Find a group of variables (called a **minimum cover** of the constraint) such that
   (a) The constraint is violated if every variable in the group equals 1 and all other variables equal 0.
   (b) But the constraint becomes satisfied if the value of *any one* of these variables is changed from 1 to 0.
3. By letting N denote the number of variables in the group, the resulting cutting plane has the form

   Sum of variables in group $\leq N - 1$.

Applying this procedure to the constraint $6x_1 + 3x_2 + 5x_3 + 2x_4 \leq 10$, we see that the group of variables $\{x_1, x_2, x_4\}$ is a *minimal cover* because

   (a) $(1, 1, 0, 1)$ violates the constraint.
   (b) But the constraint becomes satisfied if the value of *any one* of these three variables is changed from 1 to 0.

Since $N = 3$ in this case, the resulting cutting plane is $x_1 + x_2 + x_4 \leq 2$.

This same constraint also has a second minimal cover $\{x_1, x_3\}$, since $(1, 0, 1, 0)$ violates the constraint but both $(0, 0, 1, 0)$ and $(1, 0, 0, 0)$ satisfy the constraint. Therefore, $x_1 + x_3 \leq 1$ is another valid cutting plane.

The branch-and-cut approach involves generating *many* cutting planes in a similar manner before then applying clever branch-and-bound techniques. The results of including the cutting planes can be quite dramatic in tightening the LP relaxations. In some cases, the *gap* between Z for the optimal solution for the LP relaxation of the whole BIP problem and Z for this problem's optimal solution is reduced by as much as 98 percent.

Ironically, the very first algorithms developed for integer programming, including Ralph Gomory's celebrated algorithm announced in 1958, were based on cutting planes (generated in a different way), but this approach proved to be unsatisfactory in practice (except for special classes of problems). However, these algorithms relied solely on cutting planes. We now know that judiciously *combining* cutting planes and branch-and-bound techniques (along with automatic problem preprocessing) provides a powerful algorithmic approach for solving large-scale BIP problems. This is one reason that the name *branch-and-cut algorithm* has been given to this approach.

■ 11.9 THE INCORPORATION OF CONSTRAINT PROGRAMMING

No presentation of the basic ideas of integer programming is complete these days without introducing an exciting recent development—the incorporation of the techniques of *constraint programming*—that is promising to greatly expand our ability to formulate and solve integer programming models. (These same techniques also are beginning to be used in related areas of mathematical programming, especially combinatorial optimization, but we will limit our discussion to their central use in integer programming.)

The Nature of Constraint Programming

In the mid-1980s, researchers in the computer science community began to develop constraint programming by combining ideas in artificial intelligence with the development of computer programming languages. The goal was to have a flexible computer programming system that would include both *variables* and *constraints* on their values, while also allowing the description of search procedures that would generate feasible values of the variables. Each variable has a *domain* of possible values, e.g., {2, 4, 6, 8, 10}. Rather than being limited to the types of mathematical constraints used in mathematical programming, there is great flexibility in how to state the constraints. In particular, the constraints can be any of the following types.

1. Mathematical constraints, e.g., $x + y < z$.
2. Disjunctive constraints, e.g., the times of certain tasks in the problem being modeled cannot overlap.
3. Relational constraints, e.g., at least three tasks should be assigned to a certain machine.
4. Explicit constraints, e.g., although both x and y have domains {1, 2, 3, 4, 5}, (x, y) must be (1, 1), (2, 3), or (4, 5).
5. Unary constraints, e.g., z is an integer between 5 and 10.
6. Logical constraints, e.g., if x is 5, then y is between 6 and 8.

When expressing these kinds of constraints, constraint programming allows the use of various standard logic functions, such as IF, AND, OR, NOT, and so on. Excel includes many of the same logic functions. LINGO now supports all the standard logic functions and can use its global optimizer to find a globally optimal solution.

To illustrate the algorithms that constraint programming uses to generate feasible solutions, suppose that a problem has four variables—x_1, x_2, x_3, x_4—and their domains are

$$x_1 \in \{1, 2\}, x_2 \in \{1, 2\}, x_3 \in \{1, 2, 3\}, x_4 \in \{1, 2, 3, 4, 5\},$$

where the symbol $\in$ signifies that the variable on the left belongs to the set on the right. Suppose also that the constraints are

(1) All these variables must have different values,
(2) $x_1 + x_3 = 4$.

By straightforward logic, since the values of 1 and 2 must be reserved for x_1 and x_2, the first constraint immediately implies that $x_3 \in \{3\}$, which then implies that $x_4 \in \{4, 5\}$. (This process of eliminating possible values for variables is referred to as *domain reduction*.) Next, since the domain of x_3 has been changed, the process of *constraint propagation* applies the second constraint to imply that $x_1 \in \{1\}$. This again triggers the first constraint, so that

$$x_1 \in \{1\}, \quad x_2 \in \{2\}, \quad x_3 \in \{3\}, \quad x_4 \in \{4, 5\}$$

lists the only feasible solutions for the problem. This kind of *feasibility reasoning* based on alternating between the application of domain reduction and constraint propagation algorithms is a key part of constraint programming.

After the application of the constraint propagation and domain reduction algorithms to a problem, a search procedure is used to find complete feasible solutions. In the example above, since the domains of all the variables have been reduced to a single value except for x_4, the search procedure would simply try the values $x_4 = 4$ and $x_4 = 5$ to determine the complete feasible solutions for that problem. However, for a problem with many constraints and variables, the constraint propagation and domain reduction algorithms typically do not reduce the domain of each variable to a single value. It is therefore necessary to write a search procedure that will try different assignments of values to the variables. As these assignments are tried, the constraint propagation algorithm is triggered and further domain reduction occurs. The process creates a *search tree,* which is similar to the branching tree when applying the branch-and-bound technique to integer programming.

The overall process of applying constraint programming to complicated IP problems (or related problems) involves the following three steps.

1. Formulate a compact model for the problem by using a variety of constraint types (most of which do not fit the format of integer programming).
2. Efficiently find feasible solutions that satisfy all these constraints.
3. Search among these feasible solutions for an optimal solution.

The power of constraint programming lies in its great ability to perform the first two steps rather than the third, whereas the main strength of integer programming and its algorithms lie in performing the third step. Thus, constraint programming is ideally suited for a highly constrained problem that has no objective function, so the only goal is to find a feasible solution. However, it also can be extended to the third step. One method of doing so is to *enumerate* the feasible solutions and calculate the value of the objective function for each one. However, this would be extremely inefficient for problems where there are numerous feasible solutions. To circumvent this drawback, the common approach is to add a constraint that tightly bounds the objective function to values that are very near to what is anticipated for an optimal solution. For example, if the objective is to *maximize* the objective function and its value Z is anticipated to be approximately $Z = 10$ for an optimal solution, one might add the constraint that $Z \geq 9$ so that the only remaining feasible solutions to be enumerated are those that are very close to being optimal. Each time that a new best solution then is found during the search, the bound on Z can be further tightened to consider only feasible solutions that are at least as good as the current best solution.

Although this is a reasonable approach to the third step, a more attractive approach would be to integrate constraint programming and integer programming so that each is mainly used where it is strongest—steps 1 and 2 with constraint programming and step 3 with integer programming. This is part of the potential of constraint programming described next.

The Potential of Constraint Programming

In the 1990s, constraint programming features, including powerful constraint-solving algorithms, were successfully incorporated into a number of general-purpose programming languages, as well as several special-purpose programming languages. This brought computer science closer and closer to the Holy Grail of computer programming, namely, allowing the user to simply state the problem and then the computer will solve it.

As word of this exciting development began to spread beyond the computer science community, researchers in operations research began to realize the great potential of integrating constraint programming with the traditional techniques of integer programming (and other areas of mathematical programming as well). The much greater flexibility in expressing the constraints of the problem should greatly increase the ability to formulate valid models for complex problems. It also should lead to much more compact and straightforward formulations. In addition, by reducing the size of the feasible region that needs to be considered while efficiently finding solutions within this region, the constraint-solving algorithms of constraint programming might help accelerate the progress of integer programming algorithms in finding an optimal solution.

Because of their substantial differences, integrating constraint programming with integer programming is a very difficult task. Since integer programming does not recognize most of the constraints of constraint programming, this requires developing computer-implemented procedures for translating from the language of constraint programming to the language of integer programming and vice versa. Good progress is being made, but this undoubtedly will continue to be one of the most active areas of OR research for some years to come.

To illustrate the way in which constraint programming can greatly simplify the formulation of integer programming models, we now will introduce two of the most important "global constraints" of constraint programming. A **global constraint** is a constraint that succinctly expresses a global pattern in the allowable relationship between multiple variables. Therefore, a single global constraint often can replace what used to require a large number of traditional integer programming constraints while also making the model considerably more readable. To clarify the presentation, we will use very simple examples that don't require the use of constraint programming to illustrate global constraints, but these same types of constraints also can readily be used for some much more complicated problems.

The All-Different Constraint

The *all-different* global constraint simply specifies that all the variables in a given set must have different values. If $x_1, x_2, \ldots, x_n$ are the variables involved, the constraint can be written succinctly as

$$\text{all-different } (x_1, x_2, \ldots, x_n)$$

while also specifying the domains of the individual variables in the model. (These domains collectively need to include at least n different values in order to enforce the all-different constraint.)

To illustrate this constraint, consider the classical *assignment problem* presented in Sec. 8.3. Recall that this problem involves assigning n assignees to n tasks on a one-to-one basis so as to minimize the total cost of these assignments. Although the assignment problem is a particularly easy one to solve (as described in Sec. 8.4), it nicely illustrates how the all-different constraint can greatly simplify the formulation of the model.

With the traditional formulation presented in Sec. 8.3, the decision variables are the binary variables,

$$x_{ij} = \begin{cases} 1, & \text{if assignee } i \text{ performs task } j \\ 0, & \text{if not} \end{cases}$$

for $i, j = 1, 2, \ldots, n$. Ignoring the objective function for now, the functional constraints are the following.

Each assignee i is to be assigned to exactly *one* task:

$$\sum_{j=1}^{n} x_{ij} = 1 \qquad \text{for } i = 1, 2, \ldots, n.$$

Each task j is to be performed by exactly *one* assignee:

$$\sum_{i=1}^{n} x_{ij} = 1 \quad \text{for } j = 1, 2, \ldots, n.$$

Thus, there are n^2 variables and $2n$ functional constraints.

Now let us look at the much smaller model that constraint programming can provide. In this case, the variables are

$$y_i = \text{task to which assignee } i \text{ is assigned}$$

for $i = 1, 2, \ldots, n$. There are n tasks and they are numbered $1, 2, \ldots, n$, so each of the y_i variables has the domain $\{1, 2, \ldots, n\}$. Since all the assignees must be assigned different tasks, this restriction on the variables is precisely described by the single global constraint,

$$\text{all-different } (y_1, y_2, \ldots, y_n).$$

Therefore, rather than n^2 variables and $2n$ functional constraints, this complete constraint programming model (excluding the objective function) has only n variables and a *single* constraint (plus one domain for all the variables).

Now let us see how the next global constraint enables incorporating the objective function into this tiny model as well.

The Element Constraint

The *element* global constraint is most commonly used to look up a cost or profit associated with an integer variable. In particular, suppose that a variable y has domain $\{1, 2, \ldots, n\}$ and that the cost associated with each of these values is $c_1, c_2, \ldots, c_n$, respectively. Then the constraint

$$\text{element } (y, [c_1, c_2, \ldots, c_n], z)$$

constrains the variable z to equal the yth constant in the list $[c_1, c_2, \ldots, c_n]$. In other words, $z = c_y$. This variable z can now be included in the objective function to provide the cost associated with y.

To illustrate the use of the element constraint, consider the assignment problem again and let

$$c_{ij} = \text{cost of assigning assignee } i \text{ to task } j$$

for $i, j, = 1, 2, \ldots, n$. The complete constraint programming model (including the objective function for this problem is

$$\text{Minimize } Z = \sum_{i=1}^{n} z_i,$$

subject to

$$\text{element } (y_i, [c_{i1}, c_{i2}, \ldots, c_{in}], z_i) \quad \text{for } i = 1, 2, \ldots, n,$$
$$\text{all-different } (y_1, y_2, \ldots, y_n),$$
$$y_i \in \{1, 2, \ldots, n\} \quad \text{for } i = 1, 2, \ldots, n.$$

This complete model now has $2n$ variables and $(n + 1)$ constraints (plus the one domain for all the variables), which still is far smaller than the traditional integer programming formulation presented in Sec. 8.3. For example, when $n = 100$, this model has 200 variables and 101 constraints whereas the traditional integer programming model has 10,000 variables and 200 functional constraints.

As an additional example, reconsider Example 2 (Violating Proportionality) presented in Sec. 11.4. In this case, the original decision variables are

x_j = number of TV spots allocated to product j

for $j = 1, 2, 3$, where a total of five TV spots are to be allocated to the three products. However, because the profits given in Table 11.3 for different values of each x_j are not proportional to x_j, Sec. 11.4 formulates two alternative integer programming models with auxiliary binary variables for this problem. Both models are fairly complicated.

A constraint programming model that uses the element constraint is much more straightforward. For example, the profit for Product 1 given in Table 11.3 is 0, 1, 3, and 3 for $x_1 = 0, 1, 2$, and 3, respectively. Therefore, this profit is simply z_1 when the value of z_1 is given by the constraint

element $(x_1 + 1, [0, 1, 3, 3], z_1)$.

(The first component is $x_1 + 1$ instead of x_1 because $x_1 + 1 = 1, 2, 3$, or 4, and it is the value of this component that indicates the choice of position 1, 2, 3, or 4 in the list [0, 1, 3, 3].) Proceeding in the same way for the other two products, the complete model is

Maximize $Z = z_1 + z_2 + z_3$,

subject to

element $(x_1 + 1, [0, 1, 3, 3], z_1)$,
element $(x_2 + 1, [0, 0, 2, 3], z_2)$,
element $(x_3 + 1, [0, -1, 2, 4], z_3)$,
$x_1 + x_2 + x_3 = 5$,
$x_j \in \{0, 1, 2, 3\}$ for $j = 1, 2, 3$.

Now compare this model to the two integer programming models for the same problem in Sec. 11.4. Note how the use of element constraints provides a considerably more compact and transparent model.

The *all-different* and *element* constraints are but two of the various available global constraints (Selected Reference 6 describes nearly 40), but they nicely illustrate the power of constraint programming to provide a compact and readable model of a complex problem.

Current Research

Current research in integrating constraint programming and integer programming is moving along several parallel paths. For example, the most straightforward approach is to simultaneously use both a constraint programming model and an integer programming model to represent complementary parts of a problem. Thus, each relevant constraint is included in whichever model it fits or, when feasible, in both models. As a constraint programming algorithm and an integer programming algorithm are applied to the respective models, information is passed back and forth to focus the search on the feasible solutions (those that satisfy the constraints of both models).

This kind of double modeling scheme can be implemented with the Optimization Programming Language (OPL) that is incorporated into the ILOG OPL-CPLEX Development System. (ILOG is the company that provides the CPLEX optimization software that is included in your OR Courseware.) After employing the OPL modeling language, the ILOG OPL-CPLEX Development System can invoke both a constraint programming algorithm (ILOG CP Optimizer) and a mathematical programming solver (CPLEX) and then pass some information from one to the other.

Although double modeling is a good first step, the goal is to fully integrate constraint programming and integer programming so that a single hybrid model and a single algorithm can be used. It is this kind of seamless integration that will be able to fully provide the complementary strengths of both techniques. Although fully achieving this goal remains a formidable research challenge, good progress continues to be made in this direction. Selected Reference 6 describes the current state of the art in this area.

Even at this early stage, there already have been numerous successful applications of the merger of mathematical programming and constraint programming. The areas of application include network design, vehicle routing, crew rostering, the classical transportation problem with piecewise linear costs, inventory management, computer graphics, software engineering, databases, finance, engineering, and combinatorial optimization, among others. In addition, Selected Reference 2 describes how scheduling is proving to be a particularly fruitful area for the application of constraint programming. For example, because of the many complicated scheduling constraints involved, constraint programming has been used to determine the regular-season schedule for the National Football League in the United States.

These applications only begin to tap the potential of integrating constraint programming and integer programming. Further progress in completing this integration promises to open up many exciting new opportunities for important applications.

■ 11.10 CONCLUSIONS

IP problems arise frequently because some or all of the decision variables must be restricted to integer values. There also are many applications involving yes-or-no decisions (including combinatorial relationships expressible in terms of such decisions) that can be represented by binary (0–1) variables. These factors have made integer programming one of the most widely used OR techniques.

IP problems are more difficult than they would be without the integer restriction, so the algorithms available for integer programming are generally considerably less efficient than the simplex method. However, there has been tremendous progress over the past couple of decades in the ability to solve some (but not all) huge IP problems with tens or even hundreds of thousands of integer variables. This progress is due to a combination of three factors—dramatic improvements in IP algorithms, striking improvement in the linear programming algorithms used within IP algorithms, and the great speedup in computers. However, IP algorithms also will occasionally still fail to solve rather small problems (even as few as a hundred integer variables). Various characteristics of an IP problem in addition to its size, have a great influence on how readily it can be solved.

Nevertheless, size is one key factor in determining the time required to solve an IP problem, if it can be solved at all. The most important determinants of computation time for an IP algorithm are the *number of integer variables* and whether the problem has some *special structure* that can be exploited. For a fixed number of integer variables, BIP problems generally are much easier to solve than problems with general integer variables, but adding continuous variables (MIP) may not increase computation time substantially. For special types of BIP problems containing a special structure that can be exploited by a *special-purpose algorithm,* it may be possible to solve very large problems (thousands of binary variables) routinely.

Computer codes for IP algorithms now are commonly available in mathematical programming software packages. Traditionally, these algorithms usually have been based on the *branch-and-bound* technique and variations thereof.

More modern IP algorithms now use the *branch-and-cut* approach. This algorithmic approach involves combining automatic problem preprocessing, the generation of cutting planes, and clever branch-and-bound techniques. Research in this area is continuing, along with the development of sophisticated new software packages that incorporate these techniques.

The latest development in IP methodology is to begin incorporating *constraint programming*. It appears that this approach will greatly expand our ability to formulate and solve IP models.

In recent years, there has been considerable investigation into the development of algorithms (including heuristic algorithms) for integer *nonlinear* programming, and this area continues to be an active area of research. (Selected Reference 8 presents the current state of the art in this area.)

■ SELECTED REFERENCES

1. Appa, G., L. Pitsoulis, and H. P. Williams (eds.): *Handbook on Modelling for Discrete Optimization*, Springer, New York, 2006.
2. Baptiste, P., C. LePape, and W. Nuijten: *Constraint-Based Scheduling: Applying Constraint Programming to Scheduling Problems*, Kluwer Academic Publishers (now Springer), Boston, 2001.
3. Barnhart, C., P. Belobaba, and A. R. Odoni: "Applications of Operations Research in the Air Transport Industry," *Transportation Science*, **37**(4): 368–391, 2003.
4. Bixby, R. E., Z. Gu, E. Rothberg, and R. Wunderling: "Mixed Integer Programming: A Progress Report," pp.309–326 in M. Grötschel (ed.), *The Sharpest Cut: The Impact of Manfred Padberg and His Work*, MPS/SIAM Series on Optimization.
5. Hillier, F. S., and M. S. Hillier: *Introduction to Management Science: A Modeling and Case Studies Approach with Spreadsheets,* 3rd ed., McGraw-Hill/Irwin, Burr Ridge, IL, 2008, chap. 7.
6. Hooker, J. N.: *Integrated Methods for Optimization*, Springer, New York, 2007.
7. Karlof, J. K.: *Integer Programming: Theory and Practice*, CRC Press, Boca Raton, FL, 2006.
8. Li, D., and X. Sun: *Nonlinear Integer Programming*, Springer, New York, 2006.
9. Lübbecke, M. E., and J. Desrosiers: "Selected Topics in Column Generation," *Operations Research*, **53**(6): 1007–1023, November–December 2005.
10. Lustig, I., and J.-F. Puget: "Program Does Not Equal Program: Constraint Programming and Its Relationship to Mathematical Programming," *Interfaces,* **31**(6): 29–53, November–December 2001.
11. Nemhauser, G. L.: "Need and Potential for Real-Time Mixed Integer Programming," *OR/MS Today*, **34**(1): 21–22, February 2007.
12. Nemhauser, G. L., and L. A. Wolsey: *Integer and Combinatorial Optimization*, Wiley, New York, 1988, reprinted in 1999.
13. Schriver, A.: *Theory of Linear and Integer Programming*, Wiley, New York, 1986.
14. Williams, H. P.: *Model Building in Mathematical Programming*, 4th ed., Wiley, New York, 1999.
15. Wolsey, L. A.: *Integer Programming*, Wiley, New York, 1998.
16. Wolsey, L. A.: "Strong Formulations for Mixed Integer Programs: Valid Inequalities and Extended Formulations," *Mathematical Programming Series B*, **97**(1–2): 423–447, 2003.

Some Award-Winning Applications of Integer Programming:

(A link to all these articles is provided on our website, www.mhhe.com/hillier.)

A1. Abbink, E., M. Fischetti, L. Kroon, G. Timmer, and M. Vromans: "Reinventing Crew Scheduling at Netherlands Railways," *Interfaces*, **35**(5): 393–401, September–October 2005.
A2. Armacost, A. P., C. Barnhart, K. A. Ware, and A. M. Wilson: "UPS Optimizes Its Air Network," *Interfaces*, **34**(1): 15–25, January–February 2004.
A3. Bertsimas, D., C. Darnell, and R. Soucy: "Portfolio Construction Through Mixed-Integer Programming at Grantham, Mayo, Van Otterloo and Company," *Interfaces*, **29**(1): 49–66, January–February 1999.

A4. Camm, J. D., T. E. Chorman, F. A. Dill, J. R. Evans, D. J. Sweeney, and G. W. Wegryn: "Blending OR/MS, Judgment, and GIS: Restructuring P&G's Supply Chain," *Interfaces*, **27**(1): 128–142, January–February 1997.

A5. Denton, B. T., J. Forrest, and R. J. Milne: "IBM Solves a Mixed-Integer Program to Optimize Its Semiconductor Supply Chain," *Interfaces*, **36**(5): 386–399, September–October 2006.

A6. Gendron, B.: "Scheduling Employees in Quebec's Liquor Stores with Integer Programming," *Interfaces*, **35**(5): 402–410, September–October 2005.

A7. Gryffenberg, I., J. L. Lausberg, W. J. Smit, S. Uys, S. Botha, F. R. Hofmeyr, R. P. Nicolay, W. L. van der Merwe, and G. J. Wessells: "Guns or Butter: Decision Support for Determining the Size and Shape of the South African National Defense Force," *Interfaces*, **27**(1): 7–28, January–February 1997.

A8. Martin, C., D. Jones, and P. Keskinocak: "Optimizing On-Demand Aircraft Schedules for Fractional Aircraft Operators," *Interfaces*, **33**(5): 22–35. September–October 2003.

A9. Metty, T., R. Harlan, Q. Samelson, T. Moore, T. Morris, R. Sorenson, A. Scneur, O. Raskina, R. Schneur, J. Kanner, K. Potts, and J. Robbins: "Reinventing the Supplier Negotiation Process at Motorola," *Interfaces*, **35**(1), 7–23, January–February 2005.

A10. Smith, B. C., R. Darrow, J. Elieson, D. Guenther, B. V. Rao, and F. Zouaoui: "Travelocity Becomes a Travel Retailer," *Interfaces*, **37**(1): 68–81, January–February 2007.

A11. Spencer III, T., A. J. Brigandi, D. R. Dargon, and M. J. Sheehan: "AT&T's Telemarketing Site Selection System Offers Customer Support," *Interfaces*, **20**(1): 83–96, January–February 1990.

A12. Subramanian, R., R. P. Scheff, Jr., J. D. Quillinan, D. S. Wiper, and R. E. Marsten: "Coldstart: Fleet Assignment at Delta Air Lines," *Interfaces*, **24**(1): 104–120, January–February 1994.

A13. Tyagi, R., and S. Bollapragada: "SES Americom Maximizes Satellite Revenues by Optimally Configuring Transponders," *Interfaces*, **33**(5): 36–44, September–October 2003.

A14. Yu, G., M. Argüello, G. Song, S. M. McCowan, and A. White: "A New Era for Crew Recovery at Continental Airlines," *Interfaces*, **33**(1): 5–22, January–February 2003.

■ LEARNING AIDS FOR THIS CHAPTER ON OUR WEBSITE (www.mhhe.com/hillier)

Worked Examples:

Examples for Chapter 11

Demonstration Examples in OR Tutor:

Binary Integer Programming Branch-and-Bound Algorithm
Mixed Integer Programming Branch-and-Bound Algorithm

Interactive Procedures in IOR Tutorial:

Enter or Revise an Integer Programming Model
Solve Binary Integer Program Interactively
Solve Mixed Integer Program Interactively

An Excel Add-in:

Premium Solver for Education

"Ch. 11—Integer Programming" Files for Solving the Examples:

Excel Files
LINGO/LINDO File
MPL/CPLEX File

Glossary for Chapter 11

See Appendix 1 for documentation of the software.

■ PROBLEMS

The symbols to the left of some of the problems (or their parts) have the following meaning:

D: The corresponding demonstration example just listed in Learning Aids may be helpful.

I: We suggest that you use the corresponding interactive procedure just listed (the printout records your work).

C: Use the computer with any of the software options available to you (or as instructed by your instructor) to solve the problem.

An asterisk on the problem number indicates that at least a partial answer is given in the back of the book.

11.1-1. Reconsider the California Manufacturing Co. example presented in Sec. 11.1. The mayor of San Diego now has contacted the company's president to try to persuade him to build a factory and perhaps a warehouse in that city. With the tax incentives being offered the company, the president's staff estimates that the net present value of building a factory in San Diego would be $7 million and the amount of capital required to do this would be $4 million. The net present value of building a warehouse there would be $5 million and the capital required would be $3 million. (This option would be considered only if a factory also is being built there.)

The company president now wants the previous OR study revised to incorporate these new alternatives into the overall problem. The objective still is to find the feasible combination of investments that maximizes the total net present value, given that the amount of capital available for these investments is $10 million.
(a) Formulate a BIP model for this problem.
(b) Display this model on an Excel spreadsheet.
C (c) Use the computer to solve this model.

11.1-2* A young couple, Eve and Steven, want to divide their main household chores (marketing, cooking, dishwashing, and laundering) between them so that each has two tasks but the total time they spend on household duties is kept to a minimum. Their efficiencies on these tasks differ, where the time each would need to perform the task is given by the following table:

	Time Needed per Week			
	Marketing	**Cooking**	**Dishwashing**	**Laundry**
Eve	4.5 hours	7.8 hours	3.6 hours	2.9 hours
Steven	4.9 hours	7.2 hours	4.3 hours	3.1 hours

(a) Formulate a BIP model for this problem.
(b) Display this model on an Excel spreadsheet.
C (c) Use the computer to solve this model.

11.1-3. A real estate development firm, Peterson and Johnson, is considering five possible development projects. The following table shows the estimated long-run profit (net present value) that each project would generate, as well as the amount of investment required to undertake the project, in units of millions of dollars.

	Development Project				
	1	**2**	**3**	**4**	**5**
Estimated profit	1	1.8	1.6	0.8	1.4
Capital required	6	12	10	4	8

The owners of the firm, Dave Peterson and Ron Johnson, have raised $20 million of investment capital for these projects. Dave and Ron now want to select the combination of projects that will maximize their total estimated long-run profit (net present value) without investing more that $20 million.
(a) Formulate a BIP model for this problem.
(b) Display this model on an Excel spreadsheet.
C (c) Use the computer to solve this model.

11.1-4. The board of directors of General Wheels Co. is considering six large capital investments. Each investment can be made only once. These investments differ in the estimated long-run profit (net present value) that they will generate as well as in the amount of capital required, as shown by the following table (in units of millions of dollars):

	Investment Opportunity					
	1	**2**	**3**	**4**	**5**	**6**
Estimated profit	15	12	16	18	9	11
Capital required	38	33	39	45	23	27

The total amount of capital available for these investments is $100 million. Investment opportunities 1 and 2 are mutually exclusive, and so are 3 and 4. Furthermore, neither 3 nor 4 can be undertaken unless one of the first two opportunities is undertaken. There are no such restrictions on investment opportunities 5 and 6. The objective is to select the combination of capital investments that will maximize the total estimated long-run profit (net present value).
(a) Formulate a BIP model for this problem.
C (b) Use the computer to solve this model.

11.1-5. Reconsider Prob. 8.3-4, where a swim team coach needs to assign swimmers to the different legs of a 200-yard medley relay team. Formulate a BIP model for this problem. Identify the groups of mutually exclusive alternatives in this formulation.

11.1-6. Vincent Cardoza is the owner and manager of a machine shop that does custom order work. This Wednesday afternoon, he has received calls from two customers who would like to place rush orders. One is a trailer hitch company which would like some custom-made heavy-duty tow bars. The other is a mini-car-carrier company which needs some customized stabilizer bars. Both customers would like as many as possible by the end of the week (two working days). Since both products would require the use of the same two machines, Vincent needs to decide and inform the customers this afternoon about how many of each product he will agree to make over the next two days.

Each tow bar requires 3.2 hours on machine 1 and 2 hours on machine 2. Each stabilizer bar requires 2.4 hours on machine 1 and 3 hours on machine 2. Machine 1 will be available for 16 hours over the next two days and machine 2 will be available for 15 hours. The profit for each tow bar produced would be $130 and the profit for each stabilizer bar produced would be $150.

Vincent now wants to determine the mix of these production quantities that will maximize the total profit.

(a) Formulate an IP model for this problem.

(b) Use a graphical approach to solve this model.

c (c) Use the computer to solve the model.

11.1-7. Reconsider Prob. 8.2-21 involving a contractor (Susan Meyer) who needs to arrange for hauling gravel from two pits to three building sites.

Susan now needs to hire the trucks (and their drivers) to do the hauling. Each truck can only be used to haul gravel from a single pit to a single site. In addition to the hauling and gravel costs specified in Prob. 8.2-21, there now is a fixed cost of $150 associated with hiring each truck. A truck can haul 5 tons, but it is not required to go full. For each combination of pit and site, there are now two decisions to be made: the number of trucks to be used and the amount of gravel to be hauled.

(a) Formulate an MIP model for this problem.

c (b) Use the computer to solve this model.

11.2-1. Read the referenced article that fully describes the OR study summarized in the application vignette presented in Sec. 11.2. Briefly describe how integer programming was applied in this study. Then list the various financial and nonfinancial benefits that resulted from this study.

11.2-2. Select one of the actual applications of BIP by a company or governmental agency mentioned in Sec. 11.2. Read the article describing the application in the referenced issue of *Interfaces*. Write a two-page summary of the application and its benefits.

11.2-3. Select three of the actual applications of BIP by a company or governmental agency mentioned in Sec. 11.2. Read the articles describing the applications in the referenced issues of *Interfaces*. For each one, write a one-page summary of the application and its benefits.

11.3-1.* The Research and Development Division of the Progressive Company has been developing four possible new product lines. Management must now make a decision as to which of these four products actually will be produced and at what levels. Therefore, an operations research study has been requested to find the most profitable product mix.

A substantial cost is associated with beginning the production of any product, as given in the first row of the following table. Management's objective is to find the product mix that maximizes the total profit (total net revenue minus start-up costs).

	Product			
	1	**2**	**3**	**4**
Start-up cost	$50,000	$40,000	$70,000	$60,000
Marginal revenue	$ 70	$ 60	$ 90	$ 80

Let the continuous decision variables x_1, x_2, x_3, and x_4 be the production levels of products 1, 2, 3, and 4, respectively. Management has imposed the following policy constraints on these variables:

1. No more than two of the products can be produced.
2. Either product 3 or 4 can be produced only if either product 1 or 2 is produced.
3. Either $\quad 5x_1 + 3x_2 + 6x_3 + 4x_4 \leq 6{,}000$

   or $\quad 4x_1 + 6x_2 + 3x_3 + 5x_4 \leq 6{,}000$.

(a) Introduce auxiliary binary variables to formulate a mixed BIP model for this problem.

c (b) Use the computer to solve this model.

11.3-2. Suppose that a mathematical model fits linear programming except for the restriction that $|x_1 - x_2| = 0$, or 3, or 6. Show how to reformulate this restriction to fit an MIP model.

11.3-3. Suppose that a mathematical model fits linear programming except for the restrictions that

1. At least one of the following two inequalities holds:

$$3x_1 - x_2 - x_3 + x_4 \leq 12$$
$$x_1 + x_2 + x_3 + x_4 \leq 15.$$

2. At least two of the following three inequalities holds:

$$2x_1 + 5x_2 - x_3 + x_4 \leq 30$$
$$-x_1 + 3x_2 + 5x_3 + x_4 \leq 40$$
$$3x_1 - x_2 + 3x_3 - x_4 \leq 60.$$

Show how to reformulate these restrictions to fit an MIP model.

11.3-4. The Toys-R-4-U Company has developed two new toys for possible inclusion in its product line for the upcoming Christmas season. Setting up the production facilities to begin production would cost $50,000 for toy 1 and $80,000 for toy 2. Once these costs are covered, the toys would generate a unit profit of $10 for toy 1 and $15 for toy 2.

The company has two factories that are capable of producing these toys. However, to avoid doubling the start-up costs, just one factory would be used, where the choice would be based on maximizing profit. For administrative reasons, the same factory would be used for both new toys if both are produced.

Toy 1 can be produced at the rate of 50 per hour in factory 1 and 40 per hour in factory 2. Toy 2 can be produced at the rate of 40 per hour in factory 1 and 25 per hour in factory 2. Factories 1 and 2, respectively, have 500 hours and 700 hours of production time available before Christmas that could be used to produce these toys.

It is not known whether these two toys would be continued after Christmas. Therefore, the problem is to determine how many units (if any) of each new toy should be produced before Christmas to maximize the total profit.

(a) Formulate an MIP model for this problem.

C (b) Use the computer to solve this model.

11.3-5.* Northeastern Airlines is considering the purchase of new long-, medium-, and short-range jet passenger airplanes. The purchase price would be $67 million for each long-range plane, $50 million for each medium-range plane, and $35 million for each short-range plane. The board of directors has authorized a maximum commitment of $1.5 billion for these purchases. Regardless of which airplanes are purchased, air travel of all distances is expected to be sufficiently large that these planes would be utilized at essentially maximum capacity. It is estimated that the net annual profit (after capital recovery costs are subtracted) would be $4.2 million per long-range plane, $3 million per medium-range plane, and $2.3 million per short-range plane.

It is predicted that enough trained pilots will be available to the company to crew 30 new airplanes. If only short-range planes were purchased, the maintenance facilities would be able to handle 40 new planes. However, each medium-range plane is equivalent to $1\frac{1}{3}$ short-range planes, and each long-range plane is equivalent to $1\frac{2}{3}$ short-range planes in terms of their use of the maintenance facilities.

The information given here was obtained by a preliminary analysis of the problem. A more detailed analysis will be conducted subsequently. However, using the preceding data as a first approximation, management wishes to know how many planes of each type should be purchased to maximize profit.

(a) Formulate an IP model for this problem.

C (b) Use the computer to solve this problem.

(c) Use a binary representation of the variables to reformulate the IP model in part (a) as a BIP problem.

C (d) Use the computer to solve the BIP model formulated in part (c). Then use this optimal solution to identify an optimal solution for the IP model formulated in part (a).

11.3-6. Consider the two-variable IP example discussed in Sec. 11.5 and illustrated in Fig. 11.3.

(a) Use a binary representation of the variables to reformulate this model as a BIP problem.

C (b) Use the computer to solve this BIP problem. Then use this optimal solution to identify an optimal solution for the original IP model.

11.3-7. The Fly-Right Airplane Company builds small jet airplanes to sell to corporations for the use of their executives. To meet the needs of these executives, the company's customers sometimes order a custom design of the airplanes being purchased. When this occurs, a substantial start-up cost is incurred to initiate the production of these airplanes.

Fly-Right has recently received purchase requests from three customers with short deadlines. However, because the company's production facilities already are almost completely tied up filling previous orders, it will not be able to accept all three orders. Therefore, a decision now needs to be made on the number of airplanes

the company will agree to produce (if any) for each of the three customers.

The relevant data are given in the next table. The first row gives the start-up cost required to initiate the production of the airplanes for each customer. Once production is under way, the marginal net revenue (which is the purchase price minus the marginal production cost) from each airplane produced is shown in the second row. The third row gives the percentage of the available production capacity that would be used for each airplane produced. The last row indicates the maximum number of airplanes requested by each customer (but less will be accepted).

	Customer		
	1	**2**	**3**
Start-up cost	$3 million	$2 million	0
Marginal net revenue	$2 million	$3 million	$0.8 million
Capacity used per plane	20%	40%	20%
Maximum order	3 planes	2 planes	5 planes

Fly-Right now wants to determine how many airplanes to produce for each customer (if any) to maximize the company's total profit (total net revenue minus start-up costs).

(a) Formulate a model with both integer variables and binary variables for this problem.

C (b) Use the computer to solve this model.

11.4-1. Reconsider the Fly-Right Airplane Co. problem introduced in Prob. 11.3-7. A more detailed analysis of the various cost and revenue factors now has revealed that the potential profit from producing airplanes for each customer cannot be expressed simply in terms of a *start-up cost* and a fixed *marginal net revenue* per airplane produced. Instead, the profits are given by the following table.

Airplanes Produced	Profit from Customer		
	1	**2**	**3**
0	0	0	0
1	−$1 million	$1 million	$1 million
2	$2 million	$5 million	$3 million
3	$4 million		$5 million
4			$6 million
5			$7 million

(a) Formulate a BIP model for this problem that includes constraints for *mutually exclusive alternatives*.

C (b) Use the computer to solve the model formulated in part (a). Then use this optimal solution to identify the optimal number of airplanes to produce for each customer.

(c) Formulate another BIP model for this model that includes constraints for *contingent decisions*.

c **(d)** Repeat part (*b*) for the model formulated in part (*c*).

11.4-2. Reconsider the Wyndor Glass Co. problem presented in Sec. 3.1. Management now has decided that only one of the two new products should be produced, and the choice is to be made on the basis of maximizing profit. Introduce *auxiliary binary variables* to formulate an MIP model for this new version of the problem.

11.4-3.* Reconsider Prob. 3.1-11, where the management of the Omega Manufacturing Company is considering devoting excess production capacity to one or more of three products. (See the Partial Answers to Selected Problems in the back of the book for additional information about this problem.) Management now has decided to add the restriction that no more than two of the three prospective products should be produced.

(a) Introduce *auxiliary binary variables* to formulate an MIP model for this new version of the problem.

c **(b)** Use the computer to solve this model.

11.4-4. Consider the following integer nonlinear programming problem.

Maximize $Z = 4x_1^2 - x_1^3 + 10x_2^2 - x_2^4$,

subject to

$$x_1 + x_2 \le 3$$

and

$$x_1 \ge 0, \qquad x_2 \ge 0$$
$$x_1 \text{ and } x_2 \text{ are integers.}$$

This problem can be reformulated in two different ways as an equivalent pure BIP problem (with a linear objective function) with six binary variables (y_{1j} and y_{2j} for $j = 1, 2, 3$), depending on the interpretation given the binary variables.

(a) Formulate a BIP model for this problem where the binary variables have the interpretation,

$$y_{ij} = \begin{cases} 1 & \text{if } x_i = j \\ 0 & \text{otherwise.} \end{cases}$$

c **(b)** Use the computer to solve the model formulated in part (*a*), and thereby identify an optimal solution for (x_1, x_2) for the original problem.

(c) Formulate a BIP model for this problem where the binary variables have the interpretation,

$$y_{ij} = \begin{cases} 1 & \text{if } x_i \ge j \\ 0 & \text{otherwise.} \end{cases}$$

c **(d)** Use the computer to solve the model formulated in part (*c*), and thereby identify an optimal solution for (x_1, x_2) for the original problem.

11.4-5.* Consider the following special type of *shortest-path problem* (see Sec. 9.3) where the nodes are in columns and the only paths considered always move forward one column at a time.

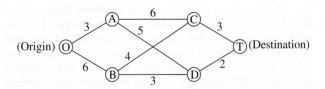

The numbers along the links represent distances, and the objective is to find the shortest path from the origin to the destination.

This problem also can be formulated as a BIP model involving both mutually exclusive alternatives and contingent decisions.

(a) Formulate this model. Identify the constraints that are for mutually exclusive alternatives and that are for contingent decisions.

c **(b)** Use the computer to solve this problem.

11.4-6. Speedy Delivery provides two-day delivery service of large parcels across the United States. Each morning at each collection center, the parcels that have arrived overnight are loaded onto several trucks for delivery throughout the area. Since the competitive battlefield in this business is speed of delivery, the parcels are divided among the trucks according to their geographical destinations to minimize the average time needed to make the deliveries.

On this particular morning, the dispatcher for the Blue River Valley Collection Center, Sharon Lofton, is hard at work. Her three drivers will be arriving in less than an hour to make the day's deliveries. There are nine parcels to be delivered, all at locations many miles apart. As usual, Sharon has loaded these locations into her computer. She is using her company's special software package, a decision support system called Dispatcher. The first thing Dispatcher does is use these locations to generate a considerable number of attractive possible routes for the individual delivery trucks. These routes are shown in the following table (where the numbers in each column indicate the order of the deliveries), along with the estimated time required to traverse the route.

Delivery Location	Attractive Possible Route									
	1	**2**	**3**	**4**	**5**	**6**	**7**	**8**	**9**	**10**
A	1				1				1	
B		2		1		2			2	2
C			3	3			3		3	
D	2						1	1		
E			2	2		3				
F		1			2					
G	3						1	2		3
H			1		3					1
I		3		4			2			
Time (in hours)	6	4	7	5	4	6	5	3	7	6

Dispatcher is an interactive system that shows these routes to Sharon for her approval or modification. (For example, the computer may not know that flooding has made a particular route infeasible.) After Sharon approves these routes as attractive possibilities with reasonable time estimates, Dispatcher next formulates and solves a BIP model for selecting three routes that minimize their total time while including each delivery location on exactly one route. This morning, Sharon does approve all the routes.
(a) Formulate this BIP model.
c (b) Use the computer to solve this model.

11.4-7. An increasing number of Americans are moving to a warmer climate when they retire. To take advantage of this trend, Sunny Skies Unlimited is undertaking a major real estate development project. The project is to develop a completely new retirement community (to be called Pilgrim Haven) that will cover several square miles. One of the decisions to be made is where to locate the two fire stations that have been allocated to the community. For planning purposes, Pilgrim Haven has been divided into five tracts, with no more than one fire station to be located in any given tract. Each station is to respond to *all* the fires that occur in the tract in which it is located as well as in the other tracts that are assigned to this station. Thus, the decisions to be made consist of (1) the tracts to receive a fire station and (2) the assignment of each of the other tracts to one of the fire stations. The objective is to minimize the overall average of the *response times* to fires.

The following table gives the average response time to a fire in each tract (the columns) if that tract is served by a station in a given tract (the rows). The bottom row gives the forecasted average number of fires that will occur in each of the tracts per day.

| Assigned Station Located in Tract | Response Times (in minutes) Fire in Tract | | | | |
	1	2	3	4	5
1	5	12	30	20	15
2	20	4	15	10	25
3	15	20	6	15	12
4	25	15	25	4	10
5	10	25	15	12	5
Average frequency of fires	2 per day	1 per day	3 per day	1 per day	3 per day

Formulate a BIP model for this problem. Identify any constraints that correspond to mutually exclusive alternatives or contingent decisions.

11.4-8. Reconsider Prob. 11.4-7. The management of Sunny Skies Unlimited now has decided that the decision on the locations of the fire stations should be based mainly on costs.

The cost of locating a fire station in a tract is $300,000 for tract 1, $350,000 for tract 2, $600,000 for tract 3, $450,000 for

tract 4, and $700,000 for tract 5. Management's objective now is the following:

> Determine which tracts should receive a station to minimize the total cost of stations while ensuring that each tract has at least one station close enough to respond to a fire in no more than 12 minutes (on the average).

In contrast to the original problem, note that the total number of fire stations is no longer fixed. Furthermore, if a tract without a station has more than one station within 12 minutes, it is no longer necessary to assign this tract to just one of these stations.
(a) Formulate a complete pure BIP model with 5 binary variables for this problem.
(b) Is this a *set covering problem?* Explain, and identify the relevant sets.
c (c) Use the computer to solve the model formulated in part (a).

11.4-9. Suppose that a state sends R persons to the U.S. House of Representatives. There are D counties in the state ($D > R$), and the state legislature wants to group these counties into R distinct electoral districts, each of which sends a delegate to Congress. The total population of the state is P, and the legislature wants to form districts whose population approximates $p = P/R$. Suppose that the appropriate legislative committee studying the electoral districting problem generates a long list of N *candidates* to be districts ($N > R$). Each of these candidates contains contiguous counties and a total population p_j ($j = 1, 2, \ldots, N$) that is acceptably close to p. Define $c_j = |p_j - p|$. Each county i ($i = 1, 2, \ldots, D$) is included in at least one candidate and typically will be included in a considerable number of candidates (in order to provide many feasible ways of selecting a set of R candidates that includes each county exactly once). Define

$$a_{ij} = \begin{cases} 1 & \text{if county } i \text{ is included in candidate } j \\ 0 & \text{if not.} \end{cases}$$

Given the values of the c_j and the a_{ij}, the objective is to select R of these N possible districts such that each county is contained in a single district and such that the largest of the associated c_j is as small as possible.
Formulate a BIP model for this problem.

11.5-1. Read the referenced article that fully describes the OR study summarized in the application vignette presented in Sec. 11.5. Briefly describe how integer programming was applied in this study. Then list the various financial and nonfinancial benefits that resulted from this study.

11.5-2.* Consider the following IP problem.

Maximize $Z = 5x_1 + x_2,$

subject to

$$\begin{aligned} -x_1 + 2x_2 &\le 4 \\ x_1 - x_2 &\le 1 \\ 4x_1 + x_2 &\le 12 \end{aligned}$$

and

$$x_1 \geq 0, \qquad x_2 \geq 0$$
x_1, x_2 are integers.

(a) Solve this problem graphically.

(b) Solve the LP relaxation graphically. Round this solution to the *nearest* integer solution and check whether it is feasible. Then enumerate *all* the rounded solutions by rounding this solution for the LP relaxation in *all* possible ways (i.e., by rounding each noninteger value both up and down). For each rounded solution, check for feasibility and, if feasible, calculate Z. Are any of these feasible rounded solutions optimal for the IP problem?

11.5-3. Follow the instructions of Prob. 11.5-2 for the following IP problem.

Maximize $Z = 220x_1 + 80x_2$,

subject to

$$5x_1 + 2x_2 \leq 16$$
$$2x_1 - x_2 \leq 4$$
$$-x_1 + 2x_2 \leq 4$$

and

$$x_1 \geq 0, \qquad x_2 \geq 0$$
x_1, x_2 are integers.

11.5-4. Follow the instructions of Prob. 11.5-2 for the following BIP problem.

Maximize $Z = 10x_1 + 25x_2$,

subject to

$$19x_1 - 6x_2 \leq 15$$
$$5x_1 + 15x_2 \leq 15$$

and

x_1, x_2 are binary.

11.5-5. Follow the instructions of Prob. 11.5-2 for the following BIP problem.

Maximize $Z = -5x_1 + 25x_2$,

subject to

$$-3x_1 + 30x_2 \leq 27$$
$$3x_1 + x_2 \leq 4$$

and

x_1, x_2 are binary.

11.5-6. Label each of the following statements as True or False, and then justify your answer by referring to specific statements in the chapter.

(a) Linear programming problems are generally considerably easier to solve than IP problems.

(b) For IP problems, the number of integer variables is generally more important in determining the computational difficulty than is the number of functional constraints.

(c) To solve an IP problem with an approximate procedure, one may apply the simplex method to the LP relaxation problem and then round each noninteger value to the nearest integer. The result will be a feasible but not necessarily optimal solution for the IP problem.

D,I **11.6-1.*** Use the BIP branch-and-bound algorithm presented in Sec. 11.6 to solve the following problem interactively.

Maximize $Z = 2x_1 - x_2 + 5x_3 - 3x_4 + 4x_5$,

subject to

$$3x_1 - 2x_2 + 7x_3 - 5x_4 + 4x_5 \leq 6$$
$$x_1 - x_2 + 2x_3 - 4x_4 + 2x_5 \leq 0$$

and

x_j is binary,   for $j = 1, 2, \ldots, 5$.

D,I **11.6-2.** Use the BIP branch-and-bound algorithm presented in Sec. 11.6 to solve the following problem interactively.

Minimize $Z = 5x_1 + 6x_2 + 7x_3 + 8x_4 + 9x_5$,

subject to

$$3x_1 - x_2 + x_3 + x_4 - 2x_5 \geq 2$$
$$x_1 + 3x_2 - x_3 - 2x_4 + x_5 \geq 0$$
$$-x_1 - x_2 + 3x_3 + x_4 + x_5 \geq 1$$

and

x_j is binary,   for $j = 1, 2, \ldots, 5$.

D,I **11.6-3.** Use the BIP branch-and-bound algorithm presented in Sec. 11.6 to solve the following problem interactively.

Maximize $Z = 3x_1 + 3x_2 + 5x_3 - 2x_4 - x_5$,

subject to

$$x_1 + 2x_2 - 3x_4 - x_5 \leq 0$$
$$-15x_1 + 30x_2 - 35x_3 + 45x_4 + 45x_5 \geq 50$$

and

x_j is binary,   for $j = 1, 2, \ldots, 5$.

D,I **11.6-4.** Reconsider Prob. 11.3-6(*a*). Use the BIP branch-and-bound algorithm presented in Sec. 11.6 to solve this BIP model interactively.

D,I **11.6-5.** Reconsider Prob. 11.4-8(*a*). Use the BIP algorithm presented in Sec. 11.6 to solve this problem interactively.

11.6-6. Consider the following statements about any pure IP problem (in maximization form) and its LP relaxation. Label each of the statements as True or False, and then justify your answer.

(a) The feasible region for the LP relaxation is a subset of the feasible region for the IP problem.

(b) If an optimal solution for the LP relaxation is an integer solution, then the optimal value of the objective function is the same for both problems.

(c) If a noninteger solution is feasible for the LP relaxation, then the nearest integer solution (rounding each variable to the nearest integer) is a feasible solution for the IP problem.

11.6-7.* Consider the assignment problem with the following cost table:

		Task				
		1	**2**	**3**	**4**	**5**
	1	39	65	69	66	57
	2	64	84	24	92	22
Assignee	3	49	50	61	31	45
	4	48	45	55	23	50
	5	59	34	30	34	18

(a) Design a branch-and-bound algorithm for solving such assignment problems by specifying how the branching, bounding, and fathoming steps would be performed. (*Hint:* For the assignees not yet assigned for the current subproblem, form the relaxation by deleting the constraints that each of these assignees must perform exactly one task.)

(b) Use this algorithm to solve this problem.

11.6-8. Five jobs need to be done on a certain machine. However, the setup time for each job depends upon which job immediately preceded it, as shown by the following table:

		Setup Time				
		Job				
		1	**2**	**3**	**4**	**5**
	None	4	5	8	9	4
	1	—	7	12	10	9
Immediately	2	6	—	10	14	11
Preceding Job	3	10	11	—	12	10
	4	7	8	15	—	7
	5	12	9	8	16	—

The objective is to schedule the *sequence* of jobs that minimizes the sum of the resulting setup times.

(a) Design a branch-and-bound algorithm for sequencing problems of this type by specifying how the branch, bound, and fathoming steps would be performed.

(b) Use this algorithm to solve this problem.

11.6-9.* Consider the following *nonlinear* BIP problem.

$$\text{Maximize} \quad Z = 80x_1 + 60x_2 + 40x_3 + 20x_4 \\ - (7x_1 + 5x_2 + 3x_3 + 2x_4)^2,$$

subject to

$$x_j \text{ is binary}, \quad \text{for } j = 1, 2, 3, 4.$$

Given the value of the first k variables $x_1, \ldots, x_k$, where $k = 0$, 1, 2, or 3, an upper bound on the value of Z that can be achieved by the corresponding feasible solutions is

$$\sum_{j=1}^{k} c_j x_j - \left(\sum_{j=1}^{k} d_j x_j \right)^2$$

$$+ \sum_{j=k+1}^{4} \max \left\{ 0, c_j - \left[\left(\sum_{i=1}^{k} d_i x_i + d_j \right)^2 - \left(\sum_{i=1}^{k} d_i x_i \right)^2 \right] \right\},$$

where $c_1 = 80$, $c_2 = 60$, $c_3 = 40$, $c_4 = 20$, $d_1 = 7$, $d_2 = 5$, $d_3 = 3$, $d_4 = 2$. Use this bound to solve the problem by the branch-and-bound technique.

11.6-10. Consider the Lagrangian relaxation described near the end of Sec. 11.6.

(a) If $\mathbf{x}$ is a feasible solution for an MIP problem, show that $\mathbf{x}$ also must be a feasible solution for the corresponding Lagrangian relaxation.

(b) If $\mathbf{x}^*$ is an optimal solution for an MIP problem, with an objective function value of Z, show that $Z \leq Z_R^*$, where Z_R^* is the optimal objective function value for the corresponding Lagrangian relaxation.

11.7-1. Read the referenced article that fully describes the OR study summarized in the application vignette presented in Sec. 11.7. Briefly describe how integer programming was applied in this study. Then list the various financial and nonfinancial benefits that resulted from this study.

11.7-2.* Consider the following IP problem.

$$\text{Maximize} \quad Z = -3x_1 + 5x_2,$$

subject to

$$5x_1 - 7x_2 \geq 3$$

and

$$x_j \leq 3 \\ x_j \geq 0 \\ x_j \text{ is integer}, \quad \text{for } j = 1, 2.$$

(a) Solve this problem graphically.

(b) Use the MIP branch-and-bound algorithm presented in Sec. 11.7 to solve this problem by hand. For each subproblem, solve its LP relaxation *graphically*.

(c) Use the binary representation for integer variables to reformulate this problem as a BIP problem.

D,I **(d)** Use the BIP branch-and-bound algorithm presented in Sec. 11.6 to solve the problem as formulated in part (*c*) interactively.

11.7-3. Follow the instructions of Prob. 11.7-2 for the following IP model.

Minimize $\quad Z = 15x_1 + 10x_2,$

subject to

$$15x_1 + 5x_2 \geq 30$$
$$10x_1 + 10x_2 \geq 30$$

and

$$x_1 \geq 0, \qquad x_2 \geq 0$$
$$x_1, x_2 \text{ are integers.}$$

11.7-4. Reconsider the IP model of Prob. 11.5-2.
(a) Use the MIP branch-and-bound algorithm presented in Sec. 11.7 to solve this problem by hand. For each subproblem, solve its LP relaxation *graphically.*
D,I **(b)** Now use the interactive procedure for this algorithm in your IOR Tutorial to solve this problem.
C **(c)** Check your answer by using an automatic procedure to solve the problem.

D,I **11.7-5.** Consider the IP example discussed in Sec. 11.5 and illustrated in Fig. 11.3. Use the MIP branch-and-bound algorithm presented in Sec. 11.7 to solve this problem interactively.

D,I **11.7-6.** Reconsider Prob. 11.3-5*a*. Use the MIP branch-and-bound algorithm presented in Sec. 11.7 to solve this IP problem interactively.

11.7-7. A machine shop makes two products. Each unit of the first product requires 3 hours on machine 1 and 2 hours on machine 2. Each unit of the second product requires 2 hours on machine 1 and 3 hours on machine 2. Machine 1 is available only 8 hours per day and machine 2 only 7 hours per day. The profit per unit sold is 16 for the first product and 10 for the second. The amount of each product produced per day must be an integral multiple of 0.25. The objective is to determine the mix of production quantities that will maximize profit.
(a) Formulate an IP model for this problem.
(b) Solve this model graphically.
(c) Use graphical analysis to apply the MIP branch-and-bound algorithm presented in Sec. 11.7 to solve this model.
D,I **(d)** Now use the interactive procedure for this algorithm in your IOR Tutorial to solve this model.
C **(e)** Check your answers in parts (*b*), (*c*), and (*d*) by using an automatic procedure to solve the model.

D,I **11.7-8.** Use the MIP branch-and-bound algorithm presented in Sec. 11.7 to solve the following MIP problem interactively.

Maximize $\quad Z = 20x_1 + 10x_2 + 25x_3 + 20x_4,$

subject to

$$x_1 + x_2 + x_3 + 2x_4 \leq 12$$
$$3x_1 + x_2 + 2x_3 + 2x_4 \leq 20$$
$$x_1 + 2x_2 + 5x_3 + 3x_4 \leq 30$$

and

$$x_j \geq 0, \qquad \text{for } j = 1, 2, 3, 4$$
$$x_j \text{ is integer,} \qquad \text{for } j = 1, 2, 3.$$

D,I **11.7-9.** Use the MIP branch-and-bound algorithm presented in Sec. 11.7 to solve the following MIP problem interactively.

Maximize $\quad Z = 3x_1 + 4x_2 + 2x_3 + x_4 + 2x_5,$

subject to

$$2x_1 - x_2 + x_3 + x_4 + x_5 \leq 3$$
$$-x_1 + 3x_2 + x_3 - x_4 - 2x_5 \leq 2$$
$$2x_1 + x_2 - x_3 + x_4 + 3x_5 \leq 1$$

and

$$x_j \geq 0, \qquad \text{for } j = 1, 2, 3, 4, 5$$
$$x_j \text{ is binary,} \qquad \text{for } j = 1, 2, 3.$$

D,I **11.7-10.** Use the MIP branch-and-bound algorithm presented in Sec. 11.7 to solve the following MIP problem interactively.

Minimize $\quad Z = 5x_1 + x_2 + x_3 + 2x_4 + 3x_5,$

subject to

$$x_2 - 5x_3 + x_4 + 2x_5 \geq -2$$
$$5x_1 - x_2 + x_5 \geq 7$$
$$x_1 + x_2 + 6x_3 + x_4 \geq 4$$

and

$$x_j \geq 0, \qquad \text{for } j = 1, 2, 3, 4, 5$$
$$x_j \text{ is integer,} \qquad \text{for } j = 1, 2, 3.$$

11.8-1.* For each of the following constraints of pure BIP problems, use the constraint to fix as many variables as possible.
(a) $4x_1 + x_2 + 3x_3 + 2x_4 \leq 2$
(b) $4x_1 - x_2 + 3x_3 + 2x_4 \leq 2$
(c) $4x_1 - x_2 + 3x_3 + 2x_4 \geq 7$

11.8-2. For each of the following constraints of pure BIP problems, use the constraint to fix as many variables as possible.
(a) $20x_1 - 7x_2 + 5x_3 \leq 10$
(b) $10x_1 - 7x_2 + 5x_3 \geq 10$
(c) $10x_1 - 7x_2 + 5x_3 \leq -1$

11.8-3. Use the following set of constraints for the *same* pure BIP problem to fix as many variables as possible. Also identify the constraints which become redundant because of the fixed variables.

$$3x_3 - x_5 + x_7 \leq 1$$
$$x_2 + x_4 + x_6 \leq 1$$
$$x_1 - 2x_5 + 2x_6 \geq 2$$
$$x_1 + x_2 - x_4 \leq 0$$

11.8-4. For each of the following constraints of pure BIP problems, identify which ones are made redundant by the binary constraints. Explain why each one is, or is not, redundant.
(a) $2x_1 + x_2 + 2x_3 \leq 5$
(b) $3x_1 - 4x_2 + 5x_3 \leq 5$
(c) $x_1 + x_2 + x_3 \geq 2$
(d) $3x_1 - x_2 - 2x_3 \geq -4$

11.8-5. In Sec. 11.8, at the end of the subsection on tightening constraints, we indicated that the constraint $4x_1 - 3x_2 + x_3 + 2x_4 \leq 5$ can be tightened to $2x_1 - 3x_2 + x_3 + 2x_4 \leq 3$ and then to $2x_1 - 2x_2 + x_3 + 2x_4 \leq 3$. Apply the procedure for tightening constraints to confirm these results.

11.8-6. Apply the procedure for *tightening constraints* to the following constraint for a pure BIP problem.

$$5x_1 - 10x_2 + 15x_3 \leq 15.$$

11.8-7. Apply the procedure for *tightening constraints* to the following constraint for a pure BIP problem.

$$x_1 - x_2 + 3x_3 + 4x_4 \geq 1.$$

11.8-8. Apply the procedure for *tightening constraints* to each of the following constraints for a pure BIP problem.
(a) $x_1 + 3x_2 - 4x_3 \leq 2$.
(b) $3x_1 - x_2 + 4x_3 \geq 1$.

11.8-9. In Sec. 11.8, a pure BIP example with the constraint, $2x_1 + 3x_2 \leq 4$, was used to illustrate the procedure for tightening constraints. Show that applying the procedure for generating cutting planes to this constraint yields the same new constraint, $x_1 + x_2 \leq 1$.

11.8-10. One of the constraints of a certain pure BIP problem is

$$x_1 + 3x_2 + 2x_3 + 4x_4 \leq 5.$$

Identify all the minimal covers for this constraint, and then give the corresponding cutting planes.

11.8-11. One of the constraints of a certain pure BIP problem is

$$25x_1 + 15x_2 + 20x_3 + 10x_4 \leq 35.$$

Identify all the minimal covers for this constraint, and then give the corresponding cutting planes.

11.8-12. Generate as many cutting planes as possible from the following constraint for a pure BIP problem.

$$3x_1 + 5x_2 + 4x_3 + 8x_4 \leq 10.$$

11.8-13. Generate as many cutting planes as possible from the following constraint for a pure BIP problem.

$$5x_1 + 3x_2 + 7x_3 + 4x_4 + 6x_5 \leq 9.$$

11.8-14. Consider the following BIP problem.

Maximize $Z = 2x_1 + 3x_2 + x_3 + 4x_4 + 3x_5$
$\qquad\qquad + 2x_6 + 2x_7 + x_8 + 3x_9,$

subject to

$$3x_2 + x_4 + x_5 \geq 3$$
$$x_1 + x_2 \leq 1$$
$$x_2 + x_4 - x_5 - x_6 \leq -1$$
$$x_2 + 2x_6 + 3x_7 + x_8 + 2x_9 \geq 4$$
$$-x_3 + 2x_5 + x_6 + 2x_7 - 2x_8 + x_9 \leq 5$$

and

all x_j binary.

Develop the tightest possible formulation of this problem by using the techniques of automatic problem reprocessing (fixing variables, deleting redundant constraints, and tightening constraints). Then use this tightened formulation to determine an optimal solution by inspection.

11.9-1. Consider the following problem.

Maximize $Z = 10x_1 + 30x_2 + 40x_3 + 30x_4,$

subject to

$x_1 \in \{2, 3\}, \quad x_2 \in \{2, 4\}, \quad x_3 \in \{3, 4\}, \quad x_4 \in \{1, 2, 3, 4\},$
all these variables must have different values,
$x_1 + x_2 + x_3 + x_4 \leq 10.$

Use the techniques of constraint programming (domain reduction, constraint propagation, a search procedure, and enumeration) to identify all the feasible solutions and then to find an optimal solution. Show your work.

11.9-2. Consider the following problem.

Maximize $Z = 5x_1 - x_1^2 + 8x_2 - x_2^2 + 10x_3 - x_3^2 + 15x_4$
$\qquad\qquad - x_4^2 + 20x_5 - x_5^2,$

subject to

$x_1 \in \{3, 6, 12\}, x_2 \in \{3, 6\}, x_3 \in \{3, 6, 9, 12\},$
$\qquad x_4 \in \{6, 12\}, x_5 \in \{9, 12, 15, 18\},$
all these variables must have different values,
$x_1 + x_3 + x_4 \leq 25.$

Use the techniques of constraint programming (domain reduction, constraint propagation, a search procedure, and enumeration) to identify all the feasible solutions and then to find an optimal solution. Show your work.

11.9-3. Consider the following problem.

Maximize $Z = 100x_1 - 3x_1^2 + 400x_2 - 5x_2^2 + 200x_3$
$\qquad\qquad - 4x_3^2 + 100x_4 - 2x_4^4,$

subject to

$x_1 \in \{25, 30\}, x_2 \in \{20, 25, 30, 35, 40, 50\},$
$\qquad x_3 \in \{20, 25, 30\}, x_4 \in \{20, 25\},$
all these variables must have different values,
$x_2 + x_3 \leq 60,$
$x_1 + x_3 \leq 50.$

Use the techniques of constraint programming (domain reduction, constraint propagation, a search procedure, and enumeration) to identify all the feasible solutions and then to find an optimal solution. Show your work.

11.9-4. Consider the Job Shop Co. example introduced in Sec. 8.3. Table 8.25 shows its formulation as an assignment problem. Use *global constraints* to formulate a compact constraint programming model for this assignment problem.

11.9-5. Consider the problem of assigning swimmers to the different legs of a medley relay team that is presented in Prob. 8.3-4. The answer in the back of the book shows the formulation of this problem as an assignment problem. Use *global constraints* to formulate a compact constraint programming model for this assignment problem.

11.9-6. Consider the problem of determining the best plan for how many days to study for each of four final examinations that is presented in Prob. 10.3-3. Formulate a compact constraint programming model for this problem.

11.9-7. Problem 10.3-2 describes how the owner of a chain of three grocery stores needs to determine how many crates of fresh strawberries should be allocated to each of the stores. Formulate a compact constraint programming model for this problem.

11.9-8. One powerful feature of constraint programming is that variables can be used as subscripts for the terms in the objective function. For example, consider the following *traveling salesman*

problem. The salesman needs to visit each of n cities (city 1, 2, . . . , n) exactly once, starting in city 1 (his home city) and returning to city 1 after completing the tour. Let c_{ij} be the distance from city i to city j for $i, j = 1, 2, . . . , n$ ($i \neq j$). The objective is to determine which route to follow so as to minimize the total distance of the tour. (As discussed further in Chap. 13, this traveling salesman problem is a famous classic OR problem with many applications that have nothing to do with salesmen.)

Letting the decision variable x_j ($j = 1, 2, . . . , n, n + 1$) denote the jth city visited by the salesman, where $x_1 = 1$ and $x_{n+1} = 1$, constraint programming allows writing the objective as

$$\text{Minimize } Z = \sum_{j=1}^{n} c_{x_j x_{j+1}}.$$

Using this objective function, formulate a complete constraint programming model for this problem.

11.10-1. From the bottom part of the selected references given at the end of the chapter, select one of these award-winning applications of integer programming. Read this article and then write a two-page summary of the application and the benefits (including nonfinancial benefits) it provided.

11.10-2. From the bottom part of the selected references given at the end of the chapter, select three of these award-winning applications of integer programming. For each one, read the article and then write a one-page summary of the application and the benefits (including nonfinancial benefits) it provided.

■ CASES

CASE 11.1   Capacity Concerns

Bentley Hamilton throws the business section of *The New York Times* onto the conference room table and watches as his associates jolt upright in their overstuffed chairs.

Mr. Hamilton wants to make a point.

He throws the front page of *The Wall Street Journal* on top of *The New York Times* and watches as his associates widen their eyes once heavy with boredom.

Mr. Hamilton wants to make a big point.

He then throws the front page of *The Financial Times* on top of the newspaper pile and watches as his associates dab the fine beads of sweat off their brows.

Mr. Hamilton wants his point indelibly etched into his associates' minds.

"I have just presented you with three leading financial newspapers carrying today's top business story," Mr. Hamilton declares in a tight, angry voice. "My dear associates, our company is going to hell in a hand basket! Shall I read you the headlines? From *The New York Times*, 'CommuniCorp stock

drops to lowest in 52 weeks.' From *The Wall Street Journal*, 'CommuniCorp loses 25 percent of the pager market in only one year.' Oh and my favorite, from *The Financial Times*, 'CommuniCorp cannot CommuniCate: CommuniCorp stock drops because of internal communications disarray.' How did our company fall into such dire straits?"

Mr. Hamilton throws a transparency showing a line sloping slightly upward onto the overhead projector. "This is a graph of our productivity over the last 12 months. As you can see from the graph, productivity in our pager production facility has increased steadily over the last year. Clearly, productivity is not the cause of our problem."

Mr. Hamilton throws a second transparency showing a line sloping steeply upward onto the overhead projector. "This is a graph of our missed or late orders over the last 12 months." Mr. Hamilton hears an audible gasp from his associates. "As you can see from the graph, our missed or late orders have increased steadily and significantly over the past 12 months. I think this trend explains why we have been losing market share, causing our stock to drop to its lowest level in 52 weeks. We have angered and lost the business of retailers, our customers who depend upon on-time deliveries to meet the demand of consumers."

"Why have we missed our delivery dates when our productivity level should have allowed us to fill all orders?" Mr. Hamilton asks. "I called several departments to ask this question."

"It turns out that we have been producing pagers for the hell of it!" Mr. Hamilton says in disbelief. "The marketing and sales departments do not communicate with the manufacturing department, so manufacturing executives do not know what pagers to produce to fill orders. The manufacturing executives want to keep the plant running, so they produce pagers regardless of whether the pagers have been ordered. Finished pagers are sent to the warehouse, but marketing and sales executives do not know the number and styles of pagers in the warehouse. They try to communicate

with warehouse executives to determine if the pagers in inventory can fill the orders, but they rarely receive answers to their questions."

Mr. Hamilton pauses and looks directly at his associates. "Ladies and gentlemen, it seems to me that we have a serious internal communications problem. I intend to correct this problem immediately. I want to begin by installing a companywide computer network to ensure that all departments have access to critical documents and are able to easily communicate with each other through e-mail. Because this intranet will represent a large change from the current communications infrastructure, I expect some bugs in the system and some resistance from employees. I therefore want to phase in the installation of the intranet."

Mr. Hamilton passes the following timeline and requirements chart to his associates (IN = Intranet).

Month 1	Month 2	Month 3	Month 4	Month 5
IN Education	Install IN in Sales	Install IN in Manufacturing	Install IN in Warehouse	Install IN in Marketing

Department	Number of Employees
Sales	60
Manufacturing	200
Warehouse	30
Marketing	75

Mr. Hamilton proceeds to explain the timeline and requirements chart. "In the first month, I do not want to bring any department onto the intranet; I simply want to disseminate information about it and get buy-in from employees. In the second month, I want to bring the sales department onto the intranet since the sales department

receives all critical information from customers. In the third month, I want to bring the manufacturing department onto the intranet. In the fourth month, I want to install the intranet at the warehouse, and in the fifth and final month, I want to bring the marketing department onto the intranet. The requirements chart under the timeline lists the number of employees requiring access to the intranet in each department."

Mr. Hamilton turns to Emily Jones, the head of Corporate Information Management. "I need your help in planning for the installation of the intranet. Specifically, the company needs to purchase servers for the internal network. Employees will connect to company servers and download information to their own desktop computers."

Type of Server	Number of Employees Server Supports	Cost of Server
Standard Intel Pentium PC	Up to 30 employees	$ 2,500
Enhanced Intel Pentium PC	Up to 80 employees	$ 5,000
SGI Workstation	Up to 200 employees	$10,000
Sun Workstation	Up to 2,000 employees	$25,000

Mr. Hamilton passes Emily the above chart detailing the types of servers available, the number of employees each server supports, and the cost of each server.

"Emily, I need you to decide what servers to purchase and when to purchase them to minimize cost and to ensure that the company possesses enough server capacity to follow the intranet implementation timeline," Mr. Hamilton says. "For example, you may decide to buy one large server during the first month to support all employees, or buy several small servers during the first month to support all employees, or buy one small server each month to support each new group of employees gaining access to the intranet."

"There are several factors that complicate your decision," Mr. Hamilton continues. "Two server manufacturers are willing to offer discounts to CommuniCorp. SGI is willing to give you a discount of 10 percent off each server purchased, but only if you purchase servers in the first or second month. Sun is willing to give you a 25 percent discount off all servers purchased in the first two months. You are also limited in the amount of money you can spend during the first month. CommuniCorp has already allocated much of the budget for the next two months, so you only have a total of $9,500 available to purchase

servers in months 1 and 2. Finally, the Manufacturing Department requires at least one of the three more powerful servers. Have your decision on my desk at the end of the week."

(a) Emily first decides to evaluate the number and type of servers to purchase on a month-to-month basis. For each month, formulate an IP model to determine which servers Emily should purchase in that month to minimize costs in that month and support the new users. How many and which types of servers should she purchase in each month? How much is the total cost of the plan?

(b) Emily realizes that she could perhaps achieve savings if she bought a larger server in the initial months to support users in the final months. She therefore decides to evaluate the number and type of servers to purchase over the entire planning period. Formulate an IP model to determine which servers Emily should purchase in which months to minimize total cost and support all new users. How many and which types of servers should she purchase in each month? How much is the total cost of the plan?

(c) Why is the answer using the first method different from that using the second method?

(d) Are there other costs that Emily is not accounting for in her problem formulation? If so, what are they?

(e) What further concerns might the various departments of CommuniCorp have regarding the intranet?

■ PREVIEWS OF ADDED CASES ON OUR WEBSITE (www.mhhe.com/hillier)

CASE 11.2 Assigning Art

Plans are being made for an exhibit of up-and-coming modern artists at the San Francisco Museum of Modern Art. A long list of possible artists, their available pieces, and the display prices for these pieces has been compiled. There also are various constraints regarding the mix of pieces that can be chosen. BIP now needs to be applied to make the selection of the pieces for the exhibit under three different scenarios.

CASE 11.3 Stocking Sets

Poor inventory management at the local warehouse for Furniture City has led to overstocking of many items and frequent shortages of some others. To begin to rectify this situation, the 20 most popular kitchen sets in Furniture City's kitchen department have just been identified. These kitchen sets are composed of up to eight features in a variety of styles, so each of these styles should be well stocked in the warehouse. However, the limited amount of warehouse space allocated to the kitchen department means that some difficult stocking decisions need to be made. After gathering the relevant data for the 20 kitchen sets, BIP now needs to be applied to determine how many of each feature and style Furniture City should stock in the local warehouse under three different scenarios.

CASE 11.4 Assigning Students to Schools, Revisited Again

As introduced in Case 4.3 and revisited in Case 6.3, the Springfield School Board needs to assign the middle school

students in the city's six residential areas to the three remaining middle schools. The new complication in that the school board has just made the decision to prohibit the splitting of residential areas among multiple schools. Therefore, since each of the six areas must be assigned to a single school, BIP now must be applied to make these assignments under the various scenarios considered in Case 4.3.

12

Nonlinear Programming

The fundamental role of linear programming in OR is accurately reflected by the fact that it is the focus of a *third* of this book. A key assumption of linear programming is that *all its functions* (objective function and constraint functions) are linear. Although this assumption essentially holds for many practical problems, it frequently does not hold. Therefore, it often is necessary to deal directly with nonlinear programming problems, so we turn our attention to this important area.

In one general form,[1] the *nonlinear programming problem* is to find $\mathbf{x} = (x_1, x_2, \ldots, x_n)$ so as to

Maximize $f(\mathbf{x})$,

subject to

$$g_i(\mathbf{x}) \leq b_i, \quad \text{for } i = 1, 2, \ldots, m,$$

and

$$\mathbf{x} \geq \mathbf{0},$$

where $f(\mathbf{x})$ and the $g_i(\mathbf{x})$ are given functions of the n decision variables.[2]

There are many different types of nonlinear programming problems, depending on the characteristics of the $f(\mathbf{x})$ and $g_i(\mathbf{x})$ functions. Different algorithms are used for the different types. For certain types where the functions have simple forms, problems can be solved relatively efficiently. For some other types, solving even small problems is a real challenge.

Because of the many types and the many algorithms, nonlinear programming is a particularly large subject. We do not have the space to survey it completely. However, we do present a few sample applications and then introduce some of the basic ideas for solving certain important types of nonlinear programming problems.

Both Appendixes 2 and 3 provide useful background for this chapter, and we recommend that you review these appendixes as you study the next few sections.

---

[1]The other *legitimate forms* correspond to those for *linear programming* listed in Sec. 3.2. Section 4.6 describes how to convert these other forms to the form given here.

[2]For simplicity, we assume throughout the chapter that *all* these functions either are *differentiable* everywhere or are *piecewise linear functions* (discussed in Secs. 12.1 and 12.8).

■ 12.1   SAMPLE APPLICATIONS

The following examples illustrate a few of the many important types of problems to which nonlinear programming has been applied.

The Product-Mix Problem with Price Elasticity

In *product-mix* problems, such as the Wyndor Glass Co. problem of Sec. 3.1, the goal is to determine the optimal mix of production levels for a firm's products, given limitations on the resources needed to produce those products, in order to maximize the firm's total profit. In some cases, there is a fixed unit profit associated with each product, so the resulting objective function will be linear. However, in many product-mix problems, certain factors introduce *nonlinearities* into the objective function.

For example, a large manufacturer may encounter *price elasticity,* whereby the amount of a product that can be sold has an inverse relationship to the price charged. Thus, the *price-demand curve* for a typical product might look like the one shown in Fig. 12.1, where $p(x)$ is the price required in order to be able to sell x units. The firm's profit from producing and selling x units of the product then would be the sales revenue, $xp(x)$, minus the production and distribution costs. Therefore, if the unit cost for producing and distributing the product is fixed at c (see the dashed line in Fig. 12.1), the firm's profit from producing and selling x units is given by the nonlinear function

$$P(x) = xp(x) - cx,$$

as plotted in Fig. 12.2. If *each* of the firm's n products has a similar profit function, say, $P_j(x_j)$ for producing and selling x_j units of product j ($j = 1, 2, \ldots, n$), then the overall objective function is

$$f(\mathbf{x}) = \sum_{j=1}^{n} P_j(x_j),$$

a sum of nonlinear functions.

Another reason that nonlinearities can arise in the objective function is the fact that the *marginal cost* of producing another unit of a given product varies with the production level. For example, the marginal cost may decrease when the production level is increased because of a *learning-curve effect* (more efficient production with more experience). On

■ **FIGURE 12.1**
Price-demand curve.

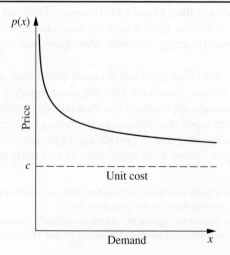

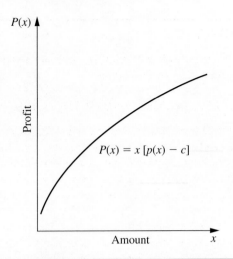

FIGURE 12.2
Profit function.

the other hand, it may increase instead, because special measures such as overtime or more expensive production facilities may be needed to increase production further.

Nonlinearities also may arise in the $g_i(\mathbf{x})$ constraint functions in a similar fashion. For example, if there is a budget constraint on total production cost, the cost function will be nonlinear if the marginal cost of production varies as just described. For constraints on the other kinds of resources, $g_i(\mathbf{x})$ will be nonlinear whenever the use of the corresponding resource is not strictly proportional to the production levels of the respective products.

The Transportation Problem with Volume Discounts on Shipping Costs

As illustrated by the P & T Company example in Sec. 8.1, a typical application of the transportation problem is to determine an optimal plan for shipping goods from various sources to various destinations, given supply and demand constraints, in order to minimize total shipping cost. It was assumed in Chap. 8 that the *cost per unit shipped* from a given source to a given destination is *fixed,* regardless of the amount shipped. In actuality, this cost may not be fixed. *Volume discounts* sometimes are available for large shipments, so that the *marginal cost* of shipping one more unit might follow a pattern like the one shown in Fig. 12.3. The resulting cost of shipping x units then is given by a *nonlinear* function $C(x)$, which is a *piecewise linear function* with slope equal to the marginal cost, like the one shown in Fig. 12.4. [The function in Fig. 12.4 consists of a line segment with slope 6.5 from (0, 0) to (0.6, 3.9), a second line segment with slope 5 from (0.6, 3.9) to (1.5, 8.4), a third line segment with slope 4 from (1.5, 8.4) to (2.7, 13.2), and a fourth line segment with slope 3 from (2.7, 13.2) to (4.5, 18.6).] Consequently, if *each* combination of source and destination has a similar shipping cost function, so that the cost of shipping x_{ij} units from source i ($i = 1, 2, \ldots, m$) to destination j ($j = 1, 2, \ldots, n$) is given by a nonlinear function $C_{ij}(x_{ij})$, then the overall objective function to be *minimized* is

$$f(\mathbf{x}) = \sum_{i=1}^{m} \sum_{j=1}^{n} C_{ij}(x_{ij}).$$

Even with this nonlinear objective function, the constraints normally are still the special linear constraints that fit the transportation problem model in Sec. 8.1.

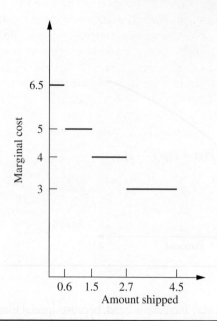

FIGURE 12.3
Marginal shipping cost.

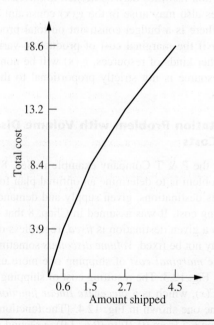

FIGURE 12.4
Shipping cost function.

Portfolio Selection with Risky Securities

It now is common practice for professional managers of large stock portfolios to use computer models based partially on nonlinear programming to guide them. Because investors are concerned about both the *expected return* (gain) and the *risk* associated with their investments, nonlinear programming is used to determine a portfolio that, under certain assumptions, provides an optimal trade-off between these two factors. This approach is based largely on path-breaking research done by Harry Markowitz and William Sharpe that helped them win the 1990 Nobel Prize in Economics.

A nonlinear programming model can be formulated for this problem as follows. Suppose that *n* stocks (securities) are being considered for inclusion in the portfolio, and let the

The **Bank Hapoalim Group** is Israel's largest banking group, providing services within Israel through a network of 327 branches, nine regional business centers, and various domestic subsidiaries. It also operates worldwide through 37 branches, offices, and subsidiaries in major financial centers in North and South America and Europe.

A major part of Bank Hapoalim's business involves providing investment advisors for its customers. To stay ahead of its competitors, management embarked on a restructuring program to provide these investment advisors with state-of-the-art methodology and technology. An OR team was formed to do this.

The team concluded that it needed to develop a flexible decision-support system for the investment advisors that could be tailored to meet the diverse needs of every customer. Each customer would be asked to provide extensive information about his or her needs, including choosing among various alternatives regarding his or her investment objectives, investment horizon, choice of an index to strive to exceed, preference with regard to liquidity and currency, etc. A series of questions also would be asked to ascertain the customer's risk-taking classification.

The natural choice of the model to drive the resulting decision-support system (called the *Opti-Money System*) was the *classical nonlinear programming model for portfolio selection* described in this section of the book, with modifications to incorporate all the information about the needs of the individual customer. This model generates an optimal weighting of 60 possible asset classes of equities and bonds in the portfolio, and the investment advisor then works with the customer to choose the specific equities and bonds within these classes.

In one recent year, the bank's investment advisors held some 133,000 consultation sessions with 63,000 customers while using this decision-support system. *The annual earnings* over benchmarks to customers who follow the investment advice provided by the system *total approximately* **US$244 million**, *while adding more than* **US$31 million** *to the bank's annual income*.

Source: M. Avriel, H. Pri-Zan, R. Meiri, and A. Peretz: "Opti-Money at Bank Hapoalim: A Model-Based Investment Decision-Support System for Individual Customers," *Interfaces,* **34**(1): 39–50, Jan.–Feb. 2004. (A link to this article is provided on our website, www.mhhe.com/hillier.)

decision variables x_j ($j = 1, 2, \ldots, n$) be the number of shares of stock j to be included. Let μ_j and σ_{jj} be the (estimated) *mean* and *variance,* respectively, of the return on each share of stock j, where σ_{jj} measures the risk of this stock. For $i = 1, 2, \ldots, n$ ($i \neq j$), let σ_{ij} be the *covariance* of the return on one share each of stock i and stock j. (Because it would be difficult to estimate all the σ_{ij} values, the usual approach is to make certain assumptions about market behavior that enable us to calculate σ_{ij} directly from σ_{ii} and σ_{jj}.) Then the expected value $R(\mathbf{x})$ and the variance $V(\mathbf{x})$ of the total return from the entire portfolio are

$$R(\mathbf{x}) = \sum_{j=1}^{n} \mu_j x_j$$

and

$$V(\mathbf{x}) = \sum_{i=1}^{n} \sum_{j=1}^{n} \sigma_{ij} x_i x_j,$$

where $V(\mathbf{x})$ measures the risk associated with the portfolio. One way to consider the trade-off between these two factors is to use $V(\mathbf{x})$ as the objective function to be minimized and then impose the constraint that $R(\mathbf{x})$ must be no smaller than the minimum acceptable expected return. The complete nonlinear programming model then would be

$$\text{Minimize} \quad V(\mathbf{x}) = \sum_{i=1}^{n} \sum_{j=1}^{n} \sigma_{ij} x_i x_j,$$

subject to

$$\sum_{j=1}^{n} \mu_j x_j \geq L$$

$$\sum_{j=1}^{n} P_j x_j \leq B$$

and

$$x_j \geq 0, \qquad \text{for } j = 1, 2, \dots, n,$$

where L is the minimum acceptable expected return, P_j is the price for each share of stock j, and B is the amount of money budgeted for the portfolio.

One drawback of this formulation is that it is relatively difficult to choose an appropriate value for L for obtaining the best trade-off between $R(\mathbf{x})$ and $V(\mathbf{x})$. Therefore, rather than stopping with one choice of L, it is common to use a *parametric* (nonlinear) programming approach to generate the optimal solution as a function of L over a wide range of values of L. The next step is to examine the values of $R(\mathbf{x})$ and $V(\mathbf{x})$ for these solutions that are optimal for some value of L and then to choose the solution that seems to give the best trade-off between these two quantities. This procedure often is referred to as generating the solutions on the *efficient frontier* of the two-dimensional graph of $(R(\mathbf{x}), V(\mathbf{x}))$ points for feasible $\mathbf{x}$. The reason is that the $(R(\mathbf{x}), V(\mathbf{x}))$ point for an optimal $\mathbf{x}$ (for some L) lies on the *frontier* (boundary) of the feasible points. Furthermore, each optimal $\mathbf{x}$ is *efficient* in the sense that no other feasible solution is at least equally good with one measure (R or V) and strictly better with the other measure (smaller V or larger R).

This application of nonlinear programming is a particularly important one. The use of nonlinear programming for portfolio optimization now lies at the center of modern financial analysis. (More broadly, the relatively new field of *financial engineering* has arisen to focus on the application of OR techniques such as nonlinear programming to various finance problems, including portfolio optimization.) As illustrated by the application vignette in this section, this kind of application of nonlinear programming is having a tremendous impact in practice. Much research also continues to be done on the properties and application of both the above model and related nonlinear programming models to sophisticated kinds of portfolio analysis.[3]

■ 12.2 GRAPHICAL ILLUSTRATION OF NONLINEAR PROGRAMMING PROBLEMS

When a nonlinear programming problem has just one or two variables, it can be represented graphically much like the Wyndor Glass Co. example for linear programming in Sec. 3.1. Because such a graphical representation gives considerable insight into the properties of optimal solutions for linear and nonlinear programming, let us look at a few examples. To highlight the difference between linear and nonlinear programming, we shall use some *nonlinear* variations of the Wyndor Glass Co. problem.

Figure 12.5 shows what happens to this problem if the only changes in the model shown in Sec. 3.1 are that both the second and the third functional constraints are replaced by the single nonlinear constraint $9x_1^2 + 5x_2^2 \leq 216$. Compare Fig. 12.5 with Fig. 3.3. The optimal solution still happens to be $(x_1, x_2) = (2, 6)$. Furthermore, it still lies on the boundary of the feasible region. However, it is *not* a corner-point feasible (CPF) solution. The optimal solution could have been a CPF solution with a different objective function (check $Z = 3x_1 + x_2$), but the fact that it need not be one means that we no longer have the tremendous simplification used in linear programming of limiting the search for an optimal solution to just the CPF solutions.

---

[3]Important recent research includes the following papers. B. I. Jacobs, K. N. Levy, and H. M. Markowitz: "Portfolio Optimization with Factors, Scenarios, and Realistic Short Positions," *Operations Research*, **53**(4): 586–599, July–Aug. 2005; A. F. Siegel and A. Woodgate: "Performance of Portfolios Optimized with Estimation Error," *Management Science*, **53**(6): 1005–1015, June 2007; H. Konno and T. Koshizuka: "Mean-Absolute Deviation Model," *IIE Transactions*, **37**(10): 893–900, Oct. 2005.

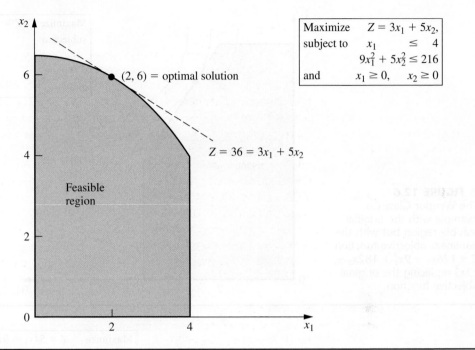

FIGURE 12.5
The Wyndor Glass Co. example with the nonlinear constraint $9x_1^2 + 5x_2^2 \le 216$ replacing the original second and third functional constraints.

Now suppose that the linear constraints of Sec. 3.1 are kept unchanged, but the objective function is made nonlinear. For example, if

$$Z = 126x_1 - 9x_1^2 + 182x_2 - 13x_2^2,$$

then the graphical representation in Fig. 12.6 indicates that the optimal solution is $x_1 = \frac{8}{3}$, $x_2 = 5$, which again lies on the boundary of the feasible region. (The value of Z for this optimal solution is $Z = 857$, so Fig. 12.6 depicts the fact that the locus of all points with $Z = 857$ intersects the feasible region at just this one point, whereas the locus of points with any larger Z does not intersect the feasible region at all.) On the other hand, if

$$Z = 54x_1 - 9x_1^2 + 78x_2 - 13x_2^2,$$

then Fig. 12.7 illustrates that the optimal solution turns out to be $(x_1, x_2) = (3, 3)$, which lies *inside* the boundary of the feasible region. (You can check that this solution is optimal by using calculus to derive it as the unconstrained global maximum; because it also satisfies the constraints, it must be optimal for the constrained problem.) Therefore, a general algorithm for solving similar problems needs to consider *all* solutions in the feasible region, not just those on the boundary.

Another complication that arises in nonlinear programming is that a *local* maximum need not be a *global* maximum (the overall optimal solution). For example, consider the function of a single variable plotted in Fig. 12.8. Over the interval $0 \le x \le 5$, this function has three local maxima—$x = 0$, $x = 2$, and $x = 4$—but only one of these—$x = 4$—is a *global maximum*. (Similarly, there are local minima at $x = 1$, 3, and 5, but only $x = 5$ is a *global minimum*.)

Nonlinear programming algorithms generally are unable to distinguish between a local maximum and a global maximum (except by finding another *better* local maximum). Therefore, it becomes crucial to know the conditions under which any local maximum is *guaranteed* to be a global maximum over the feasible region. You may recall from calculus that

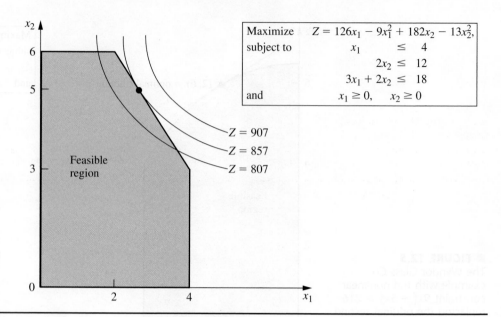

Maximize $Z = 126x_1 - 9x_1^2 + 182x_2 - 13x_2^2$,

subject to

$$x_1 \leq 4$$
$$2x_2 \leq 12$$
$$3x_1 + 2x_2 \leq 18$$

and

$$x_1 \geq 0, \quad x_2 \geq 0$$

■ **FIGURE 12.6**
The Wyndor Glass Co. example with the original feasible region but with the nonlinear objective function $Z = 126x_1 - 9x_1^2 + 182x_2 - 13x_2^2$ replacing the original objective function.

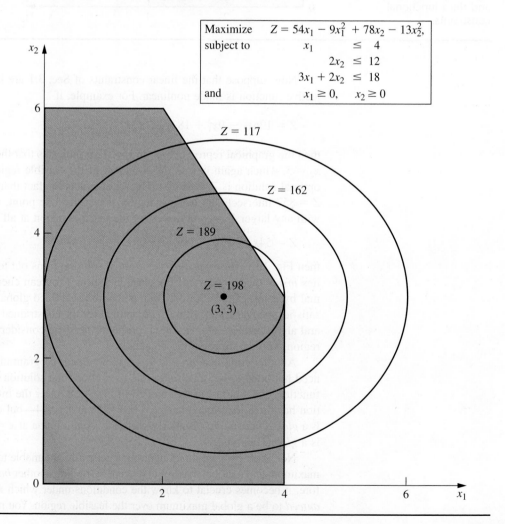

Maximize $Z = 54x_1 - 9x_1^2 + 78x_2 - 13x_2^2$,

subject to

$$x_1 \leq 4$$
$$2x_2 \leq 12$$
$$3x_1 + 2x_2 \leq 18$$

and

$$x_1 \geq 0, \quad x_2 \geq 0$$

■ **FIGURE 12.7**
The Wyndor Glass Co. example with the original feasible region but with another nonlinear objective function, $Z = 54x_1 - 9x_1^2 + 78x_2 - 13x_2^2$, replacing the original objective function.

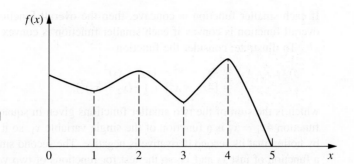

■ **FIGURE 12.8**
A function with several local maxima ($x = 0, 2, 4$), but only $x = 4$ is a global maximum.

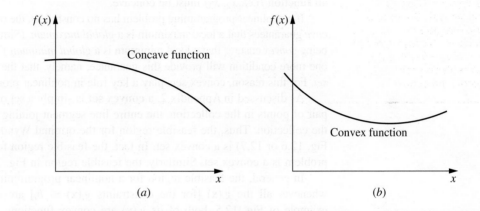

■ **FIGURE 12.9**
Examples of (a) a concave function and (b) a convex function.

(a)

(b)

when we maximize an ordinary (doubly differentiable) function of a single variable $f(x)$ without any constraints, this guarantee can be given when

$$\frac{\partial^2 f}{\partial x^2} \leq 0 \qquad \text{for all } x.$$

Such a function that is always "curving downward" (or not curving at all) is called a **concave** function.[4] Similarly, if $\leq$ is replaced by $\geq$, so that the function is always "curving upward" (or not curving at all), it is called a **convex** function.[5] (Thus, a *linear* function is both concave and convex.) See Fig. 12.9 for examples. Then note that Fig. 12.8 illustrates a function that is neither concave nor convex because it alternates between curving upward and curving downward.

Functions of multiple variables also can be characterized as concave or convex if they always curve downward or curve upward. These intuitive definitions are restated in precise terms, along with further elaboration on these concepts, in Appendix 2. (Concave and convex functions play a fundamental role in nonlinear programming, so if you are not very familiar with such functions, we suggest that you read further in Appendix 2.) Appendix 2 also provides a convenient test for checking whether a function of two variables is concave, convex, or neither.

Here is a convenient way of checking this for a function of more than two variables when the function consists of a *sum* of smaller functions of just one or two variables each.

[4]Concave functions sometimes are referred to as *concave downward*.
[5]Convex functions sometimes are referred to as *concave upward*.

If each smaller function is concave, then the overall function is concave. Similarly, the overall function is convex if each smaller function is convex.

To illustrate, consider the function

$$f(x_1, x_2, x_3) = 4x_1 - x_1^2 - (x_2 - x_3)^2$$
$$= [4x_1 - x_1^2] + [-(x_2 - x_3)^2],$$

which is the sum of the two smaller functions given in square brackets. The first smaller function $4x_1 - x_1^2$ is a function of the single variable x_1, so it can be found to be concave by noting that its second derivative is negative. The second smaller function $-(x_2 - x_3)^2$ is a function of just x_2 and x_3, so the test for functions of two variables given in Appendix 2 is applicable. In fact, Appendix 2 uses this particular function to illustrate the test and finds that the function is concave. Because both smaller functions are concave, the overall function $f(x_1, x_2, x_3)$ must be concave.

If a nonlinear programming problem has no constraints, the objective function being *concave* guarantees that a local maximum is a *global maximum.* (Similarly, the objective function being *convex* ensures that a local minimum is a *global minimum.*) If there are constraints, then one more condition will provide this guarantee, namely, that the *feasible region* is a *convex set.* For this reason, convex sets play a key role in nonlinear programming.

As discussed in Appendix 2, a **convex set** is simply a set of points such that, for each pair of points in the collection, the entire line segment joining these two points is also in the collection. Thus, the feasible region for the original Wyndor Glass Co. problem (see Fig. 12.6 or 12.7) is a convex set. In fact, the feasible region for *any* linear programming problem is a convex set. Similarly, the feasible region in Fig. 12.5 is a convex set.

In general, the feasible region for a nonlinear programming problem is a convex set whenever all the $g_i(\mathbf{x})$ [for the constraints $g_i(\mathbf{x}) \leq b_i$] are convex functions. For the example of Fig. 12.5, both of its $g_i(\mathbf{x})$ are convex functions, since $g_1(\mathbf{x}) = x_1$ (a linear function is automatically both concave and convex) and $g_2(\mathbf{x}) = 9x_1^2 + 5x_2^2$ (both $9x_1^2$ and $5x_2^2$ are convex functions so their sum is a convex function). These two convex $g_i(\mathbf{x})$ lead to the feasible region of Fig. 12.5 being a convex set.

Now let's see what happens when just one of these $g_i(\mathbf{x})$ is a concave function instead. In particular, suppose that the only changes in the original Wyndor Glass Co. example are that the second and third functional constraints are replaced by $2x_2 \leq 14$ and $8x_1 - x_1^2 + 14x_2 - x_2^2 \leq 49$. Therefore, the new $g_3(\mathbf{x}) = 8x_1 - x_1^2 + 14x_2 - x_2^2$, which is a concave function since both $8x_1 - x_1^2$ and $14x_2 - x_2^2$ are concave functions. The new feasible region shown in Fig. 12.10 is *not* a convex set. Why? Because this feasible region contains pairs of points, for example, (0, 7) and (4, 3), such that part of the line segment joining these two points is not in the feasible region. Consequently, we cannot guarantee that a local maximum is a global maximum. In fact, this example has two local maxima, (0, 7) and (4, 3), but only (0, 7) is a global maximum.

Therefore, to guarantee that a local maximum is a global maximum for a nonlinear programming problem with constraints $g_i(\mathbf{x}) \leq b_i$ ($i = 1, 2, \ldots, m$) and $\mathbf{x} \geq \mathbf{0}$, the objective function $f(\mathbf{x})$ must be a *concave* function and each $g_i(\mathbf{x})$ must be a *convex* function. Such a problem is called a *convex programming problem,* which is one of the key types of nonlinear programming problems discussed in Sec. 12.3.

12.3   TYPES OF NONLINEAR PROGRAMMING PROBLEMS

Nonlinear programming problems come in many different shapes and forms. Unlike the simplex method for linear programming, no single algorithm can solve all these different types of problems. Instead, algorithms have been developed for various individual

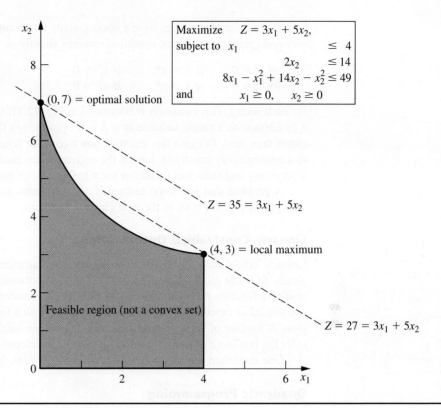

$$\text{Maximize} \quad Z = 3x_1 + 5x_2,$$
$$\text{subject to} \quad x_1 \qquad\qquad\quad \leq 4$$
$$2x_2 \qquad \leq 14$$
$$8x_1 - x_1^2 + 14x_2 - x_2^2 \leq 49$$
$$\text{and} \qquad\qquad x_1 \geq 0, \quad x_2 \geq 0$$

$(0, 7)$ = optimal solution

$Z = 35 = 3x_1 + 5x_2$

$(4, 3)$ = local maximum

Feasible region (not a convex set)

$Z = 27 = 3x_1 + 5x_2$

■ **FIGURE 12.10**
The Wyndor Glass Co. example with $2x_2 \leq 14$ and a nonlinear constraint, $8x_1 - x_1^2 + 14x_2 - x_2^2 \leq 49$, replacing the original second and third functional constraints.

classes (special types) of nonlinear programming problems. The most important classes are introduced briefly in this section. The subsequent sections then describe how some problems of these types can be solved. To simplify the discussion, we will assume throughout that the problems have been formulated (or reformulated) in the general form presented at the beginning of the chapter.

Unconstrained Optimization

Unconstrained optimization problems have *no* constraints, so the objective is simply to

$$\text{Maximize} \quad f(\mathbf{x})$$

over *all* values of $\mathbf{x} = (x_1, x_2, \ldots, x_n)$. As reviewed in Appendix 3, the *necessary* condition that a particular solution $\mathbf{x} = \mathbf{x}^*$ be optimal when $f(\mathbf{x})$ is a differentiable function is

$$\frac{\partial f}{\partial x_j} = 0 \qquad \text{at } \mathbf{x} = \mathbf{x}^*, \text{ for } j = 1, 2, \ldots, n.$$

When $f(\mathbf{x})$ is a *concave* function, this condition also is *sufficient*, so then solving for $\mathbf{x}^*$ reduces to solving the system of n equations obtained by setting the n partial derivatives equal to zero. Unfortunately, for *nonlinear* functions $f(\mathbf{x})$, these equations often are going to be *nonlinear* as well, in which case you are unlikely to be able to solve analytically for their simultaneous solution. What then? Sections 12.4 and 12.5 describe *algorithmic search procedures* for finding $\mathbf{x}^*$, first for $n = 1$ and then for $n > 1$. These procedures also play an important role in solving many of the problem types described next, where there are constraints. The reason is that many algorithms for *constrained* problems are designed so that they can focus on an *unconstrained* version of the problem during a portion of each iteration.

When a variable x_j does have a nonnegativity constraint $x_j \geq 0$, the preceding necessary and (perhaps) sufficient condition changes slightly to

$$\frac{\partial f}{\partial x_j} \begin{cases} \leq 0 & \text{at } \mathbf{x} = \mathbf{x^*}, & \text{if } x_j^* = 0 \\ = 0 & \text{at } \mathbf{x} = \mathbf{x^*}, & \text{if } x_j^* > 0 \end{cases}$$

for each such j. This condition is illustrated in Fig. 12.11, where the optimal solution for a problem with a single variable is at $x = 0$ even though the derivative there is negative rather than zero. Because this example has a concave function to be maximized subject to a nonnegativity constraint, having the derivative less than or equal to 0 at $x = 0$ is both a necessary and sufficient condition for $x = 0$ to be optimal.

A problem that has some nonnegativity constraints but no functional constraints is one special case ($m = 0$) of the next class of problems.

Linearly Constrained Optimization

Linearly constrained optimization problems are characterized by constraints that completely fit linear programming, so that *all* the $g_i(\mathbf{x})$ constraint functions are linear, but the objective function $f(\mathbf{x})$ is nonlinear. The problem is considerably simplified by having just one nonlinear function to take into account, along with a linear programming feasible region. A number of special algorithms based upon *extending* the simplex method to consider the nonlinear objective function have been developed.

One important special case, which we consider next, is quadratic programming.

Quadratic Programming

Quadratic programming problems again have linear constraints, but now the objective function $f(\mathbf{x})$ must be *quadratic*. Thus, the only difference between such a problem and a linear programming problem is that some of the terms in the objective function involve the *square* of a variable or the *product* of two variables.

■ **FIGURE 12.11**
An example that illustrates how an optimal solution can lie at a point where a derivative is negative instead of zero, because that point lies at the boundary of a nonnegativity constraint.

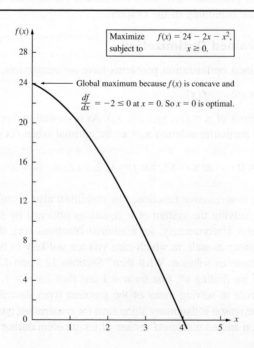

Maximize $f(x) = 24 - 2x - x^2,$
subject to $x \geq 0.$

Global maximum because $f(x)$ is concave and $\frac{df}{dx} = -2 \leq 0$ at $x = 0$. So $x = 0$ is optimal.

Many algorithms have been developed for this case under the additional assumption that $f(\mathbf{x})$ is a concave function. Section 12.7 presents an algorithm that involves a direct extension of the simplex method.

Quadratic programming is very important, partially because such formulations arise naturally in many applications. For example, the problem of portfolio selection with risky securities described in Sec. 12.1 fits into this format. However, another major reason for its importance is that a common approach to solving general linearly constrained optimization problems is to solve a sequence of quadratic programming approximations.

Convex Programming

Convex programming covers a broad class of problems that actually encompasses as special cases all the preceding types when $f(\mathbf{x})$ is a concave function to be maximized. Continuing to assume the general problem form (including maximization) presented at the beginning of the chapter, the assumptions are that

1. $f(\mathbf{x})$ is a concave function.
2. Each $g_i(\mathbf{x})$ is a convex function.

As discussed at the end of Sec. 12.2, these assumptions are enough to ensure that a local maximum is a global maximum. (If the objective were to *minimize* $f(\mathbf{x})$ instead, subject to either $g_i(\mathbf{x}) \leq b_i$ or $-g_i(\mathbf{x}) \geq b_i$ for $i = 1, 2, \ldots, m$, the first assumption would change to requiring that $f(\mathbf{x})$ must be a *convex* function, since this is what is needed to ensure that a local minimum is a global minimum.) You will see in Sec. 12.6 that the necessary and sufficient conditions for such an optimal solution are a natural generalization of the conditions just given for *unconstrained optimization* and its extension to include *nonnegativity constraints*. Section 12.9 then describes algorithmic approaches to solving convex programming problems.

Separable Programming

Separable programming is a special case of convex programming, where the one additional assumption is that

3. All the $f(\mathbf{x})$ and $g_i(\mathbf{x})$ functions are separable functions.

A **separable function** is a function where *each term* involves just a *single variable,* so that the function is separable into a sum of functions of individual variables. For example, if $f(\mathbf{x})$ is a separable function, it can be expressed as

$$f(\mathbf{x}) = \sum_{j=1}^{n} f_j(x_j),$$

where each $f_j(x_j)$ function includes only the terms involving just x_j. In the terminology of linear programming (see Sec. 3.3), separable programming problems satisfy the assumption of additivity but violate the assumption of proportionality when any of the $f_j(x_j)$ functions are nonlinear functions.

To illustrate, the objective function considered in Fig. 12.6,

$$f(x_1, x_2) = 126x_1 - 9x_1^2 + 182x_2 - 13x_2^2$$

is a separable function because it can be expressed as

$$f(x_1, x_2) = f_1(x_1) + f_2(x_2)$$

where $f_1(x_1) = 126x_1 - 9x_1^2$ and $f_2(x_2) = 182x_2 - 13x_2^2$ are each a function of a single variable—x_1 and x_2, respectively. By the same reasoning, you can verify that the objective function considered in Fig. 12.7 also is a separable function.

It is important to distinguish separable programming problems from other convex programming problems, because any such problem can be closely approximated by a linear programming problem so that the extremely efficient simplex method can be used. This approach is described in Sec. 12.8. (For simplicity, we focus there on the *linearly constrained* case where the special approach is needed only on the objective function.)

Nonconvex Programming

Nonconvex programming encompasses all nonlinear programming problems that do not satisfy the assumptions of convex programming. Now, even if you are successful in finding a *local maximum*, there is no assurance that it also will be a *global maximum*. Therefore, there is no algorithm that will find an optimal solution for all such problems. However, there do exist some algorithms that are relatively well suited for exploring various parts of the feasible region and perhaps finding a global maximum in the process. We describe this approach in Sec. 12.10. Section 12.10 also will introduce two global optimizers (available with LINGO and MPL) for finding an optimal solution for nonconvex programming problems of moderate size, as well as a search procedure that generally will find a near-optimal solution for rather large problems.

Certain specific types of nonconvex programming problems can be solved without great difficulty by special methods. Two especially important such types are discussed briefly next.

Geometric Programming

When we apply nonlinear programming to engineering design problems, as well as certain economics and statistics problems, the objective function and the constraint functions frequently take the form

$$g(\mathbf{x}) = \sum_{i=1}^{N} c_i P_i(\mathbf{x}),$$

where

$$P_i(\mathbf{x}) = x_1^{a_{i1}} x_2^{a_{i2}} \cdots x_n^{a_{in}}, \qquad \text{for } i = 1, 2, \ldots, N.$$

In such cases, the c_i and a_{ij} typically represent physical constants, and the x_j are design variables. These functions generally are neither convex nor concave, so the techniques of convex programming cannot be applied directly to these *geometric programming* problems. However, there is one important case where the problem can be transformed to an equivalent convex programming problem. This case is where *all* the c_i coefficients in each function are strictly positive, so that the functions are *generalized positive polynomials* (now called **posynomials**) and the objective function is to be minimized. The equivalent convex programming problem with decision variables $y_1, y_2, \ldots, y_n$ is then obtained by setting

$$x_j = e^{y_j}, \qquad \text{for } j = 1, 2, \ldots, n$$

throughout the original model, so now a convex programming algorithm can be applied. Alternative solution procedures also have been developed for solving these *posynomial programming* problems, as well as for geometric programming problems of other types.

Fractional Programming

Suppose that the objective function is in the form of a *fraction,* i.e., the ratio of two functions,

$$\text{Maximize} \qquad f(\mathbf{x}) = \frac{f_1(\mathbf{x})}{f_2(\mathbf{x})}.$$

Such *fractional programming* problems arise, e.g., when one is maximizing the ratio of output to person-hours expended (productivity), or profit to capital expended (rate of return), or expected value to standard deviation of some measure of performance for an investment portfolio (return/risk). Some special solution procedures have been developed for certain forms of $f_1(\mathbf{x})$ and $f_2(\mathbf{x})$.

When it can be done, the most straightforward approach to solving a fractional programming problem is to transform it to an equivalent problem of a standard type for which effective solution procedures already are available. To illustrate, suppose that $f(\mathbf{x})$ is of the *linear fractional programming* form

$$f(\mathbf{x}) = \frac{\mathbf{cx} + c_0}{\mathbf{dx} + d_0},$$

where $\mathbf{c}$ and $\mathbf{d}$ are row vectors, $\mathbf{x}$ is a column vector, and c_0 and d_0 are scalars. Also assume that the constraint functions $g_i(\mathbf{x})$ are linear, so that the constraints in matrix form are $\mathbf{Ax} \leq \mathbf{b}$ and $\mathbf{x} \geq \mathbf{0}$.

Under mild additional assumptions, we can transform the problem to an equivalent *linear programming* problem by letting

$$\mathbf{y} = \frac{\mathbf{x}}{\mathbf{dx} + d_0} \qquad \text{and} \qquad t = \frac{1}{\mathbf{dx} + d_0},$$

so that $\mathbf{x} = \mathbf{y}/t$. This result yields

Maximize $\quad Z = \mathbf{cy} + c_0 t,$

subject to

$$\mathbf{Ay} - \mathbf{b}t \leq \mathbf{0},$$
$$\mathbf{dy} + d_0 t = 1,$$

and

$$\mathbf{y} \geq \mathbf{0}, \qquad t \geq 0,$$

which can be solved by the simplex method. More generally, the same kind of transformation can be used to convert a fractional programming problem with concave $f_1(\mathbf{x})$, convex $f_2(\mathbf{x})$, and convex $g_i(\mathbf{x})$ to an equivalent convex programming problem.

The Complementarity Problem

When we deal with quadratic programming in Sec. 12.7, you will see one example of how solving certain nonlinear programming problems can be reduced to solving the complementarity problem. Given variables $w_1, w_2, \ldots, w_p$ and $z_1, z_2, \ldots, z_p$, the **complementarity problem** is to find a *feasible* solution for the set of constraints

$$\mathbf{w} = F(\mathbf{z}), \qquad \mathbf{w} \geq \mathbf{0}, \qquad \mathbf{z} \geq \mathbf{0}$$

that also satisfies the **complementarity contraint**

$$\mathbf{w}^T \mathbf{z} = 0.$$

Here, $\mathbf{w}$ and $\mathbf{z}$ are column vectors, F is a given vector-valued function, and the superscript T denotes the transpose (see Appendix 4). The problem has no objective function, so technically it is not a full-fledged nonlinear programming problem. It is called the complementarity problem because of the complementary relationships that either

$$w_i = 0 \qquad \text{or} \qquad z_i = 0 \qquad \text{(or both)} \qquad \text{for each } i = 1, 2, \ldots, p.$$

An important special case is the **linear complementarity problem,** where

$$F(\mathbf{z}) = \mathbf{q} + \mathbf{Mz},$$

where $\mathbf{q}$ is a given column vector and $\mathbf{M}$ is a given $p \times p$ matrix. Efficient algorithms have been developed for solving this problem under suitable assumptions[6] about the properties of the matrix $\mathbf{M}.$ One type involves pivoting from one basic feasible (BF) solution to the next, much like the simplex method for linear programming.

In addition to having applications in nonlinear programming, complementarity problems have applications in game theory, economic equilibrium problems, and engineering equilibrium problems.

12.4   ONE-VARIABLE UNCONSTRAINED OPTIMIZATION

We now begin discussing how to solve some of the types of problems just described by considering the simplest case—*unconstrained optimization* with just a single variable x ($n = 1$), where the differentiable function $f(x)$ to be maximized is *concave.*[7] Thus, the *necessary and sufficient condition* for a particular solution $x = x^*$ to be optimal (a global maximum) is

$$\frac{df}{dx} = 0 \qquad \text{at } x = x^*,$$

as depicted in Fig. 12.12. If this equation can be solved directly for x^*, you are done. However, if $f(x)$ is not a particularly simple function, so the derivative is not just a linear or quadratic function, you may not be able to solve the equation *analytically.* If not, a number of *search procedures* are available for solving the problem *numerically.*

The approach with any of these search procedures is to find a sequence of *trial solutions* that leads toward an optimal solution. At each iteration, you begin at the current trial solution to conduct a systematic search that culminates by identifying a new *improved*

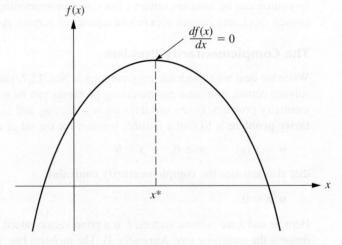

■ FIGURE 12.12
The one-variable unconstrained optimization problem when the function is concave.

[6]See R. W. Cottle, J.-S. Pang, and R. E. Stone, *The Linear Complementarity Problem,* Academic Press, Boston, 1992.
[7]See the beginning of Appendix 3 for a review of the corresponding case when $f(x)$ is not concave.

trial solution. The procedure is continued until the trial solutions have converged to an optimal solution, assuming that one exists.

We now will describe two common search procedures. The first one (the *bisection method*) was chosen because it is such an intuitive and straightforward procedure. The second one (*Newton's method*) is included because it plays a fundamental role in nonlinear programming in general.

The Bisection Method

This search procedure always can be applied when $f(x)$ is concave (so that the second derivative is negative or zero for all x) as depicted in Fig. 12.12. It also can be used for certain other functions as well. In particular, if x^* denotes the optimal solution, all that is needed[8] is that

$$\frac{df(x)}{dx} > 0 \quad \text{if } x < x^*,$$

$$\frac{df(x)}{dx} = 0 \quad \text{if } x = x^*,$$

$$\frac{df(x)}{dx} < 0 \quad \text{if } x > x^*.$$

These conditions automatically hold when $f(x)$ is concave, but they also can hold when the second derivative is positive for some (but not all) values of x.

The idea behind the bisection method is a very intuitive one, namely, that whether the slope (derivative) is positive or negative at a trial solution definitely indicates whether improvement lies immediately to the right or left, respectively. Thus, if the derivative evaluated at a particular value of x is *positive,* then x^* must be larger than this x (see Fig. 12.12), so this x becomes a *lower bound* on the trial solutions that need to be considered thereafter. Conversely, if the derivative is *negative,* then x^* must be *smaller* than this x, so x would become an *upper bound.* Therefore, after both types of bounds have been identified, each new trial solution selected between the current bounds provides a new tighter bound of one type, thereby narrowing the search further. As long as a reasonable rule is used to select each trial solution in this way, the resulting *sequence* of trial solutions must *converge* to x^*. In practice, this means continuing the sequence until the distance between the bounds is sufficiently small that the next trial solution must be within a prespecified *error tolerance* of x^*.

This entire process is summarized next, given the notation

x' = current trial solution,

$\underline{x}$ = current lower bound on x^*,

$\bar{x}$ = current upper bound on x^*,

ϵ = error tolerance for x^*.

Although there are several reasonable rules for selecting each new trial solution, the one used in the bisection method is the **midpoint rule** (traditionally called the *Bolzano search plan*), which says simply to select the midpoint between the two current bounds.

---

[8]Another possibility is that the graph of $f(x)$ is flat at the top so that x is optimal over some interval $[a, b]$. In this case, the procedure still will converge to one of these optimal solutions as long as the derivative is positive for $x < a$ and negative for $x > b$.

Summary of the Bisection Method

Initialization: Select ϵ. Find an initial $\underline{x}$ and $\overline{x}$ by inspection (or by respectively finding any value of x at which the derivative is positive and then negative). Select an initial trial solution

$$x' = \frac{\underline{x} + \overline{x}}{2}.$$

Iteration:

1. Evaluate $\dfrac{df(x)}{dx}$ at $x = x'$.

2. If $\dfrac{df(x)}{dx} \geq 0$, reset $\underline{x} = x'$.

3. If $\dfrac{df(x)}{dx} \leq 0$, reset $\overline{x} = x'$.

4. Select a new $x' = \dfrac{\underline{x} + \overline{x}}{2}$.

Stopping rule: If $\overline{x} - \underline{x} \leq 2\epsilon$, so that the new x' must be within ϵ of x^*, stop. Otherwise, perform another iteration.

We shall now illustrate the bisection method by applying it to the following example.

Example. Suppose that the function to be maximized is

$$f(x) = 12x - 3x^4 - 2x^6,$$

as plotted in Fig. 12.13. Its first two derivatives are

$$\frac{df(x)}{dx} = 12(1 - x^3 - x^5),$$

$$\frac{d^2f(x)}{dx^2} = -12(3x^2 + 5x^4).$$

■ **FIGURE 12.13**
Example for the bisection method.

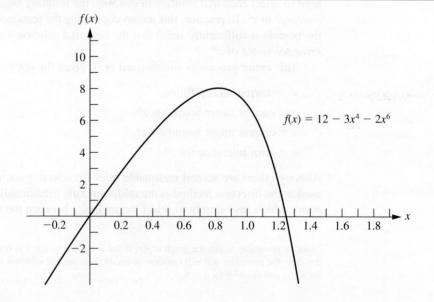

$$f(x) = 12 - 3x^4 - 2x^6$$

■ **TABLE 12.1** Application of the bisection method to the example

Iteration	$\dfrac{df(x)}{dx}$	$\underline{x}$	$\bar{x}$	New x'	$f(x')$
0	0	0	2	1	7.0000
1	−12	0	1	0.5	5.7812
2	+10.12	0.5	1	0.75	7.6948
3	+4.09	0.75	1	0.875	7.8439
4	−2.19	0.75	0.875	0.8125	7.8672
5	+1.31	0.8125	0.875	0.84375	7.8829
6	−0.34	0.8125	0.84375	0.828125	7.8815
7	+0.51	0.828125	0.84375	0.8359375	7.8839
Stop					

Because the second derivative is nonpositive everywhere, $f(x)$ is a concave function, so the bisection method can be safely applied to find its global maximum (assuming a global maximum exists).

A quick inspection of this function (without even constructing its graph as shown in Fig. 12.13) indicates that $f(x)$ is positive for small positive values of x, but it is negative for $x < 0$ or $x > 2$. Therefore, $\underline{x} = 0$ and $\bar{x} = 2$ can be used as the initial bounds, with their midpoint, $x' = 1$, as the initial trial solution. Let $\epsilon = 0.01$ be the error tolerance for x^* in the stopping rule, so the final $(\bar{x} - \underline{x}) \leq 0.02$ with the final x' at the midpoint.

Applying the bisection method then yields the sequence of results shown in Table 12.1. [This table includes both the function and derivative values for your information, where the derivative is evaluated at the trial solution generated at the *preceding* iteration. However, note that the algorithm actually doesn't need to calculate $f(x')$ at all and that it only needs to calculate the derivative far enough to determine its sign.] The conclusion is that

$$x^* \approx 0.836,$$
$$0.828125 < x^* < 0.84375.$$

Your IOR Tutorial includes an interactive procedure for executing the bisection method.

Newton's Method

Although the bisection method is an intuitive and straightforward procedure, it has the disadvantage of converging relatively slowly toward an optimal solution. Each iteration only decreases the difference between the bounds by one-half. Therefore, even with the fairly simple function being considered in Table 12.1, seven iterations were required to reduce the error tolerance for x^* to less than 0.01. Another seven iterations would be needed to reduce this error tolerance to less than 0.0001.

The basic reason for this slow convergence is that the only information about $f(x)$ being used is the value of the first derivative $f'(x)$ at the respective trial values of x. Additional helpful information can be obtained by considering the second derivative $f''(x)$ as well. This is what *Newton's method*[9] does.

---

[9]This method is due to the great 17th-century mathematician and physicist, Sir Isaac Newton. While a young student at the University of Cambridge (England), Newton took advantage of the university being closed for two years (due to the bubonic plague that devastated Europe in 1664–65) to discover the law of universal gravitation and invent calculus (among other achievements). His development of calculus led to this method.

The basic idea behind Newton's method is to approximate $f(x)$ within the neighborhood of the current trial solution by a quadratic function and then to maximize (or minimize) the approximate function exactly to obtain the new trial solution to start the next iteration. (This idea of working with a **quadratic approximation** of the objective function has since been made a key feature of many algorithms for more general kinds of nonlinear programming problems.) This approximating quadratic function is obtained by truncating the Taylor series after the second derivative term. In particular, by letting x_{i+1} be the trial solution generated at iteration i to start iteration $i + 1$ (so x_1 is the initial trial solution provided by the user to begin iteration 1), the truncated Taylor series for x_{i+1} is

$$f(x_{i+1}) \approx f(x_i) + f'(x_i)(x_{i+1} - x_i) + \frac{f''(x_i)}{2}(x_{i+1} - x_i)^2.$$

Having fixed x_i at the beginning of iteration i, note that $f(x_i)$, $f'(x_i)$, and $f''(x_i)$ also are fixed constants in this approximating function on the right. Thus, this approximating function is just a quadratic function of x_{i+1}. Furthermore, this quadratic function is such a good approximation of $f(x_{i+1})$ in the neighborhood of x_i that their values and their first and second derivatives are exactly the same when $x_{i+1} = x_i$.

This quadratic function now can be maximized in the usual way by setting its first derivative to zero and solving for x_{i+1}. (Remember that we are assuming that $f(x)$ is concave, which implies that this quadratic function is concave, so the solution when setting the first derivative to zero will be a global maximum.) This first derivative is

$$f'(x_{i+1}) \approx f'(x_i) + f''(x_i)(x_{i+1} - x_i)$$

since x_i, $f(x_i)$, $f'(x_i)$, and $f''(x_i)$ are constants. Setting the first derivative on the right to zero yields

$$f'(x_{i+1}) + f''(x_i)(x_{i+1} - x_i) = 0,$$

which directly leads algebraically to the solution,

$$x_{i+1} = x_i - \frac{f'(x_i)}{f''(x_i)}.$$

This is the key formula that is used at each iteration i to calculate the next trial solution x_{i+1} after obtaining the trial solution x_i to begin iteration i and then calculating the first and second derivatives at x_i. (The same formula is used when minimizing a convex function.)

Iterations generating new trial solutions in this way would continue until these solutions have essentially converged. One criterion for convergence is that $|x_{i+1} - x_i|$ has become sufficiently small. Another is that $f'(x)$ is sufficiently close to zero. Still another is that $|f(x_{i+1}) - f(x_i)|$ is sufficiently small. Choosing the first criterion, define ϵ as the value such that the algorithm is stopped when $|x_{i+1} - x_i| \leq \epsilon$.

Here is a complete description of the algorithm.

Summary of Newton's Method

Initialization: Select ϵ. Find an initial trial solution x_i by inspection. Set $i = 1$.

Iteration i:

1. Calculate $f'(x_i)$ and $f''(x_i)$. [Calculating $f(x_i)$ is optional.]

2. Set $x_{i+1} = x_i - \dfrac{f'(x_i)}{f''(x_i)}$.

Stopping Rule: If $|x_{i+1} - x_i| \leq \epsilon$, stop; x_{i+1} is essentially the optimal solution. Otherwise, reset $i = i + 1$ and perform another iteration.

■ **TABLE 12.2** Application of Newton's method to the example

Iteration i	x_i	$f(x_i)$	$f'(x_i)$	$f''(x_i)$	x_{i+1}
1	1	7	-12	-96	0.875
2	0.875	7.8439	-2.1940	-62.733	0.84003
3	0.84003	7.8838	-0.1325	-55.279	0.83763
4	0.83763	7.8839	-0.0006	-54.790	0.83762

Example. We now will apply Newton's method to the same example used for the bisection method. As depicted in Fig. 12.13, the function to be maximized is

$$f(x) = 12x - 3x^4 - 2x^6.$$

Thus, the formula for calculating the new trial solution (x_{i+1}) from the current one (x_i) is

$$x_{i+1} = x_i - \frac{f'(x_i)}{f''(x_i)} = x_i - \frac{12(1 - x^3 - x^5)}{-12(3x^2 + 5x^4)} = x_i + \frac{1 - x^3 - x^5}{3x^2 + 5x^4}.$$

After selecting $\epsilon = 0.00001$ and choosing $x_1 = 1$ as the initial trial solution, Table 12.2 shows the results from applying Newton's method to this example. After just four iterations, this method has converged to $x = 0.83762$ as the optimal solution with a very high degree of precision.

A comparison of this table with Table 12.1 illustrates how much more rapidly Newton's method converges than the bisection method. Nearly 20 iterations would be required for the bisection method to converge with the same degree of precision that Newton's method achieved after only four iterations.

Although this rapid convergence is fairly typical of Newton's method, its performance does vary from problem to problem. Since the method is based on using a quadratic approximation of $f(x)$, its performance is affected by the degree of accuracy of the approximation.

■ 12.5 MULTIVARIABLE UNCONSTRAINED OPTIMIZATION

Now consider the problem of maximizing a *concave* function $f(\mathbf{x})$ of *multiple* variables $\mathbf{x} = (x_1, x_2, \ldots, x_n)$ when there are no constraints on the feasible values. Suppose again that the necessary and sufficient condition for optimality, given by the system of equations obtained by setting the respective partial derivatives equal to zero (see Sec. 12.3), cannot be solved analytically, so that a numerical search procedure must be used.

As for the one-variable case, a number of search procedures are available for solving such a problem numerically. One of these (the *gradient search procedure*) is an especially important one because it identifies and uses the direction of movement from the current trial solution that maximizes the rate at which $f(\mathbf{x})$ is increased. This is one of the key ideas of nonlinear programming. Adaptations of this same idea to take constraints into account are a central feature of many algorithms for *constrained* optimization as well.

After discussing this procedure in some detail, we will briefly describe how Newton's method is extended to the multivariable case.

The Gradient Search Procedure

In Sec. 12.4, the value of the ordinary derivative was used by the bisection method to select one of just two possible directions (increase x or decrease x) in which to move from the current trial solution to the next one. The goal was to reach a point eventually where this derivative is (essentially) 0. Now, there are *innumerable* possible directions in which to move; they correspond to the possible *proportional rates* at which the respective variables can be changed. The goal is to reach a point eventually where all the partial derivatives are (essentially) 0. Therefore, a natural approach is to use the values of the *partial* derivatives to select the specific direction in which to move. This selection involves using the gradient of the objective function, as described next.

Because the objective function $f(\mathbf{x})$ is assumed to be differentiable, it possesses a gradient, denoted by $\nabla f(\mathbf{x})$, at each point $\mathbf{x}$. In particular, the **gradient** at a specific point $\mathbf{x} = \mathbf{x}'$ is the *vector* whose elements are the respective *partial derivatives* evaluated at $\mathbf{x} = \mathbf{x}'$, so that

$$\nabla f(\mathbf{x}') = \left(\frac{\partial f}{\partial x_1}, \frac{\partial f}{\partial x_2}, \cdots, \frac{\partial f}{\partial x_n} \right) \qquad \text{at } \mathbf{x} = \mathbf{x}'.$$

The significance of the gradient is that the (infinitesimal) change in $\mathbf{x}$ that *maximizes* the rate at which $f(\mathbf{x})$ increases is the change that is *proportional* to $\nabla f(\mathbf{x})$. To express this idea geometrically, the "direction" of the gradient $\nabla f(\mathbf{x}')$ is interpreted as the *direction* of the directed line segment (arrow) from the origin $(0, 0, \ldots, 0)$ to the point $(\partial f/\partial x_1, \partial f/\partial x_2, \ldots, \partial f/\partial x_n)$, where $\partial f/\partial x_j$ is evaluated at $x_j = x_j'$. Therefore, it may be said that the rate at which $f(\mathbf{x})$ increases is maximized if (infinitesimal) changes in $\mathbf{x}$ are in the *direction* of the gradient $\nabla f(\mathbf{x})$. Because the objective is to find the feasible solution maximizing $f(\mathbf{x})$, it would seem expedient to attempt to move in the direction of the gradient as much as possible.

Because the current problem has no constraints, this interpretation of the gradient suggests that an efficient search procedure should keep moving in the direction of the gradient until it (essentially) reaches an optimal solution $\mathbf{x}^*$, where $\nabla f(\mathbf{x}^*) = \mathbf{0}$. However, normally it would not be practical to change $\mathbf{x}$ *continuously* in the direction of $\nabla f(\mathbf{x})$, because this series of changes would require continuously *reevaluating* the $\partial f/\partial x_j$ and changing the direction of the path. Therefore, a better approach is to keep moving in a *fixed* direction from the current trial solution, not stopping until $f(\mathbf{x})$ stops increasing. This stopping point would be the next trial solution, so the gradient then would be recalculated to determine the new direction in which to move. With this approach, each iteration involves changing the current trial solution $\mathbf{x}'$ as follows:

Reset $\mathbf{x}' = \mathbf{x}' + t^* \nabla f(\mathbf{x}')$,

where t^* is the positive value of t that *maximizes* $f(\mathbf{x}' + t \nabla f(\mathbf{x}'))$; that is,

$$f(\mathbf{x}' + t^* \nabla f(\mathbf{x}')) = \max_{t \geq 0} f(\mathbf{x}' + t \nabla f(\mathbf{x}')).$$

[Note that $f(\mathbf{x}' + t \nabla f(\mathbf{x}'))$ is simply $f(\mathbf{x})$ where

$$x_j = x_j' + t \left(\frac{\partial f}{\partial x_j} \right)_{\mathbf{x}=\mathbf{x}'}, \qquad \text{for } j = 1, 2, \ldots, n,$$

and that these expressions for the x_j involve only constants and t, so $f(\mathbf{x})$ becomes a function of just the single variable t.] The iterations of this gradient search procedure continue until $\nabla f(\mathbf{x}) = 0$ within a small tolerance ϵ, that is, until

$$\left| \frac{\partial f}{\partial x_j} \right| \leq \epsilon \qquad \text{for } j = 1, 2, \ldots, n.\text{[10]}$$

---

[10]This stopping rule generally will provide a solution $\mathbf{x}$ that is close to an optimal solution $\mathbf{x}^*$, with a value of $f(\mathbf{x})$ that is very close to $f(\mathbf{x}^*)$. However, this cannot be guaranteed, since it is possible that the function maintains a very small positive slope ($\leq \epsilon$) over a great distance from $\mathbf{x}$ to $\mathbf{x}^*$.

An analogy may help to clarify this procedure. Suppose that you need to climb to the top of a hill. You are nearsighted, so you cannot see the top of the hill in order to walk directly in that direction. However, when you stand still, you can see the ground around your feet well enough to determine the direction in which the hill is sloping upward most sharply. You are able to walk in a straight line. While walking, you also are able to tell when you stop climbing (zero slope in your direction). Assuming that the hill is *concave,* you now can use the *gradient search procedure* for climbing to the top efficiently. This problem is a *two-variable problem,* where (x_1, x_2) represents the coordinates (ignoring height) of your current location. The function $f(x_1, x_2)$ gives the height of the hill at (x_1, x_2). You start each iteration at your current location (current trial solution) by determining the direction [in the (x_1, x_2) coordinate system] in which the hill is sloping upward most sharply (the direction of the gradient) at this point. You then begin walking in this fixed direction and continue as long as you still are climbing. You eventually stop at a new trial location (solution) when the hill becomes level in your direction, at which point you prepare to do another iteration in another direction. You continue these iterations, following a zigzag path up the hill, until you reach a trial location where the slope is essentially zero in all directions. Under the assumption that the hill [$f(x_1, x_2)$] is concave, you must then be essentially at the top of the hill.

The most difficult part of the gradient search procedure usually is to find t^*, the value of t that maximizes f in the direction of the gradient, at each iteration. Because $\mathbf{x}$ and $\nabla f(\mathbf{x})$ have fixed values for the maximization, and because $f(\mathbf{x})$ is concave, this problem should be viewed as maximizing a *concave* function of a *single variable t.* Therefore, it can be solved by the kind of search procedures for one-variable unconstrained optimization that are described in Sec. 12.4 (while considering only nonnegative values of t because of the $t \geq 0$ constraint). Alternatively, if f is a simple function, it may be possible to obtain an analytical solution by setting the derivative with respect to t equal to zero and solving.

Summary of the Gradient Search Procedure

Initialization: Select ϵ and any initial trial solution x'. Go first to the stopping rule.
Iteration:

1. Express $f(\mathbf{x}' + t \nabla f(\mathbf{x}'))$ as a function of t by setting

$$x_j = x_j' + t \left(\frac{\partial f}{\partial x_j} \right)_{\mathbf{x}=\mathbf{x}'}, \qquad \text{for } j = 1, 2, \ldots, n,$$

and then substituting these expressions into $f(\mathbf{x})$.
2. Use a search procedure for one-variable unconstrained optimization (or calculus) to find $t = t^*$ that maximizes $f(\mathbf{x}' + t \nabla f(\mathbf{x}'))$ over $t \geq 0$.
3. Reset $\mathbf{x}' = \mathbf{x}' + t^* \nabla f(\mathbf{x}')$. Then go to the stopping rule.

Stopping rule: Evaluate $\nabla f(\mathbf{x}')$ at $\mathbf{x} = \mathbf{x}'$. Check if

$$\left| \frac{\partial f}{\partial x_j} \right| \leq \epsilon \qquad \text{for all } j = 1, 2, \ldots, n.$$

If so, stop with the current $\mathbf{x}'$ as the desired approximation of an optimal solution $\mathbf{x}^*$. Otherwise, perform another iteration.

Now let us illustrate this procedure.

Example. Consider the following two-variable problem:

Maximize $\quad f(\mathbf{x}) = 2x_1x_2 + 2x_2 - x_1^2 - 2x_2^2.$

Thus,

$$\frac{\partial f}{\partial x_1} = 2x_2 - 2x_1,$$

$$\frac{\partial f}{\partial x_2} = 2x_1 + 2 - 4x_2.$$

We also can verify (see Appendix 2) that $f(\mathbf{x})$ is concave.

To begin the gradient search procedure, after choosing a suitably small value of ϵ (normally well under 0.1) suppose that $\mathbf{x} = (0, 0)$ is selected as the initial trial solution. Because the respective partial derivatives are 0 and 2 at this point, the gradient is

$$\nabla f(0, 0) = (0, 2).$$

With $\epsilon < 2$, the stopping rule then says to perform an iteration.

Iteration 1: With values of 0 and 2 for the respective partial derivatives, the first iteration begins by setting

$$x_1 = 0 + t(0) = 0,$$
$$x_2 = 0 + t(2) = 2t,$$

and then substituting these expressions into $f(\mathbf{x})$ to obtain

$$
\begin{aligned}
f(\mathbf{x}' + t\,\nabla f(\mathbf{x}')) &= f(0, 2t) \\
&= 2(0)(2t) + 2(2t) - 0^2 - 2(2t)^2 \\
&= 4t - 8t^2.
\end{aligned}
$$

Because

$$f(0, 2t^*) = \max_{t \geq 0} f(0, 2t) = \max_{t \geq 0} \{4t - 8t^2\}$$

and

$$\frac{d}{dt}(4t - 8t^2) = 4 - 16t = 0,$$

it follows that

$$t^* = \frac{1}{4},$$

so

Reset $\mathbf{x}' = (0, 0) + \frac{1}{4}(0, 2) = \left(0, \frac{1}{2}\right).$

This completes the first iteration. For this new trial solution, the gradient is

$$\nabla f\left(0, \frac{1}{2}\right) = (1, 0).$$

With $\epsilon < 1$, the stopping rule now says to perform another iteration.

Iteration 2: To begin the second iteration, use the values of 1 and 0 for the respective partial derivatives to set

$$\mathbf{x} = \left(0, \frac{1}{2}\right) + t(1, 0) = \left(t, \frac{1}{2}\right),$$

so

$$
f(\mathbf{x}' + t\,\nabla f(\mathbf{x}')) = f\left(0 + t, \frac{1}{2} + 0t\right) = f\left(t, \frac{1}{2}\right)
$$

$$
= (2t)\left(\frac{1}{2}\right) + 2\left(\frac{1}{2}\right) - t^2 - 2\left(\frac{1}{2}\right)^2
$$

$$
= t - t^2 + \frac{1}{2}.
$$

Because

$$
f\left(t^*, \frac{1}{2}\right) = \max_{t \geq 0} f\left(t, \frac{1}{2}\right) = \max_{t \geq 0}\left\{t - t^2 + \frac{1}{2}\right\}
$$

and

$$
\frac{d}{dt}\left(t - t^2 + \frac{1}{2}\right) = 1 - 2t = 0,
$$

then

$$
t^* = \frac{1}{2},
$$

so

Reset $\mathbf{x}' = \left(0, \dfrac{1}{2}\right) + \dfrac{1}{2}(1, 0) = \left(\dfrac{1}{2}, \dfrac{1}{2}\right).$

This completes the second iteration. With a typically small value of ϵ, the procedure now would continue on to several more iterations in a similar fashion. (We will forego the details.)

A nice way of organizing this work is to write out a table such as Table 12.3 which summarizes the preceding two iterations. At each iteration, the second column shows the current trial solution, and the rightmost column shows the eventual new trial solution, which then is carried down into the second column for the next iteration. The fourth column gives the expressions for the x_j in terms of t that need to be substituted into $f(\mathbf{x})$ to give the fifth column.

By continuing in this fashion, the subsequent trial solutions would be $(\frac{1}{2}, \frac{3}{4})$, $(\frac{3}{4}, \frac{3}{4})$, $(\frac{3}{4}, \frac{7}{8})$, $(\frac{7}{8}, \frac{7}{8})$, . . . , as shown in Fig. 12.14. Because these points are converging to $\mathbf{x}^* = (1, 1)$, this solution is the optimal solution, as verified by the fact that

$$
\nabla f(1, 1) = (0, 0).
$$

However, because this converging sequence of trial solutions never reaches its limit, the procedure actually will stop somewhere (depending on ϵ) slightly below $(1, 1)$ as its final approximation of $\mathbf{x}^*$.

As Fig. 12.14 suggests, the gradient search procedure zigzags to the optimal solution rather than moving in a straight line. Some modifications of the procedure have been

TABLE 12.3 Application of the gradient search procedure to the example

Iteration	$\mathbf{x}'$	$\nabla f(\mathbf{x}')$	$\mathbf{x}' + t\,\nabla f(\mathbf{x}')$	$f(\mathbf{x}' + t\,\nabla f(\mathbf{x}'))$	t^*	$\mathbf{x}' + t^*\,\nabla f(\mathbf{x}')$
1	$(0, 0)$	$(0, 2)$	$(0, 2t)$	$4t - 8t^2$	$\frac{1}{4}$	$\left(0, \frac{1}{2}\right)$
2	$\left(0, \frac{1}{2}\right)$	$(1, 0)$	$\left(t, \frac{1}{2}\right)$	$t - t^2 + \frac{1}{2}$	$\frac{1}{2}$	$\left(\frac{1}{2}, \frac{1}{2}\right)$

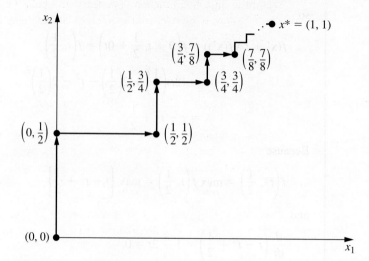

■ **FIGURE 12.14**
Illustration of the gradient search procedure when $f(x_1, x_2) = 2x_1x_2 + 2x_2 - x_1^2 - 2x_2^2$.

developed that *accelerate* movement toward the optimal solution by taking this zigzag behavior into account.

If $f(\mathbf{x})$ were *not* a concave function, the gradient search procedure still would converge to a *local* maximum. The only change in the description of the procedure for this case is that t^* now would correspond to the *first local maximum* of $f(\mathbf{x}' + t\,\nabla f(\mathbf{x}'))$ as t is increased from 0.

If the objective were to *minimize* $f(\mathbf{x})$ instead, one change in the procedure would be to move in the *opposite* direction of the gradient at each iteration. In other words, the rule for obtaining the next point would be

Reset $\mathbf{x}' = \mathbf{x}' - t^*\,\nabla f(\mathbf{x}')$.

The only other change is that t^* now would be the nonnegative value of t that *minimizes* $f(\mathbf{x}' - t\,\nabla f(\mathbf{x}'))$; that is,

$$f(\mathbf{x}' - t^*\,\nabla f(\mathbf{x}')) = \min_{t \geq 0} f(\mathbf{x}' - t\,\nabla f(\mathbf{x}')).$$

Additional examples of the application of the gradient search procedure are included in both the Worked Examples section of the book's website and your OR Tutor. The IOR Tutorial includes both an interactive procedure and an automatic procedure for applying this algorithm.

Newton's Method

Section 12.4 describes how Newton's method would be used to solve *one-variable* unconstrained optimization problems. The general version of Newton's method actually is designed to solve *multivariable* unconstrained optimization problems. The basic idea is the same as described in Sec. 12.4, namely, work with a *quadratic approximation* of the objective function $f(\mathbf{x})$, where $\mathbf{x} = (x_1, x_2, \ldots, x_n)$ in this case. This approximating quadratic function is obtained by truncating the Taylor series around the current trial solution after the second derivative term. This approximate function then is maximized (or minimized) exactly to obtain the new trial solution to start the next iteration.

When the objective function is concave and both the current trial solution $\mathbf{x}$ and its gradient $\nabla f(\mathbf{x})$ are written as *column vectors,* the solution $\mathbf{x}'$ that maximizes the approximating quadratic function has the form,

$$\mathbf{x}' = \mathbf{x} - \left[\nabla^2 f(\mathbf{x})\right]^{-1} \nabla f(\mathbf{x}),$$

where $\nabla^2 f(\mathbf{x})$ is the $n \times n$ matrix (called the *Hessian matrix*) of the second partial derivatives of $f(\mathbf{x})$ evaluated at the current trial solution $\mathbf{x}$ and $\left[\nabla^2 f(\mathbf{x})\right]^{-1}$ is the *inverse* of this Hessian matrix.

Nonlinear programming algorithms that employ Newton's method (including those that adapt it to help deal with *constrained* optimization problems) commonly approximate the inverse of the Hessian matrix in various ways. These approximations of Newton's method are referred to as **quasi-Newton methods** (or *variable metric methods*). We will comment further on the important role of these methods in nonlinear programming in Sec. 12.9.

Further description of these methods is beyond the scope of this book, but further details can be found in books devoted to nonlinear programming.

12.6 THE KARUSH-KUHN-TUCKER (KKT) CONDITIONS FOR CONSTRAINED OPTIMIZATION

We now focus on the question of how to recognize an *optimal solution* for a nonlinear programming problem (with differentiable functions). What are the necessary and (perhaps) sufficient conditions that such a solution must satisfy?

In the preceding sections we already noted these conditions for *unconstrained optimization,* as summarized in the first two rows of Table 12.4. Early in Sec. 12.3 we also gave these conditions for the slight *extension* of unconstrained optimization where the *only* constraints are nonnegativity constraints. These conditions are shown in the third row of Table 12.4. As indicated in the last row of the table, the conditions for the general case are called the **Karush-Kuhn-Tucker conditions** (or **KKT conditions**), because they were derived independently by Karush[11] and by Kuhn and Tucker.[12] Their basic result is embodied in the following theorem.

TABLE 12.4 Necessary and sufficient conditions for optimality

Problem	Necessary Conditions for Optimality	Also Sufficient If:
One-variable unconstrained	$\dfrac{df}{dx} = 0$	$f(x)$ concave
Multivariable unconstrained	$\dfrac{\partial f}{\partial x_j} = 0 \quad (j = 1, 2, \ldots, n)$	$f(\mathbf{x})$ concave
Constrained, nonnegativity constraints only	$\dfrac{\partial f}{\partial x_j} = 0 \quad (j = 1, 2, \ldots, n)$ (or ≤ 0 if $x_j = 0$)	$f(\mathbf{x})$ concave
General constrained problem	Karush-Kuhn-Tucker conditions	$f(\mathbf{x})$ concave and $g_i(\mathbf{x})$ convex $(i = 1, 2, \ldots, m)$

[11]W. Karush, "Minima of Functions of Several Variables with Inequalities as Side Conditions," M.S. thesis, Department of Mathematics, University of Chicago, 1939.

[12]H. W. Kuhn and A. W. Tucker, "Nonlinear Programming," in Jerzy Neyman (ed.), *Proceedings of the Second Berkeley Symposium,* University of California Press, Berkeley, 1951, pp. 481–492.

Theorem.   Assume that $f(\mathbf{x})$, $g_1(\mathbf{x})$, $g_2(\mathbf{x})$, . . . , $g_m(\mathbf{x})$ are *differentiable* functions satisfying certain regularity conditions.[13] Then

$$\mathbf{x}^* = (x_1^*, x_2^*, \ldots, x_n^*)$$

can be an *optimal solution* for the nonlinear programming problem only if there exist m numbers $u_1, u_2, \ldots, u_m$ such that *all* the following *KKT conditions* are satisfied:

$$
\left.
\begin{aligned}
&\textbf{1.}\;\; \frac{\partial f}{\partial x_j} - \sum_{i=1}^{m} u_i \frac{\partial g_i}{\partial x_j} \leq 0 \\[2mm]
&\textbf{2.}\;\; x_j^* \left(\frac{\partial f}{\partial x_j} - \sum_{i=1}^{m} u_i \frac{\partial g_i}{\partial x_j} \right) = 0
\end{aligned}
\right\}
\quad \text{at } \mathbf{x} = \mathbf{x}^*, \text{ for } j = 1, 2, \ldots, n.
$$

$$
\left.
\begin{aligned}
&\textbf{3.}\;\; g_i(\mathbf{x}^*) - b_i \leq 0 \\
&\textbf{4.}\;\; u_i[g_i(\mathbf{x}^*) - b_i] = 0
\end{aligned}
\right\}
\quad \text{for } i = 1, 2, \ldots, m.
$$

5. $x_j^* \geq 0$,   for $j = 1, 2, \ldots, n.$

6. $u_i \geq 0$,   for $i = 1, 2, \ldots, m.$

Note that both conditions 2 and 4 require that the product of two quantities be zero. Therefore, each of these conditions really is saying that at least one of the two quantities must be zero. Consequently, condition 4 can be combined with condition 3 to express them in another equivalent form as

$$(3, 4) \qquad g_i(\mathbf{x}^*) - b_i = 0$$
$$\text{(or} \leq 0 \;\; \text{if } u_i = 0), \qquad \text{for } i = 1, 2, \ldots, m.$$

Similarly, condition 2 can be combined with condition 1 as

$$(1, 2) \qquad \frac{\partial f}{\partial x_j} - \sum_{i=1}^{m} u_i \frac{\partial g_i}{\partial x_j} = 0$$
$$\text{(or} \leq 0 \;\; \text{if } x_j^* = 0), \qquad \text{for } j = 1, 2, \ldots, n.$$

When $m = 0$ (no functional constraints), this summation drops out and the combined condition (1, 2) reduces to the condition given in the third row of Table 12.4. Thus, for $m > 0$, each term in the summation modifies the $m = 0$ condition to incorporate the effect of the corresponding functional constraint.

In conditions 1, 2, 4, and 6, the u_i correspond to the *dual variables* of linear programming (we expand on this correspondence at the end of the section), and they have a comparable economic interpretation. However, the u_i actually arose in the mathematical derivation as *Lagrange multipliers* (discussed in Appendix 3). Conditions 3 and 5 do nothing more than ensure the feasibility of the solution. The other conditions eliminate most of the feasible solutions as possible candidates for an optimal solution.

However, note that satisfying these conditions does not guarantee that the solution is optimal. As summarized in the rightmost column of Table 12.4, certain additional *convexity* assumptions are needed to obtain this guarantee. These assumptions are spelled out in the following extension of the theorem.

Corollary.   Assume that $f(\mathbf{x})$ is a *concave* function and that $g_1(\mathbf{x})$, $g_2(\mathbf{x})$, . . . , $g_m(\mathbf{x})$ are *convex* functions (i.e., this problem is a convex programming problem), where all these functions satisfy the regularity conditions. Then $\mathbf{x}^* = (x_1^*, x_2^*, \ldots, x_n^*)$ is an *optimal solution* if and only if all the conditions of the theorem are satisfied.

---

[13]Ibid., p. 483.

Example. To illustrate the formulation and application of the *KKT conditions,* we consider the following two-variable nonlinear programming problem:

$$\text{Maximize} \quad f(\mathbf{x}) = \ln(x_1 + 1) + x_2,$$

subject to

$$2x_1 + x_2 \leq 3$$

and

$$x_1 \geq 0, \qquad x_2 \geq 0,$$

where ln denotes the natural logarithm. Thus, $m = 1$ (one functional constraint) and $g_1(\mathbf{x}) = 2x_1 + x_2$, so $g_1(\mathbf{x})$ is convex. Furthermore, it can be easily verified (see Appendix 2) that $f(\mathbf{x})$ is concave. Hence, the corollary applies, so any solution that satisfies the KKT conditions will definitely be an optimal solution. Applying the formulas given in the theorem yields the following KKT conditions for this example:

1($j = 1$). $\quad \dfrac{1}{x_1 + 1} - 2u_1 \leq 0.$

2($j = 1$). $\quad x_1\left(\dfrac{1}{x_1 + 1} - 2u_1\right) = 0.$

1($j = 2$). $\quad 1 - u_1 \leq 0.$

2($j = 2$). $\quad x_2(1 - u_1) = 0.$

3. $\qquad 2x_1 + x_2 - 3 \leq 0.$

4. $\qquad u_1(2x_1 + x_2 - 3) = 0.$

5. $\qquad x_1 \geq 0, x_2 \geq 0.$

6. $\qquad u_1 \geq 0.$

The steps in solving the KKT conditions for this particular example are outlined below.

1. $u_1 \geq 1$, from condition 1($j = 2$).
   $x_1 \geq 0$, from condition 5.

2. Therefore, $\dfrac{1}{x_1 + 1} - 2u_1 < 0.$

3. Therefore, $x_1 = 0$, from condition 2($j = 1$).

4. $u_1 \neq 0$ implies that $2x_1 + x_2 - 3 = 0$, from condition 4.

5. Steps 3 and 4 imply that $x_2 = 3$.

6. $x_2 \neq 0$ implies that $u_1 = 1$, from condition 2($j = 2$).

7. No conditions are violated by $x_1 = 0$, $x_2 = 3$, $u_1 = 1$.

Therefore, there exists a number $u_1 = 1$ such that $x_1 = 0$, $x_2 = 3$, and $u_1 = 1$ satisfy all the conditions. Consequently, $\mathbf{x}^* = (0, 3)$ is an optimal solution for this problem.

This particular problem was relatively easy to solve because the first two steps above quickly led to the remaining conclusions. It often is more difficult to see how to get started. The particular progression of steps needed to solve the KKT conditions will differ from one problem to the next. When the logic is not apparent, it is sometimes helpful to consider separately the different cases where each x_j and u_i are specified to be either equal to or greater than 0 and then trying each case until one leads to a solution.

To illustrate, suppose this approach of considering the different cases separately had been applied to the above example instead of using the logic involved in the above seven steps. For this example, eight cases need to be considered. These cases correspond to the eight combinations of $x_1 = 0$ versus $x_1 > 0$, $x_2 = 0$ versus $x_2 > 0$, and $u_1 = 0$ versus $u_1 > 0$. Each case leads to a simpler statement and analysis of the conditions. To illustrate, consider first the case shown next, where $x_1 = 0$, $x_2 = 0$, and $u_1 = 0$.

KKT Conditions for the Case $x_1 = 0$, $x_2 = 0$, $u_1 = 0$

1($j = 1$). $\dfrac{1}{0 + 1} \leq 0.$ Contradiction.

1($j = 2$). $1 - 0 \leq 0.$ Contradiction.

3. $\quad\quad 0 + 0 \leq 3.$

(All the other conditions are redundant.)

As listed below, the other three cases where $u_1 = 0$ also give immediate contradictions in a similar way, so no solution is available.

Case $x_1 = 0$, $x_2 > 0$, $u_1 = 0$ contradicts conditions 1($j = 1$), 1($j = 2$), and 2($j = 2$).
Case $x_1 > 0$, $x_2 = 0$, $u_1 = 0$ contradicts conditions 1($j = 1$), 2($j = 1$), and 1($j = 2$).
Case $x_1 > 0$, $x_2 > 0$, $u_1 = 0$ contradicts conditions 1($j = 1$), 2($j = 1$), 1($j = 2$), and 2($j = 2$).

The case $x_1 > 0$, $x_2 > 0$, $u_1 > 0$ enables one to delete these nonzero multipliers from conditions 2($j = 1$), 2($j = 2$), and 4, which then enables deletion of conditions 1($j = 1$), 1($j = 2$), and 3 as redundant, as summarized next.

KKT Conditions for the Case $x_1 > 0$, $x_2 > 0$, $u_1 > 0$

1($j = 1$). $\dfrac{1}{x_1 + 1} - 2u_1 = 0.$

2($j = 2$). $1 - u_1 = 0.$

4. $\quad\quad 2x_1 + x_2 - 3 = 0.$

(All the other conditions are redundant.)

Therefore, $u_1 = 1$, so $x_1 = -\frac{1}{2}$, which contradicts $x_1 > 0$.

Now suppose that the case $x_1 = 0$, $x_2 > 0$, $u_1 > 0$ is tried next.

KKT Conditions for the Case $x_1 = 0$, $x_2 > 0$, $u_1 > 0$

1($j = 1$). $\dfrac{1}{0 + 1} - 2u_1 = 0.$

2($j = 2$). $1 - u_1 = 0.$

4. $\quad\quad 0 + x_2 - 3 = 0.$

(All the other conditions are redundant.)

Therefore, $x_1 = 0$, $x_2 = 3$, $u_1 = 1$. Having found a solution, we know that no additional cases need be considered.

If you would like to see **another example** of using the KKT conditions to solve for an optimal solution, one is provided in the Worked Examples section of the book's website.

For problems more complicated than the above example, it may be difficult, if not essentially impossible, to derive an optimal solution *directly* from the KKT conditions. Nevertheless, these conditions still provide valuable clues as to the identity of an optimal solution, and they also permit us to check whether a proposed solution may be optimal.

There also are many valuable *indirect* applications of the KKT conditions. One of these applications arises in the *duality theory* that has been developed for nonlinear programming to parallel the duality theory for linear programming presented in Chap. 6. In particular, for any given constrained maximization problem (call it the *primal problem*), the KKT conditions can be used to define a closely associated dual problem that is a constrained minimization problem. The variables in the dual problem consist of both the Lagrange multipliers u_i ($i = 1, 2, \ldots, m$) and the primal variables x_j ($j = 1, 2, \ldots, n$).

In the special case where the primal problem is a linear programming problem, the x_j variables drop out of the dual problem and it becomes the familiar dual problem of linear programming (where the u_i variables here correspond to the y_i variables in Chap. 6). When the primal problem is a convex programming problem, it is possible to establish relationships between the primal problem and the dual problem that are similar to those for linear programming. For example, the *strong duality property* of Sec. 6.1, which states that the optimal objective function values of the two problems are equal, also holds here. Furthermore, the values of the u_i variables in an optimal solution for the dual problem can again be interpreted as *shadow prices* (see Secs. 4.7 and 6.2); i.e., they give the rate at which the optimal objective function value for the primal problem could be increased by (slightly) increasing the right-hand side of the corresponding constraint. Because duality theory for nonlinear programming is a relatively advanced topic, the interested reader is referred elsewhere for further information.[14]

You will see another indirect application of the KKT conditions in the next section.

12.7 QUADRATIC PROGRAMMING

As indicated in Sec. 12.3, the quadratic programming problem differs from the linear programming problem only in that the objective function also includes x_j^2 and $x_i x_j$ $(i \neq j)$ terms. Thus, if we use matrix notation like that introduced at the beginning of Sec. 5.2, the problem is to find $\mathbf{x}$ so as to

$$\text{Maximize} \quad f(\mathbf{x}) = \mathbf{c}\mathbf{x} - \frac{1}{2}\mathbf{x}^T\mathbf{Q}\mathbf{x},$$

subject to

$$\mathbf{A}\mathbf{x} \leq \mathbf{b} \quad \text{and} \quad \mathbf{x} \geq \mathbf{0},$$

where $\mathbf{c}$ is a row vector, $\mathbf{x}$ and $\mathbf{b}$ are column vectors, $\mathbf{Q}$ and $\mathbf{A}$ are matrices, and the superscript T denotes the transpose (see Appendix 4). The q_{ij} (elements of Q) are given constants such that $q_{ij} = q_{ji}$ (which is the reason for the factor of $\frac{1}{2}$ in the objective function). By performing the indicated vector and matrix multiplications, the objective function then is expressed in terms of these q_{ij}, the c_j (elements of $\mathbf{c}$), and the variables as follows:

$$f(\mathbf{x}) = \mathbf{c}\mathbf{x} - \frac{1}{2}\mathbf{x}^T\mathbf{Q}\mathbf{x} = \sum_{j=1}^{n} c_j x_j - \frac{1}{2}\sum_{i=1}^{n}\sum_{j=1}^{n} q_{ij} x_i x_j.$$

For each term where $i = j$ in this double summation, $x_i x_j = x_j^2$, so $-\frac{1}{2}q_{jj}$ is the coefficient of x_j^2. When $i \neq j$, then $-\frac{1}{2}(q_{ij}x_i x_j + q_{ji}x_j x_i) = -q_{ij}x_i x_j$, so $-q_{ij}$ is the total coefficient for the product of x_i and x_j.

To illustrate this notation, consider the following example of a quadratic programming problem.

$$\text{Maximize} \quad f(x_1, x_2) = 15x_1 + 30x_2 + 4x_1 x_2 - 2x_1^2 - 4x_2^2,$$

subject to

$$x_1 + 2x_2 \leq 30$$

and

$$x_1 \geq 0, \quad x_2 \geq 0.$$

---

[14]For a unified survey of various approaches to duality in nonlinear programming, see A. M. Geoffrion, "Duality in Nonlinear Programming: A Simplified Applications-Oriented Development," *SIAM Review,* **13:** 1–37, 1971.

In this case,

$$\mathbf{c} = [15 \quad 30], \qquad \mathbf{x} = \begin{bmatrix} x_1 \\ x_2 \end{bmatrix}, \qquad \mathbf{Q} = \begin{bmatrix} 4 & -4 \\ -4 & 8 \end{bmatrix},$$

$$\mathbf{A} = [1 \quad 2], \qquad \mathbf{b} = [30].$$

Note that

$$\mathbf{x}^T \mathbf{Q} \mathbf{x} = [x_1 \quad x_2] \begin{bmatrix} 4 & -4 \\ -4 & 8 \end{bmatrix} \begin{bmatrix} x_1 \\ x_2 \end{bmatrix}$$

$$= [(4x_1 - 4x_2) \quad (-4x_1 + 8x_2)] \begin{bmatrix} x_1 \\ x_2 \end{bmatrix}$$

$$= 4x_1^2 - 4x_2 x_1 - 4x_1 x_2 + 8x_2^2$$

$$= q_{11} x_1^2 + q_{21} x_2 x_1 + q_{12} x_1 x_2 + q_{22} x_2^2.$$

Multiplying through by $-\frac{1}{2}$ gives

$$-\frac{1}{2} \mathbf{x}^T \mathbf{Q} \mathbf{x} = -2x_1^2 + 4x_1 x_2 - 4x_2^2,$$

which is the nonlinear portion of the objective function for this example. Since $q_{11} = 4$ and $q_{22} = 8$, the example illustrates that $-\frac{1}{2} q_{jj}$ is the coefficient of x_j^2 in the objective function. The fact that $q_{12} = q_{21} = -4$ illustrates that both $-q_{ij}$ and $-q_{ji}$ give the total coefficient of the product of x_i and x_j.

Several algorithms have been developed for the special case of the quadratic programming problem where the objective function is a *concave* function. (A way to verify that the objective function is concave is to verify the equivalent condition that

$$\mathbf{x}^T \mathbf{Q} \mathbf{x} \geq \mathbf{0}$$

for all $\mathbf{x}$, that is, $\mathbf{Q}$ is a *positive semidefinite* matrix.) We shall describe one[15] of these algorithms, the *modified simplex method,* that has been quite popular because it requires using only the simplex method with a slight modification. The key to this approach is to construct the KKT conditions from the preceding section and then to reexpress these conditions in a convenient form that closely resembles linear programming. Therefore, before describing the algorithm, we shall develop this convenient form.

The KKT Conditions for Quadratic Programming

For concreteness, let us first consider the above example. Starting with the form given in the preceding section, its KKT conditions are the following.

1($j = 1$). $15 + 4x_2 - 4x_1 - u_1 \leq 0.$
2($j = 1$). $x_1(15 + 4x_2 - 4x_1 - u_1) = 0.$
1($j = 2$). $30 + 4x_1 - 8x_2 - 2u_1 \leq 0.$
2($j = 2$). $x_2(30 + 4x_1 - 8x_2 - 2u_1) = 0.$
3. $x_1 + 2x_2 - 30 \leq 0.$
4. $u_1(x_1 + 2x_2 - 30) = 0.$
5. $x_1 \geq 0, \qquad x_2 \geq 0.$
6. $u_1 \geq 0.$

To begin reexpressing these conditions in a more convenient form, we move the constants in conditions 1($j = 1$), 1($j = 2$), and 3 to the right-hand side and then introduce

---

[15]P. Wolfe, "The Simplex Method for Quadratic Programming," *Econometrics,* **27:** 382–398, 1959. This paper develops both a short form and a long form of the algorithm. We present a version of the *short form,* which assumes further that *either* $\mathbf{c} = \mathbf{0}$ *or* the objective function is *strictly* concave.

nonnegative *slack variables* (denoted by y_1, y_2, and v_1, respectively) to convert these inequalities to equations.

$$1(j = 1). \quad -4x_1 + 4x_2 - \quad u_1 + y_1 \qquad\qquad = -15$$
$$1(j = 2). \quad 4x_1 - 8x_2 - 2u_1 \qquad + y_2 \qquad = -30$$
$$3. \qquad\quad x_1 + 2x_2 \qquad\qquad\quad + v_1 = \quad 30$$

Note that condition $2(j = 1)$ can now be reexpressed as simply requiring that either $x_1 = 0$ or $y_1 = 0$; that is,

$$2(j = 1). \quad x_1 y_1 = 0.$$

In just the same way, conditions $2(j = 2)$ and 4 can be replaced by

$$2(j = 2). \quad x_2 y_2 = 0,$$
$$4. \qquad\quad u_1 v_1 = 0.$$

For each of these three pairs—(x_1, y_1), (x_2, y_2), (u_1, v_1)—the two variables are called **complementary variables,** because only one of the two variables can be nonzero. These new forms of conditions $2(j = 1)$, $2(j = 2)$, and 4 can be combined into one constraint,

$$x_1 y_1 + x_2 y_2 + u_1 v_1 = 0,$$

called the **complementarity constraint.**

After multiplying through the equations for conditions $1(j = 1)$ and $1(j = 2)$ by -1 to obtain nonnegative right-hand sides, we now have the desired convenient form for the entire set of conditions shown here:

$$4x_1 - 4x_2 + \quad u_1 - y_1 \qquad\qquad = 15$$
$$-4x_1 + 8x_2 + 2u_1 \qquad - y_2 \qquad = 30$$
$$x_1 + 2x_2 \qquad\qquad\quad + v_1 = 30$$
$$x_1 \geq 0, \qquad x_2 \geq 0, \qquad u_1 \geq 0, \qquad y_1 \geq 0, \qquad y_2 \geq 0, \qquad v_1 \geq 0$$
$$x_1 y_1 + x_2 y_2 + u_1 v_1 = 0$$

This form is particularly convenient because, except for the complementarity constraint, these conditions are *linear programming constraints.*

For *any* quadratic programming problem, its KKT conditions can be reduced to this same convenient form containing just linear programming constraints plus one complementarity constraint. In matrix notation again, this general form is

$$\mathbf{Qx} + \mathbf{A}^T \mathbf{u} - \mathbf{y} = \mathbf{c}^T,$$
$$\mathbf{Ax} + \mathbf{v} = \mathbf{b},$$
$$\mathbf{x} \geq \mathbf{0}, \qquad \mathbf{u} \geq \mathbf{0}, \qquad \mathbf{y} \geq \mathbf{0}, \qquad \mathbf{v} \geq \mathbf{0},$$
$$\mathbf{x}^T \mathbf{y} + \mathbf{u}^T \mathbf{v} = 0,$$

where the elements of the column vector $\mathbf{u}$ are the u_i of the preceding section and the elements of the column vectors $\mathbf{y}$ and $\mathbf{v}$ are slack variables.

Because the objective function of the original problem is assumed to be concave and because the constraint functions are linear and therefore convex, the corollary to the theorem of Sec. 12.6 applies. Thus, $\mathbf{x}$ is *optimal* if and only if there exist values of $\mathbf{y}$, $\mathbf{u}$, and $\mathbf{v}$ such that all four vectors together satisfy all these conditions. The original problem is thereby reduced to the equivalent problem of finding a *feasible solution* to these *constraints.*

It is of interest to note that this equivalent problem is one example of the *linear complementarity problem* introduced in Sec. 12.3 (see Prob. 12.3-6), and that a key constraint for the linear complementarity problem is its *complementarity constraint.*

The Modified Simplex Method

The *modified simplex method* exploits the key fact that, with the exception of the complementarity constraint, the KKT conditions in the convenient form obtained above are nothing more than linear programming constraints. Furthermore, the complementarity constraint simply implies that it is not permissible for *both* complementary variables of any pair to be (nondegenerate) basic variables (the only variables > 0) when (nondegenerate) BF solutions are considered. Therefore, the problem reduces to finding an initial BF solution to any linear programming problem that has these constraints, subject to this additional restriction on the identity of the basic variables. (This initial BF solution may be the only feasible solution in this case.)

As we discussed in Sec. 4.6, finding such an initial BF solution is relatively straightforward. In the simple case where $\mathbf{c}^T \leq \mathbf{0}$ (unlikely) and $\mathbf{b} \geq \mathbf{0}$, the initial basic variables are the elements of $\mathbf{y}$ and $\mathbf{v}$ (multiply through the first set of equations by -1), so that the desired solution is $\mathbf{x} = \mathbf{0}$, $\mathbf{u} = \mathbf{0}$, $\mathbf{y} = -\mathbf{c}^T$, $\mathbf{v} = \mathbf{b}$. Otherwise, you need to revise the problem by introducing an *artificial variable* into each of the equations where $c_j > 0$ (add the variable on the left) or $b_i < 0$ (subtract the variable on the left and then multiply through by -1) in order to use these artificial variables (call them z_1, z_2, and so on) as initial basic variables for the revised problem. (Note that this choice of initial basic variables satisfies the complementarity constraint, because as nonbasic variables $\mathbf{x} = \mathbf{0}$ and $\mathbf{u} = \mathbf{0}$ automatically.)

Next, use phase 1 of the *two-phase method* (see Sec. 4.6) to find a BF solution for the real problem; i.e., apply the simplex method (with one modification) to the following linear programming problem

$$\text{Minimize} \quad Z = \sum_j z_j,$$

subject to the linear programming constraints obtained from the KKT conditions, but with these artificial variables included.

The one modification in the simplex method is the following change in the procedure for selecting an entering basic variable.

Restricted-Entry Rule: When you are choosing an entering basic variable, exclude from consideration any nonbasic variable whose *complementary variable* already is a basic variable; the choice should be made from the other nonbasic variables according to the usual criterion for the simplex method.

This rule keeps the complementarity constraint satisfied throughout the course of the algorithm. When an optimal solution

$$\mathbf{x}^*, \mathbf{u}^*, \mathbf{y}^*, \mathbf{v}^*, z_1 = 0, \ldots, z_n = 0$$

is obtained for the phase 1 problem, $\mathbf{x}^*$ is the desired optimal solution for the original quadratic programming problem. Phase 2 of the two-phase method is not needed.

Example.   We shall now illustrate this approach on the example given at the beginning of the section. As can be verified from the results in Appendix 2 (see Prob. 12.7-1a), $f(x_1, x_2)$ is *strictly concave;* i.e.,

$$\mathbf{Q} = \begin{bmatrix} 4 & -4 \\ -4 & 8 \end{bmatrix}$$

is positive definite, so the algorithm can be applied.

The starting point for solving this example is its KKT conditions in the convenient form obtained earlier in the section. After the needed artificial variables are introduced, the linear programming problem to be addressed explicitly by the modified simplex method then is

Minimize $Z = z_1 + z_2,$

subject to

$$
\begin{aligned}
4x_1 - 4x_2 + u_1 - y_1 \qquad\qquad + z_1 \qquad &= 15 \\
-4x_1 + 8x_2 + 2u_1 \qquad - y_2 \qquad\qquad + z_2 &= 30 \\
x_1 + 2x_2 \qquad\qquad\qquad + v_1 \qquad\qquad &= 30
\end{aligned}
$$

and

$$x_1 \geq 0, \qquad x_2 \geq 0, \qquad u_1 \geq 0, \qquad y_1 \geq 0, \qquad y_2 \geq 0, \qquad v_1 \geq 0,$$
$$z_1 \geq 0, \qquad z_2 \geq 0.$$

The additional complementarity constraint

$$x_1y_1 + x_2y_2 + u_1v_1 = 0,$$

is not included explicitly, because the algorithm automatically enforces this constraint because of the *restricted-entry rule*. In particular, for each of the three pairs of complementary variables—(x_1, y_1), (x_2, y_2), (u_1, v_1)—whenever one of the two variables already is a basic variable, the other variable is *excluded* as a candidate for the entering basic variable. Remember that the only *nonzero* variables are basic variables. Because the initial set of basic variables for the linear programming problem—z_1, z_2, v_1—gives an initial BF solution that satisfies the complementarity constraint, there is no way that this constraint can be violated by any subsequent BF solution.

Table 12.5 shows the results of applying the modified simplex method to this problem. The first simplex tableau exhibits the initial system of equations *after* converting from minimizing Z to maximizing $-Z$ *and* algebraically eliminating the initial basic variables from Eq. (0), just as was done for the radiation therapy example in Sec. 4.6. The three iterations proceed just as for the regular simplex method, *except* for eliminating certain candidates for the entering basic variable because of the restricted-entry rule. In the first tableau, u_1 is eliminated as a candidate because its complementary variable (v_1) already is a basic variable (but x_2 would have been chosen anyway because $-4 < -3$). In the second tableau, both u_1 and y_2 are eliminated as candidates (because v_1 and x_2 are basic variables), so x_1 automatically is chosen as the only candidate with a negative coefficient in row 0 (whereas the *regular* simplex method would have permitted choosing *either* x_1 or u_1 because they are tied for having the largest negative coefficient). In the third tableau, both y_1 and y_2 are eliminated (because x_1 and x_2 are basic variables). However, u_1 is *not* eliminated because v_1 no longer is a basic variable, so u_1 is chosen as the entering basic variable in the usual way.

The resulting optimal solution for this phase 1 problem is $x_1 = 12$, $x_2 = 9$, $u_1 = 3$, with the rest of the variables zero. (Problem 12.7-1c asks you to verify that this solution is optimal by showing that $x_1 = 12$, $x_2 = 9$, $u_1 = 3$ satisfy the KKT conditions for the original problem when they are written in the form given in Sec. 12.6.) Therefore, the optimal solution for the quadratic programming problem (which includes only the x_1 and x_2 variables) is $(x_1, x_2) = (12, 9)$.

■ **TABLE 12.5** Application of the modified simplex method to the quadratic programming example

Iteration	Basic Variable	Eq.	Z	x_1	x_2	u_1	y_1	y_2	v_1	z_1	z_2	Right Side
0	Z	(0)	−1	0	−4	−3	1	1	0	0	0	−45
	z_1	(1)	0	4	−4	1	−1	0	0	1	0	15
	z_2	(2)	0	−4	8	2	0	−1	0	0	1	30
	v_1	(3)	0	1	2	0	0	0	1	0	0	30
1	Z	(0)	−1	−2	0	−2	1	$\frac{1}{2}$	0	0	$\frac{1}{2}$	−30
	z_1	(1)	0	2	0	2	−1	$-\frac{1}{2}$	0	1	$\frac{1}{2}$	30
	x_2	(2)	0	$-\frac{1}{2}$	1	$\frac{1}{4}$	0	$-\frac{1}{8}$	0	0	$\frac{1}{8}$	$3\frac{3}{4}$
	v_1	(3)	0	2	0	$-\frac{1}{2}$	0	$\frac{1}{4}$	1	0	$-\frac{1}{4}$	$22\frac{1}{2}$
2	Z	(0)	−1	0	0	$-\frac{5}{2}$	1	$\frac{3}{4}$	1	0	$\frac{1}{4}$	$-7\frac{1}{2}$
	z_1	(1)	0	0	0	$\frac{5}{2}$	−1	$-\frac{3}{4}$	−1	1	$\frac{3}{4}$	$7\frac{1}{2}$
	x_2	(2)	0	0	1	$\frac{1}{8}$	0	$-\frac{1}{16}$	$\frac{1}{4}$	0	$\frac{1}{16}$	$9\frac{3}{8}$
	x_1	(3)	0	1	0	$-\frac{1}{4}$	0	$\frac{1}{8}$	$\frac{1}{2}$	0	$-\frac{1}{8}$	$11\frac{1}{4}$
3	Z	(0)	−1	0	0	0	0	0	0	1	1	0
	u_1	(1)	0	0	0	1	$-\frac{2}{5}$	$-\frac{3}{10}$	$-\frac{2}{5}$	$\frac{2}{5}$	$\frac{3}{10}$	3
	x_2	(2)	0	0	1	0	$\frac{1}{20}$	$-\frac{1}{40}$	$\frac{3}{10}$	$-\frac{1}{20}$	$\frac{1}{40}$	9
	x_1	(3)	0	1	0	0	$-\frac{1}{10}$	$\frac{1}{20}$	$\frac{2}{5}$	$\frac{1}{10}$	$-\frac{1}{20}$	12

The Worked Examples section of the book's website include **another example** that illustrates the application of the modified simplex method to a quadratic programming problem. The KKT conditions also are applied to this example.

Some Software Options

Your IOR Tutorial includes an interactive procedure for the modified simplex method to help you learn this algorithm efficiently. In addition, Excel, LINGO, LINDO, and MPL/CPLEX all can solve quadratic programming problems.

The procedure for using Excel is almost the same as with linear programming. The one crucial difference is that the equation entered for the cell that contains the value of the objective function now needs to be a quadratic equation. To illustrate, consider again the example introduced at the beginning of the section, which has the objective function

$$f(x_1, x_2) = 15x_1 + 30x_2 + 4x_1x_2 - 2x_1^2 - 4x_2^2.$$

Suppose that the values of x_1 and x_2 are in cells B4 and C4 of the Excel spreadsheet, and that the value of the objective function is in cell F4. Then the equation for cell F4 needs to be

F4 = 15*B4 + 30*C4 + 4*B4*C4 − 2*(B4^2) − 4*(C4^2),

where the symbol ^2 indicates an exponent of 2. Before solving the model, you should click on the Option button and make sure that the *Assume Linear Model* option is *not* selected (since this is not a *linear* programming model).

When using MPL/CPLEX, you should set the model type to Quadratic by adding the following statement at the beginning of the model file.

OPTIONS

ModelType = Quadratic

(Alternatively, you can select the Quadratic Models option from the MPL Language option dialogue box, but then you will need to remember to change the setting when dealing with linear programming problems again.) Otherwise, the procedure is the same as with linear programming except that the expression for the objective function now is a quadratic function. Thus, for the example, the objective function would be expressed as

$$15x1 + 30x2 + 4x1*x2 - 2(x1\hat{} 2) - 4(x2\hat{} 2).$$

Nothing more needs to be done when calling CPLEX, since it will automatically recognize the model as being a quadratic programming problem.

This objective function would be expressed in this same way for a LINGO model. LINGO then will automatically call its nonlinear solver to solve the model.

In fact, the Excel, MPL/CPLEX, and LINGO/LINDO files for this chapter in your OR Courseware all demonstrate their procedures by showing the details for how these software packages set up and solve this example.

Some of these software packages also can be applied to more complicated kinds of nonlinear programming problems than quadratic programming. Although CPLEX cannot, the professional version of MPL does support some other solvers that can. The student version of MPL on the book's website includes one such solver called CONOPT (a product of ARKI Consulting) that is designed for solving convex programming problems. It can be used by adding the following statement at the beginning of the model file.

OPTIONS

ModelType = Nonlinear

Both Excel and LINGO include versatile nonlinear solvers. However, be aware that the Excel Solver is not guaranteed to find an optimal solution for complicated problems, especially nonconvex programming problems (the subject of Sec. 12.10). On the other hand, LINGO contains a *global optimizer* that will find a globally optimal solution for sufficiently small nonconvex programming problems. MPL also supports a global optimizer called LGO as one of its solvers provided on the book's website.

12.8   SEPARABLE PROGRAMMING

The preceding section showed how one class of nonlinear programming problems can be solved by an extension of the simplex method. We now consider another class, called *separable programming,* that actually can be solved by the simplex method itself, because any such problem can be approximated as closely as desired by a linear programming problem with a larger number of variables.

As indicated in Sec. 12.3, in separable programming it is assumed that the objective function $f(\mathbf{x})$ is concave, that each of the constraint functions $g_i(\mathbf{x})$ is convex, and that all these functions are separable functions (functions where each term involves just a single variable). However, to simplify the discussion, we focus here on the special case where the convex and separable $g_i(\mathbf{x})$ are, in fact, *linear functions*, just as for linear programming. (We will turn to the general case briefly at the end of this section.) Thus, only the objective function requires special treatment for this special case.

Under the preceding assumptions, the objective function can be expressed as a sum of concave functions of individual variables

$$f(\mathbf{x}) = \sum_{j=1}^{n} f_j(x_j),$$

so that each $f_j(x_j)$ has a shape[16] such as the one shown in Fig. 12.15 (either case) over the feasible range of values of x_j. Because $f(\mathbf{x})$ represents the measure of performance (say, profit) for all the activities together, $f_j(x_j)$ represents the *contribution to profit* from activity j when it is conducted at level x_j. The condition of $f(\mathbf{x})$ being separable simply implies additivity (see Sec. 3.3); i.e., there are no interactions between the activities (no cross-product terms) that affect total profit beyond their independent contributions. The assumption that each $f_j(x_j)$ is concave says that the *marginal profitability* (slope of the profit curve) either stays the same or decreases (*never* increases) as x_j is increased.

Concave profit curves occur quite frequently. For example, it may be possible to sell a limited amount of some product at a certain price, then a further amount at a lower price, and perhaps finally a further amount at a still lower price. Similarly, it may be necessary to purchase raw materials from increasingly expensive sources. In another common situation, a more expensive production process must be used (e.g., overtime rather than regular-time work) to increase the production rate beyond a certain point.

These kinds of situations can lead to either type of profit curve shown in Fig. 12.15. In case 1, the slope decreases only at certain *breakpoints*, so that $f_j(x_j)$ is a *piecewise linear function* (a sequence of connected line segments). For case 2, the slope may decrease continuously as x_j increases, so that $f_j(x_j)$ is a general concave function. Any such function can be approximated as closely as desired by a piecewise linear function, and this kind of approximation is used as needed for separable programming problems. (Figure 12.15 shows an approximating function that consists of just three line segments, but the approximation can be made even better just by introducing additional breakpoints.) This approximation is very convenient because a piecewise linear function of a single variable can be rewritten as a *linear function* of several variables, with one special restriction on the values of these variables, as described next.

Reformulation as a Linear Programming Problem

The key to rewriting a piecewise linear function as a linear function is to use a separate variable for each line segment. To illustrate, consider the piecewise linear function $f_j(x_j)$ shown in Fig. 12.15, case 1 (or the approximating piecewise linear function for case 2), which has three line segments over the feasible range of values of x_j. Introduce the three new variables x_{j1}, x_{j2}, and x_{j3} and set

$$x_j = x_{j1} + x_{j2} + x_{j3},$$

where

$$0 \le x_{j1} \le u_{j1}, \qquad 0 \le x_{j2} \le u_{j2}, \qquad 0 \le x_{j3} \le u_{j3}.$$

---

[16]$f(\mathbf{x})$ is concave if and only if *every* $f_j(x_j)$ is concave.

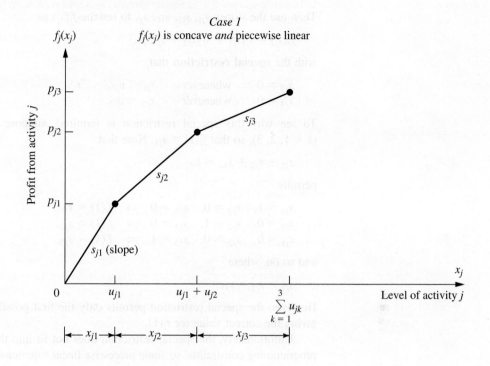

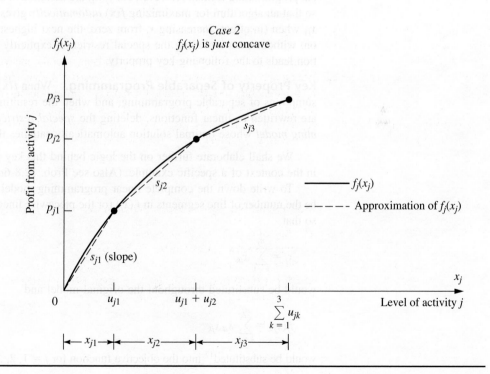

FIGURE 12.15
Shape of profit curves for separable programming.

Then use the slopes s_{j1}, s_{j2}, and s_{j3} to rewrite $f_j(x_j)$ as

$$f_j(x_j) = s_{j1}x_{j1} + s_{j2}x_{j2} + s_{j3}x_{j3},$$

with the **special restriction** that

$$
\begin{aligned}
x_{j2} &= 0 \quad &\text{whenever} \quad &x_{j1} < u_{j1}, \\
x_{j3} &= 0 \quad &\text{whenever} \quad &x_{j2} < u_{j2}.
\end{aligned}
$$

To see why this special restriction is required, suppose that $x_j = 1$, where $u_{jk} > 1$ ($k = 1, 2, 3$), so that $f_j(1) = s_{j1}$. Note that

$$x_{j1} + x_{j2} + x_{j3} = 1$$

permits

$$
\begin{aligned}
x_{j1} = 1, \quad x_{j2} = 0, \quad x_{j3} = 0 \quad &\Rightarrow \quad f_j(1) = s_{j1}, \\
x_{j1} = 0, \quad x_{j2} = 1, \quad x_{j3} = 0 \quad &\Rightarrow \quad f_j(1) = s_{j2}, \\
x_{j1} = 0, \quad x_{j2} = 0, \quad x_{j3} = 1 \quad &\Rightarrow \quad f_j(1) = s_{j3},
\end{aligned}
$$

and so on, where

$$s_{j1} > s_{j2} > s_{j3}.$$

However, the special restriction permits only the first possibility, which is the only one giving the correct value for $f_j(1)$.

Unfortunately, the special restriction does not fit into the required format for linear programming constraints, so *some* piecewise linear functions cannot be rewritten in a linear programming format. However, *our* $f_j(x_j)$ are assumed to be concave, so $s_{j1} > s_{j2} > \cdots$, so that an algorithm for maximizing $f(\mathbf{x})$ *automatically* gives the highest priority to using x_{j1} when (in effect) increasing x_j from zero, the next highest priority to using x_{j2}, and so on, without even including the special restriction explicitly in the model. This observation leads to the following key property.

Key Property of Separable Programming.   When $f(\mathbf{x})$ and the $g_i(\mathbf{x})$ satisfy the assumptions of separable programming, and when the resulting piecewise linear functions are rewritten as linear functions, deleting the *special restriction* gives a *linear programming model* whose optimal solution automatically satisfies the special restriction.

We shall elaborate further on the logic behind this key property later in this section in the context of a specific example. (Also see Prob. 12.8-6a).

To write down the complete linear programming model in the above notation, let n_j be the number of line segments in $f_j(x_j)$ (or the piecewise linear function approximating it), so that

$$x_j = \sum_{k=1}^{n_j} x_{jk}$$

would be substituted throughout the original model and

$$f_j(x_j) = \sum_{k=1}^{n_j} s_{jk}x_{jk}$$

would be substituted[17] into the objective function for $j = 1, 2, \ldots, n$. The resulting model is

$$\text{Maximize} \quad Z = \sum_{j=1}^{n} \left(\sum_{k=1}^{n_j} s_{jk}x_{jk} \right),$$

[17]If one or more of the $f_j(x_j)$ already are *linear* functions $f_j(x_j) = c_jx_j$, then $n_j = 1$ so neither of these substitutions will be made for j.

subject to

$$\sum_{j=1}^{n} a_{ij}\left(\sum_{k=1}^{n_j} x_{jk}\right) \leq b_i, \qquad \text{for } i = 1, 2, \ldots, m$$

$$x_{jk} \leq u_{jk}, \qquad \text{for } k = 1, 2, \ldots, n_j; j = 1, 2, \ldots, n$$

and

$$x_{jk} \geq 0, \qquad \text{for} \qquad k = 1, 2, \ldots, n_j; j = 1, 2, \ldots, n.$$

(The $\sum_{k=1}^{n_j} x_{jk} \geq 0$ constraints are deleted because they are ensured by the $x_{jk} \geq 0$ constraints.) If some original variable x_j has no upper bound, then $u_{jn_j} = \infty$, so the constraint involving this quantity will be deleted.

An efficient way of solving this model[18] is to use the streamlined version of the simplex method for dealing with upper bound constraints (described in Sec. 7.3). After obtaining an optimal solution for this model, you then would calculate

$$x_j = \sum_{k=1}^{n_j} x_{jk},$$

for $j = 1, 2, \ldots, n$ in order to identify an optimal solution for the original separable programming problem (or its piecewise linear approximation).

Example.   The Wyndor Glass Co. (see Sec. 3.1) has received a special order for handcrafted goods to be made in Plants 1 and 2 throughout the next 4 months. Filling this order will require borrowing certain employees from the work crews for the regular products, so the remaining workers will need to work overtime to utilize the full production capacity of the plant's machinery and equipment for these regular products. In particular, for the two new regular products discussed in Sec. 3.1, overtime will be required to utilize the last 25 percent of the production capacity available in Plant 1 for product 1 and for the last 50 percent of the capacity available in Plant 2 for product 2. The additional cost of using overtime work will reduce the profit for each unit involved from \$3 to \$2 for product 1 and from \$5 to \$1 for product 2, giving the *profit curves* of Fig. 12.16, both of which fit the form for case 1 of Fig. 12.15.

Management has decided to go ahead and use overtime work rather than hire additional workers during this temporary situation. However, it does insist that the work crew for each product be fully utilized on regular time before any overtime is used. Furthermore, it feels that the current production rates ($x_1 = 2$ for product 1 and $x_2 = 6$ for product 2) should be changed temporarily if this would improve overall profitability. Therefore, it has instructed the OR team to review products 1 and 2 again to determine the most profitable product mix during the next 4 months.

Formulation.   To refresh your memory, the linear programming model for the original Wyndor Glass Co. problem in Sec. 3.1 is

Maximize $Z = 3x_1 + 5x_2,$

subject to

$$x_1 \qquad\qquad \leq 4$$
$$2x_2 \leq 12$$
$$3x_1 + 2x_2 \leq 18$$

---

[18]For a specialized algorithm for solving this model very efficiently, see R. Fourer, "A Specialized Algorithm for Piecewise-Linear Programming III: Computational Analysis and Applications," *Mathematical Programming,* **53:** 213–235, 1992. Also see A. M. Geoffrion, "Objective Function Approximations in Mathematical Programming," *Mathematical Programming,* **13:** 23–37, 1977.

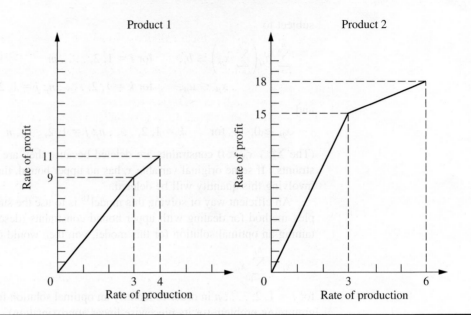

■ FIGURE 12.16
Profit data during the next 4 months for the Wyndor Glass Co.

and

$$x_1 \geq 0, \qquad x_2 \geq 0.$$

We now need to modify this model to fit the new situation described above. For this purpose, let the production rate for product 1 be $x_1 = x_{1R} + x_{1O}$, where x_{1R} is the production rate achieved on regular time and x_{1O} is the incremental production rate from using overtime. Define $x_2 = x_{2R} + x_{2O}$ in the same way for product 2. Thus, in the notation of the general linear programming model for separable programming given just before this example, $n = 2$, $n_1 = 2$, and $n_2 = 2$. Plugging the data given in Fig. 12.16 (including maximum rates of production on regular time and on overtime) into this general model gives the specific model for this application. In particular, the new linear programming problem is to determine the values of x_{1R}, x_{1O}, x_{2R}, and x_{2O} so as to

$$\text{Maximize} \qquad Z = 3x_{1R} + 2x_{1O} + 5x_{2R} + x_{2O},$$

subject to

$$x_{1R} + x_{1O} \qquad\qquad\qquad \leq 4$$
$$\qquad\qquad 2(x_{2R} + x_{2O}) \leq 12$$
$$3(x_{1R} + x_{1O}) + 2(x_{2R} + x_{2O}) \leq 18$$
$$x_{1R} \leq 3, \qquad x_{1O} \leq 1, \qquad x_{2R} \leq 3, \qquad x_{2O} \leq 3$$

and

$$x_{1R} \geq 0, \qquad x_{1O} \geq 0, \qquad x_{2R} \geq 0, \qquad x_{2O} \geq 0.$$

(Note that the upper bound constraints in the next-to-last row of the model make the first two functional constraints *redundant,* so these two functional constraints can be deleted.)

However, there is one important factor that is not taken into account explicitly in this formulation. Specifically, there is nothing in the model that requires all available regular time for a product to be fully utilized before any overtime is used for that product. In other words, it may be feasible to have $x_{1O} > 0$ even when $x_{1R} < 3$ and to have $x_{2O} > 0$

even when $x_{2R} < 3$. Such solutions would not, however, be acceptable to management. (Prohibiting such solutions is the *special restriction* discussed earlier in this section.)

Now we come to the *key property of separable programming*. Even though the model does not take this factor into account explicitly, the model does take it into account implicitly! Despite the model's having excess "feasible" solutions that actually are unacceptable, any *optimal* solution for the model is *guaranteed* to be a legitimate one that does not replace any available regular-time work with overtime work. (The reasoning here is analogous to that for the Big *M* method discussed in Sec. 4.6, where excess feasible but *nonoptimal* solutions also were allowed in the model as a matter of convenience.) Therefore, the simplex method can be safely applied to this model to find the most profitable acceptable product mix. The reasons are twofold. First, the two decision variables for each product *always* appear together as a *sum*, $x_{1R} + x_{1O}$ or $x_{2R} + x_{2O}$, in *each* functional constraint other than the upper bound constraints on individual variables. Therefore, it *always* is possible to convert an unacceptable feasible solution to an acceptable one having the same total production rates, $x_1 = x_{1R} + x_{1O}$ and $x_2 = x_{2R} + x_{2O}$, merely by replacing overtime production by regular-time production as much as possible. Second, overtime production is less profitable than regular-time production (i.e., the slope of each profit curve in Fig. 12.16 is a monotonic *decreasing* function of the rate of production), so converting an unacceptable feasible solution to an acceptable one in this way *must* increase the total rate of profit Z. Consequently, any feasible solution that uses overtime production for a product when regular-time production is still available *cannot* be optimal with respect to the model.

For example, consider the unacceptable feasible solution $x_{1R} = 1$, $x_{1O} = 1$, $x_{2R} = 1$, $x_{2O} = 3$, which yields a total rate of profit Z = 13. The acceptable way of achieving the same total production rates $x_1 = 2$ and $x_2 = 4$ is $x_{1R} = 2$, $x_{1O} = 0$, $x_{2R} = 3$, $x_{2O} = 1$. This latter solution is still feasible, but it also increases Z by $(3 - 2)(1) + (5 - 1)(2) = 9$ to a total rate of profit Z = 22.

Similarly, the optimal solution for this model turns out to be $x_{1R} = 3$, $x_{1O} = 1$, $x_{2R} = 3$, $x_{2O} = 0$, which is an acceptable feasible solution.

Another example that illustrates the application of separable programming is included in the Worked Examples section of the book's website.

Extensions

Thus far we have focused on the special case of separable programming where the only nonlinear function is the objective function $f(\mathbf{x})$. Now consider briefly the general case where the constraint functions $g_i(\mathbf{x})$ need not be linear but are convex and separable, so that each $g_i(\mathbf{x})$ can be expressed as a sum of functions of individual variables

$$g_i(\mathbf{x}) = \sum_{j=1}^{n} g_{ij}(x_j),$$

where each $g_{ij}(x_j)$ is a *convex* function. Once again, each of these new functions may be approximated as closely as desired by a *piecewise linear* function (if it is not already in that form). The one new restriction is that for each variable x_j ($j = 1, 2, \ldots, n$), all the piecewise linear approximations of the functions of this variable $[f_j(x_j), g_{1j}(x_j), \ldots, g_{mj}(x_j)]$ must have the *same* breakpoints so that the same new variables $(x_{j1}, x_{j2}, \ldots, x_{jn_j})$ can be used for all these piecewise linear functions. This formulation leads to a linear programming model just like the one given for the special case except that for each i and j, the x_{jk} variables now have different coefficients in constraint i [where these coefficients are the corresponding slopes of the piecewise linear function approximating $g_{ij}(x_j)$]. Because the

$g_{ij}(x_j)$ are required to be convex, essentially the same logic as before implies that the key property of separable programming still must hold. (See Prob. 12.8-6*b*.)

One drawback of approximating functions by piecewise linear functions as described in this section is that achieving a close approximation requires a large number of line segments (variables), whereas such a fine grid for the breakpoints is needed only in the immediate neighborhood of an optimal solution. Therefore, more sophisticated approaches that use a succession of *two-segment* piecewise linear functions have been developed[19] to obtain *successively closer approximations* within this immediate neighborhood. This kind of approach tends to be both faster and more accurate in closely approximating an optimal solution.

■ 12.9   CONVEX PROGRAMMING

We already have discussed some special cases of convex programming in Secs. 12.4 and 12.5 (unconstrained problems), 12.7 (quadratic objective function with linear constraints), and 12.8 (separable functions). You also have seen some theory for the general case (necessary and sufficient conditions for optimality) in Sec. 12.6. In this section, we briefly discuss some types of approaches used to solve the general convex programming problem [where the objective function $f(\mathbf{x})$ to be maximized is concave and the $g_i(\mathbf{x})$ constraint functions are convex], and then we present one example of an algorithm for convex programming.

There is no single standard algorithm that always is used to solve convex programming problems. Many different algorithms have been developed, each with its own advantages and disadvantages, and research continues to be active in this area. Roughly speaking, most of these algorithms fall into one of the following three categories.

The first category is **gradient algorithms,** where the gradient search procedure of Sec. 12.5 is modified in some way to keep the search path from penetrating any constraint boundary. For example, one popular gradient method is the *generalized reduced gradient* (GRG) method. The Excel Solver uses the GRG method for solving convex programming problems. (As discussed in the next section, Premium Solver also includes an Evolutionary Solver option that is well suited for dealing with *nonconvex* programming problems.)

The second category—**sequential unconstrained algorithms**—includes *penalty function* and *barrier function* methods. These algorithms convert the original constrained optimization problem to a sequence of *unconstrained optimization* problems whose optimal solutions converge to the optimal solution for the original problem. Each of these unconstrained optimization problems can be solved by the kinds of procedures described in Sec. 12.5. This conversion is accomplished by incorporating the constraints into a penalty function (or barrier function) that is subtracted from the objective function in order to impose large penalties for violating constraints (or even being near constraint boundaries). In the latter part of this section, we will describe an algorithm from the 1960s, called the **sequential unconstrained minimization technique** (or **SUMT** for short), that pioneered this category of algorithms. (SUMT also helped to motivate some of the *interior-point methods* for linear programming.)

The third category—**sequential-approximation algorithms**—includes *linear approximation* and *quadratic approximation* methods. These algorithms replace the nonlinear objective function by a succession of linear or quadratic approximations. For linearly constrained optimization problems, these approximations allow repeated application of linear or quadratic programming algorithms. This work is accompanied by other analysis that yields a sequence of solutions that converges to an optimal solution for the original problem. Although these algorithms are particularly suitable for linearly constrained optimization problems,

[19]R. R. Meyer, "Two-Segment Separable Programming," *Management Science,* **25:** 385–395, 1979.

some also can be extended to problems with nonlinear constraint functions by the use of appropriate linear approximations.

As one example of a *sequential-approximation* algorithm, we present here the **Frank-Wolfe algorithm**[20] for the case of *linearly constrained* convex programming (so the constraints are $\mathbf{Ax} \leq \mathbf{b}$ and $\mathbf{x} \geq \mathbf{0}$ in matrix form). This procedure is particularly straightforward; it combines *linear* approximations of the objective function (enabling us to use the simplex method) with a procedure for one-variable unconstrained optimization (such as described in Sec. 12.4).

A Sequential Linear Approximation Algorithm (Frank-Wolfe)

Given a feasible trial solution $\mathbf{x}'$, the linear approximation used for the objective function $f(\mathbf{x})$ is the first-order Taylor series expansion of $f(\mathbf{x})$ around $\mathbf{x} = \mathbf{x}'$, namely,

$$f(\mathbf{x}') \approx f(\mathbf{x}') + \sum_{j=1}^{n} \frac{\partial f(\mathbf{x}')}{\partial x_j}(x_j - x_j') = f(\mathbf{x}') + \nabla f(\mathbf{x}')(\mathbf{x} - \mathbf{x}'),$$

where these partial derivatives are evaluated at $\mathbf{x} = \mathbf{x}'$. Because $f(\mathbf{x}')$ and $\nabla f(\mathbf{x}')\mathbf{x}'$ have fixed values, they can be dropped to give an equivalent linear objective function

$$g(\mathbf{x}) = \nabla f(\mathbf{x}')\mathbf{x} = \sum_{j=1}^{n} c_j x_j, \qquad \text{where } c_j = \frac{\partial f(\mathbf{x})}{\partial x_j} \qquad \text{at } \mathbf{x} = \mathbf{x}'.$$

The simplex method (or the graphical procedure if $n = 2$) then is applied to the resulting linear programming problem [maximize $g(\mathbf{x})$ subject to the original constraints, $\mathbf{Ax} \leq \mathbf{b}$ and $\mathbf{x} \geq \mathbf{0}$] to find *its* optimal solution x_{LP}. Note that the linear objective function necessarily increases steadily as one moves along the line segment from $\mathbf{x}'$ to $\mathbf{x}_{LP}$ (which is on the boundary of the feasible region). However, the linear approximation may not be a particularly close one for $\mathbf{x}$ far from $\mathbf{x}'$, so the *nonlinear* objective function may not continue to increase all the way from $\mathbf{x}'$ to $\mathbf{x}_{LP}$. Therefore, rather than just accepting $\mathbf{x}_{LP}$ as the next trial solution, we choose the point that maximizes the nonlinear objective function along this line segment. This point may be found by conducting a procedure for one-variable unconstrained optimization of the kind presented in Sec. 12.4, where the one variable for purposes of this search is the fraction t of the total distance from $\mathbf{x}'$ to $\mathbf{x}_{LP}$. This point then becomes the new trial solution for initiating the next iteration of the algorithm, as just described. The sequence of trial solutions generated by repeated iterations converges to an optimal solution for the original problem, so the algorithm stops as soon as the successive trial solutions are close enough together to have essentially reached this optimal solution.

Summary of the Frank-Wolfe Algorithm

Initialization: Find a feasible initial trial solution $\mathbf{x}^{(0)}$, for example, by applying linear programming procedures to find an initial BF solution. Set $k = 1$.

Iteration k:

1. For $j = 1, 2, \ldots, n$, evaluate

$$\frac{\partial f(\mathbf{x})}{\partial x_j} \qquad \text{at } \mathbf{x} = \mathbf{x}^{(k-1)}$$

and set c_j equal to this value.

[20]M. Frank and P. Wolfe, "An Algorithm for Quadratic Programming," *Naval Research Logistics Quarterly*, **3:** 95–110, 1956. Although originally designed for quadratic programming, this algorithm is easily adapted to the case of a general concave objective function considered here.

2. Find an optimal solution $\mathbf{x}_{LP}^{(k)}$ for the following linear programming problem.

$$\text{Maximize} \quad g(\mathbf{x}) = \sum_{j=1}^{n} c_j x_j,$$

subject to

$$\mathbf{Ax} \leq \mathbf{b} \quad \text{and} \quad \mathbf{x} \geq \mathbf{0}.$$

3. For the variable t ($0 \leq t \leq 1$), set

$$h(t) = f(\mathbf{x}) \quad \text{for } \mathbf{x} = \mathbf{x}^{(k-1)} + t(\mathbf{x}_{LP}^{(k)} - \mathbf{x}^{(k-1)}),$$

so that $h(t)$ gives the value of $f(\mathbf{x})$ on the line segment between $\mathbf{x}^{(k-1)}$ (where $t = 0$) and $\mathbf{x}_{LP}^{(k)}$ (where $t = 1$). Use some procedure for one-variable unconstrained optimization (see Sec. 12.4) to maximize $h(t)$ over $0 \leq t \leq 1$, and set $\mathbf{x}^{(k)}$ equal to the corresponding $\mathbf{x}$. Go to the stopping rule.

Stopping rule: If $\mathbf{x}^{(k-1)}$ and $\mathbf{x}^{(k)}$ are sufficiently close, stop and use $\mathbf{x}^{(k)}$ (or some extrapolation of $\mathbf{x}^{(0)}, \mathbf{x}^{(1)}, \ldots, \mathbf{x}^{(k-1)}, \mathbf{x}^{(k)}$) as your estimate of an optimal solution. Otherwise, reset $k = k + 1$ and perform another iteration.

Now let us illustrate this procedure.

Example. Consider the following linearly constrained convex programming problem:

$$\text{Maximize} \quad f(\mathbf{x}) = 5x_1 - x_1^2 + 8x_2 - 2x_2^2,$$

subject to

$$3x_1 + 2x_2 \leq 6$$

and

$$x_1 \geq 0, \qquad x_2 \geq 0.$$

Note that

$$\frac{\partial f}{\partial x_1} = 5 - 2x_1, \qquad \frac{\partial f}{\partial x_2} = 8 - 4x_2,$$

so that the *unconstrained* maximum $\mathbf{x} = (\frac{5}{2}, 2)$ violates the functional constraint. Thus, more work is needed to find the *constrained* maximum.

Iteration 1: Because $\mathbf{x} = (0, 0)$ is clearly feasible (and corresponds to the initial BF solution for the linear programming constraints), let us choose it as the initial trial solution $\mathbf{x}^{(0)}$ for the Frank-Wolfe algorithm. Plugging $x_1 = 0$ and $x_2 = 0$ into the expressions for the partial derivatives gives $c_1 = 5$ and $c_2 = 8$, so that $g(\mathbf{x}) = 5x_1 + 8x_2$ is the initial linear approximation of the objective function. Graphically, solving this linear programming problem (see Fig. 12.17a) yields $\mathbf{x}_{LP}^{(1)} = (0, 3)$. For step 3 of the first iteration, the points on the line segment between $(0, 0)$ and $(0, 3)$ shown in Fig. 12.17a are expressed by

$$(x_1, x_2) = (0, 0) + t[(0, 3) - (0, 0)] \qquad \text{for } 0 \leq t \leq 1$$
$$= (0, 3t)$$

as shown in the sixth column of Table 12.6. This expression then gives

$$h(t) = f(0, 3t) = 8(3t) - 2(3t)^2$$
$$= 24t - 18t^2,$$

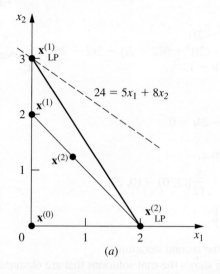 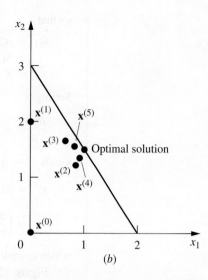

■ **FIGURE 12.17**
Illustration of the Frank-Wolfe algorithm.

(a)   (b)

■ **TABLE 12.6** Application of the Frank-Wolfe algorithm to the example

k	$\mathbf{x}^{(k-1)}$	c_1	c_2	$\mathbf{x}_{LP}^{(k)}$	$\mathbf{x}$ for $h(t)$	$h(t)$	t^*	$\mathbf{x}^{(k)}$
1	$(0, 0)$	5	8	$(0, 3)$	$(0, 3t)$	$24t - 18t^2$	$\frac{2}{3}$	$(0, 2)$
2	$(0, 2)$	5	0	$(2, 0)$	$(2t, 2 - 2t)$	$8 + 10t - 12t^2$	$\frac{5}{12}$	$\left(\frac{5}{6}, \frac{7}{6}\right)$

so that the value $t = t^*$ that maximizes $h(t)$ over $0 \leq t \leq 1$ may be obtained in this case by setting

$$\frac{dh(t)}{dt} = 24 - 36t = 0,$$

so that $t^* = \frac{2}{3}$. This result yields the next trial solution

$$\mathbf{x}^{(1)} = (0, 0) + \frac{2}{3}[(0, 3) - (0, 0)]$$
$$= (0, 2),$$

which completes the first iteration.

Iteration 2: To sketch the calculations that lead to the results in the second row of Table 12.6, note that $\mathbf{x}^{(1)} = (0, 2)$ gives

$$c_1 = 5 - 2(0) = 5,$$
$$c_2 = 8 - 4(2) = 0.$$

For the objective function $g(\mathbf{x}) = 5x_1$, graphically solving the problem over the feasible region in Fig. 12.17a gives $\mathbf{x}_{LP}^{(2)} = (2, 0)$. Therefore, the expression for the line segment between $\mathbf{x}^{(1)}$ and $\mathbf{x}_{LP}^{(2)}$ (see Fig. 12.17a) is

$$\mathbf{x} = (0, 2) + t[(2, 0) - (0, 2)]$$
$$= (2t, 2 - 2t),$$

so that

$$h(t) = f(2t, 2 - 2t)$$
$$= 5(2t) - (2t)^2 + 8(2 - 2t) - 2(2 - 2t)^2$$
$$= 8 + 10t - 12t^2.$$

Setting

$$\frac{dh(t)}{dt} = 10 - 24t = 0$$

yields $t^* = \frac{5}{12}$. Hence,

$$\mathbf{x}^{(2)} = (0, 2) + \frac{5}{12}[(2, 0) - (0, 2)]$$
$$= \left(\frac{5}{6}, \frac{7}{6}\right),$$

which completes the second iteration.

Figure 12.17b shows the trial solutions that are obtained from iterations 3, 4, and 5 as well. You can see how these trial solutions keep alternating between two trajectories that appear to intersect at approximately the point $\mathbf{x} = (1, \frac{3}{2})$. This point is, in fact, the optimal solution, as can be verified by applying the KKT conditions from Sec. 12.6.

This example illustrates a common feature of the Frank-Wolfe algorithm, namely, that the trial solutions alternate between two (or more) trajectories. When they alternate in this way, we can extrapolate the trajectories to their approximate point of intersection to estimate an optimal solution. This estimate tends to be better than using the last trial solution generated. The reason is that the trial solutions tend to converge rather slowly toward an optimal solution, so the last trial solution may still be quite far from optimal.

If you would like to see **another example** of the application of the Frank-Wolfe algorithm, one is included in the Worked Examples section of the book's website. Your OR Tutor provides **an additional example** as well. IOR Tutorial also includes an interactive procedure for this algorithm.

Some Other Algorithms

We should emphasize that the Frank-Wolfe algorithm is just one example of sequential-approximation algorithms. Many of these algorithms use *quadratic* instead of *linear* approximations at each iteration because quadratic approximations provide a considerably closer fit to the original problem and thus enable the sequence of solutions to converge considerably more rapidly toward an optimal solution than was the case in Fig. 12.17b. For this reason, even though sequential linear approximation methods such as the Frank-Wolfe algorithm are relatively straightforward to use, *sequential quadratic approximation methods* now are generally preferred in actual applications. Popular among these are the *quasi-Newton* (or *variable metric*) methods. As already mentioned in Sec. 12.5, these methods use a fast approximation of *Newton's method* and then further adapt this method to take the constraints of the problem into account. To speed up the algorithm, quasi-Newton methods compute a quadratic approximation to the curvature of a nonlinear function without explicitly calculating second (partial) derivatives. (For linearly constrained optimization problems, this nonlinear function is just the objective function; whereas with nonlinear constraints, it is the Lagrangian function described in Appendix 3.) Some quasi-Newton algorithms do not even explicitly form and solve an approximating quadratic programming problem at each iteration, but instead incorporate some of the basic ingredients of *gradient algorithms*. (See Selected Reference 2 for further details about sequential-approximation algorithms.)

We turn now from sequential-approximation algorithms to *sequential unconstrained algorithms*. As mentioned at the beginning of the section, algorithms of the latter type solve the original constrained optimization problem by instead solving a sequence of *unconstrained* optimization problems.

A particularly prominent sequential unconstrained algorithm that has been widely used since its development in the 1960s is the *sequential unconstrained minimization technique* (or *SUMT* for short).[21] There actually are two main versions of SUMT, one of which is an *exterior-point* algorithm that deals with *infeasible* solutions while using a *penalty function* to force convergence to the feasible region. We shall describe the other version, which is an *interior-point* algorithm that deals directly with *feasible* solutions while using a *barrier function* to force staying inside the feasible region. Although SUMT was originally presented as a minimization technique, we shall convert it to a maximization technique in order to be consistent with the rest of the chapter. Therefore, we continue to assume that the problem is in the form given at the beginning of the chapter and that all the functions are differentiable.

Sequential Unconstrained Minimization Technique (SUMT)

As the name implies, SUMT replaces the original problem by a *sequence* of *unconstrained* optimization problems whose solutions *converge* to a solution (local maximum) of the original problem. This approach is very attractive because unconstrained optimization problems are much easier to solve (see Sec. 12.5) than those with constraints. Each of the unconstrained problems in this sequence involves choosing a (successively smaller) strictly positive value of a scalar r and then solving for $\mathbf{x}$ so as to

Maximize $P(\mathbf{x}; r) = f(\mathbf{x}) - rB(\mathbf{x})$.

Here $B(\mathbf{x})$ is a **barrier function** that has the following properties (for $\mathbf{x}$ that are feasible for the original problem):

1. $B(\mathbf{x})$ is *small* when $\mathbf{x}$ is *far* from the boundary of the feasible region.
2. $B(\mathbf{x})$ is *large* when $\mathbf{x}$ is *close* to the boundary of the feasible region.
3. $B(\mathbf{x}) \to \infty$ as the distance from the (nearest) boundary of the feasible region $\to 0$.

Thus, by starting the search procedure with a *feasible* initial trial solution and then attempting to increase $P(\mathbf{x}; r)$, $B(\mathbf{x})$ provides a *barrier* that prevents the search from ever crossing (or even reaching) the boundary of the feasible region for the original problem.

The most common choice of $B(\mathbf{x})$ is

$$B(\mathbf{x}) = \sum_{i=1}^{m} \frac{1}{b_i - g_i(\mathbf{x})} + \sum_{j=1}^{n} \frac{1}{x_j}.$$

For feasible values of $\mathbf{x}$, note that the denominator of each term is proportional to the distance of $\mathbf{x}$ from the constraint boundary for the corresponding functional or nonnegativity constraint. Consequently, *each* term is a *boundary repulsion term* that has all the preceding three properties with respect to this particular constraint boundary. Another attractive feature of this $B(\mathbf{x})$ is that when all the assumptions of *convex programming* are satisfied, $P(\mathbf{x}; r)$ is a *concave* function.

Because $B(\mathbf{x})$ keeps the search away from the boundary of the feasible region, you probably are asking the very legitimate question: What happens if the desired solution lies there? This concern is the reason that SUMT involves solving a *sequence* of these unconstrained optimization problems for successively smaller values of r approaching zero (where the final trial solution from each one becomes the initial trial solution for the next). For example, each new r might be obtained from the preceding one by multiplying by a

[21]See Selected Reference 1.

constant θ $(0 < \theta < 1)$, where a typical value is $\theta = 0.01$. As r approaches 0, $P(\mathbf{x}; r)$ approaches $f(\mathbf{x})$, so the corresponding local maximum of $P(\mathbf{x}; r)$ converges to a local maximum of the original problem. Therefore, it is necessary to solve only enough unconstrained optimization problems to permit extrapolating their solutions to this limiting solution.

How many are enough to permit this extrapolation? When the original problem satisfies the assumptions of convex programming, useful information is available to guide us in this decision. In particular, if $\bar{x}$ is a global maximizer of $P(\mathbf{x}; r)$, then

$$f(\overline{\mathbf{x}}) \leq f(\mathbf{x}^*) \leq f(\overline{\mathbf{x}}) + rB(\overline{\mathbf{x}}),$$

where $\mathbf{x}^*$ is the (unknown) *optimal* solution for the original problem. Thus, $rB(\overline{\mathbf{x}})$ is the *maximum error* (in the value of the objective function) that can result by using $\overline{\mathbf{x}}$ to approximate $\mathbf{x}^*$, and extrapolating beyond $\overline{\mathbf{x}}$ to increase $f(\mathbf{x})$ further decreases this error. If an *error tolerance* is established in advance, then you can stop as soon as $rB(\overline{\mathbf{x}})$ is less than this quantity.

Summary of SUMT

Initialization: Identify a *feasible* initial trial solution $\mathbf{x}^{(0)}$ that is not on the boundary of the feasible region. Set $k = 1$ and choose appropriate strictly positive values for the initial r and for $\theta < 1$ (say, $r = 1$ and $\theta = 0.01$).[22]

Iteration k: Starting from $\mathbf{x}^{(k-1)}$, apply a multivariable unconstrained optimization procedure (e.g., the gradient search procedure) such as described in Sec. 12.5 to find a local maximum $\mathbf{x}^{(k)}$ of

$$P(\mathbf{x}; r) = f(\mathbf{x}) - r\left[\sum_{i=1}^{m} \frac{1}{b_i - g_i(\mathbf{x})} + \sum_{j=1}^{n} \frac{1}{x_j}\right].$$

Stopping rule: If the change from $\mathbf{x}^{(k-1)}$ to $\mathbf{x}^{(k)}$ is negligible, stop and use $\mathbf{x}^{(k)}$ (or an extrapolation of $\mathbf{x}^{(0)}, \mathbf{x}^{(1)}, \ldots, \mathbf{x}^{(k-1)}, \mathbf{x}^{(k)}$) as your estimate of a *local maximum* of the original problem. Otherwise, reset $k = k + 1$ and $r = \theta r$ and perform another iteration.

Finally, we should note that SUMT also can be extended to accommodate *equality* constraints $g_i(\mathbf{x}) = b_i$. One standard way is as follows. For each equality constraint,

$$\frac{-[b_i - g_i(\mathbf{x})]^2}{\sqrt{r}} \quad \text{replaces} \quad \frac{-r}{b_i - g_i(\mathbf{x})}$$

in the expression for $P(\mathbf{x}; r)$ given under "Summary of SUMT," and then the same procedure is used. The numerator $-[b_i - g_i(\mathbf{x})]^2$ imposes a large penalty for deviating substantially from satisfying the equality constraint, and then the denominator tremendously increases this penalty as r is decreased to a tiny amount, thereby forcing the sequence of trial solutions to converge toward a point that satisfies the constraint.

SUMT has been widely used because of its simplicity and versatility. However, numerical analysts have found that it is relatively prone to *numerical instability*, so considerable caution is advised. For further information on this issue as well as similar analyses for alternative algorithms, see Selected Reference 3.

Example. To illustrate SUMT, consider the following two-variable problem:

Maximize $\quad f(\mathbf{x}) = x_1 x_2,$

subject to

$$x_1^2 + x_2 \leq 3$$

---

[22]A reasonable criterion for choosing the initial r is one that makes $rB(\mathbf{x})$ about the same order of magnitude as $f(\mathbf{x})$ for feasible solutions $\mathbf{x}$ that are not particularly close to the boundary.

TABLE 12.7 Illustration of SUMT

k	r	$x_1^{(k)}$	$x_2^{(k)}$
0		1	1
1	1	0.90	1.36
2	10^{-2}	0.987	1.925
3	10^{-4}	0.998	1.993
		$\downarrow$	$\downarrow$
		1	2

and

$$x_1 \geq 0, \qquad x_2 \geq 0.$$

Even though $g_1(\mathbf{x}) = x_1^2 + x_2$ is convex (because each term is convex), this problem is a *nonconvex* programming problem because $f(\mathbf{x}) = x_1 x_2$ is *not* concave (see Appendix 2). However, the problem is close enough to being a convex programming problem that SUMT necessarily will still converge to an optimal solution in this case. (We will discuss nonconvex programming further, including the role of SUMT in dealing with such problems, in the next section.)

For the initialization, $(x_1, x_2) = (1, 1)$ is one obvious feasible solution that is not on the boundary of the feasible region, so we can set $\mathbf{x}^{(0)} = (1, 1)$. Reasonable choices for r and θ are $r = 1$ and $\theta = 0.01$.

For each iteration,

$$P(\mathbf{x}; r) = x_1 x_2 - r \left(\frac{1}{3 - x_1^2 - x_2} + \frac{1}{x_1} + \frac{1}{x_2} \right).$$

With $r = 1$, applying the gradient search procedure starting from $(1, 1)$ to maximize this expression eventually leads to $\mathbf{x}^{(1)} = (0.90, 1.36)$. Resetting $r = 0.01$ and restarting the gradient search procedure from $(0.90, 1.36)$ then lead to $\mathbf{x}^{(2)} = (0.983, 1.933)$. One more iteration with $r = 0.01(0.01) = 0.0001$ leads from $\mathbf{x}^{(2)}$ to $\mathbf{x}^{(3)} = (0.998, 1.994)$. This sequence of points, summarized in Table 12.7, quite clearly is converging to $(1, 2)$. Applying the KKT conditions to this solution verifies that it does indeed satisfy the necessary condition for optimality. Graphical analysis demonstrates that $(x_1, x_2) = (1, 2)$ is, in fact, a global maximum (see Prob. 12.9-13b).

For this problem, there are no local maxima other than $(x_1, x_2) = (1, 2)$, so reapplying SUMT from various feasible initial trial solutions always leads to this same solution.[23]

The Worked Examples section of the book's website provides **another example** that illustrates the application of SUMT to a convex programming problem in minimization form. You also can go to your OR Tutor to see **an additional example**. An automatic procedure for executing SUMT is included in IOR Tutorial.

Some Software Options for Convex Programming

As indicated at the end of Sec. 12.7, both Excel and LINGO can solve convex programming problems, but the student version of LINDO and CPLEX cannot except for the special case of quadratic programming (which includes the first example in this

---

[23]The technical reason is that $f(\mathbf{x})$ is a (strictly) *quasiconcave* function that shares the property of concave functions that a local maximum always is a global maximum. For further information, see M. Avriel, W. E. Diewert, S. Schaible, and I. Zang, *Generalized Concavity*, Plenum, New York, 1985.

section). Details for this example are given in the Excel and LINGO files for this chapter in your OR Courseware. The professional version of MPL supports a large number of solvers, including some that can handle convex programming. One of these, called CONOPT, is included with the student version of MPL that is on the book's website. The convex programming examples that are formulated in this chapter's MPL file have been solved with this solver after setting the model type to Nonlinear (as described at the end of Sec. 12.7).

12.10 NONCONVEX PROGRAMMING (WITH SPREADSHEETS)

The assumptions of convex programming (the function $f(\mathbf{x})$ to be maximized is *concave* and all the $g_i(\mathbf{x})$ constraint functions are *convex*) are very convenient ones, because they ensure that any *local maximum* also is a *global maximum*. (If the objective is to *minimize* $f(\mathbf{x})$ instead, then convex programming assumes that $f(\mathbf{x})$ is *convex,* and so on, which ensures that a *local minimum* also is a *global minimum*.) Unfortunately, the nonlinear programming problems that arise in practice frequently fail to satisfy these assumptions. What kind of approach can be used to deal with such *nonconvex programming* problems?

The Challenge of Solving Nonconvex Programming Problems

There is no single answer to the above question because there are so many different types of nonconvex programming problems. Some are much more difficult to solve than others. For example, a maximization problem where the objective function is nearly *convex* generally is much more difficult than one where the objective function is nearly concave. (The SUMT example in Sec. 12.9 illustrated a case where the objective function was so close to being concave that the problem could be treated as if it were a convex programming problem.) Similarly, having a feasible region that is *not* a convex set (because some of the $g_i(\mathbf{x})$ functions are not convex) generally is a major complication. Dealing with functions that are not differentiable, or perhaps not even continuous, also tends to be a major complication.

The goal of much ongoing research is to develop efficient **global optimization** procedures for finding a *globally optimal solution* for various types of nonconvex programming problems, and some progress has been made. As one example, LINDO Systems (which produces LINDO, LINGO, and What's Best) now has incorporated a global optimizer into its advanced solver that is shared by some of its software products. In particular, LINGO and What's Best have a multistart option to automatically generate a number of starting points for their nonlinear programming solver in order to quickly find a good solution. If the global option is checked, they next employ the global optimizer. The global optimizer converts a nonconvex programming problem (including even those whose formulation includes logic functions such as IF, AND, OR, and NOT) into several subproblems that are convex programming relaxations of portions of the original problem. The branch-and-bound technique then is used to exhaustively search over the subproblems. Once the procedure runs to completion, the solution found is guaranteed to be a globally optimal solution. (The other possible conclusion is that the problem has no feasible solutions.) The student version of this global optimizer is included in the version of LINGO that is provided on the book's website. However, it is limited to relatively small problems (a maximum of five nonlinear variables out of 500 variables total). The professional version of the global optimizer has successfully solved some much larger problems.

Similarly, MPL now supports a global optimizer called LGO. The student version of LGO is available to you as one of the MPL solvers provided on the book's website. LGO also can be used to solve convex programming problems.

A variety of approaches to global optimization (such as the one incorporated into LINGO described above) are being tried. We will not attempt to survey this advanced topic in any depth. (See Selected Reference 5 for some details.) We instead will begin with a simple case and then introduce a more general approach at the end of the section. We will illustrate our methodology with spreadsheets and Excel software, but other software packages also can be used.

Using the Excel Solver to Find Local Optima

We now will focus on straightforward approaches to relatively simple types of nonconvex programming problems. In particular, we will consider (maximization) problems where the objective function is nearly concave either over the entire feasible region or within major portions of the feasible region. We also will ignore the added complexity of having nonconvex constraint functions $g_i(\mathbf{x})$ by simply using linear constraints. We will begin by illustrating what can be accomplished by simply applying some algorithm for convex programming to such problems. Although any such algorithm (such as those described in Sec. 12.9) could be selected, we will use the convex programming algorithm that is employed by the Excel Solver for nonlinear programming problems.

For example, consider the following one-variable nonconvex programming problem:

$$\text{Maximize} \quad Z = 0.5x^5 - 6x^4 + 24.5x^3 - 39x^2 + 20x,$$

subject to

$$x \le 5$$
$$x \ge 0,$$

where Z represents the profit in dollars. Figure 12.18 shows a plot of the profit over the feasible region that demonstrates how highly nonconvex this function is. However, if this graph were not available, it might not be immediately clear that this is *not* a convex programming problem since a little analysis is required to verify that the objective function is not concave over the feasible region. Therefore, suppose that the Excel Solver, which is designed for solving convex programming problems, is applied to this example. Figure 12.19 demonstrates what a difficult time the Excel Solver has in attempting to cope with this

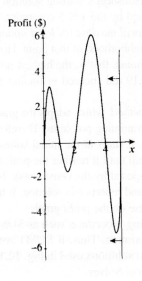

■ **FIGURE 12.19**
An example of a nonconvex programming problem (depicted in Fig. 12.18) where the Excel Solver obtains three different solutions when it starts with three different initial solutions.

	A	B	C	D	E
1	**Solver Solution**				
2	**(Starting with x=0)**				
3					
4					Maximum
5		x =	0.371	<=	5
6					
7		Profit = 0.5x⁵-6x⁴+24.5x³-39x²+20x			
8		=	$3.19		

	A	B	C	D	E
1	**Solver Solution**				
2	**(Starting with x=3)**				
3					
4					Maximum
5		x =	3.126	<=	5
6					
7		Profit = 0.5x⁵-6x⁴+24.5x³-39x²+20x			
8		=	$6.13		

	A	B	C	D	E
1	**Solver Solution**				
2	**(Starting with x=4.7)**				
3					
4					Maximum
5		x =	5.000	<=	5
6					
7		Profit = 0.5x⁵-6x⁴+24.5x³-39x²+20x			
8		=	$0.00		

	B	C
7	Profit =	
8	=	=0.5*x^5-6*x^4+24.5*x^3-39*x^2+20*x

Solver Parameters
Set Target Cell: Profit
Equal To: ● Max ○ Min ○
By Changing Cells:
x
Subject to the Constraints:
x <= Maximum

Solver Options
☐ Assume Linear Model
☑ Assume Non-Negative

Range Name	Cell
Maximum	E5
x	C5
Profit	C8

problem. The model is straightforward to formulate in a spreadsheet, with x (C5) as the changing cell and Profit (C8) as the target cell. (Note that the Solver option, Assume Linear Model, is *not* chosen in this case because this is not a linear programming model.) When $x = 0$ is entered as the initial value in the changing cell, the left spreadsheet in Fig. 12.19 shows that the solver then indicates that $x = 0.371$ is the optimal solution with Profit = \$3.19. However, if $x = 3$ is entered as the initial value instead, as in the middle spreadsheet in Fig. 12.19, Solver obtains $x = 3.126$ as the optimal solution with Profit = \$6.13. Trying still another initial value of $x = 4.7$ in the right spreadsheet, Solver now indicates an optimal solution of $x = 5$ with Profit = \$0. What is going on here?

Figure 12.18 helps to explain Solver's difficulties with this problem. Starting at $x = 0$, the profit graph does indeed climb to a peak at $x = 0.371$, as reported in the left spreadsheet of Fig. 12.19. Starting at $x = 3$ instead, the graph climbs to a peak at $x = 3.126$, which is the solution found in the middle spreadsheet. Using the right spreadsheet's starting solution of $x = 4.7$, the graph climbs until it reaches the boundary imposed by the $x \leq 5$ constraint, so $x = 5$ is the peak in that direction. These three peaks are the *local maxima* (or *local optima*) because each one is a maximum of the graph within a local neighborhood of that point. However, only the largest of these local maxima is the *global maximum,* that is, the highest point on the entire graph. Thus, the middle spreadsheet in Fig. 12.19 did succeed in finding the globally optimal solution at $x = 3.126$ with Profit = \$6.13.

The Excel Solver uses the *generalized reduced gradient method,* which adapts the gradient search method described in Sec. 12.5 to solve convex programming problems. Therefore, this algorithm can be thought of as a hill-climbing procedure. It starts at the initial solution entered into the changing cells and then begins climbing that hill until it reaches the peak (or is blocked from climbing further by reaching the boundary imposed by the constraints). The procedure terminates when it reaches this peak (or boundary) and reports this solution. It has no way of detecting whether there is a taller hill somewhere else on the profit graph.

The same thing would happen with any other hill-climbing procedure, such as SUMT (described in Sec. 12.9), that stops when it finds a local maximum. Thus, if SUMT were to be applied to this example with each of the three initial trial solutions used in Fig. 12.19, it would find the same three local maxima found by the Excel Solver.

A More Systematic Approach to Finding Local Optima

A common approach to "easy" nonconvex programming problems is to apply some algorithmic hill-climbing procedure that will stop when it finds a *local maximum* and then to restart it a number of times from a variety of initial trial solutions (either chosen randomly or as a systematic cross-section) in order to find as many distinct local maxima as possible. The best of these local maxima is then chosen for implementation. Normally, the hill-climbing procedure is one that has been designed to find a global maximum when all the assumptions of convex programming hold, but it also can operate to find a local maximum when they do not.

When employing the Excel Solver, a systematic way of applying this approach is to use the Solver Table add-in that is provided in your OR Courseware. To demonstrate, we will continue to use the spreadsheet model shown in Fig. 12.19. Figure 12.20 displays how the Solver Table is used to try six different starting points (0, 1, 2, 3, 4, and 5) as the initial trial solutions for this model by executing the following steps. In the first row of the table, enter formulas that refer to the changing cell, x (C5), and the target cell, Profit (C8). The different starting points are entered in the first column of the table (G8:G13). Then, select the entire table (G7:I13) and choose Solver Table from the Add-Ins tab (for Excel 2007) or the Tools menu (for earlier versions of Excel). The column input cell entered in the Solver Table dialogue box is the changing cell x (C5), since this is where we want the different starting points in the first column of the table to be entered. (No row input cell is entered in this dialogue box since only a column is being used to list the starting points.)

	A	B	C	D	E	F	G	H	I	J	K
1	**Using Solver Table to Try Different Starting Points**										
2											
3											
4					Maximum		Starting				
5		x =	3.126	<=	5		Point	Solution			
6							x	x*	Profit		
7		Profit = $0.5x^5-6x^4+24.5x^3-39x^2+20x$						3.126	$6.13		Select the entire
8		=	$6.13				0	0.371	$3.19		table (G7:I13),
9							1	0.371	$3.19		before choosing
10							2	3.126	$6.13		Solver Table from
11							3	3.126	$6.13		the Tools menu.
12							4	3.126	$6.13		
13							5	5.000	$0.00		

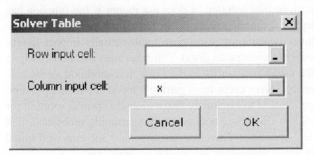

	H	I
5	Solution	
6	x*	Profit
7	=x	=Profit

Range Name	Cell
x	C5
Profit	C8

FIGURE 12.20

An application of the Solver Table (an Excel add-in provided in your OR Courseware) to the example considered in Figs. 12.18 and 12.19.

Clicking OK then causes the Solver Table to re-solve the problem for all these starting points in the first column and fill in the corresponding results (the local maximum for x and Profit referred to in the first row) in the other columns of the table.

The example has only one variable and so only one changing cell. However, the Solver Table also can be used to try multiple starting points for problems with two variables (changing cells). This is done by using the first row and first column of the table to specify different starting points for the two changing cells. Enter an equation referring to the target cell in the upper left-hand corner of the table. Select the entire table and choose Solver Table from the Add-Ins tab or Tools menu, with the two changing cells selected as the column input cell and row input cell. The Solver Table then re-solves the problem for each combination of starting points of the two changing cells and fills in the body of the table with the objective function value of the solution that is found (a local optimum) for each of these combinations. (See Sec. 6.8 for more details about setting up a two-dimensional Solver Table.)

For problems with more than two variables (changing cells), this same approach still can be used to try multiple starting points for any two of the changing cells at a time. However, this becomes a very cumbersome way of trying a broad range of starting points for all the changing cells when there are more than three or four of these cells.

Unfortunately, there generally is no guarantee of finding a globally optimal solution, no matter how many different starting points are tried. Also, if the profit graphs are not smooth (e.g., if they have discontinuities or kinks), then Solver may not even be able to find local optima. Fortunately, Excel's *Premium Solver* provides another search procedure, called *Evolutionary Solver,* to attempt to solve these somewhat more difficult nonconvex programming problems.

Evolutionary Solver

Frontline Systems, the developer of the standard Solver included with Excel, has developed Premium versions of Solver. One version of Premium Solver (Premium Solver for Education) is available in your OR courseware (but not included with standard Excel). Every version of Premium Solver, including this one, adds a search procedure called **Evolutionary Solver** in the set of tools available to search for an optimal solution for a model. The philosophy of Evolutionary Solver is based on genetics, evolution, and the survival of the fittest. Hence, this type of algorithm is sometimes called a **genetic algorithm.** We will devote Sec. 13.4 to describing how genetic algorithms operate.

Evolutionary Solver has three crucial advantages over the standard Solver (or any other convex programming algorithm) for solving nonconvex programming problems. First, the complexity of the objective function does not impact Evolutionary Solver. As long as the function can be evaluated for a given trial solution, it does not matter if the function has kinks or discontinuities or many local optima. Second, the complexity of the given constraints (including even nonconvex constraints) also doesn't substantially impact Evolutionary Solver (although the *number* of constraints does). Third, because it evaluates whole populations of trial solutions that aren't necessarily in the same neighborhood as the current best trial solution, Evolutionary Solver keeps from getting trapped at a local optimum. In fact, Evolutionary Solver is guaranteed to eventually find a globally optimal solution for any nonlinear programming problem (including nonconvex programming problems), if it is run forever (which is impractical of course). Therefore, Evolutionary Solver is well suited for dealing with many relatively small nonconvex programming problems.

On the other hand, it must be pointed out that Evolutionary Solver is not a panacea. First, it can take *much* longer than the standard Solver to find a final solution. Second, Evolutionary Solver does not perform well on models that have many constraints. Third, Evolutionary Solver is a random process, so running it again on the same model usually will yield a different final solution. Finally, the best solution found typically is not quite optimal (although it may be very close). Evolutionary Solver does not continuously move toward better solutions. Rather it is more like an intelligent search engine, trying out different random solutions. Thus, while it is quite likely to end up with a solution that is very close to optimal, it almost never returns the exact globally optimal solution on most types of nonlinear programming problems. Consequently, if often can be beneficial to run the standard Solver (GRG Nonlinear option) after the Evolutionary Solver, starting with the final solution obtained by the Evolutionary Solver, to see if this solution can be improved by searching around its neighborhood.

■ 12.11   CONCLUSIONS

Practical optimization problems frequently involve *nonlinear* behavior that must be taken into account. It is sometimes possible to *reformulate* these nonlinearities to fit into a linear programming format, as can be done for *separable programming* problems. However, it is frequently necessary to use a *nonlinear programming* formulation.

In contrast to the case of the simplex method for linear programming, there is no efficient all-purpose algorithm that can be used to solve all nonlinear programming problems. In fact, some of these problems cannot be solved in a very satisfactory manner by any method. However, considerable progress has been made for some important classes of problems, including *quadratic programming, convex programming,* and certain special types of *nonconvex programming*. A variety of algorithms that frequently perform well are available for these cases. Some of these algorithms incorporate highly efficient procedures for *unconstrained optimization* for a portion of each iteration, and some use a succession of linear or quadratic approximations to the original problem.

There has been a strong emphasis in recent years on developing high-quality, reliable *software packages* for general use in applying the best of these algorithms. For example, several powerful software packages have been developed in the Systems Optimization Laboratory at Stanford University. These packages are widely used elsewhere for solving many of the types of problems discussed in this chapter (as well as linear programming problems). The steady improvements being made in both algorithmic techniques and software now are bringing some rather large problems into the range of computational feasibility.

Research in nonlinear programming remains very active.

■ SELECTED REFERENCES

1. Fiacco, A. V., and G. P. McCormick: *Nonlinear Programming: Sequential Unconstrained Minimization Techniques,* Classics in Applied Mathematics 4, Society for Industrial and Applied Mathematics, Philadelphia, 1990. (Reprint of a classic book published in 1968.)
2. Fletcher, R.: *Practical Methods of Optimization,* 2nd ed., Wiley, New York, 2000.
3. Gill, P. E., W. Murray, and M. H. Wright: *Practical Optimization,* Academic Press, London, 1981.
4. Hillier, F. S., and M. S. Hillier: *Introduction to Management Science: A Modeling and Case Studies Approach with Spreadsheets,* 3rd ed., McGraw-Hill/Irwin, Burr Ridge, IL, 2008, chap. 8.
5. Leyffer, S., and J. More (eds.): Special Issue on Deterministic Global Optimization and Applications, *Mathematical Programming,* Series B, **103**(2), June 2005.
6. Luenberger, D., and Y. Ye: *Linear and Nonlinear Programming,* 3rd ed., Springer, New York, 2008.
7. Miller, R. E.: *Optimization: Foundations and Applications,* Wiley, New York, 1999.
8. Rardin, D.: *Optimization in Operations Research,* Prentice-Hall, Upper Saddle River, NJ, 1998.

■ LEARNING AIDS FOR THIS CHAPTER ON OUR WEBSITE (www.mhhe.com/hillier)

Worked Examples:

Examples for Chapter 12

Demonstration Examples in OR Tutor:

Gradient Search Procedure
Frank-Wolfe Algorithm
Sequential Unconstrained Minimization Technique—SUMT

Interactive Procedures in IOR Tutorial:

Interactive One-Dimensional Search Procedure
Interactive Gradient Search Procedure
Interactive Modified Simplex Method
Interactive Frank-Wolfe Algorithm

Automatic Procedures in IOR Tutorial:

Automatic Gradient Search Procedure
Sequential Unconstrained Minimization Technique—SUMT

Excel Add-ins:

Premium Solver for Education
Solver Table

"Ch. 12—Nonlinear Programming" Files for Solving the Examples:

Excel Files
LINGO/LINDO File
MPL/CPLEX/CONOPT/LGO File

Glossary for Chapter 12

See Appendix 1 for documentation of the software.

■ PROBLEMS

The symbols to the left of some of the problems (or their parts) have the following meaning:

D: The corresponding demonstration example just listed in Learning Aids may be helpful.
I: We suggest that you use the corresponding interactive routine just listed (the printout records your work).
C: Use the computer with any of the software options available to you (or as instructed by your instructor) to solve the problem.

An asterisk on the problem number indicates that at least a partial answer is given in the back of the book.

12.1-1. Read the referenced article that fully describes the OR study summarized in the application vignette presented in Sec. 12.1. Briefly describe how nonlinear programming was applied in this study. Then list the various financial and nonfinancial benefits that resulted from this study.

12.1-2. Consider the *product mix* problem described in Prob. 3.1-11. Suppose that this manufacturing firm actually encounters *price elasticity* in selling the three products, so that the profits would be different from those stated in Chap. 3. In particular, suppose that the unit costs for producing products 1, 2, and 3 are $25, $10, and $15, respectively, and that the prices required (in dollars) in order to be able to sell x_1, x_2, and x_3 units are $(35 + 100x_1^{-\frac{1}{3}})$, $(15 + 40x_2^{-\frac{1}{4}})$, and $(20 + 50x_3^{-\frac{1}{2}})$, respectively.

Formulate a nonlinear programming model for the problem of determining how many units of each product the firm should produce to maximize profit.

12.1-3. For the P & T Co. problem described in Sec. 8.1, suppose that there is a 10 percent discount in the shipping cost for all truckloads *beyond* the first 40 for each combination of cannery and warehouse. Draw figures like Figs. 12.3 and 12.4, showing the marginal cost and total cost for shipments of truckloads of peas from cannery 1 to warehouse 1. Then describe the overall nonlinear programming model for this problem.

12.1-4. A stockbroker, Richard Smith, has just received a call from his most important client, Ann Hardy. Ann has $50,000 to invest, and wants to use it to purchase two stocks. Stock 1 is a solid blue-chip security with a respectable growth potential and little risk involved. Stock 2 is much more speculative. It is being touted in two investment newsletters as having outstanding growth potential, but also is considered very risky. Ann would like a large return on her investment, but also has considerable aversion to risk. Therefore, she has instructed Richard to analyze what mix of investments in the two stocks would be appropriate for her.

Ann is used to talking in units of thousands of dollars and 1,000-share blocks of stocks. Using these units, the price per block is 20 for stock 1 and 30 for stock 2. After doing some research, Richard has made the following estimates. The expected return per block is 5 for stock 1 and 10 for stock 2. The variance of the return on each block is 4 for stock 1 and 100 for stock 2. The covariance of the return on one block each of the two stocks is 5.

Without yet assigning a specific numerical value to the minimum acceptable expected return, formulate a nonlinear programming model for this problem.

12.2-1. Reconsider Prob. 12.1-2. Verify that this problem is a convex programming problem.

12.2-2. Reconsider Prob. 12.1-4. Show that the model formulated is a convex programming problem by using the test in Appendix 2 to show that the objective function being minimized is convex.

12.2-3. Consider the variation of the Wyndor Glass Co. example represented in Fig. 12.5, where the second and third functional constraints of the original problem (see Sec. 3.1) have been replaced by $9x_1^2 + 5x_2^2 \leq 216$. Demonstrate that $(x_1, x_2) = (2, 6)$ with $Z = 36$ is indeed optimal by showing that the objective function line $36 = 3x_1 + 5x_2$ is *tangent* to this constraint boundary at $(2, 6)$. (*Hint:* Express x_2 in terms of x_1 on this boundary, and then differentiate this expression with respect to x_1 to find the slope of the boundary.)

12.2-4. Consider the variation of the Wyndor Glass Co. problem represented in Fig. 12.6, where the original objective function (see Sec. 3.1) has been replaced by $Z = 126x_1 - 9x_1^2 + 182x_2 - 13x_2^2$. Demonstrate that $(x_1, x_2) = (\frac{8}{3}, 5)$ with $Z = 857$ is indeed optimal by showing that the ellipse $857 = 126x_1 - 9x_1^2 + 182x_2 - 13x_2^2$ is *tangent* to the constraint boundary $3x_1 + 2x_2 = 18$ at $(\frac{8}{3}, 5)$. (*Hint:* Solve for x_2 in terms of x_1 for the ellipse, and then differentiate this expression with respect to x_1 to find the slope of the ellipse.)

12.2-5. Consider the following function:

$$f(x) = 240x - 300x^2 + 10x^3.$$

(a) Use the first and second derivatives to find the local maxima and local minima of $f(x)$.
(b) Use the first and second derivatives to show that $f(x)$ has neither a global maximum nor a global minimum because it is unbounded in both directions.

12.2-6. For each of the following functions, show whether it is convex, concave, or neither.
(a) $f(x) = 10x - x^2$
(b) $f(x) = x^4 + 6x^2 + 12x$
(c) $f(x) = 2x^3 - 3x^2$
(d) $f(x) = x^4 + x^2$
(e) $f(x) = x^3 + x^4$

12.2-7.* For each of the following functions, use the test given in Appendix 2 to determine whether it is convex, concave, or neither.
(a) $f(\mathbf{x}) = x_1x_2 - x_1^2 - x_2^2$
(b) $f(\mathbf{x}) = 3x_1 + 2x_1^2 + 4x_2 + x_2^2 - 2x_1x_2$
(c) $f(\mathbf{x}) = x_1^2 + 3x_1x_2 + 2x_2^2$
(d) $f(\mathbf{x}) = 20x_1 + 10x_2$
(e) $f(\mathbf{x}) = x_1x_2$

12.2-8. Consider the following function:

$$f(\mathbf{x}) = 5x_1 + 2x_2^2 + x_3^2 - 3x_3x_4 + 4x_4^2 + 2x_5^4 + x_5^2 \\ + 3x_5x_6 + 6x_6^2 + 3x_6x_7 + x_7^2.$$

Show that $f(\mathbf{x})$ is convex by expressing it as a sum of functions of one or two variables and then showing (see Appendix 2) that all these functions are convex.

12.2-9. Consider the following nonlinear programming problem:

Maximize $f(\mathbf{x}) = x_1 + x_2$,

subject to

$$x_1^2 + x_2^2 \leq 1$$

and

$$x_1 \geq 0, \qquad x_2 \geq 0.$$

(a) Verify that this is a convex programming problem.
(b) Solve this problem graphically.

12.2-10. Consider the following nonlinear programming problem:

Minimize $Z = x_1^4 + 2x_2^2$,

subject to

$$x_1^2 + x_2^2 \geq 2.$$
(No nonnegativity constraints.)

(a) Use geometric analysis to determine whether the feasible region is a convex set.
(b) Now use algebra and calculus to determine whether the feasible region is a convex set.

12.3-1. Reconsider Prob. 12.1-3. Show that this problem is a non-convex programming problem.

12.3-2. Consider the following constrained optimization problem:

Maximize $f(x) = -120x + 15x^2 - 10x^3$,

subject to

$$x \geq 0.$$

Use just the first and second derivatives of $f(x)$ to derive an optimal solution.

12.3-3. Consider the following nonlinear programming problem:

Minimize $Z = x_1^4 + 2x_1^2 + 2x_1x_2 + 4x_2^2$,

subject to

$$2x_1 + x_2 \geq 10 \\ x_1 + 2x_2 \geq 10$$

and

$$x_1 \geq 0, \qquad x_2 \geq 0.$$

(a) Of the special types of nonlinear programming problems described in Sec. 12.3, to which type or types can this particular problem be fitted? Justify your answer.
(b) Now suppose that the problem is changed slightly by replacing the nonnegativity constraints by $x_1 \geq 1$ and $x_2 \geq 1$. Convert this new problem to an equivalent problem that has just two functional constraints, two variables, and two nonnegativity constraints.

12.3-4. Consider the following geometric programming problem:

Minimize $f(\mathbf{x}) = 2x_1^{-2}x_2^{-1} + x_2^{-2}$,

subject to

$$4x_1x_2 + x_1^2x_2^2 \leq 12$$

and

$$x_1 \geq 0, \qquad x_2 \geq 0.$$

(a) Transform this problem to an equivalent convex programming problem.
(b) Use the test given in Appendix 2 to verify that the model formulated in part (*a*) is indeed a convex programming problem.

12.3-5. Consider the following linear fractional programming problem:

Maximize $f(\mathbf{x}) = \dfrac{10x_1 + 20x_2 + 10}{3x_1 + 4x_2 + 20}$,

subject to

$$x_1 + 3x_2 \leq 50$$
$$3x_1 + 2x_2 \leq 80$$

and

$$x_1 \geq 0, \qquad x_2 \geq 0.$$

(a) Transform this problem to an equivalent linear programming problem.

C (b) Use the computer to solve the model formulated in part (a). What is the resulting optimal solution for the original problem?

12.3-6. Consider the expressions in matrix notation given in Sec. 12.7 for the general form of the KKT conditions for the quadratic programming problem. Show that the problem of finding a feasible solution for these conditions is a linear complementarity problem, as introduced in Sec. 12.3, by identifying $\mathbf{w}$, $\mathbf{z}$, $\mathbf{q}$, and $\mathbf{M}$ in terms of the vectors and matrices in Sec. 12.7.

12.4-1.* Consider the following problem:

Maximize $f(x) = x^3 + 2x - 2x^2 - 0.25x^4.$

I (a) Apply the bisection method to (approximately) solve this problem. Use an error tolerance $\epsilon = 0.04$ and initial bounds $\underline{x} = 0, \bar{x} = 2.4.$

(b) Apply Newton's method, with $\epsilon = 0.001$ and $x_1 = 1.2$, to this problem.

I **12.4-2.** Use the bisection method with an error tolerance $\epsilon = 0.04$ and with the following initial bounds to interactively solve (approximately) each of the following problems.
(a) Maximize $f(x) = 6x - x^2$, with $\underline{x} = 0, \bar{x} = 4.8.$
(b) Minimize $f(x) = 6x + 7x^2 + 4x^3 + x^4$,   with $\underline{x} = -4, \bar{x} = 1.$

12.4-3. Consider the following problem:

Maximize $f(x) = 48x^5 + 42x^3 + 3.5x - 16x^6$
$\qquad\qquad - 61x^4 - 16.5x^2.$

I (a) Apply the bisection method to (approximately) solve this problem. Use an error tolerance $\epsilon = 0.08$ and initial bounds $\underline{x} = -1, \bar{x} = 4.$

(b) Apply Newton's method, with $\epsilon = 0.001$ and $x_1 = 1$, to this problem.

12.4-4. Consider the following problem:

Maximize $f(x) = 10x^3 + 60x - 2x^6 - 3x^4 - 12x^2.$

I (a) Apply the bisection method to (approximately) solve this problem. Use an error tolerance $\epsilon = 0.07$ and find appropriate initial bounds by inspection.

(b) Apply Newton's method, with $\epsilon = 0.001$ and $x_1 = 1$, to this problem.

12.4-5. Consider the following convex programming problem:

Minimize $Z = x^4 + x^2 - 4x,$

subject to

$$x \leq 2 \qquad \text{and} \qquad x \geq 0.$$

(a) Use one simple calculation *just* to check whether the optimal solution lies in the interval $0 \leq x \leq 1$ or the interval $1 \leq x \leq 2$. (Do *not* actually solve for the optimal solution in order to determine in which interval it must lie.) Explain your logic.

I (b) Use the bisection method with initial bounds $\underline{x} = 0, \bar{x} = 2$ and with an error tolerance $\epsilon = 0.02$ to interactively solve (approximately) this problem.

(c) Apply Newton's method, with $\epsilon = 0.0001$ and $x_1 = 1$, to this problem.

12.4-6. Consider the problem of maximizing a differentiable function $f(x)$ of a single unconstrained variable x. Let $\underline{x}_0$ and $\bar{x}_0$, respectively, be a valid lower bound and upper bound on the same global maximum (if one exists). Prove the following general properties of the bisection method (as presented in Sec. 12.4) for attempting to solve such a problem.

(a) Given $\underline{x}_0, \bar{x}_0$, and $\epsilon = 0$, the sequence of trial solutions selected by the *midpoint rule* must *converge* to a limiting solution. [*Hint:* First show that $\lim_{n \to \infty} (\bar{x}_n - \underline{x}_n) = 0$, where $\bar{x}_n$ and $\underline{x}_n$ are the upper and lower bounds identified at iteration n.]

(b) If $f(x)$ is concave [so that $df(x)/dx$ is a monotone decreasing function of x], then the limiting solution in part (a) must be a global maximum.

(c) If $f(x)$ is not concave everywhere, but would be concave if its domain were restricted to the interval between $\underline{x}_0$ and $\bar{x}_0$, then the limiting solution in part (a) must be a global maximum.

(d) If $f(x)$ is not concave even over the interval between $\underline{x}_0$ and $\bar{x}_0$, then the limiting solution in part (a) need not be a global maximum. (Prove this by graphically constructing a counterexample.)

(e) If $df(x)/dx < 0$ for all x, then no $\underline{x}_0$ exists. If $df(x)/dx > 0$ for all x, then no $\bar{x}_0$ exists. In either case, $f(x)$ does not possess a global maximum.

(f) If $f(x)$ is concave and $\lim_{x \to -\infty} df(x)/dx < 0$, then no $\underline{x}_0$ exists. If $f(x)$ is concave and $\lim_{x \to \infty} df(x)/dx > 0$, then no $\bar{x}_0$ exists. In either case, $f(x)$ does not possess a global maximum.

I **12.4-7.** Consider the following linearly constrained convex programming problem:

Maximize $f(\mathbf{x}) = 32x_1 + 50x_2 - 10x_2^2 + x_2^3 - x_1^4 - x_2^4,$

subject to

$$3x_1 + x_2 \leq 11$$
$$2x_1 + 5x_2 \leq 16$$

and

$$x_1 \geq 0, \qquad x_2 \geq 0.$$

Ignore the constraints and solve the resulting two *one-variable unconstrained optimization* problems. Use calculus to solve the problem involving x_1 and use the bisection method with $\epsilon = 0.001$ and initial bounds 0 and 4 to solve the problem involving x_2. Show that the resulting solution for (x_1, x_2) satisfies all of the constraints, so it is actually optimal for the original problem.

12.5-1. Consider the following unconstrained optimization problem:

Maximize $\quad f(\mathbf{x}) = 2x_1x_2 + x_2 - x_1^2 - 2x_2^2.$

D,I **(a)** Starting from the initial trial solution $(x_1, x_2) = (1, 1)$, interactively apply the gradient search procedure with $\epsilon = 0.25$ to obtain an approximate solution.

(b) Solve the system of linear equations obtained by setting $\nabla f(\mathbf{x}) = \mathbf{0}$ to obtain the exact solution.

(c) Referring to Fig 12.14 as a sample for a similar problem, draw the path of trial solutions you obtained in part (a). Then show the apparent *continuation* of this path with your best guess for the next three trial solutions [based on the pattern in part (a) and in Fig. 12.14]. Also show the exact solution from part (b) toward which this sequence of trial solutions is converging.

C **(d)** Apply the automatic routine for the gradient search procedure (with $\epsilon = 0.01$) in your IOR Tutorial to this problem.

D,I,C **12.5-2.** Starting from the initial trial solution $(x_1, x_2) = (1, 1)$, interactively apply two iterations of the gradient search procedure to begin solving the following problem, and then apply the automatic routine for this procedure (with $\epsilon = 0.01$).

Maximize $\quad f(\mathbf{x}) = 60x_1x_2 - 15x_1^2 - 80x_2^2.$

Then solve $\nabla f(\mathbf{x}) = \mathbf{0}$ directly to obtain the exact solution.

D,I,C **12.5-3.*** Starting from the initial trial solution $(x_1, x_2) = (0, 0)$, interactively apply the gradient search procedure with $\epsilon = 0.3$ to obtain an approximate solution for the following problem, and then apply the automatic routine for this procedure (with $\epsilon = 0.01$).

Maximize $\quad f(\mathbf{x}) = 8x_1 - x_1^2 - 12x_2 - 2x_2^2 + 2x_1x_2.$

Then solve $\nabla f(\mathbf{x}) = \mathbf{0}$ directly to obtain the exact solution.

D,I,C **12.5-4.** Starting from the initial trial solution $(x_1, x_2) = (0, 0)$, interactively apply two iterations of the gradient search procedure to begin solving the following problem, and then apply the automatic routine for this procedure (with $\epsilon = 0.01$).

Maximize $\quad f(\mathbf{x}) = 6x_1 + 2x_1x_2 - 2x_2 - 2x_1^2 - x_2^2.$

Then solve $\nabla f(\mathbf{x}) = \mathbf{0}$ directly to obtain the exact solution.

12.5-5. Starting from the initial trial solution $(x_1, x_2) = (0, 0)$, apply *one* iteration of the gradient search procedure to the following problem by hand:

Maximize $\quad f(\mathbf{x}) = 4x_1 + 2x_2 + x_1^2 - x_1^4 - 2x_1x_2 - x_2^2.$

To complete this iteration, approximately solve for t^* by manually applying *two* iterations of the bisection method with initial bounds $\underline{t} = 0$, $\bar{t} = 1$.

12.5-6. Consider the following unconstrained optimization problem:

Maximize $\quad f(\mathbf{x}) = 3x_1x_2 + 3x_2x_3 - x_1^2 - 6x_2^2 - x_3^2.$

(a) Describe how solving this problem can be reduced to solving a *two-variable* unconstrained optimization problem.

D,I **(b)** Starting from the initial trial solution $(x_1, x_2, x_3) = (1, 1, 1)$, interactively apply the gradient search procedure with $\epsilon = 0.05$ to solve (approximately) the two-variable problem identified in part (a).

C **(c)** Repeat part (b) with the automatic routine for this procedure (with $\epsilon = 0.005$).

D,I,C **12.5-7.*** Starting from the initial trial solution $(x_1, x_2) = (0, 0)$, interactively apply the *gradient search procedure* with $\epsilon = 1$ to solve (approximately) the following problem, and then apply the automatic routine for this procedure (with $\epsilon = 0.01$).

Maximize $\quad f(\mathbf{x}) = x_1x_2 + 3x_2 - x_1^2 - x_2^2.$

12.6-1. Reconsider the one-variable convex programming model given in Prob. 12.4-5. Use the KKT conditions to derive an optimal solution for this model.

12.6-2. Reconsider Prob. 12.2-9. Use the KKT conditions to check whether $(x_1, x_2) = (1/\sqrt{2}, 1/\sqrt{2})$ is optimal.

12.6-3.* Reconsider the model given in Prob. 12.3-3. What are the KKT conditions for this model? Use these conditions to determine whether $(x_1, x_2) = (0, 10)$ can be optimal.

12.6-4. Consider the following convex programming problem:

Maximize $\quad f(\mathbf{x}) = 12x_1 - x_1^2 + 50x_2 - x_2^2,$

subject to

$$x_1 \leq 10,$$
$$x_2 \leq 15,$$

and

$$x_1 \geq 0, \qquad x_2 \geq 0.$$

(a) Use the KKT conditions for this problem to derive an optimal solution.

(b) Decompose this problem into two separate constrained optimization problems involving just x_1 and just x_2, respectively. For each of these two problems, plot the objective function over the feasible region in order to *demonstrate* that the value of x_1 or x_2 derived in part (a) is indeed optimal. Then *prove* that this value is optimal by using just the first and second derivatives of the objective function and the constraints for the respective problems.

12.6-5. Consider the following linearly constrained optimization problem:

Maximize $\quad f(\mathbf{x}) = \ln(x_1 + 1) - x_2^2,$

subject to

$$x_1 + 2x_2 \leq 3$$

and

$$x_1 \geq 0, \qquad x_2 \geq 0,$$

where ln denotes the natural logarithm,

(a) Verify that this problem is a convex programming problem.

(b) Use the KKT conditions to derive an optimal solution.

(c) Use intuitive reasoning to demonstrate that the solution obtained in part (b) is indeed optimal.

12.6-6.* Consider the nonlinear programming problem given in Prob. 10.3-11. Determine whether $(x_1, x_2) = (1, 2)$ can be optimal by applying the KKT conditions.

12.6-7. Consider the following nonlinear programming problem:

Maximize $f(\mathbf{x}) = \dfrac{x_1}{x_2 + 1}$,

subject to

$$x_1 - x_2 \le 2$$

and

$$x_1 \ge 0, \qquad x_2 \ge 0.$$

(a) Use the KKT conditions to demonstrate that $(x_1, x_2) = (4, 2)$ is *not* optimal.
(b) Derive a solution that does satisfy the KKT conditions.
(c) Show that this problem is *not* a convex programming problem.
(d) Despite the conclusion in part (c), use *intuitive* reasoning to show that the solution obtained in part (b) is, in fact, optimal. [The theoretical reason is that $f(\mathbf{x})$ is *pseudo-concave*.]
(e) Use the fact that this problem is a linear fractional programming problem to transform it into an equivalent linear programming problem. Solve the latter problem and thereby identify the optimal solution for the original problem. (*Hint:* Use the equality constraint in the linear programming problem to substitute one of the variables out of the model, and then solve the model graphically.)

12.6-8.* Use the KKT conditions to derive an optimal solution for each of the following problems.

(a)   Maximize $f(\mathbf{x}) = x_1 + 2x_2 - x_2^3$,

subject to

$$x_1 + x_2 \le 1$$

and

$$x_1 \ge 0, \qquad x_2 \ge 0.$$

(b)   Maximize $f(\mathbf{x}) = 20x_1 + 10x_2$,

subject to

$$x_1^2 + x_2^2 \le 1$$
$$x_1 + 2x_2 \le 2$$

and

$$x_1 \ge 0, \qquad x_2 \ge 0.$$

12.6-9. What are the KKT conditions for nonlinear programming problems of the following form?

Minimize $f(\mathbf{x})$,

subject to

$$g_i(\mathbf{x}) \ge b_i, \qquad \text{for } i = 1, 2, \dots, m$$

and

$$\mathbf{x} \ge \mathbf{0}.$$

(*Hint:* Convert this form to our standard form assumed in this chapter by using the techniques presented in Sec. 4.6 and then applying the KKT conditions as given in Sec. 12.6.)

12.6-10. Consider the following nonlinear programming problem:

Minimize $Z = 2x_1^2 + x_2^2$,

subject to

$$x_1 + x_2 = 10$$

and

$$x_1 \ge 0, \qquad x_2 \ge 0.$$

(a) Of the special types of nonlinear programming problems described in Sec. 12.3, to which type or types can this particular problem be fitted? Justify your answer. (*Hint:* First convert this problem to an equivalent nonlinear programming problem that fits the form given in the second paragraph of the chapter, with $m = 2$ and $n = 2$.)
(b) Obtain the KKT conditions for this problem.
(c) Use the KKT conditions to derive an optimal solution.

12.6-11. Consider the following linearly constrained programming problem:

Minimize $f(\mathbf{x}) = x_1^3 + 4x_2^2 + 16x_3$,

subject to

$$x_1 + x_2 + x_3 = 5$$

and

$$x_1 \ge 1, \qquad x_2 \ge 1, \qquad x_3 \ge 1.$$

(a) Convert this problem to an equivalent nonlinear programming problem that fits the form given at the beginning of the chapter (second paragraph), with $m = 2$ and $n = 3$.
(b) Use the form obtained in part (a) to construct the KKT conditions for this problem.
(c) Use the KKT conditions to check whether $(x_1, x_2, x_3) = (2, 1, 2)$ is optimal.

12.6-12. Consider the following linearly constrained convex programming problem:

Minimize $Z = x_1^2 - 6x_1 + x_2^3 - 3x_2$,

subject to

$$x_1 + x_2 \le 1$$

and

$$x_1 \ge 0, \qquad x_2 \ge 0.$$

(a) Obtain the KKT conditions for this problem.
(b) Use the KKT conditions to check whether $(x_1, x_2) = (\frac{1}{2}, \frac{1}{2})$ is an optimal solution.
(c) Use the KKT conditions to derive an optimal solution.

12.6-13. Consider the following linearly constrained convex programming problem:

Maximize $f(\mathbf{x}) = 8x_1 - x_1^2 + 2x_2 + x_3$,

subject to

$$x_1 + 3x_2 + 2x_3 \leq 12$$

and

$$x_1 \geq 0, \qquad x_2 \geq 0, \qquad x_3 \geq 0.$$

(a) Use the KKT conditions to demonstrate that $(x_1, x_2, x_3) = (2, 2, 2)$ is *not* an optimal solution.

(b) Use the KKT conditions to derive an optimal solution. (*Hint:* Do some preliminary intuitive analysis to determine the most promising case regarding which variables are nonzero and which are zero.)

12.6-14. Use the KKT conditions to determine whether $(x_1, x_2, x_3) = (1, 1, 1)$ can be optimal for the following problem:

$$\text{Minimize} \qquad Z = 2x_1 + x_2^3 + x_3^2,$$

subject to

$$x_1^2 + 2x_2^2 + x_3^2 \geq 4$$

and

$$x_1 \geq 0, \qquad x_2 \geq 0, \qquad x_3 \geq 0.$$

12.6-15. Reconsider the model given in Prob. 12.2-10. What are the KKT conditions for this problem? Use these conditions to determine whether $(x_1, x_2) = (1, 1)$ can be optimal.

12.6-16. Reconsider the linearly constrained convex programming model given in Prob. 12.4-7. Use the KKT conditions to determine whether $(x_1, x_2) = (2, 2)$ can be optimal.

12.7-1. Consider the quadratic programming example presented in Sec. 12.7.

(a) Use the test given in Appendix 2 to show that the objective function is *strictly concave.*

(b) Verify that the objective function is strictly concave by demonstrating that $\mathbf{Q}$ is a *positive definite* matrix; that is, $\mathbf{x}^T\mathbf{Q}\mathbf{x} > \mathbf{0}$ for all $\mathbf{x} \neq \mathbf{0}$. (*Hint:* Reduce $\mathbf{x}^T\mathbf{Q}\mathbf{x}$ to a sum of squares.)

(c) Show that $x_1 = 12$, $x_2 = 9$, and $u_1 = 3$ satisfy the KKT conditions when they are written in the form given in Sec. 12.6.

12.7-2.* Consider the following quadratic programming problem:

$$\text{Maximize} \qquad f(\mathbf{x}) = 8x_1 - x_1^2 + 4x_2 - x_2^2,$$

subject to

$$x_1 + x_2 \leq 2$$

and

$$x_1 \geq 0, \qquad x_2 \geq 0.$$

(a) Use the KKT conditions to derive an optimal solution.

(b) Now suppose that this problem is to be solved by the modified simplex method. Formulate the linear programming problem that is to be addressed explicitly, and then identify the additional complementarity constraint that is enforced automatically by the algorithm.

I **(c)** Apply the modified simplex method to the problem as formulated in part (b).

C **(d)** Use the computer to solve the quadratic programming problem directly.

12.7-3. Consider the following quadratic programming problem:

$$\text{Maximize} \qquad f(\mathbf{x}) = 250x_1 - 25x_1^2 + 100x_2 - 100x_2^2 + 90x_1x_2,$$

subject to

$$\begin{aligned} 20x_1 + 5x_2 &\leq 90 \\ 10x_1 + 10x_2 &\leq 60 \end{aligned}$$

and

$$x_1 \geq 0, \qquad x_2 \geq 0.$$

Suppose that this problem is to be solved by the modified simplex method.

(a) Formulate the linear programming problem that is to be addressed explicitly, and then identify the additional complementarity constraint that is enforced automatically by the algorithm.

I **(b)** Apply the modified simplex method to the problem as formulated in part (a).

12.7-4. Consider the following quadratic programming problem.

$$\text{Maximize} \qquad f(\mathbf{x}) = 2x_1 + 3x_2 - x_1^2 - x_2^2,$$

subject to

$$x_1 + x_2 \leq 2$$

and

$$x_1 \geq 0, \qquad x_2 \geq 0.$$

(a) Use the KKT conditions to derive an optimal solution directly.

(b) Now suppose that this problem is to be solved by the modified simplex method. Formulate the linear programming problem that is to be addressed explicitly, and then identify the additional complementarity constraint that is enforced automatically by the algorithm.

(c) Without applying the modified simplex method, show that the solution derived in part (a) is indeed optimal ($Z = 0$) for the equivalent problem formulated in part (b).

I **(d)** Apply the modified simplex method to the problem as formulated in part (b).

C **(e)** Use the computer to solve the quadratic programming problem directly.

12.7-5. Reconsider the first quadratic programming variation of the Wyndor Glass Co. problem presented in Sec. 12.2 (see Fig. 12.6). Analyze this problem by following the instructions of parts (a), (b), and (c) of Prob. 12.7-4.

C **12.7-6.** Reconsider Prob. 12.1-4 and its quadratic programming model.

(a) Display this model [including the values of $R(\mathbf{x})$ and $V(\mathbf{x})$] on an Excel spreadsheet.

(b) Solve this model for four cases: minimum acceptable expected return = 13, 14, 15, 16.

(c) For typical probability distributions (with mean μ and variance σ^2) of the total return from the entire portfolio, the probability is fairly high (about 0.8 or 0.9) that the return will exceed $\mu - \sigma$, and the probability is extremely high (often close to 0.999) that the return will exceed $\mu - 3\sigma$. Calculate $\mu - \sigma$ and $\mu - 3\sigma$ for the four portfolios obtained in part (b). Which portfolio will give the highest μ among those that also give $\mu - \sigma \geq 0$?

12.8-1. The MFG Corporation is planning to produce and market three different products. Let x_1, x_2, and x_3 denote the number of units of the three respective products to be produced. The preliminary estimates of their potential profitability are as follows.

For the first 15 units produced of Product 1, the unit profit would be approximately $500. The unit profit would be only $60 for any additional units of Product 1. For the first 20 units produced of Product 2, the unit profit is estimated at $400. The unit profit would be $200 for each of the next 20 units and $100 for any additional units. For the first 20 units of Product 3, the unit profit would be $600. The unit profit would be $400 for each of the next 10 units and $200 for any additional units.

Certain limitations on the use of needed resources impose the following constraints on the production of the three products:

$$2x_1 + 3x_2 + 4x_3 \leq 180$$
$$3x_1 + x_2 \qquad \leq 150$$
$$x_1 \qquad + 3x_3 \leq 100.$$

Management wants to know what values of x_1, x_2 and x_3 should be chosen to maximize the total profit.
(a) Plot the profit graph for each of the three products.
(b) Use separable programming to formulate a linear programming model for this problem.
C (c) Solve the model. What is the resulting recommendation to management about the values of x_1, x_2, and x_3 to use?
(d) Now suppose that there is an additional constraint that the profit from products 1 and 2 must total at least $20,000. Use the technique presented in the "Extensions" subsection of Sec. 12.8 to add this constraint to the model formulated in part (b).
C (e) Repeat part (c) for the model formulated in part (d).

12.8-2.* The Dorwyn Company has two new products that will compete with the two new products for the Wyndor Glass Co. (described in Sec. 3.1). Using units of hundreds of dollars for the objective function, the linear programming model shown below has been formulated to determine the most profitable product mix.

Maximize $\quad Z = 4x_1 + 6x_2$,

subject to

$$x_1 + 3x_2 \leq 8$$
$$5x_1 + 2x_2 \leq 14$$

and

$$x_1 \geq 0, \qquad x_2 \geq 0.$$

However, because of the strong competition from Wyndor, Dorwyn management now realizes that the company will need to make a strong marketing effort to generate substantial sales of these products. In particular, it is estimated that achieving a production and sales rate of x_1 units of Product 1 per week will require weekly marketing costs of x_1^3 hundred dollars. The corresponding marketing costs for Product 2 are estimated to be $2x_2^2$ hundred dollars. Thus, the objective function in the model should be $Z = 4x_1 + 6x_2 - x_1^3 - 2x_2^2$.

Dorwyn management now would like to use the revised model to determine the most profitable product mix.
(a) Verify that $(x_1, x_2) = (2/\sqrt{3}, \frac{3}{2})$ is an optimal solution by applying the KKT conditions.
(b) Construct tables to show the profit data for each product when the production rate is 0, 1, 2, 3.
(c) Draw a figure like Fig. 12.15b that plots the weekly profit points for each product when the production rate is 0, 1, 2, 3. Connect the pairs of consecutive points with (dashed) line segments.
(d) Use separable programming based on this figure to formulate an approximate linear programming model for this problem.
C (e) Solve the model. What does this say to Dorwyn management about which product mix to use?

12.8-3. The B. J. Jensen Company specializes in the production of power saws and power drills for home use. Sales are relatively stable throughout the year except for a jump upward during the Christmas season. Since the production work requires considerable work and experience, the company maintains a stable employment level and then uses overtime to increase production in November. The workers also welcome this opportunity to earn extra money for the holidays.

B. J. Jensen, Jr., the current president of the company, is overseeing the production plans being made for the upcoming November. He has obtained the following data.

	Maximum Monthly Production*		Profit per Unit Produced	
	Regular Time	Overtime	Regular Time	Overtime
Power saws	12,000	8,000	$240	$80
Power drills	20,000	12,000	$160	$120

*Assuming adequate supplies of materials from the company's vendors.

However, Mr. Jensen now has learned that, in addition to the limited number of labor hours available, two other factors will limit the production levels that can be achieved this November. One is that the company's vendor for power supply units will only be able to provide 40,000 of these units for November (8,000 more than his usual monthly shipment). Each power saw and each power drill requires one of these units. Second, the vendor who supplies a key part for the gear assemblies will only be able to provide 60,000 for November (16,000 more than for other months). Each power saw requires two of these parts and each power drill requires one.

Mr. Jensen now wants to determine how many power saws and how many power drills to produce in November to maximize the company's total profit.
(a) Draw the profit graph for each of these two products.

(b) Use separable programming to formulate a linear programming model for this problem.

C **(c)** Solve the model. What does this say about how many power saws and how many power drills to produce in November?

12.8-4. Reconsider the linearly constrained convex programming model given in Prob. 12.4-7.

(a) Use the separable programming technique presented in Sec. 12.8 to formulate an approximate linear programming model for this problem. Use $x_1 = 0, 1, 2, 3$ and $x_2 = 0, 1, 2, 3$ as the breakpoints of the piecewise linear functions.

C **(b)** Use the simplex method to solve the model formulated in part (a). Then reexpress this solution in terms of the *original* variables of the problem.

12.8-5. Suppose that the separable programming technique has been applied to a certain problem (the "original problem") to convert it to the following equivalent linear programming problem:

Maximize $Z = 5x_{11} + 4x_{12} + 2x_{13} + 4x_{21} + x_{22}$,

subject to

$$3x_{11} + 3x_{12} + 3x_{13} + 2x_{21} + 2x_{22} \le 25$$
$$2x_{11} + 2x_{12} + 2x_{13} - x_{21} - x_{22} \le 10$$

and

$$0 \le x_{11} \le 2 \qquad 0 \le x_{21} \le 3$$
$$0 \le x_{12} \le 3 \qquad 0 \le x_{22} \le 1.$$
$$0 \le x_{13}$$

What was the mathematical model for the original problem? (You may define the objective function either algebraically or graphically, but express the constraints algebraically.)

12.8-6. For each of the following cases, *prove* that the key property of separable programming given in Sec. 12.8 must hold. (*Hint:* Assume that there exists an optimal solution that violates this property, and then contradict this assumption by showing that there exists a better feasible solution.)

(a) The special case of separable programming where all the $g_i(\mathbf{x})$ are linear functions.

(b) The general case of separable programming where all the functions are nonlinear functions of the designated form. [*Hint:* Think of the functional constraints as constraints on resources, where $g_{ij}(x_j)$ represents the amount of resource i used by running activity j at level x_j, and then use what the convexity assumption implies about the slopes of the approximating piece-wise linear function.]

12.8-7. The MFG Company produces a certain subassembly in each of two separate plants. These subassemblies are then brought to a third nearby plant where they are used in the production of a certain product. The peak season of demand for this product is approaching, so to maintain the production rate within a desired range, it is necessary to use temporarily some overtime in making the subassemblies. The cost per subassembly on regular time (RT) and on overtime (OT) is shown in the following table for both plants, along with the maximum number of subassemblies that can be produced on RT and on OT each day.

	Unit Cost		Capacity	
	RT	OT	RT	OT
Plant 1	$23	$38	6,000	3,000
Plant 2	$24	$36	3,000	1,500

Let x_1 and x_2 denote the total number of subassemblies produced per day at plants 1 and 2, respectively. The objective is to maximize $Z = x_1 + x_2$, subject to the constraint that the total daily cost not exceed $270,000. Note that the mathematical programming formulation of this problem (with x_1 and x_2 as decision variables) has the same form as the main case of the separable programming model described in Sec. 12.8, except that the separable functions appear in a constraint function rather than the objective function. However, the same approach can be used to reformulate the problem as a linear programming model where it is feasible to use OT even when the RT capacity at that plant is not fully used.

(a) Formulate this linear programming model.

(b) Explain why the logic of separable programming also applies here to guarantee that an optimal solution for the model formulated in part (a) never uses OT unless the RT capacity at that plant has been fully used.

12.8-8. Consider the following nonlinear programming problem:

Maximize $Z = 5x_1 + x_2$,

subject to

$$2x_1^2 + x_2 \le 13$$
$$x_1^2 + x_2 \le 9$$

and

$$x_1 \ge 0, \qquad x_2 \ge 0.$$

(a) Show that this problem is a convex programming problem.

(b) Use the separable programming technique discussed at the end of Sec. 12.8 to formulate an approximate linear programming model for this problem. Use the integers as the breakpoints of the piecewise linear function.

C **(c)** Use the computer to solve the model formulated in part (b). Then reexpress this solution in terms of the *original* variables of the problem.

12.8-9. Consider the following convex programming problem:

Maximize $Z = 32x_1 - x_1^4 + 4x_2 - x_2^2$,

subject to

$$x_1^2 + x_2^2 \le 9$$

and

$$x_1 \ge 0, \qquad x_2 \ge 0.$$

(a) Apply the separable programming technique discussed at the end of Sec. 12.8, with $x_1 = 0, 1, 2, 3$ and $x_2 = 0, 1, 2, 3$ as the breakpoint of the piecewise linear functions, to formulate an approximate linear programming model for this problem.

c **(b)** Use the computer to solve the model formulated in part (*a*). Then reexpress this solution in terms of the *original* variables of the problem.

(c) Use the KKT conditions to determine whether the solution for the original variables obtained in part (*b*) actually is optimal for the original problem (not the approximate model).

12.8-10. Reconsider the integer nonlinear programming model given in Prob. 10.3-9.

(a) Show that the objective function is not concave.

(b) Formulate an equivalent *pure binary* integer *linear* programming model for this problem as follows. Apply the separable programming technique with the feasible integers as the breakpoints of the piecewise linear functions, so that the auxiliary variables are binary variables. Then add some linear programming constraints on these binary variables to enforce the *special restriction* of separable programming. (Note that the *key property* of separable programming does not hold for this problem because the objective function is not concave.)

c **(c)** Use the computer to solve this problem as formulated in part (*b*). Then reexpress this solution in terms of the *original* variables of the problem.

D,I **12.9-1.** Reconsider the linearly constrained convex programming model given in Prob. 12.6-5. Starting from the initial trial solution $(x_1, x_2) = (0, 0)$, use one iteration of the Frank-Wolfe algorithm to obtain exactly the same solution you found in part (*b*) of Prob. 12.6-5, and then use a second iteration to verify that it is an optimal solution (because it is replicated exactly).

D,I **12.9-2.** Reconsider the linearly constrained convex programming model given in Prob. 12.6-12. Starting from the initial trial solution $(x_1, x_2) = (0, 0)$, use one iteration of the Frank-Wolfe algorithm to obtain exactly the same solution you found in part (*c*) of Prob. 12.6-12, and then use a second iteration to verify that it is an optimal solution (because it is replicated exactly). Explain why exactly the same results would be obtained on these two iterations with any other trial solution.

D,I **12.9-3.** Reconsider the linearly constrained convex programming model given in Prob. 12.6-13. Starting from the initial trial solution $(x_1, x_2, x_3) = (0, 0, 0)$, apply two iterations of the Frank-Wolfe algorithm.

D,I **12.9-4.** Consider the quadratic programming example presented in Sec. 12.7. Starting from the initial trial solution $(x_1, x_2) = (5, 5)$, apply eight iterations of the Frank-Wolfe algorithm.

12.9-5. Reconsider the quadratic programming model given in Prob. 12.7-4.

D,I **(a)** Starting from the initial trial solution $(x_1, x_2) = (0, 0)$, use the Frank-Wolfe algorithm (six iterations) to solve the problem (approximately).

(b) Show graphically how the sequence of trial solutions obtained in part (*a*) can be extrapolated to obtain a closer approximation of an optimal solution. What is your resulting estimate of this solution?

D,I **12.9-6.** Reconsider the linearly constrained convex programming model given in Prob. 12.4-7. Starting from the initial trial solution $(x_1, x_2) = (0, 0)$, use the Frank-Wolfe algorithm (four iterations) to solve this model (approximately).

D,I **12.9-7.** Consider the following linearly constrained convex programming problem:

Maximize $f(\mathbf{x}) = 3x_1x_2 + 40x_1 + 30x_2 - 4x_1^2 - x_1^4$
$- 3x_2^2 - x_2^4,$

subject to

$4x_1 + 3x_2 \leq 12$
$x_1 + 2x_2 \leq 4$

and

$x_1 \geq 0, \qquad x_2 \geq 0.$

Starting from the initial trial solution $(x_1, x_2) = (0, 0)$, apply two iterations of the Frank-Wolfe algorithm.

D,I **12.9-8.*** Consider the following linearly constrained convex programming problem:

Maximize $f(\mathbf{x}) = 3x_1 + 4x_2 - x_1^3 - x_2^2,$

subject to

$x_1 + x_2 \leq 1$

and

$x_1 \geq 0, \qquad x_2 \geq 0.$

(a) Starting from the initial trial solution $(x_1, x_2) = (\frac{1}{4}, \frac{1}{4})$, apply three iterations of the Frank-Wolfe algorithm.

(b) Use the KKT conditions to check whether the solution obtained in part (*a*) is, in fact, optimal.

12.9-9. Consider the following linearly constrained convex programming problem:

Maximize $f(\mathbf{x}) = 4x_1 - x_1^4 + 2x_2 - x_2^2,$

subject to

$4x_1 + 2x_2 \leq 5$

and

$x_1 \geq 0, \qquad x_2 \geq 0.$

(a) Starting from the initial trial solution $(x_1, x_2) = (\frac{1}{2}, \frac{1}{2})$, apply four iterations of the Frank-Wolfe algorithm.

(b) Show graphically how the sequence of trial solutions obtained in part (*a*) can be extrapolated to obtain a closer approximation of an optimal solution. What is your resulting estimate of this solution?

(c) Use the KKT conditions to check whether the solution you obtained in part (*b*) is, in fact, optimal. If not, use these conditions to derive the exact optimal solution.

12.9-10. Reconsider the linearly constrained convex programming model given in Prob. 12.9-8.

(a) If SUMT were to be applied to this problem, what would be the unconstrained function $P(\mathbf{x}; r)$ to be maximized at each iteration?

(b) Setting $r = 1$ and using $(\frac{1}{4}, \frac{1}{4})$ as the initial trial solution, manually apply one iteration of the gradient search procedure (except stop before solving for t^*) to begin maximizing the function $P(\mathbf{x}; r)$ you obtained in part (a).

D,C **(c)** Beginning with the same initial trial solution as in part (b), use the automatic procedure in your IOR Tutorial to apply SUMT to this problem with $r = 1, 10^{-2}, 10^{-4}$.

(d) Compare the final solution obtained in part (c) to the true optimal solution for Prob. 12.9-8 given in the back of the book. What is the percentage error in x_1, in x_2, and in $f(\mathbf{x})$?

12.9-11. Reconsider the linearly constrained convex programming model given in Prob. 12.9-9. Follow the instructions of parts (a), (b), and (c) of Prob. 12.9-10 for this model, except use $(x_1, x_2) = (\frac{1}{2}, \frac{1}{2})$ as the initial trial solution and use $r = 1, 10^{-2}, 10^{-4}, 10^{-6}$.

12.9-12. Reconsider the model given in Prob. 12.3-3.

(a) If SUMT were to be applied directly to this problem, what would be the unconstrained function $P(\mathbf{x}; r)$ to be *minimized* at each iteration?

(b) Setting $r = 100$ and using $(x_1, x_2) = (5, 5)$ as the initial trial solution, manually apply one iteration of the gradient search procedure (except stop before solving for t^*) to begin minimizing the function $P(\mathbf{x}; r)$ you obtained in part (a).

D,C **(c)** Beginning with the same initial trial solution as in part (b), use the automatic procedure in your IOR Tutorial to apply SUMT to this problem with $r = 100, 1, 10^{-2}, 10^{-4}$. (*Hint:* The computer routine assumes that the problem has been converted to *maximization* form with the functional constraints in $\leq$ form.)

12.9-13. Consider the example for applying SUMT given in Sec. 12.9.

(a) Show that $(x_1, x_2) = (1, 2)$ satisfies the KKT conditions.

(b) Display the feasible region graphically, and then plot the locus of points $x_1 x_2 = 2$ to demonstrate that $(x_1, x_2) = (1, 2)$ with $f(1, 2) = 2$ is, in fact, a *global maximum*.

12.9-14.* Consider the following convex programming problem:

$$\text{Maximize} \quad f(\mathbf{x}) = -2x_1 - (x_2 - 3)^2,$$

subject to

$$x_1 \geq 3 \quad \text{and} \quad x_2 \geq 3.$$

(a) If SUMT were applied to this problem, what would be the unconstrained function $P(\mathbf{x}; r)$ to be maximized at each iteration?

(b) Derive the maximizing solution of $P(\mathbf{x}; r)$ analytically, and then give this solution for $r = 1, 10^{-2}, 10^{-4}, 10^{-6}$.

D,C **(c)** Beginning with the initial trial solution $(x_1, x_2) = (4, 4)$, use the automatic procedure in your IOR Tutorial to apply SUMT to this problem with $r = 1, 10^{-2}, 10^{-4}, 10^{-6}$.

D,C **12.9-15.** Consider the following convex programming problem:

$$\text{Maximize} \quad f(\mathbf{x}) = x_1 x_2 - x_1 - x_1^2 - x_2 - x_2^2,$$

subject to

$$x_2 \geq 0.$$

Beginning with the initial trial solution $(x_1, x_2) = (1, 1)$, use the automatic procedure in your IOR Tutorial to apply SUMT to this problem with $r = 1, 10^{-2}, 10^{-4}$.

D,C **12.9-16.** Reconsider the quadratic programming model given in Prob. 12.7-4. Beginning with the initial trial solution $(x_1, x_2) = (\frac{1}{2}, \frac{1}{2})$, use the automatic procedure in your IOR Tutorial to apply SUMT to this model with $r = 1, 10^{-2}, 10^{-4}, 10^{-6}$.

D,C **12.9-17.** Reconsider the first quadratic programming variation of the Wyndor Glass Co. problem presented in Sec. 12.2 (see Fig. 12.6). Beginning with the initial trial solution $(x_1, x_2) = (2, 3)$, use the automatic procedure in your IOR Tutorial to apply SUMT to this problem with $r = 10^2, 1, 10^{-2}, 10^{-4}$.

12.9-18. Reconsider the convex programming model with an equality constraint given in Prob. 12.6-11.

(a) If SUMT were to be applied to this model, what would be the unconstrained function $P(\mathbf{x}; r)$ to be *minimized* at each iteration?

D,C **(b)** Starting from the initial trial solution $(x_1, x_2, x_3) = (\frac{3}{2}, \frac{3}{2}, 2)$, use the automatic procedure in your IOR Tutorial to apply SUMT to this model with $r = 10^{-2}, 10^{-4}, 10^{-6}, 10^{-8}$.

C **(c)** Use the standard Excel Solver to solve this problem.

C **(d)** Use Evolutionary Solver to solve this problem.

C **(e)** Use LINGO to solve this problem.

12.10-1. Consider the following nonconvex programming problem:

$$\text{Maximize} \quad f(x) = 1{,}000x - 400x^2 + 40x^3 - x^4,$$

subject to

$$x^2 + x \leq 500$$

and

$$x \geq 0.$$

(a) Identify the feasible values for x. Obtain general expressions for the first three derivatives of $f(x)$. Use this information to help you draw a rough sketch of $f(x)$ over the feasible region for x. Without calculating their values, mark the points on your graph that correspond to *local* maxima and minima.

I **(b)** Use the bisection method with $\epsilon = 0.05$ to find each of the local maxima. Use your sketch from part (a) to identify appropriate initial bounds for each of these searches. Which of the local maxima is a global maximum?

(c) Starting with $x = 3$ and $x = 15$ as the initial trial solutions, use Newton's method with $\epsilon = 0.001$ to find each of the local maxima.

D,C **(d)** Use the automatic procedure in your IOR Tutorial to apply SUMT to this problem with $r = 10^3, 10^2, 10, 1$ to find each of the local maxima. Use $x = 3$ and $x = 15$ as the initial trial solutions for these searches. Which of the local maxima is a global maximum?

C **(e)** Formulate this problem in a spreadsheet, where $f(x)$ represents profit, and then use the Solver Table to generate the solutions with the following starting points: $x = 0, 5, 10, 15, 20, 25$. Include the value of x and the profit as output cells in the Solver Table.

C **(f)** Use Evolutionary Solver to solve this problem.

c **(g)** Use the global optimizer feature of LINGO to solve this problem.

c **(h)** Use MPL and its global optimizer LGO to solve this problem.

12.10-2. Consider the following nonconvex programming problem:

Maximize $\qquad f(\mathbf{x}) = 3x_1x_2 - 2x_1^2 - x_2^2,$

subject to

$$x_1^2 + 2x_2^2 \leq 4$$
$$2x_1 - x_2 \leq 3$$
$$x_1x_2^2 + x_1^2x_2 = 2$$

and

$$x_1 \geq 0, \qquad x_2 \geq 0.$$

(a) If SUMT were to be applied to this problem, what would be the unconstrained function $P(\mathbf{x}; r)$ to be maximized at each iteration?

D,C **(b)** Starting from the initial trial solution $(x_1, x_2) = (1, 1)$, use the automatic procedure in your IOR Tutorial to apply SUMT to this problem with $r = 1, 10^{-2}, 10^{-4}$.

c **(c)** Use Evolutionary Solver to solve this problem.

c **(d)** Use the global optimizer feature of LINGO to solve this problem.

c **(e)** Use MPL and its global optimizer LGO to solve this problem.

12.10-3. Consider the following nonconvex programming problem.

Minimize $\qquad f(\mathbf{x}) = \sin 3x_1 + \cos 3x_2 + \sin(x_1 + x_2),$

subject to

$$x_1^2 - 10x_2 \geq -1$$
$$10x_1 + x_2^2 \leq 100$$

and

$$x_1 \geq 0, \qquad x_2 \geq 0.$$

(a) If SUMT were applied to this problem, what would be the unconstrained function $P(\mathbf{x}; r)$ to be minimized at each iteration?

(b) Describe how SUMT should be applied to attempt to obtain a global minimum. (Do not actually solve.)

c **(c)** Use the global optimizer feature of LINGO to solve this problem.

c **(d)** Use MPL and its global optimizer LGO to solve this problem.

c **12.10-4.** Consider the following nonconvex programming problem:

Maximize Profit $= x^5 - 13x^4 + 59x^3 - 107x^2 + 61x,$

subject to

$$0 \leq x \leq 5.$$

(a) Formulate this problem in a spreadsheet, and then use the Solver Table to solve this problem with the following starting points: $x = 0, 1, 2, 3, 4,$ and 5. Include the value of x and the profit as output cells in the Solver Table.

(b) Use Evolutionary Solver to solve this problem.

c **12.10-5.** Consider the following nonconvex programming problem:

Maximize Profit $= 100x^6 - 1{,}359x^5 + 6{,}836x^4$
$\qquad\qquad\qquad - 15{,}670x^3 + 15{,}870x^2 - 5{,}095x,$

subject to

$$0 \leq x \leq 5.$$

(a) Formulate this problem in a spreadsheet, and then use the Solver Table to solve this problem with the following starting points: $x = 0, 1, 2, 3, 4,$ and 5. Include the value of x and the profit as output cells in the Solver Table.

(b) Use Evolutionary Solver to solve this problem.

c **12.10-6.** Because of population growth, the state of Washington has been given an additional seat in the House of Representatives, making a total of 10. The state legislature, which is currently controlled by the Republicans, needs to develop a plan for redistricting the state. There are 18 major cities in the state of Washington that need to be assigned to one of the 10 congressional districts. The table below gives the numbers of registered Democrats and registered Republicans in each city. Each district must contain between 150,000 and 350,000 of these registered voters. Use Evolutionary Solver to assign each city to one of the 10 congressional districts in order to maximize the number of districts that have more registered Republicans than registered Democrats. (Hint: Use the SUMIF function.)

City	Democrats (Thousands)	Republicans (Thousands)
1	152	62
2	81	59
3	75	83
4	34	52
5	62	87
6	38	87
7	48	69
8	74	49
9	98	62
10	66	72
11	83	75
12	86	82
13	72	83
14	28	53
15	112	98
16	45	82
17	93	68
18	72	98

12.10-7. Reconsider the Wyndor Glass Co. problem introduced in Sec. 3.1.

c **(a)** Solve this problem using the standard Excel Solver.

c **(b)** Starting with an initial solution of producing 0 batches of doors and 0 batches of windows, solve this problem using Evolutionary Solver.

(c) Comment on the performance of the two approaches.

12.11-1. Consider the following problem:

Maximize $\qquad Z = 4x_1 - x_1^2 + 10x_2 - x_2^2,$

subject to

$$x_1^2 + 4x_2^2 \leq 16$$

and

$$x_1 \geq 0, \qquad x_2 \geq 0.$$

(a) Is this a convex programming problem? Answer yes or no, and then justify your answer.

(b) Can the modified simplex method be used to solve this problem? Answer yes or no, and then justify your answer (but do not actually solve.)

(c) Can the Frank-Wolfe algorithm be used to solve this problem? Answer yes or no, and then justify your answer (but do not actually solve).

(d) What are the KKT conditions for this problem? Use these conditions to determine whether $(x_1, x_2) = (1, 1)$ can be optimal.

(e) Use the separable programming technique to formulate an *approximate* linear programming model for this problem. Use the feasible integers as the breakpoints for each piecewise linear function.

C **(f)** Use the simplex method to solve the problem as formulated in part (e).

(g) Give the function $P(\mathbf{x}; r)$ to be maximized at each iteration when applying SUMT to this problem. (Do not actually solve.)

D,C **(h)** Use SUMT (the automatic procedure in your IOR Tutorial) to solve the problem as formulated in part (g). Begin with the initial trial solution $(x_1, x_2) = (2, 1)$ and use $r = 1$, 10^{-2}, 10^{-4}, 10^{-6}.

C **(i)** Formulate this problem in a spreadsheet, and then use the standard Excel Solver to solve this problem.

C **(j)** Use Evolutionary Solver to solve this problem.

C **(k)** Use LINGO to solve this problem.

■ CASES

Case 12.1   Savvy Stock Selection

Ever since the day she took her first economics class in high school, Lydia wondered about the financial practices of her parents. They worked very hard to earn enough money to live a comfortable middle-class life, but they never made their money work for them. They simply deposited their hard-earned paychecks in savings accounts earning a nominal amount of interest. (Fortunately, there always was enough money when it came time to pay her college bills.) She promised herself that when she became an adult, she would not follow the same financially conservative practices as her parents.

And Lydia kept this promise. Every morning while getting ready for work, she watches the CNN financial reports. She plays investment games on the World Wide Web, finding portfolios that maximize her return while minimizing her risk. She reads *The Wall Street Journal* and *Financial Times* with a thirst she cannot quench.

Lydia also reads the investment advice columns of the financial magazines, and she has noticed that on average, the advice of the investment advisers turns out to be very good. Therefore, she decides to follow the advice given in the latest issue of one of the magazines. In his monthly column the editor Jonathan Taylor recommends three stocks that he believes will rise far above market average. In addition, the well-known mutual fund guru Donna Carter advocates the purchase of three more stocks that she thinks will outperform the market over the next year.

BIGBELL (ticker symbol on the stock exchange: BB), one of the nation's largest telecommunications companies, trades at a price-earnings ratio well below market average. Huge investments over the last 8 months have depressed earnings considerably. However, with their new cutting edge technology, the

company is expected to significantly raise their profit margins. Taylor predicts that the stock will rise from its current price of $60 per share to $72 per share within the next year.

LOTSOFPLACE (LOP) is one of the leading hard drive manufacturers in the world. The industry recently underwent major consolidation, as fierce price wars over the last few years were followed by many competitors going bankrupt or being bought by LOTSOFPLACE and its competitors. Due to reduced competition in the hard drive market, revenues and earnings are expected to rise considerably over the next year. Taylor predicts a one-year increase of 42 percent in the stock of LOTSOFPLACE from the current price of $127 per share.

INTERNETLIFE (ILI) has survived the many ups and downs of Internet companies. With the next Internet frenzy just around the corner, Taylor expects a doubling of this company's stock price from $4 to $8 within a year.

HEALTHTOMORROW (HEAL) is a leading biotechnology company that is about to get approval for several new drugs from the Food and Drug Administration, which will help earnings to grow 20 percent over the next few years. In particular a new drug to significantly reduce the risk of heart attacks is supposed to reap huge profits. Also, due to several new great-tasting medications for children, the company has been able to build an excellent image in the media. This public relations coup will surely have positive effects for the sale of its over-the-counter medications. Carter is convinced that the stock will rise from $50 to $75 per share within a year.

QUICKY (QUI) is a fast-food chain which has been vastly expanding its network of restaurants all over the United States. Carter has followed this company closely since it went public some 15 years ago when it had only a few dozen restaurants on the west coast of the United States. Since then the company has expanded, and it now has restaurants in

■ 13.1   THE NATURE OF METAHEURISTICS

To illustrate the nature of metaheuristics, let us begin with an example of a small but modestly difficult nonlinear programming problem.

An Example: A Nonlinear Programming Problem with Multiple Local Optima

Consider the following problem.

Maximize $\quad f(x) = 12x^5 - 975x^4 + 28,000x^3 - 345,000x^2 + 1,800,000x,$

subject to

$$0 \le x \le 31.$$

Figure 13.1 graphs the objective function $f(x)$ over the feasible values of the single variable x. This plot reveals that the problem has three local optima, one at $x = 5$, another at $x = 20$, and the third at $x = 31$, where the global optimum is at $x = 20$.

The objective function $f(x)$ is sufficiently complicated that it would be difficult to determine where the global optimum lies without the benefit of viewing the plot in Fig. 13.1. Calculus could be used, but this would require solving a polynomial equation of the fourth degree (after setting the first derivative equal to zero) to determine where the critical points lie. It would even be difficult to ascertain that $f(x)$ has multiple local optima rather than just a global optimum.

This problem is an example of a *nonconvex programming* problem, a special type of nonlinear programming problem that typically has multiple local optima. Section 12.10

■ **FIGURE 13.1**
A plot of the value of the objective function over the feasible range, $0 \le x \le 31$, for the nonlinear programming example. The local optima are at $x = 5$, $x = 20$, and $x = 31$, but only $x = 20$ is a global optimum.

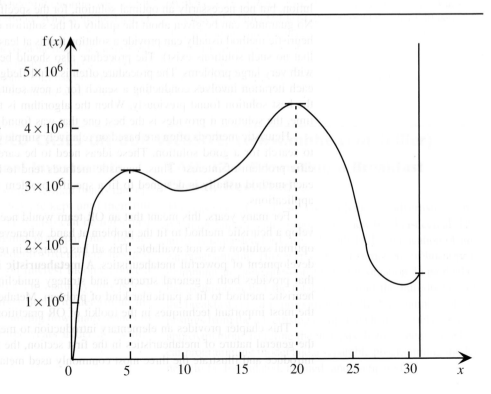

discusses nonconvex programming and even introduces a software package (Evolutionary Solver) that uses the kind of metaheuristic described in Sec. 13.4.

For nonlinear programming problems that appear to be somewhat difficult, like this one, a simple heuristic method is to conduct a **local improvement procedure.** Such a procedure starts with an initial trial solution and then, at each iteration, searches in the neighborhood of the current trial solution to find a better trial solution. This process continues until no improved solution can be found in the neighborhood of the current trial solution. Thus, this kind of procedure can be viewed as a *hill-climbing procedure* that keeps climbing higher on the plot of the objective function (assuming the objective is maximization) until it essentially reaches the top of the hill. A well-designed local improvement procedure usually will be successful in converging to a *local* optimum (the top of a hill), but it then will stop even if this local optimum is not a *global* optimum (the top of the tallest hill).

For example, the *gradient search procedure* described in Sec. 12.5 is a local improvement procedure. If it were to start with, say, $x = 0$ as the initial trial solution in Fig. 13.1, it would climb up the hill by trying successively larger values of x until it essentially reaches the top of the hill at $x = 5$, at which point it would stop. Figure 13.2 shows a typical sequence of values of $f(x)$ that would be obtained by such a local improvement procedure when starting from far down the hill.

Since the nonlinear programming example depicted in Fig. 13.1 involves only a single variable, the bisection method described in Sec. 12.4 also could be applied to this particular problem. This procedure is another example of a local improvement procedure, since each iteration starts from the current trial solution to search in its neighborhood (defined by a current lower bound and upper bound on the value of the variable) for a better solution. For example, if the search were to begin with a lower bound of $x = 0$ and an upper bound of $x = 6$ in Fig. 13.1, the sequence of trial solutions obtained by the bisection method would be $x = 3$, $x = 4.5$, $x = 5.25$, $x = 4.875$, and so forth as it converges to $x = 5$. The corresponding values of the objective function for these four trial solutions are 2.975 million, 3.286 million, 3.300 million, and 3.302 million, respectively. Thus, the second iteration provides a relatively large improvement over the first one (311,000), the third iteration gives a considerably smaller improvement (14,000), and the fourth iteration yields only a very small improvement (2000). As depicted in Fig. 13.2, this pattern is rather typical of local improvement procedures (although with some variation in the rate of convergence to the local maximum).

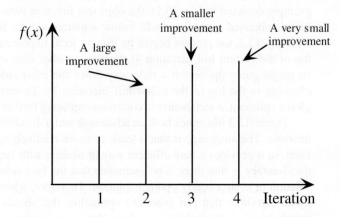

■ **FIGURE 13.2**
A typical sequence of objective function values for the solutions obtained by a local improvement procedure as it converges to a local optimum when it is applied to a maximization problem.

Just as with the gradient search procedure, this search with the bisection method would get trapped at the local optimum at $x = 5$, so it never would find the global optimum at $x = 20$. Like other local improvement procedures, both the gradient search procedure and the bisection method are designed only to keep improving on the current trial solutions within the local neighborhood of those solutions. Once they climb to the top of a hill, they must stop because they cannot climb any higher within the local neighborhood of the trial solution at the top of the hill. This illustrates the drawback of any local improvement procedure.

> **The drawback of a local improvement procedure:** When a well-designed local improvement procedure is applied to an optimization problem with multiple local optima, the procedure will converge to one local optimum and then stop. Which local optimum it finds depends on where the procedure begins the search. Thus, the procedure will find the global optimum only if it happens to begin the search in the neighborhood of this global optimum.

To try to overcome this drawback, one can restart the local improvement procedure a number of times from randomly selected initial trial solutions. Restarting from a new part of the feasible region often will lead to a new local optimum. Repeating this a number of times increases the chance that the best of the local optima obtained actually will be the global optimum. This approach works well on small problems, like the one-variable nonlinear programming example depicted in Fig. 13.1. However, it is much less successful on large problems with many variables and a complicated feasible region. When the feasible region has numerous "nooks and crannies" and restarting a local improvement procedure from only one of them will lead to the global optimum, restarting from randomly selected initial trial solutions becomes a very haphazard way to reach the global optimum.

What is needed instead is a more structured approach that uses the information being gathered to guide the search toward the global optimum. This is the role that a metaheuristic plays.

> **The nature of metaheuristics:** A metaheuristic is a general kind of solution method that orchestrates the interaction between local improvement procedures and higher level strategies to create a process that is capable of escaping from local optima and performing a robust search of a feasible region.

Thus, one key feature of a metaheuristic is its ability to escape from a local optimum. After reaching (or nearly reaching) a local optimum, different metaheuristics execute this escape in different ways. However, a common characteristic is that the trial solutions that immediately follow a local optimum are allowed to be inferior to this local optimum. Consequently, when a metaheuristic is applied to a maximization problem (such as the example depicted in Fig. 13.1), the objective function values for the sequence of trial solutions obtained typically would follow a pattern similar to that shown in Fig. 13.3. As with Fig. 13.2, the process begins by using a local improvement procedure to climb to the top of the current hill (iteration 4). However, rather than stopping there, the metaheuristic might guide the search a little way down the other side of this hill until it can start climbing to the top of the tallest hill (iteration 8). To verify that this appears to be the global optimum, a metaheuristic continues exploring further before stopping (iteration 12).

Figure 13.3 illustrates both an advantage and a disadvantage of a well-designed metaheuristic. The advantage is that it tends to move relatively quickly toward very good solutions, so it provides a very efficient way of dealing with large complicated problems. The disadvantage is that there is no guarantee that the best solution found will be an optimal solution or even a nearly optimal solution. Therefore, whenever a problem can be solved by an algorithm that can guarantee optimality, that should be done instead. The role of metaheuristics is to deal with problems that are too large and complicated to be solved by

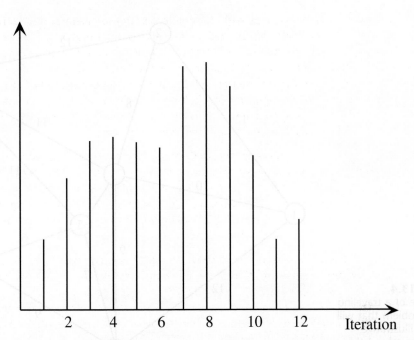

■ FIGURE 13.3
A typical sequence of objective function values for the solutions obtained by a metaheuristic as it first converges to a local optimum (iteration 4) and then escapes to converge to (hopefully) the global optimum (iteration 8) of a maximization problem before concluding its search (iteration 12).

exact algorithms. All the examples in this chapter are too small to require the use of metaheuristics, since they are intended only to illustrate in a straightforward way how metaheuristics can approach far more complicated problems.

Section 13.3 will illustrate the application of a particular metaheuristic to the nonlinear programming example depicted in Fig. 13.1. Section 13.4 then will apply another metaheuristic to the integer programming version of this same example.

Although metaheuristics sometimes are applied to difficult nonlinear programming and integer programming problems, a more common area of application is to *combinatorial optimization* problems. Our next example is of this type.

An Example: A Traveling Salesman Problem

Perhaps the most famous classic combinatorial optimization problem is called the *traveling salesman problem*. It has been given this picturesque name because it can be described in terms of a salesman (or saleswoman) who must travel to a number of cities during one tour. Starting from his (or her) home city, the salesman wishes to determine which route to follow to visit each city exactly once before returning to his home city so as to minimize the total length of the tour.

Figure 13.4 shows an example of a small traveling salesman problem with seven cities. City 1 is the salesman's home city. Therefore, starting from this city, the salesman must choose a route to visit each of the other cities exactly once before returning to city 1. The number next to each link between each pair of cities represents the distance (or cost or time) between these cities. We assume that the distance is the same in either direction. (This is referred to as a *symmetric* traveling salesman problem.) Although there commonly is a direct link between every pair of cities, we are simplifying this example by assuming that the only direct links are those shown in the figure. The objective is to determine which route will minimize the total distance that the salesman must travel.

There have been a number of applications of traveling salesman problems that have nothing to do with salesmen. For example, when a truck leaves a distribution center to deliver goods to a number of locations, the problem of determining the shortest route for

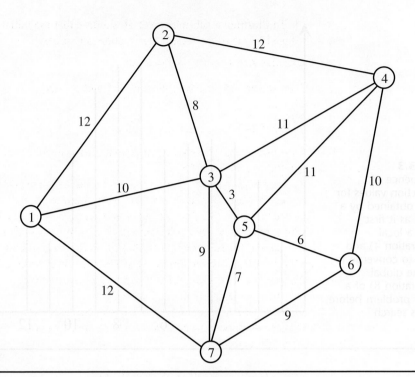

■ **FIGURE 13.4**
The example of a traveling
salesman problem that will
be used for illustrative
purposes throughout this
chapter.

doing this is a traveling salesman problem. Another example involves the manufacture of printed circuit boards for wiring chips and other components. When many holes need to be drilled into a printed circuit board, the problem of finding the most efficient drilling sequence is a traveling salesman problem.

The difficulty of traveling salesman problems increases rapidly as the number of cities increases. For a problem with n cities and a link between every pair of cities, the number of feasible routes to be considered is $(n - 1)!/2$ since there are $(n - 1)$ possibilities for the first city after the home city, $(n - 2)$ possibilities for the next city, and so forth. The denominator of 2 arises because every route has an equivalent reverse route with exactly the same distance. Thus, while a 10-city traveling salesman problem has less than 200,000 feasible solutions to be considered, a 20-city problem has roughly 10^{16} feasible solutions, while a 50-city problem has about 10^{62}.

Surprisingly, powerful algorithms based on the branch-and-cut approach introduced in Sec. 11.8 have succeeded in solving to optimality certain huge traveling salesman problems with many hundreds (or even thousands) of cities. However, because of the enormous difficulty of solving large traveling salesman problems, heuristic methods guided by metaheuristics continue to be a popular way of addressing such problems.

These heuristic methods commonly involve generating a sequence of feasible trial solutions, where each new trial solution is obtained by making a certain type of small adjustment in the current trial solution. Several methods have been suggested for how to adjust the current trial solution. Because of its ease of implementation, one popular method uses the following type of adjustment.

A **sub-tour reversal** adjusts the sequence of cities visited in the current trial solution by selecting a subsequence of the cities and simply reversing the order in which that subsequence of cities is visited. (The subsequence being reversed can consist of as few as two cities, but also can have more.)

To illustrate a sub-tour reversal, suppose that the initial trial solution for our example in Fig. 13.4 is to visit the cities in numerical order:

1-2-3-4-5-6-7-1   Distance = 69

If we select, say, the subsequence 3-4 and reverse it, we obtain the following new trial solution:

1-2-4-3-5-6-7-1   Distance = 65

Thus, this particular sub-tour reversal has succeeded in reducing the distance for the complete tour from 69 to 65.

Figure 13.5 depicts this sub-tour reversal, which leads from the initial trial solution on the left to the new trial solution on the right. The dashed lines indicate the links that are deleted from the tour (on the left) or added to the tour (on the right) by sub-tour reversal. Note that the new trial solution deletes exactly two links from the previous tour and replaces them by exactly two new links to form the new tour. This is a characteristic of any sub-tour reversal (including those where the subsequence of cities being reversed consists of more than two cities). Thus, a particular sub-tour reversal is possible only if the corresponding two new links actually exist.

This success in obtaining an improved tour by simply performing a sub-tour reversal suggests the following heuristic method for seeking a good feasible solution for any traveling salesman problem.

The Sub-Tour Reversal Algorithm

Initialization.   Start with any feasible tour as the initial trial solution.

Iteration.   For the current trial solution, consider all possible ways of performing a sub-tour reversal (except exclude the reversal of the entire tour). Select the one that provides the largest decrease in the distance traveled to be the new trial solution. (Ties may be broken arbitrarily.)

■ **FIGURE 13.5**
A sub-tour reversal that replaces the tour on the left (the initial trial solution) by the tour on the right (the new trial solution) by reversing the order in which cities 3 and 4 are visited. This sub-tour reversal results in replacing the dashed lines on the left by the dashed lines on the right as the links that are traversed in the new tour.

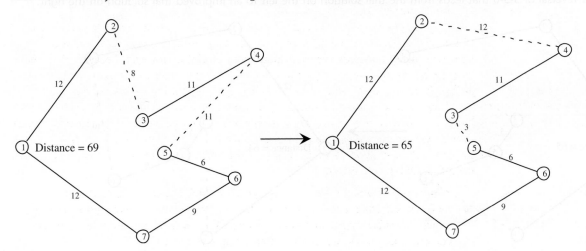

Stopping rule.   Stop when no sub-tour reversal will improve the current trial solution. Accept this solution as the final solution.

Now let us apply this algorithm to the example, starting with 1-2-3-4-5-6-7-1 as the initial trial solution. There are four possible sub-tour reversals that would improve upon this solution, as listed in the second, third, fourth, and fifth rows below.

	1-2-3-4-5-6-7-1	Distance = 69
Reverse 2-3:	1-3-2-4-5-6-7-1	Distance = 68
Reverse 3-4:	1-2-4-3-5-6-7-1	Distance = 65
Reverse 4-5:	1-2-3-5-4-6-7-1	Distance = 65
Reverse 5-6:	1-2-3-4-6-5-7-1	Distance = 66

The two solutions with Distance = 65 tie for providing the largest decrease in the distance traveled, so suppose that the first of these, 1-2-4-3-5-6-7-1 (as shown on the right side of Fig. 13.5), is chosen arbitrarily to be the next trial solution. This completes the first iteration.

The second iteration begins with the tour on the right side of Fig. 13.5 as the current trial solution. For this solution, there is only one sub-tour reversal that will provide an improvement, as listed in the second row below:

| | 1-2-4-3-5-6-7-1 | Distance = 65 |
| Reverse 3-5-6: | 1-2-4-6-5-3-7-1 | Distance = 64 |

Figure 13.6 shows this sub-tour reversal, where the entire subsequence of cities 3-5-6 on the left now is visited in reverse order (6-5-3) on the right. Thus, the tour on the right now traverses the link 4-6 instead of 4-3, as well as the link 3-7 instead of 6-7, in order to use the reverse order 6-5-3 between cities 4 and 7. This completes the second iteration.

We next try to find a sub-tour reversal that will improve upon this new trial solution. However, there is none, so the sub-tour reversal algorithm stops with this trial solution as the final solution.

Is 1-2-4-6-5-3-7-1 the optimal solution? Unfortunately, no. The optimal solution turns out to be

$$1\text{-}2\text{-}4\text{-}6\text{-}7\text{-}5\text{-}3\text{-}1 \qquad \text{Distance} = 63$$

■ **FIGURE 13.6**
The sub-tour reversal of 3-5-6 that leads from the trial solution on the left to an improved trial solution on the right.

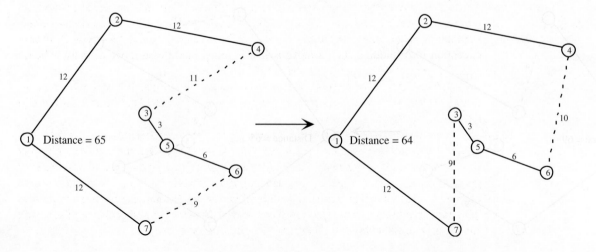

(or 1-3-5-7-6-4-2-1 by reversing the direction of this entire tour)

However, this solution cannot be reached by performing a sub-tour reversal that improves 1-2-4-6-5-3-7-1.

The sub-tour reversal algorithm is another example of a *local improvement procedure*. It improves upon the current trial solution at each iteration. When it can no longer find a better solution, it stops because the current trial solution is a local optimum. In this case, 1-2-4-6-5-3-7-1 is indeed a *local optimum* because there is no better solution within its local neighborhood that can be reached by performing a sub-tour reversal.

What is needed to provide a better chance of reaching a global optimum is to use a metaheuristic that will enable the process to escape from a local optimum. You will see how three different metaheuristics do this with this same example in the next three sections.

■ 13.2 TABU SEARCH

Tabu search is a widely used metaheuristic that uses some common-sense ideas to enable the search process to escape from a local optimum. After introducing its basic concepts, we will go through a simple example and then return to the traveling salesman example.

Basic Concepts

Any application of tabu search includes as a subroutine a *local search procedure* that seems appropriate for the problem being addressed. (A **local search procedure** operates just like a local improvement procedure except that it may not require that each new trial solution must be better than the preceding trial solution.) The process begins by using this procedure as a local *improvement* procedure in the usual way (i.e., only accepting an improved solution at each iteration) to find a local optimum. A key strategy of tabu search is that it then continues the search by allowing *non-improving moves* to the best solutions in the neighborhood of the local optimum. Once a point is reached where better solutions can be found in the neighborhood of the current trial solution, the local improvement procedure is reapplied to find a new local optimum.

Using the analogy of hill climbing, this process is sometimes referred to as the **steepest ascent/mildest descent approach** because each iteration selects the available move that goes furthest up the hill, or, when an upward move is not available, selects a move that drops least down the hill. If all goes well, the process will follow a pattern like that shown in Fig. 13.3, where a local optimum is left behind in order to climb to the global optimum.

The danger with this approach is that after moving away from a local optimum, the process will cycle right back to the same local optimum. To avoid this, a tabu search temporarily forbids moves that would return to (or perhaps toward) a solution recently visited. A **tabu list** records these forbidden moves, which are referred to as *tabu moves*. (The only exception to forbidding such a move is if it is found that a tabu move actually is better than the best feasible solution found so far.)

This use of *memory* to guide the search by using tabu lists to record some of the recent history of the search is a distinctive feature of tabu search. This feature has roots in the field of artificial intelligence.

Tabu search also can incorporate some more advanced concepts. One is *intensification,* which involves exploring a portion of the feasible region more thoroughly than usual after it has been identified as a particularly promising portion for containing very good solutions. Another concept is *diversification,* which involves forcing the search into previously unexplored areas of the feasible region. (Long-term memory is used to help implement both concepts.) However, we will focus on the basic form of tabu search summarized next without delving into these additional concepts.

Founded in 1886, **Sears, Roebuck and Company** (now commonly referred to as just **Sears**) grew to become the largest multiline retailer in the United States by the mid-20th century. It continues today to rank among the largest retailers in the world selling merchandise and services. It also provides the largest home-delivery service of furniture and appliances in the United States with over 4 million deliveries a year. Sears manages a U.S. fleet of over 1,000 delivery vehicles that includes contract carriers and Sears-owned vehicles. It also operates a U.S. fleet of about 12,500 service vehicles and the associated technicians, who make approximately 15 million on-site service calls annually to repair and install appliances and provide home improvement.

The cost of operating this huge home-delivery and home-service business runs in the *billions of dollars per year*. With many thousands of vehicles being used to make many tens of thousands of calls on customers *daily*, the efficiency of this operation has a major impact on the company's profitability.

With so many calls on customers to be made with so many vehicles, a huge number of decisions must be made *each* day. Which stops should be assigned to each vehicle's route? What should the order of the stops be (which considerably impacts the total distance and time for the route) for each vehicle? How can all these decisions be made so as to minimize total operational costs while providing satisfactory service to the customers?

It became clear that operations research was needed to address this problem. The natural formulation is as a *vehicle-routing problem with time windows* (VRPTW), for which both exact and heuristic algorithms have been developed. Unfortunately, the Sears problem is so huge that it is a very difficult combinatorial optimization problem that is beyond the reach of standard algorithms for VRPTW. Therefore, a new algorithm was developed that was based on using *tabu search* for making both the decisions on which vehicle's route serves which stops and what the sequence is of stops within a route.

The resulting new vehicle-routing-and-scheduling system, based largely on tabu search, led to *over* **$9 million** *in one-time savings* and *over* **$42 million** in *annual savings* for Sears. It also provided a number of intangible benefits, including (most importantly) *improved service to customers.*

Source: D. Weigel, and B. Cao: "Applying GIS and OR Techniques to Solve Sears Technician-Dispatching and Home-Delivery Problems," *Interfaces*, **29**(1): 112–130, Jan.–Feb. 1999. (A link to this article is provided on our website, www.mhhe.com/hillier.)

Outline of a Basic Tabu Search Algorithm

Initialization.   Start with a feasible initial trial solution.

Iteration.   Use an appropriate local search procedure to define the feasible moves into the local neighborhood of the current trial solution. Eliminate from consideration any move on the current tabu list unless that move would result in a better solution than the best trial solution found so far. Determine which of the remaining moves provides the best solution. Adopt this solution as the next trial solution, regardless of whether it is better or worse than the current trial solution. Update the tabu list to forbid cycling back to what had been the current trial solution. If the tabu list already had been full, delete the oldest member of the tabu list to provide more flexibility for future moves.

Stopping rule.   Use some stopping criterion, such as a fixed number of iterations, a fixed amount of CPU time, or a fixed number of consecutive iterations without an improvement in the best objective function value. (The latter criterion is a particularly popular one.) Also stop at any iteration where there are no feasible moves into the local neighborhood of the current trial solution. Accept the best trial solution found on any iteration as the final solution.

This outline leaves a number of questions unanswered.

1. Which local search procedure should be used?
2. How should that procedure define the *neighborhood structure* that specifies which solutions are immediate neighbors (reachable in a single iteration) of any current trial solution?
3. What is the form in which tabu moves should be represented on the tabu list?
4. Which tabu move should be added to the tabu list in each iteration?
5. How long should a tabu move be retained on the tabu list?
6. Which stopping rule should be used?

These all are important details that need to be worked out to fit the specific type of problem being addressed, as illustrated by the following examples. Tabu search only provides a general structure and strategy guidelines for developing a specific heuristic method to fit a specific situation. The selection of its parameters is a key part of developing a successful heuristic method.

The following examples illustrate the use of tabu search.

A Minimum Spanning Tree Problem with Constraints

Section 9.4 describes the minimum spanning tree problem. In brief, starting with a network that has its nodes but no links between the nodes yet, the problem is to determine which links should be inserted into the network. The objective is to minimize the total cost (or length) of the inserted links that will provide a path between every pair of nodes. For a network with n nodes, $(n - 1)$ links (with no cycles) are needed to provide a path between every pair of nodes. Such a network is referred to as a *spanning tree*.

The left-hand side of Fig. 13.7 shows a network with five nodes, where the dashed lines represent the potential links that could be inserted into the network and the number next to each dashed line represents the cost associated with inserting that particular link. Thus, the problem is to determine which four of these links (with no cycles) should be inserted into the network to minimize the total cost of these links. The right-hand side of the figure shows the desired *minimum spanning tree,* where the dark lines represent the links

■ **FIGURE 13.7**
(a) The data for a minimum spanning tree problem before choosing the links to be included in the network and (b) the optimal solution for this problem where the dark lines represent the chosen links.

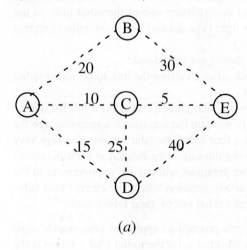

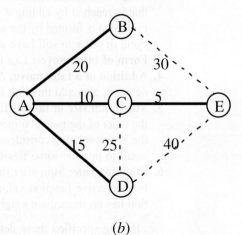

(a)

(b)

that have been inserted into the network with a total cost of 50. This optimal solution is obtained easily by applying the "greedy" algorithm presented in Sec. 9.4.

To illustrate the use of tabu search, let us now add a couple complications to this example by supposing that the following constraints also must be observed when choosing the links to include in the network.

Constraint 1: Link AD can be included only if link DE also is included.
Constraint 2: At most one of the three links—AD, CD, and AB—can be included.

Note that the previously optimal solution on the right-hand side of Fig. 13.7 violates both of these constraints because (1) link AD is included even though DE is not and (2) both AD and AB are included.

By imposing such constraints, the greedy algorithm presented in Sec. 9.4 can no longer be used to find the new optimal solution. For such a small problem, this solution probably could be found rather quickly by inspection. However, let us see how tabu search could be used on either this problem or much larger problems to search for an optimal solution.

The easiest way to take the constraints into account is to charge a huge penalty, such as the following, for violating them.

1. Charge a penalty of 100 if constraint 1 is violated.
2. Charge a penalty of 100 if two of the three links specified in constraint 2 are included. Increase this penalty to 200 if all three of the links are included.

A penalty of 100 is large enough to ensure that the constraints will not be violated for a spanning tree that minimizes the total cost, including the penalty, provided only that there exist some feasible solutions. Doubling this penalty if constraint 2 is badly violated provides an incentive for at least reducing how many of the three links are included during an iteration of the tabu search.

There are a variety of ways to answer the six questions that are needed to specify how the tabu search will be conducted. (See the list of questions that follows the outline of a basic tabu search algorithm.) Here is one straightforward way of answering the questions.

1. **Local search procedure:** At each iteration, choose the best immediate neighbor of the current trial solution that is not ruled out by its tabu status.
2. **Neighborhood structure:** An immediate neighbor of the current trial solution is one that is reached by adding a single link and then deleting one of the other links in the cycle that is formed by the addition of this link. (The deleted link must come from this cycle in order to still have a spanning tree.)
3. **Form of tabu moves:** List the links that should not be deleted.
4. **Addition of a tabu move:** At each iteration, after choosing the link to be added to the network, also add this link to the tabu list.
5. **Maximum size of tabu list:** Two. Whenever a tabu move is added to a full list, delete the older of the two tabu moves that already were on the list. (Since a spanning tree for the problem being considered only includes four links, the tabu list must be kept very small to provide some flexibility in choosing the link to be deleted at each iteration.)
6. **Stopping rule:** Stop after three consecutive iterations without an improvement in the best objective function value. (Also stop at any iteration where the current trial solution has no immediate neighbors that are not ruled out by their tabu status.)

Having specified these details, we now can proceed to apply the tabu search algorithm to the example. To get started, a reasonable choice for the initial trial solution is the optimal solution for the unconstrained version of the problem that is shown in Fig. 13.7(*b*).

Because this solution violates both of the constraints (but with the inclusion of only two of the three links specified in constraint 2), penalties of 100 need to be imposed twice. Therefore, the total cost of this solution is

$$\text{Cost} = 20 + 10 + 5 + 15 + 200 \text{ (constraint penalties)}$$
$$= 250.$$

Iteration 1. The three options for adding a link to the network in Fig. 13.7(*b*) are BE, CD, and DE. If BE were to be chosen, the cycle formed would be BE-CE-AC-AB, so the three options for deleting a link would be CE, AC, and AB. (At this point, no links have yet been added to the tabu list.) If CE were to be deleted, the change in the cost would be $30 - 5 = 25$ with no change in the constraint penalties, so the total cost would increase from 250 to 275. Similarly, if AC were to be deleted instead, the total cost would increase from 250 to $250 + (30 - 10) = 270$. However, if link AB were to be the one deleted, the link costs would change by $30 - 20 = 10$ and the constraint penalties would decrease from 200 to 100 because constraint 2 would no longer be violated, so the total cost would become $50 + 10 + 100 = 160$. These results are summarized in the first three rows of Table 13.1.

The next two rows summarize the calculations if CD were to be the link that is added to the network. In this case, the cycle created is CD-AD-AC, so AD and AC are the only options for deleting a link. AC would be a particularly bad choice because constraint 1 would still be violated (a penalty of 100), and a penalty of 200 now would need to be charged for violating constraint 2 since all three of the links specified in the constraint would be included in the network. Deleting AD instead would have the virtue of satisfying constraint 1 and not increasing the extent to which constraint 2 is violated.

The last three rows of the table show the options if DE were the added link. The cycle created by adding this link would be DE-CE-AC-AD, so CE, AC, and AD would be the options for deletion. All three would satisfy constraint 1, but deleting AD would satisfy constraint 2 as well. By completely eliminating constraint penalties, the total cost for this option would become only $50 + (40 - 15) = 75$. Since this is the smallest cost for all eight available options for moving to an immediate neighbor of the current trial solution, we choose this particular move by adding DE and deleting AD. This choice is indicated in the iteration 1 portion of Fig. 13.8 and the resulting spanning tree for beginning iteration 2 is shown to the right.

To complete the iteration, since DE was added to the network, it becomes the first link placed on the tabu list. This will prevent deleting DE next and cycling back to the trial solution that began this iteration.

■ **TABLE 13.1** The options for adding a link and deleting another link in iteration 1

Add	Delete	Cost
BE	CE	$75 + 200 = 275$
BE	AC	$70 + 200 = 270$
BE	AB	$60 + 100 = 160$
CD	AD	$60 + 100 = 160$
CD	AC	$65 + 300 = 365$
DE	CE	$85 + 100 = 185$
DE	AC	$80 + 100 = 180$
DE	AD	$75 + 0 \ \ = \ \ 75 \leftarrow$ Minimum

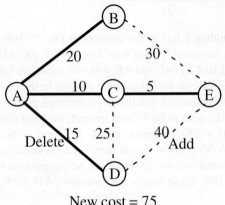

Iteration 1

Cost = 50 + 200 (constraint penalties)

New cost = 75
(Local optimum)

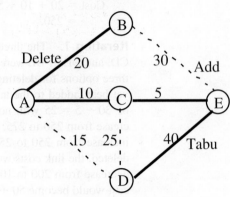

Iteration 2

Cost = 75

New cost = 85
(Escape local optimum)

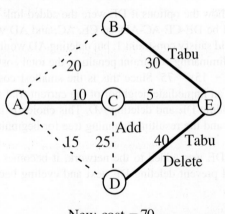

Iteration 3

Cost = 85

New cost = 70
(Override tabu status)

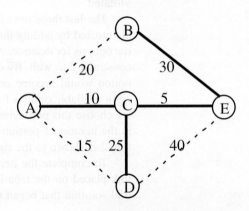

Optimal Solution

Cost = 70

Additional iterations only
find inferior solutions.

FIGURE 13.8
Application of a tabu search algorithm to the minimum spanning tree problem shown in Fig. 13.7
after also adding two constraints.

To summarize, the following decisions have been made during this first iteration.

Add link DE to the network.
Delete link AD from the network.
Add link DE to the tabu list.

Iteration 2. The upper right-hand portion of Fig. 13.8 indicates that the corresponding decisions made during iteration 2 are the following.

Add link BE to the network.
Automatically place this added link on the tabu list.
Delete link AB from the network.

Table 13.2 summarizes the calculations that led to these decisions by finding that the move in the sixth row provides the smallest cost.

The moves listed in the first and seventh rows of the table involve deleting DE, which is on the tabu list. Therefore, these moves would have been considered only if they would result in a better solution than the best trial solution found so far, which has a cost of 75. The calculation in the seventh row shows that this move would not provide a better solution. A calculation is not even needed for the first row because this move would cycle back to the preceding trial solution.

Note that the move in the sixth row is made even though it results in a new trial solution that has a larger cost (85) than for the preceding trial solution (75) that initiated iteration 2. What this means is that the preceding trial solution was a local optimum because all of its immediate neighbors (those that can be reached by making one of the moves listed in Table 13.2) have a larger cost. However, moving to the best of the immediate neighbors allows us to escape the local optimum and continue the search for the global optimum.

Before moving to iteration 3, we should interject an observation about what more advanced forms of tabu search might do here when selecting the best immediate neighbor. More general tabu search methods can change the meaning of a "best neighbor," depending on history, by using additional forms of memory to support intensification and diversification processes. As mentioned earlier, intensification focuses the search in a particularly promising region of solutions identified previously and diversification drives the search into promising new regions.

Iteration 3. The lower left-hand portion of Fig. 13.8 summarizes the decisions made during iteration 3.

Add link CD to the network.
Automatically place this added link on the tabu list.
Delete link DE from the network.

Table 13.3 shows that this move leads to the best immediate neighbor of the trial solution that initiated this iteration.

■ TABLE 13.2 The options for adding a link and deleting another link in iteration 2

Add	Delete	Cost
AD	DE*	(Tabu move)
AD	CE	85 + 100 = 185
AD	AC	80 + 100 = 180
BE	CE	100 + 0 = 100
BE	AC	95 + 0 = 95
BE	AB	85 + 0 = 85 ← Minimum
CD	DE*	60 + 100 = 160
CD	CE	95 + 100 = 195

*A tabu move. Will be considered only if it would result in a better solution than the best trial solution found previously.

■ TABLE 13.3 The options for adding a link and deleting another link in iteration 3

Add	Delete	Cost
AB	BE*	(Tabu move)
AB	CE	100 + 0 = 100
AB	AC	95 + 0 = 95
AD	DE*	60 + 100 = 160
AD	CE	95 + 0 = 95
AD	AC	90 + 0 = 90
CD	DE*	70 + 0 = 70 ← Minimum
CD	CE	105 + 0 = 105

*A tabu move. Will be considered only if it would result in a better solution than the best trial solution found previously.

An interesting feature of this move is that it is made even though it is a tabu move. The reason it is made is that, in addition to being the best immediate neighbor, it also results in a solution that is better (a cost of 70) than the best trial solution found previously (a cost of 75). This enables the tabu status of the move to be overridden. (Tabu search also can incorporate a variety of more advanced criteria for overriding tabu status.)

One more adjustment needs to be made in the tabu list before beginning the next iteration.

Delete link DE from the tabu list.

This is done for two reasons. First, the tabu list consists of links that normally should not be deleted from the network during the current iteration (with the exception noted above), but DE is no longer in the network. Second, since the size of the tabu list has been set at two and two other links (BE and CD) have been added to the list more recently, DE automatically would have been deleted from the list at this point anyway.

Continuation.   The current trial solution shown in the lower right-hand portion of Fig. 13.8 is, in fact, the optimal solution (the global optimum) for the problem. However, the tabu search algorithm has no way of knowing this, so it would continue on for a while. Iteration 4 would begin with this trial solution and with links BE and CD on the tabu list. After completing this iteration and two more, the algorithm would terminate because three consecutive iterations did not improve on the best previous objective function value (a cost of 70).

With a well-designed tabu search algorithm, the best trial solution found after the algorithm has run a modest number of iterations is likely to be a good feasible solution. It might even be an optimal solution, but no such guarantee can be given. Selecting a stopping rule that provides a relatively long run of the algorithm increases the chance of reaching the global optimum.

Having gotten our feet wet by designing and applying a tabu search algorithm to this small example, let us now apply a similar tabu search algorithm to the example of a traveling salesman problem presented in Sec. 13.1.

The Traveling Salesman Problem Example

There are some close parallels between a minimum spanning tree problem and a traveling salesman problem. In both cases, the problem is to choose which links to include in the solution. (Recall that a solution for a traveling salesman problem can be described as the sequence of links that the salesman traverses in the tour of the cities.) In both cases, the objective is to minimize the total cost or distance associated with the fixed number of links that are included in the solution. And in both cases, there is an intuitive local search procedure available that involves adding and deleting links in the current trial solution to obtain the new trial solution.

For minimum spanning tree problems, the local search procedure described in the preceding subsection involves adding and deleting only a *single* link at each iteration. The corresponding procedure described in Sec. 13.1 for traveling salesman problems involves using *sub-tour reversals* to add and delete a *pair* of links at each iteration.

Because of the close parallels between these two types of problems, the design of a tabu search algorithm for traveling salesman problems can be quite similar to the one just described for the minimum spanning problem example. In particular, using the outline of a basic tabu search algorithm presented earlier, the six questions following the outline can be answered in a similar way below.

1. **Local search algorithm:** At each iteration, choose the best immediate neighbor of the current trial solution that is not ruled out by its tabu status.
2. **Neighborhood structure:** An immediate neighbor of the current trial solution is one that is reached by making a *sub-tour reversal,* as described in Sec. 13.1 and illustrated in Fig. 13.5. Such a reversal requires adding two links and deleting two other links from the current trial solution. (We rule out a sub-tour reversal that simply reverses the direction of the tour provided by the current trial solution.)
3. **Form of tabu moves:** List the links such that a particular sub-tour reversal would be tabu if *both* links to be deleted in this reversal are on the list. (This will prevent quickly cycling back to a previous trial solution.)
4. **Addition of a tabu move:** At each iteration, after choosing the two links to be added to the current trial solution, also add these two links to the tabu list.
5. **Maximum size of tabu list:** Four (two from each of the two most recent iterations). Whenever a pair of links is added to a full list, delete the two links that already have been on the list the longest.
6. **Stopping rule:** Stop after three consecutive iterations without an improvement in the best objective function value. (Also stop at any iteration where the current trial solution has no immediate neighbors that are not ruled out by their tabu status.)

To apply this tabu search algorithm to our example (see Fig. 13.4), let us begin with the same initial trial solution, 1-2-3-4-5-6-7-1, as in Sec. 13.1. Recall how starting the sub-tour reversal algorithm (a local improvement algorithm) with this initial trial solution led in two iterations (see Figs. 13.5 and 13.6) to a local optimum at 1-2-4-6-5-3-7-1, at which point that algorithm stopped. Except for adding a tabu list, the tabu search algorithm starts off in exactly the same way, as summarized below.

Initial trial solution: 1-2-3-4-5-6-7-1   Distance = 69
Tabu list: Blank at this point.

Iteration 1: Choose to reverse 3-4 (see Fig. 13.5).
Deleted links: 2-3 and 4-5
Added links: 2-4 and 3-5
Tabu list: Links 2-4 and 3-5
New trial solution: 1-2-4-3-5-6-7-1   Distance = 65

Iteration 2: Choose to reverse 3-5-6 (see Fig. 13.6).
Deleted links: 4-3 and 6-7   (OK since not on tabu list)
Added links: 4-6 and 3-7

Tabu list: Links 2-4, 3-5, 4-6, and 3-7
New trial solution: 1-2-4-6-5-3-7-1   Distance = 64

However, rather than terminating, the tabu search algorithm now escapes from this local optimum (shown on the right side of Fig. 13.6 and the left side of Fig. 13.9) by moving next to the best immediate neighbor of the current trial solution even though its distance is longer. Considering the limited availability of links between pairs of nodes (cities) in Fig. 13.4, the current trial solution has only the two immediate neighbors listed below.

Reverse 6-5-3: 1-2-4-3-5-6-7-1   Distance = 65
Reverse 3-7: 1-2-4-6-5-7-3-1   Distance = 66

(We are ruling out reversing 2-4-6-5-3-7 to obtain 1-7-3-5-6-4-2-1 because this is simply the same tour in the opposite direction.) However, we must rule out the first of these immediate neighbors because it would require deleting links 4-6 and 3-7, which is tabu since *both* of these links are on the tabu list. (This move could still be allowed if it would improve upon the best trial solution found so far, but it does not.) Ruling out this immediate neighbor prevents us from simply cycling back to the preceding trial solution. Therefore, by default, the second of these immediate neighbors is chosen to be the next trial solution, as summarized below.

Iteration 3: Choose to reverse 3-7 (see Fig. 13.9).
Deleted links: 5-3 and 7-1
Added links: 5-7 and 3-1
Tabu list: 4-6, 3-7, 5-7, and 3-1
    (2-4 and 3-5 are now deleted from the list.)
New trial solution: 1-2-4-6-5-7-3-1   Distance = 66

The sub-tour reversal for this iteration can be seen in Fig. 13.9, where the dashed lines show the links being deleted (on the left) and added (on the right) to obtain the new trial solution. Note that one of the deleted links is 5-3 even though it was on the tabu list at the end of iteration 2. This is OK since a sub-tour reversal is tabu only if *both* of the deleted links are on the tabu list. Also note that the updated tabu list at the end of iteration 3 has

■ **FIGURE 13.9**
The sub-tour reversal of 3-7 in iteration 3 that leads from the trial solution on the left to the new trial solution on the right.

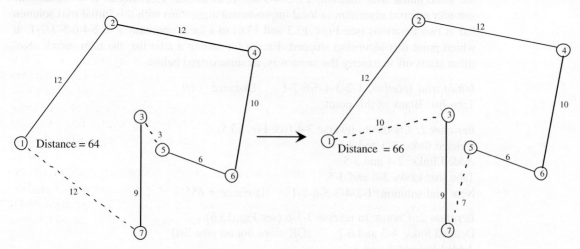

deleted the two links that had been on the list the longest (the ones added during iteration 1) since the maximum size of the tabu list has been set at four.

The new trial solution has the four immediate neighbors listed below.

Reverse 2-4-6-5-7: 1-7-5-6-4-2-3-1   Distance = 65
Reverse 6-5: 1-2-4-5-6-7-3-1   Distance = 69
Reverse 5-7: 1-2-4-6-5-7-3-1   Distance = 63
Reverse 7-3: 1-2-4-6-5-3-7-1   Distance = 64

However, the second of these immediate neighbors is tabu because *both* of the deleted links (4-6 and 5-7) are on the tabu list. The fourth immediate neighbor (which is the preceding trial solution) also is tabu for the same reason. Thus, the only viable options are the first and third immediate neighbors. Since the latter neighbor has the shorter distance, it becomes the next trial solution, as summarized below.

Iteration 4: Choose to reverse 5-7 (see Fig. 13.10).
Deleted links: 6-5 and 7-3
Added links: 6-7 and 5-3
Tabu list: 5-7, 3-1, 6-7, and 5-3
    (4-6 and 3-7 are now deleted from the list.)
New trial solution: 1-2-4-6-7-5-3-1   Distance = 63

Figure 13.10 shows this sub-tour reversal. The tour for the new trial solution on the right has a distance of only 63, which is less than for any of the preceding trial solutions. In fact, this new solution happens to be the optimal solution.

Not knowing this, the tabu search algorithm would attempt to execute more iterations. However, the only immediate neighbor of the current trial solution is the trial solution that was obtained at the preceding iteration. This would require deleting links 6-7 and 5-3, both of which are on the tabu list, so we are prevented from cycling back to the preceding trial solution. Since no other immediate neighbors are available, the stopping rule terminates the algorithm at this point with 1-2-4-6-7-5-3-1 (the best of the trial solutions) as the final solution. Although there is no guarantee that the algorithm's final solution is an optimal solution, we are fortunate that it turned out to be optimal in this case.

The metaheuristics area in your IOR Tutorial includes a procedure for applying this particular tabu search algorithm to other small traveling salesman problems.

This particular algorithm is just one example of a possible tabu search algorithm for traveling salesman problems. Various details of the algorithm could be modified in a number of reasonable ways. For example, the method typically doesn't stop when all available moves are forbidden by their tabu status, but instead just selects a "least tabu" move. Also, an important feature of general tabu search methods includes the use of multiple neighborhoods, relying on basic neighborhoods as long as they bring progress, and then including more advanced neighborhoods when the rate of finding improved solutions diminishes. The most significant additional element of tabu search is its use of intensification and diversification strategies, as mentioned earlier. But the general outline of a basic "short-term memory" tabu search approach would remain roughly the same as we have illustrated.

Both examples considered in this section fall into the category of combinatorial optimization problems involving networks. This is a particularly common area of application for tabu search algorithms. The general outline of these algorithms incorporates the principles presented in this section, but the details are worked out to fit the structure of the specific problems being considered.

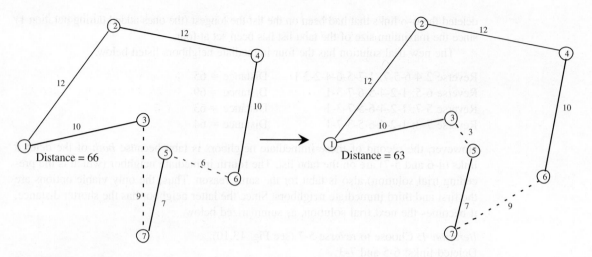

■ **FIGURE 13.10**

The sub-tour reversal of 5-7 in iteration 4 that leads from the trial solution on the left to the new trial solution on the right (which happens to be the optimal solution).

13.3   SIMULATED ANNEALING

Simulated annealing is another widely used metaheuristic that enables the search process to escape from a local optimum. To better compare and contrast it with tabu search, we will apply it to the same traveling salesman problem example before returning to the nonlinear programming example introduced in Sec. 13.1. But first, let us examine the basic concepts of simulated annealing.

Basic Concepts

Figure 13.1 in Sec. 13.1 introduced the concept that finding the global optimum of a complicated maximization problem is analogous to determining which of a number of hills is the tallest hill and then climbing to the top of that particular hill. Unfortunately, a mathematical search process does not have the benefit of keen eyesight that would enable spotting a tall hill in the distance. Instead, it is like hiking in a dense fog where the only clue for the direction to take next is how much the next step in any direction would take you up or down.

One approach, adopted into tabu search, is to climb the current hill in the steepest direction until reaching its top and then start climbing slowly downward while searching for another hill to climb. The drawback is that a lot of time (iterations) is spent climbing each hill encountered rather than searching for the tallest hill.

Instead, the approach used in simulated annealing is to focus mainly on searching for the tallest hill. Since the tallest hill can be anywhere in the feasible region, the early emphasis is on taking steps in random directions (except for rejecting some, but not all, steps that would go downward rather than upward) in order to explore as much of the feasible region as possible. Because most of the accepted steps are upward, the search will gradually gravitate toward those parts of the feasible region containing the tallest hills. Therefore, the search process gradually increases the emphasis on climbing upward by rejecting an increasing proportion of steps that go downward. Given enough time, the process often will reach and climb to the top of the tallest hill.

To be more specific, each iteration of the simulated annealing search process moves from the current trial solution to an immediate neighbor in the local neighborhood of this solution, just as for tabu search. However, the difference from tabu search lies in how an immediate neighbor is selected to be the next trial solution. Let

Z_c = objective function value for the *current* trial solution,

Z_n = objective function value for the current candidate to be the next trial solution,

T = a parameter that measures the tendency to accept the current candidate to be the next trial solution if this candidate is not an improvement on the current trial solution.

The rule for selecting which immediate neighbor will be the next trial solution is the following.

Move selection rule: Among all the immediate neighbors of the current trial solution, select one randomly to become the current candidate to be the next trial solution. Assuming the objective is *maximization* of the objective function, accept or reject this candidate to be the next trial solution as follows:

If $Z_n \geq Z_c$, always accept this candidate.

If $Z_n < Z_c$, accept the candidate with the following probability:

$$\text{Prob\{acceptance\}} = e^x \text{ where } x = \frac{Z_n - Z_c}{T}$$

(If the objective is *minimization* instead, reverse Z_n and Z_c in the above formulas.) If this candidate is rejected, repeat this process with a new randomly selected immediate neighbor of the current trial solution. (If no immediate neighbors remain, terminate the algorithm.)

Thus, if the current candidate under consideration is better than the current trial solution, it always is accepted to be the next trial solution. If it is worse, the probability of acceptance depends on how much worse it is (and on the size of T). Table 13.4 shows a sampling of these probability values, ranging from a very high probability when the current candidate is only slightly worse (relative to T) than the current trial solution to an extremely small probability when it is much worse. In other words, the move selection rule usually will accept a step that is only slightly downhill, but seldom will accept a steep downward step. Starting with a relatively large value of T (as simulated annealing does) makes the probability of acceptance relatively large, which enables the search to proceed in almost random directions. Gradually decreasing the value of T as the search continues (as simulated annealing does) gradually decreases the probability of acceptance, which increases the emphasis on mostly climbing upward. Thus, the choice of the values of T over time controls the degree of randomness in the process for allowing downward steps.

■ **TABLE 13.4** Some sample probabilities that the move selection rule will accept a downward step when the objective is maximization

$x = \dfrac{Z_n - Z_c}{T}$	Prob{acceptance} = e^x
−0.01	0.990
−0.1	0.905
−0.25	0.779
−0.5	0.607
−1	0.368
−2	0.135
−3	0.050
−4	0.018
−5	0.007

This random component, not present in basic tabu search, provides more flexibility for moving toward another part of the feasible region in the hope of finding a taller hill.

The usual method of implementing the move selection rule to determine whether a particular downward step will be accepted is to compare a **random number** between 0 and 1 to the probability of acceptance. Such a random number can be thought of as a random observation from a uniform distribution between 0 and 1. (All references to random numbers throughout the chapter will be to such random numbers.) There are a number of methods of generating these random numbers (as will be described in Sec. 20.3). For example, the Excel function RAND() generates such random numbers upon request. (The beginning of the Problems section also describes how you can use the random digits given in Table 20.3 to obtain the random numbers you will need for some of your homework problems.)

If *random number* < Prob{acceptance}, accept a downward step.
Otherwise, reject the step.

Why does simulated annealing use the particular formula for Prob{acceptance} specified by the move selection rule? The reason is that simulated annealing is based on the analogy to a *physical annealing process*. This process initially involves melting a metal or glass at a high temperature and then slowly cooling the substance until it reaches a low-energy stable state with desirable physical properties. At any given temperature T during this process, the energy level of the atoms in the substance is fluctuating but tending to decrease. A mathematical model of how the energy level fluctuates assumes that changes occur randomly except that only some of the increases are accepted. In particular, the probability of accepting an increase when the temperature is T has the same form as for Prob{acceptance} in the move selection rule for simulated annealing.

The analogy for an optimization problem in minimization form is that the energy level of the substance at the current state of the system corresponds to the objective function value at the current feasible solution of the problem. The objective of having the substance reach a stable state with an energy level that is as small as possible corresponds to having the problem reach a feasible solution with an objective function value that is as small as possible.

Just as for a physical annealing process, a key question when designing a simulated annealing algorithm for an optimization problem is to select an appropriate **temperature schedule** to use. (Because of the analogy to physical annealing, we now are referring to T in a simulated annealing algorithm as the temperature.) This schedule needs to specify the initial, relatively large value of T, as well as the subsequent progressively smaller values. It also needs to specify how many moves (iterations) should be made at each value of T. The selection of these parameters to fit the problem under consideration is a key factor in the effectiveness of the algorithm. Some preliminary experimentation can be used to guide this selection of the parameters of the algorithm. We later will specify one specific temperature schedule that seems reasonable for the two examples considered in this section, but many others could be considered as well.

With this background, we now can provide an outline of a basic simulated annealing algorithm.

Outline of a Basic Simulated Annealing Algorithm

Initialization.   Start with a feasible initial trial solution.

Iteration.   Use the *move selection rule* to select the next trial solution. (If none of the immediate neighbors of the current trial solution are accepted, the algorithm is terminated.)

Check the temperature schedule.   When the desired number of iterations have been performed at the current value of T, decrease T to the next value in the temperature schedule and resume performing iterations at this next value.

Stopping rule. When the desired number of iterations have been performed at the smallest value of T in the temperature schedule (or when none of the immediate neighbors of the current trial solution are accepted), stop. Accept the best trial solution found at any iteration (including for larger values of T) as the final solution.

Before applying this algorithm to any particular problem, a number of details need to be worked out to fit the structure of the problem.

1. How should the initial trial solution be selected?
2. What is the *neighborhood structure* that specifies which solutions are immediate neighbors (reachable in a single iteration) of any current trial solution?
3. What device should be used in the move selection rule to *randomly* select one of the immediate neighbors of the current trial solution to become the current candidate to be the next trial solution?
4. What is an appropriate temperature schedule?

We will illustrate some reasonable ways of addressing these questions in the context of applying the simulated annealing algorithm to the following two examples.

The Traveling Salesman Problem Example

We now return to the particular traveling salesman problem that was introduced in Sec. 13.1 and displayed in Fig. 13.4.

The metaheuristics area in your IOR Tutorial includes a procedure for applying the basic simulated annealing algorithm to small traveling salesman problems like this example. This procedure answers the four questions in the following way.

1. **Initial trial solution:** You may enter any feasible solution (sequence of cities on the tour), perhaps by randomly generating the sequence, but it is helpful to enter one that appears to be a good feasible solution. For the example, the feasible solution 1-2-3-4-5-6-7-1 is a reasonable choice.
2. **Neighborhood structure:** An immediate neighbor of the current trial solution is one that is reached by making a *sub-tour reversal,* as described in Sec. 13.1 and illustrated in Fig. 13.5. (However, the sub-tour reversal that simply reverses the direction of the tour provided by the current trial solution is ruled out.)
3. **Random selection of an immediate neighbor:** Selecting a sub-tour to be reversed requires selecting the slot in the current sequence of cities where the sub-tour currently begins and then the slot where the sub-tour currently ends. The beginning slot can be anywhere except the first and last slots (reserved for the home city) and the next-to-last slot. The ending slot must be somewhere after the beginning slot, excluding the last slot. (Both beginning in the second slot and ending in the next-to-last slot also is ruled out since this would simply reverse the direction of the tour.) As will be illustrated shortly, random numbers are used to give equal probabilities to selecting any of the eligible beginning slots and then any of the eligible ending slots. If this selection of the beginning and ending slots turns out to be infeasible (because the links needed to complete the sub-tour reversal are not available), this process is repeated until a feasible selection is made.
4. **Temperature schedule:** Five iterations are performed at each of five values of T (T_1, T_2, T_3, T_4, T_5) in turn, where

$T_1 = 0.2Z_c$ when Z_c is the objective function value for the initial trial solution,
$T_2 = 0.5T_1$,
$T_3 = 0.5T_2$,
$T_4 = 0.5T_3$,
$T_5 = 0.5T_4$.

This particular temperature schedule is only illustrative of what could be used. $T_1 = 0.2Z_c$ is a reasonable choice because T_1 should tend to be fairly large compared to typical values of $|Z_n - Z_c|$, which will encourage an almost random search through the feasible region to find where the search should be focused. However, by the time the value of T is reduced to T_5, almost no nonimproving moves will be accepted, so the emphasis will be on improving the value of the objective function.

When dealing with larger problems, more than five iterations probably would be performed at each value of T. Furthermore, the values of T would probably be reduced more slowly than with the temperature schedule prescribed above.

Now let us elaborate on how the random selection of an immediate neighbor is made. Suppose we are dealing with the initial trial solution of 1-2-3-4-5-6-7-1 in our example.

Initial trial solution: 1-2-3-4-5-6-7-1 $Z_c = 69$ $T_1 = 0.2Z_c = 13.8$

The sub-tour that will be reversed can begin anywhere between the second slot (currently designating city 2) and the sixth slot (currently designating city 6). These five slots can be given equal probabilities by having the following values of a random number between 0 and 1 correspond to choosing the slot indicated below.

0.0000–0.1999:   Sub-tour begins in slot 2.
0.2000–0.3999:   Sub-tour begins in slot 3.
0.4000–0.5999:   Sub-tour begins in slot 4.
0.6000–0.7999:   Sub-tour begins in slot 5.
0.8000–0.9999:   Sub-tour begins in slot 6.

Suppose that the random number generated happens to be 0.2779.

0.2779: Choose a sub-tour that begins in slot 3.

By beginning in slot 3, the sub-tour that will be reversed needs to end somewhere between slots 4 and 7. These four slots are given equal probabilities by using the following correspondence with a random number.

0.0000–0.2499:   Sub-tour ends in slot 4.
0.2500–0.4999:   Sub-tour ends in slot 5.
0.5000–0.7499:   Sub-tour ends in slot 6.
0.7500–0.9999:   Sub-tour ends in slot 7.

Suppose that the random number generated for this purpose happens to be 0.0461.

0.0461: Choose to end the sub-tour in slot 4.

Since slots 3 and 4 currently designate that cities 3 and 4 are the third and fourth cities visited in the tour, the sub-tour of cities 3-4 will be reversed.

Reverse 3-4 (see Fig. 13.5): 1-2-4-3-5-6-7-1 $Z_n = 65$

This immediate neighbor of the current (initial) trial solution becomes the current candidate to be the next trial solution. Since

$$Z_n = 65 < Z_c = 69,$$

this candidate is better than the current trial solution (remember that the objective here is to *minimize* the total distance of the tour), so this candidate is automatically accepted to be next trial solution.

This choice of a sub-tour reversal was a fortunate one because it led to a feasible solution. This does not always happen in traveling salesman problems like our example where certain pairs of cities are not directly connected by a link. For example, if the random numbers had called for reversing 2-3-4-5 to obtain the tour 1-5-4-3-2-6-7-1, Fig. 13.4

shows that this is an infeasible solution because there is no link between cities 1 and 5 as well as no link between cities 2 and 6. When this happens, new pairs of random numbers would need to be generated until a feasible solution is obtained. (A more sophisticated procedure also can be constructed to generate random numbers only for relevant links.)

To illustrate a case where the current candidate to be the next trial solution is worse than the current trial solution, suppose that the second iteration results in reversing 3-5-6 (as in Fig. 13.6) to obtain 1-2-4-6-5-3-7-1, which has a total distance of 64. Then suppose that the third iteration begins by reversing 3-7 (as in Fig. 13.9) to obtain 1-2-4-6-5-7-3-1 (which has a total distance of 66) as the current candidate to be the next trial solution. Since 1-2-4-6-5-3-7-1 (with a total distance of 64) is the current trial solution for iteration 3, we now have

$$Z_c = 64, \qquad Z_n = 66, \qquad T_1 = 13.8.$$

Therefore, since the objective here is *minimization,* the probability of accepting 1-2-4-6-5-7-3-1 as the next trial solution is

$$\text{Prob\{acceptance\}} = e^{(Z_c - Z_n)/T_1}$$
$$= e^{-2/13.8}$$
$$= 0.865.$$

If the next random number generated is less than 0.865, this candidate solution will be accepted as the next trial solution. Otherwise, it will be rejected.

Table 13.5 shows the results of using IOR Tutorial to apply the complete simulated annealing algorithm to this problem. Note that iterations 14 and 16 tie for finding the best

■ **TABLE 13.5** One application of the simulated annealing algorithm in IOR Tutorial to the traveling salesman problem example

Iteration	T	Trial Solution Obtained	Distance
0		1-2-3-4-5-6-7-1	69
1	13.8	1-3-2-4-5-6-7-1	68
2	13.8	1-2-3-4-5-6-7-1	69
3	13.8	1-3-2-4-5-6-7-1	68
4	13.8	1-3-2-4-6-5-7-1	65
5	13.8	1-2-3-4-6-5-7-1	66
6	6.9	1-2-3-4-5-6-7-1	69
7	6.9	1-3-2-4-5-6-7-1	68
8	6.9	1-2-3-4-5-6-7-1	69
9	6.9	1-2-3-5-4-6-7-1	65
10	6.9	1-2-3-4-5-6-7-1	69
11	3.45	1-2-3-4-6-5-7-1	66
12	3.45	1-3-2-4-6-5-7-1	65
13	3.45	1-3-7-5-6-4-2-1	66
14	3.45	1-3-5-7-6-4-2-1	63 ← Minimum
15	3.45	1-3-7-5-6-4-2-1	66
16	1.725	1-3-5-7-6-4-2-1	63 ← Minimum
17	1.725	1-3-7-5-6-4-2-1	66
18	1.725	1-3-2-4-6-5-7-1	65
19	1.725	1-2-3-4-6-5-7-1	66
20	1.725	1-3-2-4-6-5-7-1	65
21	0.8625	1-3-7-5-6-4-2-1	66
22	0.8625	1-3-2-4-6-5-7-1	65
23	0.8625	1-2-3-4-6-5-7-1	66
24	0.8625	1-3-2-4-6-5-7-1	65
25	0.8625	1-3-7-5-6-4-2-1	66

trial solution, 1-3-5-7-6-4-2-1 (which happens to be the optimal solution along with the equivalent tour in the reverse direction, 1-2-4-6-7-5-3-1), so this solution is accepted as the final solution. You might find it interesting to apply this software to the same problem yourself. Due to the randomness built into the algorithm, the sequence of trial solutions obtained will be different each time. Because of this feature, practitioners sometimes will reapply a simulated annealing algorithm to the same problem several times to increase the chance of finding an optimal solution. (Problem 13.3-2 asks you to do this for this same example.) The initial trial solution also may be changed each time to help facilitate a more thorough exploration of the entire feasible region.

If you would like to see **another example** of how random numbers are used to perform an iteration of the basic simulated annealing algorithm for a traveling salesman problem, one is provided in the Worked Examples section of the book's website.

Before going on to the next example, we should pause at this point to mention a couple of ways in which advanced features of tabu search can be combined fruitfully with simulated annealing. One way is by applying the *strategic oscillation* feature of tabu search to the temperature schedule of simulated annealing. Strategic oscillation adjusts the temperature schedule by decreasing the temperatures more rapidly than usual but then strategically moving the temperatures back and forth across levels where the best solutions were found. Another way involves applying the candidate-list strategies of tabu search to the move selection rule of simulated annealing. The idea here is to scan multiple neighbors to see if an improving move is found before applying the randomized rule for accepting or rejecting the current candidate to be the next trial solution. These changes have sometimes produced significant improvements.

As these ideas for applying features of tabu search to simulated annealing suggest, a *hybrid algorithm* that combines the ideas of different metaheuristics can sometimes perform better than an algorithm that is based solely on a single metaheuristic. Although we are presenting the three most commonly used metaheuristics separately in this chapter, experienced practitioners occasionally will pick and choose among the ideas of these and other metaheuristics in designing their heuristic methods.

The Nonlinear Programming Example

Now reconsider the example of a small nonlinear programming problem (only a single variable) that was introduced in Sec. 13.1. The problem is to

$$\text{Maximize} \quad f(x) = 12x^5 - 975x^4 + 28{,}000x^3 - 345{,}000x^2 + 1{,}800{,}000x,$$

subject to

$$0 \leq x \leq 31.$$

The graph of $f(x)$ in Fig. 13.1 reveals that there are local optima at $x = 5$, $x = 20$, and $x = 31$, but only $x = 20$ is a global optimum.

The metaheuristics area in IOR Tutorial includes a procedure for applying the simulated annealing algorithm to small nonlinear programming problems of the form,

$$\text{Maximize} \quad f(x_1, \ldots, x_n)$$

subject to

$$L_j \leq x_j \leq U_j, \quad \text{for } j = 1, \ldots, n,$$

where $n = 1$ or 2, and where L_j and U_j are constants $(0 \leq L_j < U_j \leq 63)$ representing the bounds on x_j. (Having relatively tight bounds on the individual variables is highly desirable for the efficiency of a simulated annealing algorithm, as well as for genetic algorithms discussed in the next section.) One or two linear functional constraints on the variables $\mathbf{x} = (x_1, \ldots, x_n)$ also can be included when $n = 2$. For the example, we have

$$n = 1, \qquad L_1 = 0, \qquad U_1 = 31,$$

with no linear functional constraints.

This procedure in IOR Tutorial designs the details of the simulated annealing algorithm for such nonlinear programming problems as follows.

1. **Initial trial solution:** You may enter any feasible solution, but it is helpful to enter one that appears to be a good feasible solution. In the absence of any clues about where the good feasible solutions might lie, it is reasonable to set each variable x_j midway between its lower bound L_j and upper bound U_j in order to start the search in the middle of the feasible region. (For this reason, $x = 15.5$ is a reasonable choice for the initial trial solution for the example.)

2. **Neighborhood structure:** Any feasible solution is considered to be an immediate neighbor of the current trial solution. However, the method described below for selecting an immediate neighbor to become the current candidate to be the next trial solution gives a preference to feasible solutions that are relatively close to the current trial solution, while still allowing for the possibility of moving to a different part of the feasible region to continue the search.

3. **Random selection of an immediate neighbor:** Set

$$\sigma_j = \frac{U_j - L_j}{6}, \qquad \text{for } j = 1, \ldots, n.$$

Then, given the current trial solution $(x_1, \ldots, x_n)$,

$$\text{reset } x_j = x_j + N(0, \sigma_j), \qquad \text{for } j = 1, \ldots, n,$$

where $N(0, \sigma_j)$ is a random observation from a *normal distribution* with mean zero and standard deviation σ_j. If this does not result in a feasible solution, then repeat this process (starting again from the current trial solution) as many times as needed to obtain a feasible solution.

4. **Temperature schedule:** As for traveling salesman problems, five iterations are performed at each of five values of T $(T_1, T_2, T_3, T_4, T_5)$ in turn, where

$T_1 = 0.2Z_c$ when Z_c is the objective function value for the initial trial solution,
$T_2 = 0.5T_1$,
$T_3 = 0.5T_2$,
$T_4 = 0.5T_3$,
$T_5 = 0.5T_4$.

The reason for setting $\sigma_j = (U_j - L_j)/6$ when selecting an immediate neighbor is that when the variable x_j is midway between L_j and U_j, any new feasible value of the variable is within three standard deviations of the current value. This gives a significant probability that the new value will move most of the way to one of its bounds even though there is a much higher probability that the new value will be relatively close to the current value. There are a number of methods for generating a random observation $N(0, \sigma_j)$ from a normal distribution (as will be discussed briefly in Sec. 20.4). For example, the Excel function, NORMINV(RAND(),0,σ_j), generates such a random observation. For your homework, here is a straightforward way of generating the random observations you need. Obtain a random number r and then use the normal table in Appendix 5 to find the value of $N(0, \sigma_j)$ such that $P\{X \leq N(0, \sigma_j)\} = r$ when X is a normal random variable with mean 0 and standard deviation σ_j.

To illustrate how the algorithm designed in this way would be applied to the example, let us start with $x = 15.5$ as the initial trial solution. Thus,

$$Z_c = f(15.5) = 3,741,121 \qquad \text{and} \qquad T_1 = 0.2Z_c = 748,224.$$

Since

$$\sigma = \frac{U - L}{6} = \frac{31 - 0}{6} = 5.167,$$

the next step is to generate a random observation $N(0, 5.167)$ from a normal distribution with mean zero and this standard deviation. To do this, we first obtain a random number, which happens to be 0.0735. Going to the normal table in Appendix 5, $P\{\text{standard normal} \leq -1.45\} = 0.0735$, so $N(0, 5.167) = -1.45(5.167) = -7.5$. The current candidate to be the next trial solution then is obtained by resetting x as

$$x = 15.5 + N(0, 5.167) = 15.5 - 7.5$$
$$= 8,$$

so that

$$Z_n = f(x) = 3,055,616.$$

Because

$$\frac{Z_n - Z_c}{T} = \frac{3,055,616 - 3,741,121}{748,224} = -0.916$$

the probability of accepting $x = 8$ as the next trial solution is

$$\text{Prob\{acceptance\}} = e^{-0.916} = 0.400.$$

Therefore, $x = 8$ will be accepted only if the corresponding random number between 0 and 1 happens to be less than 0.400. Thus, $x = 8$ is fairly likely to be rejected. (In somewhat later iterations when T is much smaller, $x = 8$ would almost certainly be rejected.) This is fortunate since Fig. 13.1 reveals that the search should focus on the portion of the feasible region between $x = 10$ and $x = 30$ in order to start climbing the tallest hill.

Table 13.6 provides the results that were obtained by using IOR Tutorial to apply the complete simulated annealing algorithm to this nonlinear programming problem. Note how the trial solutions obtained vary fairly widely over the feasible region during the early iterations, but then start approaching the top of the tallest hill more consistently during the later iterations when T has been reduced to much smaller values. Therefore, of the 25 iterations, the best trial solution of $x = 20.031$ (as compared to the optimal solution of $x = 20$) was not obtained until iteration 21.

Once again, you might find it interesting to apply this software to the same problem yourself to see what is yielded by new sequences of random numbers and random observations from normal distributions. (Problem 13.3-6 asks you to do this several times.)

■ **TABLE 13.6** One application of the simulated annealing algorithm in IOR Tutorial to the nonlinear programming example

Iteration	T	Trial Solution Obtained	f(x)
0		$x = 15.5$	3,741,121.0
1	748,224	$x = 17.557$	4,167,533.956
2	748,224	$x = 14.832$	3,590,466.203
3	748,224	$x = 17.681$	4,188,641.364
4	748,224	$x = 16.662$	3,995,966.078
5	748,224	$x = 18.444$	4,299,788.258
6	374,112	$x = 19.445$	4,386,985.033
7	374,112	$x = 21.437$	4,302,136.329
8	374,112	$x = 18.642$	4,322,687.873
9	374,112	$x = 22.432$	4,113,901.493
10	374,112	$x = 21.081$	4,345,233.403
11	187,056	$x = 20.383$	4,393,306.255
12	187,056	$x = 21.216$	4,330,358.125
13	187,056	$x = 21.354$	4,313,392.276
14	187,056	$x = 20.795$	4,370,624.01
15	187,056	$x = 18.895$	4,348,060.727
16	93,528	$x = 21.714$	4,259,787.734
17	93,528	$x = 19.463$	4,387,360.1
18	93,528	$x = 20.389$	4,393,076.988
19	93,528	$x = 19.83$	4,398,710.575
20	93,528	$x = 20.68$	4,378,591.085
21	46,764	$x = 20.031$	4,399,955.913 ← Maximum
22	46,764	$x = 20.184$	4,398,462.299
23	46,764	$x = 19.9$	4,399,551.462
24	46,764	$x = 19.677$	4,395,385.618
25	46,764	$x = 19.377$	4,383,048.039

■ 13.4 GENETIC ALGORITHMS

Genetic algorithms provide a third type of metaheuristic that is quite different from the first two. This type tends to be particularly effective at exploring various parts of the feasible region and gradually evolving toward the best feasible solutions.

After introducing the basic concepts for this type of metaheuristic, we will apply a basic genetic algorithm to the same nonlinear programming example just considered above with the additional constraint that the variable is restricted to integer values. We then will apply this approach to the same traveling salesman problem example considered in each of the preceding sections.

Basic Concepts

Just as simulated annealing is based on an analogy to a natural phenomenon (the physical annealing process), genetic algorithms are greatly influenced by another form of a natural phenomenon. In this case, the analogy is to the biological *theory of evolution* formulated by Charles Darwin in the mid-19th century. Each species of plants and animals has great individual variation. Darwin observed that those individuals with variations that impart a survival advantage through improved adaptation to the environment are most likely to survive to the next generation. This phenomenon has since been referred to as *survival of the fittest*.

The modern field of genetics provides a further explanation of this process of evolution and the *natural selection* involved in the survival of the fittest. In any species that reproduces by sexual reproduction, each offspring inherits some of the *chromosomes* from each of the two parents, where the *genes* within the chromosomes determine the individual features of the child. A child who happens to inherit the better features of the parents is slightly more likely to survive into adulthood and then become a parent who passes on some of these features to the next generation. The population tends to improve slowly over time by this process. A second factor that contributes to this process is a random, low-level mutation rate in the DNA of the chromosomes. Thus, a *mutation* occasionally occurs that changes the features of a chromosome that a child inherits from a parent. Although most mutations have no effect or are disadvantageous, some mutations provide desirable improvements. Children with desirable mutations are slightly more likely to survive and contribute to the future gene pool of the species.

These ideas transfer over to dealing with optimization problems in a rather natural way. Feasible solutions for a particular problem correspond to members of a particular species, where the fitness of each member now is measured by the value of the objective function. Rather than processing a single trial solution at a time (as with basic forms of tabu search and simulated annealing), we now work with an entire *population* of trial solutions.[1] For each iteration (generation) of a genetic algorithm, the current **population** consists of the set of trial solutions currently under consideration. These trial solutions are thought of as the currently living members of the species. Some of the youngest members of the population (including especially the fittest members) survive into adulthood and become **parents** (paired at random) who then have **children** (new trial solutions) who share some of the features (genes) of both parents. Since the fittest members of the population are more likely to become parents than others, a genetic algorithm tends to generate *improving populations* of trial solutions as it proceeds. **Mutations** occasionally occur so that certain children also can acquire features (sometimes desirable features) that are not possessed by either parent. This helps a genetic algorithm to explore a new, perhaps better part of the feasible region than previously considered. Eventually, survival of the fittest should tend to lead a genetic algorithm to a trial solution (the best of any considered) that is at least nearly optimal.

Although the analogy of the process of biological evolution defines the core of any genetic algorithm, it is not necessary to adhere rigidly to this analogy in every detail. For example, some genetic algorithms (including the one outlined below) allow the same trial solution to be a parent repeatedly over multiple generations (iterations). Thus, the analogy needs to be only a starting point for defining the details of the algorithm to best fit the problem under consideration.

Here is a rather typical outline of a genetic algorithm that we will employ for the two examples.

Outline of a Basic Genetic Algorithm

Initialization. Start with an initial population of feasible trial solutions, perhaps by generating them randomly. Evaluate the *fitness* (the value of the objective function) for each member of this current population.

Iteration. Use a random process that is biased toward the more fit members of the current population to select some of the members (an even number) to become parents. Pair up the parents randomly and then have each pair of parents give birth to two children (new *feasible* trial solutions) whose features (genes) are a random mixture of the features of the parents, except for occasional mutations. (Whenever the random mixture of features and any mutations result in an *infeasible* solution, this is a *miscarriage,* so the process of attempting

---

[1]One of the intensification strategies of tabu search also maintains a population of best solutions. The population is used to create linking paths between its members and to relaunch the search along these paths.

to give birth then is repeated until a child is born that corresponds to a *feasible* solution.) Retain the children and enough of the best members of the current population to form the new population of the same size for the next iteration. (Discard the other members of the current population.) Evaluate the fitness for each new member (the children) in the new population.

Stopping rule. Use some stopping rule, such as a fixed number of iterations, a fixed amount of CPU time, or a fixed number of consecutive iterations without any improvement in the best trial solution found so far. Use the best trial solution found on any iteration as the final solution.

Before this algorithm can be implemented the following questions need to be answered.

1. What should the population size be?
2. How should the members of the current population be selected to become parents?
3. How should the features of the children be derived from the features of the parents?
4. How should mutations be injected into the features of the children?
5. Which stopping rule should be used?

The answers to these questions depend greatly on the structure of the specific problem being addressed. The metaheuristics area in the IOR Tutorial does include two versions of the algorithm. One is for very small integer nonlinear programming problems like the example considered next. The other is for small traveling salesman problems. Both versions answer some of the questions in the same way, as described below.

1. **Population size:** Ten. (This size is reasonable for the small problems for which this software is designed, but much larger populations commonly are used for large problems.)
2. **Selection of parents:** From among the five most fit members of the population (according to the value of the objective function), select four randomly to become parents. From among the five least fit members, select two randomly to become parents. Pair up the six parents randomly to form three couples.
3. **Passage of features (genes) from parents to children:** This process is highly problem dependent and so differs for the two versions of the algorithm in the software, as described later for the two examples.
4. **Mutation rate:** The probability that an inherited feature of a child mutates into an opposite feature is set at 0.1 in the software. (Much smaller mutation rates commonly are used for large problems.)
5. **Stopping rule:** Stop after five consecutive iterations without any improvement in the best trial solution found so far.

Now we are ready to apply the algorithm to the two examples.

The Integer Version of the Nonlinear Programming Example

We return again to the small nonlinear programming problem that was introduced in Sec. 13.1 (see Fig. 13.1) and then addressed using a simulated annealing algorithm at the end of the preceding section. However, we now add the additional constraint that the problem's single variable x must have an integer value. Because the problem already has the constraint that $0 \leq x \leq 31$, this means that the problem has 32 feasible solutions, $x = 0, 1, 2, \ldots, 31$. (Having such bounds is very important for a genetic algorithm, since it reduces the search space to the relevant region.) Thus, we now are dealing with an *integer* nonlinear programming problem.

When applying a genetic algorithm, *strings of binary digits* often are used to represent the solutions of the problem. Such an *encoding* of the solutions is a particularly convenient one for the various steps of a genetic algorithm, including the process of parents giving birth to children. This encoding is easy to do for our particular problem because we simply can write each value of x in base 2. Since 31 is the maximum feasible value of x, only five binary digits are required to write any feasible value. We always will include all five binary digits even when the leading digit or digits are zeroes. Thus, for example,

$x = 3$ is   00011 in base 2,
$x = 10$ is   01010 in base 2,
$x = 25$ is   11001 in base 2.

Each of the five binary digits is referred to as one of the **genes** of the solution, where the two possible values of the binary digit describe which of two possible features is being carried in that gene to help form the overall genetic makeup. When both parents have the same feature, it will be passed down to each child (except when a mutation occurs). However, when the two parents carry opposite features on the same gene, which feature a child will inherit becomes random.

For example, suppose that the two parents are

P1:   00011 and
P2:   01010.

Since the first, third, and fourth digits agree, the children then automatically become (barring mutations)

C1:   0x01x and
C2:   0x01x,

where x indicates that this particular digit is not known yet. Random numbers are used to identify these unknown digits, where a natural correspondence is

0.0000–0.4999   corresponds to the digit being 0,
0.5000–0.9999   corresponds to the digit being 1.

For example, suppose that the next four random numbers generated are 0.7265, 0.5190, 0.0402, and 0.3639 so that the two unknown digits for the first child are both 1s and the two unknown digits for the second child are both 0s. The children then become (barring mutations)

C1:   01011 and
C2:   00010.

This particular method of generating the children from the parents is known as *uniform crossover*. It is perhaps the most intuitive of the various alternative methods that have been proposed.

We now need to consider the possibility of mutations that would affect the genetic makeup of the children.

Since the probability of a mutation in any gene (flipping the binary digit to the opposite value) has been set at 0.1 for our algorithm, we can let the random numbers

0.0000–0.0999   correspond to a mutation,
0.1000–0.9999   correspond to no mutation.

For example, suppose that in the next 10 random numbers generated, only the eighth one is less than 0.1000. This indicates that no mutation occurs in the first child, but the third gene (digit) in the second child flips its value. Therefore, the final conclusion is that the two children are

C1:   01011 and
C2:   00110.

Returning to base 10, the two parents correspond to the solutions, $x = 3$ and $x = 10$, whereas their children would have been (barring mutations) $x = 11$ and $x = 2$. However, because of the mutation, the children become $x = 11$ and $x = 6$.

For this particular example, any integer value of x such that $0 \le x \le 31$ (in base 10) is a feasible solution, so every 5-digit number in base 2 also is a feasible solution. Therefore, the above process of creating children never results in a *miscarriage* (an infeasible solution). However, if the upper bound on x were, say, $x \le 25$ instead, then miscarriages would occur occasionally. Whenever a miscarriage occurs, the solution is discarded and the entire process of creating a child is repeated until a feasible solution is obtained.

This example includes only a single variable. For a nonlinear programming problem with multiple variables, each member of the population again would use base 2 to show the value of each variable. The above process of generating children from parents then would be done in the same way one variable at a time.

Table 13.7 shows the application of the complete algorithm to this example through both the initialization step (part a of the table) and iteration 1 (part b of the table). In the initialization step, each of the members of the initial population were generated by generating five random numbers and using the correspondence between a random number and a binary digit given earlier to obtain the five binary digits in turn. The corresponding value of x in base 10 then is plugged into the objective function given at the beginning of Sec. 13.1 to evaluate the fitness of that member of the population.

The five members of the initial population that have the highest degree of fitness (in order) are members 10, 8, 4, 1, and 7. To randomly select four of these members to become parents, a random number is used to select one member to be rejected, where 0.0000–0.1999 corresponds to ejecting the first member listed (member 10), 0.2000–0.3999 corresponds to rejecting the second member, and so forth. In this case, the random number was 0.9665, so the fifth member listed (member 7) does not become a parent.

From among the five less fit members of the initial population (members 2, 1, 6, 5, and 9), random numbers now are used to select which two of these members will become parents. In this case, the random numbers were 0.5634 and 0.1270. For the first random

■ **TABLE 13.7** Application of the genetic algorithm to the integer nonlinear programming example through (a) the initialization step and (b) iteration 1

	Member	Initial Population	Value of x	Fitness
	1	0 1 1 1 1	15	3,628,125
	2	0 0 1 0 0	4	3,234,688
	3	0 1 0 0 0	8	3,055,616
	4	1 0 1 1 1	23	3,962,091
(a)	5	0 1 0 1 0	10	2,950,000
	6	0 1 0 0 1	9	2,978,613
	7	0 0 1 0 1	5	3,303,125
	8	1 0 0 1 0	18	4,239,216
	9	1 1 1 1 0	30	1,350,000
	10	1 0 1 0 1	21	4,353,187

	Member	Parents	Children	Value of x	Fitness
	10	1 0 1 0 1	0 0 1 0 1	5	3,303,125
	2	0 0 1 0 0	1 0 0 0 1	17	4,064,259
(b)	8	1 0 0 1 0	1 0 0 1 1	19	4,357,164
	4	1 0 1 1 1	1 0 1 0 0	20	4,400,000
	1	0 1 1 1 1	0 1 0 1 1	11	2,980,637
	6	0 1 0 0 1	0 1 1 1 1	15	3,628,125

number, 0.0000–0.1999 corresponds to selecting the first member listed (member 2), 0.2000–0.3999 corresponds to selecting the second member, and so forth, so the third member listed (member 6) is the one selected in this case. Since only four members (2, 1, 5, and 9) now remain for selecting the last parent, the corresponding intervals for the second random number are 0.0000–0.2499, 0.2500–0.4999, 0.5000–0.7499, and 0.7500–0.9999. Because 0.1270 falls in the first of these intervals, the first remaining member listed (member 2) is selected to be a parent.

The next step is to pair up the six parents—members 10, 8, 4, 1, 6, and 2. Let us begin by using a random number to determine the mate of the first member listed (member 10). The random number 0.8204 indicated that it should be paired up with the fifth of the other five parents listed (member 2). To pair up the next member listed (member 8), the next random number was 0.0198, which is in the interval 0.0000–0.3333, so the first of the three remaining parents listed (member 4) is chosen to be the mate of member 8. This then leaves the two remaining parents (members 1 and 6) to become the last couple.

Part (b) of Table 13.7 shows the children that were reproduced by these parents by using the process illustrated earlier in this subsection. Note that mutations occurred in the third gene of the second child and the fourth gene of the fourth child. By and large, the six children have a relatively high degree of fitness. In fact, for each pair of parents, both of the children turned out to be more fit than one of the parents. This does not always occur, but is fairly common. In the case of the second pair of parents, both of the children happen to be more fit than both parents. Fortuitously, both of these children ($x = 19$ and $x = 20$) actually are superior to *any* of the members of the preceding population given in part (a) of the table. To form the new population for the next iteration, all six children are retained along with the four most fit members of the preceding population (members 10, 8, 4, and 1).

Subsequent iterations would proceed in a similar fashion. Since we know from the discussion in Sec. 13.1 (see Fig. 13.1) that $x = 20$ (the best trial solution generated in iteration 1) actually is the optimal solution for this example, subsequent iterations would not provide any further improvement. Therefore, the stopping rule would terminate the algorithm after five more iterations and provide $x = 20$ as the final solution.

Your IOR Tutorial includes a procedure for applying this same genetic algorithm to other very small integer nonlinear programming problems. (The form and size restrictions are the same as specified in Sec. 13.3 for nonlinear programming problems.)

You might find it interesting to apply this procedure in IOR Tutorial to this same example. Because of the randomness inherent in the algorithm, different intermediate results are obtained each time that it is applied. (Problem 13.4-3 asks you to apply the algorithm to this example several times.)

Although this was a discrete example, genetic algorithms can also be applied to continuous problems such as a nonlinear programming problem without an integer constraint. In this case, the value of a continuous variable would be represented (or closely approximated) by a decimal number in base 2. For example, $x = 23\frac{5}{8}$ is 10111.10100 in base 2, and $x = 23.66$ is closely approximated by 10111.10101 in base 2. All the binary digits on both sides of the decimal point can be treated just as before to have parents reproduce children, and so forth.

The Traveling Salesman Problem Example

Sections 13.2 and 13.3 illustrated how a tabu search algorithm and a simulated annealing algorithm would be applied to the particular traveling salesman problem introduced in Sec. 13.1 (see Fig. 13.4). Now let us see how our genetic algorithm can be applied using this same example.

Rather than using binary digits in this case, we will continue to represent each solution (tour) in the natural way as a sequence of cities visited. For example, the first solution considered in Sec. 13.1 is the tour of the cities in the following order: 1-2-3-4-5-6-7-1, where city 1 is the home base where the tour must begin and end. We should point out, however, that genetic algorithms for traveling salesman problems frequently use other methods for *encoding* solutions. In general, clever methods of representing solutions (often by using strings of binary digits) can make it easier to generate children, create mutations, maintain feasibility, and so forth, in a natural way. The development of an appropriate *encoding scheme* is a key part of developing an effective genetic algorithm for any application.

A complication with this particular example is that, in a sense, it is too easy. Because of the rather limited number of links between pairs of cities in Fig. 13.4, this problem barely has 10 distinct feasible solutions if we rule out a tour that is simply a previously considered tour in the reverse direction. Therefore, it is not possible to have an initial population with 10 distinct trial solutions such that the resulting six parents then reproduce distinct children that also are distinct from the members of the initial population (including the parents).

Fortunately, a genetic algorithm can still operate reasonably well when there is a modest amount of duplication in the trial solutions in a population or in two consecutive populations. For example, even when both parents in a couple are identical, it still is possible for their children to differ from the parents because of mutations.

The genetic algorithm for traveling salesman problems in your IOR Tutorial does not do anything to avoid duplication in the trial solutions considered. Each of the 10 trial solutions in the initial population is generated in turn as follows. Starting from the home base city, random numbers are used to select the next city from among those that have a link to the home base city (cities 2, 3, and 7 in Fig. 13.4). Random numbers then are used to select the third city from among the remaining cities that have a link to the second city. This process is continued until either every city is included once in the tour (plus a return to the home base city from the last city) or a dead end is reached because there is no link from the current city to any of the remaining cities that still need to be visited. In the latter case, the entire process for generating a trial solution is restarted from the beginning with new random numbers.

Random numbers are also used to reproduce children from a pair of parents. To illustrate this process, consider the following pair of parents.

P1:   1-2-3-4-5-6-7-1
P2:   1-2-4-6-5-7-3-1

As we describe the process of generating a child from these parents, we also summarize the results in Table 13.8 to help you follow the progression.

■ **TABLE 13.8** Illustration of the process of generating a child for the traveling salesman problem example

Parent P1:	1-2-3-4-5-6-7-1		
Parent P2:	1-2-4-6-5-7-3-1		
Link	**Options**	**Random Selection**	**Tour**
1	1-2, 1-7, 1-2, 1-3	1-2	1-2
2	2-3, 2-4	2-4	1-2-4
3	4-3, 4-5, 4-6	4-3	1-2-4-3
4	3-5*, 3-7	3-5*	1-2-4-3-5
5	5-6, 5-6, 5-7	5-6	1-2-4-3-5-6
6	6-7	6-7	1-2-4-3-5-6-7
7	7-1	7-1	1-2-4-3-5-6-7-1

*A link that completes a sub-tour reversal

Ignoring the possibility of mutations for the time being, here is the main idea for how to generate a child.

Inheriting Links: Genes correspond to the links in a tour. Therefore, each of the links (genes) inherited by a child should come from one parent or the other (or both). (One other possibility described later is that a parent also can pass down a sub-tour reversal.) These links being inherited are randomly selected one at a time until a complete tour (the child) has been generated.

To start this process with the above parents, since a tour must begin in city 1, a child's initial link must come from one of the parent's links that connect city 1 to another city. For parent P1, these are links 1-2 and 1-7. (Link 1-7 qualifies since it is equivalent to take the tour in either direction.) For parent P2, the corresponding links are 1-2 (again) and 1-3. The fact that both parents have link 1-2 doubles the probability that it will be inherited by a child. Therefore, when using a random number to determine which link the child will inherit, the interval 0.0000–0.4999 (or any interval of this size) corresponds to inheriting link 1-2 whereas the intervals 0.50000–0.7499 and 0.7500–0.9999 then would correspond to the choice of link 1-7 and link 1-3, respectively. Suppose 1-2 is selected, as shown in the first row of Table 13.8. After 1-2, one parent next uses link 2-3 whereas the other uses 2-4. Therefore, in generating the child, a random choice should be made between these two options. Suppose 2-4 is selected. (See the second row of Table 13.8.) There now are three options for the link to follow 1-2-4 because the first parent uses two links (4-3 and 4-5) to connect city 4 in its tour and the second parent uses link 4-6 (link 4-2 is ignored because city 2 already is in the child's tour). When randomly selecting one of these options, suppose 4-3 is chosen to form 1-2-4-3 as the beginning of the child's tour thus far, as shown in the third row of Table 13.8.

We now come to an additional feature of this process for generating a child's tour, namely, using a *sub-tour reversal* from a parent.

Inheriting a Sub-Tour Reversal: One other possibility for a link inherited by a child is a link that is needed to complete a sub-tour reversal that the child's tour is making in a portion of a parent's tour.

To illustrate how this possibility can arise, note that the next city beyond 1-2-4-3 needs to be one of the cities not yet visited (city 5, 6, or 7), but the first parent does not have a link from city 3 to any of these other cities. The reason is that the child is using a sub-tour reversal (reversing 3-4) of this parent's tour, 1-2-3-4-5-6-7-1. Completing this sub-tour reversal requires adding the link 3-5, so this becomes one of the options for the next link in the child's tour. The other option is link 3-7 provided by the second parent (link 3-1 is not an option because city 1 must come at the very end of the tour). One of these two options is selected randomly. Suppose the choice is link 3-5, which provides 1-2-4-3-5 as the child's tour thus far, as shown in the fourth row of Table 13.8.

To continue this tour, the options for the next link are 5-6 (provided by both parents) and 5-7 (provided by the second parent). Suppose that the random choice among 5-6, 5-6, and 5-7 is 5-6, so that the tour thus far is 1-2-4-3-5-6. (See the fifth row of Table 13.8.) Since the only city not yet visited is city 7, link 6-7 is automatically added next, followed by link 7-1 to return to home base. Thus, as shown in the last row of Table 13.8, the complete tour for the child is

C1:   1-2-4-3-5-6-7-1

Figure 13.5 in Sec. 13.1 displays how closely this child resembles the first parent, since the only difference is the sub-tour reversal obtained by reversing 3-4 in the parent.

If link 5-7 had been chosen instead to follow 1-2-4-3-5, the tour would have been completed automatically as 1-2-4-3-5-7-6-1. However, there is no link 6-1 (see Fig. 13.4), so a dead end is reached at city 6. When this happens, a *miscarriage* occurs and the entire process needs to be restarted from the beginning with new random numbers until a child with a complete tour is obtained. Then this process is repeated to obtain the second child.

We now need to add one more feature—the possibility of mutations—to complete the description of the process of generating children.

> **Mutations of Inherited Links:** Whenever a particular link normally would be inherited from a parent of a child, there is a small possibility that a mutation will occur that will reject that link and instead randomly select one of the other links from the current city to another city not already on the tour, regardless of whether that link is used by either parent.

Our genetic algorithm for traveling salesman problems implemented in your IOR Tutorial uses a probability of 0.1 that a mutation will occur each time the next link in the child's tour needs to be selected. Thus, whenever the corresponding random number is less than 0.1000, the choice of the link made in the normal manner described above is rejected (if any other possible choice exists). Instead, all the other links from the current city to a city not already in the tour (including links not provided by either parent) are identified, and one of these links is randomly selected to be the next link in the tour. For example, suppose that a mutation occurs when generating the very first link for the child. Even though 1-2 had been the random choice as the first link, this link now would be rejected because of the mutation. Since city 1 also has links to cities 3 and 7 (see Fig. 13.4), either link 1-3 or link 1-7 would be randomly selected to be the first tour. (Since the parents end their tours by using one or the other of these links, this can be viewed in this case as starting the child's tour by reversing the direction of one of the parents' tours.)

We now can outline the general procedure for generating a child from a pair of parents.

Procedure for Generating a Child

1. **Initialization:** To start, designate the home base city as the *current city*.
2. **Options for the next link:** Identify all the links from the current city to another city not already in the child's tour that are used by either parent in either direction. Also, add any link that is needed to complete a sub-tour reversal that the child's tour is making in a portion of a parent's tour.
3. **Selection of the next link:** Use a random number to randomly select one of the options identified in step 2.
4. **Check for a mutation:** If the next random number is less than 0.1000, a mutation occurs and the link selected in step 3 is rejected (unless there is no other link from the current city to another city not already in the tour). If the link is rejected, identify all the other links from the current city to another city not already in the tour (including links not used by either parent). Use a random number to randomly select one of these other links.
5. **Continuation:** Add the link selected in step 3 (if no mutation occurs) or in step 4 (if a mutation occurs) to the end of the child's current incomplete tour and redesignate the city at the end of this link as the *current city*. If there still remains more than one city not included on the tour (plus the return to the home base city), return to steps 2–4 to select the next link. Otherwise, go to step 6.
6. **Completion:** With only one city remaining that has not yet been added to the child's tour, add the link from the current city to this remaining city. Then add the link from this last city back to the home base city to complete the tour for the child. However, if the needed link does not exist, a miscarriage occurs and the procedure must restart again from step 1.

This procedure is applied for each pair of parents to obtain each of their two children.

The genetic algorithm for traveling salesman problems in your IOR Tutorial incorporates this procedure for generating children as part of the overall algorithm outlined near the beginning of this section. Table 13.9 shows the results from applying this algorithm to the example through the initialization step and the first iteration of the overall algorithm. Because of the randomness built into the algorithm, its intermediate results (and perhaps the final best solution as well) will vary each time the algorithm is run to its completion. (To explore this further, Prob. 13.4-7 asks you to use your IOR Tutorial to apply the complete algorithm to this example several times.)

The fact that the example has only a relatively small number of distinct feasible solutions is reflected in the results shown in Table 13.9. Members 1, 4, 6, and 10 are identical, as are members 2, 7, and 9 (except that member 2 takes its tour in the reverse direction). Therefore, the random generation of the 10 members of the initial population resulted in only five distinct feasible solutions. Similarly, four of the six children generated (members 12, 14, 15, and 16) are identical to one of its parents (except that member 14 takes its tour in the opposite direction of its first parent). Two of the children (members 12 and 15) have a better fitness (shorter distance) than one of its parents, but neither improved upon both of its parents. None of these children provide an optimal solution (which has a distance of 63). This illustrates the fact that a genetic algorithm may require many generations (iterations) on some problems before the survival-of-the-fittest phenomenon results in clearly superior populations.

The Worked Examples section of the book's website provides **another example** of applying this genetic algorithm to a traveling salesman problem. This problem has a somewhat larger number of distinct feasible solutions than the above example, so there is a greater diversity in its initial population, the resulting parents, and their children.

■ **TABLE 13.9** One application of the genetic algorithm in IOR Tutorial to the traveling salesman problem example through (*a*) the initialization step and (*b*) iteration 1

	Member	Initial Population	Distance
	1	1-2-4-6-5-3-7-1	64
	2	1-2-3-5-4-6-7-1	65
	3	1-7-5-6-4-2-3-1	65
	4	1-2-4-6-5-3-7-1	64
(a)	5	1-3-7-5-6-4-2-1	66
	6	1-2-4-6-5-3-7-1	64
	7	1-7-6-4-5-3-2-1	65
	8	1-3-7-6-5-4-2-1	69
	9	1-7-6-4-5-3-2-1	65
	10	1-2-4-6-5-3-7-1	64

	Member	Parents	Children	Member	Distance
	1	1-2-4-6-5-3-7-1	1-2-4-5-6-7-3-1	11	69
	7	1-7-6-4-5-3-2-1	1-2-4-6-5-3-7-1	12	64
(b)	2	1-2-3-5-4-6-7-1	1-2-4-5-6-7-3-1	13	69
	6	1-2-4-6-5-3-7-1	1-7-6-4-5-3-2-1	14	65
	4	1-2-4-6-5-3-7-1	1-2-4-6-5-3-7-1	15	64
	5	1-3-7-5-6-4-2-1	1-3-7-5-6-4-2-1	16	66

■ 13.5 CONCLUSIONS

Some optimization problems (including various combinatorial optimization problems) are sufficiently complex that it may not be possible to solve for an optimal solution with the kinds of exact algorithms presented in previous chapters. In such cases, heuristic methods are commonly used to search for a good (but not necessarily optimal) feasible solution. Several metaheuristics are available that provide a general structure and strategy guidelines for designing a specific heuristic method to fit a particular problem. A key feature of these metaheuristic procedures is their ability to escape from local optima and perform a robust search of a feasible region.

This chapter has introduced the three most prominent types of metaheuristics. *Tabu search* moves from the current trial solution to the best neighboring trial solution at each iteration, much like a local improvement procedure, except that it allows a nonimproving move when an improving move is not available. It then incorporates short-term memory of the past search to encourage moving toward new parts of the feasible region rather than cycling back to previously considered solutions. In addition, it may employ intensification and diversification strategies based on long-term memory to focus the search on promising continuations. *Simulated annealing* also moves from the current trial solution to a neighboring trial solution at each iteration while occasionally allowing nonimproving moves. However, it selects the neighboring trial solution randomly and then uses the analogy to a physical annealing process to determine if this neighbor should be rejected as the next trial solution if it is not as good as the current trial solution. The third type of metaheuristic, *genetic algorithms,* works with an entire population of trial solutions at each iteration. It then uses the analogy to the biological theory of evolution, including the concept of survival of the fittest, to discard some of the trial solutions (especially the poorer ones) and replace them by some new ones. This replacement process has pairs of surviving members of the population pass on some of their features to pairs of new members just as if they were parents reproducing children.

For the sake of concreteness, we have described one basic algorithm for each metaheuristic and then adapted this algorithm to two specific types of problems (including the traveling salesman problem), using simple examples. However, many variations of each algorithm also have been developed by researchers and used by practitioners to better fit the characteristics of the complex problems being addressed. For example, literally dozens of variations of the basic genetic algorithm for traveling salesman problems presented in Sec. 13.4 (including different procedures for generating children) have been proposed, and research is continuing to determine what is most effective. (Some of the best methods for traveling salesman problems use special "k-opt" and "ejection chain" strategies that are carefully tailored to take advantage of the problem structure.) Therefore, the important lessons from this chapter are the basic concepts and intuition incorporated into each metaheuristic rather than the details of the particular algorithms presented here.

There are several other important types of metaheuristics in addition to the three that are featured in this chapter. These include, for example, ant colony optimization, scatter search, and artificial neural networks. (These suggestive names give a hint of the key idea that drives each of these metaheuristics.) Selected Reference 4 provides a thorough coverage of both these other metaheuristics and the three presented here. (Michel Gendreau and Jean-Yves Potvin are preparing a second edition to update this important reference.)

Some heuristic algorithms actually are a hybrid of different types of metaheuristics in order to combine their better features. For example, short-term tabu search (without a diversification component) is very good at finding local optima but not as good at thoroughly exploring the various parts of a feasible region to find the part containing the global optimum, whereas a genetic algorithm has the opposite characteristics. Therefore, an improved algorithm sometimes can be obtained by beginning with a genetic algorithm to try to find the tallest hills (when the objective is maximization) and then switch to a basic tabu search at the very end to climb quickly to the top of these hills. The key for designing an effective heuristic algorithm is to incorporate whatever ideas work best for the problem at hand rather than adhering rigidly to the philosophy of a particular metaheuristic.

■ SELECTED REFERENCES

1. Coello, C., D. A. Van Veldhuizen, and G. B. Lamont: *Evolutionary Algorithms for Solving Multi-Objective Problems,* Kluwer Academic Publishers (now Springer), Boston, 2002.
2. Gen, M., and R. Cheng, *Genetic Algorithms and Engineering Optimization,* Wiley, New York, 2000.
3. Glover, F.: "Tabu Search: A Tutorial," *Interfaces,* **20**(4): 74–94, July–August 1990.
4. Glover, F., and G. Kochenberger (eds.): *Handbook of Metaheuristics,* Kluwer Academic Publishers (now Springer), Boston, MA, 2003. (This reference provides a thorough coverage of all the metaheuristics considered in this chapter, as well as some other metaheuristics.)
5. Glover, F., and M. Laguna: *Tabu Search,* Kluwer Academic Publishers (now Springer), Boston, MA, 1997.
6. Gutin, G., and A. Punnen (eds.): *The Traveling Salesman Problem and Its Variations,* Kluwer Academic Publishers (now Springer), Boston, MA, 2002.
7. Haupt, R. L., and S. E. Haupt: *Practical Genetic Algorithms,* Wiley, New York, 1998.
8. Jones, D. F., S. K. Mirrazavi, and M. Tamiz: "Multiobjective Metaheuristics: An Overview of the Current State of the Art," *European Journal of Operational Research,* **137:** 1–9, 2002.
9. Laguna, M., and R. Marti: *Scatter Search: Methodology and Implementations in C,* Kluwer Academic Publishers (now Springer), Boston, 2003.
10. Michalewicz, Z., and D. B. Fogel: *How To Solve It: Modern Heuristics,* Springer, Berlin, 2002.
11. Mitchell, M.: *An Introduction to Genetic Algorithms,* MIT Press, Cambridge, MA, 1998.
12. Molina, J., M. Laguna, R. Marti, and R. Caballero: "SSPMO: A Scatter Tabu Search Procedure for Non-Linear Multiobjective Optimization," *INFORMS Journal on Computing,* **19**(1): 91–100, Winter 2007.
13. Reeves, C. R.: "Genetic Algorithms for the Operations Researcher," *INFORMS Journal on Computing,* **9:** 231–250, 1997. (Also see pp. 251–265 for commentaries on this feature article.)
14. Sarker, R., M. Mohammadian, and X. Yao (eds.): *Evolutionary Optimization,* Kluwer Academic Publishers (now Springer), Boston, MA, 2002.

■ LEARNING AIDS FOR THIS CHAPTER ON OUR WEBSITE (www.mhhe.com/hillier)

Worked Examples:

Examples for Chapter 13

Automatic Procedures in IOR Tutorial:

Tabu Search Algorithm for Traveling Salesman Problems
Simulated Annealing Algorithm for Traveling Salesman Problems

Simulated Annealing Algorithm for Nonlinear Programming Problems
Genetic Algorithm for Integer Nonlinear Programming Problems
Genetic Algorithm for Traveling Salesman Problems

Glossary for Chapter 13

See Appendix 1 for documentation of the software.

■ PROBLEMS

The symbol A to the left of some of the problems (or their parts) has the following meaning.

> A: You should use the corresponding automatic procedure in IOR Tutorial. The printout will record the results obtained at each iteration.

An asterisk on the problem number indicates that at least a partial answer is given in the back of the book.

Instructions for Obtaining Random Numbers

For each problem or its part where random numbers are needed, obtain them from the consecutive random digits in Table 20.3 in Sec. 20.3 as follows. Start from the front of the top row of the table and form *five-digit* random numbers by placing a decimal point in front of each group of five random digits (0.09656, 0.96657, etc.) in the order that you need random numbers. Always restart from the front of the top row for each new problem or its part.

13.1-1. Consider the traveling salesman problem shown below, where city 1 is the home city.

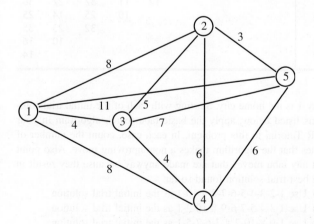

(a) List all the possible tours, except exclude those that are simply the reverse of previously listed tours. Calculate the distance of each of these tours and thereby identify the optimal tour.
(b) Starting with 1-2-3-4-5-1 as the initial trial solution, apply the sub-tour reversal algorithm to this problem.
(c) Apply the sub-tour reversal algorithm to this problem when starting with 1-2-4-3-5-1 as the initial trial solution.

(d) Apply the sub-tour reversal algorithm to this problem when starting with 1-4-2-3-5-1 as the initial trial solution.

13.1-2. Reconsider the example of a traveling salesman problem shown in Fig. 13.4.
(a) When the sub-tour reversal algorithm was applied to this problem in Sec. 13.1, the first iteration resulted in a tie for which of two sub-tour reversals (reversing 3-4 or 4-5) provided the largest decrease in the distance of the tour, so the tie was broken arbitrarily in favor of the first reversal. Determine what would have happened if the second of these reversals (reversing 4-5) had been chosen instead.
(b) Apply the sub-tour reversal algorithm to this problem when starting with 1-2-4-5-6-7-3-1 as the initial trial solution.

13.1-3. Consider the traveling salesman problem shown below, where city 1 is the home city.

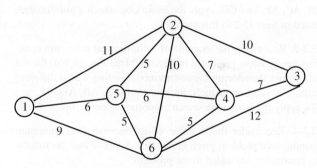

(a) List all the possible tours, except exclude those that are simply the reverse of previously listed tours. Calculate the distance of each of these tours and thereby identify the optimal solution.
(b) Starting with 1-2-3-4-5-6-1 as the initial trial solution, apply the sub-tour reversal algorithm to this problem.
(c) Apply the sub-tour reversal algorithm to this problem when starting with 1-2-5-4-3-6-1 as the initial trial solution.

13.2-1. Read the referenced article that fully describes the OR study summarized in the application vignette presented in Sec. 13.2. Briefly describe how tabu search was applied in this study. Then list the various financial and nonfinancial benefits that resulted from this study.

13.2-2.* Consider the minimum spanning tree problem depicted below, where the dashed lines represent the potential links that could be inserted into the network and the number next to each dashed line represents the cost associated with inserting that particular link.

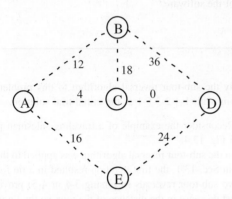

This problem also has the following two constraints:

Constraint 1: No more than one of the three links—AB, BC, and AE—can be included.

Constraint 2: Link AB can be included only if link BD also is included.

Starting with the initial trial solution where the inserted links are AB, AC, AE, and CD, apply the basic tabu search algorithm presented in Sec. 13.2 to this problem.

13.2-3. Reconsider the example of a constrained minimum spanning tree problem presented in Sec. 13.2 (see Fig. 13.7(a) for the data before introducing the constraints). Starting with a different initial trial solution, namely, the one with links AB, AD, BE, and CD, apply the basic tabu search algorithm again to this problem.

13.2-4. Reconsider the example of an unconstrained minimum spanning tree problem given in Sec. 9.4. Suppose that the following constraints are added to the problem.

Constraint 1: Either link AD or link ET must be included.

Constraint 2: At most one of the three links—AO, BC, and DE—can be included.

Starting with the optimal solution for the unconstrained problem given at the end of Sec. 9.4 as the initial trial solution, apply the basic tabu search algorithm to this problem.

13.2-5. Reconsider the traveling salesman problem shown in Prob. 13.1-1. Starting with 1-5-3-2-4-1 as the initial trial solution, apply the basic tabu search algorithm by hand to this problem.

A **13.2-6.** Consider the 8-city traveling salesman problem whose links have the associated distances shown in the following table (where a dash indicates the absence of a link).

City	2	3	4	5	6	7	8
1	14	15	—	—	—	—	17
2		13	14	20	—	—	21
3			11	21	17	9	9
4				11	10	8	20
5					15	18	—
6						9	—
7							13

City 1 is the home city. Starting with each of the initial trial solutions listed below, apply the basic tabu search algorithm in your IOR Tutorial to this problem. In each case, count the number of times that the algorithm makes a nonimproving move. Also point out any tabu moves that are made anyway because they result in the best trial solution found so far.

(a) Use 1-2-3-4-5-6-7-8-1 as the initial trial solution.

(b) Use 1-2-5-6-7-4-8-3-1 as the initial trial solution.

(c) Use 1-3-2-5-6-4-7-8-1 as the initial trial solution.

A **13.2-7.** Consider the 10-city traveling salesman problem whose links have the associated distances shown in the following table.

City	2	3	4	5	6	7	8	9	10
1	13	25	15	21	9	19	18	8	15
2		26	21	29	21	31	23	16	10
3			11	18	23	28	44	34	35
4				10	13	19	34	24	29
5					12	11	37	27	36
6						10	25	14	25
7							32	23	35
8								10	16
9									14

City 1 is the home city. Starting with each of the initial trial solutions listed below, apply the basic tabu search algorithm in your IOR Tutorial to this problem. In each case, count the number of times that the algorithm makes a nonimproving move. Also point out any tabu moves that are made anyway because they result in the best trial solution found so far.

(a) Use 1-2-3-4-5-6-7-8-9-10-1 as the initial trial solution.

(b) Use 1-3-4-5-7-6-9-8-10-2-1 as the initial trial solution.

(c) Use 1-9-8-10-2-4-3-6-7-5-1 as the initial trial solution.

13.3-1. While applying a simulated annealing algorithm to a certain problem, you have come to an iteration where the current value of T is $T = 2$ and the value of the objective function for the current trial solution is 30. This trial solution has four immediate neighbors and their objective function values are 29, 34, 31, and 24. For each of these four immediate neighbors in turn, you wish to determine the probability that the move selection rule would accept this immediate neighbor if it is randomly selected to become the current candidate to be the next trial solution.

(a) Determine this probability for each of the immediate neighbors when the objective is *maximization* of the objective function.

(b) Determine this probability for each of the immediate neighbors when the objective is *minimization* of the objective function.

A **13.3-2.** Because of its use of random numbers, a simulated annealing algorithm will provide slightly different results each time it is run. Table 13.5 shows one application of the basic simulated annealing algorithm in IOR Tutorial to the example of a traveling salesman problem depicted in Fig. 13.4. Starting with the same initial trial solution (1-2-3-4-5-6-7-1), use your IOR Tutorial to apply this same algorithm to this same example five more times. How many times does it again find the optimal solution (1-3-5-7-6-4-2-1 or, equivalently, 1-2-4-6-7-5-3-1)?

13.3-3. Reconsider the traveling salesman problem shown in Prob. 13.1-1. Using 1-4-2-3-5-1 as the initial trial solution, you are to follow the instructions below for applying the basic simulated annealing algorithm presented in Sec. 13.3 to this problem.

(a) Perform the first iteration by hand. Follow the instructions given at the beginning of the Problems section to obtain the needed random numbers. Show your work, including the use of the random numbers.

A **(b)** Use your IOR Tutorial to apply this algorithm. Observe the progress of the algorithm and record for each iteration how many (if any) candidates to be the next trial solution are rejected before one is accepted. Also count the number of iterations where a nonimproving move is accepted.

A **13.3-4.** Follow the instructions of Prob. 13.3-3 for the traveling salesman problem described in Prob. 13.2-6, using 1-2-3-4-5-6-7-8-1 as the initial trial solution.

A **13.3-5.** Follow the instructions of Prob. 13.3-3 for the traveling salesman problem described in Prob. 13.2-7, using 1-9-8-10-2-4-3-6-7-5-1 as the initial trial solution.

A **13.3-6.** Because of its use of random numbers, a simulated annealing algorithm will provide slightly different results each time it is run. Table 13.6 shows one application of the basic simulated annealing algorithm in IOR Tutorial to the nonlinear programming example introduced in Sec. 13.1. Starting with the same initial trial solution ($x = 15.5$), use your IOR Tutorial to apply this same algorithm to this same example five more times. What is the best solution found in these five applications? Is it closer to the optimal solution ($x = 20$ with $f(x) = 4,400,000$) than the best solution shown in Table 13.6?

13.3-7. Consider the following nonconvex programming problem.

$$\text{Maximize} \quad f(x) = x^3 - 60x^2 + 900x + 100,$$

subject to

$$0 \le x \le 31.$$

(a) Use the first and second derivatives of $f(x)$ to determine the critical points (along with the end points of the feasible region) where x is either a local maximum or a local minimum.

(b) Roughly plot the graph of $f(x)$ by hand over the feasible region.

(c) Using $x = 15.5$ as the initial trial solution, perform the first iteration of the basic simulated annealing algorithm presented in Sec. 13.3 by hand. Follow the instructions given at the beginning of the Problems section to obtain the needed random numbers. Show your work, including the use of the random numbers.

A **(d)** Use your IOR Tutorial to apply this algorithm, starting with $x = 15.5$ as the initial trial solution. Observe the progress of the algorithm and record for each iteration how many (if any) candidates to be the next trial solution are rejected before one is accepted. Also count the number of iterations where a nonimproving move is accepted.

13.3-8. Consider the example of a nonconvex programming problem presented in Sec. 12.10 and depicted in Fig. 12.18.

(a) Using $x = 2.5$ as the initial trial solution, perform the first iteration of the basic simulated annealing algorithm presented in Sec. 13.3 by hand. Follow the instructions given at the beginning of the Problems section to obtain the random numbers. Show your work, including the use of the random numbers.

A **(b)** Use your IOR Tutorial to apply this algorithm, starting with $x = 2.5$ as the initial trial solution. Observe the progress of the algorithm and record for each iteration how many (if any) candidates to be the next trial solution are rejected before one is accepted. Also count the number of iterations where a nonimproving move is accepted.

A **13.3-9.** Follow the instructions of Prob. 13.3-8 for the following nonconvex programming problem when starting with $x = 25$ as the initial trial solution.

$$\text{Maximize} \quad f(x) = x^6 - 140x^5 + 7000x^4 - 160,000x^3$$
$$+ 1,600,000x^2 - 5,000,000x,$$

subject to

$$0 \le x \le 50.$$

A **13.3-10.** Follow the instructions of Prob. 13.3-8 for the following nonconvex programming problem when starting with $(x_1, x_2) = (18, 25)$ as the initial trial solution.

$$\text{Maximize} \quad f(x_1, x_2) = x_1^5 - 81x_1^4 + 2330x_1^3 - 28,750x_1^2$$
$$+ 150,000x_1 + 0.5x_2^5 - 65x_2^4$$
$$+ 2950x_2^3 - 53,500x_2^2 + 305,000x_2,$$

subject to

$$x_1 + 2x_2 \le 110$$
$$3x_1 + x_2 \le 120$$

and

$$0 \le x_1 \le 36, \qquad 0 \le x_2 \le 50.$$

13.4-1. For each of the following pairs of parents, generate their two children when applying the basic genetic algorithm presented

■ **TABLE 14.1** Payoff table for the odds and evens game

Strategy		Player 2	
		1	2
Player 1	1	1	−1
	2	−1	1

from the player taking odds (player 2). If the number does not match, player 1 pays $1 to player 2. Thus, each player has two *strategies:* to show either one finger or two fingers. The resulting payoff to player 1 in dollars is shown in the *payoff table* given in Table 14.1.

In general, a two-person game is characterized by

1. The strategies of player 1.
2. The strategies of player 2.
3. The payoff table.

Before the game begins, each player knows the strategies she or he has available, the ones the opponent has available, and the payoff table. The actual play of the game consists of each player simultaneously choosing a strategy without knowing the opponent's choice.

A strategy may involve only a simple action, such as showing a certain number of fingers in the odds and evens game. On the other hand, in more complicated games involving a series of moves, a **strategy** is a predetermined rule that specifies completely how one intends to respond to each possible circumstance at each stage of the game. For example, a strategy for one side in chess would indicate how to make the next move for *every* possible position on the board, so the total number of possible strategies would be astronomical. Applications of game theory normally involve far less complicated competitive situations than chess does, but the strategies involved can be fairly complex.

The **payoff table** shows the gain (positive or negative) for player 1 that would result from each combination of strategies for the two players. It is given only for player 1 because the table for player 2 is just the negative of this one, due to the zero-sum nature of the game.

The entries in the payoff table may be in any units desired, such as dollars, provided that they accurately represent the *utility* to player 1 of the corresponding outcome. However, utility is not necessarily proportional to the amount of money (or any other commodity) when large quantities are involved. For example, $2 million (after taxes) is probably worth much less than twice as much as $1 million to a poor person. In other words, given the choice between (1) a 50 percent chance of receiving $2 million rather than nothing and (2) being sure of getting $1 million, a poor person probably would much prefer the latter. On the other hand, the outcome corresponding to an entry of 2 in a payoff table should be "worth twice as much" to player 1 as the outcome corresponding to an entry of 1. Thus, given the choice, he or she should be indifferent between a 50 percent chance of receiving the former outcome (rather than nothing) and definitely receiving the latter outcome instead.[1]

A primary objective of game theory is the development of *rational criteria* for selecting a strategy. Two key assumptions are made:

1. *Both* players are *rational*.
2. *Both* players choose their strategies solely to *promote their own welfare* (no compassion for the opponent).

---

[1]See Sec. 15.6 for a further discussion of the concept of utility.

Game theory contrasts with *decision analysis* (see Chap. 15), where the assumption is that the decision maker is playing a game with a passive opponent—nature—which chooses its strategies in some random fashion.

We shall develop the standard game theory criteria for choosing strategies by means of illustrative examples. In particular, the end of the next section describes how game theory says the odds and evens game should be played. (Problems 14.3-1, 14.4-1, and 14.5-1 also invite you to apply the techniques developed in this chapter to solve for the optimal way to play this game.) In addition, the next section presents a prototype example that illustrates the formulation of a two-person, zero-sum game and its solution in some simple situations. A more complicated variation of this game is then carried into Sec. 14.3 to develop a more general criterion. Sections 14.4 and 14.5 describe a graphical procedure and a linear programming formulation for solving such games.

14.2 SOLVING SIMPLE GAMES—A PROTOTYPE EXAMPLE

Two politicians are running against each other for the U.S. Senate. Campaign plans must now be made for the final two days, which are expected to be crucial because of the closeness of the race. Therefore, both politicians want to spend these days campaigning in two key cities, Bigtown and Megalopolis. To avoid wasting campaign time, they plan to travel at night and spend either one full day in each city or two full days in just one of the cities. However, since the necessary arrangements must be made in advance, neither politician will learn his (or her)[2] opponent's campaign schedule until after he has finalized his own. Therefore, each politician has asked his campaign manager in each of these cities to assess what the impact would be (in terms of votes won or lost) from the various possible combinations of days spent there by himself and by his opponent. He then wishes to use this information to choose his best strategy on how to use these two days.

Formulation as a Two-Person, Zero-Sum Game

To formulate this problem as a two-person, zero-sum game, we must identify the two *players* (obviously the two politicians), the *strategies* for each player, and the *payoff table*.

As the problem has been stated, each player has the following three strategies:

Strategy 1 = spend one day in each city.
Strategy 2 = spend both days in Bigtown.
Strategy 3 = spend both days in Megalopolis.

By contrast, the strategies would be more complicated in a different situation where each politician learns where his opponent will spend the first day before he finalizes his own plans for his second day. In that case, a typical strategy would be: Spend the first day in Bigtown; if the opponent also spends the first day in Bigtown, then spend the second day in Bigtown; however, if the opponent spends the first day in Megalopolis, then spend the second day in Megalopolis. There would be eight such strategies, one for each combination of the two first-day choices, the opponent's two first-day choices, and the two second-day choices.

Each entry in the payoff table for player 1 represents the *utility* to player 1 (or the negative utility to player 2) of the outcome resulting from the corresponding strategies used by the two players. From the politician's viewpoint, the objective is to *win votes,* and each additional vote (before he learns the outcome of the election) is of equal value to him. Therefore, the appropriate entries for the payoff table for politician 1 are the

---

[2]We use only *his* or only *her* in some examples and problems for ease of reading: we do not mean to imply that only men or only women are engaged in the various activities.

Variation 2 of the Example

Now suppose that the current data give Table 14.4 as the payoff table for player 1 (politician 1). This game does not have dominated strategies, so it is not obvious what the players should do. What line of reasoning does game theory say they should use?

Consider player 1. By selecting strategy 1, he could win 6 or could lose as much as 3. However, because player 2 is rational and thus will seek a strategy that will protect himself from large payoffs to player 1, it seems likely that player 1 would incur a loss by playing strategy 1. Similarly, by selecting strategy 3, player 1 could win 5, but more probably his rational opponent would avoid this loss and instead administer a loss to player 1 which could be as large as 4. On the other hand, if player 1 selects strategy 2, he is guaranteed not to lose anything and he could even win something. Therefore, because it provides the *best guarantee* (a payoff of 0), strategy 2 seems to be a "rational" choice for player 1 against his rational opponent. (This line of reasoning assumes that both players are averse to risking larger losses than necessary, in contrast to those individuals who enjoy gambling for a large payoff against long odds.)

Now consider player 2. He could lose as much as 5 or 6 by using strategy 1 or 3, but is guaranteed at least breaking even with strategy 2. Therefore, by the same reasoning of seeking the best guarantee against a rational opponent, his apparent choice is strategy 2.

If both players choose their strategy 2, the result is that both break even. Thus, in this case, neither player improves upon his best guarantee, but both also are forcing the opponent into the same position. Even when the opponent deduces a player's strategy, the opponent cannot exploit this information to improve his position. Stalemate.

The end product of this line of reasoning is that each player should play in such a way as to *minimize his maximum losses* whenever the resulting choice of strategy cannot be exploited by the opponent to then improve his position. This so-called **minimax criterion** is a standard criterion proposed by game theory for selecting a strategy. In effect, this criterion says to select a strategy that would be best even if the selection were being announced to the opponent before the opponent chooses a strategy. In terms of the payoff table, it implies that *player 1* should select the strategy whose *minimum payoff* is *largest,* whereas *player 2* should choose the one whose *maximum payoff to player 1* is the *smallest.* This criterion is illustrated in Table 14.4, where strategy 2 is identified as the *maximin strategy* for player 1 and strategy 2 is the *minimax strategy* for player 2. The resulting payoff of 0 is the value of the game, so this is a fair game.

Notice the interesting fact that the same entry in this payoff table yields both the maximin and minimax values. The reason is that this entry is both the minimum in its row and the maximum of its column. The position of any such entry is called a **saddle point.**

■ **TABLE 14.4** Payoff table for player 1 for variation 2 of the political campaign problem

Strategy		Player 2			Minimum
		1	2	3	
	1	−3	−2	6	−3
Player 1	2	2	0	2	0 ← Maximin value
	3	5	−2	−4	−4
Maximum:		5	0	6	
			↑		
			Minimax value		

The fact that this game possesses a saddle point was actually crucial in determining how it should be played. Because of the saddle point, neither player can take advantage of the opponent's strategy to improve his own position. In particular, when player 2 predicts or learns that player 1 is using strategy 2, player 2 would incur a loss instead of breaking even if he were to change from his original plan of using his strategy 2. Similarly, player 1 would only worsen his position if he were to change his plan. Thus, neither player has any motive to consider changing strategies, either to take advantage of his opponent or to prevent the opponent from taking advantage of him. Therefore, since this is a **stable solution** (also called an *equilibrium solution*), players 1 and 2 should exclusively use their maximin and minimax strategies, respectively.

As the next variation illustrates, some games do not possess a saddle point, in which case a more complicated analysis is required.

Variation 3 of the Example

Late developments in the campaign result in the final payoff table for player 1 (politician 1) given by Table 14.5. How should this game be played?

Suppose that both players attempt to apply the minimax criterion in the same way as in variation 2. Player 1 can guarantee that he will lose no more than 2 by playing strategy 1. Similarly, player 2 can guarantee that he will lose no more than 2 by playing strategy 3.

However, notice that the maximin value (-2) and the minimax value (2) do not coincide in this case. The result is that there is *no saddle point.*

What are the resulting consequences if both players plan to use the strategies just derived? It can be seen that player 1 would win 2 from player 2, which would make player 2 unhappy. Because player 2 is rational and can therefore foresee this outcome, he would then conclude that he can do much better, actually winning 2 rather than losing 2, by playing strategy 2 instead. Because player 1 is also rational, he would anticipate this switch and conclude that he can improve considerably, from -2 to 4, by changing to strategy 2. Realizing this, player 2 would then consider switching back to strategy 3 to convert a loss of 4 to a gain of 3. This possibility of a switch would cause player 1 to consider again using strategy 1, after which the whole cycle would start over again. Therefore, even though this game is being played only once, *any* tentative choice of a strategy leaves that player with a motive to consider changing strategies, either to take advantage of his opponent or to prevent the opponent from taking advantage of him.

In short, the originally suggested solution (player 1 to play strategy 1 and player 2 to play strategy 3) is an **unstable solution,** so it is necessary to develop a more satisfactory solution. But what kind of solution should it be?

TABLE 14.5 Payoff table for player 1 for variation 3 of the political campaign problem

		Player 2			
Strategy		1	2	3	Minimum
	1	0	-2	2	-2 ← Maximin value
Player 1	2	5	4	-3	-3
	3	2	3	-4	-4
Maximum:		5	4	2	
				↑	
				Minimax value	

to randomly select the pure strategy to be used from the probability distribution for the optimal mixed strategy. (Valid statistical procedures for making such a random selection are discussed in Sec. 20.4.)

Now we need to show how to find the optimal mixed strategy for each player. There are several methods of doing this. One is a graphical procedure that may be used whenever one of the players has only two (undominated) pure strategies; this approach is described in the next section. When larger games are involved, the usual method is to transform the problem to a linear programming problem that then can be solved by the simplex method on a computer; Sec. 14.5 discusses this approach.

14.4 GRAPHICAL SOLUTION PROCEDURE

Consider any game with mixed strategies such that, after dominated strategies are eliminated, one of the players has only two pure strategies. To be specific, let this player be player 1. Because her mixed strategies are (x_1, x_2) and $x_2 = 1 - x_1$, it is necessary for her to solve only for the optimal value of x_1. However, it is straightforward to plot the expected payoff as a function of x_1 for each of her opponent's pure strategies. This graph can then be used to identify the point that maximizes the minimum expected payoff. The opponent's minimax mixed strategy can also be identified from the graph.

To illustrate this procedure, consider variation 3 of the political campaign problem (see Table 14.5). Notice that the third pure strategy for player 1 is dominated by her second, so the payoff table can be reduced to the form given in Table 14.6. Therefore, for each of the pure strategies available to player 2, the expected payoff for player 1 will be:

(y_1, y_2, y_3)	**Expected Payoff**
(1, 0, 0)	$0x_1 + 5(1 - x_1) = 5 - 5x_1$
(0, 1, 0)	$-2x_1 + 4(1 - x_1) = 4 - 6x_1$
(0, 0, 1)	$2x_1 - 3(1 - x_1) = -3 + 5x_1$

Now plot these expected-payoff lines on a graph, as shown in Fig. 14.1. For any given values of x_1 and (y_1, y_2, y_3), the expected payoff will be the appropriate weighted average of the corresponding points on these three lines. In particular,

$$\text{Expected payoff for player 1} = y_1(5 - 5x_1) + y_2(4 - 6x_1) + y_3(-3 + 5x_1).$$

Remember that player 2 wants to minimize this expected payoff for player 1. Given x_1, player 2 can minimize this expected payoff by choosing the pure strategy that corresponds to the "bottom" line for that x_1 in Fig. 14.1 (either $-3 + 5x_1$ or $4 - 6x_1$, but never $5 - 5x_1$). According to the minimax (or maximin) criterion, player 1 wants to maximize

TABLE 14.6 Reduced payoff table for player 1 for variation 3 of the political campaign problem

			Player 2		
		Probability	y_1	y_2	y_3
	Probability	**Pure Strategy**	**1**	**2**	**3**
Player 1	x_1	1	0	-2	2
	$1 - x_1$	2	5	4	-3

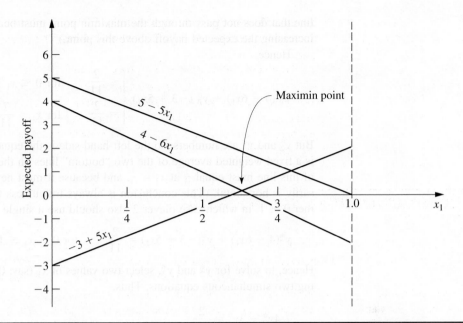

FIGURE 14.1
Graphical procedure
for solving games.

this minimum expected payoff. Consequently, player 1 should select the value of x_1 where the bottom line peaks, i.e., where the $(-3 + 5x_1)$ and $(4 - 6x_1)$ lines intersect, which yields an expected payoff of

$$\underline{v} = v = \max_{0 \le x_1 \le 1} \{\min\{-3 + 5x_1, 4 - 6x_1\}\}.$$

To solve algebraically for this optimal value of x_1 at the intersection of the two lines $-3 + 5x_1$ and $4 - 6x_1$, we set

$$-3 + 5x_1 = 4 - 6x_1,$$

which yields $x_1 = \frac{7}{11}$. Thus, $(x_1, x_2) = (\frac{7}{11}, \frac{4}{11})$ is the *optimal mixed strategy* for player 1, and

$$\underline{v} = v = -3 + 5\left(\frac{7}{11}\right) = \frac{2}{11}$$

is the value of the game.

To find the corresponding optimal mixed strategy for player 2, we now reason as follows. According to the definition of the minimax value $\bar{v}$ and the minimax theorem, the expected payoff resulting from the optimal strategy $(y_1, y_2, y_3) = (y_1^*, y_2^*, y_3^*)$ will satisfy the condition

$$y_1^*(5 - 5x_1) + y_2^*(4 - 6x_1) + y_3^*(-3 + 5x_1) \le \bar{v} = v = \frac{2}{11}$$

for all values of x_1 $(0 \le x_1 \le 1)$. Furthermore, when player 1 is playing optimally (that is, $x_1 = \frac{7}{11}$), this inequality will be an equality (by the minimax theorem), so that

$$\frac{20}{11}y_1^* + \frac{2}{11}y_2^* + \frac{2}{11}y_3^* = v = \frac{2}{11}.$$

Because (y_1, y_2, y_3) is a probability distribution, it is also known that

$$y_1^* + y_2^* + y_3^* = 1.$$

Therefore, $y_1^* = 0$ because $y_1^* > 0$ would violate the next-to-last equation; i.e., the expected payoff on the graph at $x_1 = \frac{7}{11}$ would be above the maximin point. (In general, any line that does not pass through the maximin point must be given a zero weight to avoid increasing the expected payoff above this point.)

Hence,

$$y_2^*(4 - 6x_1) + y_3^*(-3 + 5x_1) \begin{cases} \leq \dfrac{2}{11} & \text{for } 0 \leq x_1 \leq 1, \\[2mm] = \dfrac{2}{11} & \text{for } x_1 = \dfrac{7}{11}. \end{cases}$$

But y_2^* and y_3^* are numbers, so the left-hand side is the equation of a straight line, which is a fixed weighted average of the two "bottom" lines on the graph. Because the ordinate of this line must equal $\frac{2}{11}$ at $x_1 = \frac{7}{11}$, and because it must never exceed $\frac{2}{11}$, the line necessarily is horizontal. (This conclusion is always true unless the optimal value of x_1 is either 0 or 1, in which case player 2 also should use a single pure strategy.) Therefore,

$$y_2^*(4 - 6x_1) + y_3^*(-3 + 5x_1) = \frac{2}{11}, \qquad \text{for } 0 \leq x_1 \leq 1.$$

Hence, to solve for y_2^* and y_3^*, select two values of x_1 (say, 0 and 1), and solve the resulting two simultaneous equations. Thus,

$$4y_2^* - 3y_3^* = \frac{2}{11},$$

$$-2y_2^* + 2y_3^* = \frac{2}{11},$$

which has a simultaneous solution of $y_2^* = \frac{5}{11}$ and $y_3^* = \frac{6}{11}$. Therefore, the *optimal mixed strategy* for player 2 is $(y_1, y_2, y_3) = (0, \frac{5}{11}, \frac{6}{11})$.

If, in another problem, there should happen to be more than two lines passing through the maximin point, so that more than two of the y_j^* values can be greater than zero, this condition would imply that there are many ties for the optimal mixed strategy for player 2. One such strategy can then be identified by setting all but two of these y_j^* values equal to zero and solving for the remaining two in the manner just described. For the remaining two, the associated lines must have positive slope in one case and negative slope in the other.

Although this graphical procedure has been illustrated for only one particular problem, essentially the same reasoning can be used to solve any game with mixed strategies that has only two undominated pure strategies for one of the players. The Worked Examples section of the book's website provides **another example** where, in this case, it is player 2 that has only two undominated strategies, so the graphical solution procedure is applied initially from the viewpoint of that player.

■ 14.5 SOLVING BY LINEAR PROGRAMMING

Any game with mixed strategies can be solved by transforming the problem to a linear programming problem. As you will see, this transformation requires little more than applying the minimax theorem and using the definitions of the maximin value $\underline{v}$ and minimax value $\bar{v}$.

First, consider how to find the optimal mixed strategy for player 1. As indicated in Sec. 14.3,

$$\text{Expected payoff for player 1} = \sum_{i=1}^{m} \sum_{j=1}^{n} p_{ij} x_i y_j$$

and the strategy $(x_1, x_2, \ldots, x_m)$ is optimal if

$$\sum_{i=1}^{m} \sum_{j=1}^{n} p_{ij} x_i y_j \geq \underline{v} = v$$

for all opposing strategies $(y_1, y_2, \ldots, y_n)$. Thus, this inequality will need to hold, e.g., for each of the pure strategies of player 2, that is, for each of the strategies $(y_1, y_2, \ldots, y_n)$ where one $y_j = 1$ and the rest equal 0. Substituting these values into the inequality yields

$$\sum_{i=1}^{m} p_{ij} x_i \geq v \qquad \text{for } j = 1, 2, \ldots, n,$$

so that the inequality *implies* this set of n inequalities. Furthermore, this set of n inequalities *implies* the original inequality (rewritten)

$$\sum_{j=1}^{n} y_j \left(\sum_{i=1}^{m} p_{ij} x_i \right) \geq \sum_{j=1}^{n} y_j v = v,$$

since

$$\sum_{j=1}^{n} y_j = 1.$$

Because the implication goes in both directions, it follows that imposing this set of n linear inequalities is *equivalent* to requiring the original inequality to hold for all strategies $(y_1, y_2, \ldots, y_n)$. But these n inequalities are legitimate linear programming constraints, as are the additional constraints

$$x_1 + x_2 + \cdots + x_m = 1$$
$$x_i \geq 0, \qquad \text{for } i = 1, 2, \ldots, m$$

that are required to ensure that the x_i are probabilities. Therefore, any solution $(x_1, x_2, \ldots, x_m)$ that satisfies this entire set of linear programming constraints is the desired optimal mixed strategy.

Consequently, the problem of finding an optimal mixed strategy has been reduced to finding a feasible solution for a linear programming problem, which can be done as described in Chap. 4. The two remaining difficulties are that (1) v is unknown and (2) the linear programming problem has no objective function. Fortunately, both these difficulties can be resolved at one stroke by replacing the unknown constant v by the variable x_{m+1} and then *maximizing* x_{m+1}, so that x_{m+1} automatically will equal v (by definition) at the *optimal* solution for the linear programming problem!

The Linear Programming Formulation

To summarize, player 1 would find his optimal mixed strategy by using the simplex method to solve the linear programming problem:

$$\text{Maximize} \qquad x_{m+1},$$

subject to

$$p_{11}x_1 + p_{21}x_2 + \cdots + p_{m1}x_m - x_{m+1} \geq 0$$
$$p_{12}x_1 + p_{22}x_2 + \cdots + p_{m2}x_m - x_{m+1} \geq 0$$
$$\cdots\cdots\cdots\cdots\cdots\cdots\cdots\cdots\cdots\cdots\cdots\cdots\cdots\cdots$$
$$p_{1n}x_1 + p_{2n}x_2 + \cdots + p_{mn}x_m - x_{m+1} \geq 0$$
$$x_1 + x_2 + \cdots + x_m = 1$$

and

$$x_i \geq 0, \qquad \text{for } i = 1, 2, \ldots, m.$$

Note that x_{m+1} is not restricted to be nonnegative, whereas the simplex method can be applied only after *all* the variables have nonnegativity constraints. However, this matter can be easily rectified, as will be discussed shortly.

Now consider player 2. He could find his optimal mixed strategy by rewriting the payoff table as the payoff to himself rather than to player 1 and then by proceeding exactly as just described. However, it is enlightening to summarize his formulation in terms of the original payoff table. By proceeding in a way that is completely analogous to that just described, player 2 would conclude that his optimal mixed strategy is given by an optimal solution to the linear programming problem:

Minimize y_{n+1},

subject to

$$
\begin{aligned}
p_{11}y_1 + p_{12}y_2 + \cdots + p_{1n}y_n - y_{n+1} &\leq 0 \\
p_{21}y_1 + p_{22}y_2 + \cdots + p_{2n}y_n - y_{n+1} &\leq 0 \\
&\cdots\cdots\cdots\cdots\cdots\cdots\cdots\cdots\cdots\cdots\cdots \\
p_{m1}y_1 + p_{m2}y_2 + \cdots + p_{mn}y_n - y_{n+1} &\leq 0 \\
y_1 + y_2 + \cdots + y_n &= 1
\end{aligned}
$$

and

$$y_j \geq 0, \qquad \text{for } j = 1, 2, \ldots, n.$$

It is easy to show (see Prob. 14.5-6 and its hint) that this linear programming problem and the one given for player 1 are *dual* to each other in the sense described in Secs. 6.1 and 6.4. This fact has several important implications. One implication is that the optimal mixed strategies for both players can be found by solving only one of the linear programming problems because the optimal dual solution is an automatic by-product of the simplex method calculations to find the optimal primal solution. A second implication is that this brings all *duality theory* (described in Chap. 6) to bear upon the interpretation and analysis of games.

A related implication is that this provides **a simple proof of the minimax theorem.** Let x_{m+1}^* and y_{n+1}^* denote the value of x_{m+1} and y_{n+1} in the optimal solution of the respective linear programming problems. It is known from the *strong duality property* given in Sec. 6.1 that $-x_{m+1}^* = -y_{n+1}^*$, so that $x_{m+1}^* = y_{n+1}^*$. However, it is evident from the definition of $\underline{v}$ and $\bar{v}$ that $\underline{v} = x_{m+1}^*$ and $\bar{v} = y_{n+1}^*$, so it follows that $\underline{v} = \bar{v}$, as claimed by the minimax theorem.

One remaining loose end needs to be tied up, namely, what to do about x_{m+1} and y_{n+1} being unrestricted in sign in the linear programming formulations. If it is clear that $v \geq 0$ so that the optimal values of x_{m+1} and y_{n+1} are nonnegative, then it is safe to introduce nonnegativity constraints for these variables for the purpose of applying the simplex method. However, if $v < 0$, then an adjustment needs to be made. One possibility is to use the approach described in Sec. 4.6 for replacing a variable without a nonnegativity constraint by the difference of two nonnegative variables. Another is to reverse players 1 and 2 so that the payoff table would be rewritten as the payoff to the original player 2, which would make the corresponding value of v positive. A third, and the most commonly used, procedure is to add a sufficiently large fixed constant to all the entries in the payoff table that the new value of the game will be positive. (For example, setting this constant equal to the absolute value of the largest negative entry will suffice.) Because this same constant is added to every entry, this adjustment cannot alter the optimal mixed strategies in any way, so they can now be obtained in the usual manner. The indicated value of the game would be increased by the amount of the constant, but this value can be readjusted after the solution has been obtained.

Application to Variation 3 of the Political Campaign Problem

To illustrate this linear programming approach, consider again variation 3 of the political campaign problem after dominated strategy 3 for player 1 is eliminated (see Table 14.6). Because there are some negative entries in the reduced payoff table, it is unclear at the outset whether the *value* of the game v is *nonnegative* (it turns out to be). For the moment, let us assume that $v \geq 0$ and proceed without making any of the adjustments discussed in the preceding paragraph.

To write out the linear programming model for player 1 for this example, note that p_{ij} in the general model is the entry in row i and column j of Table 14.6, for $i = 1, 2$ and $j = 1, 2, 3$. The resulting model is

$$\text{Maximize} \quad x_3,$$

subject to

$$
\begin{aligned}
5x_2 - x_3 &\geq 0 \\
-2x_1 + 4x_2 - x_3 &\geq 0 \\
2x_1 - 3x_2 - x_3 &\geq 0 \\
x_1 + x_2 &= 1
\end{aligned}
$$

and

$$x_1 \geq 0, \qquad x_2 \geq 0.$$

Applying the simplex method to this linear programming problem (after adding the constraint $x_3 \geq 0$) yields $x_1^* = \frac{7}{11}$, $x_2^* = \frac{4}{11}$, $x_3^* = \frac{2}{11}$ as the optimal solution. (See Probs. 14.5-8 and 14.5-9.) Consequently, just as was found by the graphical procedure in the preceding section, the optimal mixed strategy for player 1 according to the minimax criterion is $(x_1, x_2) = (\frac{7}{11}, \frac{4}{11})$, and the value of the game is $v = x_3^* = \frac{2}{11}$. The simplex method also yields the optimal solution for the dual (given next) of this problem, namely, $y_1^* = 0$, $y_2^* = \frac{5}{11}$, $y_3^* = \frac{6}{11}$, $y_4^* = \frac{2}{11}$, so the optimal mixed strategy for player 2 is $(y_1, y_2, y_3) = (0, \frac{5}{11}, \frac{6}{11})$.

The dual of the preceding problem is just the linear programming model for player 2 (the one with variables $y_1, y_2, \ldots, y_n, y_{n+1}$) shown earlier in this section. (See Prob. 14.5-7.) By plugging in the values of p_{ij} from Table 14.6, this model is

$$\text{Minimize} \quad y_4,$$

subject to

$$
\begin{aligned}
-2y_2 + 2y_3 - y_4 &\leq 0 \\
5y_1 + 4y_2 - 3y_3 - y_4 &\leq 0 \\
y_1 + y_2 + y_3 &= 1
\end{aligned}
$$

and

$$y_1 \geq 0, \qquad y_2 \geq 0, \qquad y_3 \geq 0.$$

Applying the simplex method directly to this model (after adding the constraint $y_4 \geq 0$) yields the optimal solution: $y_1^* = 0$, $y_2^* = \frac{5}{11}$, $y_3^* = \frac{6}{11}$, $y_4^* = \frac{2}{11}$ (as well as the optimal dual solution $x_1^* = \frac{7}{11}$, $x_2^* = \frac{4}{11}$, $x_3^* = \frac{2}{11}$). Thus, the optimal mixed strategy for player 2 is $(y_1, y_2, y_3) = (0, \frac{5}{11}, \frac{6}{11})$, and the value of the game is again seen to be $v = y_4^* = \frac{2}{11}$.

Because we already had found the optimal mixed strategy for player 2 while dealing with the first model, we did not have to solve the second one. In general, you always can find optimal mixed strategies for *both* players by choosing just one of the models (either one) and then using the simplex method to solve for both an optimal solution and an optimal dual solution.

When the simplex method was applied to both of these linear programming models, a nonnegativity constraint was added that assumed that $v \geq 0$. If this assumption were violated, both models would have no feasible solutions, so the simplex method would stop quickly with this message. To avoid this risk, we could have added a positive constant, say, 3 (the absolute value of the largest negative entry), to all the entries in Table 14.6. This then would increase by 3 all the coefficients of x_1, x_2, y_1, y_2, and y_3 in the inequality constraints of the two models. (See Prob. 14.5-2.)

■ 14.6   EXTENSIONS

Although this chapter has considered only two-person, zero-sum games with a finite number of pure strategies, game theory extends far beyond this kind of game. In fact, extensive research has been done on a number of more complicated types of games, including the ones summarized in this section.

The simplest generalization is to the *two-person, constant-sum game.* In this case, the sum of the payoffs to the two players is a fixed constant (positive or negative) regardless of which combination of strategies is selected. The only difference from a two-person, zero-sum game is that, in the latter case, the constant must be zero. A nonzero constant may arise instead because, in addition to one player winning whatever the other one loses, the two players may share some reward (if the constant is positive) or some cost (if the constant is negative) for participating in the game. Adding this fixed constant does nothing to affect which strategies should be chosen. Therefore, the analysis for determining optimal strategies is exactly the same as described in this chapter for two-person, zero-sum games.

A more complicated extension is to the *n-person game,* where more than two players may participate in the game. This generalization is particularly important because, in many kinds of competitive situations, frequently more than two competitors are involved. This may occur, for example, in competition among business firms, in international diplomacy, and so forth. Unfortunately, the existing theory for such games is less satisfactory than it is for two-person games.

Another generalization is the *nonzero-sum game,* where the sum of the payoffs to the players need not be 0 (or any other fixed constant). This case reflects the fact that many competitive situations include noncompetitive aspects that contribute to the mutual advantage or mutual disadvantage of the players. For example, the advertising strategies of competing companies can affect not only how they will split the market but also the total size of the market for their competing products. However, in contrast to a constant-sum game, the size of the mutual gain (or loss) for the players depends on the combination of strategies chosen.

Because mutual gain is possible, nonzero-sum games are further classified in terms of the degree to which the players are permitted to cooperate. At one extreme is the *noncooperative game,* where there is no preplay communication between the players. At the other extreme is the *cooperative game,* where preplay discussions and binding agreements are permitted. For example, competitive situations involving trade regulations between countries, or collective bargaining between labor and management, might be formulated as cooperative games. When there are more than two players, cooperative games also allow some of or all the players to form coalitions.

Still another extension is to the class of *infinite games,* where the players have an infinite number of pure strategies available to them. These games are designed for the kind of situation where the strategy to be selected can be represented by a *continuous* decision variable. For example, this decision variable might be the time at which to take a certain action, or the proportion of one's resources to allocate to a certain activity, in a competitive situation.

However, the analysis required in these extensions beyond the two-person, zero-sum, finite game is relatively complex and will not be pursued further here. (See any of Selected References 4, 6, 7, 8, and 10 for further information.)

▇ 14.7 CONCLUSIONS

The general problem of how to make decisions in a competitive environment is a very common and important one. The fundamental contribution of game theory is that it provides a basic conceptual framework for formulating and analyzing such problems in simple situations. However, there is a considerable gap between what the theory can handle and the complexity of most competitive situations arising in practice. Therefore, the conceptual tools of game theory usually play just a supplementary role in dealing with these situations.

Because of the importance of the general problem, research is continuing with some success to extend the theory to more complex situations.

▇ SELECTED REFERENCES

1. Aumann, R. J., and S. Hart (eds.): *Handbook of Game Theory: With Application to Economics,* vols. 1, 2, and 3, North-Holland, Amsterdam, 1992, 1994, 2002.
2. Brandenburger, A., and H. Stuart: "Biform Games," *Management Science,* **53**(4): 537–549, April 2007.
3. Chatterjee, K., and W. F. Samuelson (eds.): *Game Theory and Business Applications,* Kluwer Academic Publishers (now Springer), Boston, 2001.
4. Forgó, F., J. Szép, and F. Szidarovsky: *Introduction to the Theory of Games: Concepts, Methods, Applications,* Kluwer Academic Publishers (now Springer), Boston, 1999.
5. Hohzaki, R.: "A Search Game Taking Account of Attributes of Searching Resources," *Naval Research Logistics,* **55**(1): 76–90, February 2008.
6. Mendelson, E.: *Introducing Game Theory and Its Applications,* Chapman and Hall/CRC Press, Boca Raton, FL, 2005.
7. Meyerson, R. B.: *Game Theory: Analysis of Conflict,* Harvard University Press, Cambridge, MA, 1991.
8. Owen, G.: *Game Theory,* 3d ed., Academic Press, San Diego, 1995.
9. Parthasarathy, T., B. Dutta, and A. Sen (eds.): *Game Theoretical Applications to Economics and Operations Research,* Kluwer Academic Publishers (now Springer), Boston, 1997.
10. Webb, J. N.: *Game Theory: Decisions, Interaction and Evolution,* Springer, New York, 2007.
11. Zhuang, J., and V. M. Bier: "Balancing Terrorism and Natural Disasters—Defensive Strategy with Endogenous Attacker Effort," *Operations Research,* **55**(5): 976–991, September–October 2007.

▇ LEARNING AIDS FOR THIS CHAPTER ON OUR WEBSITE (www.mhhe.com/hillier)

Worked Examples:

Examples for Chapter 14

"Ch. 14—Game Theory" Files for Solving the Examples:

Excel Files
LINGO/LINDO File
MPL/CPLEX File

Glossary for Chapter 14

See Appendix 1 for documentation of the software.

■ PROBLEMS

The symbol to the left of some of the problems (or their parts) has the following meaning.

C: Use the computer with any of the software options available to you (or as instructed by your instructor) to solve the problem.

An asterisk on the problem number indicates that at least a partial answer is given in the back of the book.

14.1-1. The labor union and management of a particular company have been negotiating a new labor contract. However, negotiations have now come to an impasse, with management making a "final" offer of a wage increase of $1.10 per hour and the union making a "final" demand of a $1.60 per hour increase. Therefore, both sides have agreed to let an impartial arbitrator set the wage increase somewhere between $1.10 and $1.60 per hour (inclusively).

The arbitrator has asked each side to submit to her a confidential proposal for a fair and economically reasonable wage increase (rounded to the nearest dime). From past experience, both sides know that this arbitrator normally accepts the proposal of the side that gives the most from its final figure. If neither side changes its final figure, or if they both give in the same amount, then the arbitrator normally compromises halfway between ($1.35 in this case). Each side now needs to determine what wage increase to propose for its own maximum advantage.

Formulate this problem as a two-person, zero-sum game.

14.1-2. Two manufacturers currently are competing for sales in two different but equally profitable product lines. In both cases the sales volume for manufacturer 2 is three times as large as that for manufacturer 1. Because of a recent technological breakthrough, both manufacturers will be making a major improvement in both products. However, they are uncertain as to what development and marketing strategy to follow.

If both product improvements are developed simultaneously, either manufacturer can have them ready for sale in 12 months. Another alternative is to have a "crash program" to develop only one product first to try to get it marketed ahead of the competition. By doing this, manufacturer 2 could have one product ready for sale in 9 months, whereas manufacturer 1 would require 10 months (because of previous commitments for its production facilities). For either manufacturer, the second product could then be ready for sale in an additional 9 months.

For either product line, if both manufacturers market their improved models simultaneously, it is estimated that manufacturer 1 would increase its share of the total future sales of this product by 8 percent of the total (from 25 to 33 percent). Similarly, manufacturer 1 would increase its share by 20, 30, and 40 percent of the total if it marketed the product sooner than manufacturer 2 by 2, 6, and 8 months, respectively. On the other hand, manufacturer 1 would lose 4, 10, 12, and 14 percent of the total if manufacturer 2 marketed it sooner by 1, 3, 7, and 10 months, respectively.

Formulate this problem as a two-person, zero-sum game, and then determine which strategy the respective manufacturers should use according to the minimax criterion.

14.1-3. Consider the following parlor game to be played between two players. Each player begins with three chips: one red, one white, and one blue. Each chip can be used only once.

To begin, each player selects one of her chips and places it on the table, concealed. Both players then uncover the chips and determine the payoff to the winning player. In particular, if both players play the same kind of chip, it is a draw; otherwise, the following table indicates the winner and how much she receives from the other player. Next, each player selects one of her two remaining chips and repeats the procedure, resulting in another payoff according to the following table. Finally, each player plays her one remaining chip, resulting in the third and final payoff.

Winning Chip	Payoff ($)
Red beats white	90
White beats blue	70
Blue beats red	50
Matching colors	0

Formulate this problem as a two-person, zero-sum game by identifying the form of the strategies and payoffs.

14.2-1. Reconsider Prob. 14.1-1.
(a) Use the concept of dominated strategies to determine the best strategy for each side.
(b) Without eliminating dominated strategies, use the minimax criterion to determine the best strategy for each side.

14.2-2.* For the game having the following payoff table, determine the optimal strategy for each player by successively eliminating dominated strategies. (Indicate the order in which you eliminated strategies.)

		Player 2		
Strategy		**1**	**2**	**3**
	1	−3	1	2
Player 1	2	1	2	1
	3	1	0	−2

14.2-3. Consider the game having the following payoff table.

		Player 2			
Strategy		**1**	**2**	**3**	**4**
	1	5	−7	−2	2
Player 1	2	−2	2	−5	5
	3	−2	5	−2	7

Determine the optimal strategy for each player by successively eliminating dominated strategies. Give a list of the dominated strategies (and the corresponding dominating strategies) in the order in which you were able to eliminate them.

14.2-4. Find the saddle point for the game having the following payoff table.

		Player 2	
Strategy	**1**	**2**	**3**
1	3	−1	3
Player 1 2	−3	1	7
3	7	3	5

Use the minimax criterion to find the best strategy for each player. Does this game have a saddle point? Is it a stable game?

14.2-5. Find the saddle point for the game having the following payoff table.

		Player 2		
Strategy	**1**	**2**	**3**	**4**
1	3	−3	−2	−4
Player 1 2	−4	−2	−1	1
3	1	−1	2	0

Use the minimax criterion to find the best strategy for each player. Does this game have a saddle point? Is it a stable game?

14.2-6. Two companies share the bulk of the market for a particular kind of product. Each is now planning its new marketing plans for the next year in an attempt to wrest some sales away from the other company. (The total sales for the product are relatively fixed, so one company can increase its sales only by winning them away from the other.) Each company is considering three possibilities: (1) better packaging of the product, (2) increased advertising, and (3) a slight reduction in price. The costs of the three alternatives are quite comparable and sufficiently large that each company will select just one. The estimated effect of each combination of alternatives on the *increased percentage of the sales* for company 1 is as follows:

		Player 2	
Strategy	**1**	**2**	**3**
1	2	3	1
Player 1 2	1	4	0
3	3	−2	−1

Each company must make its selection before learning the decision of the other company.

(a) Without eliminating dominated strategies, use the minimax (or maximin) criterion to determine the best strategy for each company.

(b) Now identify and eliminate dominated strategies as far as possible. Make a list of the dominated strategies, showing the order in which you were able to eliminate them. Then show the resulting reduced payoff table with no remaining dominated strategies.

14.2-7.* Two politicians soon will be starting their campaigns against each other for a certain political office. Each must now select the main issue she will emphasize as the theme of her campaign. Each has three advantageous issues from which to choose, but the relative effectiveness of each one would depend upon the issue chosen by the opponent. In particular, the estimated increase in the vote for politician 1 (expressed as a percentage of the total vote) resulting from each combination of issues is as follows:

		Issue for Politician 2	
	1	**2**	**3**
Issue for 1	7	−1	3
Politician 1 2	1	0	2
3	−5	−3	−1

However, because considerable staff work is required to research and formulate the issue chosen, each politician must make her own choice before learning the opponent's choice. Which issue should she choose?

For each of the situations described here, formulate this problem as a two-person, zero-sum game, and then determine which issue should be chosen by each politician according to the specified criterion.

(a) The current preferences of the voters are very uncertain, so each additional percent of votes won by one of the politicians has the same value to her. Use the minimax criterion.

(b) A reliable poll has found that the percentage of the voters currently preferring politician 1 (before the issues have been raised) lies between 45 and 50 percent. (Assume a uniform distribution over this range.) Use the concept of dominated strategies, beginning with the strategies for politician 1.

(c) Suppose that the percentage described in part (*b*) actually were 45 percent. Should politician 1 use the minimax criterion? Explain. Which issue would you recommend? Why?

14.2-8. Briefly describe what you feel are the advantages and disadvantages of the minimax criterion.

14.3-1. Consider the odds and evens game introduced in Sec. 14.1 and whose payoff table is shown in Table 14.1.

(a) Show that this game does not have a saddle point.

(b) Write an expression for the expected payoff for player 1 (the evens player) in terms of the probabilities of the two players using their respective pure strategies. Then show what this expression reduces to for the following three cases: (*i*) Player 2 definitely uses his first strategy, (*ii*) player 2 definitely uses his second strategy, (*iii*) player 2 assigns equal probabilities to using his two strategies.

(c) Repeat part (*b*) when player 1 becomes the odds player instead.

■ 15.2   DECISION MAKING WITHOUT EXPERIMENTATION

Before seeking a solution to the first Goferbroke Co. problem, we will formulate a general framework for decision making.

In general terms, the decision maker must choose an **alternative** from a set of possible decision alternatives. The set contains all the *feasible alternatives* under consideration for how to proceed with the problem of concern.

This choice of an alternative must be made in the face of uncertainty, because the outcome will be affected by random factors that are outside the control of the decision maker. These random factors determine what situation will be found at the time that the decision alternative is executed. Each of these possible situations is referred to as a possible **state of nature.**

For each combination of a decision alternative and a state of nature, the decision maker knows what the resulting payoff would be. The **payoff** is a quantitative measure of the value to the decision maker of the consequences of the outcome. For example, the payoff frequently is represented by the *net monetary gain* (profit), although other measures also can be used (as described in Sec. 15.6). If the consequences of the outcome do not become completely certain even when the state of nature is given, then the payoff becomes an *expected value* (in the statistical sense) of the measure of the consequences. A **payoff table** commonly is used to provide the payoff for each combination of an action and a state of nature.

If you previously studied game theory (Chap. 14), we should point out an interesting analogy between this decision analysis framework and the two-person, zero-sum games described in Chap. 14. The *decision maker* and *nature* can be viewed as the *two players* of such a game. The *alternatives* and the possible *states of nature* can then be viewed as the available *strategies* for these respective players, where each combination of strategies results in some *payoff* to player 1 (the decision maker). From this viewpoint, the decision analysis framework can be summarized as follows:

1. The *decision maker* needs to choose one of the *decision alternatives*.
2. *Nature* then would choose one of the possible *states of nature*.
3. Each combination of a decision alternative and state of nature would result in a *payoff*, which is given as one of the entries in a *payoff table*.
4. This payoff table should be used to find an *optimal alternative* for the decision maker according to an appropriate criterion.

Soon we will present three possibilities for this criterion, where the first one (the maximin payoff criterion) comes from game theory.

However, this analogy to two-person, zero-sum games breaks down in one important respect. In game theory, *both* players are assumed to be *rational* and choosing their strategies to *promote their own welfare*. This description still fits the decision maker, but certainly not nature. By contrast, nature now is a passive player that chooses its strategies (states of nature) in some random fashion. This change means that the game theory criterion for how to choose an optimal strategy (alternative) will not appeal to many decision makers in the current context.

One additional element needs to be added to the decision analysis framework. The decision maker generally will have some information that should be taken into account about the relative likelihood of the possible states of nature. Such information can usually be translated to a probability distribution, acting as though the state of nature is a random variable, in which case this distribution is referred to as a **prior distribution.** Prior distributions are often subjective in that they may depend upon the experience or intuition

Following the merger of Conoco Inc. and the Phillips Petroleum Company in 2002, **ConocoPhillips** became the third-largest integrated energy company in the United States with $160 billion in assets and 38,000 employees. Like any company in this industry, the management of ConocoPhillips must grapple continually with decisions about the allocation of limited investment capital across a set of risky petroleum exploration projects. These decisions have a great impact on the profitability of the company.

In the early 1990s, the then Phillips Petroleum Company became an industry leader in the application of sophisticated OR methodology to aid these decisions by developing a *decision analysis software package* called DISCOVERY. The user interface allows a geologist or engineer to model the uncertainties associated with a project and then the software interprets the inputs and constructs a decision tree that shows all the decision nodes (including opportunities to obtain additional seismic information) and the intervening event nodes. A key feature of the software is the use of an *exponential utility function* (to be introduced in Sec. 15.6) to incorporate management's attitudes about financial risk. An intuitive questionnaire is used to measure corporate risk preferences in order to determine an appropriate value of the risk tolerance parameter for this utility function.

Management uses the software to (1) evaluate *petroleum exploration projects* with a consistent risk-taking policy across the company, (2) rank projects in terms of overall preference, (3) identify the firm's appropriate level of participation in these projects, and (4) stay within budget.

Source: M. R. Walls, G. T. Morahan, and J. S. Dyer: "Decision Analysis of Exploration Opportunities in the Onshore US at Phillips Petroleum Company," *Interfaces,* **25**(6): 39–56, Nov.–Dec. 1995. (A link to this article is provided on our website, www.mhhe.com/hillier.)

of an individual. The probabilities for the respective states of nature provided by the prior distribution are called **prior probabilities.**

Formulation of the Prototype Example in This Framework

As indicated in Table 15.1, the Goferbroke Co. has two possible decision alternatives under consideration: drill for oil or sell the land. The possible states of nature are that the land contains oil and that it does not, as designated in the column headings of Table 15.1 by *oil* and *dry.* Since the consulting geologist has estimated that there is one chance in four of oil (and so three chances in four of no oil), the prior probabilities of the two states of nature are 0.25 and 0.75, respectively. Therefore, with the payoff in units of thousands of dollars of profit, the payoff table can be obtained directly from Table 15.1, as shown in Table 15.2.

We will use this payoff table next to find the optimal alternative according to each of the three criteria described below.

The Maximin Payoff Criterion

If the decision maker's problem were to be viewed as a *game against nature,* then game theory would say to choose the decision alternative according to the *minimax criterion*

TABLE 15.2 Payoff table for the decision analysis formulation of the first Goferbroke Co. problem

	State of Nature	
Alternative	Oil	Dry
1. Drill for oil	700	−100
2. Sell the land	90	90
Prior probability	0.25	0.75

■ **TABLE 15.3** Application of the maximin payoff criterion to the first Goferbroke Co. problem

Alternative	State of Nature		Minimum
	Oil	**Dry**	
1. Drill for oil	700	−100	−100
2. Sell the land	90	90	90 ← Maximin value
Prior probability	0.25	0.75	

(as described in Sec. 14.2). From the viewpoint of player 1 (the decision maker), this criterion is more aptly named the *maximin payoff criterion,* as summarized below.

> **Maximin payoff criterion:** For each possible decision alternative, find the *minimum payoff* over all possible states of nature. Next, find the *maximum* of these minimum payoffs. Choose the alternative whose minimum payoff gives this maximum.

Table 15.3 shows the application of this criterion to the prototype example. Thus, since the minimum payoff for selling (90) is larger than that for drilling (−100), the former alternative (sell the land) will be chosen.

The rationale for this criterion is that it provides the *best guarantee* of the payoff that will be obtained. Regardless of what the true state of nature turns out to be for the example, the payoff from selling the land cannot be less than 90, which provides the best available guarantee. Thus, this criterion takes the pessimistic viewpoint that, regardless of which alternative is selected, the worst state of nature for that alternative is likely to occur, so we should choose the alternative which provides the best payoff with its worst state of nature.

This rationale is quite valid when one is competing against a rational and malevolent opponent. However, this criterion is not often used in games against nature because it is an extremely conservative criterion in this context. In effect, it assumes that nature is a conscious opponent that wants to inflict as much damage as possible on the decision maker. Nature is not a malevolent opponent, and the decision maker does not need to focus solely on the worst possible payoff from each alternative. This is especially true when the worst possible payoff from an alternative comes from a relatively unlikely state of nature.

Thus, this criterion normally is of interest only to a very cautious decision maker.

The Maximum Likelihood Criterion

The next criterion focuses on the *most likely* state of nature, as summarized below.

> **Maximum likelihood criterion:** Identify the most likely state of nature (the one with the largest prior probability). For this state of nature, find the decision alternative with the maximum payoff. Choose this decision alternative.

Applying this criterion to the example, Table 15.4 indicates that the *Dry* state has the largest prior probability. In the Dry column, the sell alternative has the maximum payoff, so the choice is to sell the land.

The appeal of this criterion is that the most important state of nature is the most likely one, so the alternative chosen is the best one for this particularly important state of nature. Basing the decision on the assumption that this state of nature will occur tends to give a better chance of a favorable outcome than assuming any other state of nature.

■ **TABLE 15.4** Application of the maximum likelihood criterion to the first Goferbroke Co. problem

Alternative	State of Nature		
	Oil	Dry	
1. Drill for oil	700	−100	−100
2. Sell the land	90	90	90 ← Maximum in this column
Prior probability	0.25	0.75	

↑
Maximum

Furthermore, the criterion does not rely on questionable subjective estimates of the probabilities of the respective states of nature other than identifying the most likely state.

The major drawback of the criterion is that it completely ignores much relevant information. No state of nature is considered other than the most likely one. In a problem with many possible states of nature, the probability of the most likely one may be quite small, so focusing on just this one state of nature is quite unwarranted. Even in the example, where the prior probability of the *Dry* state is 0.75, this criterion ignores the extremely attractive payoff of 700 if the company drills and finds oil. In effect, the criterion does not permit gambling on a low-probability big payoff, no matter how attractive the gamble may be.

Bayes' Decision Rule[1]

Our third criterion, and the one commonly chosen, is *Bayes' decision rule,* described below.

Bayes' decision rule: Using the best available estimates of the probabilities of the respective states of nature (currently the prior probabilities), calculate the expected value of the payoff for each of the possible decision alternatives. Choose the decision alternative with the maximum expected payoff.

For the prototype example, these expected payoffs are calculated directly from Table 15.2 as follows:

$$E[\text{Payoff (drill)}] = 0.25(700) + 0.75(-100)$$
$$= 100.$$
$$E[\text{Payoff (sell)}] = 0.25(90) + 0.75(90)$$
$$= 90.$$

Since 100 is larger than 90, the alternative selected is to drill for oil.

Note that this choice contrasts with the selection of the sell alternative under each of the two preceding criteria.

The big advantage of Bayes' decision rule is that it incorporates all the available information, including all the payoffs and the best available estimates of the probabilities of the respective states of nature.

---

[1]The origin of this name is that this criterion is often credited to the Reverend Thomas Bayes, a nonconforming 18th-century English minister who won renown as a philosopher and mathematician. (The same basic idea has even longer roots in the field of economics.) This decision rule also is sometimes called the *expected monetary value (EMF)* criterion, although this is a misnomer for those cases where the measure of the payoff is something other than monetary value (as in Sec. 15.6).

It is sometimes argued that these estimates of the probabilities necessarily are largely subjective and so are too shaky to be trusted. There is no accurate way of predicting the future, including a future state of nature, even in probability terms. This argument has some validity. The reasonableness of the estimates of the probabilities should be assessed in each individual situation.

Nevertheless, under many circumstances, past experience and current evidence enable one to develop reasonable estimates of the probabilities. Using this information should provide better grounds for a sound decision than ignoring it. Furthermore, experimentation frequently can be conducted to improve these estimates, as described in the next section. Therefore, we will be using only Bayes' decision rule throughout the remainder of the chapter.

To assess the effect of possible inaccuracies in the prior probabilities, it often is helpful to conduct sensitivity analysis, as described below.

Sensitivity Analysis with Bayes' Decision Rule

Sensitivity analysis commonly is used with various applications of operations research to study the effect if some of the numbers included in the mathematical model are not correct. In this case, the mathematical model is represented by the payoff table shown in Table 15.2. The numbers in this table that are most questionable are the prior probabilities. We will focus the sensitivity analysis on these numbers, although a similar approach could be applied to the payoffs given in the table.

The sum of the two prior probabilities must equal 1, so increasing one of these probabilities automatically decreases the other one by the same amount, and vice versa. Goferbroke's management feels that the true chances of having oil on the tract of land are likely to lie somewhere between 15 and 35 percent. In other words, the true prior probability of having oil is likely to be in the range from 0.15 to 0.35, so the corresponding prior probability of the land being dry would range from 0.85 to 0.65.

Letting

p = prior probability of oil,

the expected payoff from drilling for any p is

$$E[\text{Payoff (drill)}] = 700p - 100(1 - p)$$
$$= 800p - 100.$$

The slanting line in Fig. 15.1 shows the plot of this expected payoff versus p. Since the payoff from selling the land would be 90 for any p, the flat line in Fig. 15.1 gives $E[\text{Payoff (sell)}]$ versus p.

The four dots in Fig. 15.1 show the expected payoff for the two decision alternatives when $p = 0.15$ or $p = 0.35$. When $p = 0.15$, the decision swings over to selling the land by a wide margin (an expected payoff of 90 versus only 20 for drilling). However, when $p = 0.35$, the decision is to drill by a wide margin (expected payoff = 180 versus only 90 for selling). Thus, the decision is very *sensitive* to p. This sensitivity analysis has revealed that it is important to do more, if possible, to develop a more precise estimate of the true value of p.

The point in Fig. 15.1 where the two lines intersect is the **crossover point** where the decision shifts from one alternative (sell the land) to the other (drill for oil) as the prior probability increases. To find this point, we set

$$E[\text{Payoff (drill)}] = E[\text{Payoff (sell)}]$$
$$800p - 100 = 90$$
$$p = \frac{190}{800} = 0.2375$$

Conclusion: Should sell the land if $p < 0.2375$.
Should drill for oil if $p > 0.2375$.

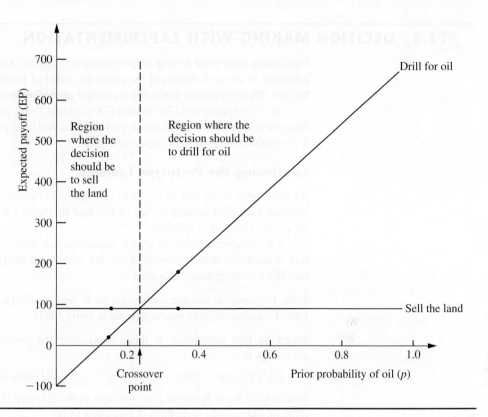

FIGURE 15.1
Graphical display of how the expected payoff for each decision alternative changes when the prior probability of oil changes for the first Goferbroke Co. problem.

Thus, when trying to refine the estimate of the true value of p, the key question is whether it is smaller or larger than 0.2375.

For other problems that have more than two decision alternatives, the same kind of analysis can be applied. The main difference is that there now would be more than two lines (one per alternative) in the graphical display corresponding to Fig. 15.1. However, the top line for any particular value of the prior probability still indicates which alternative should be chosen. With more than two lines, there might be more than one crossover point where the decision shifts from one alternative to another.

You can see **another example** of performing this kind of analysis with three decision alternatives in the Worked Examples section of the book's website. (This same example also illustrates the application of all three decision criteria considered in this section.)

For a problem with more than two possible states of nature, the most straightforward approach is to focus the sensitivity analysis on only two states at a time as described above. This again would involve investigating what happens when the prior probability of one state increases as the prior probability of the other state decreases by the same amount, holding fixed the prior probabilities of the remaining states. This procedure then can be repeated for as many other pairs of states as desired.

Practitioners sometimes use software to assist them in performing this kind of sensitivity analysis, including generating the graphs. For example, an Excel add-in in your OR Courseware called *SensIt* is designed specifically for conducting sensitivity analysis with probabilistic models such as when applying Bayes' decision rule. Complete documentation for SensIt is included on the book's website. Section 15.5 will describe and illustrate the application of SensIt.

Because the decision the Goferbroke Co. should make depends so critically on the true probability of oil, serious consideration should be given to conducting a seismic survey to estimate this probability more closely. We will explore this option in the next two sections.

■ 15.3   DECISION MAKING WITH EXPERIMENTATION

Frequently, additional testing (experimentation) can be done to improve the preliminary estimates of the probabilities of the respective states of nature provided by the prior probabilities. These improved estimates are called **posterior probabilities.**

We first update the Goferbroke Co. example to incorporate experimentation, then describe how to derive the posterior probabilities, and finally discuss how to decide whether it is worthwhile to conduct experimentation.

Continuing the Prototype Example

As mentioned at the end of Sec. 15.1, an available option before making a decision is to conduct a detailed seismic survey of the land to obtain a better estimate of the probability of oil. The cost is $30,000.

A seismic survey obtains seismic soundings that indicate whether the geological structure is favorable to the presence of oil. We will divide the possible findings of the survey into the following two categories:

USS: Unfavorable seismic soundings; oil is fairly unlikely.
FSS: Favorable seismic soundings; oil is fairly likely.

Based on past experience, if there is oil, then the probability of unfavorable seismic soundings is

$$P(\text{USS} \mid \text{State} = \text{Oil}) = 0.4, \qquad \text{so} \qquad P(\text{FSS} \mid \text{State} = \text{Oil}) = 1 - 0.4 = 0.6.$$

Similarly, if there is no oil (i.e., the true state of nature is *Dry*), then the probability of unfavorable seismic soundings is estimated to be

$$P(\text{USS} \mid \text{State} = \text{Dry}) = 0.8, \qquad \text{so} \qquad P(\text{FSS} \mid \text{State} = \text{Dry}) = 1 - 0.8 = 0.2.$$

We soon will use these data to find the posterior probabilities of the respective states of nature *given* the seismic soundings.

Posterior Probabilities

Proceeding now in general terms, we let

n = number of possible states of nature;

$P(\text{State} = \text{state } i)$ = prior probability that true state of nature is state i, for $i = 1, 2, \ldots, n$;

Finding = finding from experimentation (a random variable);

Finding j = one possible value of finding;

$P(\text{State} = \text{state } i \mid \text{Finding} = \text{finding } j)$ = posterior probability that true state of nature is state i, given that Finding = finding j, for $i = 1, 2, \ldots, n$.

The question currently being addressed is the following:

Given $P(\text{State} = \text{state } i)$ and $P(\text{Finding} = \text{finding } j \mid \text{State} = \text{state } i)$, for $i = 1, 2, \ldots, n$, what is $P(\text{State} = \text{state } i \mid \text{Finding} = \text{finding } j)$?

This question is answered by combining the following standard formulas of probability theory:

$$P(\text{State} = \text{state } i \mid \text{Finding} = \text{finding } j) = \frac{P(\text{State} = \text{state } i, \text{Finding} = \text{finding } j)}{P(\text{Finding} = \text{finding } j)}$$

The **Workers' Compensation Board (WCB) of British Columbia, Canada** is responsible for the occupational health and safety, rehabilitation, and compensation interests of this province's workers and employers. The WCB serves more than 165,000 employers who employ about 1.8 million workers in British Columbia. It spends approximately US$1 billion annually on compensation and rehabilitation.

A key factor in controlling WCB costs is to identify those short-term disability claims that pose a potentially high financial risk of converting into a *far* more expensive long-term disability claim unless there is intensive early *claim-management intervention* to provide the needed medical treatment and rehabilitation. The question was how to accurately identify these high-risk claims so as to minimize the expected total cost of claim compensation and claim-management intervention.

An OR team was formed to study this problem by *applying decision analysis*. For each of numerous categories of injury claims, based on the nature of the injury, the gender and age of the worker, etc., a *decision tree* was used to evaluate whether that category should be classified as low risk (not requiring intervention) or high risk (requiring intervention), depending on the severity of the injury. For each category, a calculation was made of the cutoff point on the critical number of short-term disability claim days paid that would trigger claim-management intervention, so as to minimize the expected cost of claim payments and claim-management intervention. A key in making this calculation was assessing the posterior probability that a claim would become a long-term disability claim, given the number of short-term disability claim days paid.

This application of decision analysis with decision trees is now *saving WCB approximately* **US \$4 million** *per year* while also enabling some injured workers to return to work sooner.

Source: E. Urbanovich, E. E. Young, M. L. Puterman, and S. O. Fattedad: "Early Detection of High-Risk Claims at the Workers' Compensation Board of British Columbia," *Interfaces*, **33**(4): 15–26, July–Aug. 2003. (A link to this article is provided on our website, www.mhhe.com/hillier.)

$$P(\text{Finding} = \text{finding } j) = \sum_{k=1}^{n} P(\text{State} = \text{state } k, \text{Finding} = \text{finding } j)$$

$$P(\text{State} = \text{state } i, \text{Finding} = \text{finding } j) =$$
$$P(\text{Finding} = \text{finding } j \mid \text{State} = \text{state } i)\, P(\text{State} = \text{state } i).$$

Therefore, for each $i = 1, 2, \ldots, n$, the desired formula for the corresponding posterior probability is

$$P(\text{State} = \text{state } i \mid \text{Finding} = \text{finding } j) =$$
$$\frac{P(\text{Finding} = \text{finding } j \mid \text{State} = \text{state } i)\, P(\text{State} = \text{state } i)}{\sum_{k=1}^{n} P(\text{Finding} = \text{finding } j \mid \text{State} = \text{state } k)\, P(\text{State} = \text{state } k)}$$

(This formula often is referred to as **Bayes' theorem** because it was developed by Thomas Bayes, the same 18th-century mathematician who is credited with developing Bayes' decision rule.)

Now let us return to the prototype example and apply this formula. If the finding of the seismic survey is unfavorable seismic soundings (USS), then the posterior probabilities are

$$P(\text{State} = \text{Oil} \mid \text{Finding} = \text{USS}) = \frac{0.4(0.25)}{0.4(0.25) + 0.8(0.75)} = \frac{1}{7},$$

$$P(\text{State} = \text{Dry} \mid \text{Finding} = \text{USS}) = 1 - \frac{1}{7} = \frac{6}{7}.$$

Similarly, if the seismic survey gives favorable seismic soundings (FSS), then

$$P(\text{State} = \text{Oil} \mid \text{Finding} = \text{FSS}) = \frac{0.6(0.25)}{0.6(0.25) + 0.2(0.75)} = \frac{1}{2},$$

$$P(\text{State} = \text{Dry} \mid \text{Finding} = \text{FSS}) = 1 - \frac{1}{2} = \frac{1}{2}.$$

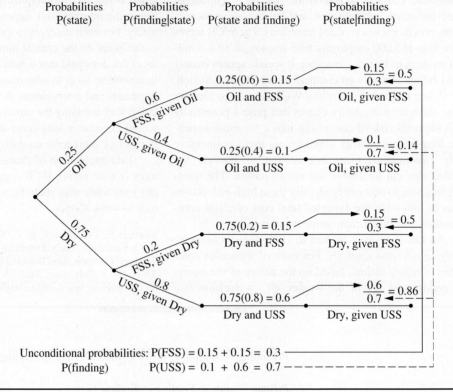

Prior Probabilities P(state)	Conditional Probabilities P(finding|state)	Joint Probabilities P(state and finding)	Posterior Probabilities P(state|finding)

FIGURE 15.2
Probability tree diagram for the full Goferbroke Co. problem showing all the probabilities leading to the calculation of each posterior probability of the state of nature given the finding of the seismic survey.

Unconditional probabilities: P(FSS) = 0.15 + 0.15 = 0.3
P(finding)   P(USS) = 0.1 + 0.6 = 0.7

The **probability tree diagram** in Fig. 15.2 shows a nice way of organizing these calculations in an intuitive manner. The prior probabilities in the first column and the conditional probabilities in the second column are part of the input data for the problem. Multiplying each probability in the first column by a probability in the second column gives the corresponding joint probability in the third column. Each joint probability then becomes the numerator in the calculation of the corresponding posterior probability in the fourth column. Cumulating the joint probabilities with the same finding (as shown at the bottom of the figure) provides the denominator for each posterior probability with this finding. (If you would like to see **another example** of using a probability tree diagram to determine the posterior probabilities, one is included in the Worked Examples section of the book's website.)

Your OR Courseware also includes an Excel template for computing these posterior probabilities, as shown in Fig. 15.3.

After these computations have been completed, Bayes' decision rule can be applied just as before, with the posterior probabilities now replacing the prior probabilities. Again, by using the payoffs (in units of thousands of dollars) from Table 15.2 and subtracting the cost of the experimentation, we obtain the results shown below.

Expected payoffs if finding is unfavorable seismic soundings (USS):

$$E[\text{Payoff (drill} \mid \text{Finding} = \text{USS})] = \frac{1}{7}(700) + \frac{6}{7}(-100) - 30$$

$$= -15.7.$$

$$E[\text{Payoff (sell} \mid \text{Finding} = \text{USS})] = \frac{1}{7}(90) + \frac{6}{7}(90) - 30$$

$$= 60.$$

	A	B	C	D	E	F	G	H
1		**Template**	**for**	**Posterior**	**Probabilities**			
2								
3		**Data:**			P(Finding | State)			
4		State of	Prior		Finding			
5		Nature	Probability	FSS	USS			
6		Oil	0.25	0.6	0.4			
7		Dry	0.75	0.2	0.8			
8								
9								
10								
11								
12		**Posterior**			P(State | Finding)			
13		**Probabilities:**			State of Nature			
14		Finding	P(Finding)	Oil	Dry			
15		FSS	0.3	0.5	0.5			
16		USS	0.7	0.14286	0.85714			
17								
18								
19								

	B	C	D
12	Posterior		P(State | Finding)
13	Probabilities:		State of Nature
14	Finding	P(Finding)	=B6
15	=D5	=SUMPRODUCT(C6:C10,D6:D10)	=C6*D6/SUMPRODUCT(C6:C10,D6:D10)
16	=E5	=SUMPRODUCT(C6:C10,E6:E10)	=C6*E6/SUMPRODUCT(C6:C10,E6:E10)
17	=F5	=SUMPRODUCT(C6:C10,F6:F10)	=C6*F6/SUMPRODUCT(C6:C10,F6:F10)
18	=G5	=SUMPRODUCT(C6:C10,G6:G10)	=C6*G6/SUMPRODUCT(C6:C10,G6:G10)
19	=H5	=SUMPRODUCT(C6:C10,H6:H10)	=C6*H6/SUMPRODUCT(C6:C10,H6:H10)

■ **FIGURE 15.3**
This *posterior probabilities* template in your OR Courseware enables efficient calculation of posterior probabilities, as illustrated here for the full Goferbroke Co. problem.

Expected payoffs if finding is favorable seismic soundings (FSS):

$$E[\text{Payoff (drill} \mid \text{Finding} = \text{FSS})] = \frac{1}{2}(700) + \frac{1}{2}(-100) - 30$$
$$= 270.$$
$$E[\text{Payoff (sell} \mid \text{Finding} = \text{FSS})] = \frac{1}{2}(90) + \frac{1}{2}(90) - 30$$
$$= 60.$$

Since the objective is to maximize the expected payoff, these results yield the optimal policy shown in Table 15.5.

However, what this analysis does not answer is whether it is worth spending $30,000 to conduct the experimentation (the seismic survey). Perhaps it would be better to forgo

■ **TABLE 15.5** The optimal policy with experimentation, under Bayes' decision
rule, for the full Goferbroke Co. problem

Finding from Seismic Survey	Optimal Alternative	Expected Payoff Excluding Cost of Survey	Expected Payoff Including Cost of Survey
USS	Sell the land	90	60
FSS	Drill for oil	300	270

this major expense and just use the optimal solution without experimentation (drill for oil, with an expected payoff of $100,000). We address this issue next.

The Value of Experimentation

Before performing any experiment, we should determine its potential value. We present two complementary methods of evaluating its potential value.

The first method assumes (unrealistically) that the experiment will remove *all* uncertainty about what the true state of nature is, and then this method makes a very quick calculation of what the resulting *improvement in the expected payoff* would be (ignoring the cost of the experiment). This quantity, called the *expected value of perfect information,* provides an *upper bound* on the potential value of the experiment. Therefore, if this upper bound is less than the cost of the experiment, the experiment definitely should be forgone.

However, if this upper bound exceeds the cost of the experiment, then the second (slower) method should be used next. This method calculates the *actual* improvement in the expected payoff (ignoring the cost of the experiment) that would result from performing the experiment. Comparing this improvement (called the *expected value of experimentation*) with the cost indicates whether the experiment should be performed.

Expected Value of Perfect Information.   Suppose now that the experiment could definitely identify what the true state of nature is, thereby providing "perfect" information. Whichever state of nature is identified, you naturally choose the action with the maximum payoff for that state. We do not know in advance which state of nature will be identified, so a calculation of the expected payoff with perfect information (ignoring the cost of the experiment) requires weighting the maximum payoff for each state of nature by the prior probability of that state of nature.

This calculation is shown at the bottom of Table 15.6 for the full Goferbroke Co. problem, where the expected value of perfect information is 242.5. Thus, if the Goferbroke Co. could learn before choosing its action whether the land contains oil, the expected payoff as of now (before acquiring this information) would be $242,500 (excluding the cost of the experiment generating the information.)

To evaluate whether the experiment should be conducted, we now use this quantity to calculate the expected value of perfect information.

The **expected value of perfect information,** abbreviated **EVPI,** is calculated as

$$\text{EVPI} = \text{expected payoff with perfect information} - \text{expected payoff without experimentation.}^{[2]}$$

TABLE 15.6 Expected payoff with perfect information for the full Goferbroke Co. problem

Alternative	State of Nature	
	Oil	Dry
1. Drill for oil	700	−100
2. Sell the land	90	90
Maximum payoff	700	90
Prior probability	0.25	0.75

Expected payoff with perfect information = 0.25(700) + 0.75(90) = 242.5

---

[2]The *value of perfect information* is a random variable equal to the payoff with perfect information *minus* the payoff without experimentation. EVPI is the expected value of this random variable.

Thus, since experimentation usually cannot provide perfect information, EVPI provides an upper bound on the expected value of experimentation.

For this same example, we found in Sec. 15.2 that the expected payoff without experimentation (under Bayes' decision rule) is 100. Therefore,

$$\text{EVPI} = 242.5 - 100 = 142.5.$$

Since 142.5 far exceeds 30, the cost of experimentation (a seismic survey), it may be worthwhile to proceed with the seismic survey. To find out for sure, we now go to the second method of evaluating the potential benefit of experimentation.

Expected Value of Experimentation. Rather than just obtain an upper bound on the *expected increase in payoff* (excluding the cost of the experiment) due to performing experimentation, we now will do somewhat more work to calculate this expected increase directly. This quantity is called the *expected value of experimentation.* (It also is sometimes called the *expected value of sample information.*)

Calculating this quantity requires first computing the expected payoff with experimentation (excluding the cost of the experiment). Obtaining this latter quantity requires doing all the work described earlier to find all the posterior probabilities, the resulting optimal policy with experimentation, and the corresponding expected payoff (excluding the cost of the experiment) for each possible finding from the experiment. Then each of these expected payoffs needs to be weighted by the probability of the corresponding finding, that is,

$$\text{Expected payoff with experimentation} = \sum_j P(\text{Finding} = \text{finding } j) \\ E[\text{payoff} \mid \text{Finding} = \text{finding } j],$$

where the summation is taken over all possible values of *j*.

For the prototype example, we have already done all the work to obtain the terms on the right side of this equation. The values of $P(\text{Finding} = \text{finding } j)$ for the two possible findings from the seismic survey—unfavorable (USS) and favorable (FSS)—were calculated at the bottom of the probability tree diagram in Fig. 15.2 as

$$P(\text{USS}) = 0.7, \qquad P(\text{FSS}) = 0.3.$$

For the optimal policy with experimentation, the corresponding expected payoff (excluding the cost of the seismic survey) for each finding was obtained in the third column of Table 15.5 as

$$E(\text{Payoff} \mid \text{Finding} = \text{USS}) = 90,$$
$$E(\text{Payoff} \mid \text{Finding} = \text{FSS}) = 300.$$

With these numbers,

$$\text{Expected payoff with experimentation} = 0.7(90) + 0.3(300)$$
$$= 153.$$

Now we are ready to calculate the expected value of experimentation.

The **expected value of experimentation,** abbreviated **EVE,** is calculated as

EVE = expected payoff with experimentation − expected payoff without experimentation.

Thus, EVE identifies the potential value of experimentation.

For the Goferbroke Co.,

$$\text{EVE} = 153 - 100 = 53.$$

Since this value exceeds 30, the cost of conducting a detailed seismic survey (in units of thousands of dollars), this experimentation should be done.

15.4 DECISION TREES

Decision trees provide a useful way of *visually displaying* the problem and then *organizing the computational work* already described in the preceding two sections. These trees are especially helpful when a *sequence of decisions* must be made.

Constructing the Decision Tree

The prototype example involves a sequence of two decisions:

1. Should a seismic survey be conducted before an action is chosen?
2. Which action (drill for oil or sell the land) should be chosen?

The corresponding decision tree (before adding numbers and performing computations) is displayed in Fig. 15.4.

The junction points in the decision tree are referred to as **nodes** (or forks), and the lines are called **branches.**

A **decision node,** represented by a square, indicates that a decision needs to be made at that point in the process. An **event node** (or chance node), represented by a circle, indicates that a random event occurs at that point.

■ **FIGURE 15.4**
The decision tree (before including any numbers) for the full Goferbroke Co. problem.

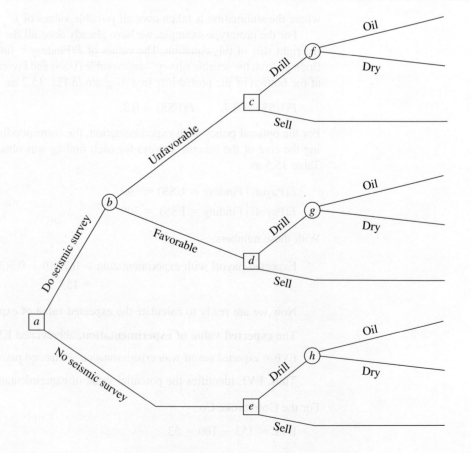

The **Westinghouse Science and Technology Center** is the Westinghouse Electric Corporation's main research and development (R&D) arm to develop new technology. The process of evaluating R&D projects to decide which ones should be initiated and then which ones should be continued as progress is made (or not made) is particularly challenging for management because of the great uncertainties and very long time horizons involved. The actual launch date for an embryonic technology may be years, even decades, removed from its inception as a modest R&D proposal to investigate the technology's potential.

As the Center came under increasing pressure to reduce costs and deliver high-impact technology quickly, the Center's controller funded an operations research project to improve this evaluation process. The OR team developed a *decision tree approach* to analyzing any R&D proposal while considering its complete sequence of key decision points. The first decision point is whether to fund the proposed embryonic project for the first year or so. If its early technical milestones are reached, the next decision point is whether to continue funding the project for some period. This may then be repeated one or more times. If the late technical milestones are reached, the next decision point is whether to prelaunch because the innovation still meets strategic business objectives. If a strategic fit is achieved, the final decision point is whether to commercialize the innovation now or to delay its launch, or to abandon it altogether. A *decision tree* with a progression of decision nodes and intervening event nodes provides a natural way of depicting and analyzing such an R&D project.

Source: R. K. Perdue, W. J. McAllister, P. V. King, and B. G. Berkey: "Valuation of R and D Projects Using Options Pricing and Decision Analysis Models," *Interfaces*, **29**(6): 57–74, Nov.–Dec. 1999. (A link to this article is provided on our website, www.mhhe.com/hillier.)

Thus, in Fig. 15.4, the first decision is represented by decision node *a*. Node *b* is an event node representing the random event of the outcome of the seismic survey. The two branches emanating from event node *b* represent the two possible outcomes of the survey. Next comes the second decision (nodes *c, d,* and *e*) with its two possible choices. If the decision is to drill for oil, then we come to another event node (nodes *f, g,* and *h*), where its two branches correspond to the two possible states of nature.

Note that the path followed from node *a* to reach any terminal branch (except the bottom one) is determined both by the decisions made and by random events that are outside the control of the decision maker. This is characteristic of problems addressed by decision analysis.

The next step in constructing the decision tree is to insert numbers into the tree as shown in Fig. 15.5. The numbers under or over the branches that are *not* in parentheses are the cash flows (in thousands of dollars) that occur at those branches. For each path through the tree from node *a* to a terminal branch, these same numbers then are added to obtain the resulting total payoff shown in boldface to the right of that branch. The last set of numbers is the probabilities of random events. In particular, since each branch emanating from an event node represents a possible random event, the probability of this event occurring from this node has been inserted in parentheses along this branch. From event node *h*, the probabilities are the *prior probabilities* of these states of nature, since no seismic survey has been conducted to obtain more information in this case. However, event nodes *f* and *g* lead out of a decision to do the seismic survey (and then to drill). Therefore, the probabilities from these event nodes are the *posterior probabilities* of the states of nature, given the finding from the seismic survey, where these numbers are given in Figs. 15.2 and 15.3. Finally, we have the two branches emanating from event node *b*. The numbers here are the probabilities of these findings from the seismic survey, Favorable (FSS) or Unfavorable (USS), as given underneath the probability tree diagram in Fig. 15.2 or in cells C15:C16 of Fig. 15.3.

Performing the Analysis

Having constructed the decision tree, including its numbers, we now are ready to analyze the problem by using the following procedure.

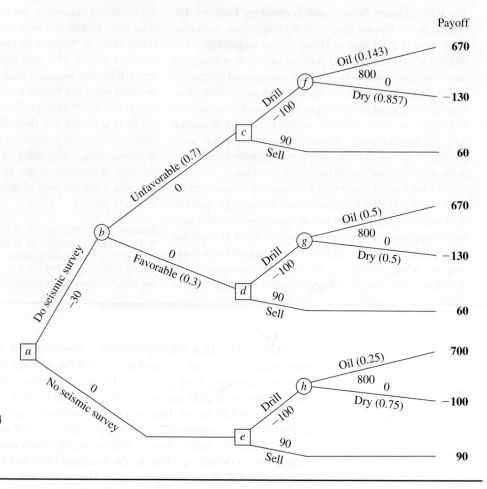

Payoff

FIGURE 15.5
The decision tree in Fig. 15.4 after adding both the probabilities of random events and the payoffs.

1. Start at the right side of the decision tree and move left one column at a time. For each column, perform either step 2 or step 3 depending upon whether the nodes in that column are event nodes or decision nodes.

2. For each event node, calculate its *expected payoff* by multiplying the expected payoff of each branch (shown in boldface to the right of the branch) by the probability of that branch and then summing these products. Record this expected payoff for each decision node in boldface next to the node, and designate this quantity as also being the expected payoff for the branch leading to this node.

3. For each decision node, compare the expected payoffs of its branches and choose the alternative whose branch has the largest expected payoff. In each case, record the choice on the decision tree by inserting a double dash as a barrier through each rejected branch.

To begin the procedure, consider the rightmost column of nodes, namely, event nodes *f*, *g*, and *h*. Applying step 2, their expected payoffs (EP) are calculated as

$$\text{EP} = \frac{1}{7}(670) + \frac{6}{7}(-130) = -15.7, \qquad \text{for node } f,$$

$$\text{EP} = \frac{1}{2}(670) + \frac{1}{2}(-130) = 270, \qquad \text{for node } g,$$

$$\text{EP} = \frac{1}{4}(700) + \frac{3}{4}(-100) = 100, \qquad \text{for node } h.$$

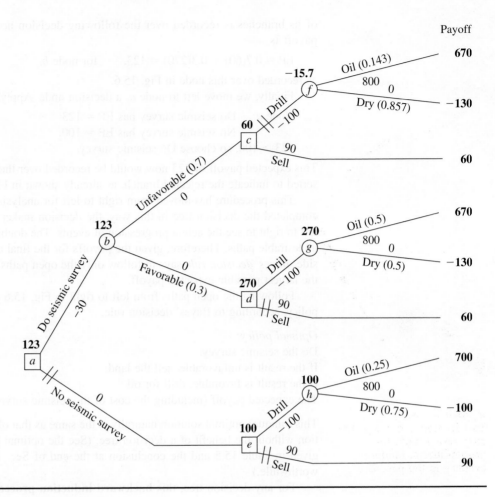

Payoff

■ FIGURE 15.6
The final decision tree that records the analysis for the full Goferbroke Co. problem when using monetary payoffs.

These expected payoffs then are placed above these nodes, as shown in Fig. 15.6.

Next, we move one column to the left, which consists of decision nodes *c, d,* and *e.* The expected payoff for a branch that leads to an event node now is recorded in boldface over that event node. Therefore, step 3 can be applied as follows.

Node *c:*   Drill alternative has EP = −15.7.
             Sell alternative has EP = 60.
60 > −15.7, so choose the Sell alternative.

Node *d:*   Drill alternative has EP = 270.
             Sell alternative has EP = 60.
270 > 60, so choose the Drill alternative.

Node *e:*   Drill alternative has EP = 100.
             Sell alternative has EP = 90.
100 > 90, so choose the Drill alternative.

The expected payoff for each chosen alternative now would be recorded in boldface over its decision node, as already shown in Fig. 15.6. The chosen alternative also is indicated by inserting a double dash as a barrier through each rejected branch.

Next, moving one more column to the left brings us to node *b.* Since this is an event node, step 2 of the procedure needs to be applied. The expected payoff for each of its branches is recorded over the following decision node. Therefore, the expected payoff is

$$EP = 0.7(60) + 0.3(270) = 123, \qquad \text{for node } b,$$

as recorded over this node in Fig. 15.6.

Finally, we move left to node a, a decision node. Applying step 3 yields

Node a: Do seismic survey has EP = 123.
            No seismic survey has EP = 100.

123 > 100, so choose Do seismic survey.

This expected payoff of 123 now would be recorded over the node, and a double dash inserted to indicate the rejected branch, as already shown in Fig. 15.6.

This procedure has moved from right to left for analysis purposes. However, having completed the decision tree in this way, the decision maker now can read the tree from left to right to see the actual progression of events. The double dashes have closed off the undesirable paths. Therefore, given the payoffs for the final outcomes shown on the right side, *Bayes' decision rule* says to follow only the open paths from left to right to achieve the largest possible expected payoff.

Following the open paths from left to right in Fig. 15.6 yields the following optimal policy, according to Bayes' decision rule.

Optimal policy:
Do the seismic survey.
If the result is unfavorable, sell the land.
If the result is favorable, drill for oil.
The expected payoff (including the cost of the seismic survey) is 123 ($123,000).

This (unique) optimal solution naturally is the same as that obtained in the preceding section without the benefit of a decision tree. (See the optimal policy with experimentation given in Table 15.5 and the conclusion at the end of Sec. 15.3 that experimentation is worthwhile.)

For any decision tree, this **backward induction procedure** always will lead to the *optimal policy* (or policies) after the probabilities are computed for the branches emanating from an event node.

Another example of solving a decision tree in this way is included in the Worked Examples section of the book's website.

■ 15.5 USING SPREADSHEETS TO PERFORM SENSITIVITY ANALYSIS ON DECISION TREES

Some helpful spreadsheet software now is available for constructing and analyzing decision trees on spreadsheets. One popular Excel add-in of this type is *TreePlan,* which is shareware developed by Professor Michael Middleton. The academic version of TreePlan (along with accompanying documentation) is included in your OR Courseware, along with Professor Middleton's companion shareware SensIt. (If you want to continue to use either software package after this course, you will need to register and pay the shareware fee.) As mentioned at the end of Sec. 15.2, SensIt is designed for conducting sensitivity analysis.

Before turning to SensIt, we will describe how TreePlan is used to create a decision tree. To simplify this discussion, we will begin by illustrating the construction of a small decision tree for the first Goferbroke Co. problem (no consideration of conducting a seismic survey) before considering the full problem.

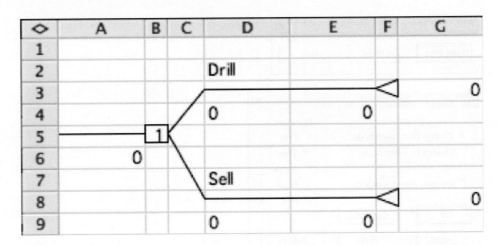

■ **FIGURE 15.7**
The default decision tree created by TreePlan by selecting Decision Tree from the Add-Ins tab or Tools menu, clicking on New Tree, and then entering the Drill and Sell labels for the two decision alternatives.

How TreePlan Constructs the Decision Tree for the First Goferbroke Co. Problem

Consider the first Goferbroke Co. problem (no seismic survey) as summarized earlier in Table 15.2. To begin creating the corresponding decision tree using TreePlan (after installing this add-in in Excel), select Decision Tree from the Add-Ins tab (for Excel 2007) or Tools menu (for earlier versions of Excel) and click on New Tree. This creates the default decision tree shown in Fig. 15.7 with a single (square) decision node with two branches. It so happens that this is exactly what is needed for the first node in the current problem. However, even if something else were needed, it is easy to make changes to a node in TreePlan. Simply select the cell containing the node (B5 in Fig. 15.7) and choose Decision Tree from the Add-Ins tab or Tools menu. This brings up a dialogue box that allows you to change the type of node (e.g., from a decision node to an event node) or add more branches.

By default, the labels for the decisions (cells D2 and D7 in Fig. 15.7) are "Decision 1," "Decision 2," and so on. These labels are changed by clicking on them and typing a new label. In Fig. 15.7, these labels have already been changed to "Drill" and "Sell," respectively.

If the decision is to drill, the next event is to learn whether or not the land contains oil. To create an event node, click on the cell containing the triangle terminal node at the end of the drill branch (cell F3 in Fig. 15.7), and choose Decision Tree from the Add-Ins tab or Tools menu. This brings up the TreePlan Acad.-Terminal Node dialogue box shown second from the top in Fig. 15.8. Choose the "Change to event node" option on the left and select the two branches option on the right and then click OK. This results in the decision tree with the nodes and branches shown in Fig. 15.9 (after replacing the default labels "Event 1" and "Event 2" with "Oil" and "Dry," respectively).

At any time, you also can click on any existing decision node (a square) or event node (a circle) and choose Decision Tree from the Add-Ins tab or Tools menu to bring up the corresponding dialogue box—"TreePlan Acad.-Decision Node" or "TreePlan Acad.-Event Node"—to make any of the modifications listed in Fig. 15.8 at that node.

Initially, each branch would show a default value of 0 for the net cash flow being generated there (the numbers appear below the branch labels: D6, D14, H4, and H9 in Fig. 15.9). Also, each of the two branches leading from the event node would display default values of 0.5 for their prior probabilities (the probabilities are just above the corresponding labels: H1 and H6 in Fig. 15.9). Therefore, you next should click on these default values and replace them with the correct numbers, namely,

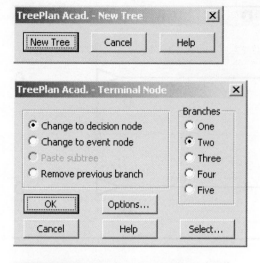

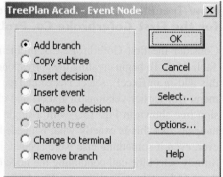

■ FIGURE 15.8
The dialogue boxes used by TreePlan for constructing a decision tree.

■ FIGURE 15.9
The decision tree constructed and solved by TreePlan for the first Goferbroke Co. problem as presented in Table 15.2, where the 1 in cell B9 indicates that the top branch (the Drill alternative) should be chosen.

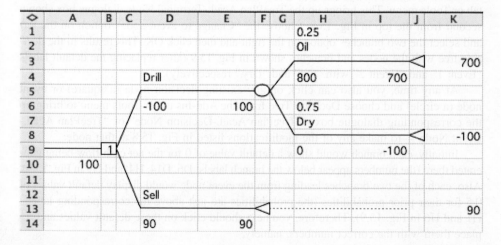

D6   = –100 (the cost of drilling is $100,000),
D14 =   90 (the profit from selling is $90,000),
H1   = 0.25 (the prior probability of oil is 0.25),
H4   = 800 (the net revenue after finding oil is $800,000),
H6   = 0.75 (the prior probability of dry is 0.75),
H9   =   0 (the net revenue after finding dry is 0),

as shown in the figure.

At each stage in constructing a decision tree, TreePlan automatically solves for the optimal policy with the current tree when using *Bayes' decision rule.* The number inside each decision node indicates which branch should be chosen (assuming the branches emanating from that node are numbered consecutively from top to bottom). Thus, for the final decision tree in Fig. 15.9, the number 1 in cell B9 specifies that the first branch (the Drill alternative) should be chosen. The number on both sides of each terminal node is the payoff if that node is reached. The number 100 in cells A10 and E6 is the *expected payoff* at those stages in the process.

We think that you will find this procedure with TreePlan quite intuitive when you execute it on a computer. If you spend considerable time with TreePlan, you also will find that it has many helpful features that haven't been described in this brief introduction.

The Decision Tree for the Full Goferbroke Co. Problem

Now consider the full Goferbroke Co. problem, where the first decision to be made is whether to conduct a seismic survey. Continuing the procedure described above, TreePlan would be used to construct and solve the decision tree shown in Fig. 15.10. Although the form is somewhat different, note that this decision tree is completely equivalent to the one in Fig. 15.6. Besides the convenience of constructing the tree directly on a spreadsheet, TreePlan also provides the key advantage of automatically solving the decision tree. Rather than relying on hand calculations as in Fig. 15.6, TreePlan instantaneously calculates all the expected payoffs at each stage of the tree, as shown next to each node, as soon as the decision tree is constructed. Instead of using double dashes, TreePlan puts a number inside each decision node indicating which branch should be chosen (assuming the branches emanating from that node are numbered consecutively from top to bottom).

Organizing the Spreadsheet to Perform Sensitivity Analysis

The end of Sec. 15.2 illustrated how sensitivity analysis can be performed on a small problem (the first Goferbroke Co. problem), where only a single decision (drill or sell) needs to be made. In that case, the analysis was quite straightforward because the expected payoff for each decision alternative could be expressed as a simple function of the model parameter (the prior probability of oil) being considered. By contrast, when a sequence of decisions needs to be made, as for the full Goferbroke Co. problem, sensitivity analysis becomes somewhat more involved. There now are more model parameters (the various costs, revenues, and probabilities) that might have sufficient uncertainty to warrant performing sensitivity analysis. Furthermore, finding the maximum expected payoff for any particular values of the model parameters now requires solving a decision tree. Therefore, using spreadsheet software such as TreePlan that automatically solves the decision tree becomes very helpful. Adding software that is specifically designed for conducting sensitivity analysis, such as SensIt, then quickly provides further insights.

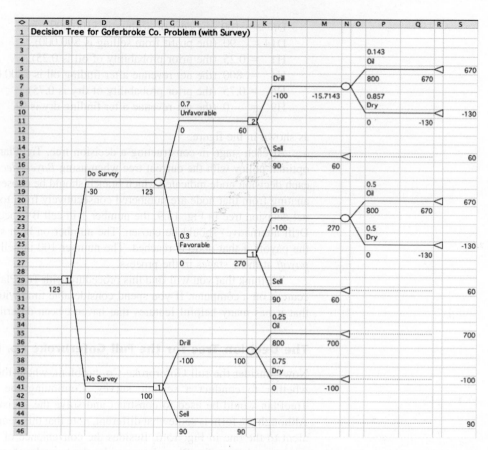

FIGURE 15.10

The decision tree constructed and solved by TreePlan for the full Goferbroke Co. problem that also considers whether to do a seismic survey.

Beginning with the spreadsheet that already contains the decision tree, the next step is to expand and organize this spreadsheet for performing sensitivity analysis. We now will illustrate this for the full Goferbroke Co. problem by starting with the spreadsheet in Fig. 15.10 that contains the decision tree constructed by TreePlan.

It is helpful to begin by consolidating the data and results into a new section, as shown on the right-hand side of Fig. 15.11. All the data cells in the decision tree now would need to make reference to the consolidated data cells (cells V4:V11), as illustrated by the formulas shown for cells P6 and P11 at the bottom of the figure. Similarly, the summarized results to the right of the decision tree make reference to the output cells within the decision tree (the decision nodes in cells B29, F41, J11, and J26, as well as the expected payoff in cell A30) by using the formulas for cells U19, V15, V26, and W19:W20 displayed at the bottom of Fig. 15.11.

The probability data in the decision tree are complicated by the fact that the posterior probabilities will need to be updated any time a change is made in any of the prior probability data. Fortunately, the template for calculating posterior probabilities (as shown in Fig. 15.3) can be used to do these calculations. The relevant portion of this template (B3:H19) has been copied (using the Copy and Paste commands in the Edit menu) to the spreadsheet in Fig. 15.11 (now appearing in U30:AA46). The data for the template refer to the probability data in the data cells PriorProbabilityOfOil (V9), ProbFSSGivenOil (V10), and ProbUSSGivenDry (V11), as shown in the formulas for cells V33:X34 at the

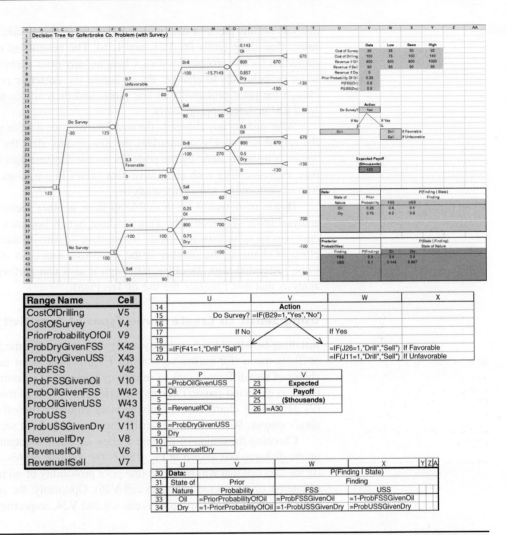

Range Name	Cell
CostOfDrilling	V5
CostOfSurvey	V4
PriorProbabilityOfOil	V9
ProbDryGivenFSS	X42
ProbDryGivenUSS	X43
ProbFSS	V42
ProbFSSGivenOil	V10
ProbOilGivenFSS	W42
ProbOilGivenUSS	W43
ProbUSS	V43
ProbUSSGivenDry	V11
RevenueIfDry	V8
RevenueIfOil	V6
RevenueIfSell	V7

■ FIGURE 15.11

In preparation for performing sensitivity analysis on the full Goferbroke problem, the data and results have been consolidated on the spreadsheet to the right of the decision tree.

bottom of Fig. 15.11. The template automatically calculates the probability of each finding and the posterior probabilities (in cells V42:X43) based on these data. The decision tree then refers to these calculated probabilities when they are needed, as shown in the formulas for cells P3:P11 in Fig. 15.11.

Consolidating the data and results offers a couple of advantages. First, it assures that each piece of data is in only one place. Each time that piece of data is needed in the decision tree, a reference is made to the single data cell. This greatly simplifies sensitivity analysis. To change a piece of data, you need to change it in only one place rather than searching through the entire tree to find and change all occurrences of that piece of data. A second advantage of consolidating the data and results is that it makes it easy for *anyone* to interpret the model. It is not necessary to understand TreePlan or how to read a decision tree in order to see what data were used in the model or what the suggested plan of action and expected payoff are.

While it takes some time and effort to consolidate the data and results, including all the necessary cross-referencing, this step is truly essential for performing sensitivity analysis. Many pieces of data are used in several places on the decision tree. For example, the revenue if Goferbroke finds oil appears in cells P6, P21, and L36. Performing sensitivity analysis on this piece of data now requires changing its value in only one place (cell V6) rather than three (cells P6, P21, and L36). The benefits of consolidation are even more important for the probability data. Changing any prior probability may cause *all* the posterior probabilities to change. By including the posterior probability template, you can change the prior probability in one place, and then all the other probabilities are calculated and updated appropriately.

After making any change in the cost data, revenue data, or probability data in Fig. 15.11, the spreadsheet nicely summarizes the new results after the actual work to obtain these results is instantly done by the posterior probability template and the decision tree. Therefore, experimenting with alternative data values in a trial-and-error manner is one useful way of performing sensitivity analysis.

However, it would be desirable to have another method of performing sensitivity analysis more systematically. This is where SensIt becomes very helpful. It provides an easy way to systematically create informative sensitivity analysis graphs that display the effect of changing the number in the respective data cells of interest. SensIt is designed to be integrated with TreePlan (although it also can perform other types of sensitivity analysis that don't require the use of TreePlan).

Using SensIt to Create Three Types of Sensitivity Analysis Graphs

Installing SensIt adds a Sensitivity Analysis menu item to the Add-Ins tab (for Excel 2007) or Tools menu (for earlier versions of Excel). This menu item has a submenu giving a choice of two different kinds of sensitivity analysis: (1) plotting a graph of a *single output* (such as expected payoff) versus a *single input* (such as the prior probability of oil) or (2) generating charts that simultaneously compare the effect of *multiple inputs* on a *single output*. We will now describe these two kinds of sensitivity analysis in turn.

Choosing the option of plotting a graph of a single output versus a single input brings up the dialogue box shown in Fig. 15.12. The top half of this dialogue box is used to specify the data cell that will be varied (the prior probability of oil in cell V9) and the output cell of interest (the expected payoff in cell V26). Optionally, the cells containing the labels for these cells may also be specified (cells U9 and V24, respectively). These labels are used to

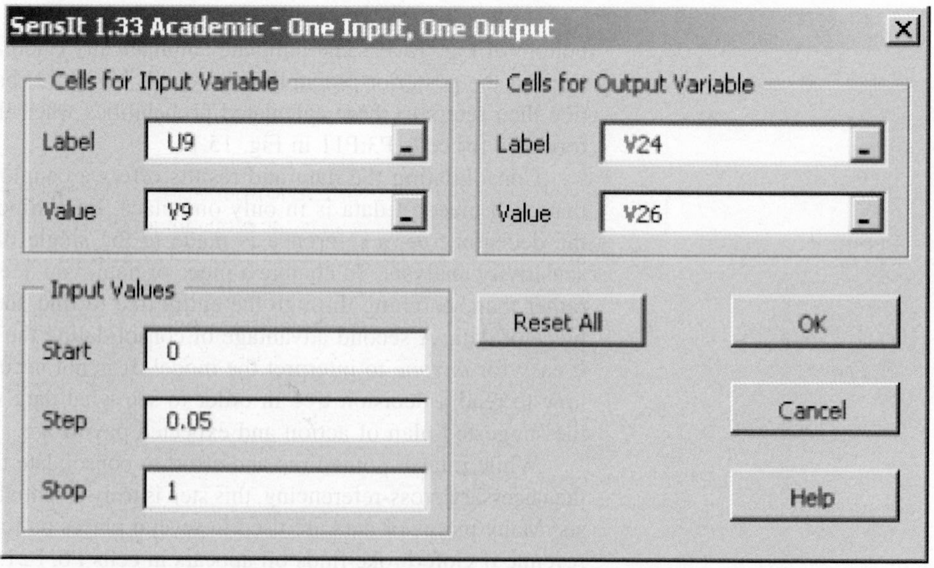

■ **FIGURE 15.12**
The dialogue box used by SensIt to plot a graph of a single output versus a single input.

SensIt 1.33 Academic - One Input, One Output	
Cells for Input Variable	**Cells for Output Variable**
Label: U9	Label: V24
Value: V9	Value: V26
Input Values	
Start: 0	Reset All / OK
Step: 0.05	Cancel
Stop: 1	Help

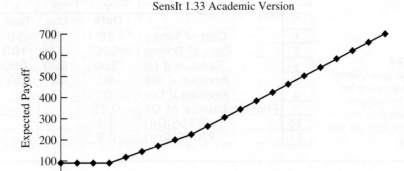

SensIt 1.33 Academic Version

■ **FIGURE 15.13**
The graph generated
by SensIt for the full
Goferbroke Co. problem
to show how the expected
payoff (when using Bayes'
decision rule) depends on
the prior probability of oil.

label the axes of the graph that is created. The bottom half of the dialogue box is used to specify the range of values to be considered for the single data cell (the prior probability of oil). In this case, all values between 0 and 1 (at intervals of 0.05) will be considered. Clicking OK then generates the graph shown in Fig. 15.13 that reveals the relationship between the prior probability of oil and the expected payoff that results from using the optimal policy given this probability.

This graph indicates that the expected payoff starts increasing when the prior probability is a little over 0.15 and then starts increasing more rapidly when this probability is around 0.3. This suggests that the optimal policy changes at roughly these values of the prior probability. To check this out, the spreadsheet in Fig. 15.11 can be used to see how the results change when the prior probability of oil is slowly increased in the vicinity of these values. This kind of trial-and-error analysis soon leads to the following conclusions about how the optimal policy depends on this probability.

Optimal Policy

Let p = Prior probability of oil.

If $p \leq 0.168$, then sell the land (no seismic survey).

If $0.169 \leq p \leq 0.308$, then do the survey: drill if favorable and sell if not.

If $p \geq 0.309$, then drill for oil (no seismic survey).

This sensitivity analysis has focused so far on investigating the effect if the true probability of finding oil is different from the original prior probability of 0.25. Similar analysis could be done with respect to the probabilities in cells V10:V11 of Fig. 15.11. However, since there is significant uncertainty about the cost and revenue data in cells V4:V7, we turn next to performing sensitivity analysis with respect to these data.

Suppose we want to investigate how the expected payoff would change if any one of the costs or revenues in cells V4:V7 were to change. This requires making some additions to the original spreadsheet (Fig. 15.11). As shown in Fig. 15.14, three columns are added for each data cell that will be varied, indicating the lowest value, base value, and highest value. Suppose that the cost of the survey and the revenue if the land is sold are fairly predictable (thus varying over a small range of 28–32 and 85–95, respectively), while the cost of drilling and the revenue if oil is struck are more variable (thus varying over a large range of 75–140 and 600–1,000, respectively).

■ **FIGURE 15.14**
Expansion of the spreadsheet in Fig. 15.11 to prepare for using SensIt to investigate the effect of changing any cost or revenue values on the expected payoff.

	U	V	W	X	Y
3		**Data**	Low	Base	High
4	Cost of Survey	30	28	30	32
5	Cost of Drilling	100	75	100	140
6	Revenue if Oil	800	600	800	1000
7	Revenue if Sell	90	85	90	95
8	Revenue if Dry	0			
9	Prior Probability of Oil	0.25			
10	P(FSS|Oil)	0.6			
11	P(USS|Dry)	0.8			

Since we want to investigate how the expected payoff would change if any one of the costs or revenues in cells V4:V7 were to change, we now have four inputs (these costs and revenues) and one output (the expected payoff). Therefore, after expanding the spreadsheet as shown in Fig. 15.14, the next step is to bring up the SensIt dialogue box for "many inputs, one output." This dialogue box (called by choosing the corresponding item in the Sensitivity Analysis menu on the Add-Ins tab for Excel 2007 or under the Tools menu for earlier versions of Excel) is shown in Fig. 15.15. It is used to specify which contiguous data cells will be varied, which output cell will be examined, and the location of the cells specifying the range (low, base, and high) for the data cells. The Step Percent box is used to specify the desired step size (as a percentage of the base value) in each input value at which the expected payoff will be recalculated until the low and high values of the input are reached. The lower right-hand side of the dialogue box gives a choice of three charts for displaying the effect of alternative values of any one of these inputs on the output. Suppose that the "single-factor spider chart" option is chosen (as shown in Fig. 15.16). Clicking OK then generates the **spider chart** shown in Fig. 15.17.

Each line in the spider chart in this figure plots the expected payoff as one of the selected data cells (V4:V7) is changed from its original value by being multiplied by the percentage indicated along the bottom of the graph. (The *cost of survey line* lies on top of the *cost of drilling line*, but is much shorter than the latter line since it only extends to

■ **FIGURE 15.15**
The dialogue box used by SensIt to simultaneously investigate the effect of changing any one of several inputs on a single output.

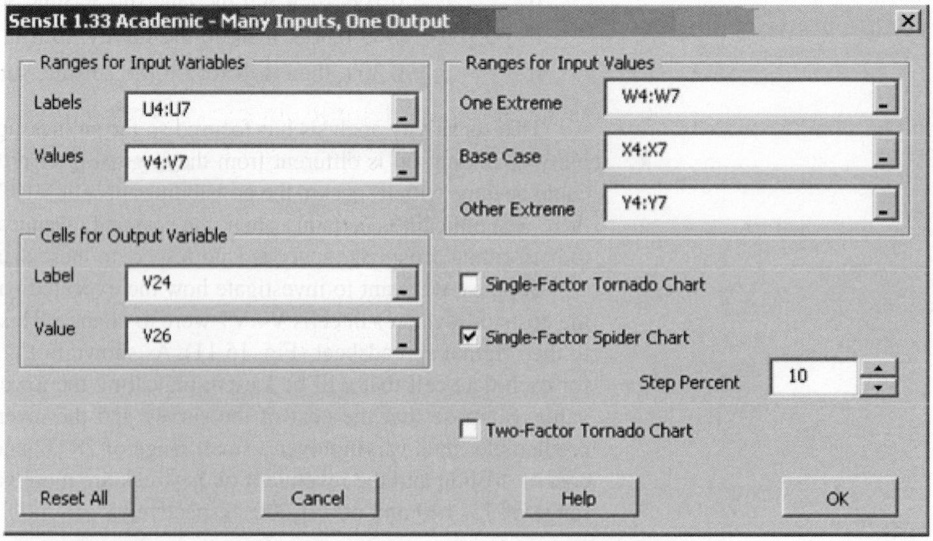

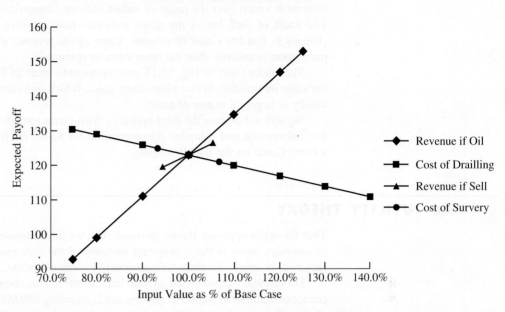

FIGURE 15.16
The spider chart generated by SensIt for the full Goferbroke Co. problem to show how the expected payoff (when using Bayes' decision rule) varies with changes in any one of the cost or revenue estimates.

93.3% on its left side and to 106.7% on its right side.) The fact that the *revenue if oil line* is the steepest one reveals that the expected payoff is particularly sensitive to the estimate of the revenue if oil is found, so any additional work on refining the estimates should focus the most attention on this one.

Now suppose that the "single-factor tornado chart" option in Fig. 15.15 is chosen instead. Clicking OK then generates the **tornado chart** shown in Fig. 15.17. Each bar in

FIGURE 15.17
The tornado chart generated SensIt for the full Goferbroke Co. problem to show how much the expected payoff (when using Bayes' decision rule) can vary over the entire range of likely values of any one of the cost or revenue estimates.

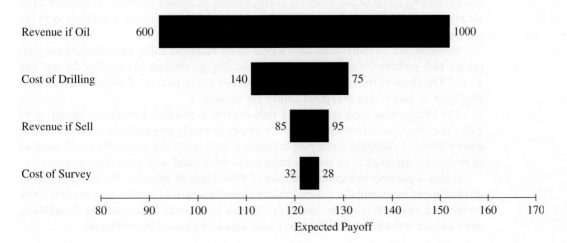

the graph shows the range of change in the expected payoff as the corresponding cost or revenue is varied over the range of values indicated numerically at the ends of each bar. The width of each bar in the graph measures how sensitive the expected payoff is to changes in that bar's cost or revenue. Once again, *revenue if oil* stands out as causing much more sensitivity than the other costs or revenues.

The spider chart in Fig. 15.16 and the tornado chart in Fig. 15.17 actually provide the same information in complementary ways. Which one conveys this information more vividly is largely a matter of taste.

We will not discuss the third option (a "two-factor tornado chart") in Fig. 15.15. Further information and complete documentation for SensIt (as for TreePlan) is provided in a Users Guide on the book's website.

■ 15.6   UTILITY THEORY

Thus far, when applying Bayes' decision rule, we have assumed that the expected payoff in *monetary terms* is the appropriate measure of the consequences of taking an action. However, in many situations this assumption is inappropriate.

For example, suppose that an individual is offered the choice of (1) accepting a 50:50 chance of winning $100,000 or nothing or (2) receiving $40,000 with certainty. Many people would prefer the $40,000 even though the expected payoff on the 50:50 chance of winning $100,000 is $50,000. A company may be unwilling to invest a large sum of money in a new product even when the expected profit is substantial if there is a risk of losing its investment and thereby becoming bankrupt. People buy insurance even though it is a poor investment from the viewpoint of the expected payoff.

Do these examples invalidate Bayes' decision rule? Fortunately, the answer is no, because there is a way of transforming *monetary values* to an appropriate scale that reflects the decision maker's preferences. This scale is called the *utility function for money.*

Utility Functions for Money

Figure 15.18 shows a typical **utility function $U(M)$ for money M.** It indicates that an individual having this utility function would value obtaining $30,000 twice as much as $10,000 and would value obtaining $100,000 twice as much as $30,000. This reflects the fact that the person's highest-priority needs would be met by the first $10,000. Having this decreasing slope of the function as the amount of money increases is referred to as having a **decreasing marginal utility for money.** Such an individual is referred to as being **risk-averse.**

However, not all individuals have a decreasing marginal utility for money. Some people are **risk seekers** instead of *risk-averse,* and they go through life looking for the "big score." The slope of their utility function *increases* as the amount of money increases, so they have an **increasing marginal utility for money.**

The intermediate case is that of a **risk-neutral** individual, who prizes money at its face value. Such an individual's utility for money is simply proportional to the amount of money involved. Although some people appear to be risk-neutral when only small amounts of money are involved, it is unusual to be truly risk-neutral with very large amounts.

It also is possible to exhibit a mixture of these kinds of behavior. For example, an individual might be essentially risk-neutral with small amounts of money, then become a risk seeker with moderate amounts, and then turn risk-averse with large amounts. In addition, one's attitude toward risk can shift over time depending upon circumstances.

An individual's attitude toward risk also may be different when dealing with one's personal finances than when making decisions on behalf of an organization. For example, managers of a business firm need to consider the company's circumstances and the collective philosophy of top management in determining the appropriate attitude toward risk when making managerial decisions.[3]

The fact that different people have different utility functions for money has an important implication for decision making in the face of uncertainty.

> When a *utility function for money* is incorporated into a decision analysis approach to a problem, this utility function must be constructed to fit the preferences and values of the decision maker involved. (The decision maker can be either a single individual or a group of people.)

The *scale* of the utility function is irrelevant. In other words, it doesn't matter whether the value of $U(M)$ at the dashed lines in Fig. 15.18 are 0.25, 0.5, 0.75, 1 (as shown) or 10,000, 20,000, 30,000, 40,000, or whatever. All the utilities can be multiplied by any positive constant without affecting which alternative course of action will

■ **FIGURE 15.18**
A typical utility function for money, where $U(M)$ is the utility of obtaining an amount of money M.

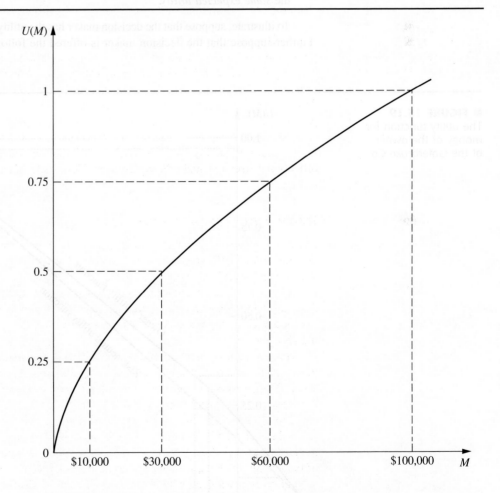

[3]For a survey of the shape of the utility function for 332 owner-managers and the impact of this shape on organizational behavior, see J. M. E. Pennings and A. Smidts, "The Shape of Utility Functions and Organizational Behavior," *Management Science,* **49:** 1251–1263, 2003.

have the largest expected utility. It also is possible to add the same constant (positive or negative) to all the utilities without affecting which course of action will have the largest expected utility.

For these reasons, we have the liberty to set the value of $U(M)$ arbitrarily for two values of M, so long as the higher monetary value has the higher utility. It is particularly convenient (although certainly not necessary) to set $U(M) = 0$ for the smallest value of M under consideration and to set $U(M) = 1$ for the largest M, as was done in Fig. 15.18. By assigning a utility of 0 to the worst outcome and a utility of 1 to the best outcome, and then determining the utilities of the other outcomes accordingly, it becomes easy to see the relative utility of each outcome along the scale from worst to best.

The key to constructing the utility function for money to fit the decision maker is the following fundamental property of utility functions.

Fundamental Property: Under the assumptions of utility theory, the decision maker's *utility function for money* has the property that the decision maker is *indifferent* between two alternative courses of action if the two alternatives have the *same expected utility*.

To illustrate, suppose that the decision maker has the utility function shown in Fig. 15.19. Further suppose that the decision maker is offered the following opportunity.

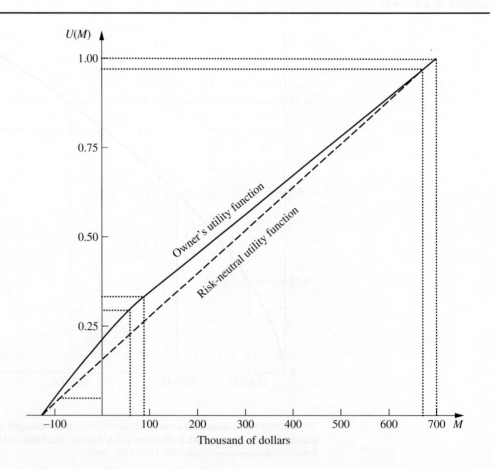

■ FIGURE 15.19
The utility function for money of the owner of the Goferbroke Co.

Offer: An opportunity to obtain either $100,000 (utility = 1) with probability p or nothing (utility = 0) with probability $(1 - p)$.

Thus,

$$E(\text{utility}) = p, \qquad \text{for this offer.}$$

Therefore, for *each* of the following three pairs of alternatives, the decision maker is indifferent between the first and second alternatives:

1. The offer with $p = 0.25$ [$E(\text{utility}) = 0.25$] or definitely obtaining $10,000 (utility = 0.25)
2. The offer with $p = 0.5$ [$E(\text{utility}) = 0.5$] or definitely obtaining $30,000 (utility = 0.5)
3. The offer with $p = 0.75$ [$E(\text{utility}) = 0.75$] or definitely obtaining $60,000 (utility = 0.75)

This example also illustrates one way in which the decision maker's utility function for money can be constructed in the first place. The decision maker would be made the same hypothetical offer to obtain either a large amount of money (for example, $100,000) with probability p or nothing. Then, for each of a few smaller amounts of money (for example, $10,000, $30,000, and $60,000), the decision maker would be asked to choose a value of p that would make him or her *indifferent* between the offer and definitely obtaining that amount of money. The utility of the smaller amount of money then is p times the utility of the large amount. This procedure, called the *equivalent lottery method* for determining utilities, is outlined below.

Equivalent Lottery Method

1. Determine the largest potential payoff, $M = maximum$, and assign it some utility, e.g., $U(maximum) = 1$.
2. Determine the smallest potential payoff, $M = minimum$, and assign it some utility smaller than in step 1, e.g., $U(minimum) = 0$.
3. To determine the utility of another potential payoff M, the decision maker is offered the following two hypothetical alternatives:

A1: Obtain a payoff of *maximum* with probability p,
   Obtain a payoff of *minimum* with probability $1 - p$.
A2: Definitely obtain a payoff of M.

Question to the decision maker: What value of p makes you *indifferent* between these two alternatives? The resulting utility of M then is

$$U(M) = p \, U(maximum) + (1 - p) \, U(minimum),$$

which simplifies to

$$U(M) = p, \quad \text{if } U(minimum) = 0, \quad U(maximum) = 1.$$

Now we are ready to summarize the basic role of utility functions in decision analysis.

When the decision maker's utility function for money is used to measure the relative worth of the various possible monetary outcomes, *Bayes' decision rule* replaces monetary payoffs by the corresponding utilities. Therefore, the optimal action (or series of actions) is the one which *maximizes the expected utility.*

Only utility functions *for money* have been discussed here. However, we should mention that utility functions can sometimes still be constructed when some of or all the important consequences of the alternative courses of action are *not* monetary. (For example, the consequences of a doctor's decision alternatives in treating a patient involve the future health of the patient.) Nevertheless, under these circumstances, it is important to incorporate such value judgments into the decision process. This is not necessarily easy, since it may require making value judgments about the relative desirability of rather intangible consequences. Nevertheless, under these circumstances, it is important to incorporate such value judgments into the decision process.

Applying Utility Theory to the Full Goferbroke Co. Problem

At the end of Sec. 15.1, we mentioned that the Goferbroke Co. was operating without much capital, so a loss of $100,000 would be quite serious. The owner of the company already has gone heavily into debt to keep going. The worst-case scenario would be to come up with $30,000 for a seismic survey and then still lose $100,000 by drilling when there is no oil. This scenario would not bankrupt the company at this point, but definitely would leave it in a precarious financial position.

On the other hand, striking oil is an exciting prospect, since earning $700,000 finally would put the company on a fairly solid financial footing.

To apply the owner's (decision maker's) *utility function for money* to the problem as described in Secs. 15.1 and 15.3, it is necessary to identify the utilities for all the possible monetary payoffs. In units of thousands of dollars, these possible payoffs and the corresponding utilities are given in Table 15.7. We now will discuss how these utilities were obtained.

As a starting point in constructing the utility function, since we have the liberty to set the value of $U(M)$ arbitrarily for two values of M (so long as the higher monetary value has the higher utility), it was convenient to set $U(-130) = 0$ and $U(700) = 1$. Then the *equivalent lottery method* was applied to determine the utility for another of the possible monetary payoffs, $M = 90$, by posing the following question to the decision maker (the owner of the Goferbroke Co.).

> Suppose you have only the following two alternatives. In units of thousands of dollars, alternative 1 is to obtain a payoff of 700 with probability p and a payoff of -130 (loss of 130) with probability $1 - p$. Alternative 2 is to definitely obtain a payoff of 90. What value of p makes you *indifferent* between these two alternatives?

> The decision maker's choice: $p = \frac{1}{3}$, so $U(90) = 0.333$.

Next, the equivalent lottery method was applied in the same way to $M = -100$. In this case, the decision maker's *point of indifference* was $p = \frac{1}{20}$, so $U(-100) = 0.05$.

At this point, a smooth curve was drawn through $U(-130)$, $U(-100)$, $U(90)$, and $U(700)$ to obtain the decision maker's *utility function for money* shown in Fig. 15.19. The values on

■ **TABLE 15.7** Utilities for the full
Goferbroke Co. problem

Monetary Payoff	Utility
−130	0
−100	0.05
60	0.30
90	0.333
670	0.97
700	1

this curve at $M = 60$ and $M = 670$ provide the corresponding utilities, $U(60) = 0.30$ and $U(670) = 0.97$, which completes the list of utilities given in the right column of Table 15.7. The shape of this curve indicates that the owner of the Goferbroke Co. is moderately *risk averse*. By contrast, the dashed line drawn at 45° in Fig. 15.19 shows what his utility function would have been if he were *risk-neutral*.

By nature, the owner of the Goferbroke Co. actually is inclined to be a risk seeker. However, the difficult financial circumstances of his company, which he badly wants to keep solvent, have forced him to adopt a moderately risk-averse stance in addressing his current decisions.

Another Approach for Estimating U(M)

The above procedure for constructing $U(M)$ asks the decision maker to repeatedly make a difficult decision about which probability would make him or her indifferent between two alternatives. Many individuals would be uncomfortable with making this kind of decision. Therefore, an alternative approach is sometimes used instead to estimate the utility function for money.

This approach is to assume that the utility function has a certain mathematical form, and then adjust this form to fit the decision maker's attitude toward risk as closely as possible. For example, one particularly popular form to assume (because of its relative simplicity) is the **exponential utility function,**

$$U(M) = R\left(1 - e^{-\frac{M}{R}}\right),$$

where R is the decision maker's *risk tolerance*. This utility function has a decreasing marginal utility for money, so it is designed to fit a *risk-averse* individual. A great aversion to risk corresponds to a small value of R (which would cause the utility function curve to bend sharply), whereas a small aversion to risk corresponds to a large value of R (which gives a much more gradual bend in the curve).

Since the owner of the Goferbroke Co. has a relatively small aversion to risk, the utility function curve in Fig. 15.19 bends quite slowly. It bends particularly slowly for the large values of M near the right side of Fig. 15.19, so the corresponding value of R in this region is approximately $R = 2000$. On the other hand, the owner becomes much more risk-averse when large losses can occur, since this now would threaten bankruptcy, so the utility function curve has considerably more curvature in this region where M has large negative values. Therefore, the corresponding value of R is considerably smaller, only about $R = 500$, in this region.

Unfortunately, it is not possible to use two different values of R for the same utility function. A drawback of the exponential utility function is that it assumes a constant aversion to risk (a fixed value of R), regardless of how much (or how little) money the decision maker currently has. This doesn't fit the Goferbroke Co. situation, since the current shortage of money makes the owner much more concerned than usual about incurring a large loss.

In other situations where the consequences of the potential losses are not as severe, assuming an exponential utility function may provide a reasonable approximation. In such a case, here is an easy (slightly approximate) way of estimating the appropriate value of R. The decision maker would be asked to choose the number R that would make him (or her) indifferent between the following two alternatives.

A_1: A 50-50 gamble where he would gain R dollars with probability 0.5 and lose $\frac{R}{2}$ dollars with probability 0.5.

A_2: Neither gain nor lose anything.

Sensitivity analysis also can become unwieldy on large problems. Although it normally is supported by the computer software, the amount of data generated can easily overwhelm an analyst or decision maker. Therefore, some graphical techniques, such as *tornado charts,* have been developed to organize the data in a readily understandable way.[5]

Other kinds of graphical techniques also are available to complement the decision tree in representing and solving decision analysis problems. One that has become quite popular is called the *influence diagram,* and researchers continue to develop others as well.[6]

Many strategic business decisions are made collectively by several members of management. One technique for group decision making is called *decision conferencing.* This is a process where the group comes together for discussions in a decision conference with the help of an analyst and a group facilitator. The facilitator works directly with the group to help it structure and focus discussions, think creatively about the problem, bring assumptions to the surface, and address the full range of issues involved. The analyst uses decision analysis to assist the group in exploring the implications of the various decision alternatives. With the assistance of a computerized group decision support system, the analyst builds and solves models on the spot, and then performs sensitivity analysis to respond to what-if questions from the group.[7]

Applications of decision analysis commonly involve a partnership between the managerial decision maker (whether an individual or a group) and an analyst (whether an individual or a team) with training in OR. Some companies do not have a staff member who is qualified to serve as the analyst. Therefore, a considerable number of management consulting firms specializing in decision analysis have been formed to fill this role.

If you would like to do more reading about the practical application of decision analysis, a good place to begin would be the November–December 1992 issue of *Interfaces.* This is a special issue devoted entirely to decision analysis and the related area of risk analysis. It includes many interesting articles, including descriptions of basic methods, sensitivity analysis, and decision conferencing. Also included are several articles on applications. Then, for a more recent perspective on the practical application of decision analysis, we suggest that you turn to Selected Reference 8. This article was the leadoff paper in the first issue of the new journal *Decision Analysis* that focuses on applied research in decision analysis. The article provides a detailed discussion of various publications that present applications of decision analysis.

■ 15.8   CONCLUSIONS

Decision analysis has become an important technique for decision making in the face of uncertainty. It is characterized by enumerating all the available decision alternatives, identifying the payoffs for all possible outcomes, and quantifying the subjective probabilities for all the possible random events. When these data are available, decision analysis becomes a powerful tool for determining an optimal course of action.

One option that can be readily incorporated into the analysis is to perform experimentation to obtain better estimates of the probabilities of the possible states of nature. Decision trees are a useful visual tool for analyzing this option or any series of decisions.

---

[5]For further information, see T. G. Eschenbach, "Spiderplots versus Tornado Diagrams for Sensitivity Analysis," *Interfaces,* **22:** 40–46, Nov.–Dec. 1992.

[6]For example, see C. Bielza and P. P. Shenoy, "A Comparison of Graphical Techniques for Asymmetric Decision Problems," *Management Science,* **45**(11): 1552–1569, Nov. 1999.

[7]For further information, see the two articles on decision conferencing in the November–December 1992 issue of *Interfaces,* where one describes an application in Australia and the other summarizes the experience of 26 decision conferences in Hungary.

Utility theory provides a way of incorporating the decision maker's attitude toward risk into the analysis.

Good software (including TreePlan and SensIt in your OR Courseware) is becoming widely available for performing decision analysis. (Selected Reference 9 provides a survey of such software.)

■ SELECTED REFERENCES

1. Bleichrodt, H., J. M. Abellan-Perpiñan, J. L. Pinto-Prades, and I. Mendez-Martinez: "Resolving Inconsistencies in Utility Measurement Under Risk: Tests of Generalizations of Expected Utility," *Management Science*, **53**(3): 469–482, March 2007.
2. Clemen, R. T.: *Making Hard Decisions: Introduction to Decision Analysis (with CD-ROM)*, 3rd ed., Duxbury Press, Pacific Grove, CA, 2006.
3. Fishburn, P. C.: "Foundations of Decision Analysis: Along the Way," *Management Science,* **35:** 387–405, 1989.
4. Fishburn, P. C.: *Nonlinear Preference and Utility Theory,* The Johns Hopkins Press, Baltimore, MD, 1988.
5. Goodwin, P., and G. Wright: *Decision Analysis for Management Judgment,* 3rd ed., Wiley, New York, 2004.
6. Hammond, J. S., R. L. Keeney, and H. Raiffa: *Smart Choices: A Practical Guide to Making Better Decisions,* Harvard Business School Press, Cambridge, MA, 1999.
7. Hillier, F. S., and M. S. Hillier: *Introduction to Management Science: A Modeling and Case Studies Approach with Spreadsheets,* 3rd ed., McGraw-Hill/Irwin, Burr Ridge, IL, 2008, chap. 9.
8. Keefer, D. L., C. W. Kirkwood, and J. L. Corner: "Perspective on Decision Analysis Applications," *Decision Analysis,* **1**(1): 4–22, 2004.
9. Maxwell, D. T.: "Software Survey: Decision Analysis," *OR/MS Today,* **33**(6): 51–61, Dec. 2006.
10. Smith, J. E., and R. L. Keeney: "Your Money or Your Life: A Prescriptive Model for Health, Safety, and Consumption Decisions," *Management Science,* **51**(9): 1309–1325, Sept. 2005.
11. Smith, J. E., and R. L. Winkler: "The Optimizer's Curse: Skepticism and Postdecision Surprise in Decision Analysis," *Management Science,* **52**(3): 311–322, March 2006.
12. Smith, J. E., and D. von Winterfeldt: "Decision Analysis in *Management Science*," *Management Science,* **50**(5): 561–574, May 2004.

■ LEARNING AIDS FOR THIS CHAPTER ON OUR WEBSITE (www.mhhe.com/hillier)

Worked Examples:

Examples for Chapter 15

"Ch. 15—Decision Analysis" Excel Files:

Template for Posterior Probabilities
TreePlan Decision Tree for First Goferbroke Co. Problem
TreePlan Decision Tree for Full Goferbroke Problem (with SensIt Graphs)

"Ch. 15—Decision Analysis" LINGO File for Selected Examples

Excel Add-Ins:

TreePlan (academic version)
SensIt (academic version)

Glossary for Chapter 15

See Appendix 1 for documentation of the software.

■ PROBLEMS

The symbols to the left of some of the problems (or their parts) have the following meaning:

T: The Excel template just listed can be helpful.
A: The corresponding Excel add-in just listed can be used.

An asterisk on the problem number indicates that at least a partial answer is given in the back of the book.

15.2-1. Read the referenced article that fully describes the OR study summarized in the application vignette presented in Sec. 15.2. Briefly describe how decision analysis was applied in this study. Then list the various financial and nonfinancial benefits that resulted from this study.

15.2-2.* Silicon Dynamics has developed a new computer chip that will enable it to begin producing and marketing a personal computer if it so desires. Alternatively, it can sell the rights to the computer chip for $15 million. If the company chooses to build computers, the profitability of the venture depends upon the company's ability to market the computer during the first year. It has sufficient access to retail outlets that it can guarantee sales of 10,000 computers. On the other hand, if this computer catches on, the company can sell 100,000 machines. For analysis purposes, these two levels of sales are taken to be the two possible outcomes of marketing the computer, but it is unclear what their prior probabilities are. If the decision is to go ahead with producing and marketing the computer, the company will produce as many chips as it finds it will be able to sell, but not more. The cost of setting up the assembly line is $6 million. The difference between the selling price and the variable cost of each computer is $600.

(a) Develop a decision analysis formulation of this problem by identifying the decision alternatives, the states of nature, and the payoff table.

(b) Develop a graph that plots the expected payoff for each of the decision alternatives versus the prior probability of selling 10,000 computers.

(c) Referring to the graph developed in part (b), use algebra to solve for the *crossover point*. Explain the significance of this point.

A (d) Develop a graph that plots the expected payoff (when using Bayes' decision rule) versus the prior probability of selling 10,000 computers.

(e) Assuming the prior probabilities of the two levels of sales are both 0.5, which decision alternative should be chosen?

15.2-3. Jean Clark is the manager of the Midtown Saveway Grocery Store. She now needs to replenish her supply of strawberries. Her regular supplier can provide as many cases as she wants. However, because these strawberries already are very ripe, she will need to sell them tomorrow and then discard any that remain unsold. Jean estimates that she will be able to sell 12, 13, 14, or 15 cases tomorrow. She can purchase the strawberries for $7 per case and sell them for $18 per case. Jean now needs to decide how many cases to purchase.

Jean has checked the store's records on daily sales of strawberries. On this basis, she estimates that the prior probabilities are 0.1, 0.3, 0.4, and 0.2 for being able to sell 12, 13, 14, and 15 cases of strawberries tomorrow.

(a) Develop a decision analysis formulation of this problem by identifying the decision alternatives, the states of nature, and the payoff table.

(b) How many cases of strawberries should Jean purchase if she uses the maximin payoff criterion?

(c) How many cases should be purchased according to the maximum likelihood criterion?

(d) How many cases should be purchased according to Bayes' decision rule?

(e) Jean thinks she has the prior probabilities just about right for selling 12 cases and selling 15 cases, but is uncertain about how to split the prior probabilities for 13 cases and 14 cases. Reapply Bayes' decision rule when the prior probabilities of 13 and 14 cases are (i) 0.2 and 0.5, (ii) 0.4 and 0.3, and (iii) 0.5 and 0.2.

15.2-4.* Warren Buffy is an enormously wealthy investor who has built his fortune through his legendary investing acumen. He currently has been offered three major investments and he would like to choose one. The first one is a *conservative investment* that would perform very well in an improving economy and only suffer a small loss in a worsening economy. The second is a *speculative investment* that would perform extremely well in an improving economy but would do very badly in a worsening economy. The third is a *countercyclical investment* that would lose some money in an improving economy but would perform well in a worsening economy.

Warren believes that there are three possible scenarios over the lives of these potential investments: (1) an improving economy, (2) a stable economy, and (3) a worsening economy. He is pessimistic about where the economy is headed, and so has assigned prior probabilities of 0.1, 0.5, and 0.4, respectively, to these three scenarios. He also estimates that his profits under these respective scenarios are those given by the following table:

	Improving Economy	Stable Economy	Worsening Economy
Conservative investment	$30 million	$ 5 million	−$10 million
Speculative investment	$40 million	$10 million	−$30 million
Countercyclical investment	−$10 million	0	$15 million
Prior probability	0.1	0.5	0.4

Which investment should Warren make under each of the following criteria?

(a) Maximin payoff criterion.

(b) Maximum likelihood criterion.

(c) Bayes' decision rule.

15.2-5. Reconsider Prob. 15.2-4. Warren Buffy decides that Bayes' decision rule is his most reliable decision criterion. He believes that 0.1 is just about right as the prior probability of an improving economy, but is quite uncertain about how to split the remaining probabilities between a stable economy and a worsening economy. Therefore, he now wishes to do sensitivity analysis with respect to these latter two prior probabilities.

(a) Reapply Bayes' decision rule when the prior probability of a stable economy is 0.3 and the prior probability of a worsening economy is 0.6.

(b) Reapply Bayes' decision rule when the prior probability of a stable economy is 0.7 and the prior probability of a worsening economy is 0.2.

(c) Graph the expected profit for each of the three investment alternatives versus the prior probability of a stable economy (with the prior probability of an improving economy fixed at 0.1). Use this graph to identify the crossover points where the decision shifts from one investment to another.

(d) Use algebra to solve for the crossover points identified in part (c).

A (e) Develop a graph that plots the expected profit (when using Bayes' decision rule) versus the prior probability of a stable economy.

15.2-6. You are given the following payoff table (in units of thousands of dollars) for a decision analysis problem:

	State of Nature		
Alternative	S_1	S_2	S_3
A_1	220	170	110
A_2	200	180	150
Prior probability	0.6	0.3	0.1

(a) Which alternative should be chosen under the maximin payoff criterion?

(b) Which alternative should be chosen under the maximum likelihood criterion?

(c) Which alternative should be chosen under Bayes' decision rule?

(d) Using Bayes' decision rule, do sensitivity analysis graphically with respect to the prior probabilities of states S_1 and S_2 (without changing the prior probability of state S_3) to determine the crossover point where the decision shifts from one alternative to the other. Then use algebra to calculate this crossover point.

(e) Repeat part (d) for the prior probabilities of states S_1 and S_3.

(f) Repeat part (d) for the prior probabilities of states S_2 and S_3.

(g) If you feel that the true probabilities of the states of nature are within 10 percent of the given prior probabilities, which alternative would you choose?

15.2-7. Dwight Moody is the manager of a large farm with 1,000 acres of arable land. For greater efficiency, Dwight always devotes the farm to growing one crop at a time. He now needs to make a decision on which one of four crops to grow during the upcoming growing season. For each of these crops, Dwight has obtained the following estimates of crop yields and net incomes per bushel under various weather conditions.

	Expected Yield, Bushels/Acre			
Weather	**Crop 1**	**Crop 2**	**Crop 3**	**Crop 4**
Dry	30	25	40	60
Moderate	50	30	35	60
Damp	60	40	35	60
Net income per bushel	$3.00	$4.50	$3.00	$1.50

After referring to historical meteorological records, Dwight also estimated the following prior probabilities for the weather during the growing season:

Dry	0.2
Moderate	0.5
Damp	0.3

(a) Develop a decision analysis formulation of this problem by identifying the decision alternatives, the states of nature, and the payoff table.

(b) Use Bayes' decision rule to determine which crop to grow.

(c) Using Bayes' decision rule, do sensitivity analysis with respect to the prior probabilities of moderate weather and damp weather (without changing the prior probability of dry weather) by re-solving when the prior probability of moderate weather is 0.2, 0.3, 0.4, and 0.6.

15.2-8.* A new type of airplane is to be purchased by the Air Force, and the number of spare engines to be ordered must be determined. The Air Force must order these spare engines in batches of five, and it can choose among only 15, 20, or 25 spares. The supplier of these engines has two plants, and the Air Force must make its decision prior to knowing which plant will be used. However, the Air Force knows from past experience that two-thirds of all types of airplane engines are produced in Plant A, and only one-third are produced in Plant B. The Air Force also knows that the number of spare engines required when production takes place at Plant A is approximated by a Poisson distribution with mean $\theta = 21$, whereas the number of spare engines required when production takes place at Plant B is approximated by a Poisson distribution with mean $\theta = 24$. The cost of a spare engine purchased now is $400,000, whereas the cost of a spare engine purchased at a later date is $900,000. Spares must always be supplied if they are demanded, and unused engines will be scrapped when the airplanes become obsolete. Holding costs and interest are to be neglected. From these data, the total costs (negative payoffs) have been computed as follows:

	State of Nature	
Alternative	$\theta = 21$	$\theta = 24$
Order 15	1.155×10^7	1.414×10^7
Order 20	1.012×10^7	1.207×10^7
Order 25	1.047×10^7	1.135×10^7

Determine the optimal alternative under Bayes' decision rule.

15.3-1. Read the referenced article that fully describes the OR study summarized in the application vignette presented in Sec. 15.3. Briefly describe how decision analysis was applied in this study. Then list the various financial and nonfinancial benefits that resulted from this study.

15.3-2.* Reconsider Prob. 15.2-2. Management of Silicon Dynamics now is considering doing full-fledged market research at a cost of $1 million to predict which of the two levels of demand is likely to occur. Previous experience indicates that such market research is correct two-thirds of the time.
(a) Find EVPI for this problem.
(b) Does the answer in part (a) indicate that it might be worthwhile to perform this market research?
(c) Develop a probability tree diagram to obtain the posterior probabilities of the two levels of demand for each of the two possible outcomes of the market research.
T (d) Use the corresponding Excel template to check your answers in part (c).
(e) Find EVE. Is it worthwhile to perform the market research?

15.3-3. You are given the following payoff table (in units of thousands of dollars) for a decision analysis problem:

	State of Nature		
Alternative	S_1	S_2	S_3
A_1	6	1	1
A_2	1	3	0
A_3	4	1	2
Prior probability	0.3	0.4	0.3

(a) According to Bayes' decision rule, which alternative should be chosen?
(b) Find EVPI.
(c) You are given the opportunity to spend $1,000 to obtain more information about which state of nature is likely to occur. Given your answer to part (b), might it be worthwhile to spend this money?

15.3-4.* Betsy Pitzer makes decisions according to Bayes' decision rule. For her current problem, Betsy has constructed the following payoff table (in units of dollars):

	State of Nature		
Alternative	S_1	S_2	S_3
A_1	50	100	-100
A_2	0	10	-10
A_3	20	40	-40
Prior probability	0.5	0.3	0.2

(a) Which alternative should Betsy choose?
(b) Find EVPI.
(c) What is the most that Betsy should consider paying to obtain more information about which state of nature will occur?

15.3-5. Using Bayes' decision rule, consider the decision analysis problem having the following payoff table (in units of thousands of dollars):

	State of Nature		
Alternative	S_1	S_2	S_3
A_1	-20	3	25
A_2	-3	5	10
A_3	4	2	15
Prior probability	0.3	0.3	0.4

(a) Which alternative should be chosen? What is the resulting expected payoff?
(b) You are offered the opportunity to obtain information which will tell you with certainty whether the first state of nature S_1 will occur. What is the maximum amount you should pay for the information? Assuming you will obtain the information, how should this information be used to choose an alternative? What is the resulting expected payoff (excluding the payment)?
(c) Now repeat part (b) if the information offered concerns S_2 instead of S_1.
(d) Now repeat part (b) if the information offered concerns S_3 instead of S_1.
(e) Now suppose that the opportunity is offered to provide information which will tell you with certainty which state of nature will occur (perfect information). What is the maximum amount you should pay for the information? Assuming you will obtain the information, how should this information be used to choose an alternative? What is the resulting expected payoff (excluding the payment)?
(f) If you have the opportunity to do some testing that will give you partial additional information (not perfect information) about the state of nature, what is the maximum amount you should consider paying for this information?

15.3-6. Reconsider the Goferbroke Co. prototype example, including its analysis in Sec. 15.3. With the help of a consulting geologist, some historical data have been obtained that provide more precise information on the likelihood of obtaining favorable seismic soundings on similar tracts of land. Specifically, when the land contains oil, favorable seismic soundings are obtained 80 percent of the time. This percentage changes to 40 percent when the land is dry.

(a) Revise Fig. 15.2 to find the new posterior probabilities.

T (b) Use the corresponding Excel template to check your answers in part (a).

(c) What is the resulting optimal policy?

15.3-7. You are given the following payoff table (in units of dollars):

	State of Nature	
Alternative	S_1	S_2
A_1	400	−100
A_2	0	100
Prior probability	0.4	0.6

You have the option of paying $100 to have research done to better predict which state of nature will occur. When the true state of nature is S_1, the research will accurately predict S_1 60 percent of the time (but will inaccurately predict S_2 40 percent of the time). When the true state of nature is S_2, the research will accurately predict S_2 80 percent of the time (but will inaccurately predict S_1 20 percent of the time).

(a) Given that the research is not done, use Bayes' decision rule to determine which decision alternative should be chosen.

(b) Find EVPI. Does this answer indicate that it might be worthwhile to do the research?

(c) Given that the research is done, find the joint probability of each of the following pairs of outcomes: (i) the state of nature is S_1 and the research predicts S_1, (ii) the state of nature is S_1 and the research predicts S_2, (iii) the state of nature is S_2 and the research predicts S_1, and (iv) the state of nature is S_2 and the research predicts S_2.

(d) Find the unconditional probability that the research predicts S_1. Also find the unconditional probability that the research predicts S_2.

(e) Given that the research is done, use your answers in parts (c) and (d) to determine the posterior probabilities of the states of nature for each of the two possible predictions of the research.

T (f) Use the corresponding Excel template to obtain the answers for part (e).

(g) Given that the research predicts S_1, use Bayes' decision rule to determine which decision alternative should be chosen and the resulting expected payoff.

(h) Repeat part (g) when the research predicts S_2.

(i) Given that research is done, what is the expected payoff when using Bayes' decision rule?

(j) Use the preceding results to determine the optimal policy regarding whether to do the research and the choice of the decision alternative.

15.3-8.* Reconsider Prob. 15.2-8. Suppose now that the Air Force knows that a similar type of engine was produced for an earlier version of the type of airplane currently under consideration. The order size for this earlier version was the same as for the current type. Furthermore, the probability distribution of the number of spare engines required, given the plant where production takes place, is believed to be the same for this earlier airplane model and the current one. The engine for the current order will be produced in the same plant as the previous model, although the Air Force does not know which of the two plants this is. The Air Force does have access to the data on the number of spares actually required for the older version, but the supplier has not revealed the production location.

(a) How much money is it worthwhile to pay for perfect information on which plant will produce these engines?

(b) Assume that the cost of the data on the old airplane model is free and that 30 spares were required. You are given that the probability of 30 spares, given a Poisson distribution with mean θ, is 0.013 for $\theta = 21$ and 0.036 for $\theta = 24$. Find the optimal action under Bayes' decision rule.

15.3-9.* Vincent Cuomo is the credit manager for the Fine Fabrics Mill. He is currently faced with the question of whether to extend $100,000 credit to a potential new customer, a dress manufacturer. Vincent has three categories for the credit-worthiness of a company: poor risk, average risk, and good risk, but he does not know which category fits this potential customer. Experience indicates that 20 percent of companies similar to this dress manufacturer are poor risks, 50 percent are average risks, and 30 percent are good risks. If credit is extended, the expected profit for poor risks is −$15,000, for average risks $10,000, and for good risks $20,000. If credit is not extended, the dress manufacturer will turn to another mill. Vincent is able to consult a credit-rating organization for a fee of $5,000 per company evaluated. For companies whose actual credit record with the mill turns out to fall into each of the three categories, the following table shows the percentages that were given each of the three possible credit evaluations by the credit-rating organization.

	Actual Credit Record		
Credit Evaluation	**Poor**	**Average**	**Good**
Poor	50%	40%	20%
Average	40%	50%	40%
Good	10%	10%	40%

(a) Develop a decision analysis formulation of this problem by identifying the decision alternatives, the states of nature, and the payoff table when the credit-rating organization is not used.

(b) Assuming the credit-rating organization is not used, use Bayes' decision rule to determine which decision alternative should be chosen.

(c) Find EVPI. Does this answer indicate that consideration should be given to using the credit-rating organization?

(d) Assume now that the credit-rating organization is used. Develop a probability tree diagram to find the posterior probabilities of the respective states of nature for each of the three possible credit evaluations of this potential customer.

т **(e)** Use the corresponding Excel template to obtain the answers for part (*d*).

(f) Determine Vincent's optimal policy.

15.3-10. An athletic league does drug testing of its athletes, 15 percent of whom use drugs. This test, however, is only 97 percent reliable. That is, a drug user will test positive with probability 0.97 and negative with probability 0.03, and a nonuser will test negative with probability 0.97 and positive with probability 0.03.

Develop a probability tree diagram to determine the posterior probability of each of the following outcomes of testing an athlete.

(a) The athlete is a drug user, given that the test is positive.

(b) The athlete is not a drug user, given that the test is positive.

(c) The athlete is a drug user, given that the test is negative.

(d) The athlete is not a drug user, given that the test is negative.

т **(e)** Use the corresponding Excel template to check your answers in the preceding parts.

15.3-11. Management of the Telemore Company is considering developing and marketing a new product. It is estimated to be twice as likely that the product would prove to be successful as unsuccessful. It it were successful, the expected profit would be $1,500,000. If unsuccessful, the expected loss would be $1,800,000. A marketing survey can be conducted at a cost of $300,000 to predict whether the product would be successful. Past experience with such surveys indicates that successful products have been predicted to be successful 80 percent of the time, whereas unsuccessful products have been predicted to be unsuccessful 70 percent of the time.

(a) Develop a decision analysis formulation of this problem by identifying the decision alternatives, the states of nature, and the payoff table when the market survey is not conducted.

(b) Assuming the market survey is not conducted, use Bayes' decision rule to determine which decision alternative should be chosen.

(c) Find EVPI. Does this answer indicate that consideration should be given to conducting the market survey?

т **(d)** Assume now that the market survey is conducted. Find the posterior probabilities of the respective states of nature for each of the two possible predictions from the market survey.

(e) Find the optimal policy regarding whether to conduct the market survey and whether to develop and market the new product.

15.3-12. The Hit-and-Miss Manufacturing Company produces items that have a probability p of being defective. These items are produced in lots of 150. Past experience indicates that p for an entire lot is either 0.05 or 0.25. Furthermore, in 80 percent of the lots produced, p equals 0.05 (so p equals 0.25 in 20 percent

of the lots). These items are then used in an assembly, and ultimately their quality is determined before the final assembly leaves the plant. Initially the company can *either* screen each item in a lot at a cost of $10 per item and replace defective items *or* use the items directly without screening. If the latter action is chosen, the cost of rework is ultimately $100 per defective item. Because screening requires scheduling of inspectors and equipment, the decision to screen or not screen must be made 2 days before the screening is to take place. However, one item can be taken from the lot and sent to a laboratory for inspection, and its quality (defective or nondefective) can be reported before the screen/no screen decision must be made. The cost of this initial inspection is $125.

(a) Develop a decision analysis formulation of this problem by identifying the decision alternatives, the states of nature, and the payoff table if the single item is not inspected in advance.

(b) Assuming the single item is not inspected in advance, use Bayes' decision rule to determine which decision alternative should be chosen.

(c) Find EVPI. Does this answer indicate that consideration should be given to inspecting the single item in advance?

т **(d)** Assume now that the single item is inspected in advance. Find the posterior probabilities of the respective states of nature for each of the two possible outcomes of this inspection.

(e) Find EVE. Is inspecting the single item worthwhile?

(f) Determine the optimal policy.

т **15.3-13.*** Consider two weighted coins. Coin 1 has a probability of 0.3 of turning up heads, and coin 2 has a probability of 0.6 of turning up heads. A coin is tossed once; the probability that coin 1 is tossed is 0.6, and the probability that coin 2 is tossed is 0.4. The decision maker uses Bayes' decision rule to decide which coin is tossed. The payoff table is as follows:

	State of Nature	
Alternative	**Coin 1 Tossed**	**Coin 2 Tossed**
Say coin 1 tossed	0	−1
Say coin 2 tossed	−1	0
Prior probability	0.6	0.4

(a) What is the optimal alternative before the coin is tossed?

(b) What is the optimal alternative after the coin is tossed if the outcome is heads? If it is tails?

15.3-14. There are two biased coins with probabilities of landing heads of 0.7 and 0.3, respectively. One coin is chosen at random (each with probability $\frac{1}{2}$) to be tossed twice. You are to receive $250 if you correctly predict how many heads will occur in two tosses.

(a) Using Bayes' decision rule, what is the optimal prediction, and what is the corresponding expected payoff?

T **(b)** Suppose now that you may observe a practice toss of the chosen coin before predicting. Use the corresponding Excel template to find the posterior probabilities for which coin is being tossed.

(c) Determine your optimal prediction after observing the practice toss. What is the resulting expected payoff?

(d) Find EVE for observing the practice toss. If you must pay $75 to observe the practice toss, what is your optimal policy?

15.4-1. Read the referenced article that fully describes the OR study summarized in the application vignette presented in Sec. 15.4. Briefly describe how decision analysis was applied in this study. Then list the various financial and nonfinancial benefits that resulted from this study.

15.4-2.* Reconsider Prob. 15.3-2. The management of Silicon Dynamics now wants to see a decision tree displaying the entire problem. Construct and solve this decision tree by hand.

15.4-3. You are given the decision tree below, where the numbers in parentheses are probabilities and the numbers on the far right are payoffs at these terminal points. Analyze this decision tree to obtain the optimal policy.

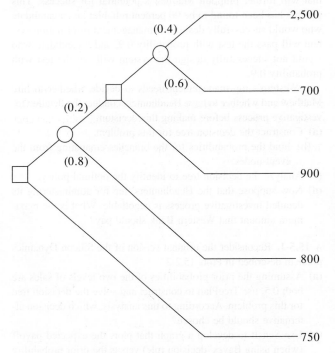

15.4-4.* The Athletic Department of Leland University is considering whether to hold an extensive campaign next year to raise funds for a new athletic field. The response to the campaign depends heavily upon the success of the football team this fall. In the past, the football team has had winning seasons 60 percent of the time. If the football team has a winning season (W) this fall, then many of the alumnae and alumni will contribute and the campaign will raise $3 million. If the team has a losing season (L),

few will contribute and the campaign will lose $2 million. If no campaign is undertaken, no costs are incurred. On September 1, just before the football season begins, the Athletic Department needs to make its decision about whether to hold the campaign next year.

(a) Develop a decision analysis formulation of this problem by identifying the decision alternatives, the states of nature, and the payoff table.

(b) According to Bayes' decision rule, should the campaign be undertaken?

(c) What is EVPI?

(d) A famous football guru, William Walsh, has offered his services to help evaluate whether the team will have a winning season. For $100,000, he will carefully evaluate the team throughout spring practice and then throughout preseason workouts. William then will provide his prediction on September 1 regarding what kind of season, W or L, the team will have. In similar situations in the past when evaluating teams that have winning seasons 50 percent of the time, his predictions have been correct 75 percent of the time. Considering that this team has more of a winning tradition, if William predicts a winning season, what is the posterior probability that the team actually will have a winning season? What is the posterior probability of a losing season? If Williams predicts a losing season instead, what is the posterior probability of a winning season? Of a losing season? Show how these answers are obtained from a probability tree diagram.

T **(e)** Use the corresponding Excel template to obtain the answers requested in part (d).

(f) Draw the decision tree for this entire problem by hand. Analyze this decision tree to determine the optimal policy regarding whether to hire William and whether to undertake the campaign.

15.4-5. The comptroller of the Macrosoft Corporation has $100 million of excess funds to invest. She has been instructed to invest the entire amount for one year in either stocks or bonds (but not both) and then to reinvest the entire fund in either stocks or bonds (but not both) for one year more. The objective is to maximize the expected monetary value of the fund at the end of the second year.

The annual rates of return on these investments depend on the economic environment, as shown in the following table:

Economic Environment	Rate of Return	
	Stocks	**Bonds**
Growth	20%	5%
Recession	−10%	10%
Depression	−50%	20%

The probabilities of growth, recession, and depression for the first year are 0.7, 0.3, and 0, respectively. If growth occurs in the first

year, these probabilities remain the same for the second year. However, if a recession occurs in the first year, these probabilities change to 0.2, 0.7, and 0.1, respectively, for the second year.
(a) Construct the decision tree for this problem by hand.
(b) Analyze the decision tree to identify the optimal policy.

15.4-6 On Monday, a certain stock closed at $10 per share. On Tuesday, you expect the stock to close at $9, $10, or $11 per share, with respective probabilities 0.3, 0.3, and 0.4. On Wednesday, you expect the stock to close 10 percent lower, unchanged, or 10 percent higher than Tuesday's close, with the following probabilities:

Today's Close	10% Lower	Unchanged	10% Higher
$ 9	0.4	0.3	0.3
$10	0.2	0.2	0.6
$11	0.1	0.2	0.7

On Tuesday, you are directed to buy 100 shares of the stock before Thursday. All purchases are made at the end of the day, at the known closing price for that day, so your only options are to buy at the end of Tuesday or at the end of Wednesday. You wish to determine the optimal strategy for whether to buy on Tuesday or defer the purchase until Wednesday, given the Tuesday closing price, to minimize the expected purchase price. Develop and evaluate a decision tree by hand for determining the optimal strategy.

15.4-7. Use the scenario given in Prob. 15.3-9.
(a) Draw and properly label the decision tree. Include all the payoffs but not the probabilities.
T (b) Find the probabilities for the branches emanating from the event nodes.
(c) Apply the backward induction procedure, and identify the resulting optimal policy.

15.4-8. Use the scenario given in Prob. 15.3.-11.
(a) Draw and properly label the decision tree. Include all the payoffs but not the probabilities.
T (b) Find the probabilities for the branches emanating from the event nodes.
(c) Apply the backward induction procedure, and identify the resulting optimal policy.

15.4-9. Use the scenario given in Prob. 15.3-12.
(a) Draw and properly label the decision tree. Include all the payoffs but not the probabilities.
T (b) Find the probabilities for the branches emanating from the event nodes.
(c) Apply the backward induction procedure, and identify the resulting optimal policy.

15.4-10. Use the scenario given in Prob. 15.3-13.
(a) Draw and properly label the decision tree. Include all the payoffs but not the probabilities.

T (b) Find the probabilities for the branches emanating from the event nodes.
(c) Apply the backward induction procedure, and identify the resulting optimal policy.

A **15.4-11.** The executive search being conducted for Western Bank by Headhunters Inc. may finally be bearing fruit. The position to be filled is a key one—Vice President for Information Processing—because this person will have responsibility for developing a state-of-the-art management information system that will link together Western's many branch banks. However, Headhunters feels they have found just the right person, Matthew Fenton, who has an excellent record in a similar position for a midsized bank in New York.

After a round of interviews, Western's president believes that Matthew has a probability of 0.75 of designing the management information system successfully. If Matthew is successful, the company will realize a profit of $4 million (net of Matthew's salary, training, recruiting costs, and expenses). If he is not successful, the company will realize a net loss of $900,000.

For an additional fee of $35,000, Headhunters will provide a detailed investigative process (including an extensive background check, a battery of academic and psychological tests, etc.) that will further pinpoint Matthew's potential for success. This process has been found to be 90 percent reliable; i.e., a candidate who would successfully design the management information system will pass the test with probability 0.9, and a candidate who would not successfully design the system will fail the test with probability 0.9.

Western's top management needs to decide whether to hire Matthew and whether to have Headhunters conduct the detailed investigative process before making this decision.
(a) Construct the decision tree for this problem.
T (b) Find the probabilities for the branches emanating from the event nodes.
(c) Analyze the decision tree to identify the optimal policy.
(d) Now suppose that the Headhunters' fee for administering its detailed investigative process is negotiable. What is the maximum amount that Western Bank should pay?

A **15.5-1.** Reconsider the original version of the Silicon Dynamics problem described in Prob. 15.2-2.
(a) Assuming the prior probabilities of the two levels of sales are both 0.5, use TreePlan to construct and solve the decision tree for this problem. According to this analysis, which decision alternative should be chosen?
(b) Use SensIt to develop a graph that plots the expected payoff (when using Bayes' decision rule) versus the prior probability of selling 10,000 computers.

A **15.5-2.** Now reconsider the expanded version of the Silicon Dynamics problem described in Probs. 15.3-2 and 15.4-2.
(a) Use TreePlan to construct and solve the decision tree for this problem.
(b) There is some uncertainty in the financial data ($15 million, $6 million, and $600) stated in Prob. 15.2.2. Each could vary

from its base value by as much as 10 percent. For each one, perform sensitivity analysis to find what would happen if its value were at either end of this range of variability (without any change in the other two pieces of data) by adjusting the values in the data cells accordingly. Then do the same for the eight cases where all these pieces of data are at one end or the other of their ranges of variability.

(c) Because of the uncertainty described in part (b), use SensIt to generate a graph that plots the expected profit over the range of variability for each piece of financial data (without any change in the other two pieces of data).

(d) Generate the corresponding spider chart and tornado chart.

A **15.5-3.** Reconsider the decision tree given in Prob. 15.4-3. Use TreePlan to construct and solve this decision tree.

A **15.5-4.** Reconsider Prob. 15.4-5. Use TreePlan to construct and solve the decision tree for this problem.

A **15.5-5.** Reconsider Prob. 15.4-6. Use TreePlan to construct and solve the decision tree for this problem.

A **15.5-6.** Jose Morales manages a large outdoor fruit stand in one of the less affluent neighborhoods of San Jose, California. To replenish his supply, Jose buys boxes of fruit early each morning from a grower south of San Jose. About 85 percent of the boxes of fruit turn out to be of satisfactory quality, but the other 15 percent are unsatisfactory. A satisfactory box contains 90 percent excellent fruit and will earn $600 profit for Jose. An unsatisfactory box contains 40 percent excellent fruit and will produce a loss of $2,000. Before Jose decides to accept a box, he is given the opportunity to sample one piece of fruit to test whether it is excellent. Based on that sample, he then has the option of rejecting the box without paying for it. Jose wonders (1) whether he should continue buying from this grower, (2) if so, whether it is worthwhile sampling just one piece of fruit from a box, and (3) if so, whether he should be accepting or rejecting the box based on the outcome of this sampling.

Use TreePlan (and the Excel template for posterior probabilities) to construct and solve the decision tree for this problem.

A **15.5-7.*** The Morton Ward Company is considering the introduction of a new product that is believed to have a 50-50 chance of being successful. One option is to try out the product in a test market, at a cost of $5 million, before making the introduction decision. Past experience shows that ultimately successful products are approved in the test market 80 percent of the time, whereas ultimately unsuccessful products are approved in the test market only 25 percent of the time. If the product is successful, the net profit to the company will be $40 million; if unsuccessful, the net loss will be $15 million.

(a) Discarding the option of trying out the product in a test market, develop a decision analysis formulation of the problem by identifying the decision alternatives, states of nature, and payoff table. Then apply Bayes' decision rule to determine the optimal decision alternative.

(b) Find EVPI.

A **(c)** Now include the option of trying out the product in a test market. Use TreePlan (and the Excel template for posterior probabilities) to construct and solve the decision tree for this problem.

A **(d)** There is some uncertainty in the stated profit and loss figures ($40 million and $15 million). Either could vary from its base by as much as 25 percent in either direction. Use TreePlan calculations to generate a graph for each that plots the expected profit over this range of variability.

A **(e)** Because of the uncertainty described in part (d), use SensIt to generate a graph that plots the expected profit over the range of variability for each of the two financial figures (without any change in the other figure).

A **(f)** Generate the corresponding spider chart and tornado chart. Interpret each one.

A **15.5-8.** Chelsea Bush is an emerging candidate for her party's nomination for President of the United States. She now is considering whether to run in the high-stakes Super Tuesday primaries. If she enters the Super Tuesday (S.T.) primaries, she and her advisers believe that she will either do well (finish first or second) or do poorly (finish third or worse) with probabilities 0.4 and 0.6, respectively. Doing well on Super Tuesday will net the candidate's campaign approximately $16 million in new contributions, whereas a poor showing will mean a loss of $10 million after numerous TV ads are paid for. Alternatively, she may choose not to run at all on Super Tuesday and incur no costs.

Chelsea's advisers realize that her chances of success on Super Tuesday may be affected by the outcome of the smaller New Hampshire (N.H.) primary occurring three weeks before Super Tuesday. Political analysts feel that the results of New Hampshire's primary are correct two-thirds of the time in predicting the results of the Super Tuesday primaries. Among Chelsea's advisers is a decision analysis expert who uses this information to calculate the following probabilities:

$P\{$Chelsea does well in S.T. primaries, given she does well in N.H.$\} = \frac{4}{7}$

$P\{$Chelsea does well in S.T. primaries, given she does poorly in N.H.$\} = \frac{1}{4}$

$P\{$Chelsea does well in N.H. primary$\} = \frac{7}{15}$

The cost of entering and campaigning in the New Hampshire primary is estimated to be $1.6 million.

Chelsea feels that her chance of winning the nomination depends largely on having substantial funds available after the Super Tuesday primaries to carry on a vigorous campaign the rest of the way. Therefore, she wants to choose the strategy (whether to run in the New Hampshire primary and then whether to run in the Super Tuesday primaries) that will maximize her expected funds after these primaries.

(a) Construct and solve the decision tree for this problem.

(b) There is some uncertainty in the estimates of a gain of $16 million or a loss of $10 million depending on the showing on Super Tuesday. Either amount could differ from this estimate by as much as 25 percent in either direction. For each of these two financial figures, perform sensitivity analysis to check how the results in part (a) would change if the value of the financial figure were at either end of this range of variability (without any change in the value of the other financial figure). Then do the same for the four cases where both financial figures are at one end or the other of their ranges of variability.

A (c) Because of the uncertainty described in part (b), use SensIt to generate a graph that plots Chelsea's expected funds after these primaries over the range of variability for each of the two financial figures (without any change in the other figure).

A (d) Generate the corresponding spider chart and tornado chart. Interpret each one.

15.6-1. Reconsider the Goferbroke Co. prototype example, including the application of utilities in Sec. 15.6. The owner now has decided that, given the company's precarious financial situation, he needs to take a much more risk-averse approach to the problem. Therefore, he has revised the utilities given in Table 15.7 as follows: $U(-130) = 0$, $U(-100) = 0.1$, $U(60) = 0.4$, $U(90) = 0.45$, $U(670) = 0.985$, and $U(700) = 1$.

(a) Analyze the revised decision tree corresponding to Fig. 15.20 by hand to obtain the new optimal policy.

A (b) Use TreePlan to construct and solve this revised decision tree.

15.6-2.* You live in an area that has a possibility of incurring a massive earthquake, so you are considering buying earthquake insurance on your home at an annual cost of $180. The probability of an earthquake damaging your home during one year is 0.001. If this happens, you estimate that the cost of the damage (fully covered by earthquake insurance) will be $160,000. Your total assets (including your home) are worth $250,000.

(a) Apply Bayes' decision rule to determine which alternative (take the insurance or not) maximizes your expected assets after one year.

(b) You now have constructed a utility function that measures how much you value having total assets worth x dollars ($x \geq 0$). This utility function is $U(x) = \sqrt{x}$. Compare the utility of reducing your total assets next year by the cost of the earthquake insurance with the expected utility next year of not taking the earthquake insurance. Should you take the insurance?

15.6-3. For your graduation present from college, your parents are offering you your choice of two alternatives. The first alternative is to give you a money gift of $19,000. The second alternative is to make an investment in your name. This investment will quickly have the following two possible outcomes:

Outcome	Probability
Receive $10,000	0.3
Receive $30,000	0.7

Your utility for receiving M thousand dollars is given by the utility function $U(M) = \sqrt{M + 6}$. Which choice should you make to maximize expected utility?

15.6-4.* Reconsider Prob. 15.6-3. You now are uncertain about what your true utility function for receiving money is, so you are in the process of constructing this utility function. So far, you have found that $U(19) = 16.7$ and $U(30) = 20$ are the utility of receiving $19,000 and $30,000, respectively. You also have concluded that you are indifferent between the two alternatives offered to you by your parents. Use this information to find $U(10)$.

15.6-5. You wish to construct your personal utility function $U(M)$ for receiving M thousand dollars. After setting $U(0) = 0$, you next set $U(1) = 1$ as your utility for receiving $1,000. You next want to find $U(10)$ and then $U(5)$.

(a) You offer yourself the following two hypothetical alternatives:

A_1: Obtain $10,000 with probability p.
Obtain 0 with probability $(1 - p)$.
A_2: Definitely obtain $1,000.

You then ask yourself the question: What value of p makes you indifferent between these two alternatives? Your answer is $p = 0.125$. Find $U(10)$.

(b) You next repeat part (a) except for changing the second alternative to definitely receiving $5,000. The value of p that makes you indifferent between these two alternatives now is $p = 0.5625$. Find $U(5)$.

(c) Repeat parts (a) and (b), but now use your personal choices for p.

15.6-6. You are given the following payoff table:

	State of Nature	
Alternative	S_1	S_2
A_1	36	49
A_2	144	0
A_3	0	81
Prior probability	p	$1 - p$

(a) Assume that your utility function for the payoffs is $U(x) = \sqrt{x}$. Plot the expected utility of each alternative versus the value of p on the same graph. For each alternative, find the range of values of p over which this alternative maximizes the expected utility.

A (b) Now assume that your utility function is the exponential utility function with a risk tolerance of $R = 50$. Use TreePlan to construct and solve the resulting decision tree in turn for $p = 0.25$, $p = 0.5$, and $p = 0.75$.

15.6-7. Dr. Switzer has a seriously ill patient but has had trouble diagnosing the specific cause of the illness. The doctor now has narrowed the cause down to two alternatives: disease *A* or disease *B*. Based on the evidence so far, she feels that the two alternatives are equally likely.

Beyond the testing already done, there is no test available to determine if the cause is disease *B*. One test is available for disease *A*, but it has two major problems. First, it is very expensive. Second, it is somewhat unreliable, giving an accurate result only 80 percent of the time. Thus, it will give a positive result (indicating disease *A*) for only 80 percent of patients who have disease *A*, whereas it will give a positive result for 20 percent of patients who actually have disease *B* instead.

Disease *B* is a very serious disease with no known treatment. It is sometimes fatal, and those who survive remain in poor health with a poor quality of life thereafter. The prognosis is similar for victims of disease *A* if it is left untreated. However, there is a fairly expensive treatment available that eliminates the danger for those with disease *A*, and it may return them to good health. Unfortunately, it is a relatively radical treatment that always leads to death if the patient actually has disease *B* instead.

The probability distribution for the prognosis for this patient is given for each case in the following table, where the column headings (after the first one) indicate the disease for the patient.

	Outcome Probabilities			
	No Treatment		Receive Treatment for Disease A	
Outcome	A	B	A	B
Die	0.2	0.5	0	1.0
Survive with poor health	0.8	0.5	0.5	0
Return to good health	0	0	0.5	0

The patient has assigned the following utilities to the possible outcomes:

Outcome	Utility
Die	0
Survive with poor health	10
Return to good health	30

In addition, these utilities should be incremented by -2 if the patient incurs the cost of the test for disease *A* and by -1 if the patient (or the patient's estate) incurs the cost of the treatment for disease *A*.

Use decision analysis with a complete decision tree to determine if the patient should undergo the test for disease *A* and then how to proceed (receive the treatment for disease *A*?) to maximize the patient's expected utility.

15.6-8. You want to choose between decision alternatives A_1 and A_2 in the following decision tree, but you are uncertain about the value of the probability p, so you need to perform sensitivity analysis of p as well.

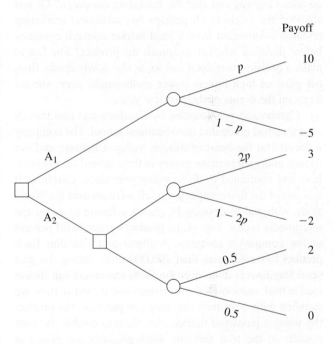

Your utility function for money (the payoff received) is

$$U(M) = \begin{cases} M^2 & \text{if } M \geq 0 \\ M & \text{if } M < 0. \end{cases}$$

(a) For $p = 0.25$, determine which alternative is optimal in the sense that it maximizes the expected utility of the payoff.

(b) Determine the range of values of the probability p ($0 \leq p \leq 0.5$) for which this same alternative remains optimal.

■ CASES

CASE 15.1   Brainy Business

While El Niño is pouring its rain on northern California, Charlotte Rothstein, CEO, major shareholder and founder of Cerebrosoft, sits in her office, contemplating the decision she faces regarding her company's newest proposed product, Brainet. This has been a particularly difficult decision. Brainet might catch on and sell very well. However, Charlotte is concerned about the risk involved. In this competitive market, marketing Brainet also could lead to substantial losses. Should she go ahead anyway and start the marketing campaign? Or just abandon the product? Or perhaps buy additional marketing research information from a local market research company before deciding whether to launch the product? She has to make a decision very soon and so, as she slowly drinks from her glass of high protein-power multivitamin juice, she reflects on the events of the past few years.

Cerebrosoft was founded by Charlotte and two friends after they had graduated from business school. The company is located in the heart of Silicon Valley. Charlotte and her friends managed to make money in their second year in business and continued to do so every year since. Cerebrosoft was one of the first companies to sell software over the World Wide Web and to develop PC-based software tools for the multimedia sector. Two of the products generate 80 percent of the company's revenues: Audiatur and Videatur. Each product has sold more than 100,000 units during the past year. Business is done over the Web: customers can download a trial version of the software, test it, and if they are satisfied with what they see, they can purchase the product (by using a password that enables them to disable the time counter in the trial version). Both products are priced at $75.95 and are exclusively sold over the Web.

Although the World Wide Web is a network of computers of different types, running different kinds of software, a standardized protocol between the computers enables them to communicate. Users can "surf" the Web and visit computers many thousand miles away, accessing information available at the site. Users can also make files available on the Web, and this is how Cerebrosoft generates its sales. Selling software over the Web eliminates many of the traditional cost factors of consumer products: packaging, storage, distribution, sales force, and so on. Instead, potential customers can download a trial version, take a look at it (that is, use the product) before its trial period expires, and then decide whether to buy it. Furthermore, Cerebrosoft can always make the most recent files available to the customer, avoiding the problem of having outdated software in the distribution pipeline.

Charlotte is interrupted in her thoughts by the arrival of Jeannie Korn. Jeannie is in charge of marketing for on-line products and Brainet has had her particular attention from the beginning. She is more than ready to provide the advice that Charlotte has requested. "Charlotte, I think we should really go ahead with Brainet. The software engineers have convinced me that the current version is robust and we want to be on the market with this as soon as possible! From the data for our product launches during the past two years we can get a rather reliable estimate of how the market will respond to the new product, don't you think? And look!" She pulls out some presentation slides. "During that time period we launched 12 new products altogether and 4 of them sold more than 30,000 units during the first 6 months alone! Even better: the last two we launched even sold more than 40,000 copies during the first two quarters!" Charlotte knows these numbers as well as Jeannie does. After all, two of these launches have been products she herself helped to develop. But she feels uneasy about this particular product launch. The company has grown rapidly during the past three years and its financial capabilities are already rather stretched. A poor product launch for Brainet would cost the company a lot of money, something that isn't available right now due to the investments Cerebrosoft has recently made.

Later in the afternoon, Charlotte meets with Reggie Ruffin, a jack-of-all-trades and the production manager. Reggie has a solid track record in his field and Charlotte wants his opinion on the Brainet project.

"Well, Charlotte, quite frankly I think that there are three main factors that are relevant to the success of this project: competition, units sold, and cost—ah, and of course our pricing. Have you decided on the price yet?"

"I am still considering which of the three strategies would be most beneficial to us. Selling for $50.00 and trying to maximize revenues—or selling for $30.00 and trying to maximize market share. Of course, there is still your third alternative; we could sell for $40.00 and try to do both."

At this point Reggie focuses on the sheet of paper in front of him. "And I still believe that the $40.00 alternative is the best one. Concerning the costs, I checked the records; basically we have to amortize the development costs we incurred for Brainet. So far we have spent $800,000 and we expect to spend another $50,000 per year for support and shipping the CDs to those who want a hardcopy on top of their downloaded software." Reggie next hands a report to Charlotte. "Here we have some data on the industry. I just received that yesterday, hot off the press. Let's see what we can learn about the industry here." He shows Charlotte some of the highlights. Reggie then agrees to compile the most relevant information contained in the report and have it ready for Charlotte the following morning. It takes him long into the night to gather the data from the pages of the report, but in the end he produces three tables, one for each of the three alternative pricing strategies. Each table shows the corresponding probability of various amounts of sales given the level of competition (high, medium, or low) that develops from other companies.

The next morning Charlotte is sipping from another power drink. Jeannie and Reggie will be in her office any moment now and, with their help, she will have to decide what to do with Brainet. Should they launch the product? If so, at what price?

When Jeannie and Reggie enter the office, Jeannie immediately bursts out: "Guys, I just spoke to our marketing research company. They say that they could do a study for us about the competitive situation for the introduction of Brainet and deliver the results within a week."

"How much do they want for the study?"

"I knew you'd ask that, Reggie. They want $10,000 and I think it's a fair deal."

At this point Charlotte steps into the conversation. "Do we have any data on the quality of the work of this marketing research company?"

"Yes, I do have some reports here. After analyzing them, I have come to the conclusion that the marketing research company is not very good in predicting the competitive environment for medium or low pricing. Therefore, we should not ask them to do the study for us if we decide on one of these two pricing strategies. However, in the case of high pricing, they do quite well: given that the competition turned out to be high, they predicted it correctly 80 percent of the time, while 15 percent of the time they predicted medium competition in that setting. Given that the competition turned out to be medium, they predicted high competition 15 percent of the time and medium competition 80 percent of the time. Finally, for the case of low competition, the numbers

TABLE 1 Probability distribution of unit sales, given a high price ($50)

Sales	Level of Competition		
	High	Medium	Low
50,000 units	0.2	0.25	0.3
30,000 units	0.25	0.3	0.35
20,000 units	0.55	0.45	0.35

TABLE 2 Probability distribution of unit sales, given a medium price ($40)

Sales	Level of Competition		
	High	Medium	Low
50,000 units	0.25	0.30	0.40
30,000 units	0.35	0.40	0.50
20,000 units	0.40	0.30	0.10

TABLE 3 Probability distribution of unit sales, given a low price ($30)

Sales	Level of Competition		
	High	Medium	Low
50,000 units	0.35	0.40	0.50
30,000 units	0.40	0.50	0.45
20,000 units	0.25	0.10	0.05

were 90 percent of the time a correct prediction, 7 percent of the time a 'medium' prediction and 3 percent of the time a 'high' prediction."

Charlotte feels that all these numbers are too much for her. "Don't we have a simple estimate of how the market will react?"

"Some prior probabilities, you mean? Sure, from our past experience, the likelihood of facing high competition is 20 percent, whereas it is 70 percent for medium competition and 10 percent for low competition," Jeannie has her numbers always ready when needed.

All that is left to do now is to sit down and make sense of all this. . . .

(a) For the initial analysis, ignore the opportunity of obtaining more information by hiring the marketing research company. Identify the decision alternatives and the states of nature. Construct the payoff table. Then formulate the decision problem in a decision tree. Clearly distinguish between decision and event nodes and include all the relevant data.
(b) What is Charlotte's decision if she uses the maximum likelihood criterion? The maximin payoff criterion?
(c) What is Charlotte's decision if she uses Bayes' decision rule?
(d) Now consider the possibility of doing the market research. Develop the corresponding decision tree. Calculate the relevant probabilities and analyze the decision tree. Should Cerebrosoft pay the $10,000 for the marketing research? What is the overall optimal policy?

■ PREVIEW OF ADDED CASES ON OUR WEBSITE (www.mhhe.com/hillier)

CASE 15.2   Smart Steering Support

The CEO of Bay Area Automobile Gadgets is contemplating whether to add a road scanning device to the company's driver support system. A series of decisions need to be made. Should basic research into the road scanning device be undertaken? If the research is successful, should the company develop the product or sell the technology? In the case of successful product development, should the company market the product or sell the product concept? Decision analysis needs to be applied to address these issues. Part of the analysis will involve using the CEO's utility function.

CASE 15.3   Who Wants to be a Millionaire?

You are a contestant on "Who Wants to be a Millionaire?" and have just answered the $250,000 question correctly. If you decide to go on to the $500,000 question and then to the $1,000,000 question, you still have the option available of using the "phone a friend" lifeline on one of the questions to improve your chances of answering correctly. You now want to use decision analysis (including a decision tree and utility theory) to decide how to proceed.

CASE 15.4   University Toys and the Engineering Professor Action Figures

University Toys has developed a series of Engineering Professor Action Figures for the local engineering school and management needs to decide how to market the dolls in the face of uncertainty about the demand. One option is to immediately ramp up for full production, advertising, and sales. Another option is to test-market the product first. A complication with this option is a rumor that a competitor is about to enter the market with a similar product. Decision analysis (including a decision tree and sensitivity analysis) now needs to be used to decide how to proceed.

16

Markov Chains

Chapter 15 focused on decision making in the face of uncertainty about *one* future event (learning the true state of nature). However, some decisions need to take into account uncertainty about *many* future events. We now begin laying the groundwork for decision making in this broader context.

In particular, this chapter presents probability models for processes that *evolve over time* in a probabilistic manner. Such processes are called *stochastic processes*. After briefly introducing general stochastic processes in the first section, the remainder of the chapter focuses on a special kind called a *Markov chain*. Markov chains have the special property that probabilities involving how the process will evolve in the future depend only on the present state of the process, and so are independent of events in the past. Many processes fit this description, so Markov chains provide an especially important kind of probability model.

For example, you will see in the next chapter that *continuous-time Markov chains* (described in Sec. 16.8) are used to formulate most of the basic models of *queueing theory*. Markov chains also provide the foundation for the study of *Markov decision models* in Chapter 19. There are a wide variety of other applications of Markov chains as well. A considerable number of books and articles present some of these applications. One is Selected Reference 4, which describes applications in such diverse areas as the classification of customers, DNA sequencing, the analysis of genetic networks, the estimation of sales demand over time, and credit rating. You also will see an application vignette in Sec. 16.2 that involves credit rating, as well as an application vignette in Sec. 16.8 that involves machine maintenance. Selected Reference 6 focuses on applications in finance and Selected Reference 3 describes applications for analyzing baseball strategy. The list goes on and on, but let us turn now to a description of stochastic processes in general and Markov chains in particular.

■ 16.1 STOCHASTIC PROCESSES

A **stochastic process** is defined to be an indexed collection of random variables $\{X_t\}$, where the index t runs through a given set T. Often T is taken to be the set of nonnegative integers, and X_t represents a measurable characteristic of interest at time t. For example, X_t might represent the inventory level of a particular product at the end of week t.

Stochastic processes are of interest for describing the behavior of a system operating over some period of time. A stochastic process often has the following structure.

The current status of the system can fall into any one of $M + 1$ mutually exclusive categories called **states.** For notational convenience, these states are labeled 0, 1, . . . , M. The random variable X_t represents the *state of the system* at time t, so its only possible values are 0, 1, . . . , M. The system is observed at particular points of time, labeled $t = 0$, 1, 2, ... Thus, the stochastic process $\{X_t\} = \{X_0, X_1, X_2, \ldots\}$ provides a mathematical representation of how the status of the physical system evolves over time.

This kind of process is referred to as being a *discrete time* stochastic process with a *finite state space.* Except for Sec. 16.8, this will be the only kind of stochastic process considered in this chapter. (Section 16.8 describes a certain *continuous time* stochastic process.)

A Weather Example

The weather in the town of Centerville can change rather quickly from day to day. However, the chances of being dry (no rain) tomorrow are somewhat larger if it is dry today than if it rains today. In particular, the probability of being dry tomorrow is **0.8** if it is dry today, but is only **0.6** if it rains today. These probabilities do not change if information about the weather before today is also taken into account.

The evolution of the weather from day to day in Centerville is a stochastic process. Starting on some initial day (labeled as day 0), the weather is observed on each day t, for $t = 0, 1, 2, \ldots$. The state of the system on day t can be either

State 0 = Day t is dry

or

State 1 = Day t has rain.

Thus, for $t = 0, 1, 2, \ldots$, the random variable X_t takes on the values,

$$X_t = \begin{cases} 0 & \text{if day } t \text{ is dry} \\ 1 & \text{if day } t \text{ has rain.} \end{cases}$$

The stochastic process $\{X_t\} = \{X_0, X_1, X_2, \ldots\}$ provides a mathematical representation of how the status of the weather in Centerville evolves over time.

An Inventory Example

Dave's Photography Store has the following inventory problem. The store stocks a particular model camera that can be ordered weekly. Let $D_1, D_2, \ldots$ represent the *demand* for this camera (the number of units that would be sold if the inventory is not depleted) during the first week, second week, . . . , respectively, so the random variable D_t (for $t = 1, 2, \ldots$) is

D_t = number of cameras that would be sold in week t if the inventory is not
depleted. (This number includes lost sales when the inventory is depleted.)

It is assumed that the D_t are independent and identically distributed random variables having a *Poisson distribution* with a mean of 1. Let X_0 represent the number of cameras on hand at the outset, X_1 the number of cameras on hand at the end of week 1, X_2 the number of cameras on hand at the end of week 2, and so on, so the random variable X_t (for $t = 0, 1, 2, \ldots$) is

X_t = number of cameras on hand at the end of week t.

Assume that $X_0 = 3$, so that week 1 begins with three cameras on hand.

$$\{X_t\} = \{X_0, X_1, X_2, \ldots\}$$

is a stochastic process where the random variable X_t represents the state of the system at time t, namely,

State at time t = number of cameras on hand at the end of week t.

As the owner of the store, Dave would like to learn more about how the status of this stochastic process evolves over time while using the current ordering policy described below.

At the end of each week t (Saturday night), the store places an order that is delivered in time for the next opening of the store on Monday. The store uses the following order policy:

If $X_t = 0$, order 3 cameras.
If $X_t > 0$, do not order any cameras.

Thus, the inventory level fluctuates between a minimum of zero cameras and a maximum of three cameras, so the possible states of the system at time t (the end of week t) are

Possible states = 0, 1, 2, or 3 cameras on hand.

Since each random variable X_t $(t = 0, 1, 2, \ldots)$ represents the state of the system at the end of week t, its only possible values are 0, 1, 2, or 3. The random variables X_t are dependent and may be evaluated iteratively by the expression

$$X_{t+1} = \begin{cases} \max\{3 - D_{t+1}, 0\} & \text{if} \quad X_t = 0 \\ \max\{X_t - D_{t+1}, 0\} & \text{if} \quad X_t \geq 1, \end{cases}$$

for $t = 0, 1, 2, \ldots$.

These examples are used for illustrative purposes throughout many of the following sections. Section 16.2 further defines the particular type of stochastic process considered in this chapter.

16.2 MARKOV CHAINS

Assumptions regarding the joint distribution of $X_0, X_1, \ldots$ are necessary to obtain analytical results. One assumption that leads to analytical tractability is that the stochastic process is a Markov chain, which has the following key property:

A stochastic process $\{X_t\}$ is said to have the **Markovian property** if $P\{X_{t+1} = j \mid X_0 = k_0,$ $X_1 = k_1, \ldots, X_{t-1} = k_{t-1}, X_t = i\} = P\{X_{t+1} = j \mid X_t = i\}$, for $t = 0, 1, \ldots$ and every sequence $i, j, k_0, k_1, \ldots, k_{t-1}$.

In words, this Markovian property says that the conditional probability of any future "event," given any past "events" and the present state $X_t = i$, is *independent* of the past events and depends only upon the present state.

A stochastic process $\{X_t\}$ $(t = 0, 1, \ldots)$ is a **Markov chain** if it has the *Markovian property*.

The conditional probabilities $P\{X_{t+1} = j \mid X_t = i\}$ for a Markov chain are called (one-step) **transition probabilities.** If, for each i and j,

$$P\{X_{t+1} = j \mid X_t = i\} = P\{X_1 = j \mid X_0 = i\}, \qquad \text{for all } t = 1, 2, \ldots,$$

then the (one-step) transition probabilities are said to be *stationary*. Thus, having **stationary transition probabilities** implies that the transition probabilities do not change over time. The existence of stationary (one-step) transition probabilities also implies that, for each i, j, and n ($n = 0, 1, 2, \ldots$),

$$P\{X_{t+n} = j \mid X_t = i\} = P\{X_n = j \mid X_0 = i\}$$

for all $t = 0, 1, \ldots$. These conditional probabilities are called **n-step transition probabilities.** To simplify notation with stationary transition probabilities, let

$$p_{ij} = P\{X_{t+1} = j \mid X_t = i\},$$
$$p_{ij}^{(n)} = P\{X_{t+n} = j \mid X_t = i\}.$$

Thus, the n-step transition probability $p_{ij}^{(n)}$ is just the conditional probability that the system will be in state j after exactly n steps (time units), given that it starts in state i at any time t. When $n = 1$, note that $p_{ij}^{(1)} = p_{ij}$.[1]

Because the $p_{ij}^{(n)}$ are conditional probabilities, they must be nonnegative, and since the process must make a transition into some state, they must satisfy the properties

$$p_{ij}^{(n)} \geq 0, \qquad \text{for all } i \text{ and } j; n = 0, 1, 2, \ldots,$$

and

$$\sum_{j=0}^{M} p_{ij}^{(n)} = 1 \qquad \text{for all } i; n = 0, 1, 2, \ldots.$$

A convenient way of showing all the n-step transition probabilities is the *n-step transition matrix*

$$
\mathbf{P}^{(n)} = \begin{array}{c} \\ 0 \\ 1 \\ \vdots \\ M \end{array}
\begin{array}{c} \text{State} \quad 0 \qquad 1 \quad \cdots \quad M \\
\begin{bmatrix}
p_{00}^{(n)} & p_{01}^{(n)} & \cdots & p_{0M}^{(n)} \\
p_{10}^{(n)} & p_{11}^{(n)} & \cdots & p_{1M}^{(n)} \\
\cdots & \cdots & \cdots & \cdots \\
p_{M0}^{(n)} & p_{M1}^{(n)} & \cdots & p_{MM}^{(n)}
\end{bmatrix}
\end{array}
$$

Note that the transition probability in a particular row and column is for the transition *from* the row state *to* the column state. When $n = 1$, we drop the superscript n and simply refer to this as the *transition matrix*.

The Markov chains to be considered in this chapter have the following properties:

1. A finite number of states.
2. Stationary transition probabilities.

We also will assume that we know the initial probabilities $P\{X_0 = i\}$ for all i.

Formulating the Weather Example as a Markov Chain

For the weather example introduced in the preceding section, recall that the evolution of the weather in Centerville from day to day has been formulated as a stochastic process $\{X_t\}$ ($t = 0, 1, 2, \ldots$) where

$$X_t = \begin{cases} 0 & \text{if day } t \text{ is dry} \\ 1 & \text{if day } t \text{ has rain.} \end{cases}$$

---

[1] For $n = 0$, $p_{ij}^{(0)}$ is just $P\{X_0 = j \mid X_0 = i\}$ and hence is 1 when $i = j$ and is 0 when $i \neq j$.

Merrill Lynch is a leading full-service financial service firm. It provides brokerage, investment, and banking services to individual retail clients and small businesses while also helping major corporations and institutions around the world raise capital. One of Merrill Lynch's affiliates, *Merrill Lynch (ML) Bank USA*, has assets of over $60 billion obtained by accepting deposits from Merrill Lynch retail customers and using these deposits to fund loans and make investments.

In 2000, ML Bank USA began to establish revolving credit lines for client companies. Within a few years, the bank had developed a portfolio of about $13 billion in credit-line commitments with over 100 institutions. Long before this point was reached, Merrill Lynch's outstanding OR group was asked to guide the management of this growing portfolio by using OR techniques to assess the *liquidity risk* (the bank's potential inability to meet its cash obligations) associated with its current and prospective credit-line commitments.

The OR group developed a *simulation model* (the topic of Chap. 20) for this purpose. However, the most important input to this model is a *Markov chain* that describes the evolution of each customer's credit rating over time. The states of the Markov chain are the various possible credit ratings (ranging from *highest investment grade to default*) that are assigned to major companies by such credit-rating agencies as Standard and Poor's and Moody's. The transition probability from state i to state j in the transition matrix for a given company is the probability that the credit-rating agency will shift its rating of the company from state i to state j from one month to the next, based on historical patterns for similar companies.

This application of operations research, including Markov chains, enabled ML Bank USA to *free up about* **$4 billion** *of liquidity for other use*, as well as to expand its portfolio of credit-line commitments by over 60 percent in less than two years. Other benefits include the ability to evaluate extreme-risk scenarios and to perform long-range planning. This outstanding work led to Merrill Lynch winning the prestigious *Wagner Prize for Excellence in Operations Research Practice* for 2004.

Source: Duffy, T., M. Hatzakis, W. Hsu, R. Labe, B. Liao, X. Luo, J. Oh, A. Setya, and L. Yang: "Merrill Lynch Improves Liquidity Risk Management for Revolving Credit Lines," *Interfaces*, **35**(5): 353–369, Sept.–Oct. 2005. (A link to this article is provided on our website, www.mhhe.com/hillier.)

$$P\{X_{t+1} = 0 \mid X_t = 0\} = 0.8,$$
$$P\{X_{t+1} = 0 \mid X_t = 1\} = 0.6.$$

Furthermore, because these probabilities do not change if information about the weather before today (day t) is also taken into account,

$$P\{X_{t+1} = 0 \mid X_0 = k_0, X_1 = k_1, \ldots, X_{t-1} = k_{t-1}, X_t = 0\} = P\{X_{t+1} = 0 \mid X_t = 0\}$$
$$P\{X_{t+1} = 0 \mid X_0 = k_0, X_1 = k_1, \ldots, X_{t-1} = k_{t-1}, X_t = 1\} = P\{X_{t+1} = 0 \mid X_t = 1\}$$

for $t = 0, 1, \ldots$ and every sequence $k_0, k_1, \ldots, k_{t-1}$. These equations also must hold if $X_{t+1} = 0$ is replaced by $X_{t+1} = 1$. (The reason is that states 0 and 1 are mutually exclusive and the only possible states, so the probabilities of the two states must sum to 1.) Therefore, the stochastic process has the *Markovian property*, so the process is a Markov chain.

Using the notation introduced in this section, the (one-step) transition probabilities are

$$p_{00} = P\{X_{t+1} = 0 \mid X_t = 0\} = 0.8,$$
$$p_{10} = P\{X_{t+1} = 0 \mid X_t = 1\} = 0.6$$

for all $t = 1, 2, \ldots$, so these are *stationary* transition probabilities. Furthermore,

$$p_{00} + p_{01} = 1, \quad \text{so} \quad p_{01} = 1 - 0.8 = 0.2,$$
$$p_{10} + p_{11} = 1, \quad \text{so} \quad p_{11} = 1 - 0.6 = 0.4.$$

Therefore, the (one-step) transition matrix is

$$\mathbf{P} = \begin{array}{c} 0 \\ 1 \end{array}\begin{bmatrix} p_{00} & p_{01} \\ p_{10} & p_{11} \end{bmatrix} = \begin{array}{c} 0 \\ 1 \end{array}\begin{bmatrix} 0.8 & 0.2 \\ 0.6 & 0.4 \end{bmatrix}$$

with column headings State 0 1 for each matrix.

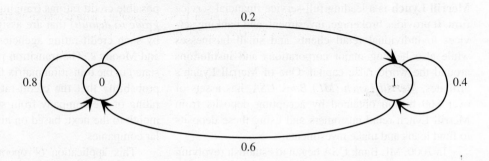

■ FIGURE 16.1
The state transition diagram for the weather example.

where these transition probabilities are for the transition *from* the row state *to* the column state. Keep in mind that state 0 means that the day is dry, whereas state 1 signifies that the day has rain, so these transition probabilities give the probability of the state the weather will be in tomorrow, given the state of the weather today.

The state transition diagram in Fig. 16.1 graphically depicts the same information provided by the transition matrix. The two nodes (circle) represent the two possible states for the weather, and the arrows show the possible transitions (including back to the same state) from one day to the next. Each of the transition probabilities is given next to the corresponding arrow.

The *n*-step transition matrices for this example will be shown in the next section.

Formulating the Inventory Example as a Markov Chain

Returning to the inventory example developed in the preceding section, recall that X_t is the number of cameras in stock at the end of week t (before ordering any more), so X_t represents the *state of the system* at time t (the end of week t). Given that the current state is $X_t = i$, the expression at the end of Sec. 16.1 indicates that X_{t+1} depends only on D_{t+1} (the demand in week $t + 1$) and X_t. Since X_{t+1} is independent of any past history of the inventory system prior to time t, the stochastic process $\{X_t\}$ $(t = 0, 1, \ldots)$ has the *Markovian property* and so is a Markov chain.

Now consider how to obtain the (one-step) transition probabilities, i.e., the elements of the (one-step) *transition matrix*

$$
\mathbf{P} = \begin{array}{c} \text{State} \\ 0 \\ 1 \\ 2 \\ 3 \end{array}
\begin{array}{cccc} 0 & 1 & 2 & 3 \end{array}
\left[\begin{array}{cccc}
p_{00} & p_{01} & p_{02} & p_{03} \\
p_{10} & p_{11} & p_{12} & p_{13} \\
p_{20} & p_{21} & p_{22} & p_{23} \\
p_{30} & p_{31} & p_{32} & p_{33}
\end{array} \right]
$$

given that D_{t+1} has a Poisson distribution with a mean of 1. Thus,

$$
P\{D_{t+1} = n\} = \frac{(1)^n e^{-1}}{n!}, \qquad \text{for } n = 0, 1, \ldots,
$$

so (to three significant digits)

$$
P\{D_{t+1} = 0\} = e^{-1} = 0.368,
$$
$$
P\{D_{t+1} = 1\} = e^{-1} = 0.368,
$$
$$
P\{D_{t+1} = 2\} = \frac{1}{2}e^{-1} = 0.184,
$$
$$
P\{D_{t+1} \geq 3\} = 1 - P\{D_{t+1} \leq 2\} = 1 - (0.368 + 0.368 + 0.184) = 0.080.
$$

For the first row of **P**, we are dealing with a transition from state $X_t = 0$ to some state X_{t+1}. As indicated at the end of Sec. 16.1,

$$X_{t+1} = \max\{3 - D_{t+1}, 0\} \quad \text{if} \quad X_t = 0.$$

Therefore, for the transition to $X_{t+1} = 3$ or $X_{t+1} = 2$ or $X_{t+1} = 1$,

$$p_{03} = P\{D_{t+1} = 0\} = 0.368,$$
$$p_{02} = P\{D_{t+1} = 1\} = 0.368,$$
$$p_{01} = P\{D_{t+1} = 2\} = 0.184.$$

A transition from $X_t = 0$ to $X_{t+1} = 0$ implies that the demand for cameras in week $t + 1$ is 3 or more after 3 cameras are added to the depleted inventory at the beginning of the week, so

$$p_{00} = P\{D_{t+1} \geq 3\} = 0.080.$$

For the other rows of **P**, the formula at the end of Sec. 16.1 for the next state is

$$X_{t+1} = \max\{X_t - D_{t+1}, 0\} \quad \text{if} \quad X_t \geq 1.$$

This implies that $X_{t+1} \leq X_t$, so $p_{12} = 0$, $p_{13} = 0$, and $p_{23} = 0$. For the other transitions,

$$p_{11} = P\{D_{t+1} = 0\} = 0.368,$$
$$p_{10} = P\{D_{t+1} \geq 1\} = 1 - P\{D_{t+1} = 0\} = 0.632,$$
$$p_{22} = P\{D_{t+1} = 0\} = 0.368,$$
$$p_{21} = P\{D_{t+1} = 1\} = 0.368,$$
$$p_{20} = P\{D_{t+1} \geq 2\} = 1 - P\{D_{t+1} \leq 1\} = 1 - (0.368 + 0.368) = 0.264.$$

For the last row of **P**, week $t + 1$ begins with 3 cameras in inventory, so the calculations for the transition probabilities are exactly the same as for the first row. Consequently, the complete transition matrix (to three significant digits) is

$$
\mathbf{P} = \begin{array}{c} \\ 0 \\ 1 \\ 2 \\ 3 \end{array}
\begin{array}{cccc} \text{State} \quad 0 & 1 & 2 & 3 \\ \left[\begin{array}{cccc} 0.080 & 0.184 & 0.368 & 0.368 \\ 0.632 & 0.368 & 0 & 0 \\ 0.264 & 0.368 & 0.368 & 0 \\ 0.080 & 0.184 & 0.368 & 0.368 \end{array}\right] \end{array}
$$

The information given by this transition matrix can also be depicted graphically with the state transition diagram in Fig. 16.2. The four possible states for the number of cameras

FIGURE 16.2
The state transition diagram for the inventory example.

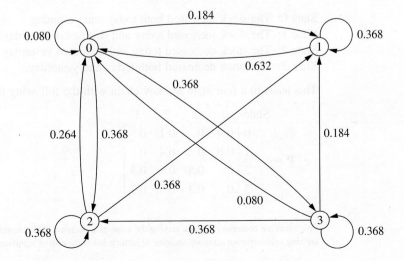

on hand at the end of a week are represented by the four nodes (circles) in the diagram. The arrows show the possible transitions from one state to another, or sometimes from a state back to itself, when the camera store goes from the end of one week to the end of the next week. The number next to each arrow gives the probability of that particular transition occurring next when the camera store is in the state at the base of the arrow.

Additional Examples of Markov Chains

A Stock Example.   Consider the following model for the value of a stock. At the end of a given day, the price is recorded. If the stock has gone up, the probability that it will go up tomorrow is **0.7.** If the stock has gone down, the probability that it will go up tomorrow is only **0.5.** (For simplicity, we will count the stock staying the same as a decrease.) This is a Markov chain, where the possible states for each day are as follows:

State 0: The stock increased on this day.
State 1: The stock decreased on this day.

The transition matrix that shows each probability of going from a particular state today to a particular state tomorrow is given by

$$\mathbf{P} = \begin{array}{c} \\ 0 \\ 1 \end{array} \begin{array}{c} \text{State} \quad 0 \qquad 1 \\ \begin{bmatrix} 0.7 & 0.3 \\ 0.5 & 0.5 \end{bmatrix} \end{array}$$

The form of the state transition diagram for this example is exactly the same as for the weather example shown in Fig. 16.1, so we will not repeat it here. The only difference is that the transition probabilities in the diagram are slightly different (0.7 replaces 0.8, 0.3 replaces 0.2, and 0.5 replaces both 0.6 and 0.4 in Fig. 16.1).

A Second Stock Example.   Suppose now that the stock market model is changed so that the stock's going up tomorrow depends upon whether it increased today *and* yesterday. In particular, if the stock has increased for the past two days, it will increase tomorrow with probability **0.9.** If the stock increased today but decreased yesterday, then it will increase tomorrow with probability **0.6.** If the stock decreased today but increased yesterday, then it will increase tomorrow with probability **0.5.** Finally, if the stock decreased for the past two days, then it will increase tomorrow with probability **0.3.** If we define the state as representing whether the stock goes up or down today, the system is no longer a Markov chain. However, we can transform the system to a Markov chain by defining the states as follows:[2]

State 0: The stock increased both today and yesterday.
State 1: The stock increased today and decreased yesterday.
State 2: The stock decreased today and increased yesterday.
State 3: The stock decreased both today and yesterday.

This leads to a four-state Markov chain with the following transition matrix:

$$\mathbf{P} = \begin{array}{c} \\ 0 \\ 1 \\ 2 \\ 3 \end{array} \begin{array}{c} \text{State} \quad 0 \quad\; 1 \quad\; 2 \quad\; 3 \\ \begin{bmatrix} 0.9 & 0 & 0.1 & 0 \\ 0.6 & 0 & 0.4 & 0 \\ 0 & 0.5 & 0 & 0.5 \\ 0 & 0.3 & 0 & 0.7 \end{bmatrix} \end{array}$$

---

[2] We again are counting the stock staying the same as a decrease. This example demonstrates that Markov chains are able to incorporate arbitrary amounts of history, but at the cost of significantly increasing the number of states.

Figure 16.3 shows the state transition diagram for this example. An interesting feature of the example revealed by both this diagram and all the values of 0 in the transition matrix is that so many of the transitions from state i to state j are impossible in one step. In other words, $p_{ij} = 0$ for 8 of the 16 entries in the transition matrix. However, check out how it always is possible to go from any state i to any state j (including $j = i$) in two steps. The same holds true for three steps, four steps, and so forth. Thus, $p_{ij}^{(n)} > 0$ for n = 2, 3, ... for all i and j.

A Gambling Example. Another example involves gambling. Suppose that a player has \$1 and with each play of the game wins \$1 with probability $p > 0$ or loses \$1 with probability $1 - p > 0$. The game ends when the player either accumulates \$3 or goes broke. This game is a Markov chain with the states representing the player's current holding of money, that is, 0, \$1, \$2, or \$3, and with the transition matrix given by

$$
\mathbf{P} = \begin{array}{c} \text{State} \\ 0 \\ 1 \\ 2 \\ 3 \end{array}
\begin{array}{cccc}
 0 & 1 & 2 & 3 \\
\begin{bmatrix}
1 & 0 & 0 & 0 \\
1 - p & 0 & p & 0 \\
0 & 1 - p & 0 & p \\
0 & 0 & 0 & 1
\end{bmatrix}
\end{array}
$$

The state transition diagram for this example is shown in Fig. 16.4. This diagram demonstrates that once the process enters either state 0 or state 3, it will stay in that state

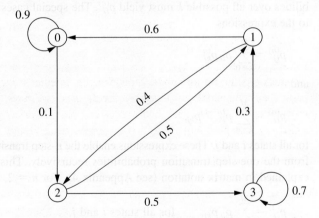

■ FIGURE 16.3
The state transition diagram for the second stock example.

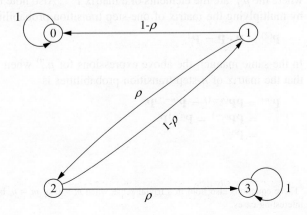

■ FIGURE 16.4
The state transition diagram for the gambling example.

forever after, since $p_{00} = 1$ and $p_{33} = 1$. States 0 and 3 are examples of what are called an **absorbing state** (a state that is never left once the process enters that state). We will focus on analyzing absorbing states in Sec. 16.7.

Note that in both the inventory and gambling examples, the numeric labeling of the states that the process reaches coincides with the physical expression of the system—i.e., actual inventory levels and the player's holding of money, respectively—whereas the numeric labeling of the states in the weather and stock examples has no physical significance.

16.3 CHAPMAN-KOLMOGOROV EQUATIONS

Section 16.2 introduced the n-step transition probability $p_{ij}^{(n)}$. The following *Chapman-Kolmogorov equations* provide a method for computing these n-step transition probabilities:

$$p_{ij}^{(n)} = \sum_{k=0}^{M} p_{ik}^{(m)} p_{kj}^{(n-m)}, \quad \text{for all } i = 0, 1, \dots, M,$$
$$j = 0, 1, \dots, M,$$
$$\text{and any } m = 1, 2, \dots, n-1,$$
$$n = m+1, m+2, \dots.^3$$

These equations point out that in going from state i to state j in n steps, the process will be in some state k after exactly m (less than n) steps. Thus, $p_{ik}^{(m)} p_{kj}^{(n-m)}$ is just the conditional probability that, given a starting point of state i, the process goes to state k after m steps and then to state j in $n-m$ steps. Therefore, summing these conditional probabilities over all possible k must yield $p_{ij}^{(n)}$. The special cases of $m = 1$ and $m = n-1$ lead to the expressions

$$p_{ij}^{(n)} = \sum_{k=0}^{M} p_{ik} p_{kj}^{(n-1)}$$

and

$$p_{ij}^{(n)} = \sum_{k=0}^{M} p_{ik}^{(n-1)} p_{kj},$$

for all states i and j. These expressions enable the n-step transition probabilities to be obtained from the one-step transition probabilities recursively. This recursive relationship is best explained in matrix notation (see Appendix 4). For $n = 2$, these expressions become

$$p_{ij}^{(2)} = \sum_{k=0}^{M} p_{ik} p_{kj}, \quad \text{for all states } i \text{ and } j,$$

where the $p_{ij}^{(2)}$ are the elements of a matrix $\mathbf{P}^{(2)}$. Also note that these elements are obtained by multiplying the matrix of one-step transition probabilities by itself; i.e.,

$$\mathbf{P}^{(2)} = \mathbf{P} \cdot \mathbf{P} = \mathbf{P}^2.$$

In the same manner, the above expressions for $p_{ij}^{(n)}$ when $m = 1$ and $m = n-1$ indicate that the matrix of n-step transition probabilities is

$$\mathbf{P}^{(n)} = \mathbf{P}\mathbf{P}^{(n-1)} = \mathbf{P}^{(n-1)}\mathbf{P}$$
$$= \mathbf{P}\mathbf{P}^{n-1} = \mathbf{P}^{n-1}\mathbf{P}$$
$$= \mathbf{P}^n.$$

---

[3]These equations also hold in a trivial sense when $m = 0$ or $m = n$, but $m = 1, 2, \dots, n-1$ are the only interesting cases.

Thus, the n-step transition probability matrix $\mathbf{P}^n$ can be obtained by computing the nth power of the one-step transition matrix $\mathbf{P}$.

n-Step Transition Matrices for the Weather Example

For the weather example introduced in Sec. 16.1, we now will use the above formulas to calculate various n-step transition matrices from the (one-step) transition matrix $\mathbf{P}$ that was obtained in Sec. 16.2. To start, the two-step transition matrix is

$$\mathbf{P}^{(2)} = \mathbf{P} \cdot \mathbf{P} = \begin{bmatrix} 0.8 & 0.2 \\ 0.6 & 0.4 \end{bmatrix} \begin{bmatrix} 0.8 & 0.2 \\ 0.6 & 0.4 \end{bmatrix} = \begin{bmatrix} 0.76 & 0.24 \\ 0.72 & 0.28 \end{bmatrix}.$$

Thus, if the weather is in state 0 (dry) on a particular day, the probability of being in state 0 two days later is 0.76 and the probability of being in state 1 (rain) then is 0.24. Similarly, if the weather is in state 1 now, the probability of being in state 0 two days later is 0.72 whereas the probability of being in state 1 then is 0.28.

The probabilities of the state of the weather three, four, or five days into the future also can be read in the same way from the three-step, four-step, and five-step transition matrices calculated to three significant digits below.

$$\mathbf{P}^{(3)} = \mathbf{P}^3 = \mathbf{P} \cdot \mathbf{P}^2 = \begin{bmatrix} 0.8 & 0.2 \\ 0.6 & 0.4 \end{bmatrix} \begin{bmatrix} 0.76 & 0.24 \\ 0.72 & 0.28 \end{bmatrix} = \begin{bmatrix} 0.752 & 0.248 \\ 0.744 & 0.256 \end{bmatrix}$$

$$\mathbf{P}^{(4)} = \mathbf{P}^4 = \mathbf{P} \cdot \mathbf{P}^3 = \begin{bmatrix} 0.8 & 0.2 \\ 0.6 & 0.4 \end{bmatrix} \begin{bmatrix} 0.752 & 0.248 \\ 0.744 & 0.256 \end{bmatrix} = \begin{bmatrix} 0.75 & 0.25 \\ 0.749 & 0.251 \end{bmatrix}$$

$$\mathbf{P}^{(5)} = \mathbf{P}^5 = \mathbf{P} \cdot \mathbf{P}^4 = \begin{bmatrix} 0.8 & 0.2 \\ 0.6 & 0.4 \end{bmatrix} \begin{bmatrix} 0.75 & 0.25 \\ 0.749 & 0.251 \end{bmatrix} = \begin{bmatrix} 0.75 & 0.25 \\ 0.75 & 0.25 \end{bmatrix}$$

Note that the five-step transition matrix has the interesting feature that the two rows have identical entries (after rounding to three significant digits). This reflects the fact that the probability of the weather being in a particular state is essentially independent of the state of the weather five days before. Thus, the probabilities in either row of this five-step transition matrix are referred to as the *steady-state probabilities* of this Markov chain.

We will expand further on the subject of the steady-state probabilities of a Markov chain, including how to derive them more directly, at the beginning of Sec. 16.5.

n-Step Transition Matrices for the Inventory Example

Returning to the inventory example included in Sec. 16.1, we now will calculate its n-step transition matrices to three decimal places for $n = 2$, 4, and 8. To start, its one-step transition matrix $\mathbf{P}$ obtained in Sec. 16.2 can be used to calculate the two-step transition matrix $\mathbf{P}^{(2)}$ as follows:

$$\mathbf{P}^{(2)} = \mathbf{P}^2 = \begin{bmatrix} 0.080 & 0.184 & 0.368 & 0.368 \\ 0.632 & 0.368 & 0 & 0 \\ 0.264 & 0.368 & 0.368 & 0 \\ 0.080 & 0.184 & 0.368 & 0.368 \end{bmatrix} \begin{bmatrix} 0.080 & 0.184 & 0.368 & 0.368 \\ 0.632 & 0.368 & 0 & 0 \\ 0.264 & 0.368 & 0.368 & 0 \\ 0.080 & 0.184 & 0.368 & 0.368 \end{bmatrix}$$

$$= \begin{bmatrix} 0.249 & 0.286 & 0.300 & 0.165 \\ 0.283 & 0.252 & 0.233 & 0.233 \\ 0.351 & 0.319 & 0.233 & 0.097 \\ 0.249 & 0.286 & 0.300 & 0.165 \end{bmatrix}.$$

For example, given that there is one camera left in stock at the end of a week, the probability is 0.283 that there will be no cameras in stock 2 weeks later, that is, $p_{10}^{(2)} = 0.283$. Similarly, given that there are two cameras left in stock at the end of a week, the probability is 0.097 that there will be three cameras in stock 2 weeks later, that is, $p_{23}^{(2)} = 0.097$.

The four-step transition matrix can also be obtained as follows:

$$\mathbf{P}^{(4)} = \mathbf{P}^4 = \mathbf{P}^{(2)} \cdot \mathbf{P}^{(2)}$$

$$= \begin{bmatrix} 0.249 & 0.286 & 0.300 & 0.165 \\ 0.283 & 0.252 & 0.233 & 0.233 \\ 0.351 & 0.319 & 0.233 & 0.097 \\ 0.249 & 0.286 & 0.300 & 0.165 \end{bmatrix} \begin{bmatrix} 0.249 & 0.286 & 0.300 & 0.165 \\ 0.283 & 0.252 & 0.233 & 0.233 \\ 0.351 & 0.319 & 0.233 & 0.097 \\ 0.249 & 0.286 & 0.300 & 0.165 \end{bmatrix}$$

$$= \begin{bmatrix} 0.289 & 0.286 & 0.261 & 0.164 \\ 0.282 & 0.285 & 0.268 & 0.166 \\ 0.284 & 0.283 & 0.263 & 0.171 \\ 0.289 & 0.286 & 0.261 & 0.164 \end{bmatrix}.$$

For example, given that there is one camera left in stock at the end of a week, the probability is 0.282 that there will be no cameras in stock 4 weeks later, that is, $p_{10}^{(4)} = 0.282$. Similarly, given that there are two cameras left in stock at the end of a week, the probability is 0.171 that there will be three cameras in stock 4 weeks later, that is, $p_{23}^{(4)} = 0.171$.

The transition probabilities for the number of cameras in stock 8 weeks from now can be read in the same way from the eight-step transition matrix calculated below.

$$\mathbf{P}^{(8)} = \mathbf{P}^8 = \mathbf{P}^{(4)} \cdot \mathbf{P}^{(4)}$$

$$= \begin{bmatrix} 0.289 & 0.286 & 0.261 & 0.164 \\ 0.282 & 0.285 & 0.268 & 0.166 \\ 0.284 & 0.283 & 0.263 & 0.171 \\ 0.289 & 0.286 & 0.261 & 0.164 \end{bmatrix} \begin{bmatrix} 0.289 & 0.286 & 0.261 & 0.164 \\ 0.282 & 0.285 & 0.268 & 0.166 \\ 0.284 & 0.283 & 0.263 & 0.171 \\ 0.289 & 0.286 & 0.261 & 0.164 \end{bmatrix}$$

State	0	1	2	3
0	0.286	0.285	0.264	0.166
1	0.286	0.285	0.264	0.166
2	0.286	0.285	0.264	0166
3	0.286	0.285	0.264	0.166

Like the five-step transition matrix for the weather example, this matrix has the interesting feature that its rows have identical entries (after rounding). The reason once again is that probabilities in any row are the *steady-state probabilities* for this Markov chain, i.e., the probabilities of the state of the system after enough time has elapsed that the initial state is no longer relevant.

Your IOR Tutorial includes a procedure for calculating $\mathbf{P}^{(n)} = \mathbf{P}^n$ for any positive integer $n \leq 99$.

Unconditional State Probabilities

Recall that one- or n-step transition probabilities are *conditional* probabilities; for example, $P\{X_n = j \mid X_0 = i\} = p_{ij}^{(n)}$. Assume that n is small enough that these conditional probabilities are not yet *steady-state* probabilities. In this case, if the *unconditional* probability $P\{X_n = j\}$ is desired, it is necessary to specify the probability distribution of the initial state, namely, $P\{X_0 = i\}$ for $i = 0, 1, \ldots, M$. Then

$$P\{X_n = j\} = P\{X_0 = 0\}\, p_{0j}^{(n)} + P\{X_0 = 1\} p_{1j}^{(n)} + \cdots + P\{X_0 = M\} p_{Mj}^{(n)}.$$

In the inventory example, it was assumed that initially there were 3 units in stock, that is, $X_0 = 3$. Thus, $P\{X_0 = 0\} = P\{X_0 = 1\} = P\{X_0 = 2\} = 0$ and $P\{X_0 = 3\} = 1$. Hence, the (unconditional) probability that there will be three cameras in stock 2 weeks after the inventory system began is $P\{X_2 = 3\} = (1)p_{33}^{(2)} = 0.165$.

■ 16.4 CLASSIFICATION OF STATES OF A MARKOV CHAIN

We have just seen near the end of the preceding section that the n-step transition probabilities for the inventory example converge to steady-state probabilities after a sufficient number of steps. However, this is not true for all Markov chains. The long-run properties of a Markov chain depend greatly on the characteristics of its states and transition matrix. To further describe the properties of Markov chains, it is necessary to present some concepts and definitions concerning these states.

State j is said to be **accessible** from state i if $p_{ij}^{(n)} > 0$ for some $n \geq 0$. (Recall that $p_{ij}^{(n)}$ is just the conditional probability of being in state j after n steps, starting in state i.) Thus, state j being accessible from state i means that it is possible for the system to enter state j eventually when it starts from state i. This is clearly true for the weather example (see Fig. 16.1) since $p_{ij} > 0$ for all i and j. In the inventory example (see Fig. 16.2), $p_{ij}^{(2)} > 0$ for all i and j, so every state is accessible from every other state. In general, a sufficient condition for *all* states to be accessible is that there exists a value of n for which $p_{ij}^{(n)} > 0$ for all i and j.

In the gambling example given at the end of Sec. 16.2 (see Fig. 16.4), state 2 is not accessible from state 3. This can be deduced from the context of the game (once the player reaches state 3, the player never leaves this state), which implies that $p_{32}^{(n)} = 0$ for all $n \geq 0$. However, even though state 2 is *not* accessible from state 3, state 3 *is* accessible from state 2 since, for $n = 1$, the transition matrix given at the end of Sec. 16.2 indicates that $p_{23} = p > 0$.

If state j is accessible from state i and state i is accessible from state j, then states i and j are said to **communicate.** In both the weather and inventory examples, all states communicate. In the gambling example, states 2 and 3 do not. (The same is true of states 1 and 3, states 1 and 0, and states 2 and 0.) In general,

1. Any state communicates with itself (because $p_{ii}^{(0)} = P\{X_0 = i \mid X_0 = i\} = 1$).
2. If state i communicates with state j, then state j communicates with state i.
3. If state i communicates with state j and state j communicates with state k, then state i communicates with state k.

Properties 1 and 2 follow from the definition of states communicating, whereas property 3 follows from the Chapman-Kolmogorov equations.

As a result of these three properties of communication, the states may be partitioned into one or more separate **classes** such that those states that communicate with each other are in the same class. (A class may consist of a single state.) If there is only one class, i.e., all the states communicate, the Markov chain is said to be **irreducible.** In both the weather and inventory examples, the Markov chain is irreducible. In both of the stock examples in Sec. 16.2, the Markov chain also is irreducible. However, the gambling example contains three classes. Observe in Fig. 16.4 how state 0 forms a class, state 3 forms a class, and states 1 and 2 form a class.

Recurrent States and Transient States

It is often useful to talk about whether a process entering a state will ever return to this state. Here is one possibility.

A state is said to be a **transient** state if, upon entering this state, the process *might never return* to this state again. Therefore, state i is transient if and only if there exists a state j

($j \neq i$) that is accessible from state i but not vice versa, that is, state i is not accessible from state j.

Thus, if state i is transient and the process visits this state, there is a positive probability (perhaps even a probability of 1) that the process will later move to state j and so will never return to state i. Consequently, a transient state will be visited only a finite number of times. To illustrate, consider the gambling example presented at the end of Sec. 16.2. Its state transition diagram shown in Fig. 16.4 indicates that both states 1 and 2 are transient states since the process will leave these states sooner or later to enter either state 0 or state 3 and then will remain in that state forever.

When starting in state i, another possibility is that the process *definitely* will return to this state.

> A state is said to be a **recurrent** state if, upon entering this state, the process *definitely will return* to this state again. Therefore, a state is recurrent if and only if it is not transient.

Since a recurrent state definitely will be revisited after each visit, it will be visited infinitely often if the process continues forever. For example, all the states in the state transition diagrams shown in Figs. 16.1, 16.2, and 16.3 are recurrent states because the process always will return to each of these states. Even for the gambling example, states 0 and 3 are recurrent states because the process will keep returning immediately to one of these states forever once the process enters that state. Note in Fig. 16.4 how the process eventually will enter either state 0 or state 3 and then will never leave that state again.

If the process enters a certain state and then stays in this state at the next step, this is considered a *return* to this state. Hence, the following kind of state is a special type of recurrent state.

> A state is said to be an **absorbing** state if, upon entering this state, the process *never will leave* this state again. Therefore, state i is an absorbing state if and only if $p_{ii} = 1$.

As just noted, both states 0 and 3 for the gambling example fit this definition, so they both are absorbing states as well as a special type of recurrent state. We will discuss absorbing states further in Sec. 16.7.

Recurrence is a class property. That is, all states in a class are either recurrent or transient. Furthermore, in a finite-state Markov chain, not all states can be transient. Therefore, all states in an irreducible finite-state Markov chain are recurrent. Indeed, one can identify an irreducible finite-state Markov chain (and therefore conclude that all states are recurrent) by showing that all states of the process communicate. It has already been pointed out that a sufficient condition for *all* states to be accessible (and therefore communicate with each other) is that there exists a value of n for which $p_{ij}^{(n)} > 0$ for all i and j. Thus, all states in the inventory example (see Fig. 16.2) are recurrent, since $p_{ij}^{(2)}$ is positive for all i and j. Similarly, both the weather example and the first stock example contain only recurrent states, since p_{ij} is positive for all i and j. By calculating $p_{ij}^{(2)}$ for all i and j in the second stock example in Sec. 16.2 (see Fig. 16.3), it follows that all states are recurrent since $p_{ij}^{(2)} > 0$ for all i and j.

As another example, suppose that a Markov chain has the following transition matrix:

$$\mathbf{P} = \begin{array}{c} \text{State} \\ 0 \\ 1 \\ 2 \\ 3 \\ 4 \end{array} \begin{array}{cccccc} 0 & 1 & 2 & 3 & 4 \\ \left[\begin{array}{ccccc} \frac{1}{4} & \frac{3}{4} & 0 & 0 & 0 \\ \frac{1}{2} & \frac{1}{2} & 0 & 0 & 0 \\ 0 & 0 & 1 & 0 & 0 \\ 0 & 0 & \frac{1}{3} & \frac{2}{3} & 0 \\ 1 & 0 & 0 & 0 & 0 \end{array}\right] \end{array}$$

Note that state 2 is an absorbing state (and hence a recurrent state) because if the process enters state 2 (row 3 of the matrix), it will never leave. State 3 is a transient state because if the process is in state 3, there is a positive probability that it will never return. The probability is $\frac{1}{3}$ that the process will go from state 3 to state 2 on the first step. Once the process is in state 2, it remains in state 2. State 4 also is a transient state because if the process starts in state 4, it immediately leaves and can never return. States 0 and 1 are recurrent states. To see this, observe from **P** that if the process starts in either of these states, it can never leave these two states. Furthermore, whenever the process moves from one of these states to the other one, it always will return to the original state eventually.

Periodicity Properties

Another useful property of Markov chains is *periodicities*. The **period** of state i is defined to be the integer t ($t > 1$) such that $p_{ii}^{(n)} = 0$ for all values of n other than $t, 2t, 3t, \ldots$ and t is the smallest integer with this property. In the gambling example (end of Section 16.2), starting in state 1, it is possible for the process to enter state 1 only at times 2, 4, . . . , so state 1 has period 2. The reason is that the player can break even (be neither winning nor losing) only at times 2, 4, . . . , which can be verified by calculating $p_{11}^{(n)}$ for all n and noting that $p_{11}^{(n)} = 0$ for n odd. You also can see in Fig. 16.4 that the process always takes two steps to return to state 1 until the process gets absorbed in either state 0 or state 3. (The same conclusion also applies to state 2.)

If there are two consecutive numbers s and $s + 1$ such that the process can be in state i at times s and $s + 1$, the state is said to have period 1 and is called an **aperiodic** state.

Just as recurrence is a class property, it can be shown that periodicity is a class property. That is, if state i in a class has period t, then all states in that class have period t. In the gambling example, state 2 also has period 2 because it is in the same class as state 1 and we noted above that state 1 has period 2.

It is possible for a Markov chain to have both a recurrent class of states and a transient class of states where the two classes have different periods greater than 1. If you would like to see a Markov chain where this occurs, **another example** of this type is provided in the Worked Examples section of the book's website.

In a finite-state Markov chain, recurrent states that are aperiodic are called **ergodic** states. A Markov chain is said to be *ergodic* if all its states are ergodic states. You will see next that a key long-run property of a Markov chain that is both irreducible and ergodic is that its n-step transition probabilities will converge to steady-state probabilities as n grows large.

■ 16.5   LONG-RUN PROPERTIES OF MARKOV CHAINS

Steady-State Probabilities

While calculating the n-step transition probabilities for both the weather and inventory examples in Sec. 16.3, we noted an interesting feature of these matrices. If n is large enough ($n = 5$ for the weather example and $n = 8$ for the inventory example), all the rows of the matrix have identical entries, so the probability that the system is in each state j no longer depends on the initial state of the system. In other words, there is a limiting probability that the system will be in each state j after a large number of transitions, and this probability is independent of the initial state. These properties of the long-run behavior of finite-state Markov chains do, in fact, hold under relatively general conditions, as summarized below.

For any irreducible ergodic Markov chain, $\lim_{n \to \infty} p_{ij}^{(n)}$ exists and is independent of i.

Furthermore,

$$\lim_{n \to \infty} p_{ij}^{(n)} = \pi_j > 0,$$

where the π_j uniquely satisfy the following **steady-state equations**

$$\pi_j = \sum_{i=0}^{M} \pi_i p_{ij}, \qquad \text{for } j = 0, 1, \ldots, M,$$

$$\sum_{j=0}^{M} \pi_j = 1.$$

If you prefer to work with a system of equations in matrix form, this system (excluding the sum = 1 equation) also can be expressed as

$$\pi = \pi \mathbf{P},$$

where $\pi = (\pi_0, \pi_1, \ldots, \pi_M)$.

The π_j are called the **steady-state probabilities** of the Markov chain. The term *steady-state* probability means that the probability of finding the process in a certain state, say j, after a large number of transitions tends to the value π_j, independent of the probability distribution of the initial state. It is important to note that the steady-state probability does *not* imply that the process settles down into one state. On the contrary, the process continues to make transitions from state to state, and at any step n the transition probability from state i to state j is still p_{ij}.

The π_j can also be interpreted as *stationary probabilities* (not to be confused with stationary transition probabilities) in the following sense. If the *initial* probability of being in state j is given by π_j (that is, $P\{X_0 = j\} = \pi_j$) for all j, then the probability of finding the process in state j at time $n = 1, 2, \ldots$ is also given by π_j (that is, $P\{X_n = j\} = \pi_j$).

Note that the steady-state equations consist of $M + 2$ equations in $M + 1$ unknowns. Because it has a unique solution, at least one equation must be redundant and can, therefore, be deleted. It cannot be the equation

$$\sum_{j=0}^{M} \pi_j = 1,$$

because $\pi_j = 0$ for all j will satisfy the other $M + 1$ equations. Furthermore, the solutions to the other $M + 1$ steady-state equations have a unique solution up to a multiplicative constant, and it is the final equation that forces the solution to be a probability distribution.

Application to the Weather Example.   The weather example introduced in Sec. 16.1 and formulated in Sec. 16.2 has only two states (dry and rain), so the above steady-state equations become

$$\pi_0 = \pi_0 p_{00} + \pi_1 p_{10},$$
$$\pi_1 = \pi_0 p_{01} + \pi_1 p_{11},$$
$$1 = \pi_0 \quad + \pi_1.$$

The intuition behind the first equation is that, in steady state, the probability of being in state 0 after the next transition must equal (1) the probability of being in state 0 now *and* then staying in state 0 after the next transition *plus* (2) the probability of being in state 1 now *and* next making the transition to state 0. The logic for the second equation is the same, except in terms of state 1. The third equation simply expresses the fact that the probabilities of these mutually exclusive states must sum to 1.

Referring to the transition probabilities given in Sec. 16.2 for this example, these equations become

$$\pi_0 = 0.8\pi_0 + 0.6\pi_1, \qquad \text{so} \qquad 0.2\pi_0 = 0.6\pi_1,$$
$$\pi_1 = 0.2\pi_0 + 0.4\pi_1, \qquad \text{so} \qquad 0.6\pi_1 = 0.2\pi_0,$$
$$1 = \pi_0 + \pi_1.$$

Note that one of the first two equations is redundant since both equations reduce to $\pi_0 = 3\pi_1$. Combining this result with the third equation immediately yields the following steady-state probabilities:

$$\pi_0 = 0.25, \qquad \pi_1 = 0.75$$

These are the same probabilities as obtained in each row of the five-step transition matrix calculated in Sec. 16.3 because five transitions proved enough to make the state probabilities essentially independent of the initial state.

Application to the Inventory Example. The inventory example introduced in Sec. 16.1 and formulated in Sec. 16.2 has four states. Therefore, in this case, the steady-state equations can be expressed as

$$\pi_0 = \pi_0 p_{00} + \pi_1 p_{10} + \pi_2 p_{20} + \pi_3 p_{30},$$
$$\pi_1 = \pi_0 p_{01} + \pi_1 p_{11} + \pi_2 p_{21} + \pi_3 p_{31},$$
$$\pi_2 = \pi_0 p_{02} + \pi_1 p_{12} + \pi_2 p_{22} + \pi_3 p_{32},$$
$$\pi_3 = \pi_0 p_{03} + \pi_1 p_{13} + \pi_2 p_{23} + \pi_3 p_{33},$$
$$1 = \pi_0 + \pi_1 + \pi_2 + \pi_3.$$

Substituting values for p_{ij} (see the transition matrix in Sec. 16.2) into these equations leads to the equations

$$\pi_0 = 0.080\pi_0 + 0.632\pi_1 + 0.264\pi_2 + 0.080\pi_3,$$
$$\pi_1 = 0.184\pi_0 + 0.368\pi_1 + 0.368\pi_2 + 0.184\pi_3,$$
$$\pi_2 = 0.368\pi_0 \qquad\quad + 0.368\pi_2 + 0.368\pi_3,$$
$$\pi_3 = 0.368\pi_0 \qquad\qquad\qquad\qquad + 0.368\pi_3,$$
$$1 = \qquad \pi_0 + \qquad \pi_1 + \qquad \pi_2 + \qquad \pi_3.$$

Solving the last four equations simultaneously provides the solution

$$\pi_0 = 0.286, \qquad \pi_1 = 0.285, \qquad \pi_2 = 0.263, \qquad \pi_3 = 0.166,$$

which is essentially the result that appears in matrix $\mathbf{P}^{(8)}$ in Sec. 16.3. Thus, after many weeks the probability of finding zero, one, two, and three cameras in stock at the end of a week tends to 0.286, 0.285, 0.263, and 0.166, respectively.

More about Steady-State Probabilities. Your IOR Tutorial includes a procedure for solving the steady-state equations to obtain the steady-state probabilities. In addition, the Worked Examples section of our website includes **another example** of applying steady-state probabilities (including using the technique described in the next subsection) to determine the best of several alternatives on a cost basis.

There are other important results concerning steady-state probabilities. In particular, if i and j are recurrent states belonging to different classes, then

$$p_{ij}^{(n)} = 0, \qquad \text{for all } n.$$

This result follows from the definition of a class.

Similarly, if j is a transient state, then

$$\lim_{n \to \infty} p_{ij}^{(n)} = 0, \qquad \text{for all } i.$$

Thus, the probability of finding the process in a transient state after a large number of transitions tends to zero.

Expected Average Cost per Unit Time

The preceding subsection dealt with irreducible finite-state Markov chains whose states were ergodic (recurrent and aperiodic). If the requirement that the states be aperiodic is relaxed, then the limit

$$\lim_{n \to \infty} p_{ij}^{(n)}$$

may not exist. To illustrate this point, consider the two-state transition matrix

$$\mathbf{P} = \begin{array}{c} \text{State} \\ 0 \\ 1 \end{array} \begin{array}{cc} 0 & 1 \\ \begin{bmatrix} 0 & 1 \\ 1 & 0 \end{bmatrix} \end{array}.$$

If the process starts in state 0 at time 0, it will be in state 0 at times 2, 4, 6, . . . and in state 1 at times 1, 3, 5, ... Thus, $p_{00}^{(n)} = 1$ if n is even and $p_{00}^{(n)} = 0$ if n is odd, so that

$$\lim_{n \to \infty} p_{00}^{(n)}$$

does not exist. However, the following limit always exists for an irreducible (finite-state) Markov chain:

$$\lim_{n \to \infty} \left(\frac{1}{n} \sum_{k=1}^{n} p_{ij}^{(k)} \right) = \pi_j,$$

where the π_j satisfy the steady-state equations given in the preceding subsection.

This result is important in computing the *long-run average cost per unit time* associated with a Markov chain. Suppose that a cost (or other penalty function) $C(X_t)$ is incurred when the process is in state X_t at time t, for $t = 0, 1, 2, . . . $. Note that $C(X_t)$ is a random variable that takes on any one of the values $C(0), C(1), . . . , C(M)$ and that the function $C(\cdot)$ is independent of t. The expected average cost incurred over the first n periods is given by

$$E\left[\frac{1}{n} \sum_{t=1}^{n} C(X_t) \right].$$

By using the result that

$$\lim_{n \to \infty} \left(\frac{1}{n} \sum_{k=1}^{n} p_{ij}^{(k)} \right) = \pi_j,$$

it can be shown that the (long-run) *expected average cost per unit time* is given by

$$\lim_{n \to \infty} E\left[\frac{1}{n} \sum_{t=1}^{n} C(X_t) \right] = \sum_{j=0}^{M} \pi_j C(j).$$

Application to the Inventory Example.   To illustrate, consider the inventory example introduced in Sec. 16.1, where the solution for the π_j was obtained in an earlier subsection. Suppose the camera store finds that a storage charge is being allocated for each camera remaining on the shelf at the end of the week. The cost is charged as follows:

$$C(x_t) = \begin{cases} 0 & \text{if} & x_t = 0 \\ 2 & \text{if} & x_t = 1 \\ 8 & \text{if} & x_t = 2 \\ 18 & \text{if} & x_t = 3 \end{cases}$$

Using the steady-state probabilities found earlier in this section, the long-run expected average storage cost per week can then be obtained from the preceding equation, i.e.,

$$\lim_{n \to \infty} E\left[\frac{1}{n} \sum_{t=1}^{n} C(X_t)\right] = 0.286(0) + 0.285(2) + 0.263(8) + 0.166(18) = 5.662.$$

Note that an alternative measure to the (long-run) expected average cost per unit time is the (long-run) *actual average cost per unit time*. It can be shown that this latter measure also is given by

$$\lim_{n \to \infty} \left[\frac{1}{n} \sum_{t=1}^{n} C(X_t)\right] = \sum_{j=0}^{M} \pi_j C(j)$$

for essentially all paths of the process. Thus, either measure leads to the same result. These results can also be used to interpret the meaning of the π_j. To do so, let

$$C(X_t) = \begin{cases} 1 & \text{if} & X_t = j \\ 0 & \text{if} & X_t \neq j. \end{cases}$$

The (long-run) expected fraction of times the system is in state j is then given by

$$\lim_{n \to \infty} E\left[\frac{1}{n} \sum_{t=1}^{n} C(X_t)\right] = \lim_{n \to \infty} E(\text{fraction of times system is in state } j) = \pi_j.$$

Similarly, π_j can also be interpreted as the (long-run) actual fraction of times that the system is in state j.

Expected Average Cost per Unit Time for Complex Cost Functions

In the preceding subsection, the cost function was based solely on the state that the process is in at time t. In many important problems encountered in practice, the cost may also depend upon some other random variable.

For example, in the inventory example introduced in Sec. 16.1, suppose that the costs to be considered are the ordering cost and the penalty cost for unsatisfied demand (storage costs are so small they will be ignored). It is reasonable to assume that the number of cameras ordered to arrive at the beginning of week t depends only upon the state of the process X_{t-1} (the number of cameras in stock) when the order is placed at the end of week $t - 1$. However, the cost of unsatisfied demand in week t will also depend upon the demand D_t. Therefore, the total cost (ordering cost plus cost of unsatisfied demand) for week t is a function of X_{t-1} and D_t, that is, $C(X_{t-1}, D_t)$.

Under the assumptions of this example, it can be shown that the (long-run) *expected average cost per unit time* is given by

$$\lim_{n \to \infty} E\left[\frac{1}{n} \sum_{t=1}^{n} C(X_{t-1}, D_t)\right] = \sum_{j=0}^{M} k(j) \, \pi_j,$$

where

$$k(j) = E[C(j, D_t)],$$

and where this latter (conditional) expectation is taken with respect to the probability distribution of the random variable D_t, given the state j. Similarly, the (long-run) actual average cost per unit time is given by

$$\lim_{n \to \infty} \left[\frac{1}{n} \sum_{t=1}^{n} C(X_{t-1}, D_t) \right] = \sum_{j=0}^{M} k(j)\pi_j.$$

Now let us assign numerical values to the two components of $C(X_{t-1}, D_t)$ in this example, namely, the ordering cost and the penalty cost for unsatisfied demand. If $z > 0$ cameras are ordered, the cost incurred is $(10 + 25z)$ dollars. If no cameras are ordered, no ordering cost is incurred. For each unit of unsatisfied demand (lost sales), there is a penalty of $50. Therefore, given the ordering policy described in Sec. 16.1, the cost in week t is given by

$$C(X_{t-1}, D_t) = \begin{cases} 10 + (25)(3) + 50 \max\{D_t - 3, 0\} & \text{if} & X_{t-1} = 0 \\ 50 \max \{D_t - X_{t-1}, 0\} & \text{if} & X_{t-1} \geq 1, \end{cases}$$

for $t = 1, 2, \ldots$. Hence,

$$C(0, D_t) = 85 + 50 \max\{D_t - 3, 0\},$$

so that

$$k(0) = E[C(0, D_t)] = 85 + 50E(\max\{D_t - 3, 0\})$$
$$= 85 + 50[P_D(4) + 2P_D(5) + 3P_D(6) + \cdots],$$

where $P_D(i)$ is the probability that the demand equals i, as given by a Poisson distribution with a mean of 1, so that $P_D(i)$ becomes negligible for i larger than about 6. Since $P_D(4) = 0.015$, $P_D(5) = 0.003$, and $P_D(6) = 0.001$, we obtain $k(0) = 86.2$. Also using $P_D(2) = 0.184$ and $P_D(3) = 0.061$, similar calculations lead to the results

$$k(1) = E[C(1, D_t)] = 50E(\max\{D_t - 1, 0\})$$
$$= 50[P_D(2) + 2P_D(3) + 3P_D(4) + \cdots]$$
$$= 18.4,$$

$$k(2) = E[C(2, D_t)] = 50E(\max\{D_t - 2, 0\})$$
$$= 50[P_D(3) + 2P_D(4) + 3P_D(5) + \cdots]$$
$$= 5.2,$$

and

$$k(3) = E[C(3, D_t)] = 50E(\max\{D_t - 3, 0\})$$
$$= 50[P_D(4) + 2P_D(5) + 3P_D(6) + \cdots]$$
$$= 1.2.$$

Thus, the (long-run) expected average cost per week is given by

$$\sum_{j=0}^{3} k(j)\pi_j = 86.2(0.286) + 18.4(0.285) + 5.2(0.263) + 1.2(0.166) = \$31.46.$$

This is the cost associated with the particular ordering policy described in Sec. 16.1. The cost of other ordering policies can be evaluated in a similar way to identify the policy that minimizes the expected average cost per week.

The results of this subsection were presented only in terms of the inventory example. However, the (nonnumerical) results still hold for other problems as long as the following conditions are satisfied:

1. $\{X_t\}$ is an irreducible (finite-state) Markov chain.
2. Associated with this Markov chain is a sequence of random variables $\{D_t\}$ which are independent and identically distributed.
3. For a fixed $m = 0, \pm 1, \pm 2, \ldots$, a cost $C(X_t, D_{t+m})$ is incurred at time t, for $t = 0, 1, 2, \ldots$.
4. The sequence $X_0, X_1, X_2, \ldots, X_t$ must be independent of D_{t+m}

In particular, if these conditions are satisfied, then

$$\lim_{n \to \infty} E\left[\frac{1}{n} \sum_{t=1}^{n} C(X_t, D_{t+m})\right] = \sum_{j=0}^{M} k(j)\pi_j,$$

where

$$k(j) = E[C(j, D_{t+m})],$$

and where this latter conditional expectation is taken with respect to the probability distribution of the random variable D_t, given the state j. Furthermore,

$$\lim_{n \to \infty} \left[\frac{1}{n} \sum_{t=1}^{n} C(X_t, D_{t+m})\right] = \sum_{j=0}^{M} k(j)\pi_j$$

for essentially all paths of the process.

■ 16.6   FIRST PASSAGE TIMES

Section 16.3 dealt with finding n-step transition probabilities from state i to state j. It is often desirable to also make probability statements about the number of transitions made by the process in going from state i to state j *for the first time*. This length of time is called the **first passage time** in going from state i to state j. When $j = i$, this first passage time is just the number of transitions until the process returns to the initial state i. In this case, the first passage time is called the **recurrence time** for state i.

To illustrate these definitions, reconsider the inventory example introduced in Sec. 16.1, where X_t is the number of cameras on hand at the end of week t, where we start with $X_0 = 3$. Suppose that it turns out that

$$X_0 = 3, \quad X_1 = 2, \quad X_2 = 1, \quad X_3 = 0, \quad X_4 = 3, \quad X_5 = 1.$$

In this case, the first passage time in going from state 3 to state 1 is 2 weeks, the first passage time in going from state 3 to state 0 is 3 weeks, and the recurrence time for state 3 is 4 weeks.

In general, the first passage times are random variables. The probability distributions associated with them depend upon the transition probabilities of the process. In particular, let $f_{ij}^{(n)}$ denote the probability that the first passage time from state i to j is equal to n. For $n > 1$, this first passage time is n if the first transition is from state i to some state k ($k \neq j$) and then the first passage time from state k to state j is $n - 1$. Therefore, these probabilities satisfy the following recursive relationships:

$$f_{ij}^{(1)} = p_{ij}^{(1)} = p_{ij},$$
$$f_{ij}^{(2)} = \sum_{k \neq j} p_{ik} f_{kj}^{(1)},$$
$$f_{ij}^{(n)} = \sum_{k \neq j} p_{ik} f_{kj}^{(n-1)}.$$

Thus, the probability of a first passage time from state i to state j in n steps can be computed recursively from the one-step transition probabilities.

In the inventory example, the probability distribution of the first passage time in going from state 3 to state 0 is obtained from these recursive relationships as follows:

$$f_{30}^{(1)} = p_{30} = 0.080,$$
$$f_{30}^{(2)} = p_{31}f_{10}^{(1)} + p_{32}f_{20}^{(1)} + p_{33}f_{30}^{(1)}$$
$$= 0.184(0.632) + 0.368(0.264) + 0.368(0.080) = 0.243,$$
$$\vdots$$

where the p_{3k} and $f_{k0}^{(1)} = p_{k0}$ are obtained from the (one-step) transition matrix given in Sec. 16.2.

For fixed i and j, the $f_{ij}^{(n)}$ are nonnegative numbers such that

$$\sum_{n=1}^{\infty} f_{ij}^{(n)} \leq 1.$$

Unfortunately, this sum may be strictly less than 1, which implies that a process initially in state i may never reach state j. When the sum does equal 1, $f_{ij}^{(n)}$ (for $n = 1, 2, \ldots$) can be considered as a probability distribution for the random variable, the first passage time.

Although obtaining $f_{ij}^{(n)}$ for all n may be tedious, it is relatively simple to obtain the expected first passage time from state i to state j. Denote this expectation by μ_{ij}, which is defined by

$$\mu_{ij} = \begin{cases} \infty & \text{if } \sum_{n=1}^{\infty} f_{ij}^{(n)} < 1 \\ \sum_{n=1}^{\infty} n f_{ij}^{(n)} & \text{if } \sum_{n=1}^{\infty} f_{ij}^{(n)} = 1. \end{cases}$$

Whenever

$$\sum_{n=1}^{\infty} f_{ij}^{(n)} = 1,$$

μ_{ij} uniquely satisfies the equation

$$\mu_{ij} = 1 + \sum_{k \neq j} p_{ik}\mu_{kj}.$$

This equation recognizes that the first transition from state i can be to either state j or to some other state k. If it is to state j, the first passage time is 1. Given that the first transition is to some state k ($k \neq j$) instead, which occurs with probability p_{ik}, the conditional expected first passage time from state i to state j is $1 + \mu_{kj}$. Combining these facts, and summing over all the possibilities for the first transition, leads directly to this equation.

For the inventory example, these equations for the μ_{ij} can be used to compute the expected time until the cameras are out of stock, given that the process is started when three cameras are available. This expected time is just the expected first passage time μ_{30}. Since all the states are recurrent, the system of equations leads to the expressions

$$\mu_{30} = 1 + p_{31}\mu_{10} + p_{32}\mu_{20} + p_{33}\mu_{30},$$
$$\mu_{20} = 1 + p_{21}\mu_{10} + p_{22}\mu_{20} + p_{23}\mu_{30},$$
$$\mu_{10} = 1 + p_{11}\mu_{10} + p_{12}\mu_{20} + p_{13}\mu_{30},$$

or

$$\mu_{30} = 1 + 0.184\mu_{10} + 0.368\mu_{20} + 0.368\mu_{30},$$
$$\mu_{20} = 1 + 0.368\mu_{10} + 0.368\mu_{20},$$
$$\mu_{10} = 1 + 0.368\mu_{10}.$$

The simultaneous solution to this system of equations is

$$\mu_{10} = 1.58 \text{ weeks},$$
$$\mu_{20} = 2.51 \text{ weeks},$$
$$\mu_{30} = 3.50 \text{ weeks},$$

so that the expected time until the cameras are out of stock is 3.50 weeks. Thus, in making these calculations for μ_{30}, we also obtain μ_{20} and μ_{10}.

For the case of μ_{ij} where $j = i$, μ_{ii} is the expected number of transitions until the process returns to the initial state i, and so is called the **expected recurrence time** for state i. After obtaining the steady-state probabilities $(\pi_0, \pi_1, \ldots, \pi_M)$ as described in the preceding section, these expected recurrence times can be calculated immediately as

$$\mu_{ii} = \frac{1}{\pi_i}, \qquad \text{for } i = 0, 1, \ldots, M.$$

Thus, for the inventory example, where $\pi_0 = 0.286$, $\pi_1 = 0.285$, $\pi_2 = 0.263$, and $\pi_3 = 0.166$, the corresponding expected recurrence times are

$$\mu_{00} = \frac{1}{\pi_0} = 3.50 \text{ weeks}, \qquad \mu_{22} = \frac{1}{\pi_2} = 3.80 \text{ weeks},$$

$$\mu_{11} = \frac{1}{\pi_1} = 3.51 \text{ weeks}, \qquad \mu_{33} = \frac{1}{\pi_3} = 6.02 \text{ weeks}.$$

■ 16.7 ABSORBING STATES

It was pointed out in Sec. 16.4 that a state k is called an *absorbing state* if $p_{kk} = 1$, so that once the chain visits k it remains there forever. If k is an absorbing state, and the process starts in state i, the probability of *ever* going to state k is called the **probability of absorption** into state k, given that the system started in state i. This probability is denoted by f_{ik}.

When there are two or more absorbing states in a Markov chain, and it is evident that the process will be absorbed into one of these states, it is desirable to find these probabilities of absorption. These probabilities can be obtained by solving a system of linear equations that considers all the possibilities for the first transition and then, given the first transition, considers the conditional probability of absorption into state k. In particular, if the state k is an absorbing state, then the set of absorption probabilities f_{ik} satisfies the system of equations

$$f_{ik} = \sum_{j=0}^{M} p_{ij} f_{jk}, \qquad \text{for } i = 0, 1, \ldots, M,$$

subject to the conditions

$$f_{kk} = 1,$$
$$f_{ik} = 0, \qquad \text{if state } i \text{ is recurrent and } i \neq k.$$

Absorption probabilities are important in random walks. A **random walk** is a Markov chain with the property that if the system is in a state i, then in a single transition the system either remains at i or moves to one of the two states immediately adjacent to i. For example, a random walk often is used as a model for situations involving gambling.

A Second Gambling Example.   To illustrate the use of absorption probabilities in a random walk, consider a gambling example similar to that presented in Sec. 16.2. However, suppose now that two players (A and B), each having \$2, agree to keep playing the game and betting \$1 at a time until one player is broke. The probability of A winning a single bet is $\frac{1}{3}$, so B wins the bet with probability $\frac{2}{3}$. The number of dollars that player A has before each bet (0, 1, 2, 3, or 4) provides the states of a Markov chain with transition matrix

$$
P = \begin{matrix} \text{State} \\ 0 \\ 1 \\ 2 \\ 3 \\ 4 \end{matrix}
\begin{matrix}
\begin{matrix} 0 & 1 & 2 & 3 & 4 \end{matrix} \\
\begin{bmatrix}
1 & 0 & 0 & 0 & 0 \\
\frac{2}{3} & 0 & \frac{1}{3} & 0 & 0 \\
0 & \frac{2}{3} & 0 & \frac{1}{3} & 0 \\
0 & 0 & \frac{2}{3} & 0 & \frac{1}{3} \\
0 & 0 & 0 & 0 & 1
\end{bmatrix}
\end{matrix}.
$$

Starting from state 2, the probability of absorption into state 0 (A losing all her money) can be obtained by solving for f_{20} from the system of equations given at the beginning of this section,

$$f_{00} = 1 \quad \text{(since state 0 is an absorbing state),}$$

$$f_{10} = \frac{2}{3} f_{00} + \frac{1}{3} f_{20},$$

$$f_{20} = \phantom{\frac{2}{3} f_{00} +} \frac{2}{3} f_{10} + \frac{1}{3} f_{30},$$

$$f_{30} = \phantom{\frac{2}{3} f_{00} + \frac{2}{3} f_{10} +} \frac{2}{3} f_{20} + \frac{1}{3} f_{40},$$

$$f_{40} = 0 \quad \text{(since state 4 is an absorbing state).}$$

This system of equations yields

$$f_{20} = \frac{2}{3}\left(\frac{2}{3} + \frac{1}{3} f_{20}\right) + \frac{1}{3}\left(\frac{2}{3} f_{20}\right) = \frac{4}{9} + \frac{4}{9} f_{20},$$

which reduces to $f_{20} = \frac{4}{5}$ as the probability of absorption into state 0.

Similarly, the probability of A finishing with \$4 (B going broke) when starting with \$2 (state 2) is obtained by solving for f_{24} from the system of equations,

$$f_{04} = 0 \quad \text{(since state 0 is an absorbing state),}$$

$$f_{14} = \frac{2}{3} f_{04} + \frac{1}{3} f_{24},$$

$$f_{24} = \phantom{\frac{2}{3} f_{04} +} \frac{2}{3} f_{14} + \frac{1}{3} f_{34},$$

$$f_{34} = \phantom{\frac{2}{3} f_{04} + \frac{2}{3} f_{14} +} \frac{2}{3} f_{24} + \frac{1}{3} f_{44},$$

$$f_{44} = 1 \quad \text{(since state 0 is an absorbing state).}$$

This yields

$$f_{24} = \frac{2}{3}\left(\frac{1}{3} f_{24}\right) + \frac{1}{3}\left(\frac{2}{3} f_{24} + \frac{1}{3}\right) = \frac{4}{9} f_{24} + \frac{1}{9},$$

so $f_{24} = \frac{1}{5}$ is the probability of absorption into state 4.

A Credit Evaluation Example.   There are many other situations where absorbing states play an important role. Consider a department store that classifies the balance of a customer's bill as fully paid (state 0), 1 to 30 days in arrears (state 1), 31 to 60 days in arrears (state 2), or bad debt (state 3). The accounts are checked *monthly* to determine the state of each customer. In general, credit is not extended and customers are expected to pay their bills promptly. Occasionally, customers miss the deadline for paying their bill. If this occurs when the balance is within 30 days in arrears, the store views the customer as being in state 1. If this occurs when the balance is between 31 and 60 days in arrears, the store views the customer as being in state 2. Customers that are more than 60 days in arrears are put into the bad-debt category (state 3), and then bills are sent to a collection agency.

After examining data over the past several years on the month by month progression of individual customers from state to state, the store has developed the following transition matrix:[4]

State \ State	0: Fully Paid	1: 1 to 30 Days in Arrears	2: 31 to 60 Days in Arrears	3: Bad Debt
0: fully paid	1	0	0	0
1: 1 to 30 days in arrears	0.7	0.2	0.1	0
2: 31 to 60 days in arrears	0.5	0.1	0.2	0.2
3: bad debt	0	0	0	1

Although each customer ends up in state 0 or 3, the store is interested in determining the probability that a customer will end up as a bad debt given that the account belongs to the 1 to 30 days in arrears state, and similarly, given that the account belongs to the 31 to 60 days in arrears state.

To obtain this information, the set of equations presented at the beginning of this section must be solved to obtain f_{13} and f_{23}. By substituting, the following two equations are obtained:

$$f_{13} = p_{10}f_{03} + p_{11}f_{13} + p_{12}f_{23} + p_{13}f_{33},$$
$$f_{23} = p_{20}f_{03} + p_{21}f_{13} + p_{22}f_{23} + p_{23}f_{33}.$$

Noting that $f_{03} = 0$ and $f_{33} = 1$, we now have two equations in two unknowns, namely,

$$(1 - p_{11})f_{13} = p_{13} + p_{12}f_{23},$$
$$(1 - p_{22})f_{23} = p_{23} + p_{21}f_{13}.$$

Substituting the values from the transition matrix leads to

$$0.8f_{13} = 0.1f_{23},$$
$$0.8f_{23} = 0.2 + 0.1f_{13},$$

and the solution is

$$f_{13} = 0.032,$$
$$f_{23} = 0.254.$$

[4]Customers who are fully paid (in state 0) and then subsequently fall into arrears on new purchases are viewed as "new" customers who start in state 1.

Thus, approximately 3 percent of the customers whose accounts are 1 to 30 days in arrears end up as bad debts, whereas about 25 percent of the customers whose accounts are 31 to 60 days in arrears end up as bad debts.

■ 16.8   CONTINUOUS TIME MARKOV CHAINS

In all the previous sections, we assumed that the time parameter t was discrete (that is, $t = 0, 1, 2, \ldots$). Such an assumption is suitable for many problems, but there are certain cases (such as for some queueing models considered in Chap. 17) where a continuous time parameter (call it t') is required, because the evolution of the process is being observed *continuously* over time. The definition of a Markov chain given in Sec. 16.2 also extends to such continuous processes. This section focuses on describing these "continuous time Markov chains" and their properties.

Formulation

As before, we label the possible **states** of the system as $0, 1, \ldots, M$. Starting at time 0 and letting the time parameter t' run continuously for $t' \geq 0$, we let the random variable $X(t')$ be the state of the system at time t'. Thus, $X(t')$ will take on one of its possible $(M + 1)$ values over some interval, $0 \leq t' < t_1$, then will jump to another value over the next interval, $t_1 \leq t' < t_2$, etc., where these transit points ($t_1, t_2, \ldots$) are random points in time (*not* necessarily integer).

Now consider the three points in time (1) $t' = r$ (where $r \geq 0$), (2) $t' = s$ (where $s > r$), and (3) $t' = s + t$ (where $t > 0$), interpreted as follows:

$t' = r$ is a past time,

$t' = s$ is the current time,

$t' = s + t$ is t time units into the future.

Therefore, the state of the system now has been observed at times $t' = s$ and $t' = r$. Label these states as

$$X(s) = i \quad \text{and} \quad X(r) = x(r).$$

Given this information, it now would be natural to seek the probability distribution of the state of the system at time $t' = s + t$. In other words, what is

$$P\{X(s + t) = j \mid X(s) = i \text{ and } X(r) = x(r)\}, \quad \text{for } j = 0, 1, \ldots, M?$$

Deriving this conditional probability often is very difficult. However, this task is considerably simplified if the stochastic process involved possesses the following key property.

A continuous time stochastic process $\{X(t'); \ t' \geq 0\}$ has the **Markovian property** if

$$P\{X(t + s) = j \mid X(s) = i \text{ and } X(r) = x(r)\} = P\{X(t + s) = j \mid X(s) = i\},$$

for all $i, j = 0, 1, \ldots, M$ and for all $r \geq 0$, $s > r$, and $t > 0$.

Note that $P\{X(t + s) = j \mid X(s) = i\}$ is a **transition probability,** just like the transition probabilities for discrete time Markov chains considered in the preceding sections, where the only difference is that t now need not be an integer.

If the transition probabilities are independent of s, so that

$$P\{X(t + s) = j \mid X(s) = i\} = P\{X(t) = j \mid X(0) = i\}$$

for all $s > 0$, they are called **stationary transition probabilities.**

Based in France, **PSA Peugeot Citroën** is one of the largest carmakers in the world. When the decision was made to introduce 25 new models between 2001 and 2004, PSA management decided to redesign its body shops so that the bodies of various car models could be assembled in each shop. An OR team was assigned the task of guiding the design process by developing tools for evaluating in advance the efficiency of the production lines for any given shop design.

The OR team developed both quick approximate methods and longer detailed methods for doing this. However, a key factor that needed to be incorporated into all the methods was the frequency with which each of the machines in the shop would go down and require repair, thereby disrupting the flow of work in the shop. The OR team used a *continuous-time Markov chain* to represent the evolution of each type of machine in going back and forth between being operational (up) and needing repair (down). Thus, the Markov chain has just two states, *up* and *down*, with some (small) transition rate for going from *up* to *down* and some (much larger) transition rate for going from *down* to *up*. The team concluded that the transition rate for going from *up* to *down* is essentially the same regardless of whether the machine currently is actually being operated or is idle, so it is not necessary to divide the *up* state into *operating* and *idle* states.

This application of operations research, including this simple, continuous-time Markov chain, had a dramatic impact on the company. By substantially improving the efficiency of the production lines in PSA body shops with minimal capital investment and no compromise in quality, it is credited with *contributing* **$130 million** *to PSA profits* (about 6.5 percent of the total profit) in the first year alone.

Source: A. Patchong, T. Lemoine, and G. Kern: "Improving Car Body Production at PSA Peugeot Citroën," *Interfaces*, **33**(1): 36–49, Jan.–Feb. 2003. (A link to this article is provided on our website, www.mhhe.com/hillier.)

To simplify notation, we shall denote these stationary transition probabilities by

$$p_{ij}(t) = P\{X(t) = j \mid X(0) = i\},$$

where $p_{ij}(t)$ is referred to as the **continuous time transition probability function.** We assume that

$$\lim_{t \to 0} p_{ij}(t) = \begin{cases} 1 & \text{if} \quad i = j \\ 0 & \text{if} \quad i \neq j. \end{cases}$$

Now we are ready to define the continuous time Markov chains to be considered in this section.

A continuous time stochastic process $\{X(t'); t' \geq 0\}$ is a **continuous time Markov chain** if it has the *Markovian property.*

We shall restrict our consideration to continuous time Markov chains with the following properties:

1. A finite number of states.
2. Stationary transition probabilities.

Some Key Random Variables

In the analysis of continuous time Markov chains, one key set of random variables is the following.

Each time the process enters state i, the amount of time it spends in that state before moving to a different state is a random variable T_i, where $i = 0, 1, \ldots, M$.

Suppose that the process enters state i at time $t' = s$. Then, for any fixed amount of time $t > 0$, note that $T_i > t$ if and only if $X(t') = i$ for all t' over the interval $s \leq t' \leq s + t$. Therefore, the Markovian property (with stationary transition probabilities) implies that

$$P\{T_i > t + s \mid T_i > s\} = P\{T_i > t\}.$$

This is a rather unusual property for a probability distribution to possess. It says that the probability distribution of the *remaining* time until the process transits out of a given state always is the same, regardless of how much time the process has already spent in that state. In effect, the random variable is memoryless; the process forgets its history. There is only one (continuous) probability distribution that possesses this property—the *exponential distribution*. The exponential distribution has a single parameter, call it q, where the mean is $1/q$ and the cumulative distribution function is

$$P\{T_i \leq t\} = 1 - e^{-qt}, \qquad \text{for } t \geq 0.$$

(We shall describe the properties of the exponential distribution in detail in Sec. 17.4.)

This result leads to an equivalent way of describing a continuous time Markov chain:

1. The random variable T_i has an exponential distribution with a mean of $1/q_i$.
2. When leaving state i, the process moves to a state j with probability p_{ij}, where the p_{ij} satisfy the conditions

$$p_{ii} = 0 \qquad \text{for all } i,$$

and

$$\sum_{j=0}^{M} p_{ij} = 1 \qquad \text{for all } i.$$

3. The next state visited after state i is independent of the time spent in state i.

Just as the one-step transition probabilities played a major role in describing discrete time Markov chains, the analogous role for a continuous time Markov chain is played by the transition intensities.

The **transition intensities** are

$$q_i = -\frac{d}{dt} p_{ii}(0) = \lim_{t \to 0} \frac{1 - p_{ii}(t)}{t}, \qquad \text{for } i = 0, 1, 2, \ldots, M,$$

and

$$q_{ij} = \frac{d}{dt} p_{ij}(0) = \lim_{t \to 0} \frac{p_{ij}(t)}{t} = q_i p_{ij}, \qquad \text{for all } j \neq i,$$

where $p_{ij}(t)$ is the *continuous time transition probability function* introduced at the beginning of the section and p_{ij} is the probability described in property 2 of the preceding paragraph. Furthermore, q_i as defined here turns out to still be the parameter of the exponential distribution for T_i as well (see property 1 of the preceding paragraph).

The intuitive interpretation of the q_i and q_{ij} is that they are *transition rates*. In particular, q_i is the *transition rate out of state i* in the sense that q_i is the expected number of times that the process leaves state i per unit of time spent in state i. (Thus, q_i is the reciprocal of the expected time that the process spends in state i per visit to state i; that is, $q_i = 1/E[T_i]$.) Similarly, q_{ij} is the *transition rate from state i to state j* in the sense that q_{ij} is the expected number of times that the process transits from state i to state j per unit of time spent in state i. Thus,

$$q_i = \sum_{j \neq i} q_{ij}.$$

Just as q_i is the parameter of the exponential distribution for T_i, each q_{ij} is the parameter of an exponential distribution for a related random variable described below.

Each time the process enters state i, the amount of time it will spend in state i before a transition to state j occurs (if a transition to some other state does not occur first) is a random variable T_{ij}, where $i, j = 0, 1, \ldots, M$ and $j \neq i$. The T_{ij} are independent random variables, where each T_{ij} has an *exponential distribution* with parameter q_{ij}, so $E[T_{ij}] = 1/q_{ij}$. The time spent in state i until a transition occurs (T_i) is the *minimum* (over $j \neq i$) of the T_{ij}. When the transition occurs, the probability that it is to state j is $p_{ij} = q_{ij}/q_i$.

Steady-State Probabilities

Just as the transition probabilities for a discrete time Markov chain satisfy the Chapman-Kolmogorov equations, the continuous time transition probability function also satisfies these equations. Therefore, for any states i and j and nonnegative numbers t and s ($0 \leq s \leq t$),

$$p_{ij}(t) = \sum_{k=0}^{M} p_{ik}(s)p_{kj}(t - s).$$

A pair of states i and j are said to *communicate* if there are times t_1 and t_2 such that $p_{ij}(t_1) > 0$ and $p_{ji}(t_2) > 0$. All states that communicate are said to form a *class*. If all states form a single class, i.e., if the Markov chain is *irreducible* (hereafter assumed), then

$$p_{ij}(t) > 0, \qquad \text{for all } t > 0 \text{ and all states } i \text{ and } j.$$

Furthermore,

$$\lim_{t \to \infty} p_{ij}(t) = \pi_j$$

always exists and is independent of the initial state of the Markov chain, for $j = 0, 1, \ldots, M$. These limiting probabilities are commonly referred to as the **steady-state probabilities** (or *stationary probabilities*) of the Markov chain.

The π_j satisfy the equations

$$\pi_j = \sum_{i=0}^{M} \pi_i p_{ij}(t), \qquad \text{for } j = 0, 1, \ldots, M \text{ and every } t \geq 0.$$

However, the following **steady-state equations** provide a more useful system of equations for solving for the steady-state probabilities:

$$\pi_j q_j = \sum_{i \neq j} \pi_i q_{ij}, \qquad \text{for } j = 0, 1, \ldots, M.$$

and

$$\sum_{j=0}^{M} \pi_j = 1.$$

The steady-state equation for state j has an intuitive interpretation. The left-hand side ($\pi_j q_j$) is the *rate* at which the process *leaves* state j, since π_j is the (steady-state) probability that the process is in state j and q_j is the transition rate out of state j given that the process is in state j. Similarly, each term on the right-hand side ($\pi_i q_{ij}$) is the *rate* at which the process *enters* state j from state i, since q_{ij} is the transition rate from state i to state j given that the process is in state i. By summing over all $i \neq j$, the entire right-hand side then gives the rate at which the process enters state j from any other state. The overall equation thereby states that the rate at which the process leaves state j must equal the rate at which the process enters state j. Thus, this equation is analogous to the conservation of flow equations encountered in many engineering and science courses.

Because each of the first $M + 1$ *steady-state equations* requires that two rates be in *balance* (equal), these equations sometimes are called the **balance equations**.

Example.   A certain shop has two identical machines that are operated continuously except when they are broken down. Because they break down fairly frequently, the top-priority assignment for a full-time maintenance person is to repair them whenever needed.

The time required to repair a machine has an exponential distribution with a mean of $\frac{1}{2}$ day. Once the repair of a machine is completed, the time until the next breakdown of that machine has an exponential distribution with a mean of 1 day. These distributions are independent.

Define the random variable $X(t')$ as

$$X(t') = \text{number of machines broken down at time } t',$$

so the possible values of $X(t')$ are 0, 1, 2. Therefore, by letting the time parameter t' run continuously from time 0, the continuous time stochastic process $\{X(t'); t' \geq 0\}$ gives the evolution of the number of machines broken down.

Because both the repair time and the time until a breakdown have exponential distributions, $\{X(t'); t' \geq 0\}$ is a *continuous time Markov chain*[5] with states 0, 1, 2. Consequently, we can use the steady-state equations given in the preceding subsection to find the steady-state probability distribution of the number of machines broken down. To do this, we need to determine all the *transition rates*, i.e., the q_i and q_{ij} for $i, j = 0, 1, 2$.

The state (number of machines broken down) increases by 1 when a breakdown occurs and decreases by 1 when a repair occurs. Since both breakdowns and repairs occur one at a time, $q_{02} = 0$ and $q_{20} = 0$. The expected repair time is $\frac{1}{2}$ day, so the rate at which repairs are completed (when any machines are broken down) is 2 per day, which implies that $q_{21} = 2$ and $q_{10} = 2$. Similarly, the expected time until a particular operational machine breaks down is 1 day, so the rate at which it breaks down (when operational) is 1 per day, which implies that $q_{12} = 1$. During times when both machines are operational, breakdowns occur at the rate of $1 + 1 = 2$ per day, so $q_{01} = 2$.

These transition rates are summarized in the rate diagram shown in Fig. 16.5. These rates now can be used to calculate the *total transition rate* out of each state.

$$q_0 = q_{01} = 2.$$
$$q_1 = q_{10} + q_{12} = 3.$$
$$q_2 = q_{21} = 2.$$

Plugging all the rates into the steady-state equations given in the preceding subsection then yields

Balance equation for state 0:	$2\pi_0 = 2\pi_1$
Balance equation for state 1:	$3\pi_1 = 2\pi_0 + 2\pi_2$
Balance equation for state 2:	$2\pi_2 = \pi_1$
Probabilities sum to 1:	$\pi_0 + \pi_1 + \pi_2 = 1$

Any one of the balance equations (say, the second) can be deleted as redundant, and the simultaneous solution of the remaining equations gives the steady-state distribution as

$$(\pi_0, \pi_1, \pi_2) = \left(\frac{2}{5}, \frac{2}{5}, \frac{1}{5}\right).$$

Thus, in the long run, both machines will be broken down simultaneously 20 percent of the time, and one machine will be broken down another 40 percent of the time.

---

[5]Proving this fact requires the use of two properties of the exponential distribution discussed in Sec. 17.4 (*lack of memory* and *the minimum of exponentials is exponential*), since these properties imply that the T_{ij} random variables introduced earlier do indeed have exponential distributions.

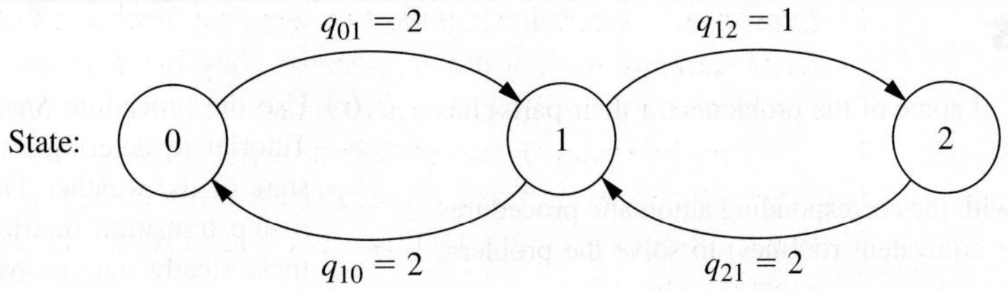

■ **FIGURE 16.5**
The rate diagram for the example of a continuous time Markov chain.

The next chapter (on queueing theory) features many more examples of continuous time Markov chains. In fact, most of the basic models of queueing theory fall into this category. The current example actually fits one of these models (the finite calling population variation of the $M/M/s$ model included in Sec. 17.6).

■ **SELECTED REFERENCES**

1. Bhat, U. N., and G. K. Miller: *Elements of Applied Stochastic Processes,* 3rd ed., Wiley, New York, 2002.
2. Bini, D., G. Latouche, and B. Meini: *Numerical Methods for Structured Markov Chains*, Oxford University Press, New York, 2005.
3. Bukiet, B., E. R. Harold, and J. L. Palacios: "A Markov Chain Approach to Baseball," *Operations Research,* **45:** 14–23, 1997.
4. Ching, W.-K., and M. K. Ng: *Markov Chains: Models, Algorithms and Applications*, Springer, New York, 2006.
5. Grassmann, W. K. (ed.): *Computational Probability,* Kluwer Academic Publishers (now Springer), Boston, MA, 2000.
6. Mamon, R. S., and R. J. Elliott (eds.): *Hidden Markov Models in Finance*, Springer, New York, 2007.
7. Resnick, S. I.: *Adventures in Stochastic Processes,* Birkhäuser, Boston, 1992.
8. Tijms, H. C.: *A First Course in Stochastic Models,* Wiley, New York, 2003.

■ **LEARNING AIDS FOR THIS CHAPTER ON OUR WEBSITE (www.mhhe.com/hillier)**

Worked Examples:

Examples for Chapter 16

Automatic Procedures in IOR Tutorial:

Enter Transition Matrix
Chapman-Kolmogorov Equations
Steady-State Probabilities

"Ch. 16—Markov Chains" LINGO File for Selected Examples

Glossary for Chapter 16

See Appendix 1 for documentation of the software.

■ PROBLEMS

The symbol to the left of some of the problems (or their parts) has the following meaning.

C: Use the computer with the corresponding automatic procedures just listed (or other equivalent routines) to solve the problem.

An asterisk on the problem number indicates that at least a partial answer is given in the back of the book.

16.2-1. Read the referenced article that fully describes the OR study summarized in the application vignette presented in Sec. 16.2. Briefly describe how a Markov chain was applied in this study. Then list the various financial and nonfinancial benefits that resulted from this study.

16.2-2. Assume that the probability of rain tomorrow is 0.5 if it is raining today, and assume that the probability of its being clear (no rain) tomorrow is 0.9 if it is clear today. Also assume that these probabilities do not change if information is also provided about the weather before today.
(a) Explain why the stated assumptions imply that the *Markovian property* holds for the evolution of the weather.
(b) Formulate the evolution of the weather as a Markov chain by defining its states and giving its (one-step) transition matrix.

16.2-3. Consider the second version of the stock market model presented as an example in Sec. 16.2. Whether the stock goes up tomorrow depends upon whether it increased today *and* yesterday. If the stock increased today and yesterday, it will increase tomorrow with probability α_1. If the stock increased today and decreased yesterday, it will increase tomorrow with probability α_2. If the stock decreased today and increased yesterday, it will increase tomorrow with probability α_3. Finally, if the stock decreased today and yesterday, it will increase tomorrow with probability α_4.
(a) Construct the (one-step) transition matrix of the Markov chain.
(b) Explain why the states used for this Markov chain cause the mathematical definition of the Markovian property to hold even though what happens in the future (tomorrow) depends upon what happened in the past (yesterday) as well as the present (today).

16.2-4. Reconsider Prob. 16.2-3. .Suppose now that whether or not the stock goes up tomorrow depends upon whether it increased today, yesterday, *and* the day before yesterday. Can this problem be formulated as a Markov chain? If so, what are the possible states? Explain why these states give the process the *Markovian property* whereas the states in Prob. 16.2-3 do not.

16.3-1. Reconsider Prob. 16.2-2.
C **(a)** Use the procedure *Chapman-Kolmogorov Equations* in your IOR Tutorial to find the *n*-step transition matrix $\mathbf{P}^{(n)}$ for $n = 2, 5, 10, 20$.
(b) The probability that it will rain today is 0.5. Use the results from part (*a*) to determine the probability that it will rain *n* days from now, for $n = 2, 5, 10, 20$.

C **(c)** Use the procedure *Steady-State Probabilities* in your IOR Tutorial to determine the steady-state probabilities of the state of the weather. Describe how the probabilities in the *n*-step transition matrices obtained in part (*a*) compare to these steady-state probabilities as *n* grows large.

16.3-2. Suppose that a communications network transmits binary digits, 0 or 1, where each digit is transmitted 10 times in succession. During each transmission, the probability is 0.995 that the digit entered will be transmitted accurately. In other words, the probability is 0.005 that the digit being transmitted will be recorded with the opposite value at the end of the transmission. For each transmission after the first one, the digit entered for transmission is the one that was recorded at the end of the preceding transmission. If X_0 denotes the binary digit entering the system, X_1 the binary digit recorded after the first transmission, X_2 the binary digit recorded after the second transmission, . . . , then $\{X_n\}$ is a Markov chain.
(a) Construct the (one-step) transition matrix.
C **(b)** Use your IOR Tutorial to find the 10-step transition matrix $\mathbf{P}^{(10)}$. Use this result to identify the probability that a digit entering the network will be recorded accurately after the last transmission.
C **(c)** Suppose that the network is redesigned to improve the probability that a single transmission will be accurate from 0.995 to 0.998. Repeat part (*b*) to find the new probability that a digit entering the network will be recorded accurately after the last transmission.

16.3-3.* A particle moves on a circle through points that have been marked 0, 1, 2, 3, 4 (in a clockwise order). The particle starts at point 0. At each step it has probability 0.5 of moving one point clockwise (0 follows 4) and 0.5 of moving one point counterclockwise. Let X_n ($n \geq 0$) denote its location on the circle after step *n*. $\{X_n\}$ is a Markov chain.
(a) Construct the (one-step) transition matrix.
C **(b)** Use your IOR Tutorial to determine the *n*-step transition matrix $\mathbf{P}^{(n)}$ for $n = 5, 10, 20, 40, 80$.
C **(c)** Use your IOR Tutorial to determine the steady-state probabilities of the state of the Markov chain. Describe how the probabilities in the *n*-step transition matrices obtained in part (*b*) compare to these steady-state probabilities as *n* grows large.

16.4-1.* Given the following (one-step) transition matrices of a Markov chain, determine the classes of the Markov chain and whether they are recurrent.

$$
\text{(a)} \ \mathbf{P} = \begin{array}{c} \\ 0 \\ 1 \\ 2 \\ 3 \end{array}
\begin{array}{c}
\begin{array}{cccc} \text{State} & \ 0 & 1 & 2 & 3 \end{array} \\
\left[\begin{array}{cccc}
0 & 0 & \frac{1}{3} & \frac{2}{3} \\
1 & 0 & 0 & 0 \\
0 & 1 & 0 & 0 \\
0 & 1 & 0 & 0
\end{array} \right]
\end{array}
$$

(b) $\mathbf{P} =$

State	0	1	2	3
0	1	0	0	0
1	0	$\frac{1}{2}$	$\frac{1}{2}$	0
2	0	$\frac{1}{2}$	$\frac{1}{2}$	0
3	$\frac{1}{2}$	0	0	$\frac{1}{2}$

16.4-2. Given each of the following (one-step) transition matrices of a Markov chain, determine the classes of the Markov chain and whether they are recurrent.

(a) $\mathbf{P} =$

State	0	1	2	3
0	0	$\frac{1}{3}$	$\frac{1}{3}$	$\frac{1}{3}$
1	$\frac{1}{3}$	0	$\frac{1}{3}$	$\frac{1}{3}$
2	$\frac{1}{3}$	$\frac{1}{3}$	0	$\frac{1}{3}$
3	$\frac{1}{3}$	$\frac{1}{3}$	$\frac{1}{3}$	0

(b) $\mathbf{P} =$

State	0	1	2
0	0	0	1
1	$\frac{1}{2}$	$\frac{1}{2}$	0
2	0	1	0

16.4-3. Given the following (one-step) transition matrix of a Markov chain, determine the classes of the Markov chain and whether they are recurrent.

$\mathbf{P} =$

State	0	1	2	3	4
0	$\frac{1}{4}$	$\frac{3}{4}$	0	0	0
1	$\frac{3}{4}$	$\frac{1}{4}$	0	0	0
2	$\frac{1}{3}$	$\frac{1}{3}$	$\frac{1}{3}$	0	0
3	0	0	0	$\frac{3}{4}$	$\frac{1}{4}$
4	0	0	0	$\frac{1}{4}$	$\frac{3}{4}$

16.4-4. Determine the period of each of the states in the Markov chain that has the following (one-step) transition matrix.

$\mathbf{P} =$

State	0	1	2	3	4	5
0	0	0	0	$\frac{2}{3}$	0	$\frac{1}{3}$
1	0	0	1	0	0	0
2	1	0	0	0	0	0
3	0	$\frac{1}{4}$	0	0	$\frac{3}{4}$	0
4	0	0	1	0	0	0
5	0	$\frac{1}{2}$	0	0	$\frac{1}{2}$	0

16.4-5. Consider the Markov chain that has the following (one-step) transition matrix.

$\mathbf{P} =$

State	0	1	2	3	4
0	0	$\frac{4}{5}$	0	$\frac{1}{5}$	0
1	$\frac{1}{4}$	0	$\frac{1}{2}$	$\frac{1}{4}$	0
2	0	$\frac{1}{2}$	0	$\frac{1}{10}$	$\frac{2}{5}$
3	0	0	0	1	0
4	$\frac{1}{3}$	0	$\frac{1}{3}$	$\frac{1}{3}$	0

(a) Determine the classes of this Markov chain and, for each class, determine whether it is recurrent or transient.

(b) For each of the classes identified in part (a), determine the period of the states in that class.

16.5-1. Reconsider Prob. 16.2-2. Suppose now that the given probabilities, 0.5 and 0.9, are replaced by arbitrary values, α and β, respectively. Solve for the *steady-state probabilities* of the state of the weather in terms of α and β.

16.5-2. A transition matrix $\mathbf{P}$ is said to be doubly stochastic if the sum over each column equals 1; that is,

$$\sum_{i=0}^{M} p_{ij} = 1, \qquad \text{for all } j.$$

If such a chain is irreducible, aperiodic, and consists of $M + 1$ states, show that

$$\pi_j = \frac{1}{M + 1}, \qquad \text{for } j = 0, 1, \ldots, M.$$

16.5-3. Reconsider Prob. 16.3-3. Use the results given in Prob. 16.5-2 to find the steady-state probabilities for this Markov chain. Then find what happens to these steady-state probabilities if, at each step, the probability of moving one point clockwise changes to 0.9 and the probability of moving one point counter-clockwise changes to 0.1.

C **16.5-4.** The leading brewery on the West Coast (labeled A) has hired an OR analyst to analyze its market position. It is particularly concerned about its major competitor (labeled B). The analyst believes that brand switching can be modeled as a Markov chain using three states, with states A and B representing customers drinking beer produced from the aforementioned breweries and state C representing all other brands. Data are taken monthly, and the analyst has constructed the following (one-step) transition matrix from past data.

	A	B	C
A	0.8	0.15	0.05
B	0.25	0.7	0.05
C	0.15	0.05	0.8

What are the steady-state market shares for the two major breweries?

16.5-5. Consider the following blood inventory problem facing a hospital. There is need for a rare blood type, namely, type AB, Rh negative blood. The demand D (in pints) over any 3-day period is given by

$$P\{D = 0\} = 0.4, \qquad P\{D = 1\} = 0.3,$$
$$P\{D = 2\} = 0.2, \qquad P\{D = 3\} = 0.1.$$

Note that the expected demand is 1 pint, since $E(D) = 0.3(1) + 0.2(2) + 0.1(3) = 1$. Suppose that there are 3 days between deliveries. The hospital proposes a policy of receiving 1 pint at each delivery and using the oldest blood first. If more blood is required than is on hand, an expensive emergency delivery is made. Blood is discarded if it is still on the shelf after 21 days. Denote the state of the system as the number of pints on hand just after a delivery. Thus, because of the discarding policy, the largest possible state is 7.

(a) Construct the (one-step) transition matrix for this Markov chain.

C (b) Find the steady-state probabilities of the state of the Markov chain.

(c) Use the results from part (b) to find the steady-state probability that a pint of blood will need to be discarded during a 3-day period. (*Hint:* Because the oldest blood is used first, a pint reaches 21 days only if the state was 7 and then $D = 0$.)

(d) Use the results from part (b) to find the steady-state probability that an emergency delivery will be needed during the 3-day period between regular deliveries.

C **16.5-6.** In the last subsection of Sec. 16.5, the (long-run) expected average cost per week (based on just ordering costs and unsatisfied demand costs) is calculated for the inventory example of Sec. 16.1. Suppose now that the ordering policy is changed to the following. Whenever the number of cameras on hand at the end of the week is 0 or 1, an order is placed that will bring this number up to 3. Otherwise, no order is placed.

Recalculate the (long-run) expected average cost per week under this new inventory policy.

16.5-7.* Consider the inventory example introduced in Sec. 16.1, but with the following change in the ordering policy. If the number of cameras on hand at the end of each week is 0 or 1, two additional cameras will be ordered. Otherwise, no ordering will take place. Assume that the storage costs are the same as given in the second subsection of Sec. 16.5.

C (a) Find the steady-state probabilities of the state of this Markov chain.

(b) Find the long-run expected average storage cost per week.

16.5-8. Consider the following inventory policy for the certain product. If the demand during a period exceeds the number of items available, this unsatisfied demand is backlogged; i.e., it is filled when the next order is received. Let Z_n ($n = 0, 1, \ldots$) denote the amount of inventory on hand minus the number of units backlogged before ordering at the end of period n ($Z_0 = 0$). If Z_n is zero or positive, no orders are backlogged. If Z_n is negative, then $-Z_n$ represents the number of backlogged units and no inventory is on hand. At the end of period n, if $Z_n < 1$, an order is placed for $2m$ units, where m is the smallest integer such that $Z_n + 2m \geq 1$. Orders are filled immediately.

Let $D_1, D_2, \ldots$, be the demand for the product in periods 1, 2, \ldots, respectively. Assume that the D_n are independent and identically distributed random variables taking on the values, 0, 1, 2, 3, 4, each with probability $\frac{1}{5}$. Let X_n denote the amount of stock on hand *after* ordering at the end of period n (where $X_0 = 2$), so that

$$X_n = \begin{cases} X_{n-1} - D_n + 2m & \text{if } X_{n-1} - D_n < 1 \\ X_{n-1} - D_n & \text{if } X_{n-1} - D_n \geq 1 \end{cases} \quad (n = 1, 2, \ldots),$$

when $\{X_n\}$ ($n = 0, 1, \ldots$) is a Markov chain. It has only two states, 1 and 2, because the only time that ordering will take place

is when $Z_n = 0, -1, -2,$ or -3, in which case 2, 2, 4, and 4 units are ordered, respectively, leaving $X_n = 2, 1, 2, 1$, respectively.

(a) Construct the (one-step) transition matrix.

(b) Use the steady-state equations to solve manually for the steady-state probabilities.

(c) Now use the result given in Prob. 16.5-2 to find the steady-state probabilities.

(d) Suppose that the ordering cost is given by $(2 + 2m)$ if an order is placed and zero otherwise. The holding cost per period is Z_n if $Z_n \geq 0$ and zero otherwise. The shortage cost per period is $-4Z_n$ if $Z_n < 0$ and zero otherwise. Find the (long-run) expected average cost per unit time.

16.5-9. An important unit consists of two components placed in parallel. The unit performs satisfactorily if one of the two components is operating. Therefore, only one component is operated at a time, but both components are kept operational (capable of being operated) as often as possible by repairing them as needed. An operating component breaks down in a given period with probability 0.2. When this occurs, the parallel component takes over, if it is operational, at the beginning of the next period. Only one component can be repaired at a time. The repair of a component starts at the beginning of the first available period and is completed at the end of the next period. Let X_t be a vector consisting of two elements U and V, where U represents the number of components that are operational at the end of period t and V represents the number of periods of repair that have been completed on components that are not yet operational. Thus, $V = 0$ if $U = 2$ or if $U = 1$ and the repair of the nonoperational component is just getting under way. Because a repair takes two periods, $V = 1$ if $U = 0$ (since then one nonoperational component is waiting to begin repair while the other one is entering its second period of repair) or if $U = 1$ and the nonoperational component is entering its second period of repair. Therefore, the state space consists of the four states (2, 0), (1, 0), (0, 1), and (1, 1). Denote these four states by 0, 1, 2, 3, respectively. $\{X_t\}$ ($t = 0, 1, \ldots$) is a Markov chain (assume that $X_0 = 0$) with the (one-step) transition matrix

$$\mathbf{P} = \begin{array}{c} \text{State} \\ 0 \\ 1 \\ 2 \\ 3 \end{array} \begin{array}{cccc} 0 & 1 & 2 & 3 \\ \begin{bmatrix} 0.8 & 0.2 & 0 & 0 \\ 0 & 0 & 0.2 & 0.8 \\ 0 & 1 & 0 & 0 \\ 0.8 & 0.2 & 0 & 0 \end{bmatrix} \end{array}.$$

C (a) What is the probability that the unit will be inoperable (because both components are down) after n periods, for $n = 2, 5, 10, 20$?

C (b) What are the steady-state probabilities of the state of this Markov chain?

(c) If it costs $30,000 per period when the unit is inoperable (both components down) and zero otherwise, what is the (long-run) expected average cost per period?

16.6-1. A computer is inspected at the end of every hour. It is found to be either working (up) or failed (down). If the computer is found to be up, the probability of its remaining up for the next hour is 0.95. If it is down, the computer is repaired, which may require more than

1 hour. Whenever the computer is down (regardless of how long it has been down), the probability of its still being down 1 hour later is 0.5.

(a) Construct the (one-step) transition matrix for this Markov chain.

(b) Use the approach described in Sec. 16.6 to find the μ_{ij} (the expected first passage time from state i to state j) for all i and j.

16.6-2. A manufacturer has a machine that, when operational at the beginning of a day, has a probability of 0.1 of breaking down sometime during the day. When this happens, the repair is done the next day and completed at the end of that day.

(a) Formulate the evolution of the status of the machine as a Markov chain by identifying three possible states at the end of each day, and then constructing the (one-step) transition matrix.

(b) Use the approach described in Sec. 16.6 to find the μ_{ij} (the expected first passage time from state i to state j) for all i and j. Use these results to identify the expected number of full days that the machine will remain operational before the next breakdown after a repair is completed.

(c) Now suppose that the machine already has gone 20 full days without a breakdown since the last repair was completed. How does the expected number of full days *hereafter* that the machine will remain operational before the next breakdown compare with the corresponding result from part (b) when the repair had just been completed? Explain.

16.6-3. Reconsider Prob. 16.6-2. Now suppose that the manufacturer keeps a spare machine that only is used when the primary machine is being repaired. During a repair day, the spare machine has a probability of 0.1 of breaking down, in which case it is repaired the next day. Denote the state of the system by (x, y), where x and y, respectively, take on the values 1 or 0 depending upon whether the primary machine (x) and the spare machine (y) are operational (value of 1) or not operational (value of 0) at the end of the day. [*Hint:* Note that $(0, 0)$ is not a possible state.]

(a) Construct the (one-step) transition matrix for this Markov chain.

(b) Find the *expected recurrence time* for the state $(1, 0)$.

16.6-4. Consider the inventory example presented in Sec. 16.1 except that demand now has the following probability distribution:

$$P\{D = 0\} = \frac{1}{4}, \qquad P\{D = 2\} = \frac{1}{4},$$

$$P\{D = 1\} = \frac{1}{2}, \qquad P\{D \geq 3\} = 0.$$

The ordering policy now is changed to ordering just 2 cameras at the end of the week if none are in stock. As before, no order is placed if there are any cameras in stock. Assume that there is one camera in stock at the time (the end of a week) the policy is instituted.

(a) Construct the (one-step) transition matrix.

C (b) Find the probability distribution of the state of this Markov chain n weeks after the new inventory policy is instituted, for $n = 2, 5, 10$.

(c) Find the μ_{ij} (the expected first passage time from state i to state j) for all i and j.

C (d) Find the steady-state probabilities of the state of this Markov chain.

(e) Assuming that the store pays a storage cost for each camera remaining on the shelf at the end of the week according to the function $C(0) = 0$, $C(1) = \$2$, and $C(2) = \$8$, find the long-run expected average storage cost per week.

16.6-5. A production process contains a machine that deteriorates rapidly in both quality and output under heavy usage, so that it is inspected at the end of each day. Immediately after inspection, the condition of the machine is noted and classified into one of four possible states:

State	Condition
0	Good as new
1	Operable—minimum deterioration
2	Operable—major deterioration
3	Inoperable and replaced by a good-as-new machine

The process can be modeled as a Markov chain with its (one-step) transition matrix **P** given by

State	0	1	2	3
0	0	$\frac{7}{8}$	$\frac{1}{16}$	$\frac{1}{16}$
1	0	$\frac{3}{4}$	$\frac{1}{8}$	$\frac{1}{8}$
2	0	0	$\frac{1}{2}$	$\frac{1}{2}$
3	1	0	0	0

C (a) Find the steady-state probabilities.

(b) If the costs of being in states 0, 1, 2, 3, are 0, \$1,000, \$3,000, and \$6,000, respectively, what is the long-run expected average cost per day?

(c) Find the *expected recurrence time* for state 0 (i.e., the expected length of time a machine can be used before it must be replaced).

16.7-1. Consider the following gambler's ruin problem. A gambler bets \$1 on each play of a game. Each time, he has a probability p of winning and probability $q = 1 - p$ of losing the dollar bet. He will continue to play until he goes broke or nets a fortune of T dollars. Let X_n denote the number of dollars possessed by the gambler after the nth play of the game. Then

$$X_{n+1} = \begin{cases} X_n + 1 & \text{with probability } p \\ X_n - 1 & \text{with probability } q = 1 - p \end{cases} \quad \text{for } 0 < X_n < T,$$

$$X_{n+1} = X_n, \qquad \text{for } X_n = 0, \text{ or } T.$$

$\{X_n\}$ is a Markov chain. The gambler starts with X_0 dollars, where X_0 is a positive integer less than T.

(a) Construct the (one-step) transition matrix of the Markov chain.

(b) Find the classes of the Markov chain.

(c) Let $T = 3$ and $p = 0.3$. Using the notation of Sec. 16.7, find $f_{10}, f_{1T}, f_{20}, f_{2T}$.

(d) Let $T = 3$ and $p = 0.7$. Find $f_{10}, f_{1T}, f_{20}, f_{2T}$.

16.7-2. A video cassette recorder manufacturer is so certain of its quality control that it is offering a complete replacement warranty if a recorder fails within 2 years. Based upon compiled data, the company has noted that only 1 percent of its recorders fail during the first year, whereas 5 percent of the recorders that survive the first year will fail during the second year. The warranty does not cover replacement recorders.

(a) Formulate the evolution of the status of a recorder as a Markov chain whose states include two absorption states that involve needing to honor the warranty or having the recorder survive the warranty period. Then construct the (one-step) transition matrix.

(b) Use the approach described in Sec. 16.7 to find the probability that the manufacturer will have to honor the warranty.

16.8-1. Read the referenced article that fully describes the OR study summarized in the application vignette presented in Sec. 16.8. Briefly describe how a continuous time Markov chain was applied in this study. Then list the various financial and nonfinancial benefits that resulted from this study.

16.8-2. Reconsider the example presented at the end of Sec. 16.8. Suppose now that a third machine, identical to the first two, has been added to the shop. The one maintenance person still must maintain all the machines.

(a) Develop the *rate diagram* for this Markov chain.

(b) Construct the *steady-state equations*.

(c) Solve these equations for the *steady-state probabilities*.

16.8-3. The state of a particular continuous time Markov chain is defined as the number of jobs currently at a certain work center, where a maximum of two jobs are allowed. Jobs arrive individually. Whenever fewer than two jobs are present, the time until the next arrival has an exponential distribution with a mean of 2 days. Jobs are processed at the work center one at a time and then leave immediately. Processing times have an exponential distribution with a mean of 1 day.

(a) Construct the *rate diagram* for this Markov chain.

(b) Write the *steady-state equations*.

(c) Solve these equations for the *steady-state probabilities*.

17

Queueing Theory

Queues (waiting lines) are a part of everyday life. We all wait in queues to buy a movie ticket, make a bank deposit, pay for groceries, mail a package, obtain food in a cafeteria, start a ride in an amusement park, etc. We have become accustomed to considerable amounts of waiting, but still get annoyed by unusually long waits.

However, having to wait is not just a petty personal annoyance. The amount of time that a nation's populace wastes by waiting in queues is a major factor in both the quality of life there and the efficiency of the nation's economy.

Great inefficiencies also occur because of other kinds of waiting than people standing in line. For example, making *machines* wait to be repaired may result in lost production. *Vehicles* (including ships and trucks) that need to wait to be unloaded may delay subsequent shipments. *Airplanes* waiting to take off or land may disrupt later travel schedules. Delays in *telecommunication* transmissions due to saturated lines may cause data glitches. Causing *manufacturing jobs* to wait to be performed may disrupt subsequent production. Delaying *service jobs* beyond their due dates may result in lost future business.

Queueing theory is the study of waiting in all these various guises. It uses *queueing models* to represent the various types of *queueing systems* (systems that involve queues of some kind) that arise in practice. Formulas for each model indicate how the corresponding queueing system should perform, including the average amount of waiting that will occur, under a variety of circumstances.

Therefore, these queueing models are very helpful for determining how to operate a queueing system in the most effective way. Providing too much service capacity to operate the system involves excessive costs. But not providing enough service capacity results in excessive waiting and all its unfortunate consequences. The models enable finding an appropriate balance between the cost of service and the amount of waiting.

After some general discussion, this chapter presents most of the more elementary queueing models and their basic results. Section 17.10 discusses how the information provided by queueing theory can be used to design queueing systems that minimize the total cost of service and waiting, and then Chap. 26 (on the book's website) elaborates considerably further on the application of queueing theory in this way.

■ 17.1 PROTOTYPE EXAMPLE

The emergency room of COUNTY HOSPITAL provides quick medical care for emergency cases brought to the hospital by ambulance or private automobile. At any hour there is always one doctor on duty in the emergency room. However, because of a growing tendency for emergency cases to use these facilities rather than go to a private physician, the hospital has been experiencing a continuing increase in the number of emergency room visits each year. As a result, it has become quite common for patients arriving during peak usage hours (the early evening) to have to wait until it is their turn to be treated by the doctor. Therefore, a proposal has been made that a second doctor should be assigned to the emergency room during these hours, so that two emergency cases can be treated simultaneously. The hospital's management engineer has been assigned to study this question.

The management engineer began by gathering the relevant historical data and then projecting these data into the next year. Recognizing that the emergency room is a queueing system, she applied several alternative queueing theory models to predict the waiting characteristics of the system with one doctor and with two doctors, as you will see in the latter sections of this chapter (see Tables 17.2 and 17.3).

■ 17.2 BASIC STRUCTURE OF QUEUEING MODELS

The Basic Queueing Process

The basic process assumed by most queueing models is the following. *Customers* requiring service are generated over time by an *input source.* These customers enter the *queueing system* and join a *queue.* At certain times, a member of the queue is selected for service by some rule known as the *queue discipline.* The required service is then performed for the customer by the *service mechanism,* after which the customer leaves the queueing system. This process is depicted in Fig. 17.1.

Many alternative assumptions can be made about the various elements of the queueing process; they are discussed next.

Input Source (Calling Population)

One characteristic of the input source is its size. The *size* is the total number of customers that might require service from time to time, i.e., the total number of distinct potential customers. This population from which arrivals come is referred to as the **calling population.** The size may be assumed to be either *infinite* or *finite* (so that the input source also is said to be either *unlimited* or *limited*). Because the calculations are far easier for the infinite case, this assumption often is made even when the actual size is some relatively

■ **FIGURE 17.1**
The basic queueing process.

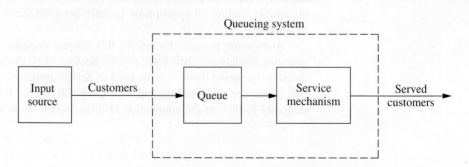

large finite number; and it should be taken to be the implicit assumption for any queueing model that does not state otherwise. The finite case is more difficult analytically because the number of customers in the queueing system affects the number of potential customers outside the system at any time. However, the finite assumption must be made if the rate at which the input source generates new customers is significantly affected by the number of customers in the queueing system.

The statistical pattern by which customers are generated over time must also be specified. The common assumption is that they are generated according to a *Poisson process;* i.e., the number of customers generated until any specific time has a Poisson distribution. As we discuss in Sec. 17.4, this case is the one where arrivals to the queueing system occur randomly but at a certain fixed mean rate, regardless of how many customers already are there (so the *size* of the input source is *infinite*). An equivalent assumption is that the probability distribution of the time between consecutive arrivals is an *exponential* distribution. (The properties of this distribution are described in Sec. 17.4.) The time between consecutive arrivals is referred to as the **interarrival time.**

Any unusual assumptions about the behavior of arriving customers must also be specified. One example is *balking,* where the customer refuses to enter the system and is lost if the queue is too long.

Queue

The queue is where customers wait *before* being served. A queue is characterized by the maximum permissible number of customers that it can contain. Queues are called *infinite* or *finite,* according to whether this number is infinite or finite. The assumption of an *infinite queue* is the standard one for most queueing models, even for situations where there actually is a (relatively large) finite upper bound on the permissible number of customers, because dealing with such an upper bound would be a complicating factor in the analysis. However, for queueing systems where this upper bound is small enough that it actually would be reached with some frequency, it becomes necessary to assume a *finite queue.*

Queue Discipline

The queue discipline refers to the order in which members of the queue are selected for service. For example, it may be first-come-first-served, random, according to some priority procedure, or some other order. First-come-first-served usually is assumed by queueing models, unless it is stated otherwise.

Service Mechanism

The service mechanism consists of one or more *service facilities,* each of which contains one or more *parallel service channels,* called **servers.** If there is more than one service facility, the customer may receive service from a sequence of these (*service channels in series*). At a given facility, the customer enters one of the parallel service channels and is completely serviced by that server. A queueing model must specify the arrangement of the facilities and the number of servers (parallel channels) at each one. Most elementary models assume one service facility with either one server or a finite number of servers.

The time elapsed from the commencement of service to its completion for a customer at a service facility is referred to as the **service time** (or *holding time*). A model of a particular queueing system must specify the probability distribution of service times for each server (and possibly for different types of customers), although it is common to assume the *same* distribution for all servers (all models in this chapter make this assumption). The service-time distribution that is most frequently assumed in practice (largely because it is far more tractable than any other) is the *exponential* distribution discussed in Sec. 17.4, and most of our models will be of this type. Other important service-time distributions are the *degenerate* distribution (constant service time) and the *Erlang* (gamma) distribution, as illustrated by models in Sec. 17.7.

An Elementary Queueing Process

As we have already suggested, queueing theory has been applied to many different types of waiting-line situations. However, the most prevalent type of situation is the following: A single waiting line (which may be empty at times) forms in the front of a single service facility, within which are stationed one or more servers. Each customer generated by an input source is serviced by one of the servers, perhaps after some waiting in the queue (waiting line). The queueing system involved is depicted in Fig. 17.2.

Notice that the queueing process in the prototype example of Sec. 17.1 is of this type. The input source generates customers in the form of emergency cases requiring medical care. The emergency room is the service facility, and the doctors are the servers.

A server need not be a single individual; it may be a group of persons, e.g., a repair crew that combines forces to perform simultaneously the required service for a customer. Furthermore, servers need not even be people. In many cases, a server can instead be a machine, a vehicle, an electronic device, etc. By the same token, the customers in the waiting line need not be people. For example, they may be items waiting for a certain operation by a given type of machine, or they may be cars waiting in front of a tollbooth.

It is not necessary that there actually be a physical waiting line forming in front of a physical structure that constitutes the service facility. The members of the queue may instead be scattered throughout an area, waiting for a server to come to them, e.g., machines waiting to be repaired. The server or group of servers assigned to a given area constitutes the service facility for that area. Queueing theory still gives the average number waiting, the average waiting time, and so on, because it is irrelevant whether the customers wait together in a group. The only essential requirement for queueing theory to be applicable is that changes in the number of customers waiting for a given service occur just as though the physical situation described in Fig. 17.2 (or a legitimate counterpart) prevailed.

Except for Sec. 17.9, all the queueing models discussed in this chapter are of the elementary type depicted in Fig. 17.2. Many of these models further assume that all

■ **FIGURE 17.2**
An elementary queueing system (each customer is indicated by a *C* and each server by an *S*).

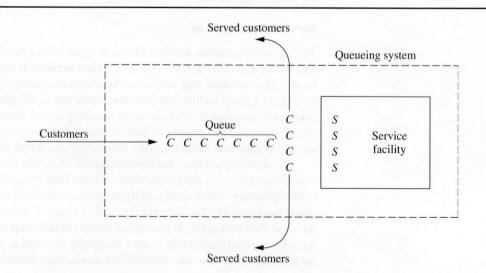

interarrival times are independent and identically distributed and that all *service times* are independent and identically distributed. Such models conventionally are labeled as follows:

Distribution of service times

— / — / — Number of servers

Distribution of interarrival times,

where M = exponential distribution (Markovian), as described in Sec. 17.4,

D = degenerate distribution (constant times), as discussed in Sec. 17.7,

E_k = Erlang distribution (shape parameter = k), as described in Sec. 17.7,

G = general distribution (any arbitrary distribution allowed),[1] as discussed in Sec. 17.7.

For example, the $M/M/s$ model discussed in Sec. 17.6 assumes that both interarrival times and service times have an exponential distribution and that the number of servers is s (any positive integer). The $M/G/1$ model discussed again in Sec. 17.7 assumes that interarrival times have an exponential distribution, but it places no restriction on what the distribution of service times must be, whereas the number of servers is restricted to be exactly 1. Various other models that fit this labeling scheme also are introduced in Sec. 17.7.

Terminology and Notation

Unless otherwise noted, the following standard terminology and notation will be used:

State of system = number of customers in queueing system.

Queue length = number of customers waiting for service to begin.

= state of system *minus* number of customers being served.

$N(t)$ = number of customers in queueing system at time t ($t \geq 0$).

$P_n(t)$ = probability of exactly n customers in queueing system at time t, given number at time 0.

s = number of servers (parallel service channels) in queueing system.

λ_n = mean arrival rate (expected number of arrivals per unit time) of new customers when n customers are in system.

μ_n = mean service rate for overall system (expected number of customers completing service per unit time) when n customers are in system. *Note:* μ_n represents *combined* rate at which all *busy* servers (those serving customers) achieve service completions.

λ, μ, ρ = see following paragraph.

When λ_n is a constant for all n, this constant is denoted by λ. When the mean service rate *per busy server* is a constant for all $n \geq 1$, this constant is denoted by μ. (In this case, $\mu_n = s\mu$ when $n \geq s$, that is, when all s servers are busy.) Under these circumstances, $1/\lambda$ and $1/\mu$ are the *expected interarrival time* and the *expected service time,* respectively. Also, $\rho = \lambda/(s\mu)$ is the **utilization factor** for the service facility, i.e., the expected fraction of

---

[1]When we refer to interarrival times, it is conventional to replace the symbol G by GI = general independent distribution.

time the individual servers are busy, because $\lambda/(s\mu)$ represents the fraction of the system's service capacity $(s\mu)$ that is being *utilized* on the average by arriving customers (λ).

Certain notation also is required to describe *steady-state* results. When a queueing system has recently begun operation, the state of the system (number of customers in the system) will be greatly affected by the initial state and by the time that has since elapsed. The system is said to be in a **transient condition.** However, after sufficient time has elapsed, the state of the system becomes essentially independent of the initial state and the elapsed time (except under unusual circumstances).[2] The system has now essentially reached a **steady-state condition,** where the probability distribution of the state of the system remains the same (the *steady-state* or *stationary* distribution) over time. Queueing theory has tended to focus largely on the steady-state condition, partially because the transient case is more difficult analytically. (Some transient results exist, but they are generally beyond the technical scope of this book.) The following notation assumes that the system is in a *steady-state condition:*

P_n = probability of exactly n customers in queueing system.

$$L = \text{expected number of customers in queueing system} = \sum_{n=0}^{\infty} nP_n.$$

$$L_q = \text{expected queue length (excludes customers being served)} = \sum_{n=s}^{\infty} (n - s)P_n.$$

$\mathcal{W}$ = waiting time in system (includes service time) for each individual customer.

$W = E(\mathcal{W})$.

$\mathcal{W}_q$ = waiting time in queue (excludes service time) for each individual customer.

$W_q = E(\mathcal{W}_q)$.

Relationships between *L*, *W*, *L_q*, and *W_q*

Assume that λ_n is a constant λ for all n. It has been proved that in a steady-state queueing process,

$$L = \lambda W.$$

(Because John D. C. Little provided the first rigorous proof, this equation sometimes is referred to as **Little's formula.**) Furthermore, the same proof also shows that

$$L_q = \lambda W_q.$$

If the λ_n are not equal, then λ can be replaced in these equations by $\bar{\lambda}$, the *average* arrival rate over the long run. (We shall show later how $\bar{\lambda}$ can be determined for some basic cases.)

Now assume that the mean service time is a constant, $1/\mu$ for all $n \geq 1$. It then follows that

$$W = W_q + \frac{1}{\mu}.$$

These relationships are extremely important because they enable all four of the fundamental quantities—L, W, L_q, and W_q—to be immediately determined as soon as

---

[2]When λ and μ are defined, these unusual circumstances are that $\rho \geq 1$, in which case the state of the system tends to grow continually larger as time goes on.

one is found analytically. This situation is fortunate because some of these quantities often are much easier to find than others when a queueing model is solved from basic principles.

17.3 EXAMPLES OF REAL QUEUEING SYSTEMS

Our description of queueing systems in Sec. 17.2 may appear relatively abstract and applicable to only rather special practical situations. On the contrary, queueing systems are surprisingly prevalent in a wide variety of contexts. To broaden your horizons on the applicability of queueing theory, we shall briefly mention various examples of real queueing systems that fall into several broad categories. We then will describe queueing systems in several prominent companies (plus one city) and the award-winning studies that were conducted to design these systems.

Some Classes of Queueing Systems

One important class of queueing systems that we all encounter in our daily lives is **commercial service systems,** where outside customers receive service from commercial organizations. Many of these involve person-to-person service at a fixed location, such as a barber shop (the barbers are the servers), bank teller service, checkout stands at a grocery store, and a cafeteria line (service channels in series). However, many others do not, such as home appliance repairs (the server travels to the customers), a vending machine (the server is a machine), and a gas station (the cars are the customers).

Another important class is **transportation service systems.** For some of these systems the vehicles are the customers, such as cars waiting at a tollbooth or traffic light (the server), a truck or ship waiting to be loaded or unloaded by a crew (the server), and airplanes waiting to land or take off from a runway (the server). (An unusual example of this kind is a parking lot, where the cars are the customers and the parking spaces are the servers, but there is no queue because arriving customers go elsewhere to park if the lot is full.) In other cases, the vehicles, such as taxicabs, fire trucks, and elevators, are the servers.

In recent years, queueing theory probably has been applied most to **internal service systems,** where the customers receiving service are *internal* to the organization. Examples include materials-handling systems, where materials-handling units (the servers) move loads (the customers); maintenance systems, where maintenance crews (the servers) repair machines (the customers); and inspection stations, where quality control inspectors (the servers) inspect items (the customers). Employee facilities and departments servicing employees also fit into this category. In addition, machines can be viewed as servers whose customers are the jobs being processed. A related example is a computer laboratory, where each computer is viewed as the server.

There is now growing recognition that queueing theory also is applicable to **social service systems.** For example, a judicial system is a queueing network, where the courts are service facilities, the judges (or panels of judges) are the servers, and the cases waiting to be tried are the customers. A legislative system is a similar queueing network, where the customers are the bills waiting to be processed. Various health-care systems also are queueing systems. You already have seen one example in Sec. 17.1 (a hospital emergency room), but you can also view ambulances, X-ray machines, and hospital beds as servers in their own queueing systems. Similarly, families waiting for low- and moderate-income housing, or other social services, can be viewed as customers in a queueing system.

completions—as *events*.) This random variable is said to have an *exponential distribution with parameter* α if its probability density function is

$$f_T(t) = \begin{cases} \alpha e^{-\alpha t} & \text{for } t \geq 0 \\ 0 & \text{for } t < 0, \end{cases}$$

as shown in Fig. 17.3. In this case, the cumulative probabilities are

$$P\{T \leq t\} = 1 - e^{-\alpha t}$$
$$P\{T > t\} = e^{-\alpha t} \qquad (t \geq 0),$$

and the expected value and variance of T are, respectively,

$$E(T) = \frac{1}{\alpha},$$

$$\text{var}(T) = \frac{1}{\alpha^2}.$$

What are the implications of assuming that T has an exponential distribution for a queueing model? To explore this question, let us examine six key properties of the exponential distribution.

Property 1: $f_T(t)$ is a strictly *decreasing* function of t ($t \geq 0$).

One consequence of Property 1 is that

$$P\{0 \leq T \leq \Delta t\} > P\{t \leq T \leq t + \Delta t\}$$

for any strictly positive values of Δt and t. [This consequence follows from the fact that these probabilities are the area under the $f_T(t)$ curve over the indicated interval of length Δt, and the average height of the curve is less for the second probability than for the first.] Therefore, it is not only possible but also relatively likely that T will take on a small value near zero. In fact,

$$P\left\{0 \leq T \leq \frac{1}{2}\frac{1}{\alpha}\right\} = 0.393$$

whereas

$$P\left\{\frac{1}{2}\frac{1}{\alpha} \leq T \leq \frac{3}{2}\frac{1}{\alpha}\right\} = 0.383,$$

so that the value T takes on is more likely to be "small" [i.e., less than half of $E(T)$] than "near" its expected value [i.e., no further away than half of $E(T)$], even though the second interval is twice as wide as the first.

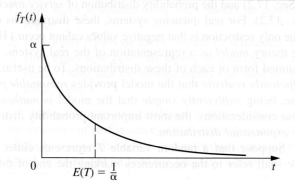

■ **FIGURE 17.3**
Probability density function for the exponential distribution.

Is this really a reasonable property for T in a queueing model? If T represents *service times*, the answer depends upon the general nature of the service involved, as discussed next.

If the service required is essentially identical for each customer, with the server always performing the same sequence of service operations, then the actual service times tend to be near the expected service time. Small deviations from the mean may occur, but usually because of only minor variations in the efficiency of the server. A small service time far below the mean is essentially impossible, because a certain minimum time is needed to perform the required service operations even when the server is working at top speed. The exponential distribution clearly does not provide a close approximation to the service-time distribution for this type of situation.

On the other hand, consider the type of situation where the specific tasks required of the server differ among customers. The broad nature of the service may be the same, but the specific type and amount of service differ. For example, this is the case in the County Hospital emergency room problem discussed in Sec. 17.1. The doctors encounter a wide variety of medical problems. In most cases, they can provide the required treatment rather quickly, but an occasional patient requires extensive care. Similarly, bank tellers and grocery store checkout clerks are other servers of this general type, where the required service is often brief but must occasionally be extensive. An exponential service-time distribution would seem quite plausible for this type of service situation.

If T represents *interarrival times,* Property 1 rules out situations where potential customers approaching the queueing system tend to postpone their entry if they see another customer entering ahead of them. On the other hand, it is entirely consistent with the common phenomenon of arrivals occurring "randomly," described by subsequent properties. Thus, when arrival times are plotted on a time line, they sometimes have the appearance of being clustered with occasional large gaps separating clusters, because of the substantial probability of small interarrival times and the small probability of large interarrival times, but such an irregular pattern is all part of true randomness.

Property 2: Lack of memory.

This property can be stated mathematically as

$$P\{T > t + \Delta t \mid T > \Delta t\} = P\{T > t\}$$

for any positive quantities t and Δt. In other words, the probability distribution of the *remaining* time until the event (arrival or service completion) occurs always is the same, regardless of how much time (Δt) already has passed. In effect, the process "forgets" its history. This surprising phenomenon occurs with the exponential distribution because

$$P\{T > t + \Delta t \mid T > \Delta t\} = \frac{P\{T > \Delta t, T > t + \Delta t\}}{P\{T > \Delta t\}}$$

$$= \frac{P\{T > t + \Delta t\}}{P\{T > \Delta t\}}$$

$$= \frac{e^{-\alpha(t + \Delta t)}}{e^{-\alpha \Delta t}}$$

$$= e^{-\alpha t}$$

$$= P\{T > t\}.$$

For *interarrival times,* this property describes the common situation where the time until the next arrival is completely uninfluenced by when the last arrival occurred. For *service times,* the property is more difficult to interpret. We should not expect it to hold in a situation where the server must perform the same fixed sequence of operations for each customer, because then a long elapsed service should imply that probably little values of Δt, this probability is essentially *proportional* to Δt, with proportionality factor α. In fact, α is the *mean rate* at which the events occur (see Property 4), so that the *expected number* of events in the interval of length Δt is *exactly* $\alpha \Delta t$. The only reason that the probability of an event's occurring differs slightly from this value is the possibility that *more than one* event will occur, which has negligible probability when Δt is small.

To see why Property 5 holds mathematically, note that the constant value of our probability (for a fixed value of $\Delta t > 0$) is just

$$P\{T \leq t + \Delta t \mid T > t\} = P\{T \leq \Delta t\}$$
$$= 1 - e^{-\alpha \Delta t},$$

for any $t \geq 0$. Therefore, because the series expansion of e^x for any exponent x is

$$e^x = 1 + x + \sum_{n=2}^{\infty} \frac{x^n}{n!},$$

it follows that

$$P\{T \leq t + \Delta t \mid T > t\} = 1 - 1 + \alpha \Delta t - \sum_{n=2}^{\infty} \frac{(-\alpha \Delta t)^n}{n!}$$
$$\approx \alpha \Delta t, \qquad \text{for small } \Delta t,\text{[3]}$$

because the summation terms become relatively negligible for sufficiently small values of $\alpha \Delta t$.

Because T can represent either interarrival or service times in queueing models, this property provides a convenient approximation of the probability that the event of interest occurs in the next small interval (Δt) of time. An analysis based on this approximation also can be made exact by taking appropriate limits as $\Delta t \to 0$.

Property 6: Unaffected by aggregation or disaggregation.

This property is relevant primarily for verifying that the *input process is Poisson.* Therefore, we shall describe it in these terms, although it also applies directly to the exponential distribution (exponential interarrival times) because of Property 4.

We first consider the aggregation (combining) of several Poisson input processes into one overall input process. In particular, suppose that there are several (n) *different* types of customers, where the customers of each type (type i) arrive according to a *Poisson input process* with parameter λ_i ($i = 1, 2, \ldots, n$). Assuming that these are *independent* Poisson processes, the property says that the *aggregate* input process (arrival of all customers without regard to type) also must be Poisson, with parameter (mean arrival rate) $\lambda = \lambda_1 + \lambda_2 + \cdots + \lambda_n$. In other words, having a Poisson process is *unaffected by aggregation.*

This part of the property follows directly from Properties 3 and 4. The latter property implies that the interarrival times for customers of type i have an exponential distribution with parameter λ_i. For this identical situation, we already discussed for Property 3 that it implies that the interarrival times for all customers also must have an exponential distribution, with parameter $\lambda = \lambda_1 + \lambda_2 + \cdots + \lambda_n$. Using Property 4 again then implies that the aggregate input process is Poisson.

The second part of Property 6 ("unaffected by disaggregation") refers to the reverse case, where the *aggregate* input process (the one obtained by combining the input processes

[3]More precisely,
$$\lim_{\Delta t \to 0} \frac{P\{T \leq t + \Delta t \mid T > t\}}{\Delta t} = \alpha.$$

for several customer types) is known to be Poisson with parameter λ, but the question now concerns the nature of the *disaggregated* input processes (the individual input processes for the individual customer types). Assuming that each arriving customer has a *fixed* probability p_i of being of type i ($i = 1, 2, \ldots, n$), with

$$\lambda_i = p_i\lambda \quad \text{and} \quad \sum_{i=1}^{n} p_i = 1,$$

the property says that the input process for customers of type i also must be Poisson with parameter λ_i. In other words, having a Poisson process is *unaffected by disaggregation.*

As one example of the usefulness of this second part of the property, consider the following situation. Indistinguishable customers arrive according to a Poisson process with parameter λ. Each arriving customer has a fixed probability p of *balking* (leaving without entering the queueing system), so the probability of entering the system is $1 - p$. Thus, there are two types of customers—those who balk and those who enter the system. The property says that each type arrives according to a Poisson process, with parameters $p\lambda$ and $(1 - p)\lambda$, respectively. Therefore, by using the latter Poisson process, queueing models that assume a Poisson input process can still be used to analyze the performance of the queueing system for those customers who enter the system.

Another example in the Worked Examples section of the books' website illustrates the application of several of the properties of the exponential distribution presented in this section.

■ 17.5   THE BIRTH-AND-DEATH PROCESS

Most elementary queueing models assume that the inputs (arriving customers) and outputs (leaving customers) of the queueing system occur according to the *birth-and-death process.* This important process in probability theory has applications in various areas. However, in the context of queueing theory, the term **birth** refers to the *arrival* of a new customer into the queueing system, and **death** refers to the *departure* of a served customer. The *state* of the system at time t ($t \geq 0$), denoted by $N(t)$, is the number of customers in the queueing system at time t. The birth-and-death process describes *probabilistically* how $N(t)$ changes as t increases. Broadly speaking, it says that *individual* births and deaths occur *randomly,* where their mean occurrence rates depend only upon the current state of the system. More precisely, the assumptions of the birth-and-death process are the following:

Assumption 1.   Given $N(t) = n$, the current probability distribution of the *remaining* time until the next *birth* (arrival) is *exponential* with parameter λ_n ($n = 0, 1, 2, \ldots$).

Assumption 2.   Given $N(t) = n$, the current probability distribution of the *remaining* time until the next *death* (service completion) is *exponential* with parameter μ_n ($n = 1, 2, \ldots$).

Assumption 3.   The random variable of assumption 1 (the remaining time until the next *birth*) and the random variable of assumption 2 (the remaining time until the next *death*) are mutually independent. The next transition in the state of the process is either

$$n \rightarrow n + 1 \quad \text{(a single birth)}$$

or

$$n \rightarrow n - 1 \quad \text{(a single death)},$$

depending on whether the former or latter random variable is smaller.

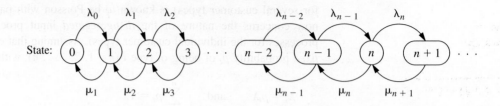

■ FIGURE 17.4
Rate diagram for the birth-
and-death process.

For a queueing system, λ_n and μ_n respectively represent the *mean arrival rate* and the *mean rate of service completions,* when there are n customers in the system. For some queueing systems, the values of the λ_n will be the same for all values of n, and the μ_n also will be the same for all n except for such small n (e.g., $n = 0$) that a server is idle. However, the λ_n and the μ_n also can vary considerably with n for some queueing systems.

For example, one of the ways in which λ_n can be different for different values of n is if potential arriving customers become increasingly likely to *balk* (refuse to enter the system) as n increases. Similarly, μ_n can be different for different n because customers in the queue become increasingly likely to *renege* (leave without being served) as the queue size increases. **Another example** in the Worked Examples section of the books' website illustrates a queueing system where both balking and reneging occur. This example then demonstrates how the general results for the birth-and-death process lead directly to various measures of performance for this queueing system.

Analysis of the Birth-and-Death Process

Because of its assumptions, the birth-and-death process is a special type of *continuous time Markov chain.* (See Sec. 16.8 for a description of continuous time Markov chains and their properties, including an introduction to the general procedure for finding steady-state probabilities that will be applied in the remainder of this section.) Queueing models that can be represented by a continuous time Markov chain are far more tractable analytically than any other.

Because Property 4 for the exponential distribution (see Sec. 17.4) implies that the λ_n and μ_n are mean rates, we can summarize these assumptions by the rate diagram shown in Fig. 17.4. The arrows in this diagram show the only possible *transitions* in the state of the system (as specified by assumption 3), and the entry for each arrow gives the mean rate for that transition (as specified by assumptions 1 and 2) when the system is in the state at the base of the arrow.

Except for a few special cases, analysis of the birth-and-death process is very difficult when the system is in a *transient* condition. Some results about the probability distribution of $N(t)$ have been obtained, but they are too complicated to be of much practical use. On the other hand, it is relatively straightforward to derive this distribution *after* the system has reached a *steady-state* condition (assuming that this condition can be reached). This derivation can be done directly from the rate diagram, as outlined next.

Consider any particular state of the system n ($n = 0, 1, 2, \ldots$). Starting at time 0, suppose that a count is made of the number of times that the process enters this state and the number of times it leaves this state, as denoted below:

$E_n(t)$ = number of times that process enters state n by time t.

$L_n(t)$ = number of times that process leaves state n by time t.

Because the two types of events (entering and leaving) must alternate, these two numbers must always either be equal or differ by just 1; that is,

$$|E_n(t) - L_n(t)| \le 1.$$

Dividing through both sides by t and then letting $t \to \infty$ gives

$$\left| \frac{E_n(t)}{t} - \frac{L_n(t)}{t} \right| \le \frac{1}{t}, \qquad \text{so} \qquad \lim_{t \to \infty} \left| \frac{E_n(t)}{t} - \frac{L_n(t)}{t} \right| = 0.$$

Dividing $E_n(t)$ and $L_n(t)$ by t gives the *actual rate* (number of events per unit time) at which these two kinds of events have occurred, and letting $t \to \infty$ then gives the *mean rate* (expected number of events per unit time):

$$\lim_{t \to \infty} \frac{E_n(t)}{t} = \text{mean rate at which process enters state } n.$$

$$\lim_{t \to \infty} \frac{L_n(t)}{t} = \text{mean rate at which process leaves state } n.$$

These results yield the following key principle:

Rate In = Rate Out Principle.   For any state of the system n ($n = 0, 1, 2, \ldots$),

mean entering rate = mean leaving rate.

The equation expressing this principle is called the **balance equation** for state n. After constructing the balance equations for all the states in terms of the *unknown* P_n probabilities, we can solve this system of equations (plus an equation stating that the probabilities must sum to 1) to find these probabilities.

To illustrate a balance equation, consider state 0. The process enter this state *only* from state 1. Thus, the steady-state probability of being in state 1 (P_1) represents the proportion of time that it would be *possible* for the process to enter state 0. Given that the process is in state 1, the mean rate of entering state 0 is μ_1. (In other words, for each cumulative unit of time that the process spends in state 1, the expected number of times that it would leave state 1 to enter state 0 is μ_1.) From any *other* state, this mean rate is 0. Therefore, the overall mean rate at which the process leaves its current state to enter state 0 (the *mean entering rate*) is

$$\mu_1 P_1 + 0(1 - P_1) = \mu_1 P_1.$$

By the same reasoning, the *mean leaving rate* must be $\lambda_0 P_0$, so the balance equation for state 0 is

$$\mu_1 P_1 = \lambda_0 P_0.$$

For every other state there are two possible transitions both into and out of the state. Therefore, each side of the balance equations for these states represents the *sum* of the mean rates for the two transitions involved. Otherwise, the reasoning is just the same as for state 0. These balance equations are summarized in Table 17.1.

Notice that the first balance equation contains two variables for which to solve (P_0 and P_1), the first two equations contain three variables (P_0, P_1, and P_2), and so on, so that there always is one "extra" variable. Therefore, the procedure in solving these equations is to solve in terms of one of the variables, the most convenient one being P_0. Thus, the first equation is used to solve for P_1 in terms of P_0; this result and the second equation are then used to solve for P_2 in terms of P_0; and so forth. At the end, the requirement that the sum of all the probabilities equal 1 can be used to evaluate P_0.

■ **TABLE 17.1** Balance equations for the birth-and-death process

State	Rate In = Rate Out
0	$\mu_1 P_1 = \lambda_0 P_0$
1	$\lambda_0 P_0 + \mu_2 P_2 = (\lambda_1 + \mu_1)P_1$
2	$\lambda_1 P_1 + \mu_3 P_3 = (\lambda_2 + \mu_2)P_2$
$\vdots$	$\vdots$
$n-1$	$\lambda_{n-2} P_{n-2} + \mu_n P_n = (\lambda_{n-1} + \mu_{n-1})P_{n-1}$
n	$\lambda_{n-1} P_{n-1} + \mu_{n+1} P_{n+1} = (\lambda_n + \mu_n)P_n$
$\vdots$	$\vdots$

Results for the Birth-and-Death Process

Applying this procedure yields the following results:

State:

0:      $P_1 = \dfrac{\lambda_0}{\mu_1}P_0$

1:      $P_2 = \dfrac{\lambda_1}{\mu_2}P_1 + \dfrac{1}{\mu_2}(\mu_1 P_1 - \lambda_0 P_0) \qquad = \dfrac{\lambda_1}{\mu_2}P_1 \qquad = \dfrac{\lambda_1 \lambda_0}{\mu_2 \mu_1}P_0$

2:      $P_3 = \dfrac{\lambda_2}{\mu_3}P_2 + \dfrac{1}{\mu_3}(\mu_2 P_2 - \lambda_1 P_1) \qquad = \dfrac{\lambda_2}{\mu_3}P_2 \qquad = \dfrac{\lambda_2 \lambda_1 \lambda_0}{\mu_3 \mu_2 \mu_1}P_0$

$\vdots$

$n-1$: $P_n = \dfrac{\lambda_{n-1}}{\mu_n}P_{n-1} + \dfrac{1}{\mu_n}(\mu_{n-1}P_{n-1} - \lambda_{n-2}P_{n-2}) = \dfrac{\lambda_{n-1}}{\mu_n}P_{n-1} = \dfrac{\lambda_{n-1}\lambda_{n-2}\cdots\lambda_0}{\mu_n \mu_{n-1}\cdots\mu_1}P_0$

n: $P_{n+1} = \dfrac{\lambda_n}{\mu_{n+1}}P_n + \dfrac{1}{\mu_{n+1}}(\mu_n P_n - \lambda_{n-1}P_{n-1}) \quad = \dfrac{\lambda_n}{\mu_{n+1}}P_n \quad = \dfrac{\lambda_n \lambda_{n-1}\cdots\lambda_0}{\mu_{n+1}\mu_n \cdots\mu_1}P_0$

$\vdots$

To simplify notation, let

$$C_n = \frac{\lambda_{n-1}\lambda_{n-2}\cdots\lambda_0}{\mu_n \mu_{n-1}\cdots\mu_1}, \qquad \text{for } n = 1, 2, \ldots,$$

and then define $C_n = 1$ for $n = 0$. Thus, the steady-state probabilities are

$$P_n = C_n P_0, \qquad \text{for } n = 0, 1, 2, \ldots.$$

The requirement that

$$\sum_{n=0}^{\infty} P_n = 1$$

implies that

$$\left(\sum_{n=0}^{\infty} C_n\right)P_0 = 1,$$

so that

$$P_0 = \left(\sum_{n=0}^{\infty} C_n\right)^{-1}.$$

When a queueing model is based on the birth-and-death process, so the state of the system n represents the number of customers in the queueing system, the key measures of performance for the queueing system (L, L_q, W, and W_q) can be obtained immediately after calculating the P_n from the above formulas. The definitions of L and L_q given in Sec. 17.2 specify that

$$L = \sum_{n=0}^{\infty} n P_n, \qquad L_q = \sum_{n=s}^{\infty} (n - s) P_n.$$

Furthermore, the relationships given at the end of Sec. 17.2 yield

$$W = \frac{L}{\lambda}, \qquad W_q = \frac{L_q}{\lambda},$$

where $\bar{\lambda}$ is the *average* arrival rate over the long run. Because λ_n is the mean arrival rate while the system is in state n ($n = 0, 1, 2, \ldots$) and P_n is the proportion of time that the system is in this state,

$$\bar{\lambda} = \sum_{n=0}^{\infty} \lambda_n P_n.$$

Several of the expressions just given involve summations with an infinite number of terms. Fortunately, these summations have analytic solutions for a number of interesting special cases,[4] as seen in the next section. Otherwise, they can be approximated by summing a finite number of terms on a computer.

These steady-state results have been derived under the assumption that the λ_n and μ_n parameters have values such that the process actually can *reach* a steady-state condition. This assumption *always* holds if $\lambda_n = 0$ for some value of n greater than the initial state, so that only a finite number of states (those less than this n) are possible. It also *always* holds when λ and μ are defined (see "Terminology and Notation" in Sec. 17.2) and $\rho = \lambda/(s\mu) < 1$. It does *not* hold if $\sum_{n=1}^{\infty} C_n = \infty$.

Section 17.6 describes several queueing models that are special cases of the birth-and-death process. Therefore, the general steady-state results just given in boxes will be used over and over again to obtain the specific steady-state results for these models.

17.6   QUEUEING MODELS BASED ON THE BIRTH-AND-DEATH PROCESS

Because each of the mean rates $\lambda_0, \lambda_1, \ldots$ and $\mu_1, \mu_2, \ldots$ for the birth-and-death process can be assigned any nonnegative value, we have great flexibility in modeling a queueing system. Probably the most widely used models in queueing theory are based directly upon this process. Because of assumptions 1 and 2 (and Property 4 for the exponential distribution), these models are said to have a **Poisson input** and **exponential service times.** The models

---

[4]These solutions are based on the following known results for the sum of any geometric series:

$$\sum_{n=0}^{N} x^n = \frac{1 - x^{N+1}}{1 - x}, \qquad \text{for any } x \neq 1,$$

$$\sum_{n=0}^{\infty} x^n = \frac{1}{1 - x}, \qquad \text{if } |x| < 1.$$

differ only in their assumptions about how the λ_n and μ_n change with n. We present three of these models in this section for three important types of queueing systems.

The *M/M/s* Model

As described in Sec. 17.2, the *M/M/s* model assumes that all *interarrival times* are independently and identically distributed according to an exponential distribution (i.e., the input process is Poisson), that all *service times* are independent and identically distributed according to another exponential distribution, and that the number of servers is s (any positive integer). Consequently, this model is just the special case of the birth-and-death process where the queueing system's *mean arrival rate* and *mean service rate per busy server* are constant (λ and μ, respectively) regardless of the state of the system. When the system has just a *single server* ($s = 1$), the implication is that the parameters for the birth-and-death process are $\lambda_n = \lambda$ ($n = 0, 1, 2, \ldots$) and $\mu_n = \mu$ ($n = 1, 2, \ldots$). The resulting rate diagram is shown in Fig. 17.5a.

However, when the system has *multiple servers* ($s > 1$), the μ_n cannot be expressed this simply, as explained below.

> **System Service Rate**: The system service rate μ_n represents the mean rate of service completions for the *overall* queueing system when there are n customers in the system. With multiple servers and $n > 1$, μ_n is *not* the same as μ, the mean service rate per busy server. Instead,
>
> $$\mu_n = n\mu \qquad \text{when } n \leq s,$$
> $$\mu_n = s\mu \qquad \text{when } n \geq s.$$

Using these formulas for μ_n, the rate diagram for the birth-and-death process shown in Fig. 17.4 reduces to the rate diagrams shown in Fig. 17.5 for the *M/M/s* model.

When $s\mu$ exceeds the mean arrival rate λ, that is, when

$$\rho = \frac{\lambda}{s\mu} < 1,$$

a queueing system fitting this model will eventually reach a steady-state condition. In this situation, the steady-state results derived in Sec. 17.5 for the general birth-and-death process are directly applicable. However, these results simplify considerably for this model and yield closed-form expressions for P_n, L, L_q, and so forth, as shown next.

■ **FIGURE 17.5**
Rate diagrams for the *M/M/s* model.

(a) *Single-server case* ($s = 1$) $\quad \lambda_n = \lambda, \qquad$ for $n = 0, 1, 2, \ldots$
$\qquad\qquad\qquad\qquad\qquad\qquad\quad \mu_n = \mu, \qquad$ for $n = 1, 2, \ldots$

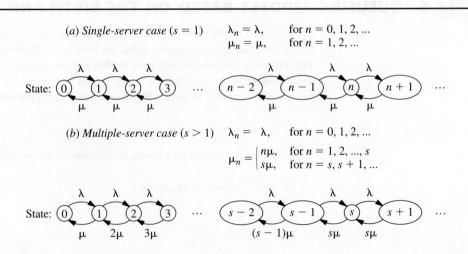

(b) *Multiple-server case* ($s > 1$) $\quad \lambda_n = \lambda, \qquad$ for $n = 0, 1, 2, \ldots$

$$\mu_n = \begin{cases} n\mu, & \text{for } n = 1, 2, \ldots, s \\ s\mu, & \text{for } n = s, s+1, \ldots \end{cases}$$

KeyCorp is a Fortune 500 company headquartered in Cleveland, Ohio. It is the thirteenth-largest bank holding company in the United States, with 19,000 employees, assets of $93 billion, and annual revenues of $6.7 billion. The company emphasizes consumer banking and has 2.4 million customers across more than 1,300 branch banks and many additional affiliate offices.

To help grow its business, KeyCorp management initiated an extensive OR study to determine how to improve customer service (defined primarily as reducing customer waiting time before beginning service) while also providing cost-effective staffing. A service-quality goal was set that at least 90 percent of the customers should have waiting times of less than 5 minutes.

The key tool in analyzing this problem was the *M/M/s queueing model*, which proved to fit this application very well. To apply this model, data were gathered that revealed that the average service time required to process a customer was a distressingly high 246 seconds. With this average service time and typical mean arrival rates, the model indicated that a 30 percent increase in the number of tellers would be needed to meet the service-quality goal. This prohibitively expensive option led

management to conclude that an extensive campaign needed to be undertaken to drastically reduce the average service time by both reengineering the customer session and providing better management of staff. Over a period of three years, this campaign led to a reduction in the average service time all the way down to 115 seconds. Frequent reapplication of the M/M/s model then revealed how the service-quality goal can be substantially surpassed while actually reducing personnel levels through improved scheduling of the personnel in the various branch banks.

The net result has been *savings of nearly* **$20 million** *per year with vastly improved service* that enables 96 percent of the customers to wait less than 5 minutes. This improvement extended throughout the company since the percentage of branch banks who meet the service-quality goal has increased from 42 percent to 94 percent. Surveys also confirm a great increase in customer satisfaction.

Source: S. K. Kotha, M. P. Barnum, and D. A. Bowen: "KeyCorp Service Excellence Management System," *Interfaces*, **26**(1): 54–74, Jan.–Feb. 1996. (A link to this article is provided on our website, www.mhhe.com/hillier.)

Results for the Single-Server Case (M/M/1). For $s = 1$, the C_n factors for the birth-and-death process reduce to

$$C_n = \left(\frac{\lambda}{\mu}\right)^n = \rho^n, \qquad \text{for } n = 0, 1, 2, \ldots$$

Therefore,

$$P_n = \rho^n P_0, \qquad \text{for } n = 0, 1, 2, \ldots,$$

where

$$P_0 = \left(\sum_{n=0}^{\infty} \rho^n\right)^{-1}$$

$$= \left(\frac{1}{1-\rho}\right)^{-1}$$

$$= 1 - \rho.$$

Thus,

$$P_n = (1 - \rho)\rho^n, \qquad \text{for } n = 0, 1, 2, \ldots.$$

Consequently,

$$L = \sum_{n=0}^{\infty} n(1 - \rho)\rho^n$$

$$= (1 - \rho)\rho \sum_{n=0}^{\infty} \frac{d}{d\rho}(\rho^n)$$

$$= (1 - \rho)\rho \frac{d}{d\rho} \left(\sum_{n=0}^{\infty} \rho^n \right)$$

$$= (1 - \rho)\rho \frac{d}{d\rho} \left(\frac{1}{1 - \rho} \right)$$

$$= \frac{\rho}{1 - \rho} = \frac{\lambda}{\mu - \lambda}.$$

Similarly,

$$L_q = \sum_{n=1}^{\infty} (n - 1)P_n$$

$$= L - 1(1 - P_0)$$

$$= \frac{\lambda^2}{\mu(\mu - \lambda)}.$$

When $\lambda \geq \mu$, so that the mean arrival rate exceeds the mean service rate, the preceding solution "blows up" (because the summation for computing P_0 diverges). For this case, the queue would "explode" and grow without bound. If the queueing system begins operation with no customers present, the server might succeed in keeping up with arriving customers over a short period of time, but this is impossible in the long run. (Even when $\lambda = \mu$, the *expected* number of customers in the queueing system slowly grows without bound over time because, even though a temporary return to no customers present always is possible, the probabilities of huge numbers of customers present become increasingly significant over time.)

Assuming again that $\lambda < \mu$, we now can derive the probability distribution of the *waiting time in the system* (so *including* service time) $\mathcal{W}$ for a random arrival when the queue discipline is first-come-first-served. If this arrival finds n customers already in the system, then the arrival will have to wait through $n + 1$ exponential service times, including his or her own. (For the customer currently being served, recall the lack-of-memory property for the exponential distribution discussed in Sec. 17.4.) Therefore, let $T_1, T_2, \ldots$ be independent service-time random variables having an exponential distribution with parameter μ, and let

$$S_{n+1} = T_1 + T_2 + \cdots + T_{n+1}, \qquad \text{for } n = 0, 1, 2, \ldots,$$

so that S_{n+1} represents the *conditional* waiting time given n customers already in the system. As discussed in Sec. 17.7, S_{n+1} is known to have an *Erlang distribution*.[5] Because the probability that the random arrival will find n customers in the system is P_n, it follows that

$$P\{\mathcal{W} > t\} = \sum_{n=0}^{\infty} P_n P\{S_{n+1} > t\},$$

which reduces after considerable manipulation (see Prob. 17.6-17) to

$$P\{\mathcal{W} > t\} = e^{-\mu(1-\rho)t}, \qquad \text{for } t \geq 0.$$

The surprising conclusion is that $\mathcal{W}$ has an *exponential* distribution with parameter $\mu(1 - \rho)$. Therefore,

$$W = E(\mathcal{W}) = \frac{1}{\mu(1 - \rho)}$$

$$= \frac{1}{\mu - \lambda}.$$

---

[5]Outside queueing theory, this distribution is known as the *gamma distribution*.

These results *include* service time in the waiting time. In some contexts (e.g., the County Hospital emergency room problem described in Sec. 17.1), the more relevant waiting time is just until service begins. Thus, consider the *waiting time in the queue* (so *excluding* service time) W_q for a random arrival when the queue discipline is first-come-first-served. If this arrival finds no customers already in the system, then the arrival is served immediately, so that

$$P\{W_q = 0\} = P_0 = 1 - \rho.$$

If this arrival finds $n > 0$ customers already there instead, then the arrival has to wait through n exponential service times until his or her own service begins, so that

$$P\{W_q > t\} = \sum_{n=1}^{\infty} P_n P\{S_n > t\}$$

$$= \sum_{n=1}^{\infty} (1 - \rho)\rho^n P\{S_n > t\}$$

$$= \rho \sum_{n=0}^{\infty} P_n P\{S_{n+1} > t\}$$

$$= \rho P\{W > t\}$$

$$= \rho e^{-\mu(1-\rho)t}, \qquad \text{for } t \geq 0.$$

Note that Wq does not quite have an exponential distribution, because $P\{W_q = 0\} > 0$. However, the *conditional* distribution of W_q, given that $W_q > 0$, does have an exponential distribution with parameter $\mu(1 - \rho)$, just as W does, because

$$P\{W_q > t \mid W_q > 0\} = \frac{P\{W_q > t\}}{P\{W_q > 0\}} = e^{-\mu(1-\rho)t}, \qquad \text{for } t \geq 0.$$

By deriving the mean of the (unconditional) distribution of W_q (or applying either $L_q = \lambda W_q$ or $W_q = W - 1/\mu$),

$$W_q = E(W_q) = \frac{\lambda}{\mu(\mu - \lambda)}.$$

If you would like to see **another example** that applies the *M/M/*1 model to determine which type of materials handling equipment a company should purchase, one is provided in the Worked Examples section of the book's website.

Results for the Multiple-Server Case ($s > 1$). When $s > 1$, the C_n factors become

$$C_n = \begin{cases} \dfrac{(\lambda/\mu)^n}{n!} & \text{for } n = 1, 2, \ldots, s \\[2ex] \dfrac{(\lambda/\mu)^s}{s!}\left(\dfrac{\lambda}{s\mu}\right)^{n-s} = \dfrac{(\lambda/\mu)^n}{s!\,s^{n-s}} & \text{for } n = s, s + 1, \ldots. \end{cases}$$

Consequently, if $\lambda < s\mu$ [so that $\rho = \lambda/(s\mu) < 1$], then

$$P_0 = 1 \Big/ \left[1 + \sum_{n=1}^{s-1} \frac{(\lambda/\mu)^n}{n!} + \frac{(\lambda/\mu)^s}{s!} \sum_{n=s}^{\infty} \left(\frac{\lambda}{s\mu}\right)^{n-s} \right]$$

$$= 1 \Big/ \left[\sum_{n=0}^{s-1} \frac{(\lambda/\mu)^n}{n!} + \frac{(\lambda/\mu)^s}{s!} \frac{1}{1 - \lambda/(s\mu)} \right],$$

where the $n = 0$ term in the last summation yields the correct value of 1 because of the convention that $n! = 1$ when $n = 0$. These C_n factors also give

$$
P_n = \begin{cases}
\dfrac{(\lambda/\mu)^n}{n!} P_0 & \text{if } 0 \le n \le s \\[2ex]
\dfrac{(\lambda/\mu)^n}{s! s^{n-s}} P_0 & \text{if } n \ge s.
\end{cases}
$$

Furthermore,

$$
\begin{aligned}
L_q &= \sum_{n=s}^{\infty} (n - s) P_n \\
&= \sum_{j=0}^{\infty} j P_{s+j} \\
&= \sum_{j=0}^{\infty} j \frac{(\lambda/\mu)^s}{s!} \rho^j P_0 \\
&= P_0 \frac{(\lambda/\mu)^s}{s!} \rho \sum_{j=0}^{\infty} \frac{d}{d\rho} (\rho^j) \\
&= P_0 \frac{(\lambda/\mu)^s}{s!} \rho \frac{d}{d\rho} \left(\sum_{j=0}^{\infty} \rho^j \right) \\
&= P_0 \frac{(\lambda/\mu)^s}{s!} \rho \frac{d}{d\rho} \left(\frac{1}{1-\rho} \right) \\
&= \frac{P_0 (\lambda/\mu)^s \rho}{s!(1-\rho)^2};
\end{aligned}
$$

$$
W_q = \frac{L_q}{\lambda};
$$

$$
W = W_q + \frac{1}{\mu};
$$

$$
L = \lambda \left(W_q + \frac{1}{\mu} \right) = L_q + \frac{\lambda}{\mu}.
$$

Figure 17.6 shows how L changes with ρ for various values of s.

The single-server method for finding the probability distribution of waiting times also can be extended to the multiple-server case. This yields[6] (for $t \ge 0$)

$$
P\{\mathcal{W} > t\} = e^{-\mu t} \left[1 + \frac{P_0(\lambda/\mu)^s}{s!(1-\rho)} \left(\frac{1 - e^{-\mu t(s-1-\lambda/\mu)}}{s - 1 - \lambda/\mu} \right) \right]
$$

and

$$
P\{\mathcal{W}_q > t\} = (1 - P\{\mathcal{W}_q = 0\}) e^{-s\mu(1-\rho)t},
$$

where

$$
P\{\mathcal{W}_q = 0\} = \sum_{n=0}^{s-1} P_n.
$$

The above formulas for the various measures of performance (including the P_n) are relatively imposing for hand calculations. However, this chapter's Excel file in your OR

---

[6]When $s - 1 - \lambda/\mu = 0$, $(1 - e^{-\mu t(s-1-\lambda/\mu)})/(s - 1 - \lambda/\mu)$ should be replaced by μt.

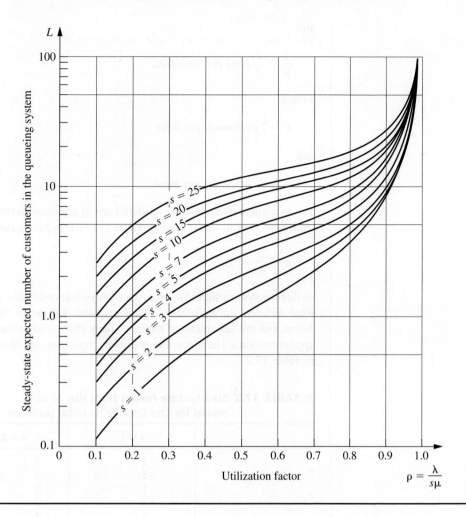

FIGURE 17.6
Values for L for the $M/M/s$ model (Sec. 17.6).

Courseware includes an Excel template that performs all these calculations simultaneously for any values of t, s, λ, and μ you want, provided that $\lambda < s\mu$.

If $\lambda \geq s\mu$, so that the mean arrival rate exceeds the maximum mean rate of service completions, then the queue grows without bound, so the preceding steady-state solutions are not applicable.

The County Hospital Example with the M/M/s Model. For the County Hospital emergency room problem (see Sec. 17.1), the management engineer has concluded that the emergency cases arrive pretty much at random (a *Poisson input process*), so that interarrival times have an exponential distribution. She also has concluded that the time spent by a doctor treating the cases approximately follows an *exponential distribution*. Therefore, she has chosen the *M/M/s* model for a preliminary study of this queueing system.

By projecting the available data for the early evening shift into next year, she estimates that patients will arrive at an *average* rate of 1 every $\frac{1}{2}$ hour. A doctor requires an average of 20 minutes to treat each patient. Thus, with one hour as the unit of time,

$$\frac{1}{\lambda} = \frac{1}{2} \text{ hour per customer}$$

and

$$\frac{1}{\mu} = \frac{1}{3} \text{ hour per customer,}$$

so that

$$\lambda = 2 \text{ customers per hour}$$

and

$$\mu = 3 \text{ customers per hour.}$$

The two alternatives being considered are to continue having just one doctor during this shift ($s = 1$) or to add a second doctor ($s = 2$). In both cases,

$$\rho = \frac{\lambda}{s\mu} < 1,$$

so that the system should approach a steady-state condition. (Actually, because λ is somewhat different during other shifts, the system will never truly reach a steady-state condition, but the management engineer feels that steady-state results will provide a good approximation.) Therefore, the preceding equations are used to obtain the results shown in Table 17.2.

■ **TABLE 17.2** Steady-state results from the *M/M/s* model for the County Hospital problem

	s = 1	s = 2
ρ	$\frac{2}{3}$	$\frac{1}{3}$
P_0	$\frac{1}{3}$	$\frac{1}{2}$
P_1	$\frac{2}{9}$	$\frac{1}{3}$
P_n for $n \geq 2$	$\frac{1}{3}\left(\frac{2}{3}\right)^n$	$\left(\frac{1}{3}\right)^n$
L_q	$\frac{4}{3}$	$\frac{1}{12}$
L	2	$\frac{3}{4}$
W_q	$\frac{2}{3}$ hour	$\frac{1}{24}$ hour
W	1 hour	$\frac{3}{8}$ hour
$P\{W_q > 0\}$	0.667	0.167
$P\left\{W_q > \frac{1}{2}\right\}$	0.404	0.022
$P\{W_q > 1\}$	0.245	0.003
$P\{W_q > t\}$	$\frac{2}{3}e^{-t}$	$\frac{1}{6}e^{-4t}$
$P\{W > t\}$	e^{-t}	$\frac{1}{2}e^{-3t}(3 - e^{-t})$

On the basis of these results, she tentatively concluded that a single doctor would be inadequate next year for providing the relatively prompt treatment needed in a hospital emergency room. You will see later (Sec. 17.8) how she checked this conclusion by applying another queueing model that provides a better representation of the real queueing system in one crucial way.

You can see **another example** of an application of the $M/M/1$ model in the Worked Examples section of the book's website, where the issue in this case is whether three employees in a fast-food restaurant should work together as one fast server or separately as three considerably slower servers.

The Finite Queue Variation of the $M/M/s$ Model (Called the $M/M/s/K$ Model)

We mentioned in the discussion of queues in Sec. 17.2 that queueing systems sometimes have a *finite queue;* i.e., the number of customers in the system is not permitted to exceed some specified number (denoted by K) so the queue capacity is $K - s$. Any customer that arrives while the queue is "full" is refused entry into the system and so leaves forever. From the viewpoint of the birth-and-death process, the mean input rate into the system becomes zero at these times. Therefore, the one modification needed in the $M/M/s$ model to introduce a finite queue is to change the λ_n parameters to

$$\lambda_n = \begin{cases} \lambda & \text{for } n = 0, 1, 2, \ldots, K - 1 \\ 0 & \text{for } n \geq K. \end{cases}$$

Because $\lambda_n = 0$ for some values of n, a queueing system that fits this model always will eventually reach a steady-state condition, even when $\rho = \lambda/s\mu \geq 1$.

This model commonly is labeled $M/M/s/K$, where the presence of the fourth symbol distinguishes it from the $M/M/s$ model. The single difference in the formulation of these two models is that K is finite for the $M/M/s/K$ model and $K = \infty$ for the $M/M/s$ model.

The usual physical interpretation for the $M/M/s/K$ model is that there is only *limited waiting room* that will accommodate a maximum of K customers in the system. For example, for the County Hospital emergency room problem, this system actually would have a finite queue if there were only K cots for the patients and if the policy were to send arriving patients to another hospital whenever there were no empty cots.

Another possible interpretation is that arriving customers will leave and "take their business elsewhere" whenever they find too many customers (K) ahead of them in the system because they are not willing to incur a long wait. This balking phenomenon is quite common in commercial service systems. However, there are other models available (e.g., see Prob. 17.5-5) that fit this interpretation even better.

The rate diagram for this model is identical to that shown in Fig. 17.5 for the $M/M/s$ model, *except* that it stops with state K.

Results for the Single-Server Case ($M/M/1/K$).

For this case,

$$C_n = \begin{cases} \left(\dfrac{\lambda}{\mu}\right)^n = \rho^n & \text{for } n = 0, 1, 2, \ldots, K \\ 0 & \text{for } n > K. \end{cases}$$

Therefore, for $\rho \neq 1$,[7]

$$P_0 = \frac{1}{\Sigma_{n=0}^{K} (\lambda/\mu)^n}$$

$$= 1 \Big/ \left[\frac{1 - (\lambda/\mu)^{K+1}}{1 - \lambda/\mu} \right]$$

$$= \frac{1 - \rho}{1 - \rho^{K+1}},$$

so that

$$P_n = \frac{1 - \rho}{1 - \rho^{K+1}} \rho^n, \qquad \text{for } n = 0, 1, 2, \ldots, K.$$

Hence,

$$L = \sum_{n=0}^{K} n P_n$$

$$= \frac{1 - \rho}{1 - \rho^{K+1}} \rho \sum_{n=0}^{K} \frac{d}{d\rho} (\rho^n)$$

$$= \frac{1 - \rho}{1 - \rho^{K+1}} \rho \frac{d}{d\rho} \left(\sum_{n=0}^{K} \rho^n \right)$$

$$= \frac{1 - \rho}{1 - \rho^{K+1}} \rho \frac{d}{d\rho} \left(\frac{1 - \rho^{K+1}}{1 - \rho} \right)$$

$$= \rho \frac{-(K+1)\rho^K + K\rho^{K+1} + 1}{(1 - \rho^{K+1})(1 - \rho)}$$

$$= \frac{\rho}{1 - \rho} - \frac{(K+1)\rho^{K+1}}{1 - \rho^{K+1}}.$$

As usual (when $s = 1$),

$$L_q = L - (1 - P_0).$$

Notice that the preceding results do not require that $\lambda < \mu$ (i.e., that $\rho < 1$).

When $\rho < 1$, it can be verified that the second term in the final expression for L converges to 0 as $K \to \infty$, so that *all* the preceding results do indeed converge to the corresponding results given earlier for the $M/M/1$ model.

The waiting-time distributions can be derived by using the same reasoning as for the $M/M/1$ model (see Prob. 17.6-28). However, no simple expressions are obtained in this case, so computer calculations are required. Fortunately, even though $L \neq \lambda W$ and $L_q \neq \lambda W_q$ for the current model because the λ_n are not equal for all n (see the end of Sec. 17.2), the *expected* waiting times for customers entering the system still can be obtained directly from the expressions given at the end of Sec. 17.5:

$$W = \frac{L}{\bar{\lambda}}, \qquad W_q = \frac{L_q}{\bar{\lambda}},$$

---

[7]If $\rho = 1$, then $P_n = 1/(K+1)$ for $n = 0, 1, 2, \ldots, K$, so that $L = K/2$.

where

$$\bar{\lambda} = \sum_{n=0}^{\infty} \lambda_n P_n$$

$$= \sum_{n=0}^{K-1} \lambda P_n$$

$$= \lambda(1 - P_K).$$

Results for the Multiple-Server Case ($s > 1$). Because this model does not allow more than K customers in the system, K is the maximum number of servers that could ever be used. Therefore, assume that $s \leq K$. In this case, C_n becomes

$$C_n = \begin{cases} \dfrac{(\lambda/\mu)^n}{n!} & \text{for } n = 0, 1, 2, \ldots, s \\[2ex] \dfrac{(\lambda/\mu)^s}{s!}\left(\dfrac{\lambda}{s\mu}\right)^{n-s} = \dfrac{(\lambda/\mu)^n}{s!\,s^{n-s}} & \text{for } n = s, s+1, \ldots, K \\[2ex] 0 & \text{for } n > K. \end{cases}$$

Hence,

$$P_n = \begin{cases} \dfrac{(\lambda/\mu)^n}{n!}P_0 & \text{for } n = 1, 2, \ldots, s \\[2ex] \dfrac{(\lambda/\mu)^n}{s!\,s^{n-s}}P_0 & \text{for } n = s, s+1, \ldots, K \\[2ex] 0 & \text{for } n > K, \end{cases}$$

where

$$P_0 = 1 \Big/ \left[\sum_{n=0}^{s} \frac{(\lambda/\mu)^n}{n!} + \frac{(\lambda/\mu)^s}{s!} \sum_{n=s+1}^{K} \left(\frac{\lambda}{s\mu}\right)^{n-s} \right].$$

(These formulas continue to use the convention that $n! = 1$ when $n = 0$.)
Adapting the derivation of L_q for the $M/M/s$ model to this case yields

$$L_q = \frac{P_0(\lambda/\mu)^s \rho}{s!(1-\rho)^2}[1 - \rho^{K-s} - (K-s)\rho^{K-s}(1-\rho)],$$

where $\rho = \lambda/(s\mu)$.[8] It can then be shown that

$$L = \sum_{n=0}^{s-1} nP_n + L_q + s\left(1 - \sum_{n=0}^{s-1} P_n\right).$$

And W and W_q are obtained from these quantities just as shown for the single-server case.

This chapter's Excel file includes an Excel template for calculating the above measures of performance (including the P_n) for this model.

One interesting special case of this model is where $K = s$ so the queue capacity is $K - s = 0$. In this case, customers who arrive when all servers are busy will leave

---

[8]If $\rho = 1$, it is necessary to apply L'Hôpital's rule twice to this expression for L_q. Otherwise, all these multiple-server results hold for all $\rho > 0$. The reason that this queueing system can reach a steady-state condition even when $\rho \geq 1$ is that $\lambda_n = 0$ for $n \geq K$, so that the number of customers in the system cannot continue to grow indefinitely.

immediately and be lost to the system. This would occur, for example, in a telephone network with s trunk lines so callers get a busy signal and hang up when all the trunk lines are busy. This kind of system (a "queueing system" with no queue) is referred to as *Erlang's loss system* because it was first studied in the early 20th century by A. K. Erlang, a Danish telephone engineer who is considered the founder of queueing theory.

It is common now for the telephone system at a call center to provide some extra trunk lines that place the caller on hold, but additional callers then get a busy signal. Such a system also fits this model, where $(K - s)$ is the number of extra trunk lines that place the caller on hold. **Another example** in the Worked Examples section of the book's website illustrates the application of this model to such a system.

The Finite Calling Population Variation of the *M/M/s* Model

Now assume that the only deviation from the *M/M/s* model is that (as defined in Sec. 17.2) the *input source* is *limited;* i.e., the size of the *calling population* is *finite.* For this case, let N denote the size of the calling population. Thus, when the number of customers in the queueing system is n ($n = 0, 1, 2, \ldots, N$), there are only $N - n$ *potential* customers remaining in the input source.

The most important application of this model has been to the machine repair problem, where one or more maintenance people are assigned the responsibility of maintaining in operational order a certain group of N machines by repairing each one that breaks down. (The example given at the end of Sec. 16.8 illustrates this application when the general procedures for solving any *continuous time Markov chain* are used rather than the specific formulas available for the birth-and-death process.) The maintenance people are considered to be individual servers in the queueing system if they work individually on different machines, whereas the entire crew is considered to be a single server if crew members work together on each machine. The machines constitute the calling population. Each one is considered to be a customer in the queueing system when it is down waiting to be repaired, whereas it is outside the queueing system while it is operational.

Note that each member of the calling population alternates between being *inside* and *outside* the queueing system. Therefore, the analog of the *M/M/s* model that fits this situation assumes that *each* member's *outside time* (i.e., the elapsed time from leaving the system until returning for the next time) has an *exponential distribution* with parameter λ. When n of the members are *inside,* and so $N - n$ members are *outside,* the current probability distribution of the *remaining* time until the next arrival to the queueing system is the distribution of the *minimum* of the *remaining outside times* for the latter $N - n$ members. Properties 2 and 3 for the exponential distribution imply that this distribution must be exponential with parameter $\lambda_n = (N - n)\lambda$. Hence, this model is just the special case of the birth-and-death process that has the rate diagram shown in Fig. 17.7.

Because $\lambda_n = 0$ for $n = N$, any queueing system that fits this model will eventually reach a steady-state condition. The available steady-state results are summarized as follows:

Results for the Single-Server Case ($s = 1$).   When $s = 1$, the C_n factors in Sec. 17.5 reduce to

$$C_n = \begin{cases} N(N - 1) \cdots (N - n + 1)\left(\dfrac{\lambda}{\mu}\right)^n = \dfrac{N!}{(N - n)!}\left(\dfrac{\lambda}{\mu}\right)^n & \text{for } n \leq N \\ 0 & \text{for } n > N, \end{cases}$$

(a) *Single-server case* ($s = 1$)

$$\lambda_n = \begin{cases} (N - n)\lambda, & \text{for } n = 0, 1, 2, \ldots, N \\ 0, & \text{for } n \geq N \end{cases}$$

$$\mu_n = \mu, \qquad \text{for } n = 1, 2, \ldots$$

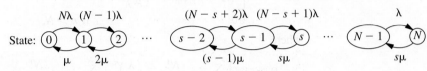

(b) *Multiple-server case* ($s > 1$)

$$\lambda_n = \begin{cases} (N - n)\lambda, & \text{for } n = 0, 1, 2, \ldots, N \\ 0, & \text{for } n \geq N \end{cases}$$

$$\mu_n = \begin{cases} n\mu, & \text{for } n = 1, 2, \ldots, s \\ s\mu, & \text{for } n = s, s + 1, \ldots \end{cases}$$

■ **FIGURE 17.7**
Rate diagrams for the finite calling population variation of the *M/M/s* model.

State diagram (b): States 0, 1, 2, ..., s−2, s−1, s, ..., N−1, N with rates $N\lambda$, $(N-1)\lambda$, ..., $(N-s+2)\lambda$, $(N-s+1)\lambda$, ..., λ on top and μ, 2μ, ..., $(s-1)\mu$, $s\mu$, ..., $s\mu$ on bottom.

for this model. Therefore, again using the convention that $n! = 1$ when $n = 0$,

$$P_0 = 1 \Big/ \sum_{n=0}^{N} \left[\frac{N!}{(N-n)!} \left(\frac{\lambda}{\mu} \right)^n \right];$$

$$P_n = \frac{N!}{(N-n)!} \left(\frac{\lambda}{\mu} \right)^n P_0, \qquad \text{if } n = 1, 2, \ldots, N;$$

$$L_q = \sum_{n=1}^{N} (n - 1) P_n,$$

which can be reduced to

$$L_q = N - \frac{\lambda + \mu}{\lambda} (1 - P_0);$$

$$L = \sum_{n=0}^{N} n P_n = L_q + 1 - P_0$$

$$= N - \frac{\mu}{\lambda} (1 - P_0).$$

Finally,

$$W = \frac{L}{\bar{\lambda}} \qquad \text{and} \qquad W_q = \frac{L_q}{\bar{\lambda}},$$

where

$$\bar{\lambda} = \sum_{n=0}^{\infty} \lambda_n P_n = \sum_{n=0}^{N} (N - n)\lambda P_n = \lambda(N - L).$$

At this point, you might find it helpful to refer back to the example at the end of Sec. 16.8, because that example completely fits this model for the single-server case. In particular, $N = 2$, $\lambda = 1$, and $\mu = 2$ for that example, so $P_0 = 0.4$, $P_1 = 0.4$, $P_2 = 0.2$, and so forth.

Results for the Multiple-Server Case ($s > 1$).   For $N \geq s > 1$,

$$
C_n = \begin{cases}
\dfrac{N!}{(N-n)!\,n!}\left(\dfrac{\lambda}{\mu}\right)^n & \text{for } n = 0, 1, 2, \ldots, s \\[3mm]
\dfrac{N!}{(N-n)!\,s!\,s^{n-s}}\left(\dfrac{\lambda}{\mu}\right)^n & \text{for } n = s, s+1, \ldots, N \\[3mm]
0 & \text{for } n > N.
\end{cases}
$$

Hence,

$$
P_n = \begin{cases}
\dfrac{N!}{(N-n)!\,n!}\left(\dfrac{\lambda}{\mu}\right)^n P_0 & \text{if } 0 \leq n \leq s \\[3mm]
\dfrac{N!}{(N-n)!\,s!\,s^{n-s}}\left(\dfrac{\lambda}{\mu}\right)^n P_0 & \text{if } s \leq n \leq N \\[3mm]
0 & \text{if } n > N,
\end{cases}
$$

where

$$
P_0 = 1 \bigg/ \left[\sum_{n=0}^{s-1} \frac{N!}{(N-n)!\,n!}\left(\frac{\lambda}{\mu}\right)^n + \sum_{n=s}^{N} \frac{N!}{(N-n)!\,s!\,s^{n-s}}\left(\frac{\lambda}{\mu}\right)^n \right].
$$

Finally,

$$
L_q = \sum_{n=s}^{N} (n-s)P_n
$$

and

$$
L = \sum_{n=0}^{s-1} nP_n + L_q + s\left(1 - \sum_{n=0}^{s-1} P_n\right),
$$

which then yield W and W_q by the same equations as in the single-server case.

This chapter's Excel files include an Excel template for performing all the above calculations.

Extensive tables of computational results also are available[9] for this model for both the single-server and multiple-server cases.

For both cases, it has been shown[10] that the preceding formulas for P_n and P_0 (and so for L_q, L, W, and W_q) *also* hold for a generalization of this model. In particular, we can *drop* the assumption that the times spent *outside* the queueing system by the members of the calling population have an *exponential distribution*, even though this takes the model outside the realm of the birth-and-death process. As long as these times are identically distributed with mean $1/\lambda$ (and the assumption of exponential service times still holds), these outside times can have *any* probability distribution!

17.7 QUEUEING MODELS INVOLVING NONEXPONENTIAL DISTRIBUTIONS

Because all the queueing theory models in the preceding section (except for one generalization) are based on the birth-and-death process, both their interarrival and service times are required to have *exponential* distributions. As discussed in Sec. 17.4, this type

[9]L. G. Peck and R. N. Hazelwood, *Finite Queueing Tables*, Wiley, New York, 1958.

[10]B. D. Bunday and R. E. Scraton, "The G/M/r Machine Interference Model," *European Journal of Operational Research*, **4**: 399–402, 1980.

of probability distribution has many convenient properties for queueing theory, but it provides a reasonable fit for only certain kinds of queueing systems. In particular, the assumption of exponential interarrival times implies that arrivals occur randomly (a Poisson input process), which is a reasonable approximation in many situations but *not* when the arrivals are carefully scheduled or regulated. Furthermore, the actual service-time distribution frequently deviates greatly from the exponential form, particularly when the service requirements of the customers are quite similar. Therefore, it is important to have available other queueing models that use alternative distributions.

Unfortunately, the mathematical analysis of queueing models with nonexponential distributions is much more difficult. However, it has been possible to obtain some useful results for a few such models. This analysis is beyond the level of this book, but in this section we shall summarize the models and describe their results.

The *M/G*/1 Model

As introduced in Sec. 17.2, the *M/G*/1 model assumes that the queueing system has a *single server* and a *Poisson input process* (exponential interarrival times) with a *fixed* mean arrival rate λ. As usual, it is assumed that the customers have *independent* service times with the *same* probability distribution. However, no restrictions are imposed on what this service-time distribution can be. In fact, it is only necessary to know (or estimate) the mean $1/\mu$ and variance σ^2 of this distribution.

Any such queueing system can eventually reach a steady-state condition if $\rho = \lambda/\mu < 1$. The readily available steady-state results[11] for this general model are the following:

$$P_0 = 1 - \rho,$$

$$L_q = \frac{\lambda^2 \sigma^2 + \rho^2}{2(1 - \rho)},$$

$$L = \rho + L_q,$$

$$W_q = \frac{L_q}{\lambda},$$

$$W = W_q + \frac{1}{\mu}.$$

Considering the complexity involved in analyzing a model that permits *any* service-time distribution, it is remarkable that such a simple formula can be obtained for L_q. This formula is one of the most important results in queueing theory because of its ease of use and the prevalence of *M/G*/1 queueing systems in practice. This equation for L_q (or its counterpart for W_q) commonly is referred to as the **Pollaczek-Khintchine formula,** named after two pioneers in the development of queueing theory who derived the formula independently in the early 1930s.

For any fixed expected service time $1/\mu$, notice that L_q, L, W_q, and W all increase as σ^2 is increased. This result is important because it indicates that the consistency of the server has a major bearing on the performance of the service facility—not just the server's average speed. This key point is illustrated in the next subsection.

When the service-time distribution is exponential, $\sigma^2 = 1/\mu^2$, and the preceding results will reduce to the corresponding results for the *M/M*/1 model given at the beginning of Sec. 17.6.

---

[11]A recursion formula also is available for calculating the probability distribution of the number of customers in the system; see A. Hordijk and H. C. Tijms, "A Simple Proof of the Equivalence of the Limiting Distribution of the Continuous-Time and the Embedded Process of the Queue Size in the *M/G*/1 Queue," *Statistica Neerlandica,* **36:** 97–100, 1976.

The complete flexibility in the service-time distribution provided by this model is extremely useful, so it is unfortunate that efforts to derive similar results for the multiple-server case have been unsuccessful. However, some multiple-server results have been obtained for the important special cases described by the following two models. (Excel templates are available in this chapter's Excel file for performing the calculations for both the $M/G/1$ model and the two models considered below when $s = 1$.)

The $M/D/s$ Model

When the service consists of essentially the same routine task to be performed for all customers, there tends to be little variation in the service time required. The $M/D/s$ model often provides a reasonable representation for this kind of situation, because it assumes that all service times actually equal some fixed *constant* (the *degenerate* service-time distribution) and that we have a *Poisson* input process with a fixed mean arrival rate λ.

When there is just a single server, the $M/D/1$ model is just the special case of the $M/G/1$ model where $\sigma^2 = 0$, so that the *Pollaczek-Khintchine formula* reduces to

$$L_q = \frac{\rho^2}{2(1 - \rho)},$$

where L, W_q, and W are obtained from L_q as just shown. Notice that these L_q and W_q are exactly *half* as large as those for the exponential service-time case of Sec. 17.6 (the $M/M/1$ model), where $\sigma^2 = 1/\mu^2$, so decreasing σ^2 can *greatly* improve the measures of performance of a queueing system.

For the multiple-server version of this model ($M/D/s$), a complicated method is available[12] for deriving the steady-state probability distribution of the number of customers in the system and its mean [assuming $\rho = \lambda/(s\mu) < 1$]. These results have been tabulated for numerous cases,[13] and the means (L) also are given graphically in Fig. 17.8.

The $M/E_k/s$ Model

The $M/D/s$ model assumes *zero* variation in the service times ($\sigma = 0$), whereas the *exponential* service-time distribution assumes a very large variation ($\sigma = 1/\mu$). Between these two rather extreme cases lies a long middle ground ($0 < \sigma < 1/\mu$), where most *actual* service-time distributions fall. Another kind of theoretical service-time distribution that fills this middle ground is the **Erlang distribution** (named after the founder of queueing theory).

The probability density function for the Erlang distribution is

$$f(t) = \frac{(\mu k)^k}{(k - 1)!} t^{k-1} e^{-k\mu t}, \qquad \text{for } t \geq 0,$$

where μ and k are strictly positive parameters of the distribution and k is further restricted to be integer. (Except for this integer restriction and the definition of the parameters, this distribution is *identical* to the *gamma distribution*.) Its mean and standard deviation are

$$\text{Mean} = \frac{1}{\mu}$$

and

$$\text{Standard deviation} = \frac{1}{\sqrt{k}} \frac{1}{\mu}.$$

[12]See N. U. Prabhu: *Queues and Inventories*, Wiley, New York, 1965, pp. 32–34; also see pp. 286–288 in Selected Reference 5.

[13]F. S. Hillier and O. S. Yu, with D. Avis, L. Fossett, F. Lo, and M. Reiman, *Queueing Tables and Graphs*, Elsevier North-Holland, New York, 1981.

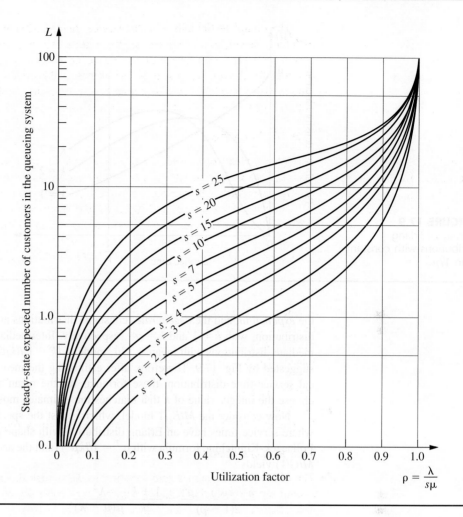

FIGURE 17.8
Values of L for the $M/D/s$ model (Sec. 17.7).

Thus, k is the parameter that specifies the degree of variability of the service times relative to the mean. It usually is referred to as the *shape parameter*.

The Erlang distribution is a very important distribution in queueing theory for two reasons. To describe the first one, suppose that $T_1, T_2, \ldots, T_k$ are k independent random variables with an identical exponential distribution whose mean is $1/(k\mu)$. Then their sum

$$T = T_1 + T_2 + \cdots + T_k$$

has an *Erlang* distribution with parameters μ and k. The discussion of the exponential distribution in Sec. 17.4 suggested that the time required to perform certain kinds of tasks might well have an exponential distribution. However, the total service required by a customer may involve the server's performing not just one specific task but a sequence of k tasks. If the respective tasks have an independent and identical exponential distribution for their duration, the total service time will have an Erlang distribution. This will be the case, e.g., if the server must perform the *same* exponential task k independent times for each customer.

The Erlang distribution also is very useful because it is a large (two-parameter) family of distributions permitting only nonnegative values. Hence, empirical service-time distributions can usually be reasonably approximated by an Erlang distribution. In fact, both

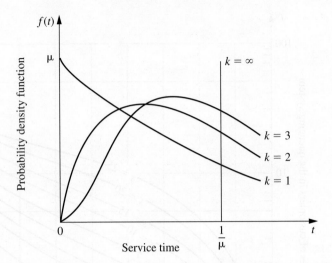

■ FIGURE 17.9
A family of Erlang
distributions with constant
mean $1/\mu$.

the *exponential* and the *degenerate* (constant) distributions are special cases of the Erlang distribution, with $k = 1$ and $k = \infty$, respectively. Intermediate values of k provide intermediate distributions with mean $= 1/\mu$, mode $= (k - 1)/(k\mu)$, and variance $= 1/(k\mu^2)$, as suggested by Fig. 17.9. Therefore, after estimating the mean and variance of an empirical service-time distribution, these formulas for the mean and variance can be used to choose the integer value of k that matches the estimates most closely.

Now consider the $M/E_k/1$ model, which is just the special case of the $M/G/1$ model where service times have an Erlang distribution with shape parameter $= k$. Applying the Pollaczek-Khintchine formula with $\sigma^2 = 1/(k\mu^2)$ (and the accompanying results given for $M/G/1$) yields

$$L_q = \frac{\lambda^2/(k\mu^2) + \rho^2}{2(1 - \rho)} = \frac{1 + k}{2k} \frac{\lambda^2}{\mu(\mu - \lambda)},$$

$$W_q = \frac{1 + k}{2k} \frac{\lambda}{\mu(\mu - \lambda)},$$

$$W = W_q + \frac{1}{\mu},$$

$$L = \lambda W.$$

With multiple servers ($M/E_k/s$), the relationship of the Erlang distribution to the exponential distribution just described can be exploited to formulate a *modified* birth-and-death process (continuous time Markov chain) in terms of individual exponential service phases (k per customer) rather than complete customers. However, it has not been possible to derive a general steady-state solution [when $\rho = \lambda/(s\mu) < 1$] for the probability distribution of the number of customers in the system as we did in Sec. 17.5. Instead, advanced theory is required to solve individual cases numerically. Once again, these results have been obtained and tabulated for numerous cases.[14] The means (L) also are given graphically in Fig. 17.10 for some cases where $s = 2$.

The Worked Examples section of the book's website includes **another example** that applies the $M/E_k/s$ model for both $s = 1$ and $s = 2$ to choose the less costly alternative.

---

[14]Ibid.

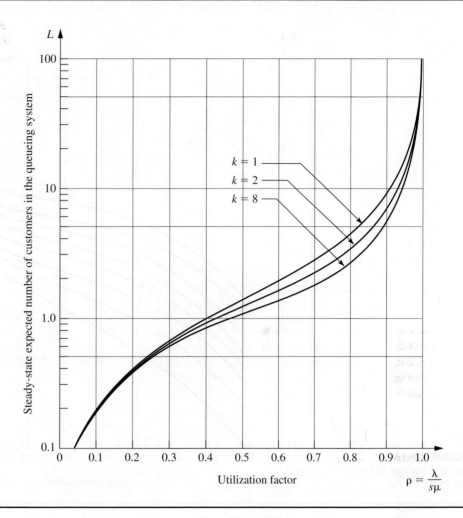

FIGURE 17.10
Values of L for the $M/E_k/2$ model (Sec. 17.7).

Models without a Poisson Input

All the queueing models presented thus far have assumed a Poisson input process (exponential interarrival times). However, this assumption is violated if the arrivals are scheduled or regulated in some way that prevents them from occurring randomly, in which case another model is needed.

As long as the service times have an exponential distribution with a fixed parameter, three such models are readily available. These models are obtained by merely *reversing* the assumed distributions of the *interarrival* and *service times* in the preceding three models. Thus, the first new model ($GI/M/s$) imposes no restriction on what the *interarrival time* distribution can be. In this case, there are some steady-state results available[15] (particularly in regard to waiting-time distributions) for both the single-server and multiple-server versions of the model, but these results are not nearly as convenient as the simple expressions given for the $M/G/1$ model. The second new model ($D/M/s$) assumes that all interarrival times equal some fixed *constant,* which would represent a queueing system where arrivals are *scheduled* at regular intervals. The third new model ($E_k/M/s$) assumes an *Erlang* interarrival time distribution, which provides a middle ground between

[15]For example, see pp. 248–260 of Selected Reference 5.

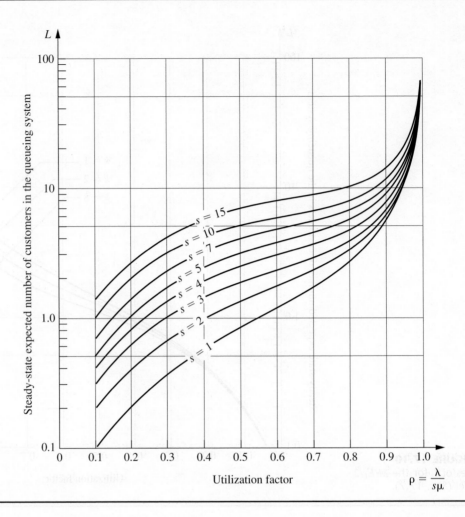

■ FIGURE 17.11
Values of L for the $D/M/s$ model (Sec. 17.7).

regularly scheduled (constant) and *completely random* (exponential) arrivals. Extensive computational results have been tabulated[16] for these latter two models, including the values of L given graphically in Figs. 17.11 and 17.12.

If neither the interarrival times nor the service times for a queueing system have an exponential distribution, then there are three additional queueing models for which computational results also are available.[17] One of these models ($E_m/E_k/s$) assumes an Erlang distribution for both these times. The other two models ($E_k/D/s$ and $D/E_k/s$) assume that one of these times has an Erlang distribution and the other time equals some fixed constant.

Other Models

Although you have seen in this section a large number of queueing models that involve nonexponential distributions, we have far from exhausted the list. For example, another distribution that occasionally is used for either interarrival times or service times is the **hyperexponential distribution.** The key characteristic of this distribution is that even though only nonnegative values are allowed, its standard deviation σ actually is larger than

---

[16]Hillier and Yu, op. cit.

[17]Ibid.

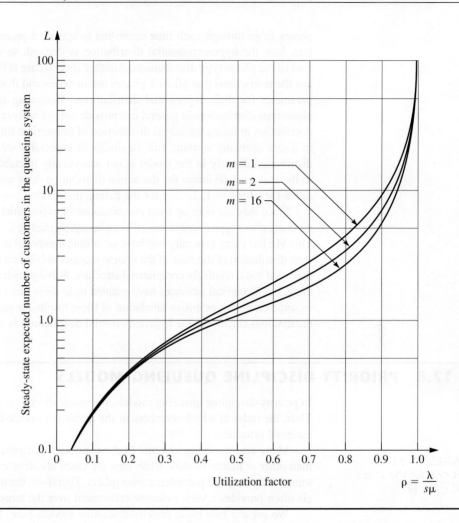

FIGURE 17.12
Values of L for the $E_k/M/2$ model (Sec. 17.7).

its mean $1/\mu$. This characteristic is in contrast to the Erlang distribution, where $\sigma < 1/\mu$ in every case except $k = 1$ (exponential distribution), which has $\sigma = 1/\mu$. To illustrate a typical situation where $\sigma > 1/\mu$ can occur, we suppose that the service involved in the queueing system is the repair of some kind of machine or vehicle. If many of the repairs turn out to be routine (small service times) but occasional repairs require an extensive overhaul (very large service times), then the standard deviation of service times will tend to be quite large relative to the mean, in which case the hyperexponential distribution may be used to represent the service-time distribution. Specifically, this distribution would assume that there are fixed probabilities, p and $(1 - p)$, for which kind of repair will occur, that the time required for each kind has an exponential distribution, but that the parameters for these two exponential distributions are different. (In general, the hyperexponential distribution is such a composite of two or more exponential distributions.)

Another family of distributions coming into general use consists of **phase-type distributions** (some of which also are called *generalized Erlangian distributions*). These distributions are obtained by breaking down the total time into a number of phases, each having an exponential distribution, where the parameters of these exponential distributions may be different and the phases may be either in series or in parallel (or both). A group of phases being *in parallel* means that the process randomly selects *one* of the phases to go through each time according to specified probabilities. This approach is, in fact, how the hyperexponential distribution is derived, so this distribution is a special case of the phase-type distributions. Another special case is the Erlang distribution, which has the restrictions that all its k phases are in series and that these phases have the *same* parameter for their exponential distributions. Removing these restrictions means that phase-type distributions in general can provide considerably more flexibility than the Erlang distribution in fitting the actual distribution of interarrival times or service times observed in a real queueing system. This flexibility is especially valuable when using the actual distribution directly in the model is not analytically tractable, and the ratio of the *mean* to the *standard deviation* for the actual distribution does not closely match the available ratios ($\sqrt{k}$ for $k = 1, 2, \ldots$) for the Erlang distribution.

Since they are built up from combinations of exponential distributions, queueing models using phase-type distributions still can be represented by a *continuous time Markov chain*. This Markov chain generally will have an infinite number of states, so solving for the steady-state distribution of the state of the system requires solving an infinite system of linear equations that has a relatively complicated structure. Solving such a system is far from a routine thing, but theoretical advances have enabled us to solve these queueing models numerically in some cases. An extensive tabulation of these results for models with various phase-type distributions (including the hyperexponential distribution) is available.[18]

17.8 PRIORITY-DISCIPLINE QUEUEING MODELS

In priority-discipline queueing models, the queue discipline is based on a *priority system*. Thus, the order in which members of the queue are selected for service is based on their assigned priorities.

Many real queueing systems fit these priority-discipline models much more closely than other available models. Rush jobs are taken ahead of other jobs, and important customers may be given precedence over others. Therefore, the use of priority-discipline models often provides a very welcome refinement over the more usual queueing models.

We present two basic priority-discipline models here. Since both models make the same assumptions, except for the nature of the priorities, we first describe the models together and then summarize their results separately.

The Models

Both models assume that there are *N priority classes* (class 1 has the highest priority and class *N* has the lowest) and that whenever a server becomes free to begin serving a new customer from the queue, the one customer selected is that member of the *highest* priority class represented in the queue who has waited longest. In other words, customers are selected to begin service in the order of their priority classes, but on a first-come-first-served basis within each priority class. A *Poisson* input process and *exponential* service times are assumed for each priority class. Except for one special case considered later, the models also make the somewhat restrictive assumption that the expected service time is the *same* for all priority classes. However, the models do permit the mean arrival rate to differ among priority classes.

The distinction between the two models is whether the priorities are *nonpreemptive* or *preemptive*. With **nonpreemptive priorities,** a customer being served cannot be ejected back into the queue (preempted) if a higher-priority customer enters the queueing system.

[18]L. P. Seelen, H. C. Tijms, and M. H. Van Hoorn, *Tables for Multi-Server Queues,* North-Holland, Amsterdam, 1985.

Therefore, once a server has begun serving a customer, the service must be completed without interruption. The first model assumes nonpreemptive priorities.

With **preemptive priorities,** the lowest-priority customer being served is *preempted* (ejected back into the queue) whenever a higher-priority customer enters the queueing system. A server is thereby freed to begin serving the new arrival immediately. (When a server does succeed in *finishing* a service, the next customer to begin receiving service is selected just as described at the beginning of this subsection, so a preempted customer normally will get back into service again and, after enough tries, will eventually finish.) Because of the lack-of-memory property of the exponential distribution (see Sec. 17.4), we do not need to worry about defining the point at which service begins when a preempted customer returns to service; the distribution of the *remaining* service time *always* is the same. (For any other service-time distribution, it becomes important to distinguish between *preemptive-resume* systems, where service for a preempted customer resumes at the point of interruption, and *preemptive-repeat* systems, where service must start at the beginning again.) The second model assumes preemptive priorities.

For both models, if the distinction between customers in different priority classes were ignored, Property 6 for the exponential distribution (see Sec. 17.4) implies that *all* customers would arrive according to a Poisson input process. Furthermore, all customers have the *same* exponential distribution for service times. Consequently, the two models actually are identical to the *M/M/s* model studied in Sec. 17.6 *except* for the order in which customers are served. Therefore, when we count just the *total* number of customers in the system, the steady-state distribution for the *M/M/s* model also applies to both models. Consequently, the formulas for L and L_q also carry over, as do the expected waiting-time results (by Little's formula) W and W_q, for a randomly selected customer. What changes is the *distribution* of waiting times, which was derived in Sec. 17.6 under the assumption of a first-come-first-served queue discipline. With a priority discipline, this distribution has a much larger *variance,* because the waiting times of customers in the highest priority classes tend to be much smaller than those under a first-come-first-served discipline, whereas the waiting times in the lowest priority classes tend to be much larger. By the same token, the breakdown of the total number of customers in the system tends to be disproportionately weighted toward the lower-priority classes. But this condition is just the reason for imposing priorities on the queueing system in the first place. We want to *improve* the *measures of performance* for each of the higher-priority classes at the expense of performance for the lower-priority classes. To determine how much improvement is being made, we need to obtain such measures as *expected waiting time in the system* and *expected number of customers in the system* for the individual priority classes. Expressions for these measures are given next for the two models in turn.

Results for the Nonpreemptive Priorities Model

Let W_k be the steady-state expected waiting time in the system (including service time) for a member of priority class k. Then

$$W_k = \frac{1}{AB_{k-1}B_k} + \frac{1}{\mu}, \qquad \text{for } k = 1, 2, \ldots, N,$$

where
$$A = s! \frac{s\mu - \lambda}{r^s} \sum_{j=0}^{s-1} \frac{r^j}{j!} + s\mu,$$

$$B_0 = 1,$$

$$B_k = 1 - \frac{\sum_{i=1}^{k} \lambda_i}{s\mu},$$

$$s = \text{number of servers,}$$

$$\mu = \text{mean service rate per busy server,}$$

$$\lambda_i = \text{mean arrival rate for priority class } i,$$

$$\lambda = \sum_{i=1}^{N} \lambda_i,$$

$$r = \frac{\lambda}{\mu}.$$

(This result assumes that

$$\sum_{i=1}^{k} \lambda_i < s\mu,$$

so that priority class k can reach a steady-state condition.) *Little's formula* still applies to individual priority classes, so L_k, the steady-state expected number of members of priority class k in the queueing system (including those being served), is

$$L_k = \lambda_k W_k, \qquad \text{for } k = 1, 2, \ldots, N.$$

To determine the expected waiting time in the queue (excluding service time) for priority class k, merely subtract $1/\mu$ from W_k; the corresponding expected queue length is again obtained by multiplying by λ_k. For the special case where $s = 1$, the expression for A reduces to $A = \mu^2/\lambda$.

An Excel template is provided in your OR Courseware for performing the above calculations.

The Worked Examples section of the book's website provides **an example** that illustrates the application of the nonpreemptive priorities model for determining how many turret lathes a factory should have when the jobs fall into three priority classes.

A Single-Server Variation of the Nonpreemptive Priorities Model

The above assumption that the expected service time $1/\mu$ is the same for all priority classes is a fairly restrictive one. In practice, this assumption sometimes is violated because of differences in the service requirements for the different priority classes.

Fortunately, for the special case of a single server, it is possible to allow different expected service times and still obtain useful results. Let $1/\mu_k$ denote the mean of the exponential service-time distribution for priority class k, so

$$\mu_k = \text{mean service rate for priority class } k, \qquad \text{for } k = 1, 2, \ldots, N.$$

Then the steady-state expected waiting time in the system for a member of priority class k is

$$W_k = \frac{a_k}{b_{k-1}b_k} + \frac{1}{\mu_k}, \qquad \text{for } k = 1, 2, \ldots, N,$$

where

$$a_k = \sum_{i=1}^{k} \frac{\lambda_i}{\mu_i^2},$$

$$b_0 = 1,$$

$$b_k = 1 - \sum_{i=1}^{k} \frac{\lambda_i}{\mu_i}.$$

This result holds as long as

$$\sum_{i=1}^{k} \frac{\lambda_i}{\mu_i} < 1,$$

which enables priority class k to reach a steady-state condition. Little's formula can be used as described above to obtain the other main measures of performance for each priority class.

Results for the Preemptive Priorities Model

For the preemptive priorities model, we need to reinstate the assumption that the expected service time is the same for all priority classes. Using the same notation as for the original nonpreemptive priorities model, having the preemption changes the *total* expected waiting time in the system (including the total service time) to

$$W_k = \frac{1/\mu}{B_{k-1}B_k}, \qquad \text{for } k = 1, 2, \ldots, N,$$

for the *single-server* case ($s = 1$). When $s > 1$, W_k can be calculated by an iterative procedure that will be illustrated soon by the County Hospital example. The L_k continue to satisfy the relationship

$$L_k = \lambda_k W_k, \qquad \text{for } k = 1, 2, \ldots, N.$$

The corresponding results for the queue (excluding customers in service) also can be obtained from W_k and L_k as just described for the case of nonpreemptive priorities. Because of the lack-of-memory property of the exponential distribution (see Sec. 17.4), preemptions do not affect the service process (occurrence of service completions) in any way. The expected total service time for any customer still is $1/\mu$.

This chapter's Excel files include an Excel template for calculating the above measures of performance for the single-server case.

The County Hospital Example with Priorities

For the County Hospital emergency room problem, the management engineer has noticed that the patients are not treated on a first-come-first-served basis. Rather, the admitting nurse seems to divide the patients into roughly three categories: (1) *critical* cases, where prompt treatment is vital for survival; (2) *serious* cases, where early treatment is important to prevent further deterioration; and (3) *stable* cases, where treatment can be delayed without adverse medical consequences. Patients are then treated in this order of priority, where those in the same category are normally taken on a first-come-first-served basis. A doctor will interrupt treatment of a patient if a new case in a higher-priority category arrives. Approximately 10 percent of the patients fall into the first category, 30 percent into the second, and 60 percent into the third. Because the more serious cases will be sent to the hospital for further care after receiving emergency treatment, the average treatment time by a doctor in the emergency room actually does not differ greatly among these categories.

The management engineer has decided to use a priority-discipline queueing model as a reasonable representation of this queueing system, where the three categories of patients constitute the three priority classes in the model. Because treatment is interrupted by the arrival of a higher-priority case, the *preemptive priorities model* is the appropriate one. Given the previously available data ($\mu = 3$ and $\lambda = 2$), the preceding percentages yield $\lambda_1 = 0.2$, $\lambda_2 = 0.6$, and $\lambda_3 = 1.2$. Table 17.3 gives the resulting expected waiting times in the queue (so *excluding* treatment time) for the respective priority classes[19] when there is one ($s = 1$) or two ($s = 2$) doctors on duty. (The corresponding results for the nonpreemptive priorities model also are given in Table 17.3 to show the effect of preempting.)

Deriving the Preemptive Priority Results. These preemptive priority results for $s = 2$ were obtained as follows. Because the waiting times for priority class 1 customers are completely unaffected by the presence of customers in the lower-priority classes, W_1

---

[19]Note that these expected times can no longer be interpreted as the expected time before treatment begins when $k > 1$, because treatment may be interrupted at least once, causing additional waiting time before service is completed.

■ TABLE 17.3 Steady-state results from the priority-discipline models for the County Hospital problem

	Preemptive Priorities		Nonpreemptive Priorities	
	$s = 1$	$s = 2$	$s = 1$	$s = 2$
A	—	—	4.5	36
B_1	0.933	—	0.933	0.967
B_2	0.733	—	0.733	0.867
B_3	0.333	—	0.333	0.667
$W_1 - \dfrac{1}{\mu}$	0.024 hour	0.00037 hour	0.238 hour	0.029 hour
$W_2 - \dfrac{1}{\mu}$	0.154 hour	0.00793 hour	0.325 hour	0.033 hour
$W_3 - \dfrac{1}{\mu}$	1.033 hours	0.06542 hour	0.889 hour	0.048 hour

will be the same for any other values of λ_2 and λ_3, including $\lambda_2 = 0$ and $\lambda_3 = 0$. Therefore, W_1 must equal W for the corresponding *one-class* model (the *M/M/s* model in Sec. 17.6) with $s = 2$, $\mu = 3$, and $\lambda = \lambda_1 = 0.2$, which yields

$$W_1 = W = 0.33370 \text{ hour}, \qquad \text{for } \lambda = 0.2$$

so

$$W_1 - \frac{1}{\mu} = 0.33370 - 0.33333 = 0.00037 \text{ hour.}$$

Now consider the first two priority classes. Again note that customers in these classes are completely unaffected by lower-priority classes (just priority class 3 in this case), which can therefore be ignored in the analysis. Let $\overline{W}_{1-2}$ be the expected waiting time in the system (so including service time) of a *random arrival* in *either* of these two classes, so the probability is $\lambda_1/(\lambda_1 + \lambda_2) = \frac{1}{4}$ that this arrival is in class 1 and $\lambda_2/(\lambda_1 + \lambda_2) = \frac{3}{4}$ that it is in class 2. Therefore,

$$\overline{W}_{1-2} = \frac{1}{4}W_1 + \frac{3}{4}W_2.$$

Furthermore, because the *expected* waiting time is the same for *any* queue discipline, $\overline{W}_{1-2}$ must also equal W for the *M/M/s* model in Sec. 17.6, with $s = 2$, $\mu = 3$, and $\lambda = \lambda_1 + \lambda_2 = 0.8$, which yields

$$\overline{W}_{1-2} = W = 0.33937 \text{ hour}, \qquad \text{for } \lambda = 0.8.$$

Combining these facts gives

$$W_2 = \frac{4}{3}\left[0.33937 - \frac{1}{4}(0.33370)\right] = 0.34126 \text{ hour.}$$

$$\left(W_2 - \frac{1}{\mu} = 0.00793 \text{ hour.}\right)$$

Finally, let $\overline{W}_{1-3}$ be the expected waiting time in the system (so including service time) for a *random arrival* in *any* of the three priority classes, so the probabilities are 0.1, 0.3, and 0.6 that it is in classes 1, 2, and 3, respectively. Therefore,

$$\overline{W}_{1-3} = 0.1W_1 + 0.3W_2 + 0.6W_3.$$

Furthermore, $\overline{W}_{1-3}$ must also equal W for the *M/M/s* model in Sec. 17.6, with $s = 2$, $\mu = 3$, and $\lambda = \lambda_1 + \lambda_2 + \lambda_3 = 2$, so that (from Table 17.2)

$$\overline{W}_{1-3} = W = 0.375 \text{ hour,} \qquad \text{for } \lambda = 2.$$

Consequently,

$$W_3 = \frac{1}{0.6}[0.375 - 0.1(0.33370) - 0.3(0.34126)]$$

$$= 0.39875 \text{ hour.}$$

$$\left(W_3 - \frac{1}{\mu} = 0.06542 \text{ hour.} \right)$$

The corresponding W_q results for the *M/M/s* model in Sec. 17.6 also could have been used in exactly the same way to derive the $W_k - 1/\mu$ quantities directly.

Conclusions. When $s = 1$, the $W_k - 1/\mu$ values in Table 17.3 for the preemptive priorities case indicate that providing just a single doctor would cause critical cases to wait about $1\frac{1}{2}$ minutes (0.024 hour) on the average, serious cases to wait more than 9 minutes, and stable cases to wait more than 1 hour. (Contrast these results with the average wait of $W_q = \frac{2}{3}$ hour for all patients that was obtained in Table 17.2 under the first-come-first-served queue discipline.) However, these values represent *statistical expectations,* so some patients have to wait considerably longer than the average for their priority class. This wait would not be tolerable for the critical and serious cases, where a few minutes can be vital. By contrast, the $s = 2$ results in Table 17.3 (preemptive priorities case) indicate that adding a second doctor would virtually eliminate waiting for all but the stable cases. Therefore, the management engineer recommended that there be two doctors on duty in the emergency room during the early evening hours next year. The board of directors for County Hospital adopted this recommendation and simultaneously raised the charge for using the emergency room!

17.9 QUEUEING NETWORKS

Thus far we have considered only queueing systems that have a *single* service facility with one or more servers. However, queueing systems encountered in OR studies are sometimes actually *queueing networks,* i.e., networks of service facilities where customers must receive service at some of or all these facilities. For example, orders being processed through a job shop must be routed through a sequence of machine groups (service facilities). It is therefore necessary to study the entire network to obtain such information as the expected total waiting time, expected number of customers in the entire system, and so forth.

Because of the importance of queueing networks, research into this area has been very active. However, this is a difficult area, so we limit ourselves to a brief introduction.

One result is of such fundamental importance for queueing networks that this finding and its implications warrant special attention here. This fundamental result is the following *equivalence property* for the *input process* of arriving customers and the *output process* of departing customers for certain queueing systems.

Equivalence property: Assume that a service facility with s servers and an infinite queue has a Poisson input with parameter λ and the same exponential service-time distribution with parameter μ for each server (the *M/M/s* model), where $s\mu > \lambda$. Then the steady-state *output* of this service facility is also a Poisson process with parameter λ.

For many decades, **General Motors Corporation (GM)** has enjoyed its position as the world's largest automotive manufacturer, before being overtaken recently by Toyota. It has manufacturing operations in 32 countries, employs over 300,000 people worldwide, and generates annual revenues of close to $200 billion. However, ever since the late 1980s, when the productivity of GM's plants ranked near the bottom in the industry, the company's market position has been steadily eroding due to ever-increasing foreign competition.

To counter this foreign competition, GM management initiated a long-term operations research project many years ago to predict and improve the throughput performance of the company's several hundred production lines throughout the world. The goal was to greatly increase the company's productivity throughout its manufacturing operations and thereby provide GM with a strategic competitive advantage.

The most important analytical tool used in this project has been a *complicated queueing model* that uses a simple single-server model as a building block. The overall model begins by considering a two-station production line where each station is modeled as a single-server queueing system with constant interarrival times and constant service times with the following exceptions. The server (commonly a machine) at each station occasionally breaks down and does not resume serving until a repair is completed. The server at the first station also shuts down when it completes a service and the buffer between the stations is full. The server at the second station shuts down when it completes a service and has not yet received a job from the first station.

The next step in the analysis is to extend this queueing model for a two-station production line to one for a production line with any number of stations. This larger queueing model then is used to analyze how production lines should be designed to maximize their throughput. (The technique of *simulation* described in Chap. 20 also is used for this purpose for relatively complex production lines.)

This application of queueing theory (and simulation), along with supporting data-collection systems, has reaped remarkable benefits for GM. According to impartial industry sources, its plants, which once were among the least productive in the industry, now rank among the very best. The resulting improvements in production throughput in over 30 vehicle plants and 10 countries *has yielded over* **$2.1 billion** *in documented savings and increased revenue.* These dramatic results led to General Motors winning the prestigious First Prize in the 2005 international competition for the Franz Edelman Award for Achievement in Operations Research and the Management Sciences.

Source: J. M. Alden, L. D. Burns, T. Costy, R. D. Hutton, C. A. Jackson, D. S. Kim, K. A. Kohls, J. H. Owen, M. A. Turnquist, and D. J. Vander Veen: "General Motors Increases Its Production Throughput," *Interfaces*, **36**(1): 6–25, Jan.–Feb. 2006. (A link to this article is provided on our website, www.mhhe.com/hillier.)

Notice that this property makes no assumption about the type of queue discipline used. Whether it is first-come-first-served, random, or even a priority discipline as in Sec. 17.8, the served customers will leave the service facility according to a Poisson process. The crucial implication of this fact for queueing networks is that if these customers must then go to another service facility for further service, this second facility *also* will have a Poisson input. With an exponential service-time distribution, the equivalence property will hold for this facility as well, which can then provide a Poisson input for a third facility, etc. We discuss the consequences for two basic kinds of networks next.

Infinite Queues in Series

Suppose that customers must all receive service at a *series* of m service facilities in a fixed sequence. Assume that each facility has an infinite queue (no limitation on the number of customers allowed in the queue), so that the series of facilities form a system of *infinite queues in series*. Assume further that the customers arrive at the first facility according to a Poisson process with parameter λ and that each facility i ($i = 1, 2, \ldots, m$) has an exponential service-time distribution with parameter μ_i for its s_i servers, where $s_i\mu_i > \lambda$. It then follows from the equivalence property that (under steady-state conditions) each service facility has a Poisson input with parameter λ. Therefore, the elementary $M/M/s$ model of Sec. 17.6 (or its priority-discipline counterparts in Sec. 17.8) can be used to analyze each service facility independently of the others!

Being able to use the $M/M/s$ model to obtain all measures of performance for each facility independently, rather than analyzing interactions between facilities, is a tremendous simplification. For example, the probability of having n customers at a given facility is given by the formula for P_n in Sec. 17.6 for the $M/M/s$ model. The *joint probability* of n_1 customers at facility 1, n_2 customers at facility 2, . . . , then, is the *product* of the individual probabilities obtained in this simple way. In particular, this joint probability can be expressed as

$$P\{(N_1, N_2, \ldots, N_m) = (n_1, n_2, \ldots, n_m)\} = P_{n_1} P_{n_2} \cdots P_{n_m}.$$

(This simple form for the solution is called the **product form solution.**) Similarly, the expected total waiting time and the expected number of customers in the entire system can be obtained by merely summing the corresponding quantities obtained at the respective facilities.

Unfortunately, the equivalence property and its implications do not hold for the case of *finite* queues discussed in Sec. 17.6. This case is actually quite important in practice, because there is often a definite limitation on the queue length in front of service facilities in networks. For example, only a small amount of buffer storage space is typically provided in front of each facility (station) in a production-line system. For such systems of finite queues in series, no simple product form solution is available. The facilities must be analyzed jointly instead, and only limited results have been obtained.

Jackson Networks

Systems of infinite queues in series are not the only queueing networks where the $M/M/s$ model can be used to analyze each service facility independently of the others. Another prominent kind of network with this property (a product form solution) is the *Jackson network*, named after the individual (James R. Jackson) who first characterized the network and showed that this property holds a few decades ago.

The characteristics of a Jackson network are the same as assumed above for the system of infinite queues in series, except now the customers visit the facilities in different orders (and may not visit them all). For each facility, its arriving customers come from *both* outside the system (according to a Poisson process) and the other facilities. These characteristics are summarized below.

A **Jackson network** is a system of m service facilities where facility i ($i = 1$, $2, \ldots, m$) has

1. An infinite queue
2. Customers arriving from outside the system according to a Poisson input process with parameter a_i
3. s_i servers with an exponential service-time distribution with parameter μ_i.

A customer leaving facility i is routed next to facility j ($j = 1, 2, \ldots, m$) with probability p_{ij} or departs the system with probability

$$q_i = 1 - \sum_{j=1}^{m} p_{ij}.$$

Any such network has the following key property.

Under steady-state conditions, each facility j ($j = 1, 2, \ldots, m$) in a Jackson network behaves as if it were an *independent M/M/s* queueing system with arrival rate

$$\lambda_j = a_j + \sum_{i=1}^{m} \lambda_i p_{ij},$$

where $s_j \mu_j > \lambda_j$.

This key property cannot be *proved* directly from the equivalence property this time (the reasoning would become circular), but its *intuitive underpinning* is still provided by the latter property. The intuitive viewpoint (not quite technically correct) is that, for each facility *i*, its input processes from the various sources (outside and other facilities) are *independent Poisson processes,* so the *aggregate* input process is Poisson with parameter λ_i (Property 6 in Sec. 17.4). The equivalence property then says that the *aggregate output* process for facility *i* must be Poisson with parameter λ_i. By disaggregating this output process (Property 6 again), the process for customers going from facility *i* to facility *j* must be Poisson with parameter $\lambda_i p_{ij}$. This process becomes one of the Poisson *input* processes for facility *j*, thereby helping to maintain the series of Poisson processes in the overall system.

The equation given for obtaining λ_j is based on the fact that λ_i is the *departure rate* as well as the arrival rate for all customers using facility *i*. Because p_{ij} is the proportion of customers departing from facility *i* who go next to facility *j*, the rate at which customers from facility *i* arrive at facility *j* is $\lambda_i p_{ij}$. Summing this product over all *i*, and then adding this sum to a_j, gives the *total arrival rate* to facility *j* from all sources.

To calculate λ_j from this equation requires knowing the λ_i for $i \neq j$, but these λ_i also are unknowns given by the corresponding equations. Therefore, the procedure is to solve *simultaneously* for $\lambda_1, \lambda_2, \ldots, \lambda_m$ by obtaining the simultaneous solution of the entire system of linear equations for λ_j for $j = 1, 2, \ldots, m$. Your IOR Tutorial includes an interactive procedure for solving for the λ_j in this way.

To illustrate these calculations, consider a Jackson network with three service facilities that have the parameters shown in Table 17.4. Plugging into the formula for λ_j for $j = 1, 2, 3$, we obtain

$$\lambda_1 = 1 \qquad\quad + 0.1\lambda_2 + 0.4\lambda_3$$
$$\lambda_2 = 4 + 0.6\lambda_1 \qquad\quad + 0.4\lambda_3$$
$$\lambda_3 = 3 + 0.3\lambda_1 + 0.3\lambda_2.$$

(Reason through each equation to see why it gives the total arrival rate to the corresponding facility.) The simultaneous solution for this system is

$$\lambda_1 = 5, \qquad \lambda_2 = 10, \qquad \lambda_3 = 7\frac{1}{2}.$$

Given this simultaneous solution, each of the three service facilities now can be analyzed *independently* by using the formulas for the *M/M/s* model given in Sec. 17.6. For example, to obtain the distribution of the number of customers $N_i = n_i$ at facility *i*, note that

$$\rho_i = \frac{\lambda_i}{s_i \mu_i} = \begin{cases} \dfrac{1}{2} & \text{for } i = 1 \\[2mm] \dfrac{1}{2} & \text{for } i = 2 \\[2mm] \dfrac{3}{4} & \text{for } i = 3. \end{cases}$$

■ **TABLE 17.4** Data for the example of a Jackson network

Facility j	s_j	μ_j	a_j	p_{ij}		
				$i = 1$	$i = 2$	$i = 3$
$j = 1$	1	10	1	0	0.1	0.4
$j = 2$	2	10	4	0.6	0	0.4
$j = 3$	1	10	3	0.3	0.3	0

Plugging these values (and the parameters in Table 17.4) into the formula for P_n gives

$$P_{n_1} = \frac{1}{2}\left(\frac{1}{2}\right)^{n_1} \quad \text{for facility 1,}$$

$$P_{n_2} = \begin{cases} \dfrac{1}{3} & \text{for } n_2 = 0 \\[2mm] \dfrac{1}{3} & \text{for } n_2 = 1 \quad \text{for facility 2,} \\[2mm] \dfrac{1}{3}\left(\dfrac{1}{2}\right)^{n_2-1} & \text{for } n_2 \geq 2 \end{cases}$$

$$P_{n_3} = \frac{1}{4}\left(\frac{3}{4}\right)^{n_3} \quad \text{for facility 3.}$$

The *joint probability* of (n_1, n_2, n_3) then is given simply by the product form solution

$$P\{(N_1, N_2, N_3) = (n_1, n_2, n_3)\} = P_{n_1}P_{n_2}P_{n_3}.$$

In a similar manner, the expected number of customers L_i at facility i can be calculated from Sec. 17.6 as

$$L_1 = 1, \qquad L_2 = \frac{4}{3}, \qquad L_3 = 3.$$

The expected *total* number of customers in the entire system then is

$$L = L_1 + L_2 + L_3 = 5\frac{1}{3}.$$

Obtaining W, the expected *total* waiting time in the system (including service times) for a customer, is a little trickier. You cannot simply add the expected waiting times at the respective facilities, because a customer does not necessarily visit each facility exactly once. However, Little's formula can still be used, where the system arrival rate λ is the sum of the arrival rates *from outside* to the facilities, $\lambda = a_1 + a_2 + a_3 = 8$. Thus,

$$W = \frac{L}{a_1 + a_2 + a_3} = \frac{2}{3}.$$

In conclusion, we should point out that there do exist other (more complicated) kinds of queueing networks where the individual service facilities can be analyzed independently from the others. In fact, finding queueing networks with a product form solution has been the Holy Grail for research on queueing networks. Some sources of additional information are Selected References 3 and 12.

■ 17.10   THE APPLICATION OF QUEUEING THEORY

Because of the wealth of information provided by queueing theory, it is widely used to guide the design (or redesign) of queueing systems. We now turn our focus to how queueing theory is applied in this way.

The most common decision that needs to be made when designing a queueing system is how many servers to provide. However, a number of other decisions also may be needed. The possible decisions include

1. Number of servers at a service facility.
2. Efficiency of the servers.
3. Number of service facilities.

4. Amount of waiting space in the queue.

5. Any priorities for different categories of customers.

The two primary considerations in making these kinds of decisions typically are (1) the cost of the service capacity provided by the queueing system and (2) the consequences of making the customers wait in the queueing system. Providing too much service capacity causes excessive costs. Providing too little causes excessive waiting. Therefore, the goal is to find an appropriate trade-off between the service cost and the amount of waiting.

Two basic approaches are available for seeking this trade-off. One is to establish one or more criteria for a satisfactory level of service in terms of how much waiting would be acceptable. For example, one possible criterion might be that the expected waiting time in the system should not exceed a certain number of minutes. Another might be that at least 95 percent of the customers should wait no longer than a certain number of minutes in the system. Similar criteria in terms of the expected number of customers in the system (or the probability distribution of this number) also could be used. The criteria also might be stated in terms of the waiting time or the number of customers in the *queue* instead of in the system. Once the criterion or criteria have been selected, it then is usually straightforward to use trial and error to find the least costly design of the queueing system that satisfies all the criteria.

The other basic approach for seeking the best trade-off involves assessing the costs associated with the consequences of making customers wait. For example, suppose that the queueing system is an *internal service system* (as described in Sec. 17.3), where the customers are the employees of a for-profit company. Making these employees wait at the queueing system causes *lost productivity,* which results in *lost profit.* This lost profit is the **waiting cost** associated with the queueing system. By expressing this waiting cost as a function of the amount of waiting, the problem of determining the best design of the queueing system can now be posed as minimizing the expected *total cost* (service cost plus waiting cost) per unit time.

We spell out this latter approach below for the problem of determining the optimal number of servers to provide.

How Many Servers Should Be Provided?

To formulate the objective function when the decision variable is the number of servers s, let

$E(\text{TC})$ = expected total cost per unit time,

$E(\text{SC})$ = expected service cost per unit time,

$E(\text{WC})$ = expected waiting cost per unit time.

Then the objective is to choose the number of servers so as to

Minimize $E(\text{TC}) = E(\text{SC}) + E(\text{WC})$.

When each server costs the same, the **service cost** is

$E(\text{SC}) = C_s s,$

where C_s is the marginal cost of a server per unit time. To evaluate WC for any value of s, note that $L = \lambda W$ gives the expected total amount of waiting in the queueing system per unit time. Therefore, when the waiting cost is proportional to the amount of waiting, this cost can be expressed as

$E(\text{WC}) = C_w L,$

where C_w is the waiting cost per unit time for each customer in the queueing system. There-fore, after estimating the constants, C_s and C_w, the goal is to choose the value of s so as to

Minimize $E(\text{TC}) = C_s s + C_w L$.

By choosing the queueing model that fits the queueing system, the value of L can be ob-tained for various values of s. Increasing s decreases L, at first rapidly and then gradually more slowly.

Figure 17.13 shows the general shape of the $E(\text{SC})$, $E(\text{WC})$, and $E(\text{TC})$ curves ver-sus the number of servers s. (For better conceptualization, we have drawn these as smooth curves even though the only feasible values of s are $s = 1, 2, \ldots$) By calculating $E(\text{TC})$ for consecutive values of s until $E(\text{TC})$ stops decreasing and starts increasing instead, it is straightforward to find the number of servers that minimizes total cost. The following example illustrates this process.

An Example

The Acme Machine Shop has a tool crib to store tools required by the shop mechanics. Two clerks run the tool crib. The clerks hand out the tools as the mechanics arrive and request them. The tools then are returned to the clerks when they are no longer needed. There have been complaints from supervisors that their mechanics have had to waste too much time waiting to be served at the tool crib, so it appears as if there should be *more* clerks. On the other hand, management is exerting pressure to reduce overhead in the plant, and this reduction would lead to *fewer* clerks. To resolve these conflicting pressures, an OR study is being conducted to determine just how many clerks the tool crib should have.

The tool crib constitutes a queueing system, with the clerks as its servers and the me-chanics as its customers. After gathering some data on interarrival times and service times, the OR team has concluded that the queueing model that fits this queueing system best is the *M/M/s* model. The estimates of the mean arrival rate λ and the mean service rate (per server) μ are

$\lambda = 120$ customers per hour,
$\mu = 80$ customers per hour,

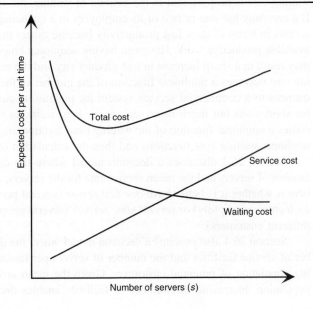

■ **FIGURE 17.13**
The shape of the expected cost curves for determining the number of servers to provide.

so the utilization factor for the two clerks is

$$\rho = \frac{\lambda}{s\mu} = \frac{120}{2(80)} = 0.75.$$

The total cost to the company of each tool crib clerk is about \$20 per hour, so $C_s = \$20$. While a mechanic is busy, the value to the company of his or her output averages about \$48 per hour, so $C_w = \$48$. Therefore, the OR team now needs to find the number of servers (tool crib clerks) s that will

Minimize $E(\text{TC}) = \$20 \, s + \$48 \, L$.

An Excel template has been provided in your OR Courseware for calculating these costs with the $M/M/s$ model. All you need to do is enter the data for the model along with the unit service cost C_s, the unit waiting cost C_w, and the number of servers s you want to try. The template then calculates $E(\text{SC})$, $E(\text{WC})$, and $E(\text{TC})$. This is illustrated in Fig. 17.14 with $s = 3$ for this example. By repeatedly entering alternative values of s, the template then can reveal which value minimizes $E(\text{TC})$ in a matter of seconds.

Table 17.5 shows the data that would be generated from this template by repeating these calculations for $s = 1, 2, 3, 4,$ and 5. Since the utilization factor for $s = 1$ is $\rho = 1.5$, a single clerk would be unable to keep up with the customers, so this option is ruled out. All larger values of s are feasible, but $s = 3$ has the smallest expected total cost. Furthermore, $s = 3$ would decrease the current expected total cost for $s = 2$ by \$61 per hour. Therefore, despite management's current drive to reduce overhead (which includes the cost of tool crib clerks), the OR team recommends that a third clerk be added to the tool crib. Note that this recommendation would decrease the utilization factor for the clerks from an already modest 0.75 all the way down to 0.5. However, because of the large improvement in the productivity of the mechanics (who are much more expensive than the clerks) through decreasing their time wasted waiting at the tool crib, management adopts the recommendation.

Other Issues

Chapter 26 on the book's website expands considerably further on the application of queueing theory, including how to deal with some other issues not considered above.

For example, the analysis displayed in Fig. 17.14 and Table 17.5 assumed that the waiting cost is proportional to the amount of waiting, but this sometimes is not the case. If a company has one or two of its employees in a queueing system, this may not be very serious in terms of their lost productivity because others may be able to handle all of the available productive work. However, having additional employees in the queueing system may result in a sharp increase in lost productivity and the resulting lost profit, so the waiting cost becomes a nonlinear function of the number in the system. Similarly, the consequences to a commercial service system for making its customers wait may be minimal for short waits but much more serious for long waits. In this case, the waiting cost becomes a nonlinear function of the waiting time. Section 26.3 describes the formulation of nonlinear waiting-cost functions and then the calculation of $E(\text{WC})$ with such functions.

Section 26.4 discusses a decision model where the decision variables are *both* the number of servers and the mean service rate for the servers. An interesting issue that arises here is whether it is better have *one fast server* (several people working together to serve each customer rapidly) or *several slow servers* (several people working separately to serve different customers).

Section 26.4 also presents a decision model where the decision variables are the number of service facilities and the number of servers per facility to provide service to a calling population of potential customers. Given the mean arrival rate for the entire calling population, increasing the number of facilities enables decreasing the mean arrival rate

	A	B	C	D	E	F	G
1		Economic Analysis of Acme Machine Shop Example					
2							
3			Data				Results
4		$\lambda =$	120	(mean arrival rate)		$L =$	1.736842105
5		$\mu =$	80	(mean service rate)		$L_q =$	0.236842105
6		$s =$	3	(# servers)			
7						$W =$	0.014473684
8		$Pr(W > t) =$	0.02581732			$W_q =$	0.001973684
9		when $t =$	0.05				
10						$\rho =$	0.5
11		$Prob(W_q > t) =$	0.00058707			n	P_n
12		when $t =$	0.05				
13						0	0.210526316
14		Economic Analysis:				1	0.315789474
15		$Cs =$	$20.00	(cost / server / unit time)		2	0.236842105
16		$Cw =$	$48.00	(waiting cost / unit time)		3	0.118421053
17						4	0.059210526
18		Cost of Service	$60.00			5	0.029605263
19		Cost of Waiting	$83.37			6	0.014802632
20		Total Cost	$143.37			7	0.007401316

	B	C
18	Cost of Service	=Cs*s
19	Cost of Waiting	=Cw*L
20	Total Cost	=CostOfService+CostOfWaiting

Range Name	Cells
CostOfService	C18
CostOfWaiting	C19
Cs	C15
Cw	C16
L	G4
s	C6
TotalCost	C20

FIGURE 17.14

This Excel template for using economic analysis to choose the number of servers with the *M/M/s* model is applied here to the Acme Machine Shop example with $s = 3$.

TABLE 17.5 Calculation of $E(TC)$ for alternative s in the Acme Machine Shop example

s	ρ	L	$E(SC) = C_s s$	$E(WC) = C_w L$	$E(TC) = E(SC) + E(WC)$
1	1.50	∞	$20	∞	∞
2	0.75	3.43	$40	$164.57	$204.57
3	0.50	1.74	$60	$83.37	$143.37
4	0.375	1.54	$80	$74.15	$154.15
5	0.30	1.51	$100	$72.41	$172.41

(workload) at each facility. The number of service facilities also affects how much time each customer will need to spend in traveling to and from the nearest facility. The waiting cost now needs to be a function of the total time lost by a customer by either waiting at a service facility or traveling to and from the facility. Therefore, Sec. 26.5 presents some travel-time models for determining the expected round-trip travel time for each customer.

◼ 17.11 CONCLUSIONS

Queueing systems are prevalent throughout society. The adequacy of these systems can have an important effect on the quality of life and productivity.

Queueing theory studies queueing systems by formulating mathematical models of their operation and then using these models to derive measures of performance. This analysis provides vital information for effectively designing queueing systems that achieve an appropriate balance between the cost of providing a service and the cost associated with waiting for that service.

This chapter presented the most basic models of queueing theory for which particularly useful results are available. However, many other interesting models could be considered if space permitted. In fact, several thousand research papers formulating and/or analyzing queueing models have already appeared in the technical literature, and many more are being published each year!

The *exponential distribution* plays a fundamental role in queueing theory for representing the distribution of interarrival and service times, because this assumption enables us to represent the queueing system as a *continuous time Markov chain.* For the same reason, *phase-type distributions* such as the *Erlang distribution,* where the total time is broken down into individual phases having an exponential distribution, are very useful. Useful analytical results have been obtained for only a relatively few queueing models making other assumptions.

Priority-discipline queueing models are useful for the common situation where some categories of customers are given priority over others for receiving service.

In another common situation, customers must receive service at several different service facilities. Models for queueing networks are gaining widespread use for such situations. This is an area of especially active ongoing research.

When no tractable model that provides a reasonable representation of the queueing system under study is available, a common approach is to obtain relevant performance data by developing a computer program for simulating the operation of the system. This technique is discussed in Chap. 20.

Section 17.10 briefly describes how queueing theory can be used to help design effective queueing systems and then Chap. 26 (on the book's website) expands considerably further on this subject.

◼ SELECTED REFERENCES

1. Asmussen, S.: *Applied Probability and Queues,* 2nd ed., Springer, New York, 2003.

2. Balsamo, S., V. de Nitto Personé, and R. Onvural: *Analysis of Queueing Networks with Blocking,* Kluwer Academic Publishers (now Springer), Boston, 2001.

3. Chen, H., and D. D. Yao: *Fundamentals of Queueing Networks: Performance, Asymptotics, and Optimization,* Springer, New York, 2001.

4. El-Taha, M., and S. Stidham, Jr.: *Sample-Path Analysis of Queueing Systems,* Kluwer Academic Publishers (now Springer), Boston, 1998.

5. Gross, D., and C. M. Harris: *Fundamentals of Queueing Theory,* 3d ed., Wiley, New York, 1998.

6. Hall, R. W. (ed.): *Patient Flow: Reducing Delay in Healthcare Delivery,* Springer, New York, 2006.
7. Hall, R. W.: *Queueing Methods: For Services and Manufacturing,* Prentice-Hall, Upper Saddle River, NJ, 1991.
8. Hassin, R., and M. Haviv: *To Queue or Not to Queue: Equilibrium Behavior in Queueing Systems,* Kluwer Academic Publishers (now Springer), Boston, 2008.
9. Hillier, F. S., and M. S. Hillier: *Introduction to Management Science: A Modeling and Case Studies Approach with Spreadsheets,* 3rd ed., McGraw-Hill/Irwin, Burr Ridge, IL, 2008, Chap. 11.
10. Papadopoulos, H. T., C. Heavy, and J. Browne: *Queueing Theory in Manufacturing Systems Analysis and Design,* Chapman Hall, London, 1993.
11. Prabhu, N. U.: *Foundations of Queueing Theory,* Kluwer Academic Publishers (now Springer), Boston, 1997.
12. Serfozo, R.: *Introduction to Stochastic Networks,* Springer, New York, 1999.
13. Stidham, S., Jr.: "Analysis, Design, and Control of Queueing Systems," *Operations Research,* **50:** 197–216, 2002.
14. Tian, N., and Z. G. Zhang: *Vacation Queueing Models: Theory and Applications,* Springer, New York, 2006.

Some Award-Winning Applications of Queueing Theory:

(A link to all these articles is provided on our website, www.mhhe.com/hillier.)

A1. Bleuel, W. H.: "Management Science's Impact on Service Strategy," *Interfaces,* **5**(1, Part 2): 4–12, November 1975.
A2. Brigandi, A. J., D. R. Dargon, M. J. Sheehan, and T. Spencer III: "AT&T's Call Processing Simulator (CAPS) Operational Design for Inbound Call Centers," *Interfaces,* **24**(1): 6–28, January–February 1994.
A3. Burman, M., S. B. Gershwin, and C. Suyematsu: "Hewlett-Packard Uses Operations Research to Improve the Design of a Printer Production Line," *Interfaces,* **28**(1): 24–36, Jan.-Feb. 1998.
A4. Quinn, P., B. Andrews, and H. Parsons: "Allocating Telecommunications Resources at L.L. Bean, Inc.," *Interfaces,* **21**(1): 75–91, January–February 1991.
A5. Ramaswami, V., D. Poole, S. Ahn, S. Byers, and A. Kaplan: "Ensuring Access to Emergency Services in the Presence of Long Internet Dial-Up Calls," *Interfaces,* **35**(5): 411–422, September–October 2005.
A6. Samuelson, D. A.: "Predictive Dialing for Outbound Telephone Call Centers," *Interfaces,* **29**(5): 66–81, September–October 1999.
A7. Swersy, A. J., L. Goldring, and E. D. Geyer, Sr.: "Improving Fire Department Productivity: Merging Fire and Emergency Medical Units in New Haven," *Interfaces,* **23**(1): 109-129, January–February 1993.
A8. Vandaele, N. J., M. R. Lambrecht, N. De Schuyter, and R. Cremmery: "Spicer Off-Highway Products Division—Brugge Improves Its Lead-Time and Scheduling Performance," *Interfaces,* **30**(1): 83–95, January–February 2000.

■ LEARNING AIDS FOR THIS CHAPTER ON OUR WEBSITE (www.mhhe.com/hillier)

Worked Examples:

Examples for Chapter 17

An Interactive Procedure in IOR Tutorial:

Jackson Network

"Ch. 17—Queueing Theory" Excel Files:

Template for $M/M/s$ Model
Template for Finite Queue Variation of $M/M/s$ Model
Template for Finite Calling Population Variation of $M/M/s$ Model
Template for $M/G/1$ Model
Template for $M/D/1$ Model
Template for $M/E_k/1$ Model
Template for Nonpreemptive Priorities Model
Template for Preemptive Priorities Model
Template for $M/M/s$ Economic Analysis of Number of Servers

"Ch. 17—Queueing Theory" LINGO File for Selected Examples

Glossary for Chapter 17

See Appendix 1 for documentation of the software.

■ PROBLEMS[20]

To the left of each of the following problems (or their parts), we have inserted a T whenever one of the templates listed above can be helpful. An asterisk on the problem number indicates that at least a partial answer is given in the back of the book.

17.2-1.* Consider a typical barber shop. Demonstrate that it is a queueing system by describing its components.

17.2-2.* Newell and Jeff are the two barbers in a barber shop they own and operate. They provide two chairs for customers who are waiting to begin a haircut, so the number of customers in the shop varies between 0 and 4. For $n = 0, 1, 2, 3, 4$, the probability P_n that exactly n customers are in the shop is $P_0 = \frac{1}{16}$, $P_1 = \frac{4}{16}$, $P_2 = \frac{6}{16}$, $P_3 = \frac{4}{16}$, $P_4 = \frac{1}{16}$.
(a) Calculate L. How would you describe the meaning of L to Newell and Jeff?
(b) For each of the possible values of the number of customers in the queueing system, specify how many customers are in the queue. Then calculate L_q. How would you describe the meaning of L_q to Newell and Jeff?
(c) Determine the expected number of customers being served.
(d) Given that an average of 4 customers per hour arrive and stay to receive a haircut, determine W and W_q. Describe these two quantities in terms meaningful to Newell and Jeff.
(e) Given that Newell and Jeff are equally fast in giving haircuts, what is the average duration of a haircut?

17.2-3. Mom-and-Pop's Grocery Store has a small adjacent parking lot with three parking spaces reserved for the store's customers. During store hours, cars enter the lot and use one of the spaces at a mean rate of 2 per hour. For $n = 0, 1, 2, 3$, the probability P_n that exactly n spaces currently are being used is $P_0 = 0.1$, $P_1 = 0.2$, $P_2 = 0.4$, $P_3 = 0.3$.

(a) Describe how this parking lot can be interpreted as being a queueing system. In particular, identify the customers and the servers. What is the service being provided? What constitutes a service time? What is the queue capacity?
(b) Determine the basic measures of performance—L, L_q, W, and W_q—for this queueing system.
(c) Use the results from part (b) to determine the average length of time that a car remains in a parking space.

17.2-4. For each of the following statements about the queue in a queueing system, label the statement as true or false and then justify your answer by referring to a specific statement in the chapter.
(a) The queue is where customers wait in the queueing system until their service is completed.
(b) Queueing models conventionally assume that the queue can hold only a limited number of customers.
(c) The most common queue discipline is first-come-first-served.

17.2-5. Midtown Bank always has two tellers on duty. Customers arrive to receive service from a teller at a mean rate of 40 per hour. A teller requires an average of 2 minutes to serve a customer. When both tellers are busy, an arriving customer joins a single line to wait for service. Experience has shown that customers wait in line an average of 1 minute before service begins.
(a) Describe why this is a queueing system.
(b) Determine the basic measures of performance—W_q, W, L_q, and L—for this queueing system. (*Hint:* We don't know the probability distributions of interarrival times and service times for this queueing system, so you will need to use the relationships between these measures of performance to help answer the question.)

---

[20]See also the end of Chap. 26 (on the book's website) for additional problems involving the application of queueing theory.

17.2-6. Explain why the utilization factor ρ for the server in a single-server queueing system must equal $1 - P_0$, where P_0 is the probability of having 0 customers in the system.

17.2-7. You are given two queueing systems, Q_1 and Q_2. The mean arrival rate, the mean service rate per busy server, and the steady-state expected number of customers for Q_2 are twice the corresponding values for Q_1. Let W_i = the steady-state expected waiting time in the system for Q_i, for $i = 1, 2$. Determine W_2/W_1.

17.2-8. Consider a single-server queueing system with *any* service-time distribution and *any* distribution of interarrival times (the *GI/G/*1 model). Use only basic definitions and the relationships given in Sec. 17.2 to verify the following general relationships:
(a) $L = L_q + (1 - P_0)$.
(b) $L = L_q + \rho$.
(c) $P_0 = 1 - \rho$.

17.2-9. Show that

$$L = \sum_{n=0}^{s-1} nP_n + L_q + s\left(1 - \sum_{n=0}^{s-1} P_n\right)$$

by using the statistical definitions of L and L_q in terms of the P_n.

17.3-1. Identify the customers and the servers in the queueing system in each of the following situations:
(a) The checkout stand in a grocery store.
(b) A fire station.
(c) The toll booth for a bridge.
(d) A bicycle repair shop.
(e) A shipping dock.
(f) A group of semiautomatic machines assigned to one operator.
(g) The materials-handling equipment in a factory area.
(h) A plumbing shop.
(i) A job shop producing custom orders.
(j) A secretarial typing pool.

17.4-1. Suppose that a queueing system has two servers, an exponential interarrival time distribution with a mean of 2 hours, and an exponential service-time distribution with a mean of 2 hours for each server. Furthermore, a customer has just arrived at 12:00 noon.
(a) What is the probability that the next arrival will come (i) before 1:00 P.M., (ii) between 1:00 and 2:00 P.M., and (iii) after 2:00 P.M.?
(b) Suppose that no additional customers arrive before 1:00 P.M. Now what is the probability that the next arrival will come between 1:00 and 2:00 P.M.?
(c) What is the probability that the number of arrivals between 1:00 and 2:00 P.M. will be (i) 0, (ii) 1, and (iii) 2 or more?
(d) Suppose that both servers are serving customers at 1:00 P.M. What is the probability that *neither* customer will have service completed (i) before 2:00 P.M., (ii) before 1:10 P.M., and (iii) before 1:01 P.M.?

17.4-2.* The jobs to be performed on a particular machine arrive according to a *Poisson* input process with a mean rate of two per hour. Suppose that the machine breaks down and will require 1 hour to be repaired. What is the probability that the number of new jobs that will arrive during this time is (a) 0, (b) 2, and (c) 5 or more?

17.4-3. The time required by a mechanic to repair a machine has an exponential distribution with a mean of 4 hours. However, a special tool would reduce this mean to 2 hours. If the mechanic repairs a machine in less than 2 hours, he is paid $100; otherwise, he is paid $80. Determine the mechanic's expected increase in pay per machine repaired if he uses the special tool.

17.4-4. A three-server queueing system has a controlled arrival process that provides customers in time to keep the servers continuously busy. Service times have an exponential distribution with mean 0.5.
  You observe the queueing system starting up with all three servers beginning service at time $t = 0$. You then note that the first completion occurs at time $t = 1$. Given this information, determine the expected amount of time after $t = 1$ until the next service completion occurs.

17.4-5. A queueing system has three servers with expected service times of 30 minutes, 20 minutes, and 15 minutes. The service times have an exponential distribution. Each server has been busy with a current customer for 10 minutes. Determine the expected remaining time until the next service completion.

17.4-6. Consider a queueing system with two types of customers. Type 1 customers arrive according to a Poisson process with a mean rate of 5 per hour. Type 2 customers also arrive according to a Poisson process with a mean rate of 5 per hour. The system has two servers, both of which serve both types of customers. For both types, service times have an exponential distribution with a mean of 10 minutes. Service is provided on a first-come-first-served basis.
(a) What is the probability distribution (including its mean) of the time between consecutive arrivals of customers of any type?
(b) When a particular type 2 customer arrives, she finds two type 1 customers there in the process of being served but no other customers in the system. What is the probability distribution (including its mean) of this type 2 customer's waiting time in the queue?

17.4-7. Consider a two-server queueing system where all service times are independent and identically distributed according to an exponential distribution with a mean of 10 minutes. Service is provided on a first-come-first-served basis. When a particular customer arrives, he finds that both servers are busy and no one is waiting in the queue.
(a) What is the probability distribution (including its mean and standard deviation) of this customer's waiting time in the queue?
(b) Determine the expected value and standard deviation of this customer's waiting time in the system.
(c) Suppose that this customer still is waiting in the queue 5 minutes after its arrival. Given this information, how does this change the expected value and the standard deviation of this customer's total waiting time in the system from the answers obtained in part (b)?

17.4-8. For each of the following statements regarding service times modeled by the exponential distribution, label the statement as true or false and then justify your answer by referring to specific statements in the chapter.

(a) The expected value and variance of the service times are always equal.

(b) The exponential distribution always provides a good approximation of the actual service-time distribution when each customer requires the same service operations.

(c) At an s-server facility, $s > 1$, with exactly s customers already in the system, a new arrival would have an expected waiting time before entering service of $1/\mu$ time units, where μ is the mean service rate for each busy server.

17.4-9. As for Property 3 of the exponential distribution, let T_1, $T_2, \ldots, T_n$ be independent exponential random variables with parameters $\alpha_1, \alpha_2, \ldots, \alpha_n$, respectively, and let $U = \min\{T_1, T_2, \ldots, T_n\}$. Show that the probability that a particular random variable T_j will turn out to be smallest of the n random variables is

$$P\{T_j = U\} = \alpha_j \bigg/ \sum_{i=1}^{n} \alpha_i, \quad \text{for } j = 1, 2, \ldots, n.$$

(*Hint*: $P\{T_j = U\} = \int_0^\infty P\{T_i > T_j \text{ for all } i \neq j \mid T_j = t\} \alpha_j e^{-\alpha_j t} dt$.)

17.5-1. Consider the birth-and-death process with all $\mu_n = 2$ ($n = 1, 2, \ldots$), $\lambda_0 = 3$, $\lambda_1 = 2$, $\lambda_2 = 1$, and $\lambda_n = 0$ for $n = 3, 4, \ldots$.

(a) Display the rate diagram.

(b) Calculate P_0, P_1, P_2, P_3, and P_n for $n = 4, 5, \ldots$.

(c) Calculate L, L_q, W, and W_q.

17.5-2. Consider a birth-and-death process with just three attainable states (0, 1, and 2), for which the steady-state probabilities are P_0, P_1, and P_2, respectively. The birth-and-death rates are summarized in the following table:

State	Birth Rate	Death Rate
0	4	—
1	2	4
2	0	6

(a) Construct the rate diagram for this birth-and-death process.

(b) Develop the balance equations.

(c) Solve these equations to find P_0, P_1, and P_2.

(d) Use the general formulas for the birth-and-death process to calculate P_0, P_1, and P_2. Also calculate L, L_q, W, and W_q.

17.5-3. Consider the birth-and-death process with the following mean rates. The birth rates are $\lambda_0 = 2$, $\lambda_1 = 3$, $\lambda_2 = 2$, $\lambda_3 = 1$, and $\lambda_n = 0$ for $n > 3$. The death rates are $\mu_1 = 3$, $\mu_2 = 4$, $\mu_3 = 1$, and $\mu_n = 2$ for $n > 4$.

(a) Construct the rate diagram for this birth-and-death process.

(b) Develop the balance equations.

(c) Solve these equations to find the steady-state probability distribution P_0, $P_1, \ldots$.

(d) Use the general formulas for the birth-and-death process to calculate P_0, $P_1, \ldots$. Also calculate L, L_q, W, and W_q.

17.5-4. Consider the birth-and-death process with all $\lambda_n = 2$ ($n = 0, 1, \ldots$), $\mu_1 = 2$, and $\mu_n = 4$ for $n = 2, 3, \ldots$.

(a) Display the rate diagram.

(b) Calculate P_0 and P_1. Then give a general expression for P_n in terms of P_0 for $n = 2, 3, \ldots$.

(c) Consider a queueing system with two servers that fits this process. What is the mean arrival rate for this queueing system? What is the mean service rate for each server when it is busy serving customers?

17.5-5.* A service station has one gasoline pump. Cars wanting gasoline arrive according to a Poisson process at a mean rate of 15 per hour. However, if the pump already is being used, these potential customers may *balk* (drive on to another service station). In particular, if there are n cars already at the service station, the probability that an arriving potential customer will balk is $n/3$ for $n = 1$, 2, 3. The time required to service a car has an exponential distribution with a mean of 4 minutes.

(a) Construct the rate diagram for this queueing system.

(b) Develop the balance equations.

(c) Solve these equations to find the steady-state probability distribution of the number of cars at the station. Verify that this solution is the same as that given by the general solution for the birth-and-death process.

(d) Find the expected waiting time (including service) for those cars that stay.

17.5-6. A maintenance person has the job of keeping two machines in working order. The amount of time that a machine works before breaking down has an exponential distribution with a mean of 10 hours. The time then spent by the maintenance person to repair the machine has an exponential distribution with a mean of 8 hours.

(a) Show that this process fits the birth-and-death process by defining the states, specifying the values of the λ_n and μ_n, and then constructing the rate diagram.

(b) Calculate the P_n.

(c) Calculate L, L_q, W, and W_q.

(d) Determine the proportion of time that the maintenance person is busy.

(e) Determine the proportion of time that any given machine is working.

(f) Refer to the nearly identical example of a *continuous time Markov chain* given at the end of Sec. 16.8. Describe the relationship between continuous time Markov chains and the birth-and-death process that enables both to be applied to this same problem.

17.5-7. Consider a single-server queueing system where interarrival times have an exponential distribution with parameter λ and service times have an exponential distribution with parameter μ. In addition, customers *renege* (leave the queueing system without being served) if their waiting time in the queue grows too large. In particular, assume that the time each customer is willing to wait in the queue before reneging has an exponential distribution with a mean of $1/\theta$.

(a) Construct the rate diagram for this queueing system.

(b) Develop the balance equations.

17.5-8.* A certain small grocery store has a single checkout stand with a full-time cashier. Customers arrive at the stand "randomly" (i.e., a Poisson input process) at a mean rate of 30 per hour. When there is only one customer at the stand, she is processed by the cashier alone, with an expected service time of 1.5 minutes. However, the stock boy has been given standard instructions that whenever there is more than one customer at the stand, he is to help the cashier by bagging the groceries. This help reduces the expected time required to process a customer to 1 minute. In both cases, the service-time distribution is exponential.

(a) Construct the rate diagram for this queueing system.

(b) What is the steady-state probability distribution of the number of customers at the checkout stand?

(c) Derive L for this system. (*Hint:* Refer to the derivation of L for the $M/M/1$ model at the beginning of Sec. 17.6.) Use this information to determine L_q, W, and W_q.

17.5-9. A department has one word-processing operator. Documents produced in the department are delivered for word processing according to a Poisson process with an expected interarrival time of 30 minutes. When the operator has just one document to process, the expected processing time is 20 minutes. When she has more than one document, then editing assistance that is available reduces the expected processing time for each document to 15 minutes. In both cases, the processing times have an exponential distribution.

(a) Construct the rate diagram for this queueing system.

(b) Find the steady-state distribution of the number of documents that the operator has received but not yet completed.

(c) Derive L for this system. (*Hint:* Refer to the derivation of L for the $M/M/1$ model at the beginning of Sec. 17.6.) Use this information to determine L_q, W, and W_q.

17.5-10. Customers arrive at a queueing system according to a Poisson process with a mean arrival rate of 2 customers per minute. The service time has an exponential distribution with a mean of 1 minute. An unlimited number of servers are available as needed so customers never wait for service to begin. Calculate the steady-state probability that exactly 1 customer is in the system.

17.5-11. Suppose that a single-server queueing system fits all the assumptions of the birth-and-death process *except* that customers always arrive in *pairs*. The mean arrival rate is 2 pairs per hour (4 customers per hour) and the mean service rate (when the server is busy) is 5 customers per hour.

(a) Construct the rate diagram for this queueing system.

(b) Develop the balance equations.

(c) For comparison purposes, display the rate diagram for the corresponding queueing system that completely fits the birth-and-death process, i.e., where customers arrive *individually* at a mean rate of 4 per hour.

17.5-12. Consider a single-server queueing system with a finite queue that can hold a maximum of 2 customers *excluding* any being served. The server can provide *batch service* to 2 customers simultaneously, where the service time has an exponential distribution with a mean of 1 unit of time regardless of the number being served. Whenever the queue is not full, customers arrive individually according to a Poisson process at a mean rate of 1 per unit of time.

(a) Assume that the server *must* serve 2 customers simultaneously. Thus, if the server is idle when only 1 customer is in the system, the server must wait for another arrival before beginning service. Formulate the queueing model as a continuous time Markov chain by defining the states and then constructing the rate diagram. Give the balance equations, but do not solve further.

(b) Now assume that the batch size for a service is 2 only if 2 customers are in the queue when the server finishes the preceding service. Thus, if the server is idle when only 1 customer is in the system, the server must serve this single customer, and any subsequent arrivals must wait in the queue until service is completed for this customer. Formulate the resulting queueing model as a continuous time Markov chain by defining the states and then constructing the rate diagram. Give the balance equations, but do not solve further.

17.5-13. Consider a queueing system that has two classes of customers, two clerks providing service, and *no queue*. Potential customers from each class arrive according to a Poisson process, with a mean arrival rate of 10 customers per hour for class 1 and 5 customers per hour for class 2, but these arrivals are lost to the system if they cannot immediately enter service.

Each customer of class 1 that enters the system will receive service from either one of the clerks that is free, where the service times have an exponential distribution with a mean of 5 minutes.

Each customer of class 2 that enters the system requires the *simultaneous use of both clerks* (the two clerks work together as a single server), where the service times have an exponential distribution with a mean of 5 minutes. Thus, an arriving customer of this kind would be lost to the system unless both clerks are free to begin service immediately.

(a) Formulate the queueing model as a continuous time Markov chain by defining the states and constructing the rate diagram.

(b) Now describe how the formulation in part (*a*) can be fitted into the format of the birth-and-death process.

(c) Use the results for the birth-and-death process to calculate the steady-state joint distribution of the number of customers of each class in the system.

(d) For each of the two classes of customers, what is the expected fraction of arrivals who are unable to enter the system?

17.6-1. Read the referenced article that fully describes the OR study summarized in the application vignette presented in Sec. 17.6. Briefly describe how queueing theory was applied in this study. Then list the various financial and nonfinancial benefits that resulted from this study.

17.6-2.* The 4M Company has a single turret lathe as a key work center on its factory floor. Jobs arrive at this work center according to a Poisson process at a mean rate of 2 per day. The processing time to perform each job has an exponential distribution with a mean of $\frac{1}{4}$ day. Because the jobs are bulky, those not being worked on are currently being stored in a room some distance from the machine. However, to save time in fetching the jobs, the production manager is proposing to add enough in-process storage space next to the turret lathe to accommodate 3 jobs in addition to the one being processed. (Excess jobs will continue to be stored temporarily in the distant room.) Under this proposal, what proportion of the time will this storage space next to the turret lathe be adequate to accommodate all waiting jobs?
(a) Use available formulas to calculate your answer.
T (b) Use the corresponding Excel template to obtain the probabilities needed to answer the question.

17.6-3. Customers arrive at a single-server queueing system according to a Poisson process at a mean rate of 30 per hour. If the server works continuously, the number of customers that can be served in an hour has a Poisson distribution with a mean of 50. Determine the proportion of time during which no one is waiting to be served.

17.6-4. Consider the $M/M/1$ model, with $\lambda < \mu$.
(a) Determine the steady-state probability that a customer's actual waiting time in the system is longer than the expected waiting time in the system, i.e., $P\{\mathcal{W} > W\}$.
(b) Determine the steady-state probability that a customer's actual waiting time in the queue is longer than the expected waiting time in the queue, i.e., $P\{\mathcal{W}_q > W_q\}$.

17.6-5. Verify the following relationships for an $M/M/1$ queueing system:

$$\lambda = \frac{(1 - P_0)^2}{W_q P_0}, \qquad \mu = \frac{1 - P_0}{W_q P_0}.$$

17.6-6. It is necessary to determine how much in-process storage space to allocate to a particular work center in a new factory. Jobs arrive at this work center according to a Poisson process with a mean rate of 4 per hour, and the time required to perform the necessary work has an exponential distribution with a mean of 0.2 hour. Whenever the waiting jobs require more in-process storage space than has been allocated, the excess jobs are stored temporarily in a less convenient location. If each job requires 1 square foot of floor space while it is in in-process storage at the work center, how much space must be provided to accommodate all waiting jobs (a) 50 percent of the time, (b) 90 percent of the time, and (c) 99 percent of the time? Derive an analytical expression to answer these three questions. *Hint:* The sum of a geometric series is

$$\sum_{n=0}^{N} x^n = \frac{1 - x^{N+1}}{1 - x}.$$

17.6-7. Consider the following statements about an $M/M/1$ queueing system and its utilization factor ρ. Label each of the statements as true or false, and then justify your answer.
(a) The probability that a customer has to wait before service begins is proportional to ρ.

(b) The expected number of customers in the system is proportional to ρ.
(c) If ρ has been increased from $\rho = 0.9$ to $\rho = 0.99$, the effect of any further increase in ρ on L, L_q, W, and W_q will be relatively small as long as $\rho < 1$.

17.6-8. Customers arrive at a single-server queueing system in accordance with a Poisson process with an expected interarrival time of 25 minutes. Service times have an exponential distribution with a mean of 30 minutes.

Label each of the following statements about this system as true or false, and then justify your answer.
(a) The server definitely will be busy forever after the first customer arrives.
(b) The queue will grow without bound.
(c) If a second server with the same service-time distribution is added, the system can reach a steady-state condition.

17.6-9. For each of the following statements about an $M/M/1$ queueing system, label the statement as true or false and then justify your answer by referring to specific statements in the chapter.
(a) The waiting time in the system has an exponential distribution.
(b) The waiting time in the queue has an exponential distribution.
(c) The conditional waiting time in the system, given the number of customers already in the system, has an Erlang (gamma) distribution.

17.6-10. The Friendly Neighbor Grocery Store has a single checkout stand with a full-time cashier. Customers arrive randomly at the stand at a mean rate of 20 per hour. The service-time distribution is exponential, with a mean of 2 minutes. This situation has resulted in occasional long lines and complaints from customers. Therefore, because there is no room for a second checkout stand, the manager is considering the alternative of hiring another person to help the cashier by bagging the groceries. This help would reduce the expected time required to process a customer to 1.5 minutes, but the distribution still would be exponential.

The manager would like to have the percentage of time that there are more than two customers at the checkout stand down below 25 percent. She also would like to have no more than 5 percent of the customers needing to wait at least 5 minutes before beginning service, or at least 7 minutes before finishing service.
(a) Use the formulas for the $M/M/1$ model to calculate L, W, W_q, L_q, P_0, P_1, and P_2 for the current mode of operation. What is the probability of having more than two customers at the checkout stand?
T (b) Use the Excel template for this model to check your answers in part (a). Also find the probability that the waiting time before beginning service exceeds 5 minutes, and the probability that the waiting time before finishing service exceeds 7 minutes.
(c) Repeat part (a) for the alternative being considered by the manager.
(d) Repeat part (b) for this alternative.
(e) Which approach should the manager use to satisfy her criteria as closely as possible?

T **17.6-11.** The Centerville International Airport has two runways, one used exclusively for takeoffs and the other exclusively for landings. Airplanes arrive in the Centerville air space to request landing instructions according to a Poisson process at a mean rate of 10 per hour. The time required for an airplane to land after receiving clearance to do so has an exponential distribution with a mean of 3 minutes, and this process must be completed before giving clearance to do so to another airplane. Airplanes awaiting clearance must circle the airport.

The Federal Aviation Administration has a number of criteria regarding the safe level of congestion of airplanes waiting to land. These criteria depend on a number of factors regarding the airport involved, such as the number of runways available for landing. For Centerville, the criteria are (1) the average number of airplanes waiting to receive clearance to land should not exceed 1, (2) 95 percent of the time, the actual number of airplanes waiting to receive clearance to land should not exceed 4, (3) for 99 percent of the airplanes, the amount of time spent circling the airport before receiving clearance to land should not exceed 30 minutes (since exceeding this amount of time often would require rerouting the plane to another airport for an emergency landing before its fuel runs out).

(a) Evaluate how well these criteria are currently being satisfied.

(b) A major airline is considering adding this airport as one of its hubs. This would increase the mean arrival rate to 15 airplanes per hour. Evaluate how well the above criteria would be satisfied if this happens.

(c) To attract additional business [including the major airline mentioned in part (b)], airport management is considering adding a second runway for landings. It is estimated that this eventually would increase the mean arrival rate to 25 airplanes per hour. Evaluate how well the above criteria would be satisfied if this happens.

T **17.6-12.** The Security & Trust Bank employs 4 tellers to serve its customers. Customers arrive according to a Poisson process at a mean rate of 2 per minute. However, business is growing and management projects that the mean arrival rate will be 3 per minute a year from now. The transaction time between the teller and customer has an exponential distribution with a mean of 1 minute.

Management has established the following guidelines for a satisfactory level of service to customers. The average number of customers waiting in line to begin service should not exceed 1. At least 95 percent of the time, the number of customers waiting in line should not exceed 5. For at least 95 percent of the customers, the time spent in line waiting to begin service should not exceed 5 minutes.

(a) Use the $M/M/s$ model to determine how well these guidelines are currently being satisfied.

(b) Evaluate how well the guidelines will be satisfied a year from now if no change is made in the number of tellers.

(c) Determine how many tellers will be needed a year from now to completely satisfy these guidelines.

17.6-13. Consider the $M/M/s$ model.

T **(a)** Suppose there is one server and the expected service time is exactly 1 minute. Compare L for the cases where the mean arrival rate is 0.5, 0.9, and 0.99 customers per minute, respectively. Do the same for L_q, W, W_q, and $P\{W > 5\}$. What

conclusions do you draw about the impact of increasing the utilization factor ρ from small values (e.g., $\rho = 0.5$) to fairly large values (e.g., $\rho = 0.9$) and then to even larger values very close to 1 (e.g., $\rho = 0.99$)?

(b) Now suppose there are two servers and the expected service time is exactly 2 minutes. Follow the instructions for part (a).

T **17.6-14.** Consider the $M/M/s$ model with a mean arrival rate of 10 customers per hour and an expected service time of 5 minutes. Use the Excel template for this model to obtain and print out the various measures of performance (with $t = 10$ and $t = 0$, respectively, for the two waiting time probabilities) when the number of servers is 1, 2, 3, 4, and 5. Then, for each of the following possible criteria for a satisfactory level of service (where the unit of time is 1 minute), use the printed results to determine how many servers are needed to satisfy this criterion.

(a) $L_q \leq 0.25$

(b) $L \leq 0.9$

(c) $W_q \leq 0.1$

(d) $W \leq 6$

(e) $P\{W_q > 0\} \leq 0.01$

(f) $P\{W > 10\} \leq 0.2$

(g) $\sum_{n=0}^{s} P_n \geq 0.95$

17.6-15. A gas station with only one gas pump employs the following policy: If a customer has to wait, the price is $3.50 per gallon; if she does not have to wait, the price is $4.00 per gallon. Customers arrive according to a Poisson process with a mean rate of 20 per hour. Service times at the pump have an exponential distribution with a mean of 2 minutes. Arriving customers always wait until they can eventually buy gasoline. Determine the expected price of gasoline per gallon.

17.6-16. You are given an $M/M/1$ queueing system with mean arrival rate λ and mean service rate μ. An arriving customer receives n dollars if n customers are already in the system. Determine the expected cost in dollars per customer.

17.6-17. Section 17.6 gives the following equations for the $M/M/1$ model:

(1) $P\{W > t\} = \sum_{n=0}^{\infty} P_n P\{S_{n+1} > t\}.$

(2) $P\{W > t\} = e^{-\mu(1-\rho)t}.$

Show that Eq. (1) reduces algebraically to Eq. (2). (*Hint:* Use differentiation, algebra, and integration.)

17.6-18. Derive W_q directly for the following cases by developing and reducing an expression analogous to Eq. (1) in Prob. 17.6-17. (*Hint:* Use the *conditional* expected waiting time in the queue given that a random arrival finds n customers already in the system.)

(a) The $M/M/1$ model

(b) The $M/M/s$ model

T **17.6-19.** Consider an $M/M/2$ queueing system with $\lambda = 3$ and $\mu = 2$. Determine the mean rate at which service completions occur during the periods when no customers are waiting in the queue.

T **17.6-20.** You are given an $M/M/2$ queueing system with $\lambda = 4$ per hour and $\mu = 6$ per hour. Determine the probability that an arriving customer will wait more than 30 minutes in the queue, given that at least 2 customers are already in the system.

17.6-21.* In the Blue Chip Life Insurance Company, the deposit and withdrawal functions associated with a certain investment product are separated between two clerks, Clara and Clarence. Deposit slips arrive randomly (a Poisson process) at Clara's desk at a mean rate of 16 per hour. Withdrawal slips arrive randomly (a Poisson process) at Clarence's desk at a mean rate of 14 per hour. The time required to process either transaction has an exponential distribution with a mean of 3 minutes. To reduce the expected waiting time in the system for both deposit slips and withdrawal slips, the actuarial department has made the following recommendations: (1) Train each clerk to handle both deposits and withdrawals, and (2) put both deposit and withdrawal slips into a single queue that is accessed by both clerks.

(a) Determine the expected waiting time in the system under current procedures for each type of slip. Then combine these results to calculate the expected waiting time in the system for a random arrival of either type of slip.

T **(b)** If the recommendations are adopted, determine the expected waiting time in the system for arriving slips.

T **(c)** Now suppose that adopting the recommendations would result in a slight increase in the expected processing time. Use the Excel template for the $M/M/s$ model to determine by trial and error the expected processing time (within 0.001 hour) that would cause the expected waiting time in the system for a random arrival to be essentially the same under current procedures and under the recommendations.

17.6-22. People's Software Company has just set up a call center to provide technical assistance on its new software package. Two technical representatives are taking the calls, where the time required by either representative to answer a customer's questions has an exponential distribution with a mean of 8 minutes. Calls are arriving according to a Poisson process at a mean rate of 10 per hour.

By next year, the mean arrival rate of calls is expected to decline to 5 per hour, so the plan is to reduce the number of technical representatives to one then.

T **(a)** Assuming that μ will continue to be 7.5 calls per hour for next year's queueing system, determine L, L_q, W, and W_q for both the current system and next year's system. For each of these four measures of performance, which system yields the smaller value?

(b) Now assume that μ will be adjustable when the number of technical representatives is reduced to one. Solve algebraically for the value of μ that would yield the same value of W as for the current system.

(c) Repeat part (b) with W_q instead of W.

17.6-23. Consider a generalization of the $M/M/1$ model where the server needs to "warm up" at the beginning of a busy period, and so serves the first customer of a busy period at a slower rate than other customers. In particular, if an arriving customer finds the server idle, the customer experiences a service time that has an exponential distribution with parameter μ_1. However, if an arriving customer finds the server busy, that customer joins the queue and subsequently experiences a service time that has an exponential distribution with parameter μ_2, where $\mu_1 < \mu_2$. Customers arrive according to a Poisson process with mean rate λ.

(a) Formulate this model as a continuous time Markov chain by defining the states and constructing the rate diagram accordingly.

(b) Develop the balance equations.

(c) Suppose that numerical values are specified for μ_1, μ_2, and λ, and that $\lambda < \mu_2$ (so that a steady-state distribution exists). Since this model has an infinite number of states, the steady-state distribution is the simultaneous solution of an infinite number of balance equations (plus the equation specifying that the sum of the probabilities equals 1). Suppose that you are unable to obtain this solution analytically, so you wish to use a computer to solve the model numerically. Considering that it is impossible to solve an infinite number of equations numerically, briefly describe what still can be done with these equations to obtain an approximation of the steady-state distribution. Under what circumstances will this approximation be essentially exact?

(d) Given that the steady-state distribution has been obtained, give explicit expressions for calculating L, L_q, W, and W_q.

(e) Given this steady-state distribution, develop an expression for $P\{\mathcal{W} > t\}$ that is analogous to Eq. (1) in Prob. 17.6-17.

17.6-24. For each of the following models, write the balance equations and show that they are satisfied by the solution given in Sec. 17.6 for the steady-state distribution of the number of customers in the system.

(a) The $M/M/1$ model.

(b) The finite queue variation of the $M/M/1$ model, with $K = 2$.

(c) The finite calling population variation of the $M/M/1$ model, with $N = 2$.

T **17.6-25.** Consider a telephone system with three lines. Calls arrive according to a Poisson process at a mean rate of 6 per hour. The duration of each call has an exponential distribution with a mean of 15 minutes. If all lines are busy, calls will be put on hold until a line becomes available.

(a) Print out the measures of performance provided by the Excel template for this queueing system (with $t = 1$ hour and $t = 0$, respectively, for the two waiting time probabilities).

(b) Use the printed result giving $P\{\mathcal{W}_q > 0\}$ to identify the steady-state probability that a call will be answered immediately (not put on hold). Then verify this probability by using the printed results for the P_n.

(c) Use the printed results to identify the steady-state probability distribution of the number of calls on hold.

(d) Print out the new measures of performance if arriving calls are lost whenever all lines are busy. Use these results to identify the steady-state probability that an arriving call is lost.

17.6-26.* Janet is planning to open a small car-wash operation, and she must decide how much space to provide for waiting cars. Janet estimates that customers would arrive randomly (i.e., a Poisson input process) with a mean rate of 1 every 4 minutes, unless the waiting area is full, in which case the arriving customers would take their cars elsewhere. The time that can be attributed to washing one car has an exponential distribution with a mean of 3 minutes. Compare the expected fraction of potential customers that will be *lost* because of inadequate waiting space if (a) 0 spaces (not including the car being washed), (b) 2 spaces, and (c) 4 spaces were provided.

17.6-27. Consider the finite queue variation of the *M/M/s* model. Derive the expression for L_q given in Sec. 17.6 for this model.

17.6-28. For the finite queue variation of the *M/M/1* model, develop an expression analogous to Eq. (1) in Prob. 17.6-17 for the following probabilities:
(a) $P\{W > t\}$.
(b) $P\{W_q > t\}$.
[*Hint:* Arrivals can occur only when the system is not full, so the probability that a random arrival finds n customers already there is $P_n/(1 - P_K)$.]

17.6-29. George is planning to open a drive-through photo-developing booth with a single service window that will be open approximately 200 hours per month in a busy commercial area. Space for a drive-through lane is available for a rental of $200 per month per car length. George needs to decide how many car lengths of space to provide for his customers.

Excluding this rental cost for the drive-through lane, George believes that he will average a profit of $4 per customer served (nothing for a drop off of film and $8 when the photographs are picked up). He also estimates that customers will arrive randomly (a Poisson process) at a mean rate of 20 per hour, although those who find the drive-through lane full will be forced to leave. Half of the customers who find the drive-through lane full wanted to drop off film, and the other half wanted to pick up their photographs. The half who wanted to drop off film will take their business elsewhere instead. The other half of the customers who find the drive-through lane full will not be lost because they will keep trying later until they can get in and pick up their photographs. George assumes that the time required to serve a customer will have an exponential distribution with a mean of 2 minutes.
T (a) Find L and the mean rate at which customers are lost when the number of car lengths of space provided is 2, 3, 4, and 5.
(b) Calculate W from L for the cases considered in part (a).
(c) Use the results from part (a) to calculate the decrease in the mean rate at which customers are lost when the number of car lengths of space provided is increased from 2 to 3, from 3 to 4, and from 4 to 5. Then calculate the increase in expected profit per hour (excluding space rental costs) for each of these three cases.
(d) Compare the increases in expected profit found in part (c) with the cost per hour of renting each car length of space. What conclusion do you draw about the number of car lengths of space that George should provide?

17.6-30. At the Forrester Manufacturing Company, one repair technician has been assigned the responsibility of maintaining three machines.

For each machine, the probability distribution of the running time before a breakdown is exponential, with a mean of 9 hours. The repair time also has an exponential distribution, with a mean of 2 hours.
(a) Which queueing model fits this queueing system?
T (b) Use this queueing model to find the probability distribution of the number of machines not running, and the mean of this distribution.
(c) Use this mean to calculate the expected time between a machine breakdown and the completion of the repair of that machine.
(d) What is the expected fraction of time that the repair technician will be busy?
T (e) As a crude approximation, assume that the calling population is infinite and that machine breakdowns occur randomly at a mean rate of 3 every 9 hours. Compare the result from part (b) with that obtained by making this approximation while using (i) the *M/M/s* model and (ii) the finite queue variation of the *M/M/s* model with $K = 3$.
T (f) Repeat part (b) when a second repair technician is made available to repair a second machine whenever more than one of these three machines require repair.

17.6-31. Reconsider the specific birth-and-death process described in Prob. 17.5-1.
(a) Identify a queueing model (and its parameter values) in Sec. 17.6 that fits this process.
T (b) Use the corresponding Excel template to obtain the answers for parts (b) and (c) of Prob. 17.5-1.

T **17.6-32.*** The Dolomite Corporation is making plans for a new factory. One department has been allocated 12 semiautomatic machines. A small number (yet to be determined) of operators will be hired to provide the machines the needed occasional servicing (loading, unloading, adjusting, setup, and so on). A decision now needs to be made on how to organize the operators to do this. Alternative 1 is to assign each operator to her own machines. Alternative 2 is to pool the operators so that any idle operator can take the next machine needing servicing. Alternative 3 is to combine the operators into a single crew that will work together on any machine needing servicing.

The running time (time between completing service and the machine's requiring service again) of each machine is expected to have an exponential distribution, with a mean of 150 minutes. The service time is assumed to have an exponential distribution, with a mean of 15 minutes (for Alternatives 1 and 2) or 15 minutes divided by the number of operators in the crew (for Alternative 3). For the department to achieve the required production rate, the machines must be running at least 89 percent of the time on average.
(a) For Alternative 1, what is the maximum number of machines that can be assigned to an operator while still achieving the required production rate? What is the resulting utilization of each operator?
(b) For Alternative 2, what is the minimum number of operators needed to achieve the required production rate? What is the resulting utilization of the operators?
(c) For Alternative 3, what is the minimum size of the crew needed to achieve the required production rate? What is the resulting utilization of the crew?

17.6-33. A shop contains three identical machines that are subject to a failure of a certain kind. Therefore, a maintenance system is provided to perform the maintenance operation (recharging) required by a failed machine. The time required by each operation has an exponential distribution with a mean of 30 minutes. However, with probability $\frac{1}{3}$, the operation must be performed a second time (with the same distribution of time) in order to bring the failed machine back to a satisfactory operational state. The maintenance system works on only one failed machine at a time, performing all the operations (one or two) required by that machine, on a first-come-first-served basis. After a machine is repaired, the time until its next failure has an exponential distribution with a mean of 3 hours.

(a) How should the states of the system be defined in order to formulate this queueing system as a continuous time Markov chain? (*Hint:* Given that a first operation is being performed on a failed machine, completing this operation *successfully* and completing it *unsuccessfully* are two separate events of interest. Then use Property 6 regarding disaggregation for the exponential distribution.)

(b) Construct the corresponding rate diagram.

(c) Develop the balance equations.

17.7-1.* Consider the $M/G/1$ model.

(a) Compare the expected waiting time in the queue if the service-time distribution is (i) exponential, (ii) constant, (iii) Erlang with the amount of variation (i.e., the standard deviation) halfway between the constant and exponential cases.

(b) What is the effect on the expected waiting time in the queue and on the expected queue length if both λ and μ are doubled and the scale of the service-time distribution is changed accordingly?

17.7-2. Consider the $M/G/1$ model with $\lambda = 0.2$ and $\mu = 0.25$.

T **(a)** Use the Excel template for this model (or hand calculations) to find the main measures of performance—L, L_q, W, W_q—for each of the following values of σ: 4, 3, 2, 1, 0.

(b) What is the ratio of L_q with $\sigma = 4$ to L_q with $\sigma = 0$? What does this say about the importance of reducing the variability of the service times?

(c) Calculate the reduction in L_q when σ is reduced from 4 to 3, from 3 to 2, from 2 to 1, and from 1 to 0. Which is the largest reduction? Which is the smallest?

(d) Use trial and error with the template to see approximately how much μ would need to be increased with $\sigma = 4$ to achieve the same L_q as with $\mu = 0.25$ and $\sigma = 0$.

17.7-3. Consider the following statements about an $M/G/1$ queueing system, where σ^2 is the variance of service times. Label each statement as true or false, and then justify your answer.

(a) Increasing σ^2 (with fixed λ and μ) will increase L_q and L, but will not change W_q and W.

(b) When choosing between a tortoise (small μ and σ^2) and a hare (large μ and σ^2) to be the server, the tortoise always wins by providing a smaller L_q.

(c) With λ and μ fixed, the value of L_q with an exponential service-time distribution is twice as large as with constant service times.

(d) Among all possible service-time distributions (with λ and μ fixed), the exponential distribution yields the largest value of L_q.

17.7-4. Marsha operates an expresso stand. Customers arrive according to a Poisson process at a mean rate of 25 per hour. The time needed by Marsha to serve a customer has an exponential distribution with a mean of 90 seconds.

(a) Use the $M/G/1$ model to find L, L_q, W, and W_q.

(b) Suppose Marsha is replaced by an expresso vending machine that requires exactly 90 seconds for each customer to operate. Find L, L_q, W, and W_q.

(c) What is the ratio of L_q in part (*b*) to L_q in part (*a*)?

T **(d)** Use trial and error with the Excel template for the $M/G/1$ model to see approximately how much Marsha would need to reduce her expected service time to achieve the same L_q as with the expresso vending machine.

17.7-5. Antonio runs a shoe repair store by himself. Customers arrive to bring a pair of shoes to be repaired according to a Poisson process at a mean rate of 1 per hour. The time Antonio requires to repair each individual shoe has an exponential distribution with a mean of 15 minutes.

(a) Consider the formulation of this queueing system where the individual shoes (not pairs of shoes) are considered to be the customers. For this formulation, construct the rate diagram and develop the balance equations, but do not solve further.

(b) Now consider the formulation of this queueing system where the pairs of shoes are considered to be the customers. Identify the specific queueing model that fits this formulation.

(c) Calculate the expected number of pairs of shoes in the shop.

(d) Calculate the expected amount of time from when a customer drops off a pair of shoes until they are repaired and ready to be picked up.

T **(e)** Use the corresponding Excel template to check your answers in parts (*c*) and (*d*).

17.7-6.* The maintenance base for Friendly Skies Airline has facilities for overhauling only one airplane engine at a time. Therefore, to return the airplanes to use as soon as possible, the policy has been to stagger the overhauling of the four engines of each airplane. In other words, only one engine is overhauled each time an airplane comes into the shop. Under this policy, airplanes have arrived according to a Poisson process at a mean rate of 1 per day. The time required for an engine overhaul (once work has begun) has an exponential distribution with a mean of $\frac{1}{2}$ day.

A proposal has been made to change the policy so that all four engines are overhauled consecutively each time an airplane comes into the shop. Although this would quadruple the expected service time, each plane would need to come to the maintenance base only one-fourth as often.

Management now needs to decide whether to continue the status quo or adopt the proposal. The objective is to minimize the average amount of flying time lost by the entire fleet per day due to engine overhauls.

(a) Compare the two alternatives with respect to the average amount of flying time lost by an airplane each time it comes to the maintenance base.

(b) Compare the two alternatives with respect to the average number of airplanes losing flying time due to being at the maintenance base.

(c) Which of these two comparisons is the appropriate one for making management's decision? Explain.

17.7-7. Reconsider Prob. 17.7-6. Management has adopted the proposal but now wants further analysis conducted of this new queueing system.

(a) How should the state of the system be defined in order to formulate the queueing model as a continuous time Markov chain?

(b) Construct the corresponding rate diagram.

17.7-8. The McAllister Company factory currently has *two* tool cribs, each with a *single* clerk, in its manufacturing area. One tool crib handles only the tools for the heavy machinery; the second one handles all other tools. However, for each crib the mechanics arrive to obtain tools at a mean rate of 18 per hour, and the expected service time is 3 minutes.

Because of complaints that the mechanics coming to the tool crib have to wait too long, it has been proposed that the two tool cribs be combined so that either clerk can handle either kind of tool as the demand arises. It is believed that the mean arrival rate to the combined two-clerk tool crib would double to 36 per hour and that the expected service time would continue to be 3 minutes. However, information is not available on the *form* of the probability distributions for interarrival and service times, so it is not clear which queueing model would be most appropriate.

Compare the status quo and the proposal with respect to the total expected number of mechanics at the tool crib(s) and the expected waiting time (including service) for each mechanic. Do this by tabulating these data for the four queueing models considered in Figs. 17.6, 17.8, 17.10, and 17.11 (use $k = 2$ when an Erlang distribution is appropriate).

17.7-9.* Consider a single-server queueing system with a Poisson input, Erlang service times, and a finite queue. In particular, suppose that $k = 2$, the mean arrival rate is 2 customers per hour, the expected service time is 0.25 hour, and the maximum permissible number of customers in the system is 2. This system can be formulated as a continuous time Markov chain by dividing each service time into two consecutive phases, each having an exponential distribution with a mean of 0.125 hour, and then defining the state of the system as (n, p), where n is the number of customers in the system ($n = 0, 1, 2$), and p indicates the phase of the customer being served ($p = 0, 1, 2$, where $p = 0$ means that no customer is being served).

(a) Construct the corresponding rate diagram. Write the balance equations, and then use these equations to solve for the steady-state distribution of the state of this Markov chain.

(b) Use the steady-state distribution obtained in part (*a*) to identify the steady-state distribution of the number of customers in the system (P_0, P_1, P_2) and the steady-state expected number of customers in the system (L).

(c) Compare the results from part (*b*) with the corresponding results when the service-time distribution is exponential.

17.7-10. Consider the $E_2/M/1$ model with $\lambda = 4$ and $\mu = 5$. This model can be formulated as a continuous time Markov chain by dividing each interarrival time into two consecutive phases, each having an exponential distribution with a mean of $1/(2\lambda) = 0.125$, and then defining the state of the system as (n, p), where n is the number of customers in the system ($n = 0, 1, 2, \ldots$) and p indicates the phase of the *next* arrival (not yet in the system) ($p = 1, 2$).

Construct the corresponding rate diagram (but do not solve further).

17.7-11. A company has one repair technician to keep a large group of machines in running order. Treating this group as an infinite calling population, individual breakdowns occur according to a Poisson process at a mean rate of 1 per hour. For each breakdown, the probability is 0.9 that only a minor repair is needed, in which case the repair time has an exponential distribution with a mean of $\frac{1}{2}$ hour. Otherwise, a major repair is needed, in which case the repair time has an exponential distribution with a mean of 5 hours. Because both of these *conditional* distributions are exponential, the *unconditional* (combined) distribution of repair times is *hyperexponential*.

(a) Compute the mean and standard deviation of this hyperexponential distribution. [*Hint:* Use the general relationships from probability theory that, for any random variable X and any pair of mutually exclusive events E_1 and E_2, $E(X) = E(X \mid E_1)P(E_1) + E(X \mid E_2)P(E_2)$ and $\text{var}(X) = E(X^2) - E(X)^2$.] Compare this standard deviation with that for an exponential distribution having this mean.

(b) What are P_0, L_q, L, W_q, and W for this queueing system?

(c) What is the conditional value of W, given that the machine involved requires major repair? A minor repair? What is the division of L between machines requiring the two types of repairs? (*Hint:* Little's formula still applies for the individual categories of machines.)

(d) How should the states of the system be defined in order to formulate this queueing system as a continuous time Markov chain? (*Hint:* Consider what additional information must be given, besides the number of machines down, for the conditional distribution of the time remaining until the next event of each kind to be exponential.)

(e) Construct the corresponding rate diagram.

17.7-12. Consider the finite queue variation of the $M/G/1$ model, where K is the maximum number of customers allowed in the system. For $n = 1, 2, \ldots$, let the random variable X_n be the number of customers in the system at the moment t_n when the nth customer has just finished being served. (Do not count the departing customer.) The times $\{t_1, t_2, \ldots\}$ are called *regeneration points*. Furthermore, $\{X_n\}$ ($n = 1, 2, \ldots$) is a discrete time Markov chain and is known as an *embedded Markov chain*. Embedded Markov chains are useful for studying the properties of continuous time stochastic processes such as for an $M/G/1$ model.

Now consider the particular special case where $K = 4$, the service time of successive customers is a fixed constant, say, 10 minutes, and the mean arrival rate is 1 every 50 minutes. Therefore, $\{X_n\}$ is an embedded Markov chain with states 0, 1, 2, 3. (Because there are never more than 4 customers in the system, there can never be more than 3 in the system at a regeneration point.) Because the system is observed at successive departures, X_n can never decrease by more than 1. Furthermore, the probabilities of transitions that result in increases in X_n are obtained directly from the Poisson distribution.

(a) Find the one-step transition matrix for the embedded Markov chain. (*Hint:* In obtaining the transition probability from state 3 to state 3, use the probability of 1 or more arrivals rather than just 1 arrival, and similarly for other transitions to state 3.)

(b) Use the corresponding routine in the Markov chains area of your IOR Tutorial to find the steady-state probabilities for the number of customers in the system at regeneration points.

(c) Compute the expected number of customers in the system at regeneration points, and compare it to the value of L for the $M/D/1$ model (with $K = \infty$) in Sec. 17.7.

17.8-1.* Southeast Airlines is a small commuter airline serving primarily the state of Florida. Their ticket counter at a certain airport is staffed by a single ticket agent. There are two separate lines—one for first-class passengers and one for coach-class passengers. When the ticket agent is ready for another customer, the next first-class passenger is served if there are any in line. If not, the next coach-class passenger is served. Service times have an exponential distribution with a mean of 3 minutes for both types of customers. During the 12 hours per day that the ticket counter is open, passengers arrive randomly at a mean rate of 2 per hour for first-class passengers and 10 per hour for coach-class passengers.

(a) What kind of queueing model fits this queueing system?

T (b) Find the main measures of performance—L, L_q, W, and W_q—for both first-class passengers and coach-class passengers.

(c) What is the expected waiting time before service begins for first-class customers as a fraction of this waiting time for coach-class customers?

(d) Determine the average number of hours per day that the ticket agent is busy.

T **17.8-2.** Consider the model with nonpreemptive priorities presented in Sec. 17.8. Suppose there are two priority classes, with $\lambda_1 = 2$ and $\lambda_2 = 3$. In designing this queueing system, you are offered the choice between the following alternatives: (1) one fast server ($\mu = 6$) and (2) two slow servers ($\mu = 3$).

Compare these alternatives with the usual four mean measures of performance (W, L, W_q, L_q) for the individual priority classes (W_1, W_2, L_1, L_2, and so forth). Which alternative is preferred if your primary concern is expected waiting time in the *system* for priority class 1 (W_1)? Which is preferred if your primary concern is expected waiting time in the *queue* for priority class 1?

17.8-3. Consider the single-server variation of the nonpreemptive priorities model presented in Sec. 17.8. Suppose there are three priority classes, with $\lambda_1 = 1$, $\lambda_2 = 1$, and $\lambda_3 = 1$. The expected service times for priority classes 1, 2, and 3 are 0.4, 0.3, and 0.2, respectively, so $\mu_1 = 2.5$, $\mu_2 = 3\frac{1}{3}$, and $\mu_3 = 5$.

(a) Calculate W_1, W_2, and W_3.

(b) Repeat part (a) when using the approximation of applying the general model for nonpreemptive priorities presented in Sec. 17.8 instead. Since this general model assumes that the expected service time is the same for all priority classes, use an expected service time of 0.3 so $\mu = 3\frac{1}{3}$. Compare the results with those obtained in part (a) and evaluate how good an approximation is provided by making this assumption.

T **17.8-4.*** A particular work center in a job shop can be represented as a single-server queueing system, where jobs arrive according to a Poisson process, with a mean rate of 8 per day. Although the arriving jobs are of three distinct types, the time required to perform any of these jobs has the same exponential distribution, with a mean of 0.1 working day. The practice has been to work on arriving jobs on a first-come-first-served basis. However, it is important that jobs of type 1 not wait very long, whereas the wait is only moderately important for jobs of type 2 and is relatively unimportant for jobs of type 3. These three types arrive with a mean rate of 2, 4, and 2 per day, respectively. Because all three types have experienced rather long delays on average, it has been proposed that the jobs be selected according to an appropriate priority discipline instead.

Compare the expected waiting time (including service) for each of the three types of jobs if the queue discipline is (a) first-come-first-served, (b) nonpreemptive priority, and (c) preemptive priority.

T **17.8-5.** Reconsider the *County Hospital* emergency room problem as analyzed in Sec. 17.8. Suppose that the definitions of the three categories of patients are tightened somewhat in order to move marginal cases into a lower category. Consequently, only 5 percent of the patients will qualify as critical cases, 20 percent as serious cases, and 75 percent as stable cases. Develop a table showing the data presented in Table 17.3 for this revised problem.

17.8-6. Reconsider the queueing system described in Prob. 17.4-6. Suppose now that type 1 customers are more important than type 2 customers. If the queue discipline were changed from first-come-first-served to a priority system with type 1 customers being given nonpreemptive priority over type 2 customers, would this increase, decrease, or keep unchanged the expected total number of customers in the system?

(a) Determine the answer without any calculations, and then present the reasoning that led to your conclusion.

T (b) Verify your conclusion in part (a) by finding the expected total number of customers in the system under each of these two queue disciplines.

17.8-7. Consider the queueing model with a preemptive priority queue discipline presented in Sec. 17.8. Suppose that $s = 1$, $N = 2$, and $(\lambda_1 + \lambda_2) < \mu$; and let P_{ij} be the steady-state probability that there are i members of the higher-priority class and j members of the lower-priority class in the queueing system ($i = 0, 1, 2, \ldots ; j = 0, 1, 2, \ldots$). Use a method analogous to that presented in Sec. 17.5 to derive a system of linear equations whose simultaneous solution is the P_{ij}. Do not actually obtain this solution.

17.9-1. Read the referenced article that fully describes the OR study summarized in the application vignette presented in Sec. 17.9. Briefly describe how queueing theory was applied in this study. Then list the various financial and nonfinancial benefits that resulted from this study.

17.9-2. Consider a queueing system with two servers, where the customers arrive from two different sources. From source 1, the customers always arrive 2 at a time, where the time between consecutive arrivals of pairs of customers has an exponential distribution with a mean of 20 minutes. Source 2 is itself a two-server queueing system, which has a Poisson input process with a mean rate of 7 customers per hour, and the service time from each of these two servers has an exponential distribution with a mean of 15 minutes. When a customer completes service at source 2, he or she immediately enters the queueing system under consideration for another type of service. In the latter queueing system, the queue discipline is preemptive priority where customers from source 1 always have preemptive priority over customers from source 2. However, service times are independent and identically distributed for both types of customers according to an exponential distribution with a mean of 6 minutes.

(a) First focus on the problem of deriving the steady-state distribution of *only* the number of source 1 customers in the queueing system under consideration. Using a continuous time Markov chain formulation, define the states and construct the rate diagram for most efficiently deriving this distribution (but do not actually derive it).

(b) Now focus on the problem of deriving the steady-state distribution of the *total* number of customers of both types in the queueing system under consideration. Using a continuous time Markov chain formulation, define the states and construct the rate diagram for most efficiently deriving this distribution (but do not actually derive it).

(c) Now focus on the problem of deriving the steady-state *joint* distribution of the number of customers of each type in the queueing system under consideration. Using a continuous time Markov chain formulation, define the states and construct the rate diagram for deriving this distribution (but do not actually derive it).

17.9-3. Consider a system of two infinite queues in series, where each of the two service facilities has a single server. All service times are independent and have an exponential distribution, with a mean of 3 minutes at facility 1 and 4 minutes at facility 2. Facility 1 has a Poisson input process with a mean rate of 10 per hour.

(a) Find the steady-state distribution of the number of customers at facility 1 and then at facility 2. Then show the product form solution for the *joint* distribution of the number at the respective facilities.

(b) What is the probability that both servers are idle?

(c) Find the expected *total* number of customers in the system and the expected *total* waiting time (including service times) for a customer.

17.9-4. Under the assumptions specified in Sec. 17.9 for a system of infinite queues in series, this kind of queueing network actually is a special case of a Jackson network. Demonstrate that this is true by describing this system as a Jackson network, including specifying the values of the a_j and the p_{ij}, given λ for this system.

17.9-5. Consider a Jackson network with three service facilities having the parameter values shown below.

Facility j	s_j	μ_j	a_j	p_{ij}		
				$i = 1$	$i = 2$	$i = 3$
$j = 1$	1	25	6	0	0.2	0.4
$j = 2$	1	30	8	0.5	0	0.3
$j = 3$	1	20	4	0.4	0.3	0

T (a) Find the total arrival rate at each of the facilities.

(b) Find the steady-state distribution of the number of customers at facility 1, facility 2, and facility 3. Then show the product form solution for the joint distribution of the number at the respective facilities.

(c) What is the probability that all the facilities have empty queues (no customers waiting to begin service)?

(d) Find the expected total number of customers in the system.

(e) Find the expected total waiting time (including service times) for a customer.

T **17.10-1.** When describing economic analysis of the number of servers to provide in a queueing system, Sec. 17.10 introduces a basic cost model where the objective is to minimize $E(\text{TC}) = C_s s + C_w L$. The purpose of this problem is to enable you to explore the effect that the relative sizes of C_s and C_w have on the optimal number of servers.

Suppose that the queueing system under consideration fits the *M/M/s* model with $\lambda = 8$ customers per hour and $\mu = 10$ customers per hour. Use the Excel template in your OR Courseware for economic analysis with the *M/M/s* model to find the optimal number of servers for each of the following cases.

(a) $C_s = \$100$ and $C_w = \$10$.

(b) $C_s = \$100$ and $C_w = \$100$.

(c) $C_s = \$10$ and $C_w = \$100$.

T **17.10-2.*** Jim McDonald, manager of the fast-food hamburger restaurant McBurger, realizes that providing fast service is a key to the success of the restaurant. Customers who have to wait very long are likely to go to one of the other fast-food restaurants in town next time. He estimates that each minute a customer has to wait in line before completing service costs him an average of 30 cents in lost future business. Therefore, he wants to be sure that enough cash registers always are open to keep waiting to a minimum. Each cash register is operated by a part-time employee who obtains the food ordered by each customer and collects the payment. The total cost for each such employee is $9 per hour.

During lunch time, customers arrive according to a Poisson process at a mean rate of 66 per hour. The time needed to serve a customer is estimated to have an exponential distribution with a mean of 2 minutes.

Determine how many cash registers Jim should have open during lunch time to minimize his expected total cost per hour.

T **17.10-3.** The Garrett-Tompkins Company provides three copy machines in its copying room for the use of its employees. However, due to recent complaints about considerable time being wasted waiting for a copier to become free, management is considering adding one or more additional copy machines.

During the 2,000 working hours per year, employees arrive at the copying room according to a Poisson process at a mean rate of 40 per hour. The time each employee needs with a copy machine is believed to have an exponential distribution with a mean of 4 minutes. The lost productivity due to an employee spending time in the copying room is estimated to cost the company an average of $40 per hour. Each copy machine is leased for $4,000 per year.

Determine how many copy machines the company should have to minimize its expected total cost per hour.

17.11-1. From the bottom part of the selected references given at the end of the chapter, select one of these award-winning applications of queueing theory. Read this article and then write a two-page summary of the application and the benefits (including nonfinancial benefits) it provided.

17.11-2. From the bottom part of the selected references given at the end of the chapter, select three of these award-winning applications of queueing theory. For each one, read the article and then write a one-page summary of the application and the benefits (including nonfinancial benefits) it provided.

■ CASES

CASE 17.1   Reducing In-Process Inventory

Jim Wells, vice-president for manufacturing of the Northern Airplane Company, is exasperated. His walk through the company's most important plant this morning has left him in a foul mood. However, he now can vent his temper at Jerry Carstairs, the plant's production manager, who has just been summoned to Jim's office.

"Jerry, I just got back from walking through the plant, and I am very upset." "What is the problem, Jim?" "Well, you know how much I have been emphasizing the need to cut down on our in-process inventory." "Yes, we've been working hard on that," responds Jerry. "Well, not hard enough!" Jim raises his voice even higher. "Do you know what I found by the presses?" "No." "Five metal sheets still waiting to be formed into wing sections. And then, right next door at the inspection station, 13 wing sections! The inspector was inspecting one of them, but the other 12 were just sitting there. You know we have a couple hundred thousand dollars tied up in each of those wing sections. So between the presses and the inspection station, we have a few million bucks worth of terribly expensive metal just sitting there. We can't have that!"

The chagrined Jerry Carstairs tries to respond. "Yes, Jim, I am well aware that that inspection station is a bottleneck. It usually isn't nearly as bad as you found it this morning, but it is a bottleneck. Much less so for the presses. You really caught us on a bad morning." "I sure hope so," retorts Jim, "but you need to prevent anything nearly this bad happening even occasionally. What do you propose to do about it?" Jerry now brightens noticeably in his response. "Well actually, I've already been working on this problem. I have a couple proposals on the table and I have asked an operations research analyst on my staff to analyze these proposals and report back with recommendations." "Great," responds Jim, "glad to see you are on top of the problem. Give this your highest priority and report back to me as soon as possible." "Will do," promises Jerry.

Here is the problem that Jerry and his OR analyst are addressing. Each of 10 identical presses is being used to form wing sections out of large sheets of specially processed metal. The sheets arrive randomly to the group of presses at a mean rate of 7 per hour. The time required by a press to form a wing section out of a metal sheet has an exponential distribution with a mean of 1 hour. When finished, the wing sections arrive randomly at an inspection station at the same mean rate as the metal sheets arrived at the presses (7 per hour). A single inspector has the full-time job of inspecting these wing sections to make sure they meet specifications. Each inspection takes her $7\frac{1}{2}$ minutes, so she can inspect 8 wing sections per hour. This inspection rate has resulted in a substantial average amount of in-process inventory at the inspection station (i.e., the average number of wing sheets waiting to complete inspection is fairly large), in addition to that already found at the group of machines.

The cost of this in-process inventory is estimated to be $8 per hour for each metal sheet at the presses or each wing section at the inspection station. Therefore, Jerry Carstairs has made two alternative proposals to reduce the average level of in-process inventory.

Proposal 1 is to use slightly less power for the presses (which would increase their average time to form a wing section to 1.2 hours), so that the inspector can keep up with their output better. This also would reduce the cost of the power for running each machine from $7.00 to $6.50 per hour. (By contrast, increasing to maximum power would increase this cost to $7.50 per hour while decreasing the average time to form a wing section to 0.8 hour.)

Proposal 2 is to substitute a certain younger inspector for this task. He is somewhat faster (albeit with some variability in his inspection times because of less experience), so he should keep up better. (His inspection time would have an Erlang distribution with a mean of 7.2 minutes and a shape parameter $k = 2$.) This inspector is in a job classification that calls for a total compensation (including benefits) of $19 per hour, whereas the current inspector is in a lower job classification where the compensation is $17 per hour. (The inspection times for each of these inspectors are typical of those in the same job classification.)

You are the OR analyst on Jerry Carstair's staff who has been asked to analyze this problem. He wants you to "use the latest OR techniques to see how much each proposal would cut down on in-process inventory and then make your recommendations."

(a) To provide a basis of comparison, begin by evaluating the status quo. Determine the expected amount of in-process inventory at the presses and at the inspection station. Then calculate the expected total cost per hour when considering all of the following: the cost of the in-process inventory, the cost of the power for runnng the presses, and the cost of the inspector.

(b) What would be the effect of proposal 1? Why? Make specific comparisons to the results from part (*a*). Explain this outcome to Jerry Carstairs.

(c) Determine the effect of proposal 2. Make specific comparisons to the results from part (*a*). Explain this outcome to Jerry Carstairs.

(d) Make your recommendations for reducing the average level of in-process inventory at the inspection station and at the group of machines. Be specific in your recommendations, and support them with quantitative analysis like that done in part (*a*). Make specific comparisons to the results from part (*a*), and cite the improvements that your recommendations would yield.

■ PREVIEW OF AN ADDED CASE ON OUR WEBSITE (www.mhhe.com/hillier)

CASE 17.2 Queueing Quandary

Many angry customers are complaining about the long waits needed to get through to a call center. It appears that more service representatives are needed to answer the calls. Another option is to train the service representatives further to enable them to answer calls more efficiently. Some possible criteria for satisfactory levels of service have been proposed. Queueing theory needs to be applied to determine how the operation of the call center should be redesigned.

Inventory Theory

"**S**orry, we're out of that item." How often have you heard that during shopping trips? In many of these cases, what you have encountered are stores that aren't doing a very good job of managing their *inventories* (stocks of goods being held for future use or sale). They aren't placing orders to replenish inventories soon enough to avoid shortages. These stores could benefit from the kinds of techniques of scientific inventory management that are described in this chapter.

It isn't just retail stores that must manage inventories. In fact, inventories pervade the business world. Maintaining inventories is necessary for any company dealing with physical products, including manufacturers, wholesalers, and retailers. For example, manufacturers need inventories of the materials required to make their products. They also need inventories of the finished products awaiting shipment. Similarly, both wholesalers and retailers need to maintain inventories of goods to be available for purchase by customers.

The annual costs associated with storing ("carrying") inventory are very large, perhaps as much as a quarter of the value of the inventory. Therefore, the costs being incurred for the storage of inventory in the United States run into the hundreds of billions of dollars annually. Reducing storage costs by avoiding unnecessarily large inventories can enhance any firm's competitiveness.

Some Japanese companies were pioneers in introducing the *just-in-time inventory system*—a system that emphasizes planning and scheduling so that the needed materials arrive "just-in-time" for their use. Huge savings are thereby achieved by reducing inventory levels to a bare minimum.

Many companies in other parts of the world also have been revamping the way in which they manage their inventories. The application of operations research techniques in this area (sometimes called *scientific inventory management*) is providing a powerful tool for gaining a competitive edge.

How do companies use operations research to improve their **inventory policy** for when and how much to replenish their inventory? They use **scientific inventory management** comprising the following steps:

1. Formulate a *mathematical model* describing the behavior of the inventory system.
2. Seek an *optimal* inventory policy with respect to this model.
3. Use a computerized *information processing system* to maintain a record of the current inventory levels.

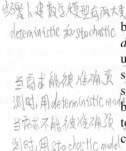

利用当前库存水平的记录，应用最优库存策略来发出何时补充库存和补充多少库存的信号

4. Using this record of current inventory levels, apply the optimal inventory policy to signal when and how much to replenish inventory.

当对建数学模型有两大类 deterministic 和 stochastic

当需求能被准确预测时，用 deterministic model 当需求不能被准确预测时，用 stochastic model

The mathematical inventory models used with this approach can be divided into two broad categories—deterministic models and stochastic models—according to the *predictability of demand* involved. The **demand** for a product in inventory is the number of units that will need to be withdrawn from inventory for some use (e.g., sales) during a specific period. If the demand in future periods can be forecast with considerable precision, it is reasonable to use an inventory policy that assumes that all forecasts will always be completely accurate. This is the case of *known demand* where a *deterministic* inventory model would be used. However, when demand cannot be predicted very well, it becomes necessary to use a *stochastic* inventory model where the demand in any period is a random variable rather than a known constant.

There are several basic considerations involved in determining an inventory policy that must be reflected in the mathematical inventory model. These are illustrated in the examples presented in the first section and then are described in general terms in Sec. 18.2. Section 18.3 develops and analyzes deterministic inventory models for situations where the inventory level is under continuous review. Section 18.4 does the same for situations where the planning is being done for a series of periods rather than continuously. Section 18.5 extends certain deterministic models to coordinate the inventories at various points along a company's supply chain. The following two sections present stochastic models, first under continuous review, and then for dealing with a perishable product over a single period. (A supplement to this chapter on the book's website introduces stochastic periodic-review models for multiple periods.) Section 18.8 then introduces a relatively new area of inventory theory, called *revenue management,* that is concerned with maximizing a company's expected revenue when dealing with the special kind of perishable product whose entire inventory must be provided to customers at a designated point in time or be lost forever. (Certain service industries, such as an airline company providing its entire inventory of seats on an particular flight at the designated time for the flight, now make extensive use of revenue management.)

■ 18.1 EXAMPLES

We present two examples in rather different contexts (a manufacturer and a wholesaler) where an inventory policy needs to be developed.

EXAMPLE 1 **Manufacturing Speakers for TV Sets**

A television manufacturing company produces its own speakers, which are used in the production of its television sets. The television sets are assembled on a continuous production line at a rate of 8,000 per month, with one speaker needed per set. The speakers are produced in batches because they do not warrant setting up a continuous production line, and relatively large quantities can be produced in a short time. Therefore, the speakers are placed into inventory until they are needed for assembly into television sets on the production line. The company is interested in determining when to produce a batch of speakers and how many speakers to produce in each batch. Several costs must be considered:

1. Each time a batch is produced, a **setup cost** of $12,000 is incurred. This cost includes the cost of "tooling up," administrative costs, record keeping, and so forth. Note that the existence of this cost argues for producing speakers in large batches.

2. The **unit production cost** of a single speaker (excluding the setup cost) is $10, independent of the batch size produced. (In general, however, the unit production cost need not be constant and may decrease with batch size.)

3. The production of speakers in large batches leads to a large inventory. The estimated **holding cost** of keeping a speaker in stock is $0.30 per month. This cost includes the cost of capital tied up in inventory. Since the money invested in inventory cannot be used in other productive ways, this cost of capital consists of the lost return (referred to as the *opportunity cost*) because alternative uses of the money must be forgone. Other components of the holding cost include the cost of leasing the storage space, the cost of insurance against loss of inventory by fire, theft, or vandalism, taxes based on the value of the inventory, and the cost of personnel who oversee and protect the inventory.

4. Company policy prohibits deliberately planning for shortages of any of its components. However, a shortage of speakers occasionally crops up, and it has been estimated that each speaker that is not available when required costs $1.10 per month. This **shortage cost** includes the extra cost of installing speakers after the television set is fully assembled otherwise, the interest lost because of the delay in receiving sales revenue, the cost of extra record keeping, and so forth.

We will develop the inventory policy for this example with the help of the first inventory model presented in Sec. 18.3.

EXAMPLE 2   **Wholesale Distribution of Bicycles**

A wholesale distributor of bicycles is having trouble with shortages of its most popular model and is currently reviewing the inventory policy for this model. The distributor purchases this model bicycle from the manufacturer monthly and then supplies it to various bicycle shops in the western United States in response to purchase orders. What the total demand from bicycle shops will be in any given month is quite uncertain. Therefore, the question is, How many bicycles should be ordered from the manufacturer for any given month, given the stock level leading into that month?

The distributor has analyzed her costs and has determined that the following are important:

1. The **ordering cost,** i.e., the cost of placing an order plus the cost of the bicycles being purchased, has two components: The administrative cost involved in placing an order is estimated as $2,000, and the actual cost of each bicycle is $350 for this wholesaler.

2. The *holding cost,* i.e., the cost of maintaining an inventory, is $10 per bicycle remaining at the end of the month. This cost represents the costs of capital tied up, warehouse space, insurance, taxes, and so on.

3. The *shortage cost* is the cost of not having a bicycle on hand when needed. This particular model is easily reordered from the manufacturer, and stores usually accept a delay in delivery. Still, although shortages are permissible, the distributor feels that she incurs a loss, which she estimates to be $150 per bicycle per month of shortage. This estimated cost takes into account the possible loss of future sales because of the loss of customer goodwill. Other components of this cost include lost interest on delayed sales revenue, and additional administrative costs associated with shortages. If some stores were to cancel orders because of delays, the lost revenues from these lost sales would need to be included in the shortage cost. Fortunately, such cancellations normally do not occur for this distributor.

We will return to a variation of this example again in Sec. 18.7.

These examples illustrate that there are two possibilities for how a firm *replenishes inventory,* depending on the situation. One possibility is that the firm *produces* the needed units itself (like the television manufacturer producing speakers). The other is that the firm *orders* the units from a supplier (like the bicycle distributor ordering bicycles from the manufacturer). Inventory models do not need to distinguish between these two ways of replenishing inventory, so we will use such terms as *producing* and *ordering* interchangeably.

Both examples deal with one specific product (speakers for a certain kind of television set or a certain bicycle model). In most inventory models, just one product is being considered at a time. All the inventory models presented in this chapter assume a single product.

Both examples indicate that there exists a trade-off between the costs involved. The next section discusses the basic cost components of inventory models for determining the optimal trade-off between these costs.

■ 18.2 COMPONENTS OF INVENTORY MODELS

Because inventory policies affect profitability, the choice among policies depends upon their relative profitability. As already seen in Examples 1 and 2, some of the costs that determine this profitability are (1) the ordering costs, (2) holding costs, and (3) shortage costs. Other relevant factors include (4) revenues, (5) salvage costs, and (6) discount rates. These six factors are described in turn below.

The **cost of ordering** an amount z (either through *purchasing* or *producing this amount*) can be represented by a function $c(z)$. The simplest form of this function is one that is directly proportional to the amount ordered, that is, $c \cdot z$, where c represents the unit price paid. Another common assumption is that $c(z)$ is composed of two parts: a term that is directly proportional to the amount ordered and a term that is a constant K for z positive and is 0 for $z = 0$. For this case,

$$c(z) = \text{cost of ordering } z \text{ units}$$
$$= \begin{cases} 0 & \text{if } z = 0 \\ K + cz & \text{if } z > 0, \end{cases}$$

where $K =$ setup cost and $c =$ unit cost.

The constant K includes the administrative cost of ordering or, when producing, the costs involved in setting up to start a production run.

There are other assumptions that can be made about the cost of ordering, but this chapter is restricted to the cases just described.

In Example 1, the speakers are produced and the setup cost for a production run is $12,000. Furthermore, each speaker costs $10, so that the *production* cost when ordering a production run of z speakers is given by

$$c(z) = 12,000 + 10z, \qquad \text{for } z > 0.$$

In Example 2, the distributor orders bicycles from the manufacturer and the *ordering* cost is given by

$$c(z) = 2,000 + 350z, \qquad \text{for } z > 0.$$

The **holding cost** (sometimes called the *storage cost*) represents all the costs associated with the storage of the inventory until it is sold or used. Included are the cost of capital tied up, space, insurance, protection, and taxes attributed to storage. The holding cost can be assessed either continuously or on a period-by-period basis. In the latter case, the cost may be a function of the maximum quantity held during a period, the average amount held, or the quantity in inventory at the end of the period. The last viewpoint is usually taken in this chapter.

In the bicycle example, the holding cost is $10 per bicycle remaining at the end of the month. In the TV speakers example, the holding cost is assessed continuously as $0.30

per speaker in inventory per month, so the average holding cost per month is $0.30 times the average number of speakers in inventory.

The **shortage cost** (sometimes called the *unsatisfied demand cost*) is incurred when the amount of the commodity required (demand) exceeds the available stock. This cost depends upon which of the following two cases applies.

In one case, called **backlogging,** the excess demand is not lost, but instead is held until it can be satisfied when the next normal delivery replenishes the inventory. For a firm incurring a temporary shortage in supplying its customers (as for the bicycle example), the shortage cost then can be interpreted as the loss of customers' goodwill and the subsequent reluctance to do business with the firm, the cost of delayed revenue, and the extra administrative costs. For a manufacturer incurring a temporary shortage in materials needed for production (such as a shortage of speakers for assembly into television sets), the shortage cost becomes the cost associated with delaying the completion of the production process.

In the second case, called **no backlogging,** if any excess of demand over available stock occurs, the firm cannot wait for the next normal delivery to meet the excess demand. Either (1) the excess demand is met by a priority shipment, or (2) it is not met at all because the orders are canceled. For situation 1, the shortage cost can be viewed as the cost of the priority shipment. For situation 2, the shortage cost is the loss of current revenue from not meeting the demand plus the cost of losing future business because of lost goodwill.[1]

Revenue may or may not be included in the model. If both the price and the demand for the product are established by the market and so are outside the control of the company, the revenue from sales (assuming demand is met) is independent of the firm's inventory policy and may be neglected. However, if revenue is neglected in the model, the *loss in revenue* must then be included in the shortage cost whenever the firm cannot meet the demand and the sale is lost. Furthermore, even in the case where demand is backlogged, the cost of the delay in revenue must also be included in the shortage cost. With these interpretations, revenue will not be considered explicitly in the remainder of this chapter.

The **salvage value** of an item is the value of a leftover item when no further inventory is desired. The salvage value represents the disposal value of the item to the firm, perhaps through a discounted sale. The negative of the salvage value is called the **salvage cost.** If there is a cost associated with the disposal of an item, the salvage cost may be positive. We assume hereafter that any salvage cost is incorporated into the *holding cost*.

Finally, the **discount rate** takes into account the time value of money. When a firm ties up capital in inventory, the firm is prevented from using this money for alternative purposes. For example, it could invest this money in secure investments, say, government bonds, and have a return on investment 1 year hence of, say, 7 percent. Thus, $1 invested today would be worth $1.07 in year 1, or alternatively, a $1 profit 1 year hence is equivalent to $\alpha = \$1/\$1.07$ today. The quantity α is known as the **discount factor.** Thus, in adding up the total profit from an inventory policy, the profit or costs 1 year hence should be multiplied by α; in 2 years hence by α^2; and so on. (Units of time other than 1 year also can be used.) The total profit calculated in this way normally is referred to as the *net present value*.

In problems having short time horizons, α may be assumed to be 1 (and thereby neglected) because the current value of $1 delivered during this short time horizon does not change very much. However, in problems having long time horizons, the discount factor must be included.

---

[1]An analysis of situation 2 is provided by E. T. Anderson, G. J. Fitzsimons, and D. Simester, "Measuring and Mitigating the Costs of Stockouts," *Management Science,* **52**(11): 1751–1763, Nov. 2006. For an analysis of whether backlogging or no backlogging provides a less costly policy under various circumstances, see B. Janakiraman, S. Seshadri, and J. G. Shanthikumar, "A Comparison of the Optimal Costs of Two Canonical Inventory Systems," *Operations Research,* **55**(5): 866–875, Sept.–Oct. 2007.

In using quantitative techniques to seek optimal inventory policies, we use the criterion of minimizing the total (expected) discounted cost. Under the assumptions that the price and demand for the product are not under the control of the company and that the lost or delayed revenue is included in the shortage penalty cost, minimizing cost is equivalent to maximizing net income. Another useful criterion is to keep the inventory policy simple, i.e., keep the rule for indicating *when to order* and *how much to order* both understandable and easy to implement. Most of the policies considered in this chapter possess this property.

As mentioned at the beginning of the chapter, inventory models are usually classified as either *deterministic* or *stochastic* according to whether the demand for a period is known or is a random variable having a known probability distribution. The production of batches of speakers in Example 1 of Sec. 18.1 illustrates deterministic demand because the speakers are used in television assemblies at a fixed rate of 8,000 per month. The bicycle shops' purchases of bicycles from the wholesale distributor in Example 2 of Sec. 18.1 illustrates random demand because the total monthly demand varies from month to month according to some probability distribution. Another component of an inventory model is the **lead time,** which is the amount of time between the placement of an order to replenish inventory (through either purchasing or producing) and the receipt of the goods into inventory. If the lead time always is the same (a *fixed* lead time), then the replenishment can be scheduled just when desired. Most models in this chapter assume that each replenishment occurs just when desired, either because the delivery is nearly instantaneous or because it is known when the replenishment will be needed and there is a fixed lead time.

Another classification refers to whether the current inventory level is being monitored continuously or periodically. In **continuous review,** an order is placed as soon as the stock level falls down to the prescribed reorder point. In **periodic review,** the inventory level is checked at discrete intervals, e.g., at the end of each week, and ordering decisions are made only at these times even if the inventory level dips below the reorder point between the preceding and current review times. (In practice, a periodic review policy can be used to approximate a continuous review policy by making the time interval sufficiently small.)

■ 18.3   DETERMINISTIC CONTINUOUS-REVIEW MODELS

The most common inventory situation faced by manufacturers, retailers, and wholesalers is that stock levels are depleted over time and then are replenished by the arrival of a batch of new units. A simple model representing this situation is the following **economic order quantity model** or, for short, the **EOQ model.** (It sometimes is also referred to as the *economic lot-size model.*)

Units of the product under consideration are assumed to be withdrawn from inventory continuously at a *known constant rate,* denoted by d; that is, the demand is d units per unit time. It is further assumed that inventory is replenished when needed by ordering (through either purchasing or producing) a batch of fixed size (Q units), where all Q units arrive simultaneously at the desired time. For the *basic EOQ model* to be presented first, the only costs to be considered are

K = setup cost for ordering one batch,

c = unit cost for producing or purchasing each unit,

h = holding cost per unit per unit of time held in inventory.

The objective is to determine when and by how much to replenish inventory so as to minimize the sum of these costs per unit time.

We assume *continuous review,* so that inventory can be replenished whenever the inventory level drops sufficiently low. We shall first assume that shortages are not allowed (but later we will relax this assumption). With the fixed demand rate, shortages can be avoided by replenishing inventory each time the inventory level drops to zero, and this also will minimize the holding cost. Figure 18.1 depicts the resulting pattern of inventory levels over time when we start at time 0 by ordering a batch of Q units in order to increase the initial inventory level from 0 to Q and then repeat this process each time the inventory level drops back down to 0.

Example 1 in Sec. 18.1 (manufacturing speakers for TV sets) fits this model and will be used to illustrate the following discussion.

The Basic EOQ Model

To summarize, in addition to the costs specified above, the basic EOQ model makes the following assumptions.

Assumptions (Basic EOQ Model).

1. A known constant *demand rate* of d units per unit time.
2. The order quantity (Q) to replenish inventory arrives all at once just when desired, namely, when the inventory level drops to 0.
3. Planned shortages are not allowed.

In regard to assumption 2, there usually is a lag between when an order is placed and when it arrives in inventory. As indicated in Sec. 18.2, the amount of time between the placement of an order and its receipt is referred to as the *lead time.* The inventory level at which the order is placed is called the **reorder point.** To satisfy assumption 2, this reorder point needs to be set at

Reorder point = (demand rate) × (lead time).

Thus, assumption 2 is implicitly assuming a *constant* lead time.

The time between consecutive replenishments of inventory (the vertical line segments in Fig. 18.1) is referred to as a *cycle.* For the speaker example, a cycle can be viewed as the time between production runs. Thus, if 24,000 speakers are produced in each production run and are used at the rate of 8,000 per month, then the cycle length is $24{,}000/8{,}000 = 3$ months. In general, the cycle length is Q/d.

The total cost per unit time T is obtained from the following components.

Production or ordering cost per cycle = $K + cQ$.

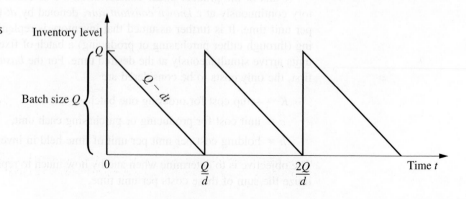

■ **FIGURE 18.1**
Diagram of inventory level as a function of time for the basic EOQ model.

The average inventory level during a cycle is $(Q + 0)/2 = Q/2$ units, and the corresponding cost is $hQ/2$ per unit time. Because the cycle length is Q/d,

$$\text{Holding cost per cycle} = \frac{hQ^2}{2d}.$$

Therefore,

$$\text{Total cost per cycle} = K + cQ + \frac{hQ^2}{2d},$$

so the total cost per unit time is

$$T = \frac{K + cQ + hQ^2/(2d)}{Q/d} = \frac{dK}{Q} + dc + \frac{hQ}{2}.$$

The value of Q, say Q^*, that minimizes T is found by setting the first derivative to zero (and noting that the second derivative is positive), which yields

$$-\frac{dK}{Q^2} + \frac{h}{2} = 0,$$

so that

$$Q^* = \sqrt{\frac{2dK}{h}},$$

which is the well-known *EOQ formula*.[2] (It also is sometimes referred to as the *square root formula*.) The corresponding *cycle time*, say t^*, is

$$t^* = \frac{Q^*}{d} = \sqrt{\frac{2K}{dh}}.$$

It is interesting to observe that Q^* and t^* change in intuitively plausible ways when a change is made in K, h, or d. As the setup cost K increases, both Q^* and t^* increase (fewer setups). When the unit holding cost h increases, both Q^* and t^* decrease (smaller inventory levels). As the demand rate d increases, Q^* increases (larger batches) but t^* decreases (more frequent setups).

These formulas for Q^* and t^* will now be applied to the speaker example. The appropriate parameter values from Sec. 18.1 are

$$K = 12,000, \qquad h = 0.30, \qquad d = 8,000,$$

so that

$$Q^* = \sqrt{\frac{(2)(8,000)(12,000)}{0.30}} = 25,298$$

and

$$t^* = \frac{25,298}{8,000} = 3.2 \text{ months.}$$

Hence, the optimal solution is to set up the production facilities to produce speakers once every 3.2 months and to produce 25,298 speakers each time. (The total cost curve is rather

---

[2] An interesting historical account of this model and formula, including a reprint of a 1913 paper that started it all, is given by D. Erlenkotter, "Ford Whitman Harris and the Economic Order Quantity Model," *Operations Research*, **38:** 937–950, 1990.

flat near this optimal value, so any similar production run that might be more convenient, say 24,000 speakers every 3 months, would be nearly optimal.)

The Worked Examples section of the book's website includes **another example** of applying the basic EOQ model when considerable sensitivity analysis also needs to be performed.

The EOQ Model with Planned Shortages

One of the banes of any inventory manager is the occurrence of an inventory shortage (sometimes referred to as a *stockout*)—demand that cannot be met currently because the inventory is depleted. This causes a variety of headaches, including dealing with unhappy customers and having extra record keeping to arrange for filling the demand later (*backorders*) when the inventory can be replenished. By assuming that planned shortages are not allowed, the basic EOQ model presented above satisfies the common desire of managers to avoid shortages as much as possible. (Nevertheless, unplanned shortages can still occur if the demand rate and deliveries do not stay on schedule.)

However, there are situations where permitting limited planned shortages makes sense from a managerial perspective. The most important requirement is that the customers generally are able and willing to accept a reasonable delay in filling their orders if need be. If so, the costs of incurring shortages described in Secs. 18.1 and 18.2 (including lost future business) should not be exorbitant. If the cost of holding inventory is high relative to these shortage costs, then lowering the average inventory level by permitting occasional brief shortages may be a sound business decision.

The **EOQ model with planned shortages** addresses this kind of situation by replacing only the third assumption of the basic EOQ model by the following new assumption.

> Planned shortages now are allowed. When a shortage occurs, the affected customers will wait for the product to become available again. Their backorders are filled immediately when the order quantity arrives to replenish inventory.

Under these assumptions, the pattern of inventory levels over time has the appearance shown in Fig. 18.2. The saw-toothed appearance is the same as in Fig. 18.1. However, now the inventory levels extend down to negative values that reflect the number of units of the product that are backordered.

Let

p = shortage cost per unit short per unit of time short,

S = inventory level just after a batch of Q units is added to inventory,

$Q - S$ = shortage in inventory just before a batch of Q units is added.

The total cost per unit time now is obtained from the following components.

Production or ordering cost per cycle = $K + cQ$.

■ **FIGURE 18.2**
Diagram of inventory level as a function of time for the EOQ model with planned shortages.

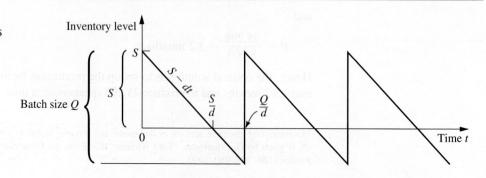

During each cycle, the inventory level is positive for a time S/d. The average inventory level *during this time* is $(S + 0)/2 = S/2$ units, and the corresponding cost is $hS/2$ per unit time. Hence,

$$\text{Holding cost per cycle} = \frac{hS}{2} \frac{S}{d} = \frac{hS^2}{2d}.$$

Similarly, shortages occur for a time $(Q - S)/d$. The average amount of shortages *during this time* is $(0 + Q - S)/2 = (Q - S)/2$ units, and the corresponding cost is $p(Q - S)/2$ per unit time. Hence,

$$\text{Shortage cost per cycle} = \frac{p(Q - S)}{2} \frac{Q - S}{d} = \frac{p(Q - S)^2}{2d}.$$

Therefore,

$$\text{Total cost per cycle} = K + cQ + \frac{hS^2}{2d} + \frac{p(Q - S)^2}{2d},$$

and the *total cost per unit time* is

$$T = \frac{K + cQ + hS^2/(2d) + p(Q-S)^2/(2d)}{Q/d}$$

$$= \frac{dK}{Q} + dc + \frac{hS^2}{2Q} + \frac{p(Q - S)^2}{2Q}.$$

In this model, there are two decision variables (S and Q), so the optimal values (S^* and Q^*) are found by setting the partial derivatives $\partial T/\partial S$ and $\partial T/\partial Q$ equal to zero. Thus,

$$\frac{\partial T}{\partial S} = \frac{hS}{Q} - \frac{p(Q - S)}{Q} = 0.$$

$$\frac{\partial T}{\partial Q} = -\frac{dK}{Q^2} - \frac{hS^2}{2Q^2} + \frac{p(Q - S)}{Q} - \frac{p(Q - S)^2}{2Q^2} = 0.$$

Solving these equations simultaneously leads to

$$S^* = \sqrt{\frac{2dK}{h}} \sqrt{\frac{p}{p + h}}, \qquad Q^* = \sqrt{\frac{2dK}{h}} \sqrt{\frac{p + h}{p}}.$$

The optimal cycle length t^* is given by

$$t^* = \frac{Q^*}{d} = \sqrt{\frac{2K}{dh}} \sqrt{\frac{p + h}{p}}.$$

The maximum shortage is

$$Q^* - S^* = \sqrt{\frac{2dK}{p}} \sqrt{\frac{h}{p + h}}.$$

In addition, from Fig. 18.2, the fraction of time that no shortage exists is given by

$$\frac{S^*/d}{Q^*/d} = \frac{p}{p + h},$$

which is independent of K.

When either p or h is made much larger than the other, the above quantities behave in intuitive ways. In particular, when $p \to \infty$ with h constant (so shortage costs dominate holding costs), $Q^* - S^* \to 0$ whereas both Q^* and t^* converge to their values for the basic EOQ model. Even though the current model permits shortages, $p \to \infty$ implies that having them is not worthwhile.

On the other hand, when $h \to \infty$ with p constant (so holding costs dominate shortage costs), $S^* \to 0$. Thus, having $h \to \infty$ makes it uneconomical to have positive inventory levels, so each new batch of Q^* units goes no further than removing the current shortage in inventory.

If planned shortages are permitted in the speaker example, the *shortage cost* is estimated in Sec. 18.1 as

$$p = 1.10.$$

As before,

$$K = 12{,}000, \qquad h = 0.30, \qquad d = 8{,}000,$$

so now

$$S^* = \sqrt{\frac{(2)(8{,}000)(12{,}000)}{0.30}} \sqrt{\frac{1.1}{1.1 + 0.3}} = 22{,}424,$$

$$Q^* = \sqrt{\frac{(2)(8{,}000)(12{,}000)}{0.30}} \sqrt{\frac{1.1 + 0.3}{1.1}} = 28{,}540,$$

and

$$t^* = \frac{28{,}540}{8{,}000} = 3.6 \text{ months.}$$

Hence, the production facilities are to be set up every 3.6 months to produce 28,540 speakers. The maximum shortage is 6,116 speakers. Note that Q^* and t^* are not very different from the no-shortage case. The reason is that p is much larger than h.

The EOQ Model with Quantity Discounts

When specifying their cost components, the preceding models have assumed that the unit cost of an item is the same regardless of the quantity in the batch. In fact, this assumption resulted in the optimal solutions being independent of this unit cost. The *EOQ model with quantity discounts* replaces this assumption by the following new assumption.

> The unit cost of an item now depends on the quantity in the batch. In particular, an incentive is provided to place a large order by replacing the unit cost for a small quantity by a smaller unit cost for every item in a larger batch, and perhaps by even smaller unit costs for even larger batches.

Otherwise, the assumptions are the same as for the basic EOQ model.

To illustrate this model, consider the TV speakers example introduced in Sec. 18.1. Suppose now that the unit cost for *every* speaker is $c_1 = \$11$ if less than 10,000 speakers are produced, $c_2 = \$10$ if production falls between 10,000 and 80,000 speakers, and $c_3 = \$9.50$ if production exceeds 80,000 speakers. What is the optimal policy? The solution to this specific problem will reveal the general method.

From the results for the basic EOQ model, the total cost per unit time T_j if the unit cost is c_j is given by

$$T_j = \frac{dK}{Q} + dc_j + \frac{hQ}{2}, \qquad \text{for } j = 1, 2, 3.$$

(This expression assumes that h is independent of the unit cost of the items, but a common small refinement would be to make h proportional to the unit cost to reflect the fact that the cost of capital tied up in inventory varies in this way.) A plot of T_j versus Q is

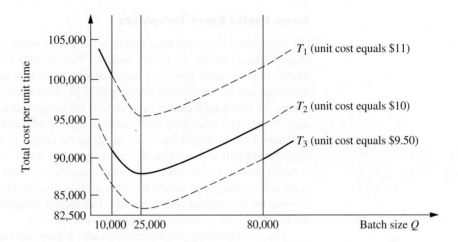

■ FIGURE 18.3
Total cost per unit time for
the speaker example with
quantity discounts.

shown in Fig. 18.3 for each j, where the solid part of each curve extends over the feasible range of values of Q for that discount category.

For each curve, the value of Q that minimizes T_j is found just as for the basic EOQ model. For $K = 12,000$, $h = 0.30$, and $d = 8,000$, this value is

$$\sqrt{\frac{(2)(8,000)(12,000)}{0.30}} = 25,298.$$

(If h were not independent of the unit cost of the items, then the minimizing value of Q would be slightly different for the different curves.) This minimizing value of Q is a feasible value for the cost function T_2. For any fixed Q, $T_2 < T_1$, so T_1 can be eliminated from further consideration. However, T_3 cannot be immediately discarded. Its minimum feasible value (which occurs at $Q = 80,000$) must be compared to T_2 evaluated at 25,298 (which is \$87,589). Because T_3 evaluated at 80,000 equals \$89,200, it is better to produce in quantities of 25,298, so this quantity is the optimal value for this set of quantity discounts.

If the quantity discount led to a unit cost of \$9 (instead of \$9.50) when production exceeded 80,000, then T_3 evaluated at 80,000 would equal \$85,200, and the optimal production quantity would become 80,000.

Although this analysis concerned a specific problem, the same approach is applicable to any similar problem. Here is a summary of the general procedure.

1. For each available unit cost c_j, use the EOQ formula for the EOQ model to calculate its optimal order quantity Q_j^*.
2. For each c_j where Q_j^* is within the feasible range of order quantities for c_j, calculate the corresponding total cost per unit time T_j.
3. For each c_j where Q_j^* is not within this feasible range, determine the order quantity Q_j that is at the endpoint of this feasible range that is closest to Q_j^*. Calculate the total cost per unit time T_j for Q_j and c_j.
4. Compare the T_j obtained for all the c_j and choose the minimum T_j. Then choose the order quantity Q_j obtained in step 2 or 3 that gives this minimum T_j.

A similar analysis can be used for other types of quantity discounts, such as incremental quantity discounts where a cost c_0 is incurred for the first q_0 units, c_1 for the next q_1 units, and so on.

Some Useful Excel Templates

For your convenience, we have included five Excel templates for the EOQ models in this chapter's Excel file on the book's website. Two of these templates are for the basic EOQ model. In both cases, you enter basic data (d, K, and h), as well as the lead time for the deliveries and the number of working days per year for the firm. The template then calculates the firm's total annual expenditures for setups and for holding costs, as well as the sum of these two costs (the *total variable cost*). It also calculates the *reorder point*—the inventory level at which the order needs to be placed to replenish inventory so the replenishment will arrive when the inventory level drops to 0. One template (the *Solver version*) enables you to enter any order quantity you want and then see what the annual costs and reorder point would be. This version also enables you to use the Excel Solver to solve for the optimal order quantity. The second template (the *analytical version*) uses the EOQ formula to obtain the optimal order quantity.

The corresponding pair of templates also is provided for the EOQ model with planned shortages. After entering the data (including the unit shortage cost p), each of these templates will obtain the various annual costs (including the annual shortage cost). With the Solver version, you can either enter trial values of the order quantity Q and maximum shortage $Q - S$ or solve for the optimal values, whereas the analytical version uses the formulas for Q^* and $Q^* - S^*$ to obtain the optimal values. The corresponding maximum inventory level S^* also is included in the results.

The final template is an analytical version for the EOQ model with quantity discounts. This template includes the refinement that the unit holding cost h is proportional to the unit cost c, so

$$h = Ic,$$

where the proportionality factor I is referred to as the *inventory holding cost rate*. Thus, the data entered includes I along with d and K. You also need to enter the number of discount categories (where the lowest-quantity category with no discount counts as one of these), as well as the unit price and range of order quantities for each of the categories. The template then finds the feasible order quantity that minimizes the total annual cost for each category, and also shows the individual annual costs (including the annual purchase cost) that would result. Using this information, the template identifies the overall optimal order quantity and the resulting total annual cost.

All these templates can be helpful for calculating a lot of information quickly after entering the basic data for the problem. However, perhaps a more important use is for performing sensitivity analysis on these data. You can immediately see how the results would change for any specific change in the data by entering the new data values in the spreadsheet. Doing this repeatedly for a variety of changes in the data is a convenient way to perform sensitivity analysis.

Observations about EOQ Models

1. If it is assumed that the unit cost of an item is constant throughout time, independent of the batch size (as with the first two EOQ models), the unit cost does not appear in the optimal solution for the batch size. This result occurs because no matter what inventory policy is used, the same number of units is required per unit time, so this cost per unit time is fixed.
2. The analysis of the EOQ models assumed that the batch size Q is constant from cycle to cycle. The resulting *optimal* batch size Q^* actually minimizes the total cost per unit time for any cycle, so the analysis shows that this constant batch size should be used from cycle to cycle even if a constant batch size is not assumed.

3. The optimal inventory level at which inventory should be replenished can never be greater than zero under these models. Waiting until the inventory level drops to zero (or less than zero when planned shortages are permitted) reduces both holding costs and the frequency of incurring the setup cost K. However, if the assumptions of *a known constant demand rate* and *the order quantity will arrive just when desired* (because of a constant lead time) are not completely satisfied, it may become prudent to plan to have some "safety stock" left when the inventory is scheduled to be replenished. This is accomplished by increasing the reorder point above that implied by the model.

4. The basic assumptions of the EOQ models are rather demanding ones. They seldom are satisfied completely in practice. For example, even when a constant demand rate is planned (as with the production line in the TV speakers example in Sec. 18.1), interruptions and variations in the demand rate still are likely to occur. It also is very difficult to satisfy the assumption that the order quantity to replenish inventory arrives just when desired. Although the schedule may call for a constant lead time, variations in the actual lead times often will occur. Fortunately, the EOQ models have been found to be robust in the sense that they generally still provide nearly optimal results even when their assumptions are only rough approximations of reality. This is a key reason why these models are so widely used in practice. However, in those cases where the assumptions are significantly violated, it is important to do some preliminary analysis to evaluate the adequacy of an EOQ model before it is used. This preliminary analysis should focus on calculating the total cost per unit time provided by the model for various order quantities and then assessing how this cost curve would change under more realistic assumptions.

Different Types of Demand for a Product

Example 2 (wholesale distribution of bicycles) introduced in Sec. 18.1 focused on managing the inventory of one model of bicycle. The demand for this product is generated by the wholesaler's customers (various retailers) who purchase these bicycles to replenish their inventories according to their own schedules. The wholesaler has no control over this demand. Because this model is sold separately from other models, its demand does not even depend on the demand for any of the company's other products. Such demand is referred to as **independent demand.**

The situation is different for the speaker example introduced in Sec. 18.1. Here, the product under consideration—television speakers—is just one component being assembled into the company's final product—television sets. Consequently, the demand for the speakers depends on the demand for the television set. The pattern of this demand for the speakers is determined internally by the production schedule that the company establishes for the television sets by adjusting the production rate for the production line producing the sets. Such demand is referred to as **dependent demand.**

The television manufacturing company produces a considerable number of products—various parts and subassemblies—that become components of the television sets. Like the speakers, these various products also are **dependent-demand products.**

Because of the dependencies and interrelationships involved, managing the inventories of dependent-demand products can be considerably more complicated than for independent-demand products. A popular technique for assisting in this task is **material requirements planning,** abbreviated as **MRP.** MRP is a computer-based system for planning, scheduling, and controlling the production of all the components of a final product. The system begins by "exploding" the product by breaking it down into all its subassemblies and then into all its individual component parts. A production schedule is then developed, using the demand and lead time for each component to determine the demand and lead time for the subsequent component in the process. In addition to a *master* *production schedule* for the final product, a *bill of materials* provides detailed information about all its components. Inventory status records give the current inventory levels, number of units on order, etc., for all the components. When more units of a component need to be ordered, the MRP system automatically generates either a purchase order to the vendor or a work order to the internal department that produces the component.[3]

The Role of Just-In-Time (JIT) Inventory Management

When the basic EOQ model was used to calculate the optimal production lot size for the speaker example, a very large quantity (25,298 speakers) was obtained. This enables having relatively infrequent setups to initiate production runs (only once every 3.2 months). However, it also causes large average inventory levels (12,649 speakers), which leads to a large total holding cost per year of over $45,000.

The basic reason for this large cost is the high setup cost of $K = \$12,000$ for each production run. The setup cost is so sizable because the production facilities need to be set up again from scratch each time. Consequently, even with less than four production runs per year, the annual setup cost is over $45,000, just like the annual holding costs.

Rather than continuing to tolerate a $12,000 setup cost each time in the future, another option for the company is to seek ways to reduce this setup cost. One possibility is to develop methods for quickly transferring machines from one use to another. Another is to dedicate a group of production facilities to the production of speakers so they would remain set up between production runs in preparation for beginning another run whenever needed.

Suppose the setup cost could be drastically reduced from $12,000 all the way down to $K = \$120$. This would reduce the optimal production lot size from 25,298 speakers down to $Q^* = 2,530$ speakers, so a new production run lasting only a brief time would be initiated more than 3 times per month. This also would reduce both the annual setup cost and the annual holding cost from over $45,000 down to only slightly over $4,500 each. By having such frequent (but inexpensive) production runs, the speakers would be produced essentially *just in time* for their assembly into television sets.

Just in time actually is a well-developed philosophy for managing inventories. A **just-in-time (JIT)** inventory system places great emphasis on reducing inventory levels to a bare minimum, and so providing the items just in time as they are needed. This philosophy was first developed in Japan, beginning with the Toyota Company in the late 1950s, and is given part of the credit for the remarkable gains in Japanese productivity through much of the late 20th century. The philosophy also has become popular in other parts of the world, including the United States, in more recent years.[4]

Although the just-in-time philosophy sometimes is misinterpreted as being incompatible with using an EOQ model (since the latter gives a large order quantity when the setup cost is large), they actually are complementary. A JIT inventory system focuses on finding ways to greatly reduce the setup costs so that the optimal order quantity will be small. Such a system also seeks ways to reduce the lead time for the delivery of an order, since this reduces the uncertainty about the number of units that will be needed when the delivery occurs. Another emphasis is on improving preventive maintenance so that the required production facilities will be available to produce the units when they are needed.

---

[3]A series of articles on pp. 32–44 of the September 1996 issue of *IIE Solutions* provides further information about MRP.

[4]For further information about applications of JIT in the United States, see R. E. White, J. N. Pearson, and J. R. Wilson, "JIT Manufacturing: A Survey of Implementations in Small and Large U.S. Manufacturing," *Management Science,* **45:** 1–15, 1999. Also see H. Chen, M. Z. Frank, and O. Q. Wu, "What Actually Happened to the Inventories of American Companies Between 1981 and 2000," *Management Science,* **51**(7): 1015–1031, July 2005.

Still another emphasis is on improving the production process to guarantee good quality. Providing just the right number of units just in time does not provide any leeway for including defective units.

In more general terms, the focus of the just-in-time philosophy is on *avoiding waste* wherever it might occur in the production process. One form of waste is unnecessary inventory. Others are unnecessarily large setup costs, unnecessarily long lead times, production facilities that are not operational when they are needed, and defective items. Minimizing these forms of waste is a key component of superior inventory management.

■ 18.4   A DETERMINISTIC PERIODIC-REVIEW MODEL

The preceding section explored the basic EOQ model and some of its variations. The results were dependent upon the assumption of a constant demand rate. When this assumption is relaxed, i.e., when the amounts that need to be withdrawn from inventory are allowed to vary from period to period, the *EOQ formula* no longer ensures a minimum-cost solution.

Consider the following periodic-review model. Planning is to be done for the next *n* periods regarding how much (if any) to produce or order to replenish inventory at the beginning of each of the periods. (The order to replenish inventory can involve either *purchasing* the units or *producing* them, but the latter case is far more common with applications of this model, so we mainly will use the terminology of *producing* the units.) The demands for the respective periods are *known* (but *not* the same in every period) and are denoted by

$$r_i = \text{demand in period } i, \qquad \text{for } i = 1, 2, \ldots, n.$$

These demands must be met on time. There is no stock on hand initially, but there is still time for a delivery at the beginning of period 1.

The costs included in this model are similar to those for the basic EOQ model:

K = setup cost for producing or purchasing any units to replenish inventory at beginning of period,

c = unit cost for producing or purchasing each unit,

h = holding cost for each unit left in inventory at end of period.

Note that this holding cost h is assessed only on inventory left at the end of a period. There also are holding costs for units that are in inventory for a portion of the period before being withdrawn to satisfy demand. However, these are *fixed* costs that are independent of the inventory policy and so are not relevant to the analysis. Only the *variable* costs that are affected by which inventory policy is chosen, such as the extra holding costs that are incurred by carrying inventory over from one period to the next, are relevant for selecting the inventory policy.

By the same reasoning, the unit cost c is an irrelevant fixed cost because, over all the time periods, all inventory policies produce the same number of units at the same cost. Therefore, c will be dropped from the analysis hereafter.

The objective is to minimize the total cost over the *n* periods. This is accomplished by ignoring the fixed costs and minimizing the total variable cost over the *n* periods, as illustrated by the following example.

An Example

An airplane manufacturer specializes in producing small airplanes. It has just received an order from a major corporation for 10 customized executive jet airplanes for the use of the corporation's upper management. The order calls for three of the airplanes to be delivered (and paid for) during the upcoming winter months (period 1), two more to be delivered during the spring (period 2), three more during the summer (period 3), and the final two during the fall (period 4).

Setting up the production facilities to meet the corporation's specifications for these airplanes requires a setup cost of $2 million. The manufacturer has the capacity to produce all 10 airplanes within a couple of months, when the winter season will be under way. However, this would necessitate holding seven of the airplanes in inventory, at a cost of $200,000 per airplane per period, until their scheduled delivery times. To reduce or eliminate these substantial holding costs, it may be worthwhile to produce a smaller number of these airplanes now and then to repeat the setup (again incurring the cost of $2 million) in some or all of the subsequent periods to produce additional small numbers. Management would like to determine the least costly production schedule for filling this order.

Thus, using the notation of the model, the demands for this particular airplane during the four upcoming periods (seasons) are

$$r_1 = 3, \qquad r_2 = 2, \qquad r_3 = 3, \qquad r_4 = 2.$$

Using units of millions of dollars, the relevant costs are

$$K = 2, \qquad h = 0.2.$$

The problem is to determine how many airplanes to produce (if any) during the beginning of each of the four periods in order to minimize the total variable cost.

The high setup cost K gives a strong incentive not to produce airplanes every period and preferably just once. However, the significant holding cost h makes it undesirable to carry a large inventory by producing the entire demand for all four periods (10 airplanes) at the beginning. Perhaps the best approach would be an intermediate strategy where airplanes are produced more than once but less than four times. For example, one such feasible solution (but not an optimal one) is depicted in Fig. 18.4, which shows the evolution of the inventory level over the next year that results from producing three airplanes at the beginning of the first period, six airplanes at the beginning of the second period, and one airplane at the beginning of the fourth period. The dots give the inventory levels after any production at the beginning of the four periods.

How can the optimal production schedule be found? For this model in general, production (or purchasing) is automatic in period 1, but a decision on whether to produce must be made for each of the other $n - 1$ periods. Therefore, one approach to solving this model is to enumerate, for each of the 2^{n-1} combinations of production decisions, the

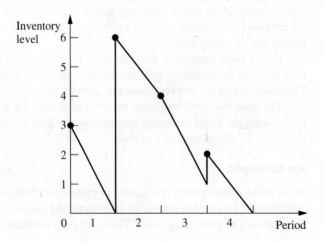

■ **FIGURE 18.4**
The inventory levels that result from one sample production schedule for the airplane example.

Inventory level

Period

possible quantities that can be produced in each period where production is to occur. This approach is rather cumbersome, even for moderate-sized n, so a more efficient method is desirable. Such a method is described next in general terms, and then we will return to finding the optimal production schedule for the example. Although the general method can be used when either producing or purchasing to replenish inventory, we now will only use the terminology of producing for definiteness.

An Algorithm

The key to developing an efficient algorithm for finding an *optimal inventory policy* (or equivalently, an *optimal production schedule*) for the above model is the following insight into the nature of an optimal policy.

> An optimal policy (production schedule) produces *only* when the inventory level is *zero*.

To illustrate why this result is true, consider the policy shown in Fig. 18.4 for the example. (Call it policy A.) Policy A violates the above characterization of an optimal policy because production occurs at the beginning of period 4 when the inventory level is *greater than zero* (namely, one airplane). However, this policy can easily be adjusted to satisfy the above characterization by simply producing one less airplane in period 2 and one more airplane in period 4. This adjusted policy (call it B) is shown by the dashed line in Fig. 18.5 wherever B differs from A (the solid line). Now note that policy B *must* have less total cost than policy A. The setup costs (and the production costs) for both policies are the same. However, the holding cost is smaller for B than for A because B has less inventory than A in periods 2 and 3 (and the same inventory in the other periods). Therefore, B is better than A, so A cannot be optimal.

This characterization of optimal policies can be used to identify policies that are not optimal. In addition, because it implies that the only choices for the amount produced at the beginning of the ith period are $0, r_i, r_i + r_{i+1}, \ldots,$ or $r_i + r_{i+1} + \cdots + r_n$, it can be exploited to obtain an efficient algorithm that is related to the *deterministic dynamic programming* approach described in Sec. 10.3.

In particular, define

C_i = total variable cost of an optimal policy for periods $i, i + 1, \ldots, n$ when period i starts with zero inventory (before producing), for $i = 1, 2, \ldots, n$.

■ FIGURE 18.5
Comparison of two inventory policies (production schedules) for the airplane example.

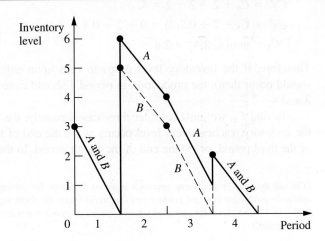

By using the dynamic programming approach of solving *backward* period by period, these C_i values can be found by first finding C_n, then finding C_{n-1}, and so on. Thus, after C_n, $C_{n-1}, \ldots, C_{i+1}$ are found, then C_i can be found from the *recursive relationship*

$$C_i = \min_{j=i, i+1, \ldots, n} \{C_{j+1} + K + h[r_{i+1} + 2r_{i+2} + 3r_{i+3} + \cdots + (j-i)r_j]\},$$

where j can be viewed as an index that denotes the (end of the) period when the inventory reaches a zero level for the first time after production at the beginning of period i. In the time interval from period i through period j, the term with coefficient h represents the total *holding cost* over this interval. When $j = n$, the term $C_{n+1} = 0$. The *minimizing value* of j indicates that if the inventory level does indeed drop to zero upon entering period i, then the production in period i should cover all demand from period i through this period j.

The algorithm for solving the model consists basically of solving for $C_n, C_{n-1}, \ldots, C_1$ in turn. For $i = 1$, the minimizing value of j then indicates that the production in period 1 should cover the demand through period j, so the second production will be in period $j + 1$. For $i = j + 1$, the new minimizing value of j identifies the time interval covered by the second production, and so forth to the end. We will illustrate this approach with the example.

The application of this algorithm is much quicker than the full dynamic programming approach.[5] As in dynamic programming, $C_n, C_{n-1}, \ldots, C_2$ must be found before C_1 is obtained. However, the number of calculations is much smaller, and the number of possible production quantities is greatly reduced.

Application of the Algorithm to the Example

Returning to the airplane example, first we consider the case of finding C_4, the cost of the optimal policy from the beginning of period 4 to the end of the planning horizon:

$$C_4 = C_5 + 2 = 0 + 2 = 2.$$

To find C_3, we must consider two cases, namely, the first time after period 3 when the inventory reaches a zero level occurs at (1) the end of the third period or (2) the end of the fourth period. In the recursive relationship for C_3, these two cases correspond to (1) $j = 3$ and (2) $j = 4$. Denote the corresponding costs (the right-hand side of the recursive relationship with this j) by $C_3^{(3)}$ and $C_3^{(4)}$, respectively. The policy associated with $C_3^{(3)}$ calls for producing only for period 3 and then following the optimal policy for period 4, whereas the policy associated with $C_3^{(4)}$ calls for producing for periods 3 and 4. The cost C_3 is then the minimum of $C_3^{(3)}$ and $C_3^{(4)}$. These cases are reflected by the policies given in Fig. 18.6.

$$C_3^{(3)} = C_4 + 2 = 2 + 2 = 4.$$
$$C_3^{(4)} = C_5 + 2 + 0.2(2) = 0 + 2 + 0.4 = 2.4.$$
$$C_3 = \min\{4, 2.4\} = 2.4.$$

Therefore, if the inventory level drops to zero upon entering period 3 (so production should occur then), the production in period 3 should cover the demand for both periods 3 and 4.

To find C_2, we must consider three cases, namely, the first time after period 2 when the inventory reaches a zero level occurs at (1) the end of the second period, (2) the end of the third period, or (3) the end of the fourth period. In the recursive relationship for C_2,

[5]The full dynamic programming approach is useful, however, for solving *generalizations* of the model (e.g., *nonlinear* production cost and holding cost functions) where the above algorithm is no longer applicable. (See Probs. 18.4-3 and 18.4-4 for examples where dynamic programming would be used to deal with generalizations of the model.)

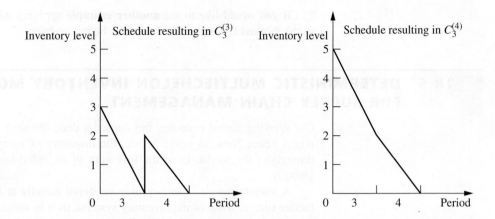

FIGURE 18.6
Alternative production schedules when production is required at the beginning of period 3 for the airplane example.

these cases correspond to (1) $j = 2$, (2) $j = 3$, and (3) $j = 4$, where the corresponding costs are $C_2^{(2)}$, $C_2^{(3)}$, and $C_2^{(4)}$, respectively. The cost C_2 is then the minimum of $C_2^{(2)}$, $C_2^{(3)}$, and $C_2^{(4)}$.

$$C_2^{(2)} = C_3 + 2 = 2.4 + 2 = 4.4.$$
$$C_2^{(3)} = C_4 + 2 + 0.2(3) = 2 + 2 + 0.6 = 4.6.$$
$$C_2^{(4)} = C_5 + 2 + 0.2[3 + 2(2)] = 0 + 2 + 1.4 = 3.4.$$
$$C_2 = \min\{4.4, 4.6, 3.4\} = 3.4.$$

Consequently, if production occurs in period 2 (because the inventory level drops to zero), this production should cover the demand for all the remaining periods.

Finally, to find C_1, we must consider four cases, namely, the first time after period 1 when the inventory reaches zero occurs at the end of (1) the first period, (2) the second period, (3) the third period, or (4) the fourth period. These cases correspond to $j = 1, 2, 3, 4$ and to the costs $C_1^{(1)}$, $C_1^{(2)}$, $C_1^{(3)}$, $C_1^{(4)}$, respectively. The cost C_1 is then the minimum of $C_1^{(1)}$, $C_1^{(2)}$, $C_1^{(3)}$, and $C_1^{(4)}$.

$$C_1^{(1)} = C_2 + 2 = 3.4 + 2 = 5.4.$$
$$C_1^{(2)} = C_3 + 2 + 0.2(2) = 2.4 + 2 + 0.4 = 4.8.$$
$$C_1^{(3)} = C_4 + 2 + 0.2[2 + 2(3)] = 2 + 2 + 1.6 = 5.6.$$
$$C_1^{(4)} = C_5 + 2 + 0.2[2 + 2(3) + 3(2)] = 0 + 2 + 2.8 = 4.8.$$
$$C_1 = \min\{5.4, 4.8, 5.6, 4.8\} = 4.8.$$

Note that $C_1^{(2)}$ and $C_1^{(4)}$ tie as the minimum, giving C_1. This means that the policies corresponding to $C_1^{(2)}$ and $C_1^{(4)}$ tie as being the optimal policies. The $C_1^{(4)}$ policy says to produce enough in period 1 to cover the demand for all four periods. The $C_1^{(2)}$ policy covers only the demand through period 2. Since the latter policy has the inventory level drop to zero at the end of period 2, the C_3 result is used next, namely, produce enough in period 3 to cover the demand for periods 3 and 4. The resulting production schedules are summarized below.

Optimal Production Schedules

1. Produce 10 airplanes in period 1.

Total variable cost = $4.8 million.

2. Produce 5 airplanes in period 1 and 5 airplanes in period 3.

Total variable cost = $4.8 million.

If you would like to see **another example** applying this algorithm, one is provided in the Worked Examples section of the book's website.

■ 18.5 DETERMINISTIC MULTIECHELON INVENTORY MODELS FOR SUPPLY CHAIN MANAGEMENT

Our growing global economy has caused a dramatic shift in inventory management in recent years. Now, as never before, the inventory of many manufacturers is scattered throughout the world. Even the inventory of an individual product may be dispersed globally.

A manufacturer's inventory may be stored initially at the point or points of manufacture (one *echelon* of the inventory system), then at national or regional warehouses (a second echelon), then at field distribution centers (a third echelon), and so on. Thus, each stage at which inventory is held in the progression through a multistage inventory system is called an **echelon** of the inventory system. Such a system with multiple echelons of inventory is referred to as a **multiechelon inventory system.** In the case of a fully integrated corporation that both manufactures its products and sells them at the retail level, its echelons will extend all the way to its retail outlets.

Some coordination is needed between the inventories of any particular product at the different echelons. Since the inventory at each echelon (except the last one) is used to replenish the inventory at the next echelon as needed, the inventory level currently needed at an echelon is affected by how soon replenishment will be needed at the various locations for the next echelon.

The analysis of multiechelon inventory systems is a major challenge. However, considerable innovative research (with roots tracing back to the middle of the 20th century) has been conducted to develop tractable multiechelon inventory models. With the growing prominence of multiechelon inventory systems, this undoubtedly will continue to be an active area of research.

Another key concept that has emerged in the global economy is that of *supply chain management.* This concept pushes the management of a multiechelon inventory system one step further by also considering what needs to happen to bring a product into the inventory system in the first place. However, as with inventory management, the main purpose still is to win the competitive battle against other companies in bringing the product to the customers as promptly as possible.

A **supply chain** is a network of facilities that procure raw materials, transform them into intermediate goods and then final products, and finally deliver the products to customers through a distribution system that includes a multiechelon inventory system. Thus, a supply chain spans procurement, manufacturing, and distribution. Since inventories are needed at all these stages, effective inventory management is one key element in managing the supply chain. To fill orders efficiently, it is necessary to understand the linkages and interrelationships of all the key elements of the supply chain. Therefore, integrated management of the supply chain has become a key success factor for some of today's leading companies.

To aid in supply chain management, multiechelon inventory models now are likely to include echelons that incorporate the early part of the supply chain as well as the echelons for the distribution of the finished product. Thus, the first echelon might be the inventory of raw materials or components that eventually will be used to produce the product. A second echelon could be the inventory of subassemblies that are produced from the raw materials or components in preparation for later assembling the subassemblies into the final product. This might then lead into the echelons for the distribution of the

An Application Vignette

finished product, starting with storage at the point or points of manufacture, then at national or regional warehouses, then at field distribution centers, and so on.

The usual objective for a multiechelon inventory model is to coordinate the inventories at the various echelons so as to minimize the total cost associated with the entire multiechelon inventory system. This is a natural objective for a fully integrated corporation that operates this entire system. It might also be a suitable objective when certain echelons are managed by either the suppliers or the customers of the company. The reason is that a key concept of supply chain management is that a company should strive to develop an informal partnership relationship with its suppliers and customers that enables them jointly to maximize their total profit. This often leads to developing mutually beneficial supply contracts that enable reducing the total cost of operating a jointly managed multiechelon inventory system.

The analysis of multiechelon inventory models tends to be considerably more complicated than those for single-facility inventory models considered elsewhere in this chapter. However, we present two relatively tractable multiechelon inventory models below that illustrate the relevant concepts.

A Model for a Serial Two-Echelon System

The simplest possible multiechelon inventory system is one where there are only two echelons and only a single installation at each echelon. Figure 18.7 depicts such a system, where the inventory at installation 1 is used to periodically replenish the inventory at installation 2. For example, installation 1 might be a factory producing a certain product with occasional production runs, and installation 2 might be the distribution center for that product. Alternatively, installation 2 might be the factory producing the product, and then installation 1 is another facility where the components needed to produce that product are themselves either produced or received from suppliers.

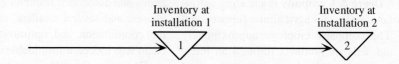

■ **FIGURE 18.7**
A serial two-echelon
inventory system.

Since the items at installation 1 and installation 2 may be different, we will refer to them as item 1 and item 2, respectively. The units of item 1 and item 2 are defined so that exactly one unit of item 1 is needed to obtain one unit of item 2. For example, if item 1 collectively consists of the components needed to produce the final product (item 2), then one set of components needed to produce one unit of the final product is defined as one unit of item 1.

The model makes the following assumptions.

Assumptions for Serial Two-Echelon Model.

1. The assumptions of the *basic EOQ model* (see Sec. 18.3) hold at installation 2. Thus, there is a known constant demand rate of d units per unit time, an order quantity of Q_2 units is placed in time to replenish inventory when the inventory level drops to zero, and planned shortages are not allowed.
2. The relevant costs at installation 2 are a *setup cost* of K_2 each time an order is placed and a *holding cost* of h_2 per unit per unit time.
3. Installation 1 uses its inventory to provide a batch of Q_2 units to installation 2 immediately each time an order is received.
4. An order quantity of Q_1 units is placed in time to replenish inventory at installation 1 before a shortage would occur.
5. Similarly to installation 2, the relevant costs at installation 1 are a *setup cost* of K_1 each time an order is placed and a *holding cost* of h_1 per unit per unit time.
6. The units increase in value when they are received and processed at installation 2, so $h_1 < h_2$.
7. The objective is to minimize the *sum* of the variable costs per unit time at the two installations. (This will be denoted by C.)

The word "immediately" in assumption 3 implies that there is essentially *zero lead time* between when installation 2 places an order for Q_2 units and installation 1 fills that order. In reality, it would be common to have a significant lead time because of the time needed for installation 1 to receive and process the order and then to transport the batch to installation 2. However, as long as the lead time is essentially fixed, this is equivalent to assuming zero lead time for modeling purposes because the order would be placed just in time to have the batch arrive when the inventory level drops to zero. For example, if the lead time is one week, the order would be placed one week before the inventory level drops to zero.

Although a zero lead time and a fixed lead time are equivalent for modeling purposes, we specifically are assuming a zero lead time because it simplifies the conceptualization of how the inventory levels at the two installations vary simultaneously over time. Figure 18.8 depicts this conceptualization. Because the assumptions of the basic EOQ model hold at installation 2, the inventory levels there vary according to the familiar saw-tooth pattern first shown in Fig. 18.1. Each time installation 2 needs to replenish its inventory, installation 1 ships Q_2 units of item 1 to installation 2. Item 1 may be identical to item 2 (as in the case of a factory shipping the final product to a distribution center). If not (as in the case of a supplier shipping the components needed to produce the final product to a factory), installation 2 immediately uses the shipment of Q_2 units of item 1 to produce Q_2 units of item 2 (the final product). The inventory at installation 2 then gets depleted at the constant

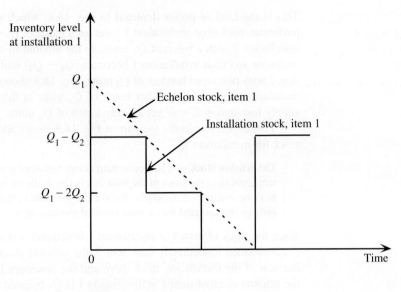

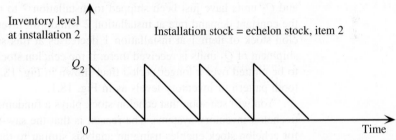

■ FIGURE 18.8
The synchronized inventory levels at the two installations when $Q_1 = 3Q_2$. The installation stock is the stock that is physically being held at the installation, whereas the echelon stock includes both the installation stock and the stock of the same item that already is downstream at the next installation (if any).

demand rate of d units per unit time until the next replenishment, which occurs just as the inventory level drops to 0.

The pattern of inventory levels over time for installation 1 is somewhat more complicated than for installation 2. Q_2 units need to be withdrawn from the inventory of installation 1 to supply installation 2 each time installation 2 needs to add Q_2 units to replenish its inventory. This necessitates replenishing the inventory of installation 1 occasionally, so an order quantity of Q_1 units is placed periodically. Using the same kind of reasoning as employed in the preceding section (including in Figs. 18.4 and 18.5), the *deterministic* nature of our model implies that installation 1 should replenish its inventory only at the instant when its inventory level is zero and it is time to make a withdrawal from the inventory in order to supply installation 2. The reasoning involves checking what would happen if installation 1 were to replenish its inventory any later or any earlier than this instant. If the replenishment were any later than this instant, installation 1 could not supply installation 2 in time to continue following the optimal inventory policy there, so this is unacceptable. If the replenishment were any earlier than this instant, installation 1 would incur the extra cost of holding this inventory until it is time to supply installation 2, so it is better to delay the replenishment at installation 1 until this instant. This leads to the following insight.

An optimal policy should have $Q_1 = nQ_2$ where n is a fixed positive integer. Furthermore, installation 1 should replenish its inventory with a batch of Q_1 units *only* when its inventory level is *zero* and it is time to supply installation 2 with a batch of Q_2 units.

This is the kind of policy depicted in Fig. 18.8, which shows the case where $n = 3$. In particular, each time installation 1 receives a batch of Q_1 units, it simultaneously supplies installation 2 with a batch of Q_2 units, so the amount of stock left on hand (called the *installation stock*) at installation 1 becomes $(Q_1 - Q_2)$ units. After later supplying installation 2 with two more batches of Q_2 units, Fig. 18.8 shows that the next cycle begins with installation 1 receiving another batch of Q_1 units at the same time as when it needs to supply installation 2 with yet another batch of Q_2 units.

The dashed line in the top part of Fig. 18.8 shows another quantity called the *echelon stock* for installation 1.

> The **echelon stock** of a particular item at any installation in a multiechelon inventory system consists of the stock of the item that is physically on hand at the installation (referred to as the *installation stock*) *plus* the stock of the same item that already is downstream (and perhaps incorporated into a more finished product) at subsequent echelons of the system.

Since the stock of item 1 at installation 1 is shipped periodically to installation 2, where it is transformed immediately into item 2, the echelon stock at installation 1 in Fig. 18.8 is the *sum* of the installation stock there and the inventory level at installation 2. At time 0, the echelon stock of item 1 at installation 1 is Q_1 because $(Q_1 - Q_2)$ units remain on hand and Q_2 units have just been shipped to installation 2 to replenish the inventory there. As the constant demand rate at installation 2 withdraws inventory there accordingly, the echelon stock of item 1 at installation 1 decreases at this same constant rate until the next shipment of Q_1 units is received there. If the echelon stock of item 1 at installation 1 were to be plotted over a longer period than shown in Fig. 18.8, you would see the same sawtooth pattern of inventory levels as in Fig. 18.1.

You will see soon that echelon stock plays a fundamental role in the analysis of multiechelon inventory systems. The reason is that the saw-tooth pattern of inventory levels for echelon stock enables using an analysis similar to that for the basic EOQ model.

Since the objective is to minimize the sum of the variable costs per unit time at the two installations, the easiest (and commonly used) approach would be to solve separately for the values of Q_2 and $Q_1 = nQ_2$ that minimize the total variable cost per unit at installation 2 and installation 1, respectively. Unfortunately, this approach overlooks (or ignores) the connections between the variable costs at the two installations. Because the batch size Q_2 for item 2 affects the pattern of inventory levels for item 1 at installation 1, optimizing Q_2 separately without considering the consequences for item 1 does not lead to an overall optimal solution.

To better understand this subtle point, it may be instructive to begin by optimizing separately at the two installations. We will do this and then demonstrate that this can lead to fairly large errors.

The Trap of Optimizing the Two Installations Separately.

Let us begin by optimizing installation 2 by itself. Since the assumptions for installation 2 fit the basic EOQ model precisely, the results presented in Sec. 18.3 for this model can be used directly. The total variable cost per unit time at this installation is

$$C_2 = \frac{dK_2}{Q_2} + \frac{h_2 Q_2}{2}.$$

(This expression for total *variable* cost differs from the one for total cost given in Sec. 18.3 for the basic EOQ model by deleting the *fixed* cost, dc, where c is the unit cost of acquiring the item.) The EOQ formula indicates that the optimal order quantity for this installation by itself is

$$Q_2^* = \sqrt{\frac{2dK_2}{h_2}},$$

so the resulting value of C_2 with $Q_2 = Q_2^*$ is

$$C_2^* = \sqrt{2dK_2h_2}.$$

Now consider installation 1 with an order quantity of $Q_1 = nQ_2$. Figure 18.8 indicates that the average inventory level of installation stock is $(n-1)Q_2/2$. Therefore, since installation 1 needs to replenish its inventory with Q_1 units every $Q_1/d = nQ_2/d$ units of time, the total variable cost per unit time at installation 1 is

$$C_1 = \frac{dK_1}{nQ_2} + \frac{h_1(n-1)Q_2}{2}.$$

To find the order quantity $Q_1 = nQ_2$ that minimizes C_1, given $Q_2 = Q_2^*$, we need to solve for the value of n that minimizes C_1. Ignoring the requirement that n be an integer, this is done by differentiating C_1 with respect to n, setting the derivative equal to zero (while noting that the second derivative is positive for positive n), and solving for n, which yields

$$n^* = \frac{1}{Q_2^*}\sqrt{\frac{2dK_1}{h_1}} = \sqrt{\frac{K_1h_2}{K_2h_1}}.$$

If n^* is an integer, then $Q_1 = n^*Q_2^*$ is the optimal order quantity for installation 1, given $Q_2 = Q_2^*$. If n^* is not an integer, then n^* needs to be rounded either up or down to an integer. The rule for doing this is the following.

Rounding Procedure for n^*

> If $n^* < 1$, choose $n = 1$.
> If $n^* > 1$, let $[n^*]$ be the largest integer $\leq n^*$, so $[n^*] \leq n^* < [n^*] + 1$, and then round as follows.
>
> If $\dfrac{n^*}{[n^*]} \leq \dfrac{[n^*]+1}{n^*}$, choose $n = [n^*]$.
>
> If $\dfrac{n^*}{[n^*]} > \dfrac{[n^*]+1}{n^*}$, choose $n = [n^*]+1$.

The formula for n^* indicates that its value depends on both K_1/K_2 and h_2/h_1. If both of these quantities are considerably greater than 1, then n^* also will be considerably greater than 1. Recall that assumption 6 of the model is that $h_1 < h_2$. This implies that h_2/h_1 exceeds 1, perhaps substantially so. The reason assumption 6 usually holds is that item 1 normally increases in value when it gets converted into item 2 (the final product) after item 1 is transferred to installation 2 (the location where the demand can be met for the final product). This means that the cost of capital tied up in each unit in inventory (usually the main component in holding costs) also will increase as the units move from installation 1 to installation 2. Similarly, if a production run needs to be set up to produce each batch at installation 1 (so K_1 is large), whereas only a relatively small administrative cost of K_2 is required for installation 2 to place each order, then K_1/K_2 will be considerably greater than 1.

The flaw in the above analysis comes in the first step when choosing the order quantity for installation 2. Rather than considering only the costs at installation 2 when doing this, the resulting costs at installation 1 also should have been taken into account. Let us turn now to the valid analysis that simultaneously considers both installations by minimizing the sum of the costs at the two locations.

Optimizing the Two Installations Simultaneously.   By adding the costs at the individual installations obtained above, the total variable cost per unit time at the two installations is

$$C = C_1 + C_2 = \left(\frac{K_1}{n} + K_2\right)\frac{d}{Q_2} + [(n-1)h_1 + h_2]\frac{Q_2}{2}.$$

The holding costs on the right have an interesting interpretation in terms of the holding costs for the *echelon stock* at the two installations. In particular, let

$e_1 = h_1$ = echelon holding cost per unit per unit time for installation 1,
$e_2 = h_2 - h_1$ = echelon holding cost per unit per unit time for installation 2.

Then the holding costs can be expressed as

$$[(n-1)h_1 + h_2]\frac{Q_2}{2} = h_1\frac{nQ_2}{2} + (h_2 - h_1)\frac{Q_2}{2}$$

$$= e_1\frac{Q_1}{2} + e_2\frac{Q_2}{2},$$

where $Q_1/2$ and $Q_2/2$ are the average inventory levels of the *echelon stock* at installations 1 and 2, respectively. (See Fig. 18.8.) The reason that $e_2 = h_2 - h_1$ rather than $e_2 = h_2$ is that $e_1Q_1/2 = h_1Q_1/2$ already includes the holding cost for the units of item 1 that are downstream at installation 2, so $e_2 = h_2 - h_1$ only needs to reflect the *value added* by converting the units of item 1 to units of item 2 at installation 2. (This concept of using echelon holding costs based on the value added at each installation will play an even more important role in our next model where there are more than two echelons.)

Using these echelon holding costs, we now have

$$C = \left(\frac{K_1}{n} + K_2\right)\frac{d}{Q_2} + (ne_1 + e_2)\frac{Q_2}{2}.$$

Differentiating with respect to Q_2, setting the derivative equal to zero (while verifying that the second derivative is positive for positive Q_2), and solving for Q_2 yields

$$Q_2^* = \sqrt{\frac{2d\left(\dfrac{K_1}{n} + K_2\right)}{ne_1 + e_2}}$$

as the optimal order quantity (given n) at installation 2. Note that this is identical to the EOQ formula for the basic EOQ model where the total setup cost is $K_1/n + K_2$ and the total unit holding cost is $ne_1 + e_2$.

Inserting this expression for Q_2^* into C and performing some algebraic simplification yields

$$C = \sqrt{2d\left(\frac{K_1}{n} + K_2\right)(ne_1 + e_2)}.$$

To solve for the optimal value of the order quantity at installation 1, $Q_1 = nQ_2^*$, we need to find the value of n that minimizes C. The usual approach for doing this would be to differentiate C with respect to n, set this derivative equal to zero, and solve for n. However, because the expression for C involves taking a square root, doing this directly is not very convenient. A more convenient approach is to get rid of the square root sign by squaring C and minimizing C^2 instead, since the value of n that minimizes C^2 also is the value that minimizes C. Therefore, we differentiate C^2 with respect to n, set this derivative equal to zero, and solve this equation for n. Since the second derivative is positive for positive n, this yields the minimizing value of n as

$$n^* = \sqrt{\frac{K_1 e_2}{K_2 e_1}}.$$

This is identical to the expression for n^* obtained in the preceding subsection except that h_1 and h_2 have been replaced here by e_1 and e_2, respectively. When n^* is not an integer, the procedure for rounding n^* to an integer also is the same as described in the preceding subsection.

Obtaining n in this way enables calculating Q_2^* with the above expression and then setting $Q_1^* = nQ_2^*$.

An Example. To illustrate these results, suppose that the parameters of the model are

$$K_1 = \$1,000, \qquad K_2 = \$100, \qquad h_1 = \$2, \qquad h_2 = \$3, \qquad d = 600.$$

Table 18.1 gives the values of Q_2^*, n^*, n (the rounded value of n^*), Q_1^*, and C^* (the resulting total variable cost per unit time) when solving in the two ways described in this section. Thus, the second column gives the results when using the imprecise approach of optimizing the two installations separately, whereas the third column uses the valid method of optimizing the two installations simultaneously.

Note that simultaneous optimization yields rather different results than separate optimization. The biggest difference is that the order quantity at installation 2 is nearly twice as large. In addition, the total variable cost C^* is nearly 3 percent smaller. With different parameter values, the error from separate optimization can sometimes lead to a considerably larger percentage difference in the total variable cost. Thus, this approach provides a pretty rough approximation. There is no reason to use it since simultaneous optimization can be performed just as readily.

A Model for a Serial Multiechelon System

We now will extend the preceding analysis to serial systems with more than two echelons. Figure 18.9 depicts this kind of system, where installation 1 has its inventory replenished periodically, then the inventory at installation 1 is used to replenish the inventory at installation 2 periodically, then installation 2 does the same for installation 3, and so on down to the final installation (installation N). Some or all of the installations might be processing centers that process the items received from the preceding installation and transform them into something closer to the finished product. Installations also are used to store items until they are ready to be moved to the next processing center or to the next storage facility that is closer to the customers for the final product. Installation N does any needed final processing and also stores the final product at a location where it can immediately meet the demand for that product on a continuous basis.

TABLE 18.1 Application of the serial two-echelon model to the example

Quantity	Separate Optimization of the Installations	Simultaneous Optimization of the Installations
Q_2^*	200	379
n^*	$\sqrt{15}$	$\sqrt{5}$
n	4	2
Q_1^*	800	758
C^*	\$1,950	\$1,897

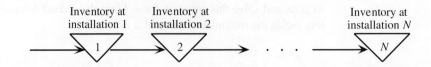

Since the items may be different at the different installations as they are being processed into something closer to the finished product, we will refer to them as item 1 while they are at installation 1, item 2 while at installation 2, and so forth. The units of the different items are defined so that exactly one unit of the item from one installation is needed to obtain one unit of the next item at the next installation.

Our model for a serial multiechelon inventory system is a direct generalization of the preceding one for a serial two-echelon inventory system, as indicated by the following assumptions for the model.

Assumptions for Serial Multiechelon Model

1. The assumptions of the basic EOQ model (see Sec. 18.3) hold at installation N. Thus, there is a known constant demand of d units per unit time, an order quantity of Q_N units is placed in time in replenish inventory when the inventory level drops to zero, and planned shortages are not allowed.
2. An order quantity of Q_1 units is placed in time to replenish inventory at installation 1 before a shortage would occur.
3. Each installation except installation N uses its inventory to periodically replenish the inventory of the next installation. Thus, installation i ($i = 1, 2, \ldots, N - 1$) provides a batch of Q_{i+1} units to installation ($i + 1$) immediately each time an order is received from installation ($i + 1$).
4. The relevant costs at each installation i ($i = 1, 2, \ldots, N$) are a *setup cost* of K_i each time an order is placed and a *holding cost* of h_i per unit per unit time.
5. The units increase in value each time they are received and processed at the next installation, so $h_1 < h_2 < \cdots < h_N$.
6. The objective is to minimize the *sum* of the variable costs per unit time at the N installations. (This will be denoted by C.)

The word "immediately" in assumption 3 implies that there is essentially zero lead time between when an installation places an order and the preceding installation fills that order, although a positive lead time that is fixed causes no complication. With zero lead time, Fig. 18.10 extends Fig. 18.8 to show how the inventory levels would vary simultaneously at the installations when there are four installations instead of only two. In this case, $Q_i = 2Q_{i+1}$ for $i = 1, 2, 3$, so each of the first three installations needs to replenish its inventory once for every two times it replenishes the inventory of the next installation. Consequently, when a complete cycle of replenishments at all four installations begins at time 0, Fig. 18.10 shows an order of Q_1 units arriving at installation 1 when the inventory level had been zero. Half of this order then is immediately used to replenish the inventory at installation 2. Installation 2 then does the same for installation 3, and installation 3 does the same for installation 4. Therefore, at time 0, some of the units that just arrived at installation 1 get transferred downstream as far as to the last installation as quickly as possible. The last installation then immediately starts using its replenished inventory of the final product to meet the demand of d units per unit time for that product.

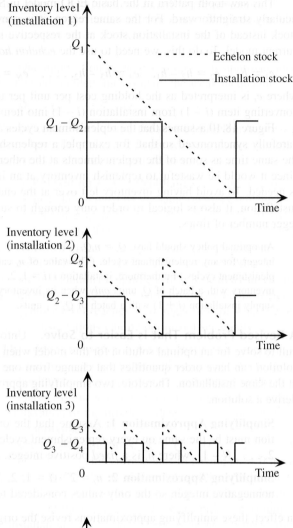

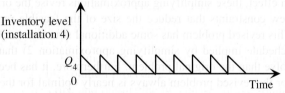

FIGURE 18.10
The synchronized inventory level at four installations ($N = 4$) when $Q_i = 2Q_{i+1}$ ($i = 1, 2, 3$), where the solid lines show the levels of the installation stock and the dashed lines do the same for the echelon stock.

Recall that the *echelon stock* at installation 1 is defined as the stock that is physically on hand there (the *installation stock*) plus the stock that already is downstream (and perhaps incorporated into a more finished product) at subsequent echelons of the inventory system. Therefore, as the dashed lines in Fig. 18.10 indicate, the echelon stock at installation 1 begins at Q_1 units at time 0 and then decreases at the rate of d units per unit time until it is time to order another batch of Q_1 units, after which the saw-tooth pattern continues. The echelon stock at installations 2 and 3 follow the same saw-tooth pattern, but with shorter cycles. The echelon stock coincides with the installation stock at installation 4, so the echelon stock again follows a saw-tooth pattern there.

(This expression for C_i assumes that the echelon inventory is replenished just as its level reaches zero, which holds for the original and revised problems, but is only an approximation for the relaxation of the problem because the lack of coordination between installations in setting order quantities tends to lead to premature replenishments.) Note that C_i is just the total variable cost per unit time for a single installation that satisfies the basic EOQ model when e_i is the relevant holding cost per unit time at the installation. Therefore, by first solving the relaxed problem, which only requires optimizing the installations separately (when using echelon holding costs instead of installation holding costs), the EOQ formula simply would be used to obtain the order quantity at each installation. It turns out that this provides a reasonable first approximation of the optimal order quantities when optimizing the installations simultaneously for the revised problem. Therefore, applying the EOQ formula in this way is the key step in phase 1 of the solution procedure. Phase 2 then applies the needed coordination between the order quantities by applying simplifying approximation 2.

When applying the EOQ formula to the respective installations, a special situation arises when $K_i/e_i < K_{i+1}/e_{i+1}$, since this would lead to $Q_i^* < Q_{i+1}^*$, which is prohibited by the relaxation of the problem. To satisfy the relaxation, which requires that $Q_i \geq Q_{i+1}$, the best that can be done is to set $Q_i = Q_{i+1}$. As described at the end of the preceding subsection, this implies that the two installations should be merged for modeling purposes.

Outline of Phase 1 (Solve the Relaxation)

1. If $\frac{K_i}{e_i} < \frac{K_{i+1}}{e_{i+1}}$ for any $i = 1, 2, \ldots, N - 1$, treat installations i and $i + 1$ as a single merged installation (for modeling purposes) with a setup cost of $K_i + K_{i+1}$ and an echelon holding cost of $e_i + e_{i+1}$ per unit per unit time. After the merger, repeat this step as needed for any other pairs of consecutive installations (which might include a merged installation). Then renumber the installations accordingly with N reset as the new total number of installations.

2. Set

$$Q_i = \sqrt{\frac{2dK_i}{e_i}}, \qquad \text{for } i = 1, 2, \ldots, N.$$

3. Set

$$C_i = \frac{dK_i}{Q_i} + \frac{e_i Q_i}{2}, \qquad \text{for } i = 1, 2, \ldots, N,$$

$$C = \sum_{i=1}^{N} C_i.$$

Phase 2 of the Solution Procedure. Phase 2 now is used to coordinate the order quantities to obtain a convenient cyclic schedule of replenishments, such as the one illustrated in Fig. 18.10. This is done mainly by rounding the order quantities obtained in phase 1 to fit the pattern prescribed in the simplifying approximations. After tentatively determining the values of $n_i = 2^{m_i}$ such that $Q_i = n_i Q_{i+1}$ in this way, the final step is to refine the value of Q_N to attempt to obtain an overall optimal solution for the revised problem.

This final step involves expressing each Q_i in terms of Q_N. In particular, given each n_i such that $Q_i = n_i Q_{i+1}$, let p_i be the product,

$$p_i = n_i n_{i+1} \cdots n_{N-1}, \qquad \text{for } i = 1, 2, \ldots, N - 1,$$

so that

$$Q_i = p_i Q_N, \qquad \text{for } i = 1, 2, \ldots, N - 1,$$

where $p_N = 1$. Therefore, the total variable cost per unit time at all the installations is

$$C = \sum_{i=1}^{N} \left[\frac{dK_i}{p_i Q_N} + \frac{e_i p_i Q_N}{2} \right].$$

Since C includes only the single order quantity Q_N, this expression also can be interpreted as the total variable cost per unit time for a *single* inventory facility that satisfies the basic EOQ model with a setup cost and unit holding cost of

$$\text{Setup cost} = \sum_{i=1}^{N} \frac{dK_i}{p_i}, \qquad \text{Unit holding cost} = \sum_{i=1}^{N} e_i p_i.$$

Hence, the value of Q_N that minimizes C is given by the EOQ formula as

$$Q_N^* = \sqrt{\frac{2d \sum_{i=1}^{N} \dfrac{K_i}{p_i}}{\sum_{i=1}^{N} e_i p_i}}.$$

Because this expression requires knowing the n_i, phase 2 begins by using the value of Q_N calculated in phase 1 as an approximation of Q_N^*, and then uses this Q_N to determine the n_i (tentatively), before using this formula to calculate Q_N^*.

Outline of Phase 2 (Solve the Revised Problem)

1. Set Q_N^* to the value of Q_N obtained in phase 1.
2. For $i = N - 1, N - 2, \ldots, 1$ in turn, do the following. Using the value of Q_i obtained in phase 1, determine the nonnegative integer value of m such that

$$2^m Q_{i+1}^* \leq Q_i < 2^{m+1} Q_{i+1}^*.$$

   If $\dfrac{Q_i}{2^m Q_{i+1}^*} \leq \dfrac{2^{m+1} Q_{i+1}^*}{Q_i},$ set $n_i = 2^m$ and $Q_i^* = n_i Q_{i+1}^*.$

   If $\dfrac{Q_i}{2^m Q_{i+1}^*} > \dfrac{2^{m+1} Q_{i+1}^*}{Q_i},$ set $n_i = 2^{m+1}$ and $Q_i^* = n_i Q_{i+1}^*.$

3. Use the values of the n_i obtained in step 2 and the above formulas for p_i and Q_N^* to calculate Q_N^*. Then use this Q_N^* to repeat step 2.[7] If none of the n_i change, use $(Q_1^*, Q_2^*, \ldots, Q_N^*)$ as the solution for the revised problem and calculate the corresponding cost C. If any of the n_i did change, repeat step 2 (starting with the current Q_N^*) and then step 3 one more time. Use the resulting solution and calculate $\overline{C}$.

This procedure provides a very good solution for the revised problem. Although the solution is not guaranteed to be optimal, it often is and, if not, it should be close. Since the revised problem is itself an approximation of the original problem, obtaining such a solution for the revised problem is very adequate for all practical purposes. Available theory guarantees that this solution will provide a good approximation of an optimal solution for the original problem.

Recall that Roundy's 98 percent approximation property guarantees that the cost of an optimal solution for the revised problem is within 2 percent of C^*, the cost of the unknown optimal solution for the original problem. In practice, this difference usually is far less

---

[7]A possible complication that would prevent repeating step 2 is if $Q_{N-1} < Q_N^*$ with this new value of Q_N^*. If this occurs, you can simply stop and use the previous value of $(Q_1^*, Q_2^*, \ldots, Q_N^*)$ as the solution for the revised problem. This same provision also applies for a subsequent attempt to repeat step 2.

than 2 percent. If the solution obtained by the above procedure is not optimal for the revised problem, Roundy's results still guarantee that its cost $\overline{C}$ is within 6 percent of C^*. Again, the actual difference in practice usually is far less than 6 percent and often is considerably less than 2 percent.

It would be nice to be able to check how close $\overline{C}$ is on any particular problem even though C^* is unknown. The relaxation of the problem provides an easy way of doing this. Because the relaxed problem does not require coordinating the inventory replenishments at the installations, the cost that is calculated for its optimal solution $\underline{C}$ is a lower bound on C^*. Furthermore, $\underline{C}$ normally is *extremely* close to C^*. Therefore, checking how close $\overline{C}$ is to $\underline{C}$ gives a conservative estimate of how close $\overline{C}$ must be to C^*, as summarized below.

> **Cost Relationships:** $\underline{C} \leq C^* \leq \overline{C}$, so $\overline{C} - C^* \leq \overline{C} - \underline{C}$, where
> $\underline{C}$ = cost of an optimal solution for the *relaxed* problem,
> C^* = cost of an (unknown) optimal solution for the *original* problem,
> $\overline{C}$ = cost of the solution obtained for the *revised* problem.

You will see in the following rather typical example that, because $\overline{C} = 1.0047\underline{C}$ for the example, it is known that $\overline{C}$ is within 0.47 percent of C^*.

An Example.   Consider a serial system with four installations that have the setup costs and unit holding costs shown in Table 18.2.

The first step in applying the model is to convert the unit holding cost h_i at each installation into the corresponding unit echelon holding cost e_i that reflects the value added at each installation. Thus,

$$e_1 = h_1 = \$0.50, \qquad e_2 = h_2 - h_1 = \$0.05,$$
$$e_3 = h_3 - h_2 = \$3, \qquad e_4 = h_4 - h_3 = \$4.$$

We now can apply step 1 of phase 1 of the solution procedure to compare each K_i/e_i with K_{i+1}/e_{i+1}.

$$\frac{K_1}{e_1} = 500, \qquad \frac{K_2}{e_2} = 120, \qquad \frac{K_3}{e_3} = 10, \qquad \frac{K_4}{e_4} = 27.5$$

These ratios decrease from left to right with the exception that

$$\frac{K_3}{e_3} = 10 < \frac{K_4}{e_4} = 27.5,$$

so we need to treat installations 3 and 4 as a single merged installation for modeling purposes. After combining their setup costs and their echelon holding costs, we now have the adjusted data shown in Table 18.3.

Using the adjusted data, Table 18.4 shows the results of applying the rest of the solution procedure to this example.

■ **TABLE 18.2** Data for the example of a four-echelon inventory system

Installation i	K_i	h_i	$d = 4,000$
1	$250	$0.50	
2	$6	$0.55	
3	$30	$3.55	
4	$110	$7.55	

■ **TABLE 18.3** Adjusted data for the four-echelon example after merging installations 3 and 4 for modeling purposes

Installation i	K_i	e_i	$d = 4{,}000$
1	$250	$0.50	
2	$6	$0.05	
3(+ 4)	$140	$7	

■ **TABLE 18.4** Results from applying the solution procedure to the four-echelon example

Installation i	Solution of Relaxed Problem Q_i		Initial Solution of Revised Problem Q_i^*		Final Solution of Revised Problem Q_i^*	
	Q_i	C_i	Q_i^*	C_i	Q_i^*	C_i
1	2,000	$1,000	1,600	$1,025	1,700	$1,013
2	980	$49	800	$50	850	$49
3(+ 4)	400	$2,800	400	$2,800	425	$2,805
		$C = $3,849		$C = $3,875		$C = $3,867

The second and third columns present the straightforward calculations from steps 2 and 3 of phase 1. For step 1 of phase 2, $Q_3 = 400$ in the second column is carried over to $Q_3^* = 400$ in the fourth column. For step 2, we find that

$$2^1 Q_3^* < Q_2 < 2^2 Q_3^*$$

since

$$2(400) = 800 < 980 < 4(400) = 1600.$$

Because

$$\frac{Q_2}{2^1 Q_3^*} = \frac{980}{800} < \frac{1600}{980} = \frac{2^2 Q_3^*}{Q_2},$$

we set $n_2 = 2^1 = 2$ and $Q_2^* = n_2 Q_3^* = 800$. Similarly, we set $n_1 = 2^1 = 2$ and $Q_1^* = n_1 Q_2^* = 1{,}600$, since

$$2(800) = 1{,}600 < 2{,}000 < 4(800) = 3{,}200 \quad \text{and} \quad \frac{2{,}000}{1{,}600} < \frac{3{,}200}{2{,}000}.$$

After calculating the corresponding C_i, the fourth and fifth columns of the table summarize these results from applying only steps 1 and 2 of phase 2.

The last two columns of the table then summarize the results from completing the solution procedure by applying step 3 of phase 2. Since $p_1 = n_1 n_2 = 4$ and $p_2 = n_2 = 2$, the formula for Q_N^* yields $Q_3^* = 425$ as the value of Q_3 that is part of the overall optimal solution for the revised problem. Repeating step 2 with this new Q_3^* again yields $n_2 = 2$ and $n_1 = 2$, so $Q_2^* = n_2 Q_3^* = 850$ and $Q_1^* = n_1 Q_2^* = 1{,}700$. Because n_2 and n_1 did not change from the first time through step 2, we indeed now have the desired solution for the revised problem, so the C_i are calculated accordingly. (This solution is, in fact, optimal for the revised problem.)

Keep in mind that the original installations 3 and 4 have been merged only for modeling purposes. They presumably will continue to be physically separate installations.

Therefore, the conclusion in the sixth column of the table that $Q_3^* = 425$ actually means that *both* installations 3 and 4 will have an order quantity of 425. As soon as installation 3 receives and processes each such order, it then will immediately transfer the entire batch to installation 4.

The bottom of the third, fifth, and seventh columns of the table show the total variable cost per unit time for the corresponding solutions. The cost C in the fifth column is 0.68 percent above $\underline{C}$ in the third column, whereas $\overline{C}$ in the seventh column is only 0.47 percent above $\underline{C}$. Since $\underline{C}$ is a lower bound on C^*, the cost of the (unknown) optimal solution for the original problem, this means that stopping after step 2 of phase 2 provided a solution that is within 0.68 percent of C^*, whereas the refinement from going on to step 3 of phase 2 improved the solution to within 0.47 percent of C^*.

Extensions of These Models

The two models presented previously in this section are both for serial inventory systems. As depicted earlier in Fig. 18.9, this restricts each installation (after the first one) to having only a single *immediate predecessor* that replenishes its inventory. By the same token, each installation (before the last one) replenishes the inventory of only a single *immediate successor.*

Many real multiechelon inventory systems are more complicated than this. An installation might have *multiple immediate successors,* such as when a factory supplies multiple warehouses or when a warehouse supplies multiple retailers. Such an inventory system is called a **distribution system.** Figure 18.11 shows a typical distribution inventory system for a particular product. In this case, this product (among others) is produced at a single factory, which sets up a quick production run each time it needs to replenish its inventory of the product. This inventory is used to supply several warehouses in different regions, replenishing their inventories of the product when needed. Each of these warehouses in turn supply several retailers within its region, replenishing their inventories of the product when needed. If each retailer has (roughly) a known constant demand

■ **FIGURE 18.11**
A typical distribution inventory system.

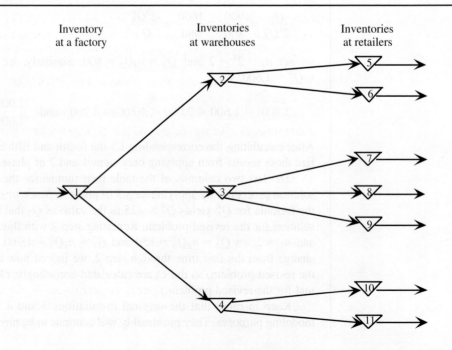

Inventory at a factory   Inventories at warehouses   Inventories at retailers

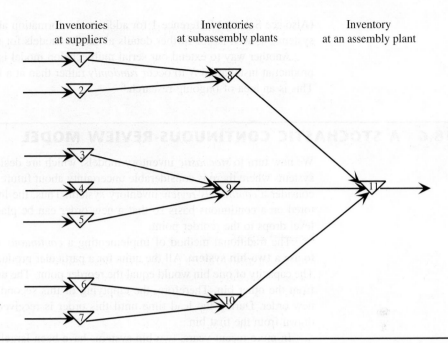

Inventories at suppliers

Inventories at subassembly plants

Inventory at an assembly plant

■ **FIGURE 18.12**
A typical assembly inventory system.

rate for the product, an extension of the serial multiechelon model can be formulated for this distribution inventory system. (We will not pursue this further.)

Another common generalization of a serial multiechelon inventory system arises when some installations have *multiple immediate predecessors,* such as when a subassembly plant receives its components from multiple suppliers or when a factory receives its subassemblies from multiple subassembly plants. Such an inventory system is called an **assembly system.** Figure 18.12 shows a typical assembly inventory system. In this case, a particular product is assembled at an assembly plant, drawing on inventories of subassemblies maintained there to assemble the product. Each of these inventories of a subassembly is replenished when needed by a plant that produces that subassembly, drawing on inventories of components maintained there to produce the subassembly. In turn, each of these inventories of a component is replenished when needed by a supplier that periodically produces this component to replenish its own inventory. Under the appropriate assumptions, another extension of the serial multiechelon model can be formulated for this assembly inventory system.

Some multiechelon inventory systems also might include both installations that have multiple immediate successors and installations that have multiple immediate predecessors. (Some installations might even fall into both categories.) Some of the greatest challenges of supply chain management come from dealing with these mixed kinds of multiechelon inventory systems. A particular challenge arises when separate organizations (e.g., suppliers, a manufacturer, and retailers) control different parts of a multiechelon inventory system, whether it be a mixed system, a distribution system, or an assembly system. In this case, a key principle of successful supply chain management is that the organizations should work together, including through the development of mutually beneficial supply contracts, to optimize the overall operation of the multiechelon inventory system.

Although the analysis of distribution systems and assembly systems presents some additional complications, the approach presented here for the serial multiechelon model (including Roundy's 98 percent approximation property) can be extended to these other kinds of multiechelon inventory systems as well. Details are provided by Selected Reference 6.

(Also see Selected Reference 1 for additional information about these kinds of inventory systems, as well as for further details about the models for serial systems.)

Another way to extend our serial multiechelon model is to allow the demand for the product at installation N to occur *randomly* rather than at a known constant demand rate. This is an area of ongoing research.[8]

18.6   A STOCHASTIC CONTINUOUS-REVIEW MODEL

We now turn to *stochastic* inventory models, which are designed for analyzing inventory systems where there is considerable uncertainty about future demands. In this section, we consider a *continuous-review* inventory system. Thus, the inventory level is being monitored on a continuous basis so that a new order can be placed as soon as the inventory level drops to the reorder point.

The traditional method of implementing a *continuous-review* inventory system was to use a **two-bin system.** All the units for a particular product would be held in two bins. The capacity of one bin would equal the reorder point. The units would first be withdrawn from the other bin. Therefore, the emptying of this second bin would trigger placing a new order. During the lead time until this order is received, units would then be withdrawn from the first bin.

In more recent years, two-bin systems have been largely replaced by **computerized inventory systems.** Each addition to inventory and each sale causing a withdrawal are recorded electronically, so that the current inventory level always is in the computer. (For example, the modern scanning devices at retail store checkout stands may both itemize your purchases and record the sales of stable products for purposes of adjusting the current inventory levels.) Therefore, the computer will trigger a new order as soon as the inventory level has dropped to the reorder point. Several excellent software packages are available from software companies for implementing such a system.

Because of the extensive use of computers for modern inventory management, continuous-review inventory systems have become increasingly prevalent for products that are sufficiently important to warrant a formal inventory policy.

A continuous-review inventory system for a particular product normally will be based on two critical numbers:

R = reorder point.
Q = order quantity.

For a manufacturer managing its finished products inventory, the order will be for a *production run* of size Q. For a wholesaler or retailer (or a manufacturer replenishing its raw materials inventory from a supplier), the order will be a *purchase order* for Q units of the product.

An inventory policy based on these two critical numbers is a simple one.

Inventory policy: Whenever the inventory level of the product drops to R units, place an order for Q more units to replenish the inventory.

Such a policy is often called a *reorder-point, order-quantity policy,* or **(R, Q) policy** for short. [Consequently, the overall model might be referred to as the (R, Q) model. Other variations of these names, such as (Q, R) policy, (Q, R) model, etc., also are sometimes used.]

[8]For example, see H. K. Shang and L.-S. Song, "Newsvendor Bounds and Heuristic for Optimal Policies in Serial Supply Chains," *Management Science,* **49**(5): 618–638, May 2003. Also see X. Chao and S. X. Zhou, "Probabilistic Solution and Bounds for Serial Inventory Systems with Discounted and Average Costs," *Naval Research Logistics,* **54**(6): 623–631, Sept. 2007.

After summarizing the model's assumptions, we will outline how R and Q can be determined.

The Assumptions of the Model

1. Each application involves a single product.
2. The inventory level is under *continuous review,* so its current value always is known.
3. An (R, Q) policy is to be used, so the only decisions to be made are to choose R and Q.
4. There is a *lead time* between when the order is placed and when the order quantity is received. This lead time can be either fixed or variable.
5. The *demand* for withdrawing units from inventory to sell them (or for any other purpose) during this lead time is uncertain. However, the probability distribution of demand is known (or at least estimated).
6. If a stockout occurs before the order is received, the excess demand is *backlogged,* so that the backorders are filled once the order arrives.
7. A fixed *setup cost* (denoted by K) is incurred each time an order is placed.
8. Except for this setup cost, the cost of the order is proportional to the order quantity Q.
9. A certain holding cost (denoted by h) is incurred for each unit in inventory per unit time.
10. When a stockout occurs, a certain shortage cost (denoted by p) is incurred for each unit backordered per unit time until the backorder is filled.

This model is closely related to the *EOQ model with planned shortages* presented in Sec. 18.3. In fact, all these assumptions also are consistent with that model, with the one key exception of assumption 5. Rather than having uncertain demand, that model assumed *known demand* with a fixed rate.

Because of the close relationship between these two models, their results should be fairly similar. The main difference is that, because of the uncertain demand for the current model, some safety stock needs to be added when setting the reorder point to provide some cushion for having well-above-average demand during the lead time. Otherwise, the trade-offs between the various cost factors are basically the same, so the order quantities from the two models should be similar.

Choosing the Order Quantity Q

The most straightforward approach to choosing Q for the current model is to simply use the formula given in Sec. 18.3 for the EOQ model with planned shortages. This formula is

$$Q = \sqrt{\frac{2dK}{h}} \sqrt{\frac{p + h}{p}},$$

where d now is the *average* demand per unit time, and where K, h, and p are defined in assumptions 7, 9, and 10, respectively.

This Q will be only an approximation of the optimal order quantity for the current model. However, no formula is available for the exact value of the optimal order quantity, so an approximation is needed. Fortunately, the approximation given above is a fairly good one.[9]

Choosing the Reorder Point R

A common approach to choosing the reorder point R is to base it on management's desired level of service to customers. Thus, the starting point is to obtain a managerial decision on service level. (Problem 18.6-3 analyzes the factors involved in this managerial decision.)

[9]For further information about the quality of this approximation, see S. Axsäter, "Using the Deterministic EOQ Formula in Stochastic Inventory Control," *Management Science,* **42:** 830–834, 1996. Also see Y.-S. Zheng, "On Properties of Stochastic Systems," *Management Science,* **38:** 87–103, 1992.

Service level can be defined in a number of different ways in this context, as outlined below.

Alternative Measures of Service Level.

1. The probability that a stockout will not occur between the time an order is placed and the order quantity is received.
2. The average number of stockouts per year.
3. The average percentage of annual demand that can be satisfied immediately (no stockout).
4. The average delay in filling backorders when a stockout occurs.
5. The overall average delay in filling orders (where the delay without a stockout is 0).

Measures 1 and 2 are closely related. For example, suppose that the order quantity Q has been set at 10 percent of the annual demand, so an average of 10 orders are placed per year. If the probability is 0.2 that a stockout *will* occur during the lead time until an order is received, then the average number of stockouts per year would be $10(0.2) = 2$.

Measures 2 and 3 also are related. For example, suppose an average of 2 stockouts occur per year and the average length of a stockout is 9 days. Since $2(9) = 18$ days of stockout per year are essentially 5 percent of the year, the average percentage of annual demand that can be satisfied immediately would be 95 percent.

In addition, measures 3, 4, and 5 are related. For example, suppose that the average percentage of annual demand that can be satisfied immediately is 95 percent and the average delay in filling backorders when a stockout occurs is 5 days. Since only 5 percent of the customers incur this delay, the overall average delay in filling orders then would be $0.05(5) = 0.25$ day per order.

A managerial decision needs to be made on the desired value of at least one of these measures of service level. After selecting one of these measures on which to focus primary attention, it is useful to explore the implications of several alternative values of this measure on some of the other measures before choosing the best alternative.

Measure 1 probably is the most convenient one to use as the primary measure, so we now will focus on this case. We will denote the desired level of service under this measure by L, so

$L =$ management's desired probability that a stockout will not occur between the time an order quantity is placed and the order quantity is received.

Using measure 1 involves working with the estimated probability distribution of the following random variable.

$D =$ demand during the lead time in filling an order.

For example, with a uniform distribution, the formula for choosing the reorder point R is a simple one.

If the probability distribution of D is a *uniform distribution* over the interval from a to b, set

$$R = a + L(b - a),$$

because then

$$P(D \leq R) = L.$$

Since the mean of this distribution is

$$E(D) = \frac{a + b}{2},$$

the amount of **safety stock** (the expected inventory level *just* before the order quantity is received) provided by the reorder point R is

$$\text{Safety stock} = R - E(D) = a + L(b - a) - \frac{a + b}{2}$$

$$= \left(L - \frac{1}{2}\right)(b - a).$$

When the demand distribution is something other than a uniform distribution, the procedure for choosing R is similar.

General Procedure for Choosing R under Service Level Measure 1.

1. Choose L.
2. Solve for R such that

$$P(D \leq R) = L.$$

For example, suppose that D has a normal distribution with mean μ and variance σ^2, as shown in Fig. 18.13. Given the value of L, the table for the normal distribution given in Appendix 5 then can be used to determine the value of R. In particular, you just need to find the value of K_{1-L} in this table and then plug into the following formula to find R.

$$R = \mu + K_{1-L}\sigma.$$

The resulting amount of safety stock is

$$\text{Safety stock} = R - \mu = K_{1-L}\sigma.$$

To illustrate, if $L = 0.75$, then $K_{1-L} = 0.675$, so

$$R = \mu + 0.675\sigma,$$

as shown in Fig. 18.13. This provides

$$\text{Safety stock} = 0.675\sigma.$$

Your OR Courseware also includes an Excel template that will calculate both the order quantity Q and the reorder point R for you. You need to enter the average demand per unit time (d), the costs (K, h, and p), and the service level based on measure 1. You also indicate whether the probability distribution of the demand during the lead time is a uniform distribution or a normal distribution. For a uniform distribution, you specify the interval over which the distribution extends by entering the lower endpoint and upper

---

■ **FIGURE 18.13**
Calculation of the reorder point R for the stochastic continuous-review model when $L = 0.75$ and the probability distribution of the demand over the lead time is a normal distribution with mean μ and standard deviation σ.

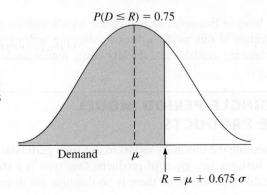

endpoint of this interval. For a normal distribution, you instead enter the mean μ and standard deviation σ of the distribution. After you provide all this information, the template immediately calculates Q and R and displays these results on the right side.

An Example

Consider once again Example 1 (manufacturing speakers for TV sets) presented in Sec. 18.1. Recall that the setup cost to produce the speakers is $K = \$12,000$, the unit holding cost is $h = \$0.30$ per speaker per month, and the unit shortage cost is $p = \$1.10$ per speaker per month.

Originally, there was a fixed demand rate of 8,000 speakers per month to be assembled into television sets being produced on a production line at this fixed rate. However, sales of the TV sets have been quite variable, so the inventory level of finished sets has fluctuated widely. To reduce inventory holding costs for finished sets, management has decided to adjust the production rate for the sets on a daily basis to better match the output with the incoming orders.

Consequently, the demand for the speakers now is quite variable. There is a *lead time* of 1 month between ordering a production run to produce speakers and having speakers ready for assembly into television sets. The demand for speakers during this lead time is a random variable D that has a normal distribution with a mean of 8,000 and a standard deviation of 2,000. To minimize the risk of disrupting the production line producing the TV sets, management has decided that the safety stock for speakers should be large enough to avoid a stockout during this lead time 95 percent of the time.

To apply the model, the order quantity for each production run of speakers should be

$$Q = \sqrt{\frac{2dK}{h}} \sqrt{\frac{p + h}{p}} = \sqrt{\frac{2(8,000)(12,000)}{0.30}} \sqrt{\frac{1.1 + 0.3}{1.1}} = 28,540.$$

This is the same order quantity that was found by the EOQ model with planned shortages in Sec. 18.3 for the previous version of this example where there was a *constant* (rather than average) demand rate of 8,000 speakers per month and planned shortages were allowed. However, the key difference from before is that safety stock now needs to be provided to counteract the variable demand. Management has chosen a service level of $L = 0.95$, so the normal table in Appendix 5 gives $K_{1-L} = 1.645$. Therefore, the reorder point should be

$$R = \mu + K_{1-L}\sigma = 8,000 + 1.645(2,000) = 11,290.$$

The resulting amount of safety stock is

Safety stock $= R - \mu = 3,290.$

The Worked Examples section of the book's website provides **another example** of the application of this model when two shipping options with different distributions for the lead time are available and the less costly option needs to be identified.

■ 18.7   A STOCHASTIC SINGLE-PERIOD MODEL FOR PERISHABLE PRODUCTS

When choosing the inventory model to use for a particular product, a distinction should be made between two types of products. One type is a **stable product,** which will remain sellable indefinitely so there is no deadline for disposing of its inventory. This is the kind of product considered in the preceding sections. The other type, by contrast, is a **perishable product,** which can be carried in inventory for only a very limited period of time before it can no longer be sold. This is the kind of product for which the single-period model (and its variations) presented in this section is designed. In particular, the single period in the model is the very limited period before the product can no longer be sold.

One example of a perishable product is a daily newspaper being sold at a newsstand. A particular day's newspaper can be carried in inventory for only a single day before it becomes outdated and needs to be replaced by the next day's newspaper. When the demand for the newspaper is a random variable (as assumed in this section), the owner of the newsstand needs to choose a daily order quantity that provides an appropriate trade-off between the potential cost of overordering (the wasted expense of ordering more newspapers than can be sold) and the potential cost of underordering (the lost profit from ordering fewer newspapers than can be sold). This section's model enables solving for the daily order quantity that would maximize the expected profit.

Because the general problem being analyzed fits this example so well, the problem is often called the **newsvendor problem.** However, it has always been recognized that the model being used is just as applicable to other perishable products as to newspapers. In fact, most of the applications have been to perishable products other than newspapers, including the examples of perishable products listed below.

Some Types of Perishable Products

As you read through the list below of various types of perishable products, think about how the inventory management of such products is analogous to a newsstand dealing with a daily newspaper since these products also cannot be sold after a single time period. All that may differ is that the length of this time period may be a week, a month, or even several months rather than just one day.

1. Periodicals, such as newspapers and magazines.
2. Flowers being sold by a florist.
3. The makings of fresh food to be prepared in a restaurant.
4. Produce, including fresh fruits and vegetables, to be sold in a grocery store.
5. Christmas trees.
6. Seasonal clothing, such as winter coats, where any goods remaining at the end of the season must be sold at highly discounted prices to clear space for the next season.
7. Seasonal greeting cards.
8. Fashion goods that will be out of style soon.
9. New cars at the end of a model year.
10. Any product that will be obsolete soon.
11. Vital spare parts that must be produced during the last production run of a certain model of a product (e.g., an airplane) for use as needed throughout the lengthy field life of that model.
12. Reservations provided by an airline for a particular flight, since the seats available on the flight can be viewed as the inventory of a perishable product (they cannot be sold after the flight has occurred).

This last type is a particularly interesting one because major airlines (and various other companies involved with transporting passengers) now are making extensive use of operations research to analyze how to maximize their revenue when dealing with this special kind of inventory. This special branch of inventory theory (commonly called *revenue management*) is the subject of the next section.

Time Inc. is the largest magazine publisher in the United States. With a portfolio of more than 125 magazines, one out of every two American adults reads a Time Inc. magazine each month.

A magazine is a good example of a perishable product, given how quickly each issue goes out of date, so the inventory model described in this section tends to fit magazines as well. From the viewpoint of Time Inc., this "newsvendor problem" for each magazine arises at three different levels—the corporate level, the wholesale level, and the retail level—but with a complication in each case that is not fully captured by the assumptions of the model. At the corporate level, a decision must be made about the number of copies of the magazine to print, but where the demand for the magazine is largely determined by negotiations with the wholesalers rather than a random variable. Similarly, each wholesaler must decide how many copies to take, but where the demand it will realize for the magazine is largely determined by negotiations with its retailers rather than a random variable. For each retailer, the demand it will realize for the magazine is indeed a random variable, but the data needed to make a reasonable estimate of the probability distribution for the random variable may not be available. (For example, if an issue of the magazine sells out before it is time for the next issue, the retailer cannot determine what the demand would have been if an adequate supply had been available.)

With the help of an OR consultant, a task force drew on *research in inventory management* to determine how to better integrate the decisions being made at the three levels. Building up from the demand at the grassroots (retail) level, OR analysis was done to make the best use of the available data to evaluate each magazine's national print order, the wholesaler allotment procedure, and the retail distribution process. Well-known solutions for formal inventory models had to be adapted so they could be implemented within the constraints of the magazine distribution channel. However, this OR study succeeded in developing a well-designed new three-echelon distribution process. The adoption of this new process has resulted in *generating incremental profits in excess of* **$3.5 million** *annually* for Time Inc.

Source: M. A. Koschat, G. L. Berk, J. A. Blatt, N. M. Kunz, M. H. LePore, and S. Blyakher: "Newsvendors Tackle the Newsvendor Problem," *Interfaces,* **33**(3): 72–84, May–June 2003. (A link to this article is provided on our website, www.mhhe.com/hillier.)

When managing the inventory of these various types of perishable products, it is occasionally necessary to deal with some considerations beyond those that will be discussed in this section. Extensive research has been conducted to extend the model to encompass these considerations, and considerable progress has been made. (Selected Reference 5 provides a literature review of this research.[10])

An Example

Refer back to Example 2 in Sec. 18.1, which involves the wholesale distribution of a particular bicycle model. There now has been a new development. The manufacturer has just informed the distributor that this model is being discontinued. To help clear out its stock, the manufacturer is offering the distributor the opportunity to make one final purchase at very favorable terms, namely, a *unit cost* of only $200 per bicycle. With these special arrangements, the distributor also would incur *no significant setup cost* to place this order.

The distributor feels that this offer provides an ideal opportunity to make one final round of sales to its customers (bicycle shops) for the upcoming Christmas season for a reduced price of only $450 per bicycle, thereby making a profit of $250 per bicycle. This will need to be a one-time sale only because this model soon will be replaced by a new model that will make it obsolete. Therefore, any bicycles not sold during this sale will become almost worthless. However, the distributor believes that she will be able to dispose of any remaining bicycles after Christmas by selling them for the nominal price of $100 each (the *salvage value*), thereby recovering half of her purchase cost. Considering

[10]More recent research includes G. Raz and E. L. Porteus, "A Fractiles Perspective to the Joint Price/Quantity Newsvendor Model," *Management Science,* **52**(11): 1764–1777, Nov. 2006.

this loss if she orders more than she can sell, as well as the lost profit if she orders fewer than can be sold, the distributor needs to decide what order quantity to submit to the manufacturer.

The administrative cost incurred by placing this special order for the Christmas season is fairly small, so this cost will be ignored until near the end of this section.

Another relevant expense is the cost of maintaining unsold bicycles in inventory until they can be disposed of after Christmas. Combining the cost of capital tied up in inventory and other storage costs, this inventory cost is estimated to be $10 per bicycle remaining in inventory after Christmas. Thus, considering the salvage value of $100 as well, the *unit holding cost* is −$90 per bicycle left in inventory at the end.

Two remaining cost components still require discussion, the shortage cost and the revenue. If the demand exceeds the supply, those customers who fail to purchase a bicycle may bear some ill will, thereby resulting in a "cost" to the distributor. This cost is the per-item quantification of the loss of goodwill times the unsatisfied demand whenever a shortage occurs. The distributor considers this cost to be negligible.

If we adopt the criterion of maximizing profit, we must include revenue in the model. Indeed, the total profit is equal to total revenue minus the costs incurred (the ordering, holding, and shortage costs). Assuming *no initial inventory,* this profit for the distributor is

> Profit = $450 × number sold by distributor
> − $200 × number purchased by distributor
> + $90 × number unsold and so disposed of for salvage value.

Let

S = number purchased by distributor
= stock (inventory) level after receiving this purchase (since there is no initial inventory)

and

D = demand by bicycle shops (a random variable),

so that

$$\min\{D, S\} = \text{number sold},$$
$$\max\{0, S - D\} = \text{number unsold}.$$

Then

$$\text{Profit} = 450 \min\{D, S\} - 200S + 90 \max\{0, S - D\}.$$

The first term also can be written as

$$450 \min\{D, S\} = 450D - 450 \max\{0, D - S\}.$$

The term $450 \max\{0, D - S\}$ represents the *lost revenue from unsatisfied demand.* This lost revenue, plus any cost of the loss of customer goodwill due to unsatisfied demand (assumed negligible in this example), will be interpreted as the *shortage cost* throughout this section.

Now note that $450D$ is independent of the inventory policy (the value of S chosen) and so can be deleted from the objective function, which leaves

$$\text{Relevant profit} = -450 \max\{0, D - S\} - 200S + 90 \max\{0, S - D\}$$

to be maximized. All the terms on the right are the *negative* of *costs,* where these costs are the *shortage cost,* the *ordering cost,* and the *holding cost* (which has a negative value here), respectively. Rather than *maximizing* the *negative* of *total cost,* we instead will do the equivalent of *minimizing*

$$\text{Total cost} = 450 \max\{0, D - S\} + 200S - 90 \max\{0, S - D\}.$$

More precisely, since total cost is a random variable (because D is a random variable), the objective adopted for the model is to *minimize the expected total cost.*

In the discussion about the interpretation of the shortage cost, we assumed that the unsatisfied demand was lost (no backlogging). If the unsatisfied demand could be met by a priority shipment, similar reasoning applies. The revenue component of net income would become the sales price of a bicycle ($450) times the demand *minus* the unit cost of the priority shipment times the unsatisfied demand whenever a shortage occurs. If our wholesale distributor could be forced to meet the unsatisfied demand by purchasing bicycles from the manufacturer for $350 each plus an air freight charge of, say, $20 each, then the appropriate shortage cost would be $370 per bicycle. (If there were any costs associated with loss of goodwill, these also would be added to this amount.)

The distributor does not know what the demand for these bicycles will be; i.e., demand D is a random variable. However, an optimal inventory policy can be obtained if information about the probability distribution of D is available. Let

$$P_D(d) = P\{D = d\}.$$

It will be assumed that $P_D(d)$ is known for all values of $d = 0, 1, 2, \ldots$.

We now are in a position to summarize the model in general terms.

The Assumptions of the Model

1. Each application involves a single perishable product.
2. Each application involves a single time period because the product cannot be sold later.
3. However, it will be possible to dispose of any units of the product remaining at the end of the period, perhaps even receiving a *salvage value* for the units.
4. There may be some initial inventory on hand going into this time period, as denoted by

   I = initial inventory.

5. The only decision to be made is the number of units to order (either through purchasing or producing) so they can be placed into inventory at the beginning of the period. Thus,

   Q = order quantity,
   S = stock (inventory) level after receiving this order
      = $I + Q$.

   Given I, it will be convenient to use S as the model's *decision variable,* which then automatically determines $Q = S - I$.
6. The *demand* for withdrawing units from inventory to sell them (or for any other purpose) during the period is a random variable D. However, the probability distribution of D is known (or at least estimated).[11]

---

[11]In practice, it commonly is necessary to estimate the probability distribution from a limited amount of past demand data. Research on how to drop assumption 6 and instead apply the available demand data directly includes R. Levi, R. O. Roundy, and D. B. Shmoys, "Provably Near-Optimal Sampling-Based Policies for Stochastic Inventory Control Models," *Mathematics of Operations Research,* **32**(4): 821–839, Nov. 2007. Also see L. Y. Chu, J. G. Shanthikumar, and Z.-J. M. Shen, "Solving Operational Statistics Via a Bayesian Analysis," *Operations Research Letters,* **36**(1): 110–116, Jan. 2008.

7. After deleting the revenue if the demand were satisfied (since this is independent of the decision S), the objective becomes to minimize the expected total cost, where the cost components are

K = setup cost for purchasing or producing the entire batch of units,

c = unit cost for purchasing or producing each unit,

h = holding cost per unit remaining at end of period (includes storage cost minus salvage value),

p = shortage cost per unit of unsatisfied demand (includes lost revenue and cost of loss of customer goodwill).

Analysis of the Model with No Initial Inventory ($I = 0$) and No Setup Cost ($K = 0$)

Before analyzing the model in its full generality, it will be instructive to begin by considering the simpler case where $I = 0$ (no initial inventory) and $K = 0$ (no setup cost).

The decision on the value of S, the amount of inventory to acquire, depends heavily on the probability distribution of demand D. More than the expected demand may be desirable, but probably less than the maximum possible demand. A trade-off is needed between (1) the risk of being short and thereby incurring shortage costs and (2) the risk of having an excess and thereby incurring wasted costs of ordering and holding excess units. This is accomplished by minimizing the expected value (in the statistical sense) of the sum of these costs.

The amount sold is given by

$$\min\{D, S\} = \begin{cases} D & \text{if } D < S \\ S & \text{if } D \geq S. \end{cases}$$

Hence, the cost incurred if the demand is D and S is stocked is given by

$$C(D, S) = cS + p \max\{0, D - S\} + h \max\{0, S - D\}.$$

Because the demand is a random variable [with probability distribution $P_D(d)$], this cost is also a random variable. The expected cost is then given by $C(S)$, where

$$C(S) = E[C(D, S)] = \sum_{d=0}^{\infty} (cS + p \max\{0, d - S\} + h \max\{0, S - d\})P_D(d)$$

$$= cS + \sum_{d=S}^{\infty} p(d - S)P_D(d) + \sum_{d=0}^{S-1} h(S - d)P_D(d).$$

The function $C(S)$ depends upon the probability distribution of D. Frequently, a representation of this probability distribution is difficult to find, particularly when the demand ranges over a large number of possible values. Hence, this *discrete random variable* is often approximated by a *continuous random variable*. Furthermore, when demand ranges over a large number of possible values, this approximation will generally yield a nearly exact value of the optimal amount of inventory to stock. In addition, when discrete demand is used, the resulting expressions may become slightly more difficult to solve analytically. Therefore, unless otherwise stated, *continuous demand* is assumed throughout the remainder of this chapter.

For this continuous random variable D, let

$f(x)$ = probability density function of D

and

$F(d)$ = cumulative distribution function (CDF) of D,

so

$$F(d) = \int_0^d f(x)\, dx.$$

When choosing an inventory level S, the CDF $F(d)$ becomes the probability that a shortage will *not* occur before the period ends. As in the preceding section, this probability is referred to as the **service level** being provided by the order quantity. The corresponding expected cost $C(S)$ is expressed as

$$C(S) = E[C(D, S)] = \int_0^\infty C(x, S) f(x)\, dx$$

$$= \int_0^\infty (cS + p \max\{0, x - S\} + h \max\{0, S - x\}) f(x)\, dx$$

$$= cS + \int_S^\infty p(x - S) f(x)\, dx + \int_0^S h(S - x) f(x)\, dx.$$

It then becomes necessary to find the value of S, say S^*, which minimizes $C(S)$. Finding a formula for S^* requires a relatively protracted and sophisticated derivation, so we will only give the answer here. However, the derivation is provided on the book's website as a supplement to this chapter for the more mathematically inclined and curious reader. (This supplement also briefly extends the model to the case where the holding costs and shortage costs are *nonlinear* instead of linear functions.)

This supplement shows that the $C(S)$ function has roughly the shape shown in Fig. 18.14, because it is a *convex* function (i.e., the second derivative is *nonnegative* everywhere). In fact, it is a *strictly convex* function (i.e., the second derivative is *strictly positive* everywhere) if $f(x) > 0$ for all $x \geq 0$. Furthermore, the first derivative becomes positive for sufficiently large S, so $C(S)$ must possess a global minimum. This global minimum is shown in Fig. 18.14 as S^*, so $S = S^*$ is the optimal inventory (stock) level to obtain when the order quantity $(Q = S^*)$ is received at the beginning of the period.

In particular, the supplement finds that the optimal inventory level S^* is that value which satisfies

$$F(S^*) = \frac{p - c}{p + h}.$$

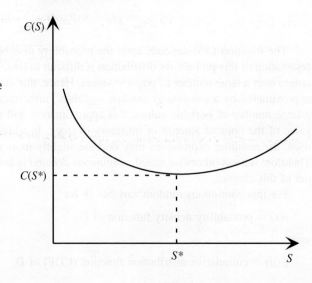

■ **FIGURE 18.14**
Graph of $C(S)$, the expected cost for the stochastic single-period model for perishable products as a function of S (the inventory level when the order quantity $Q = S - I$ is received at the beginning of the period), given that the initial inventory is $I = 0$ and the setup cost is $K = 0$.

Thus, $F(S^*)$ is the *optimal service level* and the corresponding inventory level S^* can be obtained either by solving this equation algebraically or by plotting the CDF and then identifying S^* graphically. To interpret the right-hand side of this equation, the numerator can be viewed as

$p - c =$ unit cost of underordering
    $=$ decrease in profit that results from failing to order a unit that could have been sold during the period.

Similarly,

$c + h =$ unit cost of overordering
    $=$ decrease in profit that results from ordering a unit that could not be sold during the period.

Therefore, denoting the unit cost of underordering and of overordering by C_{under} and C_{over}, respectively, this equation is specifying that

$$\text{Optimal service level} = \frac{C_{under}}{C_{under} + C_{over}}.$$

When the demand has either a uniform or an exponential distribution, an automatic procedure is available in your IOR Tutorial for calculating S^*. A similar Excel template also is included in this chapter's Excel files on the book's website.

If D is assumed to be a discrete random variable having the CDF

$$F(d) = \sum_{n=0}^{d} P_D(n),$$

a similar result is obtained. In particular, the optimal inventory level S^* is the smallest integer such that

$$F(S^*) \geq \frac{p - c}{p + h}.$$

The Worked Examples section of the book's website provides **another example** involving airline overbooking where D is a discrete random variable. The example below treats D as a continuous random variable.

Application to the Example

Returning to the bicycle example described at the beginning of this section, we assume that the demand has an exponential distribution with a mean of 10,000, so that its probability density function is

$$f(x) = \begin{cases} \dfrac{1}{10,000} e^{-x/10,000} & \text{if } x \geq 0 \\ 0 & \text{otherwise} \end{cases}$$

and the CDF is

$$F(d) = \int_{0}^{d} \frac{1}{10,000} e^{-x/10,000} \, dx = 1 - e^{-d/10,000}.$$

From the data given,

$$c = 200, \qquad p = 450, \qquad h = -90.$$

Consequently, S^* (the optimal inventory level to obtain at the outset to begin meeting the demand) is that value which satisfies

$$1 - e^{-S^*/10,000} = \frac{450 - 200}{450 - 90} = 0.69444.$$

By using the natural logarithm (denoted by ln), this equation can be solved as follows:

$$e^{-S^*/10,000} = 0.30556,$$
$$\ln e^{-S^*/10,000} = \ln 0.30556,$$
$$\frac{-S^*}{10,000} = -1.1856,$$
$$S^* = 11,856.$$

Therefore, the distributor should stock 11,856 bicycles in the Christmas season. Note that this number is slightly more than the expected demand of 10,000.

Whenever the demand has an exponential distribution with an expected value of λ, then S^* can be obtained from the relation

$$S^* = -\lambda \ln \frac{c + h}{p + h}.$$

Analysis of the Model with Initial Inventory ($I > 0$) but No Setup Cost ($K = 0$)

Now consider the case where $I > 0$, so there are already I units in inventory going into the period but prior to the receipt of the order quantity, $Q = S - I$. (For example, this case would arise for the bicycle example if the distributor begins with 500 bicycles before placing an order, so $I = 500$.) We continue to assume that $K = 0$ (no setup cost).

Let

$\overline{C}(S)$ = expected cost for the model for any value of I and K (including the current assumption that $K = 0$), given that S is the inventory level obtained when the order quantity is received at the beginning of the period,

so the objective is to choose $S \geq I$ so as to

Minimize $\overline{C}(S)$.
$S \geq I$

It will be instructive to compare $\overline{C}(S)$ with the cost function used in the preceding subsection (and plotted in Fig. 18.14),

$C(S)$ = expected cost for the model, given S, when $I = 0$ and $K = 0$.

With $K = 0$,

$$\overline{C}(S) = c(S - I) + \int_S^\infty p(x - S) f(x)\, dx + \int_0^S h(S - x) f(x)\, dx.$$

Thus, $\overline{C}(S)$ is identical to $C(S)$ except for the first term, where $C(S)$ has cS instead of $c(S - I)$. Therefore,

$$\overline{C}(S) = C(S) - cI.$$

Since I is a constant, this means that $\overline{C}(S)$ achieves its minimum at the same value of S^* as for $C(S)$, as shown in Fig. 18.14. However, since S must be constrained to $S \geq I$, if $I > S^*$, Fig. 18.14 indicates that $\overline{C}(S)$ would be minimized over $S \geq I$ by setting $S = I$ (i.e., do not place an order). This yields the following inventory policy.

Optimal Inventory Policy with $I > 0$ and $K = 0$

If $I < S^*$, order $S^* - I$ to bring the inventory level up to S^*.
If $I \geq S^*$, do not order,

where S^* again satisfies

$$F(S^*) = \frac{p - c}{p + h}.$$

Thus, in the bicycle example, if there are 500 bicycles on hand, the optimal policy is to bring the inventory level up to 11,856 bicycles (which implies ordering 11,356 additional bicycles). On the other hand, if there were 12,000 bicycles already on hand, the optimal policy would be not to order.

Analysis of the Model with a Setup Cost ($K > 0$)

Now consider the remaining version of the model where $K > 0$, so a setup cost of K is incurred for purchasing or producing the entire batch of units being ordered. (For the bicycle example, if an administrative cost of \$8,000 would be incurred to place the special order for the bicycles for the Christmas season, then $K = 8,000$.) We now will allow any value of the initial inventory, so $I \geq 0$.

With $K > 0$, the expected cost $\overline{C}(S)$, given the value of the decision variable S, is

$$\overline{C}(S) = K + c(S - I) + \int_S^\infty p(x - S)\, f(x)\, dx + \int_0^S h(S - x)\, f(x)\, dx \qquad \begin{array}{l}\text{if an order is}\\ \text{placed;}\end{array}$$

$$\overline{C}(S) = \int_S^\infty p(x - S)\, f(x)\, dx + \int_0^S h(S - x)\, f(x)\, dx \qquad\qquad \text{if do not order.}$$

Therefore, in comparison with the expected cost function $C(S)$ that is plotted in Fig. 18.14 (which assumes that $I = 0$ and $K = 0$),

$$\overline{C}(S) = K + C(S) - cI \qquad \text{if an order is placed;}$$
$$\overline{C}(I) = C(I) - cI \qquad\quad\ \text{if do not order.}$$

Because I is a constant, the cI term in both expressions can be ignored for purposes of minimizing $\overline{C}(S)$ over $S \geq I$. Consequently, the plot of $C(S)$ in Fig. 18.14 can be used to determine if an order should be placed and, if so, what value of S should be selected. This is what is done in Fig. 18.15, where s^* is the value of S such that

$$C(s^*) = K + C(S^*).$$

■ **FIGURE 18.15**
The graph of $C(S)$, the expected cost (given S) for the stochastic single-period model when $I = 0$ and $K = 0$, is being used here to determine the critical points, s^* and S^*, of the optimal inventory policy for the version of the model where $I \geq 0$ and $K > 0$.

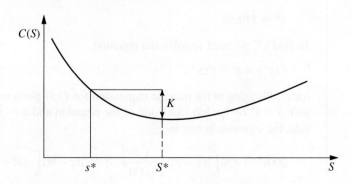

Thus,

if $I < s^*$,   then $C(S^*) < K + C(I)$,   so should order with $S = S^*$;
if $I \geq s^*$,   then $C(S) \leq K + C(I)$ for any $S \geq I$,   so should not order.

In other words, if the initial inventory I is less than s^*, then expending the setup cost K is worthwhile because bringing the inventory level up to S^* (by ordering $S - I$) will reduce the expected remaining cost by more than K when compared with not ordering. However, if $I > s^*$, then it becomes impossible to recoup the setup cost K by ordering any amount. (If $I = s^*$, incurring the setup cost K to order $S^* - s^*$ will reduce the expected remaining cost by this same amount, so there is no reason to bother ordering.) This leads to the following inventory policy.

Optimal Inventory Policy with $I \geq 0$ and $K > 0$

If $I < s^*$,   order $S^* - I$ to bring the inventory level up to S^*.
If $I^* \geq s^*$,   do not order.
(See the boxed formulas for S^* and s^* given earlier.)

When the demand has either a uniform or an exponential distribution, an automatic procedure is available in your IOR Tutorial for calculating s^* and S^*. A similar Excel template is also included in this chapter's Excel files on the book's website.

This kind of policy is referred to as an **(s, S) policy.** It has had extensive use in industry.

An (s, S) policy also is often used when applying stochastic periodic-review models to *stable products,* so multiple periods need to be considered. In this case, finding the optimal inventory policy is somewhat more complicated since the values of s and S may need to be different for different periods. The second supplement for this chapter on the book's website provides the details.

Returning to the current single-period model, we now will illustrate the calculation of the optimal inventory policy for the bicycle example when $K > 0$.

Application to the Example

Suppose that the administrative cost of placing the special order for the bicycles for the upcoming Christmas season is estimated to be $8,000. Thus, the parameters of the model now are

$$K = 8,000, \qquad c = 200, \qquad p = 450, \qquad h = -90.$$

As indicated earlier, the demand for the bicycles is assumed to have an exponential distribution with a mean of 10,000.

We found earlier for this example that

$$S^* = 11,856.$$

To find s^*, we need to solve the equation,

$$C(s^*) = K + C(S^*),$$

for s^*. Plugging twice into the expression for $C(S)$ given in the early part of this section, with $S = s^*$ on the left-hand side of the equation and $S = S^* = 11,856$ on the right-hand side, the equation becomes

$$200s^* + 450 \int_{s^*}^{\infty} (x - s^*) \frac{1}{10,000} e^{-x/10,000} dx - 90 \int_{0}^{s^*} (s^* - x) \frac{1}{10,000} e^{-x/10,000} dx$$

$$= 8{,}000 + 200(11{,}856) + 450\int_{11{,}856}^{\infty} (x - 11{,}856)\frac{1}{10{,}000}e^{-x/10{,}000}dx$$

$$- 90\int_{0}^{11{,}856} (11{,}856 - x)\frac{1}{10{,}000}e^{-x/10{,}000}dx.$$

After lengthy calculations to compute the number on the right-hand side and to reduce the left-hand side to a simpler expression in terms of s^*, this equation eventually leads to the numerical solution,

$$s^* = 10{,}674.$$

Thus, the optimal policy calls for bringing the inventory level up to $S^* = 11{,}856$ bicycles if the amount on hand is less than $s^* = 10{,}674$. Otherwise, no order is placed.

An Approximate Solution for the Optimal Policy When the Demand Has an Exponential Distribution

As this example has just illustrated, a lengthy calculation is required to solve for s^* even when the demand has a relatively straightforward distribution such as the exponential distribution. Therefore, given this demand distribution, we now will develop a close approximation to the optimal inventory policy that is easy to compute.

As described in Sec. 17.4, for an exponential distribution with a mean of $1/\alpha$, the probability density function $f(x)$ and CDF $F(x)$ are

$$\begin{aligned} f(x) &= \alpha e^{-\alpha x}, && \text{for } x \geq 0, \\ F(x) &= 1 - e^{-\alpha x}, && \text{for } x \geq 0. \end{aligned}$$

Consequently, since

$$F(S^*) = \frac{p - c}{p + h},$$

we have

$$1 - e^{-\alpha S^*} = \frac{p - c}{p + h}, \qquad \text{or} \qquad e^{-\alpha S^*} = \frac{(p + h) - (p - c)}{p + h} = \frac{h + c}{h + p},$$

so

$$S^* = \frac{1}{\alpha} \ln \frac{h + p}{h + c}$$

is the exact solution for S^*.

To begin developing an approximation for s^*, we begin with the exact equation,

$$C(s^*) = K + C(S^*).$$

Since

$$C(S) = cS + h\int_{0}^{S} (S - x)\alpha e^{-\alpha x}\, dx + p\int_{S}^{\infty} (x - S)\alpha e^{-\alpha x}\, dx$$

$$= (c + h)S + \frac{1}{\alpha}(h + p)e^{-\alpha S} - \frac{h}{\alpha}.$$

This equation becomes

$$(c + h)s^* + \frac{1}{\alpha}(h + p)e^{-\alpha s^*} - \frac{h}{\alpha} = K + (c + h)S^* + \frac{1}{\alpha}(h + p)e^{-\alpha S^*} - \frac{h}{\alpha},$$

or (by using the above result for S^*)

$$(c + h)s^* + \frac{1}{\alpha}(h + p)e^{-\alpha s^*} = K + (c + h)S^* + \frac{1}{\alpha}(c + h).$$

Although this last equation does not have a closed-form solution for $s*$, it can be solved numerically. An approximate analytical solution also can be obtained as follows. By letting

$$\Delta = S* - s*,$$

and noting that

$$e^{-\alpha S*} = \frac{h + c}{h + p},$$

the last equation yields

$$\frac{1}{\alpha}(h + p)\frac{e^{-\alpha s*}}{e^{-\alpha S*}} = \frac{K + (c + h)\Delta + \dfrac{1}{\alpha}(c + h)}{\dfrac{h + c}{h + p}},$$

which reduces to

$$e^{\alpha\Delta} = \frac{\alpha K}{c + h} + \alpha\Delta + 1.$$

If $\alpha\Delta$ is close to zero, $e^{\alpha\Delta}$ can be expanded into a Taylor series around zero. If the terms beyond the quadratic term are neglected, the result becomes

$$1 + \alpha\Delta + \frac{\alpha^2\Delta^2}{2} \cong \frac{\alpha K}{c + h} + \alpha\Delta + 1,$$

so that

$$\Delta \cong \sqrt{\frac{2K}{\alpha(c + h)}}.$$

Therefore, the desired approximation for $s*$ is

$$s* \cong S* - \sqrt{\frac{2K}{\alpha(c + h)}}.$$

Using this approximation in the bicycle example results in

$$\Delta \cong \sqrt{\frac{(2)(10,000)(8,000)}{200 - 90}} = 1,206,$$

so that

$$s* \cong 11,856 - 1,206 = 10,650,$$

which is quite close to the exact value of $s* = 10,674$.

■ 18.8   REVENUE MANAGEMENT

The beginning of the preceding section includes a list of 12 examples of perishable products. The last of these examples (reservations provided by an airline for the available inventory of seats on a particular flight) is of considerable historical interest because its early analysis led the way to a much broader and highly successful application area of operations research commonly called *revenue management*.

The starting point for revenue management was the Airline Deregulation Act of 1978, which loosened control of airline fare prices. New low-cost and charter airlines then entered the market to take advantage. Among the major airlines, American Airlines led the way in fighting back by introducing *capacity-controlled discount fares*. A limited number of discount seats were sold on various flights as needed to match or beat the fares offered by low-cost airlines, but with restrictions that included the requirement that the purchase must be made by some substantial number of days (initially 30 days) prior to departure. The usual much-larger fares would still be provided to the airline's core customer class of business travelers, who typically make their reservations well after the deadline for discount fares. (The first model in this section deals with this situation.)

Another of the oldest and most successful practices of revenue management in the airline industry has been to do *overbooking* (providing more reservations than the number of seats available on a flight, to allow for the considerable number of no-shows that usually occur). The rule of thumb in the industry is that approximately 15 percent of all seats on a flight would go unoccupied without some form of overbooking. Therefore, a large amount of additional revenue can be obtained by doing a significant amount of overbooking without incurring an undue risk of overselling a flight. However, the penalties have become substantial for denying admission to a flight for someone with a reservation, so careful analysis must be done to achieve an appropriate trade-off between the additional revenue from overbooking and the risk of incurring these penalties. (The second model in this section deals with this situation.)

When implementing revenue management, a large airline needs to process reservations for many tens of thousands of passengers flying daily. Therefore, while OR models and algorithms drive revenue management, the other essential component is sophisticated information technology. Fortunately, advances in information technology by the 1980s were providing the needed capability to automate transactions, capture and store vast amounts of data, quickly execute complex algorithms, and then implement and manage highly detailed revenue management decisions.

By 1990, the practice of revenue management at American Airlines had been refined to the point that it was generating nearly $500 million in additional revenue per year. (Selected Reference A8 tells this story.) By that time, other airlines also were scrambling to develop similar revenue management capabilities.

As a result of this history, the practice of revenue management in the airline industry today is pervasive, highly developed, and enormously effective. According to page 10 of Selected Reference 10 (the authoritative treatise on the theory and practice of revenue management), "by most estimates, the revenue gains from the use of revenue management systems are roughly comparable to many airlines' total profitability in a good year (about 4 to 5% of revenues)."

The enormous success of revenue management in the airline industry has led various other service industries with similar characteristics to develop their own revenue management systems. These industries include hotels, cruise ship lines, passenger railways, car rental companies, tour operators, theaters, and sporting venues. Revenue management also is growing in the retail industry when dealing with highly perishable products (e.g., grocery retailers), seasonal products (e.g., apparel retailers), and products that quickly become obsolete (e.g., high-tech retailers).

Achieving these outstanding results sometimes requires developing relatively complex revenue management systems with many categories of customers, fares changing over time, and so forth. The models and algorithms needed to support such systems are also relatively complex and so are beyond the scope of this book. However, to convey the general idea, we now present two basic models for elementary types of revenue management. The components of each model are described in general terms to fit any kind of company, but then the airline context is mentioned parenthetically for concreteness. Each model also is followed by an airline example.

A Model for Capacity-Controlled Discount Fares

A company has an inventory of a certain perishable product (such as the seats on an airline flight) to sell to two classes of customers (such as the leisure travelers and business travelers on the flight). The class 2 customers come first to buy single units of the product at a discounted price that is designed to help ensure that the entire inventory can be sold before the product perishes. There is a deadline for requesting the discounted price, but the company can terminate the special sale at any earlier point whenever it feels that enough has been sold. After the discounted price is no longer available, the class 1 customers begin arriving to buy single units of the product at full price. The probability distribution of the demand from class 1 customers is assumed to be known. The decision to be made is how much of the total inventory should be reserved for class 1 customers, so the discounted price would be discontinued early if the remaining inventory drops to this level before the announced deadline for the discount is reached.

The parameters (and random variable) for the model are

L = size of the inventory of the perishable product available for sale,
p_1 = price per unit paid by class 1 customers,
p_2 = price per unit paid by class 2 customers, where $p_2 < p_1$,
D = Demand by class 1 customers (a random variable),
$F(x)$ = cumulative distribution function for D, so $F(x) = P(D \leq x)$.

The decision variable is

x = inventory level that must be reserved for class 1 customers.

The key to solving for the optimal value of x, denoted by x^*, is to ask the following question and then to answer it by performing *marginal analysis*.

Question: Suppose that x units remain in inventory prior to the deadline for requesting the discounted price p_2 and a class 2 customer arrives who wishes to purchase one unit at that price. Should this request be accepted or denied?

To address the question, we need to compare the incremental revenue (or the statistical expectation of the incremental revenue) for the two options.

If accept request, incremental revenue = p_2.

If deny request, incremental revenue = $\begin{cases} 0, & \text{if } D \leq x - 1 \\ p_1, & \text{if } D \geq x \end{cases}$

so

E (incremental revenue) = $p_1 P(D \geq x)$.

Therefore, the request to make the sale to the class 2 customer should be accepted if

$$p_2 > p_1 P(D \geq x)$$

and denied otherwise. Now note that $P(D \geq x)$ decreases as x increases. Thus, if this inequality holds for a particular value of x, this value can be increased to the critical point x^* where

$$p_2 \leq p_1 P(D \geq x^*) \quad \text{and} \quad p_2 > p_1 P(D \geq x^* + 1).$$

It then follows that the optimal inventory level to reserve for class 1 customers is x^*. Equivalently, the maximum number of units that should be sold to class 2 customers before discontinuing the discounted price p_2 is $L - x^*$.

Thus far, we have assumed that the customers are buying single units of the product (such as the seats on an airline flight) so the probability distribution of D would be a discrete distribution. However, when L is large (such as the number of seats on a large airline flight), it can be much more convenient computationally to use a continuous distribution as an approximation. There also are perishable products where fractional amounts can be purchased, so continuous demand distributions would be appropriate anyway. If continuous demand distributions now are assumed, at least as an approximation, it follows from the above analysis that the optimal inventory level x^* to reserve for class 1 customers is the one that satisfies the equation,

$$p_2 = p_1 P(D > x^*).$$

Since $P(D > x^*) = 1 - P(D \leq x^*) = 1 - F(x^*)$, this equation also can be written as

$$F(x^*) = 1 - \frac{p_2}{p_1}.$$

(When a continuous distribution is being used as an approximation but x^* that solves these two equations is not an integer, x^* should be rounded *down* to an integer in order to satisfy the expressions defining the optimal integer value of x^* given at the end of the preceding paragraph.) This latter equation clearly shows that the ratio of p_2 to p_1 plays a critical role in determining the probability that the entire demand of the class 1 customers will be satisfied.

An Example Applying This Model for Capacity-Controlled Discount Fares

BLUE SKIES AIRLINES has decided to apply this model to one of its flights. This flight can accept 200 reservations for seats in the main cabin. (This number includes an allowance for overbooking because there always are some no-shows.) The flight attracts a large number of business travelers, who typically make their reservations within a few days of the flight but are willing to pay a relatively high fare of $1,000 for this flexibility. However, the substantial majority of the passengers need to be leisure travelers in order to fill up the plane. Therefore, to attract enough of these travelers, a very low discount fare of $200 is offered to passengers who make their reservations at least 14 days in advance and satisfy certain other restrictions (including no refunds).

In the terminology of the above model, the class 1 customers are the business travelers and the class 2 customers are the leisure travelers, so the parameters of the model are

$$L = 200, \quad p_1 = \$1,000, \quad p_2 = \$200.$$

Using data on the number of reservations requested by the class 1 customers for each flight in the past, it is estimated that the probability distribution of the number of reservations requested by these customers for each future flight is approximated by a normal distribution with a mean of $\mu = 60$ and standard deviation $\sigma = 20$. Thus, this is the distribution for the random variable D in the model, where $F(x)$ denotes the cumulative distribution for D. To solve for x^*, the optimal number of reservation slots to reserve for class 1 customers, we use the equation provided by the model,

$$F(x^*) = 1 - \frac{p_2}{p_1} = 1 - \frac{\$200}{\$1,000} = 0.8.$$

Using the table for a normal distribution provided by Appendix 5 yields

$$x^* = \mu + K_{0.2}\,\sigma = 60 + 0.842(20) = 76.84.$$

Since x^* actually needs to be an integer, it next is rounded *down* (as specified by the model) to the integer 76. By reserving 76 spots for customers willing to pay the fare of $1,000 for a reservation within a few days of the flight, this implies that $L - x^* = 124$ is the maximum number of reservations that should be sold at the discount fare of $200 before discontinuing this fare, even if this occurs before the deadline of 14 days prior to the flight.

An Overbooking Model

As with the preceding model, we again are dealing with a company that has an inventory of a certain perishable product (such as the seats on an airline flight) to sell to its customers. We no longer make any distinction between different classes of customers. The units in inventory become available only at a certain point in time, so each customer purchases a unit by making a nonrefundable reservation in advance to acquire the unit at the designated time. However, not all customers who make a reservation actually arrive on time to acquire their units. Those customers who fail to arrive at the designated time are referred to as *no-shows*.

Because the company anticipates that there will be a significant number of no-shows, it can increase its revenue by doing some **overbooking** (selling more reservations than the available inventory). However, care needs to be taken not to do so much overbooking that there is a substantial probability of incurring *shortages* (more demand than inventory). The reason is that there is a *shortage* cost incurred each time a customer with a reservation arrives on time to acquire a unit of inventory after the inventory has been depleted. For example, in the airline industry, a *denied-boarding cost* is incurred each time a customer with a reservation for a particular flight is *bumped* (denied admission to the flight), where this cost may include any refund of the purchase price, compensation for the inconvenience, and the cost of the loss of goodwill (lost future bookings). In some cases, this denied-boarding cost may consist instead of the compensation provided to a customer who has a seat but is willing to give it up for another customer who has been denied a seat.

The basic question addressed by this overbooking model is how much overbooking should be done so as to maximize the company's expected profit. The model makes the following assumptions.

1. The customers independently make their reservations for a unit of inventory and then have the same fixed probability of actually arriving at the designated time to acquire the unit.
2. There is a fixed net revenue obtained for each reservation that is accepted.
3. There is a fixed shortage cost incurred each time a customer with a reservation arrives on time to acquire a unit of inventory after the inventory has been depleted.

Based on these assumptions, the model has the following parameters.

p = probability that a customer who makes a reservation for a unit of inventory will actually arrive at the designated time to acquire the unit.

r = net revenue obtained for each reservation that is accepted.

s = shortage cost per unit of unsatisfied demand.

L = size of the available inventory.

The decision variable for the model is

n = number of customers that can be given a reservation for a unit of inventory,

so

$n - L$ = amount of overbooking allowed.

Given the value of n, the uncertainty is how many of the n customers with reservations for a unit of inventory will actually arrive at the designated time to acquire this unit. In other words, what is the *demand* for withdrawing units from inventory? Denote this random variable by

$D(n)$ = demand for withdrawing units from inventory.

It follows from assumption 1 that $D(n)$ has a *binomial distribution* with parameter p, so

$$P\{D(n) = d\} = \binom{n}{d} p^d (1 - p)^{n-d} = \frac{n!}{d!(n - d)!} p^d (1 - p)^{n-d},$$

where $D(n)$ has mean np and variance $np(1 - p)$.

A closely related random variable that will be important in our analysis is the *unsatisfied demand* that will occur when n customers are given a reservation. We denote this random variable by $U(n)$, so

$$U(n) = \text{unsatisfied demand} = \begin{cases} 0, & \text{if } D(n) \leq L \\ D(n) - L, & \text{if } D(n) > L \end{cases}$$

and

$$E(U(n)) = \sum_{d=L+1}^{n} (d - L)\, P\{D(n) = d\}.$$

We will be using *marginal analysis* (the analysis of the effect of increasing the value of the decision variable n by 1) to determine the optimal value of n that maximizes expected profit, so we will need to know the effect on $E(U(n))$ of increasing the value of n by 1. Starting with n reservations, the effect of adding on one more reservation is to add 1 to the unsatisfied demand only if both of two events occur. One necessary event is that the original n reservations result in depleting the entire inventory, i.e., $D(n) \geq L$, and the other required event is that the customer given the additional reservation actually will arrive at the designated time to attempt to acquire a unit of inventory. Otherwise, there is no effect on the unsatisfied demand. Consequently,

$$\Delta E(U(n)) = E(U(n + 1)) - E(U(n)) = p\, P\{D(n) \geq L\}$$

The value of $\Delta E(U(n))$ depends on the value of n since $P\{D(n) \geq L\}$, the probability of depleting the inventory, depends on n, the number of reservations. For $n < L$, $\Delta E(U(n)) = 0$, whereas $\Delta E(U(n))$ increases as n increases further since the probability of depleting the inventory increases as the number of reservations increases.

The final random variable of interest is the company's *profit* that will occur when n customers are given a reservation. We denote this random variable by $P(n)$, so

$$P(n) = \text{profit} = r\,n - s\,U(n)$$
$$E(P(n)) = r\,n - s\,E(U(n)),$$
$$\Delta E(P(n)) = E(P(n + 1)) - E(P(n)) = r - s\,\Delta E(U(n)) = r - s\,p\,P\{D(n) \geq L\}.$$

As just noted above, $\Delta E(U(n)) = 0$ for n $< L$, whereas $\Delta E(U(n))$ increases as n increases further. Therefore, $\Delta E(P(n)) > 0$ for relatively small values of n and then (assuming that $r < s\,p$) will switch to $\Delta E(P(n)) < 0$ for sufficiently large values of n. It then follows that n^*, the value of n that maximizes $E(P(n))$, is the one that satisfies

$$\Delta E(P(n^* - 1)) > 0 \quad \text{and} \quad \Delta E(P(n^*)) \leq 0,$$

or equivalently,

$$r > s \, p \, P\{D(n^* - 1) \geq L\} \quad \text{and} \quad r \leq s \, p \, P\{D(n^*) \geq L\}.$$

Since $D(n)$ has a binomial distribution, it is straightforward (albeit very tedious computationally) to solve for n^* in this way.

When L is large, it is particularly tedious to use the binomial distribution to perform these calculations. Therefore, it is common in practice to use the *normal approximation* of the binomial distribution for this application (as well as many others). In particular, the normal distribution with mean np and variance $np(1 - p)$ frequently is used as a continuous approximation of the binomial distribution with parameters n and p, since the latter distribution has this same mean and variance. With this approach, we now assume that $D(n)$ has this normal distribution and treat n as a continuous decision variable. The optimal value of n then is given approximately by the equation,

$$r = s \, p \, P\{D(n^* \geq L\}, \quad \text{i.e.,} \quad P\{D(n^*) \geq L\} = \frac{r}{sp}$$

By using the table for a normal distribution given in Appendix 5, it is straightforward to calculate n^*, as will be illustrated by the following example. If n^* is not an integer, it next should be rounded *up* to an integer in order to satisfy the expressions defining the optimal integer value of n^* given at the end of the preceding paragraph.

An Example Applying This Overbooking Model

TRANSCONTINENTAL AIRLINES has a daily flight (excluding weekends) from San Francisco to Chicago that is mainly used by business travelers. There are 150 seats available in the single cabin. The average fare per seat is $300. This is a nonrefundable fare, so no-shows forfeit the entire fare.

The company's policy is to accept 10 percent more reservations than the number of seats available on nearly all its flights, since roughly 10 percent of all its customers making reservations end up being no-shows. However, if its experience with a particular flight is much different from this, then an exception can be made and the OR group is called in to analyze what the overbooking policy should be for that particular flight. This is what has just happened regarding the daily flight from San Francisco to Chicago. Even when the full quota of 165 reservations has been reached (which happens for most of the flights), there usually has been a significant number of empty seats. While gathering its data, the OR group has discovered the reason why. Only 80 percent of the customers who make reservations for this flight actually show up to take the flight. The other 20 percent forfeit the fare (or, in most cases, allow their company to do so) because their plans have changed.

When a customer is bumped from this flight, Transcontinental Airlines arranges to put the customer on the next available flight to Chicago on another airline. The company's average cost for doing this is $200. In addition, the company gives the customer a voucher worth $400 (but would cost the company just $300) for use on a future flight. The company also feels that an additional $500 should be assessed for the intangible cost of a loss of goodwill on the part of the bumped customer. Therefore, the total cost of bumping a customer is estimated to be $1,000.

The OR group now wants to apply the overbooking model to determine how many reservations should be accepted for this flight. Using the data described above, the parameters of the model are

$$p = 0.8, \quad r = \$300, \quad s = \$1,000, \quad L = 150.$$

Because L is so large, the group decides to use the normal approximation of the binomial distribution. Therefore, this approximation of n^*, the optimal number of reservations to accept, is found by solving the equation,

$$P\{D(n^*) \geq 150\} = \frac{r}{sp} = 0.375,$$

where $D(n^*)$ has the normal distribution with mean $\mu = np = 0.8n$ and variance $\sigma^2 = np(1 - p) = 0.16n$, so $\sigma = 0.4\sqrt{n}$. Using the table for a normal distribution given in Appendix 5, since $\alpha = 0.375$ and $K_\alpha = 0.32$,

$$\frac{150 - \mu}{\sigma} = \frac{150 - 0.8n}{0.4\sqrt{n}} = 0.32,$$

which reduces to

$$0.8n + 0.128 \sqrt{n} - 150 = 0.$$

Solving for $\sqrt{n}$ in this quadratic equation yields

$$\sqrt{n} = \frac{-0.128 + \sqrt{(0.128)^2 - 4(0.8)(-150)}}{1.6} = 13.6,$$

which then gives

$$n^* = (13.6)^2 = 184.96.$$

Since x^* actually needs to be an integer, it next is rounded *up* (as specified by the model) to the integer 185.[12] The conclusion is that the number of reservations to accept for this flight should be increased from 165 to 185.

The resulting demand $D(185)$ will have a mean of $0.8(185) = 148$ and a standard deviation of $0.4 \sqrt{185} = 5.44$. Thus, Transcontinental Airlines now should be able to nearly or completely fill the 150 seats of the airplane, without an undue frequency of bumping customers, whenever the number of reservation requests reaches 185. Therefore, the new policy of increasing the number of reservations accepted from 165 to 185 should substantially increase the company's profits from this flight.

Other Models

A variety of models are used for various types of revenue management. These models frequently incorporate some of the ideas introduced in the two models presented in this section. However, the models used in practice frequently must also incorporate some additional features that are not considered in these two basic models. Here is a list of some practical considerations that may need to be taken into account.

- Different levels of service being provided (e.g., a first class cabin, a business section, and an economy section on the same airline flight).
- Different prices charged for the same service (e.g., discounts for seniors, children, students, employees, etc.).
- Different prices charged for the same service based on how much (if any) of it is refundable with an early cancellation.

[12]One step in obtaining this solution of 185 was reading the value of $K_\alpha = 0.32$ to two decimal places from the normal table. However, if interpolation is used to carry K_α to additional decimal places, the solution from the model will change to 186. Using the binomial distribution directly instead of the normal approximation also leads to a solution of 186.

- Dynamic pricing based on when the reservation is made and how well the demand is approaching the capacity.
- Varying the overbooking level based on the remaining time and expected cancellations until the service will be provided.
- Having a nonlinear shortage cost for overbooking (e.g., the first few customers may voluntarily accept modest compensation to forego the service but then it gets more costly).
- Customers buy bundles of services in combination under various terms and conditions (e.g., airline customers arranging a set of connecting flights or hotel customers staying multiple nights).
- Customers purchase multiple units (e.g., couples or families or tour groups traveling together).

Incorporating these and other practical considerations into more sophisticated models as needed is a real challenge. However, outstanding progress has been made by numerous OR researchers and practitioners. This has become one of the most exciting areas of application of operations research. Further elaboration is beyond the scope of this book, but details can be found in Selected Reference 10 and its 591 references.

■ 18.9   CONCLUSIONS

We have introduced only rather basic kinds of inventory models here, but they serve the purpose of introducing the general nature of inventory models. Furthermore, they are sufficiently accurate representations of many actual inventory situations that they frequently are useful in practice. For example, the EOQ models have been particularly widely used. These models are sometimes modified to include some type of stochastic demand, such as the stochastic continuous-review model does. The stochastic single-period model is a very convenient one for perishable products. The elementary revenue management models in Sec. 18.8 are a starting point for the sophisticated kinds of revenue management analysis that now is extensively applied in the airline industry and other service industries with similar characteristics.

In today's global economy, multiechelon inventory models (such as those introduced in Sec. 18.5) are playing an increasingly important role in helping to manage a company's supply chain.

Nevertheless, many inventory situations possess complications that are not taken into account by the models in this chapter, e.g., interactions between products or complicated types of multiechelon inventory systems. More complex models have been formulated in an attempt to fit such situations, but it is difficult to achieve both adequate realism and sufficient tractability to be useful in practice. The development of useful models for supply chain management currently is a particularly active area of research. Much research also is being conducted on developing more sophisticated revenue management models that take into account more of the complexities that arise in practice.

Continued growth is occurring in the computerization of inventory data processing, along with an accompanying growth in scientific inventory management.

■ SELECTED REFERENCES

1. Axsäter, S.: *Inventory Control,* 2nd ed., Springer, New York, 2006.
2. Bertsimas, D., and A. Thiele: "A Robust Optimization Approach to Inventory Theory," *Operations Research,* **54**(1): 150–168, January–February 2006.

3. Gallego, G., and Ö. Özer: "A New Algorithm and a New Heuristic for Serial Supply Systems," *Operations Research Letters,* **33**(4): 349–362, July 2005.

4. Harrison, T. P., H. L. Lee, and J. J. Neale (eds.): *The Practice of Supply Chain Management: Where Theory and Application Converge,* Kluwer Academic Publishers (now Springer), Boston, 2003.

5. Khouja, M.: "The Single-Period (News-Vendor) Problem: Literature Review and Suggestions for Future Research," *Omega,* **27**: 537–553, 1999.

6. Muckstadt, J., and R. Roundy: "Analysis of Multi-Stage Production Systems," pp. 59–131 in Graves, S., A. Rinnooy Kan, and P. Zipken (eds.): *Handbook in Operations Research and Management Science, Vol. 4, Logistics of Production and Inventory,* North-Holland, Amsterdam, 1993.

7. Sethi, S. P., H. Yan, and H. Zhang: *Inventory and Supply Chain Management with Forecast Updates,* Springer, New York, 2005.

8. Sherbrooke, C. C.: *Optimal Inventory Modeling of Systems: Multi-Echelon Techniques,* 2nd ed., Kluwer Academic Publishers (now Springer), Boston, 2004.

9. Simchi-Levi, D., S. D. Wu, and Z.-J. Shen (eds.): *Handbook of Quantitative Supply Chain Analysis,* Kluwer Academic Publishers (now Springer), Boston, 2004.

10. Talluri, G., and K. van Ryzin: *Theory and Practice of Yield Management,* Kluwer Academic Publishers (now Springer), Boston, 2004.

11. Tang, C. S., C.-P. Teo, and K. K. Wei (eds.): *Supply Chain Analysis: A Handbook on the Interaction of Information, System and Optimization,* Springer, New York, 2008.

12. Tayur, S., R. Ganeshan, and M. Magazine (eds.): *Quantitative Models for Supply Chain Management,* Kluwer Academic Publishers (now Springer), Boston, 1998.

13. Tiwari, V., and S. Gavirneni: "ASP, The Art and Science of Practice: Recoupling Inventory Control Research and Practice: Guidelines for Achieving Synergy," *Interfaces,* **37**(2): 176–186, March–April 2007.

14. Zipken, P. H.: *Foundations of Inventory Management,* McGraw-Hill, Boston, 2000.

Some Award-Winning Applications of Inventory Theory:

(A link to all these articles is provided on our website, www.mhhe.com/hillier.)

A1. Arntzen, B. C., G. G. Brown, T. P. Harrison, and L. L. Trafton: "Global Supply Chain Management at Digital Equipment Corporation," *Interfaces,* **25**(1): 69–93, January–February 1995.

A2. Billington, C., G. Callioni, B. Crane, J. D. Ruark, J. U. Rapp, T. White, and S. P. Willems: "Accelerating the Profitability of Hewlett-Packard's Supply Chains," *Interfaces,* **34**(1): 59–72, January–February 2004.

A3. Flowers, A. D.: "The Modernization of Merit Brass," *Interfaces,* **23**(1): 97–108, January–February 1993.

A4. Geraghty, M. K., and E. Johnson: "Revenue Management Saves National Car Rental," *Interfaces,* **27**(1): 107–127, January–February 1997.

A5. Kok, T. de, F. Janssen, J. van Doremalen, E. van Wachem, M. Clerkx, and W. Peeters: "Phillips Electronics Synchronizes Its Supply Chain to End the Bullwhip Effect," *Interfaces,* **35**(1): 37–48, January–February 2005.

A6. Lin, G., M. Ettl, S. Buckley, S. Bagchi, D. D. Yao, B. L. Naccarato, R. Allan, K. Kim, and L. Koenig: "Extended-Enterprise Supply-Chain Management at IBM Personal Systems Group and Other Divisions," *Interfaces,* **30**(1): 7–25, January–February 2000.

A7. Lyon, P., R. J. Milne, R. Orzell, and R. Rice: "Matching Assets with Demand in Supply-Chain Management at IBM Microelectronics," *Interfaces,* **31**(1): 108–124, January–February 2001.

A8. Smith, B. C., J. F. Leimkuhler, and R. M. Darrow: "Yield Management at American Airlines," *Interfaces,* **22**(1): 8–31, January–February 1992.

■ LEARNING AIDS FOR THIS CHAPTER ON OUR WEBSITE (www.mhhe.com/hillier)

Worked Examples:

Examples for Chapter 18

Automatic Procedures in IOR Tutorial:

Stochastic Single-Period Model for Perishable Products, No Setup Cost
Stochastic Single-Period Model for Perishable Products, with Setup Cost

"Ch. 18—Inventory Theory" Excel Files:

Templates for the Basic EOQ Model (a Solver Version and an Analytical Version)
Templates for the EOQ Model with Planned Shortages (a Solver Version and an Analytical Version)
Template for the EOQ Model with Quantity Discounts (Analytical Version Only)
Template for the Stochastic Continuous-Review Model
Template for the Stochastic Single-Period Model for Perishable Products, No Setup Cost
Template for the Stochastic Single-Period Model for Perishable Products, with Setup Cost

"Ch. 18—Inventory Theory" LINGO File for Selected Examples

Glossary for Chapter 18

Supplements to This Chapter

Derivation of the Optimal Policy for the Stochastic Single-Period Model for Perishable Products
Stochastic Periodic-Review Models

■ PROBLEMS

To the left of each of the following problems (or their parts), we have inserted a T whenever one of the templates listed above can be useful. An asterisk on the problem number indicates that at least a partial answer is given in the back of the book.

T **18.3-1.*** Suppose that the demand for a product is 30 units per month and the items are withdrawn at a constant rate. The setup cost each time a production run is undertaken to replenish inventory is $15. The production cost is $1 per item, and the inventory holding cost is $0.30 per item per month.
(a) Assuming shortages are not allowed, determine how often to make a production run and what size it should be.
(b) If shortages are allowed but cost $3 per item per month, determine how often to make a production run and what size it should be.

T **18.3-2.** The demand for a product is 1,000 units per week, and the items are withdrawn at a constant rate. The setup cost for placing an order to replenish inventory is $40. The unit cost of each item is $5, and the inventory holding cost is $0.10 per item per week.
(a) Assuming shortages are not allowed, determine how often to order and what size the order should be.
(b) If shortages are allowed but cost $3 per item per week, determine how often to order and what size the order should be.

18.3-3.* Tim Madsen is the purchasing agent for Computer Center, a large discount computer store. He has recently added the hottest new computer, the Power model, to the store's stock of goods. Sales of this model now are running at about 13 per week. Tim purchases

these computers directly from the manufacturer at a unit cost of $3,000, where each shipment takes half a week to arrive.

Tim routinely uses the basic EOQ model to determine the store's inventory policy for each of its more important products. For this purpose, he estimates that the annual cost of holding items in inventory is 20 percent of their purchase cost. He also estimates that the administrative cost associated with placing each order is $75.

T (a) Tim currently is using the policy of ordering 5 Power model computers at a time, where each order is timed to have the shipment arrive just about when the inventory of these computers is being depleted. Use the Solver version of the Excel template for the basic EOQ model to determine the various annual costs being incurred with this policy.

T (b) Use this same spreadsheet to generate a table that shows how these costs would change if the order quantity were changed to the following values: 5, 7, 9, . . . , 25.

T (c) Use the Solver to find the optimal order quantity.

T (d) Now use the analytical version of the Excel template for the basic EOQ model (which applies the EOQ formula directly) to find the optimal quantity. Compare the results (including the various costs) with those obtained in part (c).

(e) Verify your answer for the optimal order quantity obtained in part (d) by applying the EOQ formula by hand.

(f) With the optimal order quantity obtained above, how frequently will orders need to be placed on the average? What should the approximate inventory level be when each order is placed?

(g) How much does the optimal inventory policy reduce the total variable inventory cost per year (holding costs plus administrative costs for placing orders) for Power model computers from that for the policy described in part (*a*)? What is the percentage reduction?

18.3-4. The Blue Cab Company is the primary taxi company in the city of Maintown. It uses gasoline at the rate of 10,000 gallons per month. Because this is such a major cost, the company has made a special arrangement with the Amicable Petroleum Company to purchase a huge quantity of gasoline at a reduced price of $3.50 per gallon every few months. The cost of arranging for each order, including placing the gasoline into storage, is $2,000. The cost of holding the gasoline in storage is estimated to be $0.04 per gallon per month.

T (**a**) Use the Solver version of the Excel template for the basic EOQ model to determine the costs that would be incurred annually if the gasoline were to be ordered monthly.

T (**b**) Use this same spreadsheet to generate a table that shows how these costs would change if the number of months between orders were to be changed to the following values: 1, 2, 3, . . . , 10.

T (**c**) Use the Solver to find the optimal order quantity.

T (**d**) Now use the analytical version of the Excel template for the basic EOQ model to find the optimal order quantity. Compare the results (including the various costs) with those obtained in part (*c*).

(**e**) Verify your answer for the optimal order quantity obtained in part (*d*) by applying the EOQ formula by hand.

18.3-5. For the basic EOQ model, use the square root formula to determine how Q^* would change for each of the following changes in the costs or the demand rate. (Unless otherwise noted, consider each change by itself.)

(**a**) The setup cost is reduced to 25 percent of its original value.

(**b**) The annual demand rate becomes four times as large as its original value.

(**c**) Both changes in parts (*a*) and (*b*).

(**d**) The unit holding cost is reduced to 25 percent of its original value.

(**e**) Both changes in parts (*a*) and (*d*).

18.3-6.* Kris Lee, the owner and manager of the Quality Hardware Store, is reassessing his inventory policy for hammers. He sells an average of 50 hammers per month, so he has been placing an order to purchase 50 hammers from a wholesaler at a cost of $20 per hammer at the end of each month. However, Kris does all the ordering for the store himself and finds that this is taking a great deal of his time. He estimates that the value of his time spent in placing each order for hammers is $75.

(**a**) What would the unit holding cost for hammers need to be for Kris' current inventory policy to be optimal according to the basic EOQ model? What is this unit holding cost as a percentage of the unit acquisition cost?

T (**b**) What is the optimal order quantity if the unit holding cost actually is 20 percent of the unit acquisition cost? What is the corresponding value of TVC = total variable inventory cost per year (holding costs plus the administrative costs for placing orders)? What is TVC for the current inventory policy?

T (**c**) If the wholesaler typically delivers an order of hammers in 5 working days (out of 25 working days in an average month), what should the reorder point be (according to the basic EOQ model)?

(**d**) Kris doesn't like to incur inventory shortages of important items. Therefore, he has decided to add a safety stock of 5 hammers to safeguard against late deliveries and larger-than-usual sales. What is his new reorder point? How much does this safety stock add to TVC?

18.3-7.* Consider Example 1 (manufacturing speakers for TV sets) introduced in Sec. 18.1 and used in Sec. 18.3 to illustrate the EOQ models. Use the EOQ model with planned shortages to solve this example when the unit shortage cost is changed to $5 per speaker short per month.

T **18.3-8.** Speedy Wheels is a wholesale distributor of bicycles. Its Inventory Manager, Ricky Sapolo, is currently reviewing the inventory policy for one popular model that is selling at the rate of 500 per month. The administrative cost for placing an order for this model from the manufacturer is $1,000 and the purchase price is $400 per bicycle. The annual cost of the capital tied up in inventory is 15 percent of the value (based on purchase price) of these bicycles. The additional cost of storing the bicycles—including leasing warehouse space, insurance, taxes, and so on—is $40 per bicycle per year.

(**a**) Use the basic EOQ model to determine the optimal order quantity and the total variable inventory cost per year.

(**b**) Speedy Wheel's customers (retail outlets) generally do not object to short delays in having their orders filled. Therefore, management has agreed to a new policy of having small planned shortages occasionally to reduce the variable inventory cost. After consultations with management, Ricky estimates that the annual shortage cost (including lost future business) would be $150 times the average number of bicycles short throughout the year. Use the EOQ model with planned shortages to determine the new optimal inventory policy.

T **18.3-9.** Reconsider Prob. 18.3-3. Because of the popularity of the Power model computer, Tim Madsen has found that customers are willing to purchase a computer even when none are currently in stock as long as they can be assured that their order will be filled in a reasonable period of time. Therefore, Tim has decided to switch from the basic EOQ model to the EOQ model with planned shortages, using a shortage cost of $200 per computer short per year.

(**a**) Use the Solver version of the Excel template for the EOQ model with planned shortages (with constraints added in the Solver dialogue box that C10:C11 = integer) to find the new optimal inventory policy and its total variable inventory cost per year (TVC). What is the reduction in the value of TVC found for Prob. 18.3-3 (and given in the back of the book) when planned shortages were not allowed?

(**b**) Use this same spreadsheet to generate a table that shows how TVC and its components would change if the maximum shortage were kept the same as found in part (*a*) but the order quantity were changed to the following values: 15, 17, 19, . . . , 35.

18.5-2. Consider an inventory system that fits the model for a serial two-echelon system presented in Sec. 18.5, where $K_1 = \$25,000$, $K_2 = \$1,500$, $h_1 = \$30$, $h_2 = \$35$, and $d = 4,000$. Develop a table like Table 18.1 that shows the results from performing both separate optimization of the installations and simultaneous optimization of the installations. Then calculate the percentage increase in the total variable cost per unit time if the results from performing separate optimization were to be used instead of the results from the valid approach of performing simultaneous optimization.

18.5-3. A company soon will begin production of a new product. When this happens, an inventory system that fits the model for a serial two-echelon system presented in Sec. 18.5 will be used. At this time, there is great uncertainty about what the setup costs and holding costs will be at the two installations, as well as what the demand rate for the new product will be. Therefore, to begin making plans for the new inventory system, various combinations of possible values of the model parameters need to be checked.

Calculate Q_2^*, n^*, n, and Q_1^* for the following combinations.
(a) $(K_1, K_2) = (\$25,000, \$1,000)$, $(\$10,000, \$2,500)$, and $(\$5,000, \$5,000)$, with $h_1 = \$25$, $h_2 = \$250$, and $d = 2,500$.
(b) $(h_1, h_2) = (\$10, \$500)$, $(\$25, \$250)$, and $(\$50, \$100)$, with $K_1 = \$10,000$, $K_2 = \$2,500$, and $d = 2,500$.
(c) $d = 1,000$, $d = 2,500$, and $d = 5,000$, with $K_1 = \$10,000$, $K_2 = \$2,500$, $h_1 = \$25$, and $h_2 = \$250$.

18.5-4. A company owns both a factory to produce its products and a retail outlet to sell them. A certain new product will be sold exclusively through this retail outlet. Its inventory of this product will be replenished when needed from the factory's inventory, where an administrative and shipping cost of $200 is incurred each time this is done. The factory will replenish its own inventory of the product when needed by setting up for a quick production run. A setup cost of $5,000 is incurred each time this is done. The annual cost for holding each unit is $10 when it is held at the factory and $11 when it is held at the retail outlet. The retail outlet expects to sell 100 units of the product per month. All the assumptions of the model for a serial two-echelon system presented in Sec. 18.5 apply to the joint inventory system for the factory and retail outlet.
(a) Suppose that the factory and the retail outlet separately optimize their own inventory policies for the product. Calculate the resulting Q_2^*, n^*, n, Q_1^*, and C^*.
(b) Suppose that the company simultaneously optimizes the joint inventory policy for the factory and retail outlet for the product. Calculate the resulting Q_2^*, n^*, n, Q_1^*, and C^*.
(c) Calculate the percentage decrease in the total variable cost per unit time C^* that is achieved by using the approach described in part (b) instead of the one in part (a).

18.5-5. A company produces a certain product by assembling it at an assembly plant. All the components needed to assemble the product are purchased from a single supplier. A shipment of all the components is received from the supplier each time the assembly plant needs to replenish its inventory of the components. The company incurs a shipping cost of $500 in addition to the purchase price for the components each time this is done. Each time the supplier needs to replenish its own inventory of the components, quick production

runs are set up to produce the components. The total cost of setting up for these production runs is $50,000. The annual cost of holding each set of components is $50 when it is held by the supplier and $60 when it is held at the assembly plant. (It is higher in the latter case since there is more capital tied up in each set of components at this stage.) The assembly plant steadily produces 500 units of the product per month. All the assumptions of the model for a serial two-echelon system described in Sec. 18.5 apply to the joint inventory system for the supplier and the assembly plant.
(a) Suppose that the supplier and the assembly plant separately optimize their own inventory policies for the sets of components. Calculate the resulting Q_2^*, n^*, n, and Q_1^*. Also calculate C_1^* and C_2^*, the total variable cost per unit time for the supplier and the assembly plant, respectively, as well as $C^* = C_1^* + C_2^*$.
(b) Suppose that the supplier and the assembly plant cooperate to simultaneously optimize their joint inventory policy. Calculate the same quantities as specified in part (a) for this new inventory policy.
(c) Compare the values of C_1^*, C_2^*, and C^* obtained in parts (a) and (b). Would either organization lose money by using the joint inventory policy obtained in part (b) instead of the separate policies obtained in part (a)? If so, what financial arrangement would need to be made between these separate organizations to induce the losing organization to agree to a supply contract that follows the inventory policy obtained in part (b)? Comparing the values of C^*, what would be the total net savings for the two organizations if they can agree to follow the jointly optimal policy from part (b) instead of the separate optimal policies from part (a)?

18.5-6. Consider a three-echelon inventory system that fits the model for a serial multiechelon system presented in Sec. 18.5, where the model parameters for this particular system are given below.

Installation i	K_i	h_i	$d = 1,000$
1	$50,000	$1	
2	$2,000	$2	
3	$360	$10	

Develop a table like Table 18.4 that shows the intermediate and final results from applying the solution procedure presented in Sec. 18.5 to this inventory system. After calculating the total variable cost per unit time of the final solution, determine the maximum possible percentage by which this cost can exceed the corresponding cost for an optimal solution.

18.5-7. Follow the instructions of Prob. 18.5-6 for a five-echelon inventory model fitting the corresponding model in Sec. 18.5, where the model parameters are given below.

Installation i	K_i	h_i	$d = 1,000$
1	$125,000	$2	
2	$20,000	$10	
3	$6,000	$15	
4	$10,000	$20	
5	$250	$30	

18.5-8. Reconsider the example of a four-echelon inventory system presented in Sec. 18.5, where its model parameters are given in Table 18.2. Suppose now that the setup costs at the four installations have changed from what is given in Table 18.2, where the new values are $K_1 = \$1,000$, $K_2 = \$5$, $K_3 = \$75$, and $K_4 = \$80$. Redo the analysis presented in Sec. 18.5 for this example (as summarized in Table 18.4) with these new setup costs.

18.5-9. One of the many products produced by the Global Corporation is marketed primarily in the United States. A rough form of the product is produced in one of the corporation's plants in Asia and then is shipped to a plant in the United States for the finish work. The finished product next is sent to the corporation's distribution center in the United States. The distribution center stores the product and then uses this inventory to fill orders from various wholesalers. These sales to wholesalers remain relatively uniform throughout the year at a rate of about 10,000 units per month. The American plant uses its inventory of the finished product to send a shipment to the distribution center whenever the center needs to replenish its inventory. The associated administrative and shipping cost is about $400 per shipment. Whenever the American plant needs to replenish its inventory, the Asian plant uses its inventory of the rough product to send a shipment to the American plant, which then sets up for a quick production run to convert the rough product to a finished product. Each time this happens, the shipping cost and setup cost total about $6,000. The Asian plant replenishes its inventory of the rough product when needed by setting up for a quick production run. A setup cost of $60,000 is incurred each time this is done. The monthly cost for holding each unit is $3 at the Asian plant, $7 at the American plant, and $9 at the distribution plant. All the assumptions of the model for a serial multi-echelon system presented in Sec. 18.5 apply to the joint inventory system at the three locations for the product.

Solve this model by developing a table like Table 18.4 that shows the intermediate and final results from applying the solution procedure presented in Sec. 18.5. After calculating the total variable cost per month of the final solution, determine the maximum possible percentage by which this cost can exceed the corresponding cost for an optimal solution.

18.6-1. Henry Edsel is the owner of Honest Henry's, the largest car dealership in its part of the country. His most popular car model is the Triton, so his largest costs are those associated with ordering these cars from the factory and maintaining an inventory of Tritons on the lot. Therefore, Henry has asked his general manager, Ruby Willis, who once took a course in operations research, to use this background to develop a cost-effective policy for when to place these orders for Tritons and how many to order each time.

Ruby decides to use the stochastic continuous-review model presented in Sec. 18.6 to determine an (R, Q) policy. After some investigation, she estimates that the administrative cost for placing each order is $1,500 (a lot of paperwork is needed for ordering cars), the holding cost for each car is $3,000 per year (15 percent of the agency's purchase price of $20,000), and the shortage cost per car short is $1,000 per year (an estimated probability of $\frac{1}{3}$ of losing a car sale and its profit of about $3,000). After considering both the seriousness of incurring shortages and the high holding

cost, Ruby and Henry agree to use a 75 percent service level (a probability of 0.75 of not incurring a shortage between the time an order is placed and the delivery of the cars ordered). Based on previous experience, they also estimate that the Tritons sell at a relatively uniform rate of about 900 per year.

After an order is placed, the cars are delivered in about two-thirds of a month. Ruby's best estimate of the probability distribution of demand during the lead time before a delivery arrives is a normal distribution with a mean of 50 and a standard deviation of 15.
(a) Solve by hand for the order quantity.
(b) Use a table for the normal distribution (Appendix 5) to solve for the reorder point.
T **(c)** Use the Excel template for this model in your OR Courseware to check your answers in parts (a) and (b).
(d) Given your previous answers, how much safety stock does this inventory policy provide?
(e) This policy can lead to placing a new order before the delivery from the preceding order arrives. Indicate when this would happen.

18.6-2. One of the largest selling items in J.C. Ward's Department Store is a new model of refrigerator that is highly energy-efficient. About 80 of these refrigerators are being sold per month. It takes about a week for the store to obtain more refrigerators from a wholesaler. The demand during this time has a uniform distribution between 10 and 30. The administrative cost of placing each order is $100. For each refrigerator, the holding cost per month is $15 and the shortage cost per month is estimated to be $3.

The store's inventory manager has decided to use the stochastic continuous-review model presented in Sec. 18.6, with a service level (measure 1) of 0.8, to determine an (R, Q) policy.
(a) Solve by hand for R and Q.
T **(b)** Use the corresponding Excel template to check your answer in part (a).
(c) What will be the average number of stockouts per year with this inventory policy?

18.6-3. When using the stochastic continuous-review model presented in Sec. 18.6, a difficult managerial judgment decision needs to be made on the level of service to provide to customers. The purpose of this problem is to enable you to explore the trade-off involved in making this decision.

Assume that the measure of service level being used is $L =$ probability that a stockout will not occur during the lead time. Since management generally places a high priority on providing excellent service to customers, the temptation is to assign a very high value to L. However, this would result in providing a very large amount of safety stock, which runs counter to management's desire to eliminate unnecessary inventory. (Remember the *just-in-time philosophy* discussed in Sec. 18.3 that is heavily influencing managerial thinking today.) What is the best trade-off between providing good service and eliminating unnecessary inventory?

Assume that the probability distribution of demand during the lead time is a normal distribution with mean μ and standard deviation σ. Then the reorder point R is $R = \mu + K_{1-L}\sigma$, where K_{1-L} is obtained from Appendix 5. The amount of safety stock provided by this reorder point is $K_{1-L}\sigma$. Thus, if h denotes the holding cost for each unit held in inventory per year, the *average annual holding cost for safety stock* (denoted by C) is $C = hK_{1-L}\sigma$.

(a) Construct a table with five columns. The first column is the service level L, with values 0.5, 0.75, 0.9, 0.95, 0.99, and 0.999. The next four columns give C for four cases. Case 1 is $h = \$1$ and $\sigma = 1$. Case 2 is $h = \$100$ and $\sigma = 1$. Case 3 is $h = \$1$ and $\sigma = 100$. Case 4 is $h = \$100$ and $\sigma = 100$.

(b) Construct a second table that is based on the table obtained in part (a). The new table has five rows and the same five columns as the first table. Each entry in the new table is obtained by subtracting the corresponding entry in the first table from the entry in the next row of the first table. For example, the entries in the first column of the new table are $0.75 - 0.5 = 0.25$, $0.9 - 0.75 = 0.15$, $0.95 - 0.9 = 0.05$, $0.99 - 0.95 = 0.04$, and $0.999 - 0.99 = 0.009$. Since these entries represent increases in the service level L, each entry in the next four columns represents the increase in C that would result from increasing L by the amount shown in the first column.

(c) Based on these two tables, what advice would you give a manager who needs to make a decision on the value of L to use?

18.6-4. The preceding problem describes the factors involved in making a managerial decision on the service level L to use. It also points out that for any given values of L, h (the unit holding cost per year), and σ (the standard deviation when the demand during the lead time has a normal distribution), the average annual holding cost for the safety stock would turn out to be $C = hK_{1-L}\sigma$, where C denotes this holding cost and K_{1-L} is given in Appendix 5. Thus, the amount of variability in the demand, as measured by σ, has a major impact on this holding cost C.

The value of σ is substantially affected by the duration of the lead time. In particular, σ increases as the lead time increases. The purpose of this problem is to enable you to explore this relationship further.

To make this more concrete, suppose that the inventory system under consideration currently has the following values: $L = 0.9$, $h = \$100$, and $\sigma = 100$ with a lead time of 4 days. However, the vendor being used to replenish inventory is proposing a change in the delivery schedule that would change your lead time. You want to determine how this would change σ and C.

We assume for this inventory system (as is commonly the case) that the demands on separate days are statistically independent. In this case, the relationship between σ and the lead time is given by the formula

$$\sigma = \sqrt{d}\sigma_1,$$

where   d = number of days in the lead time,
         σ_1 = standard deviation if $d = 1$.

(a) Calculate C for the current inventory system.

(b) Determine σ_1. Then find how C would change if the lead time were reduced from 4 days to 1 day.

(c) How would C change if the lead time were doubled, from 4 days to 8 days?

(d) How long would the lead time need to be in order for C to double from its current value with a lead time of 4 days?

18.6-5. What is the effect on the amount of safety stock provided by the stochastic continuous-review model presented in Sec. 18.6 when the following change is made in the inventory system? (Consider each change independently.)

(a) The lead time is reduced to 0 (instantaneous delivery).

(b) The service level (measure 1) is decreased.

(c) The unit shortage cost is doubled.

(d) The mean of the probability distribution of demand during the lead time is increased (with no other change to the distribution).

(e) The probability distribution of demand during the lead time is a uniform distribution from a to b, but now $(b - a)$ has been doubled.

(f) The probability distribution of demand during the lead time is a normal distribution with mean μ and standard deviation σ, but now σ has been doubled.

18.6-6.* Jed Walker is the manager of Have a Cow, a hamburger restaurant in the downtown area. Jed has been purchasing all the restaurant's beef from Ground Chuck (a local supplier) but is considering switching to Chuck Wagon (a national warehouse) because its prices are lower.

Weekly demand for beef averages 500 pounds, with some variability from week to week. Jed estimates that the *annual* holding cost is 30 cents per pound of beef. When he runs out of beef, Jed is forced to buy from the grocery store next door. The high purchase cost and the hassle involved are estimated to cost him about $3 per pound of beef short. To help avoid shortages, Jed has decided to keep enough safety stock to prevent a shortage before the delivery arrives during 95 percent of the order cycles. Placing an order only requires sending a simple fax, so the administrative cost is negligible.

Have a Cow's contract with Ground Chuck is as follows: The purchase price is $1.49 per pound. A fixed cost of $25 per order is added for shipping and handling. The shipment is guaranteed to arrive within 2 days. Jed estimates that the demand for beef during this lead time has a uniform distribution from 50 to 150 pounds.

The Chuck Wagon is proposing the following terms: The beef will be priced at $1.35 per pound. The Chuck Wagon ships via refrigerated truck, and so charges additional shipping costs of $200 per order plus $0.10 per pound. The shipment time will be roughly a week, but is guaranteed not to exceed 10 days. Jed estimates that the probability distribution of demand during this lead time will be a normal distribution with a mean of 500 pounds and a standard deviation of 200 pounds.

T (a) Use the stochastic continuous-review model presented in Sec. 18.6 to obtain an (R, Q) policy for Have a Cow for each of the two alternatives of which supplier to use.

(b) Show how the reorder point is calculated for each of these two policies.

(c) Determine and compare the amount of safety stock provided by the two policies obtained in part (a).

(d) Determine and compare the average annual holding cost under these two policies.

(e) Determine and compare the average annual acquisition cost (combining purchase price and shipping cost) under these two policies.

(f) Since shortages are very infrequent, the only important costs for comparing the two suppliers are those obtained in parts (*d*) and (*e*). Add these costs for each supplier. Which supplier should be selected?

(g) Jed likes to use the beef (which he keeps in a freezer) within a month of receiving it. How would this influence his choice of supplier?

18.7-1. Read the referenced article that fully describes the OR study summarized in the application vignette presented in Sec. 18.7. Briefly describe how inventory theory was applied in this study. Then list the various financial and nonfinancial benefits that resulted from this study.

T **18.7-2.** A newspaper stand purchases newspapers for $0.55 and sells them for $0.75. The shortage cost is $0.75 per newspaper (because the dealer buys papers at retail price to satisfy shortages). The holding cost is $0.01 per newspaper left at the end of the day. The demand distribution is a uniform distribution between 50 and 75. Find the optimal number of papers to buy.

18.7-3. Freddie the newsboy runs a newstand. Because of a nearby financial services office, one of the newspapers he sells is the daily *Financial Journal*. He purchases copies of this newspaper from its distributor at the beginning of each day for $1.50 per copy, sells it for $2.50 each, and then receives a refund of $0.50 from the distributor the next morning for each unsold copy. The number of requests for this newspaper range from 15 to 18 copies per day. Freddie estimates that there are 15 requests on 40 percent of the days, 16 requests on 20 percent of the days, 17 requests on 30 percent of the days, and 18 requests on the remaining days.

(a) Use Bayes' decision rule presented in Sec. 15.2 to determine what Freddie's new order quantity should be to maximize his expected daily profit.

(b) Apply Bayes' decision rule again, but this time with the criterion of minimizing Freddie's expected daily cost of underordering or overordering.

(c) Use the stochastic single-period model for perishable products to determine Freddie's optimal order quantity.

(d) Draw the cumulative distribution function of demand and then show graphically how the model in part (*c*) finds the optimal order quantity.

18.7-4. Jennifer's Donut House serves a large variety of doughnuts, one of which is a blueberry-filled, chocolate-covered, supersized doughnut supreme with sprinkles. This is an extra large doughnut that is meant to be shared by a whole family. Since the dough requires so long to rise, preparation of these doughnuts begins at 4:00 in the morning, so a decision on how many to prepare must be made long before learning how many will be needed. The cost of the ingredients and labor required to prepare each of these doughnuts is $1. Their sale price is $3 each. Any not sold that day are sold to a local discount grocery store for $0.50. Over the last several weeks, the number of these doughnuts sold for $3 each day has been tracked. These data are summarized next.

Number Sold	Percentage of Days
0	10%
1	15%
2	20%
3	30%
4	15%
5	10%

(a) What is the unit cost of underordering? The unit cost of overordering?

(b) Use Bayes' decision rule presented in Sec. 15.2 to determine how many of these doughnuts should be prepared each day to minimize the average daily cost of underordering or overordering.

(c) After plotting the cumulative distribution function of demand, apply the stochastic single-period model for perishable products graphically to determine how many of these doughnuts to prepare each day.

(d) Given the answer in part (*c*), what will be the probability of running short of these doughnuts on any given day?

(e) Some families make a special trip to the Donut House just to buy this special doughnut. Therefore, Jennifer thinks that the cost when they run short might be greater than just the lost profit. In particular, there may be a cost for lost customer goodwill each time a customer orders this doughnut but none are available. How high would this cost have to be before they should prepare one more of these doughnuts each day than was found in part (*c*)?

18.7-5.* Swanson's Bakery is well known for producing the best fresh bread in the city, so the sales are very substantial. The daily demand for its fresh bread has a uniform distribution between 300 and 600 loaves. The bread is baked in the early morning, before the bakery opens for business, at a cost of $2 per loaf. It then is sold that day for $3 per loaf. Any bread not sold on the day it is baked is relabeled as day-old bread and sold subsequently at a discount price of $1.50 per loaf.

(a) Apply the stochastic single-period model for perishable products to determine the optimal service level.

(b) Apply this model graphically to determine the optimal number of loaves to bake each morning.

(c) With such a wide range of possible values in the demand distribution, it is difficult to draw the graph in part (*b*) carefully enough to determine the exact value of the optimal number of loaves. Use algebra to calculate this exact value.

(d) Given your answer in part (*a*), what is the probability of incurring a shortage of fresh bread on any given day?

(e) Because the bakery's bread is so popular, its customers are quite disappointed when a shortage occurs. The owner of the bakery, Ken Swanson, places high priority on keeping his customers satisfied, so he doesn't like having shortages. He feels that the analysis also should consider the loss of customer goodwill due to shortages. Since this loss of goodwill can have a negative effect on future sales, he estimates that a cost of $1.50 per loaf should be assessed each time a customer cannot purchase fresh bread because of a shortage. Determine the new optimal number of loaves to bake each day with this change. What is the new probability of incurring a shortage of fresh bread on any given day?

18.7-6. Reconsider Prob. 18.7-5. The bakery owner, Ken Swanson, now wants you to conduct a financial analysis of various inventory policies. You are to begin with the policy obtained in the first four parts of Prob. 18.7-5 (ignoring any cost for the loss of customer goodwill). As given with the answers in the back of the book, this policy is to bake 500 loaves of bread each morning, which gives a probability of incurring a shortage of $\frac{1}{3}$.

(a) For any day that a shortage *does* occur, calculate the revenue from selling fresh bread.

(b) For those days where shortages do *not* occur, use the probability distribution of demand to determine the expected number of loaves of fresh bread sold. Use this number to calculate the expected daily revenue from selling fresh bread on those days.

(c) Combine your results from parts (a) and (b) to calculate the expected daily revenue from selling fresh bread when considering *all* days.

(d) Calculate the expected daily revenue from selling day-old bread.

(e) Use the results in parts (c) and (d) to calculate the expected total daily revenue and then the expected daily profit (excluding overhead).

(f) Now consider the inventory policy of baking 600 loaves each morning, so that shortages never occur. Calculate the expected daily profit (excluding overhead) from this policy.

(g) Consider the inventory policy found in part (e) of Prob. 18.7-5. As implied by the answers in the back of the book, this policy is to bake 550 loaves each morning, which gives a probability of incurring a shortage of $\frac{1}{6}$. Since this policy is midway between the policy considered here in parts (a) to (e) and the one considered in part (f), its expected daily profit (excluding overhead and the cost of the loss of customer goodwill) also is midway between the expected daily profit for those two policies. Use this fact to determine its expected daily profit.

(h) Now consider the cost of the loss of customer goodwill for the inventory policy analyzed in part (g). Calculate the expected daily cost of the loss of customer goodwill and then the expected daily profit when considering this cost.

(i) Repeat part (h) for the inventory policy considered in parts (a) to (e).

18.7-7. Reconsider Prob. 18.7-5. The bakery owner, Ken Swanson, now has developed a new plan to decrease the size of shortages. The bread will be baked twice a day, once before the bakery opens (as before) and the other during the day after it becomes clearer what the demand for that day will be. The first baking will produce 300 loaves to cover the minimum demand for the day. The size of the second baking will be based on an estimate of the remaining demand for the day. This remaining demand is assumed to have a uniform distribution from a to b, where the values of a and b are chosen each day based on the sales so far. It is anticipated that $(b - a)$ typically will be approximately 75, as

opposed to the range of 300 for the distribution of demand in Prob. 18.7-5.

(a) Ignoring any cost of the loss of customer goodwill [as in parts (a) to (d) of Prob. 18.7-5], write a formula for how many loaves should be produced in the second baking in terms of a and b.

(b) What is the probability of still incurring a shortage of fresh bread on any given day? How should this answer compare to the corresponding probability in Prob. 18.7-5?

(c) When $b - a = 75$, what is the maximum size of a shortage that can occur? What is the maximum number of loaves of fresh bread that will not be sold? How do these answers compare to the corresponding numbers for the situation in Prob. 18.7-5 where only one (early morning) baking occurs per day?

(d) Now consider just the cost of underordering and the cost of overordering. Given your answers in part (c), how should the expected total daily cost of underordering and overordering for this new plan compare with that for the situation in Prob. 18.7-5? What does this say in general about the value of obtaining as much information as possible about what the demand will be before placing the final order for a perishable product?

(e) Repeat parts (a), (b), and (c) when including the cost of the loss of customer goodwill as in part (e) of Prob. 18.7-5.

18.7-8. Suppose that the demand D for a spare airplane part has an exponential distribution with mean 50, that is,

$$\varphi_D(\xi) = \begin{cases} \dfrac{1}{50}e^{-\xi/50} & \text{for } \xi \geq 0 \\ 0 & \text{otherwise.} \end{cases}$$

This airplane will be obsolete in 1 year, so all production of the spare part is to take place at present. The production costs now are $1,000 per item—that is, $c = 1,000$—but they become $10,000 per item if they must be supplied at later dates—that is, $p = 10,000$. The holding costs, charged on the excess after the end of the period, are $300 per item.

T (a) Determine the optimal number of spare parts to produce.

(b) Suppose that the manufacturer has 23 parts already in inventory (from a similar, but now obsolete airplane). Determine the optimal inventory policy.

(c) Suppose that p cannot be determined now, but the manufacturer wishes to order a quantity so that the probability of a shortage equals 0.1. How many units should be ordered?

(d) If the manufacturer were following an optimal policy that resulted in ordering the quantity found in part (c), what is the implied value of p?

18.7-9. Reconsider Prob. 18.6-1 involving Henry Edsel's car dealership. The current model year is almost over, but the Tritons are selling so well that the current inventory will be depleted before the end-of-year demand can be satisfied. Fortunately, there still is time to place one more order with the factory to replenish the inventory of Tritons just about when the current supply will be gone.

The general manager, Ruby Willis, now needs to decide how many Tritons to order from the factory. Each one costs $20,000. She then is able to sell them at an average price of $23,000, provided they are sold before the end of the model year. However, any of these

Tritons left at the end of the model year would then need to be sold at a special sale price of $19,500. Furthermore, Ruby estimates that the extra cost of the capital tied up by holding these cars such an unusually long time would be $500 per car, so the net revenue would be only $19,000. Since she would lose $1,000 on each of these cars left at the end of the model year, Ruby concludes that she needs to be cautious to avoid ordering too many cars, but she also wants to avoid running out of cars to sell before the end of the model year if possible. Therefore, she decides to use the stochastic single-period model for perishable products to select the order quantity. To do this, she estimates that the number of Tritons being ordered now that could be sold before the end of the model year has a normal distribution with a mean of 50 and a standard deviation of 15.

(a) Determine the optimal service level.

(b) Determine the number of Tritons that Ruby should order from the factory.

T **18.7-10.** Find the optimal ordering policy for the stochastic single-period model with a setup cost where the demand has the probability density function

$$\varphi_D(\xi) = \begin{cases} \dfrac{1}{20} & \text{for } 0 \le \xi \le 20 \\ 0 & \text{otherwise,} \end{cases}$$

and the costs are

$$\text{Holding cost} = \$1 \text{ per item,}$$
$$\text{Shortage cost} = \$3 \text{ per item,}$$
$$\text{Setup cost} = \$1.50,$$
$$\text{Production cost} = \$2 \text{ per item.}$$

Show your work, and then check your answer by using the corresponding Excel template in your OR Courseware.

T **18.7-11.** Using the approximation for finding the optimal policy for the stochastic single-period model with a setup cost when demand has an exponential distribution, find this policy when

$$\varphi_D(\xi) = \begin{cases} \dfrac{1}{25} e^{-\xi/25} & \text{for } \xi \ge 0 \\ 0 & \text{otherwise,} \end{cases}$$

and the costs are

$$\text{Holding cost} = 40 \text{ cents per item,}$$
$$\text{Shortage cost} = \$1.50 \text{ per item,}$$
$$\text{Purchase price} = \$1 \text{ per item,}$$
$$\text{Setup cost} = \$10.$$

Show your work, and then check your answer by using the corresponding Excel template in your OR Courseware.

18.8-1. Reconsider the Blue Skies Airlines example presented in Sec. 18.8. Regarding the flight under consideration, recent experience indicates that the demand for the very low discount fare of $200 is so high that it may be possible to considerably increase this fare and still usually fill up the airplane with both leisure and business travelers. Therefore, management wants to learn how the optimal number of reservation slots to reserve for class 1 customers would change if this fare were to be increased. Make this calculation for new fares of $300, $400, $500, and $600.

18.8-2. The most popular cruise offered by Luxury Cruises is a three-week cruise in the Mediterranean each July with daily ports of call at interesting tourist destinations. The ship has 1,000 cabins, so it is a challenge to fill the ship because of the high fares charged. In particular, the average regular fare for a cabin is $20,000, which is too high for many potential customers. Therefore, to help fill the ship, the company offers a special discount fare for this cruise that averages $12,000 per cabin when it announces its future cruises a year in advance. The deadline for obtaining this discount fare is 11 months before the cruise, and this discount also can be discontinued earlier at the company's discretion. Thereafter, the company uses heavy publicity to attract luxury-seeking customers who make vacation plans later and are willing to pay the regular fare averaging $20,000 per cabin. Based on past experience, it is estimated that the number of such luxury-seeking customers for this cruise has a normal distribution with a mean of 400 and a standard deviation of 100.

Use the model for capacity-controlled discount fares presented in Sec. 18.8 to determine the maximum number of cabins that should be sold at the discount fare before reserving the remaining cabins to be sold at the regular fare.

18.8-3. To help fill its seats for a particular flight, an airline offers a special nonrefundable fare of $100 for customers who make a reservation at least 21 days in advance and satisfy other restrictions. Thereafter, the fare will be $300. A total of 100 reservations will be accepted. The number of customers who have requested a reservation at full fare for this flight in the past always has been at least 31 and not more than 50. It is estimated that the integer numbers between 31 and 50 are equally likely.

Use the model for capacity-controlled discount fares to determine how many of the reservations should be reserved for customers who would pay full fare.

18.8-4. Reconsider the Transcontinental Airlines example presented in Sec. 18.8. Management has concluded that the original estimate of $500 for the intangible cost of a loss of goodwill on the part of a bumped customer is much too low and should be increased to $1,000. Use the overbooking model to determine the number of reservations that now should be accepted for this flight.

18.8-5. The management of Quality Airlines has decided to base its overbooking policy on the overbooking model presented in Sec. 18.8. This policy now needs to be applied to a new flight from Seattle to Atlanta. The airplane has 125 seats available for a nonrefundable fare of $250. However, since there commonly are a few no-shows on similar flights, the airline should accept a few more than 125 reservations. On those occasions when more than 125 arrive to take the flight, the airline will find volunteers who are willing to be put free on a later Quality Airlines flight that has available seats, in return for being given a certificate worth $500 (but that would cost the company just $300) toward any future travel on this airline. Management feels that an additional $300 should be assessed for the intangible cost of a loss of goodwill for inconveniencing these customers.

Based on previous experience with similar flights having about 125 reservations, it is estimated that the relative frequency of the number of no-shows (independent of the exact number of reservations) will be as shown below.

Number of No-Shows	Relative Frequency
0	0%
1	5
2	10
3	10
4	15
5	20
6	15
7	10
8	10
9	5

Instead of using the binomial distribution, use this distribution directly with the overbooking model to determine how much overbooking the company should do for this flight.

18.8-6. Consider the overbooking model presented in Sec. 18.8. For a specific application, suppose that the parameters of the model are $p = 0.5$, $r = \$1,000$, $s = \$5,000$, and $L = 3$. Use the binomial distribution directly (not the normal approximation) to calculate n^*, the optimal number of reservations to accept, by using trial and error.

18.8-7. The Mountain Top Hotel is a luxury hotel in a popular ski resort area. The hotel always is essentially full during winter

months, so reservations and payments must be made months in advance for week-long stays from Saturday to Saturday. Reservations can be cancelled until a month in advance, but are nonrefundable after that. The hotel has 100 rooms and the room charge for a week's stay is $3,000. Despite this high cost, the hotel's wealthy customers occasionally will forfeit this money and not show up because their plans have changed. On the average, about 10 percent of the customers with reservations are no-shows, so the hotel's management wants to do some overbooking. However, it also feels that this should be done cautiously because the consequences of turning away a customer with a reservation would be severe. These consequences include the cost of quickly arranging for alternative housing in an inferior hotel, providing a voucher for a future stay, and the intangible cost of a massive loss of goodwill on the part of the furious customer who is turned away (and surely will tell many wealthy friends about this shabby treatment). Management estimates that the cost that should be imputed to these consequences is $20,000.

Use the overbooking model presented in Sec. 18.8, including the normal approximation for the binomial distribution, to determine how much overbooking the hotel should do.

18.9-1. From the bottom part of the selected references given at the end of the chapter, select one of these award-winning applications of inventory theory. Read this article and then write a two-page summary of the application and the benefits (including nonfinancial benefits) it provided.

18.9-2. From the bottom part of the selected references given at the end of the chapter, select three of these award-winning applications of inventory theory. For each one, write a one-page summary of the application and the benefits (including nonfinancial benefits) it provided.

■ CASES

CASE 18.1   Brushing Up on Inventory Control

Robert Gates rounds the corner of the street and smiles when he sees his wife pruning rose bushes in their front yard. He slowly pulls his car into the driveway, turns off the engine, and falls into his wife's open arms.

"How was your day?" she asks.

"Great! The drugstore business could not be better!" Robert replies, "Except for the traffic coming home from work! That traffic can drive a sane man crazy! I am so tense right now. I think I will go inside and make myself a relaxing martini."

Robert enters the house and walks directly into the kitchen. He sees the mail on the kitchen counter and begins flipping through the various bills and advertisements until he comes across the new issue of *OR/MS Today*. He prepares

his drink, grabs the magazine, treads into the living room, and settles comfortably into his recliner. He has all that he wants—except for one thing. He sees the remote control lying on the top of the television. He sets his drink and magazine on the coffee table and reaches for the remote control. Now, with the remote control in one hand, the magazine in the other, and the drink on the table near him, Robert is finally the master of his domain.

Robert turns on the television and flips the channels until he finds the local news. He then opens the magazine and begins reading an article about scientific inventory management. Occasionally he glances at the television to learn the latest in business, weather, and sports.

As Robert delves deeper into the article, he becomes distracted by a commercial on television about toothbrushes. His pulse quickens slightly in fear because the commercial for Totalee toothbrushes reminds him of the dentist. The commerical concludes that the customer should buy a Totalee toothbrush because the toothbrush is Totalee revolutionary and Totalee effective. It certainly is effective; it is the most popular toothbrush on the market!

At that moment, with the inventory article and the toothbrush commercial fresh in his mind, Robert experiences a flash of brilliance. He knows how to control the inventory of Totalee toothbrushes at Nightingale Drugstore!

As the inventory control manager at Nightingale Drugstore, Robert has been experiencing problems keeping Totalee toothbrushes in stock. He has discovered that customers are very loyal to the Totalee brand name since Totalee holds a patent on the toothbrush endorsed by 9 out of 10 dentists. Customers are willing to wait for the toothbrushes to arrive at Nightingale Drugstore since the drugstore sells the toothbrushes for 20 percent less than other local stores. This demand for the toothbrushes at Nightingale means that the drugstore is often out of Totalee toothbrushes. The store is able to receive a shipment of toothbrushes several hours after an order is placed to the Totalee regional warehouse because the warehouse is only 20 miles away from the store. Nevertheless, the current inventory situation causes problems because numerous emergency orders cost the store unnecessary time and paperwork and because customers become disgruntled when they must return to the store later in the day.

Robert now knows a way to prevent the inventory problems through scientific inventory management! He grabs his coat and car keys and rushes out of the house.

As he runs to the car, his wife yells, "Honey, where are you going?"

"I'm sorry, darling," Robert yells back. "I have just discovered a way to control the inventory of a critical item at the drugstore. I am really excited because I am able to apply my industrial engineering degree to my job! I need to get the data from the store and work out the new inventory policy! I will be back before dinner!"

Because rush hour traffic has dissipated, the drive to the drugstore takes Robert no time at all. He unlocks the darkened store and heads directly to his office where he rummages through file cabinets to find demand and cost data for Totalee toothbrushes over the past year.

Aha! Just as he suspected! The demand data for the toothbrushes is almost constant across the months. Whether in winter or summer, customers have teeth to brush, and they need toothbrushes. Since a toothbrush will wear out after a few months of use, customers will always return to buy another toothbrush. The demand data shows that Nightingale Drugstore customers purchase an average of 250 Totalee toothbrushes per month (30 days).

After examining the demand data, Robert investigates the cost data. Because Nightingale Drugstore is such a good customer, Totalee charges its lowest wholesale price of only $1.25 per toothbrush. Robert spends about 20 minutes to place each order with Totalee. His salary and benefits add up to $18.75 per hour. The annual holding cost for the inventory is 12 percent of the capital tied up in the inventory of Totalee toothbrushes.

(a) Robert decides to create an inventory policy that normally fulfills all demand since he believes that stock-outs are just not worth the hassle of calming customers or the risk of losing future business. He therefore does not allow any planned shortages. Since Nightingale Drugstore receives an order several hours after it is placed, Robert makes the simplifying assumption that delivery is instantaneous. What is the optimal inventory policy under these conditions? How many Totalee toothbrushes should Robert order each time and how frequently? What is the total variable inventory cost per year with this policy?

(b) Totalee has been experiencing financial problems because the company has lost money trying to branch into producing other personal hygiene products, such as hairbrushes and dental floss. The company has therefore decided to close the warehouse located 20 miles from Nightingale Drugstore. The drugstore must now place orders with a warehouse located 350 miles away and must wait 6 days after it places an order to receive the shipment. Given this new lead time, how many Totalee toothbrushes should Robert order each time, and when should he order?

(c) Robert begins to wonder whether he would save money if he allows planned shortages to occur. Customers would wait to buy the toothbrushes from Nightingale since they have high brand loyalty and since Nightingale sells the toothbrushes for less. Even though customers would wait to purchase the Totalee toothbrush from Nightingale, they would become unhappy with the prospect of having to return to the store again for the product. Robert decides that he needs to place a dollar value on the negative ramifications from shortages. He knows that an employee would have to calm each disgruntled customer and track down the delivery date for a new shipment of Totalee toothbrushes. Robert also believes that customers would become upset with the inconvenience of shopping at Nightingale and would perhaps begin looking for another store providing better service. He estimates the costs of dealing with disgruntled customers and losing customer goodwill and future sales as $1.50 per unit short per year. Given the 6-day lead time and the shortage allowance, how many Totalee toothbrushes should Robert order each time, and when should he order? What is the maximum shortage under this optimal inventory policy? What is the total variable inventory cost per year?

(d) Robert realizes that his estimate for the shortage cost is simply that—an estimate. He realizes that employees sometimes must spend several minutes with each customer who wishes to purchase a toothbrush when none is currently available. In addition, he realizes that the cost of losing customer goodwill and future sales could vary within a wide range. He estimates that the cost of dealing with disgruntled customers and losing customer goodwill and future sales could range from 85 cents to $25 per unit short per year. What effect would changing the estimate of the unit shortage cost have on the inventory policy and total variable inventory cost per year found in part (c)?

(e) Closing warehouses has not improved Totalee's bottom line significantly, so the company has decided to institute a discount policy to encourage more sales. Totalee will charge $1.25 per toothbrush for any order of up to 500 toothbrushes, $1.15 per toothbrush for orders of more than 500 but less than 1000 toothbrushes, and $1 per toothbrush for orders of 1000 toothbrushes or more. Robert still assumes a 6-day lead time, but he does not want planned shortages to occur. Under the new discount policy, how many Totalee toothbrushes should Robert order each time, and when should he order? What is the total inventory cost (including purchase costs) per year?

■ PREVIEWS OF ADDED CASES ON OUR WEBSITE (www.mhhe.com/hillier)

CASE 18.2   TNT: Tackling Newsboy's Teaching

A young entrepreneur will be operating a firecracker stand for the Fourth of July. He has time to place only one order for the firecrackers he will sell from his stand. After obtaining the relevant financial data and some information with which to estimate the probability distribution of potential sales, he now needs to determine how many firecracker sets he should order to maximize his expected profit under different scenarios.

CASE 18.3   Jettisoning Surplus Stock

American Aerospace produces military jet engines. Frequent shortages of one critical part has been causing delays in the production of the most popular jet engine, so a new inventory policy needs to be developed for this part. There is a long lead time between when an order is placed for the part and when the order quantity is received. The demand for the part during this lead time is uncertain, but some data are available for estimating its probability distribution. In the future, the inventory level of the part will be kept under continuous review. Decisions now need to be made regarding the inventory level at which a new order should be placed and what the order quantity should be.

19

Markov Decision Processes

hapter 16 introduced *Markov chains* and their analysis. Most of the chapter was devoted to *discrete time* Markov chains, i.e., Markov chains that are observed only at discrete points in time (e.g., the end of each day) rather than continuously. Each time it is observed, the Markov chain can be in any one of a number of *states*. Given the current state, a (one-step) *transition matrix* gives the probabilities for what the state will be next time. Given this transition matrix, Chap. 16 focused on *describing the behavior* of a Markov chain, e.g., finding the steady-state probabilities for what state it is in.

Many important systems (e.g., many queueing systems) can be modeled as either a discrete time or continuous time Markov chain. It is useful to describe the behavior of such a system (as we did in Chap. 17 for queueing systems) in order to evaluate its performance. However, it may be even more useful to *design the operation* of the system so as to *optimize its performance* (as we did in Sec. 17.10 for queueing systems).

This chapter focuses on how to design the operation of a discrete time Markov chain so as to optimize its performance. Therefore, rather than passively accepting the design of the Markov chain and the corresponding fixed transition matrix, we now are being proactive. For each possible state of the Markov chain, we make a decision about which one of several alternative actions should be taken in that state. The action chosen affects the *transition probabilities* as well as both the *immediate costs* (or rewards) and *subsequent costs* (or rewards) from operating the system. We want to choose the optimal actions for the respective states when considering both immediate and subsequent costs. The decision process for doing this is referred to as a *Markov decision process*.

The first section gives a prototype example of an application of a Markov decision process. Section 19.2 formulates the basic model for these processes. The next three sections describe how to solve them.

■ 19.1 A PROTOTYPE EXAMPLE

A manufacturer has one key machine at the core of one of its production processes. Because of heavy use, the machine deteriorates rapidly in both quality and output. Therefore, at the end of each week, a thorough inspection is done that results in classifying the condition of the machine into one of four possible states:

State	Condition
0	Good as new
1	Operable—minor deterioration
2	Operable—major deterioration
3	Inoperable—output of unacceptable quality

After historical data on these inspection results are gathered, statistical analysis is done on how the state of the machine evolves from month to month. The following matrix shows the relative frequency (probability) of each possible transition from the state in one month (a row of the matrix) to the state in the following month (a column of the matrix).

State	0	1	2	3
0	0	$\frac{7}{8}$	$\frac{1}{16}$	$\frac{1}{16}$
1	0	$\frac{3}{4}$	$\frac{1}{8}$	$\frac{1}{8}$
2	0	0	$\frac{1}{2}$	$\frac{1}{2}$
3	0	0	0	1

In addition, statistical analysis has found that these transition probabilities are unaffected by also considering what the states were in prior months. This "lack-of-memory property" is the *Markovian property* described in Sec. 16.2. Therefore, for the random variable X_t, which is the state of the machine at the end of month t, it has been concluded that the stochastic process $\{X_t, t = 0, 1, 2, \ldots\}$ is a *discrete time Markov chain* whose (one-step) *transition matrix* is just the above matrix.

As the last entry in this transition matrix indicates, once the machine becomes inoperable (enters state 3), it remains inoperable. In other words, state 3 is an *absorbing state*. Leaving the machine in this state would be intolerable, since this would shut down the production process, so the machine must be replaced. (Repair is not feasible in this state.) The new machine then will start off in state 0.

The replacement process takes 1 week to complete so that production is lost for this period. The cost of the lost production (lost profit) is $2,000, and the cost of replacing the machine is $4,000, so the total cost incurred whenever the current machine enters state 3 is $6,000.

Even before the machine reaches state 3, costs may be incurred from the production of defective items. The expected costs per week from this source are as follows:

State	Expected Cost Due to Defective Items, $
0	0
1	1,000
2	3,000

We now have mentioned all the relevant costs associated with one particular *maintenance policy* (replace the machine when it becomes inoperable but do no maintenance otherwise). Under this policy, the evolution of the state of the *system* (the succession of machines) still is a Markov chain, but now with the following transition matrix:

State	0	1	2	3
0	0	$\frac{7}{8}$	$\frac{1}{16}$	$\frac{1}{16}$
1	0	$\frac{3}{4}$	$\frac{1}{8}$	$\frac{1}{8}$
2	0	0	$\frac{1}{2}$	$\frac{1}{2}$
3	1	0	0	0

To evaluate this maintenance policy, we should consider both the immediate costs incurred over the coming week (just described) and the subsequent costs that result from having the system evolve in this way. As introduced in Sec. 16.5, one such widely used measure of performance for Markov chains is the (long-run) **expected average cost per unit time**.[1]

To calculate this measure, we first derive the *steady-state probabilities* π_0, π_1, π_2, and π_3 for this Markov chain by solving the following steady-state equations:

$$\pi_0 = \pi_3,$$
$$\pi_1 = \frac{7}{8}\pi_0 + \frac{3}{4}\pi_1,$$
$$\pi_2 = \frac{1}{16}\pi_0 + \frac{1}{8}\pi_1 + \frac{1}{2}\pi_2,$$
$$\pi_3 = \frac{1}{16}\pi_0 + \frac{1}{8}\pi_1 + \frac{1}{2}\pi_2,$$
$$1 = \pi_0 + \pi_1 + \pi_2 + \pi_3.$$

(Although this system of equations is small enough to be solved by hand without great difficulty, the Steady-State Probabilities procedure in the Markov Chains area of your IOR Tutorial provides another quick way of obtaining this solution.) The simultaneous solution is

$$\pi_0 = \frac{2}{13}, \qquad \pi_1 = \frac{7}{13}, \qquad \pi_2 = \frac{2}{13}, \qquad \pi_3 = \frac{2}{13}.$$

Hence, the (long-run) expected average cost per week for this maintenance policy is

$$0\pi_0 + 1,000\pi_1 + 3,000\pi_2 + 6,000\pi_3 = \frac{25,000}{13} = \$1,923.08.$$

However, there also are other maintenance policies that should be considered and compared with this one. For example, perhaps the machine should be replaced before it reaches state 3. Another alternative is to *overhaul* the machine at a cost of $2,000. This option is not feasible in state 3 and does not improve the machine while in state 0 or 1, so it is of interest only in state 2. In this state, an overhaul would return the machine to state 1. A week is required, so another consequence is $2,000 in lost profit from lost production.

In summary, the possible decisions after each inspection are as follows:

Decision	Action	Relevant States
1	Do nothing	0, 1, 2
2	Overhaul (return system to state 1)	2
3	Replace (return system to state 0)	1, 2, 3

[1]The term *long-run* indicates that the average should be interpreted as being taken over an *extremely* long time so that the effect of the initial state disappears. As time goes to infinity, Sec. 16.5 discusses the fact that the *actual* average cost per unit time essentially always converges to the *expected* average cost per unit time.

TABLE 19.1 Cost data for the prototype example

Decision	State	Expected Cost Due to Producing Defective Items, $	Maintenance Cost, $	Cost (Lost Profit) of Lost Production, $	Total Cost per Week, $
1. Do nothing	0	0	0	0	0
	1	1,000	0	0	1,000
	2	3,000	0	0	3,000
2. Overhaul	2	0	2,000	2,000	4,000
3. Replace	1, 2, 3	0	4,000	2,000	6,000

For easy reference, Table 19.1 also summarizes the relevant costs for each decision for each state where that decision could be of interest.

What is the optimal maintenance policy? We will be addressing this question to illustrate the material in the next four sections.

19.2   A MODEL FOR MARKOV DECISION PROCESSES

The model for the Markov decision processes considered in this chapter can be summarized as follows.

1. The state i of a discrete time Markov chain is observed after each transition ($i = 0, 1, \ldots, M$).
2. After each observation, a *decision* (action) k is chosen from a set of K possible decisions ($k = 1, 2, \ldots, K$). (Some of the K decisions may not be relevant for some of the states.)
3. If decision $d_i = k$ is made in state i, an immediate *cost* is incurred that has an expected value C_{ik}.
4. The decision $d_i = k$ in state i determines what the *transition probabilities*[2] will be for the next transition from state i. Denote these transition probabilities by $p_{ij}(k)$, for $j = 0, 1, \ldots, M$.
5. A specification of the decisions for the respective states ($d_0, d_1, \ldots, d_M$) prescribes a *policy* for the Markov decision process.
6. The objective is to find an *optimal policy* according to some cost criterion which considers both immediate costs and subsequent costs that result from the future evolution of the process. One common criterion is to minimize the (long-run) *expected average cost per unit time*. (An alternative criterion is considered in Sec. 19.5.)

To relate this general description to the prototype example presented in Sec. 19.1, recall that the Markov chain being observed there represents the state (condition) of a particular machine. After each inspection of the machine, a choice is made between three possible decisions (do nothing, overhaul, or replace). The resulting immediate expected cost is shown in the rightmost column of Table 19.1 for each relevant combination of state and decision. Section 19.1 analyzed one particular policy (d_0, d_1, d_2, d_3) = (1, 1, 1, 3), where decision 1 (do nothing) is made in states 0, 1, and 2 and decision 3 (replace) is made in state 3. The resulting transition probabilities are shown in the last transition matrix given in Sec. 19.1.

Our general model qualifies to be a *Markov* decision process because it possesses the Markovian property that characterizes any Markov process. In particular, given the current state and decision, any probabilistic statement about the future of the

---

[2]The solution procedures given in the next two sections also assume that the resulting transition matrix is *irreducible*.

In 2003, **Bank One Corporation** was the sixth-largest bank in the United States. Bank One Card Services, Inc., a division of Bank One Corporation, also was the largest issuer of Visa cards in the United States, on behalf of both Bank One and several thousand marketing partners. The following year, Bank One Corporation merged with *JPMorgan Chase* under the latter name to form the third-largest banking institution in the country. *Chase* thereafter was used as the brand for its credit card services.

The credit card business is a natural application area of operations research because its success depends so directly on a careful balancing of various quantitative factors. The annual percentage rate (APR) for interest charges and the credit line of card accounts influence both card use and bank profitability. Consumers find low APR levels and high credit lines attractive. However, low APR levels may reduce bank profitability, while indiscriminate increases in credit lines increase the bank's exposure to credit loss. It is critical that these factors be balanced in different ways for different customers based on the evolving credit ratings of these customers.

With all this in mind, Bank One management asked its in-house OR group in 1999 to begin the PORTICO (portfolio control and optimization) project to evaluate approaches for improving the profitability of its credit card business. The OR group designed the PORTICO system using *Markov decision processes* to select the APR levels and credit lines for individual card holders that maximize the *net present value* of the entire portfolio of credit card customers. The group used several variables—including the credit-line level, the APR level, and some variables describing customer behavior in making payments—to determine the state into which to slot an account in any month. The transition probabilities were based on 18 months of time-series data on a random sample of 3 million credit card accounts from the bank's portfolio. The decisions to be made for each state of the Markov decision process are the APR level and credit-line level for that category of customers in the next month.

A considerable period of testing the PORTICO model verified that it would substantially increase the bank's profitability. As the actual implementation began, it was estimated that this new process would *increase annual profits by over* **$75 million**. This outstanding application of Markov decision processes led to Bank One winning the prestigious *Wagner Prize for Excellence in Operations Research Practice* for 2002.

Source: M. S. Trench, S. P. Pederson, E. T. Lau, L. Ma, H. Wang, and S. K. Nair: "Managing Credit Lines and Prices for Bank One Credit Cards," *Interfaces*, **33**(5): 4–21, Sept.–Oct. 2003. (A link to this article is provided on our website, www.mhhe .com/hillier.)

process is completely unaffected by providing any information about the history of the process. This Markovian property holds here since (1) we are dealing with a Markov chain, (2) the new transition probabilities depend on only the current state and decision, and (3) the immediate expected cost also depends on only the current state and decision.

Our description of a policy implies two convenient (but unnecessary) properties that we will assume throughout the chapter (with one exception). One property is that a policy is **stationary;** i.e., whenever the system is in state i, the rule for making the decision always is the same regardless of the value of the current time t. The second property is that a policy is **deterministic;** i.e., whenever the system is in state i, the rule for making the decision definitely chooses one particular decision. (Because of the nature of the algorithm involved, the next section considers *randomized* policies instead, where a probability distribution is used for the decision to be made.)

Using this general framework, we now return to the prototype example and find the optimal policy by enumerating and comparing all the relevant policies. In doing this, we will let R denote a specific policy and $d_i(R)$ denote the corresponding decision to be made in state i, where decisions 1, 2, and 3 are described at the end of the preceding section. Since one or more of these three decisions are the only ones that would be considered in any given state, the only possible values of $d_i(R)$ are 1, 2, or 3 for any state i.

Solving the Prototype Example by Exhaustive Enumeration

The relevant policies for the prototype example are these:

Policy	Verbal Description	$d_0(R)$	$d_1(R)$	$d_2(R)$	$d_3(R)$
R_a	Replace in state 3	1	1	1	3
R_b	Replace in state 3, overhaul in state 2	1	1	2	3
R_c	Replace in states 2 and 3	1	1	3	3
R_d	Replace in states 1, 2, and 3	1	3	3	3

Each policy results in a different transition matrix, as shown below.

R_a				
State	0	1	2	3
0	0	$\frac{7}{8}$	$\frac{1}{16}$	$\frac{1}{16}$
1	0	$\frac{3}{4}$	$\frac{1}{8}$	$\frac{1}{8}$
2	0	0	$\frac{1}{2}$	$\frac{1}{2}$
3	1	0	0	0

R_b				
State	0	1	2	3
0	0	$\frac{7}{8}$	$\frac{1}{16}$	$\frac{1}{16}$
1	0	$\frac{3}{4}$	$\frac{1}{8}$	$\frac{1}{8}$
2	0	1	0	0
3	1	0	0	0

R_c				
State	0	1	2	3
0	0	$\frac{7}{8}$	$\frac{1}{16}$	$\frac{1}{16}$
1	0	$\frac{3}{4}$	$\frac{1}{8}$	$\frac{1}{8}$
2	1	0	0	0
3	1	0	0	0

R_d				
State	0	1	2	3
0	0	$\frac{7}{8}$	$\frac{1}{16}$	$\frac{1}{16}$
1	1	0	0	0
2	1	0	0	0
3	1	0	0	0

From the rightmost column of Table 19.1, the values of C_{ik} are as follows:

		C_{ik} (in Thousands of Dollars)	
State i Decision k	1	2	3
0	0	—	—
1	1	—	6
2	3	4	6
3	—	—	6

As indicated in Sec. 16.5, the (long-run) expected average cost per unit time $E(C)$ then can be calculated from the expression

$$E(C) = \sum_{i=0}^{M} C_{ik}\pi_i,$$

where $k = d_i(R)$ for each i and $(\pi_0, \pi_1, \ldots, \pi_M)$ represents the steady-state distribution of the state of the system under the policy R being evaluated. After $(\pi_0, \pi_1, \ldots, \pi_M)$ are solved for under each of the four policies (as can be done with your IOR Tutorial), the calculation of $E(C)$ is as summarized here:

Policy	$(\pi_0, \pi_1, \pi_2, \pi_3)$	$E(C)$, in Thousands of Dollars
R_a	$\left(\dfrac{2}{13}, \dfrac{7}{13}, \dfrac{2}{13}, \dfrac{2}{13}\right)$	$\dfrac{1}{13}[2(0) + 7(1) + 2(3) + 2(6)] = \dfrac{25}{13} = \$1,923$
R_b	$\left(\dfrac{2}{21}, \dfrac{5}{7}, \dfrac{2}{21}, \dfrac{2}{21}\right)$	$\dfrac{1}{21}[2(0) + 15(1) + 2(4) + 2(6)] = \dfrac{35}{21} = \$1,667 \leftarrow$ Minimum
R_c	$\left(\dfrac{2}{11}, \dfrac{7}{11}, \dfrac{1}{11}, \dfrac{1}{11}\right)$	$\dfrac{1}{11}[2(0) + 7(1) + 1(6) + 1(6)] = \dfrac{19}{11} = \$1,727$
R_d	$\left(\dfrac{1}{2}, \dfrac{7}{16}, \dfrac{1}{32}, \dfrac{1}{32}\right)$	$\dfrac{1}{32}[16(0) + 14(6) + 1(6) + 1(6)] = \dfrac{96}{32} = \$3,000$

Thus, the optimal policy is R_b; that is, replace the machine when it is found to be in state 3, and overhaul the machine when it is found to be in state 2. The resulting (long-run) expected average cost per week is $1,667.

If you would like to go through **another small example**, one is provided in the Worked Examples section of the book's website.

Using exhaustive enumeration to find the optimal policy is appropriate for such tiny examples, where there are so few relevant policies. However, many applications have so many policies that this approach would be completely infeasible. For such cases, algorithms that can efficiently find an optimal policy are needed. The next three sections consider such algorithms.

■ 19.3 LINEAR PROGRAMMING AND OPTIMAL POLICIES

Section 19.2 described the main kind of policy (called a *stationary, deterministic* policy) that is used by Markov decision processes. We saw that any such policy R can be viewed as a rule that prescribes decision $d_i(R)$ whenever the system is in state i, for each $i = 0, 1, \ldots, M$. Thus, R is characterized by the values

$$\{d_0(R), d_1(R), \ldots, d_M(R)\}.$$

Equivalently, R can be characterized by assigning values $D_{ik} = 0$ or 1 in the matrix

$$
\begin{array}{c}
\quad\quad\quad\quad\quad \text{Decision } k \\
\quad\quad\quad 1 \quad\quad 2 \quad \cdots \quad K \\
\text{State } i \;
\begin{array}{c} 0 \\ 1 \\ \vdots \\ M \end{array}
\left[
\begin{array}{cccc}
D_{01} & D_{02} & \cdots & D_{0K} \\
D_{11} & D_{12} & \cdots & D_{1K} \\
\hline
\multicolumn{4}{c}{\cdots\cdots\cdots\cdots\cdots\cdots\cdots\cdots} \\
D_{M1} & D_{M2} & \cdots & D_{MK}
\end{array}
\right],
\end{array}
$$

where each D_{ik} ($i = 0, 1, \ldots, M$ and $k = 1, 2, \ldots, K$) is defined as

$$
D_{ik} = \begin{cases} 1 & \text{if decision } k \text{ is to be made in state } i \\ 0 & \text{otherwise.} \end{cases}
$$

Therefore, each row in the matrix must contain a single 1 with the rest of the elements 0s. For example, the optimal policy R_b for the prototype example is characterized by the matrix

$$
\text{State } i \quad
\begin{array}{c}
 \\
0 \\
1 \\
2 \\
3
\end{array}
\begin{array}{c}
\text{Decision } k \\
\begin{array}{ccc}
1 & 2 & 3
\end{array} \\
\left[
\begin{array}{ccc}
1 & 0 & 0 \\
1 & 0 & 0 \\
0 & 1 & 0 \\
0 & 0 & 1
\end{array}
\right];
\end{array}
$$

i.e., do nothing (decision 1) when the machine is in state 0 or 1, overhaul (decision 2) in state 2, and replace the machine (decision 3) when it is in state 3.

Randomized Policies

Introducing D_{ik} provides motivation for a *linear programming formulation*. It is hoped that the expected cost of a policy can be expressed as a linear function of D_{ik} or a related variable, subject to linear constraints. Unfortunately, the D_{ik} values are integers (0 or 1), and continuous variables are required for a linear programming formulation. This requirement can be handled by expanding the interpretation of a policy. The previous definition calls for making the same decision every time the system is in state i. The new interpretation of a policy will call for determining a probability distribution for the decision to be made when the system is in state i.

With this new interpretation, the D_{ik} now need to be redefined as

$$D_{ik} = P\{\text{decision} = k \mid \text{state} = i\}.$$

In other words, given that the system is in state i, variable D_{ik} is the *probability* of choosing decision k as the decision to be made. Therefore, $(D_{i1}, D_{i2}, \ldots, D_{iK})$ is the *probability distribution* for the decision to be made in state i.

This kind of policy using probability distributions is called a **randomized policy,** whereas the policy calling for $D_{ik} = 0$ or 1 is a *deterministic policy*. Randomized policies can again be characterized by the matrix

$$
\text{State } i \quad
\begin{array}{c}
 \\
0 \\
1 \\
\vdots \\
M
\end{array}
\begin{array}{c}
\text{Decision } k \\
\begin{array}{cccc}
1 & 2 & \cdots & K
\end{array} \\
\left[
\begin{array}{cccc}
D_{01} & D_{02} & \cdots & D_{0K} \\
D_{11} & D_{12} & \cdots & D_{1K} \\
\hline
\multicolumn{4}{c}{\cdots\cdots\cdots\cdots\cdots\cdots\cdots} \\
D_{M1} & D_{M2} & \cdots & D_{MK}
\end{array}
\right],
\end{array}
$$

where each row sums to 1, and now

$$0 \leq D_{ik} \leq 1.$$

To illustrate, consider a randomized policy for the prototype example given by the matrix

$$
\text{State } i \quad
\begin{array}{c}
 \\
0 \\
1 \\
2 \\
3
\end{array}
\begin{array}{c}
\text{Decision } k \\
\begin{array}{ccc}
1 & 2 & 3
\end{array} \\
\left[
\begin{array}{ccc}
1 & 0 & 0 \\
\frac{1}{2} & 0 & \frac{1}{2} \\
\frac{1}{4} & \frac{1}{4} & \frac{1}{2} \\
0 & 0 & 1
\end{array}
\right].
\end{array}
$$

This policy calls for *always* making decision 1 (do nothing) when the machine is in state 0. If it is found to be in state 1, it is left as is with probability $\frac{1}{2}$ and replaced with probability $\frac{1}{2}$, so a coin can be flipped to make the choice. If it is found to be in state 2, it is left as is with probability $\frac{1}{4}$, overhauled with probability $\frac{1}{4}$, and replaced with probability $\frac{1}{2}$.

Presumably, a random device with these probabilities (possibly a table of random numbers) can be used to make the actual decision. Finally, if the machine is found to be in state 3, it always is replaced.

By allowing randomized policies, so that the D_{ik} are continuous variables instead of integer variables, it now is possible to formulate a linear programming model for finding an optimal policy.

A Linear Programming Formulation

The convenient decision variables (denoted here by y_{ik}) for a linear programming model are defined as follows. For each $i = 0, 1, \ldots, M$ and $k = 1, 2, \ldots, K$, let y_{ik} be the steady-state unconditional probability that the system is in state i *and* decision k is made; i.e.,

$$y_{ik} = P\{\text{state} = i \text{ and decision} = k\}.$$

Each y_{ik} is closely related to the corresponding D_{ik} since, from the rules of conditional probability,

$$y_{ik} = \pi_i D_{ik},$$

where π_i is the steady-state probability that the Markov chain is in state i. Furthermore,

$$\pi_i = \sum_{k=1}^{K} y_{ik},$$

so that

$$D_{ik} = \frac{y_{ik}}{\pi_i} = \frac{y_{ik}}{\sum_{k=1}^{K} y_{ik}}.$$

There exist three sets of constraints on y_{ik}:

1. $\displaystyle\sum_{i=0}^{M} \pi_i = 1$ so that $\displaystyle\sum_{i=0}^{M} \sum_{k=1}^{K} y_{ik} = 1.$

2. From results on steady-state probabilities (see Sec. 16.5),[3]

$$\pi_j = \sum_{i=0}^{M} \pi_i p_{ij}(k)$$

so that

$$\sum_{k=1}^{K} y_{jk} = \sum_{i=0}^{M} \sum_{k=1}^{K} y_{ik} p_{ij}(k), \qquad \text{for } j = 0, 1, \ldots, M.$$

3. $y_{ik} \geq 0$,   for $i = 0, 1, \ldots, M$ and $k = 1, 2, \ldots, K$.

The long-run expected average cost per unit time is given by

$$E(C) = \sum_{i=0}^{M} \sum_{k=1}^{K} \pi_i C_{ik} D_{ik} = \sum_{i=0}^{M} \sum_{k=1}^{K} C_{ik} y_{ik}.$$

Hence, the linear programming model is to choose the y_{ik} so as to

Minimize $\displaystyle Z = \sum_{i=0}^{M} \sum_{k=1}^{K} C_{ik} y_{ik},$

---

[3]The argument k is introduced in $p_{ij}(k)$ to indicate that the appropriate transition probability depends upon the decision k.

subject to the constraints

(1) $$\sum_{i=0}^{M} \sum_{k=1}^{K} y_{ik} = 1.$$

(2) $$\sum_{k=1}^{K} y_{jk} - \sum_{i=0}^{M} \sum_{k=1}^{K} y_{ik} p_{ij}(k) = 0, \qquad \text{for } j = 0, 1, \ldots, M.$$

(3) $$y_{ik} \ge 0, \qquad \text{for } i = 0, 1, \ldots, M; k = 1, 2, \ldots, K.$$

Thus, this model has $M + 2$ functional constraints and $K(M + 1)$ decision variables. [Actually, (2) provides one *redundant* constraint, so any one of these $M + 1$ constraints can be deleted.]

Because this is a linear programming model, it can be solved by the *simplex method*. Once the y_{ik} values are obtained, each D_{ik} is found from

$$D_{ik} = \frac{y_{ik}}{\displaystyle\sum_{k=1}^{K} y_{ik}}.$$

The optimal solution obtained by the simplex method has some interesting properties. It will contain $M + 1$ basic variables $y_{ik} \ge 0$. It can be shown that $y_{ik} > 0$ for at least one $k = 1, 2, \ldots, K$, for each $i = 0, 1, \ldots, M$. Therefore, it follows that $y_{ik} > 0$ for only *one k* for each $i = 0, 1, \ldots, M$. Consequently, each $D_{ik} = 0$ or 1.

The key conclusion is that the optimal policy found by the simplex method is *deterministic* rather than randomized. Thus, allowing policies to be randomized does not help at all in improving the final policy. However, it serves an extremely useful role in this formulation by converting integer variables (the D_{ik}) to continuous variables so that linear programming (LP) can be used. (The analogy in *integer programming* is to use the *LP relaxation* so that the simplex method can be applied and then to have the *integer solutions property* hold so that the optimal solution for the LP relaxation turns out to be integer anyway.)

Solving the Prototype Example by Linear Programming

Refer to the prototype example of Sec. 19.1. The first two columns of Table 19.1 give the relevant combinations of states and decisions. Therefore, the decision variables that need to be included in the model are y_{01}, y_{11}, y_{13}, y_{21}, y_{22}, y_{23}, and y_{33}. (The general expressions given above for the model include y_{ik} for *irrelevant* combinations of states and decisions here, so these $y_{ik} = 0$ in an optimal solution, and they might as well be deleted at the outset.) The rightmost column of Table 19.1 provides the coefficients of these variables in the objective function. The transition probabilities $p_{ij}(k)$ for each relevant combination of state i and decision k also are spelled out in Sec. 19.1.

The resulting linear programming model is

Minimize $Z = 1{,}000y_{11} + 6{,}000y_{13} + 3{,}000y_{21} + 4{,}000y_{22} + 6{,}000y_{23}$
$\qquad\qquad + 6{,}000y_{33},$

subject to

$$y_{01} + y_{11} + y_{13} + y_{21} + y_{22} + y_{23} + y_{33} = 1$$
$$y_{01} - (y_{13} + y_{23} + y_{33}) = 0$$

$$y_{11} + y_{13} - \left(\frac{7}{8}y_{01} + \frac{3}{4}y_{11} + y_{22}\right) = 0$$

$$y_{21} + y_{22} + y_{23} - \left(\frac{1}{16}y_{01} + \frac{1}{8}y_{11} + \frac{1}{2}y_{21}\right) = 0$$

$$y_{33} - \left(\frac{1}{16}y_{01} + \frac{1}{8}y_{11} + \frac{1}{2}y_{21}\right) = 0$$

and

$$\text{all } y_{ik} \geq 0.$$

Applying the simplex method, we obtain the optimal solution

$$y_{01} = \frac{2}{21}, \qquad (y_{11}, y_{13}) = \left(\frac{5}{7}, 0\right), \qquad (y_{21}, y_{22}, y_{23}) = \left(0, \frac{2}{21}, 0\right), \qquad y_{33} = \frac{2}{21},$$

so

$$D_{01} = 1, \qquad (D_{11}, D_{13}) = (1, 0), \qquad (D_{21}, D_{22}, D_{23}) = (0, 1, 0), \qquad D_{33} = 1.$$

This policy calls for leaving the machine as is (decision 1) when it is in state 0 or 1, overhauling it (decision 2) when it is in state 2, and replacing it (decision 3) when it is in state 3. This is the same optimal policy found by exhaustive enumeration at the end of Sec. 19.2.

The Worked Examples section of the book's website provides **another example** of applying linear programming to obtain an optimal policy for a Markov decision process.

■ 19.4   POLICY IMPROVEMENT ALGORITHM FOR FINDING OPTIMAL POLICIES

You now have seen two methods for deriving an optimal policy for a Markov decision process: *exhaustive enumeration* and *linear programming*. Exhaustive enumeration is useful because it is both quick and straightforward for very small problems. Linear programming can be used to solve vastly larger problems, and software packages for the simplex method are very widely available.

We now present a third popular method, namely, a *policy improvement algorithm*. The key advantage of this method is that it tends to be very efficient, because it usually reaches an optimal policy in a relatively small number of iterations (far fewer than for the simplex method with a linear programming formulation).

By following the model of Sec. 19.2 and as a joint result of the current state i of the system and the decision $d_i(R) = k$ when operating under policy R, two things occur. An (expected) cost C_{ik} is incurred that depends upon only the observed state of the system and the decision made. The system moves to state j at the next observed time period, with transition probability given by $p_{ij}(k)$. If, in fact, state j influences the cost that has been incurred, then C_{ik} is calculated as follows. Let

$q_{ij}(k) =$ expected cost incurred when the system is in state i, decision k is made,
and the system evolves to state j at the next observed time period.

Then

$$C_{ik} = \sum_{j=0}^{M} q_{ij}(k) p_{ij}(k).$$

Preliminaries

Referring to the description and notation for Markov decision processes given at the beginning of Sec. 19.2, we can show that, for any given policy R, there exist values $g(R)$, $v_0(R), v_1(R), \ldots, v_M(R)$ that satisfy

$$g(R) + v_i(R) = C_{ik} + \sum_{j=0}^{M} p_{ij}(k) \, v_j(R), \qquad \text{for } i = 0, 1, 2, \ldots, M.$$

We now shall give a heuristic justification of these relationships and an interpretation for these values.

Denote by $v_i^n(R)$ the total expected cost of a system starting in state i (beginning the first observed time period) and evolving for n time periods. Then $v_i^n(R)$ has two components: C_{ik}, the cost incurred during the first observed time period, and $\sum_{j=0}^{M} p_{ij}(k) \, v_j^{n-1}(R)$, the total expected cost of the system evolving over the remaining $n - 1$ time periods. This gives the *recursive equation*

$$v_i^n(R) = C_{ik} + \sum_{j=0}^{M} p_{ij}(k) \, v_j^{n-1}(R), \qquad \text{for } i = 0, 1, 2, \ldots, M,$$

where $v_i^1(R) = C_{ik}$ for all i.

It will be useful to explore the behavior of $v_i^n(R)$ as n grows large. Recall that the (long-run) expected average cost per unit time following any policy R can be expressed as

$$g(R) = \sum_{i=0}^{M} \pi_i C_{ik},$$

which is independent of the starting state i. Hence, $v_i^n(R)$ behaves approximately as $n\, g(R)$ for large n. In fact, if we neglect small fluctuations, $v_i^n(R)$ can be expressed as the sum of two components

$$v_i^n(R) \approx n\, g(R) + v_i(R),$$

where the first component is independent of the initial state and the second is dependent upon the initial state. Thus, $v_i(R)$ can be interpreted as the effect on the total expected cost due to starting in state i. Consequently,

$$v_i^n(R) - v_j^n(R) \approx v_i(R) - v_j(R),$$

so that $v_i(R) - v_j(R)$ is a measure of the effect of starting in state i rather than state j.

Letting n grow large, we now can substitute $v_i^n(R) = n\, g(R) + v_i(R)$ and $v_j^{n-1}(R) = (n - 1)g(R) + v_j(R)$ into the *recursive equation*. This leads to the system of equations given in the opening paragraph of this subsection.

Note that this system has $M + 1$ equations with $M + 2$ unknowns, so that one of these variables may be chosen arbitrarily. By convention, $v_M(R)$ will be chosen equal to zero. Therefore, by solving the system of linear equations, we can obtain $g(R)$, the (long-run) expected average cost per unit time when policy R is followed. In principle, all policies can be enumerated and that policy which minimizes $g(R)$ can be found. However, even for a moderate number of states and decisions, this technique is cumbersome. Fortunately, there exists an algorithm that can be used to evaluate policies and find the optimal one without complete enumeration, as described next.

The Policy Improvement Algorithm

The algorithm begins by choosing an arbitrary policy R_1. It then solves the system of equations to find the values of $g(R_1), v_0(R), v_1(R), \ldots, v_{M-1}(R)$ [with $v_M(R) = 0$]. This step is called *value determination.* A better policy, denoted by R_2, is then constructed. This step is called *policy improvement.* These two steps constitute an iteration of the algorithm. Using the new policy R_2, we perform another iteration. These iterations continue until two successive iterations lead to identical policies, which signifies that the optimal policy has been obtained. The details are outlined below.

Summary of the Policy Improvement Algorithm

Initialization: Choose an arbitrary initial trial policy R_1. Set $n = 1$.
Iteration n:
Step 1: Value determination: For policy R_n, use $p_{ij}(k)$, C_{ik}, and $v_M(R_n) = 0$ to solve the system of $M + 1$ equations

$$g(R_n) = C_{ik} + \sum_{j=0}^{M} p_{ij}(k)\, v_j(R_n) - v_i(R_n), \qquad \text{for } i = 0, 1, \ldots, M,$$

for all $M + 1$ unknown values of $g(R_n), v_0(R_n), v_1(R_n), \ldots, v_{M-1}(R_n)$.
Step 2: Policy improvement: Using the current values of $v_i(R_n)$ computed for policy R_n, find the alternative policy R_{n+1} such that, for each state i, $d_i(R_{n+1}) = k$ is the decision that minimizes

$$C_{ik} + \sum_{j=0}^{M} p_{ij}(k)\, v_j(R_n) - v_i(R_n),$$

i.e., for *each* state i,

$$\underset{k=1, 2, \ldots, K}{\text{Minimize}} \quad [C_{ik} + \sum_{j=0}^{M} p_{ij}(k)\, v_j(R_n) - v_i(R_n)],$$

and then set $d_i(R_{n+1})$ equal to the minimizing value of k. This procedure defines a new policy R_{n+1}.
Optimality test: The current policy R_{n+1} is optimal if this policy is identical to policy R_n. If it is, stop. Otherwise, reset $n = n + 1$ and perform another iteration.

Two key properties of this algorithm are

1. $g(R_{n+1}) \leq g(R_n)$, for $n = 1, 2, \ldots$.
2. The algorithm terminates with an optimal policy in a finite number of iterations.[4]

Solving the Prototype Example by the Policy Improvement Algorithm

Referring to the prototype example presented in Sec. 19.1, we outline the application of the algorithm next.

Initialization. For the initial trial policy R_1, we arbitrarily choose the policy that calls for replacement of the machine (decision 3) when it is found to be in state 3, but doing nothing (decision 1) in other states. This policy, its transition matrix, and its costs are summarized next.

---

[4]This termination is guaranteed under the assumptions of the model given in Sec. 19.2, including particularly the (implicit) assumptions of a finite number of states ($M + 1$) and a finite number of decisions (K), but not necessarily for more general models. See R. Howard, *Dynamic Programming and Markov Processes*, M.I.T. Press, Cambridge, MA, 1960. Also see pp. 1291–1293 in A. F. Veinott, Jr., "On Finding Optimal Policies in Discrete Dynamic Programming with No Discounting," *Annals of Mathematical Statistics*, **37:** 1284–1294, 1966.

Policy R_1	
State	**Decision**
0	1
1	1
2	1
3	3

Transition matrix

State	0	1	2	3
0	0	$\frac{7}{8}$	$\frac{1}{16}$	$\frac{1}{16}$
1	0	$\frac{3}{4}$	$\frac{1}{8}$	$\frac{1}{8}$
2	0	0	$\frac{1}{2}$	$\frac{1}{2}$
3	1	0	0	0

Costs	
State	C_{ik}
0	0
1	1,000
2	3,000
3	6,000

Iteration 1.   With this policy, the value determination step requires solving the following four equations simultaneously for $g(R_1)$, $v_0(R_1)$, $v_1(R_1)$, and $v_2(R_1)$ [with $v_3(R_1) = 0$].

$$g(R_1) = \qquad\qquad + \frac{7}{8}v_1(R_1) + \frac{1}{16}v_2(R_1) - v_0(R_1).$$

$$g(R_1) = 1,000 \qquad\quad + \frac{3}{4}v_1(R_1) + \frac{1}{8}v_2(R_1) \; - v_1(R_1).$$

$$g(R_1) = 3,000 \qquad\qquad\qquad\quad + \frac{1}{2}v_2(R_1) \; - v_2(R_1).$$

$$g(R_1) = 6,000 + v_0(R_1).$$

The simultaneous solution is

$$g(R_1) = \frac{25,000}{13} = 1,923$$

$$v_0(R_1) = -\frac{53,000}{13} = -4,077$$

$$v_1(R_1) = -\frac{34,000}{13} = -2,615$$

$$v_2(R_1) = \frac{28,000}{13} = 2,154.$$

Step 2 (policy improvement) can now be applied. We want to find an improved policy R_2 such that decision k in state i minimizes the corresponding expression below.

State 0: $C_{0k} - p_{00}(k)(4,077) - p_{01}(k)(2,615) + p_{02}(k)(2,154) + 4,077$

State 1: $C_{1k} - p_{10}(k)(4,077) - p_{11}(k)(2,615) + p_{12}(k)(2,154) + 2,615$

State 2: $C_{2k} - p_{20}(k)(4,077) - p_{21}(k)(2,615) + p_{22}(k)(2,154) - 2,154$

State 3: $C_{3k} - p_{30}(k)(4,077) - p_{31}(k)(2,615) + p_{32}(k)(2,154).$

Actually, in state 0, the only decision allowed is decision 1 (do nothing), so no calculations are needed. Similarly, we know that decision 3 (replace) must be made in state 3. Thus, only states 1 and 2 require calculation of the values of these expressions for alternative decisions.

For state 1, the possible decisions are 1 and 3. For each one, we show below the corresponding C_{1k}, the $p_{1j}(k)$, and the resulting value of the expression.

		State 1				
Decision	C_{1k}	$p_{10}(k)$	$p_{11}(k)$	$p_{12}(k)$	$p_{13}(k)$	Value of Expression
1	1,000	0	$\frac{3}{4}$	$\frac{1}{8}$	$\frac{1}{8}$	1,923 ← Minimum
3	6,000	1	0	0	0	4,538

Since decision 1 minimizes the expression, it is chosen as the decision to be made in state 1 for policy R_2 (just as for policy R_1).

The corresponding results for state 2 are shown below for its three possible decisions.

		State 2				
Decision	C_{2k}	$p_{20}(k)$	$p_{21}(k)$	$p_{22}(k)$	$p_{23}(k)$	Value of Expression
1	3,000	0	0	$\frac{1}{2}$	$\frac{1}{2}$	1,923
2	4,000	0	1	0	0	−769 ← Minimum
3	6,000	1	0	0	0	−231

Therefore, decision 2 is chosen as the decision to be made in state 2 for policy R_2. Note that this is a change from policy R_1.

We summarize our new policy, its transition matrix, and its costs below.

Policy R_2		Transition matrix					Costs	
State	Decision	State	0	1	2	3	State	C_{ik}
0	1	0	0	$\frac{7}{8}$	$\frac{1}{16}$	$\frac{1}{16}$	0	0
1	1						1	1,000
2	2	1	0	$\frac{3}{4}$	$\frac{1}{8}$	$\frac{1}{8}$	2	4,000
3	3						3	6,000
		2	0	1	0	0		
		3	1	0	0	0		

Since this policy is not identical to policy R_1, the optimality test says to perform another iteration.

Iteration 2. For step 1 (value determination), the equations to be solved for this policy are shown below.

$$g(R_2) = \qquad\qquad + \frac{7}{8}v_1(R_2) + \frac{1}{16}v_2(R_2) - v_0(R_2).$$

$$g(R_2) = 1,000 \qquad + \frac{3}{4}v_1(R_2) + \frac{1}{8}v_2(R_2) - v_1(R_2).$$

$$g(R_2) = 4,000 \qquad + \quad v_1(R_2) \qquad\quad - v_2(R_2).$$

$$g(R_2) = 6,000 + v_0(R_2).$$

The simultaneous solution is

$$g(R_2) = \frac{5,000}{3} = 1,667$$

$$v_0(R_2) = -\frac{13,000}{3} = -4,333$$

$$v_1(R_2) = -3,000$$

$$v_2(R_2) = -\frac{2,000}{3} = -667.$$

Step 2 (policy improvement) can now be applied. For the two states with more than one possible decision, the expressions to be minimized are

State 1: $C_{1k} - p_{10}(k)(4,333) - p_{11}(k)(3,000) - p_{12}(k)(667) + 3,000$

State 2: $C_{2k} - p_{20}(k)(4,333) - p_{21}(k)(3,000) - p_{22}(k)(667) + 667.$

The first iteration provides the necessary data (the transition probabilities and C_{ik}) required for determining the new policy, except for the values of each of these expressions for each of the possible decisions. These values are

Decision	Value for State 1	Value for State 2
1	1,667	3,333
2	—	1,667
3	4,667	2,334

Since decision 1 minimizes the expression for state 1 and decision 2 minimizes the expression for state 2, our next trial policy R_3 is

Policy R_3

State	Decision
0	1
1	1
2	2
3	3

Note that policy R_3 is identical to policy R_2. Therefore, the optimality test indicates that this policy is optimal, so the algorithm is finished.

Another example illustrating the application of this algorithm is included in your OR Tutor. The Worked Examples section of the book's website provides **an additional example** as well. The IOR Tutorial also includes an *interactive* procedure for efficiently learning and applying the algorithm.

■ 19.5   DISCOUNTED COST CRITERION

Throughout this chapter, we have measured policies on the basis of their (long-run) expected average cost per unit time. We now turn to an alternative measure of performance, namely, the **expected total discounted cost.**

As first introduced in Sec. 18.2, this measure uses a *discount factor* α, where $0 < \alpha < 1$. The discount factor α can be interpreted as equal to $1/(1 + i)$, where i is the current interest rate per period. Thus, α is the *present value* of one unit of cost one period in the future. Similarly, α^m is the *present value* of one unit of cost m periods in the future.

This *discounted cost criterion* becomes preferable to the *average cost criterion* when the time periods for the Markov chain are sufficiently long that the *time value of money* should be taken into account in adding costs in future periods to the cost in the current period. Another advantage is that the discounted cost criterion can readily be adapted to dealing with a *finite-period* Markov decision process where the Markov chain will terminate after a certain number of periods.

Both the policy improvement technique and the linear programming approach still can be applied here with relatively minor adjustments from the average cost case, as we describe next. Then we will present another technique, called the *method of successive approximations*, for quickly approximating an optimal policy.

A Policy Improvement Algorithm

To derive the expressions needed for the value determination and policy improvement steps of the algorithm, we now adopt the viewpoint of *probabilistic dynamic programming* (as described in Sec. 10.4). In particular, for each state i ($i = 0, 1, \ldots, M$) of a Markov decision process operating under policy R, let $V_i^n(R)$ be the *expected total discounted cost* when the process starts in state i (beginning the first observed time period) and evolves for n time periods. Then $V_i^n(R)$ has two components: C_{ik}, the cost incurred during the first observed time period, and $\alpha \sum_{j=0}^{M} p_{ij}(k)V_j^{n-1}(R)$, the expected total discounted cost of the process evolving over the remaining $n - 1$ time periods. For each $i = 0, 1, \ldots, M$, this yields the recursive equation

$$V_i^n(R) = C_{ik} + \alpha \sum_{j=0}^{M} p_{ij}(k)V_j^{n-1}(R),$$

with $V_i^1(R) = C_{ik}$, which closely resembles the recursive relationships of probabilistic dynamic programming found in Sec. 10.4.

As n approaches infinity, this recursive equation converges to

$$V_i(R) = C_{ik} + \alpha \sum_{j=0}^{M} p_{ij}(k)V_j(R), \qquad \text{for } i = 0, 1, \ldots, M,$$

where $V_i(R)$ can now be interpreted as the expected total discounted cost when the process starts in state i and continues indefinitely. There are $M + 1$ equations and $M + 1$ unknowns, so the simultaneous solution of this system of equations yields the $V_i(R)$.

To illustrate, consider again the prototype example of Sec. 19.1. Under the average cost criterion, we found in Secs. 19.2, 19.3, and 19.4 that the optimal policy is to do nothing in states 0 and 1, overhaul in state 2, and replace in state 3. Under the discounted cost criterion, with $\alpha = 0.9$, this same policy gives the following system of equations:

$$V_0(R) = \qquad\quad + 0.9\left[\frac{7}{8}V_1(R) + \frac{1}{16}V_2(R) + \frac{1}{16}V_3(R) \right]$$

$$V_1(R) = 1,000 + 0.9\left[\frac{3}{4}V_1(R) + \frac{1}{8}V_2(R) + \frac{1}{8}V_3(R) \right]$$

$$V_2(R) = 4,000 + 0.9[\qquad V_1(R)]$$

$$V_3(R) = 6,000 + 0.9[V_0(R)].$$

The simultaneous solution is

$$V_0(R) = 14,949$$
$$V_1(R) = 16,262$$
$$V_2(R) = 18,636$$
$$V_3(R) = 19,454.$$

Thus, assuming that the system starts in state 0, the expected total discounted cost is $14,949.

This system of equations provides the expressions needed for a policy improvement algorithm. After summarizing this algorithm in general terms, we shall use it to check whether this particular policy still is optimal under the discounted cost criterion.

Summary of the Policy Improvement Algorithm (Discounted Cost Criterion)

Initialization: Choose an arbitrary initial trial policy R_1. Set $n = 1$.
Iteration n:
Step 1: Value determination: For policy R_n, use $p_{ij}(k)$ and C_{ik} to solve the system of $M + 1$ equations

$$V_i(R_n) = C_{ik} + \alpha \sum_{j=0}^{M} p_{ij}(k)V_j(R_n), \quad \text{for } i = 0, 1, \ldots, M,$$

for all $M + 1$ unknown values of $V_0(R_n), V_1(R_n), \ldots, V_M(R_n)$.
Step 2: Policy improvement: Using the current values of the $V_i(R_n)$, find the alternative policy R_{n+1} such that, for each state i, $d_i(R_{n+1}) = k$ is the decision that minimizes

$$C_{ik} + \alpha \sum_{j=0}^{M} p_{ij}(k)V_j(R_n),$$

i.e., for *each* state i,

$$\underset{k=1, 2, \ldots, K}{\text{Minimize}} \left[C_{ik} + \alpha \sum_{j=0}^{M} p_{ij}(k)V_j(R_n) \right],$$

and then set $d_i(R_{n+1})$ equal to the minimizing value of k. This procedure defines a new policy R_{n+1}.
Optimality test: The current policy R_{n+1} is optimal if this policy is identical to policy R_n. If it is, stop. Otherwise, reset $n = n + 1$ and perform another iteration.

Three key properties of this algorithm are as follows:

1. $V_i(R_{n+1}) \leq V_i(R_n)$, for $i = 0, 1, \ldots, M$ and $n = 1, 2, \ldots$.
2. The algorithm terminates with an optimal policy in a finite number of iterations.
3. The algorithm is valid without the assumption (used for the average cost case) that the Markov chain associated with every transition matrix is irreducible.

Your IOR Tutorial includes an *interactive* procedure for applying this algorithm.

Solving the Prototype Example by This Policy Improvement Algorithm. We now pick up the prototype example where we left it before summarizing the algorithm.

We already have selected the optimal policy under the average cost criterion to be our initial trial policy R_1. This policy, its transition matrix, and its costs are summarized below.

Policy R_1			Transition matrix					Costs	
State	**Decision**	**State**	**0**	**1**	**2**	**3**		**State**	C_{ik}
0	1	0	$\frac{7}{8}$	$\frac{1}{16}$	$\frac{1}{16}$			0	0
1	1	1	0	$\frac{3}{4}$	$\frac{1}{8}$	$\frac{1}{8}$		1	1,000
2	3	2	0	1	0	0		2	4,000
3	3	3	1	0	0	0		3	6,000

We also have already done step 1 (value determination) of iteration 1. This transition matrix and these costs led to the system of equations used to find $V_0(R_1) = 14,949$, $V_1(R_1) = 16,262$, $V_2(R_1) = 18,636$, and $V_3(R_1) = 19,454$.

To start step 2 (policy improvement), we only need to construct the expression to be minimized for the two states (1 and 2) with a choice of decisions.

State 1: $C_{1k} + 0.9[p_{10}(k)(14,949) + p_{11}(k)(16,262) + p_{12}(k)(18,636)$
$\qquad\qquad + p_{13}(k)(19,454)]$

State 2: $C_{2k} + 0.9[p_{20}(k)(14,949) + p_{21}(k)(16,262) + p_{22}(k)(18,636)$
$\qquad\qquad + p_{23}(k)(19,454)].$

For each of these states and their possible decisions, we show below the corresponding C_{ik}, the $p_{ij}(k)$, and the resulting value of the expression.

			State 1				
Decision	C_{1k}	$p_{10}(k)$	$p_{11}(k)$	$p_{12}(k)$	$p_{13}(k)$	**Value of Expression**	
1	1,000	0	$\frac{3}{4}$	$\frac{1}{8}$	$\frac{1}{8}$	16,262	← Minimum
3	6,000	1	0	0	0	19,454	

			State 2				
Decision	C_{2k}	$p_{20}(k)$	$p_{21}(k)$	$p_{22}(k)$	$p_{23}(k)$	**Value of Expression**	
1	3,000	0	0	$\frac{1}{2}$	$\frac{1}{2}$	20,140	
2	4,000	0	1	0	0	18,636	← Minimum
3	6,000	1	0	0	0	19,454	

Since decision 1 minimizes the expression for state 1 and decision 2 minimizes the expression for state 2, our next trial policy (R_2) is as follows:

Policy R_2	
State	**Decision**
0	1
1	1
2	2
3	3

Since this policy is identical to policy R_1, the optimality test indicates that this policy is optimal. Thus, the optimal policy under the average cost criterion also is optimal under the discounted cost criterion in this case. (This often occurs, but not always.)

Linear Programming Formulation

The linear programming formulation for the discounted cost case is similar to that for the average cost case given in Sec. 19.3. However, we no longer need the first constraint given in Sec. 19.3; but the other functional constraints do need to include the discount factor α. The other difference is that the model now contains constants β_j for $j = 0, 1, \ldots, M$. These constants must satisfy the conditions

$$\sum_{j=0}^{M} \beta_j = 1, \qquad \beta_j > 0 \qquad \text{for } j = 0, 1, \ldots, M,$$

but otherwise they can be chosen arbitrarily without affecting the optimal policy obtained from the model.

The resulting model is to choose the values of the *continuous* decision variables y_{ik} so as to

$$\text{Minimize} \qquad Z = \sum_{i=0}^{M} \sum_{k=1}^{K} C_{ik} y_{ik},$$

subject to the constraints

(1) $\displaystyle \sum_{k=1}^{K} y_{jk} - \alpha \sum_{i=0}^{M} \sum_{k=1}^{K} y_{ik} p_{ij}(k) = \beta_j,$ for $j = 0, 1, \ldots, M$,

(2) $y_{ik} \geq 0,$ for $i = 0, 1, \ldots, M; k = 1, 2, \ldots, K$.

Once the simplex method is used to obtain an optimal solution for this model, the corresponding optimal policy then is defined by

$$D_{ik} = P\{\text{decision} = k \mid \text{state} = i\} = \frac{y_{ik}}{\displaystyle\sum_{k=1}^{K} y_{ik}}.$$

The y_{ik} now can be interpreted as the *discounted* expected time of being in state i and making decision k, when the probability distribution of the *initial state* (when observations begin) is $P\{X_0 = j\} = \beta_j$ for $j = 0, 1, \ldots, M$. In other words, if

$$z_{ik}^n = P\{\text{at time } n, \text{ state} = i \text{ and decision} = k\},$$

then

$$y_{ik} = z_{ik}^0 + \alpha z_{ik}^1 + \alpha^2 z_{ik}^2 + \alpha^3 z_{ik}^3 + \cdots.$$

With the interpretation of the β_j as *initial state probabilities* (with each probability greater than zero), Z can be interpreted as the corresponding expected total discounted cost. Thus, the choice of β_j affects the optimal value of Z (but not the resulting optimal policy).

It again can be shown that the optimal policy obtained from solving the linear programming model is deterministic; that is, $D_{ik} = 0$ or 1. Furthermore, this technique is valid without the assumption (used for the average cost case) that the Markov chain associated with every transition matrix is irreducible.

Solving the Prototype Example by Linear Programming. The linear programming model for the prototype example (with $\alpha = 0.9$) is

$$\text{Minimize} \quad Z = 1{,}000y_{11} + 6{,}000y_{13} + 3{,}000y_{21} + 4{,}000y_{22} + 6{,}000y_{23} + 6{,}000y_{33},$$

subject to

$$y_{01} - 0.9(y_{13} + y_{23} + y_{33}) = \frac{1}{4}$$

$$y_{11} + y_{13} - 0.9\left(\frac{7}{8}y_{01} + \frac{3}{4}y_{11} + y_{22}\right) = \frac{1}{4}$$

$$y_{21} + y_{22} + y_{23} - 0.9\left(\frac{1}{16}y_{01} + \frac{1}{8}y_{11} + \frac{1}{2}y_{21}\right) = \frac{1}{4}$$

$$y_{33} - 0.9\left(\frac{1}{16}y_{01} + \frac{1}{8}y_{11} + \frac{1}{2}y_{21}\right) = \frac{1}{4}$$

and

$$\text{all } y_{ik} \geq 0,$$

where β_0, β_1, β_2, and β_3 are arbitrarily chosen to be $\frac{1}{4}$. By the simplex method, the optimal solution is

$$y_{01} = 1.210, \qquad (y_{11}, y_{13}) = (6.656, 0), \qquad (y_{21}, y_{22}, y_{23}) = (0, 1.067, 0),$$
$$y_{33} = 1.067,$$

so

$$D_{01} = 1, \qquad (D_{11}, D_{13}) = (1, 0), \qquad (D_{21}, D_{22}, D_{23}) = (0, 1, 0), \qquad D_{33} = 1.$$

This optimal policy is the same as that obtained earlier in this section by the policy improvement algorithm.

The value of the objective function for the optimal solution is $Z = 17{,}325$. This value is closely related to the values of the $V_i(R)$ for this optimal policy that were obtained by the policy improvement algorithm. Recall that $V_i(R)$ is interpreted as the expected total discounted cost given that the system starts in state i, and we are interpreting β_i as the probability of starting in state i. Because each β_i was chosen to equal $\frac{1}{4}$,

$$17{,}325 = \frac{1}{4}[V_0(R) + V_1(R) + V_2(R) + V_3(R)]$$

$$= \frac{1}{4}(14{,}949 + 16{,}262 + 18{,}636 + 19{,}454).$$

Finite-Period Markov Decision Processes and the Method of Successive Approximations

We now turn our attention to an approach, called the *method of successive approximations,* for *quickly* finding at least an *approximation* to an optimal policy.

We have assumed that the Markov decision process will be operating indefinitely, and we have sought an optimal policy for such a process. The basic idea of the method of successive approximations is to instead find an optimal policy for the decisions to make in the first period when the process has only n time periods to go before termination, starting with $n = 1$, then $n = 2$, then $n = 3$, and so on. As n grows large, the corresponding optimal policies will converge to an optimal policy for the infinite-period problem of interest. Thus, the policies obtained for $n = 1, 2, 3, \ldots$ provide *successive approximations* that lead to the desired optimal policy.

The reason that this approach is attractive is that we already have a quick method of finding an optimal policy when the process has only n periods to go, namely, probabilistic dynamic programming as described in Sec. 10.4.

In particular, for $i = 0, 1, \ldots, M$, let

$V_i^n = $ expected total discounted cost of following an optimal policy, given that
process starts in state i and has only n periods to go.[5]

By the *principle of optimality* for dynamic programming (see Sec. 10.2), the V_i^n are obtained from the recursive relationship

$$V_i^n = \min_k \left\{ C_{ik} + \alpha \sum_{j=0}^{M} p_{ij}(k) V_j^{n-1} \right\}, \qquad \text{for } i = 0, 1, \ldots, M.$$

The minimizing value of k provides the optimal decision to make in the first period when the process starts in state i.

To get started, with $n = 1$, all the $V_i^0 = 0$ so that

$$V_i^1 = \min_k \{ C_{ik} \}, \qquad \text{for } i = 0, 1, \ldots, M.$$

Although the method of successive approximations may not lead to an optimal policy for the infinite-period problem after only a few iterations, it has one distinct advantage over the policy improvement and linear programming techniques. It never requires solving a system of simultaneous equations, so each iteration can be performed simply and quickly.

Furthermore, if the Markov decision process actually does have just n periods to go, n iterations of this method definitely will lead to an optimal policy. (For an n-period problem, it is permissible to set $\alpha = 1$, that is, no discounting, in which case the objective is to minimize the expected total cost over n periods.)

Your IOR Tutorial includes an interactive procedure to help guide you to use this method efficiently.

Solving the Prototype Example by the Method of Successive Approximations

We again use $\alpha = 0.9$. Refer to the rightmost column of Table 19.1 at the end of Sec. 19.1 for the values of C_{ik}. Also note in the first two columns of this table that the only feasible decisions k for each state i are $k = 1$ for $i = 0$, $k = 1$ or 3 for $i = 1$, $k = 1$, 2, or 3 for $i = 2$, and $k = 3$ for $i = 3$.

For the first iteration ($n = 1$), the value obtained for each V_i^1 is shown below, along with the minimizing value of k (given in parentheses).

$$V_0^1 = \min_{k=1} \{ C_{0k} \} = 0 \qquad (k = 1)$$
$$V_1^1 = \min_{k=1,3} \{ C_{1k} \} = 1{,}000 \qquad (k = 1)$$
$$V_2^1 = \min_{k=1,2,3} \{ C_{2k} \} = 3{,}000 \qquad (k = 1)$$
$$V_3^1 = \min_{k=3} \{ C_{3k} \} = 6{,}000 \qquad (k = 3)$$

---

[5]Since we want to allow n to grow indefinitely, we are letting n be the *number of periods to go,* instead of the *number of periods from the beginning* (as in Chap. 10).

Thus, the first approximation calls for making decision 1 (do nothing) when the system is in state 0, 1, or 2. When the system is in state 3, decision 3 (replace the machine) is made.

The second iteration leads to

$$V_0^2 = 0 + 0.9\left[\frac{7}{8}(1,000) + \frac{1}{16}(3,000) + \frac{1}{16}(6,000)\right] \qquad = 1,294 \quad (k = 1)$$

$$V_1^2 = \min\left\{1,000 + 0.9\left[\frac{3}{4}(1,000) + \frac{1}{8}(3,000) + \frac{1}{8}(6,000)\right],\right.$$

$$\left. 6,000 + 0.9[1(0)]\right\} = 2,688 \quad (k = 1)$$

$$V_2^2 = \min\left\{3,000 + 0.9\left[\frac{1}{2}(3,000) + \frac{1}{2}(6,000)\right],\right.$$

$$\left. 4,000 + 0.9[1(1,000)], 6,000 + 0.9[1(0)]\right\} = 4,900 \quad (k = 2)$$

$$V_3^2 = \qquad\qquad\qquad\qquad 6,000 + 0.9[1(0)] = 6,000 \quad (k = 3).$$

where the *min* operator has been deleted from the first and fourth expressions because only one alternative for the decision is available. Thus, the second approximation calls for leaving the machine as is when it is in state 0 or 1, overhauling when it is in state 2, and replacing the machine when it is in state 3. Note that this policy is the optimal one for the infinite-period problem, as found earlier in this section by both the policy improvement algorithm and linear programming. However, the V_i^2 (the expected total discounted cost when starting in state i for the two-period problem) are not yet close to the V_i (the corresponding cost for the infinite-period problem).

The third iteration leads to

$$V_0^3 = 0 + 0.9\left[\frac{7}{8}(2,688) + \frac{1}{16}(4,900) + \frac{1}{16}(6,000)\right] \qquad = 2,730 \quad (k = 1)$$

$$V_1^3 = \min\left\{1,000 + 0.9\left[\frac{3}{4}(2,688) + \frac{1}{8}(4,900) + \frac{1}{8}(6,000)\right],\right.$$

$$\left. 6,000 + 0.9[1(1,294)]\right\} = 4,041 \quad (k = 1)$$

$$V_2^3 = \min\left\{3,000 + 0.9\left[\frac{1}{2}(4,900) + \frac{1}{2}(6,000)\right],\right.$$

$$\left. 4,000 + 0.9[1(2,688)], 6,000 + 0.9[1(1,294)]\right\} = 6,419 \quad (k = 2)$$

$$V_3^3 = \qquad\qquad\qquad\qquad 6,000 + 0.9[1(1,294)] = 7,165 \quad (k = 3).$$

Again the optimal policy for the infinite-period problem is obtained, and the costs are getting closer to those for that problem. This procedure can be continued, and V_0^n, V_1^n, V_2^n, and V_3^n will converge to 14,949, 16,262, 18,636, and 19,454, respectively.

Note that termination of the method of successive approximations after the second iteration would have resulted in an optimal policy for the infinite-period problem, although there is no way to know this fact without solving the problem by other methods.

As indicated earlier, the method of successive approximations definitely obtains an optimal policy for an n-period problem after n iterations. For this example, the first, second, and third iterations have identified the optimal immediate decision for each state if the remaining number of periods is one, two, and three, respectively.

■ 19.6 CONCLUSIONS

Markov decision processes provide a powerful tool for optimizing the performance of stochastic processes that can be modeled as a discrete time Markov chain. Applications arise in a variety of areas, such as health care, highway and bridge maintenance, inventory management, machine maintenance, cash-flow management, control of water reservoirs, forest management, control of queueing systems, and operation of communication networks. Selected References 11 and 12 provide interesting early surveys of applications. Selected Reference 10 gives an update on one that won a prestigious prize, and Selected Reference 4 describes another award-winning application. Selected References 3 and 8 include more recent information on applications.

The two primary measures of performance used are the (long-run) *expected average cost per unit time* and the *expected total discounted cost*. The latter measure requires determination of the appropriate value of a discount factor, but this measure is useful when it is important to take into account the time value of money.

The two most important methods for deriving optimal policies for Markov decision processes are *policy improvement algorithms* and *linear programming*. Under the discounted cost criterion, the *method of successive approximations* provides a quick way of approximating an optimal policy.

■ SELECTED REFERENCES

1. Altman, E.: *Constrained Markov Decision Processes*, Chapman and Hall/CRC, Boca Raton, FL, 1999.

2. Bertsekas, D. P.: *Dynamic Programming and Optimal Control*, Vol. I, 2nd ed., Athena Scientific, Belmont MA, 2000.

3. Feinberg, E. A., and A. Shwartz: *Handbook of Markov Decision Processes: Methods and Applications*, Kluwer Academic Publishers (now Springer), Boston, 2002.

4. Golabi, K., and R. Shepard: "Pontis: A System for Maintenance Optimization and Improvement of U.S. Bridge Networks," *Interfaces*, **27**(1): 71–88, January–February 1997.

5. Howard, R. A.: "Comments on the Origin and Application of Markov Decision Processes," *Operations Research*, **50**(1): 100–102, January–February 2002.

6. Hu, J., M. C. Fu, V. R. Ramezani, and S. I. Marcus: "An Evolutionary Random Policy Search Algorithm for Solving Markov Decision Processes," *INFORMS Journal on Computing*, **19**(2): 161–174, Spring 2007.

7. Puterman, M. L.: *Markov Decision Processes: Discrete Stochastic Dynamic Programming*, Wiley, New York, 1994.

8. Sennott, L. I.: *Stochastic Dynamic Programming and the Control of Queueing Systems*, Wiley, New York, 1999.

9. Smith, J. E., and K. F. McCardle: "Structural Properties of Stochastic Dynamic Programs," *Operations Research*, **50**(5): 796–809, September–October 2002.

10. Wang, K. C. P., and J. P. Zaniewski: "20/30 Hindsight: The New Pavement Optimization in the Arizona State Highway Network," *Interfaces*, **26**(3): 77–89, May–June 1996.

11. White, D. J.: "Further Real Applications of Markov Decision Processes," *Interfaces*, **18**(5): 55–61, September–October 1988.

12. White, D. J.: "Real Applications of Markov Decision Processes," *Interfaces*, **15**(6): 73–83, November–December 1985.

■ LEARNING AIDS FOR THIS CHAPTER ON OUR WEBSITE (www.mhhe.com/hillier)

Worked Examples:

Examples for Chapter 19

A Demonstration Example in OR Tutor:

Policy Improvement Algorithm—Average Cost Case

Interactive Procedures in IOR Tutorial:

Enter Markov Decision Model
Interactive Policy Improvement Algorithm—Average Cost
Interactive Policy Improvement Algorithm—Discounted Cost
Interactive Method of Successive Approximations

Automatic Procedures in IOR Tutorial (Markov Chains Area):

Enter Transition Matrix
Steady-State Probabilities

"Ch. 19—Markov Decision Proc" Files for Solving the Linear Programming Formulations:

Excel Files
LINGO/LINDO File

Glossary for Chapter 19

See Appendix 1 for documentation of the software.

■ PROBLEMS

The symbols to the left of some of the problems (or their parts) have the following meaning:

D: The demonstration example listed above may be helpful.
I: We suggest that you use the corresponding interactive procedure listed above (the printout records your work).
A: The automatic procedures listed above can be helpful.
C: Use the computer with any of the software options available to you (or as instructed by your instructor) to solve your linear programming formulation.

An asterisk on the problem number indicates that at least a partial answer is given in the back of the book.

19.2-1. Read the referenced article that fully describes the OR study summarized in the application vignette presented in Sec. 19.2. Briefly describe how Markov decision processes were applied in this study. Then list the various financial and nonfinancial benefits that resulted from this study.

19.2-2.* During any period, a potential customer arrives at a certain facility with probability $\frac{1}{2}$. If there are already two people at the facility (including the one being served), the potential customer leaves the facility immediately and never returns. However, if there is one person or less, he enters the facility and becomes an actual customer. The manager of the facility has two types of service configurations available. At the beginning of each period, a decision must be made on which configuration to use. If she uses her "slow" configuration at a cost of $3 and any customers are present during the period, one customer will be served and leave the facility with probability $\frac{3}{5}$. If she uses her "fast" configuration at a cost of $9 and any customers are present during the period, one customer will be served and leave the facility with probability $\frac{4}{5}$. The probability

D,I **19.4-3.** Use the policy improvement algorithm to find an optimal policy for Prob. 19.2-4.

D,I **19.4-4.*** Use the policy improvement algorithm to find an optimal policy for Prob. 19.2-5.

D,I **19.4-5.** Use the policy improvement algorithm to find an optimal policy for Prob. 19.2-6.

D,I **19.4-6.** Use the policy improvement algorithm to find an optimal policy for Prob. 19.2-7.

D,I **19.4-7.** Use the policy improvement algorithm to find an optimal policy for Prob. 19.2-8.

D,I **19.4-8.** Consider the blood-inventory problem presented in Prob. 16.5-5. Suppose now that the number of pints of blood delivered (on a regular delivery) can be specified at the time of delivery (instead of using the old policy of receiving 1 pint at each delivery). Thus, the number of pints delivered can be 0, 1, 2, or 3 (more than 3 pints can never be used). The cost of regular delivery is $50 per pint, while the cost of an emergency delivery is $100 per pint. Starting with the policy of taking one pint at each regular delivery if the number of pints on hand just prior to the delivery is 0, 1, or 2 pints (so there never is more than 3 pints on hand), perform two iterations of the policy improvement algorithm. (Because so few pints are kept on hand and the oldest pint always is used first, you now can ignore the remote possibility that any pints will reach 21 days on the shelf and need to be discarded.)

I **19.5-1.*** Joe wants to sell his car. He receives one offer each month and must decide immediately whether to accept the offer. Once rejected, the offer is lost. The possible offers are $600, $800, and $1,000, made with probabilities $\frac{5}{8}$, $\frac{1}{4}$, and $\frac{1}{8}$, respectively (where successive offers are independent of each other). There is a maintenance cost of $60 per month for the car. Joe is anxious to sell the car and so has chosen a discount factor of $\alpha = 0.95$.

Using the policy improvement algorithm, find a policy that minimizes the expected total discounted cost. (*Hint:* There are two actions: Accept or reject the offer. Let the state for month t be the offer in that month. Also include a state ∞, where the process goes to state ∞ whenever an offer is accepted and it remains there at a monthly cost of 0.)

19.5-2.* Reconsider Prob. 19.5-1.
(a) Formulate a linear programming model for finding an optimal policy.
C (b) Use the simplex method to solve this model. Use the resulting optimal solution to identify an optimal policy.

I **19.5-3.*** For Prob. 19.5-1, use three iterations of the method of successive approximations to approximate an optimal policy.

I **19.5-4.** The price of a certain stock is fluctuating between $10, $20, and $30 from month to month. Market analysts have

predicted that if the stock is at $10 during any month, it will be at $10 or $20 the next month, with probabilities $\frac{4}{5}$ and $\frac{1}{5}$, respectively; if the stock is at $20, it will be at $10, $20, or $30 the next month, with probabilities $\frac{1}{4}$, $\frac{1}{4}$, and $\frac{1}{2}$, respectively; and if the stock is at $30, it will be at $20 or $30 the next month, with probabilities $\frac{3}{4}$ and $\frac{1}{4}$, respectively. Given a discount factor of 0.9, use the policy improvement algorithm to determine when to sell and when to hold the stock to maximize the expected total discounted profit. (*Hint:* Include a state that is reached with probability 1 when the stock is sold and with probability 0 when the stock is held.)

19.5-5. Reconsider Prob. 19.5-4.
(a) Formulate a linear programming model for finding an optimal policy.
C (b) Use the simplex method to solve this model. Use the resulting optimal solution to identify an optimal policy.

I **19.5-6.** For Prob. 19.5-4, use three iterations of the method of successive approximations to approximate an optimal policy.

19.5-7. A chemical company produces two chemicals, denoted by C1 and C2, and only one can be produced at a time. Each month a decision is made as to which chemical to produce that month. Because the demand for each chemical is predictable, it is known that if C2 is produced this month, there is a 60 percent chance that it will also be produced again next month. Similarly, if C1 is produced this month, there is only a 30 percent chance that it will be produced again next month.

To combat the emissions of pollutants, the chemical company has two processes, process A, which is efficient in combating the pollution from the production of C2 but not from C1, and process B, which is efficient in combating the pollution from the production of C1 but not from C2. Only one process can be used at a time. The amount of pollution from the production of each chemical under each process is

	C1	C2
A	15	2
B	3	8

Unfortunately, there is a time delay in setting up the pollution control processes, so that a decision as to which process to use must be made in the month prior to the production decision. Management wants to determine a policy for when to use each pollution control process that will minimize the expected total discounted amount of all future pollution with a discount factor of $\alpha = 0.5$.
(a) Formulate this problem as a Markov decision process by identifying the states, the decisions, and the C_{ik}. Identify all the (stationary deterministic) policies.
I (b) Use the policy improvement algorithm to find an optimal policy.

19.5-8. Reconsider Prob. 19.5-7.

(a) Formulate a linear programming model for finding an optimal policy.

C **(b)** Use the simplex method to solve this model. Use the resulting optimal solution to identify an optimal policy.

I **19.5-9.** For Prob. 19.5-7, use two iterations of the method of successive approximations to approximate an optimal policy.

I **19.5-10.** Reconsider Prob. 19.5-7. Suppose now that the company will be producing either of these chemicals for only 4 more months, so a decision on which pollution control process to use 1 month hence only needs to be made three more times. Find an optimal policy for this three-period problem.

I **19.5-11.*** Reconsider the prototype example of Sec. 19.1. Suppose now that the production process using the machine under consideration will be used for only 4 more weeks. Using the discounted cost criterion with a discount factor of $\alpha = 0.9$, find the optimal policy for this four-period problem.

20

Simulation

In this final chapter, we now are ready to focus on the last of the key techniques of operations research. *Simulation* ranks very high among the most widely used of these techniques. Furthermore, because it is such a flexible, powerful, and intuitive tool, it is continuing to rapidly grow in popularity.

This technique involves using a computer to *imitate* (simulate) the operation of an entire process or system. For example, simulation is frequently used to perform risk analysis on financial processes by repeatedly imitating the evolution of the transactions involved to generate a profile of the possible outcomes. Simulation also is widely used to analyze stochastic systems that will continue operating indefinitely. For such systems, the computer randomly generates and records the occurrences of the various events that drive the system just as if it were physically operating. Because of its speed, the computer can simulate even years of operation in a matter of seconds. Recording the performance of the simulated operation of the system for a number of alternative designs or operating procedures then enables evaluating and comparing these alternatives before choosing one.

The first section describes and illustrates the essence of simulation. The following section then presents a variety of common applications of simulation. Sections 20.3 and 20.4 focus on two key tools of simulation, the generation of random numbers and the generation of random observations from probability distributions. Section 20.5 outlines the overall procedure for applying simulation. The next section describes how some simulations now can be performed efficiently on spreadsheets. One supplement to the chapter on the book's website introduces some special techniques for improving the precision of the estimates of the measures of performance of the system being simulated. A second supplement presents an innovative statistical method for analyzing the output of a simulation. A third supplement extends the spreadsheet-based approach to searching for an optimal solution for simulation models.

■ 20.1 THE ESSENCE OF SIMULATION

The technique of *simulation* has long been an important tool of the designer. For example, simulating airplane flight in a wind tunnel is standard practice when a new airplane is designed. Theoretically, the laws of physics could be used to obtain the same information about how the performance of the airplane changes as design parameters are altered, but, as a practical matter, the analysis would be too complicated to do it all. Another alternative would be to build real airplanes with alternative designs and test them in actual flight to choose the final design, but this would be far too expensive (as well as unsafe). Therefore, after some preliminary theoretical analysis is performed to develop a *rough* design, simulating flight in a wind tunnel is a vital tool for experimenting with *specific* designs. This simulation amounts to *imitating* the performance of a real airplane in a controlled environment in order to *estimate* what its actual performance will be. After a detailed design is developed in this way, a prototype model can be built and tested in actual flight to fine-tune the final design.

The Role of Simulation in Operations Research Studies

Simulation plays essentially this same role in many OR studies. However, rather than designing an airplane, the OR team is concerned with developing a design or operating procedure for some *stochastic system* (a system that evolves *probabilistically* over time). Some of these stochastic systems resemble the examples of Markov chains and queueing systems described in Chaps. 16 and 17, and others are more complicated. Rather than use a wind tunnel, the performance of the real system is *imitated* by using probability distributions to *randomly generate* various events that occur in the system. Therefore, a simulation model *synthesizes* the system by building it up component by component and event by event. Then the model *runs* the simulated system to obtain *statistical observations* of the performance of the system that result from various randomly generated events. Because the *simulation runs* typically require generating and processing a vast amount of data, these simulated statistical experiments are inevitably performed on a computer.

When simulation is used as part of an OR study, commonly it is preceded and followed by the same steps described earlier for the design of an airplane. In particular, some preliminary analysis is done first (perhaps with approximate mathematical models) to develop a rough design of the system (including its operating procedures). Then simulation is used to experiment with specific designs to estimate how well each will perform. After a detailed design is developed and selected in this way, the system probably is tested in actual use to fine-tune the final design.

To prepare for simulating a complex system, a detailed **simulation model** needs to be formulated to describe the operation of the system and how it is to be simulated. A simulation model has several basic building blocks:

1. A definition of the *state of the system* (e.g., the number of customers in a queueing system).
2. Identify the *possible states* of the system that can occur.
3. Identify the *possible events* (e.g., arrivals and service completions in a queueing system) that would change the state of the system.
4. A provision for a *simulation clock,* located at some address in the simulation program, that will record the passage of (simulated) time.
5. A method for *randomly generating the events* of the various kinds.
6. A formula for identifying *state transitions* that are generated by the various kinds of events.

Great progress is being made in developing special software (described in Sec. 20.5) for efficiently integrating the simulation model into a computer program and then performing the simulations. Nevertheless, when dealing with relatively complex systems, simulation tends to be a relatively expensive procedure. After formulating a detailed simulation model, considerable time often is required to develop and debug the computer programs needed to run the simulation. Next, many long computer runs may be needed to obtain good data on how well all the alternative designs of the system would perform. Finally, all these data (which only provide *estimates* of the performance of the alternative designs) should be carefully analyzed before drawing any final conclusions. This entire process typically takes a lot of time and effort. Therefore, simulation should not be used when a less expensive procedure is available that can provide the same (or better) information.

Simulation typically is used when the stochastic system involved is too complex to be analyzed satisfactorily by the kinds of mathematical models (e.g., queueing models) described in the preceding chapters. One of the main strengths of a mathematical model is that it abstracts the essence of the problem and reveals its underlying structure, thereby providing insight into the cause-and-effect relationships within the system. Therefore, if the modeler is able to construct a mathematical model that is both a reasonable idealization of the problem and amenable to solution, this approach usually is superior to simulation. However, many problems are too complex to permit this approach. Thus, simulation often provides the only practical approach to a problem.

Discrete-Event versus Continuous Simulation

Two broad categories of simulations are discrete-event and continuous simulations.

A **discrete-event simulation** is one where changes in the state of the system occur instantaneously at random points in time as a result of the occurrence of *discrete events*. For example, in a queueing system where the state of the system is the number of customers in the system, the discrete events that change this state are the arrival of a customer and the departure of a customer due to the completion of its service. Most applications of simulation in practice are discrete-event simulations.

A **continuous simulation** is one where changes in the state of the system occur *continuously* over time. For example, if the system of interest is an airplane in flight and its state is defined as the current position of the airplane, then the state is changing continuously over time. Some applications of continuous simulations occur in design studies of such engineering systems. Continuous simulations typically require using differential equations to describe the rate of change of the state variables. Thus, the analysis tends to be relatively complex.

By approximating continuous changes in the state of the system by occasional discrete changes, it often is possible to use a discrete-event simulation to approximate the behavior of a continuous system. This tends to greatly simplify the analysis.

This chapter focuses hereafter on discrete-event simulations. We assume this type in all subsequent references to simulation.

Now let us look at two examples to illustrate the basic ideas of simulation. These examples have been kept considerably simpler than the usual application of this technique in order to highlight the main ideas more readily. The first system is so simple, in fact, that the simulation does not even need to be performed on a computer. The second system incorporates more of the normal features of a simulation, although it, too, is simple enough to be solved analytically.

EXAMPLE 1   A Coin-Flipping Game

You are the lucky winner of a sweepstakes contest. Your prize is an all-expense-paid vacation at a major hotel in Las Vegas, including some chips for gambling in the hotel casino.

Upon entering the casino, you find that, in addition to the usual games (blackjack, roulette, etc.), they are offering an interesting new game with the following rules.

Rules of the Game

1. Each play of the game involves repeatedly flipping an unbiased coin until the *difference* between the number of heads tossed and the number of tails is 3.

2. If you decide to play the game, you are required to pay $1 for each flip of the coin. You are not allowed to quit during a play of the game.

3. You receive $8 at the end of each play of the game.

Thus, you win money if the number of flips required is fewer than 8, but you lose money if more than 8 flips are required. Here are some examples (where H denotes a head and T a tail).

HHH	3 flips.	You win $5
THTTT	5 flips.	You win $3
THHTHTHTTTT	11 flips.	You lose $3

How would you decide whether to play this game?

Many people would base this decision on *simulation,* although they probably would not call it by that name. In this case, simulation amounts to nothing more than playing the game alone many times until it becomes clear whether it is worthwhile to play for money. Half an hour spent in repeatedly flipping a coin and recording the earnings or losses that would have resulted might be sufficient. This is a true simulation because you are *imitating* the actual play of the game *without* actually winning or losing any money.

Now let us see how a computer can be used to perform this same *simulated experiment.* Although a computer cannot flip coins, it can *simulate* doing so. It accomplishes this by generating a sequence of *random observations* from a uniform distribution between 0 and 1, where these random observations are referred to as *uniform random numbers* over the interval [0, 1]. One easy way to generate these uniform random numbers is to use the **RAND()** function in Excel. For example, the lower part of Fig. 20.1 illustrates that = RAND() has been entered into cell C13 and then copied into the range C14:C62 with the Copy command. (The parentheses need to be included with this function, but nothing is inserted between them.) This causes Excel to generate the random numbers shown in cells C13:C62 of the spreadsheet. (Rows 27–56 have been hidden to save space in the figure.

The probabilities for the outcome of flipping a coin are

$$P(\text{heads}) = \frac{1}{2}, \qquad P(\text{tails}) = \frac{1}{2}.$$

Therefore, to simulate the flipping of a coin, the computer can just let *any half* of the possible random numbers correspond to *heads* and the *other half* correspond to *tails.* To be specific, we will use the following correspondence.

0.0000 to 0.4999	correspond to	*heads.*
0.5000 to 0.9999	correspond to	*tails.*

By using the formula,

= IF(RandomNumber < 0.5, "Heads", "Tails"),

in each of the column D cells in Fig. 20.1, Excel inserts Heads if the random number is less than 0.5 and inserts Tails otherwise. Consequently, the first 11 random numbers generated in column C yield the following sequence of heads (H) and tails (T):

HTTTHHHHTHHH,

	A	B	C	D	E	F	G
1	**Coin-Flipping Game**						
2							
3			Required Difference	3			
4			Cash at End of Game	$8			
5							
6				**Summary of Game**			
7			Number of Flips	11			
8			Winnings	-$3			
9							
10							
11			Random		Total	Total	
12		Flip	Number	Result	Heads	Tails	Stop?
13		1	0.6961	Heads	1	0	
14		2	0.2086	Tails	1	1	
15		3	0.1457	Tails	1	2	
16		4	0.3098	Tails	1	3	
17		5	0.6996	Heads	2	3	
18		6	0.9617	Heads	3	3	
19		7	0.6117	Heads	4	3	
20		8	0.3948	Tails	4	4	
21		9	0.7769	Heads	5	4	
22		10	0.5750	Heads	6	4	
23		11	0.6271	Heads	7	4	Stop
24		12	0.2017	Tails	7	5	NA
25		13	0.7660	Heads	8	5	NA
26		14	0.9918	Heads	9	5	NA
57		45	0.2461	Tails	23	22	NA
58		46	0.7011	Heads	24	22	NA
59		47	0.3533	Tails	24	23	NA
60		48	0.7136	Heads	25	23	NA
61		49	0.7876	Heads	26	23	NA
62		50	0.3580	Tails	26	24	NA

Range Name	Cells
CashAtEndOfGame	D4
Flip	B13:B62
NumberOfFlips	D7
RandomNumber	C13:C62
RequiredDifference	D3
Result	D13:D62
Stop?	G13:G62
TotalHeads	E13:E62
TotalTails	F13:F62
Winnings	D8

	C	D
6		**Summary of Game**
7	Number of Flips	=COUNTBLANK(Stop?)+1
8	Winnings	=CashAtEndOfGame-NumberOfFlips

	C	D	E	F
11	Random		Total	Total
12	Number	Result	Heads	Tails
13	=RAND()	=IF(RandomNumber<0.5,1,0)	=IF(Result="Heads",1,0)	=Flip-TotalHeads
14	=RAND()	=IF(RandomNumber<0.5,"Tails","Heads")	=E13+IF(Result="Heads",1,0)	=Flip-TotalHeads
15	=RAND()	=IF(RandomNumber<0.5,"Tails","Heads")	=E14+IF(Result="Heads",1,0)	=Flip-TotalHeads
16	:	:	:	:
17	:	:	:	:

	G
12	Stop?
13	
14	
15	=IF(ABS(TotalHeads-TotalTails)>=RequiredDifference,"Stop","")
16	=IF(G15="",IF(ABS(TotalHeads-TotalTails)>=RequiredDifference,"Stop",""),"NA")
17	=IF(G16="",IF(ABS(TotalHeads-TotalTails)>=RequiredDifference,"Stop",""),"NA")
18	:
19	:

■ **FIGURE 20.1**
A spreadsheet model for a simulation of the coin-flipping game (Example 1).

at which point the game stops because the number of heads (7) exceeds the number of tails (4) by 3. Cells D7 and D8 record the total number of flips (11) and resulting winnings ($8 − $11 = −$3).

The equations in the bottom part of Fig. 20.1 show the formulas that have been entered into the various cells by entering them at the top and then using the Copy command to copy them down the columns. Using these equations, the spreadsheet then records the simulation of one complete play of the game. To virtually ensure that the game will be completed, 50 flips of the coin have been simulated. Columns E and F record the cumulative number of heads and tails after each flip. The equations entered into the column G cells leave each cell blank until the difference in the numbers of heads and tails reaches 3, at which point STOP is inserted into the cell. Thereafter, NA (for Not Applicable) is inserted instead. Using the equations shown just below the spreadsheet in Fig. 20.1, cells D7 and D8 record the outcome of the simulated play of the game.

Such simulations of plays of the game can be repeated as often as desired with this spreadsheet. Each time, Excel will generate a new sequence of random numbers, and so a new sequence of heads and tails. (Excel will repeat a sequence of random numbers only if you select the range of numbers you want to repeat, copy this range with the Copy command, select Paste Special from the Edit menu, choose the Values option, and click on OK.)

Simulations normally are repeated many times to obtain a more reliable estimate of an average outcome. Therefore, this same spreadsheet has been used to generate the data table in Fig. 20.2 for 14 plays of the game. As indicated on the right-hand side of Fig. 20.2, this is done by creating a table with the column headings shown in columns J, K, and L, and then entering equations into the first row of the data table that refer to the output cells of interest in Fig. 20.1, so =NumberOfFlips is entered into cell K6 and = Winnings is entered into cell L6, while leaving cell J6 blank. The next step is to select the entire

FIGURE 20.2

A data table that records the results of performing 14 replications of a simulation with the spreadsheet in Fig. 20.1.

	I	J	K	L	M
1		Data Table for Coin-Flipping Game			
2		(14 Replications)			
3					
4			Number		
5		Play	of Flips	Winnings	
6			3	$5	
7		1	9	-$1	
8		2	5	$3	
9		3	7	$1	
10		4	11	-$3	
11		5	5	$3	
12		6	3	$5	
13		7	3	$5	
14		8	11	-$3	
15		9	7	$1	
16		10	15	-$7	
17		11	3	$5	
18		12	7	$1	
19		13	9	-$1	
20		14	5	3	
21					
22		Average	7.14	$0.86	
23					

Select the whole table (J6:L20), before choosing Table from the Data menu.

Range Name	Cell
NumberOfFlips	D7
Winnings	D8

	K	L
4	Number	
5	of Flips	Winnings
6	=NumberOfFlips	=Winnings

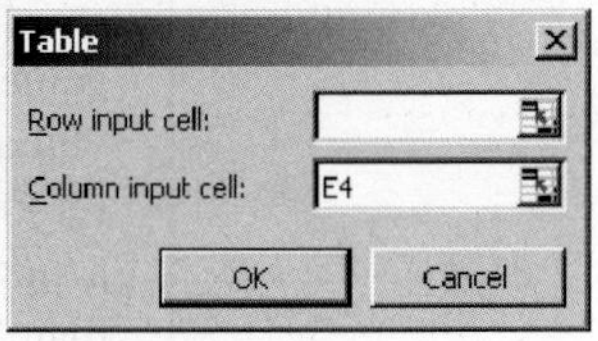

Table

Row input cell:

Column input cell: E4

OK   Cancel

	J	K	L
22	Average	=AVERAGE(K7:K20)	=AVERAGE(L7:L20)

contents of the table (cells J6:L20) and then choose Data Table from the What-If Analysis menu of the Data tab (for Excel 2007) or Table from the Data menu (for earlier versions of Excel). Finally, choose *any* blank cell (e.g., cell E4) for the column input cell and click OK. Excel then enters the numbers in the first column of the table (J7:J20) and uses the entire original spreadsheet (Fig. 20.1) in cells C13:G62 to recalculate the output cells in columns K and L for each row where *any* number is entered in row J. Entering the equations, =AVERAGE(K7:K20) or (L7:L20), into cells K22 and L22 provides the averages given in these cells.

Although this particular simulation run required using two spreadsheets—one to perform each replication of the simulation and the other to record the outcomes of the replications on a data table—we should point out that the replications of some other simulations can be performed on a single spreadsheet. This is the case whenever each replication can be performed and recorded on a single row of the spreadsheet. For example, if only a single uniform random number is needed to perform a replication, then the entire simulation run can be done and recorded by using a spreadsheet similar to Fig. 20.1.

Returning to Fig. 20.2, cell K22 shows that this sample of 14 plays of the game gives a sample average of 7.14 flips. The sample average provides an *estimate* of the true *mean* of the underlying probability distribution of the number of flips required for a play of the game. Hence, this sample average of 7.14 would seem to indicate that, on the average, you should win about $0.86 (cell L22) each time you play the game. Therefore, if you do not have a relatively high aversion to risk, it appears that you should choose to play this game, preferably a large number of times.

However, *beware!* One common error in the use of simulation is that conclusions are based on overly small samples, because statistical analysis was inadequate or totally lacking. In this case, the *sample standard deviation* is 3.67, so that the estimated *standard deviation* of the *sample average* is $3.67/\sqrt{14} \approx 0.98$. Therefore, even if it is assumed that the probability distribution of the number of flips required for a play of the game is a *normal distribution* (which is a gross assumption because the true distribution is *skewed*), any reasonable *confidence interval* for the true *mean* of this distribution would extend far above 8. Hence, a much larger sample size is required before we can draw a valid conclusion at a reasonable level of statistical significance. Unfortunately, because the standard deviation of a sample average is inversely proportional to the *square root* of the sample size, a large increase in the sample size is required to yield a relatively small increase in the precision of the estimate of the true mean. In this case, it appears that 100 simulated plays (replications) of the game *might* be adequate, depending on how close the sample average then is to 8, but 1,000 replications would be much safer.

It so happens that the true *mean* of the number of flips required for a play of this game is 9. (This mean can be found analytically, but not easily.) Thus, in the long run, you actually would average losing about $1 each time you played the game. Part of the reason that the above simulated experiment failed to draw this conclusion is that you have a small chance of a very large loss on any play of the game, but you can never win more than $5 each time. However, 14 simulated plays of the game were not enough to obtain any observations far out in the tail of the probability distribution of the amount won or lost on one play of the game. Only one simulated play gave a loss of more than $3, and that was only $7.

Figure 20.3 gives the results of running the simulation for 1,000 plays of the games (with rows 17–1000 not shown). Cell K1008 records the average number of flips as 8.97, very close to the true mean of 9. With this number of replications, the average winnings of −$0.97 in cell L1008 now provides a reliable basis for concluding that this game will

	I	J	K	L	M
1		**Data Table for Coin-Flipping Game**			
2		**(1000 Replications)**			
3					
4			Number		
5		Play	of Flips	Winnings	
6			5	$3	
7		1	3	$5	
8		2	3	$5	
9		3	7	$1	
10		4	11	-$3	
11		5	13	-$5	
12		6	7	$1	
13		7	3	$5	
14		8	7	$1	
15		9	3	$5	
16		10	9	-$1	
1001		995	5	$3	
1002		996	27	-$19	
1003		997	7	$1	
1004		998	3	$5	
1005		999	9	-$1	
1006		1000	17	-$9	
1007					
1008		Average	8.97	-$0.97	

■ **FIGURE 20.3**
This data table improves the reliability of the simulation recorded in Fig. 20.2 by performing 1,000 replications instead of only 14.

not win you money in the long run. (You can bet that the casino already has used simulation to verify this fact in advance.)

Although formally constructing a full-fledged *simulation model* was not needed to perform this simple simulation, we do so now for illustrative purposes. The *stochastic system* being simulated is the successive flipping of the coin for a play of the game. The *simulation clock* records the number of (simulated) flips t that have occurred so far. The information about the system that defines its current status, i.e., the *state of the system,* is

$N(t)$ = number of heads minus number of tails after t flips.

The *events* that change the state of the system are the flipping of a head or the flipping of a tail. The *event generation method* is the generation of a *uniform random number* over the interval [0, 1], where

0.0000 to 0.4999 ⇒ a head,
0.5000 to 0.9999 ⇒ a tail.

The *state transition formula* is

$$\text{Reset } N(t) = \begin{cases} N(t-1) + 1 & \text{if flip } t \text{ is a head} \\ N(t-1) - 1 & \text{if flip } t \text{ is a tail.} \end{cases}$$

The simulated game then ends at the first value of t where $N(t) = \pm 3$, where the resulting sampling *observation* for the simulated experiment is $8 - t$, the amount won (positive or negative) for that play of the game.

The next example will illustrate these building blocks of a simulation model for a prominent stochastic system from queueing theory.

EXAMPLE 2   An *M/M/*1 Queueing System

Consider the *M/M/*1 queueing theory model (Poisson input, exponential service times, and single server) that was discussed at the beginning of Sec. 17.6. Although this model already has been solved analytically, it will be instructive to consider how to study it by using simulation. To be specific, suppose that the values of the *mean arrival rate* λ and *mean service rate* μ are

$$\lambda = 3 \text{ per hour}, \qquad \mu = 5 \text{ per hour}.$$

To summarize the physical operation of the system, arriving customers enter the queue, eventually are served, and then leave. Thus, it is necessary for the simulation model to describe and synchronize the arrival of customers and the serving of customers.

Starting at time 0, the simulation clock records the amount of (simulated) time t that has transpired so far during the simulation run. The information about the queueing system that defines its current status, i.e., the state of the system, is

$N(t)$ = number of customers in system at time t.

The events that change the state of the system are the *arrival* of a customer or a *service completion* for the customer currently in service (if any). We shall describe the event generation method a little later. The state transition formula is

$$\text{Reset } N(t) = \begin{cases} N(t) + 1 & \text{if arrival occurs at time } t \\ N(t) - 1 & \text{if service completion occurs at time } t. \end{cases}$$

There are two basic methods used for advancing the simulation clock and recording the operation of the system. We did not distinguish between these methods for Example 1 because they actually coincide for that simple situation. However, we now describe and illustrate these two **time advance methods** (fixed-time incrementing and next-event incrementing) in turn.

With the **fixed-time incrementing** time advance method, the following two-step procedure is used repeatedly.

Summary of Fixed-Time Incrementing

1. *Advance time* by a small *fixed amount*.
2. *Update the system* by determining what events occurred during the elapsed time interval and what the resulting state of the system is. Also record desired information about the performance of the system.

For the queueing theory model under consideration, only two types of events can occur during each of these elapsed time intervals, namely, one or more *arrivals* and one or more *service completions*. Furthermore, the probability of two or more arrivals or of two or more service completions during an interval is negligible for this model if the interval is relatively short. Thus, the only two possible events during such an interval that need to be investigated are the arrival of one customer and the service completion for one customer. Each of these events has a known probability.

To illustrate, let us use 0.1 hour (6 minutes) as the small fixed amount by which the clock is advanced each time. (Normally, a considerably smaller time interval would be used to render negligible the probability of multiple arrivals or multiple service completions, but this choice will create more action for illustrative purposes.) Because both interarrival times and service times have an exponential distribution, the probability P_A that a time interval of 0.1 hour will include an *arrival* is

$$P_A = 1 - e^{-3/10} = 0.259,$$

and the probability P_D that it will include a *departure* (service completion), given that a customer was being served at the beginning of the interval, is

$$P_D = 1 - e^{-5/10} = 0.393.$$

To randomly generate either kind of event according to these probabilities, the approach is similar to that in Example 1. The computer again is used to generate a *uniform random number* over the interval [0, 1], that is, a random observation from the *uniform distribution* between 0 and 1. If we denote this uniform random number by r_A,

$r_A < 0.259 \Rightarrow$ arrival occurred,
$r_A \geq 0.259 \Rightarrow$ arrival did not occur.

Similarly, with *another* uniform random number r_D,

$r_D < 0.393 \Rightarrow$ departure occurred,
$r_D \geq 0.393 \Rightarrow$ departure did not occur,

given that a customer was being served at the beginning of the time interval. With no customer in service then (i.e., no customers in the system), it is assumed that no departure can occur during the interval even if an arrival does occur.

Table 20.1 shows the result of using this approach for 10 iterations of the *fixed-time incrementing* procedure, starting with no customers in the system and using time units of minutes.

Step 2 of the procedure (updating the system) includes recording the desired measures of performance about the aggregate behavior of the system during this time interval. For example, it could record the *number of customers* in the queueing system and the *waiting time* of any customer who just completed his or her wait. If it is sufficient to estimate only the mean rather than the probability distribution of each of these random variables, the computer will merely add the value (if any) at the end of the current time interval to a cumulative sum. The sample averages will be obtained after the simulation run is completed by dividing these sums by the sample sizes involved, namely, the total number of time intervals and the total number of customers, respectively.

To illustrate this estimating procedure, suppose that the simulation run in Table 20.1 were being used to estimate W, the steady-state expected waiting time of a customer in the queueing system (including service). Two customers arrived during this simulation run, one during the first time interval and the other during the seventh one, and each remained in

■ **TABLE 20.1** Fixed-time incrementing applied to Example 2

t, time (min)	N(t)	r_A	Arrival in Interval?	r_D	Departure in Interval?
0	0				
6	1	0.096	Yes	—	
12	1	0.569	No	0.665	No
18	1	0.764	No	0.842	No
24	0	0.492	No	0.224	Yes
30	0	0.950	No	—	
36	0	0.610	No	—	
42	1	0.145	Yes	—	
48	1	0.484	No	0.552	No
54	1	0.350	No	0.590	No
60	0	0.430	No	0.041	Yes

the system for three time intervals. Therefore, since the duration of each time interval is 0.1 hour, the estimate of W is

$$\text{Est}\{W\} = \frac{3 + 3}{2} (0.1 \text{ hour}) = 0.3 \text{ hour}.$$

This is, of course, only an extremely rough estimate, based on a sample size of only two. (Using the formula for W given in Sec. 17.6, its true value is $W = 1/(\mu - \lambda) = 0.5$ hour.) A much, much larger sample size normally would be used.

Another deficiency with using only Table 20.1 is that this simulation run started with no customers in the system, which causes the initial observations of waiting times to tend to be somewhat smaller than the expected value when the system is in a steady-state condition. Since the goal is to estimate the *steady-state* expected waiting time, it is important to run the simulation for some time without collecting data until it is believed that the simulated system has essentially reached a steady-state condition. (The second supplement to this chapter on the book's website describes a special method for circumventing this problem.) This initial period waiting to essentially reach a steady-state condition before collecting data is called the **warm-up period.**

Next-event incrementing differs from fixed-time incrementing in that the simulation clock is incremented by a *variable* amount rather than by a fixed amount each time. This variable amount is the time from the event that has just occurred until the *next event* of any kind occurs; i.e., the clock jumps from event to event. A summary follows.

Summary of Next-Event Incrementing

1. *Advance time* to the time of the *next event* of any kind.
2. *Update the system* by determining its new state that results from this event and by randomly generating the time until the next occurrence of any event type that can occur from this state (if not previously generated). Also record desired information about the performance of the system.

For this example the computer needs to keep track of two future events, namely, the next arrival and the next service completion (if a customer currently is being served). These times are obtained by taking a random observation from the probability distribution of interarrival and service times, respectively. As before, the computer takes such a random observation by generating and using a random number. (This technique will be discussed in Sec. 20.4.) Thus, each time an arrival or service completion occurs, the computer determines how long it will be until the next time this event will occur, adds this time to the current clock time, and then stores this sum in a computer file. (If the service completion leaves no customers in the system, then the generation of the time until the next service completion is postponed until the next arrival occurs.) To determine which event will occur next, the computer finds the minimum of the clock times stored in the file. To expedite the bookkeeping involved, simulation programming languages provide a "timing routine" that determines the occurrence time and type of the next event, advances time, and transfers control to the appropriate subprogram for the event type.

Table 20.2 shows the result of applying this approach through five iterations of the next-event incrementing procedure, starting with no customers in the system and using time units of minutes. For later reference, we include the *uniform random numbers* r_A and r_D used to generate the interarrival times and service times, respectively, by the method to be described in Sec. 20.4. These r_A and r_D are the same as those used in Table 20.1 in order to provide a truer comparison between the two time advance mechanisms.

The Excel files for this chapter in your OR Courseware include an automatic procedure, called **Queueing Simulator,** for applying the next-event incrementing procedure

■ **TABLE 20.2** Next-event incrementing applied to Example 2

t, time (min)	N(t)	r_A	Next Interarrival Time	r_D	Next Service Time	Next Arrival	Next Departure	Next Event
0	0	0.096	2.019	—	—	2.019	—	Arrival
2.019	1	0.569	16.833	0.665	13.123	18.852	15.142	Departure
15.142	0	—	—	—	—	18.852	—	Arrival
18.852	1	0.764	28.878	0.842	22.142	47.730	40.994	Departure
40.994	0	—	—	—	—	47.730	—	Arrival
47.730	1							

to various kinds of queueing systems. (This software is a good example of *discrete-event simulation software* that is widely used for applying simulation.) Queueing Simulator allows the queueing system to have either a single server or multiple servers. Several options (exponential, Erlang, degenerate, uniform, or translated exponential) are available for the probability distributions of interarrival times and service times. Figure 20.4 shows the input and output (in units of hours) from applying Queueing Simulator to the current example for a simulation run with 10,000 customer arrivals. Using the notation for various measures of performance for queueing systems introduced in Sec. 17.2, column *F* gives the estimate of each of these measures provided by the simulation run. [Using the formulas given in Sec. 17.6 for an *M/M/*1 queueing system, the true values of these measures are $L = 1.5$, $L_q = 0.9$, $W = 0.5$, $W_q = 0.3$, $P_0 = 0.4$, and $P_n = 0.4(0.6)^n$.] Columns *G* and *H* show the corresponding 95 percent confidence interval for each of these measures. Note that these confidence intervals are somewhat wider than might have been expected after such a long simulation run. In general, surprisingly long simulation runs are required to obtain relatively precise estimates (narrow confidence intervals) for the measures of performance for a queueing system (or for most stochastic systems).

■ **FIGURE 20.4**
The output obtained by using the Queueing Simulator that is included in this chapter's Excel files to perform a simulation of Example 2 over a period of 10,000 customer arrivals.

	A	B	C	D	E	F	G	H
1		Queueing Simulator						
2								
3			Data				Results	
4		Number of Servers =	1			Point	95% Confidence Interval	
5						Estimate	Low	High
6		Interarrival Times			L =	1.418286281	1.320246685	1.516325877
7		Distribution =	Exponential		L_q =	0.820371314	0.734901398	0.905841229
8		Mean =	0.333333333		W =	0.475627484	0.447222041	0.504032927
9					W_q =	0.275114516	0.248998719	0.301230313
10								
11		Service Times			P_0 =	0.402085033	0.386200645	0.417969421
12		Distribution =	Exponential		P_1 =	0.244395195	0.236088826	0.252701564
13		Mean =	0.2		P_2 =	0.145351997	0.138638859	0.152065136
14					P_3 =	0.09046104	0.084038151	0.096883929
15					P_4 =	0.052988644	0.047272227	0.05870506
16		Length of Simulation Run			P_5 =	0.030234667	0.025540066	0.034929268
17		Number of Arrivals =	10,000		P_6 =	0.015582175	0.012223063	0.018941288
18					P_7 =	0.008315125	0.005760629	0.010869622
19					P_8 =	0.004584301	0.002657593	0.006511009
20					P_9 =	0.00271883	0.001266236	0.004171425
21		**Run Simulation**			P_{10} =	0.001392827	0.000427267	0.002358388

The next-event incrementing procedure is considerably better suited for this example and similar stochastic systems than the fixed-time incrementing procedure. Next-event incrementing requires fewer iterations to cover the same amount of simulated time, and it generates a precise schedule for the evolution of the system rather than a rough approximation.

The next-event incrementing procedure will be illustrated again in the second supplement to this chapter on the book's website in the context of a full statistical experiment for estimating certain measures of performance for another queueing system. That supplement also describes the statistical method that is used by Queueing Simulator to obtain its point estimates and confidence intervals.

Several pertinent questions about how to conduct a simulation study of this type still remain to be answered. These answers are presented in a broader context in subsequent sections.

More Examples in Your OR Courseware

Simulation examples are easier to understand when they can be *observed in action,* rather than just talked about on a printed page. Therefore, the simulation area of your IOR Tutorial includes an automatic procedure called "Animation of a Queueing System" that shows a simulation where you actually observe the customers entering and leaving a queueing system. Thus, viewing this animation illustrates the sequence of events that the next-event incrementing procedure would generate during the simulation of a queueing system. In addition, the simulation area of your OR Tutor includes **two demonstration examples** that should be viewed at this time.

Both demonstration examples involve a bank that plans to open up a new branch office. The questions address how many teller windows to provide and then how many tellers to have on duty at the outset. Therefore, the system being studied is a *queueing system.* However, in contrast to the *M/M/*1 queueing system just considered in Example 2, this queueing system is too complicated to be solved analytically. This system has multiple servers (tellers), and the probability distributions of interarrival times and service times do not fit the standard models of queueing theory. Furthermore, in the second demonstration, it has been decided that one class of customers (merchants) needs to be given nonpreemptive priority over other customers, but the probability distributions for this class are different from those for other customers. These complications are typical of those that can be readily incorporated into a simulation study.

In both demonstrations, you will be able to see customers arrive and served customers leave as well as the next-event incrementing procedure being applied simultaneously to the simulation run.

The demonstrations also introduce you to an *interactive procedure* called "Interactively Simulate Queueing Problem" in your IOR Tutorial that you should find very helpful in dealing with some of the problems at the end of this chapter.

■ 20.2 SOME COMMON TYPES OF APPLICATIONS OF SIMULATION

Simulation is an exceptionally versatile technique. It can be used (with varying degrees of difficulty) to investigate virtually any kind of stochastic system. This versatility has made simulation the most widely used OR technique for studies dealing with such systems, and its popularity is continuing to increase.

Because of the tremendous diversity of its applications, it is impossible to enumerate all the specific areas in which simulation has been used. However, we will briefly describe here some particularly important categories of applications.

The first three categories concern types of stochastic systems considered in detail in other chapters. It is common to use the kinds of mathematical models described in those chapters to analyze simplified versions of the system and then to apply simulation to refine the results.

Design and Operation of Queueing Systems

Section 17.3 gives many examples of commonly encountered queueing systems that illustrate how such systems pervade many areas of society. Many mathematical models are available (including those presented in Chap. 17) for analyzing relatively simple types of queueing systems. Unfortunately, these models can only provide rough approximations at best of more complicated queueing systems. However, simulation is well suited for dealing with even very complicated queueing systems, so many of its applications fall into this category.

The two demonstration examples of simulation in your OR Tutor (both dealing with how much teller service to provide a bank's customers) are of this type. Because queueing applications of simulation are so pervasive, your OR Courseware includes an automatic procedure called *Queueing Simulator* (illustrated earlier in Fig. 20.4) for simulating queueing systems. (As already pointed out in the preceding section, this special procedure is provided in one of this chapter's Excel files.)

Among the award-winning applications of queueing models presented in Sec. 17.3, one of these also made heavy use of simulation. This was an application that involved AT&T developing a PC-based system to help its business customers design or redesign their call centers, resulting in more than $750 million in annual profit for these customers. This application of simulation is described further in the application vignette presented in Sec. 20.5.

Managing Inventory Systems

Sections 18.6 and 18.7 present models for the management of simple kinds of inventory systems when the products involved have uncertain demand. However, inventory systems that arise in practice often have complications that are not taken into account by these particular models. Although other mathematical models sometimes can help analyze these more complicated systems, simulation often plays a key role as well.

As one example, an article in the April 1996 issue of *OR/MS Today* describes an OR study of this kind that was done for the *IBM PC Company* in Europe. Facing unrelenting pressure from increasingly agile and aggressive competitors, the company had to find a way to greatly improve its performance in quickly filling customer orders. The OR team analyzed how to do this by simulating various redesigns of the company's entire *supply chain* (the network of facilities that spans procurement, manufacturing, and distribution, including all the inventories accumulated along the way). This led to major changes in the design and operation of the supply chain (including its inventory systems) that greatly improved the company's competitive position. Direct cost savings of $40 million per year also were achieved.

Section 20.6 will illustrate the application of simulation to a relatively simple kind of inventory system.

Estimating the Probability of Completing a Project by the Deadline

One of the key concerns of a project manager is whether his or her team will be able to complete the project by the deadline. Section 22.4 (on the book's website) describes how the PERT three-estimate approach can be used to obtain a rough estimate of the probability of meeting the deadline with the current project plan. That section also describes

An Application Vignette

Since its founding in 1914, **Merrill Lynch** has been a leading full-service financial service firm that strives to bring Wall Street to Main Street by making financial markets accessible to everyone. It employs a highly trained sales force of over 15,000 financial advisors throughout the United States and operates in 36 countries. A Fortune 100 company with net revenues of $26 billion in 2005, it manages client assets that total over $1.7 trillion.

Faced with increasing competition from discount brokerage firms and electronic brokerage firms, a task force was formed in late 1998 to recommend a product or service response to the marketplace challenge. Merrill Lynch's strong operations research group was charged with doing the detailed analysis of two potential new pricing options for clients. One option would replace charging for trades individually by charging a fixed percentage of a client's assets at Merrill Lynch and then allowing an unlimited number of free trades and complete access to a financial advisor. The other option would allow self-directed investors to invest online directly for a fixed low fee per trade without consulting a financial advisor.

The great challenge facing the OR group was to determine a "sweet spot" for the prices for these options that would be likely to grow the firm's business and increase its revenues while minimizing the risk of losing revenue instead. A key tool in attacking this problem proved to be *simulation*. To undertake a *major simulation study*, the group assembled and evaluated an extensive volume of data on the assets and trading activity of the firm's five million clients. For each segment of the client base, a careful analysis was done of its offer-adoption behavior by using managerial judgment, market research, and experience with clients. With this input, the group then formulated and ran a *simulation model* with various pricing scenarios to identify the pricing sweet spot.

The implementation of these results had a profound impact on Merrill Lynch's competitive position, restoring it to a leadership role in the industry. Instead of continuing to lose ground to the fierce new competition, *client assets managed by the company had increased by* **$22 billion** *and its incremental revenue reached* **$80 million** *within 18 months*. The CEO of Merrill Lynch called the new strategy "the most important decision we as a firm have made (in the last 20 years)." This enormously successful application of simulation led to Merrill Lynch winning the prestigious First Prize in the 2001 international competition for the Franz Edelman Award for Achievement in Operations Research and the Management Sciences.

Source: S. Altschuler, D. Batavia, J. Bennett, R. Labe, B. Liao, R. Nigam, and J. Oh: "Pricing Analysis for Merrill Lynch Integrated Choice," *Interfaces*, **32**(1): 5–19, Jan.–Feb. 2002. (A link to this article is provided on our website, www.mhhe.com/hillier.)

---

three simplifying approximations made by this approach to be able to estimate this probability. Unfortunately, because of these approximations, the resulting estimate always is overly optimistic, and sometimes by a considerable amount.

Consequently, it is becoming increasingly common now to use simulation to obtain a better estimate of this probability. This involves generating random observations from the probability distributions of the duration of the various activities in the projects. By using the project network, it then is straightforward to simulate when each activity begins and ends, and so when the project finishes. By repeating this simulation thousands of times (in one computer run), a very good estimate can be obtained of the probability of meeting the deadline.

A detailed illustration of this particular kind of application can be found in Sec. 28.2 on the book's website.

Design and Operation of Manufacturing Systems

Surveys consistently show that a large proportion of the applications of simulation involve manufacturing systems. Many of these systems can be viewed as a queueing system of some kind (e.g., a queueing system where the machines are the servers and the jobs to be processed are the customers). However, various complications inherent in these systems (e.g., occasional machine breakdowns, defective items needing to be reworked, and multiple types of jobs) go beyond the scope of the usual queueing models. Such complications can be handled readily by simulation.

Here are a few examples of the kinds of questions that might be addressed.

1. How many machines of each type should be provided?
2. How many materials-handling units of each type should be provided?
3. Considering their due dates for completion of the entire production process, what rule should be used to choose the order in which the jobs currently at a machine should be processed?
4. What are realistic due dates for jobs?
5. What will be the bottleneck operations in a new production process as currently designed?
6. What will be the throughput (production rate) of a new production process?

Selected Reference A1 describes an award-winning application of this last type. General Motors Corporation was so successful in applying simulation to predict and improve the throughput performance of its production lines that it both increased revenue and saved over $2.1 billion in 30 vehicle plants and 10 countries.

Design and Operation of Distribution Systems

Any major manufacturing corporation needs an efficient *distribution system* for distributing its goods from its factories and warehouses to its customers. There are many uncertainties involved in the operation of such a system. When will vehicles become available for shipping the goods? How long will a shipment take? What will be the demands of the various customers? By generating random observations from the relevant probability distributions, simulation can readily deal with these kinds of uncertainties. Thus, it is used quite often to test various possibilities for improving the design and operation of these systems.

One award-winning application of this kind is described in the January–February 1991 issue of *Interfaces. Reynolds Metal Company* spends over $250 million annually to deliver its products and receive raw materials. Shipments are made by truck, rail, ship, and air across a network of well over a hundred shipping locations including plants, warehouses, and suppliers. A combination of mixed binary integer programming (Chap. 11) and simulation was used to design a new distribution system with central dispatching. The new system both improved on-time delivery of shipments and reduced annual freight costs by over $7 million.

Financial Risk Analysis

Financial risk analysis was one of the earliest application areas of simulation, and it continues to be a very active area. For example, consider the evaluation of a proposed capital investment with uncertain future cash flows. By generating random observations from the probability distributions for the cash flow in each of the respective time periods (and considering relationships between time periods), simulation can generate thousands of scenarios for how the investment will turn out. This provides a *probability distribution* of the return (e.g., net present value) from the investment. This distribution (sometimes called the *risk profile*) enables management to assess the risk involved in making the investment.

A similar approach enables analyzing the risk associated with investing in various securities, including the more exotic financial instruments such as puts, calls, futures, stock options, etc.

Section 28.4 on the book's website provides a detailed example of using simulation for financial risk analysis.

Health Care Applications

Health care is another area where, like the evaluation of risky investments, analyzing future uncertainties is central to current decision making. However, rather than dealing with uncertain future cash flows, the uncertainties now involve such things as the evolution of human diseases.

Here are a few examples of the kinds of simulations that have been performed to guide the design of health care systems.

1. Simulating the use of hospital resources when treating patients with coronary heart disease.
2. Simulating health expenditures under alternative insurance plans.
3. Simulating the cost and effectiveness of screening for the early detection of a disease.
4. Simulating the use of the complex of surgical services at a medical center.
5. Simulating the timing and location of calls for ambulance services.
6. Simulating the matching of donated kidneys with transplant recipients.
7. Simulating the operation of an emergency room.

Applications to Other Service Industries

Like health care, other service industries also have proved to be fertile fields for the application of simulation. These industries include government services, banking, hotel management, restaurants, educational institutions, disaster planning, the military, amusement parks, and many others. In many cases, the systems being simulated are, in fact, queueing systems of some type.

Selected Reference A5 describes an award-winning application in this category. The *United States Postal Service* had identified *automation technology* as the only way it would be able to handle its increasing mail volume while remaining price competitive and satisfying service goals. Extensive planning over several years was required to convert to a largely automated system that would meet these goals. The backbone of the analysis leading to the adopted plan was performed with a comprehensive simulation model called META (model for evaluating technology alternatives). This model was first applied extensively at the national level, and then it was moved down to the local level for detailed planning. The resulting plan required a cumulative capital investment of $12 billion, but also was projected to achieve labor savings of over $4 billion per year. Another consequence of this highly successful application of simulation was that the value of OR tools now is recognized at the highest levels of the Postal Service. Operations research techniques continue to be used by the planning staff both at headquarters and in the field divisions.

Military Applications

There is probably no other sector of society where simulation is used as extensively as in the military. The military reliance on simulation to perform war gaming actually traces back several centuries and the U.S. military academics have included war gaming in their curriculum from their inception. However, the advent of powerful computers has led to a phenomenal growth in the military use of simulation, especially in the U.S. Department of Defense. War gaming to simulate military operations is now routinely used to plan future military operations, update military doctrine, and train officers. Simulation also is widely used to help make military procurement decisions.

New Applications

More new innovative applications of simulation are being made each year. Many of these applications are first announced publicly at the annual Winter Simulation Conference, held each December in some U.S. city. Since its beginning in 1967, this conference has been an institution in the simulation field. It now is attended by nearly a thousand participants, divided roughly equally between academics and practitioners. Hundreds of papers are presented to announce both methodological advances and new innovative applications.

20.3 GENERATION OF RANDOM NUMBERS

As the examples in Sec. 20.1 demonstrated, implementing a simulation model requires random numbers to obtain random observations from probability distributions. One method for generating such random numbers is to use a physical device such as a spinning disk or an electronic randomizer. Several tables of random numbers have been generated in this way, including one containing 1 million random digits, published by the Rand Corporation. An excerpt from the Rand table is given in Table 20.3.

Physical devices now have been replaced by the computer as the primary source for generating random numbers. For example, we pointed out in Sec. 20.1 that Excel uses the RAND() function for this purpose. Many other software packages also have the capability of generating random numbers whenever needed during a simulation run.

Characteristics of Random Numbers

The procedure used by a computer to obtain random numbers is called a *random number generator.*

> A **random number generator** is an algorithm that produces sequences of numbers that follow a specified probability distribution and possess the appearance of randomness.

The reference to *sequences of numbers* means that the algorithm produces many random numbers in a serial manner. Although an individual user may need only a few of the numbers,

■ **TABLE 20.3** Table of random digits

09656	96657	64842	49222	49506	10145	48455	23505	90430	04180
24712	55799	60857	73479	33581	17360	30406	05842	72044	90764
07202	96341	23699	76171	79126	04512	15426	15980	88898	06358
84575	46820	54083	43918	46989	05379	70682	43081	66171	38942
38144	87037	46626	70529	27918	34191	98668	33482	43998	75733
48048	56349	01986	29814	69800	91609	65374	22928	09704	59343
41936	58566	31276	19952	01352	18834	99596	09302	20087	19063
73391	94006	03822	81845	76158	41352	40596	14325	27020	17546
57580	08954	73554	28698	29022	11568	35668	59906	39557	27217
92646	41113	91411	56215	69302	86419	61224	41936	56939	27816
07118	12707	35622	81485	73354	49800	60805	05648	28898	60933
57842	57831	24130	75408	83784	64307	91620	40810	06539	70387
65078	44981	81009	33697	98324	46928	34198	96032	98426	77488
04294	96120	67629	55265	26248	40602	25566	12520	89785	93932
48381	06807	43775	09708	73199	53406	02910	83292	59249	18597
00459	62045	19249	67095	22752	24636	16965	91836	00582	46721
38824	81681	33323	64086	55970	04849	24819	20749	51711	86173
91465	22232	02907	01050	07121	53536	71070	26916	47620	01619
50874	00807	77751	73952	03073	69063	16894	85570	81746	07568
26644	75871	15618	50310	72610	66205	82640	86205	73453	90232

Source: Reproduced with permission from The Rand Corporation, *A Million Random Digits with 100,000 Normal Deviates.* Copyright, The Free Press, Glencoe, IL, 1955, top of p. 182.

generally the algorithm must be capable of producing many numbers. *Probability distribution* implies that a probability statement can be associated with the occurrence of each number produced by the algorithm.

We shall reserve the term **random number** to mean a random observation from some form of a *uniform distribution,* so that all possible numbers are *equally likely.* When we are interested in some other probability distribution (as in the next section), we shall refer to *random observations* from that distribution.

Random numbers can be divided into two main categories, random integer numbers and uniform random numbers, defined as follows:

A **random integer number** is a random observation from a *discretized uniform distribution* over some range $\underline{n}, \underline{n} + 1, \ldots, \overline{n}$. The probabilities for this distribution are

$$P(\underline{n}) = P(\underline{n} + 1) = \cdots = P(\overline{n}) = \frac{1}{\overline{n} - \underline{n} + 1}.$$

Usually, $\underline{n} = 0$ or 1, and these are convenient values for most applications. (If $\underline{n}$ has another value, then subtracting either $\underline{n}$ or $\underline{n} - 1$ from the random integer number changes the lower end of the range to either 0 or 1.)

A **uniform random number** is a random observation from a (continuous) *uniform distribution* over some interval $[a, b]$. The probability density function of this uniform distribution is

$$f(x) = \begin{cases} \dfrac{1}{b - a} & \text{if } a \leq x \leq b \\ 0 & \text{otherwise.} \end{cases}$$

When a and b are not specified, they are assumed to be $a = 0$ and $b = 1$.

The random numbers initially generated by a computer usually are random integer numbers. However, if desired, these numbers can immediately be converted to a uniform random number as follows:

For a given *random integer number* in the range 0 to $\overline{n}$, dividing this number by $\overline{n}$ yields (approximately) a *uniform random number.* (If $\overline{n}$ is small, this approximation should be improved by adding $\frac{1}{2}$ to the random integer number and then dividing by $\overline{n} + 1$ instead.)

This is the usual method used for generating uniform random numbers. With the huge values of $\overline{n}$ commonly used, it is an essentially exact method.

Strictly speaking, the numbers generated by the computer should not be called random numbers because they are predictable and reproducible (which sometimes is advantageous), given the random number generator being used. Therefore, they are sometimes given the name **pseudo-random numbers.** However, the important point is that they satisfactorily play the role of random numbers in the simulation if the method used to generate them is valid.

Various relatively sophisticated statistical procedures have been proposed for testing whether a generated sequence of numbers has an acceptable appearance of randomness. Basically the requirements are that each successive number in the sequence have an equal probability of taking on any one of the possible values and that it be statistically independent of the other numbers in the sequence.

Congruential Methods for Random Number Generation

There are a number of random number generators available, of which the most popular are the *congruential methods* (additive, multiplicative, and mixed). The mixed congruential method includes features of the other two, so we shall discuss it first.

The **mixed congruential method** generates a *sequence* of random integer numbers over the range from 0 to $m - 1$. The method always calculates the next random number from the last one obtained, given an initial random number x_0, called the **seed,** which may be obtained from some published source such as the Rand table. In particular, it calculates the $(n + 1)$st random number x_{n+1} from the nth random number x_n by using the recurrence relation

$$x_{n+1} \equiv (ax_n + c)(\text{modulo } m),$$

where a, c, and m are positive integers ($a < m$, $c < m$). This mathematical notation signifies that x_{n+1} is the *remainder* when $ax_n + c$ is divided by m. Thus, the *possible* values of x_{n+1} are $0, 1, \ldots, m - 1$, so that m represents the desired number of *different* values that could be generated for the random numbers.

To illustrate, suppose that $m = 8$, $a = 5$, $c = 7$, and $x_0 = 4$. The resulting sequence of random numbers is calculated in Table 20.4. (The sequence is not continued further because it would just begin repeating the numbers in the same order.) Note that this sequence includes each of the eight possible numbers exactly once. This property is a necessary one for a sequence of *random* integer numbers, but it does not occur with some choices of a and c. (Try $a = 4$, $c = 7$, and $x_0 = 3$.) Fortunately, there are rules available for choosing values of a and c that will guarantee this property. (There are no restrictions on the seed x_0 because it affects only where the sequence begins and not the progression of numbers.)

The number of consecutive numbers in a sequence before it begins repeating itself is referred to as the **cycle length.** Thus, the cycle length in the example is 8. The *maximum* cycle length is m, so the only values of a and c considered are those that yield this maximum cycle length.

Table 20.5 illustrates the conversion of random integer numbers to uniform random numbers. The left column gives the random integer numbers obtained in the rightmost column of Table 20.4. The right column gives the corresponding uniform random numbers from the formula

$$\text{Uniform random number} = \frac{\text{random integer number} + \dfrac{1}{2}}{m}.$$

■ **TABLE 20.4 Illustration of the mixed congruential method**

n	x_n	$5x_n + 7$	$(5x_n + 7)/8$	x_{n+1}
0	4	27	$3 + \dfrac{3}{8}$	3
1	3	22	$2 + \dfrac{6}{8}$	6
2	6	37	$4 + \dfrac{5}{8}$	5
3	5	32	$4 + \dfrac{0}{8}$	0
4	0	7	$0 + \dfrac{7}{8}$	7
5	7	42	$5 + \dfrac{2}{8}$	2
6	2	17	$2 + \dfrac{1}{8}$	1
7	1	12	$1 + \dfrac{4}{8}$	4

■ TABLE 20.5 Converting random integer numbers to uniform random numbers

Random Integer Number	Uniform Random Number
3	0.4375
6	0.8125
5	0.6875
0	0.0625
7	0.9375
2	0.3125
1	0.1875
4	0.5625

Note that each of these uniform random numbers lies at the midpoint of one of the eight equal-sized intervals 0 to 0.125, 0.125 to 0.25, . . . , 0.875 to 1. The small value of $m = 8$ does not enable us to obtain other values over the interval [0, 1], so we are obtaining fairly rough approximations of real uniform random numbers. In practice, *far* larger values of m generally are used.

The Worked Examples section of the book's website includes **another example** of applying the mixed congruential method with a relatively small value of $m(m = 16)$ and then converting the resulting random integer numbers to uniform random numbers. This example then explores the problems that arise from using such a small value of m.

For a binary computer with a word size of b bits, the usual choice for m is $m = 2^b$; this is the total number of nonnegative integers that can be expressed within the capacity of the word size. (Any undesired integers that arise in the sequence of random numbers are just not used.) With this choice of m, we can ensure that each possible number occurs exactly once before any number is repeated by selecting any of the values $a = 1, 5, 9, 13, \ldots$ and $c = 1, 3, 5, 7, \ldots$. For a decimal computer with a word size of d digits, the usual choice for m is $m = 10^d$, and the same property is ensured by selecting any of the values $a = 1, 21, 41, 61, \ldots$ and $c = 1, 3, 7, 9, 11, 13, 17, 19, \ldots$ (that is, all positive *odd* integers *except* those ending with the digit 5). The specific selection can be made on the basis of the *serial correlation* between successively generated numbers, which differs considerably among these alternatives.[1]

Occasionally, random integer numbers with only a relatively small number of digits are desired. For example, suppose that only three digits are desired, so that the possible values can be expressed as 000, 001, . . . , 999. In such a case, the usual procedure still is to use $m = 2^b$ or $m = 10^d$, so that an extremely large number of random integer numbers can be generated before the sequence starts repeating itself. However, except for purposes of calculating the next random integer number in this sequence, all but three digits of each number generated would be discarded to obtain the desired three-digit random integer number. One convention is to take the *last* three digits (i.e., the three trailing digits).

The **multiplicative congruential method** is just the special case of the mixed congruential method where $c = 0$. The **additive congruential method** also is similar, but it sets $a = 1$ and replaces c by some random number preceding x_n in the sequence, for example, x_{n-1} (so that more than one seed is required to start calculating the sequence).

The mixed congruential method provides tremendous flexibility in choosing a particular random number generator (a specific combination of values of a, c, and m). However, great care needs to be taken in choosing the random number generator because most

---

[1]See R. R. Coveyou, "Serial Correlation in the Generation of Pseudo-Random Numbers," *Journal of the Association of Computing Machinery,* **7:** 72–74, 1960.

combinations of values of a, c, and m lead to undesirable properties (e.g., a cycle length less than m). When researchers identify attractive random number generators, extensive testing is done to find any flaws, and this might lead to a better random number generator. For example, several years ago, $m = 2^{31}$ was considered an attractive choice, but experts now consider it unacceptable and are instead recommending that certain much larger numbers, including specific values of m near 2^{191}, should be used.[2]

20.4 GENERATION OF RANDOM OBSERVATIONS FROM A PROBABILITY DISTRIBUTION

Given a sequence of random numbers, how can one generate a sequence of random observations from a given probability distribution? Several different approaches are available, depending on the nature of the distribution.

Simple Discrete Distributions

For some simple discrete distributions, a sequence of random *integer* numbers can be used to generate random observations in a straightforward way. Merely allocate the possible values of a random number to the various outcomes in the probability distribution in direct proportion to the respective probabilities of those outcomes.

For Example 1 in Sec. 20.1, where flips of a coin are being simulated, the possible outcomes of one flip are a head or a tail, where each outcome has a probability of $\frac{1}{2}$. Therefore, rather than using uniform random numbers (as was done in Sec. 20.1), it would have been sufficient to use *random digits* to generate the outcomes. Five of the ten possible values of a random digit (say, 0, 1, 2, 3, 4) would be assigned an association with a head and the other five (say, 5, 6, 7, 8, 9) a tail.

As another example, consider the probability distribution of the outcome of a throw of two dice. It is known that the probability of throwing a 2 is $\frac{1}{36}$ (as is the probability of throwing a 12), the probability of throwing a 3 is $\frac{2}{36}$, and so on. Therefore, $\frac{1}{36}$ of the possible values of a random integer number should be associated with throwing a 2, $\frac{2}{36}$ of the values with throwing a 3, and so forth. Thus, if two-digit random integer numbers are being used, 72 of the 100 values will be selected for consideration, so that a random integer number will be rejected if it takes on any one of the other 28 values. Then 2 of the 72 possible values (say, 00 and 01) will be assigned an association with throwing a 2, four of them (say, 02, 03, 04, and 05) will be assigned an association with throwing a 3, and so on.

Using random *integer* numbers in this kind of way is convenient when they either are being drawn from a table of random numbers or are being generated directly by a congruential method. However, when performing the simulation on a computer, it usually is more convenient to have the computer generate *uniform* random numbers and then use them in the corresponding way. All the subsequent methods for generating random observations use uniform random numbers.

The Inverse Transformation Method

For more complicated distributions, whether discrete or continuous, the *inverse transformation method* can sometimes be used to generate random observations. Letting X be the random variable involved, we denote the cumulative distribution function by

$$F(x) = P\{X \le x\}.$$

[2]For recommendations on the choice of the random number generator, see P. L'Ecuyer, R. Simard, E. J. Chen, and W. D. Kelton, "An Object-Oriented Random-Number Package with Many Long Streams and Substreams," *Operations Research*, **50**: 1073–1075, 2002.

Generating each observation then requires the following two steps.

Summary of Inverse Transformation Method

1. Generate a *uniform random number r* between 0 and 1.
2. Set $F(x) = r$ and solve for x, which then is the desired random observation from the probability distribution.

This procedure is illustrated in Fig. 20.5 for the case where $F(x)$ is plotted graphically and the uniform random number r happens to be 0.5269.

Although the graphical procedure illustrated by Fig. 20.5 is convenient if the simulation is done manually, the computer must revert to some alternative approach. For *discrete* distributions, a *table lookup approach* can be taken by constructing a table that gives a "range" (jump) in the value of $F(x)$ for each possible value of $X = x$. Excel provides a convenient VLOOKUP function to implement this approach when performing a simulation on a spreadsheet.

To illustrate how this function works, suppose that a company is simulating the *maintenance program* for its machines. The time between breakdowns of one of these machines always is 4, 5, or 6 days, where these times occur with probabilities 0.25, 0.5, and 0.25, respectively. The first step in simulating these breakdowns is to create the table shown in Fig. 20.6 somewhere in the spreadsheet. Note that each number in the second column gives the cumulative probability *prior* to the number of days in the third column. The second and third columns (below the column headings) constitute the "lookup table." The VLOOKUP function has three arguments. The first argument gives the address of the cell that is providing the uniform random number being used. The second argument identifies the range of cell addresses for the lookup table. The third argument indicates which column of the lookup table (the second and third columns in Fig. 20.6) provides the random

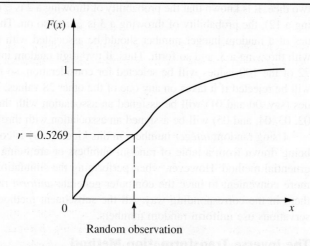

■ **FIGURE 20.5**
Illustration of the inverse transformation method for obtaining a random observation from a given probability distribution.

■ **FIGURE 20.6**
The table that would be constructed in a spreadsheet for using Excel's VLOOKUP function to implement the inverse transformation method for the maintenance program example.

Distribution of time between breakdowns

Probability	Cumulative	Number of Days
0.25	0	4
0.5	0.25	5
0.25	0.75	6

observation, so this argument equals 2 in this case. The VLOOKUP function with these three arguments is entered as the equation for each cell in the spreadsheet where a random observation from the distribution is to be entered.

For certain *continuous* distributions, the inverse transformation method can be implemented on a computer by first solving the equation $F(x) = r$ *analytically* for x. **An example** in the Worked Examples section of the book's website illustrates this approach (after first applying the graphical approach).

We also illustrate this approach next with the exponential distribution.

Exponential and Erlang Distributions

As indicated in Sec. 17.4, the cumulative distribution function for the **exponential distribution** is

$$F(x) = 1 - e^{-\alpha x}, \qquad \text{for } x \geq 0,$$

where $1/\alpha$ is the mean of the distribution. Setting $F(x) = r$ thereby yields

$$1 - e^{-\alpha x} = r,$$

so that

$$e^{-\alpha x} = 1 - r.$$

Therefore, taking the natural logarithm of both sides gives

$$\ln e^{-\alpha x} = \ln (1 - r),$$

so that

$$-\alpha x = \ln (1 - r),$$

which yields

$$x = \frac{\ln (1 - r)}{-\alpha} \ .$$

Now note that $1 - r$ is itself a uniform random number. Therefore, to save a subtraction, it is common in practice simply to use the *original* uniform random number r directly in place of $1 - r$. This gives

$$\text{Random observation} = \frac{\ln r}{-\alpha}$$

as the desired random observation from the exponential distribution.

This direct application of the inverse transformation method provides the most straightforward way of generating random observations from an exponential distribution. (More complicated techniques also have been developed for this distribution[3] that are faster for a computer than calculating a logarithm.)

A natural extension of this procedure for the exponential distribution also can be used to generate a random observation from an **Erlang** (gamma) **distribution** (see Sec. 17.7). The sum of k independent exponential random variables, each with mean $1/(k\alpha)$, has the Erlang distribution with shape parameter k and mean $1/\alpha$. Therefore, given a sequence of k uniform random numbers between 0 and 1, say, $r_1, r_2, \ldots, r_k$, the desired random observation from the Erlang distribution is

$$x = \sum_{i=1}^{k} \frac{\ln r_i}{-k\alpha},$$

---

[3]For example, see J. H. Ahrens and V. Dieter, "Efficient Table-Free Sampling Methods for Exponential, Cauchy, and Normal Distributions," *Communications of the ACM*, **31:** 1330–1337, 1988.

which reduces to

$$x = -\frac{1}{k\alpha} \ln \left[\prod_{i=1}^{k} r_i \right],$$

where Π denotes multiplication.

Normal and Chi-Square Distributions

A particularly simple (but inefficient) technique for generating a random observation from a **normal distribution** is obtained by applying the *central limit theorem*. Because a uniform random number has a *uniform distribution* from 0 to 1, it has mean $\frac{1}{2}$ and standard deviation $1/\sqrt{12}$. Therefore, this theorem implies that the sum of n uniform random numbers has approximately a normal distribution with mean $n/2$ and standard deviation $\sqrt{n/12}$. Thus, if $r_1, r_2, \ldots, r_n$ are a sample of uniform random numbers, then

$$x = \frac{\sigma}{\sqrt{n/12}} \sum_{i=1}^{n} r_i + \mu - \frac{n}{2} \frac{\sigma}{\sqrt{n/12}}$$

is a random observation from an approximately normal distribution with mean μ and standard deviation σ. This approximation is an excellent one (except in the tails of the distribution), even with small values of n. Thus, values of n from 5 to 10 may be adequate; $n = 12$ also is a convenient value, because it eliminates the square root terms from the preceding expression.

Since tables of the normal distribution are widely available (e.g., see Appendix 5), another simple method to generate a close approximation of a random observation is to use such a table to implement the inverse transformation method directly. This is fairly convenient when you are generating a few random observations by hand, but less so for computer implementation since it requires storing a large table and then using a table lookup.

Various *exact* techniques for generating random observations from a normal distribution have also been developed.[4] These exact techniques are sufficiently fast that, in practice, they generally are used instead of the approximate methods described above. A routine for one of these techniques usually is already incorporated into a software package with simulation capabilities. For example, Excel uses the function, NORMINV(RAND(), μ, σ), to generate a random observation from a normal distribution with mean μ and standard deviation σ.

A simple method for handling the **chi-square distribution** is to use the fact that it is obtained by summing squares of standardized normal random variables. Thus, if $y_1, y_2, \ldots, y_n$ are n random observations from a normal distribution with mean 0 and standard deviation 1, then

$$x = \sum_{i=1}^{n} y_i^2$$

is a random observation from a chi-square distribution with n degrees of freedom.

The Acceptance-Rejection Method

For many continuous distributions, it is not feasible to apply the inverse transformation method because $x = F^{-1}(r)$ cannot be computed (or at least computed efficiently). Therefore, several other types of methods have been developed to generate random observations

---

[4]See again the reference cited in footnote 3 on the preceding page.

from such distributions. Frequently, these methods are considerably faster than the inverse transformation method even when the latter method can be used. To provide some notion of the approach for these alternative methods, we now illustrate one called the **acceptance-rejection method** on a simple example.

Consider the *triangular distribution* having the probability density function

$$f(x) = \begin{cases} x & \text{if } 0 \leq x \leq 1 \\ 1 - (x - 1) & \text{if } 1 \leq x \leq 2 \\ 0 & \text{otherwise.} \end{cases}$$

The acceptance-rejection method uses the following two steps (perhaps repeatedly) to generate a random observation.

1. Generate a uniform random number r_1 between 0 and 1, and set $x = 2r_1$ (so that the range of possible values of x is 0 to 2).
2. Accept x with

$$\text{Probability} = \begin{cases} x & \text{if } 0 \leq x \leq 1 \\ 1 - (x - 1) & \text{if } 1 \leq x \leq 2, \end{cases}$$

to be the desired random observation [since this probability equals $f(x)$]. *Otherwise, reject x and repeat the two steps.*

To randomly generate the event of accepting (or rejecting) x according to this probability, the method implements step 2 as follows:

2. Generate a uniform random number r_2 between 0 and 1.

Accept x if $r_2 \leq f(x)$.
Reject x if $r_2 > f(x)$.

If x is rejected, repeat the two steps.

Because $x = 2r_1$ is being accepted with a probability $= f(x)$, the probability distribution of *accepted values* has $f(x)$ as its density function, so accepted values are valid *random observations* from $f(x)$.

We were fortunate in this example that the *largest* value of $f(x)$ for any x was exactly 1. If this largest value were $L \neq 1$ instead, then r_2 would be multiplied by L in step 2. With this adjustment, the method is easily extended to other probability density functions over a finite interval, and similar concepts can be used over an infinite interval as well.

■ 20.5 OUTLINE OF A MAJOR SIMULATION STUDY

Thus far, this chapter has focused mainly on the *process* of performing a simulation and some applications from doing so. We now place this material into broader perspective by briefly outlining all the typical steps involved in a major operations research study that is based on applying simulation. (Nearly the same steps also apply when the study is applying other operations research techniques instead.)

Step 1: Formulate the Problem and Plan the Study
The operations research team needs to begin by meeting with management to address the following kinds of questions.

1. What is the problem that management wants studied?
2. What are the overall objectives for the study?
3. What specific issues should be addressed?

Call centers have been one of the fastest-growing industries worldwide for many years. In the United States alone, many hundreds of thousands of businesses use call centers located around the world to enable customers to place an order simply by making a free telephone call to an 800 number.

The 800-network market is a lucrative one for telecommunication companies, so they are happy to sell the needed technology to their business customers and then to help these customers design efficient call centers. **AT&T** was the pioneer in developing and marketing this service to its customers. Its approach was to develop a highly flexible and sophisticated *simulation model*, called the *Call Processing Simulator (CAPS)*, that enables its customers to study various scenarios for how to design and operate their call centers.

CAPS contains four modules. The *call generation module* generates incoming calls arriving randomly, with mean arrival rates varying over the course of the day. The *network module* simulates how an incoming call can be answered immediately or placed on hold or receive a busy signal, where the latter cases can result in either the caller persevering until getting through or giving up and taking his or her business elsewhere. The *automatic call distribution module* simulates how AT&T's automatic call distribution system equitably distributes calls to available agents. The *call service module* simulates agents serving calls and then doing any necessary follow-up work.

The development and refinement of CAPS over a period of many years carefully followed the steps of a major simulation study described in Sec. 20.5. This meticulous approach has paid off big-time for AT&T. The company has completed as many as 2,000 CAPS studies per year for its business customers, helping it increase, protect, and *regain more than* **$1 billion** in an $8 billion 800-network market. This also has *generated more than* **$750 million** *in annual profit* for AT&T's business customers who received CAPS studies. This sophisticated application of simulation led to AT&T winning the prestigious First Prize in the 1993 international competition for the Franz Edelman Award for Achievement in Operations Research and the Management Sciences.

Source: A. J. Brigandi, D. R. Dargon, M. J. Sheehan, and T. Spencer III: "AT&T's Call Processing Simulator (CAPS) Operational Design for Inbound Call Centers," *Interfaces*, **24**(1): 6–28, Jan.–Feb. 1994. (A link to this article is provided on our website, www.mhhe.com/hillier.)

4. What kinds of alternative system configurations should be considered?
5. What measures of performance of the system are of interest to management?
6. What are the time constraints for performing the study?

In addition, the team also will meet with engineers and operational personnel to learn the details of just how the system would operate. (The team generally will also include one or more members with a first-hand knowledge of the system.)

Step 2: Collect the Data and Formulate the Simulation Model

The types of data needed depend on the nature of the system to be simulated. For example, key pieces of data for a queueing system would be the distribution of *interarrival times* and the distribution of *service times*. For most other cases as well, it is the *probability distributions* of the relevant quantities that are needed. Generally, it will only be possible to *estimate* these distributions, but it is important to do so. In order to generate representative scenarios of how a system will perform, it is essential for simulation to generate *random observations* from these distributions rather than simply using averages.

A simulation model often is formulated in terms of a *flow diagram* that links together the various components of the system. Operating rules are given for each component, including the probability distributions that control when events will occur there.

Step 3: Check the Accuracy of the Simulation Model

Before constructing a computer program, the OR team should engage the people most intimately familiar with how the system will operate in checking the accuracy of the simulation model. This often is done by performing a structured walk-through of the conceptual model, using an overhead projector, before an audience of all the key people. Typically at such meetings, several erroneous model assumptions will be discovered and corrected, a few new assumptions will be added, and some issues will be resolved about how much detail is needed in the various parts of the model.

Step 4: Select the Software and Construct a Computer Program

There are several major classes of software used for simulations. One is *spreadsheet software.* Example 1 in Sec. 20.1 illustrated how Excel is able to perform some basic simulations on a spreadsheet. In addition, some excellent Excel add-ins now are available to enhance this kind of spreadsheet modeling. The next section focuses on the use of these add-ins.

Other classes of software for simulations are intended for more extensive applications where it is no longer convenient to use spreadsheet software. One such class is a *general-purpose programming language,* such as C, FORTRAN, BASIC, etc. Such languages (and their predecessors) often were used in the early history of the field because of their great flexibility for programming any sort of simulation. However, because of the considerable programming time required, they are not used nearly as much now.

Many commercial software packages that don't use spreadsheets also have been developed specifically to perform simulations. Historically, these simulation software packages have been classified into two categories, general-purpose simulation languages and application-oriented simulators. *General-purpose simulation languages* provide many of the features needed to program any simulation model efficiently. *Application-oriented simulators* (or just *simulators* for short) are designed for simulating fairly specific types of systems. However, as time has gone on, the distinction between these two categories has become increasingly blurred. General-purpose simulation languages now may include some special features that make them almost as well suited as simulators for certain specific kinds of applications. Conversely, today's simulators tend to include more flexibility then they previously had for dealing with a broader class of systems.

Another way of categorizing simulation software packages is by whether they use an event-scheduling approach or a process approach to discrete-event simulation modeling. The *event-scheduling approach* closely follows the *next-event incrementing* time advance method described in Sec. 20.1. The *process approach* still uses next-event incrementing in the background but focuses the modeling instead on describing the processes that generate the events. Most contemporary simulation software packages now use the process approach.

It has become increasingly common for simulation software packages to include **animation** capabilities for displaying simulations in action. In an animation, key elements of a system are represented in a computer display by icons that change shape, color, or position when there is a change in the state of the simulation system. The major reason for the popularity of animation is its ability to communicate the essence of a simulation model (or of a simulation run) to managers and other key personnel.

Because of the growing importance of simulation, there now are approximately 50 software companies marketing simulation software packages. Selected Reference 12 provides a survey of these packages. (*OR/MS Today* updates this survey every two years.)

Step 5: Test the Validity of the Simulation Model

After the computer program has been constructed and debugged, the next key step is to test whether the simulation model incorporated into the program is providing valid results for the system it is representing. Specifically, will the measures of performance for the real system be closely approximated by the values of these measures generated by the simulation model?

In some cases, a mathematical model may be available to provide results for a simple version of the system. If so, these results also should be compared with the simulation results.

When no real data are available to compare with simulation results, one possibility is to conduct a *field test* to collect such data. This would involve constructing a small prototype of some version of the proposed system and placing it into operation.

Another useful validation test is to have knowledgeable operational personnel check the creditability of how the simulation results change as the configuration of the simulated system is changed. Watching animations of simulation runs also is a useful way of checking the validity of the simulation model.

Step 6: Plan the Simulations to Be Performed

At this point, you need to begin making decisions on which system configurations to simulate. This often is an evolutionary process, where the initial results for a range of configurations help you to hone in on which specific configurations warrant detailed investigation.

Decisions also need to be made now on some statistical issues. One such issue (unless using the special technique described in the second supplement to this chapter on the book's website) is the *length of the warm-up period* while waiting for the system to essentially reach a steady-state condition before starting to collect data. Preliminary simulation runs often are used to analyze this issue. Since systems frequently require a surprisingly long time to essentially reach a steady-state condition, it is helpful to select *starting conditions* for a simulated system that appear to be roughly representative of steady-state conditions in order to reduce this required time as much as possible.

Another key statistical issue is the *length of the simulation run* following the warm-up period for each system configuration being simulated. Keep in mind that simulation does not produce *exact* values for the measures of performance of a system. Instead, each simulation run can be viewed as a *statistical experiment* that is generating *statistical observations* of the performance of the simulated system. These observations are used to produce *statistical estimates* of the measures of performance. Increasing the length of a run increases the precision of these estimates. (The first supplement to this chapter on the book's website also describes special *variance-reducing techniques* that can sometimes be used to increase the precision of these estimates.)

The statistical theory for designing statistical experiments conducted through simulation is little different than for experiments conducted by directly observing the performance of a physical system.[5] Therefore, the inclusion of a professional statistician (or at least an experienced simulation analyst with a strong statistical background) on the OR team can be invaluable at this step.

Step 7: Conduct the Simulation Runs and Analyze the Results

The output from the simulation runs now provides statistical estimates of the desired measures of performance for each system configuration of interest. In addition to a *point estimate* of each measure, a *confidence interval* normally should be obtained to indicate the range of likely values of the measure (just as was done for Example 2 in Sec. 20.1). The second supplement to this chapter on the book's website describes one method for doing this.[6]

---

[5]For details about the relevant statistical theory for applying simulation, see Chaps. 9–12 in Selected Reference 10. Also see Selected References 8 and 9 for authoritative treatises on the design and analysis of simulation experiments.

[6]See pp. 530–531 in Selected Reference 10 for alternative methods.

These results might immediately indicate that one system configuration is clearly superior to the others. More often, they will identify the few strong candidates to be the best one. In the latter case, some longer simulation runs would be conducted to better compare these candidates.[7] Additional runs also might be used to fine-tune the details of what appears to be the best configuration.

Step 8: Present Recommendations to Management

After completing its analysis, the OR team needs to present its recommendations to management. This usually would be done through both a written report and a formal oral presentation to the managers responsible for making the decisions regarding the system under study.

The report and presentation should summarize how the study was conducted, including documentation of the validation of the simulation model. A demonstration of the *animation* of a simulation run might be included to better convey the simulation process and add credibility. Numerical results that provide the rationale for the recommendations need to be included.

Management usually involves the OR team further in the initial implementation of the new system, including the indoctrination of the affected personnel.

■ 20.6   PERFORMING SIMULATIONS ON SPREADSHEETS

Section 20.5 outlines the typical steps involved in major simulation studies of complex systems, including the use of general simulation languages or specialized simulators that are needed to study most such systems efficiently. However, not all simulation studies are nearly that involved. In fact, when studying relatively simple systems, it is sometimes possible to run the needed simulations quickly and easily on spreadsheets. In particular, whenever a spreadsheet model can be formulated to analyze a system without taking uncertainties into account (except through sensitivity analysis), it usually is possible to extend the model to use simulation to consider the effect of the uncertainties. Therefore, we now will focus on these simpler cases where spreadsheets can be used to perform the simulations effectively.

As illustrated by Example 1 in Sec. 20.1, the standard Excel package has some basic simulation capabilities, including the ability to generate uniform random numbers and to generate random observations from some probability distributions. An exciting subsequent advancement has been the development of powerful Excel add-ins that greatly extend these capabilities. One of these add-ins is *Crystal Ball,* developed by Decisioneering, Inc. (now Oracle). In addition to its strong functionality for performing simulations, the Professional Edition of Crystal Ball also includes two other modules. One is CB Predictor for generating forecasts from time-series data, as described and illustrated in Chapter 27 (a supplementary chapter on the book's website). The other is OptQuest, which enhances Crystal Ball by using its output from a series of simulation runs to automatically search for an optimal solution for a simulation model, as described in the third supplement to this chapter on the book's website.

Some of the other simulation add-ins are available as shareware. One is RiskSim, developed by Professor Michael Middleton. We have provided the academic version of RiskSim for you in your OR Courseware. Although not as elaborate or powerful as Crystal Ball, RiskSim is easy to use and is well documented on the book's website. (If you want to continue to use

---

[7]Methodology for using simulation to attempt to identify the best system configuration is referred to as *simulation optimization*. This is a very active area of current research. For example, see Selected References 7, 13, and 4.

it after this course, you should register and pay the shareware fee.) Like any Excel add-ins, any of these simulation add-ins need to be installed before they will show up in Excel.

This section focuses on the functionality of Crystal Ball to illustrate what can be done with simulation add-ins. You can practice using Crystal Ball yourself by going to its website (currently www.decisioneering.com/downloadform.html) to download this software for a temporary trial period (currently 30 days). Your school (like many others) may also have a site license for this popular software package.

We have included end-of-chapter problems for this section that are well suited for using Crystal Ball. RiskSim on the book's website also can be used for these problems.

Business spreadsheets typically include some *input cells* that display key data (e.g., the various costs associated with producing or marketing a product) and one or more *output cells* that show measures of performance (e.g., the profit from producing or marketing the product). The user writes Excel equations to link the inputs to the outputs so that the output cells will show the values that correspond to the values that are entered into the input cells. In some cases, there will be uncertainty about what the correct values for the input cells will turn out to be. Sensitivity analysis can be used to check how the outputs change as the values for the input cells change. However, if there is considerable uncertainty about the values of some input cells, a more systematic approach to analyzing the effect of the uncertainty would be helpful. This is where simulation enters the picture.

With simulation, instead of entering a single number in an input cell where there is uncertainty, a *probability distribution* that describes the uncertainty is entered instead. By generating a *random observation* from the probability distribution for each such input cell, the spreadsheet can calculate the output values in the usual way. This is called a **trial** by Crystal Ball. By running the number of trials specified by the user (typically hundreds or thousands), the simulation thereby generates the same number of random observations of the output values. The Crystal Ball program records all this information and then gives you the choice of printing out detailed statistics in tabular or graphical form (or both) that roughly shows the underlying *probability distribution* of the output values. A summary of the results also includes estimates of the mean and standard deviation of this distribution.

Now let us go through an example in detail to illustrate this process.

An Inventory Management Example—Freddie the Newsboy's Problem

Consider the following problem being faced by a newsboy named Freddie. One of the daily newspapers that Freddie sells from his newsstand is the *Financial Journal*. A distributor brings the day's copies of the *Financial Journal* to the newsstand early each morning. Any copies unsold at the end of the day are returned to the distributor the next morning. However, to encourage ordering a large number of copies, the distributor does give a small refund for unsold copies.

Here are Freddie's cost figures.

Freddie pays $1.50 per copy delivered.
Freddie sells it at $2.50 per copy.
Freddie's refund is $0.50 per unsold copy.

Partially because of the refund, Freddie always has taken a plentiful supply. However, he has become concerned about paying so much for copies that then have to be returned unsold, particularly since this has been occurring nearly every day. He now thinks he might be better off by ordering only a minimal number of copies and saving this extra cost.

To investigate this further, he has compiled the following record of his daily sales.

Freddie sells anywhere between 40 and 70 copies inclusively on any given day. The frequency of the numbers between 40 and 70 are roughly equal.

The decision that Freddie needs to make is the number of copies to order per day from the distributor. His objective is to maximize his average daily profit.

You may recognize this problem as an example of the *newsvendor problem* discussed in Sec. 18.7. Thus, the *stochastic one-period inventory model for perishable products* (with no setup cost) presented there can be used to solve this problem. However, for illustrative purposes, we now will show how simulation can be used to analyze this simple inventory system in the same way that it analyzes more complex inventory systems that are beyond the reach of available inventory models.

A Spreadsheet Model for This Problem

Figure 20.7 shows a spreadsheet model for this problem. Given the data cells C4:C6, the decision variable is the order quantity to be entered in cell C9. (The number 60 has been entered arbitrarily in this figure as a first guess of a reasonable value.) The bottom of the figure shows the equations used to calculate the output cells C14:C16. These output cells are then used to calculate the output cell Profit (C18).

■ FIGURE 20.7
A spreadsheet model for applying simulation to the example that involves Freddie the newsboy. The assumption cell is SimulatedDemand (C12), the forecast cell is Profit (C18), and the decision variable is OrderQuantity (C9).

	A	B	C	D	E	F
1		**Freddie the Newsboy**				
2						
3			**Data**			
4		Unit Sale Price	$2.50			
5		Unit Purchase Cost	$1.50			
6		Unit Salvage Value	$0.50			
7						
8			**Decision Variable**			
9		Order Quantity	60			
10						
11			**Simulation**		Minimum	Maximum
12		Demand	55	*Discrete Uniform*	40	70
13						
14		Sales Revenue	$137.50			
15		Purchasing Cost	$90.00			
16		Salvage Value	$2.50			
17						
18		Profit	$50.00			

	B	C
14	Sales Revenue	=UnitSalePrice*MIN(OrderQuantity,Demand)
15	Purchasing Cost	=UnitPurchaseCost*OrderQuantity
16	Salvage Value	=UnitSalvageValue*MAX(OrderQuantity-Demand,0)
17		
18	Profit	=SalesRevenue-PurchasingCost+SalvageValue

Range Name	Cell
Demand	C12
OrderQuantity	C9
Profit	C18
PurchasingCost	C15
SalesRevenue	C14
SalvageValue	C16
UnitPurchaseCost	C5
UnitSalePrice	C4
UnitSalvageValue	C6

The only uncertain input quantity in this spreadsheet is the day's demand in cell C12. This quantity can be anywhere between 40 and 70 inclusively. Since the frequency of the integer numbers between 40 and 70 are about the same, the probability distribution of the day's demand can reasonably be assumed to be a *discrete uniform distribution* between 40 and 70, as indicated in cells D12:F12. Rather than enter a single number permanently into SimulatedDemand (C12), what Crystal Ball will do is to enter this probability distribution into this cell. (Before turning to Crystal Ball, an arbitrary number 55 has been entered temporarily into this cell in Fig. 20.7.) By using Crystal Ball to generate a *random observation* from this probability distribution, the spreadsheet can calculate the output cells in the usual way to complete one trial. By running the number of trials specified by the user (typically hundreds or thousands), the simulation thereby generates the same number of random observations of the values in the output cells. Crystal Ball records this information for the output cell(s) of particular interest (Freddie's daily profit) and then, at the end, displays it in a variety of convenient forms that reveal an estimate of the underlying probability distribution of Freddie's daily profit. (More about this later.)

The Application of Crystal Ball

Four steps are needed to use the spreadsheet in Fig. 20.7 to perform the simulation with Crystal Ball.

1. Define the random input cells.
2. Define the output cells to forecast.
3. Set the run preferences.
4. Run the simulation.

We now describe each of these four steps in turn.

Define the Random Input Cells.   A random input cell is an input cell that has a random value (such as the daily demand for the *Financial Journal*), so an assumed probability distribution needs to be entered into the cell instead of permanently entering a single number. The only random input cell in Fig. 20.7 is Demand (C12). Crystal Ball refers to each such random input cell as an **assumption cell.**

The following procedure is used to define an assumption cell.

Procedure for Defining an Assumption Cell

1. Select the cell by clicking on it.
2. If the cell does not already contain a value, enter *any* number into the cell.
3. Click on the Define Assumption button (⬜) in the Crystal Ball tab (for Excel 2007) or toolbar (for earlier versions of Excel).
4. Select a probability distribution to enter into the cell by clicking on this distribution in the Distribution Gallery shown in Fig. 20.8.
5. Click on OK (or double click on the distribution) to bring up a dialogue box for the selected distribution.
6. Use this dialogue box to enter the parameters for the distribution, preferably by referring to the cells in the spreadsheet that contain the values of these parameters. If desired, a name also can be entered for the assumption cell. (If the cell already has a name next to it or above it on the spreadsheet, that name will appear in the dialogue box.)
7. Click on OK.

The **Distribution Gallery** mentioned in step 4 provides a wide variety of 21 probability distributions from which to choose. Figure 20.8 displays six basic distributions, but 15 more also are available by clicking on the All button. (When there is uncertainty about which continuous distribution provides the best fit to historical data, Crystal Ball provides

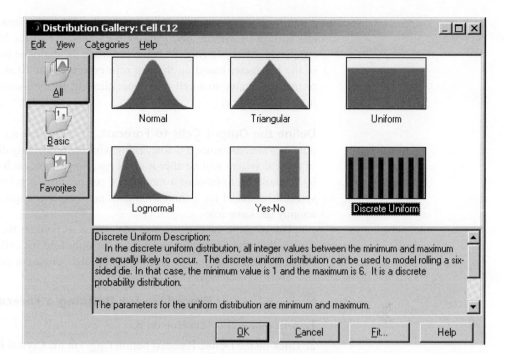

■ **FIGURE 20.8**
The Crystal Ball Distribution Gallery dialogue box showing the basic distributions. In addition to the 6 distributions displayed here, 15 more distributions can be accessed by clicking on the All button.

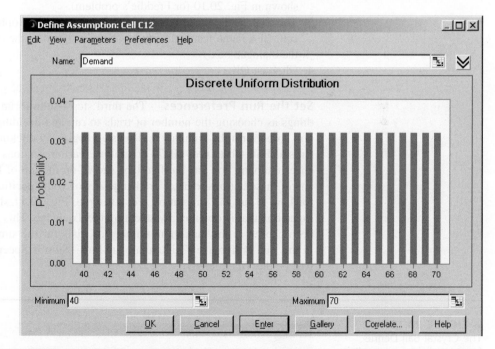

■ **FIGURE 20.9**
The Crystal Ball Discrete Uniform Distribution dialogue box. It is being used here to enter a discrete uniform distribution with the parameters 40(=E12) and 70(=F12) into the assumption cell Demand (C12) in the spreadsheet model in Fig. 20.7.

a procedure to choose an appropriate distribution. This procedure is described in Sec. 28.6 on the book's website.)

In Freddie's case, double clicking on the discrete uniform distribution in the Distribution Gallery brings up the Discrete Uniform Distribution dialogue box shown in Fig. 20.9, which is used to enter the parameters of the distribution. For each of the parameters (Minimum and Maximum), we refer to the data cells in E12 and F12 on the spreadsheet by typing the formulas =E12 and =F12 for Minimum and Maximum, respectively. After entering the cell references, the dialogue box will show the actual value of the parameter based on the cell reference (40 and 70 as shown in Fig. 20.9). To see or make a change to a cell reference, clicking on the parameter will show the underlying cell reference.

Define the Output Cells to Forecast.   Crystal Ball refers to the output of a simulations as a *forecast,* since it is forecasting what the probability distribution of the performance of the real system will be after it starts operating. Thus, each output cell that is being used by a simulation to forecast a measure of performance is referred to as a **forecast cell.** The spreadsheet model for a simulation does not include a target cell, but a forecast cell plays roughly the same role.

The measure of performance of interest to Freddie the newsboy is his daily profit from selling the *Financial Journal,* so the only forecast cell in Fig. 20.7 is Profit (C18). The following procedure is used to define such an output cell as a forecast cell.

Procedure for Defining a Forecast Cell

1. Select the cell by clicking on it.
2. Click on the Define Forecast button (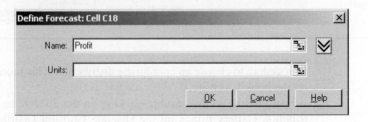) in the Crystal Ball tab (Excel 2007) or toolbar (earlier versions of Excel), which brings up the Define Forecast dialogue box (as shown in Fig. 20.10 for Freddie's problem).
3. This dialogue box can be used to define a name and (optionally) units for the forecast cell. (If a range name already has been assigned to the cell, that name will appear in the dialogue box.)
4. Click on OK.

Set the Run Preferences.   The third step—setting run preferences—refers to such things as choosing the number of trials to run and deciding on other options regarding how to perform the simulation. This step begins by clicking on Run Preferences in the Crystal Ball tab (Excel 2007) or toolbar (earlier versions of Excel). The Run Preferences dialogue box has the five tabs shown on the top of Fig. 20.11. You can click on any of these buttons to enter or change any of your specifications controlled by that tab for how to run the simulation. For example, Fig. 20.11 shows the version of the dialogue box that is obtained by selecting the Trials tab. This figure indicates that 500 has been chosen as the maximum number of trials for the simulation. (The second option in the Run Preferences Trials dialogue box—Stop if Specified Precision is Reached—will be described later.)

■ **FIGURE 20.10**
The Crystal Ball Define Forecast dialogue box. It is being used here to define the forecast cell Profit (C18) in the spreadsheet model in Fig. 20.7.

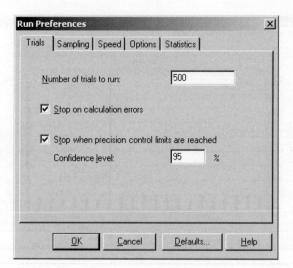

■ FIGURE 20.11
The Crystal Ball Run
Preferences dialogue box
after selecting the Trials tab.

Run the Simulation.   At this point, the stage is set to begin running the simulation. To start, you only need to click on the Start Simulation button (▷). However, if a simulation has been run previously, you should first click on the Reset Simulation button (◁◁) to reset the simulation before starting a new one.

Once started, a forecast window displays the results of the simulation as it runs. Figure 20.12 shows the forecast for Profit (Freddie's daily profit from selling the *Financial Journal*) after all 500 trials have been completed. The default view of the forecast is the frequency chart shown on the left side of the figure. The height of the vertical lines in the frequency chart indicates the relative frequency of the various profit values that were obtained during the simulation run. For example, consider the tall vertical line at $60. The right-hand side of the chart indicates a frequency of about 175 there, which means that about 175 of the 500 trials led to a profit of $60. Thus, the left-hand side of the chart indicates that the estimated probability of a profit of $60 is 175/500 = 0.35. This is the profit that results whenever the demand equals or exceeds the order quantity of 60. The remainder of the time, the profit was scattered fairly evenly between $20 and $60. These profit values correspond to trials where the demand was between 40 and 60 units, with lower profit values corresponding to demands closer to 40 and higher profit values corresponding to demands closer to 60. The mean of the 500 profit values is $45.94, as indicated by the *mean line* at this point.

The statistics table in Fig. 20.12 is obtained by choosing Statistics from the View menu. These statistics summarize the outcome of the 500 trials of the simulation. These 500 trials provide a sample of 500 random observations from the underlying probability distribution of Freddie's daily profit. The most interesting statistics about this sample provided by the table include the *mean* of $45.94, the *median* of $50.00 (indicating that $50 was the middle profit value from the 500 trials when listing the profits from smallest to largest), the *mode* of $60 (meaning that this was the profit value that occurred most frequently), and the *standard deviation* of $13.91. The information near the bottom of the table regarding the *minimum* and *maximum* profit values also is particularly useful.

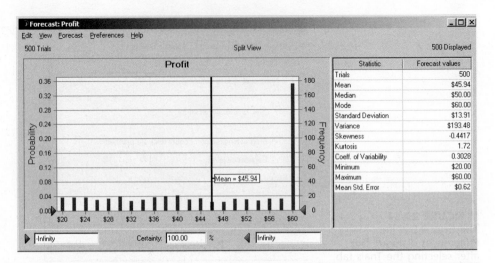

■ **FIGURE 20.12**
The frequency chart and statistics table provided by Crystal Ball to summarize the results of running the simulation model in Fig. 20.7 for the example that involves Freddie the newsboy.

Which of these statistics in Fig. 20.12 are particularly relevant really depends on what Freddie wants to achieve. The mean usually is the most important since, despite the wide fluctuations in the daily profits, the average daily profit will converge to the mean as time goes on. Therefore, multiplying the mean by the number of days that the newsstand will be open during the year gives (very closely) what the total annual profit from selling the *Financial Journal* will be, which is a very relevant quantity to want to maximize. However, if Freddie is an individual who focuses much more on the present than the future, then the median and mode might be of considerable interest to him. If he considers a profit of $50 to be a good day and his goal is to achieve a good day at least half the time, then he will want the median to be at least $50 (as it is). If he gains particular satisfaction out of achieving the maximum possible profit of $60 (given an order quantity of 60), then he will want to make sure that this will happen more often than any other specific profit (as indicated by the mode of $60). On the other hand, if Freddie is risk averse and so is particularly concerned with avoiding bad days (profits far below the mean) as much as possible, then he would have a special interest in having a relatively small standard deviation and a relatively large minimum.

Keep in mind that the statistics in Fig. 20.12 are based on using an order quantity of 60, whereas the objective is to determine the best order quantity. If Freddie has a particularly strong interest in more than one of the statistics, one approach would be to rerun the simulation model in Fig. 20.12 with various order quantities and then let Freddie choose the one whose set of statistics he likes best. In most situations, however, the mean will be the one statistic of special interest. In this case, the objective is to determine the order quantity that maximizes the mean. (We will assume this objective hereafter.) After estimating the optimal order quantity according to this objective, Freddie then should be shown the corresponding frequency chart and statistics table (and perhaps other information described subsequently as well) to make sure that everything else is satisfactory with this order quantity.

In addition to the frequency chart and statistics table presented in Fig. 20.12, the View menu provides some other useful ways of displaying the results of a simulation run, including a percentiles table, a cumulative chart, and a reverse cumulative chart. These alternative displays are shown in a split view in Fig. 20.13. The percentiles table

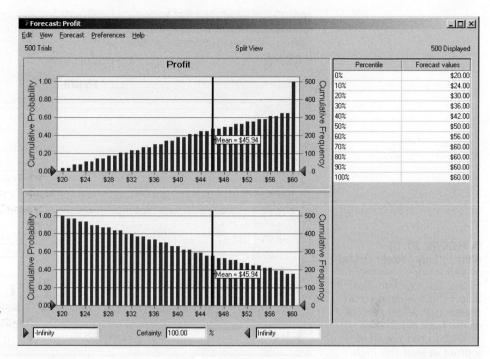

■ FIGURE 20.13
Three more forms in which Crystal Ball displays the results of running the simulation model in Fig. 20.7 for the example that involves Freddie the newsboy.

is based on listing the profit values generated by the 500 trials from smallest to largest, dividing this list into 10 equal parts (50 values in each), and then recording the value at the end of each part. Thus, the value 10 percent through the list is $24, the value 20 percent through the list is $30, and so forth. (For example, the intuitive interpretation of the 10 percent percentile of $24 is that 10 percent of the trials have profit values less than or equal to $24 and the other 90 percent of the trials have profit values greater than or equal to $24, so $24 is the dividing line between the smallest 10 percent of the values and the largest 90 percent.) The cumulative chart on the top left of Fig. 20.13 provides similar (but more detailed) information about this same list of the smallest-to-largest profit values. The horizontal axis shows the entire range of values from the smallest possible profit value ($20) to the largest possible profit value ($60). For each value in this range, the chart cumulates the number of actual profits generated by the 500 trials that are less than or equal to that value. This number equals the frequency shown on the right or, when divided by the number of trials, the probability shown on the left. The reverse cumulative chart on the bottom left of Fig. 20.13 is constructed in the same way as the cumulative chart except for the following crucial difference. For each value in the range from $20 to $60, the reverse cumulative chart cumulates the number of actual profits generated by the 500 trials that are *greater* than or equal to that value.

Figure 20.14 illustrates another of the many ways provided by Crystal Ball for extracting helpful information from the results of a simulation run. Freddie the newsboy feels that he has had a reasonably satisfactory day if he obtains a profit of at least $40 from selling the *Financial Journal*. Therefore, he would like to know the percentage of days that he could expect to achieve this much profit if he were to adopt the order quantity currently being analyzed (60). An estimate of this percentage (65.80 percent) is shown in the Certainty box below the frequency chart in Fig. 20.14. Crystal Ball can provide this

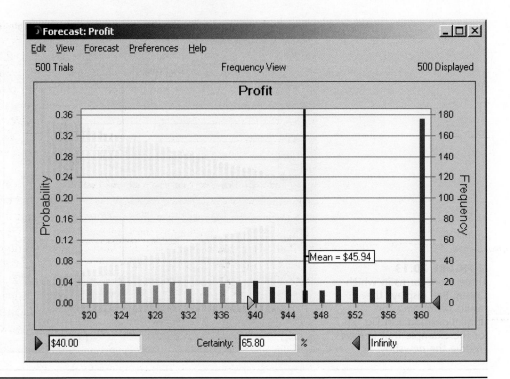

■ FIGURE 20.14
After setting a lower bound of $40 for desirable profit values, the Certainty box below this frequency chart reveals that 65.80 percent of the trials in Freddie's simulation run provided a profit at least this high.

percentage in two ways. First, the user can drag the triangle on the left just under the chart (originally at $20 in Fig. 20.12) to the right until it was at $40 (as in Fig. 20.14). Alternatively, $40 can be typed directly into the box in the lower left-hand corner. If desired, the probability of obtaining a profit between any two values also could be estimated immediately by dragging the two triangles to those values.

How Accurate Are the Simulation Results?

An important number provided by Fig. 20.12 is the mean of $45.94. This number was calculated as the *average* of the 500 random observations from the underlying probability distribution of Freddie's daily profit that were generated by the 500 trials. This *sample average* of $45.94 thereby provides an *estimate* of the *true mean* of this distribution. However, the true mean might deviate somewhat from $45.94. How accurate can we expect this estimate to be?

The answer to this key question is provided by the *mean standard error* of $0.62 given at the bottom of the statistics table in Fig. 20.12. The mean standard error is calculated as $s/\sqrt{n}$, where s is the sample standard deviation and n is the number of trials. It is an estimate of the standard deviation of the sample average, so the sample average is within one mean standard error of the true mean most of the time. In other words, the true mean can readily deviate from the sample mean by any amount up to the mean standard error, but most of the time (approximately 68 percent of the time), it will not deviate by more than that. Thus, the interval from $45.94 − $0.62 = $45.32 to $45.94 + $0.62 = $46.56 is a 68 percent *confidence interval* for the true mean. Similarly, a larger confidence interval can be obtained by using an appropriate multiple of the mean standard error to subtract from the sample mean and then to add to the sample mean. For example, the appropriate multiple for a 95 percent confidence interval is 1.965, so such a confidence interval ranges from $45.94 − 1.965($0.62) = $44.72 to $45.94 + 1.965($0.62) = $47.16. (This multiple of 1.965 will change slightly if the number of trials is different from 500.) Therefore, it is very likely that the true mean is somewhere between $44.72 and $47.16.

If greater precision is required, the mean standard error normally can be reduced by increasing the number of trials in the simulation run. However, the reduction tends to be small unless the number of trials is increased substantially. For example, cutting the mean standard error in half requires approximately quadrupling the number of trials. Thus, a surprisingly large number of trials may be required to obtain the desired degree of precision.

Since the number of trials required to obtain the desired degree of accuracy cannot be predicted very well in advance of the simulation run, the temptation is to specify an extremely large number of trials. This specified number might turn out to be many times as large as necessary and thereby cause an excessively long computer run. Fortunately, Crystal Ball has a special method of precision control for stopping the simulation run early, as soon as the desired precision has been reached. This method is triggered by choosing the option ("Stop when precision control limits are reached") in the Run Preferences Trials dialogue box shown in Fig. 20.11. The specified precision is entered in the Expanded Define Forecast dialogue box displayed in Fig. 20.15. (This dialogue box is brought up by clicking on the More button (⌄) in the Define Forecast dialogue box shown in Fig. 20.10.) Figure 20.15 indicates that the precision control is being applied to the mean (but not to the standard deviation or to a specified percentile). The run preferences in Fig. 20.11 indicate that a 95 percent confidence interval is being used. The width of half of the confidence interval, measured from its midpoint to either end, is considered to be the precision that has been achieved. The desired precision can be specified in either absolute terms (using the same units as for the confidence interval) or in relative terms (expressed as a percentage of the midpoint of the confidence interval).

Figure 20.15 indicates that the decision was made to specify the desired precision in absolute terms as $1. The 95 percent confidence interval for the mean after 500 trials was found to be $45.94 plus-or-minus $1.22, so $1.22 is the precision that was achieved after all these trials. Crystal Ball also calculates the confidence interval (and so the current precision) periodically to check whether the current precision is under $1, in which case the run would be stopped. However, this never happened, so Crystal Ball allowed the simulation to run until the maximum number of trials (500) was reached.

To obtain the desired precision, the simulation would need to be restarted to generate additional trials. This is done by entering a larger number (such as 5,000) for the maximum number of trials (including the 500 already obtained) in the Run Preferences dialogue box (shown in Fig. 20.11) and then clicking on the Start Simulation button (▷). Figure 20.16 shows the results from doing this. The first row indicates that the desired precision was obtained after only 500 additional trials, for a total of 1000 trials. (The default value for the frequency of checking the precision is every 500 trials, so the precision of $1 actually was reached somewhere between 500 and 1000 trials.) Because of the additional trials, some of the statistics have changed slightly from those given in Fig. 20.12. For example, the best estimate of the mean now is $46.46, with a precision of $0.85. Thus, it is very likely (95 percent confidence) that the true value of the mean is within $0.85 of $46.46.

95 percent confidence interval: $45.61 ≤ Mean ≤ $47.31

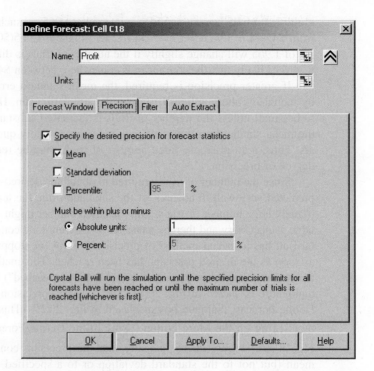

■ **FIGURE 20.15**
This Expanded Define Forecast dialogue box is being used to specify how much precision is desired in Freddie's simulation run.

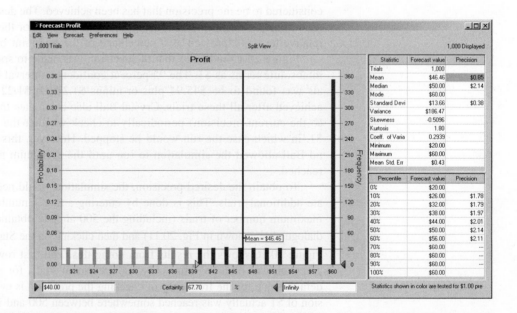

■ **FIGURE 20.16**
The results obtained after continuing Freddie's simulation run until the precision specified in Fig. 20.15 has been achieved.

The precision also is given for the current estimates of the median and the standard deviation, as well as for the estimates of the percentiles given in the percentiles table. Therefore, a 95 percent confidence interval also can be calculated for each of these quantities by adding and subtracting its precision from its estimate.

Application of the Decision Table Tool

The results presented in Figs. 20.12 and 20.16 were from a simulation run that fixed Freddie's daily order quantity at 60 copies of the *Financial Journal* (as indicated in cell C9 of the spreadsheet in Fig. 20.7). Freddie wanted this order quantity tried first because it seems to provide a reasonable compromise between being able to fully meet the demand on many days (about two-thirds of them) and often not having many unsold copies on those days. However, the results obtained do not reveal whether 60 is the *optimal* order quantity that would maximize his average daily profit. Many more simulation runs with other order quantities will be needed to determine (or at least estimate) the optimal order quantity.

Fortunately, Crystal Ball provides a special feature called the **Decision Table tool** that systematically applies simulation to identify at least an approximation of an optimal solution for problems with only one or two decision variables. Freddie's problem has only a single decision variable, OrderQuantity (C9) in the spreadsheet model of Fig. 20.7, so we now will apply this tool.

An intuitive approach for searching for an optimal solution would be to use trial and error. Try different values of the decision variable(s), run a simulation for each, and see which one provides the best estimate of the chosen measure of performance. This is what the Decision Table tool does, but it does it in a systematic way. Its dialogue boxes enable you to quickly specify what you want to do. Then, after you click one button, all the desired simulations are run and the results soon are displayed in the Decision Table. If desired, you also can view some charts, including a *trend chart,* that provide additional details about the results.

If you have previously used either an Excel data table or the Solver Table that is included in your OR Courseware for performing sensitivity analysis systematically, the Decision Table works in much the same way. In particular, the layout for a Decision Table with either one or two decision variables is similar to that for either a one-dimensional or two-dimensional Solver Table (introduced in Sec. 6.8). Two is the maximum number of decision variables that can be varied simultaneously in a Decision Table.

Since the number of copies that Freddie's customers want to purchase varies widely from day to day (anywhere from 40 to 70 copies), it would seem sensible to begin by trying a sampling of possible order quantities, say, 40, 45, 50, 55, 60, 65, and 70. To do this with the Decision Table tool, the first step is to define the decision variable being investigated, namely, OrderQuantity (C9) in Fig. 20.7, by using the following procedure.

Procedure for Defining a Decision Variable

1. Select the cell containing the decision variable by clicking on it.
2. If the cell does not already contain a value, enter *any* number into the cell.
3. Click on the Define Decision button (⊕) on the Crystal Ball tab or toolbar, which brings up the Define Decision Variable dialogue box (as shown in Fig. 20.17 for Freddie's problem).
4. Enter the lower limit and the upper limit of the range of values to be simulated for the decision variable.
5. Click on either Continuous or Discrete to define whether the decision variable is continuous or discrete.
6. If Discrete is selected in step 5, use the Step box to specify the difference between successive possible values (not just those to be simulated) of the decision variable. (The default value is 1.)
7. Click on OK.

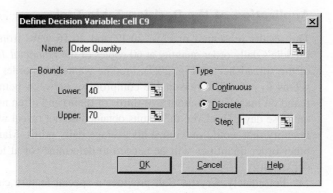

■ **FIGURE 20.17**
This Define Decision Variable dialogue box specifies the characteristics of the decision variable OrderQuantity (C9) in the simulation model in Fig. 20.7 for the example that involves Freddie the newsboy.

Figure 20.17 shows the application of this procedure to Freddie's problem. Since simulations will be run for order quantities ranging from 40 to 70, these limits for the range have been entered on the left. The order quantity can have any integer value within this range, so this is indicated on the right.

Now we are ready to choose Decision Table from the Crystal Ball Tools menu. This brings up the sequence of three dialogue boxes shown in Fig. 20.18.

The Step 1 dialogue box is used to choose one of the forecast cells listed there to be the target cell for the Decision Table. Freddie's spreadsheet model in Fig. 20.7 has only one forecast cell, Profit (C18), so select it and then click on the Next button.

■ **FIGURE 20.18**
To prepare for generating a Decision Table, these three dialogue boxes specify (1) which forecast cell will be the target cell, (2) which one or two decision variables will be varied, and (3) the running options. The choices made here are for the example that involves Freddie the newsboy.

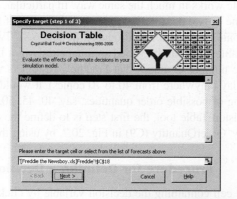
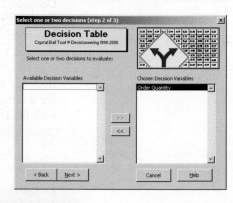
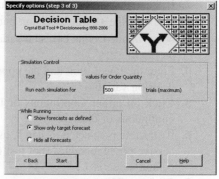

Initially, the left side of the Step 2 dialogue box includes a list of all the cells that have been defined as decision variables. This consists of the single decision variable, OrderQuantity (C9), for Freddie's problem. The purpose of this dialogue box is to choose which one or two decision variables to vary for the Decision Table. This is done by selecting these decision variables on the left side and then clicking on the double right arrows (>>) between the two boxes, which brings these decision variables to the right side. Figure 20.19 shows the result of doing this with Freddie's decision variable.

The Step 3 dialogue box is used to specify the options for the Decision Table. The first entry box records the number of values of the decision variable for which simulations will be run. Crystal Ball then distributes the values evenly over the range of values specified in the Define Decision Variable dialogue box (Fig. 20.17). For Freddie's problem, the range of values is 40 to 70, so entering 7 into the first entry box in the Step 3 dialogue box results in choosing 40, 45, 50, 55, 60, 65, and 70 as the seven values of the order quantity for which simulations will be run. After selecting the run size for each simulation and specifying what you want to see while the simulations are running, the last step is to click the Start button.

After Crystal Ball runs the simulations, the Decision Table is created in a new spreadsheet as shown in Figure 20.19. For each of the order quantities shown at the top, row 2 gives the mean of the values of the target cell, Profit (C18), obtained in all the trials of that simulation run. Cells D2:F2 reveal that an order quantity of 55 achieved the largest mean profit of $47.49, while order quantities of 50 and 60 essentially tied for the second largest mean profit.

The sharp drop off in mean profits on both sides of these order quantities virtually guarantees that the optimal order quantity lies between 50 and 60 (and probably close to 55). To pin this down better, the logical next step would be to generate another Decision Table that considers all integer order quantities between 50 and 60. You are asked to do this in Problem 20.6-6. (The third supplement to this chapter on the book's website will use the OptQuest module of Crystal Ball to pin down the optimal order quantity in another way.)

The upper left-hand corner of the Decision Table provides three options for obtaining more detailed information about the results of the simulation runs for the cells that you select. One option is to view the forecast chart of interest, such as a frequency chart or cumulative chart, by choosing a forecast cell in row 2 and then clicking on the Forecast Charts button. Another option is to see the results of two or more simulations runs together. This is done by selecting a set of forecast cells, say, cells E2:F2 in Fig. 20.19, and then clicking on the Overlay Chart button. The resulting overlay chart in shown in

■ **FIGURE 20.19**
The Decision Table for Freddie's problem.

A	B	C	D	E	F	G	H
Trend Chart / Overlay Chart / Forecast Charts	Order Quantity (40)	Order Quantity (45)	Order Quantity (50)	Order Quantity (55)	Order Quantity (60)	Order Quantity (65)	Order Quantity (70)
2	$40.00	$44.17	$46.66	$47.49	$46.64	$44.14	$39.97
3	1	2	3	4	5	6	7

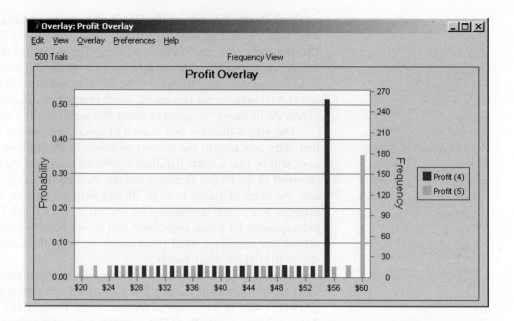

■ **FIGUER 20.20**
The overlay chart that
compares the frequency
distributions for order
quantities of 55 and 60
for Freddie's problem.

Fig. 20.20. The dark lines show the frequency chart for cell E2 (an order quantity of 55) while the light lines do the same for cell F2 (an order quantity of 60), so the results for these two cases can be compared side by side. (On a color monitor, you will see different colors used to distinguish between the different cases.)

The third option is to select all the forecast cells of interest (cells B2:H2 in Fig. 20.19) and then click on the Trend Chart button. This generates an interesting chart, called the *trend chart,* shown in Fig. 20.21. The key points along the horizontal axis are the seven vertical grid lines that correspond to the seven cases (order quantities of 40, 45, . . . , 70) for which the simulations were run. The vertical axis gives the profit values obtained in the trials of these simulation runs. The bands in the chart summarize information about the frequency distribution of the profit values from each simulation run. (On a color monitor, the bands appear in color—light blue for the center band, red for the adjacent pair of bands, green for the next pair, and dark blue for the outer pair of bands.) These bands are centered on the *medians* of the frequency distributions. In other words, the center of the middle band (the lightest one) gives the profit value such that half of the trials gave a larger value and half gave a smaller value. This middle band contains the middle 10 percent of the profit values (so 45 percent are on each side of the band). Similarly, the middle three bands contain the middle 25 percent of the profit values, the middle five bands contain the middle 50 percent of the profit values, and all seven bands contain the middle 90 percent of the profit values. (These percentages are listed to the right of the trend chart.) Thus, 5 percent of the profit values generated in the trials of each simulation run lie above the top band and 5 percent lie below the bottom band.

The trend chart received its name because it shows the trends graphically as the value of the decision variable (the order quantity in this case) increases. In Fig. 20.21, for example, consider the middle band (which gets hidden in the narrow part of the chart on the left). In going from the third-order quantity (50) to the fourth one (55), the middle band trends upward, but then it trends downward thereafter. Thus, the median value

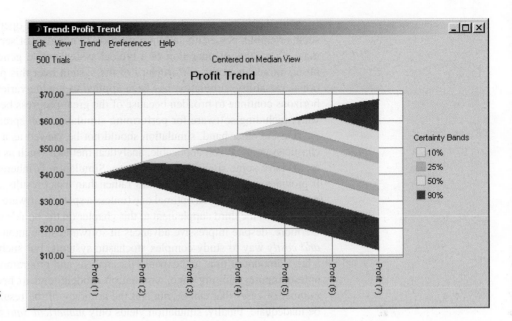

FIGUER 20.21
The trend chart that shows the trend in the range of various portions of the frequency distribution as the order quantity increases for Freddie's problem.

of the profit values generated in the respective simulation runs increases as the order quantity increases until the median reaches its peak at an order quantity of 55, after which the median trends downward. Similarly, most of the other bands also trend downward as the order quantity increases above 55. This suggests that an order quantity of 55 is a particularly attractive one in terms of its entire frequency distribution and not just its mean value. The fact that the trend chart spreads out as it moves to the right provides the further insight that the variability of the profit values increase as the order quantity is increased. Although the largest order quantities provide some chance of particularly high profits on occasional days, they also can lead to an unusually low profit on any given day. This risk profile may be relevant to Freddie if he is concerned about the variability of his daily profits.

If you would like to read more about how to perform simulations on spreadsheets with Crystal Ball, Chap. 28 on the book's website provides several additional examples and further details. These examples include applications to contract bidding, project management, cash flow management, financial risk analysis, and revenue management.

20.7 CONCLUSIONS

Simulation is a widely used tool for estimating the performance of complex stochastic systems if contemplated designs or operating policies are to be used.

We have focused in this chapter on the use of simulation for predicting the *steady-state* behavior of systems whose states change only at discrete points in time. However, by having a series of runs begin with the prescribed *starting conditions,* we can also use simulation to describe the *transient* behavior of a proposed system. Furthermore, if we use differential equations, simulation can be applied to systems whose states change *continuously* with time.

Simulation is one of the most popular techniques of operations research because it is such a flexible, powerful, and intuitive tool. In a matter of seconds or minutes, it can simulate even years of operation of a typical system while generating a series of statistical observations about the performance of the system over this period. Because of its exceptional versatility, simulation has been applied to a wide variety of areas. Furthermore, its horizons continue to broaden because of the great progress being made in simulation software, including software for performing simulations on spreadsheets.

On the other hand, simulation should not be viewed as a panacea when studying stochastic systems. When applicable, analytical methods (such as those presented in Chaps. 15 to 19) have some significant advantages. Simulation is inherently an imprecise technique. It provides only *statistical estimates* rather than exact results, and it *compares alternatives* rather than generating an optimal one (unless a special software package such as OptQuest—described in the third supplement to this chapter on the book's website—is being used). Furthermore, despite impressive advances in software, simulation still can be a relatively *slow and costly* way to study complex stochastic systems. For such systems, it usually requires a large amount of time and expense for analysis and programming, in addition to considerable computer running time. Simulation models tend to become unwieldy, so that the number of cases that can be run and the accuracy of the results obtained often turn out to be inadequate. Finally, simulation yields only *numerical data* about the performance of the system, so that it provides no additional insight into the cause-and-effect relationships within the system except for the clues that can be gleaned from these numbers (and from the analysis required to construct the simulation model). Therefore, it is very expensive to conduct a sensitivity analysis of the parameter values assumed by the model. The only possible way would be to conduct new series of simulation runs with different parameter values, which would tend to provide relatively little information at a relatively high cost.

For all these reasons, analytical methods (when available) and simulation have important complementary roles for studying stochastic systems. An analytical method is well suited for doing at least preliminary analysis, for examining cause-and-effect relationships, for doing some rough optimization, and for conducting sensitivity analysis. When the mathematical model for the analytical method does not capture all the important features of the stochastic system, simulation is well suited for incorporating all these features and then obtaining detailed information about the measures of performance of the few leading candidates for the final system configuration.

Simulation provides a way of *experimenting* with proposed systems or policies without actually implementing them. Sound statistical theory should be used in designing these experiments. Surprisingly long simulation runs often are needed to obtain *statistically significant* results. However, *variance-reducing techniques* (described in the first supplement to this chapter on the book's website) occasionally can be very helpful in reducing the length of the runs needed.

Several tactical problems arise when we apply traditional statistical estimation procedures to simulated experiments. These problems include prescribing appropriate *starting conditions,* determining how long a *warm-up period* is needed to essentially reach a steady-state condition, and dealing with *statistically dependent* observations. These problems can be eliminated by using the *regenerative method* of statistical analysis (described in the second supplement to this chapter on the book's website). However, there are some restrictions on when this method can be applied.

Simulation unquestionably has a very important place in the theory and practice of OR. It is an invaluable tool for use on those problems where analytical techniques are inadequate, and its usage is continuing to grow.

■ SELECTED REFERENCES

1. Argon, N. T., and S. Andradóttir: "Replicated Batch Means for Steady-State Simulations," *Naval Research Logistics*, **53**(6): 508–524, September 2006.

2. Asmussen, S., and P. W. Glynn: *Stochastic Simulation*, Springer, New York, 2007.

3. Banks, J., J. S. Carson, II, B. L. Nelson, and D. M. Nicol: *Discrete-Event System Simulation,* 4th ed., Prentice-Hall, Upper Saddle River, NJ, 2005.

4. Branke, J., S. E. Chick, and C. Schmidt: "Selecting a Selection Procedure," *Management Science*, **53**(12): 1916–1932, December 2007.

5. del Castillo, E.: *Process Optimization: A Statistical Approach*, Springer, New York, 2007.

6. Fishman, G. S.: *Discrete-Event Simulation: Modeling, Programming, and Analysis,* Springer, New York, 2001.

7. Fu, M. C.: "Optimization for Simulation: Theory vs. Practice," *INFORMS Journal on Computing,* **14**(3): 192–215, Summer 2002.

8. Kleijnen, J. P. C.: *Design and Analysis of Simulation Experiments*, Springer, New York, 2008.

9. Kleijnen, J. P. C., S. M. Sanchez, T. W. Lucas, and T. M. Cioppa: "State-of-the-Art Review: A User's Guide to the Brave New World of Designing Simulation Experiments," *INFORMS Journal on Computing*, **17**(3): 263–289, Summer 2005.

10. Law, A. M., and W. D. Kelton: *Simulation Modeling and Analysis,* 3rd ed., McGraw-Hill, New York, 2000.

11. Nance, R. E., and R. G. Sargent: "Perspectives on the Evolution of Simulation," *Operations Research,* **50**(1): 161–172, January–February 2002.

12. Swain, J.: "Software Survey: New Frontiers in Simulation," *OR/MS Today*, **34**(5): 32–43, October 2007.

13. Tekin, E., and I. Sabuncuoglu: "Simulation Optimization: A Comprehensive Review on Theory and Applications," *IIE Transactions*, **36**(11): 1067–1081, November 2004.

14. Whitt, W.: "Planning Queueing Simulations," *Management Science,* **35**(11): 1341–1366, November 1989.

Some Award-Winning Applications of Simulation:

(A link to all these articles is provided on our website, www.mhhe.com/hillier.)

A1. Alden, J. M., L. D. Burns, T. Costy, R. D. Hutton, C. A. Jackson, D. S. Kim, K. A. Kohls, J. H. Owen, M. A. Turnquist, and D. J. Vander Veen: "General Motors Increases Its Production Throughput," *Interfaces,* **36**(1): 6–25, January–February 2006.

A2. Barabba, V., C. Huber, F. Cooke, N. Pudar, J. Smith, and M. Paich: "A Multimethod Approach for Creating New Business Models: The General Motors OnStar Project," *Interfaces,* **32**(1): 20–34, January–February 2002.

A3. Beis, D. A., P. Loucopoulos, Y. Pyrgiotis, and K. G. Zografos: "PLATO Helps Athens Win Gold: Olympic Games Knowledge Modeling for Organizational Change and Resource Management," *Interfaces,* **36**(1): 26–42, January–February 2006.

A4. Brinkley, P. A., D. Stepto, K. R. Haag, J. Folger, K. Wang, K. Liou, and W. D. Carr: "Nortel Redefines Factory Information Technology: An OR-Driven Approach," *Interfaces,* **28**(1): 37–52, January–February 1998.

A5. Cebry, M. E., A. H. DeSilva, and F. J. DiLisio: "Management Science in Automating Postal Operations: Facility and Equipment Planning in the United States Postal Service," *Interfaces,* **22**(1): 110–130, January–February 1992.

A6. Duffy, T., M. Hatzakis, W. Hsu, R. Labe, B. Liao, X. Luo, J. Oh, A. Setya, and L. Yang: "Merrill Lynch Improves Liquidity Risk Management for Revolving Credit Lines," *Interfaces*, **35**(5): 353–369, September–October 2005.

A7. Hueter, J., and W. Swart: "An Integrated Labor-Management System for Taco Bell," *Interfaces,* **28**(1): 75–91, January–February 1998.

A8. Larson, R. C., M. F. Cahn, and M. C. Shell: "Improving the New York City Arrest-to-Arraignment System," *Interfaces,* **23**(1): 76–96, January–February 1993.

A9. Mulvey, J. M., G. Gould, and C. Morgan: "An Asset and Liability Management System for Towers Perrin-Tillinghast," *Interfaces,* **30**(1): 96–114, January–February 2000.

A10. Pfeil, G., R. Holcomb, C. T. Muir, and S. Taj: "Visteon's Sterling Plant Uses Simulation-Based Decision Support in Training, Operations, and Planning," *Interfaces,* **30**(1): 115–133, January–February 2000.

■ LEARNING AIDS FOR THIS CHAPTER ON OUR WEBSITE (www.mhhe.com/hillier)

Worked Examples:

Examples for Chapter 20

Demonstration Examples in OR Tutor:

Simulating a Basic Queueing System
Simulating a Queueing System with Priorities

An Automatic Procedure in IOR Tutorial:

Animation of a Queueing System

Interactive Procedures in IOR Tutorial:

Enter Queueing Problem
Interactively Simulate Queueing Problem

"Ch. 20—Simulation" Excel Files:

Spreadsheet Examples
Queueing Simulator

Excel Add-In:

RiskSim (academic version)

Glossary for Chapter 20

Supplements to This Chapter:

Variance-Reducing Techniques
Regenerative Method of Statistical Analysis
Optimizing with OptQuest

See Appendix 1 for documentation of the software.

■ PROBLEMS

The symbols to the left of some of the problems (or their parts) have the following meaning:

D: The demonstration examples for this chapter may be helpful.
I: We suggest that you use the interactive procedures listed in Learning Aids (the printout records your work).
E: Use Excel.
A: Use an Excel simulation add-in, such as RiskSim or Crystal Ball.
Q: Use the Queueing Simulator.
R: Use *three-digit* uniform random numbers (0.096, 0.569, etc.) that are obtained from the consecutive random digits in Table 20.3, starting from the front of the top row, to do each problem part.

20.1-1.* Use the uniform random numbers in cells C13:C18 of Fig. 20.1 to generate six random observations for each of the following situations.
(a) Throwing an unbiased coin.
(b) A baseball pitcher who throws a strike 60 percent of the time and a ball 40 percent of the time.
(c) The color of a traffic light found by a randomly arriving car when it is green 40 percent of the time, yellow 10 percent of the time, and red 50 percent of the time.

20.1-2. The weather can be considered a stochastic system, because it evolves in a probabilistic manner from one day to the next. Suppose for a certain location that this probabilistic evolution satisfies the following description:

The probability of rain tomorrow is 0.6 if it is raining today. The probability of its being clear (no rain) tomorrow is 0.8 if it is clear today.

(a) Use the uniform random numbers in cells C17:C26 of Fig. 20.1 to simulate the evolution of the weather for 10 days, beginning the day after a clear day.
E **(b)** Now use a computer with the uniform random numbers generated by Excel to perform the simulation requested in part (*a*) on a spreadsheet.

20.1-3. Jessica Williams, manager of Kitchen Appliances for the Midtown Department Store, feels that her inventory levels of stoves have been running higher than necessary. Before revising the inventory policy for stoves, she records the number sold each day over a period of 25 days, as summarized below.

Number sold	2	3	4	5	6
Number of days	4	7	8	5	1

(a) Use these data to estimate the probability distribution of daily sales.
(b) Calculate the mean of the distribution obtained in part (*a*).

(c) Describe how uniform random numbers can be used to simulate daily sales.
(d) Use the uniform random numbers 0.4476, 0.9713, and 0.0629 to simulate daily sales over 3 days. Compare the average with the mean obtained in part (*b*).
E **(e)** Formulate a spreadsheet model for performing a simulation of the daily sales. Perform 300 replications and obtain the average of the sales over the 300 simulated days.

20.1-4. The William Graham Entertainment Company will be opening a new box office where customers can come to make ticket purchases in advance for the many entertainment events being held in the area. Simulation is being used to analyze whether to have one or two clerks on duty at the box office.

While simulating the beginning of a day at the box office, the first customer arrives 5 minutes after it opens and then the interarrival times for the next four customers (in order) are 3 minutes, 9 minutes, 1 minute, and 4 minutes, after which there is a long delay until the next customer arrives. The service times for these first five customers (in order) are 8 minutes, 6 minutes, 2 minutes, 4 minutes, and 7 minutes.

(a) For the alternative of a single clerk, plot a graph that shows the evolution of the number of customers at the box office over this period.
(b) Use this figure to estimate the usual measures of performance—L, L_q, W, W_q, and the P_n (as defined in Sec. 17.2)—for this queueing system.
(c) Repeat part (*a*) for the alternative of two clerks.
(d) Repeat part (*b*) for the alternative of two clerks.

20.1-5. Consider the $M/M/1$ queueing theory model that was discussed in Sec. 17.6 and Example 2, Sec. 20.1. Suppose that the mean arrival rate is 10 per hour, the mean service rate is 12 per hour, and you are required to estimate the expected waiting time before service begins by using simulation.
R **(a)** Starting with the system empty, use next-event incrementing to perform the simulation by hand until two service completions have occurred.
R **(b)** Starting with the system empty, use fixed-time incrementing (with 2 minutes as the time unit) to perform the simulation by hand until two service completions have occurred.
D,I **(c)** Use the interactive procedure for simulation in your IOR Tutorial (which incorporates next-event incrementing) to interactively execute a simulation run until 20 service completions have occurred.
Q **(d)** Use the Queueing Simulator to execute a simulation run with 10,000 customer arrivals.
E **(e)** Use the Excel template for this model in the Excel files for Chap. 17 to obtain the usual measures of performance for this queueing system. Then compare these exact results with the corresponding point estimates and 95 percent confidence intervals obtained from the simulation run in part (*d*). Identify any measure whose exact result falls outside the 95 percent confidence interval.

20.1-6. The Rustbelt Manufacturing Company employs a maintenance crew to repair its machines as needed. Management now wants a simulation study done to analyze what the size of the crew should be, where the crew sizes under consideration are 2, 3, and 4. The time required by the crew to repair a machine has a uniform distribution over the interval from 0 to twice the mean, where the mean depends on the crew size. The mean is 4 hours with two crew members, 3 hours with three crew members, and 2 hours with four crew members. The time between breakdowns of some machine has an exponential distribution with a mean of 5 hours. When a machine breaks down and so requires repair, management wants its average waiting time before repair begins to be no more than 3 hours. Management also wants the crew size to be no larger than necessary to achieve this.

(a) Develop a simulation model for this problem by describing its basic building blocks listed in Sec. 20.1 as they would be applied to this situation.

R (b) Consider the case of a crew size of 2. Starting with one machine needing repair, where this repair is starting just now, use next-event incrementing to perform the simulation by hand for 20 hours of simulated time.

R (c) Repeat part (b), but this time with fixed-time incrementing (with 1 hour as the time unit).

D,I (d) Use the interactive procedure for simulation in your IOR Tutorial (which incorporates next-event incrementing) to interactively execute a simulation run over a period of 10 breakdowns for each of the three crew sizes under consideration.

Q (e) Use the Queueing Simulator to simulate this system over a period of 10,000 breakdowns for each of the three crew sizes.

(f) Use the $M/G/1$ queueing model presented in Sec. 17.7 to obtain the expected waiting time W_q analytically for each of the three crew sizes. (You can either calculate W_q by hand or use the template for this model in the Excel files for Chap. 17.) Which crew size should be used?

20.1-7. While performing a simulation of a single-server queueing system, the number of customers in the system is 0 for the first 10 minutes, 1 for the next 17 minutes, 2 for the next 24 minutes, 1 for the next 15 minutes, 2 for the next 16 minutes, and 1 for the next 18 minutes. After this total of 100 minutes, the number becomes 0 again. Based on these results for the first 100 minutes, perform the following analysis (using the notation for queueing models introduced in Sec. 17.2).

(a) Plot a graph showing the evolution of the number of customers in the system over these 100 minutes.

(b) Develop estimates of P_0, P_1, P_2, P_3.

(c) Develop estimates of L and L_q.

(d) Develop estimates of W and W_q.

20.1-8. View the first demonstration example (*Simulating a Basic Queueing System*) in the simulation area of your OR Tutor.

D,I (a) Enter this *same problem* into the interactive procedure for simulation in your IOR Tutorial. Interactively execute a simulation run for 20 minutes of simulated time.

Q (b) Use the Queueing Simulator with 5,000 customer arrivals to estimate the usual measures of performance for this queueing system under the current plan to provide two tellers.

Q (c) Repeat part (b) if three tellers were to be provided.

Q (d) Now perform some sensitivity analysis by checking the effect if the level of business turns out to be even higher than projected. In particular, assume that the average time between customer arrivals turns out to be only 0.9 minute instead of 1.0 minute. Evaluate the alternatives of two tellers and three tellers under this assumption.

(e) Suppose *you* were the manager of this bank. Use your simulation results as the basis for a managerial decision on how many tellers to provide. Justify your answer.

D,I **20.1-9.** View the second demonstration example (*Simulating a Queueing System with Priorities*) in the simulation area of your OR Tutor. Then enter this *same problem* into the interactive procedure for simulation in your IOR Tutorial. Interactively execute a simulation run for 20 minutes of simulated time.

20.1-10.* Hugh's Repair Shop specializes in repairing German and Japanese cars. The shop has two mechanics. One mechanic works on only German cars and the other mechanic works on only Japanese cars. In either case, the time required to repair a car has an exponential distribution with a mean of 0.2 day. The shop's business has been steadily increasing, especially for German cars. Hugh projects that, by next year, German cars will arrive randomly to be repaired at a mean rate of 4 per day, so the time between arrivals will have an exponential distribution with a mean of 0.25 day. The mean arrival rate for Japanese cars is projected to be 2 per day, so the distribution of interarrival times will be exponential with a mean of 0.5 day.

For either kind of car, Hugh would like the expected waiting time in the shop before the repair is completed to be no more than 0.5 day.

(a) Formulate a simulation model for performing a simulation to estimate what the expected waiting time until repair is completed will be next year for either kind of car.

D,I (b) Considering only German cars, use the interactive procedure for simulation in your IOR Tutorial to interactively perform this simulation over a period of 10 arrivals of German cars.

Q (c) Use the Queueing Simulator to perform this simulation for German cars over a period of 10,000 car arrivals.

Q (d) Repeat part (c) for Japanese cars.

D,I (e) Hugh is considering hiring a second mechanic who specializes in German cars so that two such cars can be repaired simultaneously. (Only one mechanic works on any one car.) Repeat part (b) for this option.

Q (f) Use the Queueing Simulator with 10,000 arrivals of German cars to evaluate the option described in part (e).

Q (g) Another option is to train the two current mechanics to work on either kind of car. This would increase the expected repair time by 10 percent, from 0.2 day to 0.22 day. Use the Queueing Simulator with 20,000 arrivals of cars of either kind to evaluate this option.

(h) Because both the interarrival-time and service-time distributions are exponential, the $M/M/1$ and $M/M/s$ queueing models introduced in Sec. 17.6 can be used to evaluate all the above options analytically. Use these models to determine W, the expected waiting time until repair is completed, for each of the cases considered in parts (c), (d), (f), and (g). (You can either calculate W by hand or use the template for the $M/M/s$ model in the Excel files for Chap. 17.) For each case, compare the estimate of W obtained by simulation with the analytical value. What does this say about the number of car arrivals that should be included in the simulation?

(i) Based on the above results, which option would you select if you were Hugh? Why?

20.1-11. Vistaprint produces monitors and printers for computers. In the past, only some of them were inspected on a sampling basis. However, the new plan is that they all will be inspected before they are released. Under this plan, the monitors and printers will be brought to the inspection station one at a time as they are completed. For monitors, the interarrival time will have a uniform distribution between 10 and 20 minutes. For printers, the interarrival time will be a constant 15 minutes.

The inspection station has two inspectors. One inspector works on only monitors and the other one only inspects printers. In either case, the inspection time has an exponential distribution with a mean of 10 minutes.

Before beginning the new plan, management wants an evaluation made of how long the monitors and printers will be held up waiting at the inspection station.

(a) Formulate a simulation model for performing a simulation to estimate the expected waiting times (both before beginning inspection and after completing inspection) for either the monitors or the printers.

D,I (b) Considering only the monitors, use the interactive procedure for simulation in your IOR Tutorial to interactively perform this simulation over a period of 10 arrivals of monitors.

D,I (c) Repeat part (b) for the printers.

Q (d) Use the Queueing Simulator to repeat parts (b) and (c) with 10,000 arrivals in each case.

Q (e) Management is considering the option of providing new inspection equipment to the inspectors. This equipment would not change the expected time to perform an inspection but it would decrease the variability of the times. In particular, for either product, the inspection time would have an Erlang distribution with a mean of 10 minutes and shape parameter $k = 4$. Use the Queueing Simulator to repeat part (d) under this option. Compare the results with those obtained in part (d).

20.2-1. Read the referenced article that fully describes the OR study summarized in the application vignette presented in Sec. 20.2. Briefly describe how simulation was applied in this study. Then list the various financial and nonfinancial benefits that resulted from this study.

20.2-2. Section 20.2 introduced actual applications of simulation that are described in Selected References A1 and A5. Select one of these applications and read the corresponding article. Write a two-page summary of the application and the benefits it provided.

20.3-1.* Use the mixed congruential method to generate the following sequences of random numbers.

(a) A sequence of 10 *one-digit* random integer numbers such that $x_{n+1} \equiv (x_n + 3)$ (modulo 10) and $x_0 = 2$

(b) A sequence of eight random integer numbers between 0 and 7 such that $x_{n+1} \equiv (5x_n + 1)$ (modulo 8) and $x_0 = 1$

(c) A sequence of five *two-digit* random integer numbers such that $x_{n+1} \equiv (61x_n + 27)$ (modulo 100) and $x_0 = 10$

20.3-2. Reconsider Prob. 20.3-1. Suppose now that you want to convert these random integer numbers to (approximate) uniform random numbers. For each of the three parts, give a formula for this conversion that makes the approximation as close as possible.

20.3-3. Use the mixed congruential method to generate a sequence of five *two-digit* random integer numbers such that $x_{n+1} \equiv (11x_n + 23)$ (modulo 100) and $x_0 = 52$.

20.3-4. Use the mixed congruential method to generate a sequence of three *three-digit* random integer numbers such that $x_{n+1} \equiv (201x_n + 503)$ (modulo 1,000) and $x_0 = 485$.

20.3-5. You need to generate five uniform random numbers.

(a) Prepare to do this by using the mixed congruential method to generate a sequence of five random integer numbers between 0 and 31 such that $x_{n+1} \equiv (13x_n + 15)$ (modulo 32) and $x_0 = 14$.

(b) Convert these random integer numbers to uniform random numbers as closely as possible.

20.3-6. You are given the *multiplicative congruential generator* $x_0 = 1$ and $x_{n+1} \equiv 7x_n$ (modulo 13) for $n = 0, 1, 2, \ldots$.

(a) Calculate x_n for $n = 1, 2, \ldots, 12$.

(b) How often does each integer between 1 and 12 appear in the sequence generated in part (a)?

(c) Without performing additional calculations, indicate how x_{13}, $x_{14}, \ldots$ will compare with $x_1, x_2, \ldots$.

20.4-1. Reconsider the coin flipping game introduced in Sec. 20.1 and analyzed with simulation in Figs. 20.1, 20.2, and 20.3.

(a) Simulate one play of this game by repeatedly flipping your own coin until the game ends. Record your results in the format shown in columns B, D, E, F, and G of Fig. 20.1. How much would you have won or lost if this had been a real play of the game?

E (b) Revise the spreadsheet model in Fig. 20.1 by using Excel's VLOOKUP function instead of the IF function to generate each simulated flip of the coin. Then perform a simulation of one play of the game.

E (c) Use this revised spreadsheet model to generate a data table with 14 replications like Fig. 20.2.

E (d) Repeat part (c) with 1,000 replications (like Fig. 20.3).

20.4-2.* Apply the inverse transformation method as indicated next to generate three random observations from the uniform distribution between -10 and 40 by using the following uniform random numbers: 0.0965, 0.5692, 0.6658.

(a) Apply this method graphically.

(b) Apply this method algebraically.

(c) Write the equation that Excel would use to generate each such random observation.

R **20.4-3.** Obtaining uniform random numbers as instructed at the beginning of the Problems section, generate three random observations from each of the following probability distributions.

(a) The uniform distribution from 25 to 75.

(b) The distribution whose probability density function is

$$f(x) = \begin{cases} \dfrac{1}{4}(x+1)^3 & \text{if } -1 \le x \le 1 \\ 0 & \text{otherwise.} \end{cases}$$

(c) The distribution whose probability density function is

$$f(x) = \begin{cases} \dfrac{1}{200}(x-40) & \text{if } 40 \le x \le 60 \\ 0 & \text{otherwise.} \end{cases}$$

R **20.4-4.** Obtaining uniform random numbers as instructed at the beginning of the Problems section, generate three random observations from each of the following probability distributions.

(a) The random variable X has $P\{X = 0\} = \frac{1}{2}$. Given $X \ne 0$, it has a uniform distribution between -5 and 15.

(b) The distribution whose probability density function is

$$f(x) = \begin{cases} x - 1 & \text{if } 1 \le x \le 2 \\ 3 - x & \text{if } 2 \le x \le 3. \end{cases}$$

(c) The geometric distribution with parameter $p = \frac{1}{3}$, so that

$$P\{X = k\} = \begin{cases} \dfrac{1}{3}\left(\dfrac{2}{3}\right)^{k-1} & \text{if } k = 1, 2, \dots \\ 0 & \text{otherwise.} \end{cases}$$

20.4-5. Each time an unbiased coin is flipped three times, the probability of getting 0, 1, 2, and 3 heads is $\frac{1}{8}, \frac{3}{8}, \frac{3}{8}$, and $\frac{1}{8}$, respectively. Therefore, with eight groups of three flips each, *on the average,* one group will yield 0 heads, three groups will yield 1 head, three groups will yield 2 heads, and one group will yield 3 heads.

(a) Using your own coin, flip it 24 times divided into eight groups of three flips each, and record the number of groups with 0 head, with 1 head, with 2 heads, and with 3 heads.

(b) Obtaining uniform random numbers as instructed at the beginning of the Problems section, simulate the flips specified in part (*a*) and record the information indicated in part (*a*).

E **(c)** Formulate a spreadsheet model for performing a simulation of three flips of the coin and recording the number of heads. Perform one replication of this simulation.

E **(d)** Use this spreadsheet to generate a data table with 8 replications of the simulation. Compare this frequency distribution of the number of heads with the probability distribution of the number of heads with three flips.

E **(e)** Repeat part (*d*) with 800 replications.

20.4-6.* The game of craps requires the player to throw two dice one or more times until a decision has been reached as to whether he (or she) wins or loses. He wins if the first throw results in a

sum of 7 or 11 or, alternatively, if the first sum is 4, 5, 6, 8, 9, or 10 and the same sum reappears before a sum of 7 has appeared. Conversely, he loses if the first throw results in a sum of 2, 3, or 12 or, alternatively, if the first sum is 4, 5, 6, 8, 9, or 10 and a sum of 7 appears before the first sum reappears.

E **(a)** Formulate a spreadsheet model for performing a simulation of the throw of two dice. Perform one replication.

E **(b)** Perform 25 replications of this simulation.

(c) Trace through these 25 replications to determine both the number of times the simulated player would have won the game of craps and the number of losses when each play starts with the next throw after the previous play ends. Use this information to calculate a preliminary estimate of the probability of winning a single play of the game.

(d) For a large number of plays of the game, the proportion of wins has *approximately* a normal distribution with mean = 0.493 and standard deviation = $0.5\sqrt{n}$. Use this information to calculate the number of simulated plays that would be required to have a probability of at least 0.95 that the proportion of wins will be less than 0.5.

R **20.4-7.** Obtaining uniform random numbers as instructed at the beginning of the Problems section, use the inverse transformation method and the table of the normal distribution given in Appendix 5 (with linear interpolation between values in the table) to generate 10 random observations (to three decimal places) from a normal distribution with mean = 1 and variance = 4. Then calculate the sample average of these random observations.

R **20.4-8.** Obtaining uniform random numbers as instructed at the beginning of the Problems section, generate three random observations (approximately) from a normal distribution with mean = 5 and standard deviation = 10.

(a) Do this by applying the central limit theorem, using three uniform random numbers to generate each random observation.

(b) Now do this by using the table for the normal distribution given in Appendix 5 and applying the inverse transformation method.

R **20.4-9.** Obtaining uniform random numbers as instructed at the beginning of the Problems section, generate four random observations (approximately) from a normal distribution with mean = 0 and standard deviation = 1.

(a) Do this by applying the central limit theorem, using three uniform random numbers to generate each random observation.

(b) Now do this by using the table for the normal distribution given in Appendix 5 and applying the inverse transformation method.

(c) Use your random observations from parts (*a*) and (*b*) to generate random observations from a chi-square distribution with 2 degrees of freedom.

R **20.4-10.** Obtaining uniform random numbers as instructed at the beginning of the Problems section, generate two random observations from each of the following probability distributions.

(a) The exponential distribution with mean = 10

(b) The Erlang distribution with mean = 10 and shape parameter $k = 2$ (that is, standard deviation = $2\sqrt{2}$)

(c) The normal distribution with mean = 10 and standard deviation = $2\sqrt{2}$. (Use the central limit theorem and $n = 6$ for each observation.)

20.4-11. Richard Collins, manager and owner of Richard's Tire Service, wishes to use simulation to analyze the operation of his shop. One of the activities to be included in the simulation is the installation of automobile tires (including balancing the tires). Richard estimates that the cumulative distribution function (CDF) of the probability distribution of the time (in minutes) required to install a tire has the graph shown below.

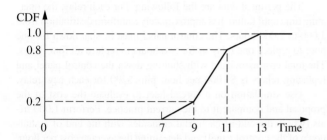

(a) Use the inverse transformation method to generate five random observations from this distribution when using the following five uniform random numbers: 0.2655, 0.3472, 0.0248, 0.9205, 0.6130.
(b) Use a nested IF function to write an equation that Excel can use to generate each random observation from this distribution.

R **20.4-12.** Obtaining uniform random numbers as instructed at the beginning of the Problems section, generate four random observations from an exponential distribution with mean = 20. Then use these four observations to generate one random observation from an Erlang distribution with mean = 4 and shape parameter $k = 4$.

20.4-13. Let $r_1, r_2, \ldots, r_n$ be uniform random numbers. Define $x_i = -\ln r_i$ and $y_i = -\ln (1 - r_i)$, for $i = 1, 2, \ldots, n$, and $z = \sum_{i=1}^{n} x_i$. Label each of the following statements as true or false, and then justify your answer.
(a) The numbers $x_1, x_2, \ldots, x_n$ and $y_1, y_2, \ldots, y_n$ are random observations from the same exponential distribution.
(b) The average of $x_1, x_2, \ldots, x_n$ is equal to the average of $y_1, y_2, \ldots, y_n$.
(c) z is a random observation from an Erlang (gamma) distribution.

20.4-14. Consider the discrete random variable X that is uniformly distributed (equal probabilities) on the set $\{1, 2, \ldots, 8\}$. You wish to generate a series of random observations x_i ($i = 1, 2, \ldots$) of X. The following three proposals have been made for doing this. For each one, analyze whether it is a valid method and, if not, how it can be adjusted to become a valid method.
(a) Proposal 1: Generate uniform random numbers r_i ($i = 1, 2, \ldots$), and then set $x_i = n$, where n is the integer satisfying $n/8 \leq r_i < (n + 1)/8$.

(b) Proposal 2: Generate uniform random numbers r_i ($i = 1, 2, \ldots$), and then set x_i equal to the greatest integer less than or equal to $1 + 8r_i$.
(c) Proposal 3: Generate x_i from the mixed congruential generator $x_{n+1} \equiv (5x_n + 7)$ (modulo 8), with starting value $x_0 = 4$.

R **20.4-15.** Obtaining uniform random numbers as instructed at the beginning of the Problems section, use the acceptance-rejection method to generate three random observations from the triangular distribution used to illustrate this method in Sec. 20.4.

R **20.4-16.** Obtaining uniform random numbers as instructed at the beginning of the Problems section, use the acceptance-rejection method to generate three random observations from the probability density function

$$f(x) = \begin{cases} \dfrac{1}{50}(x - 10) & \text{if } 10 \leq x \leq 20 \\ 0 & \text{otherwise.} \end{cases}$$

R **20.4-17.** An insurance company insures four large risks. The number of losses for each risk is independent and identically distributed on the points $\{0, 1, 2\}$ with probabilities 0.7, 0.2, and 0.1, respectively. The size of an individual loss has the following cumulative distribution function:

$$F(x) = \begin{cases} \dfrac{\sqrt{x}}{20} & \text{if } 0 \leq x \leq 100 \\ \dfrac{x}{200} & \text{if } 100 < x \leq 200 \\ 1 & \text{if } x > 200. \end{cases}$$

Obtaining uniform random numbers as instructed at the beginning of the Problems section, perform a simulation experiment twice of the total loss generated by the four large risks.

20.4-18. A company provides its three employees with health insurance under a group plan. For each employee, the probability of incurring medical expenses during a year is 0.9, so the number of employees incurring medical expenses during a year has a binomial distribution with $p = 0.9$ and $n = 3$. Given that an employee incurs medical expenses during a year, the total amount for the year has the distribution $100 with probability 0.9 or $10,000 with probability 0.1. The company has a $5,000 deductible clause with the insurance company so that each year the insurance company pays the total medical expenses for the group in excess of $5,000. Use the uniform random numbers 0.01 and 0.20, in the order given, to generate the number of claims based on a binomial distribution for each of 2 years. Use the following uniform random numbers, in the order given, to generate the amount of each claim: 0.80, 0.95, 0.70, 0.96, 0.54, 0.01. Calculate the total amount that the insurance company pays for 2 years.

20.5-1. Read the referenced article that fully describes the OR study summarized in the application vignette presented in Sec. 20.5. Briefly describe how simulation was applied in this study. Then list the various financial and nonfinancial benefits that resulted from this study.

A **20.6-1.** The results from a simulation run are inherently random. This problem will demonstrate this fact and investigate the impact of the number of trials on this randomness. Consider the example involving Freddie the newsboy that was introduced in Sec. 20.6. The spreadsheet model is available in this chapter's Excel files on the book's website. When using Crystal Ball, make sure that the "Use Same Sequence of Random Numbers" option is *not* checked and that the Monte-Carlo Sampling Method is selected in the Sampling tab of Run Preferences. Use an order quantity of 60.

(a) Set the number of trials to 100 in Run Preferences and run the simulation of Freddie's problem five times. Note the mean profit for each simulation run.

(b) Repeat part (*a*) except set the number of trials to 1,000 in Run Preferences.

(c) Compare the results from part (*a*) and part (*b*) and comment on any differences.

A **20.6-2.** The Aberdeen Development Corporation (ADC) is reconsidering the Aberdeen Resort Hotel project. It would be located on the picturesque banks of Grays Harbor and have its own championship-level golf course.

The cost to purchase the land would be $1 million, payable now. Construction costs would be approximately $2 million, payable at the end of the year 1. However, the construction costs are uncertain. These costs could be up to 20 percent higher or lower than the estimate of $2 million. Assume that the construction costs would follow a triangular distribution.

ADC is very uncertain about the annual operating profits (or losses) that would be generated once the hotel is constructed. Its best estimate for the annual operating profit that would be generated in years 2, 3, 4, and 5 is $700,000. Due to the great uncertainty, the estimate of the standard deviation of the annual operating profit in each year also is $700,00. Assume that the yearly profits are statistically independent and follow the normal distribution.

After year 5, ADC plans to sell the hotel. The selling price is likely to be somewhere between $4 and $8 million (assume a uniform distribution). ADC uses a 10 percent discount rate for calculating net present value. (For purposes of this calculation, assume that each year's profits are received at year end.) Perform 1,000 trials of a simulation of this project on a spreadsheet.

(a) What is the mean net present value (NPV) of the project? (*Hint:* The NPV(rate, cash stream) function in Excel returns the NPV of a stream of cash flows assumed to start one year from now. For example, NPV(10%, C5:F5) returns the NPV at a 10 percent discount rate when C5 is a cash flow at the end of year 1, D5 at the end of year 2, E5 at the end of year 3, and F5 at the end of year 4.)

(b) What is the estimated probability that the project will yield an NPV greater than $2 million?

(c) ADC also is concerned about cash flow in years 2, 3, 4, and 5. Generate a forecast of the distribution of the *minimum* annual operating profit (undiscounted) earned in any of the four years. What is the mean value of the minimum annual operating profit over the four years?

(d) What is the probability that the annual operating profit will be at least $0 in all four years of operation?

A **20.6-3.** The Avery Co. factory has been having a maintenance problem with the control panel for one of its production processes. This control panel contains four identical electromechanical relays that have been the cause of the trouble. The problem is that the relays fail fairly frequently, thereby forcing the control panel (and the production process it controls) to be shut down while a replacement is made. The current practice is to replace the relays only when they fail. The average total cost of doing this has been $3.19 per hour. To attempt to reduce this cost, a proposal has been made to replace all four relays whenever any one of them fails to reduce the frequency with which the control panel must be shut down. Would this actually reduce the cost?

The pertinent data are the following. For each relay, the operating time until failure has approximately a uniform distribution from 1,000 to 2,000 hours. The control panel must be shut down for one hour to replace one relay or for two hours to replace all four relays. The total cost associated with shutting down the control panel and replacing relays is $1,000 per hour plus $200 for each new relay.

Use simulation on a spreadsheet to evaluate the cost of the proposal and compare it to the current practice. Perform 1,000 trials (where the end of each trial coincides with the end of a shutdown of the control panel) and determine the average cost per hour.

A **20.6-4.** For one new product to be produced by the Aplus Company, bushings will need to be drilled into a metal block and cylindrical shafts inserted into the bushings. The shafts are required to have a radius of at least 1.0000 inch, but the radius should be as little larger than this as possible. With the proposed production process for producing the shafts, the probability distribution of the radius of a shaft has a triangular distribution with a minimum of 1.0000 inch, a most likely value of 1.0010 inches, and a maximum value of 1.0020 inches. With the proposed method of drilling the bushings, the probability distribution of the radius of a bushing has a normal distribution with a mean of 1.0020 inches and a standard deviation of 0.0010 inch. The clearance between a bushing and a shaft is the difference in their radii. Because they are selected at random, there occasionally is interference (i.e., negative clearance) between a bushing and a shaft to be mated.

Management is concerned about the disruption in the production of the new product that would be caused by this occasional interference. Perhaps the production processes for the shafts and bushings should be improved (at considerable cost) to lessen the chance of interference. To evaluate the need for such improvements, management has asked you to determine how frequently interference would occur with the currently proposed production processes.

Estimate the probability of interference by performing 500 trials of a simulation on a spreadsheet.

A **20.6-5.** Reconsider Prob. 20.4-6 involving the game of craps. Now the objective is to estimate the probability of winning a play of this game. If the probability is greater than 0.5, you will want to go to Las Vegas to play the game numerous times until you eventually win a considerable amount of money. However, if the probability is less than 0.5, you will stay home.

You have decided to perform simulation on a spreadsheet to estimate this probability. Perform the number of trials (plays of the game) indicated below *twice*.

(a) 100 trials.

(b) 1,000 trials.

(c) 10,000 trials.

(d) The true probability is 0.493. What conclusion do you draw from the above simulation runs about the number of trials that appears to be needed to give reasonable assurance of obtaining an estimate that is within 0.007 of the true probability?

A **20.6-6.** Consider the example involving Freddie the newsboy that was introduced in Sec. 20.6. The spreadsheet model is available in this chapter's Excel files on the book's website. The Decision Table generated in Sec. 20.6 (see Fig. 20.19) for Freddie's problem suggests that 55 is the best order quantity, but this table only considered order quantities that were a multiple of 5. Refine

the search by generating a Decision Table for Freddie's problem that considers all integer order quantities between 50 and 60.

20.7-1. From the bottom part of the Selected References given at the end of the chapter, select one of these award-winning applications of simulation. Read this article and then write a two-page summary of the application and the benefits (including nonfinancial benefits) it provided.

20.7-2. From the bottom part of the Selected References given at the end of the chapter, select three of these award-winning applications of simulation. For each one, read the article and then write a one-page summary of the application and the benefits (including nonfinancial benefits) it provided.

■ CASES

CASE 20.1   Reducing In-Process Inventory, Revisited

Reconsider case 17.1. The current and proposed queueing systems in this case were to be analyzed with the help of queueing models to determine how to reduce in-process inventory as much as possible. However, these same queueing systems also can be effectively analyzed by applying simulation with the help of the Queueing Simulator in your OR Courseware.

Use simulation to perform all the analysis requested in this case.

CASE 20.2   Action Adventures

The Adventure Toys Company manufactures a popular line of action figures and distributes them to toy stores at the wholesale price of $10 per unit. Demand for the action figures is seasonal, with the highest sales occurring before Christmas and during the spring. The lowest sales occur during the summer and winter (post-Christmas) months.

Each month the monthly "base" sales follow a normal distribution with mean equal to the previous month's actual "base" sales and with a standard deviation of 500 units. The actual sales in any month are the monthly base sales multiplied by the seasonality factor for the month, as shown in the table below. Base sales in December 2009 were 6,000, with actual sales equal to $(1.18)(6,000) = 7,080$. It is now January 1, 2010.

Month	Seasonality Factor	Month	Seasonality Factor
January	0.79	July	0.74
February	0.88	August	0.98
March	0.95	September	1.06
April	1.05	October	1.10
May	1.09	November	1.16
June	0.84	December	1.18

Cash sales typically account for about 40 percent of monthly sales, but this figure has been as low as 28 percent and as high as 48 percent in some months. The remainder of the sales are made on a 30-day interest-free credit basis, with full payment received one month after delivery. In December 2009, 42 percent of sales were cash sales and 58 percent were on credit.

The production costs depend upon the labor and material costs. The plastics required to manufacture the action figures fluctuate in price from month to month, depending on market conditions. Because of these fluctuations, production costs can be anywhere from $6 to $8 per unit. In addition to these variable production costs, the company incurs a fixed cost of $15,000 per month for manufacturing the action figures. The company assembles the products to order. When a batch of a particular action figure is ordered, it is immediately manufactured and shipped within a couple days.

The company utilizes eight molding machines to mold the action figures. These machines occasionally break down and require a $5,000 replacement part. Each machine requires a replacement part with a 10 percent probability each month.

The company has a policy of maintaining a minimum cash balance of at least $20,000 at the end of each month. The balance at the end of December 2009 (or equivalently, at the beginning of January 2010) is $25,000. If required, the company will take out a short-term (1 month) loan to cover expenses and maintain the minimum balance. The loans must be paid back the following month with interest (using the current month's loan interest rate). For example, if March's annual interest rate is 6 percent (so 0.5 percent per month) and a $1,000 loan is taken out in March, then $1,005 is due in April. However, a new loan can be taken out each month.

Any balance remaining at the end of a month (including the minimum balance) is carried forward to the following month, and also earns savings interest. For example, if the ending balance in March is $20,000, and March's savings interest is 3 percent per annum (so 0.25 percent per month), then $50 of savings interest is earned in April.

Both the loan interest rate and the savings interest rate are set monthly based upon the Prime rate. The loan interest rate is set at Prime + 2 percent, while the savings interest rate is set at Prime − 2 percent. However, the loan interest rate is capped at (can't exceed) 9 percent and the savings interest rate will never drop below 2 percent.

The Prime rate in December 2009 was 5 percent per annum. This rate depends upon the whims of the Federal Reserve Board. In particular, for each month there is a 70 percent chance it will stay unchanged, a 10 percent chance it will increase by 25 basis points (0.25 percent), a 10 percent chance it will decrease by 25 basis points, a 5 percent chance it will increase by 50 basis points, and a 5 percent chance it will decrease by 50 basis points.

(a) Formulate a simulation model on a spreadsheet to track the company's cash flows from month to month. Indicate the probability distributions (both the type and the parameters) for the assumption cells directly on the spreadsheet. Simulate 1,000 trials for the year 2010, and paste your results in the spreadsheet.

(b) Adventure Toys management wants information about what the company's net worth might be at the end of 2010, including the likelihood that the net worth will exceed zero. (The net worth is defined here as the ending cash balance *plus* savings interest and account receivables *minus* any loans and interest due.) Display the results of your simulation run from part (*a*) in the various forms that you think would be helpful to management in analyzing this issue.

(c) Arrangements need to be made to obtain a specific credit limit from the bank for the short-term loans that might be needed during 2010. Therefore, Adventure Toys management also would like information regarding the size of the largest short-term loan that might be needed during 2010. Display the results of your simulation run from part (*a*) in the various forms that you think would be helpful to management in analyzing this issue.

■ PREVIEWS OF ADDED CASES ON OUR WEBSITE (www.mhhe.com/hillier)

CASE 20.3 Planning Planers

A factory's planer department has had a difficult time keeping up with its workload, which has seriously disrupted the production schedule for subsequent operations. At times, the work pours in and a big backlog builds up. Then there might be a long pause when not much comes in, so the planers stand idle part of the time. Three separate proposals have been made to relive the bottleneck in the planer department: (1) obtain one additional planer, (2) eliminate the variability of the interarrival times of the jobs, and (3) reduce the variability of the time required to perform the jobs. Any one or any combination of these proposals can be adopted. With the help of the Queueing Simulator, simulation is to be used to determine what should be done so as to minimize the expected total cost per hour.

CASE 20.4 Pricing under Pressure

A client of a large investment bank is interested in purchasing a European call option for a certain stock that provides him with the right to purchase the stock at a fixed price 12 weeks from today. The client then would exercise this option in 12 weeks only if this fixed price is less than the market price of the stock at that time. The bank now needs to determine what price should be charged for the call option. This price should be the mean value of the option in 12 weeks. Based on a random walk model of how a stock price evolves from week to week, simulation is to be used to estimate this mean value. To start, the various elements of a simulation model need to be carefully formulated.

Documentation for the OR Courseware

You will find a wealth of software resources on the book's website (www.mhhe.com/hillier). The entire software package is called *OR Courseware*.

The individual software packages are discussed briefly below.

OR TUTOR

OR Tutor is a Web document consisting of a set of HTML pages that often contain JavaScript. Any browser that supports JavaScript can be used. It can be viewed with either an IBM-compatible PC or a Macintosh.

This resource has been designed to be your personal tutor by illustrating and illuminating key concepts in an interactive manner. It contains 16 *demonstration examples* that supplement the examples in the book in ways that cannot be duplicated on the printed page. Each one vividly demonstrates one of the algorithms or concepts of OR in action. Most combine an *algebraic description* of each step with a *geometric display* of what is happening. Some of these geometric displays become quite dynamic, with moving points or moving lines, to demonstrate the evolution of the algorithm. The demonstration examples also are integrated with the book, using the same notation and terminology, with references to material in the book, etc. Students find them an enjoyable and effective learning aid.

IOR TUTORIAL

Another key tutorial feature of the OR Courseware is a software package called *Interactive Operations Research Tutorial,* or *IOR Tutorial* for short. A product of Accelet Corporation, it has been designed specifically for use with this book. Innovative tutorial features are employed to make the process of learning the algorithms in the book as efficient and enjoyable as possible. It is implemented in Java 2, so it can operate on any platform.

IOR Tutorial features a large number of *interactive procedures* for the various topic areas covered in the book. Each of these interactive procedures enables you to *interactively execute* one of the algorithms of OR. While viewing all relevant information on the computer screen, you make the decision on how the next step of the algorithm should be performed, and then the computer does all the necessary number crunching to execute that step. When a previous mistake is discovered, the procedure allows you to quickly backtrack to correct the mistake. To get you started properly, the computer points out any mistake made on the first iteration (where possible). When done, you can print out all the work performed to turn in for homework.

In our judgment, these interactive procedures provide the "right" way in this computer age for students to do homework designed to help them learn the algorithms of OR. The procedures enable you to focus on concepts rather than mindless number crunching, thereby making the learning process far more efficient and effective as well as stimulating. They also point you in the right direction, including organizing the work to be done. However, the procedures do not do the thinking for you. As in any good homework assignment, you are allowed to make mistakes (and to learn from those mistakes), so that hard thinking will need to be done to try to stay on the right path. We have been careful in designing the division of labor between the computer and the student to provide an efficient, complete learning process.

991

Once you have learned the logic of a particular algorithm with the help of an interactive procedure, you will want to be able to apply the algorithm quickly with an automatic procedure thereafter. Such a procedure is provided by one or more of the software packages discussed below for most of the algorithms described in this book. However, for certain algorithms that are not included in these commercial packages (as well as a few that are), we have provided special automatic procedures in IOR Tutorial. These procedures are designed only for solving the textbook-size problems in the book.

EXCEL FILES

The OR Courseware includes separate Excel files for nearly every chapter in this book. The files for each chapter typically include several spreadsheets that will help you formulate and solve the various kinds of models described in the chapter. Two types of spreadsheets are included. First, each time an example is presented that can be solved using Excel, the complete spreadsheet formulation and solution is given in that chapter's Excel files. This provides a convenient reference, or even useful templates, when you set up spreadsheets to solve similar problems with the Excel Solver (or the Premium Solver discussed in the next subsection). Second, for many of the models in the book, template files are provided that already include all the equations necessary to solve the model. You simply enter the data for the model and the solution is immediately calculated.

EXCEL ADD-INS

Four Excel add-ins are included in OR Courseware. One is *Premium Solver for Education,* which is a more powerful version of the standard Solver in Excel and also adds Evolutionary Solver discussed in Sec. 12.10. See the book's website for instructions for how to download this add-in from the website (www.solver.com) of the developer (Frontline Systems Inc.), using both the textbook code (HLITOR) and a course code that needs to be obtained by the instructor (following instructions on our website).

Three other Excel add-ins are academic versions of *SensIt* (introduced in Sec. 15.5), *TreePlan* (introduced in Sec. 15.5), and *RiskSim* (introduced in Sec. 20.6). All are shareware developed by Professor Michael R. Middleton for Windows and Macintosh. Documentation is included on the book's website for all three add-ins. Since this software is shareware, those desiring to use it after the course should register and pay the shareware fee.

As with any Excel add-in, each of these add-ins needs to be installed in Excel before it is operational. (The same is true for the standard Excel Solver.) Installation instructions are included in the OR Courseware.

MPL/CPLEX

As discussed at length in Secs. 3.6 and 4.8, MPL is a state-of-the-art modeling language and its prime solver CPLEX is a particularly prominent and powerful solver. Several other powerful solvers (described in the next paragraph) also are available with MPL. The student version of the latest releases of MPL, CPLEX, and these other solvers is included in the OR Courseware. Although this student version is limited to *much* smaller problems than the massive linear, integer, and quadratic programming problems commonly solved in practice by the full version, it still can handle *far* larger problems than any you will encounter in this book.

The book's website provides an extensive MPL tutorial and documentation, as well as MPL/CPLEX formulations and solutions for virtually every example in the book to which they can be applied. Also included in the OR Courseware is the student version of OptiMax 2000, which enables fully integrating MPL models into Excel and solving with CPLEX. In addition, the convex programming solver CONOPT, the global oprimizer LGO, the linear and integer programming solver CoinMP, the linear, integer, and quadratic programming solver LINDO, and the stochastic solver BendX are included in MPL for solving such problems.

The website for further exploring MPL and its solvers is www.maximalsoftware.com.

LINGO/LINDO FILES

This book also features the popular modeling language LINGO (see especially the end of Sec. 3.7, the supplements to Chap. 3, and Appendix 4.1), including the traditional LINDO syntax subset (see Sec. 4.8 and Appendix 4.1). A student version of LINGO (with the LINDO subset) is included in the OR Courseware. Updated student versions of LINGO/LINDO (as well as the companion spreadsheet solver *What's Best*) also can be downloaded from the website, www.lindo.com.

The OR Courseware includes extensive LINGO/LINDO files or (when LINDO is not relevant) LINGO files for many of the chapters. Each file provides the LINGO and LINDO models and solutions for the various examples in the chapter to which they can be applied. The book's website also provides LINGO and LINDO tutorials.

UPDATES

The software world evolves very rapidly during the lifetime of one edition of a textbook. We believe that the documentation provided in this appendix is accurate at the time of this writing, but changes inevitably will occur as time passes.

You can visit the book's website, www.mhhe.com/hillier, for information about software updates.

2

Convexity

As introduced in Chap. 12, the concept of *convexity* is frequently used in OR work, especially in the area of non-linear programming. Therefore, we further introduce the properties of convex or concave functions and convex sets here.

CONVEX OR CONCAVE FUNCTIONS OF A SINGLE VARIABLE

We begin with definitions.

> **Definitions:** A *function* of a single variable $f(x)$ is a **convex function** if, for *each* pair of values of x, say, x' and x'' ($x' < x''$),
>
> $$f[\lambda x'' + (1 - \lambda)x'] \le \lambda f(x'') + (1 - \lambda)f(x')$$
>
> for all values of λ such that $0 < \lambda < 1$. It is a **strictly convex function** if $\le$ can be replaced by $<$. It is a **concave function** (or a **strictly concave function**) if this statement holds when $\le$ is replaced by $\ge$ (or by $>$).

This definition of a convex function has an enlightening geometric interpretation. Consider the graph of the function $f(x)$ drawn as a function of x, as illustrated in Fig. A2.1 for a function $f(x)$ that decreases for $x < 1$, is constant for $1 \le x \le 2$, and increases for $x > 2$. Then $[x', f(x')]$ and $[x'', f(x'')]$ are two points on the graph of $f(x)$, and $[\lambda x'' + (1 - \lambda)x', \lambda f(x'') + (1 - \lambda)f(x')]$ represents the various points on the line segment between these two points (but excluding these endpoints) when $0 < \lambda < 1$. Thus, the $\le$ inequality in the definition indicates that this line segment lies entirely above or on the graph of the function, as in Fig. A2.1. Therefore, $f(x)$ is *convex* if, for *each* pair of points on the graph of $f(x)$, the line segment joining these two points lies entirely above or on the graph of $f(x)$.

For example, the particular choice of x' and x'' shown in Fig. A2.1 results in the entire line segment (except the two endpoints) lying *above* the graph of $f(x)$. This also occurs for other choices of x' and x'' where either $x' < 1$ or $x'' > 2$ (or both). If $1 \le x' < x'' \le 2$, then the entire line segment lies *on* the graph of $f(x)$. Therefore, this $f(x)$ is convex.

This geometric interpretation indicates that $f(x)$ is convex if it only "bends upward" whenever it bends at all. (This condition is sometimes referred to as *concave upward,* as opposed to *concave downward* for a concave function.) To be more precise, if $f(x)$ possesses a second derivative everywhere, then $f(x)$ is convex if and only if $d^2f(x)/dx^2 \ge 0$ for all possible values of x.

The definitions of a *strictly convex function,* a *concave function,* and a *strictly concave function* also have analogous geometric interpretations. These interpretations are summarized below in terms of the second derivative of the function, which provides a convenient test of the status of the function.

> **Convexity test for a function of a single variable:** Consider any function of a single variable $f(x)$ that possesses a second derivative at all possible values of x. Then $f(x)$ is
>
> 1. *Convex* if and only if $\dfrac{d^2f(x)}{dx^2} \ge 0$ for all possible values of x
> 2. *Strictly convex* if and only if $\dfrac{d^2f(x)}{dx^2} > 0$ for all possible values of x
> 3. *Concave* if and only if $\dfrac{d^2f(x)}{dx^2} \le 0$ for all possible values of x
> 4. *Strictly concave* if and only if $\dfrac{d^2f(x)}{dx^2} < 0$ for all possible values of x

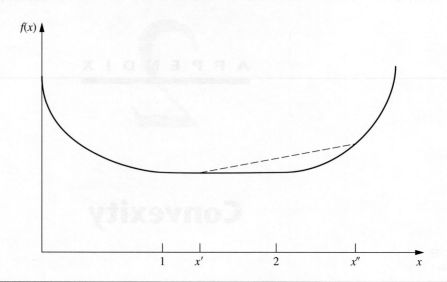

■ FIGURE A2.1
A convex function.

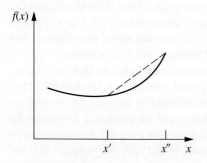

■ FIGURE A2.2
A strictly convex function.

Note that a strictly convex function also is convex, but a convex function is *not* strictly convex if the second derivative equals zero for some values of *x*. Similarly, a strictly concave function is concave, but the reverse need not be true.

Figures A2.1 to A2.6 show examples that illustrate these definitions and this convexity test.

Applying this test to the function in Fig. A2.1, we see that as *x* is increased, the slope (first derivative) either increases (for $0 \leq x < 1$ and $x > 2$) or remains constant (for $1 \leq x_1 \leq 2$). Therefore, the second derivative always is nonnegative, which verifies that the function is convex. However, it is *not* strictly convex because the second derivative equals zero for $1 \leq x \leq 2$.

However, the function in Fig. A2.2 is strictly convex because its slope always is increasing so its second derivative always is greater than zero.

The piecewise linear function shown in Fig. A2.3 changes its slope at $x = 1$. Consequently, it does not possess

a first or second derivative at this point, so the convexity test cannot be fully applied. (The fact that the second derivative equals zero for $0 \leq x < 1$ and $x > 1$ makes the function eligible to be either convex or concave, depending upon its behavior at $x = 1$.) Applying the definition of a concave function, we see that if $0 < x' < 1$ and $x'' > 1$ (as shown in Fig. A2.3), then the entire line segment joining $[x', f(x')]$ and $[x'', f(x'')]$ lies *below* the graph of $f(x)$, except for the two endpoints of the line segment. If either $0 \leq x' < x'' \leq 1$ or $1 \leq x' < x''$, then the entire line segment lies *on* the graph of $f(x)$. Therefore, $f(x)$ is concave (but *not* strictly concave).

The function in Fig. A2.4 is strictly concave because its second derivative always is less than zero.

As illustrated in Fig. A2.5, any linear function has its second derivative equal to zero everywhere and so is both convex and concave.

The function in Fig. A2.6 is *neither* convex nor concave because as *x* increases, the slope fluctuates between decreasing and increasing so the second derivative fluctuates between being negative and positive.

CONVEX OR CONCAVE FUNCTIONS OF SEVERAL VARIABLES

The concept of a convex or concave function of a single variable also generalizes to functions of more than one variable. Thus, if $f(x)$ is replaced by $f(x_1, x_2, \ldots, x_n)$, the definition still applies if *x* is replaced everywhere by $(x_1, x_2, \ldots, x_n)$. Similarly, the corresponding geometric interpretation is still valid after generalization of the concepts of *points* and *line segments*. Thus, just as a particular value of (x, y) is interpreted as a point in two-dimensional space, each possible

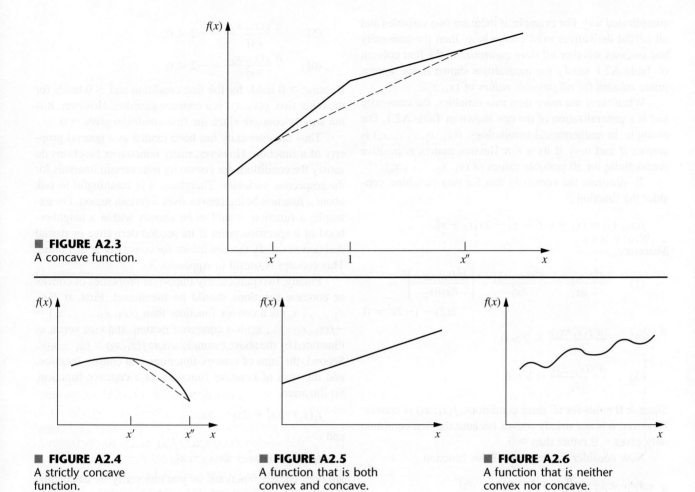

■ FIGURE A2.3
A concave function.

■ FIGURE A2.4
A strictly concave
function.

■ FIGURE A2.5
A function that is both
convex and concave.

■ FIGURE A2.6
A function that is neither
convex nor concave.

value of $(x_1, x_2, \ldots, x_m)$ may be thought of as a point in m-dimensional (Euclidean) space. By letting $m = n + 1$, the points on the graph of $f(x_1, x_2, \ldots, x_n)$ become the possible values of $[x_1, x_2, \ldots, x_n, f(x_1, x_2, \ldots, x_n)]$. Another point, $(x_1, x_2, \ldots, x_n, x_{n+1})$, is said to lie above, on, or below the graph of $f(x_1, x_2, \ldots, x_n)$, according to whether x_{n+1} is larger, equal to, or smaller than $f(x_1, x_2, \ldots, x_n)$, respectively.

Definition: The **line segment** joining any two points $(x_1', x_2', \ldots, x_m')$ and $(x_1'', x_2'', \ldots, x_m'')$ is the collection of points

$$(x_1, x_2, \ldots, x_m) = [\lambda x_1'' + (1 - \lambda)x_1', \lambda x_2'' + (1 - \lambda)x_2', \ldots, \lambda x_m'' + (1 - \lambda)x_m']$$

such that $0 \le \lambda \le 1$.

Thus, a line segment in m-dimensional space is a direct generalization of a line segment in two-dimensional space. For example, if

$$(x_1', x_2') = (2, 6), \qquad (x_1'', x_2'') = (3, 4),$$

then the line segment joining them is the collection of points

$$(x_1, x_2) = [3\lambda + 2(1 - \lambda), 4\lambda + 6(1 - \lambda)],$$

where $0 \le \lambda \le 1$.

Definition: $f(x_1, x_2, \ldots, x_n)$ is a **convex function** if, for each pair of points on the graph of $f(x_1, x_2, \ldots, x_n)$, the line segment joining these two points lies entirely above or on the graph of $f(x_1, x_2, \ldots, x_n)$. It is a **strictly convex function** if this line segment actually lies entirely above this graph except at the endpoints of the line segment. **Concave functions** and **strictly concave functions** are defined in exactly the same way, except that *above* is replaced by *below*.

Just as the second derivative can be used (when it exists everywhere) to check whether a function of a single variable is convex, so second partial derivatives can be used to check functions of several variables, although in a more complicated way. For example, if there are two variables and all partial derivatives exist everywhere, then the convexity test assesses whether *all three quantities* in the first column of Table A2.1 satisfy the inequalities shown in the appropriate column for *all possible values* of (x_1, x_2).

When there are more than two variables, the convexity test is a generalization of the one shown in Table A2.1. For example, in mathematical terminology, $f(x_1, x_2, \ldots, x_n)$ is convex if and only if its $n \times n$ Hessian matrix is positive semidefinite for all possible values of $(x_1, x_2, \ldots, x_n)$.

To illustrate the convexity test for two variables, consider the function

$$f(x_1, x_2) = (x_1 - x_2)^2 = x_1^2 - 2x_1x_2 + x_2^2.$$

Therefore,

(1) $\dfrac{\partial^2 f(x_1, x_2)}{\partial x_1^2} \dfrac{\partial^2 f(x_1, x_2)}{\partial x_2^2} - \left[\dfrac{\partial^2 f(x_1, x_2)}{\partial x_1 \partial x_2}\right]^2 =$

$$2(2) - (-2)^2 = 0,$$

(2) $\dfrac{\partial^2 f(x_1, x_2)}{\partial x_1^2} = 2 > 0,$

(3) $\dfrac{\partial^2 f(x_1, x_2)}{\partial x_2^2} = 2 > 0.$

Since ≥ 0 holds for all three conditions, $f(x_1, x_2)$ is convex. However, it is *not* strictly convex because the first condition only gives $= 0$ rather than > 0.

Now consider the negative of this function

$$g(x_1, x_2) = -f(x_1, x_2) = -(x_1 - x_2)^2$$
$$= -x_1^2 + 2x_1x_2 - x_2^2.$$

In this case,

(4) $\dfrac{\partial^2 g(x_1, x_2)}{\partial x_1^2} \dfrac{\partial^2 g(x_1, x_2)}{\partial x_2^2} - \left[\dfrac{\partial^2 g(x_1, x_2)}{\partial x_1 \partial x_2}\right]^2 =$

$$-2(-2) - 2^2 = 0,$$

(5) $\dfrac{\partial^2 g(x_1, x_2)}{\partial x_1^2} = -2 < 0,$

(6) $\dfrac{\partial^2 g(x_1, x_2)}{\partial x_2^2} = -2 < 0.$

Because ≥ 0 holds for the first condition and ≤ 0 holds for the other two, $g(x_1, x_2)$ is a concave function. However, it is *not* strictly concave since the first condition gives $= 0$.

Thus far, convexity has been treated as a general property of a function. However, many nonconvex functions do satisfy the conditions for convexity over certain intervals for the respective variables. Therefore, it is meaningful to talk about a function being convex over a certain region. For example, a function is said to be convex within a neighborhood of a specified point if its second derivative or partial derivatives satisfy the conditions for convexity at that point. This concept is useful in Appendix 3.

Finally, two particularly important properties of convex or concave functions should be mentioned. First, if $f(x_1, x_2, \ldots, x_n)$ is a convex function, then $g(x_1, x_2, \ldots, x_n) = -f(x_1, x_2, \ldots, x_n)$ is a concave function, and vice versa, as illustrated by the above example where $f(x_1, x_2) = (x_1 - x_2)^2$. Second, the sum of convex functions is a convex function, and the sum of concave functions is a concave function. To illustrate,

$$f_1(x_1) = x_1^4 + 2x_1^2 - 5x_1$$

and

$$f_2(x_1, x_2) = x_1^2 + 2x_1x_2 + x_2^2$$

are both convex functions, as you can verify by calculating their second derivatives. Therefore, the sum of these functions

$$f(x_1, x_2) = x_1^4 + 3x_1^2 - 5x_1 + 2x_1x_2 + x_2^2$$

is a convex function, whereas its negative

$$g(x_1, x_2) = -x_1^4 - 3x_1^2 + 5x_1 - 2x_1x_2 - x_2^2,$$

is a concave function.

■ **TABLE A2.1 Convexity test for a function of two variables**

Quantity	Convex	Strictly Convex	Concave	Strictly Concave
$\dfrac{\partial^2 f(x_1, x_2)}{\partial x_1^2} \dfrac{\partial^2 f(x_1, x_2)}{\partial x_2^2} - \left[\dfrac{\partial^2 f(x_1, x_2)}{\partial x_1 \partial x_2}\right]^2$	≥ 0	> 0	≥ 0	> 0
$\dfrac{\partial^2 f(x_1, x_2)}{\partial x_1^2}$	≥ 0	> 0	≤ 0	< 0
$\dfrac{\partial^2 f(x_1, x_2)}{\partial x_2^2}$	≥ 0	> 0	≤ 0	< 0
Values of (x_1, x_2)	All possible values			

CONVEX SETS

The concept of a convex function leads quite naturally to the related concept of a **convex set.** Thus, if $f(x_1, x_2, \ldots, x_n)$ is a convex function, then the collection of points that lie above or on the graph of $f(x_1, x_2, \ldots, x_n)$ forms a convex set. Similarly, the collection of points that lie below or on the graph of a concave function is a convex set. These cases are illustrated in Figs. A2.7 and A2.8 for the case of a single independent variable. Furthermore, convex sets have the important property that, for any given group of convex sets, the collection of points that lie in all of them (i.e., the intersection of these convex sets) is also a convex set. Therefore, the collection of points that lie both above or on a convex function and below or on a concave function is a convex set, as illustrated in Fig. A2.9. Thus, convex sets may be viewed intuitively as a collection of points whose bottom boundary is a convex function and whose top boundary is a concave function.

Although describing convex sets in terms of convex and concave functions may be helpful for developing intuition about their nature, their actual definition has nothing to do (directly) with such functions.

Definition: A **convex set** is a collection of points such that, for each pair of points in the collection, the entire line segment joining these two points is also in the collection.

The distinction between nonconvex sets and convex sets is illustrated in Figs. A2.10 and A2.11. Thus, the set of points shown in Fig. A2.10 is not a convex set because there exist many pairs of these points, for example, (1, 2) and (2, 1), such that the line segment between them does not lie entirely within the set. This is not the case for the set in Fig. A2.11, which is convex.

In conclusion, we introduce the useful concept of an extreme point of a convex set.

Definition: An **extreme point** of a convex set is a point in the set that does not lie on any line segment that joins two other points in the set.

Thus, the extreme points of the convex set in Fig. A2.11 are (0, 0), (0, 2), (1, 2), (2, 1), (1, 0), and all the infinite number of points on the boundary between (2, 1) and (1, 0). If this particular boundary were a line segment instead, then the set would have only the five listed extreme points.

■ **FIGURE A2.7**
Example of a convex set determined by a convex function.

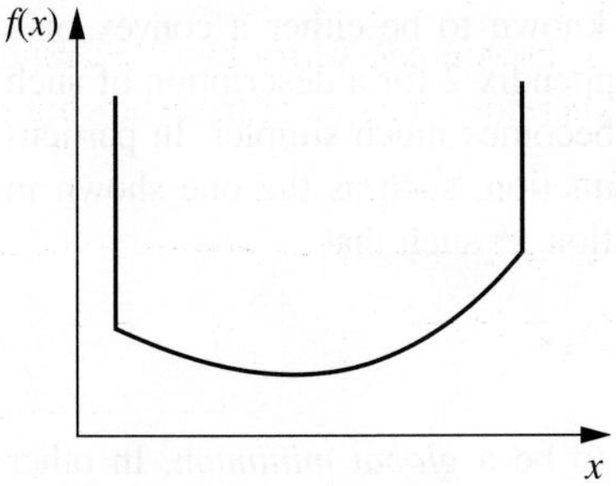

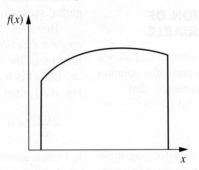

■ **FIGURE A2.8**
Example of a convex set determined by a concave function.

■ **FIGURE A2.9**
Example of a convex set determined by both convex and concave functions.

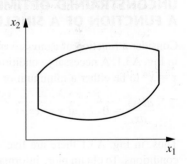

■ **FIGURE A2.10**
Example of a set that is not convex.

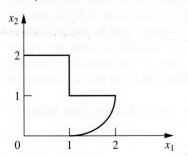

■ **FIGURE A2.11**
Example of a convex set.

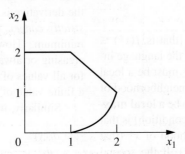

Classical Optimization Methods

This appendix reviews the classical methods of calculus for finding a solution that maximizes or minimizes (1) a function of a single variable, (2) a function of several variables, and (3) a function of several variables subject to equality constraints on the values of these variables. It is assumed that the functions considered possess continuous first and second derivatives and partial derivatives everywhere. Some of the concepts discussed next have been introduced briefly in Secs. 12.2 and 12.3.

UNCONSTRAINED OPTIMIZATION OF A FUNCTION OF A SINGLE VARIABLE

Consider a function of a single variable, such as that shown in Fig. A3.1. A necessary condition for a particular solution $x = x^*$ to be either a minimum or a maximum is that

$$\frac{df(x)}{dx} = 0 \qquad \text{at } x = x^*.$$

Thus, in Fig. A3.1 there are five solutions satisfying these conditions. To obtain more information about these five **critical points,** it is necessary to examine the second derivative. Thus, if

$$\frac{d^2f(x)}{dx^2} > 0 \qquad \text{at } x = x^*,$$

then x^* must be at least a **local minimum** [that is, $f(x^*) \leq f(x)$ for all x sufficiently close to x^*]. Using the language introduced in Appendix 2, we can say that x^* must be a local minimum if $f(x)$ is *strictly convex* within a neighborhood of x^*. Similarly, a sufficient condition for x^* to be a **local maximum** (given that it satisfies the necessary condition) is that $f(x)$ be *strictly concave* within a neighborhood of x^* (that is, the second derivative is *negative* at x^*). If the second

derivative is zero, the issue is not resolved (the point may even be an *inflection point*), and it is necessary to examine higher derivatives.

To find a **global minimum** [i.e., a solution x^* such that $f(x^*) \leq f(x)$ for all x], it is necessary to compare the local minima and identify the one that yields the smallest value of $f(x)$. If this value is less than $f(x)$ as $x \to -\infty$ and as $x \to +\infty$ (or at the endpoints of the function, if it is defined only over a finite interval), then this point is a global minimum. Such a point is shown in Fig. A3.1, along with the **global maximum,** which is identified in an analogous way.

However, if $f(x)$ is known to be either a convex or a concave function (see Appendix 2 for a description of such functions), the analysis becomes much simpler. In particular, if $f(x)$ is a *convex* function, such as the one shown in Fig. A2.1, then any solution x^* such that

$$\frac{df(x)}{dx} = 0 \qquad \text{at } x = x^*$$

is known automatically to be a *global minimum*. In other words, this condition is not only a *necessary* but also a *sufficient* condition for a global minimum of a convex function. This solution need not be unique, since there could be a tie for the global minimum over a single interval where the derivative is zero. On the other hand, if $f(x)$ actually is *strictly convex,* then this solution must be the only global minimum. (However, if the function is either always decreasing or always increasing, so the derivative is nonzero for all values of x, then there will be no global minimum at a finite value of x.)

Similarly, if $f(x)$ is a *concave* function, then having

$$\frac{df(x)}{dx} = 0 \qquad \text{at } x = x^*$$

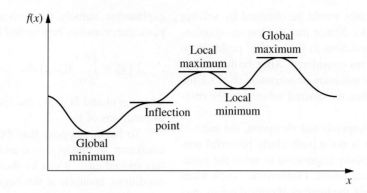

FIGURE A3.1
A function having several maxima and minima.

becomes both a *necessary* and *sufficient* condition for x^* to be a *global maximum*.

UNCONSTRAINED OPTIMIZATION OF A FUNCTION OF SEVERAL VARIABLES

The analysis for an unconstrained function of several variables $f(\mathbf{x})$, where $\mathbf{x} = (x_1, x_2, \ldots, x_n)$, is similar. Thus, a *necessary* condition for a solution $\mathbf{x} = \mathbf{x}^*$ to be either a minimum or a maximum is that

$$\frac{\partial f(\mathbf{x})}{\partial x_j} = 0 \qquad \text{at } \mathbf{x} = \mathbf{x}^*, \quad \text{for } j = 1, 2, \ldots, n.$$

After the critical points that satisfy this condition are identified, each such point is then classified as a local minimum or maximum if the function is *strictly convex* or *strictly concave,* respectively, within a neighborhood of the point. (Additional analysis is required if the function is neither.) The *global minimum* and *maximum* would be found by comparing the local minima and maxima and then checking the value of the function as some of the variables approach $-\infty$ or $+\infty$. However, if the function is known to be *convex* or *concave,* then a critical point must be a *global minimum* or a *global maximum,* respectively.

CONSTRAINED OPTIMIZATION WITH EQUALITY CONSTRAINTS

Now consider the problem of finding the *minimum* or *maximum* of the function $f(\mathbf{x})$, subject to the restriction that $\mathbf{x}$ must satisfy all the equations

$$g_1(\mathbf{x}) = b_1$$
$$g_2(\mathbf{x}) = b_2$$
$$\vdots$$
$$g_m(\mathbf{x}) = b_m,$$

where $m < n$. For example, if $n = 2$ and $m = 1$, the problem might be

Maximize $\qquad f(x_1, x_2) = x_1^2 + 2x_2,$

subject to

$$g(x_1, x_2) = x_1^2 + x_2^2 = 1.$$

In this case, (x_1, x_2) is restricted to be on the circle of radius 1, whose center is at the origin, so that the goal is to find the point on this circle that yields the largest value of $f(x_1, x_2)$. This example will be solved after a general approach to the problem is outlined.

A classical method of dealing with this problem is the **method of Lagrange multipliers.** This procedure begins by formulating the **Lagrangian function**

$$h(\mathbf{x}, \boldsymbol{\lambda}) = f(\mathbf{x}) - \sum_{i=1}^{m} \lambda_i [g_i(\mathbf{x}) - b_i],$$

where the new variables $\boldsymbol{\lambda} = (\lambda_1, \lambda_2, \ldots, \lambda_m)$ are called *Lagrange multipliers.* Notice the key fact that for the *feasible* values of $\mathbf{x}$,

$$g_i(\mathbf{x}) - b_i = 0, \qquad \text{for all } i,$$

so $h(\mathbf{x}, \boldsymbol{\lambda}) = f(x)$. Therefore, it can be shown that if $(\mathbf{x}, \boldsymbol{\lambda}) = (\mathbf{x}^*, \boldsymbol{\lambda}^*)$ is a *local* or *global minimum* or *maximum* for the unconstrained function $h(\mathbf{x}, \boldsymbol{\lambda})$, then $\mathbf{x}^*$ is a corresponding *critical point* for the original problem. As a result, the method now reduces to analyzing $h(\mathbf{x}, \boldsymbol{\lambda})$ by the procedure just described for unconstrained optimization. Thus, the $n + m$ partial derivatives would be set equal to zero

$$\frac{\partial h}{\partial x_j} = \frac{\partial f}{\partial x_j} - \sum_{i=1}^{m} \lambda_i \frac{\partial g_i}{\partial x_j} = 0, \qquad \text{for } j = 1, 2, \ldots, n,$$

$$\frac{\partial h}{\partial \lambda_i} = -g_i(\mathbf{x}) + b_i = 0, \qquad \text{for } i = 1, 2, \ldots, m,$$

and then the critical points would be obtained by solving these equations for $(\mathbf{x}, \boldsymbol{\lambda})$. Notice that the last m equations are equivalent to the constraints in the original problem, so only feasible solutions are considered. After further analysis to identify the *global minimum* or *maximum* of $h(\cdot)$, the resulting value of $\mathbf{x}$ is then the desired solution to the original problem.

From a practical computational viewpoint, the method of Lagrange multipliers is not a particularly powerful procedure. It is often essentially impossible to solve the equations to obtain the critical points. Furthermore, even when the points can be obtained, the number of critical points may be so large (often infinite) that it is impractical to attempt to identify a global minimum or maximum. However, for certain types of small problems, this method can sometimes be used successfully.

To illustrate, consider the example introduced earlier. In this case,

$$h(x_1, x_2) = x_1^2 + 2x_2 - \lambda(x_1^2 + x_2^2 - 1),$$

so that

$$\frac{\partial h}{\partial x_1} = 2x_1 - 2\lambda x_1 = 0,$$

$$\frac{\partial h}{\partial x_2} = 2 - 2\lambda x_2 = 0,$$

$$\frac{\partial h}{\partial \lambda} = -(x_1^2 + x_2^2 - 1) = 0.$$

The first equation implies that either $\lambda = 1$ or $x_1 = 0$. If $\lambda = 1$, then the other two equations imply that $x_2 = 1$ and $x_1 = 0$. If $x_1 = 0$, then the third equation implies that $x_2 = \pm 1$. Therefore, the two critical points for the original problem are $(x_1, x_2) = (0, 1)$ *and* $(0, -1)$. Thus, it is apparent that these points are the global maximum and minimum, respectively.

THE DERIVATIVE OF A DEFINITE INTEGRAL

In presenting the classical optimization methods just described, we have assumed that you are already familiar with derivatives and how to obtain them. However, there is a special case of importance in OR work that warrants additional

explanation, namely, the derivative of a definite integral. In particular, consider how to find the derivative of the function

$$F(y) = \int_{g(y)}^{h(y)} f(x, y)\, dx,$$

where $g(y)$ and $h(y)$ are the limits of integration expressed as functions of y.

To begin, suppose that these limits of integration are constants, so that $g(y) = a$ and $h(y) = b$, respectively. For this special case, it can be shown that, given the regularity conditions assumed at the beginning of this appendix, the derivative is

$$\frac{d}{dy} \int_a^b f(x, y)\, dx = \int_a^b \frac{\partial f(x, y)}{\partial y}\, dx.$$

For example, if $f(x, y) = e^{-xy}$, $a = 0$, and $b = \infty$, then

$$\frac{d}{dy} \int_0^\infty e^{-xy}\, dx = \int_0^\infty (-x)e^{-xy}\, dx = -\frac{1}{y^2}$$

at any positive value of y. Thus, the intuitive procedure of interchanging the order of differentiation and integration is valid for this case.

However, finding the derivative becomes a little more complicated than this when the limits of integration are functions. In particular,

$$\frac{d}{dy} \int_{g(y)}^{h(y)} f(x, y)\, dx = \int_{g(y)}^{h(y)} \frac{\partial f(x, y)}{\partial y}\, dx$$
$$+ f(h(y), y)\frac{dh(y)}{dy} - f(g(y), y)\frac{dg(y)}{dy},$$

where $f(h(y), y)$ is obtained by writing out $f(x, y)$ and then replacing x by $h(y)$ wherever it appears, and similarly for $f(g(y), y)$. To illustrate, if $f(x, y) = x^2 y^3$, $g(y) = y$, and $h(y) = 2y$, then

$$\frac{d}{dy} \int_y^{2y} x^2 y^3\, dx = \int_y^{2y} 3x^2 y^2\, dx + (2y)^2 y^3(2) - y^2 y^3(1)$$
$$= 14y^5$$

at any positive value of y.

4

Matrices and Matrix Operations

A **matrix** is a rectangular array of numbers. For example,

$$\mathbf{A} = \begin{bmatrix} 2 & 5 \\ 3 & 0 \\ 1 & 1 \end{bmatrix}$$

is a 3×2 matrix (where 3×2 is said "3 by 2") because it is a rectangular array of numbers with three rows and two columns. (Matrices are denoted in this book by **boldface capital letters.**) The numbers in the rectangular array are called the **elements** of the matrix. For example,

$$\mathbf{B} = \begin{bmatrix} 1 & 2.4 & 0 & \sqrt{3} \\ -4 & 2 & -1 & 15 \end{bmatrix}$$

is a 2×4 matrix whose elements are 1, 2.4, 0, $\sqrt{3}$, -4, 2, -1, and 15. Thus, in more general terms,

$$\mathbf{A} = \begin{bmatrix} a_{11} & a_{12} & \cdots & a_{1n} \\ a_{21} & a_{22} & \cdots & a_{2n} \\ \multicolumn{4}{c}{\cdots\cdots\cdots\cdots\cdots\cdots} \\ a_{m1} & a_{m2} & \cdots & a_{mn} \end{bmatrix} = \| a_{ij} \|$$

is an $m \times n$ matrix, where $a_{11}, \ldots, a_{mn}$ represent the numbers that are the elements of this matrix; $\| a_{ij} \|$ is shorthand notation for identifying the matrix whose element in row i and column j is a_{ij} for every $i = 1, 2, \ldots, m$ and $j = 1, 2, \ldots, n$.

MATRIX OPERATIONS

Because matrices do not possess a numerical value, they cannot be added, multiplied, and so on as if they were individual numbers. However, it is sometimes desirable to perform certain manipulations on arrays of numbers. Therefore, rules have been developed for performing operations on matrices that are analogous to arithmetic operations. To describe these, let $\mathbf{A} = \| a_{ij} \|$ and $\mathbf{B} = \| b_{ij} \|$ be two matrices having the same number of rows and the same number of columns. (We shall change this restriction on the size of $\mathbf{A}$ and $\mathbf{B}$ later when discussing matrix multiplication.)

Matrices $\mathbf{A}$ and $\mathbf{B}$ are said to be *equal* ($\mathbf{A} = \mathbf{B}$) if and only if *all* the corresponding elements are equal ($a_{ij} = b_{ij}$ for all i and j).

The operation of *multiplying a matrix by a number* (denote this number by k) is performed by multiplying each element of the matrix by k, so that

$$k\mathbf{A} = \| ka_{ij} \|.$$

For example,

$$3\begin{bmatrix} 1 & \frac{1}{3} & 2 \\ 5 & 0 & -3 \end{bmatrix} = \begin{bmatrix} 3 & 1 & 6 \\ 15 & 0 & -9 \end{bmatrix}.$$

To add two matrices $\mathbf{A}$ and $\mathbf{B}$, simply add the corresponding elements, so that

$$\mathbf{A} + \mathbf{B} = \| a_{ij} + b_{ij} \|.$$

To illustrate,

$$\begin{bmatrix} 5 & 3 \\ 1 & 6 \end{bmatrix} + \begin{bmatrix} 2 & 0 \\ 3 & 1 \end{bmatrix} = \begin{bmatrix} 7 & 3 \\ 4 & 7 \end{bmatrix}.$$

Similarly, *subtraction* is done as follows:

$$\mathbf{A} - \mathbf{B} = \mathbf{A} + (-1)\mathbf{B},$$

so that

$$\mathbf{A} - \mathbf{B} = \| a_{ij} - b_{ij} \|.$$

For example,

$$\begin{bmatrix} 5 & 3 \\ 1 & 6 \end{bmatrix} - \begin{bmatrix} 2 & 0 \\ 3 & 1 \end{bmatrix} = \begin{bmatrix} 3 & 3 \\ -2 & 5 \end{bmatrix}.$$

Note that, with the exception of multiplication by a number, all the preceding operations are defined only when the two matrices involved are the same size. However, all of these operations are straightforward because they involve performing only the same comparison or arithmetic operation on the corresponding elements of the matrices.

There exists one additional elementary operation that has not been defined—**matrix multiplication**—but it is considerably more complicated. To find the element in row i, column j of the matrix resulting from multiplying matrix **A** times matrix **B**, it is necessary to multiply each element in row i of **A** by the corresponding element in column j of **B** and then to add these products. To do this element-by-element multiplication, we need the following restriction on the sizes of **A** and **B**:

Matrix multiplication **AB** is defined if and only if the *number of columns* of **A** equals the *number of rows* of **B**.

Thus, if **A** is an $m \times n$ matrix and **B** is an $n \times s$ matrix, then their product is

$$\mathbf{AB} = \left\| \sum_{k=1}^{n} a_{ik} b_{kj} \right\|,$$

where this product is an $m \times s$ matrix. However, if **A** is an $m \times n$ matrix and **B** is an $r \times s$ matrix, where $n \neq r$, then **AB** is not defined.

To illustrate matrix multiplication,

$$\begin{bmatrix} 1 & 2 \\ 4 & 0 \\ 2 & 3 \end{bmatrix} \begin{bmatrix} 3 & 1 \\ 2 & 5 \end{bmatrix} = \begin{bmatrix} 1(3)+2(2) & 1(1)+2(5) \\ 4(3)+0(2) & 4(1)+0(5) \\ 2(3)+3(2) & 2(1)+3(5) \end{bmatrix}$$

$$= \begin{bmatrix} 7 & 11 \\ 12 & 4 \\ 12 & 17 \end{bmatrix}.$$

On the other hand, if one attempts to multiply these matrices in the reverse order, the resulting product

$$\begin{bmatrix} 3 & 1 \\ 2 & 5 \end{bmatrix} \begin{bmatrix} 1 & 2 \\ 4 & 0 \\ 2 & 3 \end{bmatrix}$$

is not even defined.

Even when both **AB** and **BA** are defined,

$$\mathbf{AB} \neq \mathbf{BA}$$

in general. Thus, *matrix multiplication* should be viewed as a specially designed operation whose properties are quite different from those of *arithmetic multiplication*. To understand why this special definition was adopted, consider the following system of equations:

$$\begin{aligned} 2x_1 - \ x_2 + 5x_3 + \ x_4 &= 20 \\ x_1 + 5x_2 + 4x_3 + 5x_4 &= 30 \\ 3x_1 + \ x_2 - 6x_3 + 2x_4 &= 20. \end{aligned}$$

Rather than write out these equations as shown here, they can be written much more concisely in matrix form as

$$\mathbf{Ax} = \mathbf{b},$$

where

$$\mathbf{A} = \begin{bmatrix} 2 & -1 & 5 & 1 \\ 1 & 5 & 4 & 5 \\ 3 & 1 & -6 & 2 \end{bmatrix}, \quad \mathbf{x} = \begin{bmatrix} x_1 \\ x_2 \\ x_3 \\ x_4 \end{bmatrix}, \quad \mathbf{b} = \begin{bmatrix} 20 \\ 30 \\ 20 \end{bmatrix}.$$

It is this kind of multiplication for which matrix multiplication is designed.

Carefully note that *matrix division* is *not* defined.

Although the matrix operations described here do not possess certain of the properties of arithmetic operations, they do satisfy these laws

$$\begin{aligned} \mathbf{A} + \mathbf{B} &= \mathbf{B} + \mathbf{A}, \\ (\mathbf{A} + \mathbf{B}) + \mathbf{C} &= \mathbf{A} + (\mathbf{B} + \mathbf{C}), \\ \mathbf{A}(\mathbf{B} + \mathbf{C}) &= \mathbf{AB} + \mathbf{AC}, \\ \mathbf{A}(\mathbf{BC}) &= (\mathbf{AB})\mathbf{C}, \end{aligned}$$

when the relative sizes of these matrices are such that the indicated operations are defined.

Another type of matrix operation, which has no arithmetic analog, is the **transpose operation.** This operation involves nothing more than interchanging the rows and columns of the matrix, which is frequently useful for performing the multiplication operation in the desired way. Thus, for any matrix $\mathbf{A} = \| a_{ij} \|$, its transpose $\mathbf{A}^T$ is

$$\mathbf{A}^T = \| a_{ji} \|.$$

For example, if

$$\mathbf{A} = \begin{bmatrix} 2 & 5 \\ 1 & 3 \\ 4 & 0 \end{bmatrix},$$

then

$$\mathbf{A}^T = \begin{bmatrix} 2 & 1 & 4 \\ 5 & 3 & 0 \end{bmatrix}.$$

SPECIAL KINDS OF MATRICES

In arithmetic, 0 and 1 play a special role. There also exist special matrices that play a similar role in matrix theory. In particular, the matrix that is analogous to 1 is the **identity matrix I,** which is a *square* matrix whose elements are 0s except for 1s along the main diagonal. Thus,

$$I = \begin{bmatrix} 1 & 0 & 0 & \cdots & 0 \\ 0 & 1 & 0 & \cdots & 0 \\ 0 & 0 & 1 & \cdots & 0 \\ \multicolumn{5}{c}{\cdots\cdots\cdots\cdots\cdots} \\ 0 & 0 & 0 & \cdots & 1 \end{bmatrix}$$

The number of rows or columns of I can be specified as desired. The analogy of I to 1 follows from the fact that for any matrix A,

$$IA = A = AI,$$

where I is assigned the appropriate number of rows and columns in each case for the multiplication operation to be defined.

Similarly, the matrix that is analogous to 0 is the **null matrix 0,** which is a matrix of any size whose elements are *all* 0s. Thus,

$$0 = \begin{bmatrix} 0 & 0 & \cdots & 0 \\ 0 & 0 & \cdots & 0 \\ \multicolumn{4}{c}{\cdots\cdots\cdots\cdots} \\ 0 & 0 & \cdots & 0 \end{bmatrix}$$

Therefore, for any matrix A,

$$A + 0 = A, \qquad A - A = 0, \qquad \text{and}$$
$$0A = 0 = A0,$$

where 0 is the appropriate size in each case for the operations to be defined.

On certain occasions, it is useful to partition a matrix into several smaller matrices, called **submatrices.** For example, one possible way of partitioning a 3×4 matrix would be

$$A = \begin{bmatrix} a_{11} & a_{12} & a_{13} & a_{14} \\ a_{21} & a_{22} & a_{23} & a_{24} \\ a_{31} & a_{32} & a_{33} & a_{34} \end{bmatrix} = \begin{bmatrix} a_{11} & A_{12} \\ A_{21} & A_{22} \end{bmatrix},$$

where

$$A_{12} = [a_{12}, \quad a_{13}, \quad a_{14}], \qquad A_{21} = \begin{bmatrix} a_{21} \\ a_{31} \end{bmatrix},$$

$$A_{22} = \begin{bmatrix} a_{22} & a_{23} & a_{24} \\ a_{32} & a_{33} & a_{34} \end{bmatrix}$$

all are submatrices. Rather than perform operations element by element on such partitioned matrices, we can do them in terms of the submatrices, provided the partitionings are such that the operations are defined. For example, if B is a partitioned 4×1 matrix such that

$$B = \begin{bmatrix} b_1 \\ b_2 \\ b_3 \\ b_4 \end{bmatrix} = \begin{bmatrix} b_1 \\ B_2 \end{bmatrix},$$

then

$$AB = \begin{bmatrix} a_{11}b_1 + A_{12}B_2 \\ A_{21}b_1 + A_{22}B_2 \end{bmatrix}.$$

VECTORS

A special kind of matrix that plays an important role in matrix theory is the kind that has either a *single row* or a *single column*. Such matrices are often referred to as **vectors.** Thus,

$$x = [x_1, x_2, \ldots, x_n]$$

is a **row vector,** and

$$x = \begin{bmatrix} x_1 \\ x_2 \\ \vdots \\ x_n \end{bmatrix}$$

is a **column vector.** (Vectors are denoted in this book by **boldface lowercase letters.**) These vectors also are sometimes called *n-vectors* to indicate that they have n elements. For example,

$$x = [1, 4, -2, \tfrac{1}{3}, 7]$$

is a 5-vector.

A **null vector 0** is either a row vector or a column vector whose elements are *all* 0s, that is,

$$0 = [0, 0, \ldots, 0] \qquad \text{or} \qquad 0 = \begin{bmatrix} 0 \\ 0 \\ \vdots \\ 0 \end{bmatrix}.$$

(Although the same symbol 0 is used for either kind of *null vector,* as well as for a *null matrix,* the context normally will identify which it is.)

One reason vectors play an important role in matrix theory is that any $m \times n$ matrix can be partitioned into either m row vectors or n column vectors, and important properties of the matrix can be analyzed in terms of these vectors. To amplify, consider a set of n-vectors $x_1, x_2, \ldots, x_m$ of the same type (i.e., they are either all row vectors or all column vectors).

Definition: A set of vectors $x_1, x_2, \ldots, x_m$ is said to be **linearly dependent** if there exist m numbers (denoted by $c_1, c_2, \ldots, c_m$), some of which are not zero, such that

$$c_1 x_1 + c_2 x_2 + \cdots + c_m x_m = 0.$$

Otherwise, the set is said to be **linearly independent.**

APPENDIX 5

Table for a Normal Distribution

TABLE A5.1 Areas under the normal curve from K_α to ∞

$$P\{\text{standard normal} > K_\alpha\} = \int_{K_\alpha}^{\infty} \frac{1}{\sqrt{2\pi}}\, e^{-x^2/2}\, dx = \alpha$$

K_α	.00	.01	.02	.03	.04	.05	.06	.07	.08	.09
0.0	.5000	.4960	.4920	.4880	.4840	.4801	.4761	.4721	.4681	.4641
0.1	.4602	.4562	.4522	.4483	.4443	.4404	.4364	.4325	.4286	.4247
0.2	.4207	.4168	.4129	.4090	.4052	.4013	.3974	.3936	.3897	.3859
0.3	.3821	.3783	.3745	.3707	.3669	.3632	.3594	.3557	.3520	.3483
0.4	.3446	.3409	.3372	.3336	.3300	.3264	.3228	.3192	.3156	.3121
0.5	.3085	.3050	.3015	.2981	.2946	.2912	.2877	.2843	.2810	.2776
0.6	.2743	.2709	.2676	.2643	.2611	.2578	.2546	.2514	.2483	.2451
0.7	.2420	.2389	.2358	.2327	.2296	.2266	.2236	.2206	.2177	.2148
0.8	.2119	.2090	.2061	.2033	.2005	.1977	.1949	.1922	.1894	.1867
0.9	.1841	.1814	.1788	.1762	.1736	.1711	.1685	.1660	.1635	.1611
1.0	.1587	.1562	.1539	.1515	.1492	.1469	.1446	.1423	.1401	.1379
1.1	.1357	.1335	.1314	.1292	.1271	.1251	.1230	.1210	.1190	.1170
1.2	.1151	.1131	.1112	.1093	.1075	.1056	.1038	.1020	.1003	.0985
1.3	.0968	.0951	.0934	.0918	.0901	.0885	.0869	.0853	.0838	.0823
1.4	.0808	.0793	.0778	.0764	.0749	.0735	.0721	.0708	.0694	.0681
1.5	.0668	.0655	.0643	.0630	.0618	.0606	.0594	.0582	.0571	.0559
1.6	.0548	.0537	.0526	.0516	.0505	.0495	.0485	.0475	.0465	.0455
1.7	.0446	.0436	.0427	.0418	.0409	.0401	.0392	.0384	.0375	.0367
1.8	.0359	.0351	.0344	.0336	.0329	.0322	.0314	.0307	.0301	.0294
1.9	.0287	.0281	.0274	.0268	.0262	.0256	.0250	.0244	.0239	.0233
2.0	.0228	.0222	.0217	.0212	.0207	.0202	.0197	.0192	.0188	.0183
2.1	.0179	.0174	.0170	.0166	.0162	.0158	.0154	.0150	.0146	.0143
2.2	.0139	.0136	.0132	.0129	.0125	.0122	.0119	.0116	.0113	.0110
2.3	.0107	.0104	.0102	.00990	.00964	.00939	.00914	.00889	.00866	.00842
2.4	.00820	.00798	.00776	.00755	.00734	.00714	.00695	.00676	.00657	.00639
2.5	.00621	.00604	.00587	.00570	.00554	.00539	.00523	.00508	.00494	.00480
2.6	.00466	.00453	.00440	.00427	.00415	.00402	.00391	.00379	.00368	.00357
2.7	.00347	.00336	.00326	.00317	.00307	.00298	.00289	.00280	.00272	.00264
2.8	.00256	.00248	.00240	.00233	.00226	.00219	.00212	.00205	.00199	.00193
2.9	.00187	.00181	.00175	.00169	.00164	.00159	.00154	.00149	.00144	.00139

K_α	.0	.1	.2	.3	.4	.5	.6	.7	.8	.9
3	.00135	$.0^3968$	$.0^3687$	$.0^3483$	$.0^3337$	$.0^3233$	$.0^3159$	$.0^3108$	$.0^4723$	$.0^4481$
4	$.0^4317$	$.0^4207$	$.0^4133$	$.0^5854$	$.0^5541$	$.0^5340$	$.0^5211$	$.0^5130$	$.0^6793$	$.0^6479$
5	$.0^6287$	$.0^6170$	$.0^7996$	$.0^7579$	$.0^7333$	$.0^7190$	$.0^7107$	$.0^8599$	$.0^8332$	$.0^8182$
6	$.0^9987$	$.0^9530$	$.0^9282$	$.0^9149$	$.0^{10}777$	$.0^{10}402$	$.0^{10}206$	$.0^{10}104$	$.0^{11}523$	$.0^{11}260$

Source: F. E. Croxton, *Tables of Areas in Two Tails and in One Tail of the Normal Curve.* Copyright 1949 by Prentice-Hall, Inc., Englewood Cliffs, NJ.

PARTIAL ANSWERS TO SELECTED PROBLEMS

CHAPTER 3

3.1-2. (a)

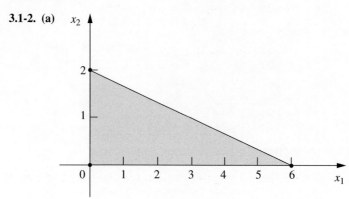

3.1-5. $(x_1, x_2) = (13, 5)$; $Z = 31$.

3.1-11. (b) $(x_1, x_2, x_3) = (26.19, 54.76, 20)$; $Z = 2,904.76$.

3.2-3. (b) Maximize $\quad Z = 4,500x_1 + 4,500x_2$,

subject to

$$
\begin{aligned}
x_1 \qquad\qquad &\leq \quad 1 \\
x_2 &\leq \quad 1 \\
5,000x_1 + 4,000x_2 &\leq 6,000 \\
400x_1 + \quad 500x_2 &\leq \quad 600
\end{aligned}
$$

and

$$x_1 \geq 0, \qquad x_2 \geq 0.$$

3.4-3. (a) *Proportionality*: OK since it is implied that a fixed fraction of the radiation dosage at a given entry point is absorbed by a given area.

Additivity: OK since it is stated that the radiation absorption from multiple beams is additive.

Divisibility: OK since beam strength can be any fractional level.

Certainty: Due to the complicated analysis required to estimate the data on radiation absorption in different tissue types, there is considerable uncertainty about the data, so sensitivity analysis should be used.

3.4-12. (b) From Factory 1, ship 200 units to Customer 2 and 200 units to Customer 3.
From Factory 2, ship 300 units to Customer 1 and 200 units to Customer 3.

3.4-13. (c) $Z = \$152,880$; $A_1 = 60,000$; $A_3 = 84,000$; $D_5 = 117,600$. All other decision variables are 0.

3.4-15. (b) Each optimal solution has $Z = \$13,330$.

3.5-2. (c, e)

Resource	Resource Usage per Unit of Each Activity		Totals		Resource Available
	Activity 1	Activity 2			
1	2	1	10	$\leq$	10
2	3	3	20	$\leq$	20
3	2	4	20	$\leq$	20
Unit Profit	20	30	$166.67		
Solution	3.333	3.333			

3.5-5. (a) Minimize $\quad Z = 84C + 72T + 60A,$

subject to

$$90C + 20T + 40A \geq 200$$
$$30C + 80T + 60A \geq 180$$
$$10C + 20T + 60A \geq 150$$

and

$$C \geq 0, \qquad T \geq 0, \qquad A \geq 0.$$

CHAPTER 4

4.1-4. (a) The corner-point solutions that are *feasible* are $(0, 0)$, $(0, 1)$, $(\frac{1}{4}, 1)$, $(\frac{2}{3}, \frac{2}{3})$, $(1, \frac{1}{4})$, and $(1, 0)$.

4.3-4. $(x_1, x_2, x_3) = (0, 10, 6\frac{2}{3})$; $Z = 70$.

4.6-1. (a, c) $(x_1, x_2) = (2, 1)$; $Z = 7$.

4.6-3. (a, c, e) $(x_1, x_2, x_3) = (\frac{4}{5}, \frac{9}{5}, 0)$; $Z = 7$.

4.6-9. (a, b, d) $(x_1, x_2, x_3) = (0, 15, 15)$; $Z = 90$.
(c) For both the Big M method and the two-phase method, only the final tableau represents a feasible solution for the real problem.

4.6-13. (a, c) $(x_1, x_2) = (-\frac{8}{7}, \frac{18}{7})$; $Z = \frac{80}{7}$.

4.7-5. (a) $(x_1, x_2, x_3) = (0, 1, 3)$; $Z = 7$.
(b) $y_1^* = \frac{1}{2}, y_2^* = \frac{5}{2}, y_3^* = 0$. These are the marginal values of resources 1, 2, and 3, respectively.

CHAPTER 5

5.1-1. (a) $(x_1, x_2) = (2, 2)$ is optimal. Other CPF solutions are $(0, 0)$, $(3, 0)$, and $(0, 3)$.

5.1-12. $(x_1, x_2, x_3) = (0, 15, 15)$ is optimal.

5.2-2. $(x_1, x_2, x_3, x_4, x_5) = (0, 5, 0, \frac{5}{2}, 0)$; $Z = 50$.

5.3-1. (a) Right side is $Z = 8$, $x_2 = 14$, $x_6 = 5$, $x_3 = 11$.
(b) $x_1 = 0, 2x_1 - 2x_2 + 3x_3 = 5, x_1 + x_2 - x_3 = 3$.

CHAPTER 6

6.1-1. (a) Minimize $\quad W = 15y_1 + 12y_2 + 45y_3,$

subject to

$$-y_1 + y_2 + 5y_3 \geq 10$$
$$2y_1 + y_2 + 3y_3 \geq 20$$

and

$$y_1 \geq 0, \qquad y_2 \geq 0, \qquad y_3 \geq 0.$$

6.3-1. (c)

Complementary Basic Solutions

Primal Problem				Dual Problem	
Basic Solution	Feasible?	Z = W	Feasible?	Basic Solution	
(0, 0, 20, 10)	Yes	0	No	(0, 0, −6, −8)	
(4, 0, 0, 6)	Yes	24	No	$\left(1\frac{1}{5}, 0, 0, -5\frac{3}{5}\right)$	
(0, 5, 10, 0)	Yes	40	No	(0, 4, −2, 0)	
$\left(2\frac{1}{2}, 3\frac{3}{4}, 0, 0\right)$	Yes and optimal	45	Yes and optimal	$\left(\frac{1}{2}, 3\frac{1}{2}, 0, 0\right)$	
(10, 0, −30, 0)	No	60	Yes	(0, 6, 0, 4)	
(0, 10, 0, −10)	No	80	Yes	(4, 0, 14, 0)	

6.3-7. (c) Basic variables are x_1 and x_2. The other variables are nonbasic.

(e) $x_1 + 3x_2 + 2x_3 + 3x_4 + x_5 = 6$, $4x_1 + 6x_2 + 5x_3 + 7x_4 + x_5 = 15$, $x_3 = 0$, $x_4 = 0$, $x_5 = 0$.
Optimal CPF solution is $(x_1, x_2, x_3, x_4, x_5) = (\frac{3}{2}, \frac{3}{2}, 0, 0, 0)$.

6.4-3. Maximize $W = 8y_1 + 6y_2$,

subject to

$$\begin{aligned} y_1 + 3y_2 &\leq 2 \\ 4y_1 + 2y_2 &\leq 3 \\ 2y_1 \quad &\leq 1 \end{aligned}$$

and

$$y_1 \geq 0, \qquad y_2 \geq 0.$$

6.4-8. (a) Minimize $W = 120y_1 + 80y_2 + 100y_3$,

subject to

$$\begin{aligned} y_2 - 3y_3 &= -1 \\ 3y_1 - y_2 + y_3 &= 2 \\ y_1 - 4y_2 + 2y_3 &= 1 \end{aligned}$$

and

$$y_1 \geq 0, \qquad y_2 \geq 0, \qquad y_3 \geq 0.$$

6.6-1. (d) Not optimal, since $2y_1 + 3y_2 \geq 3$ is violated for $y_1^* = \frac{1}{5}$, $y_2^* = \frac{3}{5}$.
(f) Not optimal, since $3y_1 + 2y_2 \geq 2$ is violated for $y_1^* = \frac{1}{5}$, $y_2^* = \frac{3}{5}$.

6.7-1.

Part	New Basic Solution $(x_1, x_2, x_3, x_4, x_5)$	Feasible?	Optimal?
(a)	(0, 30, 0, 0, −30)	No	No
(b)	(0, 20, 0, 0, −10)	No	No
(c)	(0, 10, 0, 0, 60)	Yes	Yes
(d)	(0, 20, 0, 0, 10)	Yes	Yes
(e)	(0, 20, 0, 0, 10)	Yes	Yes
(f)	(0, 10, 0, 0, 40)	Yes	No
(g)	(0, 20, 0, 0, 10)	Yes	Yes
(h)	(0, 20, 0, 0, 10, $x_6 = -10$)	No	No
(i)	(0, 20, 0, 0, 0)	Yes	Yes

6.7-3. $-10 \le \theta \le \frac{10}{9}$

6.7-12. **(a)** $b_1 \ge 2,\ 6 \le b_2 \le 18,\ 12 \le b_3 \le 24$
(b) $0 \le c_1 \le \frac{15}{2},\ c_2 \ge 2$

6.8-4. **(f)** The allowable range for the unit profit from producing toys is $2.50 to $5.00. The corresponding range for producing subassemblies is $-$3.00 to $-$1.50.

6.8-6. **(f)** For part (*a*), the change is within the allowable increase of $10, so the optimal solution does not change. For part (*b*), the change is outside the allowable decrease of $5, so the optimal solution might change. For part (*c*), the sum of the percentages of the allowable changes is 250 percent, so the 100 percent rule for simultaneous changes in objective function coefficients indicates that the optimal solution might change.

CHAPTER 7

7.1-2. $(x_1, x_2, x_3) = (\frac{2}{3}, 2, 0)$ with $Z = \frac{22}{3}$ is optimal.

7.1-6. **(a)** The new optimal solution is $(x_1, x_2, x_3, x_4, x_5) = (0, 0, 9, 3, 0)$ with $Z = 117$.

7.2-1. **(a, b)**

Range of θ	Optimal Solution	$Z(\theta)$
$0 \le \theta \le 2$	$(x_1, x_2) = (0, 5)$	$120 - 10\theta$
$2 \le \theta \le 8$	$(x_1, x_2) = \left(\dfrac{10}{3}, \dfrac{10}{3}\right)$	$\dfrac{320 - 10\theta}{3}$
$8 \le \theta$	$(x_1, x_2) = (5, 0)$	$40 + 5\theta$

7.2-4.

	Optimal Solution		
Range of θ	x_1	x_2	$Z(\theta)$
$0 \le \theta \le 1$	$10 + 2\theta$	$10 + 2\theta$	$30 + 6\theta$
$1 \le \theta \le 5$	$10 + 2\theta$	$15 - 3\theta$	$35 + \theta$
$5 \le \theta \le 25$	$25 - \theta$	0	$50 - 2\theta$

7.3-3. $(x_1, x_2, x_3) = (1, 3, 1)$ with $Z = 8$ is optimal.

CHAPTER 8

8.1-3. **(b)**

		Destination			
		Today	Tomorrow	Dummy	Supply
Source	Dick	3.0	2.7	0	5
	Harry	2.9	2.8	0	4
	Demand	3	4	2	

8.2-2. **(a)** Basic variables: $x_{11} = 4,\ x_{12} = 0,\ x_{22} = 4,\ x_{23} = 2,\ x_{24} = 0,\ x_{34} = 5,\ x_{35} = 1,\ x_{45} = 0$; $Z = 53$.
(b) Basic variables: $x_{11} = 4,\ x_{23} = 2,\ x_{25} = 4,\ x_{31} = 0,\ x_{32} = 0,\ x_{34} = 5,\ x_{35} = 1,\ x_{42} = 4$; $Z = 45$.
(c) Basic variables: $x_{11} = 4,\ x_{23} = 2,\ x_{25} = 4,\ x_{32} = 0,\ x_{34} = 5,\ x_{35} = 1,\ x_{41} = 0,\ x_{42} = 4$; $Z = 45$.

8.2-7. **(a)** $x_{11} = 3,\ x_{12} = 2,\ x_{22} = 1,\ x_{23} = 1,\ x_{33} = 1,\ x_{34} = 2$; three iterations to reach optimality.
(b, c) $x_{11} = 3,\ x_{12} = 0,\ x_{13} = 0,\ x_{14} = 2,\ x_{23} = 2,\ x_{32} = 3$; already optimal.

8.2-10. $x_{11} = 10,\ x_{12} = 15,\ x_{22} = 0,\ x_{23} = 5,\ x_{25} = 30,\ x_{33} = 20,\ x_{34} = 10,\ x_{44} = 10$; cost $=$ $77.30. Also have other tied optimal solutions.

11.5-2. (a) $(x_1, x_2) = (2, 3)$ is optimal.
(b) None of the feasible rounded solutions are optimal for the integer programming problem.

11.6-1. $(x_1, x_2, x_3, x_4, x_5) = (0, 0, 1, 1, 1)$, with $Z = 6$.

11.6-7. (b)

Task	1	2	3	4	5
Assignee	1	3	2	4	5

11.6-9. $(x_1, x_2, x_3, x_4) = (0, 1, 1, 0)$, with $Z = 36$.

11.7-2. (a, b) $(x_1, x_2) = (2, 1)$ is optimal.

11.8-1. (a) $x_1 = 0, x_3 = 0$

CHAPTER 12

12.2-7. (a) Concave.

12.4-1. (a) Approximate solution = 1.0125.

12.5-3. Exact solution is $(x_1, x_2) = (2, -2)$.

12.5-7. (a) Approximate solution is $(x_1, x_2) = (0.75, 1.5)$.

12.6-3.

$$-4x_1^3 - 4x_1 - 2x_2 + 2u_1 + u_2 = 0 \quad \text{(or } \leq 0 \text{ if } x_1 = 0).$$
$$-2x_1 - 8x_2 + u_1 + 2u_2 = 0 \quad \text{(or } \leq 0 \text{ if } x_2 = 0).$$
$$-2x_1 - x_2 + 10 = 0 \quad \text{(or } \leq 0 \text{ if } u_1 = 0).$$
$$-x_1 - 2x_2 + 10 = 0 \quad \text{(or } \leq 0 \text{ if } u_2 = 0).$$
$$x_1 \geq 0, \quad x_2 \geq 0, \quad u_1 \geq 0, \quad u_2 \geq 0.$$

12.6-6. $(x_1, x_2) = (1, 2)$ cannot be optimal.

12.6-8. (a) $(x_1, x_2) = (1 - 3^{-1/2}, 3^{-1/2})$.

12.7-2. (a) $(x_1, x_2) = (2, 0)$ is optimal.
(b) Minimize $Z = z_1 + z_2$,

subject to

$$2x_1 + u_1 - y_1 + z_1 = 8$$
$$2x_2 + u_1 - y_2 + z_2 = 4$$
$$x_1 + x_2 + v_1 = 2$$
$$x_1 \geq 0, \quad x_2 \geq 0, \quad u_1 \geq 0, \quad y_1 \geq 0, \quad y_2 \geq 0, \quad v_1 \geq 0, \quad z_1 \geq 0,$$
$$z_2 \geq 0.$$

12.8-2. (b) Maximize $Z = 3x_{11} - 3x_{12} - 15x_{13} + 4x_{21} - 4x_{23}$,

subject to

$$x_{11} + x_{12} + x_{13} + 3x_{21} + 3x_{22} + 3x_{23} \leq 8$$
$$5x_{11} + 5x_{12} + 5x_{13} + 2x_{21} + 2x_{22} + 2x_{23} \leq 14$$

and

$$0 \leq x_{ij} \leq 1, \quad \text{for } i = 1, 2, 3; j = 1, 2, 3.$$

12.9-8. (a) $(x_1, x_2) = \left(\dfrac{1}{3}, \dfrac{2}{3}\right)$.

12.9-14. (a) $P(x; r) = -2x_1 - (x_2 - 3)^2 - r\left(\dfrac{1}{x_1 - 3} + \dfrac{1}{x_2 - 3}\right)$.

(b) $(x_1, x_2) = \left[3 + \left(\dfrac{r}{2}\right)^{1/2}, 3 + \left(\dfrac{r}{2}\right)^{1/3}\right]$

CHAPTER 13

13.2-2. The best solution found has links AC, BC, CD, and DE.

13.4-2. (a) For the first child, the options for the first link are 1-2, 1-8, 1-5, and 1-4 so the random numbers 0.09656 and 0.96657 say to choose link 1-2 and no mutation occurs. The options for the second link then are 2-3, 2-8, and 2-4, and so forth. A mutation occurs with the fifth link. The complete first child is 1-2-8-5-6-4-7-3-1.

CHAPTER 14

14.2-2. Player 1: strategy 2; player 2: strategy 1.

14.2-7. (a) Politician 1: issue 2; politician 2: issue 2.
(b) Politician 1: issue 1; politician 2: issue 2.

14.4-4. $(x_1, x_2) = (\frac{2}{5}, \frac{3}{5})$; $(y_1, y_2, y_3) = (\frac{1}{5}, 0, \frac{4}{5})$; $v = \frac{8}{5}$.

14.5-3. (a) Maximize x_4,

subject to

$$5x_1 + 2x_2 + 3x_3 - x_4 \geq 0$$
$$4x_2 + 2x_3 - x_4 \geq 0$$
$$3x_1 + 3x_2 \qquad - x_4 \geq 0$$
$$x_1 + 2x_2 + 4x_3 - x_4 \geq 0$$
$$x_1 + x_2 + x_3 \qquad = 1$$

and

$$x_1 \geq 0, \quad x_2 \geq 0, \quad x_3 \geq 0, \quad x_4 \geq 0.$$

CHAPTER 15

15.2-2. (a)

	State of Nature	
Alternative	Sell 10,000	Sell 100,000
Build Computers	0	54
Sell Rights	15	15

(c) Let p = prior probability of selling 10,000. They should build when $p \leq 0.722$, and sell when $p > 0.722$.

15.2-4. (c) Warren should make the countercyclical investment.

15.2-8. Order 25.

15.3-2. (a) EVPI = EP (with perfect info) − EP (without more info) = 34.5 − 27 = $7.5 million.
(d)

Data:		P (Finding | State)	
State of Nature	Prior Probability	Finding	
		Sell 10,000	Sell 100,000
Sell 10,000	0.5	0.666666667	0.333333333
Sell 100,000	0.5	0.333333333	0.666666667

Posterior Probabilities:		P (State | Finding)	
		State of Nature	
Finding	**P (Finding)**	**Sell 10,000**	**Sell 100,000**
Sell 10,000	0.5	0.666666667	0.333333333
Sell 100,000	0.5	0.333333333	0.666666667

15.3-4. (b) EVPI = EP (with perfect info) − EP (without more info) = 53 − 35 = $18
(c) Betsy should consider spending up to $18 to obtain more information.

15.3-8. (a) Up to $230,000
(b) Order 25.

15.3-9. (a)

	State of Nature		
Alternative	**Poor Risk**	**Average Risk**	**Good Risk**
Extend Credit	−15,000	10,000	20,000
Don't Extend Credit	0	0	0
Prior Probabilities	0.2	0.5	0.3

(c) EVPI = EP (with perfect info) − EP (without more info) = 11,000 − 8,000 = $3,000. This indicates that the credit-rating organization should not be used.

15.3-13. (a) Guess coin 1.
(b) Heads: coin 2; tails: coin 1.

15.4-2. The optimal policy is to do no market research and build the computers.

15.4-4. (c) EVPI = EP (with perfect info) − EP (without more info) = 1.8 − 1 = $800,000
(d)

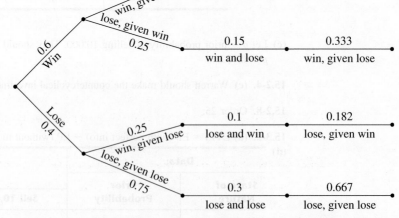

Prior Probabilities P (state)	Conditional Probabilities P (finding|state)	Joint Probabilities P (state and finding)	Posterior Probabilities P (state|finding)

(f) Leland University should hire William. If he predicts a winning season then they should hold the campaign. If he predicts a losing season then they should not hold the campaign.

15.5-7. (a) Choose to introduce the new product (expected payoff is $12.5 million).
(b) $7.5 million.
(c) The optimal policy is not to test but to introduce the new product.
(f) Both charts indicate that the expected payoff is sensitive to both parameters, but is somewhat more sensitive to changes in the profit if successful than to changes in the loss if unsuccessful.

15.6-2. (a) Choose not to buy insurance (expected payoff is $249,840).
(b) U(insurance) = 499.82
U(no insurance) = 499.8
Optimal policy is to buy insurance.

15.6-4. $U(10) = 9$

CHAPTER 16

16.3-3. (c) $\pi_0 = \pi_1 = \pi_2 = \pi_3 = \pi_4 = \frac{1}{5}$.

16.4-1. (a) All states belong to the same recurrent class.

16.5-7. (a) $\pi_0 = 0.182$, $\pi_1 = 0.285$, $\pi_2 = 0.368$, $\pi_3 = 0.165$.
(b) 6.50

CHAPTER 17

17.2-1. Input source: population having hair; customers: customers needing haircuts; and so forth for the queue, queue discipline, and service mechanism.

17.2-2. (b) $L_q = 0.375$
(d) $W - W_q = 24.375$ minutes

17.4-2. (c) 0.0527

17.5-5. (a) State:

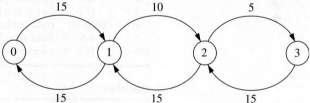

(c) $P_0 = \frac{9}{26}$, $P_1 = \frac{9}{26}$, $P_2 = \frac{3}{13}$, $P_3 = \frac{1}{13}$.
(d) $W = 0.11$ hour.

17.5-8. (b) $P_0 = \frac{2}{5}$, $P_n = (\frac{3}{5})(\frac{1}{2})^n$
(c) $L = \frac{6}{5}$, $L_q = \frac{3}{5}$, $W = \frac{1}{25}$, $W_q = \frac{1}{50}$

17.6-2. (a) $P_0 + P_1 + P_2 + P_3 + P_4 = 0.96875$ or 97 percent of the time.

17.6-21. (a) Combined expected waiting time = 0.211
(c) An expected process time of 3.43 minutes would cause the expected waiting times to be the same for the two procedures.

17.6-26. (a) 0.429

17.6-32. (a) three machines
(b) three operators

17.7-1. (a) W_q (exponential) $= 2W_q$ (constant) $= \frac{8}{5}W_q$ (Erlang).
(b) W_q (new) $= \frac{1}{2} W_q$ (old) and L_q (new) $= L_q$ (old) for all distributions.

17.7-6. (a, b) Under the current policy an airplane loses 1 day of flying time as opposed to 3.25 days under the proposed policy.
Under the current policy 1 airplane is losing flying time per day as opposed to 0.8125 airplane.

17.7-9.

Service Distribution	P_0	P_1	P_2	L
Erlang	0.561	0.316	0.123	0.561
Exponential	0.571	0.286	0.143	0.571

17.8-1. (a) This system is an example of a nonpreemptive priority queueing system.

(c) $\dfrac{W_q \text{ for first-class passengers}}{W_q \text{ for coach-class passengers}} = \dfrac{0.033}{0.083} = 0.4$

17.8-4. (a) $W = \frac{1}{2}$
(b) $W_1 = 0.20$, $W_2 = 0.35$, $W_3 = 1.10$
(c) $W_1 = 0.125$, $W_2 = 0.3125$, $W_3 = 1.250$

17.10-2. 4 cash registers

CHAPTER 18

18.3-1. (a) $t = 1.83$, $Q = 54.77$
(b) $t = 1.91$, $Q = 57.45$, $S = 52.22$

18.3-3. (a) Data

$d =$	676	(demand/year)
$K =$	$75	(setup cost)
$h =$	$600.00	(unit holding cost)
$L =$	3.5	(lead time in days)
$WD =$	365	(working days/year)

Results

Reorder point =	6.5
Annual setup cost =	$10,140
Annual holding cost =	$ 1,500
Total variable cost =	$11,640

Decision

$Q =$	5	(order quantity)

(d) Data

$d =$	676	(demand/year)
$K =$	$75	(setup cost)
$h =$	$600	(unit holding cost)
$L =$	3.5	(lead time in days)
$WD =$	365	(working days/year)

Results

Reorder point =	6.48
Annual setup cost =	$3,900
Annual holding cost =	$3,900
Total variable cost =	$7,800

Decision

$Q =$	13	(order quantity)

The results are the same as those obtained in part (c).

(f) Number of orders per year $= 52$

$ROP = 6.5 -$ inventory level when each order is placed

(g) The optimal policy reduces the total variable inventory cost by \$3,840 per year, which is a 33 percent reduction.

18.3-6. (a) $h = \$3$ per month which is 15 percent of the acquisition cost.

(c) Reorder point is 10.

(d) $ROP = 5$ hammers, which adds \$20 to his TVC (5 hammers $\times$ \$4 holding cost).

18.3-7. $t = 3.26$, $Q = 26,046$, $S = 24,572$

18.3-12. (a) Optimal $Q = 500$

18.4-4. Produce 3 units in period 1 and 4 units in period 3.

18.6-6. (b) Ground Chuck: $R = 145$.

Chuck Wagon: $R = 829$.

(c) Ground Chuck: safety stock $= 45$.

Chuck Wagon: safety stock $= 329$.

(f) Ground Chuck: \$39,378.71.

Chuck Wagon: \$41,958.61.

Jed should choose Ground Chuck as their supplier.

(g) If Jed would like to use the beef within a month of receiving it, then Ground Chuck is the better choice. The order quantity with Ground Chuck is roughly 1 month's supply, whereas with Chuck Wagon the optimal order quantity is roughly 3 month's supply.

18.7-5. (a) Optimal service level $= 0.667$

(c) $Q^* = 500$

(d) The probability of running short is 0.333.

(e) Optimal service level $= 0.833$

CHAPTER 19

19.2-2. (c) Use slow service when no customers or one customer is present and fast service when two customers are present.

19.2-3. (a) The possible states of the car are dented and not dented.

(c) When the car is not dented, park it on the street in one space. When the car is dented, get it repaired.

19.2-5. (c) State 0: attempt ace; state 1: attempt lob.

19.3-2. (a) Minimize $\quad Z = 4.5y_{02} + 5y_{03} + 50y_{14} + 9y_{15}$,

subject to

$$y_{01} + y_{02} + y_{03} + y_{14} + y_{15} = 1$$

$$y_{01} + y_{02} + y_{03} - \left(\frac{9}{10}y_{01} + \frac{49}{50}y_{02} + y_{03} + y_{14}\right) = 0$$

$$y_{14} + y_{15} - \left(\frac{1}{10}y_{01} + \frac{1}{50}y_{02} + y_{15}\right) = 0$$

and

$$\text{all } y_{ik} \geq 0.$$

19.3-4. (a) Minimize $\quad Z = -\frac{1}{8}y_{01} + \frac{7}{24}y_{02} + \frac{1}{2}y_{11} + \frac{5}{12}y_{12}$,

subject to

$$y_{01} + y_{02} - \left(\frac{3}{8}y_{01} + y_{11} + \frac{7}{8}y_{02} + y_{12}\right) = 0$$

$$y_{11} + y_{12} - \left(\frac{5}{8}y_{01} + \frac{1}{8}y_{02}\right) = 0$$

$$y_{01} + y_{02} + y_{11} + y_{12} = 1$$

and

$$y_{ik} \geq 0 \qquad \text{for } i = 0, 1; k = 1, 2.$$

19.4-2. Car not dented: park it on the street in one space. Car dented: repair it.

19.4-4. State 0: attempt ace. State 1: attempt lob.

19.5-1. Reject $600 offer, accept any of the other two.

19.5-2. (a) Minimize $\quad Z = 60(y_{01} + y_{11} + y_{21}) - 600y_{02} - 800y_{12} - 1,000y_{22}$,

subject to

$$y_{01} + y_{02} - (0.95)\left(\frac{5}{8}\right)(y_{01} + y_{11} + y_{21}) = \frac{5}{8}$$

$$y_{11} + y_{12} - (0.95)\left(\frac{1}{4}\right)(y_{01} + y_{11} + y_{21}) = \frac{1}{4}$$

$$y_{21} + y_{22} - (0.95)\left(\frac{1}{8}\right)(y_{01} + y_{11} + y_{21}) = \frac{1}{8}$$

and

$$y_{ik} \geq 0 \qquad \text{for } i = 0, 1, 2; k = 1, 2.$$

19.5-3. After three iterations, approximation is, in fact, the optimal policy given for Prob. 19.5-1.

19.5-11. In periods 1 to 3: Do nothing when the machine is in state 0 or 1; overhaul when machine is in state 2; and replace when machine is in state 3. In period 4: Do nothing when machine is in state 0, 1, or 2; replace when machine is in state 3.

CHAPTER 20

20.1-1. (b) Let the numbers 0.0000 to 0.5999 correspond to strikes and the numbers 0.6000 to 0.9999 correspond to balls. The random observations for pitches are 0.7520 = ball, 0.4184 = strike, 0.4189 = strike, 0.5982 = strike, 0.9559 = ball, and 0.1403 = strike.

20.1-10. (b) Use $\lambda = 4$ and $\mu = 5$.

(i) Answers will vary. The option of training the two current mechanics significantly decreases the waiting time for German cars, without a significant impact on the wait for Japanese cars, and does so without the added cost of a third mechanic. Adding a third mechanic lowers the average wait for German cars even more, but comes at an added cost for the third mechanic.

20.3-1. (a) 5, 8, 1, 4, 7, 0, 3, 6, 9, 2

20.4-2. (b) $F(x) = 0.0965$ when $x = -5.18$
$F(x) = 0.5692$ when $x = 18.46$
$F(x) = 0.6658$ when $x = 23.29$

20.4-6. **(a)** Here is a sample replication.

Summary of Results:

Win? (1 = Yes, 0 = No)	0
Number of Tosses =	3

Simulated Tosses

Toss	Die 1	Die 2	Sum
1	4	2	6
2	3	2	5
3	6	1	7
4	5	2	7
5	4	4	8
6	1	4	5
7	2	6	8

Results

Win?	Lose?	Continue?
0	0	Yes
0	0	Yes
0	1	No
NA	NA	No
NA	NA	No
NA	NA	No
NA	NA	No

SUBJECT INDEX